Chemical Reagents for Protein Modification

Fourth Edition

Chemical Reagents for Protein Modification

Fourth Edition

Roger L. Lundblad

Chapel Hill, North Carolina, USA

CRC Press
Taylor & Francis Group
Boca Raton London New York

CRC Press is an imprint of the
Taylor & Francis Group, an **informa** business

CRC Press
Taylor & Francis Group
6000 Broken Sound Parkway NW, Suite 300
Boca Raton, FL 33487-2742

First issued in paperback 2020

ISBN-13: 978-1-4665-7190-7 (hbk)
ISBN-13: 978-0-367-65916-5 (pbk)

Library of Congress Cataloging-in-Publication Data

Lundblad, Roger L.
  Chemical reagents for protein modification / Roger L. Lundblad. -- Fourth edition.
    pages cm
  Includes bibliographical references and index.
    ISBN 978-1-4665-7190-7 (alk. paper)
  1. Proteins--Chemical modification. 2. Chemical tests and reagents. I. Title.

QP551.L88 2013
572'.6--dc23   2013045155

Visit the Taylor & Francis Web site at
http://www.taylorandfrancis.com

and the CRC Press Web site at
http://www.crcpress.com

*This work is dedicated to the memory of my mother,
Doris Ruth Peterson Lundblad, who never understood what
I wrote but told me that it did help her to sleep.*

Contents

Preface

This is the fourth edition of a work first published in 1983 in collaboration with Dr. Claudia Noyes. There have been two editions in the intervening years, which have provided a modest update. The current work has undergone major revision and has emerged far larger than when I started this work. I have tried to provide an even more encyclopedic coverage of a field that is not nearly as popular as it was when the first work appeared in 1983. On the other hand, times do change and chemical modification may again have the popularity of 40 years ago (see Chargaff, 1970). As with current editions, I continue to be embarrassed by what I missed in early work and apologize in advance to those investigators who I have yet again missed in the current work. As a word of caution, care must be taken with specificity. Mass spectrometry does provide a way of identifying modification, which, in the past, would have been silent. Finally, I cannot recall the source, perhaps Efraim Racker, who commented, "no reagent is as specific as when it is discovered."

REFERENCE

Chargaff, E., Triviality in science: A brief meditation on fashions, *Perspect. Biol. Med.* Spring, 324–333, 1970.

Acknowledgments

I acknowledge the persistence of Barbara Norwitz in getting this work started and the help of Jill Jurgensen in turning a box of stuff into a useful volume. I also thank Professor Bryce Plapp of the University of Iowa for his continuing support of the thermodynamically challenged.

Author

Roger L. Lundblad is a native of San Francisco, California. He earned his undergraduate education at Pacific Lutheran University and his PhD in biochemistry from the University of Washington. After his postdoctoral work in the laboratories of Stanford Moore and William Stein at The Rockefeller University, he joined the faculty of the University of North Carolina at Chapel Hill. He then joined the Hyland Division of Baxter Healthcare in 1990. Currently, Dr. Lundblad works as an independent consultant at Chapel Hill, North Carolina, and writes on biotechnological issues. He is an adjunct professor of pathology at the University of North Carolina at Chapel Hill.

1 Introduction to the Chemical Modification of Proteins

This is the fourth edition of this work and, as with past editions, it is intended to provide a comprehensive review of reagents used for the chemical modification of proteins. There is a considerable history of the chemical modification of proteins, which is largely beyond the scope of the present volume. There will be occasional mention of the older literature in the current work when necessary but the serious reader is directed to several excellent reviews.[1-4] While the technology may be dated, the data are useful and can be very instructive. The three-volume work by J. Leyden Webb[5] is of considerable value in the current discipline of systems biology[6] as he discusses early work that was performed in whole animals and selected tissues. A minimum amount of effort is expended on the functional consequences of modification except when such would provide insight into the chemistry of reaction.

The use of the chemical modification of proteins has evolved over the past 80 years benefiting from advances in analytical, physical, and organic chemistry. However, some questions remain. Mudd and Joffe[7] modified antibodies with formaldehyde to determine the importance of basic amino acids in antibody function. Other early work used formaldehyde for the analysis of lysine in proteins and the role of lysine in the acid–base behavior of proteins.[8] Some 20 years later, Liener and Wada[9] published work on the chemical modification of soy bean hemagglutinin where a variety of reagents including acetic anhydride, carbobenzoxy chloride, 2,4-dinitrobenzenesulfonic acid, sodium nitrite, iodine, iodoacetamide, p-chloromercuribenzoate, and methanol/HCl (esterification) were used to show the importance of amino groups and tyrosyl residues in the hemagglutination activity. It was mentioned that methods were not available for the modification of arginine, histidine, or tryptophan. The assay methods were laborious but the conclusions valid. It is important to recognize the value of the early literature; the lack of sophisticated analysis does not invalidate the observed events just as inability to obtain an electronic copy does not imply nonexistence of the information. The early work on the chemical modification of proteins was focused toward the study of protein function with increasing emphasis on enzymatic activity. In addition to these fundamental activities, there was also early interest in the application of chemical modification for the production of protein therapeutic products[10] and use of proteins as plastics[11,12] and adhesives.[13] There was a marked increase in the activity of chemical modification in the 1960s, and the work in this area became increasingly sophisticated and the reader is directed to some later reviews in this area[14-18] as well as several volumes of *Methods in Enzymology*.[19-21] The basic information on solution protein chemistry obtained from 1960 through 1980 established the basis for present work in areas of proteomics, structural biology, and chemical biology.

Proteomics as a discipline can be said to have replaced classical protein chemistry[22-24] driven by advances in mass spectrometry (MS) and other analytical technologies but much of the work in proteomics has been enabled by earlier work in classical protein chemistry.[23] Isotope-coded affinity tags (ICAT) (Figure 1.1) were developed by Aebersold and colleagues.[25-27] The chemistry used in ICAT technology is based on earlier work (Figure 1.1) on the alkylation of cysteine residues in proteins (Chapter 2). ICAT enabled the relatively specific introduction of a deuterium-labeled moiety on the sulfhydryl groups of a protein. The use of a chemically identical modifying reagent not containing deuterium allows the comparison of protein expression.[28] The presence of a biotin moiety permits the isolation of modified peptides. Subsequent work has refined this technique[29-31],

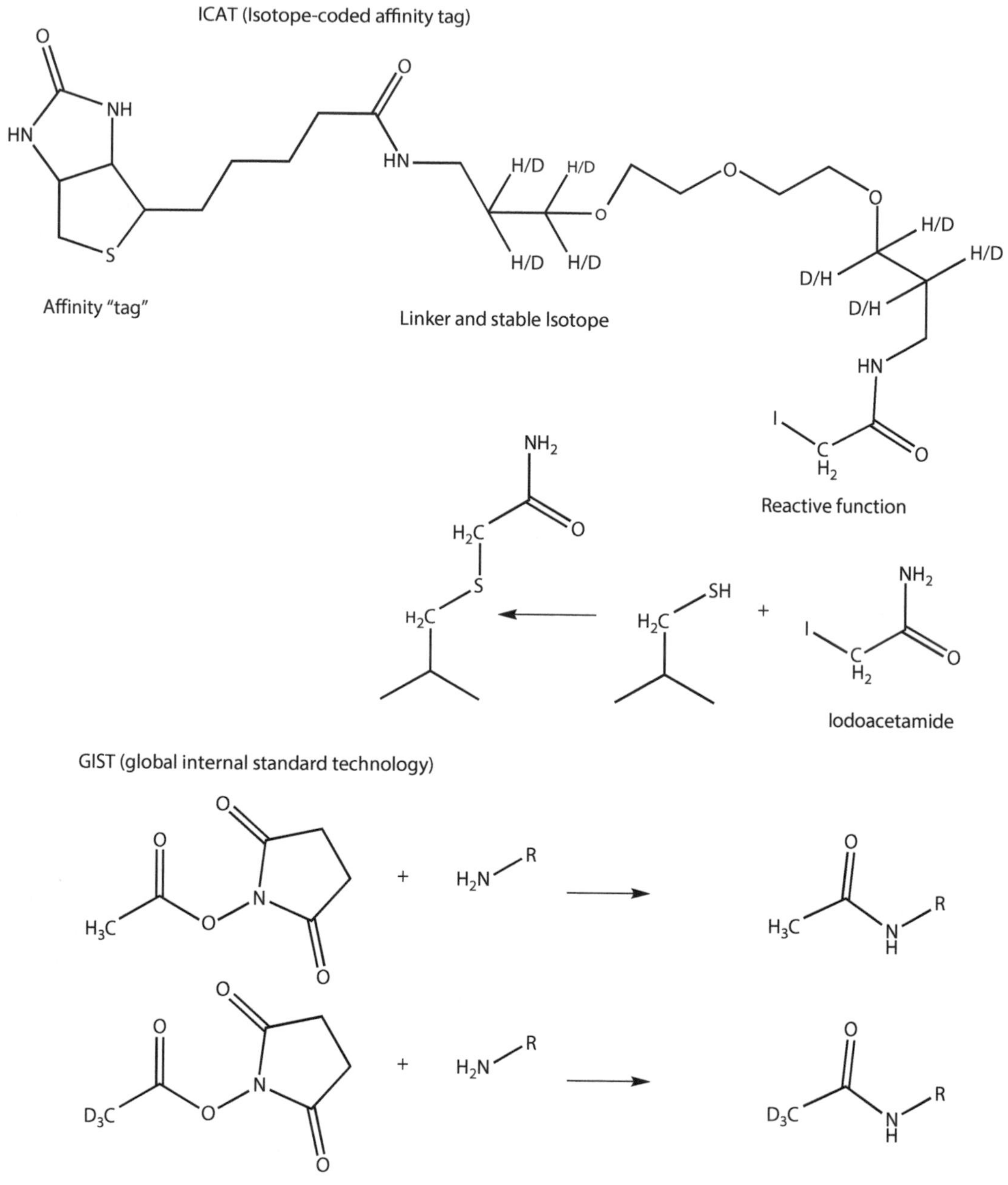

FIGURE 1.1 ICAT and Global Internal Standard Technology (GIST) for proteomic research. The use of these reagents for the determination of differential protein expression is based on the inclusion of *light* or *heavy* isotopes. See Gygi, S.P., Rist, B., Gerber, S.A., Turecek, F., Gels, M.H., and Aebersold, R., Quantitative analysis of complex protein mixtures using isotope-coded affinity tags, *Nat. Biotechnol.* 17, 994–999, 1999; Chakraborty, A. and Regnier, F.E., Global internal standard technology for comparative proteomics, *J. Chromatogr. A* 949, 173–184, 2002; Sebastino, R., Cirreria, A., Lapadula, M., and Righetti, P.G., A new deuterated alkylating agent for quantitative proteomics, *Rapid Commun. Mass Spectrom.* 17, 2380–2386, 2003; Kuyama, H., Watanabe, M., Todo, C., Ando, E., Tanaka, K., and Nishimura, O., An approach to quantitative proteome analysis by labeling tryptophan residues, *Rapid Commun. Mass Spectrom.* 17, 1642–1650, 2003.

and reagents that target residues other than cysteine have been developed.[32–34] The development of a reagent with an acid-labile link to a resin permits the facile purification of peptides.[35] A related approach uses the incorporation of stable isotope-labeled amino acids in cell culture (SILAC).[36]

Other examples of site-specific chemical modification in proteomics include the use of fluorogenic reagents for the modification of lysine residues in proteins providing high sensitivity in the analysis by 2-D capillary electrophoresis,[37] the selective modification of thiophorylated peptides at low pH with an derivative of iodoacetic acid,[38] and the oxidation of methionine residues with hydrogen peroxide.[39] The in situ oxidation of methionine[40] was used to identify methionine peptides in an earlier version of 2D electrophoresis (diagonal electrophoresis).[41,42] It has been possible to modify 3-nitrotyrosine residues in complex mixtures by first reducing the protein mixture with dithiothreitol followed by alkylation with iodoacetic acid. The nitrotyrosine residues are then reduced to the aminotyrosine derivative with sodium dithionite and alkylated with a sulfosuccinimidyl derivative.[43] The presence of a biotin tag allows the isolation of the peptides containing the modified tyrosine residues by affinity chromatography on an avidin-based matrix. This is another example of the application of previous work[44,45] in classical solution protein chemistry to proteomics.

Structural biology is concerned with the three-dimensional structural and the interactions of complex biological macromolecules with emphasis on proteins and nucleic acids and the organization of these polymers into larger structures such as membranes and chromatin. Examples of protein chemical modification techniques that have been useful in structural biology include membrane probes,[46] surface residue oxidation,[47] and covalent cross-linking.[48]

Chemical biology has been defined as "...application of chemistry to the study of molecular events in biological systems...."[49] Morrison and Weiss[50] report that chemical biology can been defined as the use of chemistry to understand biology and the use of biology to advance chemistry. In an otherwise excellent consideration of origins of chemical Morrison and Weiss[50] seem to have missed the use of the term chemical biology in a title by Florkin in 1960.[51] Notwithstanding the difficulty that I have in separating chemical biology from biological chemistry/chemistry, chemical biology is here. Chemical biology uses much from the work on the chemical modification of proteins[52] as does chemical proteomics.[53] Chemical proteomics is focused on defining the molecular basis of drug action.[54]

Chemical modification of proteins has been used to produce biotherapeutics.[55–60] Recent work on the use of chemical modification in biopharmaceutical development has focused on two general applications. The first is modification of biopharmaceuticals to extend circulatory half-life and perhaps reduce immunogenicity, while the second application uses chemical modification to produce protein conjugates. The modification of proteins and peptides with poly(ethylene) glycol (PEG) (Figure 1.2) is frequently used in the manufacture of biopharmaceuticals. There have been numerous articles on the use of PEG in the past decade with development work occurring during this time as PEGylation became a mature technology.[61–71] Successful modification of therapeutic proteins and peptides with PEG is associated with an extension of circulatory half-life and reduced or eliminated immunogenicity. It is thought that these properties might arise from the physical blocking of the therapeutic from immunological surveillance and catabolic recognition but it is likely that increased size results in decreased renal clearance. Reaction of *activated* PEG molecules with proteins can occur at cysteine residues which can be inserted by mutagenesis and at primary amino groups; reaction has been demonstrated to occur at other amino acid residues.[67] Protein engineering has been used to introduce a free cysteine residue in an immunotoxin which can be subsequently modified with PEG to yield a monosubstituted derivative.[68] Reaction at pH values below neutrality can result in preferential reaction at N-terminal α-amino groups rather than ε-amino groups (see Table 1.1).[69,70] The use of cleavable linkers has permitted the design of PEG-protein conjugates with half-lives depending on in vivo rate of cleavage of the linkers between the PEG moiety and the protein.[71] There has been continued work in this area[72,73] complicated by the work showing an increased level of antibodies against PEG[74] that may involve the terminal methoxy function.[75] The function

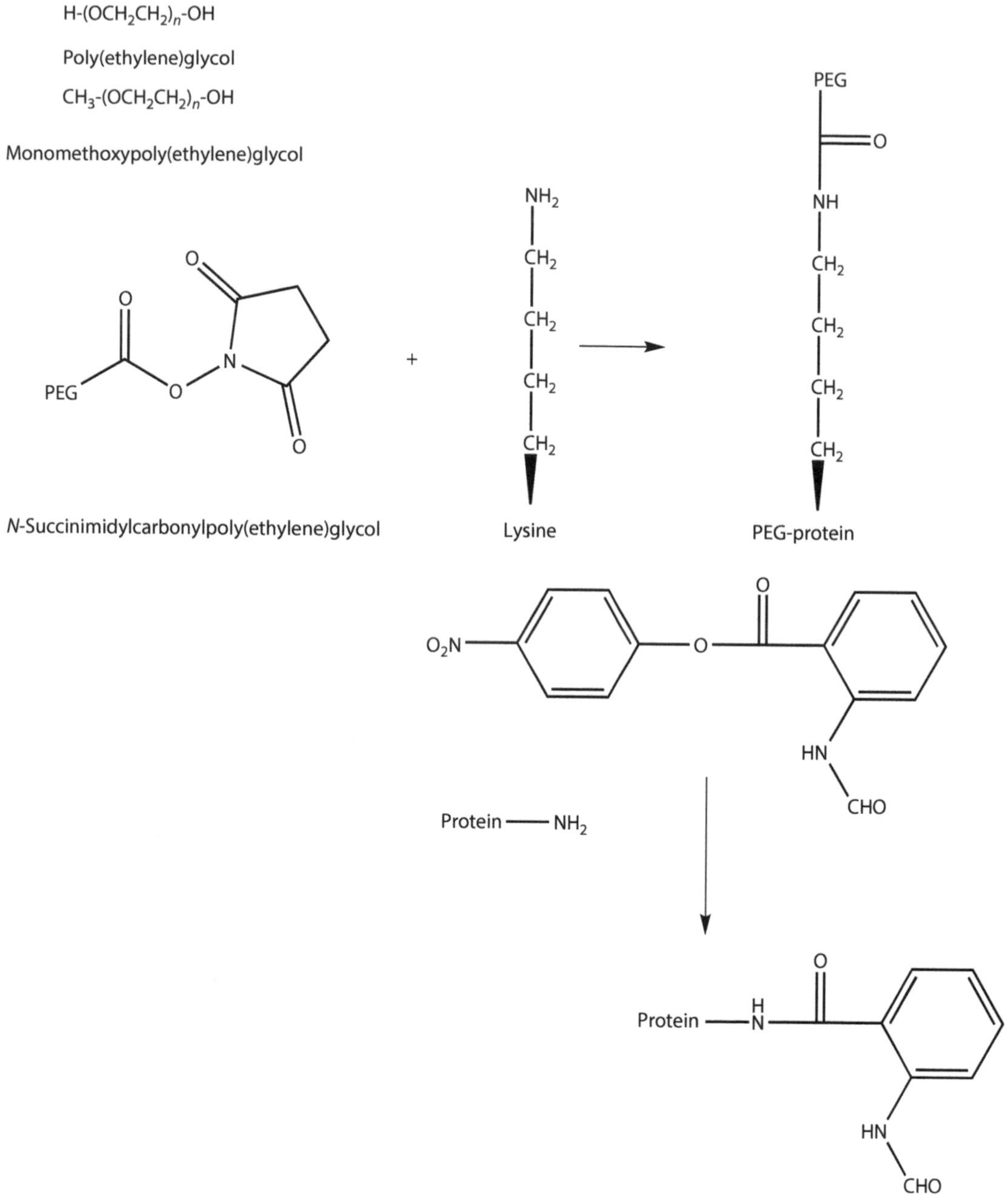
H-(OCH2CH2)n-OH
Poly(ethylene)glycol
CH3-(OCH2CH2)n-OH
Monomethoxypoly(ethylene)glycol
PEG
O
N
O
O
O
+
NH2
CH2
CH2
CH2
CH2
PEG
O
NH
CH2
CH2
CH2
CH2
N-Succinimidylcarbonylpoly(ethylene)glycol
Lysine
PEG-protein
O2N
O
O
HN
CHO
Protein NH2
Protein
H
N
O
HN
CHO

FIGURE 1.2 Modification of protein with PEG and glycosylation. Shown are the structures of PEG, the monomethyl derivative, and the reaction with proteins (Ryan, S.M., Mantovani, G., Wang, X. et al., Advances in PEGylation of important biotech molecules: Delivery aspects, *Expert Opin. Drug Deliv.* 5, 371–383, 2008). Shown at the bottom is a procedure for the preparation of a neoglycoprotein using a *p*-nitrophenyl ester anthranilate carbohydrate where the anthranilate moiety is coupled to the reducing sugar by reductive amination (Luyai, A., Lasanajak, Y., Smith, D.F. et al., Facile preparation of fluorescent neoglycoproteins using *p*-nitrophenyl anthranilate as a heterobifunctional linker, *Bioconjug. Chem.* 20, 1618–1624, 2009).

TABLE 1.1
Dissociation of Ionizable Groups in Proteins

Potential Nucleophile	pKa[a]	pKa[b]
γ-Carboxyl (glutamic acid)	4.25	4.2 ± 0.9
β-Carboxyl (aspartic acid)	3.65	3.5 ± 1.2
α-Carboxyl (isoleucine)	2.36	3.3 ± 0.8
Sulfhydryl (cysteine)	10.46	6.8 ± 2.7
α-Amino (isoleucine)	9.68	7.7 ± 0.5
Phenolic hydroxyl (tyrosine)	10.13	10.3 ± 1.2
ε-Amino (lysine)	10.79	10.5 ± 1.1
Imidazole (histidine)	6.00	6.6 ± 1.0
Guanidino (arginine)	12.48	—
Serine (hydroxyl)	—	13.06 ± 0.5[c]

[a] Taken from Mooz, E.D., Data on the naturally occurring amino acids, in *Practical Handbook of Biochemistry and Molecular Biology*, ed. G.D. Fasman, CRC Press, Boca Raton, FL, 1989; Also see Dawson, R.M.C., Elliott, D.C., and Jones, K.M., *Data for Biochemical Research*, Oxford University Press, Oxford, U.K., 1969.

[b] Taken from Grimsley, G.R., A summary of the measured pK values of the ionizable groups in folded proteins, *Protein Sci.* 18, 247–251, 2009.

[c] Determined for *N*-acetylserineamide. See Bruice, T.C., Fife, T.H., Bruno, J.J., and Brandon, N.E., Hydroxyl group catalysis. II. The reactivity of the hydroxyl group of serine. The nucleophilicity of alcohols and the ease of hydrolysis of their acetyl esters as related to their pKa, *Biochemistry* 1, 7–12, 1962.

of PEGylation is compared with that of protein glycosylation.[76,77] Other protein therapeutics that utilized site-specific chemical modification in the manufacturing process included various protein conjugates. Coupling of peptides to albumin has been described as improving circulatory half-life in the case of glucagon-like peptide 1.[78,79] A variety of monoclonal antibody conjugates have also proved useful.[80–85] Urokinase was coupled to pulmonary surfactant protein using a heterobifunctional cross-linker, *S*-sulfosuccinimidyl-4-(*p*-maleimidophenyl)butyrate.[86] Cross-linking of proteins with glutaraldehyde, PEG diacrylate, and formaldehyde is done for the preparation of hydrogels used extensively as biomaterials.[87–91] Carbamylation with potassium cyanate is used in the manufacturing process of some allergoids.[92–95] Allergoids are also produced by the modification of allergens with formaldehyde and glutaraldehyde with most recent work using glutaraldehyde.[96–98]

Chemical modification is used for the selective fragmentation of proteins for the determination of the primary structure of proteins, the preparation of large fragments for characterization by MS, and the chemical synthesis of proteins. This includes reagents such as cyanogen bromide for the cleavage of specific peptide bonds, citraconic anhydride for the reversible blocking of lysine residues restricting tryptic cleavage to arginine residues,[99] and the reversible blocking of arginine residues with 1,2-cyclohexanedione restricting tryptic cleavage to lysine residues (see Chapter 15).

Cyanogen bromide has been used to obtain large fragments of mass spectrometric analysis of proteins.[100] Cyanogen bromide fragments have also been used in the manufacture of semisynthetic proteins via native chemical ligation (Figure 1.3).[101] A 223-residue hybrid of *Streptomyces griseus* trypsin was synthesized by native chemical ligation from chemically synthesized amino-terminal

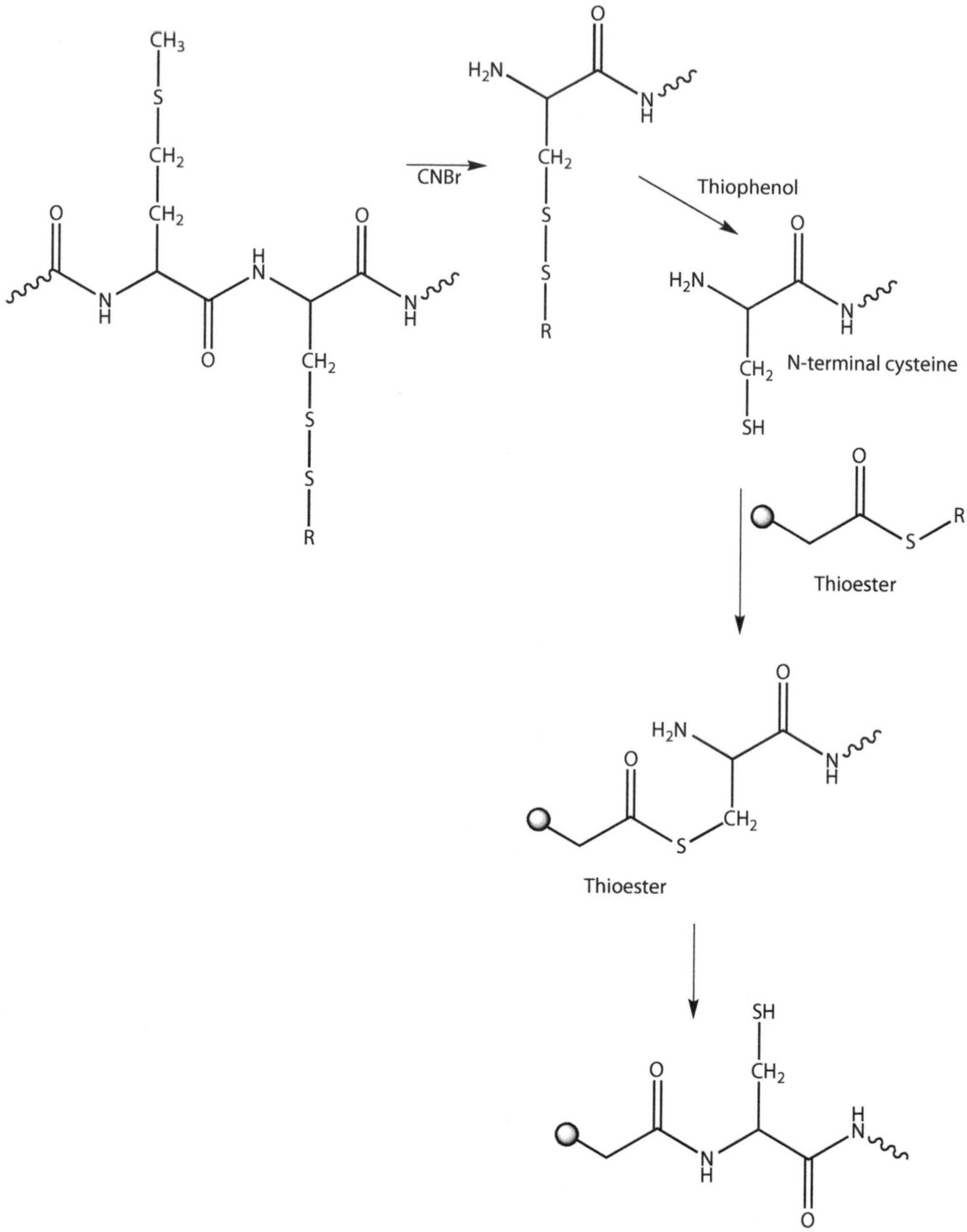

FIGURE 1.3 Cyanogen bromide cleavage products and native chemical ligation. Shown is the cyanogen bromide cleavage of a protein and the combination of the N-terminal cysteine fragment with a C-terminal thioester protein fragment to form a semisynthetic protein. In this scheme, the recombinant protein containing an engineered Met-Cys is expressed as an inclusion body that is renatured in the presence of oxidized glutathione resulting in an *S*-glutathionyl protein that was subjected to cyanogen bromide cleavage in the presence of thiophenol resulting in a C-terminal cysteine, which then can undergo native chemical ligation with a thioester peptide derivative to form a hybrid protein (Pál, G., Santamaria, F., Kossiakoff, A.A., and Lu, W., The first semi-synthetic serine protease made by native chemical ligation, *Protein Expr. Purif.* 29, 185–192, 2003). The cleavage of a Met-Cys bond without protecting the cysteine has been demonstrated (Richardson, J.P. and MacMillan, D., Optimisation of chemical protein cleavage for erythropoietin semi-synthesis using native chemical ligation, *Org. Biomol. Chem.* 6, 3977–3982, 2008). Care was taken with respect to solvent and exclusion of oxygen.

fragment and a larger C-terminal fragment obtained by cyanogen bromide cleavage.[102] The *S. griseus* trypsin was engineered to include a Met57-Cys58 bond, expressed in *Escherichia coli* and obtained as an inclusion body which was refolded in the presence of oxidized glutathione providing a derivative with five *S*-gluthionyl-cysteinyl residues. Cyanogen bromide cleavage was performed resulting a C-terminal derivative containing the five *S*-glutathionyl residues. Native chemical ligation with the thioester derivative was allowed to proceed in the presence of thiophenol yielding a hybrid protein with the expected properties; the *S*-glutathionyl groups were removed by the thiophenol during the ligation process. MacMillan and Arham[103] obtained fragment for native chemical ligation by cleavage of a Met-Cys bond with cyanogen bromide in either 80% formic acid or 0.3 MHCl/8 M urea following by treatment with 1 mM dithiothreitol. Protection of the cysteine residues with oxidative sulfitolysis (Chapter 7) prior to reaction with cyanogen bromide did not result in cleavage under the same conditions as used for the unmodified protein. It is noted that Smith and Kyte[104] reported the reversible modification of cysteine with cyanogen bromide. This short discussion on cyanogen bromide illustrates the point that prior knowledge does not dictate future outcome (you can perform a cyanogen bromide cleavage with free cysteine present) and you need to understand all products (you may have to reduce the products to recover the free cysteine).

One of the several goals of this chapter is the introduction of some basic concepts of organic chemistry (Figure 1.4) that provide the basis for the chemical modification of the various functional groups on proteins. Most chemical modification reactions of proteins are S_N2 reactions (substitution, nucleophilic, bimolecular). There are a much smaller number of S_N1 (substitution, nucleophilic, unimolecular) reactions and free radical–mediated reactions. An example is provided by the β-elimination of phosphorylated serine residues to form dehydroalanine. Michael addition chemistry is then used to add a photolabile group, which is then used to fragment the polypeptide chain.[108]

Reactivity of functional groups in biopolymers depends in part on the nucleophilicity of the specific functional group, although steric issues and solvent environment have an important role. There are several older sources that are well worth examination for more information in this area.[109,110] So, let me start with a question from a work in 1987, "…What is a nucleophile."[111] A consideration of various articles suggests that, with biopolymers, the definition is based on the kinetic data for a substitution or displacement reaction[112]; a more practical definition may be the possession of a pair of electrons that can form a new bond with another molecule.[113] The kinetic data may yield conclusion that are based on extrinsic conditions, such as solvent, as well as intrinsic nucleophilicity since intrinsic nucleophilicity can be enhanced.[111] Understanding intrinsic nucleophilicity can be challenging[114] prompting the study of gas-phase reactions to avoid solvent issues.[115] An electrophile can accept a pair of electrons. Nucleophiles can be Lewis bases and electrophiles are Lewis acids.[116,117] The concept of hard and soft bases and acids[116] provides insight into intrinsic nucleophilicity.[117] For example, a sulfur center nucleophile, a "soft" nucleophile, reacts more rapidly with an alkylating agent (soft electrophile) than does an oxygen center nucleophile, while an acylating agent such as acetic anhydride is a harder electrophile and the advantage of the sulfur center nucleophile is reduced.[117] Hard metal ions such as Mg^{2+} prefer binding to oxygen while soft metal ions such as Cu^{2+} prefer sulfur.[102]

Solvation has a potent effect on nucleophilic substitution reactions.[117,118–122] In the case of proteins, the solvent effect would include the effect of bulk solution on completely exposed residues and the effect of local microenvironment on all functional groups. Banks[121] reported on the effect of solvent on nucleophilic substitution reaction of halogen-substituted aziridine with ammonia. The solvents evaluated included water, tetrahydrofuran, and acetonitrile; reactions proceeded more rapidly in aqueous solvent. Another example is provided by a study showing that the order of reactivity of halides in the displacement of tosylate from *n*-hexyl tosylate in MeOH (I > Br > Cl) is reversed in dimethyl sulfoxide (DMSO).[123] Mendel Friedman[124] reported on the effect of DMSO on the modification of functional groups in proteins and model compounds. DMSO accelerates the rate of reaction of acrylonitrile with dipeptides; DMSO also accelerates the rate of reaction of acrylonitrile with polylysine, lysozyme, or bovine serum albumin. The first-order plot (log A_t/A_o) of the extent

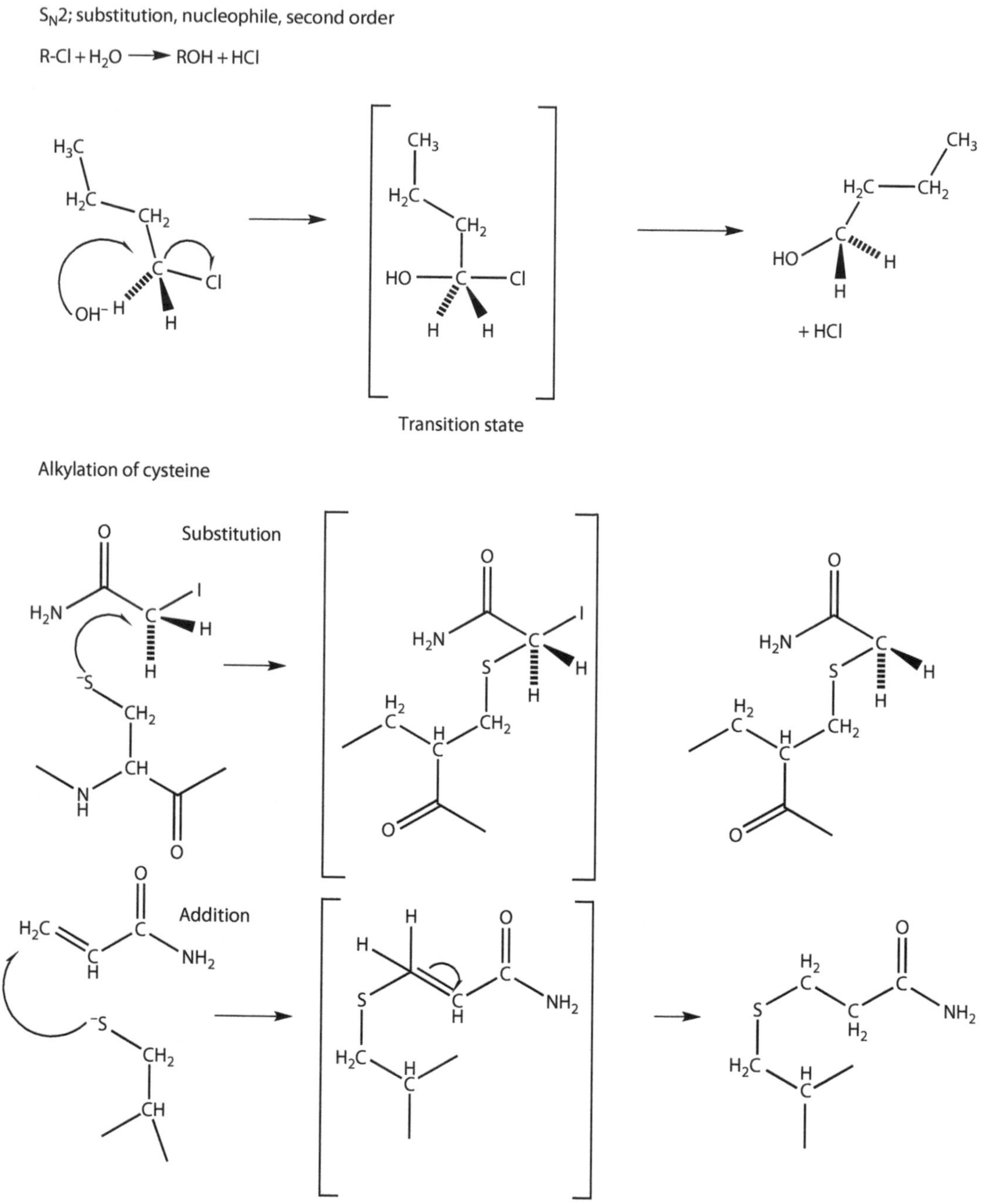

S_N2; substitution, nucleophile, second order

$R\text{-Cl} + H_2O \longrightarrow ROH + HCl$

Transition state

Alkylation of cysteine

FIGURE 1.4 Basic chemical reactions for chemical modification. At the top, shown is a classical nucleophilic substitution reaction, base-catalyzed hydrolysis of an alkyl halide to yield an alcohol showing the formation of a transition state. In the following are shown two S_N2 reactions of cysteine, nucleophilic substitution with iodoacetamide and nucleophilic addition with a double bond Michael addition as with a maleimide. The rate of a nucleophilic substitution is controlled, in part, by the nucleofugality of the leaving group.

of reaction of acrylonitrile with proteins in the presence of DMSO is no longer linear after 3 h of reaction (approximately 75% modification) (pH 8.4). This suggests that 25% of the amino groups are unavailable for reaction (this is consistent with later work as discussed in Chapter 13). Friedman suggests that the rate enhancement observed with DMSO is due to increased nucleophilicity of accessible amino groups in peptides and proteins. It is noted that DMSO has proved useful as solvent in other chemical modification reactions.[125–127] Landini and Mai[128] observed that the second-order rate constant for the S_N2 substitution reaction of n-hexyl methane sulfonate by various inorganic anions does vary with solvent. While there was little difference in the rate constant for iodide in MeOH and DMSO (5.8×10^4 M^{-1} s^{-1} vs. 5.3×10^4 M^{-1} s^{-1}), the value for chloride was 0.9×10 M^{-1} s^{-1} in MeOH and 36×10 M^{-1} s^{-1} in DMSO. Abrams has examined the effect of solvent on nucleophilic substitution and suggests the major effect is on the transition state.[129] Solvent environment/local electrostatic potential also influences functional group pKa and, hence, nucleophilicity, and the effect of local environment on the reactivity of functional groups is discussed in more detail later. Brotzel and Mayr[130] studied the reaction of stabilized diarylcarbenium ions with various amino acids. Two observations are worth noting. These investigators describe a nucleophilicity factor, N, as being the determinant in the reaction of functional groups in peptides and proteins. Thus, while the pKa values of the primary amine groups studied varied by approximately 4 units, reactivity (second-order rate constant) varied only by a factor of 4. These investigators also reported that the thiolate anion was more reactive than amino groups by a factor of 10^4.

Proteins are the most complex of the biological polymers reflecting the diversity of the various monomer units; excluding posttranslational modification, there are 20 naturally occurring amino acids found in proteins (18 L-amino acids, one imino acid and glycine); posttranslational modifications include glycosylation, phosphorylation, methylation, acetylation, hydroxylation, sulfation, and the attachment of C-terminal GPI anchors.[131,132] The individual amino acids vary considerably in nucleophilic character as defined by reactivity. Some such as those with aliphatic side chains such as leucine and isoleucine are essentially unreactive except for free radical insertion. Other amino acids with functional groups can vary considerably in reactivity; for example cysteine is very reactive while serine is much less reactive. Seven of the 20 amino acids (lysine, histidine, arginine, tyrosine, tryptophan, aspartic acid, and glutamic acid) have functional side chains that are subject to facile modification; serine and threonine can be modified with chemical reagents but with more difficulty (unless as with serine residues at enzyme active sites). Five of these residues (usually) carry a charge at physiological pH and modification of three residues, lysine, aspartic acid, and glutamic acid, can change the charge and properties of a protein.[132–136] The modification of arginine can also change protein surface charge but is not pursued as frequently. In addition, modification can be accomplished at the C-terminal carboxyl and amino-terminal amino groups. The acid dissociation constants for *typical* amino acid functional groups are presented in Table 1.1 and deserve a comment.[136–138] In particular, note the difference in the values for the sulfhydryl group of cysteine where the *older* value is 10.46, while the more recent value is 6.8 ± 2.7.[138] The latter value[138] is an average value for a cysteine residue in a protein with a range from 2.5 to 11.1; the value for cysteine in an alanine pentapeptide is 8.6. The ionization of a specific functional group in a protein is influenced by intrinsic pKa of the specific functional group and the effect of the local electrostatic potential.[139–141] The intrinsic pKa is the pKa of the functional group transferred from bulk solution into a protein with no interaction with other functional groups in a protein. In the case of cysteine mentioned earlier, the low pKa value is usually associated with a residue involved in catalytic function.[143] The pKa for the thiol function of selenocysteine is 5.3 compared to 8.6 for cysteine.[144,145]

The physical accessibility of a chemical reagent to a functional group in a protein can depend on the exposure of the functional group to bulk solvent and on the presence of surrounding amino acids that can either attract or repulse a reagent. Examples are provided later on the modification of lysine with, for example, pyridoxal-5-phosphate in Chapter 13; in this case, the presence of a negative charge on the reagent increases the specificity of the reagent. Amino acid analysis permits the differentiation of globular proteins from outer membrane proteins[146] and from connective tissue proteins such as elastin and collagen; elastin and collagen contain disproportionate amounts of protein and glycine as well as unique residues such as hydroxyproline and hydroxylysine.[147,148] It is a bit difficult to make generalizations about the amino composition of proteins, but some amino acids such as histidine, tryptophan, and methionine are usually present at low concentration, while alanine and valine are present at higher concenrations.[146,149] The accessibility of amino acid residues to a solvent is variable[150,151], and efforts are made to use compositional data to predict solution behavior based on residue exposure.[152–156] I would remiss if I did acknowledge the contributions[157,158] of the late Fred Richards at Yale University to the concept of surface and buried residues in proteins. Richards and his colleagues were also responsible for the discovery of the limited proteolysis of bovine pancreatic ribonuclease resulting in the formation of S-protein and S-peptide.[159,160] Arthur Lesk has written an excellent book[161] on protein structure that provides a lucid summary for accessible and buried surface area. Miller and coworkers[162] evaluated the solvent-accessible surface residues in 46 monomer proteins. The majority of exposed surface area is provided by hydrophobic amino acids (58%) with lesser contribution from polar (24%) and charged amino acids (19%); interior residues (buried) are 58% hydrophobic and 39% polar but only 4% charged. There is an asymmetric distribution of accessible residues[163] consistent with the existence of hydrophobic residues at domain interface regions[163,164] and the existence of polar regions such as anion-binding exosites[165] important for regulatory protease function[166] or sequences rich in basic amino acids such as found in histones[167,168] important for binding to nucleic acids.[169] The principle of *exposed* and *buried* amino acid residues as a determinant of chemical reactivity was an intense area of discussion some 40 years ago. The intensity of discussion is not really captured in the literature. The reader is directed to an excellent review[4] by Glazer in 1976 for a perspective on the relations of *exposed*, *buried*, and partially buried (sluggishly reacting).

Organic (apolar solvents) can be used to mimic the hydrophobic environment of proteins. As an example, consider the effect of the addition of an organic solvent, ethyl alcohol, on the pKa of acetic acid. In 100% H_2O, acetic acid has a pKa of 4.70. The addition of 80% ethyl alcohol results in an increase of the pKa to 6.9. In 100% ethyl alcohol, the pKa of acetic acid is 10.3 (Table 1.2). Thus, the

---

TABLE 1.2

Solvent Effects on Apparent pKa Values for Amino Acids and Related Compounds[a]

Functional Group	ΔpKa		
	86% EtOH	65% EtOH	20% Dioxane
CH_3COOH	+2.24	+1.19	+0.37
Alanine-COOH	+1.79	+1.19	+0.23
Alanine-αNH_3^+	+0.13	+0.30	−0.05
Lys-COOH	+1.73	+1.05	+0.10
Lys-αNH_3^+	+0.18	+0.05	−0.15
Lys-εNH_3^+	+0.14	0.00	−0.20
Arg-COOH	+1.12	+1.32	+0.11
Arg-αNH_3^+	−0.01	+0.36	−0.10
Arg-GuanidinoNH$_3^+$	+0.25	+1.52	+0.55

[a] See Frohliger, J.O., Gartska, R.A., Irwin, H.H., and Steward, O.W., Determination of ionization constants of monobasic acids in ethanol-water solvents by direct potentiometry, *Anal. Chem.* 40, 1400–1411, 1963; Frohliger, J.O., Dziedzic, J.E., and Steward, O.W., Simplified spectrophotometric determination of acid dissociation constants, *Anal. Chem.* 42, 1189–1191, 1970.

pKa of a buried aspartyl or glutamyl residue will be increased when compared to a surface residue. In cases of amines, the pKa is lowered with the addition of organic solvent[170] showing the preference for an uncharged species (see Table 1.2). Considering the importance of this information, it is surprising that there are no more studies in this area. Some 70 years ago, Richardson[171] concluded that lowering the dielectric constant decreases the acidity (increases the pKa of carboxylic acids with little effect on the dissociation of protonated amino groups). These observations were confirmed by Duggan and Schmidt.[172] The increase in the pKa of carboxyl groups in organic solvents has a favorable effect on transpeptidation reactions[173,174] where the carboxyl groups are required to be protonated. While it may be a bit of an oversimplification, it is useful to understand that an uncharged group is favored in a hydrophobic environment, so the pKa of an acid is increased while the pKa for dissociation of a conjugate acid such as the ammonium form of the ε-amino group of lysine would be decreased. In support of this concept, a study by García-Moreno and workers[175] demonstrated that substitution mutagenesis replacing Val66 in staphylococcal nuclease (a *buried* residue) with a lysine (V66K) resulted in a residue with a pKa of ≤6.38. This study used the changes in the ionization constant of a *buried* residues from the value in water as a means to estimate the effective dielectric constant. These investigators also provide a listing of residues in other protein with perturbed pKa values. There are a number of other studies on the perturbation of the pKa values of functional groups in proteins.[176–182] Finally, while it is possible to make a generalization that being defined as *buried* can result in a decrease in the pKa for a lysine residue while the pKa for a dicarboxylic acid residue increases, there are exceptions where a *buried* lysine at position 38 in staphylococcal nuclease has a normal or slightly elevated pKa while aspartic acid or glutamic acid at the same position have elevated pKa values (7.0 and 7.2, respectively).[182] This group has further pursued the issue of internal glutamic acid residues[183] and lysine residues[184] in staphylococcal nuclease. Isom and coworkers[150]* used substitution mutagenesis to insert 25 glutamic acid residues at various points in the staphylococcal nuclease amino acid sequence. Two of the glutamic acid residues had a normal pKa (4.5) while the remaining 23 had elevated values ranging from 5.2 to 9.4. In subsequent work,[184] Isom and coworkers used the same technical approach to evaluate inserted lysine residues. Nineteen of the inserted lysine residues had a pKa lower than the normal value of 10.4; the values ranged from 5.3 to 9.3 with six inserted lysine residues having a normal pKa value. Further work by this group[185] showed that arginine residues inserted in staphylococcal protease did not demonstrate a change in pKa. The changes in the pKa values for glutamic acid residues and lysine residues are somewhat unusual in that such perturbations are more frequently associated with amino acid residues involved in catalysis.[176]

Other factors that can influence the pKa of a functional group include hydrogen bonding with an adjacent functional group, the direct electrostatic effect of the presence of a charged group in the immediate vicinity of a potential nucleophile, and direct steric effects on the availability of a given functional group (see aforementioned discussion of differential reaction of carbon atoms in hexoses). There are several useful articles on the effect of hydrogen bonding on functional group reactivity,[186–188] and the reader is directed to an excellent article by Taylor and Kennard for a general discussion of hydrogen-bond geometry[189] and the referenced discussion by Glusker.[190] Oda and coworkers[191] used 2D nuclear magnetic resonance (NMR) technology to measure the pKa values of carboxyl groups in *Escherichia coli* ribonuclease HI. The values for aspartic acid residues ranged from <2 to 6.1, while those for glutamic acid residues ranged from 3.2 to 4.5. The pKa value for Asp10 was 6.1, while that for Asp70 was measured to be 2.6; these are two active-site residues, one of which serves as a proton donor while the other is a proton acceptor and influences the respective pKa values. A similar situation exists for pepsin where the two active-site carboxyl groups have pKa values of 1.57 and 5.02.[192] The perturbation of residues involved in catalysis has been mentioned earlier.[177] Webb and coworkers[193] have measured the pKa values for hen egg white lysozyme and found values as low as 1.2 for an aspartic acid residue and values of 2.6 and 5.5 for two glutamic acid

---

* Reaction rate can also be enhanced by increasing local concentration as with the use of affinity reagents.

residues. Dyson and coworkers[194] evaluated the effect of neighboring residues on the chemical properties of two buried residues in *E. coli* thioredoxin, Asp26 and Lys57. The pKa of Asp26 is raised from 7.5 to 9.4 in a mutant where Lys57 is substituted with a methionine (K57M). There are two cysteine residues in the enzyme active site, C32 and C35 with pKa values of 7.1 and 9.9, respectively. The pKa value of C32 is raised to 8.0 in the K57M mutant as well as in mutant where Asp26 is changed to Ala (D26A) and cannot be distinguished from C35. Andersson and coworkers[195] designed several 42-residue polypeptides that folded into a hairpin structure with a helix-loop-helix motif. The reactivity of lysine and ornithine residues was evaluated using acylation with *p*-nitrophenyl fumarate (pH 8.0). Ornithine at position 34 in the base polypeptide (KA-I) was approximately threefold more reactive than the lysine at 19; lysine residues at positions 10, 15, and 33 were not modified. Replacement of Lys15 with an alanine (K15A) resulted in increased modification at Orn34. These investigators suggest that replacing Lys15 with alanine has the effect of raising the pKa of Lys19 decreasing reactivity. Another example of the effect of a neighboring group on the reaction of a specific amino acid residue is provided by the comparison of the rates of modification of the active-site cysteinyl residue by chloroacetic acid and chloroacetamide in papain.[196,197] A rigorous evaluation of the effect of pH and ionic strength on the reaction of papain with chloroacetic acid and chloroacetamide demonstrated the importance of a neighboring imidazolium group in enhancing the rate of reaction at low pH. Similar results had been reported earlier by Gerwin[198] for the essential cysteine residues in streptococcal proteinase. The essence of the experimental observations is that the plot of the pH dependence of the second-order rate constant for the reaction with chloroacetic acid is bell shaped with an optimum at about pH 6.0, while that of chloroacetamide is S shaped, approaching a maximal rate of reaction at pH 10.0. Gerwin demonstrated that the reaction of chloroacetic acid and chloroacetamide with reduced glutathione did not demonstrate this difference in pH dependence. The decrease in the pKa of an active-site lysine in phosphonoacetaldehyde hydrolase to 9.3,[199] as shown by reaction with 2,4-dinitrophenyl acetate, is suggested to be a result of positively charged environment and/or proximity to the amino terminal of a helix. Remote sites have also been shown to influence the reactivity of histidine residues in ribonuclease A.[200]

However, it is possible to be misled by amino acid sequence alone. Acetoacetate decarboxylase provides an example of a very reasonable explanation for enhanced active-site functional group reactivity based on amino acid sequence being trumped by the crystal structure. The pKa of the lysine residues at the active site of acetoacetate decarboxylase is close to 6 where the normal pKa is between 10 and 11. O'Leary and Westheimer[201] showed that acetoacetate decarboxylase was inactivated by either acetic anhydride or 2,4-dinitrophenyl acetate; the inactivation was associated with the acetylation of 1 mol of lysine per mole of enzyme. This lysine residue was identified as the same lysine residue (Lys115) forming a Schiff base with substrate. Subsequent work by Schmidt and Westheimer[202] on the reaction of acetoacetate decarboxylase with 2,4-dinitrophenyl propionate demonstrated the active lysine residue had a pKa of 5.9. It was suggested that the low pKa was a result of proximity to the Lys116 ammonium group; this suggestion was supported by subsequent work by Highbarger and coworkers.[203] Mutation of Lys116 to arginine (K116R) had reduced enzymatic activity but the pKa of Lys115 was determined to be 5.95, similar to wild type. Mutation of Lys116 to cysteine (K116C) also had reduced enzymatic activity but the pKa of Lys115 was determined to be >9.2; the reaction of K116C with 2-bromoethylammonium bromide resulting in the formation of *S*-2-aminoethylcysteine (Chapter 2)[204–206] resulted in an increase in enzymatic activity and a return of the pKa of Lys115 to 5.9. The development of a crystal structure for acetoacetate decarboxylase[207] showed that, rather than an influence of a neighboring amino group, the low pKa of Lys115 was due to its location in a hydrophobic environment. This suggestion has been supported by subsequent analyses from other research groups.[208,209]

Partitioning between bulk solution and local microenvironment can contribute to chemical reactivity in proteins by concentrating the reagent at the site of reaction. The attractive effect of adjacent residues such as the histidine residues in papain and streptococcal proteinase is an example of this as cited earlier. This partitioning process can cause a *selective* increase (or decrease) in reagent concentration in the vicinity of a potentially reactive species. This can result in the demonstration of saturation kinetics in what should be an uncomplicated second-order reaction. An excellent example is provided by the studies of the reaction of iodoacetate with serine transhydroxymethylase.[210] Schirch and coworkers[210] showed that the holoenzyme did not react with iodoacetate; removal of the cofactor pyridoxal-5'-phosphate forming the apoenzyme which did react with iodoacetate. The enzyme was not modified with iodoacetamide. It was suggested that cationic residue near to the modified cysteine is responsible for the observed kinetics. It should be noted that iodoacetamide was developed as a membrane-permeable alternative for iodoacetate.[5]

Affinity labeling[211] is another example of the partitioning of a reagent between bulk solution and the microenvironment surrounding the modified residue on the protein. Here, the partitioning is driven by the structural relationship of a substrate or inhibitor and the chemical reagent. Thus, tosyl-phenylalanine chloromethyl ketone (TPCK; L-1-tosylamido-2-phenylethyl chloromethyl ketone) was designed on the basis of relationship to chymotrypsin substrates[212], while tosyl-lysyl chloromethyl ketone (TLCK; 1-chloro-3-tosylamido-7-amino-2-heptanone) was designed after trypsin substrates.[213] Affinity labeling has continued to be used with somewhat more complex derivatives.[214,215] Affinity labeling has also moved into a new application referred to as activity-based affinity probes.[216–220]

A final consideration in the role of reagent partitioning in protein modification is the distribution of an organic reagent such as tetranitromethane (TNM) between bulk solution (polar) and the protein microenvironment (nonpolar). This concept is much like my use of a separatory funnel in the organic laboratory at Pacific Lutheran University to extract the desired product from the reaction mixture. The concept is similar to the log P value used to characterize pharmaceutical products.[221] Usually, an organic solvent such as benzene, which is not miscible with water, is used in an aqueous reaction; a reagent that is poorly soluble in water might be expected to move to the interior of a protein that is a relatively nonpolar environment. While this is an attractive concept and there are some reagents that are poorly soluble in water, there are not many actual demonstrations of this effect. It was not unreasonable that TNM provided a useful example of the effect of partitioning on functional group reactivity as it is poorly soluble in water and soluble in organic solvents, and there is considerable information on the use of this compound to modify proteins (see Chapters 4 and Chapter 11). TNM was originally proposed as a reagent to modify exposed tyrosyl residues in proteins.[222,223] Subsequent work suggested that there might be a preference for reaction with buried tyrosyl residues in proteins.[224] Subsequent work on the modification of tyrosine residues in proteins has provided mixed results with respect to whether exposed or buried tyrosine residues are more susceptible to reaction with TNM.[225,226] The discovery of nitric oxide and peroxynitrite changed the conversation of protein nitration from protein structure studies to biological regulation.[227,228] As an aside, the author was pursuing the nitration of thrombin and other coagulation proteases with TNM in the early 1970s and was strongly advised by an NIH advisory committee not to pursue the nitration of tyrosine as there was no future in this derivative. The reaction of TNM with cytochrome c has been an interesting study in the concept of exposed and buried residues. Horse heart cytochrome c has four tyrosine residues, Tyr48, Tyr67, Tyr74, and Tyr97. The crystal structure has been determined[229,230], and Tyr48 and Tyr are considered buried residues with the hydroxyl groups pointing toward the heme group while Try74 and Tyr97 are partially buried/exposed residues.[229] So, from one perspective, Tyr48 and Tyr67 are in a nonpolar environment and should be preferentially modified by TNM; on the other hand, since these residues are buried, Tyr48 and Tyr97 should have an elevated pKa and be less reactive than the surface residues, which would be unprotonated before the buried residues.[231] Now, let us consider the experimental data. First, Skov and coworkers[232] modified horse heart cytochrome c with TNM (fourfold molar excess over tyrosine; 0.05 M Tris, pH 8.4); two of the four tyrosine residues, Tyr48 and Tyr67, were modified consistent with the modification of residues in a nonpolar environment. Sokolovsky and coworkers[233] modified cytochrome c with TNM in 0.1 M Tris-0.1 M KCl, pH 8.0; a 40-fold molar excess resulted in preferential modification of Tyr67 (80%); modification was also found (at a much lesser level) at Tyr48 and Tyr74 (20%).

In more recent work, Batthyány and coworkers[234] reported that Tyr67 and Tyr74 were modified in cytochrome c by TNM (40-fold molar excess, 0.1 M Tris-0.1 M KCl, pH 8.0, 23°C). Species with one modified tyrosine, either Tyr67 or Tyr74, were obtained in a chromatographic analysis; a dini-trated species with modification at Tyr67 and Tyr74 was also obtained. Batthyány and coworkers[234] also observed that peroxynitrite readily modified Tyr97 and Tyr74; under more rigorous conditions, all four tyrosine residues were modified by peroxynitrite with dinitration and trinitration observed. Tyrosine 48 was the least susceptible to modification with peroxynitrite. Casewell and Spiro[235] used ultraviolet (UV) resonance Raman spectroscopy to measure tyrosine nitration by TNM in cyto-chrome c. These investigators showed that two tyrosine residues were modified with a 50-fold molar excess of reagent at pH 8.0 (Tris) at 0°C. These investigators did not identify the modified residues. Finally, Šantrůček and coworkers[236] used matrix-assisted laser desorption/ionization (MALDI)–time-of-flight (TOF) MS to characterize cytochrome c modified with TNM (50 mM NH_4HCO_3, pH 8.0, 23°C). The modification of Try67 (said to have zero solvent exposure) was modified at a molar equivalent amount of TNM; the modification of Try48 was shown with a 5-fold molar excess of TNM, while Tyr97 was nitrated at a 10-fold molar excess of TNM. The modification of Tyr74 was not observed; the iodination of Tyr74 was observed as very low level of reagent (I_s/KI) as was modification of Tyr67. It is possible to conclude that the majority of observations support the hypothesis that TNM would preferentially modify a buried tyrosine residue and would not modify a solvent-exposed tyrosine residue. However, this is not absolute and reactivity may also depend on protein history and whether the reaction mixture was stirred or not stirred; there may also be solvent differences and pH effects; Sokolovsky and coworkers[233] observed that there was an increase in the extent of nitration of cytochrome c by an eightfold molar excess of TNM with stirring. In deference to the legend of James Bond, it is not clear that shaken was an option. The relation of TNM and peroxynitrite as reagents for nitration is discussed in greater detail in Chapter 4. The nitration of cytochrome c by peroxynitrite is considered to have physiological consequences.[237]

The first edition of this work was published in 1984 in collaboration with Dr. Claudia Noyes. At that time, amino acid analyzer, UV-VIS spectrophotometer, gel electrophoresis, and amino acid sequencer were the major tools of the protein chemist. Dr. Noyes managed the sequencer opera-tion in our laboratories at the University of North Carolina at Chapel Hill, which was one of the best with one of her signature accomplishments, the establishment of the site of mutation in factor $IX_{Chapel Hill}$.[238] Today, mutations are identified by nucleic acid sequence analysis, and more sophisti-cated tools have been developed for the study of proteins. The goals and purposes of chemical modi-fication of proteins have also changed. There is a small interest in the use of chemical modification to determine the importance of functional groups. There is a larger interest in the use of chemical modification in more complex systems such as chemical biology[239] and in the manufacture of pro-tein conjugates.[18,240] Thus, while studies on the chemical modification of proteins no longer have the attention of past years, it is hoped to capture the fundamental chemistry for current work. The authors feel that this is important for the use of chemical modification in more complex systems.

The reaction pattern of a given reagent with free amino acids or amino acid derivatives does not necessarily provide the basis for reaction with such amino acid residues in protein. The reader is directed to the information on alkylation in Chapter 2 and on acylation in Chapter 3 for examples of such differences in reactivity. Furthermore, the reaction pattern of a given reagent with one protein cannot necessarily be extrapolated to all proteins. Although somewhat dated, the material com-piled by Shibata some years ago[241] tabulates the differences in reaction of different proteins with various reagents. As we will see in subsequent chapters, a reagent considered specific for a specific functional group can, in fact, react with another functional group. Examples include the reaction of N-ethylmaleimide with protein amino groups,[242] the reaction of diisopropylphosphorofluoridate[243] or phenylmethylsulfonyl fluoride (PMSF) with tyrosine,[244] and the reaction of iodoacetic acid with glutamic acid.[245] The use of MS to characterize chemical modification should eliminate conclu-sions based on assumptions.[246] Indeed, some of the following discussion is irrelevant considering the potential of MS—however, not all workers used MS and MS is not necessarily useful for the characterization of chemical modification in whole cells. Finally, even with major advances in MS, its application to intact proteins is still a challenge[246,247] but significant progress has been made with monoclonal antibodies.[248,249]

The characterization of the chemical modification of a protein, even with MS, can be a challenging proposition. MS does provide the potential advantage of establishing specificity of modification but is not necessarily useful with establishing stoichiometry. MS could establish that modification had occurred at a unique residue.[250–254] The recent work by Anjum and coworkers[254] deserves some comment as it is relevant to the current application of chemical modification in chemical biology. While TPCK was developed as a reagent for the modification of the active-site histidine of α-chymotrypsin (see earlier), it was subsequently demonstrated to be promiscuous and modified cysteine residues in other proteins.[255] Anjum and coworkers[254] showed that the antitumor activity of TPCK was associated with the inhibition of protein kinases. L-1-tosylamido-2-phenylethyl chloromethyl ketone was added to serum-starved HEK293 cells that were then stimulated with either EGF or insulin. Analysis of the TPCK treated cells showed inhibition of several protein kinases (RSK, S6K, Akt, and MSK1) based lack of phosphorylation of their respective substrates. Transfected protein kinase RSK1 was isolated from TPCK-treated HEK293 cells by immunoprecipitation. The purified protein was further purified by sodium dodecyl sulfate polyacrylamide gel electrophoresis (SDS-PAGE), digested in the gel by trypsin followed by mass spectrometric analysis. This procedure demonstrated the modification of Cys241 in RSK1; the analysis of S6K1, AKT, and MSK1 also showed the modification of a cysteine residue. There are several studies that have focused directly on the use of MS to evaluate the chemical modification of proteins. Fligge and coworkers[251] used electrospray MS to study the modification of arginine residues in human angiotensin II with 1,2-cyclohexanedione and the rate of the tryptic digestion of bovine neurotensin. These investigators also used electrospray MS to evaluate the iodination of tyrosine residues in human angiotensin II. Other work has focused on the characterization of protein conjugates.[250,256]

The classical protein chemistry approach first established the extent and specificity of modification. Specificity of modification can frequently be established by the stability of modification. As an example, with acetylation (see Chapter 3), the modification of the ε-amino group of lysine would be stable to mild base but the O-acetylation of tyrosine (or serine/threonine) would be labile. This issue is discussed with each individual reaction in the following chapters. Amino acid analysis was used for the evaluation of the specificity of chemical modification of proteins; today, MS would provide the necessary information (see earlier). The extent of modification can be determined by radioisotope incorporation, by spectroscopy when a chromogenic or fluorogenic reagent is used, and by amino acid analysis; in the case of amino acid analysis, success is frequently dependent on the formation of a unique product such as S-carboxymethylcysteine or carboxymethylhistidine.

The establishment of the specificity of modification is easy when there is one mole residue modified per mole of protein and there is a single site of modification. The issue is a bit more complicated when modification is not stoichiometric, and there are several moles modification per mole of protein. It should be noted that there can be an issue even when there is apparent stoichiometric modification. An early study[257] on the modification of the lysine groups of insulin with acetyl-N-hydroxysuccinimide as a function of pH and reagent concentration provides an excellent example that remains of current value. Lindsay and Shall[257] modified bovine or porcine insulin with N-hydroxysuccinimide acetate at either pH 6.9 or 8.5 using either an equimolar or threefold molar excess of reagent. There are three primary amino groups, N-terminal glycine, N-terminal phenylalanine, and the ε-amino group of B-29 lysine that could react with N-hydroxysuccinimide acetate. The several reaction products could be separated on DEAE-Sephadex A-50. Two monosubstituted derivatives representing modification of the N-terminal amino groups of glycine or phenylalanine were obtained a low molar excess at pH 6.9; raising the reaction pH to 8.5 decreased the amount of the phenylalanine derivative and an acetyllysine derivative is produced. Reaction with a threefold molar excess at pH 8.5 produces two diacetyl derivatives (N-acetylphenylalanine and N-acetylglycine and N^ε-acetyllysine and N-acetylglycine). There was no evidence for a disubstituted derivative with *N*-acetyl phenylalanine and *N*-acetyllysine. Analysis of this data suggested that the order of reactivity is glycine > phenylalanine ≈ lysine at pH 8.5 while the order of reactivity at pH 6.9 is glycine ≈ phenylalanine > lysine. Wallace and coworkers[258] showed that the reaction of 2,4,6-trinitrobenzenesulfonic acid at pH 8.5 resulted in the modification of the *N*-terminal glycine, while earlier studies showed preferential reaction at the ε-amino group of LyrB29 with 2,4-dinitrophenyl sulfonate at pH 11.5 (0.1 M sodium carbonate).[259–261] A better-known example is the inactivation of bovine pancreatic ribonuclease with iodoacetate (pH 5.5), which results in two derivatives, 1-carboxymethylhistidine119-RNAse and 3-carboxylmethylhistidine[12]-RNAse that could be separated from each other and native RNAse by isocratic chromatography on IRC-50 ion exchange resin in 0.2 M phosphate, pH 6.47.[262] The reaction of iodoacetamide (pH 5.5) results in the formation of 3-carboxamidomethylhistidine[12]-RNAse and a fully active *S*-carboxamide methylmethionine derivative.[263] Frequently, however, while there is good evidence that multiple modified species are obtained as a result of the reaction, it is not possible to separate apparently uniquely modified species. In the reaction of TNM with thrombin,[264] apparent stoichiometry of inactivation was obtained with equivalent modification of two separate tyrosine residues (Tyr71 and Tyr85 in the B chain); it was not possible to separate these derivatives.

Establishing the stoichiometry of modification should be a relatively straightforward process. First, the molar quantity of modified residue is established by analysis. This could be spectrophotometry as, for example, with the trinitrophenylation of primary amino groups, the nitration of tyrosine with TNM, or the alkylation of tryptophan with 2-hydroxy-5-nitrobenzyl bromide. Prior to 1990, amino analysis would have been the method to determine the modification of a residue by, for example, photooxidation of histidine and the oxidation of the indole ring of tryptophan with *N*-bromosuccinimide or the appearance of a modified residue such as with *S*-carboxymethylcysteine or N^1- or N^3-carboxymethylhistidine. Advances in technology enabled the use of MS for the identification of chemically modified residues.[265,266] When spectral change or radiolabel incorporation is used to establish stoichiometry, mass spectrometric or amino acid analysis should be performed to determine that there is not a reaction with another amino acid. For example, the extent of oxidation of tryptophan by *N*-bromosuccinimide can be determined by UV spectrometry or fluorescence but amino acid analysis or mass spectrometric analysis is required to determine if modification has also occurred with another amino acid such as histidine or methionine. MS would be the method of choice in 2013 but amino acid analysis could still be useful.[267] Other techniques such as Raman spectroscopy,[268–272] near-infrared spectrophotometry,[273–275] electrophoresis,[276–280] and small-angle x-ray scattering (SAXS)[281–284] are proving useful. Some of these techniques such as near-infrared spectroscopy and Raman spectroscopy will be more useful when applied to more complex systems. However, reaction with 2,4,6-trinitrobenzenesulfonic acid was recently used to measure lysine modification in the chemical crosslinking of collagen to gold nanoclusters.[285]

More frequently, there is the situation where there are several moles of a given residue modified per mole of protein but there is a reason to suspect stoichiometric chemical modification. In some of these situations as discussed earlier, it is possible to fractionate the protein into uniquely modified species. Assessing the stoichiometry of modification from the functional consequences of such modification with any degree of comfort is a far more difficult proposition. First, there must be a clear, unambiguous signal that can be effectively measured. This is not as easy as it sounds and becomes a larger issue with more complex problems. If there is a defined catalytic event to measure, such as the tryptic hydrolysis of tosyl-L-arginine methyl ester, the modification of the active-site histidine or serine results in the loss of activity with a total loss of activity with the modification of one mole of His57 (chymotrypsinogen numbering). Now take thrombin that, like trypsin, hydrolyzes tosyl-L-arginine methyl ester, but also clots fibrinogen. Similar to trypsin, both fibrinogen clotting activity and the esterase activity are inactivated by the modification of the active-site serine or histidine; however, modification with TNM[264] markedly affects fibrinogen clotting activity with little effect on the esterase activity. The situation becomes even more complicated when, for example, with glutamate dehydrogenase, the modification of a tyrosine residue with TNM results in the loss of allosterism.[286,287] The work by Price and Radda[286] also discussed the issue of nonhomogeneity of products from chemical modification. They did have evidence to suggest the nitration of a unique tyrosine residues but recognized the limitations of spectrometric analysis; Piszkiewicz and coworkers[287] did identify Try412 as a residue in glutamate dehydrogenase that was modified by TNM.

In a situation where there are clearly multiple sites of reaction that can be distinguished by analytical techniques, the approach advanced by Ray and Koshland is useful.[288] This analysis is based on establishing a relationship between the rate of loss of biological activity and the rate of modification of a single residue. A similar approach has been advanced by Tsou[289–291] that is based on establishing a relationship between the number of residues modified and the change in biological activity. Horiike and McCormick[292] have explored the approach of relating changes in activity to the extent of chemical modification. These investigators state that the original concepts that form the basis of this approach are sound, but that extrapolation from a plot of activity remaining versus residues modified is not necessarily sound. Such extrapolation is only valid if the *nonessential* residues react much slower (rate at least 10 times slower). Given a situation where all residues within a given group are equally reactive toward the reagent in question, the number of essential residues obtained from such a plot is correct only when the total number of residues is equal to the number of essential residues that is, in turn, equal to 1.0. However, it is important to emphasize that this approach is useful when there is a difference in the rate of reaction of an *essential* residue or residues and all other residues in that class as is the example in the modification of histidyl residues with diethylpyrocarbonate in lactate dehydrogenase[293,294] and pyridoxamine-5′-phosphate oxidase.[295] The reader is also referred to the work of Rakitzis[296] for another discussion of the study of the kinetics of protein modification. There has been continuing citation to this work in the 20 years since the publication of this chapter.[297–302] Rakitzis[296] discusses the issue of protein heterogeneity in terms of multiple conformational states resulting from ligand binding or protein instability including dissociation of oligomers into monomers, protein modification cooperativity where the reactions are in the form of a catenary, reagent instability such as that seen with diethylpyrocarbonate, and saturation as first described by Kitz and Wilson.[303] While saturation kinetics is observed (and expected) with affinity-labeling reagents,[211] it is a bit surprising to observe such kinetics with simple reagents such as cyanate.[304] Rakitzis[296] also discusses the thermodynamics of the chemical modification of protein. There are several other studies on the thermodynamics of the chemical modification of proteins that are worth to consider.[305–308] Lennette and Plapp[305] compared the reaction of bromoacetate with the histidine residues in bovine pancreatic ribonuclease with the reaction of bromoacetate with histidine hydantoin. The rate of reaction of bromoacetate with N-3 of His12 is 120 times more rapid than the corresponding reaction with the N-3 (tele-N) of histidine hydantoin, while the reaction of N-1 (*pros* N) of His119 is 1000 times more rapid than that of the corresponding nitrogen in histidine hydantoin. The pH-independent rate constant for the combined (pseudobiomolecular) reaction with RNAse is some 440 times greater than the reaction of bromoacetate with histidine hydantoin. Lennette and Plapp[305] were able to show that the difference in reaction rate between the enzyme and model compound reflected a decrease in enthalpy rather than an increase in entropy suggesting the importance of increased inherent reactivity of the two histidine residues. While a decrease in enthalpy may not always be a driving factor in the reaction of chemical reagent with a protein functional group,[307] there are other examples such as the reaction of chloroacetamide with ficin as compared to mercaptoethanol where the decrease in enthalpy is striking.[308] Lennette and Plapp[305] also observed that acetate inhibited the reaction of bromoacetate with bovine pancreatic ribonuclease. These investigators suggested that acetate inhibited the carboxymethylation reaction by binding to the enzyme in a similar manner to bromoacetate. This is yet another reminder to beware of the effects of buffer on reactions.[309] In particular, Tris(tris-(hydroxymethyl)aminomethane, has been demonstrated to function as modest nucleophile.[310–313]

The study of reaction rate was of more importance some years ago where the relationship of reaction rate and residue modification was used to define functional groups involved in enzyme catalysis when less specific reagents such as acetic anhydride (Chapter 3) were used to modify a protein. It can be argued that the advent of site-specific mutagenesis provided another approach to the identification of functional groups important in the activity of a protein[314], and this technology is used more frequently than chemical modification. There is no question as to the value of site-specific mutagenesis, but there is still a value for using the rate of chemical modification in determining functional group reactivity in proteins as demonstrated by the identification of exposed residues.[315,316]

A study of variables affecting the reaction of diethylpyrocarbonate[317] with an aminopeptidase[318] isolated from Pronase[319–321] provides an interesting example of the value of chemical modification in defining the role of specific amino acid residues in catalysis. The aminopeptides have a mass of 25–30 kDa and are stimulated by the presence of calcium ions; however, calcium ions do not appear to be an obligate requirement. A careful study of the pH dependence of enzymatic activity (k_{cat} and k_{cat}/K_m) suggested the importance of groups with pKa 6.9 and 9.9. Enzymatic activity was inhibited by the reaction with diethylpyrocarbonate (50 mM phosphate, pH 6.0, 25°C). When enzymatic activity of the modified enzyme was measured in the absence of calcium ions, complete inactivation was associated with the modification of 1 mol of histidine/mole protein; the measurement of activity in the presence of calcium ions showed that complete inactivation was associated with the modification of 2 mol of histidine per mole of protein. Further analysis showed that the loss of enzymatic activity was associated with a decrease in k_{cat} with no change in K_m. At the point where there is total loss of enzymatic activity measured in the absence of calcium ions, 40% of activity is retained when measured in the presence of calcium ions. The study of the rate of reaction of diethylpyrocarbonate with the aminopeptidase in the absence of calcium ions shows the very rapid modification of 1 mol of histidine/mole of protein with the much slower modification of a second mole of histidine/mole of protein; in the presence of calcium ions, only the rapid modification is observed. The measurement of decarbethoxylation (A_{240}) in the presence of hydroxylamine (see Chapter 14) showed restoration of 1 mol of histidine/mole protein after 4 h of incubation; approximately 60% of activity was recovered when measured in the presence of calcium ions with no activity recovered when measured in the absence of calcium ions. It is suggested that the two histidine residues are involved in catalysis with one histidine binding calcium ions quite tightly. I was unable to find further work on this particular enzyme but Wilk and coworkers[222] reported that several histidyl residues were important for both structural stability and catalytic function in aspartyl aminopeptidase.[322] Lim and Turner[323] showed the modification of two histidine residues in porcine kidney aminopeptidase P with diethylpyrocarbonate with the total loss of activity.

This study explores the problem of competing nucleophiles at an enzyme active site. In the case of the aminopeptidase discussed earlier, the nucleophiles were two histidine residues. The modification of the two histidine residues in pancreatic ribonuclease is another situation where there are two *identical* nucleophiles.[262] In both of these situations, local environmental factors influenced residue reactivity. However, there are other situations where a highly reactive nucleophile, such as a cysteine, precludes modification of a less reactive nucleophile such as histidine. This was the situation faced by Teh-Yung (Darrell) Liu working on streptococcal proteinase at the Rockefeller University in New York. There was a cysteine residue at the active site of streptococcal proteinase, and in homology with the serine proteinases, it was reasonable to assume that there was also a histidine residue. While it is possible to modify the histidine at the active site of serine proteinase as the imidazolium ring is a more powerful nucleophile than the serine hydroxyl group, this was not a possibility with the histidine residue in streptococcal proteinase. Liu used sodium tetrathionate (Chapter 7) to reversibly block the sulfhydryl group at the active site of streptococcal proteinase permitting the modification of a histidine residue with α-N-bromoacetylarginine methyl ester.[324] The use of a positively charged reagent took advantage of the negative charge introduced with the S-sulfenylsulfonate derivative at the enzyme active site; the reaction did not occur with either iodoacetate or iodoacetamide. Murachi and coworkers[325] used tetrathionate modification of a cysteine in stem bromelain to prevent photosensitized oxidation permitting the identification of histidine residue as a primary target. In work on streptococcal proteinase discussed earlier, Gerwin[198] showed that there was a marked difference in the pH dependence for the reaction of chloroacetate and chloroacetamide with the active-site cysteine residue. Chloroacetate showed a bell-shaped pH dependence curve while chloroacetamide had a sigmoidal pH dependence. It was suggested that a histidine residue was responsible for the observed pH dependence of chloroacetate. These studies showed that, as the presence of the free thiol precludes the modification of the active-site histidine in streptococcal proteinase, the presence of a histidine residue influences the alkylation of the active-site cysteine.

The majority of the studies on the chemical modification of proteins have focused on enzymes. The functional characterization of chemical modification is relatively straightforward with the use of a peptide chloromethyl ketone that is assumed to occur at the active-site histidine in a serine protease.[326] However, even these specific reagents can be promiscuous in functional effect.[215,254,255,327]

The characterization of the modified protein can provide a significant challenge even with stoichiometric modification. A more difficult problem is encountered with a modified protein with fractional activity.[264,328–330] The most critical aspect in the characterization of the modified protein is the method used to determine activity. The rigorous determination of binding constants and kinetic constants is absolutely essential; the reporting of percent change in activity is clearly inadequate. The reader is directed to several classic works in this area[331,332] as well as more recent expositions in this area.[333–337] For the reader who, like the author, is somewhat challenged by physically biochemistry, the consideration of some more basic information[338–340] will be useful. Finally, the reader is directed to an excellent review by Plapp.[314] While this discussion is directed toward the use of site-specific mutagenesis for the study of enzymes, much of the content is equally applicable to the characterization of chemically modified proteins. Particular consideration should be given to the section on kinetics with emphasis on the importance of V/K (catalytic efficiency) for the evaluation of the effect of a modification on catalytic activity and the discussion on the importance of understanding that K_m is not necessarily a measure of affinity. Evaluation contribution of individual residues to the overall catalytic process is also discussed. The reader is referred to other review articles for consideration of this latter issue.[341,342] This type of analysis would markedly increase the value of studies where several different reagents are used for the chemical modification of a protein.

The characterization of a partially modified biological polymer is also of importance in the characterization of biopharmaceuticals. Frequently, there is some activity lost in the transition from active pharmaceutical ingredient to final drug product and during the storage of the drug product over a period of time. It is critical to understand the change in activity and the nature of the chemical modification. For example, does the loss of 10% of the activity mean the loss of 10% of product or is there 100% of product with 90% activity. The careful use of enzyme kinetics and binding assays can resolve these issues. The issue of the nature of the chemical modification can be assessed by MS. Most biopharmaceuticals contain a protein or protein derivative as the active pharmaceutical ingredient (API) and there is interest in the deamidation of asparagine/formation of isoaspartic acid and methionine oxidation[343–345] although a modification such as glycation is also a possibility.[346] The study of the rate of chemical modification is of substantial value in establishing the stability of biopharmaceutical products.[347,348]

Chemical modification can also be used to assess conformational change in proteins.[349,350] Surface mapping with free radicals or short-lived species such as acetic anhydride has proved useful. The reader is referred to the individual chapters and a recent review.[350]

Chemical modification can also be used to enhance the analytical capabilities of MS. Examples of this include carbamylation of lysine to improve the accuracy of molecular weight determination by SDS-PAGE.[351] Oxidation[352] or guanidation[353] of peptides has been used to improve analysis by MS. O-Methylisourea has been used to convert lysine to homoarginine,[354,355] which improves mass spectrometric analysis of peptides.[356,357] While not absolute, evidence suggests that guanidation occurs preferentially at the ε-amino groups of lysine,[357] making it possible to modify the free amino-terminal amino acid in a peptide or protein with ITRAQ (isobaric tag for relative and absolute quantitation) reagents prior to tryptic digestion permitting the identification of new N-terminal amino acids.[358]

Dolnik and Gurske[351] showed that carbamylation neutralized the charge on lysine residues permitting more effective binding of a cationic detergent, cetyltrimethylammonium bromide. As an aside, mutation in amino acid sequence can cause anomalous behavior on SDS electrophoresis.[280,359,360]

Performic acid oxidation has been used to expand the mass distribution of tryptic peptides.[361] Expanding the mass distribution should provide more peptides with unique mass characteristics for mass spectrometric analysis. Other methods of oxidation such as electrochemical oxidation have been proposed.[362,363] Roeser and coworkers[364] also suggest that oxidation products can be used for the attachment of probes.

Chemical modification for characterization of function of purified proteins is not as popular as it was some 30 years ago. However, the technology built over the past 60 years is of considerable value in the emerging disciplines of proteomics, chemical biology, structural biology, and chemical proteomics.

REFERENCES

1. Putnam, F., The chemical modification of proteins, in *The Proteins*, Vol. 1, Part B, eds. H. Neurath and K. Bailey, pp. 893–972, Academic Press, New York, 1953.
2. Glazer, A.N., Chemical modification of proteins, *Ann. Rev. Biochem.* 39, 101–130, 1970.
3. Cohen, L.A., Chemical modification of proteins as a probe of structure and function, in *The Enzymes*, 3rd edn., Vol. 1, ed. P.D. Boyer, pp. 147–211, Academic Press, New York, 1970.
4. Glazer, A.N., The chemical modification of proteins by group-specific and site-specific reagents, in *The Proteins*, 3rd edn., Vol. II, eds. H. Neurath and R.L. Hill, Chapter 1, pp. 2–103, Academic Press, New York, 1976.
5. Webb, J.L., *Enzyme and Metabolic Inhibitors*, Academic Press, New York, 1963.
6. Zadran, S. and Levine, R.D., Perspectives in metabolic engineering: Understanding cellular regulation toward the control of metabolic routes, *Appl. Biochem. Biotechnol.* 169, 55–65, 2013.
7. Mudd, S. and Joffe, E.W., The modification of antibodies by formaldehyde, *J. Gen. Physiol.* 16, 947–960, 1933.
8. Kekwick, R.A. and Cannan, R.K., The effect of formaldehyde on the hydrogen ion dissociation curve of egg albumin, *Biochem. J.* 30, 235–240, 1936.
9. Liener, I.E. and Wada, S., Chemical modification of the soy bean hemagglutinin, *J. Biol. Chem.* 222, 695–704, 1950.
10. Kosower, E.M., The therapeutic possibilities arising from the chemical modification of proteins, *Proc. Natl. Acad. Sci. USA* 53, 897–501, 1965.
11. Brother, G.H. and McKinney, L.L., Protein plastics from soy bean products. Action of handening or tanning agents on protein material, *J. Ind. Eng. Chem.* 30, 1236–1240, 1938.
12. Pinner, S.H., The cross linking of protein plastics, *Br. Plastics Moulded Products Trader* 23, 157–162, 1950.
13. Coffman, J.R., Chemical modification of proteins, US Patent 2562534, 1951.
14. Means, G.R. and Feeney, R.F., *Chemical Modification of Proteins*, Holden-Day, San Francisco, CA, 1971.
15. Lundblad, R.L. and Noyes, C.M., *Chemical Reagents for the Modification of Proteins*, CRC Press, Boca Raton, FL, 1984.
16. Jollès, P. and Jörnvall, H. (eds.), *Proteomics in Functional Genomics: Protein Structure Analysis*, Birkhäuser, Basel, Switzerland, 2000.
17. Shuford, C.M. and Muddiman, D.C., Capitalizing on the hydrophobic bias of electrospray ionization through chemical modification in mass spectrometry-based proteomics, *Expert Rev. Proteomics* 8, 317–323, 2011.
18. Stephanopoulos, N. and Francis, M.B., Choosing an effective protein bioconjugation strategy, *Nat. Chem. Biol.* 7, 876–884, 2011.
19. Hirs, C.H.W. (ed.), *Methods in Enzymology*, Vol. 11, Academic Press, New York, 1967.
20. Hirs, C.H.W. and Timasheff, S.N. (eds.), *Methods in Enzymology*, Vol. 25, Academic Press, New York, 1972.

21. Hirs, C.H.W. and Timasheff, S.N. (eds.), *Methods in Enzymology*, Vol. 47, Academic Press, New York, 1977.
22. Bradshaw, R.A. and Burlingame, A.L., From proteins to proteomics, *IUBMB Life* 57, 267–272, 2005.
23. Lundblad, R.L., *The Evolution from Protein Chemistry to Proteomics*, CRC Press, Boca Raton, FL, 2006.
24. Bergeron, J.J. and Bradshaw, R.A., What has proteomics accomplished? *Mol. Cell. Proteomics* 6, 1824–1826, 2007.
25. Gygi, S.P., Rist, B., Gerber, S.A. et al., Quantitative analysis of complex protein mixtures using isotope-coded affinity tags, *Nat. Biotechnol.* 17, 994–999, 1999.
26. Griffin, T.J., Goodless, D.R., and Aebersold, R., Advances in proteome analysis by mass spectrometry, *Curr. Opin. Biotechnol.* 12, 607–612, 2001.
27. Tao, W.A. and Aebersold, R., Advances in quantitative proteomics via stable isotope tagging and mass spectrometry, *Curr. Opin. Biotechnol.* 14, 110–118, 2003.
28. Smolka, M., Zhou, H., and Aebersold, R., Quantitative protein profiling using two-dimensional gel electrophoresis, isotope-coded affinity tag labeling, and mass spectrometry, *Mol. Cell. Proteomics* 1, 19–29, 2002.
29. Smolka, M.B., Zhou, H., Purkayastha, S., and Aebersold, R., Optimization of the isotope-coded affinity tag-labeling procedure for quantitative proteome analysis, *Anal. Biochem.* 297, 25–31, 2001.
30. Zhang, R., Scioma, C.S., Wang, S., and Regnier, F.E., Fractionation of isotopically labeled peptides in quantitative proteomics, *Anal. Chem.* 73, 5142–5149, 2001.
31. Hansen, K.C., Schmitt-Ulms, G., Chalkley, R.J. et al., Mass spectrometric analysis of protein mixtures at low levels using cleavable ^{13}C-isotope-coded affinity tag and multidimensional chromatography, *Mol. Cell. Proteomics* 2, 229–314, 2003.
32. Goshe, M.B., Conrads, T.P., Panisko, E.A. et al., Phosphoprotein isotope-coded affinity tag approach for isolating and quantitating phosphopeptides in proteome-wide analyses, *Anal. Chem.* 73, 2578–2586, 2001.
33. Kuyama, H., Watanabe, M., Toda, C. et al., An approach to quantitative proteome analysis by labeling tryptophan residues, *Rapid Commun. Mass Spectrom.* 17, 1642–1650, 2003.
34. Goshe, M.B. and Smith, R.D., Stable isotope-coded proteomic mass spectrometry, *Curr. Opin. Biotechnol.* 14, 101–109, 2003.
35. Qiu, Y., Sousa, E.A., Hewick, R.M., and Wang, J.H., Acid-labile isotope-coded extractants: A class of reagents for quantitative mass spectrometric analysis of complex protein mixtures, *Anal. Chem.* 74, 4969–4979, 2002.
36. Ong, S.-E., Blagoev, B., Kratchmarova, I. et al., Stable isotope labeling by amino acids in cell culture, SILAC, as a simple and accurate approach to expression proteomics, *Mol. Cell. Proteomics* 1, 376–386, 2002.
37. Michels, D.A., Hu, S., Schoenherr, R.M. et al., Fully automated two-dimensional capillary electrophoresis for high sensitivity protein analysis, *Mol. Cell. Proteomics* 1, 69–74, 2002.
38. Kwon, S.W., Kim, S.C., Jaunbergs, J. et al., Selective enrichment of thiophosphorylated polypeptides as a tool for the analysis of protein phosphorylation, *Mol. Cell. Proteomics* 2, 242–247, 2003.
39. Gevaert, K., Van Damme, J., Goethals, M. et al., Chromatographic isolation of methionine-containing peptides for gel-free proteome analysis *Mol. Cell. Proteomics* 1, 896–903, 2002.
40. Spande, T.F., Witkop, B., Degani, Y., and Patchornik, A., Selective cleavage and modification of peptides and proteins. *Adv. Protein Chem.* 24, 97–260, 1970.
41. Brown, J.R. and Hartley, B.S., Location of disulphide bridges by diagonal paper electrophoresis. The disulphide bridges of bovine chymotrypsinogen A, *Biochem. J.* 101, 214–228, 1966.
42. Tang, J. and Hartley, B.S., A diagonal electrophoretic method for selective purification of methionine peptides, *Biochem. J.* 102, 593–599, 1967.
43. Nikov, G., Bhat, V., Wishnok, J.S., and Tannenbaum, S.R., Analysis of nitrated proteins by nitrotyrosine-specific affinity probes and mass spectrometry, *Anal. Biochem.* 320, 214–222, 2003.
44. Sokolovsky, M., Riordan, J.F., and Vallee, B.L., Conversion of 3-nitrotyrosine to 3-aminotyrosine in peptides and proteins, *Biochem. Biophys. Res. Commun.* 27, 20–25, 1967.
45. Scherrer, P. and Stoeckenius, W., Selective nitration of tyrosines-26 and -64 in bacteriorhodopsin with tetranitromethane, *Biochemistry* 23, 6195–6202, 1984.
46. Farrens, D.L., What site-directed labeling studies tell us about the mechanism of rhodopsin activation and G-protein binding, *Photochem. Photobiol. Sci.* 9, 1466–1474, 2010.
47. Pan, Y., Ruan, X., Valvano, M.A., and Konerman, L., Validation of membrane protein topology models by oxidative labeling and mass spectrometry, *J. Am. Soc. Mass Spectrom.* 23, 889–898, 2012.

48. Moen, R.J., Thomas, D.D., and Klein, J.C., Conformationally trapping the actin-binding cleft of myosin with a bifunctional spin label, *J. Biol. Chem.* 288, 3016–3024, 2013.

49. Begley, T.P., Chemical biology: An educational challenge for chemistry departments, *Nat. Chem. Biol.* 1, 236–238, 2005.

50. Morrison, K.J. and Weiss, G.A., The origins of chemical biology, *Nat. Chem. Biol.* 2, 3–6, 2006.

51. Florkin, M. (trans. T. Wood), *Unity and Diversity in Biochemistry: An Introduction to Chemical Biology*, Pergamon Press, Oxford, U.K., 1960.

52. Takaoka, Y., Ojida, A., and Hamachi, I., Protein organic chemistry and applications for labeling and engineering in live-cell systems, *Angew. Chem. Int. Ed. Engl.* 52, 4088–4106, 2013.

53. Li, X., Foley, E.A., Molloy, K.R. et al., Quantitative chemical proteomics approach to identify post-translational modification-mediated protein-protein interactions, *J. Am. Chem. Soc.* 134, 1982–1985, 2012.

54. Bantascheff, M. and Drewes, G., Chemoproteomic approaches to drug target identification and drug profiling, *Bioorg. Med. Chem.* 20, 1973–1978, 2012.

55. Smith, R.A., Dewdney, J.M., Fears, R., and Poste, G., Chemical derivatization of therapeutic proteins, *Trends Biotechnol.* 11, 397–403, 1993.

56. Pozansky, M.J., Soluble enzyme-albumin conjugates: New possibilities for enzyme replacement therapy, *Methods Enzymol.* 137, 566–574, 1988.

57. Sharifi, J., Khawli, L.A., Hornick, J.L., and Epstein, A.L., Improving monoclonal antibody pharmacokinetics via chemical modification, *Q. J. Nucl. Med.* 42, 242–249, 1998.

58. Awwad, M., Strome, P.G., Gilman, S.C., and Axelrod, H.R., Modification of monoclonal antibody carbohydrates by oxidation, conjugation, or deoxymannojirimycin does not interfere with antibody effector functions, *Cancer Immunol. Immunother.* 38, 23–30, 1994.

59. Brader, M.L., Sukumar, M., Pekar, A.H. et al., Hybrid insulin cocrystals for controlled release delivery, *Nat. Biotechnol.* 20, 800–804, 2002.

60. Lundblad, R.L. and Bradshaw, R.A., Application of site-specific chemical modification in the manufacture of biopharmaceuticals: I. An overview, *Biotechnol. Appl. Biochem.* 26, 143–151, 1997.

61. Kozlowski, A. and Harris, J.M., Improvements in. protein PEGylation: Pegylated interferons for treatment of hepatitis C, *J. Control. Release* 72, 217–224, 2001.

62. Harrington, K.J., Mubasher, M., and Peters, A.M., Polyethylene glycol in the design of tumor-targeting radiolabelled macromolecules—Lessons from liposomes and monoclonal antibodies, *Q. J. Nucl. Med.* 46, 171–180, 2002.

63. Chapman, A.P., PEGylated antibodies and antibody fragments for improved therapy: A review, *Adv. Drug Deliv. Rev.* 54, 531–545, 2002.

64. Pasut, G. and Veronese, F.M., PEG conjugates in clinical development or use as anticancer agents: An overview, *Adv. Drug Deliv. Rev.* 61, 1177–1188, 2009.

65. Harris, J.M. and Chess, R.B., Effect of pegylation on pharmaceuticals, *Nat. Rev. Drug Discov.* 2, 214–221, 2003.

66. Lipovsek, D., Adnectins: Engineered target-binding protein therapeutics, *Protein Eng. Des. Sel.* 24, 3–9, 2011.

67. Wylie, D.C., Voloch, M., Lee, S. et al., Carboxyalkylated histidine is a pH-dependent product of pegylation with SC-PEG, *Pharm. Res.* 18, 1354–1360, 2001.

68. Tsutsumi, Y., Onda, M., Nagata, S. et al., Site-specific chemical modification with polyethylene glycol of recombinant immunotoxin anti-Tac(Fv)-PE38 (LMB-2) improves antitumor activity and reduces animal toxicity and Immunogenicity, *Proc. Natl. Acad. Sci. USA* 97, 8548–8553, 2000.

69. Kerwin, B.A., Chang, B.S., Gegg, C.V. et al., Interactions between PEG and type I soluble tumor necrosis factor receptor: Modulation by pH and by PEGylation at the N terminus, *Protein Sci.* 11, 1825–1833, 2002.

70. Lee, H., Jang, I.H., Ryu, S.H., and Park, T.G., N-Terminal site-specific mono-PEGylation of epidermal growth factor, *Pharm. Res.* 20, 818–825, 2003.

71. Greenwald, R.B., Yang, K., Zhao, H. et al., Controlled release of proteins from their poly(ethylene glycol) conjugates: Drug delivery systems employing 1,6-elimination, *Bioconjug. Chem.* 14, 395–403, 2003.

72. Ryan, S.M., Mantovani, G., Wang, X. et al., Advances in PEGylation of important biotech molecules: Delivery aspects, *Expert Opin. Drug Deliv.* 5, 371–383, 2008.

73. Delaittre, G., Justribó-Hernández, G., Nolte, R.J., and Cornelissen, J.J., Amine-reactive PEGylated nanoparticles for potential bioconjugation, *Macramol. Rapid Commun.* 32, 19–24, 2011.

74. Garay, R.P., El-Gewely, R., Armstrong, J.K. et al., Antibodies against polyethylene glycol in healthy subjects and in patients treated with PEG-conjugated agents, *Expert Opin. Drug Deliv.* 9, 1319–1323, 2012.

127. Wehr, N.B. and Levine, R.L., Quantitation of protein carbamylation by dot blot, *Anal. Biochem.* 423, 241–245, 2012.
128. Landini, D. and Maia, A., Anion nucleophilicity in ionic liquids: A comparison with traditional molecular solvents of different polarity, *Tetrahedron Lett.* 46, 3961–3963, 2005.
129. Abraham, M.H., Solvent effect on reaction rate, *Pure Appl. Chem.* 57, 1055–1064, 1985.
130. Brotzel, F. and Mayr, H., Nucleophilicities of amino acid and peptides, *Org. Biomol. Chem.* 5, 3814–3820, 2007.
131. Walsh, C.T., Garneau-Tsodikova, S., and Gatto, G.J., Jr. Protein posttranslational modifications: The chemistry of proteome diversifications, *Angew. Chem. Int. Ed.* 44, 7342–7372, 2005.
132. Patthy, L., *Protein Evolution*, 2nd edn., Blackwell, Oxford, U.K., 2008.
133. Cassini, G., Illy, S., and Pileni, M.P., Chemically modified proteins solubilized in AOT reverse micelles. Influence of protein charges on intermicellar interactions, *Chem. Phys. Lett.* 221, 205–212, 1994.
134. Franco, T.T., Andrews, A.T., and Asenjo, J.A., Conservative chemical modification of proteins to study the effects of a single protein property on partitioning in aqueous two-phase systems, *Biotechnol. Bioeng.* 49, 290–299, 1996.
135. Zschörnig, O., Paasche, G., Thieme, C. et al., Modulation of lysozyme charge influences interaction with phospholipid vesicles, *Colloids. Surf. B Biointerfaces* 42, 69–73, 2005.
136. Thurkill, R.L., Grimsley, G.R., Scholtz, J.M., and Pace, C.N., pK values of the ionizable groups of proteins, *Protein Sci.* 15, 1214–1218, 2006.
137. Pace, C.N., Grimsley, G.R., and Scholtz, J.M., Protein ionizable groups: pK values and their contribution to protein stability and solubility, *J. Biol. Chem.* 284, 13285–13289, 2009.
138. Grimsley, G.R., Scholtz, J.M., and Pace, C.N., A summary of the measured pK values of the ionizable groups in folded proteins, *Protein Sci.* 28, 247–251, 2009.
139. Drummond, C.J., Grieser, F., and Healy, T.W., Acid-base equilibria in aqueous micellar solutions, *J. Chem. Soc. Faraday Trans. I* 85, 521–535, 1989.
140. Allewell, N.M., Oberoi, H., Hariharan, M., and LiCata, V.J., Electrostatic effects of in proteins: Experimental and computational approaches, in *Proteins*, Vol. 2, ed. G. Allen, JAI Press, London, U.K., 1997.
141. Juffer, A.H., Argos, P., and Vogel, H.J., Calculating acid-dissociation constants of proteins using the boundary element method, *J. Phys. Chem. B* 101, 7664–7673, 1997.
142. Davies, M.N., Toseland, C.P., Moss, D.P., and Flower, D.R., Benchmarking pKa prediction, *BMC Biochem.* 7, 18, 2006.
143. Nelson, N.J., Day, A.E., Zeng, B.-B. et al., Isotope-coded, iodoacetamide-based reagent to determine individual cysteine pK_a values by matrix-assisted laser desorption/ionization time-of-flight mass spectrometry, *Anal. Biochem.* 375, 187–195, 2008.
144. Huber, R.E. and Criddle, R.S., Comparison of the chemical properties of selenocysteine and selenocystine with their sulfur analogs. *Arch. Biochem. Biophys.* 122, 164–173, 1967.
145. Weissjohann, L.A., Schneider, A., Abbas, M. et al., Selenium in chemistry and biochemistry in comparison to sulfur, *Biol. Chem.* 388, 997–1006, 2007.
146. Gromiha, M.M. and Suwa, M., A simple statistical method for discriminating outer membrane proteins with better accuracy, *Bioinformatics* 21, 961–968, 1995.
147. Piez, K.A. and Gross, J., The amino acid composition and morphology of some invertebrate and vertebrate collagens, *Biochim. Biophys. Acta* 34, 24–39, 1959.
148. Maestro, M.M., Turnay, J., Olmo, N. et al., Biochemical and mechanical behavior of ostrich pericardium as a new biomaterial, *Acta Biomater.* 2, 213–219, 2006.
149. Nunn, B.L. and Keil, R.G., Size distribution and amino acid chemistry of base-extractable proteins from Washington coast sediment, *Biogeochemistry* 75, 177–200, 2005.
150. Gromiha, M.M. and Ahmad, S., Role of solvent accessibility in structure based drug design, *Curr. Comput. Aided Drug Des.* 1, 223–235, 2005.
151. Bernadó, P., Blackledge, M., and Sancho, J., Sequence-specific solvent accessibilities of protein residues in unfolded protein ensembles, *Biophys. J.* 91, 4536–4543, 2006.
152. Ponnuswamy, P.K., Muthasamy, R., and Manavalan, P., Amino acid composition and thermal stability of proteins, *Int. J. Biol. Macromol.* 4, 186–190, 1982.
153. Gekko, K. and Hasegawa, Y., Compressibility-structure relationship of globular proteins, *Biochemistry* 25, 6563–6571, 1986.
154. Gromiha, M.M., Oobatake, M., and Sarai, A., Important amino acid properties for enhanced thermostability from mesophilic to thermophilic proteins, *Biophys. Chem.* 82, 51–67, 1999.
155. Wang, C.-H. and Damodaran, S., Thermal gelation of globular proteins: Weight-average molecular weight dependence of gel strength, *J. Agric. Food Chem.* 38, 1157–1164, 1990.

156. Lienqueo, M.E., Mahn, A., Navarro, G. et al., New approaches for predicting protein retention time in hydrophobic interaction chromatography, *J. Mol. Recognit.* 19, 260–269, 2006.

157. Richards, R.M. and Richmond, T., Solvents, interfaces and protein structure, *Ciba Found. Symp.* (60), 23–45, 1977.

158. Richards, F.M., Areas, volumes, packing and protein structure, *Annu. Rev. Biophys. Bioeng.* 6, 151–176, 1977.

159. Kalman, S.M., Linderstrøm-Lang, K., Ottesen, M., and Richards, F.M., Degradation of ribonuclease by subtilisin, *Biochim. Biophys. Acta* 16, 297–299, 1955.

160. Richards, F.M., On the enzymic activity of subtilisin-modified ribonuclease, *Proc. Natl. Acad. Sci. USA* 44, 162–166, 1958.

161. Lesk, A.M., *Introduction to Protein Architecture*, Oxford University Press, Oxford, U.K., 2004.

162. Miller, S., Janin, J., Lesk, A.M., and Clothia C., Interior and surface of monomeric proteins, *J. Mol. Biol.* 196, 641–656, 1987.

163. Argos, P., An investigation of protein and domain interfaces, *Protein Eng.* 2, 101–113, 1988.

164. Keskin, O., Ma, B., and Nussinov, R., Hot regions in protein-protein interactions: The organization and contribution of structurally conserved hot spot residues, *J. Mol. Biol.* 345, 1281–1294, 2005.

165. Sheehan, J.P. and Sadler, J.E., Molecular mapping of the heparin-binding exosite of thrombin, *Proc. Natl. Acad. Sci. USA* 91, 5518–5522, 1994.

166. Wang, M., Zajicek, J., Geiger, J.H. et al., Solution structure of the complex of VEK-30 and plasminogen kringle 2, *J. Struct. Biol.* 169, 349–359, 2010.

167. Ohe, Y., Hayashi, H., and Iwai, K., Human spleen histone H2B. Isolation and amino acid sequence, *J. Biochem.* 85, 615–624, 1979.

168. Rill, R.L. and Oosterhof, D.K., *Staphylococcus aureus* protease. A probe of exposed nonbasic histone sequences in nucleosomes, *J. Biol. Chem.* 256, 12687–12691, 1981.

169. Clark, R.J. and Felsenfeld, G., Association of arginine-rich histones with G-C-rich regions of DNA in chromatin, *Nat. New Biol.* 240, 226–229, 1972.

170. Crowhurst, L., Llewellyn Lancaster, N., Pérez Arlandis, J.M., and Welton, T., Manipulating solute nucleophilicity with room temperature ionic liquids, *J. Am. Chem. Soc.* 126, 11549–11555, 2004.

171. Richardson, G.M., The principle of formaldehyde, alcohol, and acetone titrations. With a discussion of the proof and implication of the zwitterionic conception, *Proc. R. Soc. B (London)* 115, 121–141, 1934.

172. Duggan, E.L. and Schmidt, C.L.A., The dissociation of certain amino acids in dioxane-water mixtures, *Arch. Biochem.* 1, 453–471, 1943.

173. Canova-Davis, E. and Carpenter, F.H., Semisynthesis of insulin: Specific activation of arginine carboxyl group of the B chain of desoctapeptide—(B23–36)-insulin (Bovine), *Biochemistry* 20, 7053–7058, 1981.

174. Canova-Davis, E., Kessler, T.J., and Ling, V.T., Transpeptidation during the analytical proteolysis of proteins, *Anal. Biochem.* 196, 39–45, 1991.

175. García-Moreno, B., Dwyer, J.J., GIttis, A.G. et al., Experimental measurement of the effective dielectric in the hydrophobic core of a protein, *Biophys. Chem.* 64, 211–224, 1997.

176. Harris, T.K. and Turner, G.J., Structural basis of perturbed pKa values of catalytic groups in enzyme active sites, *IUBMB Life* 53, 85–98, 2002.

177. Mehler, E.L., Fuxreiter, M., Simon, L., and Garcia-Moreno, E.B., The role of hydrophobic microenvironments in modulating pK(a) shifts in proteins, *Proteins Struct. Funct. Genet.* 48, 293–292, 2002.

178. Edgecomb, S.P. and Murphy, K.P., Variability of the pKa of histidine side-chains correlates with burial within proteins, *Proteins Struct. Funct. Genet.* 49, 1–6, 2002.

179. Kim, J., Mao, J., and Gunner, M.R., Are acidic and basic groups in buried proteins predicted to be ionized, *J. Mol. Biol.* 348, 1283–1298, 2005.

180. Thurkill, R.L., Grimsley, G.R., Scholtz, M., and Pace, C.N., Hydrogen bonding markedly reduces the pK of buried carboxyl groups in proteins, *J. Mol. Biol.* 362, 594–604, 2006.

181. Castanada, C.A., Fitch, C.A., Majumdar, A. et al., Molecular determinants of the pK(a) values of Asp and Glu residues in staphylococcal nuclease, *Proteins Struct. Funct. Bioinf.* 77, 570–588, 2009.

182. Harms, M.M., Castanada, C.A., Schlessman, J.L. et al., The pK(a) values of acidic and basic residues buried at the same internal location in a proteins are governed by different factors, *J. Mol. Biol.* 389, 34–47, 2009.

183. Isom, D.G., Castanada, C.A., Velu, P.D. et al., Charges in the hydrophobic interior of proteins, *Proc. Natl. Acad. Sci. USA* 107, 16096–16100, 2010.

184. Isom, D.G., Castanada, C.A., Cannon, B.R. et al., Large shifts in pK_a values of lysine residues buried inside a protein, *Proc. Natl. Acad. Sci. USA* 108, 5260–5265, 2011.

185. Harris, M.J., Schlessman, J.L., Sue, G.R., and Garcia-Moreno, B., Arginine residues at internal positions in a protein are always charged, *Proc. Natl. Acad. Sci. USA* 108, 18954–18959, 2011.

186. Sadekov, I.D., Minkin, V.I., and Lutskii, A.E., Intramolecular hydrogen bonding and reactivity of organic compounds, *Upspekhi Khimi* 39, 380–311, 1970.

187. Szedja, W., Effects of internal hydrogen bonding on the reactivity of the hydroxyl-group in esterification under phase transfer conditions, *J. Chem. Soc. Chem. Commun.* (5), 380–411, 1981.

188. Yamauchi, K., Hosokawa, T., and Kinoshita, M., Possible effect of hydrogen bonding on methylation of pyrimidine and pyridone nucleosides, *J. Chem. Soc. Perkin Trans. 1* 1, 13–15, 1989.

189. Taylor, R. and Kennard, O., Hydrogen-bond geometry in organic crystals, *Acc. Chem. Res.* 17, 320–326, 1984.

190. Glusker, J.P., The binding of ions to proteins, in *Protein*, Vol. 2, ed. G. Allen, Chapter 4. JAI Press, London, U.K., 1999.

191. Oda, Y., Yamazaki, T., Nagayama, K. et al., Individual ionization constants of all the carboxyl groups in ribonuclease HI from *Escherichia coli* determined by NMR, *Biochemistry* 33, 5275–5284, 1994.

192. Lin, Y., Fusek, M., Lin, X. et al., pH Dependence of kinetic parameters of pepsin, rhizopuspepsin, and their active-site hydrogen bond mutants, *J. Biol. Chem.* 267, 18413–18418, 1992.

193. Webb, H., Tynan-Connolly, B.M., Lee, G.M. et al., Remeasuring HEWL pK_a values by NMR spectroscopy: Methods, analysis, accuracy, and implications for theoretical pK_a calculations, *Proteins* 79, 685–702, 2011.

194. Dyson, H.J., Jeng, M.R., Tennant, L.L. et al., Effects of buried charged groups on cysteine thiol ionization and reactivity in *Escherichia coli* thioredoxin: Structural and functional characterization of mutants of Asp 26 and Lys 57, *Biochemistry* 36, 2622–2636, 1997.

195. Andersson, L.K., Caspersson, M., and Baltzer, L., Control of lysine reactivity in four-helix bundle proteins by site-selective pK_a depression: Expanding the versatility of proteins by postsynthetic functionalization, *Chem. Eur. J.* 8, 3687–3697, 2002.

196. Chaiken, I.M. and Smith, E.L., Reaction of chloroacetamide with the sulfhydryl groups of papain, *J. Biol. Chem.* 244, 5087–5094, 1969.

197. Chaiken, I.M. and Smith, E.L., Reaction of the sulfhydryl group of papain with chloroacetic acid, *J. Biol. Chem.* 244, 5095–5099, 1969.

198. Gerwin, B.I., Properties of the single sulfhydryl group of streptococcal proteinase. A comparison of the rates of alkylation by chloroacetic acid and chloroacetamide, *J. Biol. Chem.* 242, 451–456, 1967.

199. Zhang, G., Mazurkie, A.S., Dunaway-Mariano, D., and Allen, K.N., Kinetic evidence for a substrate-induced fit in phosphonoacetaldehyde hydrolase catalysis, *Biochemistry* 41, 13370–13377, 2002.

200. Fisher, B.M., Schultz, L.W., and Raines, R.T., Coulombic effects of remote subsites on the active site of ribonuclease A, *Biochemistry* 37, 17386–17401, 1998.

201. O'Leary, M.H. and Westheimer, F.H., Acetoacetate decarboxylase. Selective acetylation of the enzyme, *J. Biol. Chem.* 7, 913–919, 1968.

202. Schmidt, D.E., Jr. and Westheimer, F.H., pK of the lysine amino group at the active site of acetoacetate decarboxylase, *Biochemistry* 10, 1249–1253, 1971.

203. Highbarger, L.A., Gerlt, J.A., and Kenyon, G.L., Mechanism of the reaction catalyzed by acetoacetate decarboxylase. Importance of lysine 116 in determining the pK_a of active-site lysine 115, *Biochemistry* 35, 41–46, 1996.

204. Okazaki, K., Yamada, H., and Imoto, T., A convenient *S*-aminoethylation of cysteinyl residues in reduced proteins, *Anal. Biochem.* 149, 516–520, 1985.

205. Thevis, M., Ogorzalek Loo, R.R., and Loo, J.A., In-gel derivatization of proteins by cysteine-specific cleavages and their analyses by mass spectrometry, *J. Proteome Res.* 2, 163–172, 2003.

206. Marincean, S., Rabago Smith, M., Beltz, L., and Borhan, B., Selectivity of labeled bromoethylamine for protein alkylation, *J. Mol. Model.* 18, 4547–4556, 2012.

207. Ho, M.C., Ménétret, J.F., Tsurata, H., and Allen, K.N., The origin of the electrostatic perturbation in acetoacetate decarboxylase, *Nature* 459, 393–397, 2009.

208. Gerit, J.A., Acetoacetate decarboxylase: Hydrophobics, not electrostatics, *Nat. Chem. Biol.* 5, 454–455, 2009.

209. Ishikita, H., Origin of the p*K*a shift of the catalytic lysine in acetoacetate decarboxylase, *FEBS Lett.* 584, 364–3468, 2010.

210. Schirch, L., Slagel, S., Barra, D. et al., Evidence for a sulfhydryl group at the active site of serine transhydroxymethylase, *J. Biol. Chem.* 255, 2986–2989, 1980.

211. Plapp, B.V., Application of affinity labeling for studying structure and function in enzymes, *Methods Enzymol.* 87, 469–499, 1982.

212. Schoellman, G. and Shaw, E., A new method for labelling the active center of chymotrypsin, *Biochem. Biophys. Res. Commun.* 7, 36–40, 1962.
213. Petra, P.H., Cohen, W., and Shaw, E.N., Isolation and characterization of the alkylated histidine from TLCK inhibited trypsin, *Biochem. Biophys. Res. Commun.* 21, 612–628, 1965.
214. Lundblad, R.L., Bergstrom, J., De Vrekder, R. et al., Measurement of active coagulation factors in Autoplex-T® with colorimetric active site specific assay technology, *Thromb. Haemost.* 80, 811–815, 1998.
215. Lawrence, C.P. and Chow, S.C., Suppression of human T cell proliferation by the caspase inhibitors, z-VAD-FMK and z-IETD-FMK is independent of their caspase inhibition properties, *Toxicol. Appl. Pharmacol.* 265, 103–112, 2012.
216. Binda, O., Boyce, M., Rush, J.S. et al., A chemical method for labeling lysine methyltransferase substrates, *ChemBioChem* 12, 330–334, 2011.
217. Chalker, M.J., Bernades, G.J., and Davis, B.G., A "tag-and-modify" approach to site-selective protein modification, *Acc. Chem. Res.* 44, 730–741, 2011.
218. Chen, Y.X., Triola, G., and Waldmann, H., Bioorthogonal chemistry for site-specific labeling and surface immobilization of proteins, *Acc. Chem. Res.* 44, 762–773, 2011.
219. Rowland, M.M., Rostic, H.E., Gong, D. et al., Phosphatidylinositol 3,4,5-triphosphate activity probes for the labeling and proteomic characterization of protein binding partners, *Biochemistry* 50, 11143–11161, 2011.
220. Geurink, P.P., Prely, L.M., van der Marel, G.A. et al., Photoaffinity labeling in activity-based protein profiling, *Top. Curr. Chem.* 324, 85–113, 2012.
221. Mannhold, R., Poda, G.I., Ostermann, C., and Tetko, I.V., Calculation of molecular lipophilicity: State-of-the-art and comparison of log P methods on more than 96,000 compounds, *J. Pharm. Sci.* 98, 861–893, 2009.
222. Sokolovsky, M., Riordan, J.F., and Vallee, B.L., Tetranitromethane. A reagent for nitration of tyrosyl residues in proteins, *Biochemistry* 5, 3582–3589, 1966.
223. Riordan, J.F., Sokolovsky, M., and Vallee, B.L., Environmentally sensitive tyrosyl residues. Nitration with tetranitromethane, *Biochemistry* 6, 358–361, 1967.
224. Myers, B., II and Glazer, A.N., Spectroscopic studies of the exposure of tyrosine residues in proteins with special reference to the subtilisins, *J. Biol. Chem.* 246, 412–419, 1971.
225. Preills, J.P., Dolmans, M., Leonis, J., and Brew, K., Nitration of tyrosyl residues in human α-lactalbumin. Effect on lactose synthase specific activity, *Eur. J. Biochem.* 60, 533–539, 1975.
226. Cuatrecasas, P., Fuchs, S., and Anfinsen, C.B., Tyrosyl residues in the active site of staphylococcal nuclease. Modifications by tetranitromethane, *J. Biol. Chem.* 243, 4787–4798, 1968.
227. Castro, L., Demicheli, V., Tórtora, V., and Radi, R., Mitochondrial protein tyrosine nitration, *Free Radic. Res.* 45, 37–52, 2011.
228. Wiseman, D.A. and Thurmond, D.C., The good and the bad effects of cysteine *S*-nitrosylation and tyrosine nitration upon insulin exocytosis: A balancing act, *Curr. Diabetes Rev.* 8, 303–315, 2012.
229. Dickerson, R.E., Takano, T., Eisenberg, D. et al., Ferricytochrome c. I. General features of the horse and bonito proteins at 2.8 Ang. resolution, *J. Biol. Chem.* 246, 1511–1535, 1971.
230. Bushnell, G.W., Louie, G.V., and Brayer, G.D., High-resolution three-dimensional structure of horse heart cytochrome *c*, *J. Mol. Biol.* 214, 585–595, 1990.
231. Bruice, T.C., Gregory, M.J., and Salters, S.L., Reaction of tetranitromethane. I. Kinetics and mechanism of nitration of phenols by tetranitromethane, *J. Am. Chem. Soc.* 90, 1612–1619, 1968.
232. Skov, K., Hofmann, T., and Williams, G.R., The nitration of cytochrome c, *Can. J. Biochem.* 47, 750, 1969.
233. Sokolovsky, M., Aviram, I., and Schejter, A., Nitrocytochrome *c*. I. Structure and enzymic properties, *Biochemistry* 9n, 5113–5118, 1970.
234. Battghyány, C., Souza, J.M., Durán, R. et al., Time course and site (s) of cytochrome c tyrosine nitration by peroxynitrite, *Biochemistry* 44, 8038–8046, 2005.
235. Caswell, D.S. and Spiro, T.G., Tyrosine and tryptophan modification monitored by ultraviolet resonance Raman spectroscopy, *Biochim. Biophys. Acta* 873, 73–78, 1986.
236. Šantrůček, J., Strohälm, M., Kadlčik, V. et al., Tyrosine residues modification studied by MALDI-TOF mass spectrometry, *Biochem. Biophys. Res. Commun.* 323, 1151–1156, 2004.
237. Souza, J.M., Castro, L., Cassina, A.M. et al., Nitrocytochrome c: Synthesis, purification, and functional studies, *Methods Enzymol.* 441, 197–215, 2008.
238. Noyes, C.M., Griffith, M.J., Roberts, H.R., and Lundblad, R.L., Identification of the molecular defect in factor IX Chapel Hill: Substitution of histidine for arginine at position 145, *Proc. Natl. Acad. Sci. USA* 80, 4200–4202, 1983.

239. Daggett, K.A. and Sakmar, T.P., Site-specific in vitro and in vivo incorporation of molecular probes to study G protein-coupled receptors, *Curr. Opin. Chem. Biol.* 15, 392–398, 2011.

240. González-Valdez, J., Rito-Palomares, M., and Benavides, J., Advances and trends in the design, analysis, and characterization of polymer-protein conjugates for "PEGylaided" bioprocesses, *Anal. Bioanal. Chem.* 403, 2225–2235, 2012.

241. Shibata, K., Chemical assessibility and environment of amino acid residues in native proteins, in *New Techniques in Amino Acids, Peptide and Protein Analysis*, Chapter 9, eds. A. Niederwieser and G. Pataki, pp. 341–385, Ann Arbor Science Publishers, Ann Arbor, MI, 1971.

242. Levert, K.L., Lloyd, R.B., and Waldrop, G.L., Do cysteine 230 and lysine 238 of biotin carboxylase play a role in the activation of biotin? *Biochemistry* 39, 4122–4128, 2000.

243. Chaiken, I.M. and Smith, E.L., Reaction of a specific tyrosine residue of papain with diisopropylfluorophosphate, *J. Biol. Chem.* 244, 4247–4250, 1960.

244. De Vendittis, E., Urshy, T., Rullo, R. et al., Phenylmethylsulfonyl fluoride inactivates an archaeal superoxide dismutase by chemical modification of a specific tyrosine residue. Cloning, sequencing and expression of the gene coding for *Sulfolobus solfataricus* superoxide dismutase, *Eur. J. Biochem.* 268, 1794–1801, 2001.

245. Hashimoto, J. and Takahashi, K., Chemical modifications of ribonuclease U1, *J. Biochem.* 81, 1175–1180, 1977.

246. Collier, T.S. and Muddiman, D.C., Analytical strategies for the glocal quantification of intact proteins, *Amino Acids* 43, 1109–1117, 2012.

247. Sikanen, T., Aura, S., Franssila, S. et al., Microchip capillary electrophoresis-electrospray ionization-mass spectrometry of intact proteins using uncoated Ormocomp microchips, *Anal. Chim. Acta* 711, 69–76, 2012.

248. Rosati, S., Rose, R.J., Thompson, N.J. et al., Exploring an orbitrap analyzer for the characterization of intact antibodies by native mass spectrometry, *Angew. Chem. Int. Ed. Engl.* 51, 12992–12996, 2012.

249. Thompson, N.J., Rosati, S., Rose, R.J., and Heck, A.J. The impact of mass spectrometry on the study of intact antibodies from post-translational modifications to structural analysis, *Chem. Commun. (Camb.)* 49, 538–548, 2013.

250. Bennett, K.L., Smith, S.V., Lambrecht, R.M. et al., Rapid characterization of chemically-modified proteins by electrospray mass spectrometry, *Bioconjug. Chem.* 7, 16–22, 1996.

251. Fligge, T.A., Kast, J., Burns, K., and Przybylski, M., Direct monitoring of protein-chemical reactions by utilizing nanoelectrospray mass spectrometry, *J. Am. Soc. Mass Spectrom.* 10, 112–118, 1999.

252. Fenaille, F., Guy, P.A., and Tabet, J.C., Study of protein modification by 4-hydroxy-2-nonenol and other short chain aldehydes analyzed by electrospray ionization tandem mass spectrometry, *J. Am. Soc. Mass Spectrom.* 14, 215–226, 2003.

253. Woods, A.G., Sokolowska, I., and Darie, C.C., Identification of consistent alkylation of cysteine-less peptides in a proteomics experiment, *Biochem. Biophys. Res. Commun.* 419, 305–308, 2012.

254. Anjum, R., Pae, E., Blenis, J., and Ballif, B.A., TPCK inhibits AGC kinases by direct activation loop adduction at phenylalanine-directed cysteine residues, *FEBS Lett.* 586, 3471–3476, 2012.

255. Jonák, J., Petersen, T.E., Clark, B.F., and Rychlik, I., *N*-Tosyl-L-phenylalanylchloromethane reacts with cysteine 81 in the molecule of elongation factor Tu from *Escherichia coli*, *FEBS Lett.* 150, 485–488, 1982.

256. Skinner, J.P., Chi, L., Ozeata, P.F. et al., Introduction of the mass spread function for characterization of protein conjugates, *Anal. Chem.* 84, 1172–1177, 2012.

257. Lindsay, D.G. and Shall, S., The acetylation of insulin, *Biochem. J.* 121, 737–745, 1971.

258. Wallace, G.R., McLeod, A., and Chain, B.M., Chromatographic analysis of the trinitrophenyl derivatives of insulin, *J. Chromatogr. B Biomed. Appl.* 427, 239–246, 1988.

259. Li, C.H., Preparation and properties of dinitrophenyl-NH(ε)-insulin, *Nature* 175, 1402, 1956.

260. Keck, K., Grossberg, A.L., and Pressman, D., Antibodies of restricted heterogeneity induced by DNP-insulin, *Immunochemistry* 10, 331–335, 1973.

261. Keck, K., Ir gene control of carrier recognition. I. Immunogenicity of bovine insulin derivatives, *Eur. J. Immunol.* 5, 801–807, 1975.

262. Crestfield, A.M., Stein, W.H., and Moore, S., Alkylation and identification of the histidine residues at the active site of ribonuclease, *J. Biol. Chem.* 238, 2413–2419, 1963.

263. Fruchter, R.G. and Crestfield, A.M., The specific alkylation by iodoacetamide of histidine-12 in the active site of ribonuclease, *J. Biol. Chem.* 242, 5807–5812, 1967.

264. Lundblad, R.L., Noyes, C.M., Featherstone, G.L., Harrison, J.H., and Jenzano, J.W., The reaction of bovine alpha-thrombin with tetranitromethane. Characterization of the modified protein, *J. Biol. Chem.* 263, 3729–3734, 1988.

265. Silberring, J. and Nyberg, F., Analysis of tyrosine- and methionine-containing neuropeptides by fass atom bombardment mass spectrometry, *J. Chromatogr.* 562, 459–467, 1991.

266. Silberring, J. and Nyberg, F., Application of photodiode array detection and fast atom bombardment mass spectrometry for the identification of the arginine residues in neuropeptides, *Biomed. Chromatogr.* 5, 240–247, 1991.

267. Takahashi, T., Hiramoto, S., Wato, S. et al., Identification of essential amino acid residues of an α-amylase inhibitor from *Phaseolus vulgaris* white kidney beans, *J. Biochem.* 126, 838–844, 1999.

268. Zhao, Y., Ma, C.-Y., Yuen, S.-N., and Phillips, D.L., Study of succinylated food proteins by Raman spectroscopy, *J. Food Agric. Chem.* 52, 1815–1823, 2004.

269. Knee, K.M., Roden, C.K., Flory, M.R., and Mukerji, I., The role β93 Cys in the inhibition of Hb S fiber formation, *Biophys. Chem.* 127, 181–193, 2007.

270. Saha, A. and Yakovlev, V.V., Structural changes of human serum albumin in response to a low concentration of heavy ions, *J. Biophotonics* 2, 670–677, 2010.

271. Kodali, V.K., Scrimgeour, J., Kim, S. et al., Nonperturbative chemical modification of graphene for protein micropatterning, *Langmuir* 27, 863–865, 2011.

272. Budhavaram, N.K. and Barone, J.R., Quantifying amino acid and protein substitution using Raman spectroscopy, *J. Raman Spectrosc.* 42, 355, 362, 2011.

273. Peydecastaing, J., Bras, J., Vaca-Garcia, C. et al., NIR study of chemically modified cellulosic biopolymers, *Mol. Cryst. Liq. Cryst.* 448, 717–724, 2006.

274. Gelles, J. and Chan, S.I., Chemical modification of the Cua center in cytochrome c oxidase by sodium-*p*-(hydroxymercuri)benzoate, *Biochemistry* 24, 3963–3972, 1985.

275. Liebherr, R.B., Soukka, T., Wolfbeis, O.S., and Gorris, H.H., Maleimide activation of photon upconverting nanoparticles for bioconjugation, *Nanotechnology* 23, 485103, 2012.

276. Tong, J., Luxenhofer, R., Yi, X. et al., Protein modification with amphiphilic block copoly(2-oxazoline) as a new platform for enhanced cellular delivery, *Mol. Pharm.* 7, 984–992, 2010.

277. Wang, S., Ionescu, R., Peekhaus, N. et al., Separations of post-translational modifications antibodies by exploiting subtle conformational changes under mildly acidic conditions, *J. Chromatogr.* 1217, 6496–6502, 2010.

278. Boswell, C.A., Tesar, D.B., Mukhyala, K. et al., Effects of charge on antibody tissue distribution and pharmacokinetics, *Bioconjug. Chem.* 21, 2153–2163, 2010.

279. Garcia-Giménez, J.L., Ledesma, A.M. et al., Histone carbonylation occurs in proliferating cells, *Free Radic. Biol. Med.* 52, 1453–1464, 2012.

280. Shi, Y., Mowery, R.A., Ashley, J. et al., Abnormal SDS-PAGE migration of cytosolic proteins can identify domains and mechanisms that control surfactant binding, *Protein Sci.* 21, 1197–1209, 2012.

281. Tarabout, C., Roux, S., Gobeaux, F. et al., Control of peptide nanotube diameter by chemical modifications of an aromatic residue involved in a single close contact, *Proc. Natl. Acad. Sci. USA* 108, 7679–7684, 2011.

282. Frisman, I., Shachaf, Y., Seliktar, D., and Blanco-Peled, H., Stimulus-responsive hydrogels made from biosynthetic fibrinogen conjugates for tissue engineering: Structural characterization, *Langmuir* 27, 6977–6986, 2011.

283. Schroer, M.A., Markgraf, J., Wieland, D.C. et al., Nonlinear pressure dependence of the interaction potential of dense protein solutions, *Phys. Rev. Lett.* 106(17), 178102, 2011.

284. Nygaard, J., Munch, H.K., Thulstrup, P.W. et al., Metal ion controlled self-assembly of a chemically reengineered protein drug studied by small-angle X-ray scattering, *Langmuir* 28, 12159–12170, 2012.

285. Castaneda, L., Valle, J., Yang, N. et al., Collagen cross-linking with Au nanoparticles, *Biomacromolecules* 9, 3383–3388, 2008.

286. Price, N.C. and Radda, G.K., Desensitization of glutamate dehydrogenase by reaction of tyrosine residues, *Biochem. J.* 114, 419–427, 1969.

287. Piskiewicz, D., Landon, M., and Smith, E.L., Bovine glutamate dehydrogenase. Loss of allosteric inhibition by guanosine triphosphate and nitration of tyrosine 412, *J. Biol. Chem.* 246, 1324–1329, 1973.

288. Ray, W.J., Jr. and Koshland, D.E., Jr., A method for characterizing the type and numbers of groups involved in enzyme action. *J. Biol. Chem.* 236, 1973–1979, 1961.

289. Tsou, C.-L., Relation between modification of functional groups of proteins and their biological activity. I. A graphical method for the determination of the number and type of essential groups, *Sci. Sinica* 11, 1535–1558, 1962.

290. Tsou, C.-L., Kinetics of substrate reaction during irreversible modification of enzyme activity, *Adv. Enzymol.* 61, 381–436, 1988.

291. Zhou, J.-M., Liu, C., and Tsou, C.-L., Kinetics of trypsin inhibition by its specific inhibitors, *Biochemistry* 28, 1070–1076, 1989.
292. Horiike, K. and McCormick, D.B., Correlations between biological activity and the number of functional groups chemically modified, *J. Theor. Biol.* 79, 403–414, 1979.
293. Holbrook, J.J. and Ingram, V.A., Ionic properties of an essential histidine residue in pig heart lactate dehydrogenase, *Biochem. J.* 131, 729–738, 1973.
294. Bloxham, D.P., The chemical reactivity of the histidine-195 residue in lactate dehydrogenase thiomethyl-ated at the cysteine-165 residue, *Biochem. J.* 193, 93–97, 1981.
295. Horiike, K., Tsuge, H., and McCormick, D.B., Evidence for an essential histidyl residue at the active site of pyridoxamine (pyridoxine)-5′-phosphate oxidase from rabbit liver, *J. Biol. Chem.* 254, 6638–6643, 1979.
296. Rakitzis, E.T., Kinetics of protein modification reactions, *Biochem. J.* 217, 341–351, 1984.
297. Rakitzis, E.T., Kinetic analysis of regeneration by dilution of a covalently modified protein, *Biochem. J.* 268, 669–670, 1990.
298. Page, M.G.P., The reaction of cephalosporins with penicillin-binding protein 1bγ from *Escherichia coli*, *Biochim. Biophys. Acta* 1205, 199–206, 1994.
299. Dubus, A., Normark, S., Kania, M., and Page, M.G.P., Role of Asparagine 152 in catalysis of β-lactam hydrolysis by *Escherichia coli* ampC β-lactamase studied by site-directed mutagenesis, *Biochemistry* 34, 7757–7764, 1995.
300. Yang, S.J., Jiang, S.S., Tzeng, C.M. et al., Involvement of tyrosine residue in the inhibition of plant vacu-olar H⁺-pyrophosphatase by tetranitromethane, *Biochim. Biophys. Acta* 1294, 89–97, 1996.
301. Chu, C.L., Hsiao, Y.Y., Chen, C.H. et al., Inhibition of plant vacuolar H⁺-ATPase by diethylpyrocarbon-ate, *Biochim. Biophys. Acta* 1506, 12–22, 2001.
302. Hsiao, Y.Y., Van, R.C., Hung, H.H. et al., Diethylpyrocarbonate inhibition of vacuolar H⁺-pyrophosphatase possibly involves a histidine residue, *J. Protein Chem.* 21, 51–58, 2002.
303. Kitz, R. and Wilson, I.B., Esters of methanesulfonic acid as irreversible inhibitors of acetylcholine ester-ase, *J. Biol. Chem.* 237, 3245–3249, 1962.
304. Shen, W.C. and Colman, R.F., Cyanate modification of essential lysine residues of the diphosphopyridine nucleotide-specific isocitrate dehydrogenase of pig heart, *J. Biol. Chem.* 250, 2973–2978, 1975.
305. Lennette, E.P. and Plapp, B.V., Transition-state analysis of the facilitated alkylation of ribonuclease A by bromoacetate, *Biochemistry* 18, 3938–3946, 1979.
306. Pincus, M.R., Hummel, C.F., Brandtrauf, P.W., and Carty, R.P., Enthalpic and entropic determinants for the specificity of alkylation of the histidine-12 residue of ribonuclease-A by 4 bromoacetamido nucleo-side affinity labels and bromoacetamide, *Int. J. Pept. Protein Res.* 36, 56–66, 1990.
307. Rakitzis, E.T., Utilization of the free-energy of the reversible binding of protein and modifying agent towards the rate-enhancement of protein covalent modification, *Biochem. J.* 269, 835–838, 1990.
308. Whitaker, J.R. and Lee, L.S., Ficin- and papain-catalyzed reactions. Effect of temperature on reactivity of the essential sulfhydryl group of ficin in the presence and absence of competitive inhibitors, *Arch. Biochem. Biophys.* 148, 208–216, 1972.
309. Lundblad, R.L., Buffers, in *Biochemistry and Molecular Biology Compendium*, Chapter 6, pp. 349–353, CRC Press/Taylor & Francis, Boca Raton, FL, 2007.
310. Bizzozero, Q.A., Bixler, H.A., Davis, J.D. et al., Chemical deacylation reduces the adhesive properties of proteolipid protein and leads to decompaction of the myelin sheath, *J. Neurochem.* 76, 1129–1141, 2001.
311. Burcham, P.C., Fontaine, F.R., Petersen, D.R., and Pyke, S.M., Reactivity with Tris(hydroxymethyl)ami-nomethane confounds immunodetection of acrolein-adducted proteins, *Chem. Res. Toxicol.* 16, 1196–1201, 2003.
312. Peroza, E.A. and Freisinger, E., Tris is a non-innocent buffer during intein-mediated protein cleavage, *Protein Expr. Purif.* 57, 217–225, 2008.
313. Gab, J., John, H., Melzer, M., and Blum, M.M., Stable adducts of nerve agents sarin, soman and cyclosa-rin with TRIS, TES and related buffer compounds—Characterization by LC-ESI-MS/MS and NMR and implications for analytical chemistry, *J. Chromatogr. B Anal. Technol. Biomed. Life Sci.* 878, 1382–1390, 2010.
314. Plapp, B.V., Site-directed mutagenesis: A tool for studying enzyme catalysis, *Methods Enzymol.* 249, 91–119, 1995.
315. Gau, B.C., Chen, H., Zhang, Y., and Gross, M.L., Sulfate radical anion as a new reagent for fast photo-chemical oxidation of proteins, *Anal. Chem.* 82, 7821–7827, 2010.
316. Gong, B., Ramos, A., Vázquez-Fernández, E. et al., Probing structural differences between PrP(C) and PrP(Sc) by surface nitration and acetylation: Evidence of conformational change in the C-terminus, *Biochemistry* 50, 4963–4972, 2011.

317. Yang, S.-H., Wu, C.-H., and Lin, W.-Y., Chemical modification of aminopeptidase isolated from Pronase, *Biochem. J.* 302, 595–600, 1994.

318. Vosbeck, K.D., Greenberg, B.D., Ochoa, M.S. et al., Proteolytic enzymes of the K-1 strain of *Streptomyces griseus* obtained from a commercial preparation (Pronase). Effect of pH, metal ions, an amino acids on aminopeptidase activity, *J. Biol. Chem.* 253, 257–260, 1973.

319. Narahashi, Y. and Yanagita, M., Studies on proteolytic enzymes (Pronase) of *Streptomyces griseus* K-1. I. Nature and properties of the proteolytic enzyme system, *J. Biochem.* 62, 633–641, 1967.

320. Narahashi, Y., Shibuya, K., and Yanagita, M., Studies on proteolytic enzymes (Pronase) of *Streptomyces griseus* K-1. II. Separation of exo- and endopeptidases of Pronase, *J. Biochem.* 64, 427–437, 1968.

321. Trop, M. and Birk, Y., The trypsin-like enzyme *Streptomyces griseus* (Pronase), *Biochem. J.* 109, 475–476, 1968.

322. Wilk, S., Wilk, E., and Magnusson, R.P., Identification of histidine residues important in the catalysis and structure of aspartyl aminopeptidase, *Arch. Biochem. Biophys.* 407, 176–183, 2002.

323. Lim, J. and Turner, A.J., Chemical modification of porcine kidney aminopeptidase P indicates the involvement of two critical histidine residues, *FEBS Lett.* 381, 188–190, 1996.

324. Liu, T.-Y., Demonstration of the presence of a histidine residue at the active site of streptococcal proteinase, *J. Biol. Chem.* 247, 4029–4032, 1967.

325. Murachi, T., Tsudzuki, T., and Okmura, K., Photosensitized inactivation of stem bromelain. Oxidation of histidine, methionine, and tryptophan residues, *Biochemistry* 14, 249–255, 1975.

326. Coggins, J.R., Kray, W., and Shaw, E., Affinity labelling of proteinase with tryptic specificity by peptides with C-terminal lysine chloromethyl ketone, *Biochem. J.* 137, 579–585, 1974.

327. Herzog, C., Yang, C., Holmes, A., and Kaushal, G.P., zVAD-fmk prevents cisplatin-induced cleavage of autophagy proteins but impairs autophagic flux and worsen renal function, *Am. J. Physiol. Renal Physiol.* 303, F1239–F1250, 2012.

328. Levy, H.M., Leber, P.D., and Ryan, E.M., Inactivation of myosin by 2,4-dinitrophenol and protection by adenosine triphosphate and other phosphate compounds, *J. Biol. Chem.* 238, 3654–3659, 1963.

329. Grouselle, M. and Pudlis, J., Chemical studies on yeast hexokinase. Specific modification of a single tyrosyl residue with 1-ethyl-3-(3-dimethylaminopropyl) carbodiimide, *Eur. J. Biochem.* 74, 471–480, 1977.

330. Makinen, K.K., Mäkinen, P.L., Wilkes, S.H. et al., Chemical modification of *Aeromonas* aminopeptidase. Evidence for the involvement of tyrosyl and carboxyl groups in the activity of the enzyme, *Eur. J. Biochem.* 128, 257–265, 1982.

331. Dixon, M. and Webb, E.C., *Enzymes*, 3rd edn., Academic Press, New York, 1979.

332. Siegal, I.H., *Enzyme Kinetics: Behavior and Analysis of Rapid Equilibrium and Steady-State Enzyme Systems*, Wiley-Interscience, New York, 1975.

333. D.L. Purich,(ed.), *Contemporary Enzyme Kinetics and Mechanism*, Academic Press, New York, 1983.

334. Northrup, D.B., Rethinking fundamentals of enzyme action, *Adv. Enzymol.* 73, 25–55, 1999.

335. Wang, J., Araki, T., Matsuoka, M., and Ogawa, T., A graphical method of analyzing pH dependence of enzyme activity, *Biochim. Biophys. Acta* 1435, 177–183, 1999.

336. Ragin, O., Gruaz-Guyon, A., and Barbet, J., Equilibrium expert: An add-in to Microsoft Excel for multiple binding equilibrium simulations and parameter estimations, *Anal. Biochem.* 310, 1–14, 2002.

337. Liao, F., Tian, K.C., Yang, X. et al., Kinetic substrate quantification by fitting the enzyme reaction curve to the integrated Michaelis-Menten equation, *Anal. Bioanal. Chem.* 375, 756–762, 2003.

338. Brey, W.S., *Physical Chemistry and Its Biological Applications*, Academic Press, New York, 1978.

339. Sheehan, D., *Physical Biochemistry: Principles and Applications*, John Wiley & Sons Ltd., Chichester, U.K., 2000.

340. Price, N.C. et al., *Physical Chemistry for Biochemists*, 3rd edn., Oxford University Press, Oxford, U.K., 2001.

341. Peracchi, A., Enzyme catalysis: Removing chemically 'essential' residues by site-directed mutagenesis, *Trends Biochem. Sci.* 26, 497–503, 2001.

342. Addington, T.A., Mertz, R.W., Siegel, J.B. et al., Janus: Prediction and ranking of mutations required for functional interconversion of enzymes, *J. Mol. Biol.* 428, 1378–1389, 2013.

343. Liu, M., Cheetham, J., Chauchon, N. et al., Protein isoaspartate methyltransferase-mediated ^{18}O-labeling of isoaspartic acid for mass spectrometry analysis, *Anal. Chem.* 84, 1056–1062, 2012.

344. Luo, Q., Joubert, M.K., Stevenson, R. et al. Chemical modifications in therapeutic protein aggregates generated under different stress conditions, *J. Biol. Chem.* 286, 25134–25144, 2011.

345. Boyd, D., Kaschak, T., and Yan, B., HIC resolution of an IgG1 with an oxidized Trp, in a complementarity determining region, *J. Chromatogr. B Anal. Technol. Biomed. Life Sci.* 879, 955–960, 2011.

346. Gandhi, S., Ren, D., Xiao, G. et al., Elucidation of degradants in acidic peak of cation exchange chromatography in an IgG1 monoclonal antibody formed on long-term storage in a liquid formulation, *Pharm. Res.* 29, 209–224, 2012.

347. Shimura, K., Hoshima, M., Komiya, K. et al., Estimation of deamidation rates of major deamidation sites in a Fab fragment of mouse IgG1-κ by capillary isoelectric focusing of mutated Fab fragment, *Anal. Chem.* 85, 1705–1710, 2013.

348. Demgl, S., WEhmer, M., Hesse, F. et al., Aggregation and chemical modification of monoclonal antibodies under upstream processing conditions, *Pharm. Res.* 30, 1380–1399, 2013.

349. Carven, G.J. and Stern, L.J., Probing the ligand-induced conformational change in HLA-DR1 by selective chemical modification and mass spectrometric mapping, *Biochemistry* 44, 13625–13637, 2005.

350. Lundblad, R.L., *Approaches to Conformational Analysis of Biopharmaceuticals*, CRC/Taylor & Francis, Boca Raton, FL, 2010.

351. Dolnik, V. and Gurske, W.A., Chemical modification of proteins to improve the accuracy of their relative molecular mass determination by electrophoresis, *Electrophoresis* 32, 2893–2897, 2011.

352. Williams, B.J., Russell, W.K., and Russell, D.H., High-throughput method for on-target performic acid oxidation of MALDI-deposited samples, *J. Mass Spectrom.* 45, 157–166, 2010.

353. Brancia, F.L., Montgomergy, H., Tanaka, K., and Kumashiro, S., Guanidation labeling derivatization strategy for global characterization of peptide mixtures by liquid chromatography matrix-assisted laser desorption/ionization mass spectrometry, *Anal. Chem.* 76, 2748–2755, 2004.

354. Kimmel, J.R., Guanidation of proteins, *Methods Enzymol.* 11, 584–589, 1967.

355. Markland, F.S., Bacharach, A.D., Weber, B.H. et al., Chemical modification of yeast-3-phosphoglycerate kinase, *J. Biol. Chem.* 250, 1361–1370, 1975.

356. Bunk, D.M. and MacFarlane, R.D., Derivatization to enhance sequence-specific fragmentation of peptides and proteins, *Int. J. Mass Spectrom. Ion Process* 126, 123–136, 1993.

357. Wordwood, S., Mohammed, S., Christen, I.M. et al., Guanidation chemistry for qualitative and quantitative proteomics, *Res. Commun. Mass Spectrom.* 20, 3245–3256, 2006.

358. Ericksson, M., Mori, L., Jinwei, I. et al., Identification of proteolytic cleavage sites by quantitative proteomics, *J. Proteome Res.* 6, 2850–2856, 2007.

359. Armstrong, D.J. and Roman, A., The anomalous electrophoretic behavior of the human papillomavirus type 16 E7 protein is due the high content of acidic amino acid residues, *Biochem. Biophys. Res. Commun.* 192, 1380–1387, 1993.

360. Lock, R.A., Zhang, Q.Y., Berry, A.M., and Paton, J.C., Sequence variation in the *Streptococcus pneumoniae* pneumolysin gene affecting haemolytic activity and electrophoretic mobility of the toxin, *Microb. Pathog.* 21, 71–83, 1996.

361. Matthieson, R., Bauw, G., and Welinder, K.G., Use of performic acid oxidation to expand the mass distribution of tryptic peptides, *Anal. Chem.* 76, 6840–6850, 2004.

362. Basile, F. and Hauser, N., Rapid online nonenzymatic protein digestion combining microwave heating acid hydrolysis and electrochemical oxidation, *Anal. Chem.* 83, 359–367, 2011.

363. Roeser, J., Bischoff, R., Bruins, A.P., and Permentier, H.P., Oxidative protein labeling in mass-spectrometry-based proteomics, *Anal. Bioanal. Chem.* 397, 3441–3455, 2010.

364. Roeser, J., Alting, N.F., Permentier, H.P. et al., Chemical labeling of electrochemically cleaved peptides, *Rapid Commun. Mass Spectrom.* 27, 546–552, 2011.

2 Alkylating Agents

An alkyl group is defined as "the residue left when a hydrogen atom is removed from an aliphatic hydrocarbon...."[1,*] Alkylation is defined as "the introduction of an alkyl group into a hydrocarbon chain or aromatic ring."[1] An alkyl halide is then a "molecule consisting of an alkyl group bonded to one or more halides."[2] Within the realm of protein chemistry, the definition of alkylation is somewhat broader including the reaction of α-haloalkanoic acid derivatives such as chloroacetate but not that of acetyl chloride, which is an acylating agent (Figure 2.1).

Most alkylating reagents react with the target functional group by an S_N2 mechanism (see Chapter 1). The S_N2 reaction can occur with a neutral nucleophile such as an unprotonated amino group but is more rapid with an anionic nucleophile such as the thiolate anion. Alkylation occurs with the reaction of some of the more classical affinity-labeling reagents (Figure 2.2) such as TLCK trypsin (1-chloro-3-tosylamido-7-amino-2-heptanone, TLCK)[3], where modification occurs at an active-site histidine. Hence, the chlorine derivative is selected to reduce reactivity such that specific binding would be the dominant factor in the modification of a specific residue. As shown in Figure 2.2, the rate of reaction of the α-carbonyl haloalkyl derivatives increases with the complexity of the reagent. As seen below (Table 2.1), chloromethane is essentially unreactive toward cysteine, a more reactive nucleophile than histidine, in the absence of enzymatic coupling (including biotransformation to the aldehyde).[4] The work of Pincus and coworkers[5] provides an example of the power of specific binding on residue modification. These investigators developed 2'(3')-O-bromoacetyluridine (Figure 2.3) for the modification of the active-site histidine in pancreatic ribonuclease A. This affinity label reacted with pancreatic ribonuclease A at a rate approximately 3000 times (4.03×10^{-2} M^{-1} s^{-1}) faster than with free histidine (1.3×10^{-5} M^{-1} s^{-1}). It is recognized that this is not a direct comparison but appears to be useful as defined by these authors. The point is that the observed rate of inactivation is greatly increased by the increased local concentration of reagent resulting from specific binding. Some proteins demonstrate unexpected specificity in reaction with the peptide chloromethylketones.[6–8] Kupfer and coworkers showed that the catalytic subunit of cyclic AMP-dependent protein kinase was inactivated by either TLCK or TPCK; chloroacetic acid or chloroacetamide had no effect. This study showed that modification with TLCK occurred at the sulfhydryl group of a cysteine residue, and it is suggested that TLKC is an inhibitor based on the binding of the basic lysine residue. These investigators warned about interpreting results with affinity labels in crude system, and specificity should be supported by additional work. In other studies, Solomon and coworkers[8] reported that TLCK (IC$_{50}$ = 1 mM) was more effective than TPCK (IC$_{50}$ = 9 mM) in the inhibition of protein kinase C; these investigators suggest, as do Kupfer and coworkers,[6] that there is specific binding of the lysyl side chain of TLCK. These investigators also demonstrated saturation kinetics consistent with affinity labeling.[9]

The chemistry and reactions of some of the various alkylating agents are described in Tables 2.1 through 2.9. There are some alkylating agents that are not listed as they are rarely used to modify proteins in vitro but have major importance in in vivo situations. One such is the glycation of proteins that involve the nonenzymatic reaction of reducing sugars such as glucose and similar metabolites with an aldehyde function with the ε-amino group of lysine (Figure 2.4). While the process of glycation is of great physiologic importance, glycation is not useful for chemical modification. *Click* chemistry as introduced by Barry Sharpless and colleagues[10] is a very clever concept (Figure 2.5)

---

* There is a third edition of this reference, which is, unfortunately, out of print. http://www.penguin.co.uk/nf/Book/BookDisplay/0,9780140514452,00.html.

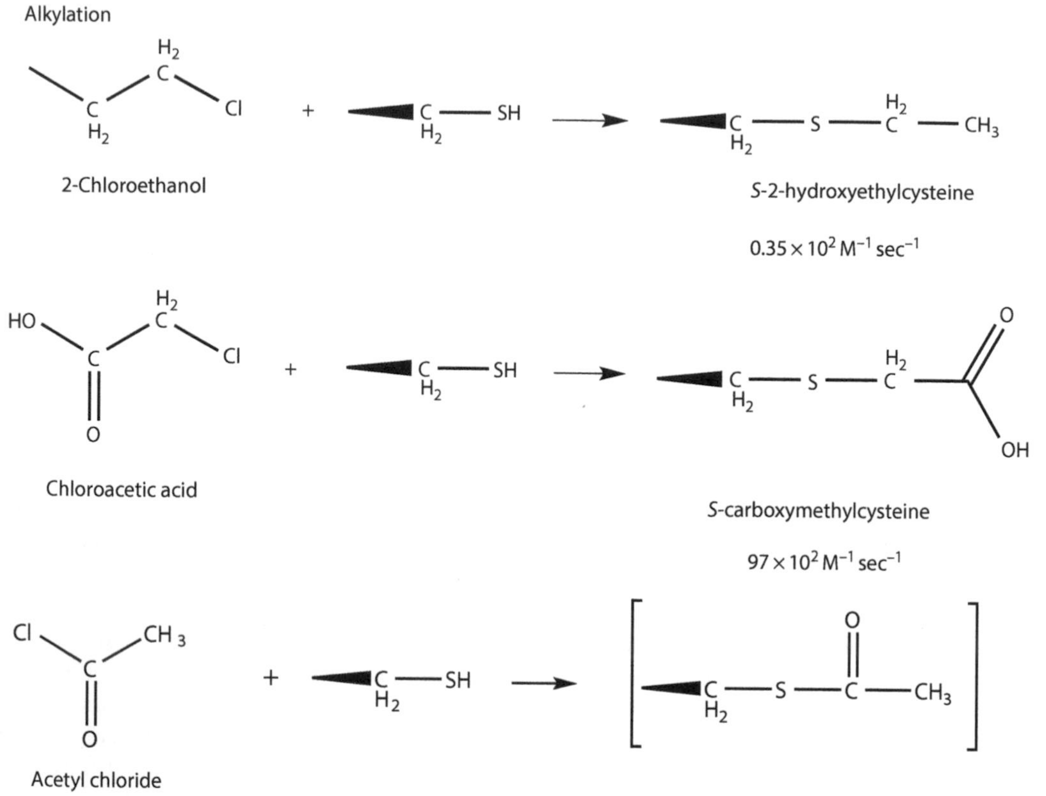

Alkylation

2-Chloroethanol

S-2-hydroxyethylcysteine

$0.35 \times 10^2 \, M^{-1} \, sec^{-1}$

Chloroacetic acid

S-carboxymethylcysteine

$97 \times 10^2 \, M^{-1} \, sec^{-1}$

Acetyl chloride

FIGURE 2.1   Alkylation of nucleophiles. A comparison of the reaction of two alkylating agents with cysteine. The data are taken from Dahl, K.H. and McKinley-McKee, J.S., The reactivity of affinity labels. A kinetic study of the reaction of alkyl halides with thiolate anions—A model reaction for protein alkylation, *Bioorg. Chem.* 10, 329–341, 1981. The data shown is the estimated pH-independent rate constants at 37°C. Also shown is the reaction of acetyl chloride with the sulfhydryl group of cysteine to yield an unstable thioester derivative.

that will likely have value in chemical biology with in vivo incorporation of alkynes and other reactive groups into cellular proteins.[11,12] These are mostly alkylation reactions although concept is driven by the application of simple chemistry with stoichiometric yields. Ning and coworkers[13] have been successful in introducing the cyclooctyne function into proteins by chemical coupling (Figure 2.5). Quinones have also been shown to alkylate cysteinyl residues (Figure 2.6) in proteins, but the reactions are complex and complicated by the redox reaction of this class of compounds where quinones can oxidize cysteine to cystine.[14–17] The term *adductomics*[18,19] has been advanced to describe the study of exposure to electrophiles, most of which lead to alkylation of proteins and nucleic acids. In particular, the alkylation of the cysteine residues in human serum albumin is used as a biomarker of exposure to environmental electrophiles.[19]

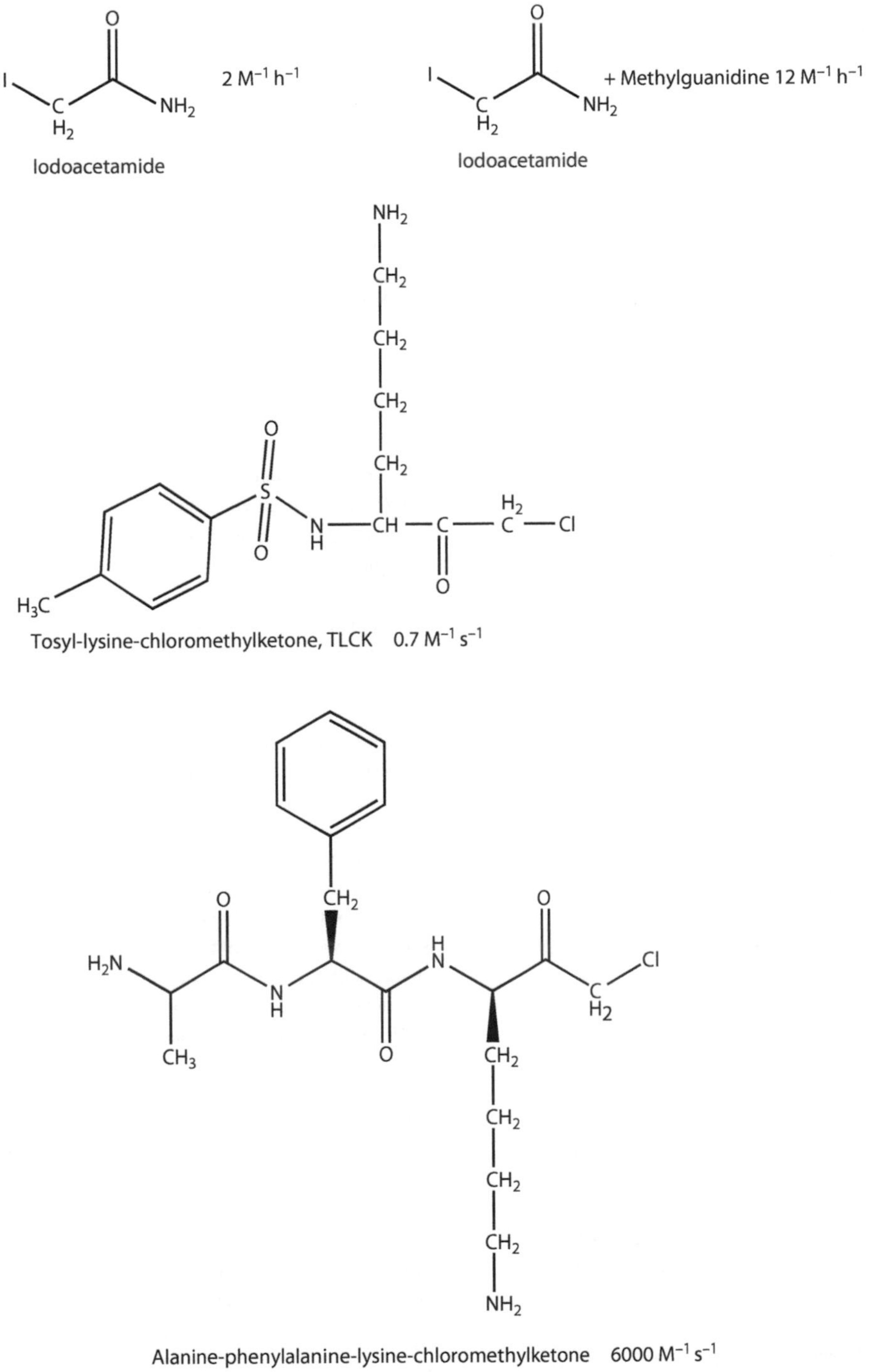

FIGURE 2.2 The basis of affinity labels. Shown is the structure of iodoacetamide with the rate constants for reaction with trypsin in the presence and absence of methylguanidine (Ingami, T., The alkylation of the active site of trypsin with iodoacetamide in the presence of alkylguanidines, *J. Biol. Chem.* 240, PC3453–PC3455, 1965), the structure of 1-chloro-3-tosylamido-7-amino-2-heptanone (TLCK), and a tripeptide chloromethylketone with the rates of reaction with trypsin (Collen, D., Lijnen, H.R., De Cock, F. et al., Kinetics properties of tripeptide lysyl chloromethylketone and lysyl *p*-nitroanilide derivatives towards trypsin-like serine proteinases, *Biochim. Biophys. Acta* 165, 158–166, 1980).

TABLE 2.1

Reactions of Alkyl Halides, Alkyl Sulfonyl

Compound	Chemistry	Reactions	References
Methyl chloride (chloromethane)	CH_3Cl (CAS 74-87-5)	• Biotransformation to formaldehyde.[a] • Bioconjugation to cysteine in glutathione.[b]	2.1.1–2.1.5
Methyl bromide (bromomethane)	CH_3Br	• Monomethyl and dimethyl derivatives of α-amino groups. • S-methyl cysteine. • N_1- and N_3-methylhistidine. • ε-Methylamino lysine. • β-Methyl aspartic. • O-Methyltyrosine.	2.1.6
Iodomethane (methyl iodide)	CH_3I[c]	• ε-Methylamino lysine. • S-methyl cysteine. • N_1- and N_3-methylhistidine.	2.1.7–2.1.21
Ethyl chloride[a,b]	CH_3CH_2Cl	S-ethylcysteine (in glutathione).	2.1.22–2.1.24
1,2-dichloroethane[d]	$ClCH_2CH_2Cl$	S-(2-chloroethyl)-cysteine (in glutathione). S,S'-ethylene-bis-glutathione.	2.1.25–2.1.33
1,2-dibromoethane[e]	$BrCH_2H_2Br$	S-bromoethylcysteine (in glutathione).	2.1.34–2.1.36
Dibromoketones	Dibromoketones have been used to cross-link functional residues in proteins.		2.1.37, 2.1.38
Bromomethyl benzene (α-bromotoluene; benzyl bromide)[f,g]	$BrCH_2C_6H_5$	Reaction with cysteine and methionine.[h] Substantial reaction has also been reported with histidine and tyrosine. There was no detectable reaction with tryptophan. Majority of recent work has used bis, tris, and tetrakis derivatives. Reaction rate data in Table 2.2.	2.1.39–2.1.45
2-Hydroxy-5-nitrobenzyl bromide		Selective modification of tryptophan under neutral or acidic conditions.[i] Multiple substitution of tryptophan with HNB has been observed as has the modification of cysteine and methionine. The reaction of histidine with HNB in a protein has been observed.	2.1.46–2.1.51
1,1′-thiobis (2-chloroethane) (mustard gas; mustards)	$ClCH_2CH_2SCH_2CH_2Cl$	Reaction with a number of amino acids including lysine, histidine, N-*terminal amino*, cysteine, and carboxyl groups.[j]	2.1.52–2.1.60
2-Chloro-N-(2-chloroethyl)-N-methylamine [nitrogen mustard, bis- (β-chloroethyl) methylamine; mechlorethamine	$ClCH_2CH_2N(CH_3)$ CH_2CH_2Cl	Not as reactive as the sulfur mustards but with a similar specificity. Has therapeutic use as antineoplastic agent.[k]	2.1.61–2.1.69
Dimethyl sulfate (DMS)	$CH_3SO_4CH_3$	The ability of DMS to alkylate DNA has been extensively studied. There is some use of DMS for the synthesis of N-methyl amino acids. The reaction of DMS with proteins is relatively nonspecific with modification of tyrosine, amino groups, and cysteine being reported.[l]	2.1.70–2.1.77

TABLE 2.1 (continued)
Reactions of Alkyl Halides, Alkyl Sulfonyl

Compound	Chemistry	Reactions	References
Methyl-*p*-nitrobenzene sulfonate		Reaction with histidine and cysteine residues in protein.[m]	2.1.78–2.1.85
Methyl methanesulfonate		Primary reaction is methylation of nucleic acids. Methylation of cysteine, lysine, and histidine in proteins is a secondary reaction.[n]	2.1.86–2.1.89
1-Methyl-1-nitrosourea		Methylation of nucleic acids. Methylation of proteins and/or peptides has been reported. Carbamylation of amino groups has been observed.[o]	2.1.86–2.1.89
2-Bromoethylamine	$BrCH_2CH_2NH_2$	Reaction with sulfhydryl groups, α-amino groups, and histidine. Used frequently to generate γ-thialysine from cysteine residues in proteins.[p]	2.1.90–2.1.99

[a] Chlorinated hydrocarbons are mostly unreactive toward functional groups in proteins unless they undergo biotransformation to a reactive derivative such as formaldehyde.[2.1.1] As the alkyl chain increases, reaction is more difficult even with halogenated derivatives that would be more reactive such as bromine or iodine.

[b] Two pathways have been suggested for ethyl chloride. The first is the P-450-mediated conversion to acetaldehyde that can be subsequently converted to ethanol or acetate; the second involved the conjugation with the cysteine residue in glutathione in a reaction mediated by glutathione transferase. A similar process has been proposed for ethyl bromide. Both ethyl chloride and presumably ethyl bromide can undergo a slow nonenzymatic conjugation reaction with the cysteine residue in glutathione. Both are essentially unreactive and require relatively high levels for genotoxicity.

[c] Iodomethane is a liquid (mp −66.5°C; bp 42.5°C) that can be used in the gas phase to modify proteins.[2.1.7–2.1.14] Reaction in the gas phase likely occurs through the S_N2 mechanism seen for solution chemistry.[2.1.15,2.1.16] Davis and coworkers[2.1.14] observe that an advantage of the gas-phase reaction of iodomethane enables the preparation of radiolabeled protein with high specific activity. It is also possible to modify lyophilized proteins with iodomethane in the gas phase since the reaction with protein nucleophiles occurs the absence of water.[7–9] Reaction proteins with iodomethane in the gas phase can occur at lysine, imidazole and tyrosine. Esterification of unprotonated carboxyl groups with iodomethane can also be detected; specific formation of methyl esters in lyophilized proteins can be observed in vacuo with gaseous methanol/HCl.[2.1.10] Iodomethane was used to modify the N-terminal amino group of glycopeptides to obtain the quaternary amine derivative for MS.[2.1.17] In this case, the sodiated peptide (sodium salt of carboxyl group) was prepared by ion exchange chromatography (AG50W-X8; preconditioned with NaOH) before modification. The deuterated iodomethane (CD_3I) was used to improve resolution on MS. Also of note in this study was the microwave treatment of proteins prior to enzymatic hydrolysis that reduced the time required for hydrolysis. Iodomethane can also be used to prepare strong anion-exchange matrices by the formation of quaternary amines from tertiary amines on the surface of epoxy-based monoliths.[2.1.18] Permethylation of glycans using ^{13}C-labeled iodomethane has been used for the relative quantitation of glycans.[2.1.19] Permethylation with iodomethane was also used to solubilize a protein in DMSO.[2.1.20] It should be noted that methyl iodide was one of the first reagents used to methylate proteins.[2.1.21,2.1.22]

[d] As with the monohaloalkanes, the dihalo derivatives are quite unreactive with sulfhydryl groups in proteins. Geminal haloalkanes (e.g., dibromomethane; 1,1′-dibromoethane) are even less reactive. 1,2-Dichloroethane can be converted to 2-chloroethanol and subsequently metabolized to chloroacetaldehyde and chloroacetic acid. 1,2-Dichloroethane can also be conjugated to the cysteine residue in glutathione to yield the *S*-(2-chloroethyl)cysteine derivative. *S*-(2-chloroethyl) cysteine. The *S*-(2-chloroethyl) derivative of either cysteine or reduced glutathione can subsequently react with

(continued)

TABLE 2.1 (continued)
Reactions of Alkyl Halides, Alkyl Sulfonyl

sulfur or nitrogen nucleophile in proteins via an episulfone intermediate (Figure 2.7); reaction can also occur with N^7 in 2-deoxyguanine. With S-(2-chloroethyl)glutathione, the relative rates of reaction are glutathione (1) > cystyr (0.78) ≫ histyr (0.05) > 2-deoxyguanine (0.02). Despite the extensive in vivo studies, the chemistry of protein modification by S-(2-chloroethyl)glutathione or S-(2-chloroethyl)cysteine is not clear. Studies with hemoglobin did not establish a site of modification, while studies with protein disulfide isomerase showed the modification of two cysteine residues. The studies with protein disulfide isomerase suggested other sites that were not identified. The S-(2-chloroethyl) derivatives are frequently referred to as half-mustards and are highly reactive through the aforementioned episulfonium ion. Neutralization of the potential toxicity of the half-mustard derivatives can be accomplished via reactive nucleophile species such as 2,4-dithiopurine.

e   In analogy with other haloalkanes, 1,2-dibromoethane is metabolized by the P-450 system to form bromoacetaldehyde or conjugated to glutathione to form S-bromoethyl cysteine in a reaction catalyzed by glutathione transferase. Presumably, the same in vivo reactions occur as described for 1,2-dichloroethane. There are more studies on 1,2-dichloroethane than on other dihaloethanes as reflection of greater commercial use. However, there is reason to suspect the dibromo derivatives as being more reactive than the dichloroderivatives.[2.1.36] The literature on 1,2-diiodoethane is extremely limited. Dibromoacetone (dibromopropanone) has been used as a cross-linking agent for proximal residues at enzyme active sites such as cysteine and histidine at the active site of glyceraldehyde-3-phosphate dehydrogenase[2.1.37] and for proximal cysteinyl residues in fatty acid synthase dimers.[2.1.38]

f   Bromomethyl benzene (Figure 2.8) is a highly reactive, poorly soluble reagent that has had limited but successful use in reaction with cysteine or methionine.[2.1.39–2.1.42] The rate of hydrolysis of reagent ($t_{1/2} ≈ 75$ min at 25°C) can be a significant factor in determining rate constants.[2.1.40,2.1.42] More recent work has focused on the use of bis-, tris-, and tetrakis-bromomethyl derivatives of benzene to modify cysteine residues in a phage-expressed peptide to generate an oriented and potentially discontinuous scaffold for display.[2.1.43–2.1.45]

g   The preferred term is monobromomethyl benzene but additional references were found using benzyl bromide as the search term. It was of interest that benzyl bromide that is the older term was also used in more recent work. This provides an example of the necessity to use multiple databases for search; for older work, it is recommended to go to the print source.

h   Bromomethyl benzene (benzyl bromide) was used originally for the modification of methionine in proteins[2.1.40,2.1.41], although it is somewhat more reactive with cysteine. Rogers and coworkers observed that at pH 6.8, the rate of reaction bromomethyl benzene is twice that of reaction with methionine; reaction with histidine and tryptophan is negligible. The lack of reaction of bromomethylbenzene with tryptophan is of interest as placing a hydroxyl group on the benzene ring *ortho* to the methylenebromide group increases both reactivity and specificity resulting in the modification of tryptophan.[2.1.46] Aleksic and coworkers[2.1.42] have evaluated the reaction of bromomethyl benzene (benzyl bromide) with a panel of single nucleophile peptides as part of project on evaluating skin-sensitizing reagents. They found highest reactivity with cysteine at pH 7.4 with lesser, but strong, reaction with tyrosine and lysine (both at pH 10) and lesser reaction with histidine at pH 7.4; little, if any, reaction was observed with arginine (pH 10). Methionine peptides were not evaluated. Benzyl bromide is used as a protecting group for the hydroxyl group in tyrosine during peptide synthesis. A fluorescent derivative has been prepared[2.1.41] seeking to take advantage of the small number of methionine residues in proteins. While methionine was successfully modified with this reagent, other nucleophiles also reacted with the reagent; replacement of methionine with selenomethionine with a decrease in reagent concentration and shorter reaction improved specificity. See Table 2.2 for rate constants for the reaction of benzyl bromide iodoacetate or iodoacetamide with methionine residues.

i   HNB (Figure 2.9) was allowed to react at pH 3.0 with a mixture of amino acids.[2.1.46,2.1.47] There was total modification of tryptophan, 20% modification of cysteine, and approximately 7% formation of methionine sulfoxide. Reaction at neutral pH did not change reactivity, while at pH 11.5, there was greater than 90% modification of cysteine and 40%–60% modification of tyrosine. HNB is subject to rapid hydrolysis in water ($t_{1/2} < 30$ s). Multiple substitution of tryptophan by HNB has been reported by several groups.[2.1.48,2.1.49] Reaction of HNB does occur with thiouracil and thiouridine at pH 10 to yield the S-alkylated derivative.[2.1.50] The reaction of HNB with histidine resulting in alkylation at the nitrogen residues in the imidazole ring in a protein has been reported.[2.1.51]

j   The sulfur mustards (Figure 2.10) are a highly reactive group of chemicals developed for the purpose of warfare and are optimized for damage.[2.1.52, 2.1.53] These reagents are of no particular value for protein modification, but there has been some interesting work.[2.1.54–2.1.56] The modification of Cys34 in albumin has been used as a biomarker for exposure to mustard-type reagents.[2.1.57]

TABLE 2.1 (continued)
Reactions of Alkyl Halides, Alkyl Sulfonyl

k Nitrogen mustards (Figure 2.11) were developed as with the sulfur mustards as a chemical warfare agent. The mustards were developed early in the last century.[2.1.61] There was an intense effort both in the United States and in Europe to understand the chemistry of both the nitrogen mustards and the sulfur mustards. This work also established sites of modification in proteins and mechanism of action.[2.1.62,2.1.63] Some of this work was done in the laboratories of Max Bergmann on the 5th floor of Flexner Hall at what was the Rockefeller Institute for Medical Research and today is the Rockefeller University. The late William H. Stein, a Nobel laureate with Stanford Moore, was involved in the studies that established the nitrogen mustard modification of various amino acid residues in protein. I had the honor of working with Professor Stein and he had some interesting stories on the challenges of working with the toxic agents. A great portion of the work, which was all classified during the period of WWII, was published in 1946. However, much detail is missing but can be retrieved by spending time at the Library of Congress. Another example of this early work was the successful use of mustards in the treatment of blood diseases such as Hodgkin's disease and lymphoma which was also published in 1946. The authors on this study[2.1.64] went on to become the leaders of clinical hematology in the latter part of the twentieth century. Recent work has provided considerable understanding of the mechanism of action of nitrogen mustards.[2.1.66–2.1.69]

l DMS (Figure 2.12) has had limited use for the modification of proteins with far more extensive use in mapping nucleic acids. Eyem and colleagues[2.1.73] explored the reaction of DMS with cysteine and cysteinyl residues in proteins. They noted the optimal pH was 7–8 and decreased with increasing pH differing from the pH dependence of other alkylation reactions at sulfhydryl groups. DMS is nonspecific[2.1.72] and has been used to measure tyrosine exposure on conformational change.[2.1.74] Cleavage at S-methylcysteine residues with cyanogen bromide has been reported.[2.1.75] The solvent conditions for the reaction of S-methylcysteine with cyanogen bromide can be manipulated to drive the conversion to serine with a resulting N,O acyl shift for use in the synthesis of glycopeptides.[2.1.76] In these studies, the S-methylation of cysteine was accomplished with methyl-p-nitrobenzene sulfonate (Figure 2.12).[2.1.77]

m The reaction of methyl-p-nitrobenzene sulfonate with histidine (Figure 2.13) appears to depend on presence in an enzyme active site,[2.1.78,2.1.79,2.1.82] while no such requirement exists for the modification of cysteine residues as reaction can take place with a denatured protein.[2.1.77,2.1.80] In some cases with histidine,[2.1.84] there is the formation of an intermediate kinetically similar to a Michaelis complex prior to the covalent modification reaction. Finally, methyl-p-nitrobenzenesulfonate has been used as a model compound in the study of nucleophilic reactions.[2.1.84,2.1.85]

n Methane methylsulfonate (MMS) (Figure 2.14) is used more frequently for the methylation of bases in DNA and RNA. MMS is used infrequently for the modification of proteins. MMS is relatively nonspecific and can modify amino groups, the imidazole ring, and cysteine in proteins. There is limited information on the chemistry of the reaction. MMS is frequently confused with methyl methanethiosulfonate, a reagent used for the modification of cysteine/cystine residues to form mixed disulfides (see Chapter 7).

o 1-Methyl-1-nitrosourea is another reagent that is used primarily for the methylation of nucleic acids (Figure 2.14). While the pattern of methylation of nucleic acids by 1-methyl-1-nitrosourea and DMS is similar, there are marked differences in the reaction of the two reagents with proteins.[2.1.86–2.1.89]

p 2-Bromoethylamine (CAS 2576-47-8) was introduced to replace ethyleneimine[2.1.90] for the modification of cysteinyl residues in proteins yielding γ-thialysine providing additional sites for peptide bond cleavage by trypsin. Reaction is thought to occur predominantly by an S_N2 mechanism via an aziridinium intermediate resembling ethyleneimine (Figure 2.15).[2.1.91,2.1.92] Specificity is optimized by reaction at alkaline pH.[2.1.14] It is primarily used to generate site for trypsin cleavage in proteins[2.1.93,2.1.94] and for the process of chemical rescue when a functionally useful lysine residue is converted to a cysteine residue by mutagenesis resulting in inactivation; activity is restored by the use of 2-bromoethylamine to generate a γ-thialysine residue.[2.1.95,2.1.96] Bromoethylamine is also used to create kidney damage in animal model systems.[2.1.97–2.1.99]

References to Table 2.1

2.1.1. Ristau, C., Bolt, H.M., and Vangala, R.R., Formation and repair of DNA lesions in kidneys of male mice after acute exposure to methyl chloride, *Arch. Toxicol.* 64, 254–256, 1990.

2.1.2. Dekant, W., Frischman, C., and Speerschneider, P., Sex, organ and species specific bioactivation of chloromethane by cytochrome P4502E1, *Xenobiotica* 25, 1259–1265, 1995.

2.1.3. Graves, R.J., Callander, R.D., and Green, T., The role of formaldehyde and S-chloromethylglutathione in the bacterial mutagenicity of methyl chloride, *Mutat. Res.* 320, 235–243, 1994.

2.1.4. Kornbrust, D.J. and Bus, J.S., The role of glutathione and cytochrome P-450 in the metabolism of methyl chloride, *Toxicol. Appl. Pharmacol.* 67, 246–256, 1983.

(continued)

TABLE 2.1 (continued)
Reactions of Alkyl Halides, Alkyl Sulfonyl

2.1.5. Jäger, R., Peter, H., Sterzel, W., and Bolt, H.M., Biochemical effects of methyl chloride in relation to its tumorigenicity, *J. Cancer Res. Clin. Oncol.* 114, 64–70, 1988.

2.1.6. Ferranti, P., Sannolo, N., Mamone, G. et al., Structural characterization by mass spectrometry of hemoglobin adducts formed after in vivo exposure to methyl bromide, *Carcinogenesis* 17, 2661–2671, 2006.

2.1.7. Altamirano, M.M., Mulliert, G., and Calcagno, M., Sulfhydryl groups of glucosamine-6-phosphate isomerase deaminase from *Escherichia coli*, *Arch. Biochem. Biophys.* 258, 95–100, 1987.

2.1.8. Taralp, A. and Kaplan, H., Chemical modification of lyophilized proteins in nonaqueous environments, *J. Protein Chem.* 16, 183–193, 1997.

2.1.9. Vakos, H.T., Kaplan, H., Black, B. et al., Use of the pH memory effect in lyophilized proteins to achieve preferential methylation of alpha-amino groups, *J. Protein Chem.* 19, 231–237, 2000.

2.1.10. Vakos, H.T., Black, B., Dawson, B. et al., In vacuo esterification of carboxyl groups in lyophilized proteins, *J. Protein Chem.* 20, 521–531, 2001.

2.1.11. Hoang, V.M., Conrads, T.P., Veenstra, T.D. et al., Quantitative proteomics employing primary amine affinity tags, *J. Biomol. Tech.* 14, 216–223, 2003.

2.1.12. Laremore, T.N., Weber, T.N., and Choma, C.T., An evaluation of the utility of in vacuo methylation for mass-spectrometry-based analyses of peptides, *Rapid Commun. Mass Spectrom.* 19, 2045–2054, 2005.

2.1.13. Liu, X., Chan, K., Chu, I.K., and Li, J., Microwave-assisted nonspecific proteolytic digestion and control methylation for glycomics applications, *Carbohydr. Res.* 343, 2870–2877, 2008.

2.1.14. Davis, L.D., Spencer, W.J., Pham, V.T. et al., ^{14}C radiolabeling of proteins to monitor biodistribution of ingested proteins, *Anal. Biochem.* 410, 57–61, 2011.

2.1.15. Hagghu, M., Irani, M., and Gholami, M.R., Theoretical study of kinetics and mechanism of reactions of hydroxylamine and amineoxide anion with methyl iodide in gas and aqueous phases, *Prog. React. Kinet. Mech.* 32, 29–50, 2007.

2.1.16. Garver, J.M., Eyet, N., Villano, S.M. et al., Mechanistic investigation of S_N2 dominated gas phase alkyl iodide reactions, *Int. J. Mass Spectrom.* 30, 151–158, 2011.

2.1.17. Liu, X., Chan, X., Chu, I.K., and Li, J., Microwave-assisted nonspecific proteolytic digestion and controlled methylation for glycomics applications, *Carbohydr. Res.* 343, 2870–2877, 2008.

2.1.18. Dinh, N.P., Cam, Q.M., Nguyen, A.M. et al., Functionalization of epoxy-based monoliths for ion exchange chromatography of proteins, *J. Sep. Sci.* 32, 2556–2564, 2009.

2.1.19. Alvarez-Manilla, G., Warren, N.L., Abney, T. et al., Tools for glycomics: Relative quantitation of glycans by isotopic permethylation using $^{13}CH_3I$, *Glycobiology* 17,677–687, 2007.

2.1.20. Vauhkonen, M., Kinnunen, P.K., and Rauvala, H., Solubilization of proteins in dimethyl sulfoxide by permethylation: Application to structural studies of apolipoprotein B, *Anal. Biochem.* 148, 357–364, 1985.

2.1.21. Blackburn, S., Carter, E.G.H., and Phillips, H., The methylation of wool with methyl sulfate and methyl halides, *Biochem. J.* 35, 627–639, 1941.

2.1.22. Fedtke, N., Certa, H., Ebert, R., and Wiegand, H.-J., Species differences in the biotransformation of ethyl chloride. II. GSH-dependent metabolism, *Arch. Toxicol.* 68, 217–223, 1994.

2.1.23. Holder, J.W., Analysis of chloroethane toxicity and carcinogenicity including a comparison with bromoethane, *Toxicol. Ind. Health* 24, 655–675, 2008.

2.1.24. Landry, T.D., Ayres, J.A., Johnson, K.A., and Wall, J.M., Ethyl chloride: A two-week inhalation study and effects on liver non-protein sulfhydryl concentrations, *Fund. Appl. Toxicol.* 2, 230–234, 1982.

2.1.25. Gwinn, M.R., Johns, D.O., Bateson, T.F., and Guyton, K.Z., A review of the genotoxicity of 1,2-dichloroethane (EDC), *Mutat. Res. Rev. Mutat. Res.* 727, 42–53, 2011.

2.1.26. Erve, J.L.L., Deinzer, M.L., and Reed, D.J., Reaction of human hemoglobin toward the alkylating agent *S*-(2-chloroethyl)glutathione, *J. Toxicol. Environ. Health* 49, 127–143, 1996.

2.1.27. Guengerich, F.P., Mason, P.S., Stott, W.I. et al., Roles of 2-haloethylene oxides and 2-haloacetaldehydes derived from vinyl bromide and vinyl chloride in irreversible binding to proteins and DNA, *Cancer Res.* 41, 4391–4398, 1981.

2.1.28. Guengerich, F.P., Crawford, W.M., Jr., and Watanabe, P.G., Activation of vinyl chloride to covalently bound metabolites: Role of 2-chloroethylene oxide and 2-chloroacetaldehyde, *Biochemistry* 18, 5177–5182, 1979.

2.1.29. Jean, P.A. and Reed, D.J., In vitro dipeptide, nucleoside and glutathione alkylation by *S*-(2-chloroethyl)glutathione and *S*-(2-chloroethyl)-L-cysteine, *Chem. Res. Toxicol.* 2, 455–460, 1989.

TABLE 2.1 (continued)
Reactions of Alkyl Halides, Alkyl Sulfonyl

2.1.30. Smit, N.A., Zefirov, N.S., Bodrikov, I.V., and Kirmer, M.Z., Episulfonium ions: Myths and realities, *Acc. Chem. Res.* 12, 282–288, 1978.

2.1.31. Webb, W.W., Elfana, A.A., Webster, K.D. et al., Role for a episulfonium ion in *S*-(2-chloroethyl)-DL-cysteine-induced cytotoxicity and its reaction with glutathione, *Biochemistry* 26, 3017–3023, 1987.

2.1.32. Kaetzel, R.S., Stapels, M.D.M., Barofsky, D.F., and Reed, D.J., Alkylation of protein disulfide isomerase by the episulfonium ion derived from the glutathione conjugate of 1,2-dichloroethane and mass spectrometric characterization of the adducts, *Arch. Biochem. Biophys.* 423, 136–147, 2004.

2.1.33. Liu, J., Powell, K.L., Thames, H.D., and MacLeod, M.C., Detoxication of sulfur half-mustards by nucleophilic scavengers: Robust activity of thiopurines, *Chem. Res. Toxicol.* 23, 488–496, 2010.

2.1.34. Albano, E., Poli, G., Tomasi, A. et al., Toxicity of 1,2-dibromoethane in isolated hepatocytes: Role of lipid peroxidation, *Chem. Biol. Interact.* 50, 255–265, 1984.

2.1.35. Thomas, C., Will, Y., Schoenberg, S.L. et al., Conjugative metabolism of 1,2-dibromoethane in mitochondria: Disruption of oxidative phosphorylation and alkylation of mitochondrial DNA, *Biochem. Pharmacol.* 61, 595–603, 2001.

2.1.36. Watanbe, K., Liberman, R.G., Skipper, P.L. et al., Analysis of DNA adducts formed in vivo in rats and mice from 1,2-dibromoethane, 1,2-dichloroethane, dibromoethane, and dichloromethane using HPLC/accelerator mass spectrometry and relevance to risk estimates, *Chem. Res. Toxicol.* 20, 1594–1600, 2007.

2.1.37. Moore, J., Jr. and Fenselau, A., Reaction of glyceraldehyde-3-phosphate dehydrogenase with dibromoacetone, *Biochemistry* 11, 3753–3762, 1972.

2.1.38. Witkowski, A., Joshi, A.K., Rangan, V.S. et al., Dibromopropanone cross-linking of the phosphopantetheine and active-site cysteine thiols of the fatty acid synthase can occur both inter- and intrasubunit. Reevaluation of the side-by-side, antiparallel subunit model, *J. Biol. Chem.* 274, 11557–11563, 1999.

2.1.39. Schramm, H.J. and Lawson, W.B., The active center of chymotrypsin. II. Modification of methionine residues in chymotrypsin with a simple benzene derivative, *Hoppe-Seyler's Zeit. Physiol. Chem.* 332, 97–100, 1963.

2.1.40. Rogers, G.A., Shaltiel, N., and Boyer, P.D., Facile alkylation of methionine by benzyl bromide and demonstration of fumarase inactivation accompanied by alkylation of a methionine residue, *J. Biol. Chem.* 251, 5711–5717, 1976.

2.1.41. Lang, S., Spratt, D.E., Guillemette, J.G., and Palmer, M., Selective labeling of selenomethionine residues in proteins with a fluorescent derivative of benzyl bromide, *Anal. Biochem.* 359, 253–259, 2006.

2.1.42. Aleksic, M., Thain, E., Roger, D. et al., Reactivity profiling: Covalent modification of single nucleophile peptides for skin sensitization risk assessment, *Toxicol. Sci.* 108, 401–411, 2009.

2.1.43. Timmerman, P., Beld, J., Puijk, W.C., and Meloen, R.H., Rapid and quantitative cyclization of multiple peptide loops onto synthetic scaffolds for structural mimicry of protein surfaces, *ChemBioChem* 6, 821–824, 2005.

2.1.44. Heinis, C., Rutherford, T., Freund, S., and Winter, G., Phage-encoded combinatorial chemical libraries based on bicyclic peptides, *Nat. Chem. Biol.* 5, 502–507, 2009.

2.1.45. Smeenk, L.E., Dailly, N., Hiemstra, H. et al., Synthesis of water-soluble scaffolds for peptide cyclization, labeling, and ligation, *Org. Lett.* 14, 1194–1197, 2012.

2.1.46. Horton, H.R. and Koshland, D.E., Jr., A highly reactive colored reagent with selectivity for the tryptophan residue in proteins. 2-Hydroxy-5-nitrobenzyl bromide, *J. Am. Chem. Soc.* 87, 1126–1132, 1965.

2.1.47. Loudon, G.M. and Koshland, D.E., Jr., The chemistry of a reported group: 2-Hydroxy-5-nitrobenzyl bromide, *J. Biol. Chem.* 245, 2247–2254, 1970.

2.1.48. Lundblad, R.L. and Noyes, C.M., Observations on the reaction of 2-hydroxy-5-nitrobenzyl bromide with a peptide-bound tryptophanyl residue, *Anal. Biochem.* 136, 93–100, 1984.

2.1.49. Strohalm, M., Kodicek, M., and Pecher, M., Tryptophan modification by 2-hydroxy-5-nitrobenzyl bromide studies by MALDI-TOF mass spectrometry, *Biochem. Biophys. Res. Commun.* 312, 811–816, 2003.

2.1.50. Sato, E. and Kanaoka, Y., Reaction of thiouracil and thiouridine with 2-hydroxy-5-nitrobenzyl bromide, *Biochim. Biophys. Acta* 232, 213–216, 1971.

2.1.51. Barman, T.E., The chemistry of the reaction of 2-hydroxy-5-nitrobenzyl bromide with his-32 of α-lactalbumin, *Eur. J. Biochem.* 83, 2247–2254, 1970.

2.1.52. Ghabili, K., Agutter, P.S., Ghanei, M. et al., Sulfur mustard toxicity: History, chemistry, pharmacokinetics, and pharmacodynamics, *Crit. Rev. Toxicol.* 41, 384–403, 2011.

2.1.53. du Vigneaud, V., Stevens, C.M., McDuffie, H.F., Jr. et al., Reactions of mustard-type vesicants with α-amino acids, *J. Am. Chem. Soc.* 70, 1620–1624, 1948.

(continued)

TABLE 2.1 (continued)
Reactions of Alkyl Halides, Alkyl Sulfonyl

2.1.54. Black, R.M., Clarke, R.J., Harrison, J.M., and Read, R.W., Biological fate of sulphur mustard: Identification of valine and histidine adducts in haemoglobin from casualties of sulphur mustard poisoning, *Xenobiotica* 27, 499–512, 1997.

2.1.55. Hambrook, J.L., Howells, D.J., and Schock, C., Biological fate of sulphur mustard (1,1′-thiobis(2-chloroethane)): Uptake, distribution and retention of ^{35}S in skin and in blood after cutaneous application of ^{35}S-sulphur mustard in rat and in comparison with human blood in vitro, *Xenotiotica* 23, 537–561, 1993.

2.1.56. Black, A.T., Hayden, P.J., Casillas, R.P., Heck, D.E. et al., Expression of proliferative and inflammatory markers in a full-thickness human skin equivalent following exposure to the model sulfur vesicant, 2-chloroethyl sulfide, *Toxicol. Appl. Pharmacol.* 249, 178–187, 2010.

2.1.57. Yeo, T.H., Ho, M.L., and Loke, W.K., Development of a liquid chromatography-multiple reaction monitoring procedure for concurrent verification of exposure to different forms of mustard agents, *J. Anal. Toxicol.* 32, 51–56, 2008.

2.1.58. Smith, J.R., Capcio, B.R., Korte, W.D. et al., Analysis for plasma protein biomarkers following an accidental human exposure to sulfur mustard, *J. Anal. Toxicol.* 32, 17–24, 2008.

2.1.59. Kroening, K.K., Richardson, D.D., Afton, S., and Caruso, J.A., Screening hydrolysis products of sulfur mustard agents by high-performance liquid chromatography with inductively coupled plasma mass spectrometry detection, *Anal. Bioanal. Chem.* 393, 1949–1956, 2009.

2.1.60. Dell'Amico, E., Bernasconi, S., Cavalca, L. et al., New insight into the biodegradation of thiodiglycol, the hydrolysis product as Yperite (sulfur mustard gas), *J. Appl. Microbiol.* 106, 1111–1121, 2009.

2.1.61. Sartori, M., *The War Gases. Chemistry and Analysis*, Van Nostrand and Company, New York, 1939.

2.1.62. Golumbic, C., Fruton, J.S., and Bergmann, M., Chemical reactions of the nitrogen mustard gases; the transformations of methyl-bis(β-chloroethyl)amine in water, *J. Org. Chem.* 11, 518–535, 1946.

2.1.63. Fruton, J.S., Stein, W.H., and Bergman, M., Chemical reactions of the nitrogen mustard gases: The reactions of the nitrogen mustard gases with protein constituents, *J. Org. Chem.* 11, 559–570, 1946.

2.1.64. Goodman, L.S., Wintrobe, M.M., Dameshek, W. et al., Nitrogen mustard therapy; use of methyl-bis (beta-chloroethyl) amine hydrochloride and tris(beta-chloroethyl)amine hydrochloride for Hodgkin's disease, lymphosarcoma, leukemia and certain allied and miscellaneous disorders, *J. Am. Med. Assoc.* 132, 126–132, 1946.

2.1.65. Williamson, C.E. and Witten, B., Reaction mechanism of some aromatic nitrogen mustards, *Cancer Res.* 27, 33–38, 1967.

2.1.66. Fung, L.W., Ho, C., Roth, E.F., Jr., and Nagel, R.L., The alkylation of hemoglobin S by nitrogen mustard. High resolution proton nuclear magnetic resonance studies, *J. Biol. Chem.* 250, 4786–4789, 1975.

2.1.67. Wu, H.C. and Bayley, H., Single-molecule detection of nitrogen mustards by covalent reaction with a protein nanopore, *J. Am. Chem. Soc.* 130, 6813–6819, 2008.

2.1.68. Polavarapu, A., Baik, M.-H., Stubblefield, S.G.W. et al., The mechanism of guanine alkylation by nitrogen mustards—A computational study, *J. Org. Chem.* 77(14), 5914–5921, 2012.

2.1.69. Lindahl, L.M., Fenger-Grøn, M., and Iversen, L., Topical nitrogen mustard therapy in patients with Langerhans cell histiocytosis, *Br. J. Dermatol.* 166, 642–645, 2012.

2.1.70. Prashad, M., Har, D., Hu, B. et al., An efficient and practical N-methylation of amino acid derivatives, *Org. Lett.* 5, 125–128, 2003.

2.1.71. Biron, E., Chatterjee, J., and Kessler, H., Optimized selective N-methylation of peptides on solid supports, *J. Pept. Sci.* 12, 213–219, 2006.

2.1.72. Moore, G., Chemical modification of ribosomes with dimethyl sulfate. Probe to the structural organization of ribosomal proteins and RNA, *Can. J. Biochem.* 53, 328–337, 1975.

2.1.73. Eyem, J., Sjödahl, J., and Sjöquist, J., S-Methylation of cysteine residues in peptides and proteins with dimethylsulfate, *Anal. Biochem.* 74, 359–368, 1976.

2.1.74. Ghélis, C., Transient conformational states in proteins followed by differential labeling, *Biophys. J.* 32, 503–514, 1980.

2.1.75. Gross, E. and Morell, J.L., The reaction of cyanogen bromide with S-methylcysteine: Fragmentation of the peptide 14–29 of bovine pancreatic ribonuclease A, *Biochem. Biophys. Res. Commun.* 59, 1145–1150, 1974.

2.1.76. Okamoto, R., Souma, S., and Kajihara, Y., Efficient substitution reaction from cysteine to the serine residue of glycosylated polypeptide: Repetitive peptide segment ligation strategy and the synthesis of glycosylated tetracontapeptide having acid labile sialyl-T(N) antigens, *J. Org. Chem.* 74, 2494–2501, 2009.

2.1.77. Heinrikson, R.L., The selective S-methylation of sulfhydryl groups in proteins and peptides with methyl-*p*-nitrobenzenesulfonate, *J. Biol. Chem.* 246, 4090–4096, 1971.

TABLE 2.1 (continued)
Reactions of Alkyl Halides, Alkyl Sulfonyl

2.1.78. Nakagawa, Y. and Bender, M.L., Modification of alpha-chymotrypsin by methyl-*p*-nitrobenzenesulfonate, *J. Am. Chem. Soc.* 91, 1566–1567, 1969.

2.1.79. Nakagawa, Y. and Bender, M.L., Modification of alpha-chymotrypsin by methyl-*p*-nitrobenzene sulfonate, *Biochemistry* 9, 259–267, 1970.

2.1.80. Hunziker, P.E., Cysteine modification of metallothionein, *Methods Enzymol.* 205, 399–400, 1991.

2.1.81. Glick, B.R., The chemical modification of *Escherichia coli* ribosomes with methyl-*p*-nitrobenzenesulfonate. Evidence for the involvement of a histidine residue in the functioning of the ribosomal peptidyl transferase, *Can. J. Biochem.* 58, 1345–1347, 1984.

2.1.82. Swenson, R.P., Williams, C.H., Jr., and Massey, V., Methylation of the active center histidine 217 in *D*-amino acid oxidase by methyl-*p*-nitrobenzenesulfonate, *J. Biol. Chem.* 259, 5585–5590, 1984.

2.1.83. Marcus, J.P. and Dekker, E.E., Identification of a second active site residue in *Escherichia coli* L-threonine dehydrogenase: Methylation of histidine-90 with methyl-*p*-nitrobenzenesulfonate, *Arch. Biochem. Biophys.* 316, 413–420, 1995.

2.1.84. Kurz, J.L. and Lu, J.Y.W., Pressure dependence of rate of nucleophilic attack by water in aqueous solution. Hydrolysis of methyl-*p*-nitrobenzenesulfonate and ethyl trichloroacetate, *J. Phys. Chem.* 87, 1444–1448, 1983.

2.1.85. Hayaki, S., Kido, K., Sato, H., and Sakaki, S., Ab initio study on S_N2 reaction of methyl-*p*-nitrobenzenesulfonate and chloride anion in [mmim][PF_6], *Phys. Chem. Chem. Phys.* 12, 1822–1826, 2010.

2.1.86. Paik, W.K., Dimaria, P., Kim, S. et al., Alkylation of protein by methyl methanesulfonate and 1-methy-1-nitrosourea in vitro, *Cancer Lett.* 23, 9–17, 1984.

2.1.87. Boffa, L.C. and Bolognesi, C., Nuclear proteins damage by alkylating agents with different degrees of carcinogenicity, *Chem. Biol. Interact.* 55, 235–245, 1985.

2.1.88. Trézl, L., Park, K.S., Kim, S., and Paik, W.K., Studies on in vitro S-methylation of naturally occurring thiol compounds with *N*-methyl-*N*-nitrosourea and methyl methanesulfonate, *Environ. Res.* 43, 417–426, 1987.

2.1.89. Zhang, F., Bartels, M.J., Pottenger, L.H., and Gollapudi, B.B., Differential adduction of proteins vs. deoxynucleosides by methyl methanesulfonate and 1-methyl-1-nitrosourea in vitro, *Rapid Commun. Mass Spectrom.* 19, 438–448, 2005.

2.1.90. Okazaki, K., Yamada, H., and Imoto, T., A convenient *S*-2-aminoethylation of cysteinyl residues in reduced proteins, *Anal. Biochem.* 149, 516–520, 1985.

2.1.91. Hopkins, C.E., Hernandez, G., Lee, J.P., and Tolan, D.R., Aminoethylation in model peptides reveals conditions for maximizing thiol specificity, *Arch. Biochem. Biophys.* 443, 1–10, 2005.

2.1.92. Marincean, S., Smith, M.R., Beltz, L., and Borhan, B., Selectivity of labeled bromoethylamine for protein alkylation, *J. Mol. Model.* 18(9), 4547–4556, 2012.

2.1.93. Thevis, M., Ogorzalek Loo, R.R., and Loo, J.A., In-gel derivatization of proteins for cysteine-specific cleavages and their analysis by mass spectrometry, *J. Proteome Res.* 2, 163–172, 2003.

2.1.94. Rehulková, H., Marchetti-Deschmann, M., Pittenauer, E. et al., Improved identification of hordeins by cysteine alkylation with 2-bromoethylamine, SDS-PAGE and subsequent in-gel tryptic digestion, *J. Mass Spectrom.* 44, 1613–1621, 2009.

2.1.95. Hopkins, C.E., O'Conner, P.B., Allen, K.N. et al., Chemical-modification rescue assessed by mass spectrometry demonstrates that gamma-thia-lysine yield the same activity as lysine in aldolase, *Protein Sci.* 11, 1591–1599, 2002.

2.1.96. Woodyer, R., Wheatley, J.L., Relyea, H.A. et al., Site-directed mutagenesis of active site residues of phosphite dehydrogenase, *Biochemistry* 44, 4765–4774, 2005.

2.1.97. Sabatini, S., Pathophysiology of drug-induced papillary necrosis, *Fundam. Appl. Toxicol.* 4, 909–921, 1984.

2.1.98. Anderson, W.P., Woods, R.L., Thomas, C.J. et al., Renal medullary antihypertensive mechanisms, *Clin. Exp. Pharmacol. Physiol.* 22, S426–S429, 1995.

2.1.99. Sasaki, D., Yamada, A., Umeno, H. et al., Comparison of the course of biomarker changes and kidney injury in a rat model of drug-induced acute kidney injury, *Biomarkers* 16, 553–566, 2011.

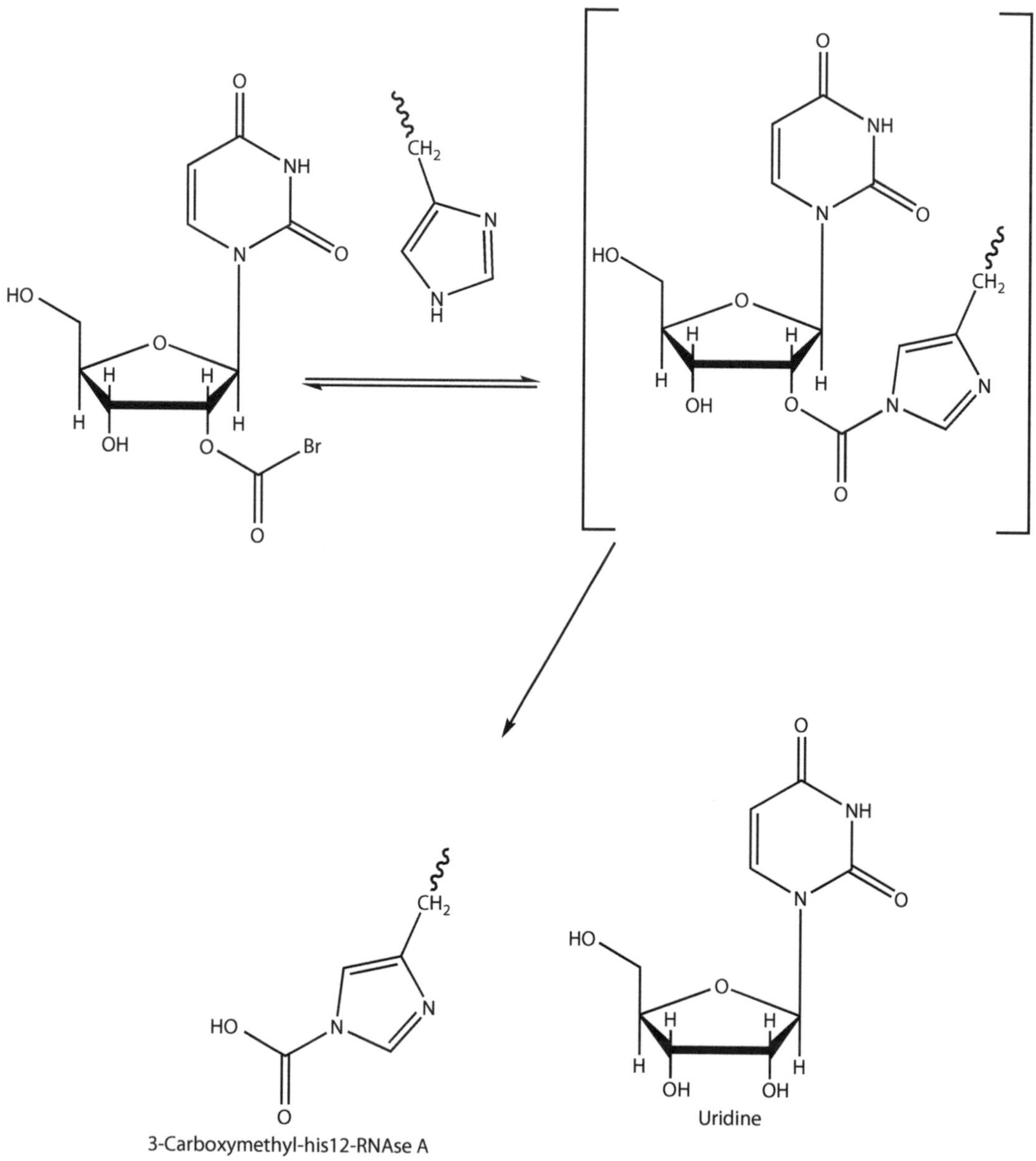

FIGURE 2.3 The reaction of an affinity label with pancreatic ribonuclease A. Shown is the structure of 2′(3′)-*O*-bromoacetyluridine and reactions with histidine. Relevant kinetic information is also presented. The reaction with RNAse is characterized by the reversible formation of an intermediate that rapidly decomposes to form 3-carboxymethyl histidine. (See Pincus, M., Thi, L.L., and Carty, R.P., The kinetics and specificity of the reaction of 2′(3′)-*O*-bromoacetyluridine with bovine pancreatic ribonuclease A, *Biochemistry* 13, 3533–3661, 1975; Hummel, C.F., Pincus, M.R., Brandt-Raf, P.W. et al., Reaction of (Bromoacetamido)nucleoside affinity labels with ribonuclease A: Evidence for steric control of reaction specificity and alkylation rate, *Biochemistry* 26, 135–146, 1987.)

TABLE 2.2
Rate Constants for the Reaction of Bromomethyl Benzene and Other Alkylating Agents with Amino Acids and Amino Acid Derivatives[a]

Amino Acid	Reagent/Conditions	Second-Order Rate Constants ($M^{-1} min^{-1}$)
His	BMB[b]/pH 7.3, 50 mM phosphate, $I = 0.5$[c]/25°C	0.12
Met	BMB/pH 6.8, 50 mM phosphate/25°C	3.84
Met	BMB/pH 6.8, 50 mM phosphate, $I = 0.5$[c]/25°C	5.05
Met	BMB/pH 6.8, 50 mM phosphate, $I = 0.5$/25°C with 6.0 M urea	5.9
N-Ac-L-Cys	BMB/pH 6.8, 50 mM phosphate, $I = 0.5$/25°C	3.1
GSH[d]	BMB/pH 6.8, 50 mM phosphate, $I = 0.5$/25°C	9.5
GSH	BMB/pH 9.5, 50 mM carbonate, $I = 0.5$/25°C	1×10^{3}[e]
Met[f]	Iodoacetate/pH 6.2/40°C	0.16
Met[f]	Iodoacetamide/pH 5.3/40°C	0.04

[a] Data adapted from Rogers, G.A., Shaltiel, N., and Boyer, P.D., Facile alkylation of methionine by benzyl bromide and demonstration of fumarase inactivation accompanied by alkylation of a methionine residue, *J. Biol. Chem.* 251, 5711–5717, 1976.

[b] BMB, bromomethyl benzene.

[c] Ionic strength adjusted with KCl.

[d] GSH, glutathione.

[e] Increase in rate of reaction at pH 9.5 emphasizes importance of thiolate anion in the reaction with haloalkane.

[f] These data were taken from Stark, G.R. and Stein, W.H., Alkylation of the methionine residues of ribonuclease in 8 M urea, *J. Biol. Chem.* 239, 3755–2761, 1964.

TABLE 2.3

α-Halo Carbonyl Compounds

Compound	Chemistry	Reactions	References
Chloroacetic acid[a]	ClCH$_2$COOH	Modification of cysteine; other nucleophiles with difficulty.[b]	2.3.1–2.3.13
Bromoacetic acid	BrCH$_2$COOH	Modification of cysteine, methionine, histidine, and glutamic acid (γ-carboxyl).[c] More reactive than chloroacetate.	2.3.14–2.3.19
Iodoacetic acid	ICH$_2$COOH	Modification of cysteine, methionine, histidine, and glutamic acid (γ-carboxyl).[d] More reactive than either chloroacetate or bromoacetate. Iodoacetate (and iodoacetamide) is used to carboxymethylate proteins after reduction in proteomic analysis.	2.3.20–2.3.48
Chloroacetamide	ClCH$_2$CONH$_2$	Relatively unreactive with most nucleophiles.[e]	2.3.49–2.3.53
Bromoacetamide	BrCH$_2$CONH$_2$	More reactive than chloroacetamide with reaction at histidine and methionine residues. Not extensively used for protein modification.[f]	2.3.54–2.3.62
Iodoacetamide	ICH$_2$CONH$_2$	More reactive than the other haloalkyl functions. Reported modification of cysteine, methionine, histidine, and lysine. Glutamate(γ-carboxyl), α-amino groups.[g]	2.3.63, 2.3.64
p-Bromophenacyl bromide		Modification of carboxyl groups, amino groups, and sulfhydryl groups.[h]	2.3.65–2.3.73
Bromoketo acid generated from suicide substrate		Cysteine alkylation.[i]	2.3.74–2.3.76

[a] The haloacetyl function (Figure 2.16) serves as base for the development of a variety of site-directed modifications in proteins using affinity labeling. These are more usefully cited in chapters on the modification of specific amino acid residues in proteins. One example, however, is provided by N^ε-haloacetyl-L-lysine derivatives of methotrexate (Figure 2.17). ICATs may also use a haloalkyl function (Figure 2.17).[2.3.1] The order of reactivity of haloacetate protein nucleophiles was established in early work by Korman and Clarke in 1956.[2.3.2] In studies on the reaction of chloroacetic acid, bromoacetic acid, and iodoacetic acid with tyrosine and histidine, the order of reactivity was chloroacetic < bromoacetic ≈ iodoacetic. There was modification of the phenolic hydroxyl, α-amino (tyrosine), and the imidazole ring of histidine. The modification of the phenolic hydroxyl group was not observed with chloroacetate. These reactions were performed at pH 9.0 in the presence of magnesium oxide. The fluoroderivatives are essentially unreactive toward nucleophiles found in proteins. Fluoroacetate is used as a rodenticide with the toxicity based on in vivo incorporation into fluorocitrate that is potent inhibitor of aconitase in the TCA cycle.[2.3.3] Haloacetic acids are used (illegally) for the preservation of alcoholic beverages.[2.3.4] A bacterium that degrades haloacetate derivatives has been isolated.[2.3.5]

[b] As with the haloalkanes (Table 2.1), chloroacetic acid (and chloroacetamide) reacts slowly with nucleophiles in proteins (Table 2.4). The low reactivity of the chloroacetyl function is an advantage with affinity labels where nonspecific binding of inhibitor is to be avoided. The low reactivity of the chloroacetate and chloroacetamide is also useful in studies with highly reactive functional groups in proteins when use of the more highly reactive iodoacetyl derivatives is not useful.[2.3.6] Chloroacetic acid was used for the preparation of one of the first ion-exchange celluloses developed for protein purification[2.3.7] and is currently used in the preparation of derivatized starches celluloses and dextran-based hydrogels.[2.3.8–2.3.11] Sawdust has been modified with chloroacetic acid in a process to develop biodegradable plastics.[2.3.12] Yu and coworkers[2.3.13] have advanced a novel using chloroacetyl–coenzyme A to introduce a chloroacetyl function on the epsilon-amino group of a lysine residue in protein. These investigators selected the chloroderivative on lower reactivity compared to other haloacetyl functions.

[c] Bromoacetic acid was the first haloacetyl derivative introduced for use in the study of bromine derivatives by Steinauer in 1974.[2.3.14] Bromoacetate can be used to modify methionine residues in proteins[2.3.15–2.3.18] in addition to reactions at cysteine and histidine. In the case of ribonuclease, the modification of pancreatic ribonuclease A is facilitated by the formation of a transition-state intermediate.[2.3.19] Earlier studies on ribonuclease showed that modification of lysine residue with bromoacetate occurred at pH 8.5, while only His12 and His119 were modified at pH 5.5.

TABLE 2.3 (continued)
α-Halo Carbonyl Compounds

d   Iodoacetate and iodoacetamide are the most reactive haloacetyl-based compounds (Table 2.4; Chapter 7) and have a long history of use in the modification of cysteinyl, histidyl, and methionyl residues in proteins.[2.3.20–2.3.25] The modification of histidine occurs at an enzyme active site in the absence of sulfhydryl groups.[2.3.26] The modification of lysine residues and α-amino groups has been reported as a secondary reaction largely dependent on solvent conditions; in the case of bovine pancreatic ribonuclease A, lysine modification occurs at pH 9.5–10.[2.3.22] The modification of the γ-carboxyl group of glutamic acid has been reported.[2.3.27,2.3.28] Iodoacetate is frequently used to demonstrate the importance of a cysteinyl residue in the function of a protein.[2.3.29–2.3.31] One of the more interesting early studies[2.3.32] reported that iodoacetate prevented thermal coagulation of serum proteins, while iodoacetamide had no effect; cysteine promoted aggregation. The ability of iodoacetate to inhibit the thermal coagulation of plasma (iodoacetate index) was advanced as a diagnostic test some 60 years ago[2.3.33–2.3.35] but apparently failed to gain traction. The early work on iodoacetate in whole animal, tissue, and cell studies has been reviewed by Webb.[2.3.36] Iodoacetate is still actively used to generate an animal model of osteoarthritis.[2.3.37–2.3.42] Iodoacetate has a long history of use to carboxymethylate proteins after reduction in the studies of proteins. The early use was focused on the preparation of forms of proteins for structural analysis.[2.3.43] More recently, reduction and carboxymethylation/carboxamidomethylation are used in proteomic studies.[2.3.44–2.3.48]

e   Chloroacetamide was used to identify the most reactive cysteine residue in tubulin.[2.3.49] This use was based on the low reactivity of this haloacetamide as mentioned earlier for chloroacetic acid.[2.3.6] The haloacetamide and neutral reagents are therefore not subject to interaction with ionizing groups at or near the protein nucleophile. Chaiken and Smith demonstrated that while the inactivation of papain with chloroacetic acid showed a bell-shaped pH dependence with a maximum rate at approximately pH 6.0, modification with chloroacetamide showed a sigmoidal curve reaching a plateau at approximately pH 11.[2.3.50,2.3.51] Gerwin had obtained similar results in earlier work on streptococcal proteinase.[2.3.6] The results on the two sulfhydryl proteinases are interpreted as demonstrating the importance of a histidine residue in the reaction with chloroacetate. This would suggest that the haloacid was functioning an affinity label.[2.3.52,2.3.53]

f   Bromoacetamido group has been used more frequently as the functional portion of an affinity label.[2.3.54–2.3.57] There was considerable early use of bromoacetamide to study ion channels in intact cells and tissues.[2.3.58–2.3.61] (See the following iodoacetamide comment for rationale for use in intact cells/tissues). Bromoacetyl bromide (Figure 2.16) has been used to couple a chelating agent to a monoclonal antibody containing an engineered cysteine residue.[2.3.62]

g   Iodoacetamide was introduced as a more lipid-soluble derivative of iodoacetate for transport across membranes.[2.3.63] Much of the early work with iodoacetate and related haloacetyl derivatives was performed in intact animals and tissue/cell systems that was reviewed by J Leyden Webb in 1973.[2.3.36] Iodoacetamide is also used in the reduction and carboxymethylation of proteins.[2.3.64]

h   p-Bromophenacyl bromide (Figure 2.18) reacts with a variety of functional groups in proteins. This reagent is best known for reaction with phospholipase A serving as a unique differentiating characteristic for this class of enzymes.[2.3.65–2.3.70] p-Bromophenacyl bromide also inactivates pepsin by modification of a carboxyl group.[2.3.71–2.3.73]

i   A bromoacetyl function is generated from a bromoenol lactone (BEL) (Figure 2.19). Phospholipase A2β hydrolyzes the lactone-generating enol, which can under tautomerization generate a bromoacetyl function.[2.3.74,2.3.75] The BELs were developed as inhibitors for chymotrypsin.[2.3.76]

References to Table 2.3

2.3.1.   Rosowsky, A., Solan, V.C., Forsch, R.A. et al., Methotrexate analogues. 30. Dihydrofolate reductase inhibition and in vitro tumor cell growth inhibition by $N^ε$-(haloacetyl)-L-lysine and $N^δ$-(haloactetyl)-L-ornithine analogues and an acivicin analogue of methotrexate, *J. Med. Chem.* 30, 1463–1469, 1987.

2.3.2.   Korman, S. and Clarke, H.T., Carboxymethylamino acids and peptides, *J. Biol. Chem.* 221, 113–131, 1956.

2.3.3.   Goncharov, N.V., Jenkins, R.O., and Radilov, A.S., Toxicology of fluoroacetate: A review with possible directions for therapy research, *J. Appl. Toxicol.* 26, 148–161, 2006.

2.3.4.   Cardador, M.J. and Gllego, M., Development of a method for the quantitation of chloro-, bromo-, and iodoacetic acids in alcoholic beverages, *J. Agric. Food Chem.* 60, 725–730, 2012.

2.3.5.   Horisaki, T., Yoshida, E., Sumiya, K. et al., Isolation and characterization of monochloroacetic acid-degrading bacteria, *J. Gen. Appl. Microbiol.* 57, 277–284, 2011.

2.3.6.   Gerwin, B.I., Properties of the single sulfhydryl group of Streptococcal proteinase. A comparison of the rate of alkylation by chloroacetic acid and chloroacetamide, *J. Biol. Chem.* 242, 451–456, 1967.

(*continued*)

TABLE 2.3 (continued)
α-Halo Carbonyl Compounds

2.3.7. Peterson, E.A. and Sober, H.A., Chromatography of proteins. I. Cellulose ion-exchange adsorbents, *J. Am. Chem. Soc.* 78, 751–755, 1956.

2.3.8. Yanli, W., Wenyuan, G., and Xia, L., Carboxymethyl Chinese yam starch: Synthesis, characterization and influence of reaction parameters, *Carbohydr. Res.* 344, 1764–1769, 2009.

2.3.9. Björses, K., Faxälv, L., Montan, C. et al., In vitro and in vivo evaluation of chemically modified degradable starch microspheres for topical hemostasis, *Acta Biomater.* 7, 2558–2565, 2011.

2.3.10. Peschel, D., Zhang, K., Aggarwal, N. et al., Synthesis of novel celluloses derivatives and investigation of their mitogenic activity in the presence and absence of FGF1, *Acta Biomater.* 6, 2116–2125, 2010.

2.3.11. Sun, G., Shen, Y.I., Ho, C.C. et al., Functional groups affect physical and biological properties of dextran-based hydrogels, *J. Biomed. Mater. Res. A* 93, 1080–1090, 2010.

2.3.12. Schilling, C.H., Tomasik, P., Karpovich, D.S. et al., Preliminary studies on converting agricultural waste into biodegradable plastics. Part III. Sawdust, *J. Polym. Environ.* 13, 177–182, 2005.

2.3.13. Yu, M., Sorio de Carvalho, L.P., Sun, G., and Blanchard, J.S., Activity-based substrate profiling for Gcn5-related N-acetyltransferases: The use of chloroacetyl-coenzyme A to identify protein substrates, *J. Am. Chem. Soc.* 128, 15356–15357.

2.3.14. Steinauer, E., Untersuchungen über die physiologische Wirkung der Bromprärate, *Arch. Pathol. Anat.* 59, 65–133, 1874.

2.3.15. Glick, D.M., Goren, H.J., and Barnard, E.A., Concurrent bromoacetate reaction at histidine and methionine residues in ribonuclease, *Biochem. J.* 102, 7C–10C, 1967.

2.3.16. Goren, H.J., Glick, D.M., and Barnard, E.A., Analysis of carboxymethylated residues in proteins by an isotopic method, and its application to the bromoacetate-ribonuclease reaction, *Arch. Biochem. Biophys.* 126, 607–623, 1968.

2.3.17. Nigen, A.M., Keim, P., Marshall, R.C. et al., Carbon 13 nuclear magnetic resonance spectroscopy of myoglobins carboxymethylated with enriched [2–13C]bromoacetate, *J. Biol. Chem.* 248, 3724–3732, 1973.

2.3.18. Eakin, R.T., Morgan, L.O., and Matwiyoff, N.A., Carbon-13 nuclear magnetic resonance spectroscopy of [2–$^{13+C}$] carboxymethylcytochrome c, *Biochemistry* 14, 4538–4543, 1975.

2.3.19. Lennette, E.P. and Plapp, B.V., Transition-state analysis of the facilitated alkylation of ribonuclease A by bromoacetate, *Biochemistry* 18, 3938–3946, 1979.

2.3.20. Moore, J., Jr. and Fenselau, A., Selective cysteine modification in glyceraldehyde-3-phosphate dehydrogenase, *Biochemistry* 11, 3762–3770, 1972.

2.3.21. Lowbridge, J. and Fruton, J.S., Studies on the extended active site of papain. *J. Biol. Chem.* 249, 6754–6761, 1974.

2.3.22. Gundlach, H.G., Stein, W.H., and Moore, S., The nature of the amino acid residues involved in the inactivation of ribonuclease by iodoacetate, *J. Biol. Chem.* 234, 1754–1760, 1959.

2.3.23. Gundlach, H.G., Moore, S., and Stein, W.H., The reaction of iodoacetate with methionine, *J. Biol. Chem.* 234, 1761–1764, 1959.

2.3.24. Stark, G.R. and Stein, W.H., Alkylation of the methionine residues of ribonuclease in 8 M urea, *J. Biol. Chem.* 239, 3755–3761, 1964.

2.3.25. Cummings, J.G., Lau, S.M., Powell, P.J., and Thorpe, C., Reductive half-reaction in medium-chain acyl-CoA dehydrogenase: Modulation of internal equilibrium by carboxymethylation of a specific methionine residues, *Biochemistry* 31, 8523–8529, 1992.

2.3.26. Liu, T.Y., Demonstration of the presence of a histidine residue at the active site of streptococcal proteinase, *J. Biol. Chem.* 242, 4029–4032, 1967.

2.3.27. Hashimoto, J. and Takahashi, K., Chemical modifications of ribonuclease U1, *J. Biochem.* 81, 1175–1180, 1977.

2.3.28. Yoshida, H. and Hanazawa, H., Carboxymethylation of an active site glutamic acid residue of ribonuclease F1 iodoacetate, *Biochimie* 71, 687–692, 1989.

2.3.29. Nagy, L., Johnson, B.R., Hauschka, P., and Szabo, S., Characterization of proteases and protease inhibitors in the rat stomach, *Am. J. Physiol.* 272, G1151–G1158, 1997.

2.3.30. Sárkány, Z., Skern, T., and Polgár, L., Characterization of the active site thiol group of rhinovirus 2A proteinase, *FEBS Lett.* 481, 289–292, 2000.

2.3.31. Kamińska, J., Wiśniewska, A., and Kościelak, J., Chemical modifications of α1,6-fucosyltransferase define amino acid residues of catalytic importance, *Biochimie* 85, 303–310, 2003.

2.3.32. Huggins, C. and Jensen, E.V., Thermal coagulation of serum proteins; the effects of iodoacetate, iodoacetamide, and thiol compounds on coagulation, *J. Biol. Chem.* 179, 645–654, 1949.

TABLE 2.3 (continued)
α-Halo Carbonyl Compounds

2.3.33. Homburger, F., Pfeiffer, P.H., Page, O. et al., Evaluation of diagnostic tests for cancer; inhibition of thermal coagulation of serum by iodoacetic acid, the Huggins-Miller-Jensen test, *Cancer* 3, 15–25, 1950.

2.3.34. Glass, G.B., Boyd, L.J., and Dworecki, I.J., Thermal coagulation point and iodoacetate index as detectors of abnormal serum proteins and underlying illness, *Proc. Soc. Exp. Biol. Med.* 76, 10–15, 1951.

2.3.35. Ellerbrook, L.D., Meek, E.C., and Lippincott, S.W., Studies of various tests for malignant neoplastic diseases. VI. Tests for the least coagulable serum protein and the iodoacetate index, *J. Natl. Cancer Inst.* 12, 49–89, 1951.

2.3.36. Webb, J.L., Iodoacetate and iodoacetamide, in *Enzyme and Metabolic Inhibitors*, Vol. III, Chapter 1, pp. 1–283, Academic Press, New York, 1968.

2.3.37. van der Kraan, P.M., Vitters, E.L., van de Putte, L.B., and van Den Berg, W.B., Development of osteoarthritic lesions in mice by "metabolic" and "mechanical" alterations in the knee joints, *Am. J. Pathol.* 135, 1001–1014, 1989.

2.3.38. Cledes, G., Felizardo, R., Foucart, J.M., and Carpentier, P., Validation of a chemical osteoarthritis model in rabbit temporomandibular joint: A compliment to biomechanical models, *Int. J. Oral Maxillofac. Surg.* 35, 1026–1033, 2006.

2.3.39. Pomonis, J.D., Boulet, J.M., Gottshall, S.L. et al., Development and pharmacological characterization of a rat model of osteoarthritis, *Pain* 114, 339–346, 2005.

2.3.40. Ameye, L.G. and Young, M.F., Animal models of osteoarthritis: Lessons learned while seeking the "Holy Grail", *Curr. Opin. Rheumatol.* 18, 537–547, 2006.

2.3.41. Ahmed, A.S., Li, J., Erlandsson-Harris, H. et al., Suppression of pain and joint destruction by inhibition of the proteasome system in experimental osteoarthritis, *Pain* 153, 18–26, 2012.

2.3.42. Nagase, H., Kumakura, S., and Shimada, K., Establishment of a novel objective and quantitative method to assess pain-related behavior in monosodium iodoacetate-induced osteoarthritis in rat knee, *J. Pharmacol. Toxicol. Methods* 65, 29–36, 2012.

2.3.43. Crestfield, A.M., Moore, S., and Stein, W.H., The preparation and enzymatic hydrolysis of reduced and *S*-carboxymethylated proteins, *J. Biol. Chem.* 238, 622–627, 1963.

2.3.44. Boja, E.S. and Fales, H.M., Overalkylation of a protein digest with iodoacetamide, *Anal. Chem.* 73, 3576–3582, 2001.

2.3.45. Herbert, B., Galvani, M., Hamdan, M. et al., Reduction and alkylation of proteins in preparation of two-dimensional map analysis: Why, when, and how?, *Electrophoresis* 22, 2046–2057, 2001.

2.3.46. Galvani, M., Hamdan, M., Herbert, B., and Righetti, P.G., Alkylation kinetics of proteins in preparation of two-dimensional maps: A matrix assisted laser desorption/ionization-mass spectrometry investigation, *Electrophoresis* 22, 2058–2065, 2001.

2.3.47. Galvani, M., Rovatti, L., Hamdan, M. et al., Protein alkylation in the presence/absence of thiourea in proteome analysis: A matrix-assisted laser desorption/ionization-time-of-flight-mass spectrometry investigation, *Electrophoresis* 22, 2066–2074, 2001.

2.3.48. Nielsen, M.L., Vermeulen, M., Bonaldi, T. et al., Iodoacetamide-induced artifact mimics ubiquitination in mass spectrometry, *Nat. Methods* 5, 459–460, 2008.

2.3.49. Britto, P.J., Knipling, L., and Wolff, J., The local electrostatic environment determines cysteine reactivity of tubulin, *J. Biol. Chem.* 277, 29018–29027, 2002.

2.3.50. Chaiken, I.M. and Smith, E.L., Reaction of chloroacetamide with the sulfhydryl group of papain, *J. Biol. Chem.* 244, 5087–5094, 1969.

2.3.51. Chaiken, I.M. and Smith, E.L., Reaction of the sulfhydryl group of papain with chloroacetic acid, *J. Biol. Chem.* 244, 5095–5099, 1969.

2.3.52. Plapp, B.V., Mechanism of carboxymethylation of bovine pancreatic nucleases by haloacetates and tosylglycolate, *J. Biol. Chem.* 248, 4896–4900, 1973.

2.3.53. Dahl, K.H. and McKinley-McKee, J.S., Enzymatic catalysis in the affinity labeling of liver alcohol dehydrogenase with haloacids, *Eur. J. Biochem.* 118, 507–513, 1981.

2.3.54. Aliau, S., Delettre, G., Mattras, H. et al., Steroidal affinity labels of the estrogen receptor alpha. 4. Electrophilic 11_β-aryl derivatives of estradiol, *J. Med. Chem.* 43, 613–628, 2000.

2.3.55. Ferry, G., Ubeaud, C., Mozo, J. et al., New substrate analogues of human serotonin N-acetyltransferase produce in situ specific and potent inhibitors, *Eur. J. Biochem.* 271, 418–428, 2004.

2.3.56. Thomas, J.M. and Perrin, D.M., Active site labeling of G8 in the hairpin ribozyme: Implications for structure and mechanism, *J. Am. Chem. Soc.* 128, 16540–16545, 2006.

(continued)

TABLE 2.3 (continued)
α-Halo Carbonyl Compounds

2.3.57. Peacock, H., Bachu, R., and Beal, P.A., Covalent stabilization of a small molecule-RNA complex, *Bioorg. Chem. Chem. Lett.* 21, 5002–5005, 2011.

2.3.58. Oxford, G.S., Wu, C.H., and Narahashi, T., Removal of sodium channel inactivation in squid giant axons by n-bromoacetamide, *J. Gen. Physiol.* 71, 227–247, 1978.

2.3.59. Wang, G.K., Modification of sodium channel inactivation in single myelinated nerve fibers by methionine-reactive chemicals, *Biophys. J.* 46, 121–124, 1984.

2.3.60. Horn, R., Vandenberg, C.A., and Lange, K., Statistical analysis of single sodium channels. Effects of *N*-bromoacetamide, *Biophys. J.* 45, 323–335, 1984.

2.3.61. Patlak, J. and Horn, R., Effect of *N*-bromoacetamide on single sodium channel currents in excised membrane patches, *J. Gen. Physiol.* 79, 333–351, 1982.

2.3.62. Tinianow, J.N., Gill, H.S., Ogasawara, A. et al., Site-specifically ^{89}Zr-labeled monoclonal antibodies for ImmunoPET, *Nucl. Med. Biol.* 37, 289–297, 2010.

2.3.63. Goddard, D.R., The reversible heat activation induced germination and increased respiration in the ascospores of *Neurospora tetrasperma*, *J. Gen. Physiol.* 19, 45–60, 1935–1936.

2.3.64. Krüger, R., Hung, C.W., Edelson-Averbukh, M., and Lehmann, W.D., Iodoacetamide-alkylated methionine can mimic neutral loss of phosphoric acid from phosphopeptides as exemplified by nano-electrospray ionization quadrupole time-of-flight parent ion scanning, *Rapid Commun. Mass Spectrom.* 19, 1709–1716, 2005.

2.3.65. Mitchell, S.M., Poyser, N.L., and Wilson, N.H., Effect of *p*-bromophenacyl bromide, an inhibitor of phospholipase A2, on arachidonic acid release and prostaglandin synthesis by the guinea-pig uterus in vitro, *Br. J. Pharmacol.* 59, 107–113, 1977.

2.3.66. Roberts, M.F., Deems, R.A., Mincey, T.C. et al., Chemical modification of the histidine residue in phospholipase A2 (*Naja naja naja*). A case of half-site reactivity, *J. Biol. Chem.* 252, 2405–2411, 1977.

2.3.67. Shaw, J.O., Roberts, M.F., Ulevitch, R.J. et al., Phospholipase A2 contamination of cobra venom factor preparations. Biologic role in complement-dependent in vivo reaction and inactivation with *p*-bromophenacyl bromide, *Am. J. Pathol.* 91, 517–530, 1978.

2.3.68. Boegerman, S.C., Deems, R.A., and Dennis, E.A., Phospholipid binding and the activation of group IA secreted phospholipase A2, *Biochemistry* 43, 3907–3916, 2004.

2.3.69. Marchi-Salvador, D.P., Fernandes, C.A., Silveira, L.B. et al., Crystal structure of a phospholipase A(2) homolog complexed with *p*-bromophenacyl bromide reveals important structural changes associated with the inhibition of myotoxic activity, *Biochim. Biophys. Acta* 1794, 1583–1590, 2009.

2.3.70. Harada, N., Zhao, J., Kurihara, H. et al., Stimulation of Fc$_\gamma$RI on primary sensory neurons increases insulin-like growth factor-I production, thereby reducing reperfusion-induced renal injury in mice, *J. Immunol.* 185, 1303–1310, 2010.

2.3.71. Erlanger, B.F., Vratsanos, S.M., Wassermann, N., and Cooper, A.G., A chemical investigation of the active center of pepsin, *Biochem. Biophys. Res. Commun.* 23, 243–245, 1966.

2.3.72. Kageyama, T. and Takahashi, K., Pepsinogens and pepsins from gastric mucosa of Japanese monkey. Purification and characterization, *J. Biochem.* 79, 455–468, 1976.

2.3.73. Tanji, M., Kageyama, T., and Takahashi, K., Tuna pepsinogens and pepsins. Purification, characterization, and amino-terminal sequences, *Eur. J. Biochem.* 177, 251–259, 1988.

2.3.74. Song, H., Ramanadham, S., Bao, S. et al., A bromoenol lactone suicide substrate inactivates group VIA phospholipase A2 by generating a diffusible bromomethyl keto acid that alkylated cysteine thiols, *Biochemistry* 45, 1061–1073, 2006.

2.3.75. Song, H., Wohltmann, M., Tan, M. et al., Group IVA PLA2 (iPLA2β) is activated upstream of p38 mitogen-activated protein kinase (MAPK) in pancreatic islet β-cell signaling, *J. Biol. Chem.* 287, 5528–5541, 2012.

2.3.76. Daniels, S.B., Cooney, E., Sofia, M.J. et al., Haloenol lactones. Potent enzyme-activated irreversible inhibitors for α-chymotrypsin, *J. Biol. Chem.* 258, 15046–15053, 1983.

TABLE 2.4
Rate Constants for the Reaction of Haloacetates and Haloacetamides with Nucleophiles

Reagent	Nucleophile/Experimental Conditions	Rate (M^{-1} s^{-1})	Reference
2-Chloroethanol	Cysteine/1.0 M diethanolamine–HCl, pH 9.0, 37°C	0.0002	2.4.1
Chloroacetate	GSH[a]/in a wide-range buffer (pH 9.0) of acetate/phosphate/boric acid,[b] 25°C	0.009[c]	2.4.2
Chloroacetate	L-cysteine, pH 6.0 (pH stat with 0.3 M KCl for ionic strength), 25°C	0.0008	2.4.3
Chloroacetate	2-Mercaptoethanol/in a wide-range buffer (pH 8.4) of acetate/phosphate/boric acid, 25°C	0.028	2.4.4
Chloroacetate	Cysteine/in a wide-range buffer (pH 8.4) of acetate/phosphate/boric acid, 25°C	0.011	2.4.4
Chloroacetamide	2-Mercaptoethanol/in a wide-range buffer (pH 8.4) of acetate/phosphate/boric acid, 25°C	0.169	2.4.4
Chloroacetamide	Cysteine/in a wide-range buffer (pH 8.4) of acetate/phosphate/boric acid, 25°C	0.061	
Chloroacetamide	2-Mercaptoethanol (pH stat), 30°C in 0.1 M KCl	0.0086 (pH 7.0) 0.05 (pH 8.0) 0.11 (pH 9.0)	2.4.5
Chloroacetamide	Cysteine (pH stat), 30°C in 0.1 M KCl	0.0067 (pH 7.0) 0.042 (pH 8.0) 0.097 (pH 9.0)	2.4.5
Chloroacetamide	GSH, pH 9.0, 30°C	0.135[c]	2.4.6
2-Bromoethanol	Cysteine/1.0 M diethanolamine–HCl, pH 9.0, 37°C	0.0025	2.4.1
Bromoethane	N-Acetyl cysteine/0.5 M borate, pH 9.5, 25°C	0.00056	2.4.7
2-Bromoethanol	N-Acetyl cysteine/0.5 M borate, pH 9.5, 25°C	0.000011	2.4.7
2-Bromoethylamine	Ac-CysTrpArg-amide/MOPS-TAPS-AMPSO, 21°C	0.015 (pH 8.4)[c] 0.0058 (pH 9.2)[c] 0.001 (pH 10)[c,d]	2.4.8
Bromoacetate	Cysteine/1.0 M diethanolamine–HCl, pH 9.0, 37°C	2.9	2.4.1
Bromoacetate	2-Mercaptoethanol/in a wide-range buffer (pH 8.4) of acetate/phosphate/boric acid, 25°C	1.97	2.4.4
Bromoacetate	Cysteine/in a wide-range buffer (pH 8.4) of acetate/phosphate/boric acid, 25°C	1.03	2.4.4
Bromoacetate	2-Mercaptoethanol/in a wide-range buffer (pH 8.4) of acetate/phosphate/boric acid, 25°C	8.79	2.4.4
Bromoacetate	Histidine/0.1 M acetate, pH 5.5, 25°C	0.0000086[e]	2.4.9
Bromoacetate	Histidine hydantoin, pH 7.72, 25°C	0.000059	2.4.10
Bromoacetamide	Cysteine/in a wide-range buffer (pH 8.4) of acetate/phosphate/boric acid, 25°C	5.71	2.4.4
Iodoacetate	Cysteine/1.0 M diethanolamine–HCl, pH 9.0, 37°C	5.5	2.4.1
Iodoacetate	N-Acetyl cysteine/1.0 M diethanolamine–HCl, pH 9.0, 37°C	1.25	2.4.1
Iodoacetate	2-Mercaptoethanol/1.0 M diethanolamine–HCl, pH 9.0, 37°C	1.4	2.4.1
Iodoacetate	GSH/1.0 M diethanolamine–HCl, pH 9.0, 37°C	4.5	2.4.1
Iodoacetate	Cystamine/1.0 M diethanolamine–HCl, pH 9.0, 37°C	6.4	2.4.1
Iodoacetate	2-Mercaptoethanol/0.5 M Triethanolamine, pH 8.55, 23.5°C	0.134	2.4.11

(continued)

TABLE 2.4 (continued)
Rate Constants for the Reaction of Haloacetates and Haloacetamides with Nucleophiles

Reagent	Nucleophile/Experimental Conditions	Rate (M^{-1} s^{-1})	Reference
Iodoacetate	2-Mercaptoethanol/0.5 M Triethanolamine, pH 7.55, 23.5	0.026	2.4.11
Iodoacetate	Methionine/pH 6.2, 40°C[f]	0.0026	2.4.12
Iodoacetamide	Methionine/pH 5.3, 40°C[f]	0.0007	2.4.12
Iodoacetate	Methionine/pH 3.7 with 8.0 M urea, 40°C[f]	0.0023	2.4.12
Iodoacetate	Methionine/pH 5.8 with 8.0 M urea, 40°C[f]	0.0034	2.4.12
Iodoacetamide	Methionine/pH 4.0 with 8.0 M urea, 40°C[f]	0.0010	2.4.12
Iodoacetamide	Methionine/pH 7.0 with 8.0 M urea, 40°C[f]	0.0009	2.4.12

[a]   GSH, reduced glutathione.

[b]   The wide-range buffer was developed for kinetic studies on streptococcal proteinase (Gerwin, B.I., Stein, W.H., and Moore, S., On the specificity of streptococcal proteinase, *J. Biol. Chem.* 241, 3331–3339, 1966). Experiments performed on the modification of streptococcal proteinase with chloroacetamide used this buffer; substitution of a triethanolamine buffer at equal ionic strength gave similar results ruling out possible specific buffer effects.

[c]   Estimated from graphic data at pH 9.0. An apparent limit value of 0.016 M^{-1} s^{-1} was observed at pH 10.

[d]   Specificity of reaction with cysteine is maximized at alkaline pH.

[e]   The rate of reaction of bromoacetate with histidine-119 in RNAse under the same reaction conditions is 0.0184 M^{-1} s^{-1}.

[f]   The reactions were performed in the absence of buffer as protons are not released or consumed during the reaction with the thioether sulfur. The authors noted that carboxylic acid buffers reacted with the alkylating reagents.

References to Table 2.4

2.4.1.   Dahl, K.H. and McKinley-McKee, J.S., The reactivity of affinity labels: A kinetics study of the reaction of alkyl halides with thiolate anions—A model reaction for protein alkylation, *Bioorg. Chem.* 10, 329–341, 1981.

2.4.2.   Gerwin, B.I., Properties of the single sulfhydryl group of Streptococcal proteinase, *J. Biol. Chem.* 242, 451–456, 1967.

2.4.3.   Sluyterman, L.A.Æ., The rate-limiting reaction in papain action as derived from the reaction of the enzyme with chloroacetic acid, *Biochim. Biophys. Acta* 151, 178–187, 1968.

2.4.4.   Wandinger, A. and Creighton, D.J., Solvent isotope effects on the rates of alkylation of thiolamine models of papain, *FEBS Lett.* 116, 116–121, 1980.

2.4.5.   Lindley, H., A study of the kinetics of the reaction between thiol compounds and chloroacetamide, *Biochem. J.* 74, 577–584, 1960.

2.4.6.   Lindley, H., The reaction of thiol compounds and chloroacetamide 2. The reaction between chloroacetamide and cysteine peptides, *Biochem. J.* 82, 418–425, 1962.

2.4.7.   Schindler, J.F. and Viola, R.E., Conversion of cysteinyl residues to unnatural amino acid analogs. Examination in a model system, *J. Protein Chem.* 15, 737–742, 1996.

2.4.8.   Hopkins, C.E., Hernandez, G., Lee, J.P., and Tolan, D.R., Aminoethylation in model peptides reveals conditions for maximizing thiol specificity, *Arch. Biochem. Biophys.* 441, 1–10, 2005.

2.4.9.   Heinrikson, R.L., Stein, W.H., Crestfield, A.M., and Moore, S., The reactivities of the histidine residues at the active site of ribonuclease toward halo acids of different structure, *J. Biol. Chem.* 240, 2921–2934, 1965.

2.4.10. Lennette, E.P. and Plapp, B.V., Kinetics of carboxymethylation of histidine hydantoin, *Biochemistry* 18, 3933–3938, 1979.

2.4.11. Dahl, K.H. and McKinley-McKee, J.S., Enzymatic catalysis in the affinity labelling of liver alcohol dehydrogenase with haloacetate, *Eur. J. Biochem.* 118, 507–513, 1981.

2.4.12. Stark, G.R. and Stein, W.H., Alkylation of the methionine residues of ribonuclease in 8 M urea, *J. Biol. Chem.* 239, 3755–3761, 1964.

TABLE 2.5
Halobimanes[a]

Compound	Chemistry	Reactions	References
Monochlorobimane		Reaction with sulfhydryl groups to yield fluorescent derivatives. The majority of use is the measurement of glutathione in cells.[b]	2.5.1–2.5.11
Monobromobimane (MBB)		Reaction with sulfhydryl groups to yield fluorescent derivatives. MBB is used for modification of cysteine residues in proteins as well as for cellular glutathione.[c]	2.5.12–2.5.16
Dibromobimane		Cross-linking of cysteine residues in proteins.[d]	2.5.17, 2.5.18

[a] The bimanes (Figure 2.20) were first synthesized in 1978 by Kosower and coworkers.[2.5.1] This was observed by the scientists during their research. The bimanes were identified as a highly fluorescent by-product in the synthesis of 2-octadecynoic acid. There has been some use in the modification of cysteine residues in proteins but more extensive use in the measurement of thiol groups in cells[2.5.2] and thiol groups in solution.[2.5.3,2.5.4] The in vivo measurement of glutathione with monochlorobimane or MBB uses the glutathione-*S*-transferase.[2.5.5,2.5.6]

[b] Monochlorobimane is used to measure glutathione levels inside cells as a measure of cell viability.[2.5.5,2.5.7–2.5.11]

[c] MBB has been used to modify cysteine residues in proteins.[2.5.12–2.5.14] Hu and coworkers[2.5.15] showed that MBB was an affinity label for certain isozymes of rat liver glutathione-*S*-transferase in the absence of the substrate, glutathione. MBB does react with tris(2-carboxyethyl)phosphine to form a fluorescent product that might interfere with the assay of thiols.[2.5.16]

[d] Dibromobimane has been used as a cross-linking agent for thiols in proteins.[2.5.17] The close spacing of the reactive methylenebromides permits the use of dibromobimane[2.5.18] as a molecular ruler.[2.5.19–2.5.20]

References to Table 2.5

2.5.1. Kosower, E.M., Pazhenchevsky, B., and Hershowitz, E., 1,5-Diazabicylco[3.3.0]octadienediones (9,10-dioxabimanes). Strongly fluorescent Syn isomers, *J. Am. Chem. Soc.* 100, 6516–6518, 1978.

2.5.2. Sen, C.K., Roy, S., and Packer, L., Flow cytometric determination of cellular thiols, *Methods Enzymol.* 299, 247–258, 1999.

2.5.3. Fahey, R.C. and Newton, G.L., Determination of low-molecular-weight thiols using monobromobimane fluorescent labeling and high-performance liquid chromatography, *Methods Enzymol.* 143, 85–96, 1987.

2.5.4. Shen, X., Peter, E.A., Bir, S. et al., Analytical measurement of discrete hydrogen sulfide pools in biological specimens, *Free Radic. Biol. Med.* 52, 2276–2283, 2012.

2.5.5. Cook, J.A., Iype, S.N., and Mitchell, J.B., Differential specificity of monochlorobimane for isozymes of human and rodent glutathione S-transferases, *Cancer Res.* 51, 1606–1612, 1991.

2.5.6. Meyer, A.J., May, M.J., and Fricker, M., Quantitative in vivo measurement of glutathione in *Arabidopsis cells*, *Plant J.* 27, 67–78, 2001.

2.5.7. Shenker, B.J., Mayro, J.S., Rooney, C. et al., Immunotoxic effects of mercuric compounds on human lymphocytes and monocytes. IV. Alterations in cellular glutathione content, *Immunopharmacol. Immunotoxicol.* 15, 273–290, 1993.

2.5.8. Hadley, D.W. and Chow, S., Evaluation of methods for measuring cellular glutathione content using flow cytometry, *Cytometry* 15, 349–358, 1994.

2.5.9. van der Ven, A.J.A.M., Mier, P., Peters, W.H.M. et al., Monochlorobimane does not selectively label glutathione in peripheral blood mononuclear cells, *Anal. Biochem.* 217, 41–47, 1994.

2.5.10. Raychaudhuri, S.P., Raychaudhuri, S.K., Atkuri, K.R. et al., Nerve growth factor: A key local regulator in the pathogenesis of inflammatory arthritis, *Arthritis Rheum.* 63, 3243–3252, 2011.

2.5.11. Uno, K., Okuno, K., Kato, T. et al., Pre-operative intracellular glutathione levels of peripheral monocytes as a biomarker to predict survival of colorectal cancer patients, *Cancer Immunol. Immunother.* 59, 1457–1465, 2010.

2.5.12. Donald, L.J., Crane, B.R., Anderson, D.H., and Duckworth, H.W., The role of cysteine 206 in allosteric inhibition of *Escherichia coli* citrate synthase. Studies by chemical modification, site-directed mutagenesis, and fluorine-19 NMR, *J. Biol. Chem.* 266, 20709–20713, 1991.

(*continued*)

TABLE 2.5 (continued)
Halobimanes[a]

2.5.13. Wu, M.-X., Filley, S.J., Xiong, J. et al., A cysteine in the C-terminal region of alanyl-tRNA synthetase is important for aminoacylation activity, *Biochemistry* 33, 12260–12266, 1994.

2.5.14. Kim, Y.J., Pannell, L.K., and Sackett, D.L., Mass spectrometric measurement of differential reactivity of cysteine to localize protein-ligand binding sites, *Anal. Biochem.* 332, 376–383, 2004.

2.5.15. Hu, L., Borleske, B.L., and Colman, R.F., Probing the active site of α-class rat liver glutathione S-transferases using affinity labeling by monobromobimane, *Protein Sci.* 6, 43–52, 1997.

2.5.16. Graham, D.E., Harich, K.C., and White, R.H., Reductive dehalogenation of monobromobimane by tris(2-carboxyethyl)phosphine, *Anal. Biochem.* 318, 325–328, 2003.

2.5.17. Bhattacharjee, H. and Rosen, B.P., Spatial proximity of Cys113, Cys172, and Cys442 in the metalloactivation domain of the ArsA ATPase, *J. Biol. Chem.* 271, 24465–24470, 1996.

2.5.18. Petrotchenko, E.V., Xiao, K., Cable, J. et al., BiPS, a photocleavable, isotopically coded, fluorescent cross-linker for structural proteomics, *Mol. Cell. Proteomics* 8, 273–286, 2009.

2.5.19. Sinz, A. and Wang, K., Mapping protein interfaces with a fluorogenic cross-linker and mass spectrometry: Application to nebulin-calmodulin complexes, *Biochemistry* 40, 7903–7913, 2001.

2.5.20. Yang, X.-C., Torres, M.P., Marzluff, W.F., and Dominski, Z., Three proteins of the U7-specific Sm ring function as a molecular ruler to determine the site of 2′-end processing in mammalian histone pre-mRNA, *Mol. Cell. Biol.* 29, 4045–4056, 2009.

TABLE 2.6
Reaction of Carbonyl Compounds with Proteins[a]

Compound	Chemistry	Reactions	References
Alkyl and aromatic aldehydes[b]	Reductive alkylation (formation of Schiff base that is reduced to form a stable alkyl derivative).[b] Acetaldehyde has been reported to form some more complex adducts similar to those formed with formaldehyde.[c]	Modification of ε-amino groups and α-amino groups in proteins. Reductive alkylation has been used to couple therapeutic products such as PEG.	2.6.16–2.6.41
Formaldehyde/ paraformaldehyde (methanal; CAS 50-00-0)	Modification of protein nucleophiles and cross-linking reaction.[d]	Amino groups, histidine residues, and sulfhydryl groups.	2.6.42–2.6.66
Glutaraldehyde	Protein cross-linking agent used for tissue preservation and vaccine manufacture.[e]	Primary reaction with ε-amino groups of lysine.	2.6.67–2.6.83
Ketones	Ketones can be used to modify proteins.[f]	Acetone can modify amino groups in proteins. Ketone functions (carbonyl groups) are generated in proteins during oxidation.	2.6.84–2.6.88

[a] Carbonyl compounds have the potential to react with a variety of nucleophiles in proteins. Most work is based on the reaction of primary amines and secondary amines (Figure 2.21). It is suggested that aldehydes and ketones can also react with sulfhydryl groups to yield mercaptals.[2.6.1] There is evidence to support the reaction of formaldehyde with cysteine,[2.6.2–2.6.4] and the evidence that supports the formation of a stable product in protein is less than convincing.[2.6.5] The bulk of the work is focused on reaction with other amino acid residues although it is suggested that a methylene bond can be formed between cysteine and another nucleophile in a protein.[2.6.5] As discussed previously with the haloalkanes, formaldehyde can be generated from methyl chloride. Other aldehydes and ketones are generated in vivo and do react with proteins. One of the most notable is 4-HNE (Figure 2.22),[2.6.6–2.6.9] and there is substantial literature of 4-HNE and other carbonyl derivatives arising during oxidative stress.[2.6.10–2.6.15] As a practical matter, formaldehyde is usually supplied a 37% (wt/wt; approximately 12.3 M) solution in water with methanol added for stability. Acetaldehyde (CAS 75-07-0; ethanal)) is miscible with water. As would be expected, propionaldehyde and butyraldehyde are less soluble and more toxic and have seen very little use in the modification of proteins.[2.6.15]

[b] The seminal work on reductive alkylation (Figure 2.23) was published by Means and Feeney[2.6.16] in 1968 and used sodium borohydride to stabilize the Schiff base formed with the carbonyl compound and the primary amine. Friedman and coworkers replaced sodium borohydride with sodium cyanoborohydride in 1974.[2.6.17] Sodium cyanoborohydride is still the reagent of choice for this modification reaction.[2.6.18–2.6.20] The N-terminal amino group is also subject to reductive alkylation. Huang and coworkers[2.6.21] were able to obtain specific coupling of an aldehyde derivative of PEG to recombinant human fibroblast growth factor 2 bound to heparin–Sepharose®. Binding to heparin influences the reactivity of lysine residues in fibroblast growth factor.[2.6.22] Thus, Huang and coworkers obtained a superior therapeutic product through reductive alkylation with the butyraldehyde PEG derivative.[2.6.21] It should be noted that Calvete and colleagues[2.6.23] modified bovine seminal plasma protein PDC-109 bound to heparin–Sepharose with acetic anhydride or 1,2-cyclohexanedione in defining the amino residues in the heparin-binding region of this protein. The propionaldehyde derivative of PEG has also been used for coupling to a protein.[2.6.24] Shang and coworkers[2.6.25] used a variation of solid-phase coupling of PEG to proteins. These investigators inserted the propionaldehyde derivative of PEG into a membrane and then circulated the protein across the membrane. A monosubstituted product could be eluted from the membrane. There is a report of a PEG acetaldehyde that could be used for coupling to biopolymers.[2.6.26] Bentley and coworkers avoided the polymerization problem by generating the acetaldehyde hydrate in situ by hydrolysis of the diacetyl derivative. Aldehyde-based tags have been used to identify N-homocysteinylation in proteins.[2.6.27]

(continued)

TABLE 2.6 (continued)
Reaction of Carbonyl Compounds with Proteins[a]

[c]　Acetaldehyde is a metabolite in catabolism of ethanol, and there has been interest in the nature of the reaction between acetaldehyde and cellular constituents including protein.[2.6.28–2.6.33] There is limited information on the chemistry of the adduct formed between acetaldehyde and proteins.[2.6.34,2.6.35] There has been far more interest in the action of acetaldehyde on DNA.[2.6.36,2.6.37] Acetaldehyde is one of the several reactive derivatives formed during degradation of poly(ethylene) glycol.[2.6.38–2.6.40] Acetaldehyde and formaldehyde are reported as being present in DMSO and 1,2-propanediol, fluids used as cryoprotectants.[2.6.41]

[d]　While formaldehyde can be used under controlled conditions to yield defined methylated products that are used for NMR and proteomic analysis[2.6.42–2.6.45] and for the preparation of dimethyl derivatives of amine-containing metabolites,[2.6.46] reaction at higher concentrations and higher temperature results in a complex series of reactions (Figure 2.24). Formaldehyde is used for tissue fixation for histology[2.6.47–2.6.52] and for the preparation of toxoids[2.6.53–2.6.56] and vaccines.[2.6.57–2.6.60] Formaldehyde has been used for the manufacture of allergoids,[2.6.61] and formaldehyde-fixed platelets have been evaluated as a therapeutic opportunity.[2.6.62] The use of formaldehyde for tissue fixation worked well for classical histology but was less suited for the molecular diagnostic era. It has been a bit surprising, at least to me, that methods have been developed for the retrieval of antigen for immunohistochemical analyses and DNA for PCR and related technologies.[2.6.49,2.6.63–2.6.66]

[e]　Glutaraldehyde is widely used for a variety of purposes[2.6.67–2.6.73] from the fixation of animal heart valves for use in cardiovascular surgery to the preparation of tissues for electron microscopy and thus is *understood* in the medical device and diagnostic community. Glutaraldehyde treatment reduces or eliminates the antigenicity of cardiovascular tissue, permitting its use as an implant material. It is used as histological preservative where the masking of antigens can be reversed by heat treatment.[2.6.73–2.6.75] Glutaraldehyde is used to produce allergoids (Chapter 1, p.5) but the immunological response to modified proteins can be variable.[2.6.76–2.6.78] It is of equal interest that although glutaraldehyde is widely used, the chemistry is still poorly understood. First, glutaraldehyde exists largely as a polymer in solution or as a hydrate (Figure 2.25).[2.6.79,2.6.80] Second, although glutaraldehyde has been used extensively for the cross-linking of proteins, the chemistry of the reaction is only poorly understood. The presence of terminal aldehyde functional groups (α, ω) would suggest that cross-linking occurs via reaction with primary amino groups, most likely the ε-amino groups of lysine residues. The initial product of the reaction should be Schiff bases. This reaction should be reversible and it is not, suggesting that the cross-linking reaction occurs via a different mechanism(s) (Figure 2.25).[2.6.80–2.6.83]

[f]　The reaction of acetone with oxytocin was a subject of some interest, and an interesting structure (Figure 2.26) of the acetone adduct with oxytocin was solved in 1968.[2.6.84] More recent studies have used deuterated acetone as a stable-isotope label for quantitative proteomics.[2.6.85] Ketones can modify C-terminal amino acids via reaction with oxyamine derivative.[2.6.86] Ketones are formed in proteins via oxidation[2.6.87] and could then react with nucleophiles in other proteins. Finally, diketones such as 2,3-butanedione can modify arginine residues in proteins.[2.6.88]

References to Table 2.6

2.6.1.　Cecil, R., Intramolecular bonds in proteins. I. The role of sulfur in proteins, in *The Proteins*, 2nd edn., Vol. 1, Chapter 5, ed. H. Neurath, pp. 379–476, Academic Press, New York, 1963.

2.6.2.　Ratner, S. and Clarke, H.T., The action of formaldehyde upon cysteine, *J. Am. Chem.* Soc. 59, 200–206, 1937.

2.6.3.　Kallen, R.G., Equilibria for the reaction of cysteine and derivatives with formaldehyde and protons, *J. Am. Chem. Soc.* 93, 6227–6235, 1971.

2.6.4.　Wlodek, L., The reaction of sulfhydryl groups with carbonyl compounds, *Acta Biochim. Pol.* 35, 307–317, 1988.

2.6.5.　Toews, J., Rogalski, J.C., Clark, T.J., and Kast, J., Mass spectrometric identification of formaldehyde-induced peptide modifications under in vivo protein cross-linking conditions, *Anal. Chim. Acta* 618, 168–183, 2008.

2.6.6.　Maier, C.S., Chavez, J., Wang, J., and Wu, J., Protein adducts of aldehydic lipid peroxidation product identification and characterization of protein adducts using an aldehyde/keto-reactive probe in combination with mass spectrometry, *Methods Enzymol.* 473, 305–330, 2010.

2.6.7.　Gueraud, F., Atalay, M., Bresgen, N. et al., Chemistry and biochemistry of lipid peroxidation products, *Free Radic. Res.* 44, 1098–1124, 2010.

2.6.8.　Bartsch, H., Arab, K., and Nair, J., Biomarkers for hazard identification in humans, *Environ. Health* 10(Suppl 1), S11, 2011.

2.6.9.　Fritz, K.S. and Petersen, D.R., Exploring the biology of lipid peroxidation-derived protein carbonylation, *Chem. Res. Toxicol.* 24, 1411–1419, 2011.

TABLE 2.6 (continued)
Reaction of Carbonyl Compounds with Proteins[a]

2.6.10. Irazusta, V., Moreno-Cermeño, A., Cabiscol, E. et al., Proteomic strategies for the analysis of carbonyl groups on proteins, *Curr. Protein Pept. Sci.* 11(8), 652–658, 2010.

2.6.11. Medina-Navarro, R., Nieto-Aguilar, R., and Alvares-Aguilar, C., Protein conjugated with aldehydes derived from lipid peroxidation as an independent parameter of the carbonyl stress in kidney damage, *Lipids Health Dis.* 10, 201, 2011.

2.6.12. Pillon, N.J., Croze, M.L., and Vella, R.E., The lipid peroxidation by-product 4-hydroxy-2-nonenal (4-HNE) induces insulin resistance in skeletal muscle through carbonyl and oxidative, *Endocrinology* 153, 2099–2111, 2012.

2.6.13. Vásquez-Garzón, V.R., Rouimi, P. et al., Evaluation of three simple direct or indirect carbonyl detection methods for characterization of oxidative modification of proteins, *Toxicol. Mech. Methods* 22, 296–304, 2012.

2.6.14. Sabuncuoğlu, S., Öztas, Y., Cetinkaya, D.U. et al., Oxidative proteins damage with carbonyl levels and nitrotyrosine expression after chemotherapy in bone marrow transplantation patients, *Pharmacology* 89, 283–286, 2012.

2.6.15. Mano, J., Reactive carbonyl species: Their production from lipid peroxides, action in environmental stress and the detoxification mechanism, *Plant Physiol. Biochem.* 50, 90–97, 2012.

2.6.16. Means, G.E. and Feeney, R.E., Reductive alkylation of amino groups in proteins, *Biochemistry* 7, 2192–2201, 1968.

2.6.17. Friedman, M., Williams, L.D., and Masri, M.S., Reductive alkylation of proteins with aromatic aldehydes and sodium cyanoborohydride, *Int. J. Pept. Protein Res.* 6, 183–185, 1974.

2.6.18. Allart, B., Lehtolainen, P., Yla-Herttuala, S. et al., A stable bis-allyloxycarbonyl biotin aldehyde derivative for biotinylation via reductive alkylation: Application to the synthesis of a biotinylated doxorubicin derivative, *Bioconjug. Chem.* 14, 187–194, 2003.

2.6.19. Stefanowicz, P., Kijewska, M., Kapczyńska, K., and Szewczuk, Z., Methods of the site-selective solid phase synthesis of peptide-derived Amadori products, *Amino Acids* 38, 881–889, 2010.

2.6.20. Kurita, Y. and Isogai, A., Reductive N-alkylation of chitosan with acetone and levulinic acid in aqueous media, *Int. J. Biol. Macromol.* 47, 184–189, 2010.

2.6.21. Huang, Z., Ye, C., Liu, Z. et al., Solid-phase N-terminus PEGylation of recombinant human fibroblast growth factor 2 on heparin-sepharose column, *Bioconjug. Chem.* 23, 74–750, 2012.

2.6.22. Ori, A., Free, P., Courty, J. et al., Identification of heparin-binding sites in proteins by selective labeling, *Mol. Cell. Proteomics* 8, 2256–2265, 2009.

2.6.23. Calvete, J.J., Campanero-Rhodes, M.A., Raida, M., and Sanz, L., Characterization of the conformational and quaternary structure-dependent heparin-binding region of bovine seminal plasma protein PDC-109. *FEBS Lett.* 444, 260–264, 1999.

2.6.24. Maleki, A., Najafabadi, A.R., Roohvand, F. et al., Evaluation of bioactivity and pharmacokinetic characteristics of PEGylated *P. pastoris*-expressed erythropoietin, *Drug Deliv.* 18, 570–577, 2011.

2.6.25. Shang, X., Yu, D., and Ghosh, R., Integrated solid-phase synthesis and purification of PEGylated protein, *Biomacromolecules* 12, 2772–2779, 2011.

2.6.26. Bentley, M.D., Roberts, M.J., and Harris, J.M., Reductive amination using poly(ethylene glycol) acetaldehyde hydrate generated in situ: Applications to chitosan and lysozyme, *J. Pharm. Sci.* 87, 1446–1449, 1998.

2.6.27. Zang, T., Dai, S., Chen D. et al., Chemical methods for the detection of protein N-homocysteinylation via selective reactions with aldehydes, *Anal. Chem.* 81, 9065–9071, 2009.

2.6.28. Sabol, D.A., Basista, M.H., Brecher, A.S. et al., Coagulation protein function VIII: Diametric effects of acetaldehyde on factor VII and factor IX function, *Dig. Dis. Sci.* 44, 2564–2567, 1999.

2.6.29. Lakatos, A., Jobst, K., Juricskay, Z., and Kalasz, V., The effect of ethanol on histone glycation in diabetic rats, *Alcohol Alcohol.* 35, 145–147, 2000.

2.6.30. Latvala, J., Melkko, J., Pakkilia, S. et al., Assays for acetaldehyde-derived adducts in blood proteins based on antibodies against acetaldehyde/lipoprotein condensates, *Alcohol. Clin. Exp. Res.* 25, 1648–1653, 2001.

2.6.31. Upadhya, S.C. and Ravindranath, V., Detection and localization of protein-acetaldehyde adducts in rat brain after chronic ethanol treatment, *Alcohol. Clin. Exp. Res.* 26, 856–863, 2002.

2.6.32. Reichardt, P., Schreiber, A., Wichmann, G. et al., Identification and quantification of in vitro adduct formation between protein reactive xenobiotics and a lysine containing model peptide, *Environ. Toxicol.* 18, 29–36, 2003.

2.6.33. Bootorabi, F., Janis, J., Valjakka, J. et al., Modification of carbonic anhydrase II with acetaldehyde, the first metabolite of ethanol, leads to decreased enzyme activity, *BMC Biochem.* 9, 32, 2008.

(continued)

TABLE 2.6 (continued)
Reaction of Carbonyl Compounds with Proteins[a]

2.6.34. Hoffman, T., Meyer, R.J., Sorrell, M.F., and Tuma, D.J., Reaction of acetaldehyde with proteins: Formation of stable fluorescent adducts, *Alcohol. Clin. Exp. Res.* 17, 69–74, 1993.

2.6.35. Moncada, C. and Israel, Y., Generation of acetate and production of ethyl-lysine in the reaction of acetaldehyde plus serum albumin, *Alcohol* 17, 87–91, 1998.

2.6.36. Balbo, S., Meng, L., Bliss, R.L. et al., Kinetics of DNA adduct formation in the oral cavity after drinking alcohol, *Cancer Epidemiol. Biomarkers Prev.* 21, 601–608, 2012.

2.6.37. Hori, K., Miyamoto, S., Yukawa, Y. et al., Stability of acetaldehyde-derived DNA adduct in vitro, *Biochem. Biophys. Res. Commun.* 423, 642–646, 2012.

2.6.38. Dwyer, D.F. and Tiedje, J.M., Metabolism of polyethylene glycol by two anaerobic bacteria, *Desulfovibrio desulfuricans* and a *Bacteroides* sp., *Appl. Environ. Microbiol.* 52, 852–856, 1986.

2.6.39. Frings, J., Schramm, E., and Schink, B., Enzymes involved in anaerobic polyethylene glycol degradation by *Pelobacter venetianus* and *Bacteroides* strain PG1, *Appl. Environ. Microbiol.* 58, 2164–2167, 1992.

2.6.40. Hemenway, J.N., Carvalho, T.C., Rao, V.M. et al., Formation of reactive impurities in aqueous and neat polyethylene glycol 400 and effects of antioxidants and oxidation inducers, *J. Pharm. Sci.* 101(9), 3305–3318, 2012.

2.6.41. Legge, M. and Byers, M.S., Varying amounts of different aldehydes present in the cryoprotectants dimethyl sulphoxide and 1,2-propanediol, *Cryobiology* 64, 297–300, 2012.

2.6.42. Abraham, S.J., Kobayashi, T., Solaro, R.J., and Gaponenko, V., Differences in lysine pKa values may be used to improve NMR signal dispersion in reductively methylated proteins, *J. Biomol. NMR* 43, 239–246, 2009.

2.6.43. Sledz, P., Zheng, H., Hurzyn, K. et al., New surface contacts formed upon reductive lysine methylation: Improving the probability of protein crystallization, *Protein Sci.* 19, 1395–1404, 2010.

2.6.44. Murphy, J.P., Kong, F., Pinto, D.M., and Wang-Pruski, G., Relative quantitative proteomic analysis reveals wound response proteins correlated with after-cooking darkening, *Proteomics* 10, 4258–4269, 2010.

2.6.45. She, Y.M., Rosu-Myles, M., Walrond, L., and Cyr, T.D., Quantification of protein isoforms in mesenchymal stem cells by reductive dimethylation of lysines in intact proteins, *Proteomics* 12, 369–379, 2012.

2.6.46. Guo, K., Ji, C., and Li, L., Stable-isotope dimethylation labeling combined with LC-ESI MS for quantification of amine-containing metabolites in biological samples, *Anal. Chem.* 79, 8631–8638, 2007.

2.6.47. Hopwood, D., Fixatives and fixation: A review, *Histochem. J.* 1, 323–360, 1969.

2.6.48. Leong, A.S. and Gilham, P.N., The effects of progressive formaldehyde fixation on the preservation of tissue antigens, *Pathology* 21, 266–268, 1989.

2.6.49. Matsuda, K.M., Chung, J.Y., and Hewitt, S.M., Histo-proteomic profiling of formalin-fixed, paraffin-embedded tissue, *Expert Rev. Proteomics* 7, 227–237, 2010.

2.6.50. Ralton, L.D. and Murray, G.I., The use of formalin fixed wax embedded tissue for proteomic analysis, *J. Clin. Pathol.* 64, 297–302, 2011.

2.6.51. Engel, K.B. and Moore, H.M., Effects of preanalytical variables on the detection of proteins by immunohistochemistry in formalin-fixed, paraffin-embedded tissue, *Arch. Pathol. Lab. Med.* 135, 537–543, 2011.

2.6.52. Tanca, A., Pagnozzi, D., and Addis, M.F., Setting proteins free: Progresses and achievements in proteomics of formalin-fixed, paraffin-embedded tissues, *Proteomics Clin. Appl.* 6, 7–21, 2012.

2.6.53. Petre, J., Pizzza, M., Nencioni, L. et al., The reaction of bacterial toxins with formaldehyde and its use for antigen stabilization, *Dev. Biol. Stand.* 87, 125–134, 1996.

2.6.54. Bolgiano, B., Fowler, S., Turner, K. et al., Monitoring of diphtheria, pertussis and tetanus toxoids by circular dichroism, fluorescence spectroscopy and size-exclusion chromatography, *Dev. Biol. (Basel)* 100, 51–59, 2000.

2.6.55. Inić-Kanada, A. and Stojanović, M., Tetanus toxoid purification: Chromatographic procedures as an alternative to ammonium-sulphate precipitation, *J. Chromatogr. B Anal. Technol. Biomed. Life Sci.* 879, 2213–2219, 2011.

2.6.56. Keller, J.E., Characterization of new formalin-detoxified botulinum neurotoxin toxoids, *Clin. Vaccine Immunol.* 15, 1374–1379, 2008.

2.6.57. Salk, D. and Salk, J., Vaccinology of poliomyelitis, *Vaccine* 2, 59–74, 1984.

2.6.58. Brown, F., Review of accidents caused by incomplete inactivation of viruses, *Dev. Biol. Stand.* 81, 103–107, 1993.

2.6.59. Rappuoli, R., Toxin inactivation and antigen stabilization: Two different uses of formaldehyde, *Vaccine* 12, 579–581, 1994.

2.6.60. Ohno, T., Autologous formalin-fixed tumor vaccine, *Curr. Pharm. Des.* 11, 1181–1188, 2005.

2.6.61. Dormann, D., Ebner, C., Jarman, E.R. et al., Responses of human birch pollen allergen-reactive T cells to chemically modified allergens (allergoids), *Clin. Exp. Allergy* 28, 1374–1383, 1998.

TABLE 2.6 (continued)
Reaction of Carbonyl Compounds with Proteins[a]

2.6.62. Fischer, T.H., Wolberg, A.S., Bode, A.P., and Nichols, T.C., The interaction of factor VIIa with rehydrated, lyophilized platelets, *Platelets* 19, 182–191, 2008.

2.6.63. Nirmalan, N.J., Harnden, P., Selby, P.J., and Banks, R.E., Mining the archival formalin-fixed paraffin-embedded tissue proteome: Opportunities and challenges, *Mol. Biosyst.* 4, 712–720, 2008.

2.6.64. Grizzle, W.E., Special symposium: Fixation and tissue processing models, *Biotech. Histochem.* 84, 185–193, 2009.

2.6.65. With, C.M., Evers, D.L., and Mason, J.T., Regulatory and ethical issues on the utilization of FFPE tissues in research, *Methods Mol. Biol.* 724, 1–21, 2011.

2.6.66. Klopfleisch, R., Weiss, A.T., and Gruber, A.D., Excavation of a buried treasure—DNA, mRNA, miRNA and protein analysis in formalin fixed, paraffin embedded tissues, *Histol. Histopathol.* 26, 797–810, 2011.

2.6.67. Goetz, W.A., Lim, H.S., Lansac, E. et al., A temporarily stented, autologous pericardial aortic valve prosthesis, *J. Heart Valve Dis.* 11, 696–702, 2002.

2.6.68. Nimni, M.E., Glutaraldehyde fixation revisited, *J. Long Term Eff. Med. Implants* 11, 151–161, 2001.

2.6.69. Neethling, W.M.L., Hodge, A.J., and Glancy, R., Glutaraldehyde-fixed kangaroo aortic wall tissue: Histology, crosslink stability and calcification potential, *J. Biomed. Mater. Res.* 66B, 356–363, 2003.

2.6.70. Wengerter, K. and Dardik, H., Biological vascular grafts, *Semin. Vasc. Surg.* 12, 46–51, 1999.

2.6.71. Chen, R.H. and Adams, D.H., Decreased porcine valve antigenicity with in vitro culture, *Ann. Thorac. Surg.* 71(5 Suppl), S393–S395, 2001.

2.6.72. Hopwood, D., Fixation and fixatives, in *Theory and Practice of Histological Techniques*, eds. J.D. Bancroft and A. Stevens, 4th edn., Chapter 2, pp. 23–45, Churchill Livingstone, New York, 1996.

2.6.73. Jamur, M.C., Faraco, C.D., Lunardi, L.O. et al., Microwave fixation improves antigenicity of glutaraldehyde-sensitive antigens while preserving ultrastructural detail, *J. Histochem. Cytochem.* 43, 307–311, 1995.

2.6.74. Brorson, S.H., Fixative-dependent increase in immunogold labeling following antigen retrieval on acrylic and epoxy sections, *Biotech. Histochem.* 74, 248–260, 1999.

2.6.75. Yamashita, S. and Okada, Y., Application of heat-induced antigen retrieval to aldehyde-fixed fresh frozen sections, *J. Histochem. Cytochem.* 53, 1421–1432, 2005.

2.6.76. Zipeto, D., Matucci, A., Ripamonte, C. et al., Induction of human immunodeficiency virus neutralizing antibodies using fusion complexes, *Microbes Infect.* 8, 1424–1433, 2006.

2.6.77. Zhu, X., Chu, W., Wang, T. et al., Variations in dominant antigen determinants of glutaraldehyde polymerized human, bovine, and porcine hemoglobin, *Artif. Cells Blood Substit. Immobil. Biotechnol.* 35, 518–532, 2007.

2.6.78. Wu, K.J., Wang, C.Y., and Lu, H.K., Effect of glutaraldehyde on the humoral immunogenicity and structure of porcine dermal collagen membranes, *Arch. Oral. Biol.* 49, 305–311, 2004.

2.6.79. Kawahara, J.-I., Ohmori, T., Ohkubo, T., Hattori, S., and Kawamura, M., The structure of glutaraldehyde in aqueous solution determined by ultraviolet adsorption and light scattering. *Anal. Biochem.* 201, 94–98, 1992.

2.6.80. Migneault, I., Dartiguenave, C., Bertrand, M.J., and Waldron, K.C., Glutaraldehyde: Behavior in aqueous solution, reaction with proteins, and application to enzyme crosslinking, *Biotechniques* 37, 790–802, 2004.

2.6.81. Richards, F.H. and Knowles, J.R., Glutaraldehyde as a protein cross-linking reagent. *J. Mol. Biol.* 37, 231–233, 1968.

2.6.82. Walt, D.R. and Agayn, V.I., The chemistry of enzyme and protein immobilization with glutaraldehyde, *TRAC* 13, 425–430, 1994.

2.6.83. Wine, Y., Cohen-Hadar, N., Freeman, A., and Frolow, F., Elucidation of the mechanism and end products of glutaraldehyde crosslinking reaction by x-ray structure analysis, *Biotechnol. Bioeng.* 98, 711–718, 2007.

2.6.84. Hruby, V.J., Yamashiro, D., and du Vigneaud, V., The structure of acetone-oxytocin with studies on the reaction of acetone with various peptides, *J. Am. Chem. Soc.* 90, 7106–7110, 1968.

2.6.85. Zhai, J., Liu, X., Huang, Z., and Zhu, H., RABA (reductive methylation by acetone): A novel stable isotope labeling approach for quantitative proteomics, *J. Am. Soc. Mass Spectrom.* 20, 1366–1377, 2009.

2.6.86. Yi, L., Sun, H., Wu, Y.-W. et al., A highly efficient strategy for modification of proteins at the C-terminus, *Angew. Chem. Int. Ed.* 49, 9417–9421, 2010.

2.6.87. Mirzaei, M. and Regnier, F., Enrichment of carbonylated peptides using Girard P reagent and strong cation exchange chromatography, *Anal. Chem.* 78, 770–778, 2006.

2.6.88. Yankeelov, J.A., Jr., Modification of arginine by diketones, *Methods Enzymol.* 25, 566–579, 1972.

TABLE 2.7
Michael Addition Reagents

Compound	Chemistry and Reactions	References
Vinyl sulfones	Reaction with nucleophiles in proteins. Reaction with nucleophiles in proteins such as lysine and cysteine via Michael addition. Divinyl sulfone has been used as a cross-linking agent.[a] Vinyl sulfones are extremely reactive and have been designated as toxic chemicals.	2.7.1–2.7.18
Acrylamide and related unsaturated propane derivatives	Acrylamide, and related compounds, is an unsaturated organic that behaves as an electrophile and reacts with nucleophilic residues in proteins. Originally considered an unwanted consequence of polyacrylamide electrophoresis, modification with acrylamide is used in the specific modification of proteins. As with other electrophilic functions, acrylamide can be used as means for attaching larger molecules.[b]	2.7.19–2.7.27
4-Hydroxy-2-nonenal (4-HNE)	Lipid oxidation product which can react with nucleophilic groups, primarily sulfhydryl, in proteins.[c]	2.7.28–2.7.38
Acrolein (propenal)	This compound and some related chemicals are reactive toward nucleophilic centers in proteins. Most of the concern is directed in environmental toxicology although some derivatives have proved useful.[d]	2.7.39–2.7.44
N-alkyl maleimides	N-alkyl maleimides such as N-ethylmaleimide are remarkably specific for reaction with cysteinyl residues in proteins.[e] Some comparative rates of reaction of maleimide with sulfhydryl groups are given in Table 2.8.	2.7.45–2.7.58
Squaric acid diester	An electrophilic reagent that can be substituted by a nucleophile to yield a disubstituted product that could be a cross-link. It is a nucleophilic substitution at a vinyl carbon.[f]	2.7.59–2.7.65

[a] Vinyl sulfones (Figure 2.26) are better known as cross-linking/conjugation reagents for various polymers[2.7.1–2.7.5] and as a functional group for coupling macromolecules to an agarose or similar matrix.[2.7.7–2.7.13] Friedman and Finley[2.7.14] reported on the use of ethyl vinyl sulfone for the measurement of lysine in food products, while a latter study[2.7.15] focused on the use of alkyl vinyl sulfones for the measurement of sulfhydryl groups in wool. Modification of lysine, cysteine, and histidine was observed in the various proteins with sulfhydryl groups being the most reactive. The vinyl sulfone functional group has been incorporated into active-site-directed reagent for deubiquitinating enzymes[2.7.16] and for a deneddylase of the ULP family[2.7.17] to several peptide active-site-directed reagents. 4-Nitrostyrene has been used as a reagent to modify reduced wool proteins[2.7.18]; the modified wool products were orange.

[b] Acrylamide modifies sulfhydryl groups in proteins (Figure 2.27). Originally considered to be an unwanted complication of the electrophoresis of reduced proteins in acrylamide gel systems,[2.7.19] modification with acrylamide is now considered to be a useful site-specific modification of cysteine in proteins.[2.7.20–2.7.26] Some of these studies[2.7.23] focus on the molecular mechanism for acrylamide toxicity. The acrylamide moiety is used as the functional group in some protein kinase inhibitors.[2.7.27]

[c] 4-Hydroxy-2-nonenal (4-HNE) is a product produced from the oxidation of lipids that modifies sulfhydryl groups (Figure 2.22) and other residues in proteins[2.7.28–2.7.32] such as histidine in proteins. 4-Hydroxy-2-nonenal is somewhat nonspecific as it can modify nucleophiles such as lysine by Schiff base formation and sulfhydryl groups by Michael addition; histidine is also subject to modification by 4-HNE via Michael addition. 4-Hydroxy-2-nonenal is somewhat unstable[2.7.33] with the stability suggested to reflect reactivity with Michael addition partners or hydrogen sulfide.[2.7.34] There has been work on the effect of membranes in stabilizing 4-HNE after its formation prior to transfer to either extracellular or intracellular space.[35]

Analysis of 4-HNE adducts with protein-bound cysteine (SIRT3) by MS provided some interesting chemistry and a warning to the analytical biochemist. Fritz and coworkers[2.7.36] observed that analysis of 4-HNE adducts with collision-induced dissociation resulted in the loss of 4-HNE from the protein adduct by retro-Michael addition.[2.7.37,2.7.38] Stabilization of the 4-HNE adduct with sodium borohydride prevented the retro-Michael addition under the conditions of collision-induced dissociation.

TABLE 2.7 (continued)
Michael Addition Reagents

^d Acrolein is an environmental pollutant as well as an endogenous metabolite[2.7.40] similar thus to 4-HNE. Acrolein reacts with nucleophilic sites in proteins including cysteine, lysine, and histidine yielding a variety of products (Figure 2.28).[2.7.41–2.7.43] The adduct products also are reactive, and Furuhata and coworkers[2.7.44] used an acrolein-modified protein to prepare a conjugate with glutathione, which was used as an antigen in the preparation of a monoclonal antibody against glutathione.

^e The maleimide functional group (Figure 2.29) is used for the direct modification of cysteine in proteins (N-ethylmaleimide) and for the introduction of probes into proteins.[2.7.45–2.7.49] This reaction is a Michael addition, which is a reaction between a nucleophile (thiolate anion) and an olefin (the maleimide ring). Bednar has examined the chemistry of the reaction of N-ethylmaleimide with cysteine and other thiols in some detail.[2.7.50] Bednar[2.7.50] also reports data for the decomposition of NEM in several buffers, and this information should be considered for the determination of truly accurate kinetic data. This reaction can be followed by the decrease in absorbance at 300 nm, the absorbance maximum of N-ethylmaleimide. The extinction coefficient of N-ethylmaleimide is 620 M^{-1} cm^{-1} at 302 nm.[2.7.47] The spectrophotometric assay is not sensitive, and the modification is usually monitored by the incorporation of radiolabeled reagent. The alkylation product (an S-succinimidyl cysteine) is stable and can be determined by amino acid analysis following acid hydrolysis. Although the reagent is reasonably specific for cysteine, reaction with other nucleophiles must be considered.[2.7.51] The reaction of N-ethylmaleimide with sulfhydryl compounds is quite rapid (Table 2.8). N-ethylmaleimide-sensitive factor (NSF) is a term used to identify ATPase involved in a variety of cell funtions.[2.7.52–2.7.55] Maleimide chemistry is also used to manufacture carbohydrate microarrays. In one study, a thiol-functionalized carbohydrate is coupled to a maleimide matrix,[2.7.56] while the other study[2.7.57] used a maleimide-functionalized carbohydrate to couple to a thiol matrix. Maleimides are compatible with phosphorothioate chemistry; no addition to phosphorothioate esters was observed.[2.7.58]

^f Squaric acid diethyl ester (1,2-diethoxycyclobutene-3.4-dione) (Figure 2.30) has been used for the production of neoglycoproteins.[2.7.59,2.7.60] The use of this reagent has the advantage in that the coupling reaction can be monitored by UV spectroscopy.[2.7.59] Other investigators have used MS combined with the ProteinChip® system to monitor oligosaccharide coupling using squaric acid diester. Squaric acid diester has been used for the preparation of polysaccharide–protein conjugates in the development of vaccines against polysaccharide epitopes.[2.7.61] Wurm and coworkers[2.7.62] noted that the coupling of hapten/polysaccharide to the squaric acid diethyl ester occurs at neutral pH[2.7.63] and the monosubstituted derivative is relatively unreactive until taken to pH 9.0 for coupling to protein.[2.7.64] Hou and coworkers[2.7.63] also noted the methyl ester was as effective as the diethyl derivative. Wurm[2.7.62] and coworkers have an excellent discussion of the experimental variables for the use of squaric acid diester in bioconjugate formation. Iijima[2.7.64] reported on the reaction of methyl squarate and dimethyl squarate with bovine pancreatic ribonuclease A and some low molecular weight amines. Dimethyl squarate was more reactive than methyl squarate, and, at pH 7–8, the α-amino group was more reactive than the ε-amino group. In a similar reaction, Zhang and Bello[2.7.65] reported the coupling of colchicine to melittin (Figure 2.25), which has a lysine with a pKa of 6.5.

References to Table 2.7

2.7.1. Norrish, R.G.W. and Brookman, E.F., The formation and structure of polymers of two insoluble cross-linked type, *Proc. R. Soc. Lond. Ser. A* 163, 205–220, 1937.

2.7.2. Shafir, G., Galperin, A., and Margel, S., Synthesis and characterization of recombinant factor VIIa-conjugated magnetic iron oxide nanoparticles for hemophilia treatment, *J. Biomed. Mater. Res. A* 91, 1056–1064, 2009.

2.7.3. Gotte, G., Laurents, D.V., and Libonati, M., Three-dimensional domain-swapped oligomers of ribonuclease A: Identification of a fifth tetramer, pentamers and hexamers and detection of trace heptameric, octomeric and nonameric species, *Biochim. Biophys. Acta* 1764, 44–54, 2006.

2.7.4. Ibrahim, S., Kang, Q.K., and Ramamurthi, A., The impact of hyaluronic acid oligomer content on physical, mechanical, and biologic properties of divinyl sulfone-crosslinked hyaluronic acid hydrogels, *J. Biomed. Mater. Res. A* 94, 355–370, 2010.

2.7.5. Chawla, K., Yu, T.B., Liao, S.W., and Guan, Z., Biodegradable and biocompatible synthetic saccharide-peptide hydrogels for three-dimensional stem cell culture, *Biomacromolecules* 12, 560–567, 2011.

2.7.6. Li, L., Crow, D., Turatti, F. et al., Site-specific conjugation of monodispersed DOTA-PEGn to a thiolated diabody reveals the effect of increasing peg size on kidney clearance and tumor uptake with improved 64-copper PET imaging, *Bioconjug. Chem.* 22, 709–716, 2011.

(continued)

TABLE 2.7 (continued)
Michael Addition Reagents

2.7.7. Lihme, A., Schafer-Nielsen, C., Lasen, K.P. et al., Divinylsulphone-activated agarose. Formation of stable and non-leaking affinity matrices by immobilization of immunoglobulins and other proteins, *J. Chromatogr.* 376, 299–305, 1986

2.7.8. Berna, P.P., Berna, N., Porath, J., and Oscarsson, S., Comparison of the protein adsorption selectivity of salt-promoted agarose-based adsorbents. Hydrophobic, thiophilic and electron donor-acceptor adsorbents, *J. Chromatogr. A* 800, 151–159, 1998.

2.7.9. Frydlová, J., Kucerová, Z., and Tichá, M., Interaction of pepsin with aromatic amino acids and their derivatives immobilized to Sepharose, *J. Chromatogr. B Anal. Technol. Biomed. Life Sci.* 863, 133–140, 2008.

2.7.10. Morales-Sanfrutos, J., Lopez-Jaramillo, J., Ortega-Muñoz, M. et al., Vinyl sulfone: A versatile function for simple bioconjugation and immobilization, *Org. Biomol. Chem.* 8, 667–675, 2010.

2.7.11. Cheng, F., Shang, J., and Ratner, D.M., A versatile method for functionalizing surfaces with bioactive glycans, *Bioconjug. Chem.* 22, 50–57, 2011.

2.7.12. Lopez-Jaramillo, J., Ortega-Muñoz, M., Megia-Fernandez, A. et al., Vinyl sulfone functionalization: A feasible approach for the study of the lectin-carbohydrate interactions, *Bioconjug. Chem.* 23, 846–855, 2012.

2.7.13. Yu, A., Shang, J., Cheng, F. et al., Biofunctional paper via covalent modification of cellulose, *Langmuir* 28(30), 11265–11273, 2012.

2.7.14. Friedman, M. and Finley, J.W., Reactions of proteins with ethyl vinyl sulfone, *Int. J. Pept. Protein Res.* 7, 481–486, 1975.

2.7.15. Masri, M.S. and Friedman, M., Protein reactions with methyl and ethyl vinyl sulfones, *J. Protein Chem.* 7, 49–54, 1988.

2.7.16. Borodovsky, A., Kessler, B.M., Casagrande, R. et al., A novel active site-directed probe specific for deubiquitylating enzymes reveals proteasome association of USP14, *EMBO J.* 20, 5187–5196, 2001.

2.7.17. Gan-Erdene, T., Nagamalleswari, K., Yin, L. et al., Identification and characterization of DEN1, a deneddylase of the ULP family, *J. Biol. Chem.* 278, 28892–28900, 2003.

2.7.18. Masri, M.S., Windle, J., and Friedman, M., *p*-Nitrophenol: New alkylating agent for sulfhydryl groups in reduced soluble peptides and keratin, *Biochem. Biophys. Res. Commun.* 47, 1408–1413, 1972.

2.7.19. Yan, J.X., Keet, W.C., Herbert, B.R. et al., Identification and quantitation of cysteine in proteins separated by gel electrophoresis, *J. Chromatogr.* 813, 187–200, 1998.

2.7.20. Bordini, E., Hamdan, M., and Righetti, P.G., Probing acrylamide alkylation sites in cysteine-free proteins by matrix-assisted laser desorption/ionization time-of-flight. *Rapid Commun. Mass Spectrom.* 14, 840–848, 2000.

2.7.21. Mineki, R. et al., In situ alkylation with acrylamide for identification of cysteinyl residues in proteins during one- and two-dimensional sodium dodecyl sulphate-polyacrylamide gel electrophoresis. *Proteomics* 2, 1672–1681, 2002.

2.7.22. Cahill, M.A. et al., Analysis of relative isotopologue abundances for quantitative profiling of complex protein mixtures labelled with acrylamide/D-3-acrylamide alkylation tag system. *Rapid Commun. Mass Spectrom.* 17, 1283–1290, 2003.

2.7.23. LoPachin, R.M., Molecular mechanisms of the conjugated α,β-unsaturated carbonyl derivatives: Relevance to neurotoxicity and neurodegenerative diseases, *Toxicol. Sci.* 104, 235–249, 2008.

2.7.24. Lü, Z.R., Zou, H.C., Park, S.J. et al., The effects of acrylamide on brain creatine kinase: Inhibition kinetics and computational docking simulation, *Int. J. Biol. Macromol.* 44, 128–132, 2009.

2.7.25. Ghahghaei, A., Rekas, A., Carver, J.A., and Augusteyn, R.C., Structure/function studies on dogfish α-crystallin, *Mol. Vis.* 15, 2411–2420, 2009.

2.7.26. Sciandrelio, G., Mauro, M., Cardonna, F. et al., Acrylamide catalytically inhibits topoisomerase II in V79 cells, *Toxic. In Vitro* 24, 830–834, 2010.

2.7.27. Garuti, L., Roberti, M., and Bottegoni, G., Irreversible protein kinase inhibitors, *Curr. Med. Chem.* 18, 2981–2994, 2011.

2.7.28. Stevens, S.M., Jr., Rauniyar, N., and Prokai, L., Rapid characterization of covalent modifications to rat brain mitochondrial proteins after ex vivo exposure to 4-hydroxy-2-nonenal by liquid chromatography-tandem mass spectrometry using data-dependent and neutral-driven MS3 acquisition, *J. Mass Spectrom.* 42, 1599–1605, 2007.

2.7.29. Zhu, X., Gallogly, M.M., Mieyal, J.J. et al., Covalent cross-linking of glutathione and carnosine to proteins by 4-oxy-2-nonenal, *Chem. Res. Toxicol.* 22, 1050–1059, 2009.

TABLE 2.7 (continued)
Michael Addition Reagents

2.7.30. Wakita, C., Maeshima, T., Yamazaki, A. et al., Stereochemical configuration of 4-hydroxy-2-nonenal-cysteine adducts and their stereoselective formation in a redox-regulated protein, *J. Biol. Chem.* 284, 28810–28822, 2009.

2.7.31. Rauniyar, N. and Prokai, L., Detection and identification of 4-hydroxy-2-nonenal Schiff-base adducts along with products of Michael addition using data-dependent neutral loss-driven MS3 acquisition: Method evaluation through an in vitro study on cytochrome c oxidase modifications, *Proteomics* 9, 5188–5193, 2009.

2.7.32. Chavez, J., Chung, W.G., Miranda, C.L. et al., Site-specific protein adducts of 4-hydroxy-2(E)-nonenal in human THP-1 monocytic cells: Protein carbonylation is diminished by ascorbic acid, *Chem. Res. Toxicol.* 23, 37–47, 2010.

2.7.33. Spies-Martin, D., Sommerburg, O., Langhams, C.-D., and Leichsenring, M., Measurement of 4-hydroxynonenal in small volume blood plasma samples: Modification of a gas chromatographic -mass spectrometric method for clinical settings, *J. Chromatogr. B* 774, 231–239, 2002.

2.7.34. Schreier, S.M., Muellner, M.K., Steinkellner, H. et al., Hydrogen sulfide scavenges the cytotoxic lipid oxidation product 4-HNE, *Neurotox. Res.* 17, 249–256, 2010.

2.7.35. Vazdar, M., Jurkiewicz, P., Hof, M. et al., Behavior of 4-hydroxynonenal in phospholipid membranes, *J. Phys. Chem. B* 116, 6411–6415, 2012.

2.7.36. Fritz, K.S., Kellersberger, K.A., Gomez, J.D., and Petersen, D.R., 4-HNE adduct stability characterized by collision-induced dissociation and electron transfer dissociation mass spectrometry, *Chem. Res. Toxicol.* 25, 965–970, 2012.

2.7.37. Rao, H.S.P. and Jothiligam, S., Studies on NaI/DMSO induced *retro*-Michael addition (RMA) reactions on some 1,5-dicarbonyl compounds, *J. Chem. Sci.* 117, 27–32, 2005.

2.7.38. Dragojevic, S., Sunji, V., Bencetic-Mihaljevic, V. et al., Determination of aqueous stability and degradation products of series of coumarin dimers, *J. Pharm. Biomed. Anal.* 54, 37–47, 2011.

2.7.39. Logue, J.M., Price, P.N., Sherman, M.H., and Singer, B.C., A method to estimate the chronic health impact of air pollutants in U.S. residences, *Environ. Health Perspect.* 120, 216–222, 2012.

2.7.40. Mano, J., Reactive carbonyl species: Their production from lipid peroxides, action in environmental stress, and the detoxification mechanism, *Plant Physiol. Biochem.* 59, 90–97, 2012.

2.7.41. Hashmi, M., Vamvakas, S., and Anders, M.W., Bioactivation mechanism of S-(3-oxopropyl)-N-acetyl-L-cysteine, the mercapturic acid of acrolein, *Chem. Res. Toxicol.* 5, 360–365, 1992.

2.7.42. LoPachin, R.M., Gavin, T., Petersen, D.R., and Barber, D.S., Molecular mechanisms of 4-hydroxy-2-nonenal and acrolein toxicity: Nucleophile targets and adduct formation, *Chem. Res. Toxicol.* 22, 1499–1508, 2009.

2.7.43. Maeshima, T., Honda, K., Chikazawa, M. et al., Quantitative analysis of acrolein-specific adducts generated during lipid peroxidation—Modification of proteins in vitro: Identification of N^{ϵ}-(3-propanal)histidine as the major adduct, *Chem. Res. Toxicol.* 25(7), 1384–1392, 2012.

2.7.44. Furuhata, A., Hondo, K., Shibata, T. et al., Monoclonal antibody against protein-bound glutathione: Use of glutathione conjugate of acrolein-modified proteins as an immunogen, *Chem. Res. Toxicol.* 25(7), 1393–1401, 2012.

2.7.45. Gregory, J.D., The stability of N-ethylmaleimide and its reaction with sulfhydryl groups, *J. Am. Chem. Soc.* 77, 3922–3923, 1955.

2.7.46. Hill, B.G., Reily, C., and Oh, J.Y., Methods for the determination and quantitation of the reactive thiol proteome, *Free Radic. Biol. Med.* 47, 675–683, 2009.

2.7.47. Leslie, J., Spectral shifts in the reaction of N-ethylmaleimide with proteins, *Anal. Biochem.* 10, 162–167, 1965.

2.7.48. Gorin, G., Martic, P.A., and Doughty, G., Kinetics of the reaction of N-ethylmaleimide with cysteine and some congeners, *Arch. Biochem. Biophys.* 115, 593–597, 1966.

2.7.49. Schelté, P.S., Boeckler, C., Frisch, B., and Schuber, F., Differential reactivity of maleimide and bromoacetyl functions with thiols: Application to the preparation of liposomal diepitope constructs, *Bioconjug. Chem.* 11, 118–123, 2000.

2.7.50. Bednar, R.A., Reactivity and pH dependence of thiol conjugation to N-ethylmaleimide: Detection of a conformational change in chalcone isomerase, *Biochemistry* 29, 3684–3690, 1990.

2.7.51. Smyth, D.G., Blumenfeld, O.O., and Konigsberg, W., Reaction of N-ethylmaleimide with peptides and amino acids, *Biochem. J.* 91, 589–595, 1964.

(continued)

TABLE 2.7 (continued)
Michael Addition Reagents

2.7.52. Zhao, C., Slevin, J.T., and Whiteheart, S.W., Cellular functions of NSF: Not just SNAPs and SNAREs, *FEBS Lett.* 581, 2140–2149, 2007.

2.7.53. Hussain, S. and Davanger, S., The discovery of the soluble *N*-ethylmaleimide-sensitive factor attachment protein receptor complex and the molecular recognition of synaptic vesicle transmitter release: The 2010 Kavli in neuroscience, *Neuroscience* 190, 12–20, 2011.

2.7.54. Zhao, C., Smith, E.C., and Whiteheart, S.W., Requirements for the catalytic cycle of the *N*-ethylmaleimide-sensitive factor (NSF), *Biochim. Biophys. Acta* 1823, 159–171, 2012.

2.7.55. Huang, S.P. and Craft, C.M., Potential cellular function of *N*-ethylmaleimide sensitive factor in the photoreceptor, *Adv. Exp. Med. Biol.* 723, 791–797, 2012.

2.7.56. Scurr, D.J., Horlacher, T., Oberli, M.A. et al., Surface characterization of carbohydrate microarrays, *Langmuir* 26, 17143–17155, 2010.

2.7.57. Lee, M.R., Park, S., and Shin, I., Carbohydrate microarrays for enzymatic reactions and quantification of binding affinities for glycan-protein interactions, *Methods Mol. Biol.* 808, 103–116, 2012.

2.7.58. Sánchez, A., Pedroso, E., and Grandas, A., Conjugation reactions involving maleimides and phosphorothioate oligonucleotides, *Bioconjug. Chem.* 23, 300–307, 2012.

2.7.59. Tietze, L.F., Schröter, C., Gabius, S. et al., Conjugation of *p*-aminophenyl glycosides with squaric acid diester to a carrier protein and the use of neoglycoprotein in the histochemical detection of lectins, *Bioconjug. Chem.* 2, 148–153, 1991.

2.7.60. Chernyak, A., Karavanov, A., Ogawa, Y., and Kováč, P., Conjugating oligosaccharides to proteins by squaric acid diester chemistry: Rapid monitoring of the progress of conjugation, and recovery of used ligand, *Carbohydr. Res.* 330, 479–486, 2001.

2.7.61. Xu, P., Alam, M.M., Kalsy, A. et al., Simple, direct conjugation of bacterial O-SP-core antigens to proteins: Development of cholera conjugate vaccines, *Bioconjug. Chem.* 22, 2179–2185, 2011.

2.7.62. Wurm, F., Dingels, C., Frey, H., and Klok, H.-A., Squaric acid mediated synthesis and biological activity of a library of linear and hyperbranched poly(glycerol)-protein conjugates, *Biomacromolecules* 13, 1161–1173, 2012.

2.7.63. Hou, S.-J., Saksena, R., and Kováč, P., Preparation of glycoconjugates by dialkyl squarate chemistry revisited, *Carbohydr. Res.* 343, 196–210, 2008.

2.7.64. Iijima, H., The reaction of methyl squarate and dimethyl squarate with ribonuclease and model compounds, *Biochem. Int.* 19, 353–359, 1989.

2.7.65. Zhang, S. and Bello, J., Reaction of colchicine with peptide amino groups, *Biopolymers* 31, 1241–1242, 1991.

TABLE 2.8
Rates of Nucleophile Modification by Maleimides

Reagent	Nucleophile/Experimental Conditions	Rate (M^{-1} s^{-1})	Reference
2-Mercaptoethanol	N-ethylmaleimide (1.0 M KCl, 0.4 mM ethylenediaminetetraacetic acid [EDTA])	11×10^6 M^{-1} min^{-1a}	2.8.1
2-Mercaptoethanol-amine	N-ethylmaleimide (1.0 M KCl, 0.4 mM EDTA)	4×10^6 M^{-1} min^{-1a}	2.8.1
L-cysteine	N-ethylmaleimide (1.0 M KCl, 0.4 mM EDTA)	1.1×10^6 M^{-1} min^{-1a}	2.8.1
Cysteine	N-ethylmaleimide (18 mM acetate, 0.11 M NaCl, pH 4.78, 25°C)	8.43 M^{-1} s^{-1b}	2.8.2

[a] pH-independent rate constant (see Jones, J.G., Otieno, S., Barnard, E.A., and Bhargava, A.K., Essential and nonessential thiols of yeast hexokinase reactions with iodoacetate and iodoacetamide, *Biochemistry* 14, 2376–2403, 1975).

[b] This value was obtained from a least squares fit of data obtained with 1.0 mM N-ethylmaleimide and 0.7 mM cysteine; a value of 8.55 was obtained by least squares fit of data obtained at 0.5 mM N-ethylmaleimide and 0.35 mM cysteine.

References to Table 2.8
2.8.1. Bednar, R.A., Reactivity and pH dependence of thiol conjugation to N-ethylmaleimide—Detection of a conformational changes in chalcone isomerase, *Biochemistry* 29, 3684–3690, 1990.

2.8.2. Gorin, G., Martic, P.A., and Doughty, G., Kinetics of the reaction of N-ethylmaleimide with cysteine and some congeners, *Arch. Biochem. Biophys.* 115, 593–597, 1966.

TABLE 2.9
Ethyleneimine and Ethylene Oxide

Compound	Chemistry Reactions	Reference
Ethyleneimine	Historic use in modifying cysteine residues to form γ-thialysine residues in proteins. Replaced with 2-bromoethylamine (Table 2.1).[a]	2.9.1–2.9.8
Ethylene oxide	Ethylene oxide is best known as a chemical that exists in a gas form under ambient conditions and is used for in situ sterilization of medical devices.[b]	2.9.9–2.9.14

[a] Ethyleneimine was used to modify cysteine residues to produce S-aminoethyl cysteine, which added an additional trypsin cleavage site to proteins for structural analysis.[2.9.1–2.9.3] Ethyleneimine was also used to identify functional cysteine residues in proteins.[2.9.4] Toxicity issue eliminated its use for this purpose, and ethyleneimine has been replaced by 2-bromoethylamine passes through an ethyleneimine intermediate structure prior to reaction.[2.9.5] As a result of high positive charge density, poly(ethyleneimine) is used as a vehicle for nucleic acid delivery.[2.9.6–2.9.8]

[b] Ethylene oxide is used for in situ sterilization of medical devices.[2.9.9] Ethylene oxide can modify proteins and nucleic acids and is considered toxic.[2.9.10,2.9.11] Green fluorescent protein has been suggested as an indicator for the removal of ethylene oxide.[2.9.10] The reaction of ethylene glycol with plasma albumin resulted in the modification of arginine, cysteine, histidine, lysine, methionine, and tyrosine.[2.9.11] The reaction results in the formation of hydroxyethyl derivatives (Figure 2.15). Unless the ethylene oxide is removed from devices used in the collection or transfer of blood or blood products or in devices such as indwelling catheters, reaction with blood proteins can occur[2.9.12] resulting in the formation of neoantigens and immunological sequelae.[2.9.13,2.9.14]

References to Table 2.9

2.9.1.   Raftery, M.A. and Cole, R.D., On the aminoethylation of proteins, *J. Biol. Chem.* 241, 3457–3461, 1966.

2.9.2.   Givol, D., The cleavage of rabbit immunoglobulin G by trypsin after mild reduction and aminoethylation, *Biochem. J.* 104, 39C–40C, 1967.

2.9.3.   Schroeder, W.A., Shelton, J.R., and Robertson, B., Modification of methionyl residues during aminoethylation, *Biochim. Biophys. Acta* 147, 590–592, 1967.

2.9.4.   Whitney, P.L., Powell, J.T., and Sanford, G.L., Oxidation and chemical modification of lung β-galactosidase-specific lectin, *Biochem. J.* 238, 683–689, 1986.

2.9.5.   Hopkins, C.E., Hernandez, G., Lee, J.P., and Tolan, D.R., Aminoethylation in model peptides reveals conditions for maximizing thiol specificity, *Arch. Biochem. Biophys.* 443, 1–10, 2005.

2.9.6.   Zeng, Q., Han, J., Zhao, D. et al., Protection of adenovirus from neutralizing antibody by cationic PEG derivative ionically linked to adenovirus, *Int. J. Nanomed.* 7, 985–997, 2012.

2.9.7.   Endres, T., Zhend, M., Beck-Broichsitter, M. et al., Optimising the self-assembly of siRNA loaded PEG-PCL-lPEI nano-carriers employing different preparation techniques, *J. Control. Release* 160, 583–591, 2012.

2.9.8.   Jäger, M., Schubert, S., Ochrimenko, S. et al., Branched and linear poly(ethylene imine)-based conjugates: Synthetic modification, characterization, and application, *Chem. Soc. Rev.* 41, 4755–4767, 2012.

2.9.9.   Mendes, G.C., Brandão, T.R., and Silva, C.L., Ethylene oxide sterilization of medical devices: A review, *Am. J. Infect. Control* 35, 574–581, 2007.

2.9.10. Dias, F.N., Ishii, M., Nagaroto, S.L. et al., Sterilization of medical devices by ethylene oxide, determination of the dissipation of residues, and use of Green Fluorescent Protein as an indicator of process control, *J. Biomed. Mater. Res. B Appl. Biomater.* 91, 626–630, 2009.

2.9.11. Starbuck, W.C. and Busch, H., Hydroxyethylation of amino acids in plasma albumin with ethylene oxide, *Biochim. Biophys. Acta* 78, 594–605, 1963.

2.9.12. Tock, R.M. and Chen, Y.C., Aeration of medical plastics, *J. Biomed. Mater. Res.* 8, 69–80, 1974.

2.9.13. Maurer, P.H., Immunologic studies with ethylene oxide-treated human serum, *J. Exp. Med.* 113, 1029–1040, 1961.

2.9.14. Grammer, L.C., Shaughnessy, M.A., Paterson, B.F., and Patterson, R., Characterization of an antigen in acute anaphylactic dialysis reactions: Ethylene oxide-altered human serum albumin, *J. Allergy Clin. Immunol.* 76, 670–675, 1985.

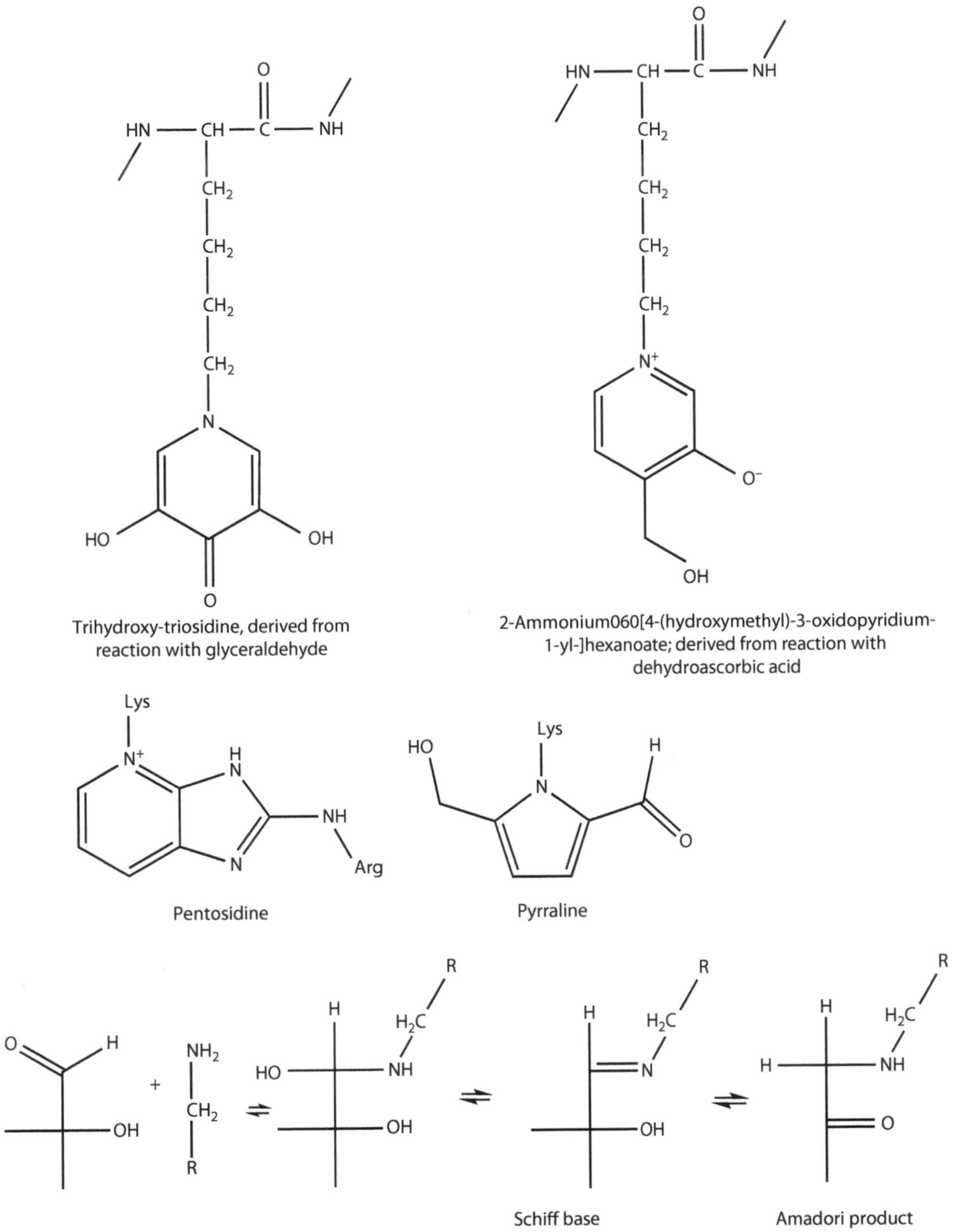

FIGURE 2.4 The modification of proteins by the process of glycation. This process can result in the formation of complex structures known as advanced glycation end products (AGEs). (See Miyata, T., Taneda, S., Kawai, R. et al., Identification of pentosidine as a native structure for advanced glycation end products in β_2-microglobulin-containing amyloid fibrils in patients with dialysis-related amyloidosis, *Proc. Natl. Acad. Sci. USA* 93, 2353–2358, 1996; Madian, A.G. and Regnier, F.E., Proteomic identification of carbonylated proteins and their oxidation sites, *J. Proteome Res.* 9, 3766–2780, 2010).

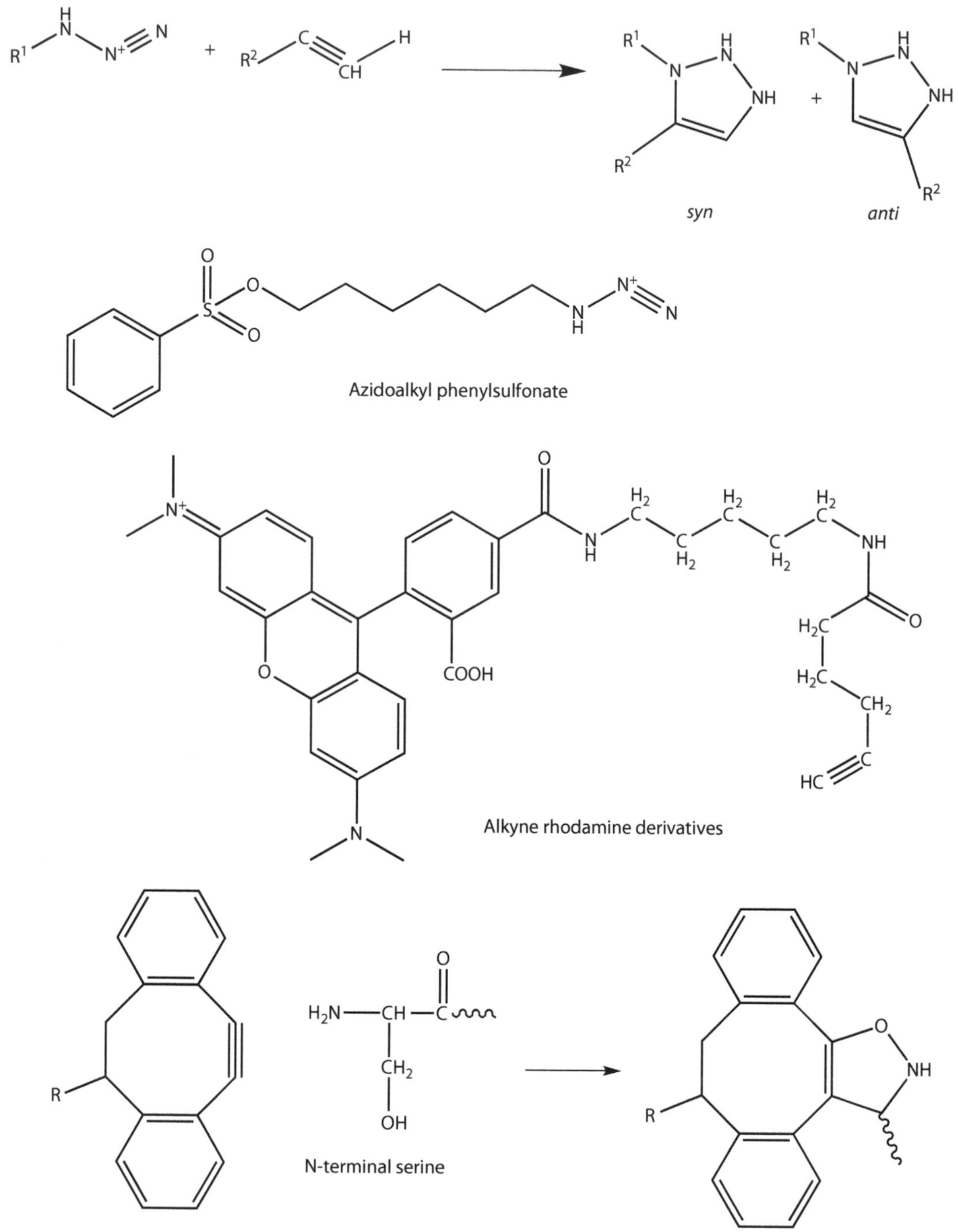

Azidoalkyl phenylsulfonate

Alkyne rhodamine derivatives

N-terminal serine

FIGURE 2.5 Click chemistry. Azide- and alkyne-based chemical probes for proteomic profiling. Note that these reagents can be bifunctional and useful for the study of protein–protein interaction (Breinbauer, R. and Kohn, M., Azide-alkyne coupling: A powerful reaction for bioconjugate chemistry, *Chem Bio Chem* 4, 1147–1149, 2003). Shown at the bottom is a cyclooctyne that can be used to modify a protein via alkyne–nitrone cycloaddition (Ning, X., Temming, R.P., Dommerholt, J. et al., Protein modification by strain-promoted alkyne-nitrone cycloaddition, *Angew. Chem. Int. Ed.* 49, 3065–3068, 2010). This reaction is dependent on the presence of an N-terminal serine that can be modified to an isoxazoline that could couple with the alkyne function.

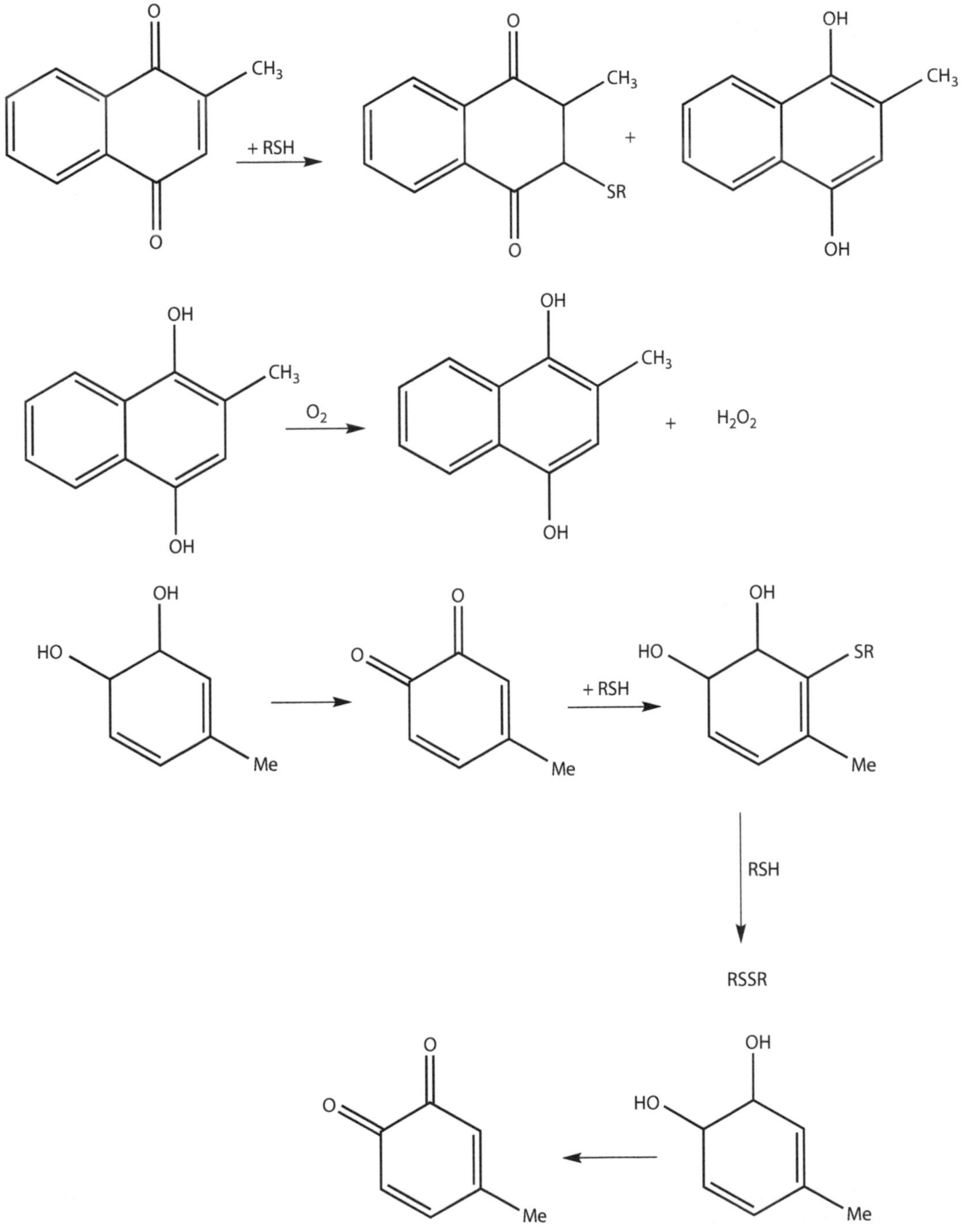

FIGURE 2.6 Reaction of quinones with nucleophiles in proteins. Shown at the top is the reaction of mena-dione with a sulfhydryl group (Nickerson, W.J., Falcone, G., and Strauss, G., Studies on quinone-thioethers. I. Mechanism of formation and properties of thiodione, *Biochemistry* 2, 537–543, 1963). This chemistry can be used for the assay of glutathione using redox indicators (Pacsiai-Ong, E.J., McCarley, R.L., Wang, W., and Strongin, R.M., Electrochemical detection of glutathione using redox indicators, *Anal. Chem.* 76, 7577–7581, 2006). At the bottom is shown a mechanism for the participation of quinones in disulfide formation (Jongberg, S., Lund, M.N., Waterhouse, A.I., and Skibsted, L.H., 4-Methylcatechol inhibits protein oxidation in meat but not disulfide formation, *J. Agric. Food Chem.* 58, 10329–10335, 2011).

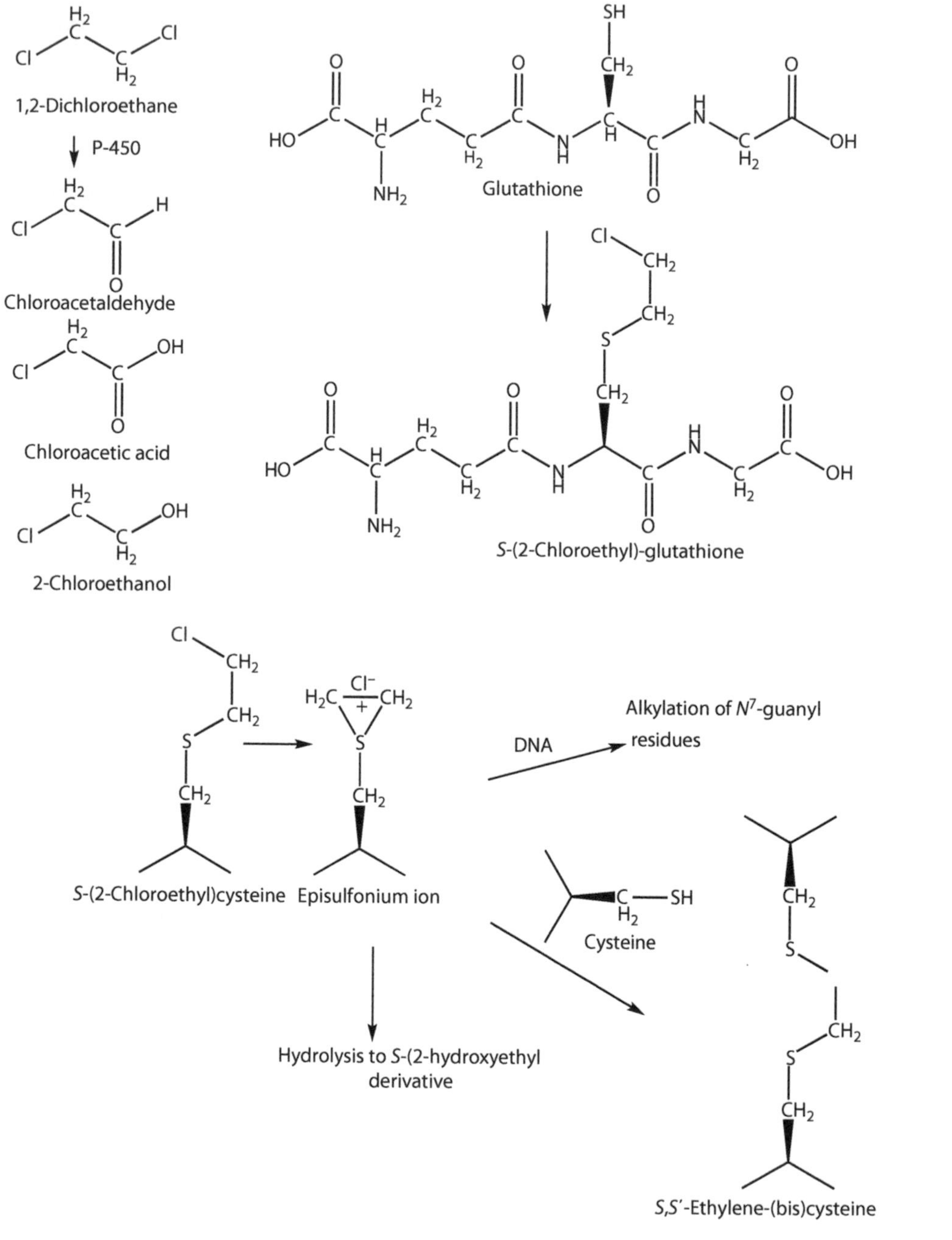

FIGURE 2.7 The reactions of 1,2-dichloroethane. Shown on the left are the oxidative reactions (P-450) of 1,2-chloroethane to various derivatives that result in the intermediate formation of glycolic acid leading to carbon dioxide. Not shown are the possible reactions of the aldehyde derivative with various protein and nucleic acid nucleophiles. Shown on the left is the substitution reaction of 1,2-dichloroethane with the cysteine residue in glutathione that results in the formation of S-(2-chloroethyl)glutathione that reacts with protein or nucleic acid nucleophiles via an episulfonium intermediate or undergoes hydrolysis to form the hydroxyethyl derivative. (Adapted from Jean, P.A. and Reed, D.J., In vitro dipeptide, nucleoside, and glutathione alkylation by S-(2-chloroethyl)glutathione and S-(2-chloroethyl)-L-cysteine, *Chem. Res. Toxicol.* 2, 455–460, 1989; Smit, N.A., Zefirov, N.S., Bodrikov, I.F., and Kirmer, N.Z., Episulfonium ions: Myths and realities, *Acc. Chem. Res.* 12, 282–288, 1978; Gwinn, M.R., Johns, D.O., Bateson, T.F., and Guyton, K.Z., A review of the genotoxicity of 1,2-dichloroethane (EDC), *Mut. Res. Rev. Mutat. Res.* 727, 42–53, 2011).

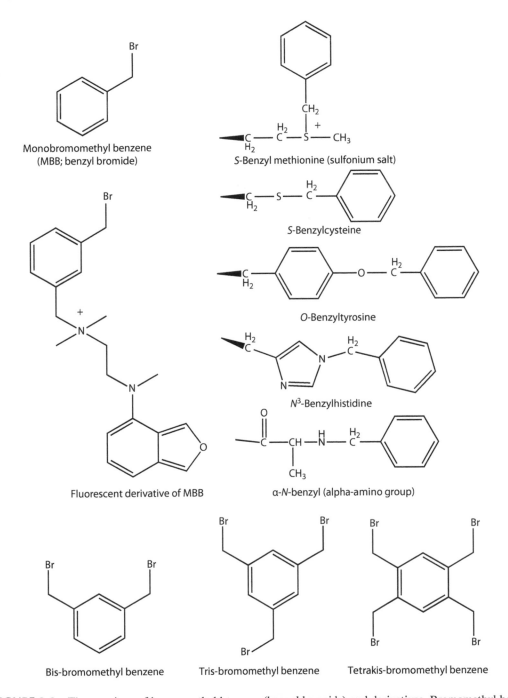

FIGURE 2.8 The reactions of bromomethyl benzene (benzyl bromide) and derivatives. Bromomethyl benzene forms stable, covalent adducts with most nucleophiles in proteins. The reactions with methionine and sulfhydryl groups are then characterized (see Rogers, G.A., Shaltiel, N., and Boyer, P.D., Facile alkylation of methionine by benzyl bromide and demonstration of fumarase inactivation accompanied by alkylation of a methionine residue, *J. Biol. Chem.* 251, 5711–5717, 1976). The bis, tris, and tetrakis derivatives have been used to form stable peptide loop–binding domains from cysteine-containing peptides (see Smeenk, L.E., Dailly, N., Hiemstra, H. et al., Synthesis of water-soluble scaffolds for peptide cyclization, labeling, and ligation, *Org. Lett.* 14, 1194–1197, 2012). A fluorescent derivative of MBB has been prepared (Lang, S., Spratt, D.E., Guillemette, J.G., and Palmer, M., Selective labeling of selenomethionine residues in proteins with a fluorescent derivative of benzyl bromide, *Anal. Biochem.* 359, 253–259, 2006).

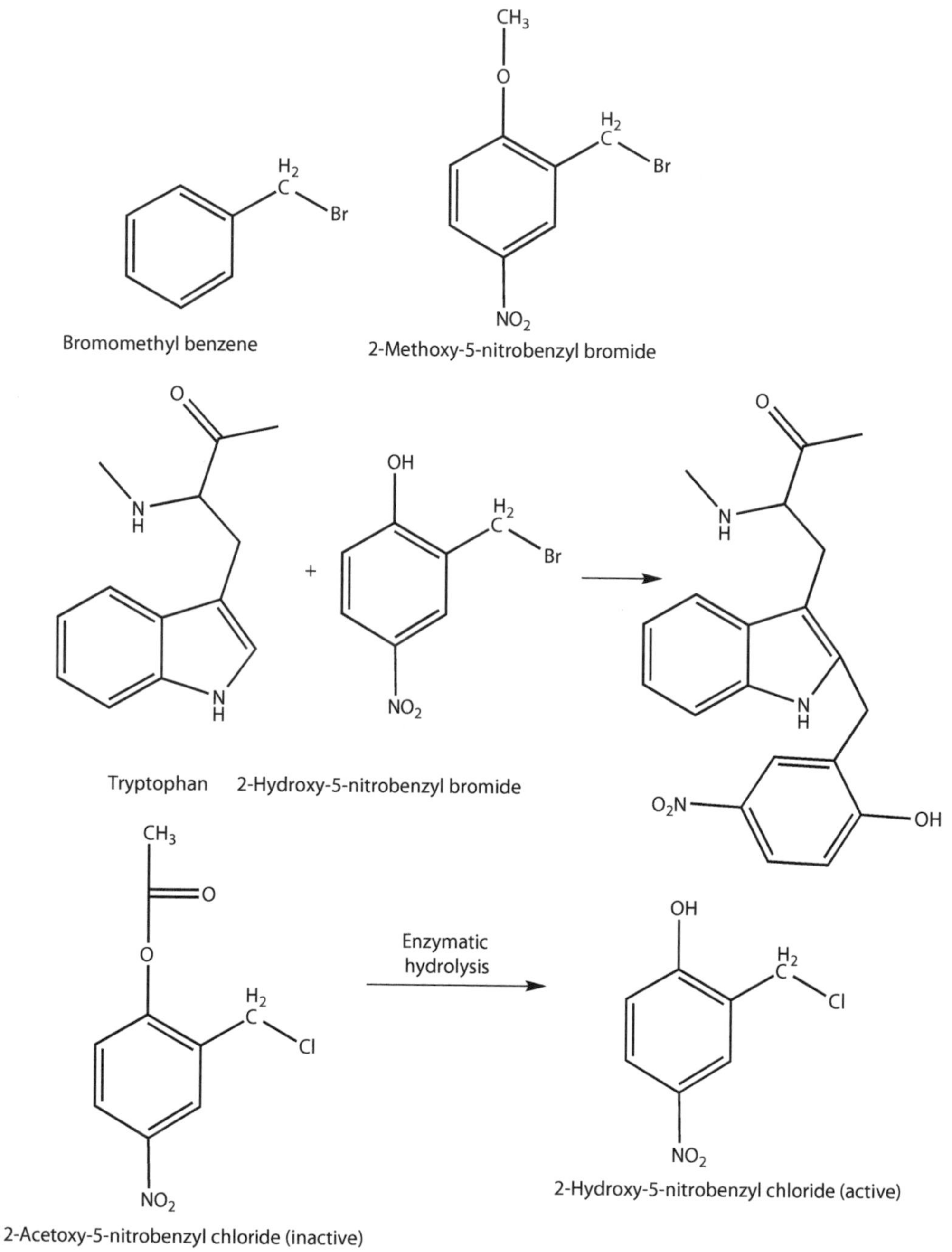

Bromomethyl benzene

2-Methoxy-5-nitrobenzyl bromide

Tryptophan   2-Hydroxy-5-nitrobenzyl bromide

O_2N ———— OH

2-Acetoxy-5-nitrobenzyl chloride (inactive)

Enzymatic hydrolysis

2-Hydroxy-5-nitrobenzyl chloride (active)

FIGURE 2.9   Reactions of HNB. Shown is the structure of HNB that was developed as a *reporter* group with specificity for tryptophan (Horton, H.R. and Koshland, D.E., Jr., A highly reactive colored reagent with selectivity for the tryptophan residue in proteins. 2-Hydroxy-5-nitrobenzyl bromide, *J. Am. Chem. Soc.* 87, 1126–1132, 1965). This reagent is sensitive to changes in pH. 2-Methoxy-5-nitrobenzyl bromide is not sensitive to pH but is sensitive to changes in solvent polarity (Horton, H.R., Kelly, H., and Koshland, D.E., Jr., Environmentally sensitive protein reagents. 2-Methoxy-5-nitrobenzyl bromide, *J. Biol. Chem.* 240, 722–724, 1965). Shown at the bottom is the structure of 2-acetoxy-5-nitrobenzyl chloride that is inactive but on hydrolysis by chymotrypsin releases the active 2-hydroxy-5-nitrobenzyl chloride (Horton, H.R. and Young, G., 2-Acetoxy-5-nitrobenzyl chloride. A reagent designed to introduce a reporter group near the active site of chymotrypsin, *Biochim. Biophys. Acta* 194, 272–278, 1969).

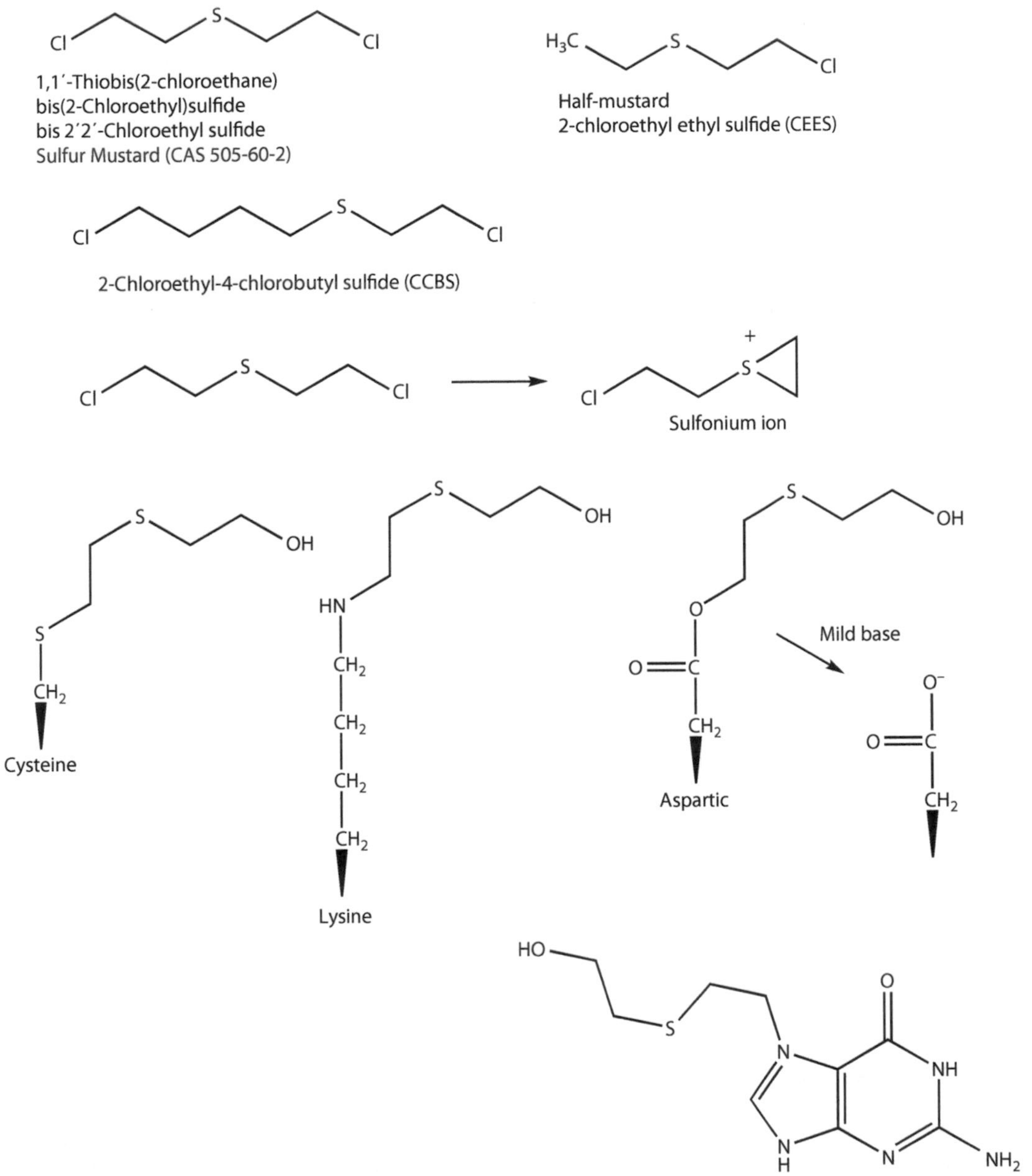

FIGURE 2.10 Structure and reactions of sulfur mustards. Shown are the structures of sulfur mustard, half-mustard, and another sulfur mustard derivative and the reaction to form a cyclic sulfonium ion (Wang, Q.-Q., Begum, R.A., Day, V.W., and Bowman-James, K., Sulfur, oxygen, and nitrogen mustards: Stability and reactivity, *Org. Biomol. Chem.* 10, 8786–8793, 2012). Also shown are the reaction products with cysteine, lysine, and aspartic acid sharing the hydroxyethylthioethyl derivative. The ester derivative formed with aspartic (and glutamic acid) is labile in base forming thiodiglycol (Lawrence, R.J., Smith, J.R., Boyd, B.L., and Capacio, B.R., Improvements in the methodology of monitoring sulfur mustard exposure by gas chromatography-mass spectrometry analysis of cleaved and derivatized blood protein adducts, *J. Anal. Toxicol.* 32, 31–36, 2008). Shown at the bottom is the reaction product obtained with the guanine base in DNA (Fidder, A., Moes, G.W.H., Scheffer, A.G. et al., Synthesis, characterization, and quantitation of the major adducts formed between sulfur mustard and DNA of calf thymus and human blood, *Chem. Res. Toxicol.* 7, 199–204, 1994). Other products are formed with DNA including a guanine dimer (not shown).

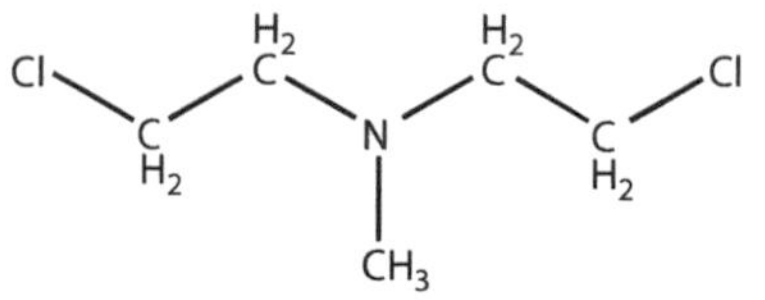

2-Chloro-*N*-(2-chloroethyl)-*N*-methylamine (CAS 51-75-2)

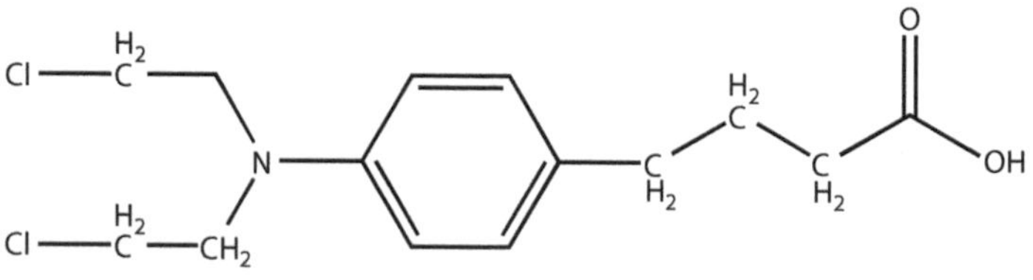

Chlorambucil (CAS 305-03-3)
(4-[bis(2-chloroethyl)amino]benzenebutanoic acid)

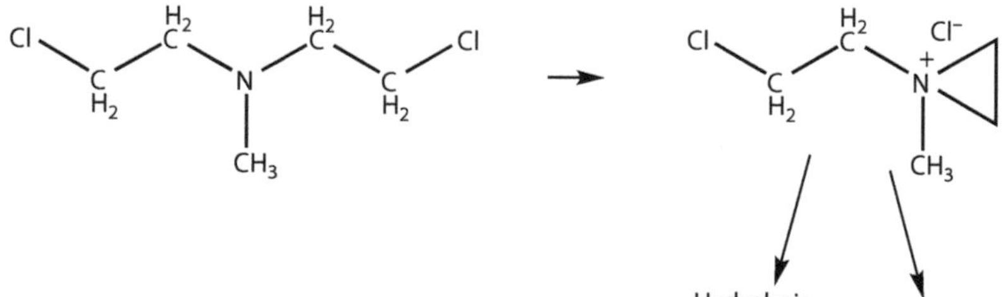

Hydrolysis

Reaction with
nucleophiles

FIGURE 2.11 Structure and reactions of nitrogen mustards. Shown is the structure of the original nitrogen mustard (2-chloro-*N*-(2-chloroethyl)-*N*-methylethanamine; mechlorethamine) and a derivative drug, chlorambucil. The reaction of nitrogen mustard is an S_N1 reaction involving the generation of an aziridinium intermediate (Polavarapu, A., Baik, M.-H., Stubblefield, S.G.W. et al., The mechanism of guanine alkylation by nitrogen mustards—A computational study, *J. Org. Chem.* 77(14), 5914–5921, 2012). The reaction products are similar to those described for sulfur mustards in Figure 2.10 (Wang, Q.-Q., Begum, R.A., Day, V.W., and Bowman-James, K., Sulfur, oxygen, and nitrogen mustards: Stability and reactivity, *Org. Biomol. Chem.* 10, 8786–8793, 2012).

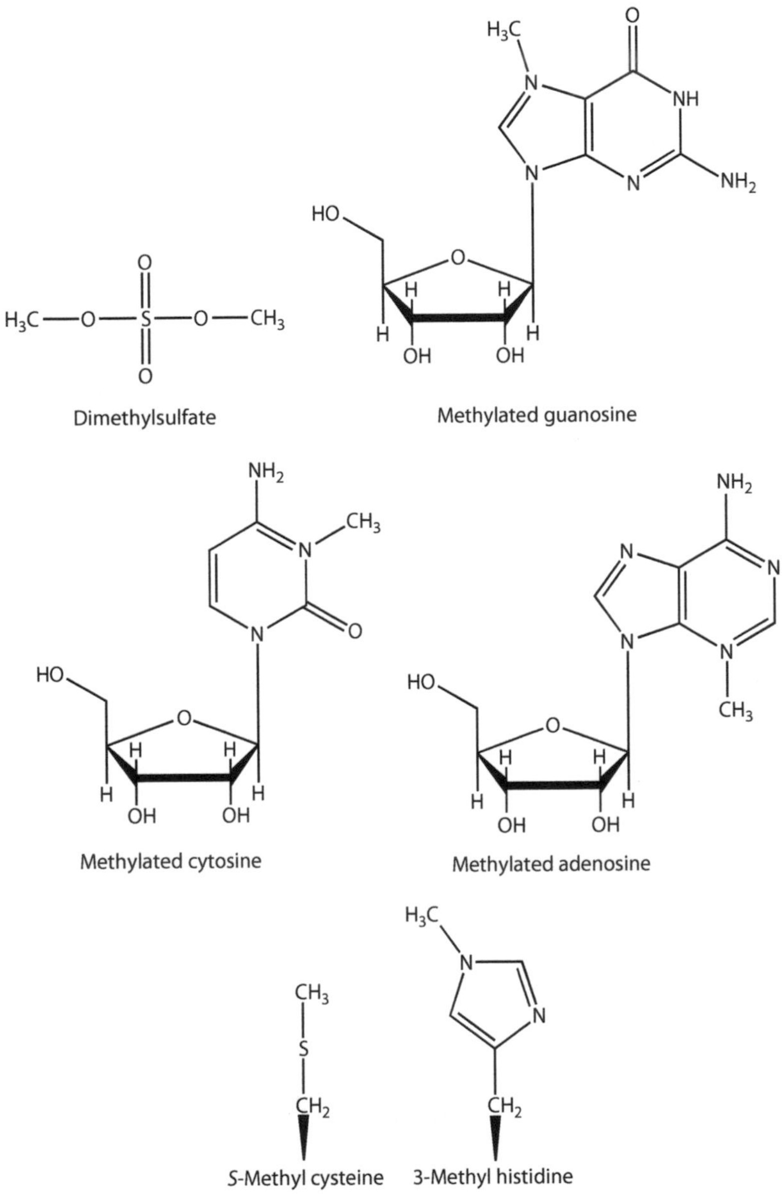

FIGURE 2.12 Reactions of DMS. DMS is primarily recognized as an alkylating agent for DNA with the products as shown. DMS will alkylate protein with methylation of cysteine and histidine as major products; there is minor modification of lysine and arginine (Boffa, L.C., Bolognesi, C., and Mariani, M.R., Specific targets of alkylating agents in nuclear proteins of cultured hepatocytes, *Mutat. Res.* 190, 119–123, 1987).

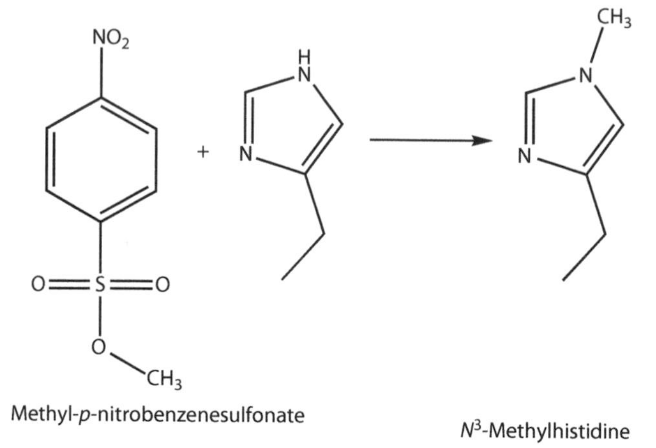

Methyl-*p*-nitrobenzenesulfonate

*N*³-Methylhistidine

FIGURE 2.13   Reactions of methyl-*p*-nitrobenzenesulfonate. Marcus, J.P. and Dekker, E.E., Identification of a second active site residue in Escherichia coli ʟ-threonine dehydrogenase: Methylation of histidine-90 with methyl *p*-nitrobenzenesulfonate. *Arch. Biochem. Biophys.* 316(1), 413–420, January 10, 1995.

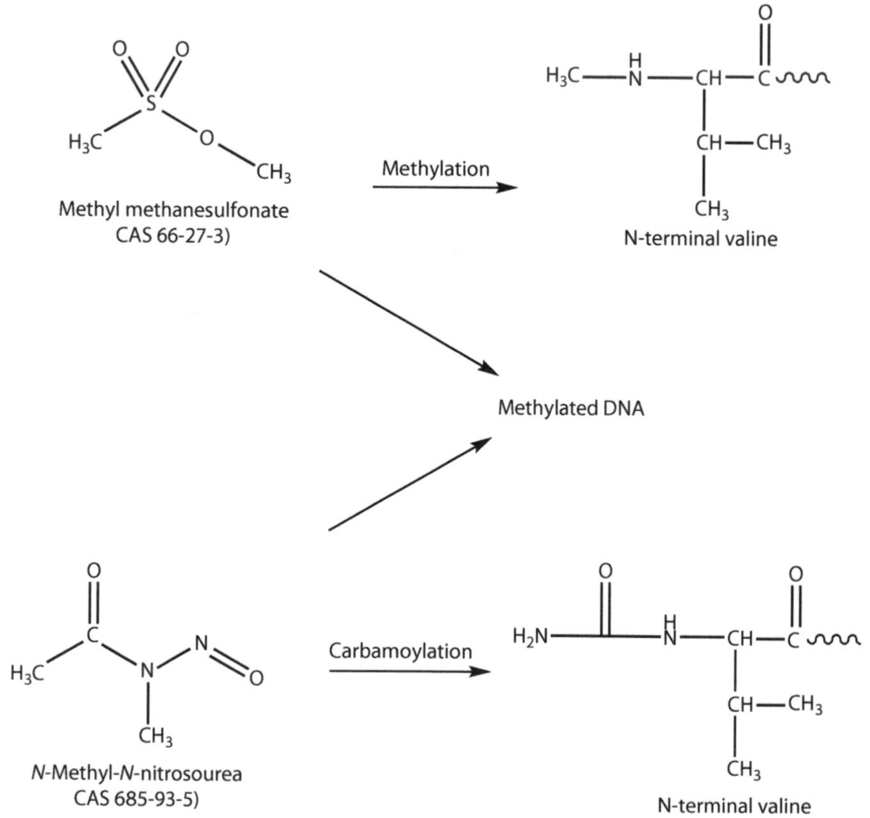

Methyl methanesulfonate
CAS 66-27-3)

Methylation

N-terminal valine

Methylated DNA

N-Methyl-*N*-nitrosourea
CAS 685-93-5)

Carbamoylation

N-terminal valine

FIGURE 2.14   Reactions of methyl methanesulfonate and 1-methyl-1-nitrosourea. Both of these reagents methylate DNA bases (Wyatt, M.D. and Pittman, D.L., Methylating agents and DNA repair responses: Methylated bases and sources of strand breaks, *Chem. Res. Toxicol.* 19, 1580–1594, 2006). Methyl methanesulfonate methylates proteins, while 1-methyl-1-nitrosourea carbamylates proteins (Zhang, F., Bartels, M.J., Pottenger, L.H., and Gollapudi, B.B., Differential adduction of proteins vs. deoxynucleosides by methyl methanesulfonate and 1-methyl-1-nitrosourea in vitro, *Rapid Commun. Mass Spectrom.* 19, 439–448, 2005).

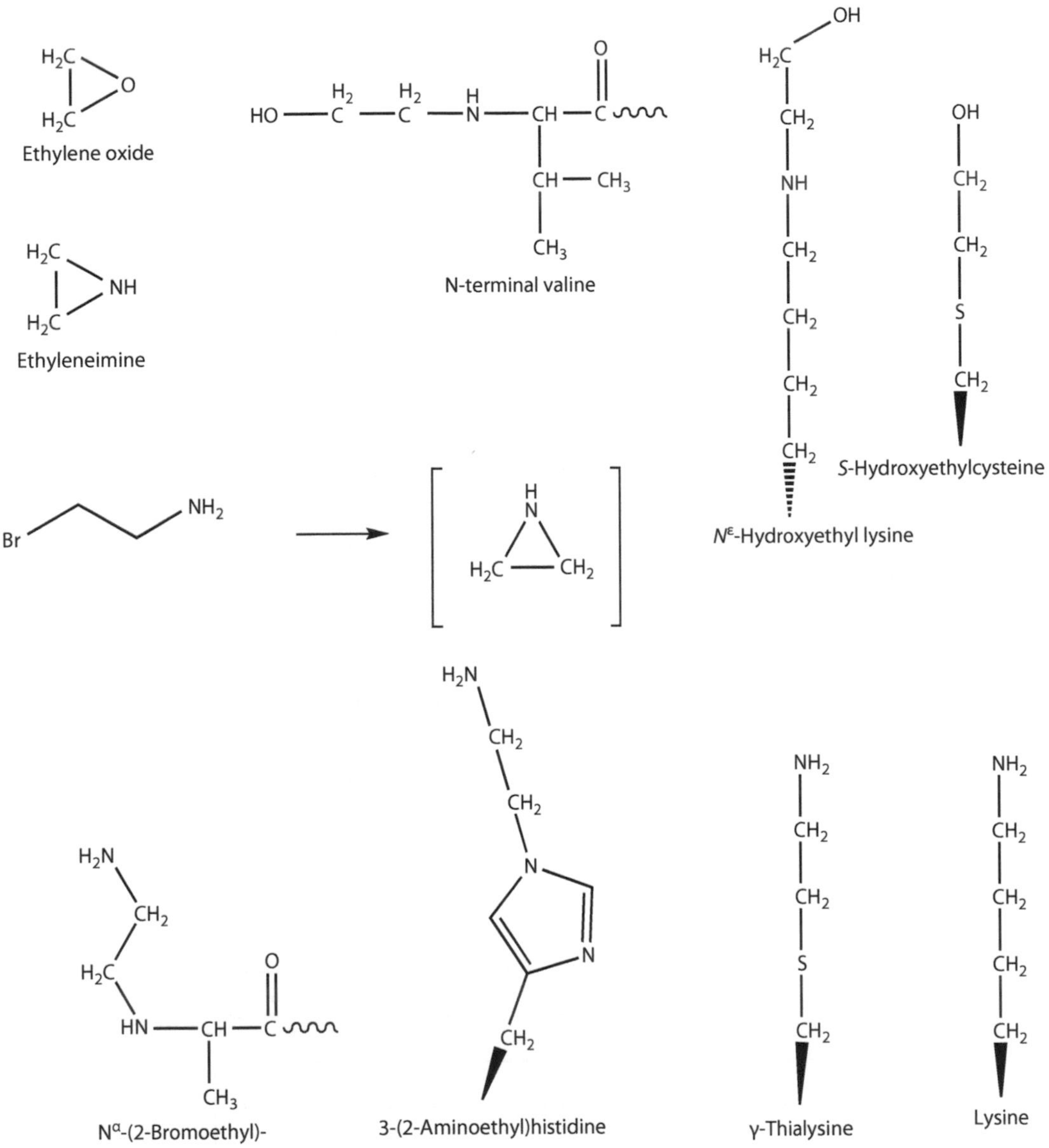

FIGURE 2.15 Reaction of 2-bromoethylamine with amino acid residues. 2-Bromoethylamine was introduced to replace ethyleneimine that is also shown in the figure. 2-Bromoethylamine has been shown to modify cysteine ≫ histidine ≫ α-amino groups; reaction has also been suggested to occur at methionine residues and with carboxyl groups (Hopkins, C.E., Hernandez, G., Lee, J.P., and Tolan, D.R., Aminoethylation in model peptides reveals conditions for maximizing thiol specificity, *Arch. Biochem. Biophys.* 443, 1–10, 2005). Ethylene oxide modifies amines, histidine, and cysteine in proteins (Starbuck, W.C. and Busch, H., Hydroxyethylation of amino acids in plasma albumin with ethylene oxide, *Biochim. Biophys. Acta* 78, 594–605, 1963; Bolt, H.M., Peter, H.Y., and Föst, U., Analysis of macromolecular ethylene oxide adducts, *Int. Arch. Occup. Environ. Health* 60, 141–144, 1988; van Welie, R.T., van Dijck, R.G., Vermeulen, N.P. et al., Mercapturic acids, protein adducts, and DNA adducts as biomarkers of electrophilic chemicals, *Crit. Rev. Toxicol.* 22, 271–306, 1992; Törnqvist, M. and Kautiainen, A., Adducted proteins for identification of endogenous electrophiles, *Environ. Health Perspect.* 99, 39–44, 1993). N-terminal valine is shown as a model since it is in hemoglobin and modification is used as a biomarker (Boysesn, G., Georgieva, N.I., Upton, P.B. et al., N-Terminal globin adducts as biomarkers for formation of butadiene derived epoxides, *Chem. Biol. Interact.* 166, 84–92, 2007).

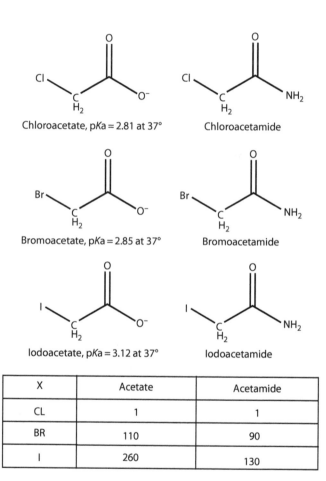

Chloroacetate, pKa = 2.81 at 37°   Chloroacetamide

Bromoacetate, pKa = 2.85 at 37°   Bromoacetamide

Iodoacetate, pKa = 3.12 at 37°   Iodoacetamide

X	Acetate	Acetamide
CL	1	1
BR	110	90
I	260	130

FIGURE 2.16 Haloacetate and haloacetamides. Shown are the structures of the various haloacetate and haloacetamides. Also shown are the relative rates of reaction of the various haloacetates and haloacetamides with the sulfhydryl function of thiodiglycol (Hellstrom, N., Das reaktionsvermügen der mercaptidogruppe. II, *Zeitf. Physiol. Chem. A* 163, 33–52, 1932).

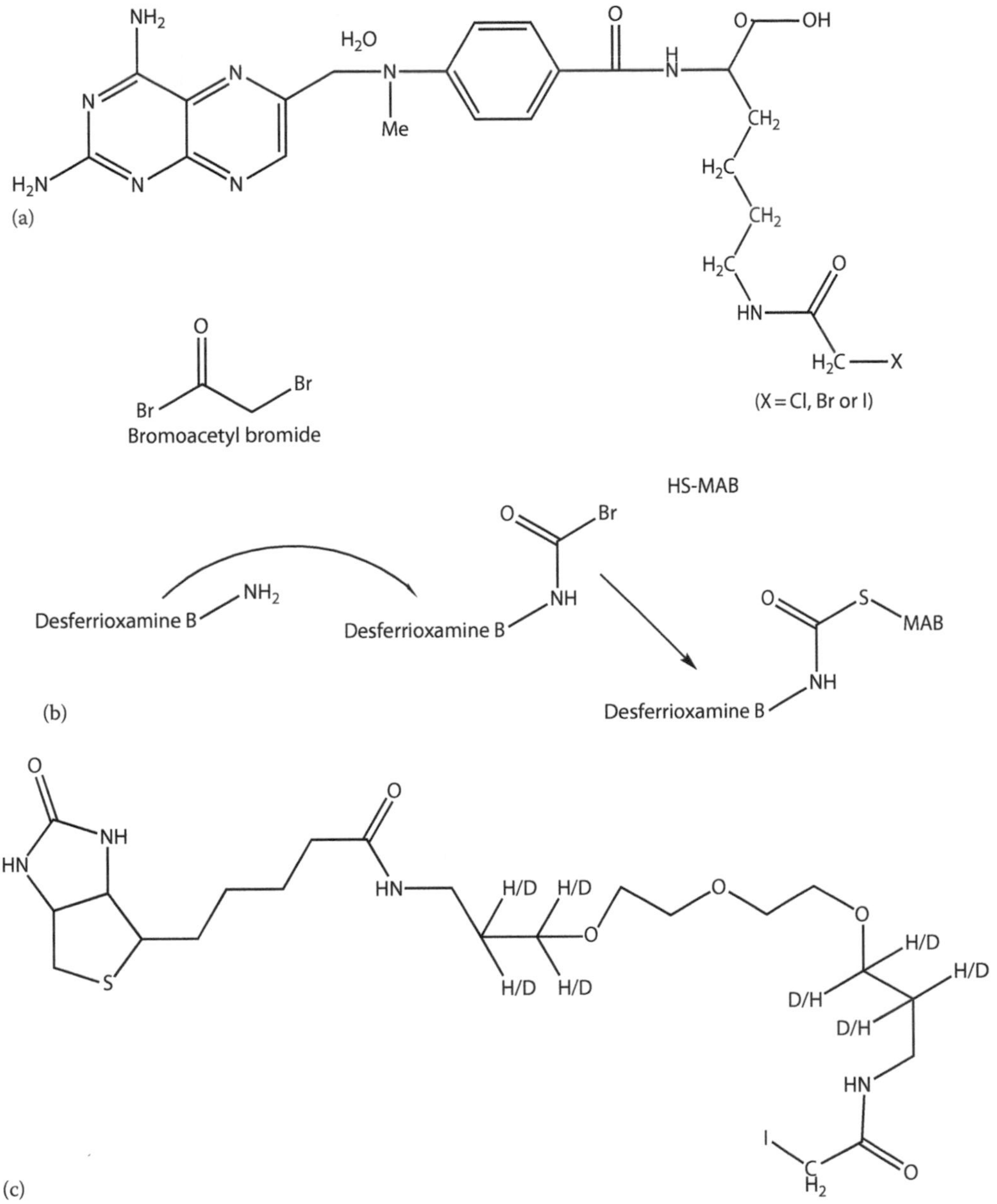

FIGURE 2.17 Utility of the haloalkyl function. The haloalkyl function is used to attach a variety of molecules to proteins. The coupling is usually accomplished through a cysteinyl residue in the target protein. The examples shown in this figure are only several of many applications. Shown is a methotrexate analog (a) using a haloacetyl function (Rosowsky, A., Solan, V.C., Forsch, R.A. et al., Methotrexate analogues. 30. Dihydrofolate reductase inhibition and in vitro tumor growth inhibition by N^{ε}-(haloacetyl)-L-lysine and N^{δ}-(haloacetyl)-L-ornithine analogues and an acivicin analogue of methotrexate, *J. Med. Chem.* 30, 1463–1469, 1987). The lysine derivative is shown in this illustration. Also shown (b) is the bromoacetyl derivative of desferrioxamine B that is coupled to a monoclonal antibody mutated to contain a single cysteine (thio-trastuzumab). The resulting derivative is used in positron emission tomography as an imaging agent (Tinianow, J.N., Gill, H.S., Ogasawara, A. et al., Site-specifically ^{89}Zr-labeled monoclonal antibodies for ImmunoPET, *Nucl. Med. Biol.* 37, 289–297, 2010). The lower illustration (c) shows an ICAT used in proteomic analysis (c.f. Gygi, S.P., Rist, B., Gerber, S.A. et al., Quantitative analysis of complex protein mixtures using isotope-coded affinity tags, *Nat. Biotechnol.* 17, 994–999, 1999).

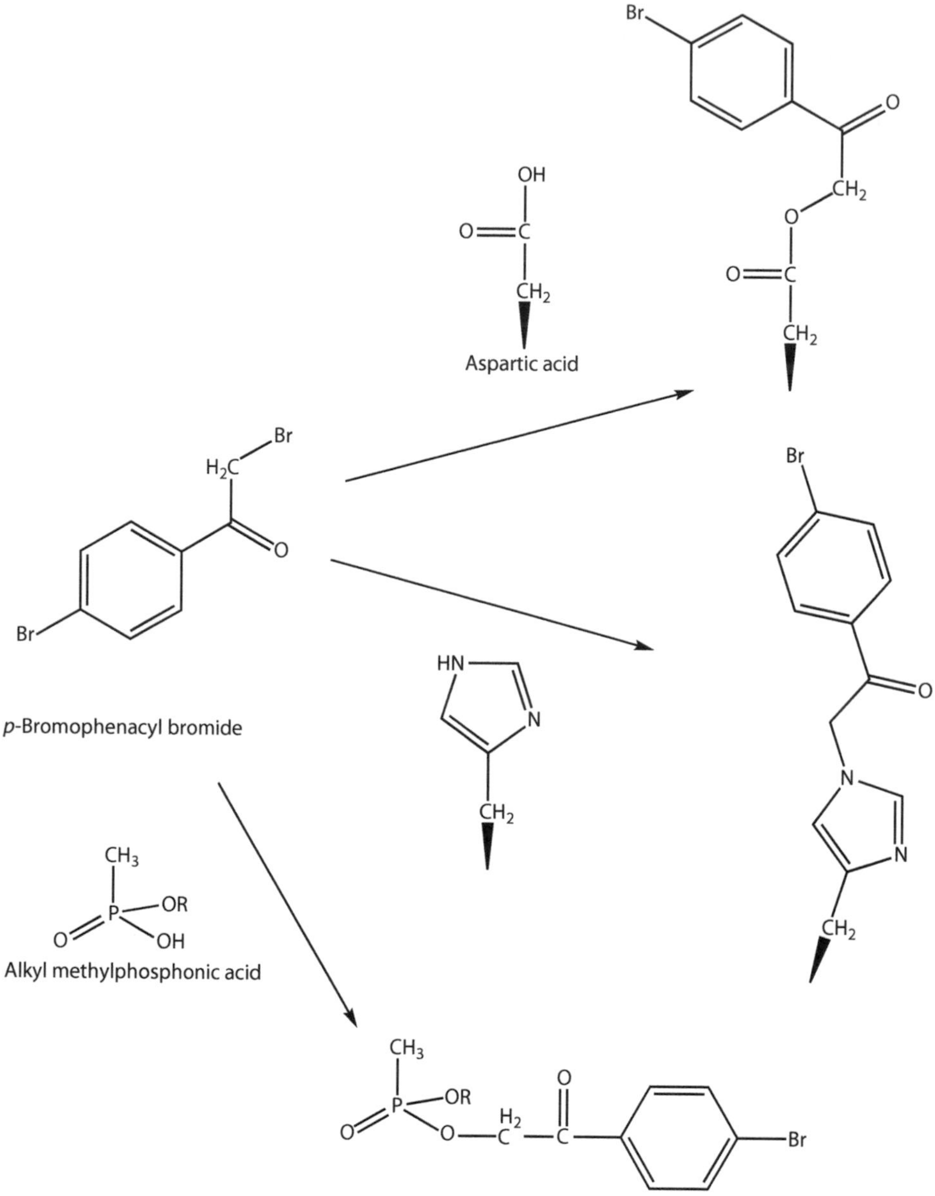

FIGURE 2.18 The reactions of *p*-bromophenacyl bromide. *p*-Bromophenacyl bromide reacts with the β-carboxyl group of aspartic acid residues in proteins as well as the imidazole group of histidine. *p*-Bromophenacyl bromide is also used to prepare derivatives of carboxylic acid and thiol compounds before HPLC analysis (Liu, H.Y., Ding, L., Yu, Y. et al., Comparison of three derivatization reagents for the simultaneous determination of highly hydrophilic pyrimidine antitumor agents in human plasma by LC-MS/MS, *J. Chromatogr. B Anal. Technol. Biomed. Life Sci.* 893–894, 49–56, 2012). Inactivation with *p*-bromophenacyl bromide is considered a characteristic of phospholipase A2 activity (Lane, J., O'Leary, M.A., and Isbister, G.K., Coagulant effects of black snake (*Pseudechis* spp.) venoms and in vitro efficacy of commercial antivenom, *Toxicon* 58, 239–246, 2011). Shown at the bottom is the use of *p*-bromophenacyl bromide for preparing derivatives for liquid chromatography/mass spectrometry (LC/MS) analysis (Katagi, M., Tatsuno, M., NIshikawa, M., and Tsuchihashi, H., On-line solid-phase extraction liquid chromatography-continuous flow frit fast atom bombardment mass spectrometric and tandem mass spectrometric determination of hydrolysis products of nerve agents alkyl methylphosphonic acids by *p*-bromophenacyl derivatization, *J. Chromatogr. A* 833, 169–179, 1999).

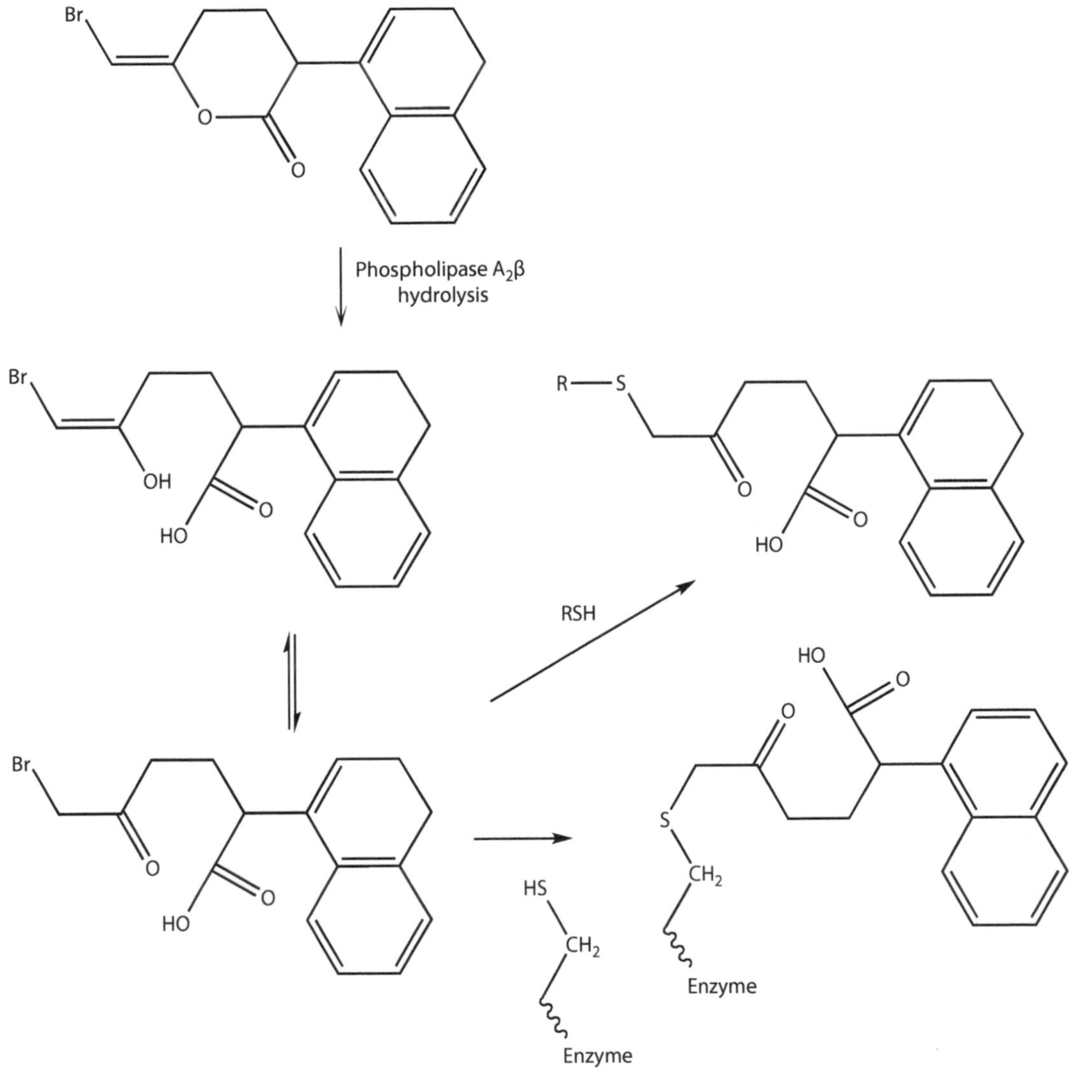

FIGURE 2.19 Bromoenol Lactone suicide substrate. Shown is the structure of (*E*)-6-(bromomethylene) tetrahydro-3-(1-naphthalenyl)-2H-pyran-2-one (BEL) which was developed as a suicide substrate for calcium-independent phospholipase A2 (Hazen, S.L., Zupan, L.A., Weiss, R.H. et al., Suicide inhibition of canine myocardial cytosolic calcium-independent phospholipase A2). Mechanism-based discrimination between calcium-dependent and-independent phospholipase A2. *J. Biol. Chem.* 266, 7227–7232, 1991. The bromo-methylcarbonyl compound then reacts with cysteinyl residues on the enzyme or can diffuse to react with thiol-containing compounds distant from the enzyme (Song, H., Ramanadham, S., Bao, S. et al., A bromoenol lactone suicide substrate inactivates group VIA phospholipase A2 by generating a diffusible bromomethyl keto acid that alkylates cysteine thiols, *Biochemistry* 45, 1061–1073, 2006).

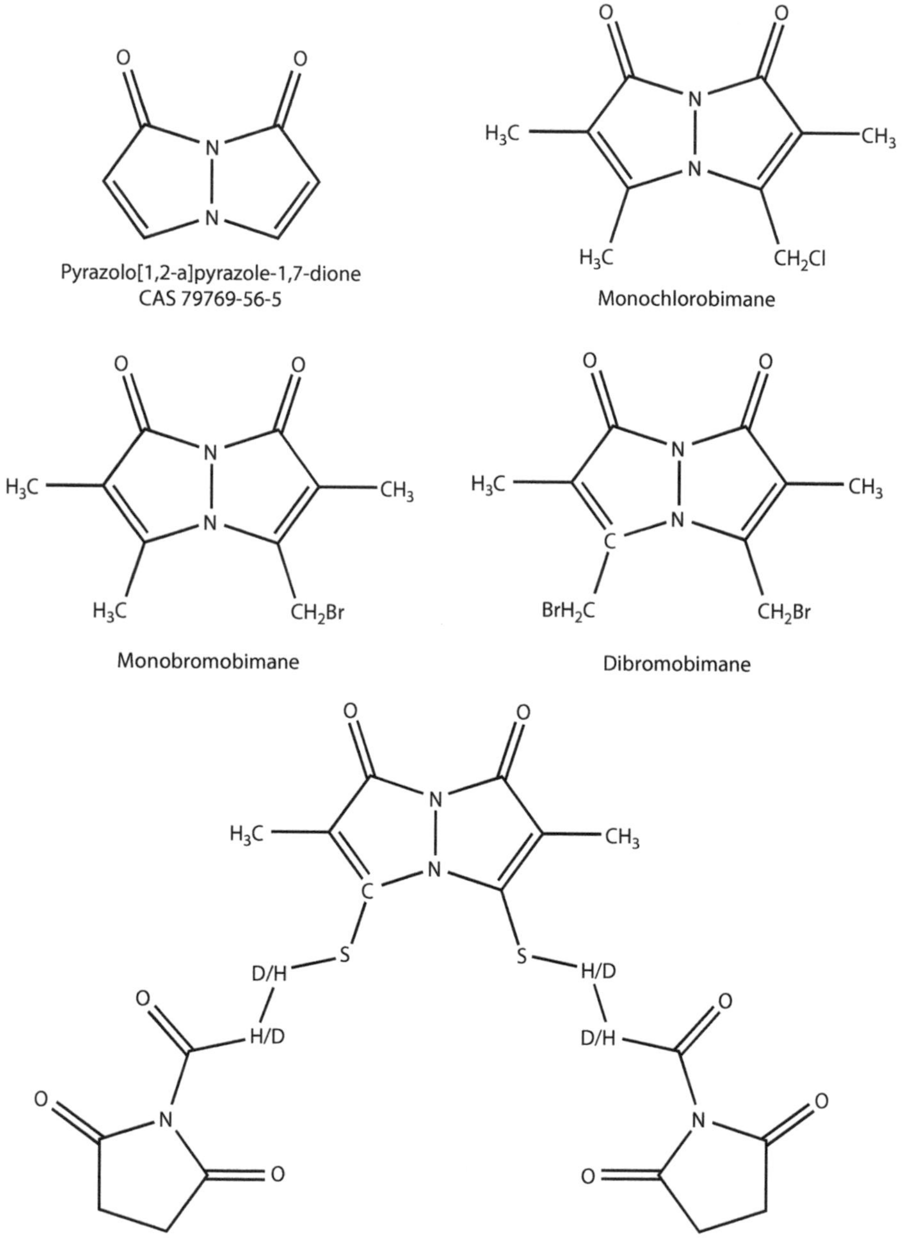

FIGURE 2.20 The structures and reactions of halobimanes. Shown are the structures of the various haloalkanes and parent bimane. Also shown is the structure of bimane bisthiopropionic acid *N*-succinimidyl ester (BiPs) (Petrotchenko, E.V., Xiao, K., Cable, J. et al., BiPS, a photocleavable, isotopically coded, fluorescent cross-linker for structural proteomics, *Mol. Cell. Proteomics* 8, 273–286, 2009).

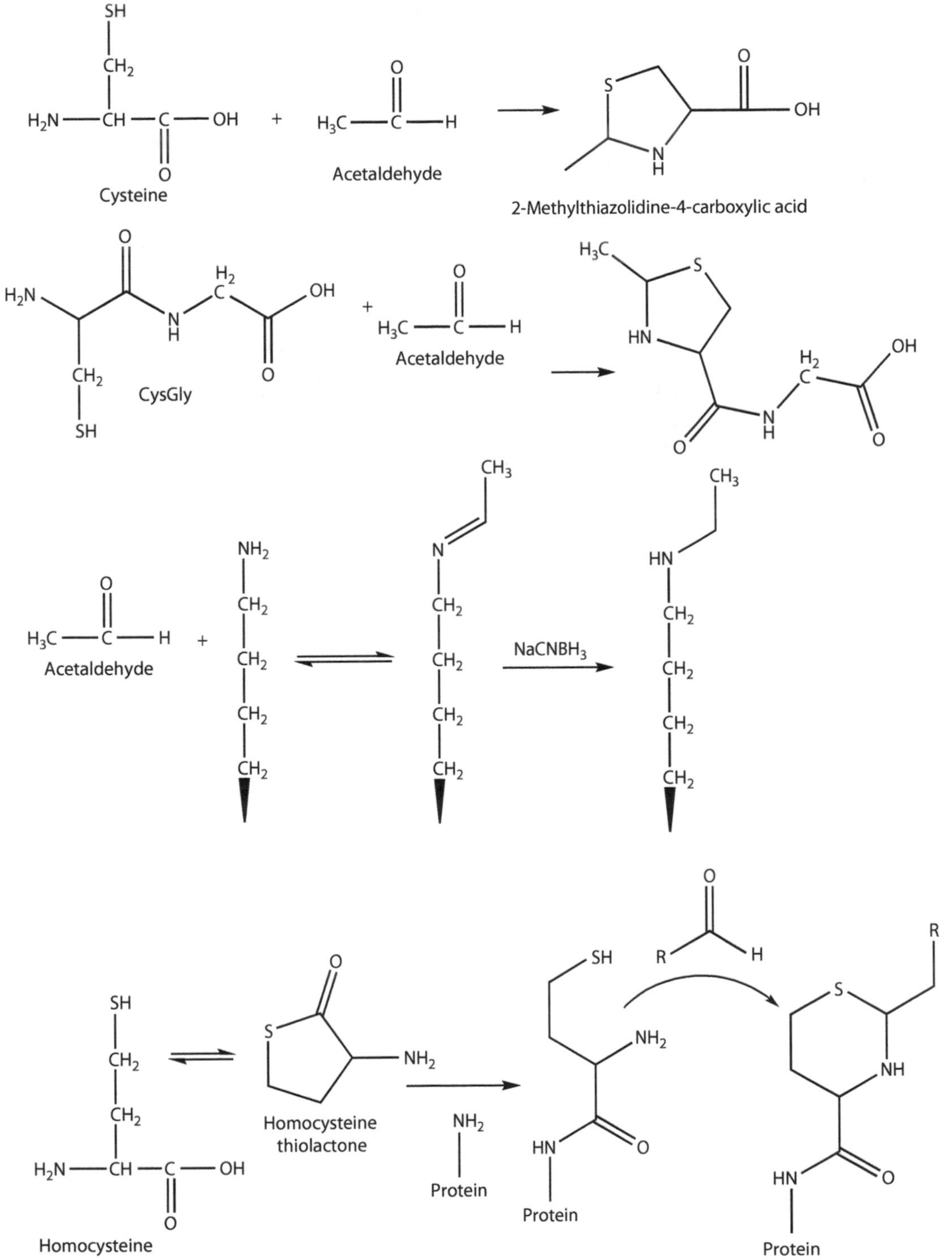

FIGURE 2.21

(*continued*)

FIGURE 2.21 (continued)   Reaction of carbonyl compounds with proteins. Shown is the reaction of acet-aldehyde with cysteine to form 2-methylthiazolidine-4-carboxylic acid (Reischl, R.J., Ricker, W., Keller, T. et al., Occurrence of 2-methylthiazolidine-4-carboxylic acid, a condensation product of cysteine and acetal-dehyde, in human blood as a consequence of ethanol consumption, *Anal. Bioanal. Chem.* 404, 1779–1787, 2012). *N*-acetylation (not shown) of the thiazolidine ring is required to provide stability for analysis. A simi-lar reaction product is obtained with the reaction of acetaldehyde with the dipeptide degradation product obtained from glutathione, cysteinylglycine, by the action of γ-glutamyltransferase (Anni, H., Pristatsky, P., and Israel, Y., Binding of acetaldehyde to a glutathione metabolite: Mass spectrometric characterization of an acetaldehyde-cysteinylglycine conjugate, *Alcohol. Clin. Exp. Res.* 27, 1613–1621, 2003). Acetaldehyde does not form a stable product with the cysteinyl residue in intact glutathione. Acetaldehyde forms stable and unstable Schiff base reaction products with the ε-amino group of lysine and N-terminal amino groups (Braun, K.P., Cody, R.B., Jr., Jones, D.R., and Peterson, C.M., A structural assignment for a stable acetaldehyde-lysine adduct, *J. Biol. Chem.* 270, 11263–11266, 1995; Pietrzak, E.R., Shanley, B.C., and Coon, P.A., Antibodies made against a formaldehyde-protein adduct cross react with an acetaldehyde-protein-adduct. Implications for the origin of antibodies in human serum which recognize acetaldehyde-protein adducts, *Alcohol Alcohol.* 30, 373–378, 1995). The Schiff base products can be stabilized by reduction with sodium cyanoborohydride (Tuma, D.J., Jennett, R.B., and Sorrell, M.F., The interaction of acetaldehyde with tubulin, *Ann. N.Y. Acad. Sci.* 492, 277–286, 1987). Shown at the bottom is the use of aldehyde-based tags to identify *N*-homocysteinylation in proteins. The *R* could be a fluorescent probe or other useful function such as biotin. Homocysteine is formed in a number of disease states including Alzheimer's disease and forms homocysteine thiolactone that can react with proteins and is detected by reaction with aldehydes (Zang, T., Dai, S., Chen, D., Lee, B.W.K. et al., Chemical methods for the detection of protein *N*-homocysteinylation via selective reactions with aldehydes, *Anal. Chem.* 81, 9065–9071, 2009).

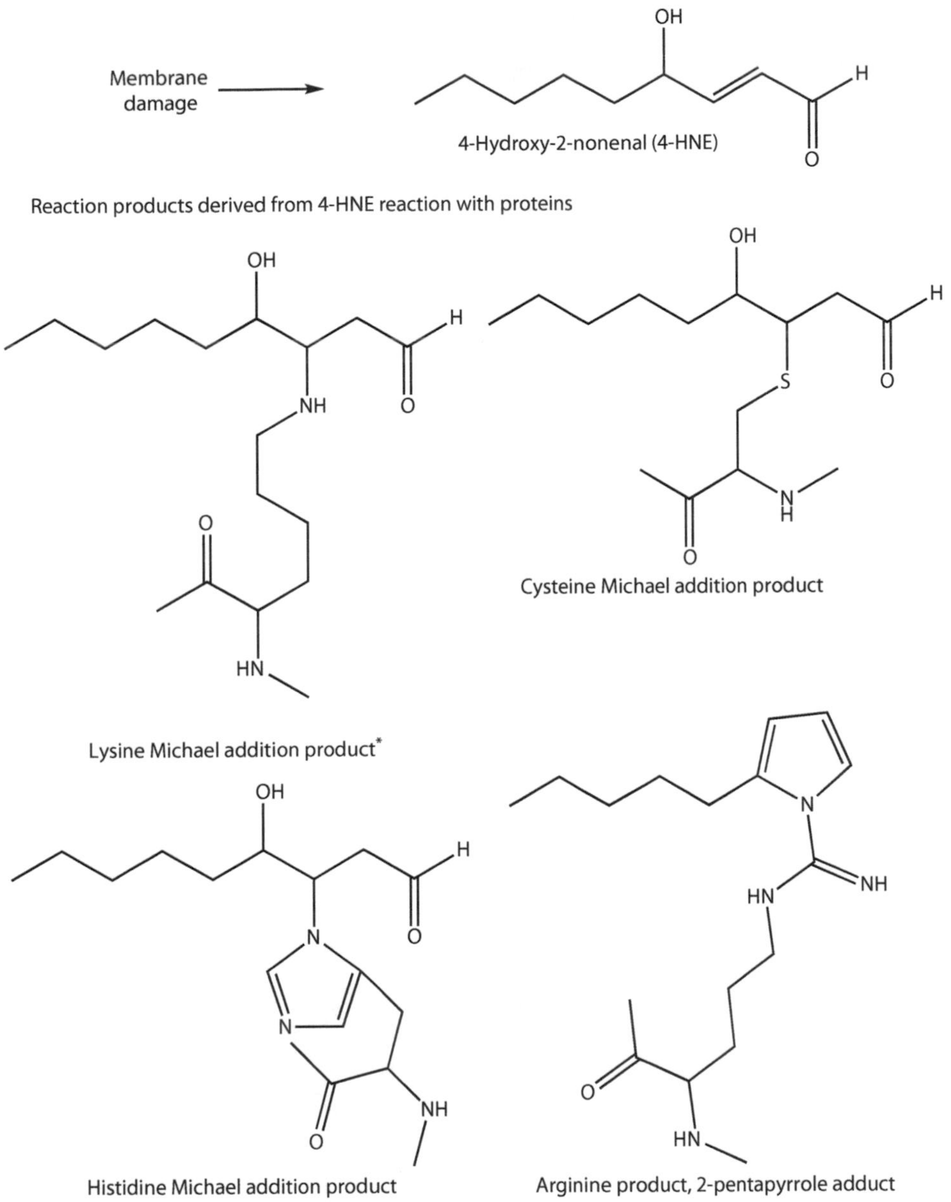

FIGURE 2.22 The reaction of 4-hydroxy-2-nonenal with amino acid residues in proteins. The presence of a double bond and an aldehyde function in 4-HNE permits the formation of a Michael addition product and a Schiff base. (See Tsai, L., Szweda, P.A., Vinogradova, O., and Szweda, L.I., Structural characterization and immunochemical detection of a fluorophore derived from 4-hydroxy-2-nonenal and lysine, *Proc. Natl. Acad. Sci. USA* 95, 7975–7980, 1998; Rauniyar, N. and Prokal, L., Detection and identification of 4-hydroxy-2-nonenal Schiff-base adducts along with products of Michael addition using data-dependent neutral loss-driven MS³ acquisition: Method evaluation through an in vitro study on cytochrome c oxidase modifications, *Proteomics* 9, 5188–5193, 2009). The reaction of 4-HNE with lysine (and other nucleophiles) is complex as shown in the bottom. A cyclic fluorophore can be formed from the product of reaction of 4-HNE with two lysine residues; if Schiff base is reduced, a stable imine is formed without formation of the cyclic fluorophore (Fenaille, F., Guy, P.A., and Tabet, J.-C., Study of protein modification by 4-hydroxy-2-nonenal and other short chain aldehydes analyzed by electrospray ionization tandem mass spectrometry, *J. Am. Soc. Mass Spectrom.* 14, 215–226, 2003); Isom, A.L., Modification of cytochrome c by 4-hydroxy-2-nonenal: Evidence for histidine, lysine, and arginine-aldehyde adducts, *J. Am. Soc. Mass Spectrom.* 15, 1136–1147, 2004).

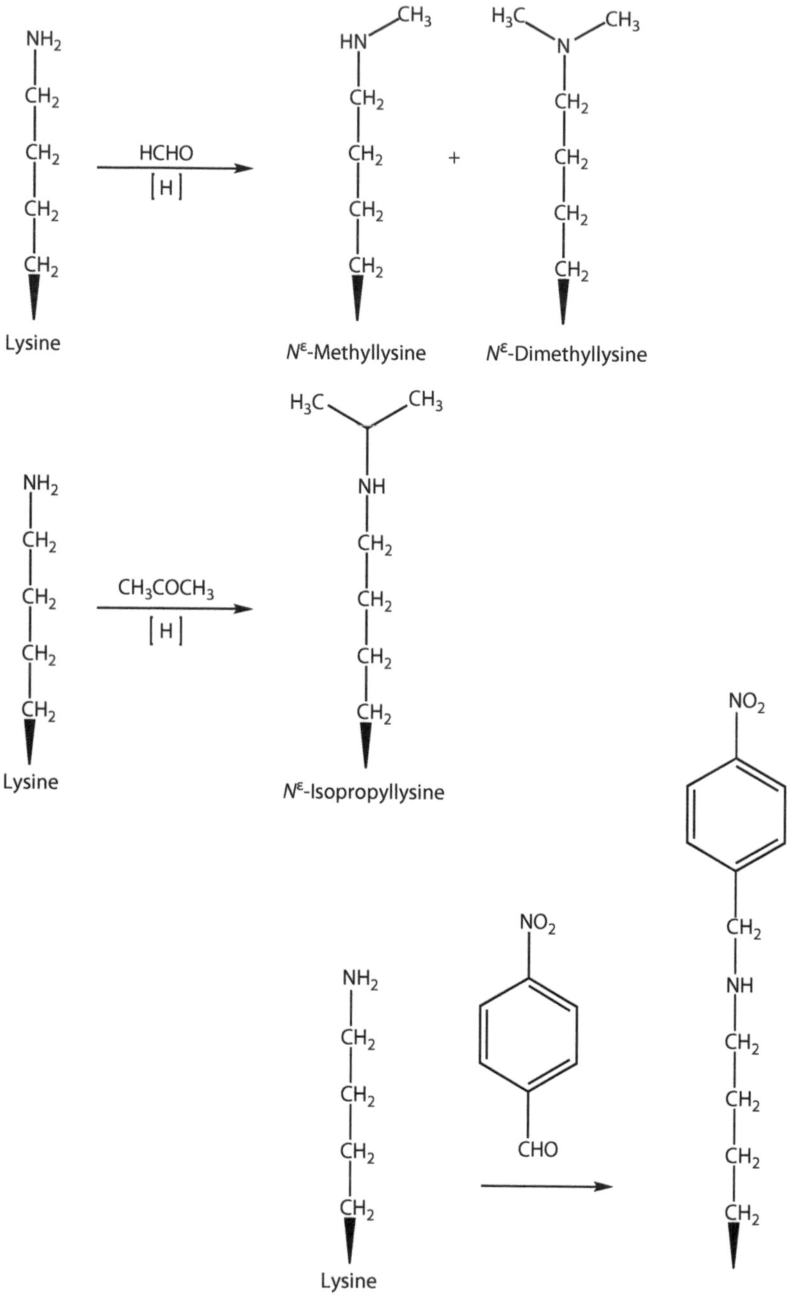

FIGURE 2.23 Reductive alkylation of lysine residues in proteins. Shown is the production of the N^ε-monomethyl and N^ε-dimethyl derivatives of lysine and the N^ε-isopropyl derivative of lysine. The methyl derivatives are obtained from initial reaction with formaldehyde or acetone, respectively, with lysine followed by reduction of the Schiff base intermediate with sodium borohydride (Means, G.E. and Feeney, R.E., Reductive alkylation of amino groups in proteins, *Biochemistry* 7, 2192–2201, 1968). The preparation of aromatic alkyl derivatives is shown at the bottom with the reaction of 4-nitrobenzaldehyde with lysine (Friedman, M., Williams, L.D., and Masri, M.S., Reductive alkylation of proteins with aromatic aldehydes and sodium cyanoborohydride, *Int. J. Pept. Protein Res.* 6, 183–185, 1974). These investigators also reported the superior behavior of sodium cyanoborohydride as a reducing agent in the process of reductive alkylation.

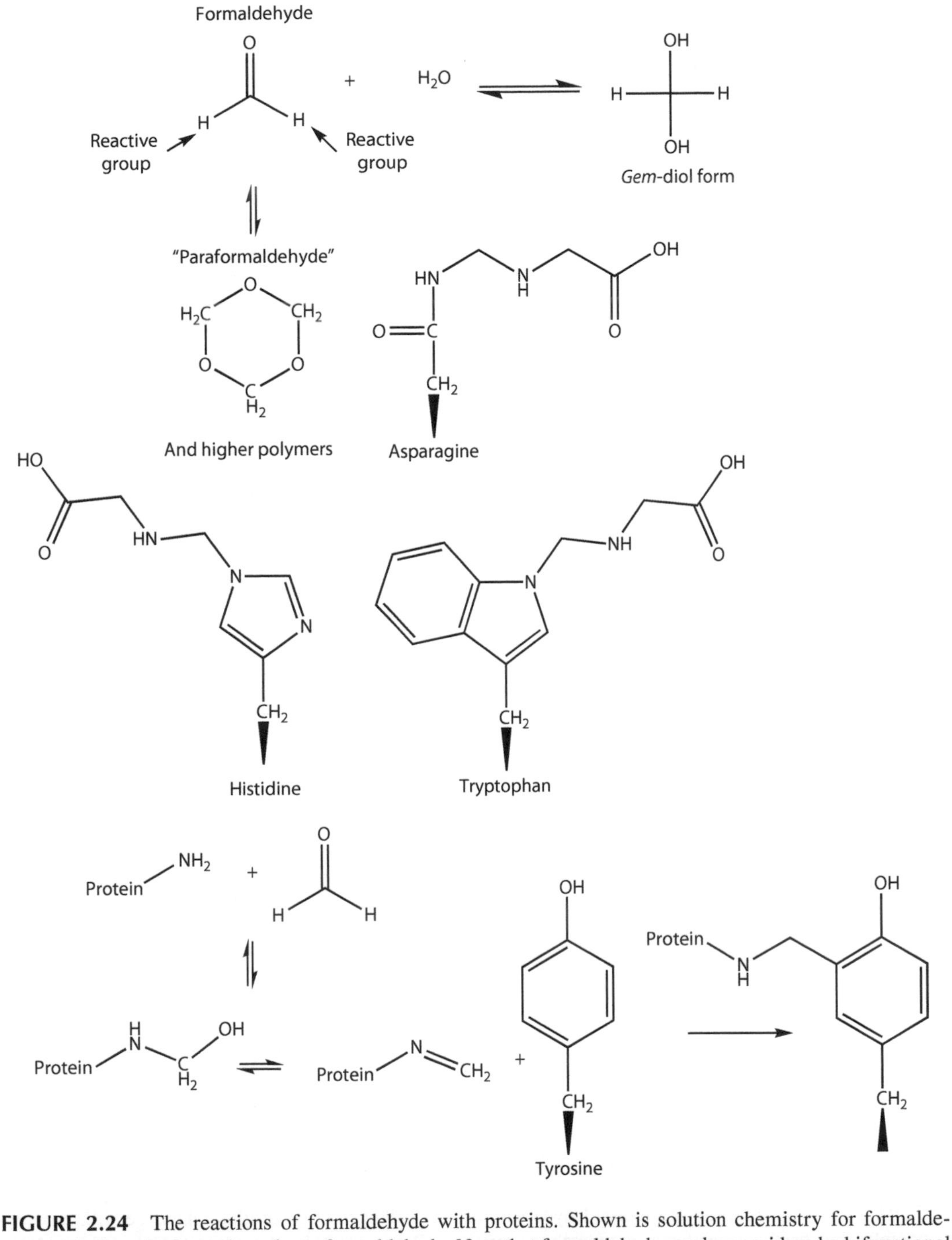

FIGURE 2.24 The reactions of formaldehyde with proteins. Shown is solution chemistry for formaldehyde including the formation of paraformaldehyde. Note that formaldehyde can be considered a bifunctional reagent (Sutherland, B.W., Toews, J., and Kast, J., Utility of formaldehyde cross-linking and mass spectrometry in the study of protein-protein interactions, *J. Mass Spectrom.* 43, 699–715, 2008). Shown below are several suggested products from the reaction of glycine and formaldehyde with the indicated amino acid residue (Metz, B., Kersten, G.F.A., Hoogerhout, P. et al., Identification of formaldehyde-induced modifications in proteins. Reactions with model peptides, *J. Biol. Chem.* 272, 6235–6243, 2004). At the bottom is one of several potential mechanisms for the cross-linking of proteins with formaldehyde (Metz, B., Kersten, G.F.A., Baart, G.J.E. et al., Identification of formaldehyde-induced modifications in proteins: Reactions with insulin, *Bioconjug. Chem.* 17, 815–822, 2006).

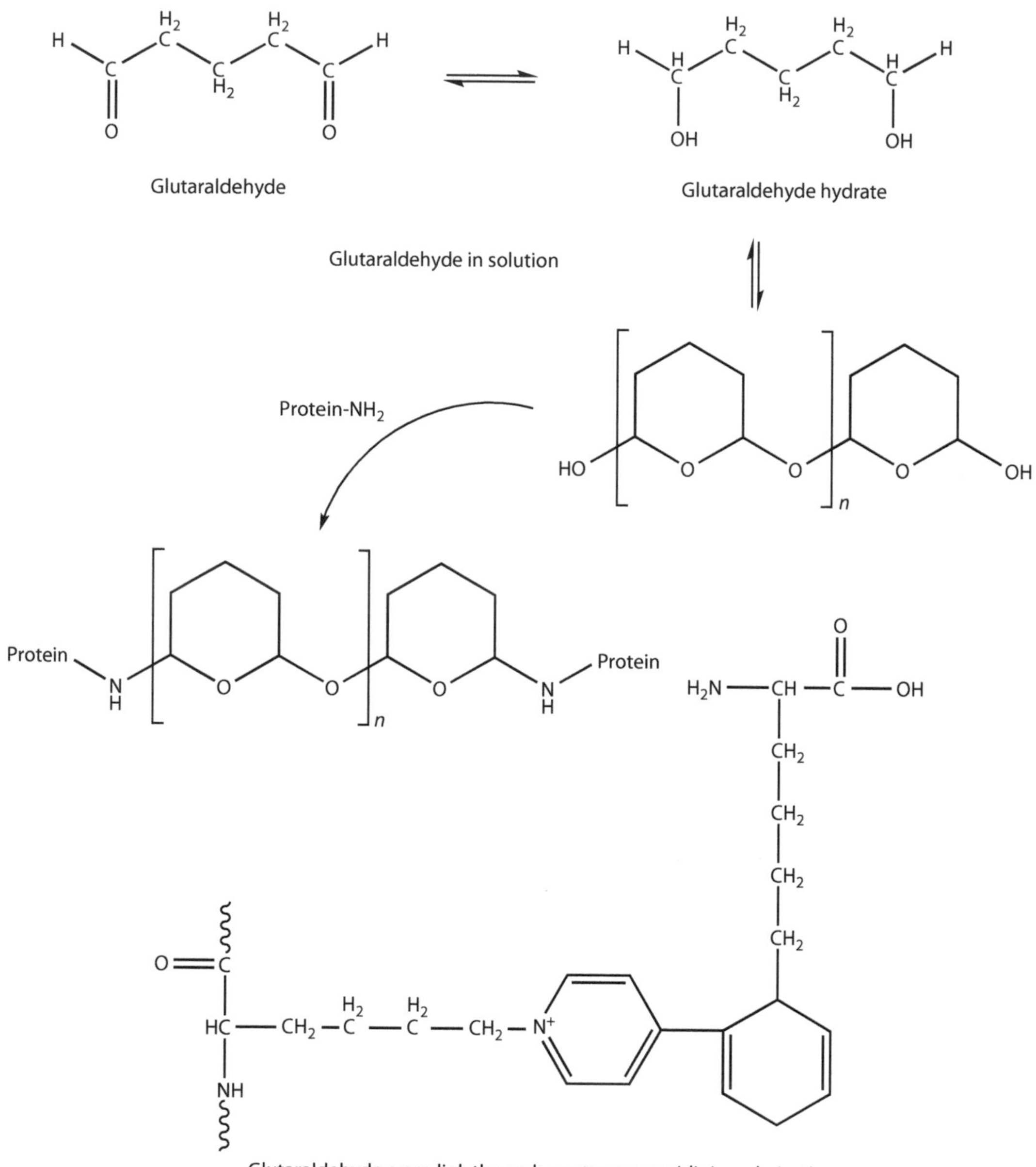

FIGURE 2.25 The reaction of glutaraldehyde with proteins. The reaction of glutaraldehyde with proteins is complex but useful even if not completely understood (Bejugam, N.K., Gayakwad, S.G., Uddin, A.N., and D'Souza, M.J., Microencapsulation of protein into biodegradable matrix: A smart solution cross-linking technique, *J. Microencapsul.* 30(3), 274–282, 2012). The scheme shown in this figure was advanced by Walt and Agayn (Walt, D.R. and Agayn, V.I., The chemistry of enzyme and protein immobilization with glutaraldehyde, *TRAC* 13, 425–430, 1994). See also Richards, F.M. and Knowles, J.R., Glutaraldehyde as a protein cross-linkage reagent, *J. Mol. Biol.* 37, 231–233, 1968.

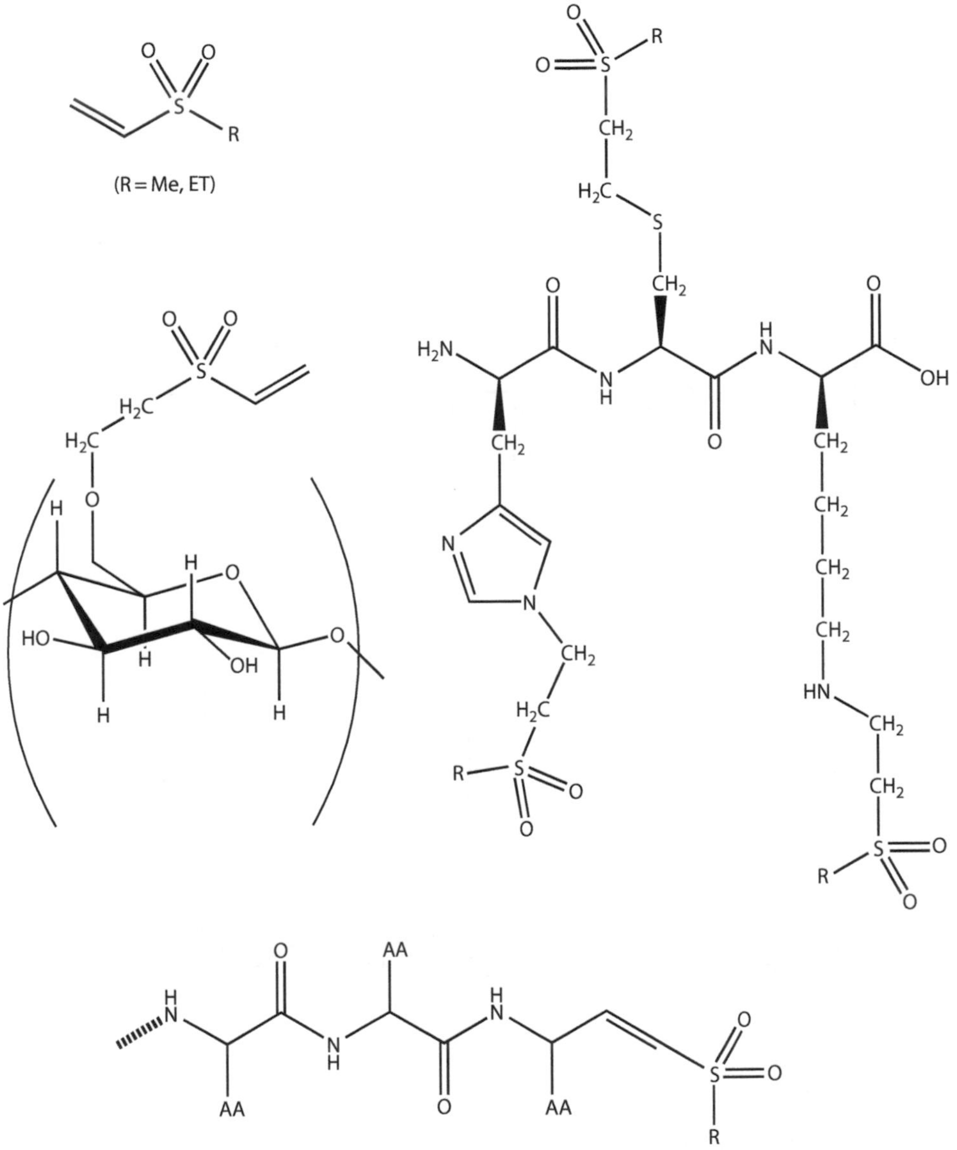

FIGURE 2.26 Vinyl sulfones. Shown is the structure of methyl/ethyl vinyl sulfone and the most common modification products obtained on reaction with peptides and proteins (Masri, M.S. and Friedman, M., Protein reactions with methyl and ethyl vinyl sulfones, *J. Protein Chem.* 7, 49–54, 1988). Also shown is a vinyl sulfone derivative of paper that could be used for patterning biomolecules using ink-jet printer technology (Yu, A., Shang, J., Cheng, F. et al., Biofunctional paper via covalent modification of cellulose, *Langmuir* 28(30), 11265–11273, 2012). A peptide vinyl sulfone inhibitor is shown at the bottom of the figure (Bogyo, M., Screening for selective small molecule inhibitors of the proteasome using activity-based probes, *Methods Enzymol.* 399, 609–622, 2005).

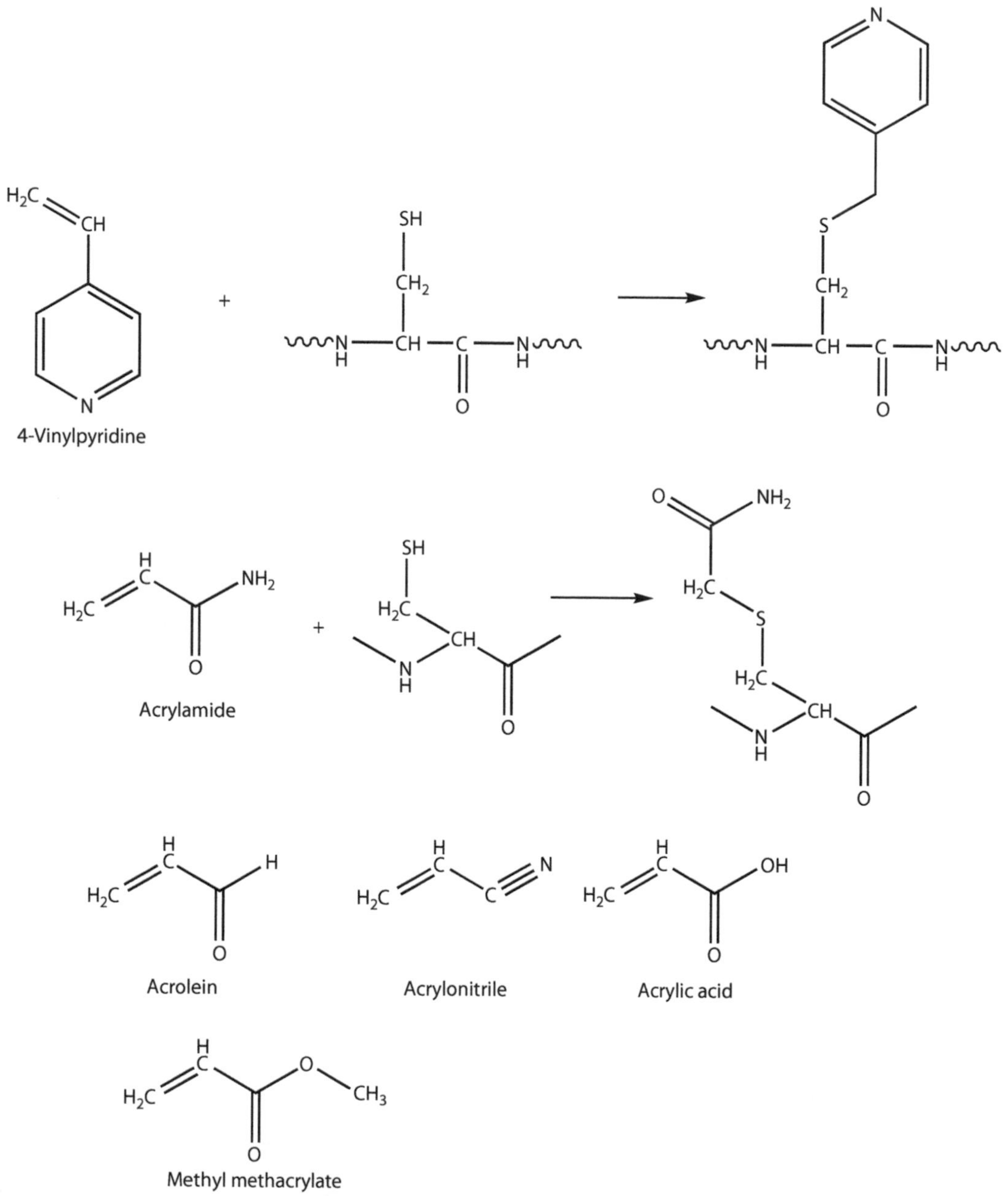

FIGURE 2.27 Vinyl pyridine, acrylamide, acrolein, acrylonitrile, acrylic acid, and methyl acrylate. These compounds react with cysteine residues in proteins via Michael addition reactions.

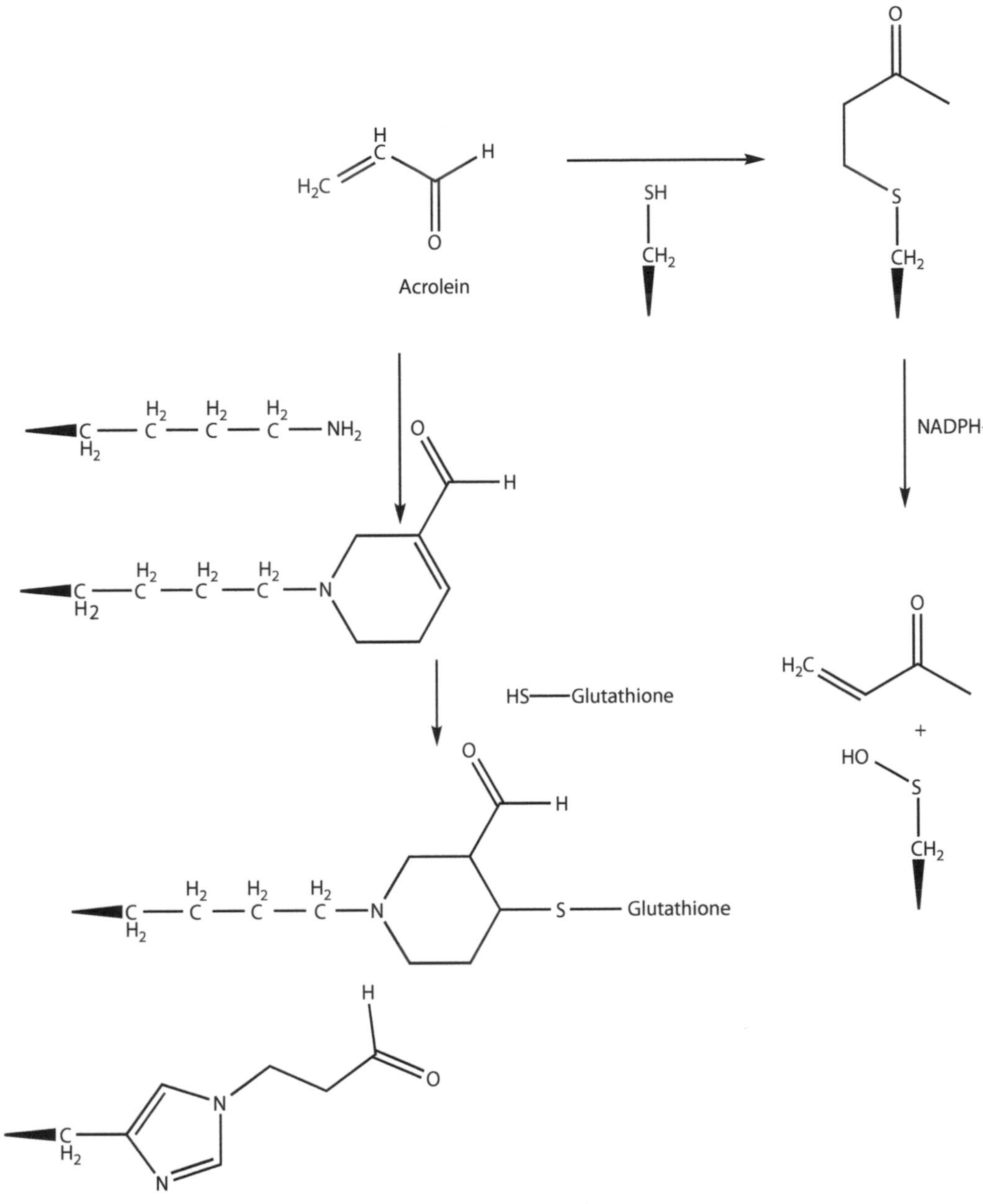

FIGURE 2.28 Reaction of acrolein with proteins. Shown is the reaction of acrolein with lysine to form N^ε-(3-formyl-3,4-dehydropiperidino)lysine and the subsequent formation of an adduct with glutathione. The product of the reaction of acrolein with histidine [N^τ-(3-propanal)histidine] (see Furuhata, A., Honda, K., Shibata, T. et al., Monoclonal antibody against protein-bound glutathione: Use of glutathione conjugate of acrolein-modified proteins as an immunogen, *Chem. Res. Toxicol.* 25(7), 1393–1401, 2012; Maeshima, T., Honda, K., Chicazawa, M. et al., Quantitative analysis of acrolein-specific adducts generated during lipid peroxidation—Modification of proteins in vitro: Identification of N^τ-(3-propanal)histidine as the major adduct, *Chem. Res. Toxicol.* 25(7), 1384–1392, 2012). Shown on the right side of the figure is a *bioactivation* mechanism for the reaction product [S-(3-oxopropyl)-N-acetyl-L-cysteine] to generate oxidized cysteine and acrolein (Hashmi, M., Vamvakas, S., and Anders, M.W., Bioactivation mechanism of S-(3-oxopropyl)-N-acetyl-L-cysteine, the mercapturic acid of acrolein, *Chem. Res. Toxicol.* 5, 360–365, 1992).

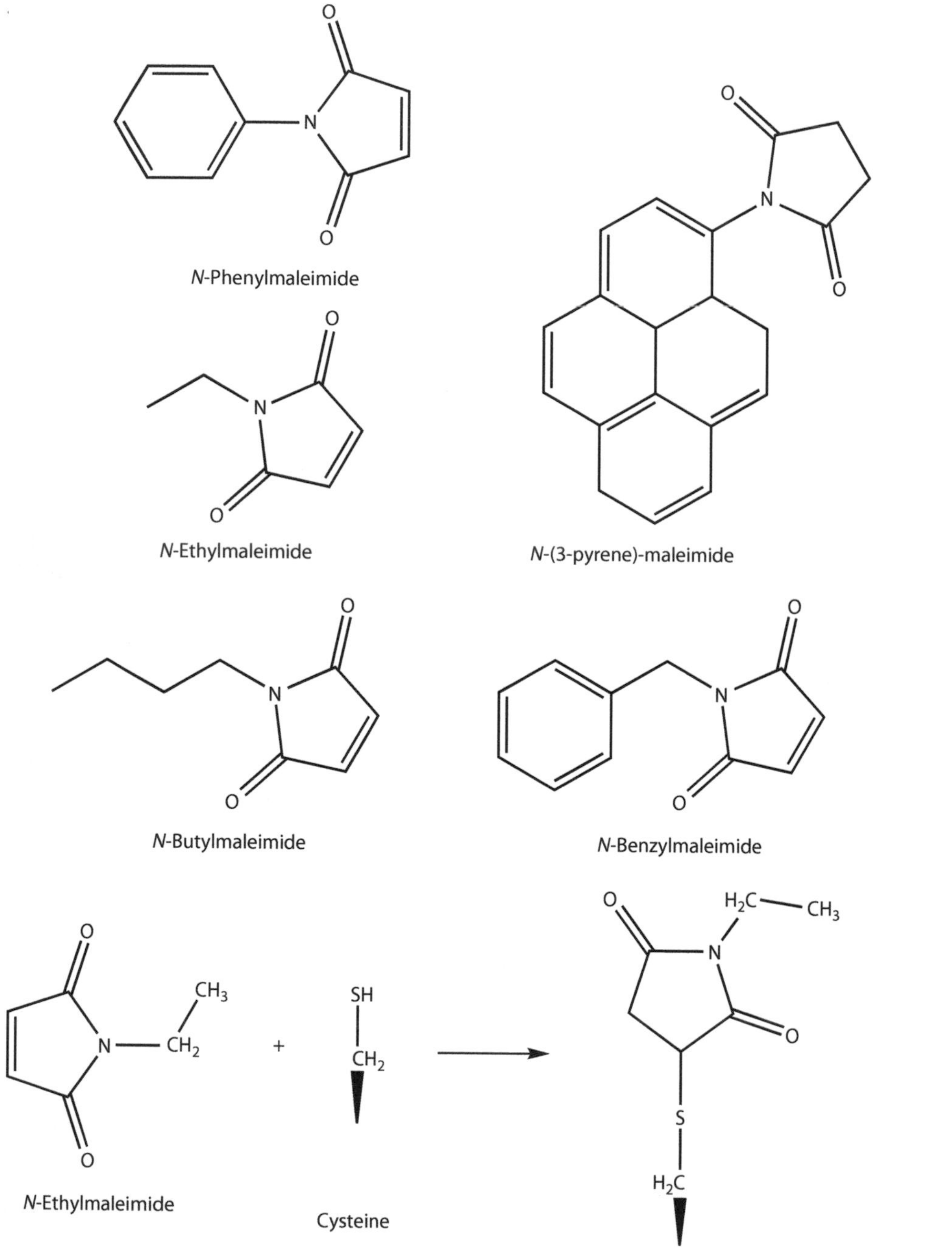

FIGURE 2.29 The reaction of *N*-ethylmaleimide with cysteinyl residue. Also shown are some hydrophobic maleimides that are used in membrane studies.

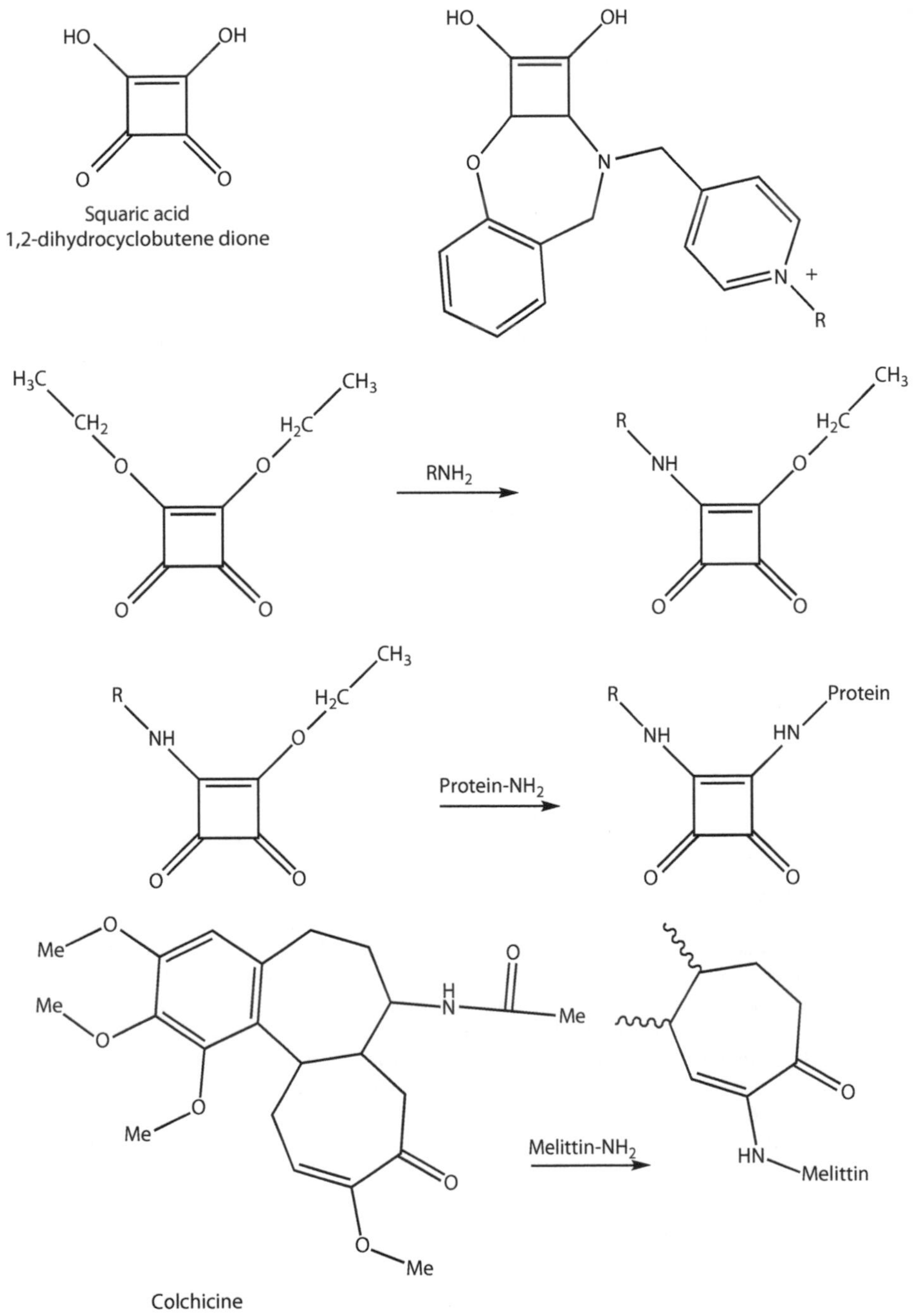

FIGURE 2.30 The reaction of vinylog of carboxylic acids with nucleophiles. Shown is the structure of the diethyl ester of squaric acid (1,2-dihydroxycyclobutene 3,4-dione) and the reaction process for bioconjugation a partner to a protein (Chernyak, A., Karavanov, A., Ogawa, Y., and Kováč, P., Conjugating oligosaccharides to proteins by squaric acid diester chemistry: Rapid monitoring of the progress of conjugation, and recovery of used ligand, *Carbohydr. Res.* 330, 479–486, 2001). Also shown is the reaction of colchicine with an amine (Zhang, S. and Bello, J., Reaction of colchicine with peptide amino groups, *Biopolymers* 31, 1241–1242, 1991). Shown at the upper right is the structure of a new reagent proposed for coupling to lysine or cysteine via squarate chemistry (Cui, D., Prashar, D., Sejwal, P., and Luk, Y.Y., Water-driven ligations using cyclic amino squarates: A class of useful S_N1-like reactions, *Chem. Commun.* 47, 1348–1350, 2011).

REFERENCES

1. Sharp, D.W.A. (ed.), *Dictionary of Chemistry*, 2nd edn., Penguin Books, London, U.K., 1990.
2. *Oxford Dictionary of the English Language*, Oxford University Press, Oxford, U.K., 2012. http://www.oup.com.
3. Shaw, E., Mares-Guia, M., and Cohen, W., Evidence of an active-center histidine in trypsin through use of a specific reagent, 1-chloro-3-tosylamido-7-amino-2-heptanone, the chloromethyl ketone derived from N^α-tosyl-L-lysine, *Biochemistry* 4, 2219–2224, 1965.
4. Redford-Ellis, M. and Gowenlock, A.M., Reaction of chloromethane with human blood, *Acta Pharmacol. Toxicol.* 30, 36–48, 1971.
5. Pincus, M., Thi, L.L., and Carty, R.P., The kinetics and specificity of the reaction of 2′(3′)-*O*-bromoacetyluridine with bovine pancreatic ribonuclease A, *Biochemistry* 14, 3653–3661, 1975.
6. Kupfer, A., Gani, V., Jiménez, J.S., and Shaltiel, S., Affinity labeling of the catalytic subunit of cyclic AMP-dependent protein kinase by N^α-tosyl-L-lysine chloromethyl ketone, *Proc. Natl. Acad. Sci. USA* 76, 3073–3077, 1979.
7. Jonák, J., Petersen, T.E., Clark, B.F. et al., *N*-Tosyl-L-phenylalanylchloromethane reacts with cysteine 81 in the molecule of elongation factor Tu from *Escherichia coli*, *FEBS Lett.* 150, 485–488, 1982.
8. Solomon, D.H., O'Brian, C.A., and Weinstein, I.B., *N*-α-Tosyl-L-lysine chloromethyl ketone and *N*-α-L-phenylalanine chloromethyl ketone inhibit protein kinase C, *FEBS Lett.* 190, 342–344, 1985.
9. Plapp, B.V., Application of affinity labeling for studying structure and function of enzymes, *Methods Enzymol.* 87, 469–499, 1982.
10. Kolb, H.C. and Sharpless, K.B., Click chemistry: Diverse chemical function from a few good reactions, *Angew. Chem. Int. Ed. Engl.* 40, 2004–2021, 2001.
11. Johnson, J.A., Lu, Y.Y., Van Deventer, J.A., and Tirrell, D.A., Residue-specific incorporation of non-canonical amino acids into proteins: Recent developments and applications, *Curr. Opin. Chem. Biol.* 14, 774–780, 2010.
12. van Hest, J.C. and van Delft, F.L., Protein modification by strain-promoted alkyne-azide cycloaddition, *ChemBioChem* 12, 1309–1312, 2011.
13. Ning, X., Temming, R.P., Doomerholt, J. et al., Protein modification by strain-promoted alkyne-nitrone cycloaddition, *Angew. Chem. Int. Ed. Engl.* 49, 3065–3068, 2010.
14. Chi, B.K., Albrecht, D., Gronau, K. et al., The redox-sensing regulator YodB senses quinones and diamide via a thiol-disulfide switch in *Bacillus subtilis*, *Proteomics* 10, 3155–3164, 2010.
15. Hu, W., Tedesco, S., McDonagh, B. et al., Selection of thiol and disulfide-containing proteins of *Escherichia coli* on activated thiol-Sepharose, *Anal. Biochem.* 398, 245–253, 2010.
16. Jongberg, S., Lund, M.N., Waterhouse, A.I., and Skibsted, L.H., 4-Methylcatechol inhibits protein oxidation in meat but not disulfide formation, *J. Food Agric. Chem.* 59, 10329–10335, 2011.
17. Samuni, A. and Goldstein, S., Redox properties and thiol reactivity of geldanamycin and its analogues in aqueous solutions, *J. Phys. Chem. B* 116, 6404–6410, 2012.
18. Spilsberg, B., Rudberget, T., Johannessen, L.E. et al., Detection of food-derived damaged nucleosides with possible adverse effects on human health using a global adductomics approach, *J. Agric. Food Chem.* 58, 6370–6375, 2010.
19. Rappaport, S.M., Li, H., Grigoryan, H. et al., Adductomics: Characterizing exposures to reactive electrophiles, *Toxicol. Lett.* 213, 83–90, 2012.

3 Acylating Agents

The acylation of amino acids involves the transfer of an acyl group (carbonyl group attached to an alkyl or aryl group) from an active donor such as a carboxylic acid anhydride (e.g., acetic anhydride), a carboxylic acid halide (e.g., acetyl chloride), or an active ester such as p-nitrophenyl acetate or N-hydroxysuccinimidyl acetate. This reaction can be described as nucleophilic acyl addition involving the carbonyl carbon. This mechanism is also used in the N-myristoylation of proteins.[1] Other acylating agents include N-acetylimidazole, 2-(acetoxy)-benzoic acid (aspirin), and the thioesters. The acylation of the ε-amino group of lysine or the α-amino group is stable under physiological conditions of temperature and pH, while the acylation of tyrosine can be reversed by base or hydroxylamine. The acylation of an amino group forms a peptide/isopeptide bond which is cleaved under conditions of acid hydrolysis used for the preparation of amino acid analysis. Acylation can also occur at the serine hydroxyl, threonine hydroxyl, the thiol group of cysteine, and the imidazole ring of histidine. The stability of these modifications is variable. The acylation of cysteine in proteins by palmitic acid that can occur by both enzymatic and nonenzymatic pathways[2,3] can yield relatively stable products that are reversed in vivo by protein thioesterases.[3,4] Acylation of the indole ring of tryptophan by acetic anhydride or acetyl chloride is accomplished in anhydrous trifluoroacetic acid.[5] I could not find evidence for the acylation of the indole ring of tryptophan in proteins. The acetylation of the guanidino nitrogen of arginine was accomplished with octanoic acid in fuming sulfuric acid (oleum)* at 60°C.[6] Aldehydes and ketones differ from other carbonyl compounds such as acetyl chloride or acetic anhydride in that these compounds do not have a leaving group. As mentioned earlier, an acylating agent such as acetic anhydride forms a stable amide bond with an amino group, while an aldehyde forms a Schiff base product that is a reversible reaction; the Schiff base may be stabilized by reduction.

The acylation of proteins with long-chain fatty acids was mentioned earlier. The acylation of proteins with long-chain fatty acids can be an important posttranslational modification as shown in the coupling of myristic acid to amino-terminal glycine residue[7,8] and coupling of palmitic acid to cysteine.[9,10] The reversible coupling of palmitic acid to proteins is critical to transport and intracellular signaling.[11,12] The reversible acetylation of lysine residues in proteins is also an important regulatory process.[13–17] The process of in vivo acylation of proteins is likely more heterogeneous than originally described.[18–20] The in vivo acylation and deacetylation of proteins is an enzymatic process involving acetyl coenzyme A.[21] Thioesters are intermediates in native chemical ligation (Figure 3.6) and thioester peptides have been used in the synthesis of novel protein.[22–25]

Organic acid chlorides, such as acetyl chloride, are acyl halides and highly reactive.[26] The reactivity is similar to that observed for the alkyl halides. Thus, the rate of hydrolysis of acetyl fluoride in water/acetone (25/75) has a relative rate of 2.2×10^2 (benzenesulfonyl fluoride has a rate of 1 in H_2O/acetone; 1/1); the relative rate of hydrolysis of acetyl chloride in water/acetone (25/75) is 1.7×10^7, while the relative rate for the hydrolysis of dimethylcarbamoyl chloride in water[26] is 1.1×10^5. The high reactivity of the organic acid chlorides such as acetyl chloride provides little opportunity for selective modification; in fact, there is little application of acyl chlorides for the modification of

---

* As one who has used fuming sulfuric acid, I would suggest that these are rigorous conditions unlikely to have application to any protein. However, it is noted that complete denaturation of the stalk domain in the hyperthermophilic archaebacterium *Staphylothermus marinus* requires 70% sulfuric acid or fuming trifluoromethanesulfonic acid.[29]

proteins. Specificity of modification can be achieved with active-site-directed reagents such as *N, N*-diphenylcarbamoyl chloride that modifies the active-site serine in chymotrypsin.[27]

Some examples of important acylating agents are shown in Table 3.1. Some second-order rate constants for the reaction of various acylating agents with nucleophiles are given in Table 3.2. It is observed that the reaction of acetic anhydride with amines is much more rapid than other reagents. Data for the reaction of acetyl chloride would be expected to be more rapid than acetic anhydride based on rates of hydrolysis in water.[28]

TABLE 3.1

Some Acylating Agents Used for the Modification of Proteins

Reagent	Chemistry and Reactions	References
Acetic anhydride	Acetic anhydride and other organic acid anhydrides are members of group of chemicals that can transfer an acyl group to a nucleophile such as lysine (nucleophilic acyl transfer) (Figure 3.1).[3.1.1] While the primary reaction of acetic anhydride is with amino groups (N-terminal and ε-amino group of lysine), reaction may occur at hydroxyl groups and the imidazole ring of histidine; these later reactions are transient not yielding stable products. There may be a transient reaction at thiol groups but no evidence to support such reaction in a protein. Thiol esters such as mercaptoethyl acetate can be prepared and fatty acid acyl chlorides are used to reaction with cysteine residues in proteins (see acetyl chloride below). A variety of analytical methods are available for determining the extent of reactions including an antibody against ε-amino group of lysine.[a] Early work on a macroscale used acetic anhydride in half-saturated sodium acetate[b] at 0°C–4°C. Current work uses either a Tris buffer at pH 8.0 or a pH-stat method at pH 8.0. The inclusion of glycerol has been reported to improve the specificity of modification by competing for the modification of hydroxyl groups.[3.1.2] Acetic anhydride has been used for the assessment of protein conformation. Acetic anhydride was used for trace labeling of proteins as a method for measuring lysine reactivity.[c] While the term trace labeling is no longer used, the technique frequently described as differential labeling continues.[3.1.3–3.1.10] Reaction with acetic anhydride, acetyl chloride, or ketene (see Ketene below) results in charge neutralization of amino groups. Acetic anhydride can be used in the gas phase for the acylation of lyophilized proteins.[3.1.11] Acetic anhydride and deuterated acetic anhydride are used in proteomic analysis[3.1.12–3.1.14] in a technical approach referred to as stable isotope labeling.[d]	3.1.1–3.1.14
Ketene	Ketene (CAS 463-51-4) (ethenone, carbomethene, $CH_2=C=O$) is an early acetylating agent for proteins (Figure 3.2).[3.1.15–3.1.17] Derivatives of ketene are still used for the modification of proteins.[3.1.18] An alkyne-functionalized ketene has been developed and selective modification of the amino-terminal amine achieved by reaction at pH 6.3.[3.1.19]	3.1.15–3.1.19
Acetyl chloride and other alkyl halides including thioacetyl halides	Acetyl chloride[3.1.20–3.1.22] (Figure 3.1) is not frequently used for proteins but is used for fatty alcohols and carbohydrates. Acid chloride derivative of fatty acids are used to modify proteins. However, the synthesis of S-acetyl-N-acetyl cysteine methyl ester has been reported for use as substrate for cholinesterases. It is suggested that there are mechanistic differences in the reaction of acetyl chloride and thioacetyl chloride with ammonia.[3.1.23] Acetyl chloride is used for the modification of carbohydrate hydroxyl groups.[3.1.24,3.1.25] There is considerable use of long-chain alkanoic chlorides for the modification (lipidation) of proteins.[3.1.26–3.1.29] Oxalyl chloride was used to modify amine groups on silica amine beads.[3.1.30] Acetyl chloride was not effective as the silica amine beads are said to be soluble in the reagent.	3.1.20–3.1.30
Succinic anhydride	The reaction of succinic anhydride (Figure 3.3) with proteins is similar to that of acetic anhydride.[3.1.31] Succinic anhydride is easier to work with than acetic anhydride as it is a nonvolatile solid. The modification of amines with succinic anhydride converts a positive charge to a negative charge; this results in a net charge change of −2 (loss of positive charge combined with addition of negative charge).	3.1.14, 3.1.31–3.1.38

(*continued*)

TABLE 3.1 (continued)
Some Acylating Agents Used for the Modification of Proteins

Reagent	Chemistry and Reactions	References
	Succinylation has been useful in the solubilization of proteins.[3.1.31,3.1.32] n-Alkenyl derivatives of succinic anhydride such as 2-octenylsuccinic anhydride have been shown to act as a plasticizer in ethyl cellulose films.[3.1.33] There is a similar enhancement in the properties of starch and starch acetate coatings.[3.1.33,3.1.34] Modification with 2-octenylsuccinic anhydride improved the hemostatic effectiveness of starch microspheres.[3.1.35] Succinic anhydride has also been used for the modification of eucalyptus wood biomass (ball-milled wood)[3.1.36] or wood–plastic composites[3.1.37] to increase hydrophilicity. Succinic anhydride and deuterated succinic anhydride have been used for the analysis of neuropeptides by mass spectrometry[3.1.14] and for the determination of the site of protein modification by electrophilic lipids such as 4-hydroxy-2-nonenal.[3.1.38] In the latter case, the peptides obtained from tryptic hydrolysis are modified with *light* and *heavy* succinic anhydride[d] and then captured by affinity chromatography on hydrazine matrices.	
Citraconic anhydride	Citraconic anhydride (Figure 3.3) (methylmaleic anhydride) has a reactivity profile similar to succinic anhydride in that modification of lysine residues results in a net charge change of −2 (loss of positive charge combined with addition of a negative charge). Modification with citraconic anhydride is reversible.[3.1.39] While considered specific for the modification of lysine residues in proteins (modification of N-terminal amino groups is a possibility), modification of sulfhydryl groups via Michael addition at the double bond has been reported.[3.1.40] The reversible nature of the modification has permitted the use of *arg tricks* where tryptic cleavage can be limited to arginine residues yielding peptides that can be subjected to the removal of the citraconyl group.[3.1.41,3.1.42] More recently, citraconic anhydride has been used for antigen retrieval from formaldehyde- and glutaraldehyde-fixed tissues.[3.1.43–3.1.46]	3.1.39–3.1.46
N-Hydroxysuccinimide esters	A variety of acyl groups can be used as N-hydroxysuccinimide esters (Figure 3.4) including acetyl (N-hydroxysuccinimide acetate, also referred to as acetoxy derivative).[3.1.47] N-Hydroxysuccinimide acetate (acetoxysuccinimide) and the deuterated form have been used in quantitative proteomics.[3.1.48,3.1.49] N-Hydroxysuccinimide esters are used to modify amine groups in peptides and proteins, but also to modify hydroxyl functions; the hydroxyl derivatives can be removed with hydroxylamine[3.1.50] or by boiling.[3.1.51] The N-hydroxysuccinimide function can be used in the preparation of either homobifunctional reagents[3.1.52] or heterobifunctional reagents for protein cross-linking.[3.1.53] (Figure 3.4). These cross-linking reagents can be used to study protein–protein interactions:[3.1.52] for the coupling of a cytotoxic drug to an antibody,[3.1.53] for the coupling of a glycated peptide to ovalbumin used as an immunogen,[3.1.54] as well as for the coupling of a functional group to a solid-phase matrix.[3.1.30] N-Hydroxysuccinimide is also used in combination with water-soluble carbodiimide for coupling lysine residues with carboxyl groups in proteins.[3.1.55–3.1.58] In this reaction, the protein carboxyl group is converted to an N-hydroxysuccinimidyl derivative (Figure 3.4). It has been demonstrated that there is a difference in specificity of modification of peptide cations depending on the reaction in solution phase versus gas phase.[3.1.59] There was preferential modification	3.1.47–3.1.59

TABLE 3.1 (continued)

Some Acylating Agents Used for the Modification of Proteins

Reagent	Chemistry and Reactions	References
	of N-terminal amino groups at pH 5.0 in solution with N-sulfohydroxysuccinimide acetate, while there is preferential modification of the ε-amino group of lysine in the gas phase (there was no modification of the amino-terminal amine with N-sulfohydroxysuccinimide acetate but modification was observed with bis(succinimidyl)subearate).	
Lactones and lactams	β-Lactones (Figure 3.5) are of current interest as inhibitors of bacterial virulence factor.[3.1.60] β-Lactones react preferentially with sulfhydryl groups[3.1.61] and have been used to prepare acylated carrier proteins.[3.1.62] β-Lactams have also been used for the acylation of carrier proteins.[3.1.63] β-Thiolactones are more reactive than lactones;[3.1.61] homocysteine thiolactone has been demonstrated to modify lysine residues in proteins.[3.1.64] There has been recent interest in the use of β-lactams and β-lactones in chemical biology.[3.1.65–3.1.67] This has included the use of γ-butyrolactones.[3.1.66] There is more discussion of these compounds in Chapter 25.	3.1.60–3.1.67
Thioesters	Thioesters are more *active* than oxyesters[3.1.68,3.1.69] and do have a critical role in native chemical ligation (Figure 3.6).[3.1.70,3.1.71] While there are no direct comparisons, it would appear that reaction with sulfhydryl groups is preferred to reaction with amines or hydroxyl groups.[3.1.72] As a side note, Tris buffer interferes with intein-mediated protein cleavage by forming an amide adduct.[3.1.73,e] Acetyl coenzyme A is an example of the use of thioester reactivity in metabolism.[3.1.74] The acylation of amine groups by CoA and adenylate thioesters of cholic acid has been reported.[3.1.75] The hydrazine group is a very reactive nucleophile for reaction with a thioester.[3.1.76] This observation supported the development of a bifunctional reagent containing an α-hydrazino acetamido and azido group that would react with the thioester function of an intein. Rate constants for the reaction of acetyl-CoA, palmitoyl-CoA, and p-nitrophenyl thioacetate with low molecular weight thiols and a large number of peptides containing cysteine have been reported (Table 3.2).[3.1.77] The thiolate form is the reactive species with the various thioesters. There is another study on the reaction of thioesters with amines[3.1.78] and on the relative rates of thiol exchange and hydrolysis of thioesters.[3.1.79]	3.1.68–3.1.79
Dithioesters	Dithioesters are relatively specific for aliphatic amines forming a thioamide linkage (Figure 3.7). While relatively easy to synthesize, storage is a problem.[3.1.80,3.1.81] Dithioesters have been used in polymers having biological application. S-Thiobenzoyl-2-thiopropionate was used as a chain transfer agent in RAFT polymerization to prepare a heparin-like polymer for immobilization of peptide growth factors (e.g., vascular endothelial growth factor, basic fibroblast growth factor).[3.1.82] Following polymerization of the dithioester was reduced to the thiol group with butyl amine permitting coupling to a surface plasmon resonance surface. The growth factors are bound by electrostatic interaction with a 4-styrene sulfonate matrix; heparin inhibits the binding of the growth factors. Alternatively, the dithioester function can be reduced with butyl amine in the presence of vinyl sulfone creating a semitelechelic polymer with Michael acceptor function (Figure 3.7).[3.1.83,3.1.84]	3.1.80–3.1.84

(continued)

TABLE 3.1 (continued)
Some Acylating Agents Used for the Modification of Proteins

Reagent	Chemistry and Reactions	References
Phthalic anhydride and tetrahydrophthalic anhydride	Phthalic anhydride (Figure 3.8) is used in the chemical industry, and there is interest in pulmonary toxicology.[3.1.85] Phthalic anhydride-modified hemoglobin is suggested for use as a biological marker for exposure.[3.1.86] Phthalic anhydride reacts with hemoglobin to form both phthalimides and phthalimide derivatives; the chemistry of the formation of these adducts has been previously described.[3.1.87,3.1.88] The study on hemoglobin[3.1.86] also showed differences in the reaction sites for phthalic anhydride and hexahydrophthalic anhydride on human hemoglobin. There has been limited use of phthalic anhydride for the modification of proteins. Phthalic anhydride was used to modify horseradish peroxidase to improve thermal stability.[3.1.89] 3-Hydroxyphthalic anhydride was used to modify bovine β-lactoglobulin providing derivatives that blocked the CD4 receptor for HIV virus.[3.1.90] Tetrahydrophthalic anhydride is used for the reversible modification of lysine residues in proteins permitting alteration of proteolysis[3.1.91,3.1.92] and for the dissociation and solubilization of membrane proteins.[3.1.93,3.1.94] Early work has provided information on several reversible protecting groups including phthalic acid derivatives.[3.1.95]	3.1.85–3.1.95
Mixed acid anhydrides	Formally, a mixed acid anhydride is an asymmetrical anhydride composed of two dissimilar carboxylate acids as opposed to a symmetrical anhydride (Figure 3.9). It also refers to amino acid adenylates for the aminoacylation of tRNA.[3.1.96] The acylation of ε-amino groups of lysine in histone H3 by a bile acyl adenylate has been reported.[3.1.97] Acetyl phosphate is a metabolic intermediate; there was no reported use for acetylation of proteins; there was one reference when a protein was phosphorylated by acetyl phosphate.[3.1.98] Methyl acetyl phosphate (Figure 3.9) was used to modify an anion-binding site in hemoglobin.[3.1.99] It was shown to have antisickling activity with sickle cell hemoglobin[3.1.100] by increasing the solubility of this protein.[3.1.101] Methyl acetyl phosphate is selective for the modification of two lysine residues in hemoglobin[3.1.99]; the modification of a lysine residue in phenylalanine dehydrogenase has also been demonstrated.[3.1.102] More recently, methyl acetate phosphate was used to modify the anion-binding site in human interleukin-1 receptor antagonist[3.1.103] establishing the structural basis for subsequent protein engineering to reduce aggregation. The modification of other functional groups has not been reported.	3.1.96–3.1.103
N-Acetylimidazole	N-Acetylimidazole (Figure 3.10) is more reactive than the thioesters described earlier.[f] N-Acetylimidazole is used as source of the acetyl function in organic synthesis[3.1.104,3.1.105] and in model systems for enzyme reactions.[3.1.106,3.1.107] N-substituted imidazoles with other acyl functions have proved useful.[3.1.108–3.1.110] Some of these acylimidazoles have therapeutic potential.[3.1.110,3.1.111] The majority of interest in N-acetylimidazole has focused on the modification of tyrosine residues in proteins (see Chapter 17).	3.1.104–3.1.111
Aspirin [2-(acetoxy) benzoic acid]	2-(Acetoxy)benzoic acid (Figure 3.10) has been demonstrated to acetylate lysine residues and histidine residues in proteins.[3.1.112] The acetylation of albumin was observed over 40 years ago.[3.1.113–3.1.115] Subsequently, it was established that the antithrombotic action of aspirin was due to the inactivation of an enzyme responsible for the synthesis of prostaglandins that caused platelet aggregation.[3.1.116,3.1.117] Aspirin does affect other	3.1.112–3.1.121

TABLE 3.1 (continued)
Some Acylating Agents Used for the Modification of Proteins

Reagent	Chemistry and Reactions	References
	systems such as the endothelial synthesis of nitric oxide.[3.1.118,3.1.119] *N*-Acetylimidazole was observed to be slightly more potent than aspirin in the inhibition of inducible nitric oxide synthase.[3.1.118] There is also considerable interest in other in vivo effects of protein acetylation by aspirin.[3.1.120,3.1.121]	
p-Nitrophenyl acetate	*p*-Nitrophenyl acetate (Figure 3.10) is best known as a nonspecific substrate for esterase activity.[3.1.122,3.1.123] Albumin has well-documented esterase activity toward *p*-nitrophenyl acetate involving specific tyrosine residue.[3.1.124–3.1.130] *p*-Nitrophenyl acetate also modifies lysine residues in albumin[3.1.131] and ubiquitin.[3.1.132] Acetylation of serine, threonine, and *N*-terminal aspartic acid was also observed.[3.1.131]	3.1.122–3.1.132
Diethylpyrocarbonate	Diethylpyrocarbonate (ethoxyformic anhydride) is a reagent that is used to modify nucleic acids (Figure 3.11).[3.1.133–3.1.138] There is use of diethylpyrocarbonate for the inhibition of ribonuclease permitting isolation of RNA from tissues.[3.1.139–3.1.141] Diethylpyrocarbonate reacts with histidine residues in proteins; there is also reaction with lysine and tyrosine that can be distinguished from histidine modification (Figure 3.12).[3.1.142–3.1.144] Modification at tyrosine results in a change in ultraviolet absorbance while the modification of lysine is not reversed by hydroxylamine;[3.1.145] modification of histidine and tyrosine is reversed by hydroxylamine.[3.1.146,3.1.147]	3.1.133–3.1.147

[a] Antibody to ε-amino group of lysine (Nagaraj, R.H., Nahomi, R.B., Shanthakumar, S. et al., Acetylation of αA-crystallin in the human lens: Effects on structure and chaperone function, *Biochim. Biophys. Acta* 1822, 120–129, 2012).

[b] The pH of a 0.1 M sodium acetate solution at 25°C is 8.9, thus promoting deprotonation of amino groups. The sodium acetate/acetic acid also serves as a catalyst for the acetylation reaction (Green, R.W., Ang, K.P., and Lam, L.C., Acetylation of collagen, *Biochem. J.* 54, 181–187, 1953). Sodium acetate was included as a catalyst for the acetylation of starch with acetic anhydride in densified carbon dioxide (supercritical carbon dioxide) (Muljana, H., Picchioni, F., Knez, Z. et al., Insights in starch acetylation in sub- and supercritical CO_2. *Carbohydr. Res.* 346, 1224–1231, 2011; Muljana, H., Picchione, F., Heeres, H.J, and Janssen, L.P.B.M., Green starch conversions: Studies on starch acetylation in densified CO_2, *Carbohydr. Polym.* 82, 653–662, 2010).

[c] Trace labeling is a technique using acetic anhydride to measure the reactivity of individual lysine residues in a protein (Giedroc, D.P., Sinha, S.K., Brew, K., and Puett, D., Differential trace labeling of calmodulin: investigation of binding sites and conformational states by individual lysine reactivities. Effects of β-endorphin, trifluoperazine, and ethylene glycol bis(β-aminoethyl ether)-*N,N,N′,N′*-tetraacetic acid, *J. Biol. Chem.* 260, 13406–13413, 1985).

[d] Stable isotope labeling is a method for comparing samples such as cell extracts (experimental and control) (Altelaar, A.F., Frese, C.K., Preisinger, C. et al., Benchmarking stable isotope labeling based quantitative proteomics, *J. Proteomics* 88, 14–26, 2012). There is more discussion of stable isotope labeling in Chapter 25.

[e] Tris can be considered a nucleophile and can interfere with a number of reactions (e.g., Burcham, P.C., Fontaine, F.R., Petersen, D.R., and Pyke, S.M., Reactivity with Tris(hydroxymethyl)aminomethane confounds immunodetection of acrolein-adducted proteins, *Chem. Res. Toxicol.* 16, 1196–1201, 2003; Gab, J., John, H., Melzer, M., and Blum, M.M., Stable adducts of nerve agents sarin, soman and cyclosarin with TRIS, TES and related buffer compounds-Characterization by LC-ESI-MS/MS and implications for analytical chemistry, *J. Chromatogr. B Anal. Technol. Biomed. Life Sci.* 878, 17–18, 2010).

[f] The free energy of hydrolysis ($-\Delta G°$ at pH 7.0 is 1.84 kJ for acetyl thiol esters while the value is 3.17 kJ for acetylimidazole (Jencks, W.P., Cordes, S., and Carriuolo, J., The free energy of thiol ester hydrolysis, *J. Biol. Chem.* 235, 3608–3614, 1960).

(*continued*)

TABLE 3.1 (continued)

Some Acylating Agents Used for the Modification of Proteins

References to Table 3.1

3.1.1.   Uraki, Z., Terminiello, L., Bier, M., and Nord, F.F., Mechanism of enzyme action. LXIII. Specificity of acetylation of proteins with [acetic]anhydride-C14, *Arch. Biochem. Biophys.* 69, 644–652, 1957.

3.1.2.   Fojo, A.T., Reuben, P.M., Whitney, P.L., and Awad, W.M., Jr., Effect of glycerol of protein acetylation by acetic anhydride, *Arch. Biochem. Biophys.* 240, 43–50, 1985.

3.1.3.   Gong, B., Ramos, A., and Várquez-Fernández, E., Probing structural differences between PrP(C) and PrP(Sc) by surface nitration and acetylation: Evidence of conformational change in the C-terminus, *Biochemistry* 50, 4963–4972, 2011.

3.1.4.   Nahomi, R.B., Oya-Ito, T., and Nagaraj, R.H., The combined effect of acetylation on the chaperone and anti-apoptotic functions of human α-crystallin, *Biochem. Biophys. Acta* 1832, 195–203, 2012.

3.1.5.   Vadvalkar, S.S., Baily, C.N., Matsuzaki, S. et al., Metabolic inflexibility and protein lysine acetylation in heart mitochondria of a chronic model of type 1 diabetes, *Biochem. J.* 449(1), 253–261, 2012.

3.1.6.   Yadav, S.P., Brew, K., Majercik, M.H., and Puett, D., A label selection approach to assess the role of individual amino groups in human choriogonadotropin receptor binding, *J. Biol. Chem.* 269, 3991–3998, 1994.

3.1.7.   Yadav, S.P., Brew, K., and Puett, D., Holoprotein formation of human chorionic gonadotropin: Differential trace labeling with acetic anhydride, *Mol. Endocrinol.* 8, 1547–1558, 1994.

3.1.8.   Chen, J., Smith, D.L., and Griep, M.A., The role of the 6 lysines and the terminal amine of *Escherichia coli* single-strand binding protein in its binding of single-stranded DNA, *Protein Sci.* 7, 1781–1788, 1998.

3.1.9.   Buechler, J.A., Vedvick, T.A., and Taylor, S.S., Differential labeling of the catalytic subunit of cAMP-dependent protein kinase with acetic anhydride: Substrate-induced conformational changes, *Biochemistry* 28, 3018–3024, 1989.

3.1.10. Hochleitner, E.O., Borchers, C., Parker, C. et al., Characterization of a discontinuous epitope of the human immunodeficiency virus (HIV) core protein p24 by epitope excision and differential chemical modification followed by mass spectrometric peptide mapping analysis, *Protein Sci.* 9, 487–496, 2000.

3.1.11. Taralp, A. and Kaplan, H., Chemical modification of lyophilized proteins in nonaqueous environments, *J. Protein Chem.* 16, 183–193, 1997.

3.1.12. Che, F.Y., Eippper, B.A., Mains, R.E., and Fricker, L.D., Quantitative peptidomics of pituitary glands from mice deficient in copper transport, *Cell Mol. Biol.* (Noisy-le-grand) 49, 713–722, 2003.

3.1.13. Shetty, V., Nickens, Z., Shah, P. et al., Investigation of sialylation aberration in *N*-linked glycopeptides by lectin and tandem labeling (LTL) quantitative proteomics, *Anal. Chem.* 82, 9201–9210, 2010.

3.1.14. Hou, X., Xie, F., and Sweedler, J.V., Relative quantitation of neuropeptides over a thousand-fold concentration range, *J. Am. Soc. Mass Spectrom.* 23, 2083–2093, 2012.

3.1.15. Neuberger, A., CXC. Note on the action of acetylating agents on amino-acids, *Biochem. J.* 32, 1452–1456, 1938.

3.1.16. Porter, R.R., The unreactive groups of proteins, *Biochim. Biophys. Acta* 2, 105–112, 1948.

3.1.17. Rovery, M. and Desnuelle, P., Sur l'acétylation des groupes phénol protéiques par le cétène, *Biochim. Biophys. Acta* 2, 514–521, 1948.

3.1.18. Pedone, E. and Brocchini, S., Synthesis of two photolabile poly(ethylene glycol) derivatives for protein conjugation, *React. Funct. Polym.* 66, 167–176, 2006.

3.1.19. Chan, A. O.-Y., Ho, C.-M., Chong, H.-C. et al., Modification of N-terminal groups of peptides and proteins using ketenes, *J. Am. Chem. Soc.* 134, 2589–2598, 2012.

3.1.20. Palling, D.J. and Jencks, W.P., Nucleophilic reactivity toward acetyl chloride in water, *J. Am. Chem. Soc.* 106, 4889–4876, 1984.

3.1.21. *Vogel's Textbook of Practical Organic Chemistry*, eds. B.S. Funiss, A.J. Hannaford, V. Rogers, P.W.G. Smith, and A.R. Tatchell, Longmans/Wiley, New York, 1987.

3.1.22. Riordan, J.F. and Vallee, B.L., Acetylation, *Methods Enzymol.* 11, 565–570, 1967.

3.1.23. Han, I.S., Kim, C.K., Sohn, C.K. et al., Comparative studies on the reactions of acetyl and thioacetyl halides with NH$_3$ in the gas phase and in aqueous solution: A theoretical study, *J. Phys. Chem. A* 115, 1364–1370, 2011.

3.1.24. Zhou, Y., Ramström, O., and Dong, H., Organosilicon-mediated regioselective acetylation of carbohydrates, *Chem. Commun.* 48, 5370–5372, 2012.

3.1.25. Shelma, R. and Sharma, C.P., Acyl modified chitosan derivatives for oral delivery of insulin and curcumin, *J. Mater. Sci. Mater. Med.* 21, 2133–2140, 2010.

TABLE 3.1 (continued)
Some Acylating Agents Used for the Modification of Proteins

3.1.26. Méndez Altolín, E.J., González Canavachiolo, V.L., and Sierra Pérez, R., Gas chromatographic determination of high molecular weight alcohols from policosanol in omega-3 fish oil by acylation with acetyl chloride, *J. AOAC Int.* 91, 1013–1019, 2008.

3.1.27. Gunstone, F., Reactions of the carboxyl group, in *Fatty Acid and Lipid Chemistry*, Chapter 8, pp. 201–222, Blackie Academic/Chapman and Hall, London, U.K., 1991.

3.1.28. Kocevar, N., Obermajer, N., and Kreft, S., Membrane permeability of acylated cystatin depends on the fatty acyl chain length, *Chem. Biol. Drug Des.* 72, 217–224, 2008.

3.1.29. Plou, F.J. and Ballesteros, A., Acylation of subtilisin with long chain acyl residues affects its activity and thermostability in aqueous medium, *FEBS Lett.* 339, 200–204, 1994.

3.1.30. Gautam, S. and Loh, K.-C., Immobilization of hydrophobic peptidic ligands to hydrophilic chromatographic matrix: A preconcentration approach, *Anal. Biochem.* 423, 202–209, 2012.

3.1.31. Klotz, I.M., Succinylation, *Methods Enzymol.* 11, 575–589, 1967.

3.1.32. Muszynska, G. and Riordan, J.F., Chemical modification of carboxypeptidase A crystals. Nitration of tyrosine-248, *Biochemistry* 15, 46–51, 1976.

3.1.33. Tarvainen, M., Sutinen, R., Peltonen, S. et al., Enhanced film-forming properties for ethyl cellulose and starch acetate using *n*-alkenyl succinic anhydrides as novel plasticizers, *Eur. J. Pharm. Sci.* 19, 363–371, 2003.

3.1.34. Baranauskiene, R., Bylaite, E., Zukauskaite, J., and Venskutonis, R.P., Flavor retention of peppermint (*Mentha piperita* L.) essential oil spray-dried in modified starches during encapsulation and storage, *J. Agric. Food Chem.* 55, 3027–3036, 2007.

3.1.35. Björses, K., Faxälv, L., Montan, C. et al., In vitro and in vivo evaluation of chemically modified degradable starch microspheres for topical haemostasis, *Acta Biomater.* 7, 2558–2565, 2011.

3.1.36. Zhang, A.P., Liu, C.F., Sun, R.C. et al., Homogeneous acylation of eucalyptus wood at room temperature in dimethyl sulfoxide/*N*-methylimidazole, *Bioresour. Technol.* 125, 328–331, 2012.

3.1.37. Burgstaller, C. and Stadlbauer, W., Influence of chemical modification on the water uptake properties of wood plastic composites, *J. Biobased Mater. Bioenerg.* 6, 380–387, 2012.

3.1.38. Han, B., Hare, M., Wickramasekara, S. et al., A comparative "bottom up" proteomics strategy for the site-specific identification and quantification of protein modifications by electrophilic lipids, *J. Proteomics* 75, 5724–5733, 2012.

3.1.39. Takahashi, K., Specific modification of arginine residues in proteins with ninhydrin, *J. Biochem.* 80, 1173–1176, 1976.

3.1.40. Brinegar, A.C. and Kinsella, J.E., Reversible modification of lysine in β-lactoglobulin using citraconic anhydride. Effects on sulfhydryl groups, *Int. J. Pept. Protein Res.* 18, 18–25, 1981.

3.1.41. Takahashi, K., The structure and function of ribonuclease T1. XXIII. Inactivation of ribonuclease T1 by reversible blocking of amino groups with *cis*-aconitic anhydride and related dicarboxylic acid anhydrides, *J. Biochem.* 81, 641–646, 1977.

3.1.42. Kadlčik, V., Strohalm, M., and Kodíček, M., Citraconylation—A simple method for high protein sequence coverage in MALDI-TOF mass spectrometry, *Biochem. Biophys. Res. Commun.* 305, 1091–1093, 2003.

3.1.43. Namimatsu, S., Ghazizadeh, M., and Sugisaki, Y., Reversing the effects of formalin fixation with citraconic anhydride and heat: A universal antigen retrieval method, *J. Histochem. Cytochem.* 53, 3–11, 2005.

3.1.44. Dai, W., Sato, S., Ishizaki, M. et al., A new antigen retrieval method using citraconic anhydride for immunoelectron microscopy: Localization of surfactant pro-protein C (proSP-C) in type II alveolar epithelial cells, *J. Submicrosc. Cytol. Pathol.* 36, 219–224, 2004.

3.1.45. Leong, A.S. and Haffajee, Z., Citraconic anhydride: A new antigen retrieval solution, *Pathology* 42, 77–81, 2010.

3.1.46. Moriguchi, K., Mitamura, Y., Iwami, J. et al., Energy filtering transmission electron microscopy immunocytochemistry and antigen retrieval of surface layer proteins from *Tannerella forsythensis* using microwave or autoclave heating with citraconic anhydride, *Biotech. Histochem.* 87, 485–493, 2012.

3.1.47. Scholten, A., Visser, N.F.C., van den Heuvel, R.H.H., and Heck, A.J.R., Analysis of protein-protein interaction surfaces using a combination of efficient lysine acetylation and nanoLC-MALDI-MS/MS applied to the E9: Im9 bacteriotoxin-immunity protein complex, *J. Am. Soc. Mass Spectrom.* 17, 983–994, 2006.

3.1.48. Ji, J., Chakraborty, A., Geng, M. et al., Strategy for qualitative and quantitative analysis in proteomics based on signature peptides, *J. Chromatogr. B* 745, 197–210, 2000.

3.1.49. Zhang, R., Sioma, C.S., Wang, S., and Regnier, F.E., Fractionation of isotopically labeled peptides in quantitative proteomics, *Anal. Chem.* 73, 5142–5149, 2001.

(*continued*)

TABLE 3.1 (continued)
Some Acylating Agents Used for the Modification of Proteins

3.1.50. Miller, B.T., Collins, T.J., Rogers, M.E., and Kurosky, A., Peptide biotinylation with amine-reactive esters: Differential side chain reactivity, *Peptides* 18, 1585–1595, 1997.

3.1.51. Abello, N., Kersitjens, H.A.M., Postma, D.S., and Bischoff, R., Selective acylation of primary amines in peptides and proteins, *J. Proteome Res.* 6, 4770–4776, 2007.

3.1.52. Fritzsche, R., Ihling, C.H., Götze. M. et al., Optimizing the enrichment of cross-linked products for mass spectrometric protein analysis, *Rapid Commun. Mass Spectrom.* 26, 653–658, 2012.

3.1.53. Zhao, R.Y., Wilhelm, S.D., Audette, C. et al., Synthesis and evaluation of hydrophilic linkers for antibody-Maytansinoid conjugates, *J. Med. Chem.* 54, 3606–3623, 2011.

3.1.54. McKillop, A.M., McCluskey, J.T., Boyd, A.C. et al., Production and characterization of specific antibodies for evaluation of glycated insulin in plasma and biological tissues, *J. Endocrinol.* 167, 153–163, 2000.

3.1.55. Grabarek, Z. and Gergely, J., Zero-length crosslinking procedure with the use of active esters, *Anal. Biochem.* 185, 131–135, 1990.

3.1.56. Debord, J.D. and Lyon, L.A., On the unusual stability of succinimidyl esters in pNIPAm-AAc microgels, *Bioconjug. Chem.* 18, 601–604, 2007.

3.1.57. Wang, C., Yan, Q., Liu, H.B. et al. Different EDC/NHS activation mechanisms between PAA and PMAA brushes and the following amidation reactions, *Langmuir* 27, 12058–12068, 2011.

3.1.58. Everaerts, F., Torrianni, M., Hendriks, M., and Feijen, J., Biomechanical properties of carbodiimide crosslinked collagen: Influence of the formation of ester crosslinks, *J. Biomed. Mater. Res. A* 85, 547–555, 2008.

3.1.59. Mentinova, M., Barefoot, N.Z., and McLuckey, S.A., Solution versus gas-phase modification of peptide cations with NHS-ester reagents, *J. Am. Soc. Mass Spectrom.* 23, 282–289, 2012.

3.1.60. Böttcher, T. and Sieber, S.A., Structurally refined β-lactones as potent inhibitors of devastating bacterial virulence factors, *ChemBioChem* 10, 663–666, 2009.

3.1.61. Noel, A., Delpech, B., and Crich, D., Comparison of the reactivity of β-thiolactones and β-lactones toward ring-opening by thiols and amines, *Org. Biomol. Chem.* 10, 6480–6483, 2012.

3.1.62. Amoroso, J.W., Borkeley, L.S., Prasad, G., and Schnarr, N.A., Direct acylation of carrier proteins with functionalized β-lactones, *Org. Lett.* 12, 2330–2333, 2010.

3.1.63. Prasad, G., Amoroso, J.W., Borketey, L.S., and Schnarr, N.A., *N*-Activated β-lactams as versatile reagents for acyl carrier protein labeling, *Org. Biomol. Chem.* 10, 1992–2002, 2012.

3.1.64. Stroylova, Y.Y., Chobert, J.M., Muronetz, V.I. et al., *N*-Homocysteinylation of ovine prion protein induces amyloid-like transformation, *Arch. Biochem. Biophys.* 526, 29–37, 2012.

3.1.65. Marqués-López, W. and Christmann, M., β-Lactones through catalytic asymmetric heterodimerization of ketenes, *Angew. Chem. Int. Ed.* 51, 8696–8698, 2012.

3.1.66. Kunzmann, M.H., Staub, I., Böttcher, T., and Sieber, S.A., Protein reactivity of natural product-derived γ-butyrolactones, *Biochemistry* 50, 910–916, 2011.

3.1.67. Kluge, A.F. and Petter, R.C., Acylating drugs: Redesigning natural covalent inhibitors, *Curr. Opin. Chem. Biol.* 14, 421–427, 2010.

3.1.68. Jencks, W.P., Cordes, S., and Carriugolo, J., The free energy of thiol ester hydrolysis, *J. Biol. Chem.* 235, 3608–3614, 1960.

3.1.69. Torchinsky, Y.M., *Sulfur in Proteins*, Pergamon Press, Oxford, U.K., 1961.

3.1.70. Snydor, J.R., Mariano, M., Sideris, S., and Nock, S., Establishment of intein-mediated protein ligation under denaturing conditions: C-terminal labeling of a single-chain antibody for biochip screening, *Bioconjug. Chem.* 13, 707–712, 2002.

3.1.71. Anderson, S., Surfaces for immobilization of *N*-terminal cysteine derivatives via native chemical ligation, *Langmuir* 24, 13962–13968, 2008.

3.1.72. Sang, S.L.W. and Silvius, J.R., Novel thioester reagents afford efficient and specific *S*-acylation of unprotected peptides under mild conditions in aqueous solution, *J. Pept. Res.* 66, 169–180, 2005.

3.1.73. Peroza, E.A. and Freisinger, E., Tris is a non-innocent buffer during intein-mediated protein cleavage, *Protein Expr. Purif.* 57, 217–225, 2008.

3.1.74. Kursula, P., Ojala, J., Lambeir, A.M., and Wierenga, R.K., The catalytic cycle of biosynthetic thiolase: A conformational journey of an acetyl group through four binding modes and two oxanion holes, *Biochemistry* 41, 15543–15556, 2002.

TABLE 3.1 (continued)
Some Acylating Agents Used for the Modification of Proteins

3.1.75. Mitamura, K., Aoyama, E., Sakai, T. et al., Characterization of non-enzymatic acylation of amino or thiol groups of bionucleophiles by the acyl-adenylate or acyl-CoA thioesters of cholic acid, *Anal. Bioanal. Chem.* 400, 2253–2259, 2011.

3.1.76. Kalia, J. and Raines, R.T., Reactivity of intein thioesters: Appending a functional group to a protein, *ChemBioChem* 7, 1375–1383, 2006.

3.1.77. Bizzozero, O.A., Bixler, H.A., and Pastuszyn, A., Structural determinants influencing the reaction of cysteine-containing peptides with palmitoyl-coenzyme A and other thioesters, *Biochim. Biophys. Acta* 1545, 278–288, 2001.

3.1.78. Castro, E.A., Kinetics and mechanism of the aminolysis of thioesters and thiocarbonates in solution, *Pure Appl. Chem.* 81, 685–696, 2009.

3.1.79. Bracher, P.J., Snyder, P.W., Bohall, B.R., and Whitesides, G.M., The relative rates of thiol-thioester exchange and hydrolysis for alkyl and aryl thioalkanoates in water, *Orig. Life Evol. Biosph.* 41, 399–412, 2011.

3.1.80. Levesque, G., Arsène, P., Fanneau-Bellenger, V., and Pham, T.-N., Protein thioacylation. 1. Reagents design and synthesis, *Biomacromolecules* 1, 387–399, 2000.

3.1.81. Levesque, G., Arsène, P., Fanneau-Bellenger, V., and Pham, T.-N., Protein thioacylation. 2. Reagent stability in aqueous media and thioacylation kinetics, *Biomacromolecules* 1, 400–406, 2000.

3.1.82. Christman, K.L., Vázquez-Dorbatt, V., Schopf, E. et al., Nanoscale growth factor patterns by immobilization on a heparin-mimicking polymer, *J. Am. Chem. Soc.* 130, 16585–16591, 2008.

3.1.83. Grover, G.N., Steevens, N.S., Alconcel, N.M. et al., Trapping of thiol-terminated acrylate polymers with divinyl sulfone to generate well-defined semitelechelic Michael acceptor polymers, *Macromolecules* 42, 7657–7663, 2009.

3.1.84. Roth, P.J., Jochum, F.D., Zentel, R., and Theato, P., Synthesis of hetero-telechelic α,ω bio-functionalized polymers, *Biomacromolecules* 11, 238–244, 2010.

3.1.85. Fukuyamam, T., Tajima, Y., Ueda, H. et al., A method for measuring mouse respiratory allergic reaction to low-dose chemical exposure to allergens: An environmental chemical of uncertain allergenicity, a typical contact allergen and a non-sensitizing irritant, *Toxicol. Lett.* 195, 35–43, 2010.

3.1.86. Jeppsson, M.C., Jönsson, B.A.G., Kristiansson, M., and Lindh, C.H., Identification of covalent binding sites of phthalic anhydride in human hemoglobin, *Chem. Res. Toxicol.* 21, 2156–2163, 2008.

3.1.87. Verbicky, J.W., Jr. and Williams, L., Thermolysis of *N*-alkyl-substituted phthalmic acids. Steric inhibition of imide formation, *J. Org. Chem.* 46, 175–177, 1981.

3.1.88. Reichardt, P., Schreiber, A., Wichmann, G. et al., Identification and quantification of *in vitro* adduct formation between protein reactive xenobiotics and a lysine-containing model peptide, *Environ. Toxicol.* 18, 29–36, 2003.

3.1.89. O'Brien, A.M., Smith, A.T., and Ó'Fágáin, C., Effects of phthalic anhydride modification on horseradish peroxidase stability and activity, *Biotechnol. Bioeng.* 21, 233–240, 2003.

3.1.90. Neurath, A.R., Jiang, S., Strick, N. et al., Bovine β-lactoglobulin modified by 3-hydroxyphthalic anhydride blocks the CD4 cell receptor for HIV, *Nat. Med.* 2, 230–234, 1996.

3.1.91. Swaney, J.B., Selective proteolytic digestion as a method for the modification of human HDL3 structure, *J. Lipid Res.* 24, 245–252, 1983.

3.1.92. Wearne, S.J., Factor Xa cleavage of fusion proteins. Elimination of non-proteolytic cleavage by reversible acylation, *FEBS Lett.* 263, 23–26, 1990.

3.1.93. Howlett, G.J. and Wardrop, A.J., Dissociation and reconstitution of human erythrocyte membrane proteins using 3,4,5,6-tetrahydrophthalic anhydride, *Arch. Biochem. Biophys.* 188, 429–433, 1978.

3.1.94. Palacián, E., González, P.F., Piñero, M., and Hernández, F., Dicarboxylic acid anhydrides as dissociating agent of protein-containing structures, *Mol. Cell Biochem.* 97, 101–111, 1990.

3.1.95. Kimura, Y., Zaitsu, K., Motomura, Y., and Ohkura, Y., A practical reagent for reversible amino-protection of insulin, 3,4,5,6-tetrahydrophthalic anhydride, *Biol. Pharm. Bull.* 17, 881–885, 1994.

3.1.96. Tavekova, S. and Kluger, R., Biomimetic aminoacylation of ribonucleotides and RNA with aminoacyl phosphate esters and lanthanum salts, *J. Am. Chem. Soc.* 129, 15848–15854, 2007.

3.1.97. Mano, N., Kasuga, K., Kobayashi, N., and Goto, J., A nonenzymatic modification of the amino-terminal domain of histone H3 by bile acid acyl adenylate, *J. Biol. Chem.* 279, 55034–55034, 2004.

3.1.98. Sinha, A., Gupta, S., Bhutani, S. et al., PhoP-PhoP interaction at adjacent PhoP binding sites is influenced by protein phosphorylation, *J. Bacteriol.* 190, 1317–1328, 2008.

(continued)

TABLE 3.1 (continued)
Some Acylating Agents Used for the Modification of Proteins

3.1.99. Ueno, H., Pospischil, M.A., and Manning, J.M., Methyl acetyl phosphate as a covalent probe for anion-binding sites in human and bovine hemoglobin, *J. Biol. Chem.* 264, 12344–12351, 1989.

3.1.100. Ueno, H., Yatco, E., Benjamin, L.J., and Manning, J.M., Effects of methyl acetyl phosphate, a covalent antisickling agent, on the density of sickle erythrocytes, *J. Lab. Clin. Med.* 120, 152–158, 1992.

3.1.101. Xu, A.S., Labotka, R.J., and London, R.E., Acetylation of human hemoglobin by methyl acetylphosphate. Evidence of broad regio-selectively revealed by NMR studies, *J. Biol. Chem.* 274, 26629–2632, 1999.

3.1.102. Kataoka, K., Tanizawa, K., Fukui, T. et al., Identification of active site lysine residues of phenylalanine dehydrogenase by chemical modification with methyl acetyl phosphate combined with site-directed mutagenesis, *J. Biochem.* 116, 1370–1376, 1994.

3.1.103. Raibekas, A.A., Bures, E.J., Siska, C.C. et al., Anion binding and controlled aggregation of human interleukin-1 receptor antagonist, *Biochemistry* 44, 9871–9878, 2005.

3.1.104. Markley, L.D. and Dorman, L.C., Comparative study of terminating agents for use in solid-phase peptide synthesis, *Tetrahedron Lett.* May(21), 1787–1790, 1970.

3.1.105. Kise, N., Agui, S., Morimoto, S., and Ueda, N., Electroreductive acylation of aromatic ketones with acylimidazoles, *J. Org. Chem.* 70, 9407–9410, 2005.

3.1.106. Alzoubi, B.M., Liehr, G., and van Eldik, R., Mechanistic elucidation of the substitution behavior of alkyl cobaloximes in water and methanol as solvents, *Inorg. Chem.* 43, 6093–6100, 2004.

3.1.107. Roy, S., George, C.B., and Ratner, M.A., Catalysis by a zinc-porphyrin-based metal-organic framework from theory to computational design, *J. Phys. Chem. C* 116, 23494–23502, 2012.

3.1.108. Kogan, R.L. and Fife, T.H., Influence of hydrophobic and steric effects in the acyl group on acylation of α-chymotrypsin by *N*-acylimidazoles, *Biochemistry* 24, 2610–2614, 1985.

3.1.109. Wang, Y. and Chen, Y., Reactions of *N*-acylimidazole with nucleosides and nucleotides, *Heterocycles* 28, 593–601, 1989.

3.1.110. Honda, T., Dinkova-Kostova, A.T., Davie, E. et al., Synthesis and biological evaluation of 1-[2-cyano-3,12-dioxooleana-1,9(11)0dien-28-oyl]-4-ethynylimidazole. A novel and highly potent anti-inflammatory and cytoprotective agent, *Bioorg. Med. Chem. Lett.* 21, 2188–2191, 2011.

3.1.111. Santos, R.C., Salvador, J.A.R., Marin, S., and Cascante, M., Novel semisynthetic derivatives of betulin and betulinic acid with cytotoxic activity, *Bioorg. Med. Chem.* 17, 6241–6250, 2009.

3.1.112. Macdonald, J.M., Haas, A.L., and London, R.E., Novel mechanism of surface catalysis of protein adduct formation, NMR studies of the acetylation of ubiquitin, *J. Biol. Chem.* 275, 31908–31913, 2000.

3.1.113. Hawkins, D., Pinckard, R.N., and Farr, R.S., Acetylation of human serum albumin by acetylsalicylic acid, *Science* 160, 780–781, 1968.

3.1.114. Hawkins, D., Pinkard, R.N., Crawford, I.P., and Farr, R.S., Structural changes in human serum albumin induced by ingestion of acetylsalicylic acid, *J. Clin. Invest.* 48, 536–542, 1969.

3.1.115. Pinkard, R.N., Hawkins, D., and Farr, R.S., The inhibitory effect of salicylate on the acetylation of human albumin by acetylsalicylic acid, *Arthritis Rheum.* 13, 361–368, 1970.

3.1.116. Roth, G.J. and Majerus, P.W., The mechanism of the effect of aspirin on human platelets. I. Acetylation of a particulate fraction protein, *J. Clin. Invest.* 56, 624–632, 1975.

3.1.117. Roth, G.J., Stanford, N., and Majerus, P.W., Acetylation of prostaglandin synthase by aspirin, *Proc. Natl. Acad. Sci. USA* 72, 3073–3076, 1975.

3.1.118. Amin, A.R., Vyas, P., Attur, M. et al., The mode of action of aspirin-like drugs: Effect on inducible nitric oxide synthase, *Proc. Natl. Acad. Sci. USA* 92, 7926–7930, 1995.

3.1.119. Jung, S.-B., Kim, C.-S., Naqvi, A. et al., Histone deacetylase 3 antagonizes aspirin-stimulated endothelial nitric oxide production by reversing lysine acetylation of endothelial nitric oxide synthase, *Circ. Res.* 107, 877–887, 2010.

3.1.120. Marimulthu, S., Chivukula, R.S.V., Alfonso, L.F. et al., Aspirin acetylates multiple cellular proteins in HCT-116 colon cancer cells: Identification of novel targets, *Int. J. Oncol.* 39, 1273–1283, 2011.

3.1.121. Svensson, J., Bergman, A.-C., Adamson, U. et al., Acetylation and glycation of fibrinogen in vitro at specific lysine residues in a concentration dependent manner: A mass spectrometric and isotope labeling study, *Biochem. Biophys. Res. Commun.* 421, 335–342, 2012.

3.1.122. Koitka, M., Höchel, J., Obst, D. et al., Determination of rat serum esterase activities by an HPLC method using *S*-acetylthiocholine iodide and *p*-nitrophenyl acetate, *Anal. Biochem.* 381, 113–122, 2008.

3.1.123. Wheelock, C.E., Philips, B.M., Anderson, B.S. et al., Applications of carboxylesterase activity in environmental monitoring and toxicity identification evaluations (TIEs), *Rev. Environ. Contam. Toxicol.* 195, 117–178, 2008.

TABLE 3.1 (continued)
Some Acylating Agents Used for the Modification of Proteins

3.1.124. Means, G.E. and Bender, M.L., Acetylation of human serum albumin by p-nitrophenyl acetate, *Biochemistry* 14, 4989–4994, 1975.

3.1.125. Noctor, T.A. and Wainer, I.W., The in situ acetylation of an immobilized human serum albumin chiral stationary phase for high-performance liquid chromatography in the examination of drug-protein phenomena, *Pharm. Res.* 9, 480–484, 1992.

3.1.126. Sakurai, Y., Ma, S.F., Watanabe, H. et al., Esterase-like activity of serum albumin: Characterization of its structural chemistry using p-nitrophenyl esters as substrates, *Pharm. Res.* 21, 285–292, 2004.

3.1.127. Matsushita, S., Isima, Y., Chuang, V.T. et al., Functional analysis of recombinant human serum albumin domains for pharmaceutical applications, *Pharm. Res.* 21, 1924–1932, 2004.

3.1.128. Masson, P., Fremont, M.T., Darvesh, S. et al., Aryl acylamidase activity of human serum albumin with o-nitrotrifluoroacetanilide as the substrate, *J. Enzyme Inhib. Med. Chem.* 22, 463–469, 2007.

3.1.129. Suji, G., Khedkar, S.A., Singh, S.K. et al., Binding of lipoic acid induces conformational change and appearance of a new binding site in methylglyoxal modified serum albumin, *Protein J.* 27, 205–214, 2008.

3.1.130. Ascenzi, P., Gioia, M., Fanali, G. et al., Pseudo-enzymatic hydrolysis of 4-nitrophenyl acetate by human serum albumin: pH-dependence of rates of individual steps, *Biochem. Biophys. Res. Commun.* 424, 451–455, 2012.

3.1.131. Lockridge, O., Xue, W., Gaydess, A. et al., Pseudo-esterase activity of human albumin: Slow turnover on tyrosine 411 and stable acetylation of 82 residues including 59 lysines, *J. Biol. Chem.* 283, 22582–22590, 2008.

3.1.132. Jabush, J.R. and Deutsch, H.F., Location of lysines acetylated in ubiquitin reacted with p-nitrophenyl acetate, *Arch. Biochem. Biophys.* 238, 170–177, 1985.

3.1.133. Leonard, N.J., McDonald, J.J., Henderson, R.E. et al., Reaction of diethyl pyrocarbonate with nucleic acid components. Adenosine, *Biochemistry* 10, 3335–3342, 1971.

3.1.134. Kondorosi, A., Svab, Z., Solymosy, F., and Fedorcsák, I., Effect of diethyl pyrocarbonate on the biological activity of deoxyribonucleic acids isolated from bacteriophages, *J. Gen. Virol.* 16, 373–380, 1972.

3.1.135. Ehrenfeld, E., Interaction of diethylpyrocarbonate with poliovirus double-stranded RNA, *Biochem. Biophys. Res. Commun.* 56, 214–219, 1974.

3.1.136. Henderson, R.E., Kirkegaard, L.H., and Leonard, N.J., Reaction of diethyl pyrocarbonate with nucleic acid components. Adenosine-containing nucleotides and dinucleoside phosphates, *Biochim. Biophys. Acta* 294, 356–364, 1973.

3.1.137. Nielsen, P.E., Chemical and photochemical probing of DNA complexes, *J. Mol. Recognit.* 3, 1–25, 1990.

3.1.138. Bailly, C. and Waring, M.J., Diethylpyrocarbonate and osmium tetroxide as probes for drug-induced changes in DNA conformation in vitro, *Methods Mol. Biol.* 90, 51–79, 1997.

3.1.139. Wiener, S.L., Wiener, R., Urivetsky, M., and Meilman, E., Inhibition of ribonuclease by diethyl pyrocarbonate and other methods, *Biochim. Biophys. Acta* 259, 378–385, 1972.

3.1.140. Diez, C., Bertsch, G., and Simm, A., Isolation of full-size mRNA from cells sorted by flow cytometry, *J. Biochem. Biophys. Methods* 40, 69–80, 1999.

3.1.141. Nadkarni, M.A., Martin, F.E., Hunter, N., and Jacques, N.A., Methods for optimizing DNA extraction before quantifying oral bacterial numbers by real-time PCR, *FEMS Microbiol. Lett.* 296, 45–51, 2009.

3.1.142. Abe, K. and Anan, F.K., The chemical modification of beef liver catalase. V. Ethoxyformylation of histidine and tyrosine residues of catalase with diethylpyrocarbonate, *J. Biochem.* 80, 229–237, 1976.

3.1.143. Chang, L.H. and Tam, M.F., Site-directed mutagenesis and chemical modification of histidine residue on an α-class chick liver glutathione S-transferase CL 3-3. Histidines are not needed for the activity of the enzyme and diethyl-pyrocarbonate modifies both histidine and lysine residues, *Eur. J. Biochem.* 211, 805–811, 1993.

3.1.144. Hnízda, A., Santrůcek, J., Sanda, M. et al., Reactivity of histidine and lysine side-chains with diethylpyrocarbonate— A method to identify surface exposed residues in proteins, *J. Biochem. Biophys. Methods* 70, 1091–1097, 2008.

3.1.145. Secundo, F., Carrea, G., D'Arrigo, P., and Servi, S., Evidence for an essential lysyl residue in phospholipase D from *Streptomyces* sp. by modification with diethyl pyrocarbonate and pyridoxal-5-phosphate, *Biochemistry* 35, 9631–9636, 1996.

3.1.146. Dage, J.L., Sun, H., and Halsall, H.B., Determination of diethylpyrocarbonate-modified amino acid residues in α_1-acid glycoprotein by high-performance liquid chromatography electrospray ionization-mass spectrometry and matrix-assisted laser desorption/ionization time-of-flight mass spectrometry, *Anal. Biochem.* 257, 176–185, 1998.

3.1.147. Miles, E.W., Modification of histidyl residues in proteins by diethylpyrocarbonate, *Methods Enzymol.* 47, 431–442, 1977.

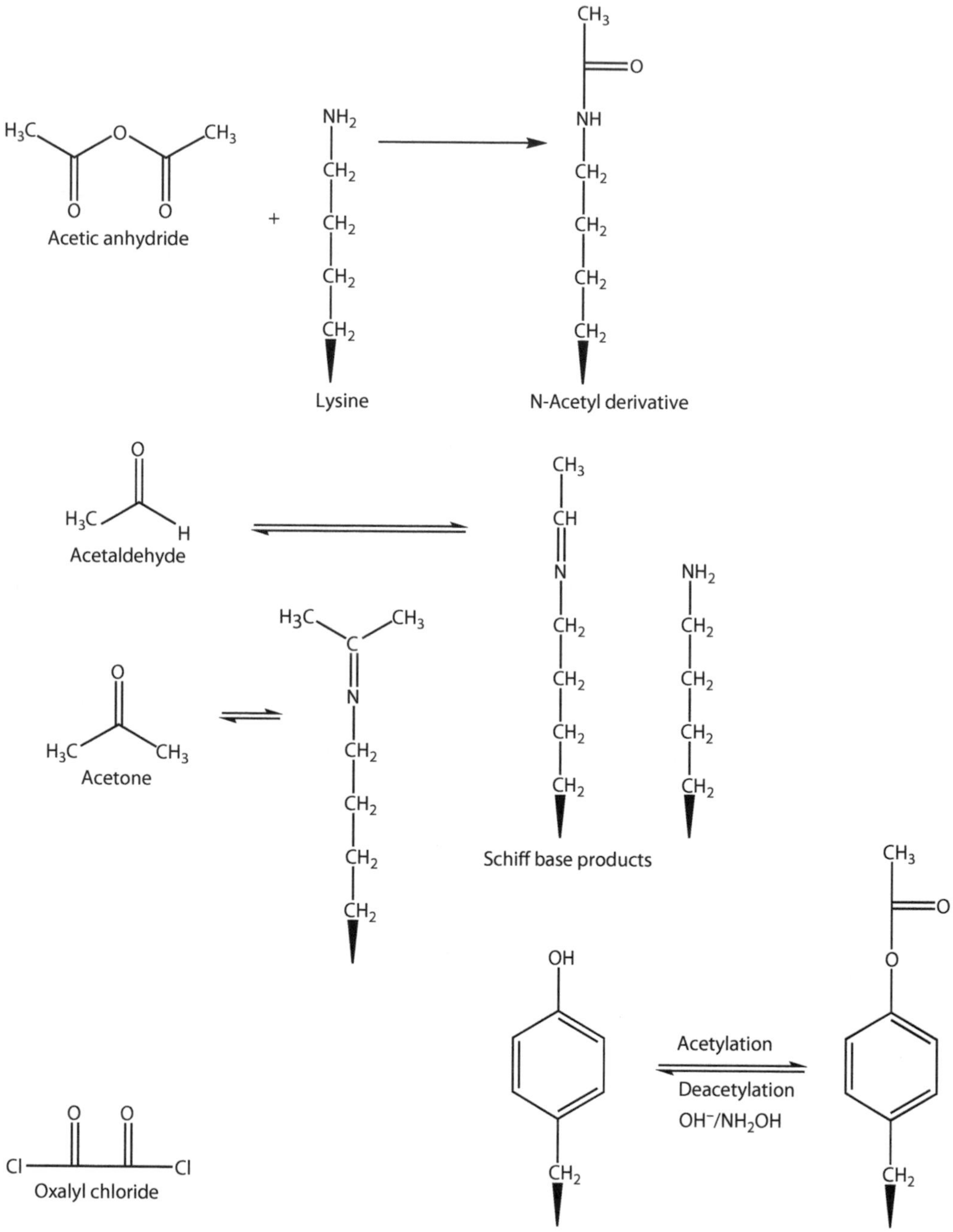

FIGURE 3.1 Reaction of acetic anhydride and acetyl chloride with groups. Shown are acetic anhydride and acetyl chloride. The commonality of the mechanism of action of the two acylating agents is the ability to participate in nucleophilic acyl addition reactions, that is, the transfer of the acetyl group to a protein nucleophile. Acetylated amino groups (amide groups) are stable, while the ester derivatives formed with hydroxyl groups (tyrosine is shown) in proteins are labile; acetylated hydroxyl groups in polysaccharides tend to be much more stable. Acetyl chloride has been used for the acetylation of polysaccharides (Xie, H., King, A., Kilpelainen, I. et al., Thorough chemical modification of wood-based lignocellulosic materials in ionic liquids, *Biomacromolecules* 8, 3740–3748, 2007): There is far more use of acetic anhydride/sodium acetate (Muljana, H., Picchioni, F., Knez, Z. et al., Insights in starch acetylation in sub- and supercritical CO_2, *Carbohydr. Res.* 346, 1224–1231, 2011). Oxalyl chloride has been used once; it is a very toxic material.

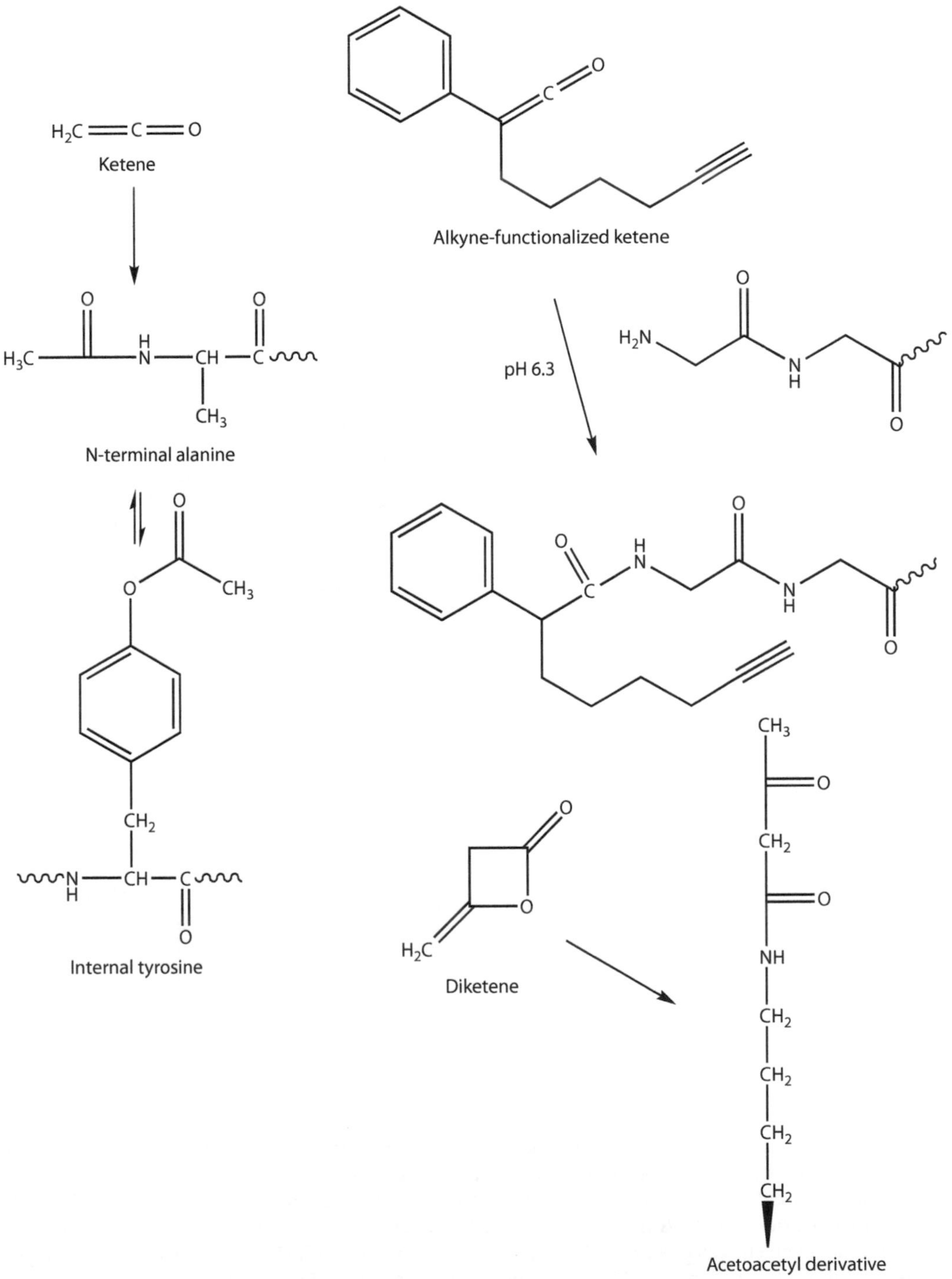

FIGURE 3.2 Structure and reactions of some ketenes. Shown is the structure of ketene (Neuberger, A., CXC. Note on the action of acetylating agents on amino-acids, *Biochem. J.* 32, 1452–1456, 1938) and a recently developed alkyne-functionalized ketene (Chan, A.O.-Y., Ho, C.-M., Chong, H.-C. et al., Modification of N-terminal α-amino groups of peptides and proteins using ketenes, *J. Am. Chem. Soc.* 134, 2589–2598, 2012). Greater than 99% specificity of modification of amino-terminal amino acid is achieved by reaction at pH 6.3 for many but not all peptide sequences where there was significant modification of lysine.

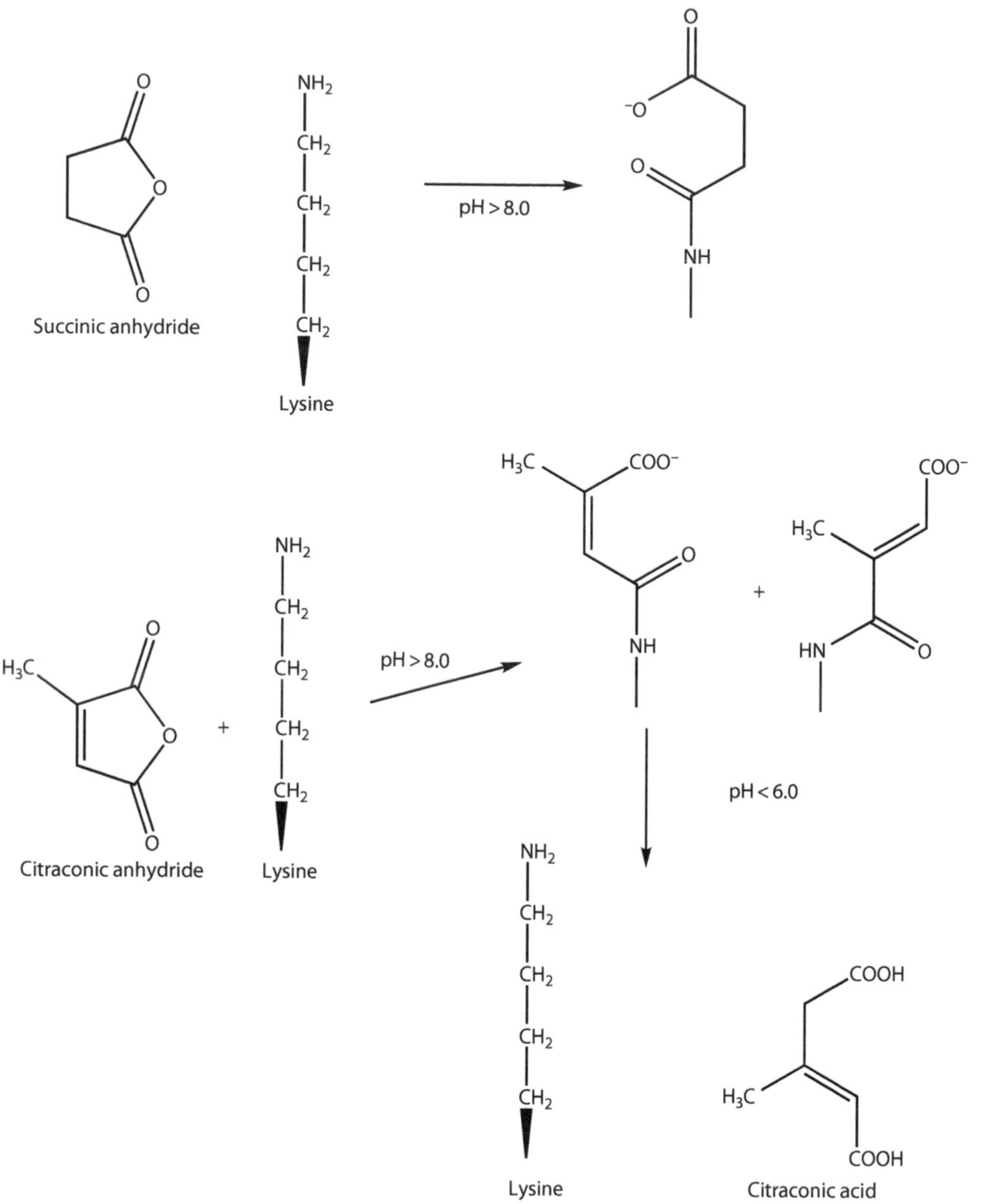

FIGURE 3.3 Succinylation and citraconylation. Succinic anhydride preferentially reacts with amino groups in proteins although reaction at hydroxyl groups and sulfhydryl groups is a possibility (Klotz, I.M., Succinylation, *Methods Enzymol.* 11, 575–589, 1967; Klapper, M.H. and Klotz, I.M., Acylation with dicarboxylic acid anhydrides, *Methods Enzymol.* 25, 531–536, 1972). The reaction product of succinic anhydride with amines is relatively stable but is labile under the condition of acid hydrolysis for amino acid analysis. The succinyl derivative is stable to MS and succinic anhydride is used in stable isotope labeling (Che, F.Y. and Fricker, L.D., Quantitative peptidomics of mouse pituitary: Comparison of different stable isotopic tags, *J. Mass Spectrom.* 40, 238–249, 2005). Also shown is the reaction of citraconic anhydride (Atassi, M.Z. and Habeeb, A.F.S.A., Reaction of proteins with citraconic anhydride, *Methods Enzymol.* 25, 546–553, 1972). The reaction of citraconic anhydride with proteins is similar to succinic anhydride with the difference in the stability of the product. Citraconylated amines are stable at alkaline pH but are unstable at acid pH (Son, Y.J., Kim, C.K., Kim, Y.B. et al., Effects of citraconylation on enzymatic modification of human proinsulin using trypsin and carboxypeptidase B, *Biotechnol. Prog.* 25, 1064–1070, 2009).

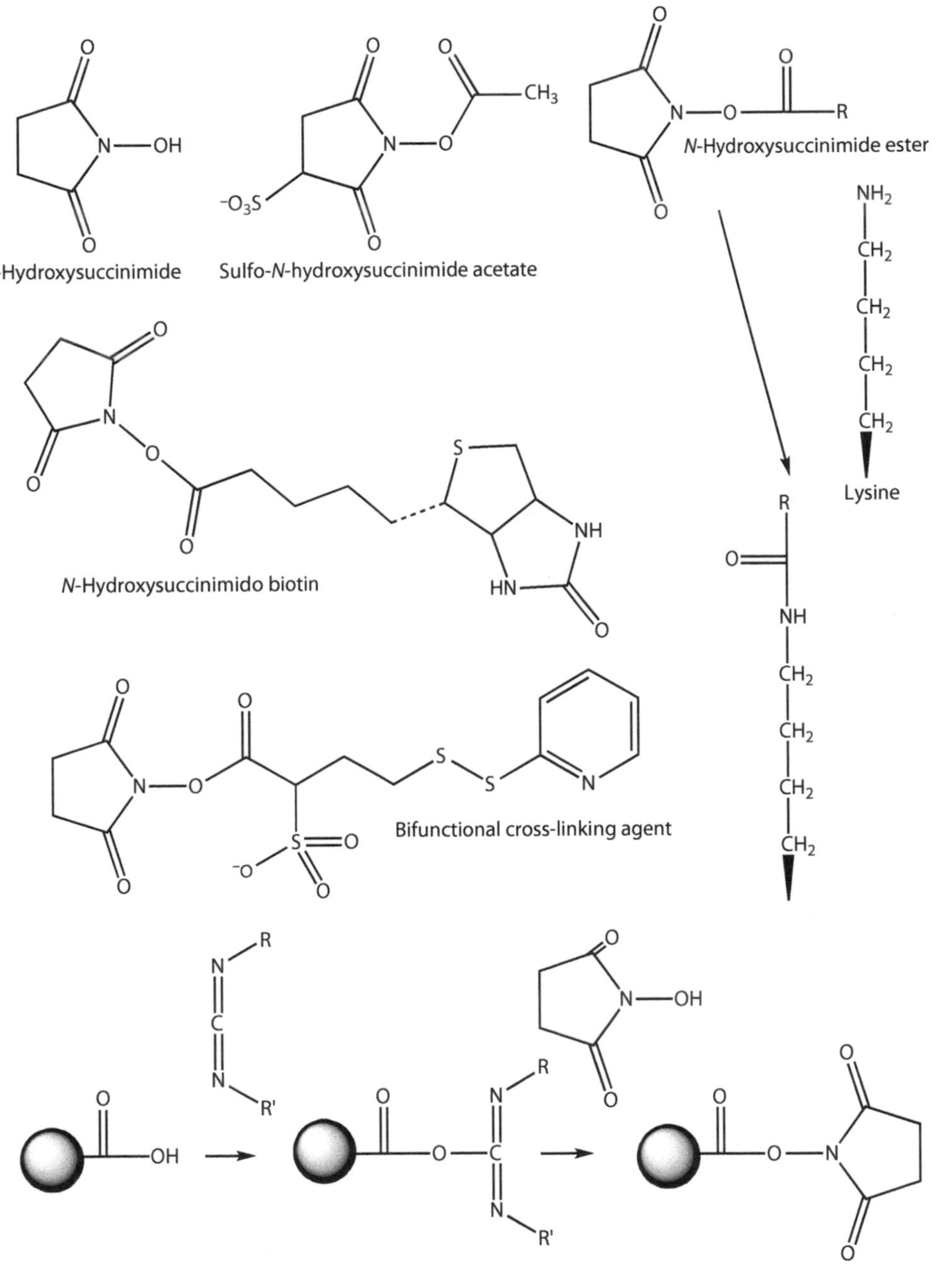

FIGURE 3.4 *N*-Hydroxysuccinimide esters and protein acylation. Shown are the structure of *N*-hydroxysuccinimide and sulfo-*N*-hydroxysuccinimide and the reaction of an *N*-hydroxysuccinimide ester with lysine to form an *N*ᵉ-acyl lysine derivative. Shown below is the structure of the *N*-hydroxysuccinimide derivative of biotin used to attach biotin to proteins (Ori, A., Free, P., Courty, J. et al., Identification of heparin-binding sites in proteins by selective labeling, *Mol. Cell. Proteomics* 8, 2256–2265, 2009). Shown below that is a bifunctional cross-linking reagent containing an *N*-hydroxysuccinimidyl function and a dithiopyridine function (Zhao, R.Y., Wilhelm, S.D., Audette, C. et al., Synthesis and evaluation of hydrophilic linkers for antibody-maytansinoid conjugates, *J. Med. Chem.* 54, 3606–2623, 2011). Shown at the bottom is the conversion of a carboxyl group to *N*-hydroxysuccinimidyl derivative that can then be coupled to a lysine residue forming an isopeptide bond (Grabarek, Z. and Gergely, J., Zero-length crosslinking procedure with the use of active esters, *Anal. Biochem.* 185, 131–135, 1990).

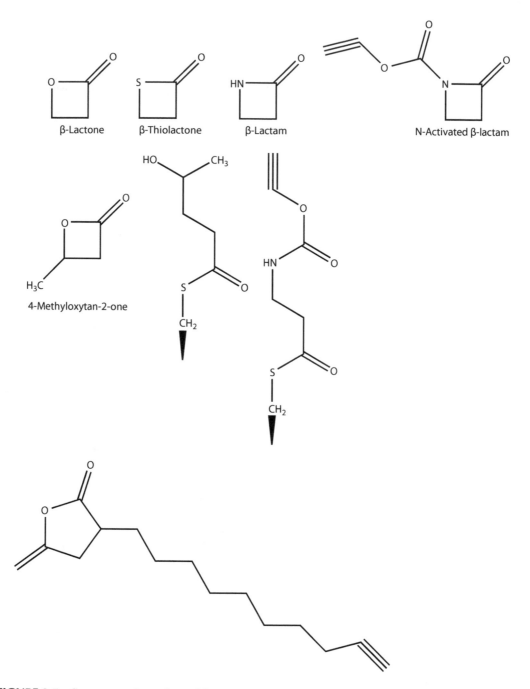

FIGURE 3.5 Structure and reactions of β-lactones, β-lactams, γ-butyrolactone, and derivative compounds. Shown at the top are generic structures for β-lactones, β-thiolactones, β-lactams, and N-activated β-lactams. Below is the reaction product obtained with a β-lactone (4-methyloxytan-2-one) and a sulfhydryl group (Amoroso, J.W., Borketsky, L.S., Prasad, G., and Schnarr, N.A., Direct acylation of carrier proteins with functionalized β-lactones, *Org. Lett.* 12, 2330–2333, 2010.). To the right is the product obtained on the reaction of cysteine with prop-2-yn-1-yl-2-oxoazetidine-1-carboxylate (Prasad, G., Amoroso, J.W., Borketey, L.S., and Schnarr, N.A., *N*-Activated β-lactams as versatile reagents for acyl carrier protein labeling, *Org. Biomol. Chem.* 10, 1992–2002, 2012). At the bottom are some derivatives of γ-butyrolactone (Kunzmann, M.H., Staub, I., Böttcher, T., and Sieber, S.A., Protein reactivity of natural product derived γ-butyrolactones, *Biochemistry* 50, 910–916, 2011).

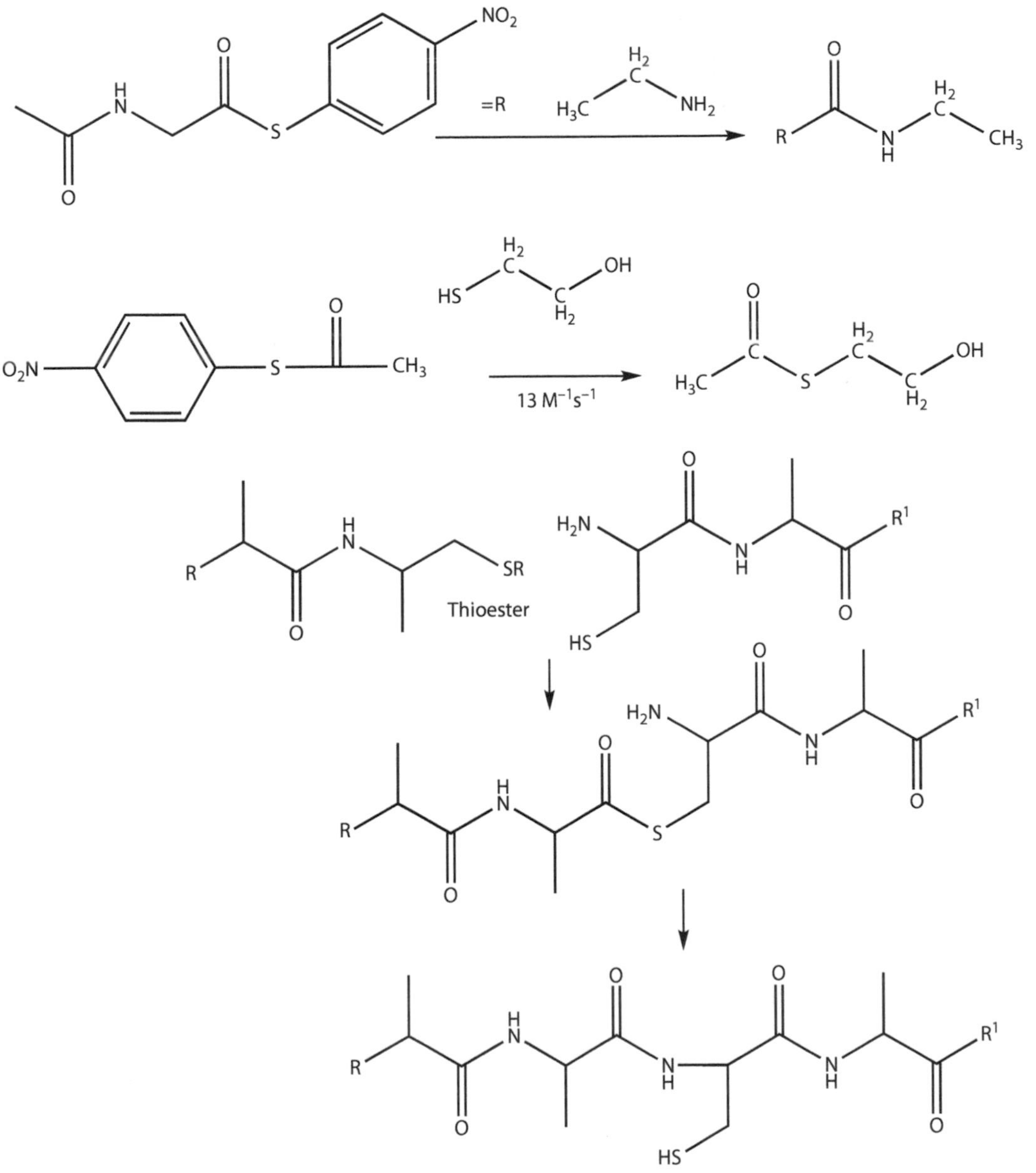

FIGURE 3.6 Preparation and reactions of thioesters. Shown is the preparation of thioester. Shown is the reaction of a thioester with a nitrogen nucleophile; reaction is much more rapid with the hydrazide (Kalla, J. and Raines, R.T., Reactivity of intein thioesters: Appending a functional group to a protein, *ChemBioChem* 7, 1375–1383, 2006). Below is the reaction of *p*-nitrophenyl thioacetate with 2-mercaptoethanol. As shown, the second-order rate constant is 13 M^{-1} s^{-1} (pH 7.5/25°C) for this reaction; the rate for the reaction with the thiol in the peptide, YCWL, was 114 M^{-1} s^{-1} (Bizzozero, O.A., Bixler, H.A., and Pastuszyn, A., Structural determinants influencing the reaction of cysteine-containing peptides with palmitoyl-coenzyme A and other thioesters, *Biochim. Biophys. Acta* 1545, 278–288, 2001). The process of native chemical ligation is shown at the bottom, which involves a thioester to form a peptide bond (Anderson, S., Surfaces for immobilization of N-terminal cysteine derivatives via native chemical ligation, *Langmuir* 24, 13962–13968, 2008; Ziao, J., Hamilton, B.S., and Tolbert, T.J., Synthesis of N-terminally linked protein and peptide dimers by native chemical ligation, *Bioconjug. Chem.* 21, 1943–1947, 2010.). Peptide thioesters are prepared for use in native chemical ligation (Hackenberger, C.P., The reduction of oxidized methionine residues in peptide thioesters with NH$_4$I-Me$_2$S, *Org. Biomol. Chem.* 4, 2291–2295, 2006.).

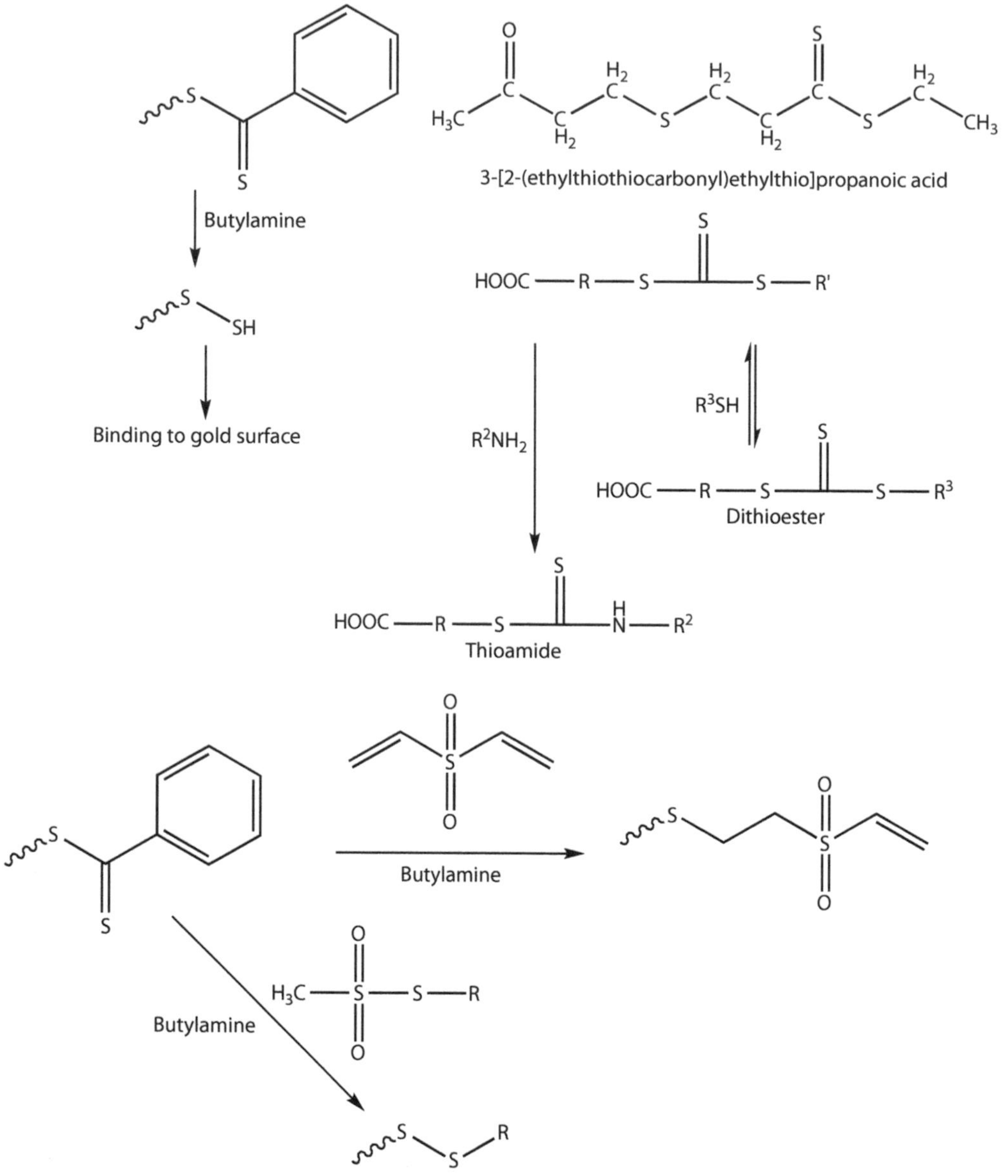

FIGURE 3.7 Structure and reaction of dithioesters. Shown are the reaction of a dithioester to form a thio-amide (irreversible) and the reversible reaction with a thiol group (Levesque, G., Arsene, P., Fanneau-bellenger, V., and Pham, T.-N., Protein thioacylation: 2. Reagent stability in aqueous media and thioacylation kinetics, *Biomacromolecules* 1, 400–406, 2000). Shown on the left is the reduction of a terminal dithioester function resulting from a reversible addition–fragmentation chain transfer (RAFT) polymerization forming a thiol that can then bind to a gold surface for surface plasmon resonance analysis (Christman, K.L., Vásquez-Dorbatt, V., Schopf, E. et al., Nanoscale growth factor patterns by immobilization on a heparin-mimicking polymer, *J. Am. Chem. Soc.* 130, 16585–16591, 2008). At the bottom is shown the conversion of the terminal dithioester function into a Michael acceptor via reaction with divinyl sulfone in the presence of butyl amine (Grover, G.N., Alconcel, S.N.S., Matsumoto, N.M., and Maynard, H.D., Trapping of thiol-terminated acrylate polymers with divinyl sulfone to generate well-defined semitelechelic Michael acceptor polymers, *Macromolecules* 42, 7657–7663, 2009), or the reaction with a methylthiosulfonate derivative (e.g., *N*-biotinylaminoethyl meth-anethiosulfonate) to form a disulfide linked derivative (Roth, P.J., Jochum, F.D., Zentel, R., and Theato, P., Synthesis of hetero-telechelic α,ω bio-functionalized polymers, *Biomacromolecules* 11, 238–244, 2010).

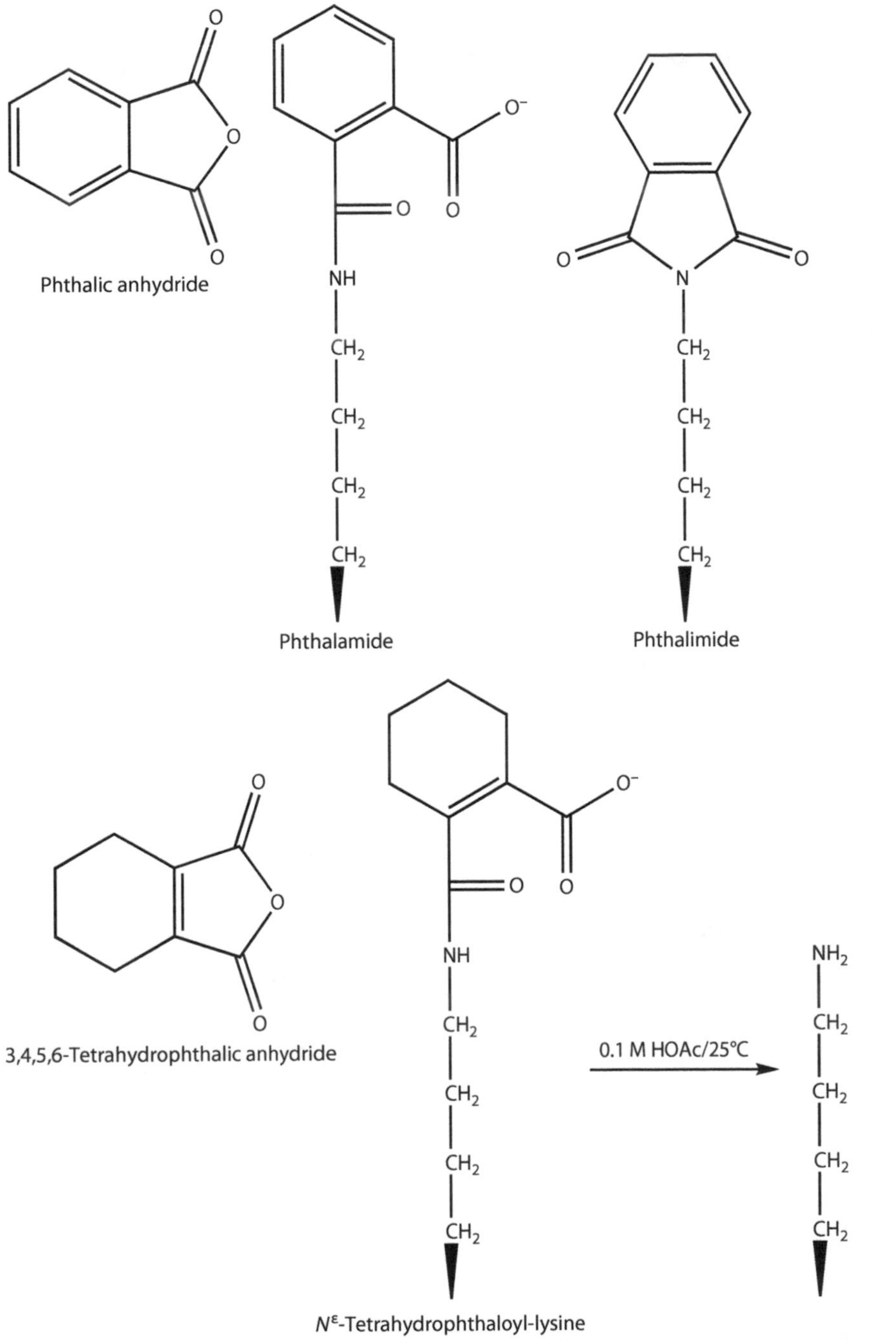

FIGURE 3.8 The structure and reaction of phthalic anhydride, tetrahydrophthalic anhydride. Shown is the structure of phthalic anhydride and the reaction products obtained with lysine. It is likely that the phthalamide is the initial and major product that can rearrange to form the phthalimide (Verbicky, J.W., Jr. and Williams, L., Thermolysis of *N*-alkyl-substituted phthalmic acids. Steric inhibition of imide formation, *J. Org. Chem.* 46, 175–177, 1981). The structure of 3,4,5,6-tetrahydrophthalic anhydride is shown as the tetrahydrophthalimide product with lysine. This reaction can be reversed by mild acid (Kimura, Y., Zaitsu, K., Motomura, Y., and Ohkura, Y., A practical reagent for reversible amino-protection of insulin, 3,4,5,6-tetrahydrophthalic anhydride, *Biol. Pharm. Bull.* 17, 881–885, 1994).

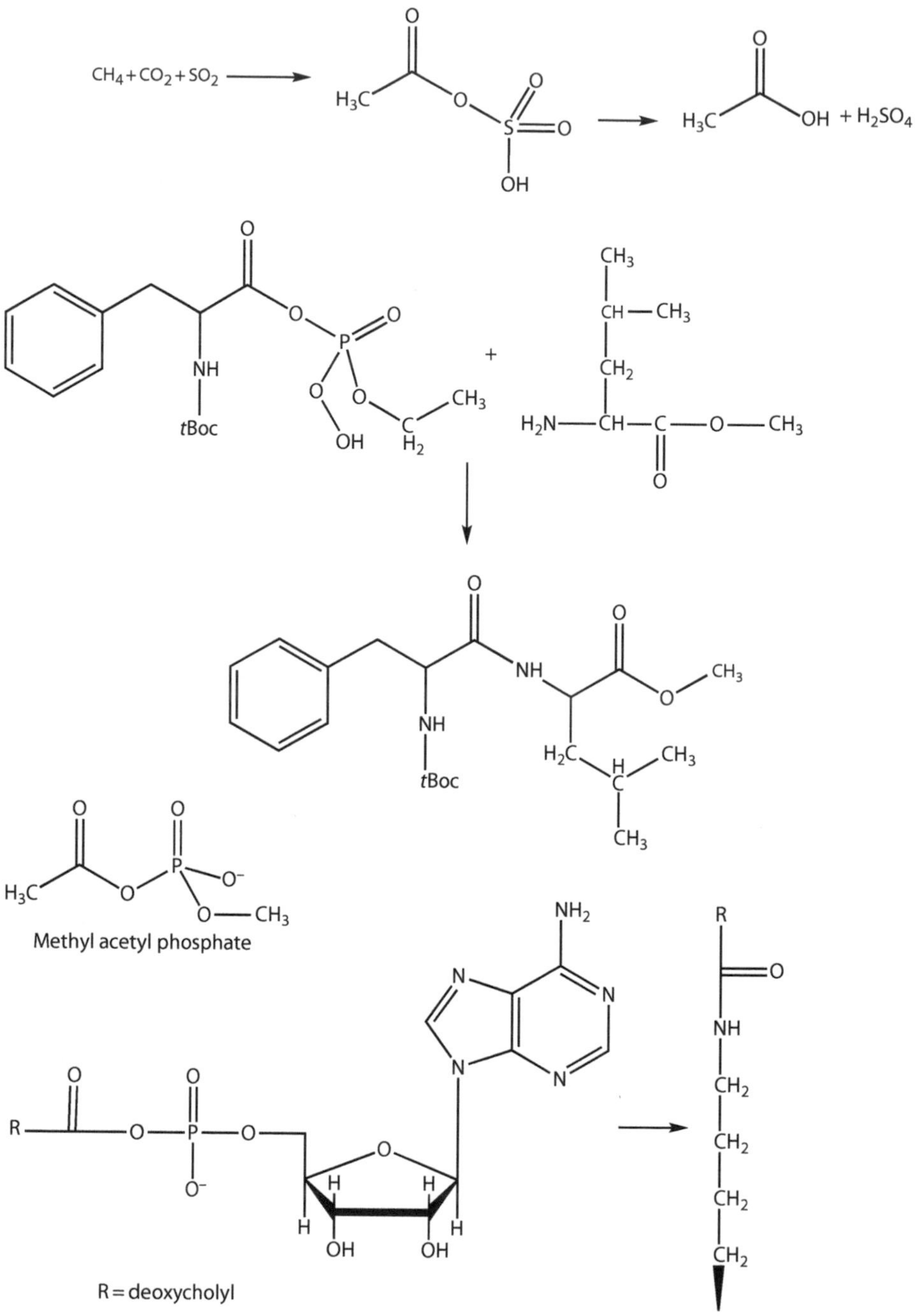

FIGURE 3.9 The structure and reactions of mixed acid anhydrides. Shown is the synthesis of acetic acid from methane via a mixed anhydride between acetic acid and sulfuric acid (Zerella, M., Mukhopadhyay, S., and Bell, A.T., Synthesis of mixed acid anhydrides from methane and carbon dioxide in acid solvents, *Org. Lett.* 5, 3913–3196, 2003). This is a method of producing acetic acid from methane. Shown below that reaction is the formation of a peptide bond between an aminoacyl phosphate ester and leucine methyl ester (Dhiman, R.S., Opinska, L.G., and Kluger, R., Biomimetic peptide bond formation in water with aminoacyl phosphate esters, *Org. Biomol. Chem.* 9, 5645–5647, 2011). At the bottom is the reaction of a bile acid acyl adenylate with a lysine residue (see Mano, N., Kasuga, K., Kobayashi, N., and Goto, J., A nonenzymatic modification of the amino-terminal domain of histone H3 by bile acid acyl adenylate, *J. Biol. Chem.* 279, 55034–55041, 2004). R is deoxycholyl and *t*Boc is *t*-butyloxycarbonyl.

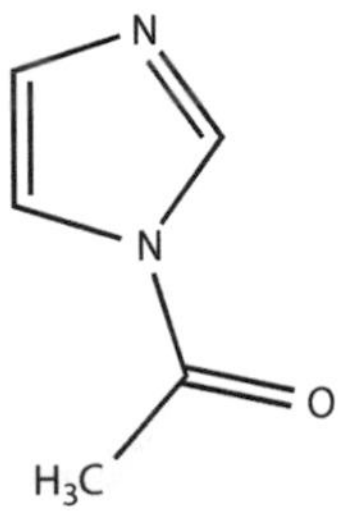

N-Acetylimidazole

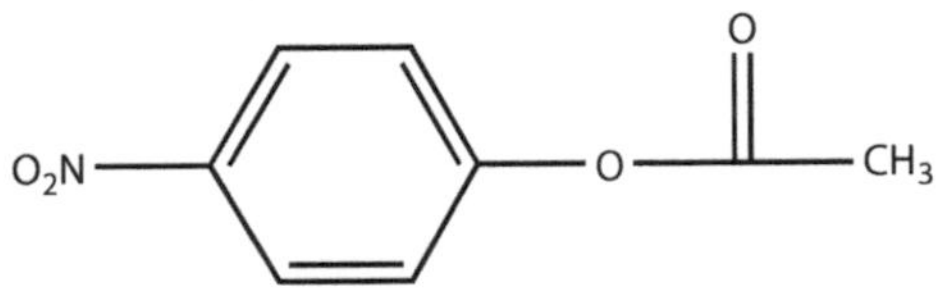

p-Nitrophenyl acetate

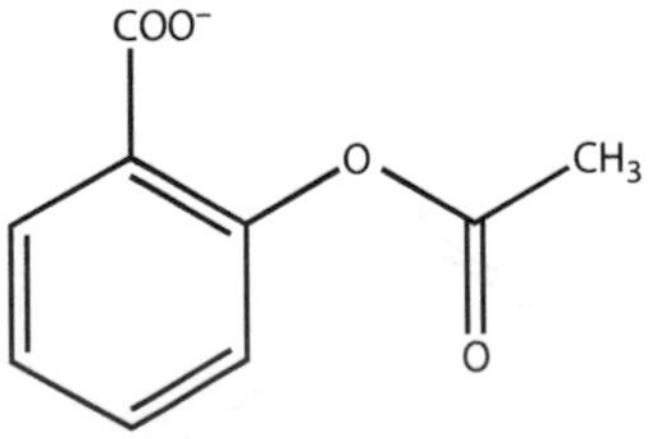

Aspirin
(2-acetoxy-benzoic acid)

FIGURE 3.10 *N*-Acetylimidazole, 2-(ethoxy)benzoic acid (aspirin), and *p*-nitrophenyl acetate. These three reagents modify various nucleophilic residues by acetylation of proteins under mild conditions. Only *N*-acetylimidazole is used routinely for the modification of proteins where it is reasonably, but not completely, specific for tyrosine (Basu, S. and Kirley, T.L., Identification of a tyrosine residue responsible for *N*-acetylimidazole-induced increase of activity of ecto-nucleoside triphosphate diphosphohydrolase 3, *Purinergic Signal.* 1, 271–280, 2005).

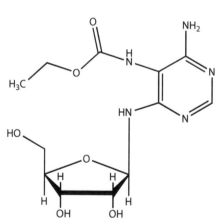

Diethylpyrocarbonate

5(4)-N-carbethoxyaminoimidazole-
4(5)-N'-carbethoxycarboxamidine

2-Amino-5-carbethoxyamino-4-hydroxy-6-
ribofuranosylaminopyrimidine

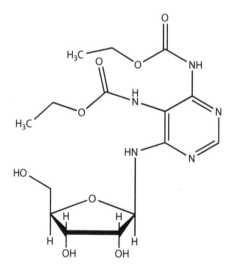

4,5-Dicarbethoxyamino-6-N-ribofuranosyl-
aminopyrimidine

4-Amino-5-carbethoxamino-6-N-ribofuranosyl-
aminopyrimidine

FIGURE 3.11 Reaction of diethylpyrocarbonate with nucleic acids. Shown is 5(4)-*N*-carbethoxyaminoimidazole-4(5)-*N'*-carbethoxycarboxamidine obtained from the reaction of adenine and diethylpyrocarbonate (Leonard, N.J., McDonald, J.J., and Reichmann, M.E., Reaction of diethyl pyrocarbonate with nucleic acid components, I. Adenine, *Proc. Natl. Acad. Sci. USA* 67, 93–98, 1970). The reaction of adenosine yields different products arising from the opening of the imidazole ring (Leonard, N.J., McDonald, J.J., Henderson, R.E.L., and Reichmann, M.E., Reaction of diethyl pyrocarbonate with nucleic acid components. Adenosine, *Biochemistry* 10, 3335–3342, 1971), which are shown at the bottom of the figure. 2-Amino-5-carbethoxyamino-4-hydroxy-6-*N*-ribofuranosylaminopyrimidine is derived from the reaction of diethylpyrocarbonate from guanine (Vincze, A., Henderson, R.E.I., McDonald, J.J., and Leonard, N.J., Reaction of diethyl pyrocarbonate with nucleic acid components. Bases and nucleosides derived from guanine, cytosine, and uracil, *J. Am. Chem. Soc.* 95, 2677–2682, 1973).

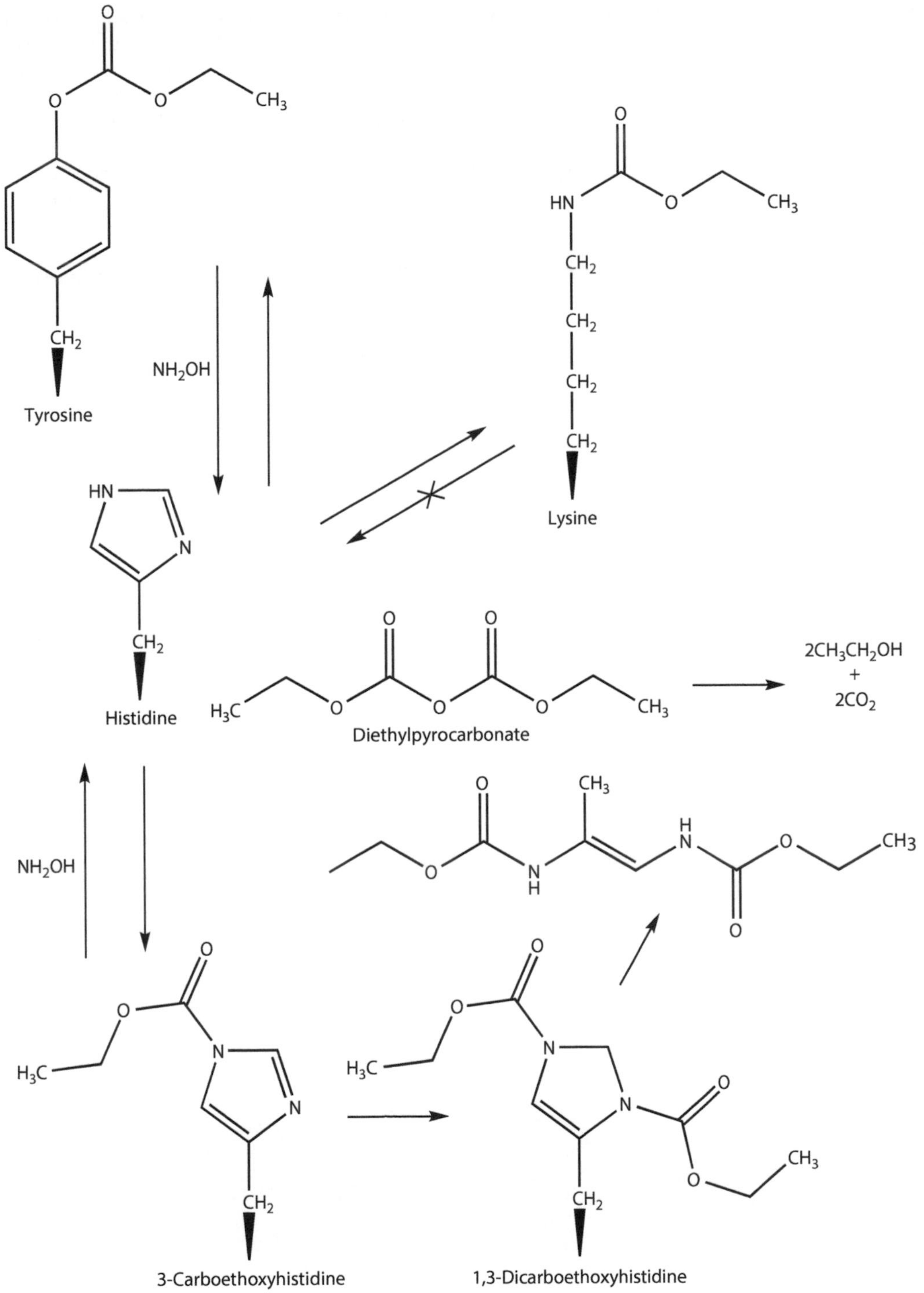

FIGURE 3.12

(*continued*)

FIGURE 3.12 (continued)   The reaction of diethylpyrocarbonate with histidine and other amino acid residues. Shown is the reaction of diethylpyrocarbonate with histidine to yield the monosubstituted and disubstituted product. Acylation of the monosubstituted product is reversed by base or hydroxylamine while the disubstituted product decomposes with rupture of the imidazole ring (Miles, E.W., Modification of histidyl residues in proteins by diethylpyrocarbonate, *Methods Enzymol.* 27, 431–442, 1967). There is also reaction with the phenolic hydroxyl of tyrosine that is reversed by hydroxylamine (Dage, J.L., Sun, H., and Halsall, H.B., Determination of diethylpyrocarbonate-modified amino acid residues in α_1-acid glycoprotein by high-performance liquid chromatography electrospray ionization-mass spectrometry and matrix-assisted laser desorption/ionization time-of-flight mass spectrometry, *Anal. Biochem.* 257, 176–185, 1998). Modification at ε-amino groups and α-amino groups is not reversed by hydroxylamine (Secundo, F., Carrea, G., D'Arrigo, P., and Servi, S., Evidence for an essential lysyl residue in phospholipase D from *Streptomyces* sp. by modification with diethylpyrocarbonate and pyridoxal-5-phosphate, *Biochemistry* 35, 9631–9636, 1996) and may be more complex than the simple acylation step shown in this figure (Wolf, B., Lesnaw, J.A., and Reichmann, M.E., A mechanism of the irreversible inactivation of bovine pancreatic ribonuclease by diethylpyrocarbonate, *Eur. J. Biochem.* 13, 519–525, 1970).

TABLE 3.2
Rate Constants for the Reaction of Acylating Agents with Nucleophiles

Acylating Agent	Nucleophile/Conditions	Rate (M^{-1} s^{-1})	Reference
Acetic anhydride[a]	L-Phenylalanine/2%(V/V)acetonitrile –0.1 KCl/10°C[b]	7.3×10^2	3.2.1
Acetic anhydride	Valine16 in porcine elastase (amino-terminal amino acid/10°C/0.1 M KCl	3.2×10	3.2.1
Acetic anhydride	Lysine87 in porcine elastase (amino-terminal amino acid/10°C/0.1 M KCl	2.2×10^3	3.2.1
Acetic anhydride	Lysine 224 in porcine elastase (amino-terminal amino acid/10°C/0.1 M KCl	1.8×10^3	3.2.1
Acetic anhydride	3-Methylbutylamine/25°C/pH stat	1×10^4	3.2.2
Acetic anhydride	Cyclohexylamine/25°C/pH stat	2.1×10^3	3.2.2
Acetic anhydride	Piperidine/25°C/pH stat	5.16×10^4	3.2.2
Acetyl chloride	Ethylamine/0.06 M NH_2OH-0.2 M phosphate, pH 11.07/22°C	8.3×10^4	3.2.3
Diethylpyrocarbonate	N-Acetylhistidine/50 mM acetate, pH 6.0, 25°C	2.16^c	3.2.4
Diethylpyrocarbonate	N-Acetylhistidine/20°C/phosphate buffer	24^b	3.2.5
Diethylpyrocarbonate	Essential histidine in 20°C/phosphate buffer	216^b	3.2.5
Palmitoyl-CoA	Cysteine ethyl ester/60 mM MOPS-0.2 M KCl, pH 7.5/25°C	0.23	3.2.6
Acetyl-CoA	Cysteine ethyl ester/60 mM MOPS-0.2 M KCl, pH 7.5/25°C	3.9	3.2.6
p-Nitrophenyl thioacetate	Cysteine ethyl ester/60 mM MOPS-0.2 M KCl, pH 7.5/25°C	39	3.2.6
Palmitoyl-CoA	2-Mercaptoethanol/60 mM MOPS-0.2 M KCl, pH 7.5/25°C	0.10	3.2.6
Acetyl-CoA	2-Mercaptoethanol/60 mM MOPS-0.2 M KCl, pH 7.5/25°C	1.1	3.2.6
p-Nitrophenyl thioacetate	2-Mercaptoethanol/60 mM MOPS-0.2 M KCl, pH 7.5/25°C	13	3.2.6
Palmitoyl-CoA	IRYCWLRR/60 mM MOPS-0.2 M KCl, pH 7.5/25°C	55	3.2.6
Acetyl-CoA	IRYCWLRR/60 mM MOPS-0.2 M KCl, pH 7.5/25°C	34.4	3.2.6
p-Nitrophenyl thioacetate	IRYCWLRR/60 mM MOPS-0.2 M KCl, pH 7.5/25°C	200	3.2.6
Palmitoyl-CoA	AcIDYCWLDD/60 mM MOPS-0.2 M KCl, pH 7.5/25°C	6.1	3.2.6
Acetyl-CoA	AcIDYCWLDD/60 mM MOPS-0.2 M KCl, pH 7.5/25°C	4.1	3.2.6
p-Nitrophenyl thioacetate	AcIDYCWLDD/60 mM MOPS-0.2 M KCl, pH 7.5/25°C	52	3.2.6
p-Nitrophenyl acetate	α-Acetyllysine/20°C/Tris buffer	$1.6 \times 10^{-8,b}$	3.2.7
p-Nitrophenyl acetate	ε-Acetyllysine/20°C/Tris buffer	$28.6 \times 10^{-8,b}$	3.2.7
Sulfo-N-hydroxysuccinimide acetate	Rubisco[d]/50 mM Mops-10 mM $MgCl_2$-10 mM $NaHCO_3$, pH 8.0/25°C	20.7^e	3.2.8
Sulfo-N-hydroxysuccinimide AMCA[f]	Rubisco[d]/50 mM Mops-10 mM $MgCl_2$-10 mM $NaHCO_3$, pH 8.0/25°C	239^e	3.2.8

(continued)

TABLE 3.2 (continued)
Rate Constants for the Reaction of Acylating Agents with Nucleophiles

Acylating Agent	Nucleophile/Conditions	Rate (M^{-1} s^{-1})	Reference
Z-Ser-hydroxysuccinimide ester[g]	p-Anisidine in DMSO/25°C	0	3.2.9
Z-Ile-N-hydroxysuccinimide ester[h]	p-Anisidine/DMSO/25°C	6.59×10^{-6}	3.2.9
Acetyl-N-hydroxysuccinimide ester	N-α-Acetyllysine/2°C/pH=pKa (10.53)	197	3.2.10
Acetyl-N-hydroxysuccinimide ester	N-ϵ-Acetyllysine/2°C/pH=pKa (8.95)	10.7	3.2.10
Phenylacetyl-N-hydroxysuccinimide ester	N-α-Acetyllysine/2°C/pH=pKa (10.53)	372	3.2.10
Phenylacetyl-N-hydroxysuccinimide ester	N-ϵ-Acetyllysine/2°C/pH=pKa (8.95)	10.7	3.2.10

[a] Acetic anhydride is subject to hydrolysis, and measurement of rate based on consumption of reagent must be considered in the determination of reaction rate with a nucleophile such as an amine. The amine may also catalyze the hydrolysis of the reagent (King, I.F., Rathore, R., Lam, J.Y.L. et al., pH optimization of nucleophilic reactions in water, *J. Am. Chem. Soc.* 114, 3028–3033, 1992).

[b] pH-independent second-order rate constant (rate constant for unprotonated amino group).

[c] Diethylpyrocarbonate is unstable in aqueous solution and kinetic studies must be corrected for hydrolysis (Figure 3.12) (Dickenson, C.J. and Dickinson, F.M., The role of an essential histidine residue of yeast alcohol dehydrogenase, *Eur. J. Biochem.* 52, 595–603, 1975).

[d] Rubisco (ribulose-1,5-bisphosphate carboxylase/oxygenase) activity.

[e] Rate is measured by loss of enzyme activity resulting from reaction at Lys247.

[f] N-Hydroxysuccinimide 7-amino-4-methylcoumarin-3-acetate.

[g] N-Benzyloxycarbonyl-serine ester of N-hydroxysuccinimide.

[h] N-Benzyloxycarbonyl-isoleucine ester of N-hydroxysuccinimide.

References to Table 3.2

3.2.1. Kaplan, H., Stevenson, K.J., and Hartley, B.S., Competitive labelling: A method for determining the reactivity of individual groups in proteins, *Biochem. J.* 174, 289–299, 1971.

3.2.2. King, J.F., Guo, Z.R., and Klassen, D.F., Nucleophilic attack vs. general base assisted hydrolysis in the reactions of acetic anhydride with primary and secondary amines. pH-yield studies in the recognition and assessment of the nucleophilic and general base reactions, *J. Org. Chem.* 59, 1095–1101, 1994.

3.2.3. Palling, D.J. and Jencks, W.P., Nucleophilic reactivity toward acetyl chloride in water, *J. Am. Chem. Soc.* 106, 4869–4876, 1984.

3.2.4. Topham, C.M. and Dalziel, K., Chemical modification of sheep-liver 6-phosphogluconate dehydrogenase by diethylpyrocarbonate. Evidence for an essential histidine residue, *Eur. J. Biochem.* 155, 87–94, 1986.

3.2.5. Holbrook, J.J. and Ingram, V.A., Ionic properties of an essential histidine residue in pig heart lactate dehydrogenase, *Biochem. J.* 131, 729–738, 1973.

3.2.6. Bizzozero, O.A., Bixler, H.A., and Pastuszyn, A., Structural determinants influencing the reaction of cysteine-containing peptides with palmitoyl-coenzyme A and other thioesters, *Biochim. Biophys. Acta* 1545, 278–288, 2001.

3.2.7. Kristol, D.S., Krautheim, P., Stanley, S., and Parker, R.C., The reaction of p-nitrophenyl acetate with lysine and lysine derivatives, *Bioorg. Chem.* 4, 299–304, 1975.

3.2.8. Salvucci, M.E., Covalent modification of a highly reactive and essential lysine residue of ribulose-1.5-bisphosphate carboxylase/oxygenase activase, *Plant Physiol.* 103, 501–508, 1993.

3.2.9. Stefanowicz, P. and Siemion, I.Z., Reactivity of N-hydroxysuccinimide esters, *Polish J. Chem.* 66, 111–118, 1992.

3.2.10. Hirata, H., Yamashina, T., and Higuchi, K., Chemistry of succinimido esters. XVI. Kinetic studies on the N-arylacetylation of amino by N-succinimidyl arylacetate in aqueous solution, *Yukagaku* (*J. Jpn. Oil Chem. Soc.*) 38, 9–17, 1989.

REFERENCES

1. Bhatnagar, R.S., Fütterer, K., Waksman, G., and Gordon, J.I., The structure of myristoyl-CoA: Protein N-myristoyltransferase, *Biochim. Biophys. Acta* 1441, 162–172, 1999.
2. Linder, M.E. and Deschenes, R.J., Palmitoylation: Policing protein stability and traffic, *Nat. Rev. Mol. Cell Biol.* 8, 74–84, 2007.

3. Yang, W., Di Vizio, D., Kirchner, M. et al., Proteome scale characterization of human S-acylated proteins in lipid raft-enriched and non-raft membranes, *Mol. Cell. Proteomics* 9, 54–70, 2010.

4. Strecker, H.J., Thioesterases for acyl and aminoacyl mercaptans, *J. Am. Chem. Soc.* 76, 3354–3355, 1954.

5. Previero, A., Prota, G., and Coletti-Previero, M.A., C-acylation of the tryptophan indole ring and its usefulness in protein chemistry, *Biochim. Biophys. Acta* 285, 269–278, 1972.

6. Kravchenya, N.A., Sokolov, K.N., Sokolova, T.N., and Sokolov, N.K., Regioselective N-acylation of arginine in oleum, *Pharm. Chem. J.* 46, 376–377, 2012 (translated from *Khimiko-Farmatsevticheskii Zhurnal* 46(6), 49–50, 2012).

7. Ferri, N., Paoletti, R., and Corsini, A., Lipid-modified proteins as biomarkers for cardiovascular disease: A review, *Biomarkers* 10, 219–237, 2005.

8. Martin, D.D., Beauchamp, E., and Berthiaume, L.G., Post-translational myristoylation: Fat matters in cellular life and death, *Biochemie* 93, 18–31, 2011.

9. Greaves, J., Prescott, G.R., Gorlecku, O.A., and Chamberlain, L.H., The fat controller: Role of palmitoylation in intracellular protein trafficking and targeting to membrane and microdomains, *Mol. Membr. Biol.* 16, 67–79, 2009.

10. Alcart-Ramos, C., Valero, R.A., and Rodriguez-Crespo, I., Protein palmitoylation and subcellular trafficking, *Biochim. Biophys. Acta* 1808, 2981–2994, 2011.

11. James, G. and Olson, E.N., Fatty acylated proteins as components of intracellular signaling pathways, *Biochemistry* 29, 2623–2634, 1990.

12. Triola, G., Waldmann, H., and Hedberg, C., Chemical biology of lipidated proteins, *ACS Chem. Biol.* 7, 87–99, 2012.

13. Mckinsey, T.A., Therapeutic potential for HDAC inhibitors in the heart, *Annu. Rev. Pharmacol. Toxicol.* 52, 303–319, 2012.

14. Leggatt, G.R. and Gabrielli, B., Histone deacetylase inhibitors in the generation of the anti-tumour immune response, *Immunol. Cell Biol.* 90, 33–38, 2012.

15. Zhang, H., Xiao, Y., Zhu, Z. et al., Immune regulation by histone deacetylases: A focus on the alteration of FOXP3 activity, *Immunol. Cell Biol.* 90, 95–100, 2012.

16. Shubassi, G., Robert, T., Vanoli, F. et al., Acetylation: A novel link between double-strand break repair and autophagy, *Cancer Res.* 72, 1332–1335, 2012.

17. Waluk, D.P., Sucharski, F., Sipos, L. et al., Reversible lysine acetylation regulates activity of human glycine N-acetyltransferase-like 2 (hGLYATL2), *J. Biol. Chem.* 287, 16158–16167, 2012.

18. Bradshaw, H.B., Rimmerman, N., Hu, S.S. et al., Novel endogenous N-acyl glycines identification and characterization, *Vitam. Horm.* 81, 191–205, 2009.

19. Waluk, D.P., Schultz, N., and Hunt, M.C., Identification of glycine N-acyltransferase-like 2 (GLYATL2) as a transferase that produces N-acyl glycines in humans, *FASEB J.* 24, 2795–2803, 2010.

20. Mitamura, K., Aoyama, E., Sakai, T. et al., Characterization of non-enzymatic acylation of amino or thiol groups of bionucleophiles by the acyl-adenylate or acyl-CoA thioester of cholic acid, *Anal. Bioanal. Chem.* 400, 2253–2259, 2011.

21. Mellert, H.S. and McMahon, S.B., Biochemical pathways that regulate acetyltransferase and deacetylase activity in mammalian cells, *Trends Biochem. Sci.* 34, 571–578, 2009.

22. Raibut, L., Ollivier, N., and Melnyk, O., Sequential native peptide ligation strategies for total chemical protein synthesis, *Chem. Soc. Rev.* 41, 7001–7015, 2012.

23. Dendane, N., Melnyk, O., Xu, T. et al., Direct characterization of native chemical ligation of peptides on silicon nanowires, *Langmuir* 28, 13336–13344, 2012.

24. van de Langemheen, H., Brouwer, A.J., Kemmink, J. et al., Synthesis of cyclic peptides containing a thioester handle for native chemical ligation, *J. Org. Chem.* 77, 10058–10064, 2012.

25. Phillips, J.J., Millership, C., and Main, E.R., Fibrous nanostructures from the self-assembly of designed repeat protein modules, *Angew. Chem. Int. Ed. Engl.* 51, 13132–13135, 2012.

26. Cohen, S., Biological reactions of carbonyl halides, in *The Chemistry of Acyl Halides*, Chapter 14, ed. S. Patel, pp. 313–348, Interscience/Wiley, London, U.K., 1972.

27. Erlanger, B.F., Cooper, A.G., and Cohen, W., The inactivation of chymotrypsin by diphenylcarbamyl chloride and its reactivation by nucleophilic agents, *Biochemistry* 5, 190–196, 1966.

28. King, I.F., Rathore, R., Lam, J.Y.L. et al., pH Optimization of nucleophilic reactions in water, *J. Am. Chem. Soc.* 114, 3028–3033, 1992.

29. Peters, J., Nitsch, M., Kühlmorgen, B. et al., Tetrabrachion: A filamentous archaebacterial surface protein assembly of unusual structure and extreme stability, *J. Mol. Biol.* 245, 385–401, 1995.

4 Nitration and Nitrosylation

The nitration and nitrosylation of functional groups in proteins (Figure 4.1) is of considerable interest as the *in vivo* formation of 3-nitrotyrosine is used as a biomarker for free radical activity/biological oxidations[1-4] and S-nitrosylation is viewed as a regulatory process.[5-8] Nitrosylation and nitration can occur in the same target proteins, but there is some suggestion that there might be exclusivity or reciprocity of modification[9-12] such as observed for the posttranslational modification of O-linked N-acetylglucosamine and O-phosphate in the carboxyl terminal domain of RNA polymerase II.[13] Gross and Hattori[9] suggest the possibility of an electrostatic interaction between the aromatic ring of a target tyrosine and the thiol group of cysteine. Most of these studies on nitrosylation or nitration are concerned with the action of nitric oxide, but a detailed discussion of these physiological reactions is beyond the scope of the current work. There are a number of useful references for those who wish to pursue these reactions in further detail.[10,12,14-16] For the purposes of the current work, the term "nitration" is meant to describe the covalent attachment of a nitro group (NO_2) to a carbon atom, while nitrosylation is considered to describe the covalent attachment of a nitric oxide group (NO) to form a nitroso derivative, most often an S-nitroso derivative of cysteine. Nitration and nitrosation are both posttranslational modifications; nitration is irreversible while nitrosation is similar to sulfenylation, sulfhydration, and S-thiolation in that the reaction is usually reversible.[17] Nitrosylation and nitrosation are equivalent terms both describing a reaction in which a nitroso compound is formed[18] by the introduction of a nitrosyl group (NO) into a molecule by reaction with nitric oxide radical or a derived form such as nitrogen trioxide or peroxynitrite. While nitrosation and nitrosylation are considered descriptors of the same chemical phenomena, nitrosylation is used more frequently, and based on my experience, literature searches should use both terms to assure complete coverage of this dynamic area.

The S-nitrosylation reaction of cysteine residues is a chemical modification considered similar to oxidation as a regulatory process[19-22] similar to cysteine sulfenic acid[23] and sulfhydration.[24] Sulfhydration is the process of hydrogen sulfide modification of the cysteine thiol group to form a persulfide; hydrogen sulfide has been compared to nitric oxide as a regulatory factor.[25] The formation of S-nitrosocysteine in proteins can proceed to the formation of cystine[26] as observed with cysteine sulfenic acid.[27] The formation of disulfide and mixed disulfide bonds following S-nitrosylation is discussed in the succeeding text. In vivo nitrosylation occurs through nitric oxide, originally described as endothelial-derived relaxing factor[28] derived from arginine. The reaction of nitric oxide to form S-nitrosocysteine in proteins can occur through various mechanisms including reaction of the nitric oxide radical, formation of nitrogen trioxide, peroxynitrite formation followed by homolytic cleavage to form nitric oxide, and nitrous acid reaction in acidic regions.[29-35] There is a suggestion that the oxidation of nitric oxide to form nitrogen trioxide is favored in a hydrophobic environment.[31,35] The nitrosylation of proteins is thought to be mediated through the action of S-nitrosoglutathione through a process of transnitrosylation.[6,36,37] Most protein S-nitrosothiols are thought to be reduced by intracellular thiols to yield the parent thiol[38,39], although some protein S-nitrosothiol derivatives have unusual stability.[40] Paige and coworkers[40] suggested that there is a conformational change in the target protein after S-nitrosylation. In support of this concept, Schreiter and coworkers[41] used crystallography to show that S-nitrosylation caused a reversible conformational change in blackfin tuna myoglobin.

S-Nitrosation can generally be reversed by exogenous thiol to yield the parent cysteine residue; other reagents can be used for reduction as described in the following text for the biotin switch assay. Thioredoxin can reduce S-nitrosocysteine in proteins[42] and is thought to be an important part

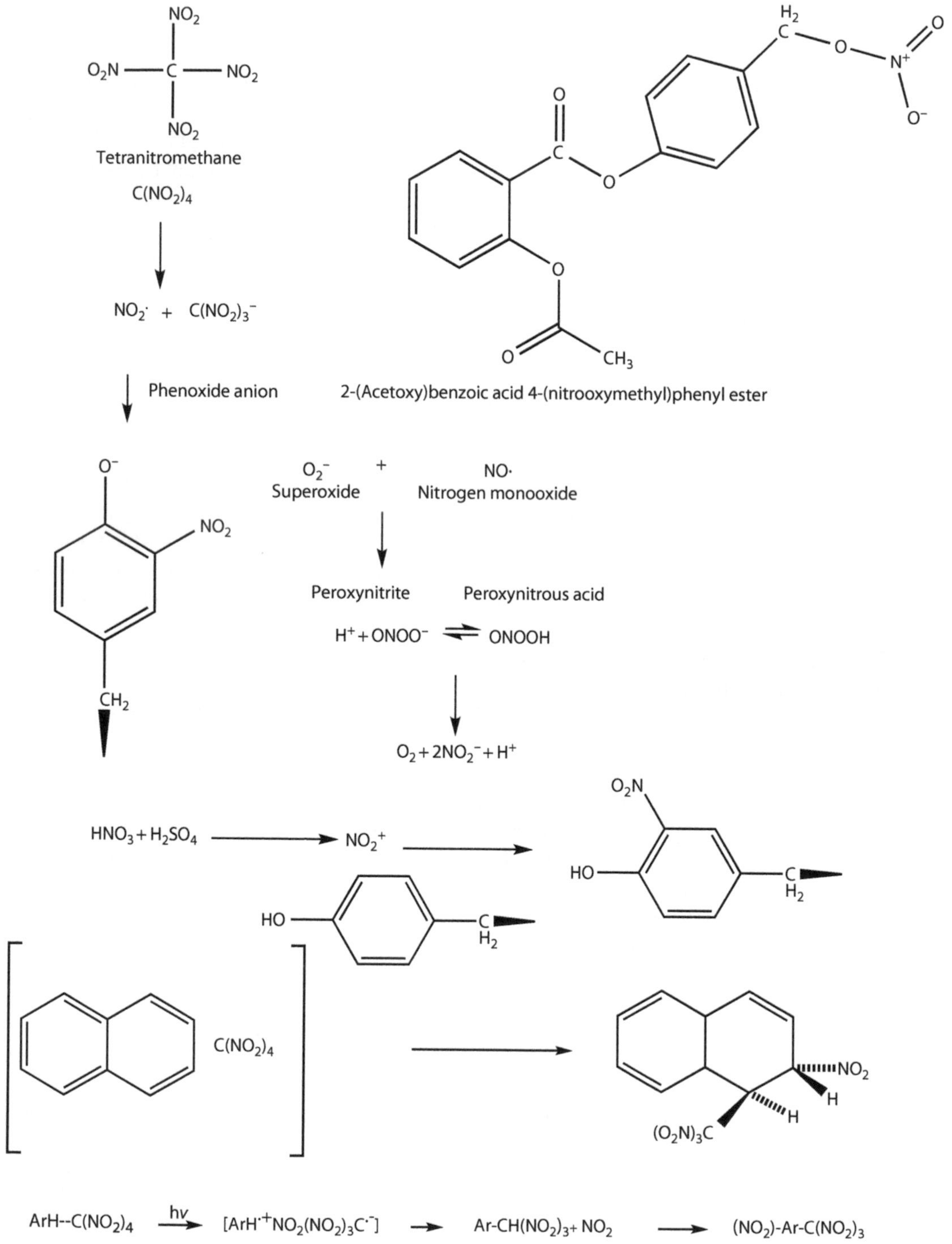

FIGURE 4.1 (continued)

of the redox regulatory system; *S*-nitrosoglutathione reductase has been identified as an important constituent of the redox regulatory system.[43-47] This enzyme was originally described as a formaldehyde dehyrogenase.[48-50] *S*-Nitrosylation of proteins has not been used to study cysteine residues such as described for other reagents such as *N*-ethylmaleimide or the alkyl methanethiosulfonates.

A consideration of the data suggests that the *S*-nitrosylation reaction is preferable with the thiolate anion[51,52], although the reaction clearly also occurs with the thiol considering that *S*-nitrosocysteine was prepared by the reaction of sodium nitrite in dilute sulfuric acid[53] and model peptides[54] and nitrosylated with sodium nitrite in dilute hydrochloric acid. The nitrosylating species likely differs from basic to acid conditions, with the nitrosonium ion the active species in acid. The study by Taldone and coworkers[54] used several different techniques to nitrosylate cysteine residues in different peptides and did not observe an effect of sequence; the use of nitrous acid, nitric oxide in gas phase, *S*-nitroso-*N*-acetylpenicillinamine (SNAP), and *S*-nitrosoglutathione was evaluated as nitric oxide donors, and the most satisfactory results were obtained with *S*-nitrosoglutathione. Morakinyo and coworkers[55] performed a careful study of the nitrosylation of cystamine. At acid concentration less than nitrite concentration (low acid), nitrous is the nitrosylation reagent, while at acid concentrations higher than nitrite concentration (high acid), the nitrosonium cation (NO[+]) is the nitrosylation reagent. The reaction of thiols with sodium nitrite in dilute mineral acid is considered the method of choice for preparing low molecular weight thiols.[56,57] Clancy and Abramson[58] described a solid-phase method for the preparation of *S*-nitrosoglutathione. 2,2-Dithiobis(ethylamine) was coupled of a matrix via an amide linkage (BioRad Affigel® 10) and reduced with dithiothreitol to yield a free sulfhydryl group of the matrix. This was converted to an *S*-nitrosyl derivative with nitrous acid provided a matrix that was red in color. The *S*-nitrosothiol matrix could convert cysteine or glutathione to the corresponding *S*-nitroso derivative. Meyer and coworkers[59] used glutathione transferase to synthesize *S*-nitrosoglutathione from alkyl nitrite such as amyl nitrite or *t*-butylnitrite and glutathione.

A variety of reagents (Figure 4.2) can be used to affect the in vitro *S*-nitrosylation of proteins including *S*-nitrosoglutathione,[26,60-65] *S*-nitrosocysteine,[66-68] Angeli's salt,[63-66,69] SNAP,[65,70] and nitric oxide

FIGURE 4.1 (continued) Some nitrating reagents and their reactions. At the top left is TNM and the suggested decomposition to form nitrogen dioxide and the trinitromethide radical in the nitration of tyrosine. The nitration of tyrosine likely proceeds via the phenoxide radical with nitration *ortho* to the hydroxyl groups with the trinitromethide radical proceeding to the trinitromethane anion (Bruice, T.C., Gregory, M.J., and Walters, S.L., Reactions of tetranitromethane. I. Kinetics and mechanisms of nitration of phenols by tetranitromethane, *J. Am. Chem. Soc.* 90, 1612–1619, 1968). Not shown are other products of the reaction including Plummerer's ketone (Grossi, L., Evidence of an electron-transfer mechanism in the peroxynitrite-mediated oxidation of 4-alkylphenols and tyrosine, *J. Org. Chem.* 68, 6349–6353, 2003). At the top right is the structure of nitrated aspirin derivative (Kashfi, K. and Rigas, B., Molecular targets of nitric-oxide-donating aspirin in cancer, *Biochem. Soc. Trans.* 33, 701–704, 2005). At the center is a representation of the formation of peroxynitrite (Radi, R., Protein tyrosine nitration: Biochemical mechanisms and structural basis of functional effects, *Accounts Chem. Res.* 46, 550–559, 2013). Shown below is the nitration of tyrosine by nitric acid in the presence of sulfuric acid proceeding through the formation of nitronium ion in electrophilic aromatic addition (Ross, D.S., Kuhlman, K.E., and Malhotra, R., Studies on aromatic nitration 2. Nitrogen 14 NMR study of the nitric acid/nitronium equilibrium in aqueous sulfuric acid, *J. Am. Chem. Soc.* 105, 4299–4302, 1982). Shown at the bottom is the photochemical reaction of an aromatic compound with tetranitromethane. The mechanism shown involves the initial formation of a complex between the aromatic compound and tetranitromethane undergoing photolysis resulting in the formation of a complex between nitrogen dioxide, an aromatic radical cation, and a trinitromethide radical anion (Eberson, L. and Hartshorn, M.P., The formation and reactions of adducts from the photochemical reactions of aromatic compounds with tetranitromethane and other X-NO$_2$ reagents, *Aust. J. Chem.* 51, 1061–1081, 1998). This complex then undergoes reaction resulting in the eventual formation of an aromatic derivative with a nitro function and a trimethyl nitro function (Calvert, J.L., Eberson, L., Hartshorn, M.P. et al., Photochemical nitration by tetranitromethane. XVI. The regiochemistry of adduct formation in the photochemical reaction of 1,8-dimethylnaphthalene and tetranitromethane: Thermal 1,3-dipolar nitro addition reactions, *Aust. J. Chem.* 47, 1211–1222, 1994).

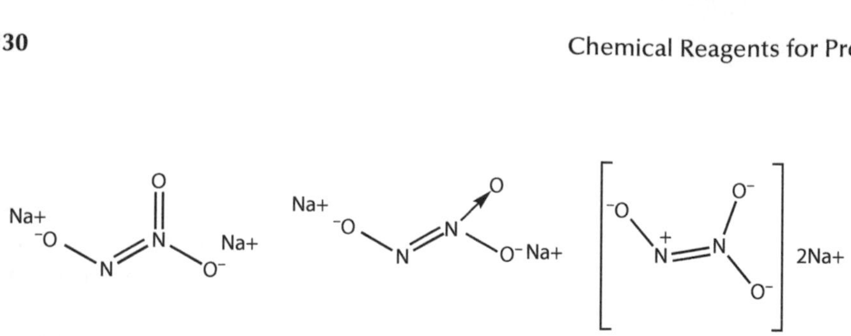

Angeli's salt (sodium α-oxyhyponitrite)

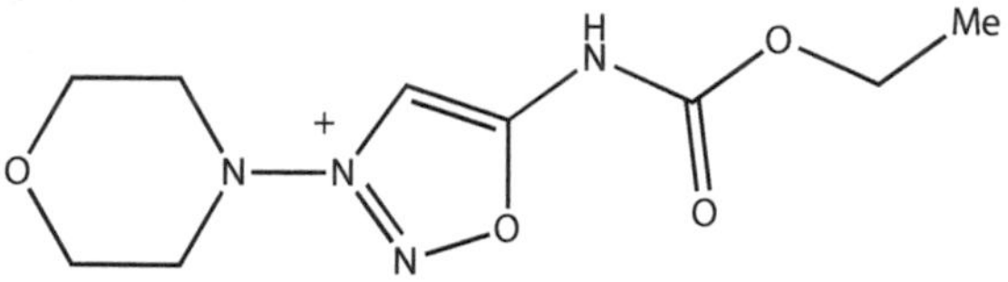

3-Morpholinosydnonimine ethyl ester
Molsidomine SIN-10

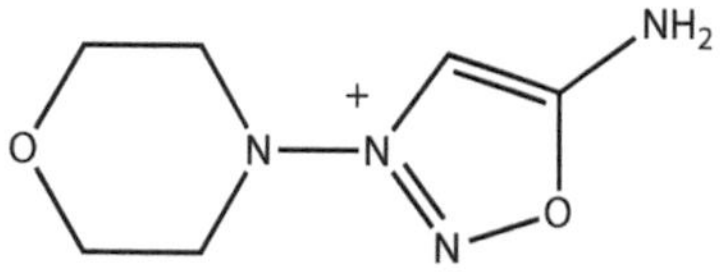

3-Morpholinosydnonimine; linsidomine; SIN-1

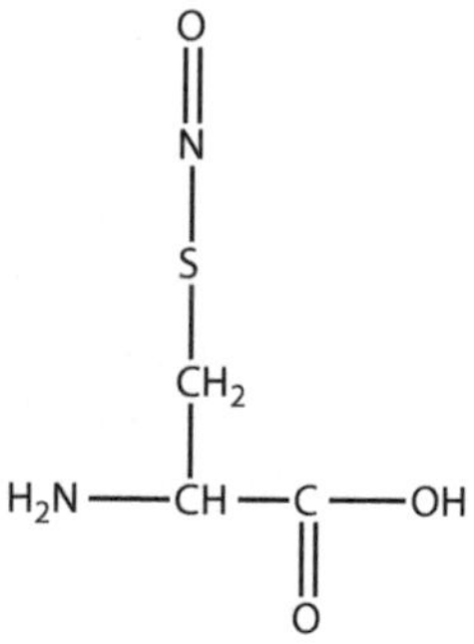

S-Nitrosocysteine

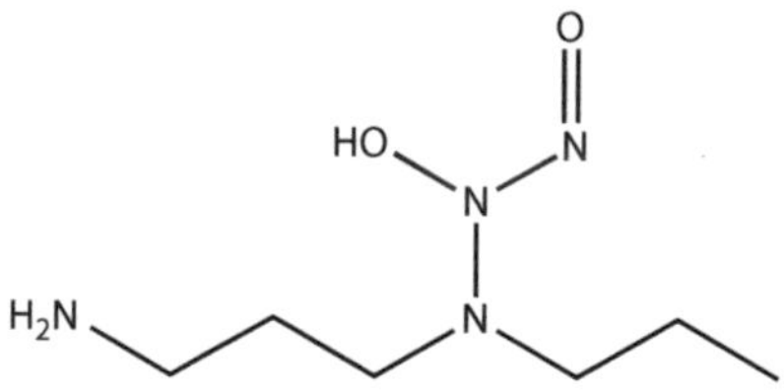

Propylamine propylamine NONOate
[(Z)-1-[N-(3-aminopropyl)-N-(n-propylamine
diazen-1-ium-1,2-diolate]; papanonoate

$NaNO_2$

pH < 4

$3HNO_2 \rightleftharpoons H_3O^+ + NO_3^- + 2NO$

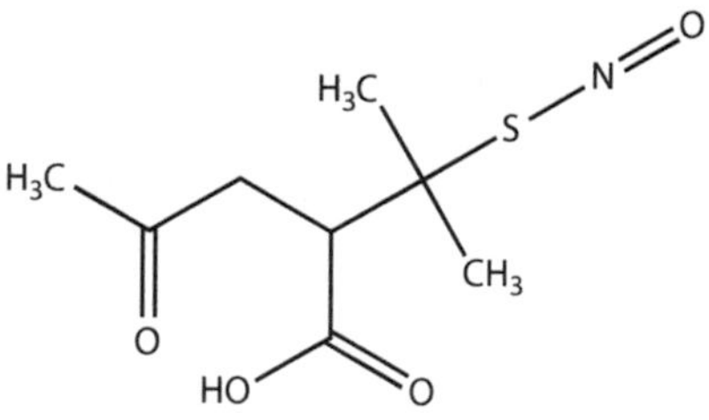

S-Nitroso-N-acetylpenicillamine (SNAP)

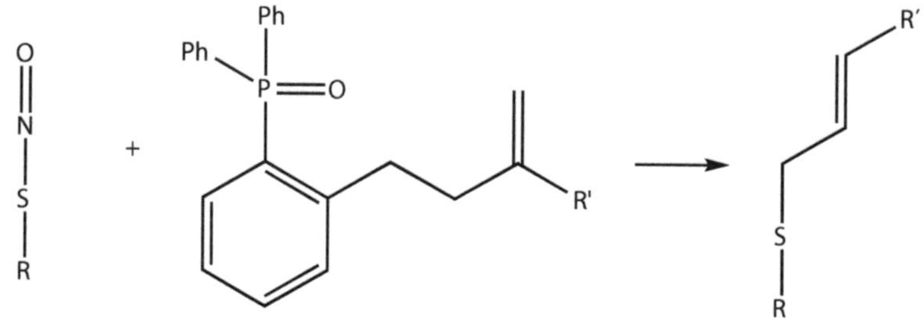

FIGURE 4.2 (continued)

in the gas phase for modification of a crystalline protein.[71] Depending on reagent, reaction conditions, and protein target, the use of some of these reagents, such as Angeli's salt, results in modification beyond S-nitrosylation such as disulfide bond formation and carbonyl formation.[64,65] Väänänen and coworkers[64] observed that Angeli's salt, papanonoate(papa nonoate; propylamine propylamine NONOate), SIN-1(linsidomine), and S-nitrosoglutathione inactivated papain. Carbonyl formation was seen with Angeli's salt, SIN-1, and papanonoate but not with S-nitrosoglutathione. Inactivation of papain with either Angeli's salt or SIN-1 was partially reversed with dithiothreitol, while inactivation with either papanonoate or S-nitrosoglutathione was fully reversed. The authors suggest that reaction with S-nitrosoglutathione or papanonoate produced an S-nitrosocysteine and/or a mixed disulfide, reaction with SIN-1 produced S-nitrosocysteine and sulfinic acid/sulfonic acid, and reaction with Angeli's salt produced a mixed disulfide and/or sulfinamide. The use of acidified nitrite (Figure 4.2) has been used topically to treat fungus in nail, and such use is accompanied by nitrosylation of cysteine.[72] The antimicrobial/antifungal activity of acidified nitrite has been recognized for some time[73] and is of continued interest for antimicrobial use.[74–76] Acidified nitrous acid has also been useful for the diazotization of amino groups in reagents for coupling to proteins.[77]

Given the role of S-nitrosylation in regulation, there has been considerable interest in factors that influence the specificity of modification. Stamler and coworkers[78] suggested that the presence of an

FIGURE 4.2 (continued) Nitrosylation and nitrosylation reagents. Shown at the top are three structural depictions of Angeli's salt (Hunt, H.R., Cox, J.R., Jr., and Ray, J.D., The heat of formation of crystalline sodium α-oxyhyponitrite, the structure of aqueous α-oxyhyponitrite ion, *Inorg. Chem.* 1, 938–941, 1962; Hughes, M.N. and Nicklin, H.G., The chemistry of peroxonitrites. Part II. Copper(II)-catalyzed reaction between hydroxylamine and peroxonitrite in alkali, *J. Chem. Soc. A.* 925–928, 1970; Nakagawa, H., Controlled release of HNO from chemical donors for biological applications, *J. Inorg. Biochem.* 118, 187–190, 2013). Shown below is the structure of 3-morpholinosydnonimine(linsidomine). This is frequently referred to as SIN-1 in the literature. It has in vitro use as nitric oxide donor (Polte, T., Oberle, S., and Schröder, H., The nitric oxide donor SIN-1 protects endothelial cells from tumor necrosis factor-α-mediated cytotoxicity: Possible role for cyclic GMP and heme oxygenase, *J. Mol. Cell. Cardiol.* 29, 3305–3310, 1997) although it will oxidize cysteine and nitrate tyrosine in addition to a nitrosylation reaction (Daiber, A., Daub, S., Bachschmid, M. et al., Protein tyrosine nitration and thiol oxidation by peroxynitrite-strategies to prevent these oxidative modifications, *Int. J. Mol. Sci.* 14, 7542–7570, 2013). 3-Morpholinosydnonimine (SIN-1) is used with varying degrees of success as a nitric oxide donor (Wegner, H.E., Knispel, H.H., Meier, T. et al., Nitric oxide donor, linsidomine chlorohydrate (SIN-1), in the diagnosis and treatment of erectile dysfunction: Critical appraisal and review of the literature, *Int. Urol. Nephrol.* 27, 621–628, 1995; Messin, R., Fenyvesi, T., Carrerr-Bruhwyler, F. et al., A pilot double-blind randomized placebo-controlled study of molsidomine 16 mg once-a-day in patients suffering from stable angina pectoris: Correlation between efficacy and over time plasma concentrations, *Eur. J. Clin. Pharmacol.* 59, 227–232, 2003). Shown below to the right is propylamine propylamine NONOate [(Z)-1-[N-(aminopropyl)-N-(n-propyl)amino diazen-1,2-diolate] commonly referred to as papanonoate or papa nonoate, which is an effective source of nitric oxide (Nicolay, J.P., Liebig, G., Niemoeller, O.M. et al., Inhibition of suicidal erythrocyte death by nitric oxide, *Pflufers Arch.* 456, 293–305, 2008; Rivera-Tirado, E., López-Casillas, M., and Wesdemiotis, C., Characterization of diazeniumdiolate nitric oxide donors (NONOates) by electrospray ionization mass spectrometry, *Rapid Commun. Mass Spectrom.* 25, 3581–3586, 2011). At the bottom left is S-nitrosocysteine (Riego, J.A., Broniowska, K.A., Kettenhofen, N.J., and Hogg, N., Activation and inhibition of soluble guanyl cyclase by S-nitrosocysteine: Involvement of amino acid transport system L, *Free Radic. Biol. Med.* 47, 269–274, 2009; Ratanatwanate, C., Chyao, A., and Balkus, K.J., Jr., S-Nitrosocysteine-decorated PbS QDs/TiO$_2$ nanotubes for enhanced production of singlet oxygen, *J. Am. Chem. Soc.* 133, 3492–3497, 2011; Hickok, J.R., Vasudevan, D., Thatcher, G.R., and Thomas, D.D., Is S-nitrosocysteine a true surrogate for nitric oxide?, *Antioxid. Redox Signal.* 17, 962–968, 2011). S-Nitrosoglutathione is not shown as the S-nitrosocysteine residue would be the active source of nitric oxide. To the right below is SNAP (Stasko, N.A., Fischer, T.H., and Schoenfisch, M.H., S-Nitrosothiol-modified dendrimers as nitric oxide delivery vehicles, *Biomacromolecules* 9, 834–841, 2008; Um, H.C., Jang, J.H., Kim, D.H. et al., Nitric oxide activates Nrf2 through S-nitrosylation of Keap1 in PC12 cells, *Nitric Oxide* 25, 161–168, 2011). Shown at the bottom is a method for converting an S-nitroso thiol to a stable thioether (Zhang, D., Devarie-Baez, N.O., Pan, J. et al., One-pot thioether formation from S-nitrosothiols, *Org. Lett.* 12, 5674–5676, 2010).

acidic amino acid (glutamic or aspartic acid) following a cysteine residue in the primary structure enhanced the nitrosylation of that cysteine residue. This concept was extended by other investigators[79,80] but other investigators have shown that there are other structural factors, which may have an important role in the susceptibility of a cysteine residue to nitrosylation.[81] Ascenzi and coworkers[79] do suggest that microenvironment rather than specific sequence elements is of importance in the susceptibility of a specific cysteine residue to nitrosylation. Thus, in this situation, and others, such as the histidine and lysine residues in pancreatic ribonuclease or the active site residues in serine proteases, proximity in primary structure is not critical for interaction. Foster and coworkers[82] also showed that protein structure, nitric oxide donor stereochemistry, and allosteric effectors all influence the site of S-nitrosylation. Recent computational analysis by Talipov and Timerghazin[83] has validated the importance of the protein microenvironment for the stability of S-nitrosothiols emphasizing the importance of interactions between the S-nitroso group and polar residues in proteins. It is apparent that there is considerable variability[40,86-88] in the stability of S-nitroso groups in proteins resulting from local structural factors. Zhang and coworkers[87] have described a reaction that converts the S-nitroso function to a stable thioether function (Figure 4.2). Giustarini and coworkers[88] reported that papain, creatine phosphokinase, and glyceraldehyde-3-phosphate dehydrogenase were both S-nitrosylated and S-thiolated by S-nitrosoglutathione, while only S-nitrosylation was observed on the reaction of S-nitrosoglutathione with bovine serum albumin, yeast alcohol dehydrogenase, or rabbit muscle skeletal actin. The modified proteins were treated with 0.5 mM glutathione to evaluate the intracellular stability of the various derivative proteins. With the exception of actin, the various nitrosylated proteins were rapidly denitrosylated in the presence of reduced glutathione. The S-glutathionyl modification was more stable than the S-nitroso derivatives to reduction by glutathione. Sadidi and coworkers[89] reported that either peroxynitrite or nitrogen dioxide inactivated tyrosine hydroxylase and the inactivation was associated with the nitration of tyrosine residues. However, if the reaction with either peroxynitrite or nitrogen dioxide is performed in the presence of cysteine or glutathione, inactivation is observed but with nitration of tyrosine; rather, S-thiolation occurs, which is reversed by dithiothreitol.

The S-nitrosothiols have varying stabilities; the $t_{1/2}$ for S-nitrosocysteine in 10 mM sodium phosphate, pH 7.4, at 37°C is 120 s, 70 min for S-nitrosohomocysteine, and 5.5 h for S-nitrosoglutathione.[57] Metal ions have a major effect on the stability of S-nitrosothiol. Hydrolysis cleavage of S-nitrosothiols in the presence of mercuric ions provided the basis for an early assay for thiols,[53] which was used for the measurement of S-nitrosylation[90] but has been replaced by biotin switch technologies.[91] Park[56] reported a second-order rate constant of 3×10^{-4} M^{-1} s^{-1} for decomposition of S-nitrosoglutathione in 0.01 M phosphate, pH 7.4/37°C; the second-order rate constant for S-nitrosocysteine was 0.11 M^{-1} s^{-1} and 0.3 M^{-1} s^{-1} for S-nitrosocystamine. Park[56] also reported that S-nitrosoglutathione reacted with cysteine to yield oxidized glutathione, cystine, and the mixed disulfide between cysteine and glutathione. A rate constant was not provided for this reaction(s), which was described as extremely fast. Park also showed that S-nitrosoglutathione inactivated yeast alcohol dehydrogenase in a rapid reaction; the inactivation was reversed with 2-mercaptoethanol.

A technique described as *biotin switch* technique (BST) (Figure 4.3) has been described for the identification of S-nitrosocysteinyl residues in proteins.[92-97] Jaffrey and coworkers[92] introduced this analytical approach in 2001. The approach used the modification of available sulfhydryl groups in S-nitrosylated proteins with S-methyl methanethiosulfonate (MMTS) (see Chapter 7); ascorbate is then used to reduce the S-nitrosocysteine to cysteine, which is then modified via disulfide exchange with N-[6-(biotinamido)hexyl]-3'-(2'-pyridyldithio)propionamide (Figure 4.4). The biotin-labeled proteins were then isolated by affinity chromatography on streptavidin–agarose. The replacement of biotin-containing reagent with a tandem mass tag reagent (Figure 4.4)[98] in the biotin switch assay permits quantitative determination[99] of S-nitrosylation. The use of a tandem mass tag technology[99] permits quantitative analysis of S-nitrosylation. There have also been other approaches to the identification of S-nitrosylated proteins including the capture of proteins denitrosylated by ascorbate[100] using a matrix-bound pyridyl disulfide.[101] In a related approach, Foster and coworkers[82] nitrosylated a

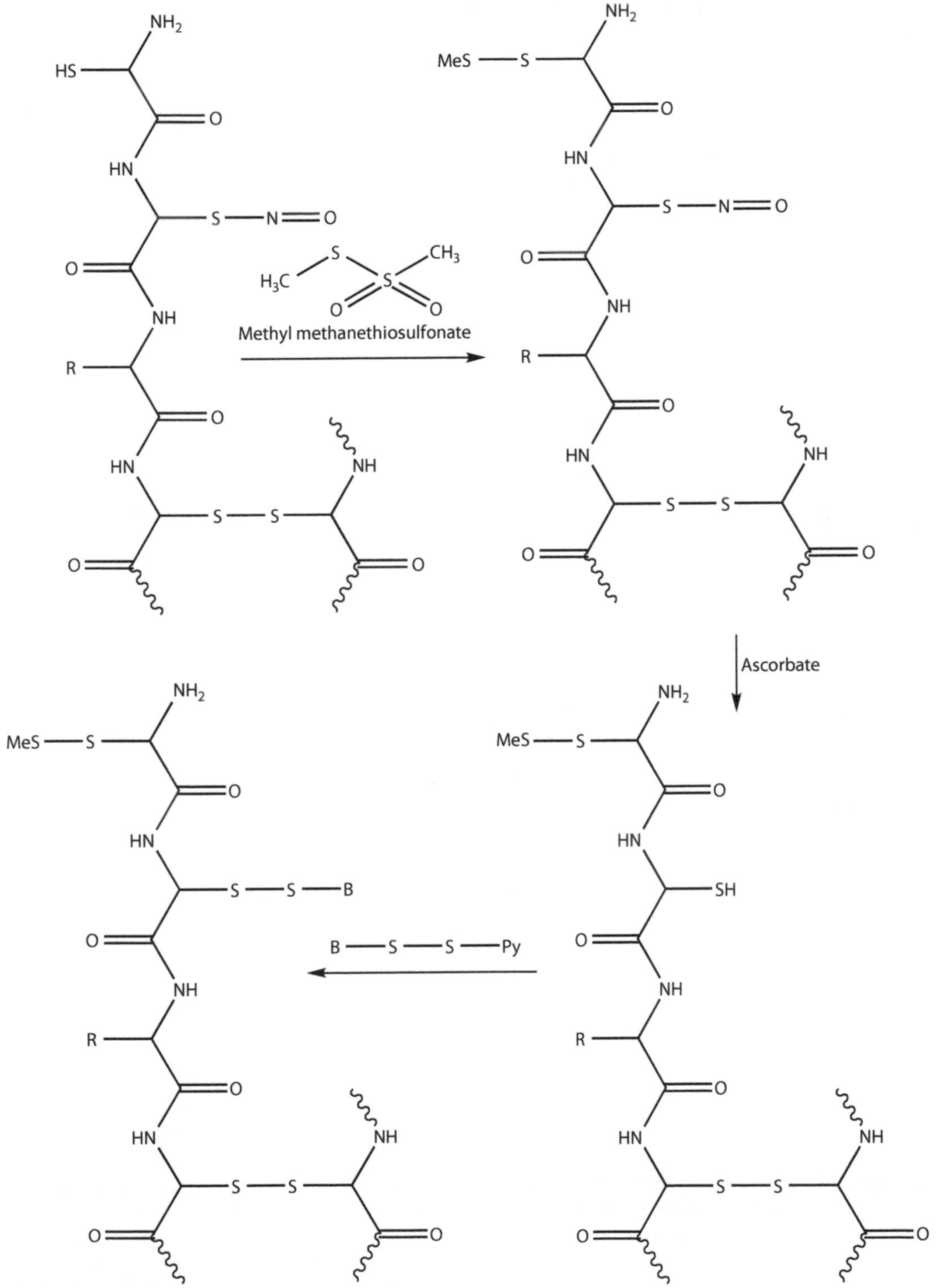

FIGURE 4.3

(continued)

FIGURE 4.3 (continued) BST for identifying protein *S*-nitrosylation. Shown is the method for the labeling of *S*-nitrosocysteine residues in proteins. The free sulfhydryl groups in *S*-nitrosylated proteins are blocked with methyl methanethiosulfonate. The *S*-nitrosyl groups are reduced with ascorbate to yield a free sulfhydryl group; in this scheme, ascorbate is suggested not to reduce disulfide bonds. The newly liberated sulfhydryl is modified with *N*-[6-(biotinamido)hexyl]-3′-(2′-pyridyldithio)propionamide(B-S-S-Py) to yield a biotin-labeled protein, which can be isolated by affinity chromatography on streptavidin–agarose or other separation technology including microplates (see Jaffrey, S.R., Erdjument-Bromage, H., Ferris, C.D. et al., Protein *S*-nitrosylation: A physiological signal for neuronal nitric oxide, *Nat. Cell Biol.* 3, 193–197, 2001; Uehara, T. and Nishiya, T., Screening systems for the identification of *S*-nitrosylated proteins, *Nitric Oxide* 25, 108–111, 2011).

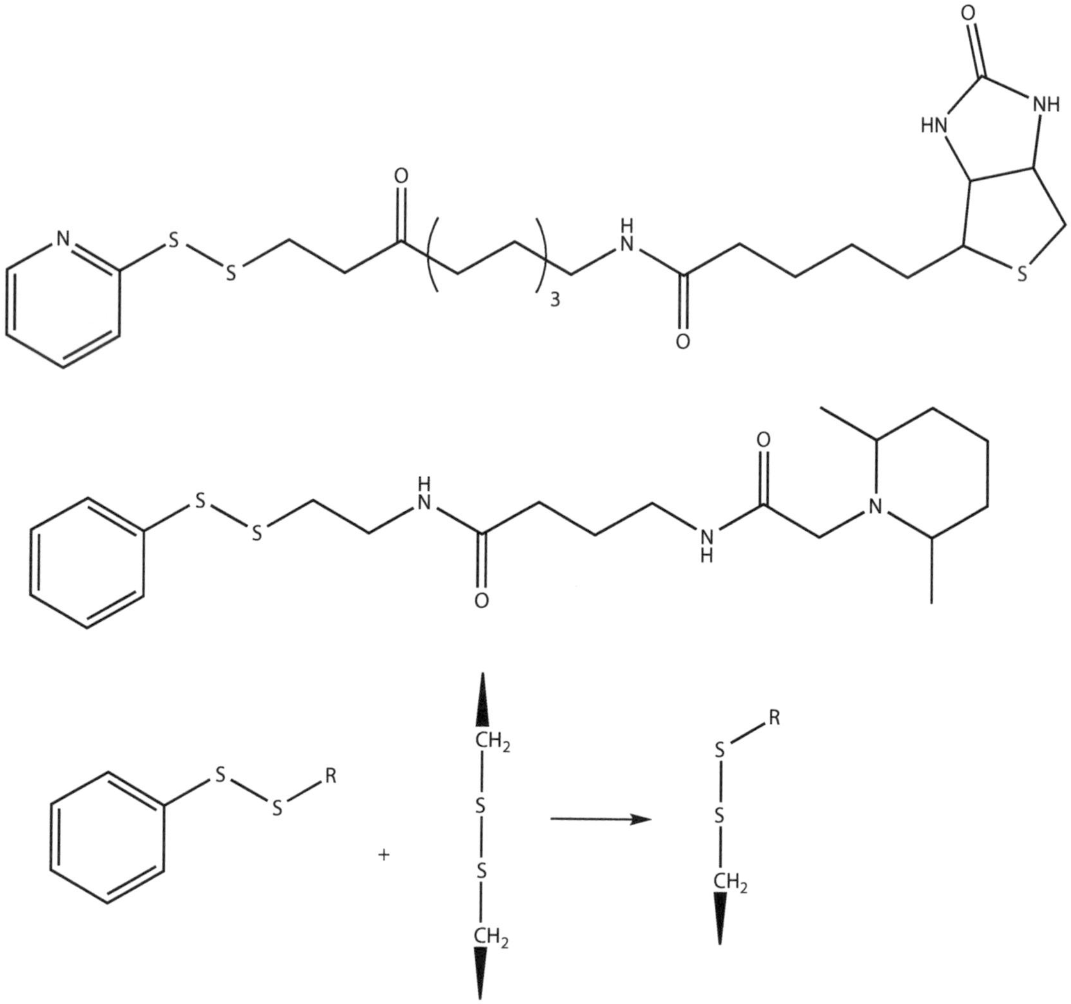

FIGURE 4.4 Reagent used for the BST. Shown are several tagging reagents used for the detection of *S*-nitrosylated proteins. *N*-[6-(biotinamido)hexyl]-3′-(2′-pyridyldithio)propionamide(B–S–S–Py) is shown at the top. This reagent was used in the original description of the biotin switch assay (Jaffrey, S.R., Erdjument-Bromage, H., Ferris, C.D. et al., Protein *S*-nitrosylation: A physiological signal for neuronal nitric oxide, *Nature Cell Biol.* 3, 193–197, 2001). A tandem mass tag reagent for labeling sulfhydryl groups that were nitrosylated is shown below (Murray, C.I., Uhrigshardt, H., O'Meally, R.N. et al., Identification and quantification of *S*-nitrosylation by cysteine reactive tandem mass tag switch assay, *Mol. Cell. Proteomics* 11, M111.013441, 2012). The reaction with the dithiopyridyl function is shown below the tandem mass tag; R indicates either the biotin-containing moiety or the mass tag moiety.

protein microarray with either *S*-nitrosocysteine or *S*-nitrosoglutathione (both prepared by the reaction of sodium nitrite in 0.5 M HCl). The microarray is then reduced with ascorbate, reacted with methyl methane thiosulfonate, and subsequently with *N*-[6-(biotinamido)hexyl]-3′-(2′-pyridyldithio) propionamide; the biotinylated proteins were detected by an antibody to biotin. The system was validated with a commercial protein microarray (Invitrogen Protoarray® for kinase substrate identification). There was some concern that the ascorbate would reduce disulfide bonds providing a spurious signal from the *S*-nitrosylated protein resulting from sulfhydryl groups not associated with *S*-nitrosylation.[102] Forrester and coworkers[103] subsequently showed that sunlight affected the ascorbic acid reduction of nitrosylated proteins resulting in artifactual signals. These investigators argue that the mechanism of ascorbate reduction involving a transnitrosylation reaction yielding *O*-nitroascorbate confers specificity to the reduction of *S*-nitrosothiols. The issue of ascorbate reduction of disulfides was not entirely settled by this rather nice piece of chemistry. Giustarini and coworkers[104] showed that ascorbate could reduce disulfide bonds in some low molecular weight disulfide and some protein mixed disulfides but no protein disulfides. These investigators did show a large effect of sunlight on the rate of loss of *S*-nitrosothiol in bovine serum albumin. Ascorbate as a reducing agent is discussed in greater detail in Chapter 6. Despite these concerns about specificity, ascorbate continues to be used for the biotin switch assay.[105] Komatsubara and coworkers[105] used proteomic technique to evaluate the effect of 1-methyl-4-phenylpyridinium (MPP+) on *S*-nitrosylation in neuroblastoma cells. In addition to 2D electrophoresis with immunoblotting, *S*-nitrosylated proteins were reduced with ascorbate and subsequently isolated with a pyridyl disulfide matrix as described by Forrester and coworkers.[100] While not widely used, this variation on the biotin switch assay has been referred to as SNO-RAC as an acronym. There are a number of approaches to replace ascorbate including sinapinic acid,[106] glutaredoxin-1,[91] triphenylphosphine (TPP),[5] and, most recently, methylhydrazine.[107] The BST takes advantage of the relative susceptibility of the *S*-nitroso group to reduction. However, as noted by Forrester and coworkers,[103] the direct one-electron reduction of *S*-nitrosothiols by ascorbate is not a thermodynamically favorable reaction based on earlier studies based on the earlier work of Hou and coworkers on the electrochemistry of *S*-nitrosothiol reduction.[108] More recent studies by Peng and Meyerhoff[109] may require a further evaluation of whether the ascorbate reduction is, in fact, thermodynamically reasonable. Regardless, Forrester and coworkers[103] did propose a novel mechanism for the reduction of *S*-nitrosothiols by ascorbate involving the nitrosylation of ascorbate as an intermediate in the process. The author notes from his personal experiences that the chemistry of ascorbate can be perverse. Other work by Wang and coworkers[110] showed that *contaminating* copper was required for the ascorbate reductions of *S*-nitrosothiols. Wang and coworkers[110] also used *N*-ethylmaleimide instead of methyl methanethiosulfonate for blocking free sulfhydryl groups in nitrosylated proteins and a cyanine dye for detection of the free sulfhydryl obtained from reduction of the *S*-nitrosothiol. The biotin switch approach has been used to define the *S*-nitroso proteome.[111,112] Huang and coworkers[111] added SNAP to endothelial cells and used the BST to identify *S*-nitrosylated proteins by 2D electrophoresis (the electrophoresis was performed in the absence of reducing agent to preserve the biotin labeling). At least 89 nitrosylated proteins were detected by Western blot analysis (streptavidin–horseradish peroxidase). It was possible to excise 28 positive spots from the gel and identified by LC/tandem mass spectrometry (MS-MS). Chen and coworkers[112] used iodoacetamide instead of methyl methanethiosulfonate for blocking sulfhydryl groups that were not nitrosylated, used an iodoacetyl derivative of biotin (PEO-iodoacetyl-biotin) that permitted irreversible modification of the cysteinyl residue after ascorbate reduction, and reduced disulfide bonds prior to proteolysis to obtain peptides for affinity purification/LC-MS-MS. A related approach used labeling with a fluorophore after ascorbate reduction.[113] Notwithstanding the lack of understanding on the chemistry of the biotin switch, it is noted that the biotin assay switch concept has been used for the identification of cysteine sulfenic acid[114] and persulfide modifications[115] in proteins.

While it is unlikely that *S*-nitrosylation of proteins will be a useful technique, such as those described in Chapter 7, for modification of cysteine residues, the chemistry is of considerable interest because of the physiological importance of the modification. The study of the physiological importance of S-nitrosylation has provided some interesting information. N-Acetylcysteine is used therapeutically for a variety of conditions with a varied degree of success.[116–119] There is poor (and that is being generous) understanding of the mechanism(s) of action of N-acetylcysteine, but it is likely that the antioxidant potential would be an important part of any mechanism of action. To this end, Samuni and coworkers[120] have reviewed the reaction of a number of substances suggested to have a role in oxidative stress. The rate of reaction of nitroxyl (HNO) with N-acetylcysteine (5×10^5 M^{-1} s^{-1}) was more rapid than that of peroxynitrite (4.2×10^2 M^{-1} s^{-1}) or hydrogen peroxide ($\sim 5 \times 10^{-1}$ M^{-1} s^{-1}) but slower than nitrogen dioxide radical ($\cdot$NO$_2$) ($\sim 10^7$ M^{-1} s^{-1}). Whiteside and coworkers[121] reported on the formation of the S-nitroso derivative of an aminothiol drug, Ethyol® (WR-1065; [N-mercaptoethyl]-1-3-diaminopropane) and comparison of the properties of this compound with related S-nitrosothiol compounds. The stability of the various derivatives in phosphate-buffered saline (PBS) (pH 7.4) at 25°C was determined by measured decrease in absorbance at 334 nm. A $t_{1/2}$ of 14.5 h was found for S-nitroso-N-acetylcysteine, while a $t_{1/2}$ of 8.74 h was found for the S-nitrosoglutathione, 1.51 h for WR-1065, and 0.64 h for S-nitrosocysteine. The various S-nitrosothiol derivatives were evaluated for inactivation of cathepsin H; the S-nitroso derivative of WR-1065 was the most potent inactivator (8.33×10^{-1} M^{-1} min^{-1}), while S-nitrosoglutathione was less effective (9.2×10^{-2} M^{-1} min^{-1}). Yu and coworkers[122] reported that S-nitrosoglutathione and SNAP (S-nitrosopenicillamine) were effective in the intracellular nitrosylation of PTEN (protein with sequence homology to tensin) but only in the presence of cysteine. S-nitrosocysteine was effective without added cysteine. It was suggested that S-nitrosoglutathione and SNAP could not pass through the membrane; transnitrosylation of exogenous cysteine was required to produce S-nitrosocysteine, which could pass through the membrane. S-Nitrosocysteine can readily cross a cell membrane[122] as there is a specific transport mechanism.[123–125] Yu and workers[122] also showed the PTEN was inactivated by either S-nitrosocysteine, S-nitrosoglutathione, or SNAP (25°C/10 min/20 mM Tris, pH 6.7); the inactivation was reversed with dithiothreitol. The S-nitrosylation reaction was also associated with a loss of immunoreactivity (Western blot) with a monoclonal antibody directed toward the C-terminal region of the protein; as with enzymatic activity, immunoreactivity was recovered on reaction with dithiothreitol. There was only slight loss of immunoreactivity with nitrosylation when a polyclonal antibody was used in the Western blot. Oxidation with hydrogen peroxide also resulted in a loss of enzymatic activity but no change in immunoreactivity. It has been demonstrated that S-nitrosylation of PTEN affects Cys83 while Cys124 is modified with H$_2$O$_2$ resulting in disulfide bond formation with Cys71.[126]

The S-nitrosocysteine residues are reactive similar to cysteine sulfenic acid in readily forming internal disulfide bonds and mixed disulfide bonds.[17,55,117,127–131] Cheng and coworkers[132] modified the two nonconserved cysteine residues in glypican-1 with sodium nitroprusside in the presence of cupric ions (CuCl$_2$). Glypican-1 is a proteoglycan substituted with heparan sulfate and attached to the cell surface via a glycosylphosphatidyl inositol.[133] The glypicans are thought to have an important role in cell growth including tumor growth.[134,135] Cheng and coworkers[132] reported a reaction between the S-nitrosylated cysteine and an unsubstituted amino group of a glucosamine residue in the heparan sulfate chain resulting in diazotization and chain cleavage (Figure 4.5). These investigators also suggest that the proximity of the glucosamine amino group promotes ionization of the thiol group to the thiolate anion, which is considered more susceptible to nitrosylation. This effect of neighboring group on the ionization is seen with other proteins[136,137] and also discussed in Chapter 7. Ding and coworkers[138] have previously reported the nitrosylation of glypican-1 cysteinyl residues in the presence of cupric ions; in the process of nitrosylation, cupric ions are reduced to cuprous ions. The nitric oxide donor is not established as the nitric oxide is derived in situ from arginine. Treatment of the S-nitrosylated protein with ascorbate releases nitric oxide, which resulted in the cleavage of the heparan sulfate chain; however, the mechanism of cleavage was not understood. The role of copper ions in S-nitrosylation was referenced earlier in the work of Wang and coworkers[110] who noted a requirement for cuprous ion (CuCl) for the ascorbate reduction of S-nitrosylated human serum albumin. The role of copper ions, cupric and cuprous, in the formation and decomposition

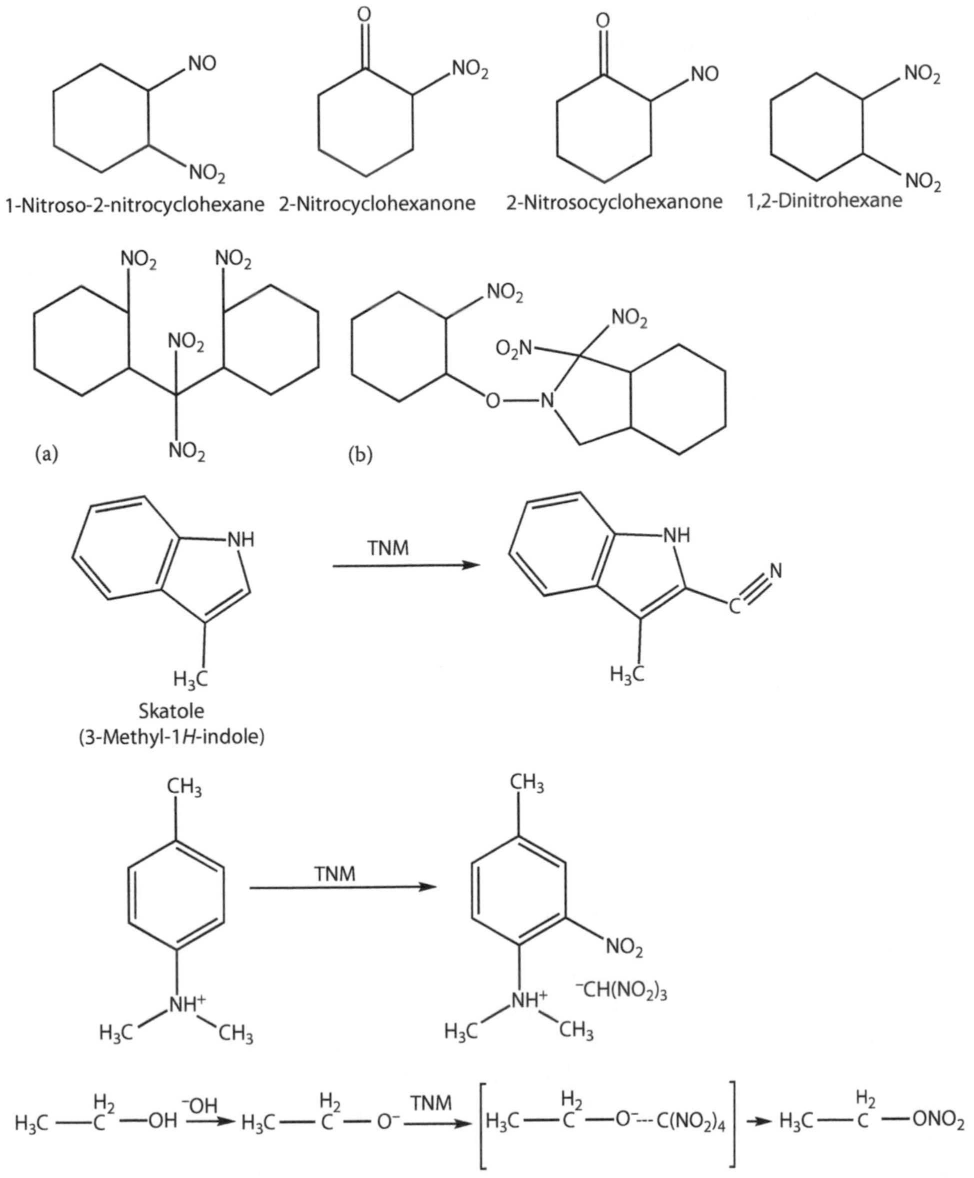

FIGURE 4.5 Some adducts obtained from the reaction of tetranitromethane with olefins. Shown at the top are structures of some derivatives obtained from the reaction of TNM with cyclohexene and more complex product (a) (Bradshaw, R.W., The reaction of tetranitromethane with olefin, *Tetrahedron Lett.* (46), 5711–5716, 1966). Shown at the right is an alternative structure (b) for a product of reaction with TNM and cyclohexene (Torssell, K., The structure of the tetranitromethane-olefin adduct, *Acta Chem. Scand.* 21, 1392–1393, 1967). Shown below those structures is the product of the reaction of tetranitromethane and skatole (Spande, T.F., Fontana, A., and Witkop, B., An unusual reaction of skatole with tetranitromethane, *J. Am. Chem. Soc.* 91, 6199–9200, 1969). Shown below is the reaction of 4-dimethylaminotoluidine to form 3-nitro-4-dimethylaminotoluidine (Schmidt, E. and Fischer, H., Zur kenntnis des tetranitro-methans. II. Mitteilung: Tetranitro-methan als nitrierungsmittel (I), *Ber. Dtsch. Chem. Ges.* 53, 1528–1537, 1920; Isaacs, N.S. and Abed, O.H., The mechanism of aromatic nitration by tetranitromethane, *Tetrahedron Lett.* 23, 2799–2802, 1982). At the bottom is shown a mechanism for the reaction of TNM with ethoxide ion to yield ethyl nitrate. The proposed mechanism involves the formation of a radical anion–cation pair (Walters, S.L. and Bruice, T.C., Reaction of tetranitromethane. II. The kinetics and products for the reactions of tetranitromethane with inorganic ions and alcohols, *J. Am. Chem. Soc.* 93, 2269–2282, 1971).

of nitric oxide donors and *S*-nitrosothiol products has been of interest for some time. Williams[57] observed that micromolar concentrations of cupric ions accelerated the rate of decomposition of SNAP. The product of the decomposition of reaction was a disulfide. Williams also suggested that cupric ion was reduced to cuprous ion in the reaction but recycled to cupric ion. Ding and coworkers[138] presented a mechanism where the reduction of cupric ions to cuprous ions is required for the formation of *S*-nitrosothiol from nitric oxide. It is important to note the low concentration of cupric ion required for the effect on SNAP composition.[57] Gu and Lewis[139] extended understanding on the effect of pH and metal ions on the decomposition of *S*-nitrosothiols using *S*-nitrosocysteine. Maximum rate of decomposition was observed at neutral pH with increased stability at either acid or basic pH. The rate of decomposition was much more rapid in a low-purity phosphate buffer (≥98%) than in a high-purity (≥99.99%) buffer. The addition of a chelating agent (Desferal®) to the low-purity buffer decreased the rate of decomposition, while the addition of 20 ppm ferrous ion increased the rate of decomposition in the high-purity buffer. Gu and Lewis[139] also observed that the rate of decomposition of *S*-nitrosocysteine was more rapid in a citric acid–Tris buffer than in a glycine buffer and ascribed the difference to metal ion contamination. Wang and coworkers[110] have also addressed the issue of metal ion contamination. They observed that the inclusion of EDTA in their biotin switch assay system reduced biotin labeling while, as noted earlier, the inclusion of cuprous ion increased biotin labeling. They also noted that treated buffer with Chelex® resin also decreased biotinylation. Dicks and coworkers[140] had earlier shown the Cu^{+1}-mediated decomposition of *S*-nitrosothiols varied with the structure of the *S*-nitrosothiol and the effect of added thiol. A theoretical study by Toubin and coworkers[141] supported a role for cuprous ion in the decomposition of *S*-nitrosothiols. While cuprous ions may promote decomposition of *S*-nitrosothiols, there are, together with the studies cited earlier, data to support a role for cuprous ion in the *S*-nitrosylation of proteins by nitric oxide donors.[142,143]

NITRATION

The chemical nitration of proteins dates at least to 1771 with the use of nitric acid (acid of nitre) to dye silk.[144] Early work on the nitration of proteins used a mixture of nitric acid and sulfuric acid[145] or nitric acid alone.[146] Johnson and Kohmann[145] reviewed the early history of the nitration or proteins including the origin of the term, xanthoprotein, to describe the product of the nitration of proteins. As shown in Figure 4.1, the sulfuric acid-catalyzed electrophilic nitration of tyrosine by nitric acid also provides the 3′-derivative (*o*-nitrophenolic). The structure of 3′-nitrotyrosine obtained with nitric acid was established by Johnson and Kohmann in 1915[145]; these investigators later established the structure of 3,5-dinitrotyrosine.[147] Wormall referred to proteins obtained with nitric acid as xanthoproteins, while nitrated proteins obtained with TNM were referred to as nitrated proteins. Wormall[146] noted that new epitopes were found in both his nitrated proteins and xanthoproteins. Xanthoproteins were used as substrates for the detection of proteolytic enzymes.[148,149] The reaction of nitric acid with proteins to yield a yellow color, the xanthoproteic/xanthoprotein reaction, has been used for the estimation of protein.[150–153] As with the other nitrating agents discussed in the succeeding text, the yellow color from the treatment of proteins with nitric acid is a product of tyrosine nitration.[145,146,154–156] The reader is directed to a review on nitration mechanisms by nitric acid.[157,158]

TNM is the reagent used mostly for the in vitro nitration of proteins.[159–161] Peroxynitrite, derived from nitric oxide, has been identified as in vivo reagent for nitration and nitrosylation of proteins[162]; peroxynitrite can be used for in vitro studies.[163–166] Both TNM and peroxynitrite can oxidize cysteine thiol groups[163,164] as well as nitrating tyrosine. In addition, peroxynitrite can undergo homolytic cleavage yielding superoxide and nitric oxide[167]; hydroxyl radical is also a decomposition product of peroxynitrite together with carbonate and nitrogen dioxide radicals.[168] Peroxynitrite at acid pH (peroxynitrous acid) reacts preferentially with thiol groups to yield the *S*-nitroso derivative proceeding to the disulfide, while reaction at alkaline pH yields various oxidation and degradation products including a disulfide.[167,169,170] TNM also oxidizes sulfhydryl groups[171]; it is possible to protect sulfhydryl groups prior to reaction with TNM.[172,173] While the oxidative reactions observed with peroxynitrite are likely of physiological interest, the focus of the current section is nitration. Tyrosine and tryptophan are the two sites of nitration in proteins, and there is considerable discussion of the modification in Chapters 11 (tyrosine) and 12 (tryptophan). Thus, the remaining part of this chapter will focus first on the chemistry of TNM and finally on the chemistry of peroxynitrite [oxidoperoxonitrate (1-)]/peroxynitrous acid (hydrogen oxoperoxonitrate).

There was early work on the use of TNM for nitration of organic compounds such as ethyl nitrate from potassium ethylate[174] as well as the generation of yellow color from reaction with aromatic hydrocarbons.[175] Early work on TNM recognized the value (and danger) of TNM as an explosive.[176,177] Other early work on TNM noted the reaction with unsaturated aliphatic compounds with interesting results.[178–183] The early work was qualitative in nature[178–181] in reporting the development of a yellow color on reaction of TNM with various unsaturated organics including carvene (1-methyl-4-prop-1-en-2ylcyclohexene) and trimethylbutylene.[161–164] Lagercrantz and Yhland[182] reported that electron spin resonance detected radicals in mixtures of olefins and TNM in the presence of light. Bradshaw[183] examined the reaction of TNM with cyclohexene and obtained support for 1-nitroso-2-nitrocyclohexane as a product (Figure 4.5). Spande and coworkers[184] reported that the reaction of TNM with skatole yielded 2-cyano-3-methylindole. A chromogenic reaction of TNM with organic compounds has been used to determine the presence of unsaturation.[185] Atassi reported that TNM reacted with the vinyl groups of heme resulting in a nitrated derivative (Figure 4.1).[186] There is some suggestion that heme can also be nitrated by nitrite or peroxynitrite.[187] The nitration of heme by acidified sodium nitrite has been reported by other investigators.[188] Heme also appears to have a role in peroxynitrite nitration of proteins.[189]

The current use of TNM with proteins dates back to the work by Wormall in 1930,[146] which, in turn, was based on earlier observations on the nitration of various organic compounds with TNM by Schmidt and Fischer in 1920.[190] Tetranitromethane (MW 296) is relatively insoluble (immiscible; log $P = -2.05$[191,192]) and is a liquid (density 1.64[20]) at 25°C (MP 13.8°C) that can be prepared by the reaction of acetic anhydride and nitric acid.[176,193,194] TNM is usually introduced in aqueous reactions as an ethanolic solution; neat TNM is approximately 8.4 M so it is no surprise that stock solutions in ethanol are 0.84 M representing a 1:10 dilution. A consideration of the literature (see Chapter 11) has not revealed mention of a solubility problem in reaction mixtures including one study with 1.65 mM.[195] It is not uncommon to wash the neat TNM with water to remove acid resulting from the decomposition of TNM to form nitric acid and nitroform.[196,197] Early work on the decomposition of TNM in base showed that the major products were nitric acid and trinitromethane (nitroform) (Figure 4.1).[190] Trinitromethane was also a product of the reaction of TNM with potassium ethoxylate to yield ethyl nitrate. In the absence of contrary information, it is accepted that the reaction of TNM with tyrosine results in 3-nitrotyrosine and trinitromethane. Bruice and coworkers[198] studied the reaction of TNM with several phenols and proposed a mechanism for nitration of phenols that involved the generation of a nitrogen dioxide and nitroform anion with reaction of nitrogen dioxide with the phenyl ring. As part of their mechanism, Bruice and coworkers[207] propose the formation of a charge-transfer complex between TNM and the phenoxide ion prior to generation of nitrogen dioxide (a free radical) and trinitromethane (nitroform). Substitution of the phenyl ring by nitrogen dioxide was *ortho* except with phenol where substitution was both *ortho* and *para*. The yield of trinitromethane anion was approximately 100% in the reaction of TNM with the various phenol derivatives. The yield of nitrated product was approximately 30% for *p*-cresol and *p*-chlorophenol and 20% for *p*-cyanophenol. Reaction of TNM with phenol provided 8% yield for *o*-nitrophenol and 13% yield for *p*-nitrophenol. A nitrated product was not obtained in the reaction of TNM and 2,4,6-trimethylphenol, but there was 100% recovery of trinitromethane and nitrite. These investigators note that measurement of trinitromethane (nitroformate) product cannot serve as measurement of protein nitration without validation. Also, the absorbance maximum of nitroformate confounds the spectral measurement of nitrophenol formation unless the nitroformate is removed. The photochemical reaction of TNM and aromatic compounds has also been shown to involve complex formation between the aromatic compound and TNM.[199–205] Eberson and Hartshorn[201] reviewed their work on the reaction of TNM with various aromatic compounds in 1998. They suggest that TNM and an aromatic compound (e.g., naphthalene) form a complex that when irradiated yields an aromatic cationic radical and a tetranitromethane anion radical. The tetranitromethane anion radical fragments to yield nitrogen dioxide and a trinitromethide anion, which then form a *triad* complex with the aromatic cationic radical. The trinitromethide anion reacts with aromatic cationic radical forming an intermediate, which then reacts with nitric oxide forming nitro/trinitromethyl adducts (Figure 4.1 bottom). There is considerable interest in the photochemistry of the reaction of TNM with aromatic compounds.[203–206] Other observations of interest include the observation that nitrogen dioxide nitrates aerosolized protein.[207] Capellos and coworkers[208] observed the formation of nitrogen dioxide and nitroformate anion on pulse radiolysis at 248 nm.

The reaction of TNM to form trinitromethane is also used to detect superoxide radical by measurement of nitroformate anion [$C(NO_2)_3^-$] at 350 nm.[209] The spectral measurement of nitroformate was used to determine the extinction coefficient of the hydrated electron at 578 nm.[210] In this study, pulse radiolysis was used to produce hydrated electrons, which then reduced TNM to trinitromethane; trinitromethane [$HC(NO_2)_3$] is considered a strong acid and readily dissociates to nitroformate [trinitromethane anion; $C(NO_2)_3^-$].[211] Rabani and coworkers[210] compared the initial absorption at 578 nm to the final absorbance at 350 nm (nitroformate) to obtain an extinction coefficient (10.6×10^3 M^{-1} cm^{-1}) for the hydrated electron. The reaction of hydrated electrons with TNM has been studied by other investigators[212] who demonstrated that the rate of reaction is rapid (~10^{10} M^{-1} s^{-1}) permitting the reduction of TNM to be used for the measurement of hydrated electrons.[213–215] Christen and Riordan[216] reported that TNM reacted with carbanions. Malhotra and Dwivedi[217] subsequently showed that isocitrate lyase from germinating castor seed endosperm catalyzed the reaction of succinate or isocitrate with TNM in a reaction product in trinitromethane anion presumably through reaction with a carbanion derived from the succinate/isocitrate substrate. In a related study, Jewett and Bruice[218] described the reaction of TNM with a variety of *pseudo acids* such as acetonitrile (methyl cyanide), dimethylsulfone, acetone, and nitroethane. The rate of reaction is dependent on the pKa of the pseudo acid, and the rapid rate (10^{10} M^{-1} min^{-1}) is interpreted to suggest the involvement of a radical species. In other work, Walters and Bruice[219] also reported results from the reaction of TNM with alcohols and inorganic ions. With straight-chain aliphatic alcohols, such as ethanol, the reaction is suggested to proceed via the ethoxide to a complex between TNM and ethoxide resulting in ethylnitrate. There was no reaction with trifluoroethanol or N-acetylglycinamide. Reaction with N-acetylserineamide was not enhanced by a base, and analysis of products showed approximately 90% trinitromethane and 60%–100% nitrite. The failure of base to enhance the reaction suggests that the alkoxide ion is not the reactive species. While there may be other reaction of TNM with protein such as peptide backbone, reaction with hydroxyl groups of serine or threonine is not supported by the results obtained by Walters and Bruice.[219] The reaction of TNM with t-butyl alcohol gave anomalous results. The rate of reaction (measured by decrease in TNM absorbance at 350 nm) was more rapid than that with other alkoxides, and the yield of trinitromethane was approximately 50% and the yield of nitrite was close to 200%.

It is clear from a consideration earlier that the reaction of TNM with proteins can be complex and form reaction products with functional sites other than tyrosine in proteins. However, it is also clear if MS allows a more careful analysis of modified proteins than possible during the time of major use of TNM for the modification of proteins.[220]

There is considerably more current work on peroxynitrite [$^-$ONOO; oxoperoxonitrate(1$^-$)] than TNM but has seen much less use in organic chemistry. While oxoperoxonitrate(1$^-$) is the more accurate chemical name, peroxynitrite will be used in the current work as will be the use of nitric oxide for NO· instead of nitrogen monoxide or oxidonitrogen(·).[221,222] Peroxynitrite is formed in vivo from nitric oxide and superoxide anion[223] (Figures 4.1 and 4.5); peroxynitrite may also be formed from nitroxyl anion and superoxide[224] and perhaps other sources (Figure 4.5). For the purposes of in vitro investigation, peroxynitrite can be synthesized by a number of methods[224–233] as well as being available from several commercial sources. Wang and Deen[231] developed a clever device for sustaining a continuous flow of peroxynitrite from the in situ reaction of nitric oxide and superoxide anion. Other examples of the in situ generation of peroxynitrite include the formation of peroxynitrite from sodium nitrite and hydrogen peroxide in the presence of myeloperoxidase[234]; by the action of xanthine oxidase with nitrite, molecular oxygen, and a reducing agent such as pterin[235]; from nitrate generated from PaPaNONOate and xanthine oxidase with pterin/xanthine oxidase as the source of superoxide anion[236]; or from nitrite and oxygen with xanthine and NADH.[237] A number of investigators have used a micropump for delivery of peroxynitrite to the reaction vessel. Peroxynitrous acid (hydrogen oxoperoxynitrate; HONOO) has a pKa of 6.8 and decomposes rapidly at acid pH ($t_{1/2} \leq 1$ s at pH 7.0/37°C).[223,238] Peroxynitrite is more stable at basic pH ($t_{1/2} = 45.6$ min at pH 12/25°C).[239] Hughes and Nicklin[238] also reported increased stability of peroxynitrite (perinitrite) at alkaline pH with a $t_{1/2}$ of 7–8 min at pH ≥ 11. Hughes and Nicklin[238] also reported that borate anions enhanced the rate of peroxynitrite decomposition; peroxynitrate is considered to decompose at acid pH by isomerization to nitrate[223,238]; decomposition at alkaline pH may be some more complex where homolytic cleavage to nitrogen dioxide and hydroxyl radical is suggested.[240] Factors such as carbon dioxide (p. 145; Table 11.4) and nitrogen trioxide (formed from nitric oxide and nitrogen dioxide)[241] can influence the decomposition of peroxynitrite; nitrogen trioxide does not react with peroxynitrous acid. Peroxynitrite in base can be stored at −80°C for months; it may also be stored in the dry form with somewhat less success.[242] The concentration of peroxynitrite may be determined by UV spectroscopy[243]; $\varepsilon_{305\,nm} = 1.70$ mM^{-1} cm^{-1}. The extinction coefficient has increased over the years as the purity of preparations of peroxynitrite has been improving.[238] The pKa for peroxynitrous acid has been reported to be dependent on solvent; a pKa of 6.55 was reported a low buffer concentration (1 mM phosphate, 20°C–22°C) and a pKa of 7.33 in 0.65 M phosphate (20°C–22°C).[244] The study of peroxynitrite as well as other active nitrogen species can be complicated by side reactions with solvent components.[245] Tris was observed to reduce the oxidation of tryptophan[246] and inner mitochondrial membrane components[247] by peroxynitrite. Molina and coworkers[248] have examined the effect of pH and buffer components on the decomposition of peroxynitrite. These investigators note that homolytic reaction of peroxonitrous acid to yield hydroxyl radical and nitrogen dioxide may be a minor pathway and the reaction with carbonate (this chapter, p. 145; Table 11.4) is more significant. For the current discussion, Molina and coworkers[248] observed that Tris accelerates peroxynitrite decomposition while phosphate appears to stabilize peroxynitrite. These investigators also observed that the rate of decomposition has decreased with increasing purity of peroxynitrite. As comment, Molina and coworkers[248] discussed the decomposition of peroxynitrous acid as an isomerization reaction resulting in the formation of nitric acid (Figure 4.6). It has been reported that the isomerization of peroxynitrite is enhanced by the heme–albumin–Fe (III) complex preventing the nitration of free tyrosine; this isomerization is prevented by the presence of ibuprofen, which enhances the nitration of free tyrosine by peroxynitrite.[249] The use of HEPES [4-(2-hydroxyethyl)-1-piperazinesulfonic acid] or MOPS [3-(N-morpholino)propanesulfonic acid] in the presence of copper (I) ions released nitric oxide.[250] Other studies[251] have shown that either peroxynitrite or SIN-1 (3-morpholinosyndnonimine N-ethylcarbamide) produced hydrogen peroxide. This group also reported the formation of piperazine radical derived by oxidation of HEPES. There are several reports on the cis–trans isomerization of peroxynitrite (Figure 4.6)[252–254] and some suggestion of increasing amounts of the cis-form at higher pH; it is also suggested that the cis-form is more stable than the trans-form.[252,254] However, there is one report[253] suggesting that the cis- and trans-forms have comparable stability and are in rapid equilibrium. Mannitol and ethanol have been reported to stabilize peroxynitrite through the formation of complexes.[255] The presence of ethyl nitrate in mice has been reported after the administration of ethanol, and it is suggested that ethyl nitrite is formed from the reaction of ethanol and peroxynitrite (Figure 4.6).[256] It has been shown that ethanol inhibits DNA strand breakage by peroxynitrite (SIN-1).[257]

Peroxynitrite can oxidize or nitrate proteins and as a chemical modification reagent, at least ex vivo, appears to be more reactive and less selective in reaction with proteins.[258] The reaction of

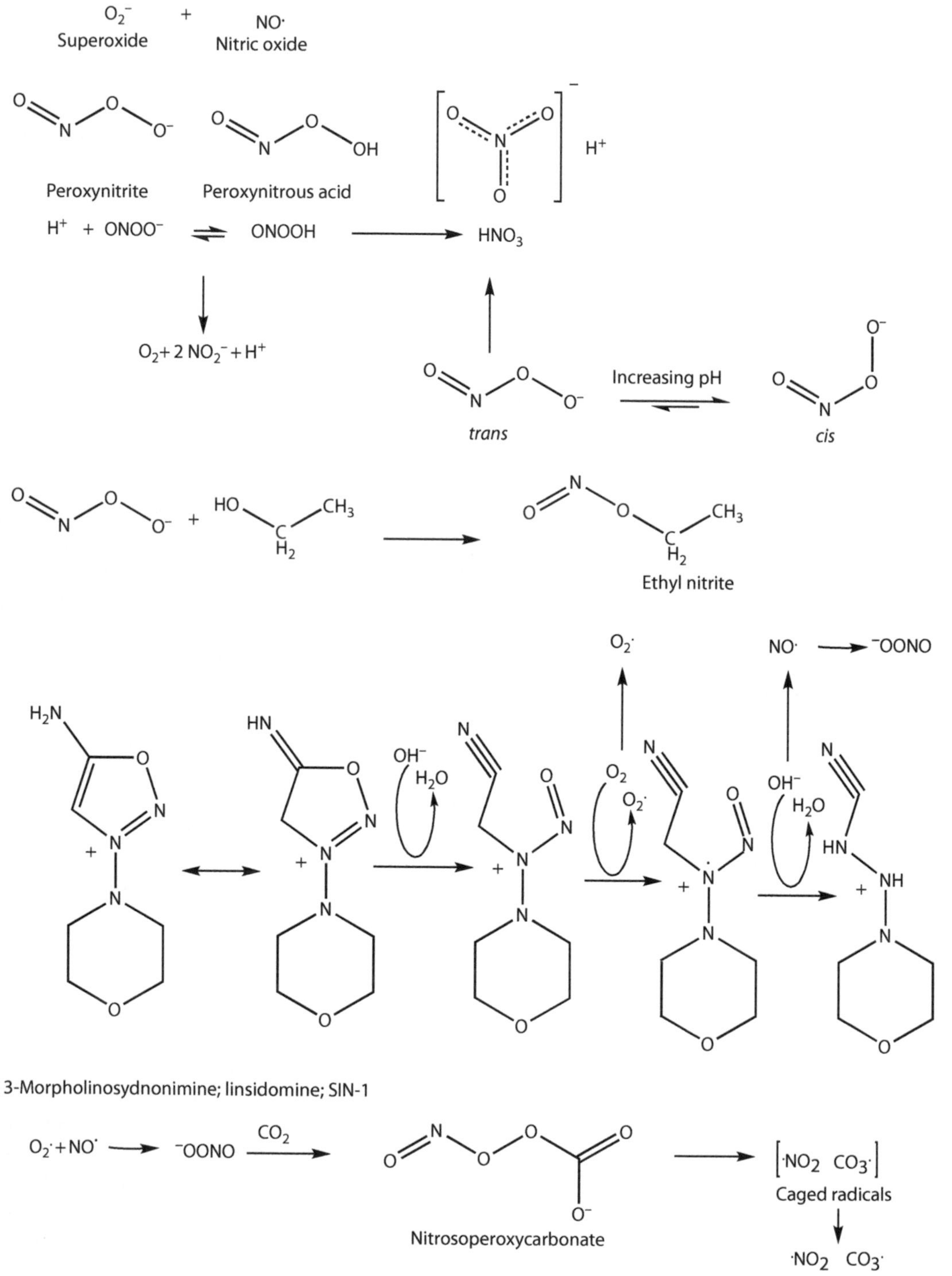

FIGURE 4.6 (continued)

peroxynitrite with functional groups on proteins is complex[224] including oxidation of cysteine,[259] methionine,[260–262] and tryptophan[246,263–265]; nitrosylation of cysteine[30]; and nitration of tryptophan[266] and tyrosine.[267] Peshenko and Shichi[259] observed that peroxynitrite oxidized the active center cysteine in bovine 1-cys peroxiredoxin more slowly than hydrogen peroxide but more rapidly than t-butyl hydroperoxide. When the concentration of peroxynitrite was equimolar or less, cysteine sulfenic acid is the product; with higher concentrations of peroxynitrite, the oxidation proceeds to cysteine sulfinic acid, then cysteine sulfonic acid. Activity could be recovered with dithiothreitol consistent with the reduction of cysteine sulfenic acid to cysteine; use of 2-mercaptoethanol as a reducing agent did not restore active but a mixed disulfide between cysteine and 2-mercaptoethanol was formed as neither free cysteine or cysteine sulfenic acid could be detected after reaction with 2-mercaptoethanol. Perrin and Koppenol[261] showed that the rate of reaction of peroxynitrous acid (pH 6.6) with either methionine or N-acetylmethionine was several orders of magnitude ($\sim 10^3$ M^{-1} s^{-1}) faster than the value at pH 8.6 (0.2 M^{-1} s^{-1} for methionine, 10 M^{-1} s^{-1} for N-acetylmethionine). Peroxynitrite does oxidize tryptophanyl residues in proteins[246,263,264]; nitration of tryptophanyl residues also occurs and is promoted by the presence of carbonate.[264,265] Pollet and coworkers[246] also reported a reaction of Tris buffer with peroxynitrite. The nitration of tyrosine by peroxynitrite has been known for some time.[267] Alvarez and Radi[268] have reviewed the reaction of peroxynitrite with various amino acids. The various amino acid derivatives described for the reaction of peroxynitrite with proteins are shown in Figure 4.7.

The rate of reaction of peroxynitrite with proteins is rapid with second-order rate constants varying from 10^4 to 10^6 M^{-1} s^{-1}. More rapid rates of reaction are seen with heme proteins such as myeloperoxidase.[269] The inactivation of tryptophan hydroxylase with peroxynitrite[270] is associated with the oxidation of a cysteinyl residue. A rate constant of 3.4×10^4 M^{-1} s^{-1}($25°C$, pH 7.4) was reported for the inactivation reaction. The rate of formation of compound II in myeloperoxidase with peroxynitrite[18] is more rapid (6.2×10^6 M^{-1} s^{-1}; pH 7.2, 12°C). The rate of formation of compound II was observed to decrease with increasing pH (2.5×10^5 M^{-1} s^{-1} at pH 8.9, 12°C) suggesting the participation of peroxynitrous acid rather than peroxynitrite. The formation of compound II (ferryl iron, FeIV) from the resting state (ferric iron, FeIII) proceeds without the observed formation of compound I. For those who, like the author, are actively involved in heme chemistry, the review by Battistuzzi and coworkers[271] is very useful.

FIGURE 4.6 (continued) Chemistry of peroxynitrite and peroxynitrous acid. Shown are the formation of peroxynitrite [oxidoperoxynitrate(-1)] from superoxide anion and nitric oxide and the isomerization of peroxynitrite to form nitrate (Molina, C., Kissner, R., and Koppenol, W.H., Decomposition of peroxynitrite: Influence of pH and buffer, *Dalton Trans.* 42, 9898–9905, 2013). Also shown is the *trans/cis* isomerization of peroxynitrite with suggestion that the *cis* form is the more stable conformation (Crow, J.P., Spruell, C., Chen, J. et al., On the pH-dependent yield of hydroxyl radical products from peroxynitrite, *Free Radic. Biol. Med.* 16, 331–338, 1994; Liang, B. and Andrews, L., Infrared spectra of *cis*- and *trans*-peroxynitrite anion, OONO-, in solid argon, *J. Am. Chem. Soc.* 123, 9848–9854, 2001). Shown below is the suggested reaction between peroxynitrite and ethanol resulting in ethyl nitrite (Deng, X.-S., Bludeau, P., and Deitrich, R.A., Formation of ethyl nitrite in vivo after ethanol administration, *Alcohol* 34, 217–223, 2004). Below this is a depiction of the process of peroxynitrite formation from SIN-1. The decomposition of SIN-1 results in the sequential and equimolar amounts of nitric oxide and superoxide, which rapidly form peroxynitrite (~ 10 M^{-1} s^{-1}) (Hodges, G.R., Marwaha, J., Paul, T., and Ingold, K.U., A novel procedure for generating both nitric oxide and superoxide in situ from chemical sources at any chosen mole ratio. First application: Tyrosine oxidation and a comparison with preformed peroxynitrite, *Chem. Res. Toxicol.* 13, 1287–1293, 2000; Stevens, J.F., MIranda, C.L., Frei, B., and Buhler, D.R., Inhibition of peroxynitrite-mediated LDL oxidation by prenylated flavonoids: The α,β-unsaturated keto functionality of 2′-hydroxychalcones as a novel antioxidant pharmacophore, *Chem. Res. Toxicol.* 16, 1277–1286, 2003). At the bottom is the reaction of peroxynitrite with carbon dioxide (Squadrito, G.L. and Pryor, W.A., Mapping the reaction of peroxynitrite with CO_2: Energetics, reactive species, and biological implications, *Chem. Res. Toxicol.* 15, 885–895, 2002; Papina, A.A. and Koppenol, W.H., Two pathways of carbon dioxide catalyzed oxidative coupling of phenol by peroxynitrite, *Chem. Res. Toxicol.* 19, 382–391, 2006).

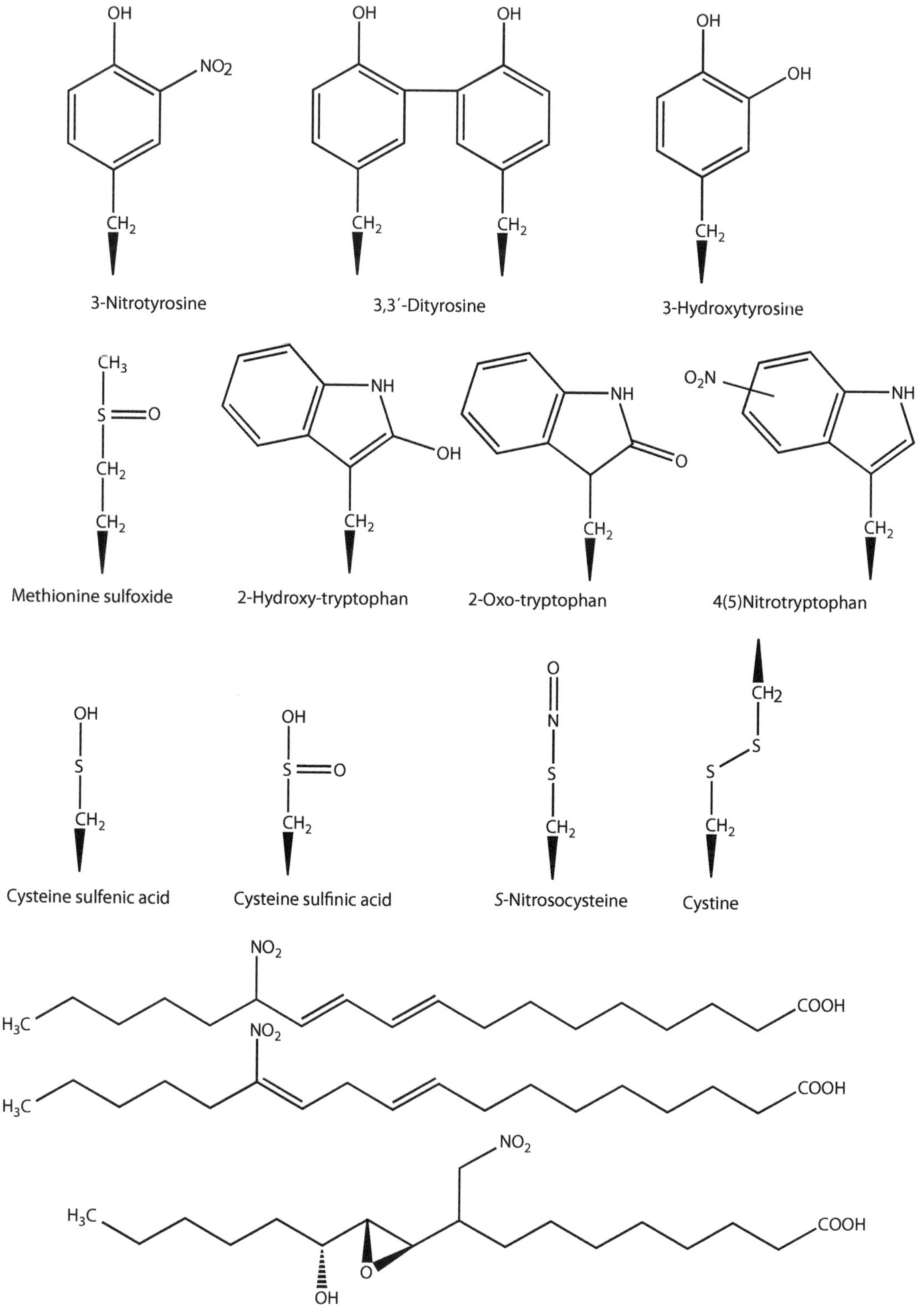

FIGURE 4.7 (continued)

Alvarez and Radi[268] also tabulated other second-order rate constants for the direct reaction of peroxynitrite with various amino acids from various studies. The rate reported for tryptophan oxidation is slow ($37\ M^{-1}\ s^{-1}$); oxidation of methionine ($2.3 \times 10^2\ M^{-1}\ s^{-1}$) and cysteine ($4.5 \times 10^3\ M^{-1}\ s^{-1}$) was more rapid. Gundaydin and Houk[272] have suggested that the nitration of p-methylphenol (as a model for a tyrosyl residue) by either peroxynitrous acid or nitrosoperoxycarbonate proceeds via the formation of nitrogen dioxide (Figure 4.6). These investigators argue that decomposition of either peroxynitrous acid or nitrosoperoxycarbonate is requisite for the nitration of tyrosine and that there is no direct reaction of either reagent with tyrosine. It would also appear that, in the absence of carbonate, nitration of tyrosine with peroxynitrite occurs at alkaline pH. The nitration of tyrosine with TNM also has a preference for alkaline pH while oxidation of cysteine with TNM occurs equally well at either pH 6 or pH 8.[270]

Radi and coworkers[273] first noted the effect of bicarbonate on peroxynitrite in studies on the oxidation of luminol in 1993. Somewhat later, Lymar and Hurst[274] described the reaction between peroxynitrite and carbon dioxide/carbonate to form a highly reactive complex ($ONOOCO_2^-$). Tien and coworkers[275] performed a systematic study of the effect of carbon dioxide on peroxynitrite modification of proteins. There is little effect of carbon dioxide on carbonyl formation (± 1.3 mM CO_2 less than 0.1 carbonyl/subunit) on glutamine synthetase with peroxynitrite at pH 7.5, while there is significant carbonyl formation (0.6 carbonyl per subunit) in the absence of carbon dioxide at pH 8.6 (phosphate); there was less than 0.1 carbonyl/subunit in the presence of 1.3 mM CO_2^-. The presence of CO_2 also suppressed the oxidation of methionine. There was little effect of CO_2 on the nitration of tyrosine at pH 8.6 but a major enhancement of nitration at pH 7.5 when compared to phosphate buffer. It is suggested that the formation of a complex between carbonate/carbon dioxide and peroxynitrate results in the stability of the peroxynitrite. Protein nitration may also be enhanced by heme with nitric oxide and superoxide[276] or nitrite and hydrogen peroxide.[189] It should be emphasized that carbon dioxide, not bicarbonate/carbonate, is the molecular species that reacts with peroxynitrite to form 1-carboxylato-2-nitrosodioxidane (nitrosoperoxycarbonate),[242,277,278] which undergoes homolytic cleavage to yield two radicals, nitrogen dioxide, and carbonate radical (Figure 4.6).

There is some evidence to suggest that regardless of stability issues, the species responsible for the nitration reaction is derived from peroxynitrous acid (hydrogen oxidoperoxidonitrate). It has been suggested that the active species is nitrogen dioxide, but other mechanisms have also been suggested (Figure 4.6).[272] I want to emphasize again that peroxynitrite/peroxynitrous acid is a product

FIGURE 4.7 (continued) Derivative compounds obtained by the action of peroxynitrite. The reaction of peroxynitrite with amino acids/peptides/proteins results in the modification of a number of amino acids some of which are shown (see Alvarez, B., Ferrer-Sueta, G., Freeman, B.A., and Radi, R., Kinetics of peroxynitrite reaction with amino acids and human serum albumin, *J. Biol. Chem.* 274, 842–848, 1999; Alvarez, B. and Radi, R., Peroxynitrite reactivity with amino acids and proteins, *Amino Acids* 25, 295–311, 2003). At the bottom are shown products of the reaction of peroxynitrite with linoleic acid [(Z,Z)-9,12-octadecadienoic acid]. First, there is the product of the reaction of peroxynitrite with linoleic acid resulting in the formation of a nitrated derivative. The 10-nitroisomer is shown; nitration does occur at other sites (Baker, P.R.S., Schopfer, F.J., Sweeney, S., and Freeman, B.A., Red cell membrane and plasma linoleic acid nitration products: Synthesis, clinical identification, and quantitation, *Proc. Natl. Acad. Sci. USA* 101, 11577–11582, 2004). Shown below is the product of a Michael addition of the nitrated linoleic acid to a protein sulfhydryl group (Batthyany, C., Schopfer, F.J., Baker, P.R.S. et al., Reversible post-translational modification of proteins by nitrated fatty acids *in vivo*, *J. Biol. Chem.* 281, 20450–20463, 2006). Not shown is the reversal of the cysteine modification by reduced glutathione. At the bottom of the figure is a derivative formed in the nitration of linoleic acid (O'Donnell, V.B., Eiserich, J.P., Chumley, P.H. et al., Nitration of unsaturated fatty acids by nitric oxide-derived reactive nitrogen species peroxynitrite, nitrous acid, nitrogen dioxide, and nitronium ion, *Chem. Res. Toxicol.* 12, 83–92, 1999) and an epoxy derivative formed by nitration of a hydroxylated metabolite of linoleic acid (Manini, P., Camera, E., Picardo, M. et al., Biomimetic nitration of the linoleic acid metabolite 13-hydroxyoctadecadienoic acid: Isolation and spectral characterization of novel chain-rearranged epoxy nitro derivatives, *Chem. Phys. Lipids* 151, 51–61, 2008).

of nature and not of synthetic organic chemistry such as TNM. It is an unstable compound and most likely is the product of in situ production and use. In this regard, it has been suggested that the nitration of tyrosine occurs in a hydrophobic environment.[279,280] Zhang and coworkers prepared a 23-mer peptide with a single tyrosyl residue at position 4 (Y-4), position 8 (Y-8), or position 12 (Y-12), which were incorporated into a liposome preparation (1,2-dilauroyl-*sn*-glycero-3-phosphatidylcholine). The nitration of these peptides with peroxynitrite was compared to that of tyrosine in solution; the peroxynitrite in 0.01 M NaOH was delivered by an infusion pump. There was a modest effect of depth (nitration of Y-12 > nitration of Y-4) on nitration with peroxynitrite (0.1 M phosphate, pH 7.4). There was 20%–40% more nitration (50 µM peroxynitrite) of the peptides in the liposomes than for tyrosine in solution. The extent of nitration of tyrosine and peptide Y-8 was evaluated as a function of peroxynitrite concentration. At 50 µM peroxynitrite, there is slightly more nitration of peptide Y-8 than free tyrosine, while at 300 µM peroxynitrite concentration, there is approximately twice as much nitration of tyrosine than that observed for peptide Y-8. The peroxynitrite was infused by pump until the final concentration of reagent was obtained. Control experiments showed that the peroxynitrite stock solution was stable over the course of the experiment. The use of SIN-1 as the source of peroxynitrite (SIN-1 decomposes to form nitric oxide and superoxide anion, which then form peroxynitrite) yielded different results in that penetration depth of the peptide resulted in more extensive nitration (Y-12 > Y-8 > Y-4); nitration of free tyrosine by SIN-1 yielded less than 1 µM nitration with at least 7 µM nitration for Y-4 and approximately 11 µM nitration for Y-12. The use of myeloperoxidase/nitrite/hydrogen peroxide resulted in more nitration of Y-4 (approximately 80 µM) and greatly reduced nitration of Y-12 (approximately 10 µM); there was approximately 30 µM nitration of free tyrosine. Accepting the hypothesis that a hydrophobic environment promotes nitration of tyrosine by peroxynitrite, the data obtained by Zhang and coworkers[279,280] with SIN-1 and peroxynitrite slowly delivered by a pump are consistent with this hypothesis; data obtained with peroxynitrite produced by myeloperoxidase does not fit the model. It would be further useful to the model if SIN-1 is hydrophobic. I could not find octanol/water partition data for either peroxynitrite or SIN-1 nor could I find solubility data. I did find solubility data for molsidomine,[281] which argue that SIN-1 is water soluble and not be expected to partition into the liposome. However, there is a problem with the aforementioned argument in that the peroxynitrite would be formed on the surface on the liposome and diffuse into the liposome. If it is assumed that nitrogen dioxide ($\cdot NO_2$) is the molecular species responsible for the nitration of tyrosine, it has been reported that nitrogen dioxide has an octanol–water partitioning coefficient of 2.7[282] and would be expected to easily diffuse into the liposome. There are issues with the delivery of peroxynitrite into in vitro reactions. Kim and coworkers[283] evaluated the effect of peroxynitrate dose and rate of reaction dosing on DNA modification. The studies involved a comparison of a bolus administration of peroxynitrite, infusion of peroxynitrite via pump, and in situ generation via decomposition of SIN-1. DNA damage was assessed by a plasmid nicking assay and a mutagenicity assay. The plasmid nicking assay measures the physical damage to DNA by destruction of bases and strand breakage. The number of DNA strand breaks was greatest with bolus administration, lowest with infusion by pump and intermediate, but closer to bolus administration with SIN-1. A similar ordering of effects was observed with mutation frequency with the highest frequency seen with bolus administration and least with SIN-1. Somewhat different results were obtained by Crow and coworkers[284] on the reaction of yeast alcohol dehydrogenase with peroxynitrite where similar results were obtained with either bolus administration or infusion. The results of Shao and coworkers[285] add to the complexity of understanding the factor influencing peroxynitrite nitration of protein. These investigators[285] observed the nitration of Tyr192 in apolipoprotein A-I by peroxynitrite. Peroxynitrite was either a bolus administration or generated in situ by myeloperoxidase/hydrogen peroxide/nitrite. Electronic paramagnetic resonance (EPR) measurements showed that Tyr192 is in a polar environment. When apolipoprotein A-I is incorporated into HDL, nitration of Tyr192 is decreased, and it suggested that this decrease is a result of decreased exposure of Tyr192 and a more hydrophobic environment. Shao and coworkers[285] showed that six other tyrosyl residues were modified in apolipoprotein A-1 (10%–20% 3-nitrotyrosine) with either bolus peroxynitrite or myeloperoxidase/hydrogen peroxide/nitrite. Analysis of the results suggested that bolus peroxynitrite was more selective in the nitration of tyrosine than the myeloperoxidase system. Shao and coworkers[285] also examined the nitration of several heptapeptides containing basic amino acid residues and tyrosine. They concluded that the presence of lysyl residues did not influence tyrosine nitration with either system. In an earlier review, Ischiropoulos[286] had suggested a potential role for acidic amino acid residues in determining the specificity of tyrosine nitration with peroxynitrite. Ischiropoulos[286] also suggested the importance of surface exposure and lack of steric hindrance as factors in promoting tyrosine nitration. Other studies have also suggested the importance of local environmental factors, which would lower tyrosyl hydroxyl pKa and facilitate nitration.[287,288] Souza and coworkers[288] studied the nitration of bovine pancreatic ribonuclease A (RNAse A), lysozyme, and phospholipase A_2 with bolus peroxynitrate (in two portions) in the presence of sodium bicarbonate (100 mM potassium phosphate–25 mM sodium bicarbonate, pH 7.4 with 0.1 mM diethyltriaminepentaacetic acid). Souza and workers[288] showed that there was one major site of nitration (Tyr115) in RNAse with two lesser sites of modification, two sites of modification in lysozyme (Tyr20 and Try 23) that appeared to be mutually exclusive, and one major site of modification in phospholipase A_2 with one minor site of modification. Hypochlorous acid with nitrite did not result in nitration of RNAse, myeloperoxidase/hydrogen peroxide/nitrite modified Tyr115, and SIN-1 also nitrated Tyr115 in RNAse. Souza and coworkers[288] also studied the reaction of bolus peroxynitrite with a mixture of RNAse, lysozyme, and phospholipase A_2; phospholipase A_2 was modified to a greater extent (55%) than either lysozyme (23%) or RNAse (22%). If myeloperoxidase/hydrogen peroxide/nitrite was used to nitrate the mixture of the three proteins, RNAse was the major nitrated product. These investigators do suggest that the local environment including exposure was important in the susceptibility of a tyrosine residue to nitration.

Diethylaminepentaacetic acid (DPTA) is frequently included in reactions using bolus addition of peroxynitrite to chelate transition metal ions such as iron and copper, which promote decomposition of peroxynitrite[289,290] EDTA has been reported to inhibit the Cu(II)-catalyzed decomposition of peroxynitrite at alkaline pH (9.8–11.1).[290] Beckman and coworkers[291] have reported that the Fe(III)–EDTA complex catalyzed the nitration of 4-hydroxyphenylacetic acid, while the complex of Fe(III) with DPTA was inactive and the complex of Fe(III) with desferrioxamine was inhibitory reflecting a direct interaction of desferrioxamine with peroxynitrite.[289] As an aside, the copper (II)-catalyzed nitration (hydrogen peroxide/nitrite) of tyrosine has been reported.[292] Copper (II) has also been reported to catalyze the nitrosylation of propofol with peroxynitrite.[293] Kohnen and coworkers[293] also reported that Cu(I) and Cu(II) accelerated the decomposition of peroxynitrite.

The chemistry of peroxynitrite creates difficulty in understanding the factors responsible for specificity of tyrosine nitration in proteins or, for that matter, determining whether there is nitration or oxidation. Similar issues in the prediction of residue modification are present with TNM (Chapter 11 and earlier comments in the current chapter). Data presented by Zhang and coworkers[279,280] would suggest the importance of a hydrophobic environment, while the data obtained from the reaction of peroxynitrite with apolipoprotein A-1[285] would argue for a polar environment as would the data of Souza and coworkers[288] on the reaction of peroxynitrite with several proteins. More recent work by Seeley and Stevens[294] suggests that the nitration of a tyrosine residue is enhanced by the presence of a basic and/or an uncharged (e.g., serine) amino acid residue. Bayden and coworkers[295] argue to the importance of either basic or acidic amino acid residue in proximity to the susceptible tyrosine residue. These investigators propose the importance of hydrogen bonding to the tyrosine hydroxyl as enhancing susceptibility to nitration. These investigators also note the importance of tyrosine exposure and the ability to accommodate a nitro group. It is obvious to me that the use of peroxynitrite as a reagent is not trivial. It is also noted that some of the studies cited earlier present data with very low levels of modification (2%–10%). The use of peroxynitrite either as bolus, delivered by infusion, or as produced in situ by myeloperoxidase/hydrogen peroxide/nitrite or SIN-1 is complicated by reagent stability. A further complication, likely related to the source of reagent, is the observation that different results may be obtained from different sources of peroxynitrite.

Finally, peroxynitrite, while reactive, is also nonspecific. In addition to proteins, peroxynitrite does oxidize[296,297] and nitrate[298,299] guanine as well as break DNA strands.[257,283,298,300,301] Peroxynitrite has been reported to degrade extracellular matrix[302] and hyaluronan.[303–305]

There are a variety of other observations that are of interest in the chemistry of peroxynitrite. Peroxynitrite can nitrate unsaturated fatty acids such as linoleic acid,[306,307] which in turn can reversibly modify proteins by Michael addition (Figure 4.7).[308] Peroxynitrite has been reported to nitrate, nitrosylate, and hydroxylate phenols (Figure 4.7),[278,309,310] which is suggested to have a protective, antioxidant function.[311–315] Other protective agents include reducing compounds such as glutathione[316] and ebelsen.[317] There has been considerable interest in the therapeutic control of peroxynitrite.[318–321]

REFERENCES

1. Kumar, Y., Liang, C., Bo, Z. et al., Serum proteome and cytokine analysis in a longitudinal cohort of adults with primary dengue infection reveals predictive markers of DHF, *PLoS Negl. Trop. Dis.* 6, e1887, 2012.
2. Namiduru, E.S., Namiduru, M., Tarakçioğlu, M. et al., Levels of malondialdehyde, myeloperoxidase and nitrotyrosine in patients with chronic viral hepatitis B and C, *Adv. Clin. Exp. Med.* 21, 47–53, 2012.
3. Sharov, V.S., Pal, R., Dermina, E.S. et al., Fluorogenic tagging of protein 3-nitrotyrosine with 4-(aminomethyl)benzene sulfonate is tissue: A useful alternative to immunohistochemistry for fluorescence microscopy imaging of protein nitration, *Free Radic. Biol. Med.* 53, 1877–1885, 2012.
4. Misko, T.P., Radabaugh, M.R., Highkin, M. et al., Characterization of nitrotyrosine as a biomarker for arthritis and joint injury, *Osteoarthritis Cartilage* 21, 151–156, 2013.
5. Li, S., Wang, H., Xian, M., and Whorton, A.R., Identification of protein nitrosothiols using phosphine-mediated selective reduction, *Nitric Oxide* 26, 20–26, 2012.
6. Broniowska, K.A., Diers, A.R., and Hogg, N., S-Nitrosoglutathione, *Biochim. Biophys. Acta* 1830, 3173–3181, 2013.
7. Couturier, J., Chibani, K., Jacquot, J.P., and Rouhier, N., Cysteine-based redox regulation and signaling in plants, *Front. Plant Sci.* 4, 105, 2013.
8. Kolesnik, B., Palten, K., Schrammel, A. et al., Efficient nitrosation of glutathione by nitric oxide, *Free Radic. Biol. Med.,* 63, 51–64, 2013.
9. Gross, S.S. and Hattori, Y., Room for both tyrosine nitration and cysteine nitrosylation to regulate NFκB activity, but perhaps only one modification at a time, *Cardiovasc. Res.* 67, 747–748, 2005.
10. Penna, C., Perrelli, M.G., Tullio, F. et al., Post-ischemic early acidosis in cardiac postconditioning modifies the activity of antioxidant enzymes, reduces nitration, and favors protein S-nitrosylation, *Pflugers Arch.* 462, 219–233, 2011.
11. Burgoyne, J.R., Rudyk, O., Moyr, M., and Eaton, P., Nitrosative protein oxidation is modulated by early endotoxemia, *Nitric Oxide* 25, 118–124, 2011.
12. Wiseman, D.A. and Thurmond, D.C., The good and bad effects of cysteine S-nitrosylation and tyrosine nitration upon insulin exocytosis: A balancing act, *Curr. Diabetes Rev.* 8, 303–315, 2012.
13. Comer, F.I. and Hart, G.W., Reciprocity between O-GlcNAc and O-phosphate on the carboxyl terminal domain of RNA polymerase II, *Biochemistry* 40, 7845–7852, 2001.
14. Ckless, K., Lampert, A., Reiss, J. et al., Inhibition of arginase activity enhances inflammation in mice with allergic airway disease, in association with increases in protein S-nitrosylation and tyrosine nitration, *J. Immunol.* 181, 4255–4264, 2008.
15. Ytterberg, A.J. and Jensen, O.N., Modification-specific proteomics in plant biology, *J. Proteomics* 73, 2249–2266, 2010.
16. Martínez-Ruiz, A., Cadenas, S., and Lamas, S., Nitric oxide signalling: Classical, less classical, and nonclassical mechanisms, *Free Radic. Biol. Med.* 51, 17–29, 2011.
17. Vazquez-Torres, A., Redox active sensors of oxidative and nitrosative stress, *Antioxid. Redox Signal.* 17, 1201–1214, 2013.
18. *Oxford English Dictionary*, Oxford, U.K., 2013.
19. Liu, S., Kawai, K., Tyurin, V.A. et al., Nitric oxide-dependent pro-oxidant and pro-apoptotic effect of metallothioneins in HL-60 cells challenged with cupric nitrilotriacetate, *Biochem. J.* 354, 397–406, 2001.
20. Tyurina, Y.Y., Basova, L.V., Konduru, N.V. et al., Nitrosative stress inhibits the aminophospholipid translocase resulting in phosphatidylserine externalization and macrophage engulfment: Implications for the resolution of inflammation, *J. Biol. Chem.* 282, 8498–8509, 2007.

21. Hashemy, S.I., Johansson, C., Carsten, B. et al., Oxidation and *S*-nitrosation of cysteines in human cytosolic and mitochondrial glutaredoxins: Effects of structure and activity, *J. Biol. Chem.* 282, 14428–14436, 2007.

22. Hare, J.M., Beigi, F., and Tziiomalos, K., Nitric oxide and cardiobiology-methods for intact hearts and isolated myocytes, *Methods Enzymol.* 441, 369–392, 2008.

23. Sabens Liedhegner, E.A., Gao, X.H., and Mieyal, J.J., Mechanisms of altered redox regulation in neurodegenerative diseases—Focus on *S*-glutathionylation, *Antioxid. Redox Signal.* 16, 543–566, 2012.

24. Paul, B.D. and Snyder, S.H., H_2S signalling through protein sulfhydration and beyond, *Nat. Rev. Mol. Cell Biol.* 13, 499–507, 2012.

25. Wang, R., Hydrogen sulfide: A new EDRF, *Kidney Int.* 76, 700–704, 2009.

26. Lu, X.M., Tompkins, R.G., and Fischman, A.J., Nitric oxide activates intradomain disulfide bond formation in the kinase loop of Akt1/PKBα after burn injury, *Int. J. Mol. Med.* 31, 740–750, 2013.

27. Rehder, D.S. and Borges, C.R., Cysteine sulfenic acid as an intermediate in disulfide bond formation and nonenzymatic protein folding, *Biochemistry* 49, 7748–7755, 2010.

28. Fleming, I. and Busse, R., NO: The primary EDRF, *J. Mol. Cell. Cardiol.* 31, 5–14, 1999.

29. Goldstein, S. and Czapski, G., Mechanism of the nitrosation of thiols and amines by oxygenated ·NO solutions: The nature of the nitrosyl intermediates, *J. Am. Chem. Soc.* 118, 3419–3425, 1996.

30. Viner, R.I., Williams, T.D., Schöneich, P., Peroxynitrite modification of protein thiols: Oxidation, nitrosylation, and *S*-glutathiolation of functionally important cysteine residue(s) in the sarcoplasmic reticulum Ca-ATPase, *Biochemistry* 38, 12408–12415, 1999.

31. Nedospasov, A., Rafikov, R., Beda, N., and Nudler, E., An autocatalytic mechanism of protein nitrosylation, *Proc. Natl. Acad. Sci. USA* 97, 13543–13548, 2000.

32. Ullrich, V. and Kissner, R., Redox signaling: Bioorganic chemistry at its best, *J. Inorg. Biochem.* 100, 2079–2089, 2006.

33. McCollister, B.D., Myers, J.T., Jones-Carson, J. et al., N_2O_3 enhances the nitrosative potential of IFNγ-primed macrophages in response to salmonella, *Immunobiology* 212, 759–769, 2007.

34. Coupe, P.J. and Williams, D.L.H., Formation of peroxynitrite at high pH and the generation of *S*-nitrosothiols from thiols and peroxynitrous acid in acid solution, *J. Chem. Soc. Perkin Trans.* 2, 1595–1599, 2001.

35. Möller, M.N., Li, Q., Lancaster, J.R., Jr., and Denicola, A., Acceleration of nitric oxide autoxidation and nitrosation by membranes, *IUBMB Life* 59, 243–248, 2007.

36. Tyurin, V.A., Tyurina, Y.Y., Liu, S.-X. et al., Quantitation of *S*-nitrosothiols in cells and biological fluids, *Methods Enzymol.* 352, 347–360, 2002.

37. Landino, L.M., Protein thiol modification by peroxynitrite anion and nitric oxide donors, *Methods Enzymol.* 440, 95–109, 2008.

38. Ferrini, M.E., Simons, B.J., Bassett, D.J. et al., *S*-Nitrosoglutathione reductase inhibition regulates allergen-induced lung inflammation and airway hyperreactivity, *PLoS One* 8(7), e70351, 2013.

39. Zhang, Y. and Hogg, N., *S*-Nitrosothiols: Cellular formation and transport, *Free Radic. Biol. Med.* 38, 831–838, 2005.

40. Paige, J.S., Xu, G., Stancevic, B., and Jaffrey, S.R., Nitrosothiol reactivity profiling identifies *S*-nitrosylated proteins with unexpected stability, *Chem. Biol.* 15, 1307–1316, 2008.

41. Schreiter, E.R., Rodriguez, M.M., Weichsel, A. et al., *S*-Nitrosylation-induced conformational change in blackfin tuna myoglobin, *J. Biol. Chem.* 282, 19773–19780, 2007.

42. Sengupta, R. and Holmgren, A., The role of thioredoxin in the regulation of cellular processes by *S*-nitrosylation, *Biochim. Biophys. Acta* 1820, 689–700, 2012.

43. Benhar, M., Forrester, M.T., and Stamler, J.S., Protein denitrosylation: Enzymatic mechanisms and cellular functions, *Nat. Rev. Mol. Cell Biol.* 10, 721–732, 2009.

44. Beigi, F., Gonzalez, D.R., Minhas, K.M. et al., Dynamic denitrosylation via *S*-nitroglutathione reductase regulates cardiovascular function, *Proc. Natl. Acad. Sci. USA* 109, 4314–4319, 2012.

45. Foster, M.W., Yang, Z., Gooden, D.M. et al., Proteomic characterization of the cellular response to nitrosative stress mediated by *S*-nitrosoglutathione reductase inhibition, *J. Proteome Res.* 11, 2480–2491, 2012.

46. Ozawa, K., Tsumoto, H., Wei, W. et al., Proteomic analysis of the role of *S*-nitrosoglutathione reductase in lipopolysaccharide-challenged mice, *Proteomics* 12, 2024–2035, 2012.

47. Sips, P.Y., Irie, T., Zou, L. et al., Reduction of cardiomyocyte *S*-nitrosylation by *S*-nitrosoglutathione reductase protects against sepsis-induced myocardial depression, *Am. J. Physiol. Heart Circ. Physiol.* 304, H1134–H1146, 2013.

48. van Ophem, P.W., Van Beeumen, J., and Duine, J.A., NAD-linked, factor-dependent formaldehyde dehydrogenase or trimeric zinc-containing, long-chain alcohol dehydrogenase from *Amycolatopsis methanolica*, *Eur. J. Biochem.* 206, 511–518, 1992.

49. Fernández, M.R., Biosca, J.A., and Parés, X., S-Nitrosoglutathione reductase activity of human and yeast glutathione-dependent formaldehyde dehydrogenase and its nuclear and cytoplasmic localisation, *Cell. Mol. Life Sci.* 60, 1013–1018, 2003.

50. Sakamoto, A., Ueda, M., and Morikawa, H., Arabidopsis glutathione-dependent formaldehyde dehydrogenase is an S-nitrosoglutathione reductase, *FEBS Lett.* 515, 20–24, 2002.

51. Benoit, R.M. and Auer, M., A direct way of redox sensing, *RNA Biol.* 8, 18–23, 2011.

52. Nakamura, T. and Lipton, S.A., Emerging role of protein-protein transnitrosylation in cell signaling pathways, *Antioxid. Redox Signal.* 18, 239–249, 2013.

53. Saville, B., A scheme for the colorimetric determination of microgram amounts of thiols, *Analyst* 83, 670–672, 1958.

54. Taldone, F.S., Tummala, M., Goldstein, E.J. et al., Studying the S-nitrosylation of model peptides and eNOS protein by mass spectrometry, *Nitric Oxide* 13, 176–187, 2005.

55. Morakinyo, M.K., Chipinda, I., Hettick, J. et al., Detailed mechanistic investigation into the S-nitrosation of cysteamine, *Can. J. Chem.* 90, 724–738, 2012.

56. Park, J.-W., Reaction of S-nitrosoglutathione with sulfhydryl groups in proteins, *Biochem. Biophys. Res. Commun.* 152, 916–920, 1988.

57. Williams, D.L.H., S-Nitrosothiols and role of metal ions in decomposition to nitric oxide, *Methods Enzymol.* 268, 299–308, 1996.

58. Clancy, R.M. and Abramson, S.B., Novel synthesis of S-nitrosoglutathione and degradation by human neutrophils, *Anal. Biochem.* 204, 365–371, 1992.

59. Meyer, D.J., Kramer, H., and Ketterer, B., Human glutathione transferase catalysis of the formation of S-nitrosoglutathione from organic nitrites plus glutathione, *FEBS Lett.* 351, 427–428, 1994.

60. King, M., Gildemeister, O., Gaston, B., and Mannick, J.B., Assessment of S-nitrosothiols on diaminofluorescein gels, *Anal. Biochem.* 346, 69–76, 2005.

61. Xu, L., Han, C., Lim, K., and Wu, T., Activation of cytosolic phospholipase $A_2\alpha$ through nitric oxide-induced S-nitrosylation: Involvement of inducible nitric-oxide synthase and cyclooxygenase-2, *J. Biol. Chem.* 283, 3077–3087, 2008.

62. Tsang, A.H.K., Lee, Y.-I., Ko, H.S. et al., S-Nitrosylation of XIAP compromises neuronal survival in Parkinson's disease, *Proc. Natl. Acad. Sci. USA* 106, 4900–4905, 2009.

63. Angeli, A., Über das Nitrohydroxylamin, *Chem. Central-Blatt* (15), 799, 1896.

64. Väänänen, A.J., Kankuri, E., and Rauhala, P., Nitric oxide-related species-induced protein oxidation: Reversible, irreversible, and protective effects on enzyme function of papain, *Free Radic. Biol. Med.* 38, 1102–1111, 2005.

65. Miranda, K.M., Katori, T., Torres de Holding, C.L. et al., Comparison of the NO and HNO donating properties of diazeniumdiolates: Primary amine adducts release HNO *in vivo*, *J. Med. Chem.* 48, 8220–8228, 2005.

66. Landino, L.M., Koumas, M.T., Mason, C.E., and Alston, J.A., Modification of tubulin cysteines by nitric oxide and nitrosyl donors alters tubulin polymerization activity, *Chem. Res. Toxicol.* 20, 1693–1700, 2007.

67. Marshall, H.E. and Stamler, J.S., Inhibition of NF-κB by S-nitrosylation, *Biochemistry* 40, 1688–1693, 2001.

68. Gu, Z., Kaul, M., Yan, B. et al., S-Nitrosylation of matrix metalloproteinases: Signaling pathway to neuronal cell death, *Science* 297, 1186–1190, 2002.

69. Liu, M., Hou, J., Huang, X. et al., Site-specific proteomics approach for study protein S-nitrosylation, *Anal. Chem.* 82, 7160–7168, 2010.

70. Ckless, K., Reynaert, N.L., Taatjes, D.J. et al., In situ detection and visualization of S-nitrosylated proteins following chemical derivatization: Identification of RNA GTPase as a target for S-nitrosylation, *Nitric Oxide* 11, 216–227, 2004.

71. Chan, N.-L., Rogers, P.H., and Arnone, A., Crystal structure of the S-nitroso form of liganded human hemoglobin, *Biochemistry* 37, 16459–16464, 1998.

72. Finnen, M.J., Hennessy, A., McLean, S. et al., Topical application of acidified nitrite to the nail renders it antifungal and causes nitrosation of cysteine groups in the nail plate, *Br. J. Dermatol.* 157, 494–500, 2007.

73. Bancroft, K., Grant, I.F., and Alexander, M., Toxicity of NO_2: Effect of nitrate on microbial activity in an acid soil, *Appl. Environ. Microbiol.* 38, 940–944, 1979.

74. Ormerod, A.D., Shah, A.A., Li, H. et al., An observational prospective study of topical acidified nitrite for killing methicillin-resistant *Staphylococcus aureus* (MRSA) in contaminated wounds, *BMC Res. Notes* 4, 458, 2011.

75. Dave, R.N., Joshi, H.M., and Venugopalan, V.P., Biomedical evaluation of a novel nitrogen oxides releasing wound dressing, *J. Mater. Sci. Mater. Med.* 23, 3097–3106, 2012.

76. Opländer, C., Müller, T., Bachin, M. et al., Characterization of novel nitrite-based nitric oxide generating delivery systems for topical dermal application, *Nitric Oxide* 28, 24–32, 2013.

77. Fujiwara, K., Matsumoto, N., Masuyama, Y. et al., New hapten-protein conjugation method using *N*-(*m*-aminobenzyloxy)succinimide as a two-level heterobifunctional agent: Thyrotropin-releasing hormone as a model peptide without free amino or carboxyl groups, *J. Immunol. Methods* 175, 123–129, 1994.

78. Stamler, J.S., Toone, E.J., Lipton, S.A., and Sucher, N.J., (S)NO signals: Translocation, regulation, and a consensus motif, *Neuron* 18, 691–696, 1997.

79. Ascenzi, P., Calasanti, M., Persichini, T. et al., Re-evaluation of amino acid sequence and structural consensus rules for cysteine-nitric oxide reactivity, *Biol. Chem.* 381, 623–628, 2000.

80. Osburn, S., O'Hair, R.A.J., Black, S.M., and Ryzhov, V., Post-translational modification in the gas phase: Mechanism of cysteine *S*-nitrosylation via ion-molecule reactions, *Rapid Commun. Mass Spectrom.* 25, 3216–3222, 2011.

81. Tummala, M., Ryzhov, V., Ravi, K., and Black S.M., Identification of the cysteine nitrosylation sites in human endothelial nitric oxide synthase, *DNA Cell Biol.* 27, 25–33, 2008.

82. Foster, M.W., Forrester, M.T., and Stamler, J.S., A protein microarray-based analyses of *S*-nitrosylation, *Proc. Natl. Acad. Sci. USA* 106, 18948–18953, 2009.

83. Talipov, M.R. and Timerghazin, Q.K., Protein control of *S*-nitrosothiol reactivity: Interplay of antagonistic resonance structures, *J. Phys. Chem.* 117, 1827–1837, 2013.

84. Nogueria, L., Figueiredo-Freitas, C., Casimiro-Lopes, G. et al., Myosin is reversibly inhibited by *S*-nitrosylation, *Biochem. J.* 424, 221–231, 2009.

85. Smith, B.C. and Marletta, M.A., Mechanisms of *S*-nitrosothiol formation and selectivity in nitric oxide signaling, *Curr. Opin. Chem. Biol.* 16, 498–506, 2012.

86. Talipov, M.R. and Timerghazin, Q.K., Protein control of *S*-nitrosothiol reactivity: Interplay of antagonistic resonance structure, *J. Phys. Chem. B.* 117, 1827–1837, 2013.

87. Zhang, D.-H., Devarie-Baez, N.O., Pan, J. et al., One-pot thioether from *S*-nitrosothiols, *Org. Lett.* 12, 5674–5676, 2010.

88. Giustarini, D., Milzani, A., Aldini, G. et al., *S*-Nitrosation versus *S*-glutathionylation of protein sulfhydryl groups by *S*-nitrosoglutathione, *Antioxid. Redox Signal.* 7, 930–939, 2005.

89. Sadidi, M., Geddes, T.J., and Kuhn, D.M., *S*-Thiolation of tyrosine hydroxylase by reactive nitrogen species in the presence of cysteine or glutathione, *Antioxid. Redox Signal.* 7, 863–869, 2005.

90. Hirayama, A., Noronha-Dutra, A.A., Gordge, M.P. et al., *S*-Nitrosothiols are stored by platelets and released during platelet-neutrophil interactions, *Nitric Oxide* 3, 95–104, 1999.

91. Aesif, S.W., Janssen-Heininger, Y.M., and Reynaert, N.L., Protocols for the detection of *S*-glutathionylated and *S*-nitrosylated proteins *in situ*, *Methods Enzymol.* 474, 289–296, 2010.

92. Jaffrey, S.R., Erdjument-Bromage, H., Ferris, C.D. et al., Protein S-nitrosylation: A physiological signal for neuronal nitric oxide, *Nat. Cell Biol.* 3, 193–197, 2001.

93. Jaffrey, S.R. and Snyder, S.H., The biotin switch method for the detection of *S*-nitrosylated proteins, *Sci. STKE* 2001(86), 11, 2001.

94. Torta, F., Elviri, L., and Bachi, A., Direct and indirect detection methods for the analysis of *S*-nitrosylated peptides and proteins, *Methods Enzymol.* 473, 265–280, 2010.

95. Burgoyne, J.R. and Eaton, P., A rapid approach for the detection, quantification, and discovery of novel sulfenic acid or *S*-nitrosothiol modified proteins using a biotin-switch method, *Methods Enzymol.* 473, 281–303, 2010.

96. Uehara, T. and Nishiya, T., Screening systems for the identification of *S*-nitrosylated proteins, *Nitric Oxide* 25, 108–111, 2011.

97. Astier, J., Rasul, S., Koen, E. et al., *S*-Nitrosylation: An emerging post-translational protein modification in plants, *Plant Sci.* 181, 527–533, 2011.

98. Murray, C.I., Uhrigshardt, H., O'Meally, R.N. et al., Identification and quantification of *S*-nitrosylation by cysteine reactive tandem mass tag switch assay, *Mol. Cell. Proteomics* 11, M111.01344, 2012.

99. Hung, C.W. and Tholey, A., Tandem mass tag protein labelling for top-down identification and quantification, *Anal. Chem.* 84, 161–170, 2012.

100. Forrester, M.T., Thompson J.W., Foster, M.W. et al., Proteomic analysis of *S*-nitrosylation and denitrosylation by resin-assisted capture, *Nat. Biotechnol.* 27, 557–559, 2009.

101. Egorov, T.A., Svenson, A., Rydén, L., and Carlsson, J., A rapid and specific method for isolation of thiol-containing peptides from large proteins by thiol-disulfide exchange on a solid support, *Proc. Natl. Acad. Sci. USA* 72, 3029–3033, 1975.

102. Landino, L.M., Koumas, M.T., Mason, C.E., and Alston, J.A., Ascorbic acid reduction of microtubule protein disulfides and its relevance to protein *S*-nitrosylation assays, *Biochem. Biophys. Res. Commun.* 340, 347–352, 2006.

103. Forrester, M.T., Foster, M.W., and Stamler, J.S., Assessment and application of the biotin switch technique for examining protein S-nitrosylation under conditions of pharmacologically induced oxidative stress, *J. Biol. Chem.* 282, 13977–13983, 2007.

104. Giustarini, D., Dalle-Donne, I., Colombo, R. et al., Is ascorbate able to reduce disulfide bridges? A cautionary note, *Nitric Oxide* 19, 252–258, 2008.

105. Komatsubara, A.T., Asano, T., Tsumoto, H. et al., Proteomic analysis of S-nitrosylation induced by 1-methyl-4-phenylpyridinium (MPP+), *Proteome Sci.* 10, 74, 2012.

106. Kallakunta, V.M., Staruch, A., and Matus, B., Sinapinic acid can replace ascorbate in the biotin switch assay, *Biochim. Biophys. Acta* 1800, 23–30, 2010.

107. Wiesweg, M., Berchner-Pfannschmidt, U., Fandrey, J. et al., Rocket fuel for the quantification of S-nitrosothiols. High specific reduction of S-nitrosothiols to thiols by methylhydrazine, *Free Radic. Res.* 47, 104–115, 2013.

108. Hou, Y., Wang, J., Arias, F. et al., Electrochemical studies of S-nitrosothiols, *Bioorg. Med. Chem. Lett.* 8, 3065–3070, 1998.

109. Peng, B. and Meyerhoff, M.E., Reexamination of the direct electrochemical reduction of S-nitrosothiols, *Electroanalysis* 25, 914–921, 2013.

110. Wang, X., Kettenhofen, N.J., Shiva, S. et al., Copper dependence of the biotin switch assay: Modified assay for measuring cellular and blood nitrosylated proteins, *Free Radic. Biol. Med.* 44, 1362–1372, 2008.

111. Huang, B., Liao, C.L., Lin, Y.P. et al., S-Nitrosoproteome in endothelial cells revealed by a modified biotin switch approach coupled with Western blot-based two-dimensional gel electrophoresis, *J. Proteome Res.* 8, 4835–4843, 2009.

112. Chen, Y.J., Ku, W.C., Lin, P.Y. et al., S-Alkylating labeling strategy for site-specific identification of the S-nitrosoproteome, *J. Proteome Res.* 9, 6417–6439, 2010.

113. Wiktorowicz, J.E., Stafford, S., Rea, H. et al., Quantification of cysteinyl S-nitrosylation by fluorescence in unbiased proteomic studies, *Biochemistry* 50, 5601–5714, 2011.

114. Burns, R.N. and Moniri, N.H., Agonist- and hydrogen peroxide-mediated oxidation of the β2 adrenergic receptor: Evidence of receptor S-sulfenylation as detected by a modified biotin-switch assay, *J. Pharmacol. Exp. Ther.* 339, 914–921, 2011.

115. Pan, J. and Carroll, K.S., Persulfide reactivity in the detection of protein S-sulfhydration, *ACS Chem. Biol.* 1110–1016, 2013.

116. Millea, P.I., N-Acetylcysteine: Multiple clinical applications, *Am. Fam. Physician* 80, 265–169, 2009.

117. Jegatheeswaran, S. and Siriwardena, A.K., Experimental and clinical evidence for modification of hepatic ischaemia-reperfusion injury by N-acetylcysteine during major liver surgery, *HPB(Oxford)* 13, 71–78, 2011.

118. Szakmany, T., Hauser, B., and Radermacher, P., N-Acetylcysteine for sepsis and systemic inflammatory response in adults, *Cochrane Database Syst. Rev.* 9, CD006616, 2012.

119. Berk, M., Malhi, G.S., Gray, L.J., and Dean, O.M., The promise of N-acetylcysteine in neuropsychiatry, *Trends Pharmacol. Sci.* 34, 167–177, 2013.

120. Samuni, Y., Goldstein, S., Dean, O.M., and Berk, M., The chemistry and biological activities of N-acetylcysteine, *Biochim. Biophys. Acta* 1830, 4117–4129, 2013.

121. Whiteside, W.M., Sears, D.N., Young, P.R., and Rubin, D.B., Properties of selected S-nitrosothiols compared to nitrosylated WR-1065, *Radiat. Res.* 157, 578–588, 2002.

122. Yu, C.-Y., Li, S., and Whorton, A.R., Redox regulation of PTEM by S-nitrosothiols, *Mol. Pharmacol.* 68, 847–854, 2005.

123. Zhang, Y. and Hogg, N., The mechanism of transmembrane S-nitrosothiol transport, *Proc. Natl. Acad. Sci. USA*, 101, 7891–7896, 2004.

124. Li, S. and Whorton, A.R., Identification of stereoselective transporters for S-nitroso-L-cysteine: Role of LAT1 and LAT2 in biological activity of S-nitrosothiols, *J. Biol. Chem.* 280, 20102–20110, 2005.

125. Sandmann, J., Schwedhelm, K.S., and Tsikas, D., Specific transport of S-nitrosocysteine in human red blood cells: Implications of formation of S-nitrosothiols and transport of NO bioactivity within the vasculature, *FEBS Lett.* 579, 4119–4124, 2005.

126. Numajrir, N., Takasawa, K., Nishiya, T. et al., On-off system for P13-kinase-Akt signaling through S-nitrosylation on phosphatase with sequence homology to tensin (PTEN), *Proc. Natl. Acad. Sci. USA* 108, 10349–10354, 2011.

127. McCarthy, S.M., Bove, P.F., Matthews, D.E. et al., Nitric oxide regulation of MMP-9 activation and its relationship to modifications of the cysteine switch, *Biochemistry* 47, 5832–5840, 2008.

128. Engelman, R., Weisman-Shomer, P., Ziv, T. et al., Multilevel regulation of 2-cys peroxiredoxin reaction cycle by *S*-nitrosylation, *J. Biol. Chem.* 288, 11312–11324, 2013.
129. Mieyal, J.J., Gallogly, M.M., Qanungo, S. et al., Molecular mechanisms and clinical implications of reversible protein *S*-glutathionylation, *Antioxid. Redox Signal.* 10, 1941–1988, 2008.
130. Zee, R.S., Vos, C.B., Pimentel, D.R. et al., Redox regulation of sirtuin-1 by *S*-glutathionylation, *Antioxid. Redox Signal.* 13, 1023–1032, 2010.
131. Morakinya, M.K., Stanzia, R.M., and Simoyi, R.H., Modulation of homocysteine toxicity by *S*-nitrosylation formation: A mechanistic approach, *J. Phys. Chem. B* 114, 9894–9904, 2010.
132. Cheng, F., Svensson, G., Fransson, L.-A., and Mani, K., Non-conserved, *S*-nitrosylated cysteine in glypican-1 react with *N*-unsubstituted glucosamines in heparin sulfate and catalytic deaminative cleavage, *Glycobiology* 22, 1480–1486, 2012.
133. De Cat, B. and David, G., Developmental roles of the glypicans, *Semin. Cell Dev. Biol.* 12, 117–125, 2001.
134. Filmus, J., Glypicans in growth control and cancer, *Glycobiology* 11, 19R–23R, 2001.
135. Iozzo, R.V. and Sanderson, R.D., Proteoglycans in cancer biology, tumor microenvironments and angiogenesis, *J. Cell Mol. Med.* 15, 1013–1031, 2011.
136. Creighton, T.E., Role of the environment in the refolding of reduced pancreatic trypsin inhibitor, *J. Mol. Biol.* 144, 521–550, 1980.
137. Dyson, H.J., Jen, M.-F., Tennant, L.L. et al., Effects of buried charged groups of cysteine thiol ionization and reactivity in *Escherichia coli* thioredoxin: Structure and functional characterization of mutants of Asp 26 and Lys 57, *Biochemistry* 36, 2622–2636, 1997.
138. Ding, K., Mani, K., Cheng, F. et al., Copper-dependent autocleavage of glypican-1 heparan sulfate by nitric oxide derived from intrinsic nitrosothiols, *J. Biol. Chem.* 277, 33353–33360, 2002.
139. Gu, J. and Lewis, R.S., Effect of pH and metal ions on the decomposition rate of *S*-nitrosocysteine, *J. Biomed. Eng.* 35, 1554–1560, 2007.
140. Dicks, A.P., Belosa, P.H., and Williams, D.L.H., Decomposition of *S*-nitrosothiols: The effect of added thiols, *J. Chem. Soc. Perkins Trans.* 2, 1429–1434, 1997.
141. Toubin, C., Yeung, D.Y.-H., English, A.M., and Peslherbe, G.H., Theoretical evidence that Cu+ complexation promotes degradation of *S*-nitrosothiols, *J. Am. Chem. Soc.* 124, 14816–14817, 2002.
142. Tao, L. and English, A.M., Mechanism of *S*-nitrosylation of recombinant human brain calbindin D28K, *Biochemistry* 42, 3326–3334, 2003.
143. Ogulener, N. and Ergun, Y., Neocuproine inhibits the decomposition of endogenous *S*-nitrosothiol by ultraviolet irradiation in the mouse gastric fundus, *Eur. J. Pharmacol.* 485, 269–274, 2004.
144. Woulfe, P., Experiments to shew the nature of Aurum Mosaicum, *Philos. Trans.* 61, 114–130, 1771.
145. Johnson, T.B. and Kohmann, E.F., Studies on nitrated proteins. I. The determination of the structure of nitrotyrosine, *J. Am. Chem. Soc.* 37, 1863–1884, 1915.
146. Wormall, A., Immunological specificity of chemically altered proteins. Halogenated and nitrated proteins, *J. Exp. Med.* 5, 295–317, 1930.
147. Johnson, T.B. and Kohmann, E.F., Nitrated proteins. II. Synthesis of 3,5-dinitrotyrosine, *J. Am. Chem. Soc.* 37, 2164–2170, 1915.
148. Pechmann, E.v., Über enzymatische hydrolyse von xanthoproteinen und deren verwendung zur colorimetrischen bestimmung proteolytischer fermente, *Biochem. Zeit.* 321, 248–260, 1950.
149. Noack, R., Koldovský, O., Friedrich, M. et al., Proteolytic and peptidase activities of the jejunum and ileum of the rat during postnatal development, *Biochem. J.* 100, 775–778, 1966.
150. Osborne, T.B. and Campbell, G.F., Conglutin and vitellin, *J. Am. Chem. Soc.* 18, 609–623, 1896.
151. Buruiana, L.M., Queiques propriétés spécifques de la substance hyaline de O.Hammersten, *Naturwissenschaften* 44, 41–42, 1957.
152. Larsson, L., Sörbo, B., Tiselius, H.G., and Ohman, S., A method for quantitative wet chemical analysis of urinary calculi, *Clin. Chim. Acta* 140, 9–20, 1984.
153. Marcone, M.F., Characterization of the edible bird's nest the "Caviar of the East," *Food Res. Int.* 38, 1125–1134, 2005.
154. Cordier, D., Grasset, L., and Ville, A., Nitration of tyrosine in fibrous protein silk, *Process Biochem.* 19, 225–227, 1984.
155. Bible, K.C., Boerner, S.A., and Kaufmann, S.H., A one-step method for protein estimation in biological sample: Nitration of tyrosine in nitric acid, *Anal. Biochem.* 267, 217–221, 1999.
156. Boerner, S.A., Lee, Y.K., Kaufmann, S.H., and Bible, K.C., The nitric acid method for protein estimation in biological samples, in *Protein Protocols Handbook*, 2nd edn., ed. J.M. Walker, Chapter 6, pp. 31–40, Humana Press, Totowa, NJ, 2002.

157. Ridd, J.H., The range of radical processes in nitration by nitric acid, *Chem. Soc. Rev.* 20, 149–165, 1991.

158. Freire de Queiroz, J., Walkimar de M. Carmeiro, J., and Sabino, A.A., Electrophilic aromatic nitration: Understanding its mechanism and substituent effects, *J. Org. Chem.* 71, 6192–6203, 2006.

159. Fujisawa, Y., Kato, K., and Giulivi, C., Nitration of tyrosine residues 368 and 345 in the β-subunit elicits FoF1-ATPase activity loss, *Biochem. J.* 423, 219–231, 2009.

160. Ghesquière, B., Colaert, N., Helsens, K. et al., In vitro and in vivo protein-bound tyrosine nitration characterized by diagonal chromatography, *Mol. Cell. Proteomics* 8, 2642–2652, 2009.

161. Zhang, Y., Yang, H., and Pöschl, U., Analysis of nitrated proteins and tryptic peptides by HPLC-chip-MS/MS: Site-specific quantification, nitration degree, and reactivity of tyrosine residues, *Anal. Bioanal. Chem.* 399, 459–471, 2011.

162. Nitric oxide. Part D. Nitric oxide, detection, mitochondria and cell function, and peroxynitrite reactions, *Methods Enzymology*, Vol. 359, ed. E. Cadenas and L. Packer, Academic Press, New York, 2002.

163. Vana, L., Kanaan, N.M., Hakala, K. et al., Peroxynitrite-induced nitrative and oxidative modifications alter tau filament formation, *Biochemistry* 50, 1203–1212, 2011.

164. Guinga-Cagmat, J.D., Stevens, S.M., Jr., Ratliff, M.V. et al., Identification of tyrosine nitration in UCH-L1 and GAPDH, *Electrophoresis* 32, 1692–1705, 2011.

165. Corpas, F.J., Leterrier, M., Begara-Morales, J.C. et al., Inhibition of peroxisomal hydroxypyruvate reductase (HPR1) by tyrosine nitration, *Biochim. Biophys. Acta* 1830, 4981–4989, 2013.

166. Díaz-Moreno, I., Nieto, P.M., Del Conte, R. et al., A non-damaging method to analyze the configuration and dynamics of nitrotyrosines in proteins, *Chemistry* 18, 3872–3878, 2012.

167. Radi, R., Beckman, J.S., Bush, K.M., and Freeman, B.A., Peroxynitrite oxidation of sulfhydryl. The cytotoxic potential of superoxide and nitric oxide, *J. Biol. Chem.* 266, 4244–4250, 1991.

168. Wardman, P., Methods to measure the reactivity of peroxynitrite-derived oxidants toward reduced fluoresceins and rhodamines, *Methods Enzymol.* 441, 261–282, 2008.

169. Grossi, L., Montevecchi, P.C., and Strazzari, S., S-Nitrosothiol and disulfide formation through peroxynitrite-promoted oxidation of thiols. *Eur. J. Org. Chem.* 2001, 131–135, 2001.

170. Qiojano, C., Alvarez, B., Gatti, R.M. et al., Pathways of peroxynitrite oxidation of thiol groups, *Biochem. J.* 322, 167–173, 1997.

171. Lane, R.S. and Dekker, E.E., Oxidation of sulfhydryl groups of bovine liver 2-keto-4-hydroxyglutarate aldolase by tetranitromethane, *Biochemistry* 11, 3295–3303, 1972.

172. Hsieh, W.T. and Matthews, K.S., Tetranitromethane modification of the tyrosine residues of the lactose repressor, *J. Biol. Chem.* 256, 4856–4862, 1981.

173. Prozororvski, V., Krook, M., Atrian, S. et al., Identification of reactive tyrosine residues in cysteine-reactive dehydrogenases, *FEBS Lett.* 304, 46–50, 1992.

174. Hantzsch, A. and Rinckenberger, A., Ueber nitroform, *Ber. Dtsch. Chem. Ges.* 37, 628–641, 1899.

175. Werner, A., Aur Frage nach den Beziehungen zwischen Farbe und Konstitution, *Ber. Dtsch. Chem. Ges.* 42, 4324–4328, 1910.

176. Chattaway, F.D., Simple method of preparing tetranitromethane, *J. Chem. Soc. Trans.* 97, 2099–2102, 1910.

177. Menzies, A.W.C., The vapor pressure of tetranitro-methane, *J. Am. Chem. Soc.* 41, 1336–1337, 1919.

178. Clarke, H.T., Macbeth, A.K., and Stewart, A.W., Colours produced by tetranitromethane with compounds containing elements capable of showing change in valency, *Proc. Chem. Soc.* 138, 161–162, 1913.

179. Harper, E.M. and Macbeth, A.K., IX. Colorations produced by some organic nitro-compounds with special reference to tetranitromethane, *J. Chem. Soc. Trans.* 107, 87–96, 1915.

180. Macbeth, A.K., CXCVII. Colorations produced by some organic nitro-compounds with special reference to tetranitromethane—Part II, *J. Chem. Soc. Trans.* 107, 1824–1827, 1915.

181. Schmidt, E., Schuacer, R., Bajen, W., and Wagnzr, A., [Tetranitromethane. V. Tetranitromethane as a nitrating agent. II], *Ber. Dtsch. Chem. Ges. B*, 55B, 1751–1759, 1922.

182. Lagercrantz, C. and Yhland, M., Light-induced free-radicals in solutions of some unsaturated compounds and tetranitromethane, *Acta. Chem. Scand.* 16, 1807–1809, 1962.

183. Bradshaw, R.W., The reaction of tetranitromethane with olefins, *Tetrahedron Lett.* 7, 5711–5716, 1966.

184. Spande, T.F., Fontana, A., and Witkop, B., An unusual reaction of skatole with tetranitromethane, *J. Am. Chem. Soc.* 91, 6199–6200, 1969.

185. Heibronner, E., Zur tetranitromethane-probe, *Helv. Chim. Acta* 36, 1121–1124, 1953.

186. Atassi, M.Z., Nitration of the vinyl groups of ferriheme, *Biochim. Biophys. Acta* 177, 663–665, 1969.

187. Yi, J., Thomas, L.M., and Musayev, F.N., Crystallographic trapping of heme loss intermediates during the nitrite-induced degradation of human hemoglobin, *Biochemistry* 50, 8323–8332, 2011.

188. Bonnett, R. and Martin, R.A., Interaction of nitrite with hemes and related compounds, *IARC Sci. Publ.* 14(Enivon.N-nitroso compd.anal.form.proc.work, 1975), 487–493, 1975.

189. Bian, K., Gao, Z., Weisbrodt, N., and Muran, F., The nature of heme/iron-induced protein tyrosine nitrations, *Proc. Natl. Acad. Sci. USA* 100, 5712–5717, 2003.

190. Schmidt, E. and Fischer, H., Zur kenntnis des tetranitro-methans. II. Mitteilung: Tetranitro-methan als nitrierungsmittel (I), *Ber. Dtsch. Chem. Ges. Abteilung B*, 53B, 1529–1537, 1920.

191. *Handbook of Data on Common Organic Compounds*, Vol. II, ed. D.R. Lide and G.W.A. Milne, CRC Press, Boca Raton, FL, 1995.

192. *Handbook of Physical Properties of Organic Chemicals*, ed. P.H. Howard and W.M. Meylan, Lewis Publishers/CRC Press, Boca Raton, FL, 1997.

193. Liang, P., Tetranitromethane, in *Organic Syntheses*, Collective Vol. 3, ed. E.C. Horning, p. 803, John Wiley, Hoboken, NJ, 1955.

194. Feiser, L.F. and Feiser, M., *Reagents for Organic Synthesis*, Vol. 1, p. 1147, Wiley Interscience, New York, 1967.

195. Trent, M.S., Wosham, L.M.S., and Ernst-Fonberg, M., HlyC, the internal protein acyltransferase that activates hemolysin toxin: The role of conserved tyrosine and arginine in enzymatic activity as probed by chemical modification and site-directed mutagenesis, *Biochemistry* 38, 8831–8838, 1999.

196. Riordan, J.F. and Vallee, B.L., Nitration with tetranitromethane, *Methods Enzymol.* 25, 515–521, 1972.

197. Dabbous, M.K., Seif, M., and Brinkley, E.C., The action of tetranitromethane on acid-soluble tropocollagen, *Biochem. Biophys. Res. Commun.* 48, 1586–1592, 1972.

198. Bruice, T.C., Gregory, M.J., and Walters, S.L., Reactions of tetranitromethane. I. Kinetics and mechanisms of nitration of phenols by tetranitromethane, *J. Am. Chem. Soc.* 90, 1612–1619, 1968.

199. Calvert, J.L., Eberson, L., Hartshorn, M.P. et al., Photochemical nitration by tetranitromethane. XVI. The regiochemistry of adduct formation in the photochemical reaction of 1,8-dimethylnaphthalene and tetranitromethane; thermal 1,3-dipolar nitro addition reactions, *Aust. J. Chem.* 47, 1211–1222, 1994.

200. Eberson, L., Hartshorn, M.P., and Persson, O., Photochemical nitration by tetranitromethane. Part XLII. Photolysis of some 4-methoxystyrene derivatives with tetranitromethane, *Acta Chem. Scand.* 52, 745–750, 1998.

201. Eberson, L. and Hartshorn, M.P., The formation and reactions of adducts from the photochemical reactions of aromatic compounds with tetranitromethane and other X-NO$_2$ reagents, *Aust. J. Chem.* 51, 1061–1081, 1998.

202. Schürmann, K.S. and Lehnig, M., Mechanistic studies of the photochemical nitration of phenols, 1,2-dimethoxybenzene and anisole with tetranitromethane by ^{15}N CIDNP, *Appl. Magn. Reson.* 18, 375–384, 2000.

203. Rasmussen, M., Aakesson, E., Eberson, L., and Sundstromem, V., Ultrafast formation of tetranitromethanide (C(NO$_2$)$_3^-$) by photoinduced dissociative electron transfer and subsequent ion pair coupling reaction in acetonitrile and dichloromethane, *J. Phys. Chem. B* 105, 2027–2035, 2001.

204. Naqvi, K.R. and Melo, T.B., Reduction of tetranitromethane by electronically excited aromatics in acetonitrile: Spectra and molar absorption coefficients of radical cations of anthracene, phenanthrene, and pyrene, *Chem. Phys. Lett.* 428, 83–87, 2006.

205. Pulatti, M., Arguello, J.E., and Penenory, A.B., Photochemical electron-transfer generation of arylthiirane radical cations with tetranitromethane and chloranil—Some novel observations, *Eur. J. Org. Chem.* 4528–4536, 2006.

206. Issacs, N.S. and Abed, O.H., The mechanism of aromatic nitration by tetranitromethane, *Tetrahedron Lett.* 23, 2799–2802, 1982.

207. Shiraiwa, M., Selzie, K., Yang, H. et al., Multiphase chemical kinetics of the nitration of aerosolized protein by ozone and nitrogen dioxide, *Environ. Sci. Technol.* 46, 6672–6680, 2012.

208. Capellos, C., Iyer, S., Liang, Y., and Gamms, L.A., Transient species and product formation from electronically excited tetranitromethane, *J. Chem. Soc. Faraday Trans.* 82, 2195–2206, 1986.

209. Shao, J., Geacintov, N.E., and Shafirovich, V., Oxidative modification of guanine bases initiated by oxyl radicals derived from photolysis of azo compounds, *J. Phys. Chem. B* 114, 6685–6692, 2010.

210. Rabani, J., Mulac, W.A., and Matheson, M.S., The pulse radiolysis of tetranitromethane, *J. Phys. Chem.* 69, 53–70, 1965.

211. Chaudhri, S.A. and Asmus, K.-D., Dissociation equilibrium of nitroform in polar solvents studied by pulse radiolysis, *J. Chem. Soc. Faraday Trans.* 1972(1), 385–392, 1972.

212. Kadhum, A.A.H. and Salmon, G.A., Reactivity of solvated electron in tetrahydrofuran, *J. Chem. Soc. Faraday Trans.* 82, 2521–2530, 1986.

213. Johnson, D.W. and Salmon, G.A., The yield and extinction coefficient of the solvated electron in methanol: Pulse radiolysis of nitrobenzene and tetranitromethane solutions, *Can. J. Chem.* 55, 2030–2043, 1977.

214. Gardes-Albert, M., Jore, D., Abedinzadeh, Z. et al., Reduction of tetranitromethane by primary species formed in radiolysis of water by heavy Ar18+ ions, *J. Chim. Phys. Phys.-Chim. Biol.* 93, 103–110, 1996.
215. Janik, I., Ulanski, P., and Rosiak, J.M., Pulse radiolysis of poly(vinyl methyl ether) in aqueous solution: Formation and structure of primary radicals, *Nucl. Instrum. Methods Phys. Res. B* 15, 318–323, 1999.
216. Christen, P. and Riordan, J.F., Spectrophotometric determination of carbanions with tetranitromethane, *Anal. Chim. Acta* 51, 47–52, 1970.
217. Malhotra, O.P. and Dwivedi, U.N., Formation of enzyme-bound carbanion-intermediate in the isocitrate lyase-catalyzed reaction: Enzymatic reaction of tetranitromethane with substrates and its dependence on effector, pH, and metal ions, *Arch. Biochem. Biophys.* 250, 238–248, 1986.
218. Jewett, S.W. and Bruice, T.C., Reactions of tetranitromethane. Mechanisms of the reaction of tetranitromethane with pseudo acids, *Biochemistry* 11, 3338–3350, 1972.
219. Walters, S.L. and Bruice, T.C., Reactions of tetranitromethane. II. The kinetics and products for the reaction of tetranitromethane with inorganic ions and alcohols, *J. Am. Chem. Soc.* 93, 2269–2282, 1971.
220. Li, B., Held, J.M., Schilling, B. et al., Confident identification of 3-nitrotyrosine modifications in mass spectral data across multiple mass spectrometry platforms, *J. Proteomics* 74, 2510–2521, 2011.
221. Koppenol, W.H. and Trayham, J.B., Say NO to nitric oxide: Nomenclature for nitrogen- and oxygen-containing compounds, *Methods Enzymol.* 268, 3–7, 1996.
222. Koppenol, W.H., NO nomenclature?, *Nitric Oxide Biol. Chem.* 6, 96–100, 2002.
223. Pryor, W.A. and Squadrito, G.L., The chemistry of peroxynitrite: A product of the reaction of nitric oxide with superoxide, *Am. J. Physiol.* 268, L699–L722, 1995.
224. Klotz, L.-O. and Sies, H., Reversible conversion of nitrosyl anion to nitric oxide, *Methods Enzymol.* 349, 101–106, 2002.
225. Uppu, R.M., Squadrito, G.L., Caeto, R., and Pryor, W.A., Selecting the most appropriate synthesis of peroxynitrite, *Methods Enzymol.* 269, 285–295, 1996.
226. Uppu, R.M. and Pryor, W.A., Synthesis of peroxynitrite in a two-phase system isoamyl nitrite and hydrogen peroxide, *Anal. Biochem.* 236, 242–249, 1996.
227. Koppenol, W.H., Kissner, R., and Beckman, J.S., Synthesis of peroxynitrite: To go with the flow or on solid grounds?, *Methods Enzymol.* 269, 296–302, 1996.
228. Uppu, R.M., Squadrito, G.L., Cueto, R., and Pryor, W.A., Synthesis of peroxynitrite by azide-ozone reaction, *Methods Enzymol.* 269, 311–321, 1996.
229. Uppu, R.M. and Pryor, W.A., Biphasic synthesis of high concentrations of peroxynitrite using water-insoluble alkyl nitrite and hydrogen peroxide, *Methods Enzymol.* 269, 322–329, 1996.
230. Saha, A., Goldstein, S., Cabelli, D., and Czapski, G., Determination of optimal conditions for synthesis of peroxynitrite by mixing acidified hydrogen peroxide with nitrite, *Free Radic. Biol. Med.* 24, 653–659, 1998.
231. Wang, C. and Deen, W.M., Peroxynitrite delivery methods for toxicity studies, *Chem. Res. Toxicol.* 17, 32–44, 2004.
232. Robinson, K.M. and Beckman, J.S., Synthesis of peroxynitrite from nitrite and hydrogen peroxide, *Methods Enzymol.* 396, 207–214, 2005.
233. Uppu, R.M., Synthesis of peroxynitrite using isoamyl nitrite and hydrogen peroxide in a homogeneous solvent system, *Anal. Biochem.* 354, 165–168, 2006.
234. Sampson, J.B., Ye, Y., Rosen, H., and Beckman, J.S., Myeloperoxidase and horseradish peroxidase catalyze tyrosine nitration in proteins from nitrite and hydrogen peroxide, *Arch. Biochem. Biophys.* 356, 207–213, 1998.
235. Godber, B.L., Doel, J.J., Durgan, J. et al., A new route to peroxynitrite: A role for xanthine oxidoreductase, *FEBS Lett.* 475, 93–96, 2000.
236. Sawa, T., Akaike, T., and Maeda, H., Tyrosine nitration by peroxynitrite formed from nitric oxide and superoxide generated by xanthine oxidase, *J. Biol. Chem.* 275, 32467–32474, 2000.
237. MIllar, T.M., Peroxynitrite formation from the simultaneous reduction of nitrite and oxygen by xanthine oxidase, *FEBS Lett.* 562, 129–133, 2004.
238. Hughes, M.N. and Nicklin, H.G., The chemistry of perinitrite. Part I. Kinetics of decomposition of perinitrous acid, *J. Chem. Soc.* (A), 450–452, 1968.
239. Priyam, A., Bhattarcharya, S.C., and Saba, A., Volatile interface of biological oxidant and luminescent CdTe quantum dots: Implications in nanodiagnostics, *Phys. Chem. Chem. Phys.* 11, 520–527, 2009.
240. Kirsch, M., Korth, H.-G., Wensing, A. et al., Product formation and kinetic simulations in the pH range 1–14 account for a free-radical mechanism of peroxynitrite decomposition, *Arch. Biochem. Biophys.* 418, 133–150, 2003.

241. Goldstein, S., Czapski, G., Lind, J., and Merénye, G., Effect of ·NO on the decomposition of peroxynitrite reaction of N_2O_3 with ONOO⁻, *Chem. Res. Toxicol.* 12, 132–136, 1999.

242. Kissner, R., Beckman, J.S., and Koppenol, W.H., Peroxynitrite studied by stopped-flow spectroscopy, *Methods Enzymol.* 301, 342–352, 1999.

243. Bohle, D.S., Glassbrenner, P.A., and Hansert, B., Synthesis of pure tetramethylammonium peroxynitrite, *Methods Enzymol.* 269, 302–311, 1996.

244. Kissner, R., Nauser, T., Bugnon, P. et al., Formation and properties of peroxynitrite as studied by laser flash photolysis, high-pressure stopped flow technique and pulse radiolysis, *Chem. Res. Toxicol.* 10, 1285–1292, 1997.

245. Keynes, R.G., Griffith, C., and Garthwaite, J., Superoxide-dependent consumption of nitric oxide in biological media may confound in vitro experiments, *Biochem. J.* 369, 399–406, 2003.

246. Pollet, E., Martinez, A., Methan, B. et al., Role of tryptophan oxidation in peroxynitrite-dependent protein chemiluminescence, *Arch. Biochem. Biophys.* 349, 74–80, 1998.

247. Gadelha, F.R., Thomson, L., Fagian, M.M. et al., Ca^{2+}-independent permeabilization of the inner mitochondrial membrane by peroxynitrite is mediated by membrane protein thiol cross-linking and lipid peroxidation, *Arch. Biochem. Biophys.* 345, 243–250, 1997.

248. Molina, C., Kissner, R., and Koppenol, W.H., Decomposition kinetics of peroxynitrite: Influence of pH and buffer, *Dalton Trans.* 42, 9898–9905, 2013.

249. Ascenzi, P., di Masi, A., Coletta, M. et al., Ibuprofen impairs allosterically peroxynitrite isomerization by ferric human serum heme-albumin, *J. Biol. Chem.* 284, 31006–31017, 2009.

250. Schmidt, K., Pfeiffer, S., and Mayer, B., Reaction of peroxynitrite with HEPES or MOPS results in the formation of nitric oxide donors, *Free Radic. Biol. Med.* 24, 859–862, 1998.

251. Kirsch, M., Lomonosova, E.E., Korth, H.-G. et al., Hydrogen peroxide formation by reaction of peroxynitrite with HEPES and related tertiary amines. Implications for a general mechanism, *J. Biol. Chem.* 273, 12716–12724, 1998.

252. Crow, J.P., Spruell, C., Chen, J. et al., On the pH-dependent yield of hydroxyl radical products from peroxynitrite, *Free Radic. Biol. Med.* 16, 331–338, 1994.

253. Symons, M.C.R., *Cis-* and *Trans-*conformations for peroxynitrite anions, *J. Inorg. Biochem.* 78, 299–301, 2000.

254. Lang, B. and Andrews, L., Infrared spectra of *cis-* and *trans-*peroxynitrite anion, OONO-, in solid argon, *J. Am. Chem. Soc.* 123, 9848–9854, 2001.

255. Alvarez, B., Ferrer-Sueta, G., and Radi, R., Slowing of peroxynitrite decomposition in the presence of mannitol and ethanol, *Free Radic. Biol. Med.* 24, 1331–1337, 1998.

256. Deng, X.-S., Bludeau, P., and Deitrich, R.A., Formation of ethyl nitrite after ethanol administration, *Alcohol* 34, 217–223, 2004.

257. Cao, Z. and Li, Y., Potent inhibition of peroxynitrite-induced DNA strand breakage by ethanol: Possible implications for ethanol-mediated cardiovascular protection, *Pharmacol. Res.* 50, 13–19, 2004.

258. Cohen, S., Biological reactions of carbonyl halides, in *The Chemistry of Acyl Halides*, ed. S. Patel, Chapter 10, pp. 313–348, Interscience/Wiley, London, U.K., 1972.

259. Peshenko, I.V. and Shichi, H., Oxidation of active center cysteine of bovine 1-cys peroxiredoxin to the cysteine sulfenic acid form by peroxide and peroxynitrite, *Free Radic. Biol. Med.* 31, 292–303, 2001.

260. Pryor, W.A., Jin, X., and Squadrito, G.L., One- and two-electron oxidations of methionine by peroxynitrite, *Proc. Natl. Acad. Sci. USA* 91, 11173–11177, 1994.

261. Perrin, D. and Koppenol, W.H., The quantitative oxidation of methionine to methionine sulfoxide by peroxynitrite, *Arch. Biochem. Biophys.* 377, 266–272, 2000.

262. McLean, S., Bowman, L.A.H., Sanguinetti, G. et al., Peroxynitrite toxicity in *Escherichia coli* K12 elicits expression of oxidative stress responses and protein nitration and nitrosylation, *J. Biol. Chem.* 281, 20724–20731, 2010.

263. Kato, Y., Kawaskishi, S., Aoki, T. et al., Oxidative modification of tryptophan residues exposed to peroxynitrite, *Biochem. Biophys. Res. Commun.* 234, 82–84, 1997.

264. Yamakura, F., Matsumoto, T., Fujimura, T. et al., Modification of a single tryptophan residue in human Cu,Zn-superoxide dismutase by peroxynitrite in the presence of bicarbonate, *Biochim. Biophys. Acta* 1548, 38–46, 2001.

265. Yamakura F., Matsumoto, T., Ikeda, K. et al., Nitrated and oxidized products of a single tryptophan residue in human Cu,Zn-superoxide dismutase treated with either peroxynitrite-carbon dioxide or myeloperoxidase-hydrogen peroxide-nitrite, *J. Biochem.* 138, 57–69, 2005.

266. Kawasaki, H., Ikeda, K., Shigenaga, A. et al., Mass spectrometric identification of tryptophan nitration sites on proteins in peroxynitrate-treated lysates from PC12 cells, *Free Radic. Biol. Med.* 50, 419–427, 2011.

267. Ischiropoulos, H., Zhu, L., Chen, J. et al., Peroxynitrite-mediated tyrosine nitration catalyzed by super-oxide dismutase, *Arch. Biochem. Biophys.* 298, 431–437, 1992.

268. Alvarez, B. and Radi, R., Peroxynitrite reactivity with amino acids and proteins, *Amino Acids* 25, 295–311, 2003.

269. Floris, R., Piersma, S.R., Yang, G. et al., Interaction of myeloperoxidase with peroxynitrite. A comparison with lactoperoxidase, horseradish peroxidase and catalase, *Eur. J. Biochem.* 215, 767–775, 1993.

270. Kuhn, D.M. and Geddes, T.J., Peroxynitrite inactivates tryptophan hydroxylase via sulfhydryl oxidation. Coincident nitration of enzyme tyrosyl residues has minimal impact on catalytic activity, *J. Biol. Chem.* 274, 29726–29732, 1999.

271. Battistuzzi, G., Bellei, M., Bartolotti, C.A., and Sole, M., Redox properties of heme peroxidases, *Arch. Biochem. Biophys.* 500, 21–36, 2010.

272. Gundaydin, H. and Houk, K.N., Mechanisms of peroxynitrite-mediated nitration of tyrosine, *Chem. Res. Toxicol.* 22, 894–898, 2009.

273. Radi, R., Cosgrove, T.P., Beckman, J.S., and Freeman, B.A., Peroxynitrite-induced luminol chemilumi-nescence, *Biochem. J.* 290, 51–57, 1993.

274. Lymar, S.V. and Hurst, J.K., Rapid reaction between peroxonitrite and carbon dioxide: Implications for biological activity, *J. Am. Chem. Soc.* 117, 8867–8868, 1995.

275. Tien, M., Berlett, B.S., Levine, R.L. et al., Peroxynitrite-mediated modification of proteins at physiological carbon dioxide concentration: pH dependence of carbonyl formation, tyrosine nitration, and methionine oxidation, *Proc. Natl. Acad. Sci. USA* 96, 7809–7814, 1999.

276. Thomas, D.D., Espey, M.G., Vitek, M.P. et al., Protein nitration is mediated by heme and free metals through Fenton-type chemistry: An alternative to the NO/O_2^- reaction, *Proc. Natl. Acad. Sci. USA* 99, 12691–12696, 2002.

277. Squadrito, G.L. and Pryor, W.A., Mapping the reaction of peroxynitrite with CO_2: Energetics, reactive species, and biological implications, *Chem. Res. Toxicol.* 15, 885–895, 2002.

278. Papina, A.A. and Koppenol, W.H., Two pathways of carbon dioxide catalyzed oxidative coupling of phenol by peroxynitrite, *Chem. Res. Toxicol.* 19, 382–391, 2006.

279. Zhang, H., Joseph, J., Felix, J. et al., Nitration and oxidation of a hydrophobic tyrosine probe by peroxynitrite in membranes: Comparison with nitration and oxidation of tyrosine by peroxynitrite in aqueous solution, *Biochemistry* 40, 7675–7686, 2001.

280. Zhang, H., Bhargava, K., Keszler, A. et al., Transmembrane nitration of hydrophobic tyrosyl peptides. Localization, characterization, mechanism of nitration, and biological implications, *J. Biol. Chem.* 278, 8969–8978, 2003.

281. Buyuidiev, R., Michallova, V., and Titeva, St., Equilibrium solubility and stability of aqueous solutions of molsidomine, *Farmatsiya* (Sofia, Bulgaria) 52, 3–7, 2005.

282. Signorelli, S., Möller, M.N., Coitiño, E.L., and Denicola, A., Nitrogen dioxide solubility and permeation in lipid membranes, *Arch. Biochem. Biophys.* 512, 190–196, 2011.

283. Kim, M.Y., Dong, M., and Dedon, P.C., Effects of peroxynitrite dose and dose rate on DNA damage and mutation in the *supF* shuttle vector, *Chem. Res. Toxicol.* 18, 76–86, 2005.

284. Crow, J.P., Beckman, J.S., and McCord, J.M., Sensitivity of the essential zinc-thiolate moiety of yeast alcohol dehydrogenase to hypochlorite and peroxynitrite, *Biochemistry* 34, 3544–3552, 1995.

285. Shao, B., Bergt, C., Fui, X. et al., Tyrosine 192 in apolipoprotein A-1 is the major site of nitration and chlorination by myeloperoxidase, but only chlorination, markedly impairs ABCA1-dependent cholesterol transport, *J. Biol. Chem.* 280, 5983–5993, 2005.

286. Ischiropoulos, H., Biological selectivity and functional aspects of protein tyrosine nitration, *Biochem. Biophys. Res. Commun.* 305, 776–783, 2003.

287. Reiter, C.D., Teng, R.J., and Beckman, J.S., Superoxide reacts with nitric oxide to nitrate tyrosine at physiological pH via peroxynitrite, *J. Biol. Chem.* 275, 32460–32466, 2000.

288. Souza, J.M., Daikhin, E., Yudkoff, M. et al., Factors determining the selectivity of protein tyrosine nitration, *Arch. Biochem. Biophys.* 371, 169–178, 1995.

289. Beckman, J.S., Beckman, T.W., Chen, J. et al., Apparent hydroxyl radical production by peroxynitrite: Implications for endothelial injury from nitric oxide and superoxide, *Proc. Natl. Acad. Sci. USA* 87, 1620–1624, 1990.

290. Babich, O.A. and Gould, E.S., Electron transfer 151. Decomposition of peroxynitrite as catalyzed by copper (II), *Res. Chem. Intermed.* 28, 575–583, 2002.

291. Beckman, J.S., Ischiropoulos, H., Zhu, L. et al., Kinetics of superoxide dismutase- and iron-catalyzed nitration of phenolics by peroxynitrite, *Arch. Biochem. Biophys.* 298, 435–445, 1992.

292. Qiao, L., Lu, Y., Liu, B., and Girault, H.H., Copper-catalyzed tyrosine-nitration, *J. Am. Chem. Soc.* 133, 19823–19833, 2011.

293. Kohnen, S., Halusiak, E., Mouithys-Mickalad, A. et al., Catalytic activation of copper(II) salts on the reaction of peroxynitrite with propofol in alkaline solution, *Nitric Oxide* 12, 252–260, 2005.

294. Seeley, K.W. and Stevens, S.M., Jr., Investigation of local structural effects on peroxynitrite-mediated tyrosine nitration using targeted mass spectrometry, *J. Proteomics* 75, 1691–1700, 2012.

295. Bayden, A.S., Yakovlev, V.A., Graves, P.R. et al., Factors influencing protein tyrosine nitration-structure-based proteomics models, *Free Radic. Biol. Med.* 50, 749–762, 2011.

296. Yu, H., Venkatarangan, L., Wishnok, J.S., and Tannenbaum, S.R., Quantitation of four guanine oxidation products from reaction of DNA with varying doses of peroxynitrite, *Chem. Res. Toxicol.* 18, 1849–1857, 2005.

297. Cui, L., Ye, W., Prestwich, E.G. et al., Comparative analysis of four oxidized guanine lesions from reactions of DNA with peroxynitrite, singlet oxygen, and γ-radiation, *Chem. Res. Toxicol.* 26, 195–202, 2013.

298. Niles, J.C., Wishnok, J.S., and Tannenbaum, S.R., Peroxynitrite-induced oxidation and nitration products of guanine and 8-oxoguanine: Structures and mechanisms of product formation, *Nitric Oxide* 14, 109–121, 2006.

299. Sawa, T., Tatermichi, M., Akaike, T. et al., Analysis of urinary 8-nitroguanine, a marker of nitrative nucleic acid damage, by high-performance liquid chromatography-electrochemical detection coupled with immunoaffinity purification: Association with cigarette smoking, *Free Radic. Biol. Med.* 40, 711–720, 2006.

300. Salgo, M.G., Stone, K., Squadrito, G.L. et al., Peroxynitrite causes DNA nicks in plasmid pBR322, *Biochem. Biophys. Res. Commun.* 210, 1025–1030, 1985.

301. Zhou, X., Liberman, R.G., Skipper, P.L. et al., Quantification of DNA strand breaks and abasic sites by oxime derivatization and accelerator mass spectrometry: Application to gamma-radiation and peroxynitrite, *Anal. Biochem.* 343, 84–92, 2005.

302. Kennett, E.C. and Davies, M.J., Degradation of extracellular matrix by peroxynitrite/peroxynitrous acid, *Free Radic. Biol. Med.* 45, 716–725, 2008.

303. Al-Assaf, S., Navaratnam, S., Parsons, B.J., and Phillips, G.O., Chain scission of hyaluronan by peroxynitrite, *Arch. Biochem. Biophys.* 411, 73–82, 2003.

304. Stankovska, M., Hrabarova, E., Valachova, K. et al., The degradative action of peroxynitrite on high-molecular-weight hyaluronan, *Neuroendocrinology* 27(Suppl. 2), 31–34, 2006.

305. Kennett, E.C. and Davies, M.J., Degradation of matrix glycosaminoglycans by peroxynitrite/peroxynitrous acid: Evidence for a hydroxyl-radical-like mechanism, *Free Radic. Biol. Med.* 42, 1278–1289, 2007.

306. O'Donnell, V.B., Elserich, J.P., Chumley, P.H. et al., Nitration of unsaturated fatty acids nitric oxide derived reactive nitrogen species peroxynitrite, nitrous acid, nitrogen dioxide, and nitronium ion, *Chem. Res. Toxicol.* 12, 83–92, 1999.

307. Manini, P., Camera, E., Picardo, M. et al., Biomimetic nitration of the linoleic acid metabolite 13-hydroxyoctadienoic acid: Isolation and spectral characterization of novel chain-rearranged epoxy nitro derivatives, *Chem. Phys. Lipids* 151, 51–61, 2008.

308. Batthyany, C., Schopfer, F.J., Baker, P.R.S. et al., Reversible post-translational modification of proteins by nitrated fatty acids *in vivo*, *J. Biol. Chem.* 281, 20450–20463, 2006.

309. Daiber, A., Mehl, M., and Ullrich, V., New aspects in the reaction mechanism of phenol with peroxynitrite: The role of phenoxy radicals, *Nitric Oxide* 2, 259–269, 1998.

310. Uppu, R.M., Lemercier, J.-N., Squadrito, G.L. et al., Nitrosation by peroxynitrite: Use of phenol as a probe, *Arch. Biochem. Biophys.* 358, 1–16, 1998.

311. Halliwell, B., Zhao, K., and Whitman, M., Nitric oxide and peroxynitrite: The ugly, the uglier and the not so good: A personal view of recent controversies, *Free Radic. Res.* 31, 651–659, 1999.

312. Boveris, A., Alvarez, S., Arnaiz, S.L., and Valdez, L.B., Peroxynitrite scavenging by mitochondrial reductants and plant polyphenols, in *Handbook of Antioxidants*, 2nd edn., ed. E. Cadenas and L. Packer, pp. 351–369, Marcel Dekker, New York, 2002.

313. Bhattacharyya, J., Biswas, S., and Datta, A.G., Mode of action of endotoxin: Role of free radicals and antioxidants, *Curr. Med. Chem.* 11, 359–368, 2004.

314. Huang, D., Ou, B., and Prior, R.L., The chemistry behind antioxidant capacity assays, *J. Agric. Food Chem.* 53, 1841–1856, 2005.

315. Uppu, R.M., Nossarman, B.D., Greco, A.J. et al., Cardiovascular effects of peroxynitrite, *Clin. Exp. Pharmacol. Physiol.* 34, 933–937, 2007.

316. Sims, N.R., Nilsson, M., and Muyderman, H., Mitochondrial glutathione: A modulator of brain cell death, *J. Bioenerg. Biomembr.* 36, 329–333, 2004.
317. Arteel, G.E., Briviba, K., and Sies, H., Protection against peroxynitrite, *FEBS Lett.* 445, 226–230, 1999.
318. Muscoli, C., Cuzzocrea, S., Ndengele, M.M. et al., Therapeutic manipulation of peroxynitrite attenuates the development of opiate-induced antinociceptive tolerance in mice, *J. Clin. Invest.* 117, 3530–3539, 2007.
319. Reboucas, J.S., Spasojević, I., and Batinić-Haberle, I., Quality of potent Mn porphyrin-based SOD mimics and peroxynitrite scavengers for pre-clinical mechanistic/therapeutic purposes, *J. Pharm. Biomed. Anal.* 48, 1046–1049, 2008.
320. Chen, A.F., Chen, D.D., Daiber, A. et al., Free radical biology of the cardiovascular system, *Clin. Sci. (Lond.)* 123, 73–91, 2012.
321. Batinić-Haberle, I., Spasojević, I., Tse, H.M. et al., Design of Mn porphyrins for treating oxidative stress injuries and their redox-based regulation of cellular transcriptional activities, *Amino Acids* 42, 95–113, 2012.

5 Oxidation

There are several reagents that can be used for the oxidation of proteins, which can be either a specific or nonspecific procedure. There are some reagents such as *t*-butyl peroxide that have considerable specificity, while there are other reagents such as hypochlorous acid or oxygen radicals that are relatively nonspecific. Specificity of protein oxidation can be influenced by protein sequence[1,2] as well as protein conformation.[3] This latter consideration is of importance for the use of hydroxyl radicals for protein surface mapping.[4] It is observed that, as with other chemical modifications, there can be conformational change secondary to the modification.[5,6]

The in vivo oxidation of proteins can be an important component of redox systems.[7] An example is provided by the methionine–methionine sulfoxide pair and the role of methionine sulfoxide reductase.[8,9] Another example is provided with oxidation and cell signaling.[10–15] Oxidation of lipids[13] can create reactive species such as 2-hydroxy-4-nonenal (HNE) (Figure 5.1), which can react with protein nucleophiles including cysteine, lysine, and histidine (see Chapter 2).[16]

The *in vitro* oxidation of proteins can be an undesirable consequence of the manufacture of biopharmaceuticals.[17,18] The work reported by Lam and coworkers[17] merits comment. The observation that formulation of a Fab antibody fragment with polysorbate 20 (Tween 20) resulted in oxidation of a specific tryptophanyl residue (Trp50) is somewhat unusual. Investigation showed that hydrogen peroxide did not oxidize the tryptophan residue, but replacement of His31 with Asn eliminated the oxidation of the tryptophan residue. It is suggested that peroxide derived from the polysorbate reacted with a metal ion bound to the histidine generating free radical, which then reacted with Trp50. In a conceptually related study, Mozziconacci and coworkers[19] observed that a methionine residue acted as a mediator in the oxidation of a tyrosine residue in Met-enkephalin by hydroxyl radical. Hydroxyl radical was formed by pulse radiolysis of water resulting in the formation of a hydroxylsulfuryl radical or sulfur radical cation followed by the transfer of an electron to a tyrosine residue with concomitant recovery of the methionine residue. In the hydroxyl radical action on Leu-enkephalin, there is direct action resulting in the formation of dihydroxyphenylalanine and tyrosine.

The remedy to the problem of peroxides is to assure that lots of polysorbate and other excipients are free of contaminating materials such as peroxides and aldehydes.[18,20–23] Rigorous quality control of all excipient materials is essential for product quality. The potential impurities in polysorbate 80 are a complex mixture of volatile and nonvolatile aldehydes and peroxides. The presence of impurities such as peroxides in excipients used in the formulation of biopharmaceuticals is of importance as lyophilization will not remove all potentially reactive substances; furthermore, the oxidation reactions can take place at low temperature.[24] While methionine is considered to be the amino acid residue in proteins most sensitive to peroxides, modification may also occur at other amino acid residues including cysteine, lysine, and tryptophan.[24,25] The problem with using the combination of sophisticated chromatographic techniques and mass spectrometry is confounded by challenges in sample management.[26,27] Aggregation has also been advanced as a surrogate assay for biological activity.[28] Interest in the aggregation of therapeutic proteins is based on the relationship between aggregation and immunogenicity.[29,30] Hermeling and colleagues concluded that aggregation is critical for immunogenicity, but oxidation does not necessarily result in immunogenicity.[31] It is recognized that the relationship between methionine oxidation and aggregation is complex.[32] It should be noted that there are potential issues with accelerated stability studies yielding data that are not necessarily indicative of stability at lower temperatures.[33] Oxidation of a single methionine residue in human von Willebrand factor (Met1606) is considered a risk factor for thrombosis.[34] The measurement of

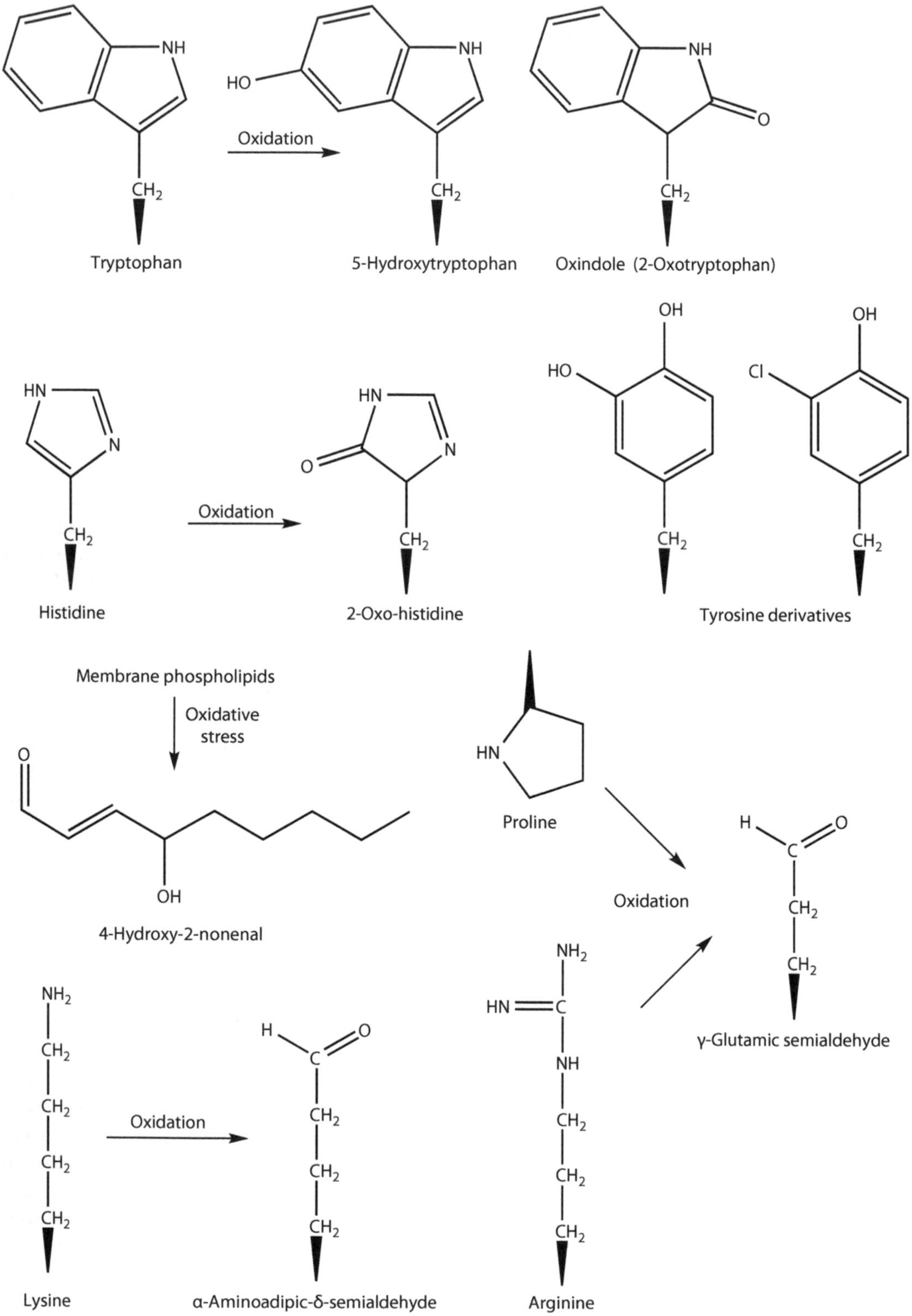

FIGURE 5.1 (continued)

oxidized methionine continues to prove to be challenging. A spectral method has been developed[35] using Fourier transform infrared spectroscopy (FTIR) but requires substantial amounts of protein. Immunoassays have been used to identify modified amino residues such as 3-nitrotyrosine in proteins.[36] An immunoassay for methionine sulfoxide has been reported[37] but has not been verified.[38]

Interest in the role of oxidative stress in disease processes has contributed to the study of the molecular processes involved in in vivo modification of proteins, nucleic acids, and carbohydrates.[39,40] Oxidative modification can also result from environmental sources [41,42] and from exposure to light.[43] It is a major understatement to acknowledge that the advent of mass spectrometry has greatly enhanced the study of oxidative modifications of proteins.[44–46] There are several recent articles on the photooxidation of proteins.[47,48] Pattison and coworkers[48] distinguish direct damage from either ultraviolet B (UVB) (288–320 nm) or ultraviolet A (UVA) (320–400 nm) radiation to tryptophanyl, tyrosyl, histidyl, and cystinyl residues in proteins; indirect damage from singlet oxygen would include cysteinyl and methionyl residues in addition to the four amino acids subject to direct damage. Photooxidation is of importance to the textile industry[49] and to the stability of protein pharmaceuticals.[50] Heavy-chain fibroin, the main protein component of silk thread, is rich in tyrosine. Approximately 25% of the 277 tyrosine residues in heavy chain were subject to photooxidation, and most were in a –GAGY– sequence pattern.[49]

Finally, while largely outside of the primary thrust of the current work, there is considerable interest in the development of biomarkers for the measurement of oxidative stress.[51–53] A variety of derivatives can be obtained from the oxidation of amino acid residues in proteins and some are shown in Figure 5.1. These derivatives can be obtained from the action of hydroxy radicals, peroxynitrous acid, peroxy acids, photochemical reactions, and other substances, some of which are described in the following.

Photooxidation has a long history of use for the modification of proteins but has been frustrated by specificity issues and the emergence of other chemical approaches. Dyes, which can absorb incident radiation resulting in the formation of singlet oxygen such as rose bengal (Figure 5.2), have been used to modify proteins. The ability of dyes such as indocyanine green to absorb incident light and transfer energy to proteins such as albumin is of importance for tissue welding.[54]

Rose bengal dye was introduced by Westhead[55] to complement methylene blue in photooxidation with improved specificity.[55,56] Photooxidation with rose bengal was used to identify the protein(s) at the peptidyl transferase site of a bacterial ribosomal subunit.[57,58] Histidine was the only amino acid modified under the reaction conditions used by these investigators. With the exception of EF-G-GTP-binding activity, the loss of biological activity is most closely related to the *fast* histidine loss. In subsequent experiments, methylene blue dye (Eastman, dye content 91%) was used.[59] Peptidyl transferase activity was lost at a more rapid rate in the presence of methylene blue than rose bengal, but data are not presented regarding any differences in residues modified or whether amino acid residues other than histidine are modified in the presence of methylene blue. Other investigators have also explored the effects of photooxidation on peptidyl transferase activity in *Escherichia coli*

FIGURE 5.1 (continued) Oxidation processes and products important for protein chemistry. Shown are several of the possible structures for tryptophan oxidation products that can be oxidized to other degradation products, the oxidation of histidine to 2-oxohistidine with further oxidation to various degradation products, and some tyrosine oxidation products. Lipids are also subject to oxidation with the formation of derivatives such as 4-hydroxy-2-nonenal, which can modify proteins by a Michael addition (see Chapter 2) or by the formation of a Schiff base (see Chapter 2). Shown at the bottom is the oxidation of lysine, arginine, and protein to form carbonyl derivatives. The process of protein oxidation is complex with multiple sites of modification and multiple products (see G. Aldini, K.-J. Yeum, E. Niki, and R.M. Russell, (eds.), *Biomarkers for Antioxidant Defense and Oxidative Damage, Principles and Practical Applications*, Wiley-Blackwell, Ames, IA, 2010; Froelich, J.M. and Reid, G.E., The origin and control of ex vivo oxidative peptide modifications prior to mass spectrometry analysis, *Proteomics* 8, 1334–1345, 2008; Roeser, J., Bischoff, R., Bruins, A.P., and Permentier, H.P., Oxidative protein labeling in mass-spectrometry-based proteomics, *Anal. Bioanal. Chem.* 397, 3441–3455, 2010).

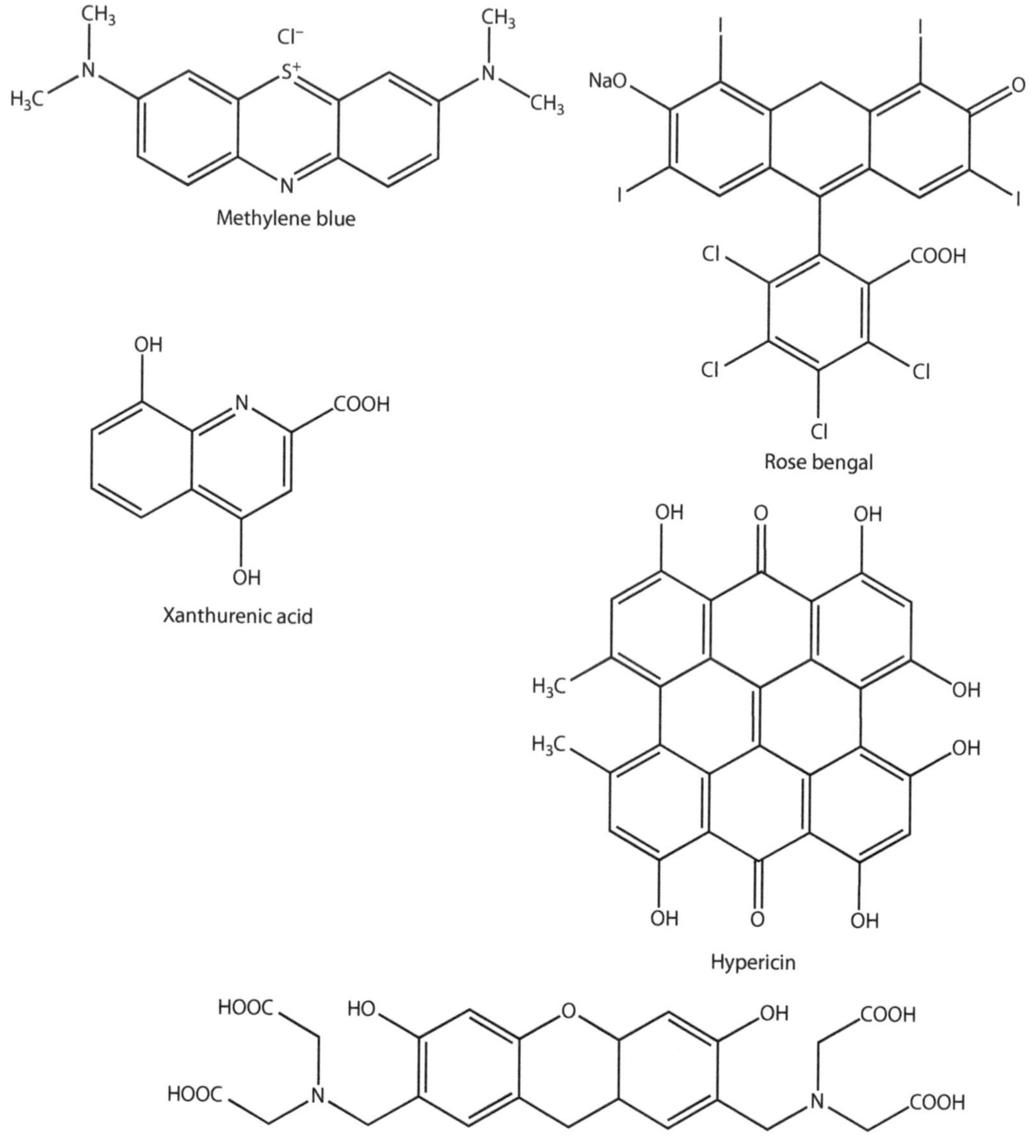

FIGURE 5.2 Structures of some dyes used for photooxidation of proteins. Shown are the structures for methylene blue (Silman, I., Roth, E., Paz, A. et al., The specific interaction of the photosensitizer methylene blue with acetylcholinesterase provides a model system for studying the molecular consequences of photodynamic therapy, *Chem. Biol. Interact.* 203, 63–66, 2012), rose bengal (Suryo Rahmanto, A., Pattison, D.I., and Davies, M.J., Photo-oxidation induced inactivation of the selenium-containing protective enzymes thioredoxin reductase and glutathione peroxidase, *Free Radic. Biol. Med.* 53, 1308–1316, 2012), xanthurenic acid (Roberts, J.E., FInley, E.L., Patat, S.A., and Schey, K.L., Photooxidation of lens proteins with xanthurenic acid: A putative chromophore for cataractogenesis, *Photochem. Photobiol.* 74, 740–744, 2001), hypericin (Kubin, A., Alth, G., Jindra, R. et al. Wavelength-dependent photoresponse of biological and aqueous model systems using the photodynamic plant pigment hypericin, *J. Photochem. Photobiol. B* 36, 103–108, 1996), and calcein (Beghetto, C., Renken, C., Eriksson, O. et al., Implications of the generation of reactive oxygen species by photoactivated calcein for mitochondrial studies, *Eur. J. Biochem.* 267, 5585–5592, 2000) that have been used for photooxidation of proteins. These varying substances share the mechanism of generating singlet oxygen, which then oxidizes amino acid residues such as histidine and tryptophan. Rose bengal is the most widely used of the various photooxidation dyes.

ribosomes.[60] These experiments were performed in 0.030 M Tris, 0.020 M $MgCl_2$, 0.220 KCl, and pH 7.5 (9 mg ribosomes in 0.300 mL) with either eosin or rose bengal as the photooxidation agent. Irradiation was performed at 0°C–4°C using a 500 W slide projector (26 cm from condenser lens to sample) for 20 min. Photooxidation has also been used to study the role of histidine residues in polypeptide chain elongation factor Tu from *E. coli*.[61] The reaction is performed in 0.05 M Tris, 0.010 M Hg $(OAc)_2$, 0.005 M 2-mercaptoethanol, 10% glycerol, and pH 7.9. Irradiation is performed at 0°C–4°C with gentle stirring using a 375 W tungsten lamp at a distance of 15 cm. A glass plate was placed in the light beam to eliminate UV irradiation. The rose bengal dye is removed after 5–30 min from the reaction by chromatography on DEAE-Sephadex A-25 or A-50 equilibrated with 0.050 M Tris, pH 7.9, 0.010 M Mg $(OAc)_2$, 0.005 M 2-mercaptoethanol, and 10% glycerol. Amino acid analysis after acid hydrolysis (6 N HCl, 22 h, 110°C, or 4 M methanesulfonic acid, 0.2% 2-aminoethylindol, 115°C, 24 h for the determination of tryptophan) demonstrated that only histidine is modified (approximately 5 out of 10 residues are modified; only one residue is modified in the presence of guanosine diphosphate). Photooxidation with methylene blue (25 mM Tris, pH 7.9, 0.05% methylene blue; 8°C) abolished placental anticoagulant protein activity with loss only of histidine residues (based on amino acid analysis).[62] Diol dehydrase was inactivated with first-order kinetics by photooxidation in the presence of either rose bengal or methylene blue.[63] In these experiments, the substitution of a helium atmosphere markedly decreased the rate of enzyme inactivation. There was a difference in the pH dependence of the photooxidation reaction performed in the presence of rose bengal (optimum pH 6.2) as compared to the reaction in the presence of methylene blue (optimum >pH 8.0). The pH dependence for the rose bengal reaction was suggested to reflect the charge status of this compound.

Despite problems with residue specificity, photooxidation continues to be of value as a method to modify histidine residues in proteins. A histidine residue was suggested to be at the active site of Ehrlich cell plasma membrane NADH–ferricyanide oxidoreductase on the basis of inactivation by either diethyl pyrocarbonate or photooxidation with rose bengal.[64] Metal-catalyzed photooxidation of His21 in human growth hormone has been reported.[65] His21, together with His19 and Glu174, forms a cation-binding site in this protein. The photooxidation of histidine in the presence of tetrasulfated aluminum phthalocyanin[66] is more effective at low fluence (100 W m^{-2}) when compared to irradiation at 500 W m^{-2}. Hypericin (Figure 5.2) (active ingredient of St. John's wort) has been demonstrated to act as a sensitizer in the photooxidation of α-crystallin from calf lenses.[67] Modification of histidine, methionine, and tryptophan was observed under these experimental conditions (10 mM NH_4CO_3, pH 7.0, >300 nm, 24 W m^{-2}) by mass spectrometric analysis. Photooxidation of histidine has been observed in the presence of calcein.[68] Calcein is a fluorescent probe (Figure 5.2) used in studies of cell viability, migration, and proliferation.[69,70] Xanthurenic acid acts as a photosensitizer for the photooxidation of cytosolic lens protein.[71] Analysis by mass spectrometry demonstrated the site-specific modification of histidine and tryptophan.

Oxidation is the process of removal of electrons, for example, the conversion of cuprous (Cu^{+1}) to cupric (Cu^{+2}) or ferrous Fe^{+2} to ferric (Fe^{+3}). With organic molecules, oxidation is also the addition of oxygen as in conversion of ethanol to acetaldehyde. The addition of a carbonyl function is commonly observed with oxidation of proteins[72] such as with the formation of α-aminoadipic semialdehyde[73] or the oxidation of sulfur in cysteine, cystine, and methionine.[74] The reader is directed to a useful discussion of oxidation (and free radicals) for a more thorough discussion of the chemistry of oxidation and reduction.[75]

Oxidation can be accomplished with a variety of agents including peroxy compounds such as peroxides, peracids such as performic acid, and peroxynitrite, single oxygen donors such as periodate and iodosobenzoate,[76,77] or hydroxy radicals. The reader is referred to a review of oxidants by Webb,[78] which describes early work in this area. It is of interest that iodosobenzoate was the most frequently used oxidant in 1966.[79] Mutation of cysteine to aspartic acid is a mimic of oxidation.[80,81]

Table 5.1 describes some of the peroxide reagents used for the modification of proteins. This category is of greatest interest for several reasons. One is the physiological role of hydrogen peroxide

TABLE 5.1
Peroxides and Protein Oxidation

Compound	Chemistry and Reactions	References
Hydrogen peroxide	Primary action on methionine to methionine sulfoxide[a,b] with secondary reaction at cysteine[c,d] with conversion to cysteine sulfenic acid intermediate to cystine; oxidation to cysteic acid has also been reported. Oxidation also at tryptophan[e] and histidine.[f] Peptide bond cleavage can occur with H_2O_2 at alkaline pH or elevated temperature.[g] It is of interest that H_2O_2 oxidation of protein can occur at temperatures as low as -80°C; minimal oxidation occurs at 4°C and 22°C.[h]	5.1.1–5.1.5
t-Butyl hydroperoxide	Primary action on methionine to methionine sulfoxide[i] although oxidation to cysteine to cysteine sulfenic acid[j] (and then to cystine) and oxidation of tryptophan[k] also occurs.	5.1.6–5.1.12
Benzoyl peroxide	A milder oxidizing agent than hydrogen peroxide with current food processing for the removal of color. There are solubility issues with benzoyl peroxide precluding work at concentrations greater than 250 μM. The reversible inactivation of protein kinase C has been reported, suggesting that the product of oxidation is cysteine sulfenic acid/cystine. Benzoyl peroxide is recognized as a safe ingredient in topical applications and is used in topical acne preparations.	5.1.13–5.1.18

[a] Methionine sulfoxide is the usual product; conversion to methionine sulfone usually requires more vigorous reaction conditions (30% H_2O_2/5% acetic acid, 23°C).

[b] The author is disappointed to see many authors still using room temperature to define reaction temperature; when an author uses room temperature, I arbitrarily assign the value of 23°C.

[c] The reaction of hydrogen peroxide with cysteine shows pH dependence with a low rate at pH 4.0 (approximately 9×10^{-4} M^{-1} s^{-1} at 25°) to a maximal rate at pH 10 (10 M^{-1} s^{-1}).[5.1.1] As a result, H_2O_2 preferentially oxidized methionine at low pH with preferential oxidation of cystine to cysteine sulfenic acid at pH 10.

[d] Cysteine sulfenic acid is the primary product of the H_2O_2 oxidation of cysteine to cysteic acid, as with conversion of methionine to methionine sulfone occurs under rigorous conditions (30% H_2O_2/5% acetic acid, 23°C).

[e] Direct oxidation of tryptophan by hydrogen peroxide occurs with some difficulty (100°C, 2 h, pH 8.5).[5.1.3]

[f] Histidine oxidation to 2-imidazolone (pH 7.4) was not observed with H_2O_2 alone but was observed with H_2O_2 in the presence of $CuSO_4$ or ascorbate with $CuSO_4$ in the absence of H_2O_2.[5.1.4] The susceptibility of a residue to oxidation with H_2O_2 is dependent on amino acid sequence.[5.1.1]

[g] Peptide cleavage is observed under alkaline conditions (0.1 M NaOD) with H_2O_2 anion (HOO⁻) (Gómez-Reyes, B. and Yatsimirky, A.K., Kinetics of amide and peptide cleavage by alkaline hydrogen peroxide, *Org. Lett.* 5, 4831–4834, 2003) and cleavage of peptide bond in a lipase B (0.2–2 M H_2O_2 at 40°C) (Törnvall, U., Hedstrom, M., Schillén, K., and Hatti-Kaul, R., Structural and chemical changes in *Pseudozyma antarctica* lipase B on exposure to hydrogen peroxide, *Biochimie* 92, 1867–1875, 2010).

[h] These experiments were performed with apomyoglobin to assess conditions used for protein footprinting with H_2O_2. It is suggested that H_2O_2 oxidation is more effective in an ice lattice than in solution at 4°C.[5.1.5]

[i] t-Butyl hydroperoxide has a long history of use in inducing oxidative stress and appears to be more effective than hydrogen peroxide.[5.1.6,5.1.7] Keck[5.1.8] introduced t-butyl hydroperoxide for the specific modification of methionine in proteins. This specificity has been demonstrated in a human monoclonal antibody (pH 5.2, citrate/phosphate with mannitol polysorbate).[5.1.9]

[j] The oxidation of cysteinyl residues in human albumin resulting in the formation of albumin dimer has been reported.[5.1.10] The oxidation of a single cysteine in actin (Cys374) (solvent conditions not reported, 25°C, 16 h) has been reported.[5.1.11] At the concentration used for cysteine modification in these studies, there was no modification of methionine.

[k] The modification of tryptophan with t-butyl hydroperoxide in recombinant antibodies (pH 5.0, 23°C, 24 h) has been reported.[5.1.12] The oxidation of Met253 is more extensive than Trp32; the modification of Met83 is less than the modification of Trp32.

TABLE 5.1 (continued)
Peroxides and Protein Oxidation

References to Table 5.1

5.1.1. Froelich, J.M. and Reid, G.E., The origin and control of ex vivo oxidative peptide modifications prior to mass spectrometry analysis, *Proteomics* 8, 1334–1345, 2008.

5.1.2. Luo, D., Smith, S.W., and Anderson, B.D., Kinetics and mechanism of the reaction of cysteine and hydrogen peroxide in aqueous solution, *J. Pharm. Sci.* 94, 304–316, 2005.

5.1.3. Kell, G. and Steinhart, H., Oxidation of tryptophan by hydrogen peroxide in model systems, *J. Food Sci.* 55, 1120–1123, 1990.

5.1.4. Rubino, J.T., Chenkin, M.P., Keller, M. et al., A comparison of methionine, histidine and cysteine in copper (I)-binding peptides reveals differences relevant to copper uptake by organisms in diverse environments, *Metallomics* 3, 61–73, 2011.

5.1.5. Hambly, D.M. and Gross, M.L., Cold chemical oxidation of proteins, *Anal. Chem.* 81, 7235–7242, 2009.

5.1.6. Kihlman, B.A., Experimentally induced chromosome aberrations in plants. I. The production of chromosome aberrations by cyanide and other heavy metal complexing agents, *J. Biophys. Biochem. Cytol.* 3, 363–380, 1957.

5.1.7. Paetzke-Brunner, I., Wieland, O.H., and Feil, G., Activation of the pyruvate dehydrogenase complex in isolated fat cell mitochondria by hydrogen peroxide and t-butyl hydroperoxide, *FEBS Lett.* 122, 29–32, 1980.

5.1.8. Keck, R.G., The use of t-butyl hydroperoxide as a probe for methionine oxidation in proteins, *Anal. Biochem.* 236, 56–62, 1996.

5.1.9. Chumsae, C., Gaza-Buseco, G., Sun, J., and Liu, H., Comparison of methionine oxidation in thermal stability and chemically stressed samples of a fully human monoclonal antibody, *J. Chromatogr. B* 850, 285–294, 2007.

5.1.10. Ogasawara, Y., Namai, T., Togawa, T., and Ischii, K., Formation of albumin dimers induced by exposure to peroxides in human plasma: A possible biomarker for oxidative stress, *Biochem. Biophys. Res. Commun.* 340, 353–358, 2006.

5.1.11. DalleDonne, I., Milzani, A., and Colombo, R., The *tert*-butyl hydroperoxide-induced oxidation of actin Cys-374 is coupled with structural changes in distant regions of the protein, *Biochemistry* 38, 12471–12480, 1999.

5.1.12. Hensel, M., Steurer, R., Fichtl, J. et al., Identification of potential sites for tryptophan oxidation in recombinant antibodies using *tert*-butylhydroperoxide and quantitative LC-MS, *PLoS* 6, e17708, 2011.

5.1.13. Finley, J.W., Wheeler, E.L., and Witt, S.C., Oxidation of glutathione by hydrogen peroxide and other oxidizing agents, *J. Agric. Food Chem.* 29, 404–407, 1977.

5.1.14. Gopalakrishna, R., Gundimeda, U., Anderson, W.B. et al., Tumor promoter benzoyl peroxide induces sulfhydryl oxidation in protein kinase C: Its reversibility is related to the cellular resistance to peroxide-induced cytotoxicity, *Arch. Biochem. Biophys.* 363, 246–258, 1999.

5.1.15. Valacchi, G., Rimbach, G., Saliou, C. et al., Effect of benzoyl peroxide on antioxidant status, NF-κB and interleukin-1α gene expression in human keratinocytes, *Toxicology* 165, 225–234, 2001.

5.1.16. Food and Drug Administration, HHS, Classification of benzoyl peroxide as safe and effective and revision of labeling to drug facts format; topical acne drug products for over-the-counter human use; final rule, *Fed. Regist.* 75, 9767–9777, 2010.

5.1.17. Listiyani, M.A., Campbell, R.E., Miracle, R.E. et al., Effect of temperature and bleaching agent on bleaching of liquid Cheddar whey, *J. Dairy Sci.* 95, 36–49, 2012.

5.1.18. Jervis, S., Campbell, R., Wojciechowski, K.L. et al., Effect of bleaching whey on sensory and functional properties of 80% whey protein concentrate, *J. Dairy Sci.* 95, 2848–2862, 2012.

in redox systems,[82–85] while the other is the role of peroxides derived from excipients[86–89] used in the manufacture of therapeutic proteins as mentioned earlier. Table 5.2 describes a variety of oxidizing agents. Perchloric acid is an oxidizing agent but is used primarily as a solvent to extract histones and high-mobility group proteins.[90–95] Performic acid oxidized cysteinyl residues to cysteic acid[96,97] and methionine to methionine sulfone[98]; there is also oxidation of tyrosine, tryptophan,

TABLE 5.2
Some Oxidizing Agents and Their Action on Proteins

Compound	Chemistry and Reactions	References
Performic acid	HCOOOH; considered to be a strong oxidizing agent with potential for explosion.[a] Performic acid is used for the oxidation of cysteine and cystine to cysteic acid in proteins; there is also modification of methionine and tryptophan.	5.2.1–5.2.7
Peroxynitrite[b]	Peroxynitrite is derived from the reaction of superoxide radical with nitric oxide and is in equilibrium with peroxynitrous acid. Peroxynitrous acid is very unstable, while peroxynitrite is stable in solution or as a solid. Peroxynitrite is an oxidizing agent with modification occurring at cysteine, methionine, and tryptophan residues.[c] Peroxynitrite also reacts with transition metal centers in proteins. Peroxynitrite can react with cysteine to form S-nitrosocysteine and with tyrosine to yield 3-nitrotyrosine.	5.2.8–5.2.11
Iodosobenzoate	A mild oxidizing agent with modest use. Iodosobenzoate was developed for the determination of cysteine in proteins and subsequently for the modification of cysteine in proteins. Iodosobenzoate will oxidize tryptophan and has been used for the cleavage of tryptophan peptide bonds in proteins. o-Iodoxybenzoate is formed by disproportionation of o-iodosobenzoate and would appear to be a more potent oxidizing agent converting cysteine to cysteic acid (cysteine sulfonic acid).	5.2.12–5.2.17
Iodine	I_2; a mild oxidizing agent that converts cysteine to cysteine sulfenic acid via the intermediate formation of cysteine iodide; the cysteine sulfenic acid can then form a disulfide with a suitable partner; there is some work suggesting that iodine oxidation of cysteine under more vigorous conditions can result in the formation of cysteine sulfinic acid and cysteine sulfonic acid. Iodine is used with a varying degree of success for forming disulfide bonds in peptide synthesis. There can also be reaction with histidine and tyrosine to form iodinated derivatives.	5.2.18–5.2.24
Permanganate	Permanganate, usually the potassium salt, converts cysteine to cysteic acid. There is evidence for the oxidation of other amino acids including methionine, tyrosine, and tryptophan; it should be noted that these oxidation products are poorly characterized.	5.2.25–5.2.27
Periodic acid	Periodate is best known for oxidation of vicinal diols in glycoproteins and carbohydrates to aldehydes, which can be coupled to a suitable acceptor. Oxidation of N-terminal serine and threonine to produce an aldehyde function that can be used for bioconjugation is also observed. Periodate oxidation of peptide mixtures derived by proteolysis provides derivatives that can be coupled to a solid-phase hydrazide matrix, thus providing a method for enriching glycopeptides for proteomic analysis. The conditions for periodate oxidation must be carefully developed and the product evaluated as there is the potential for significant side reaction with potential modification of methionine, tyrosine, and cysteine.	5.2.28–5.2.37
HOCl	HOCl (hypochlorous acid; CAS 7790-92-3) is an in vivo product of the action of myeloperoxidase in neutrophils and serves an antibacterial function.[e] HOCl is a strong oxidizing agent but a weak acid (pKa=7.5). Only available in solution. Investigators have used myeloperoxidase to produce hypochlorous acid from chloride ion and hydrogen peroxide.[f] Reagent-grade sodium hypochlorite is available and used for in vitro experiments.[g] Sodium hypochlorite is commercial bleach.	5.2.38–5.2.44

[a] There are anecdotal reports of explosions associated with the use of performic acid (Weingartsofer-Olmos, A. and Giguere, P.A., Performic acid explosion, *Chem. Eng. News* 30, 3041, 1952; Oehischlager, A. et al., Performic acid explosion, *Chem. Eng. News* 65(15), 2, 1987). Performic acid should be used without incident (Weisenberger, G.A. and Vogt, P.F., Safety notables: Information from the literature, *Org. Proc. Res. Dev.* 11, 1087–1090, 2007).

TABLE 5.2 (continued)
Some Oxidizing Agents and Their Action on Proteins

[b] The half-life for the decomposition of peroxynitrous acid at pH 7.0 (37°C) is approximately 1 s (Pryor, W.A. and Squadrito, G.L., The chemistry of peroxynitrite a product from the reaction of nitric oxide with superoxide, *Am. J. Physiol.* 268, L699–L722, 1995). Peroxynitrite is reasonably stable at alkaline pH (Goldstein, S. and Merenyl, G., The chemistry of peroxynitrite: Implications for biological activity, *Methods Enzymol.* 436, 49–61, 2008). Sodium peroxynitrite is available in 0.3 M sodium hydroxide solution (reported to be stable for 6 months and -80°C). https://www.caymanchem.com/app/template/Product.vm/catalog/81565/a/z.

[c] Amino acid oxidations by peroxynitrite are second-order reactions while the modifications of histidine, tyrosine, and phenylalanine are first-order reactions, which involve species hydroxyl radical and nitrogen dioxide (nitrite radical). Second-order rate constants have been reported for the reaction of peroxynitrite with cysteine ($\sim 10^3$ M^{-1} s^{-1}), methionine (10 M^{-1} s^{-1}), and tryptophan (37 M^{-1} s^{-1}) (Alvarez, B. and Radi, R., Peroxynitrite reactivity with amino acids and proteins, *Amino Acids* 25, 295–311, 2003).

[d] The reaction of peroxynitrite with tyrosine is discussed in Chapter 17.

[e] There was some interest in the use of hypochlorous acid as a topical antibacterial agent (Smith, J.L., Drennan, A.M., Rettie, T., and Campbell, W., Experimental observations on the antiseptic acid of hypochlorous acid and its application to wound treatment, *Brit. Med. J.* 2(2847), 129–136, 1915; Robson, M.C., Payne, W.G., Ko, F. et al., Hypochlorous acid as a potential wound care agent: Part II: Stabilized hypochlorous acid: Its role in decreasing tissue bacterial bioburden and overcoming the inhibition of infection on wound healing, *J. Burns Wounds* 6:e6, 2007). Most works with hypochlorite as an antibacterial agent used the sodium salt, which is the active component of commercial bleach (Coetzee, E., Whitelaw, A., Kahn, D., and Rode, H., The use of topical, unbuffered sodium hypochlorite in the management of burn wound infection, *Burns* 4, 529–533, 2012; Parirokh, M., Jaladi, S., Haghdoost, A.A., and Abbott, P.V., Comparison of the effect of various irrigants on apically extruded debris after root canal preparation, *J. Endod.* 38, 196–199, 2012).

[f] Hypochlorous acid can be generated in situ for the modification of proteins (Carr, A.C. and Winterbourn, C.C., Oxidation of neutrophil glutathione and protein thiols by myeloperoxidase-derived hypochlorous acid, *Biochem. J.* 327, 275–281, 1997; Hazen, S.L., Hsu, F.F., Gaut, J.P. et al., Modification of proteins and lipids by myeloperoxidase, *Methods Enzymol.* 300, 88–105, 1999).

[g] Sodium hypochlorite is available in solution from a variety of vendors. Solutions of hypochlorite may be standardized by spectrometric methods (Volf, I., Roth, A., Cooper, J. et al., Hypochlorite modified LDL are a stronger agonist for platelets that copper oxidized LDL, *FEBS Lett.* 483, 155–159, 2000; Summers, F.A., Forsman, Q.A., and Hawkins, C.L., Identification of proteins susceptible to thiol oxidation in endothelial cells exposed to hypochlorous acid and N-chloramines, *Biochem. Biophys. Res. Commun.* 425, 157–161, 2012) using a molar extinction coefficient of 350 cm^{-1} at 292 nm for the OCl$^-$ anion (Morris, J.C., The acid ionization constant of HOCl from 5 to 35°, *J. Phys. Chem.* 70, 3798–3805, 1966).

References to Table 5.2

5.2.1. Hirs, C.H.W., Stein, W.H., and Moore, S., Peptides obtained by chymotryptic hydrolysis of performic acid oxidized ribonuclease: A partial structural formula for the oxidized protein, *J. Biol. Chem.* 221, 151–169, 1956.

5.2.2. Hirs, C.H.W., Performic acid oxidation of proteins, *Methods Enzymol.* 11, 197–199, 1967.

5.2.3. Simpson, R.J., Performic acid oxidation of proteins, *CSH Protoc.* 2007, pdb.prot4696, 2007.

5.2.4. Matthiesen, R., Bauw, G., and Welinder, K.G., Use of performic acid oxidation to expand the mass distribution of tryptic peptides, *Anal. Chem.* 76, 6848–6852, 2004.

5.2.5. Dai, J., Wang, J., Zhang, Y. et al., Enrichment and identification of cysteine-containing peptides from tryptic digests of performic oxidized proteins by strong cation exchange LC and MALDI-TOF/TOF-MS, *Anal. Chem.* 77, 7594–7604, 2005.

5.2.6. Yamaguchi, M., Nakayama, D., Shima, K. et al., Selective isolation of N-terminal peptides from proteins and their *de novo* sequencing by matrix-assisted laser desorption/ionization time-of-flight mass spectrometry without regard to unblocking or blocking of N-terminal amino acids, *Rapid Commun. Mass Spectrom.* 22, 3313–3319, 2008.

5.2.7. Cao, R., Liu, Y., Chen, P. et al., Improvement of hydrophobic integral membrane protein identification by mild performic acid oxidation-assisted digestion, *Anal. Biochem.* 407, 196–204, 2010.

5.2.8. Alvarez, B. and Radi, R., Peroxynitrite reactivity with amino acids and proteins, *Amino Acids* 25, 295–311, 2003.

(continued)

TABLE 5.2 (continued)
Some Oxidizing Agents and Their Action on Proteins

5.2.9. Ferrer-Sueta, G. and Radi, R., Chemical biology of peroxynitrite: Kinetics, diffusion, and radicals, *ACS Chem. Biol.* 4, 161–177, 2009.

5.2.10. Calcerrada, P., Peluffo, G., and Radi, R., Nitric oxide-derived oxidants with a focus on peroxynitrite: Molecular targets, cellular responses and therapeutic implications, *Curr. Pharm. Des.* 17, 3905–3932, 2011.

5.2.11. Koppenol, W.H., Bounds, P.L., Nauser, T. et al., Peroxynitrous acid: Controversy and consensus surrounding an enigmatic oxidant, *Dalton Trans.* 41, 13779–13787, 2012.

5.2.12. Parker, D.J. and Allison, W.S., The mechanism of inactivation of glyceraldehyde 3-phosphate dehydrogenase by tetrathionate, *o*-iodosobenzoate, and iodine monochloride, *J. Biol. Chem.* 244, 180–189, 1969.

5.2.13. Andreo, C.S., Ravizzini, R.A., and Vallejos, R.H., Sulphydryl groups in photosynthetic energy conservation. V. Localization of the new disulfide bridges formed by *o*-iodosobenzoate in coupling factor of spinach chloroplasts, *Biochim. Biophys. Acta* 547, 370–379, 1979.

5.2.14. Verma, K.K., Determination of ascorbic acid with *o*-iodosobenzoate: Analysis of mixtures of ascorbic acid with methionine and cysteine or glutathione, *Talanta* 29, 41–45, 1982.

5.2.15. Rizk, M.S., Belal, F., and Eid, M.M., 2-Iodoxybenzoate as a titrant for the determination of some pharmaceutically-important thiol compounds, *Acta Pharm. Hung.* 63, 13–18, 1993.

5.2.16. Duhe, R.J. Nielsen, M.D., Dittman, A.H. et al., Oxidation of critical cysteine residues of type I adenylyl cyclase by *o*-iodosobenzoate or nitric oxide reversibly inhibits stimulation by calcium and calmodulin, *J. Biol. Chem.* 269, 7290–7296, 1994.

5.2.17. Smith, J.K., Patil, C.N., Patlolla, S. et al., Identification of a redox-sensitive switch within the JAK2 catalytic domain, *Free Radic. Biol. Med.* 52, 1101–1110, 2012.

5.2.18. Simonsen, C.A., The oxidation of cysteine with iodine: Formation of a sulfinic acid, *J. Biol. Chem.* 101, 36–42, 1933.

5.2.19. Anson, M.L., The sulfhydryl groups of egg albumin, *J. Gen. Physiol.* 24, 399–421, 1941.

5.2.20. Trundle, D. and Cunningham, L.W., Iodine oxidation of the sulfhydryl groups of creatine kinase, *Biochemistry* 8, 1919–1925, 1969.

5.2.21. Kellenberger, C., Hietter, H., and Luu, B., Regioselective formation of the three disulfide bonds of a 35-residue insect peptide, *Pept. Res.* 8, 321–327, 1995.

5.2.22. Söll, R. and Beck-Sickinger, A.G., On the synthesis of orexin A: A novel one-step procedure to obtain peptides with two intramolecular disulphide bonds, *J. Pept. Sci.* 6, 387–397, 2000.

5.2.23. Muttenthaler, M., Ramos, Y.G., Feytens, D. et al., *p*-Nitrobenzyl protection for cysteine and selenocysteine: A more stable alternative to the acetamidomethyl group, *Biopolymers* 94, 423–432, 2010.

5.2.24. Ready, K.M., Kumari, Y.B., Mallikharjunasarma, D. et al., Large scale solid phase synthesis of peptide drugs: Use of commercial anion exchange resins as quenching agent for removal of iodine during sulphide bond formation, *Int. J. Pept.* 2012, 323907, 2012.

5.2.25. Khan, Z., Ahmed Al-Thabaiti, S., Yousif, O.A. et al., MnO_2 nanostructures of different morphologies from amino acids-MnO_4-reactions in aqueous solutions, *Colloids Surf. B Biointerfaces* 81, 381–384, 2010.

5.2.26. Xiao, C., Chen, J., Liu, B. et al., Sensitive and selective electrochemical sensing of L-cysteine based on a caterpillar-like manganese dioxide-carbon nanocomposite, *Phys. Chem. Chem. Phys.* 13, 1568–1574, 2011.

5.2.27. Haas, W.J., Sizer, I.W., and Loofbourow, J.R., The effect of permanganate on the ultraviolet absorption spectra of aromatic amino acids and proteins, *Biochim. Biophys. Acta* 6, 601–605, 1951.

5.2.28. Suzuki, N., Quesenberry, M.S., and Wang, J.K., Efficient mobilization of proteins by modification of plate surfaces with polystyrene derivatives, *Anal. Biochem.* 247, 412–416, 1997.

5.2.29. Hage, D.S., Wolfe, C.A., and Oates, M.R., Development of a kinetic model to describe the effective rate of antibody oxidation by periodate, *Bioconjug. Chem.* 8, 914–920, 1997.

5.2.30. Behrens, C.R., Hooker, J.M., Obermeyer, A.C. et al., Rapid chemoselective bioconjugation through oxidative coupling of anilines and aminophenols, *J. Am. Chem. Soc.* 133, 16398–16401, 2011.

5.2.31. Geoghegan, K.F. and Stroh, J.G., Site-directed conjugation of nonpeptide groups to peptides and proteins via periodate oxidation of a 2-amino alcohol. Application to modification at *N*-terminal serine, *Bioconjug. Chem.* 3, 138–146, 1992.

5.2.32. Mikolajczyk, S.D., Meyer, D.L., Starling, J.J. et al., High yield, site-specific coupling of *N*-terminally modified β-lactamase to a proteolytically derived single sulfhydryl murine Fab′. *Bioconjug. Chem.* 5, 636–646, 1994.

TABLE 5.2 (continued)
Some Oxidizing Agents and Their Action on Proteins

5.2.33. Rose, K., Chen, J., Dragovic, M. et al., New cyclization reaction at the amino terminus of peptides and proteins, *Bioconjug. Chem.* 10, 1038–1043, 1999.

5.2.34. Cohen, A.M., Kostyleva, R., Chisholm, K.A., and Pinto, D.M., Iodination on tyrosine residues during oxidation with sodium periodate in solid phase extraction of N-linked glycopeptides, *J. Am. Soc. Mass Spectrom.* 23, 68–75, 2012.

5.2.35. Wolschner, C., Giese, A., Kretzschmar, H.A. et al., Design of anti- and pro-aggregation variants to assess the effects of methionine oxidation in human prion protein, *Proc. Natl. Acad. Sci. USA* 106, 7756–7761, 2009.

5.2.36. Clamp, J.R. and Hough, L., The periodate oxidation of amino acids with reference to studies on glycoproteins, *Biochem. J.* 94, 17–24, 1965.

5.2.37. Chiari, M., Ettori, C., Righetti, P.G. et al., Oxidation of cysteine to cysteic acid in proteins by peroxyacids, as monitored by immobilized pH gradients, *Electrophoresis* 12, 376–377, 1991.

5.2.38. Brickman, C.M., Atherton, J.P., and Kantor, N.L., A new simplified method for oxidation of human complement component C2, *J. Immunol. Methods* 132, 157–164, 1990.

5.2.39. Glaser, C.B., Morser, J., Clarke, J.H. et al., Oxidation of a specific methionine in thrombomodulin by activated neutrophil products blocks cofactor activity. A potential rapid mechanism for modulation of coagulation, *J. Clin. Invest.* 90, 2565–2573, 1992.

5.2.40. Hamann, M., Zhang, T., Hendrich, S., and Thomas, J.A., Quantitation of protein sulfinic and sulfonic acid, irreversibly oxidized protein cysteine sites in cellular proteins, *Methods Enzymol.* 348, 146–156, 2002.

5.2.41. Hawkins, C.L., Pattison, D.I., and Davies, M.J., Hypochlorite-induced oxidation of amino acids, peptides, and proteins, *Amino Acids* 25, 259–274, 2003.

5.2.42. Pitt, A.R. and Spickett, C.M., Mass spectrometric analysis of HOCl- and free-radical-induced damage to lipids and proteins, *Biochem. Soc. Trans.* 36, 1077–1082, 2008.

5.2.43. Nakamura, M., Shishido, N., Nunomura, A. et al., Specific reaction of Met 35 in amyloid beta peptide with hypochlorous acid, *Free Radic. Res.* 44, 734–741, 2010.

5.2.44. Saha, S., Graessler, J., Schwarz, P.E. et al., Modified high-density lipoprotein modulates aldosterone release through scavenger receptors via extra cellular signal-regulated kinase and Janus kinase-dependent pathways, *Mol. Cell. Biochem.* 366, 1–10, 2012.

and other amino acids including peptide bond cleavages.[98,99] Performic acid has been used for the oxidation of proteins prior to analysis by mass spectrometry. Peroxynitrite (Figure 5.3) is a relatively recently described in vivo oxidant. There is some use of peroxynitrite for the in vitro nitration of tyrosine (Chapters 4 and 11). Peroxynitrite can oxidize cysteine, methionine, and tryptophan in second-order reactions.[100] Peroxynitrite is also involved in the oxidation of lipids and redox regulation.[101–103] Iodosobenzoate is an oxidizing agent and also cleaves tryptophan-containing peptide bonds[104,105]; iodosobenzoate also undergoes a disproportionation reaction forming *o*-iodoxybenzoic acid (Figure 5.4), which degrades tyrosyl residues.[106] Esters of 2-iodoxybenzoic are hypervalent oxidizing agents (periodinanes*).[107] Early work on iodosobenzoate and iodoxybenzoate demonstrated an effect on in vitro immune cells[108] and subsequently in in vivo immune responses.[109] Somewhat later, Kersley and Simpson[110] reported on a clinical trial of calcium *o*-iodoxybenzoate for rheumatoid arthritis, which was unsuccessful. Iodine oxidizes cysteine to cystine via the intermediate formation of cysteine sulfenyl iodide, which undergoes hydrolysis to form cysteine sulfenic acid leading to a disulfide (Figure 5.5).[111] While most cysteine sulfenyl iodides are unstable, the cysteine sulfenyl iodide in α-lactoglobulin was demonstrated to be relatively stable.[112] Iodine has considerable use in forming disulfide bonds in peptide synthesis.[113]

Potassium permanganate (Figure 5.5) had some early use in the oxidation of cysteine residues in proteins[114] and is used in histochemistry for the oxidation of cysteine to cysteic acid.[115–117] Permanganate also resulted in the oxidation of tryptophan and tyrosine.[118] Permanganate is a strong

---

* Periodinanes are hypervalent iodine derivatives (Figure 5.3).

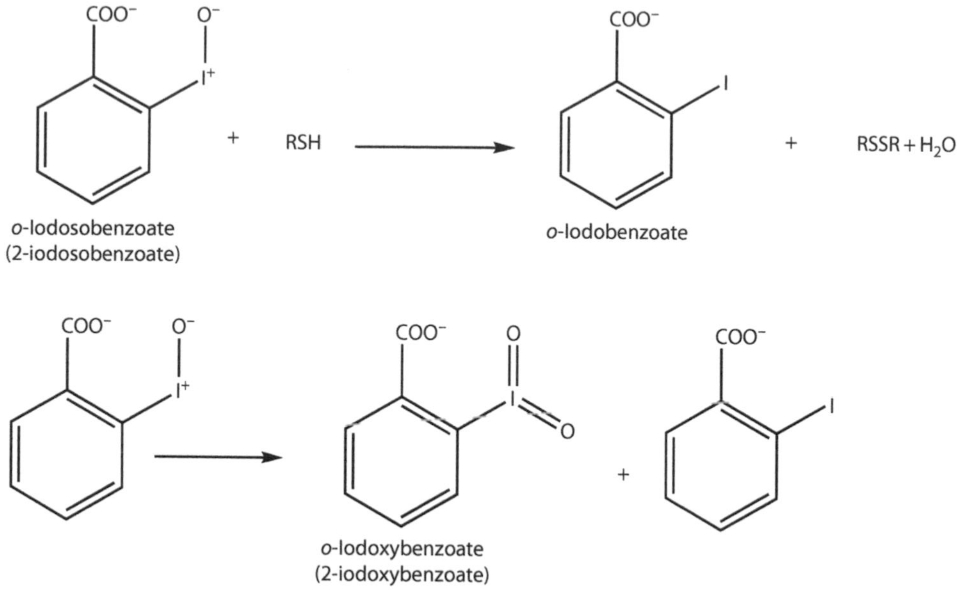

COO⁻ O⁻

o-Iodosobenzoate
(2-iodosobenzoate)

+ RSH ⟶

COO⁻

o-Iodobenzoate

+ RSSR + H₂O

COO⁻ O⁻

⟶

COO⁻ O

o-Iodoxybenzoate
(2-iodoxybenzoate)

+

COO⁻

FIGURE 5.3   The formation and oxidative reactions of peroxynitrite. Shown is the formation of peroxynitrite from nitric oxide radical and superoxide radical anion, which is in equilibrium with peroxynitrous acid. Peroxynitrous acid is very unstable with the formation of nitrate either directly or through the formation of hydroxyl radical and nitrogen dioxide radical. Cysteine, methionine, and tryptophan are the primary targets for oxidation by peroxynitrite in second-order reactions (see Alvarez, B. and Radl, R., Peroxynitrite reactivity with amino acids and proteins, *Amino Acids* 25, 295–311, 2003). It has been suggested that the reaction of peroxynitrite with amino acids involves the initial formation of a complex with peroxynitrous acid (Koppenol, W.H., Bounds, P.L., Nauser, T. et al., Peroxynitrous acid: Controversy and consensus surrounding an enigmatic oxidant, *Dalton Trans.* 41, 13779–13787, 2012).

oxidizing agent where cysteine is converted to cysteic acid (cysteine sulfonic acid).[114] Permanganate is a structural analogue of phosphate thus serving as an affinity label for the modification of enzymes.[119,120] In one case, the inactivation is based on the oxidation of cysteine,[119] while with *E. coli* alkaline phosphate, inactivation may involve the formation of a manganate ester at the enzyme active site.[120] Current use of permanganate is for the oxidation of pyrimidine bases in DNA[121] in the process of DNA footprinting.[122]

Periodic acid (Figure 5.6) is known for its use in the oxidation of vicinal hydroxyl groups in carbohydrates to produce aldehyde function, which can then be coupled with amines.[123,124] Periodic acid is also a critical component of the periodic acid–Schiff (PAS) stain used for the detection of glycoproteins and polysaccharides.[125,126] This approach has been modified for binding of a hydrazide dye to the oxidized glycoprotein[127] although care must be taken concerning side reactions occurring after periodate oxidation.[128] Periodic acid oxidation of glycopeptides and subsequent coupling to a hydrazide matrix has proved useful for the isolation of glycopeptides.[129] The periodate oxidation–hydrazide chemistry has also proved useful for the attachment of protein.[130] A recent report from Kirkeby[131] shows a change in immunohistochemistry in tissue sections with periodate oxidation suggesting the destruction of some antigenic sites and the exposure of some cryptic epitopes. Kristiansen and coworkers[132] showed that the limited (1%–20%) periodate oxidation of polysaccharides may result in derivatives with altered physical and chemical properties.

There are other chemical oxidizing agents that are mostly of in vivo interest. Hypochlorous acid is formed in vivo through the action of myeloperoxidase[133] and is a relatively nonspecific oxidizing and chlorinating agent for proteins.[134–138] There are some examples of site-specific modification in proteins with hypochlorous acid including cysteine[139,140] and methionine.[141,142] Hypochlorous acid

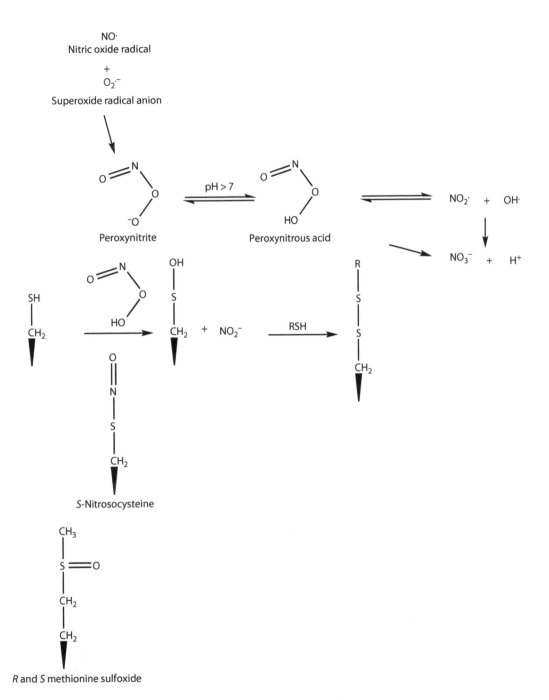

FIGURE 5.4 Reactions of *o*-iodosobenzoate (2-iodosobenzoate). Shown is the oxidation of a sulfhydryl by *o*-iodosobenzoate, which was an early method for the determination of cysteine in proteins (Hellerman, L., Chinard, F.P., and Ramsdell, P.A., *o*-Iodosobenzoic acid, a reagent for the estimation of cysteine, glutathione, and the substituent sulfhydryl groups of certain proteins, *J. Am. Chem. Soc.* 63, 2551–2553, 1941). Also shown is the disproportionation of *o*-iodosobenzoate to form *o*-iodoxybenzoate and *o*-iodobenzoate.

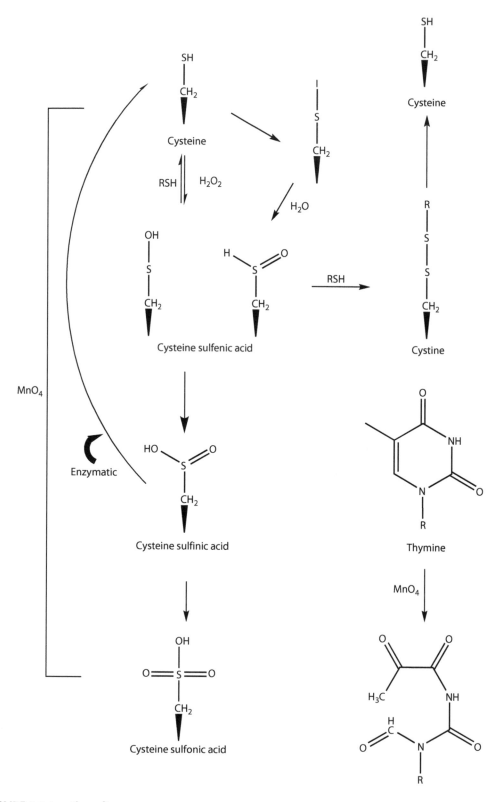

FIGURE 5.5 (continued)

produced cysteine sulfinic acid, cysteine sulfonic acid, and a disulfide derivative from a synthetic peptide mimicking the cysteine switch domain, which is suggested to regulate the matrix metalloproteinase (MMP) activity.[139] Reaction of MMP-7 with hypochlorous acid resulted in the activation of the enzyme and the formation of cysteine sulfinic acid.[139] Reaction of hypochlorous acid with recombinant human myoglobin[140] results in the formation of cystine and cysteic acid; cysteine sulfenic acid and cysteine sulfinic acid were not detected. There was some modification of methionine at higher concentrations of hypochlorous acid. Modification of a mutant protein where the target cysteine (Cys110) was replaced with an alanine (Cys100Ala) allowed the modification of tryptophan and tyrosine. Nagy and Ashby[143] reported that cysteine sulfenic acid is formed from cysteine in alkaline solution with hypobromous or hypochlorous acid. Cysteine sulfenic acid either proceeds to cystine or to the formation of cysteine sulfinic acid. Hypochlorous acid converts an internal methionine to methionine sulfoxide,[142] while an N-terminal methionine is converted to dehydromethionine (Figure 5.7), which is suggested as a biomarker for oxidative stress.[141] Nakamura and coworkers[142] observed that the rate of oxidation of N-acetylmethionine by hypochlorous acid (7×10^6 M^{-1} s^{-1}) is slightly faster than that observed for a methionine residue in amyloid beta peptide (Aβ25-35) (1.4×10^6 M^{-1} s^{-1}). These rates are markedly slower than those reported for the oxidation of methionine by peroxynitrite (1.6×10^3 M^{-1} s^{-1}).[100] Hypochlorous acid can lead to chlorination of target substances, lipids and proteins, as well as oxidation (Figure 5.7).[135,144,145] The oxidative modifications induced by hypochlorous acid are very similar to those induced by free radicals.

FIGURE 5.5 (continued) Oxidation reactions of hydrogen peroxide, iodine, and permanganate. Shown is the oxidation of cysteine by iodine to cysteine sulfenic acid with subsequent formation of a disulfide (Trundle, D. and Cunningham, L.W., Iodine oxidation of the sulfhydryl groups of creatine kinase, *Biochemistry* 8, 1919–1925, 1969). Also shown is the oxidation of cysteine to cysteine sulfenic acid by hydrogen peroxide. Cysteine sulfenic acid and cystine can be reduced to cysteine. Permanganate (KMnO$_4$) oxidizes thiols to sulfonic acids (Yano, M. and Hayatsu, H., Permanganate oxidation of 4-thiouracil derivatives. Isolation and properties of I-substituted 2-pyrimidone 4-sulfonates, *Biochim. Biophys. Acta* 199, 303–315, 1970). Permanganate is better known for the oxidation of pyrimidine bases (Hayatsu, H. and Ukita, T., The selective degradation of pyrimidines in nucleic acids by permanganate oxidation, *Biochem. Biophys. Res. Commun.* 29, 556–561, 1971).

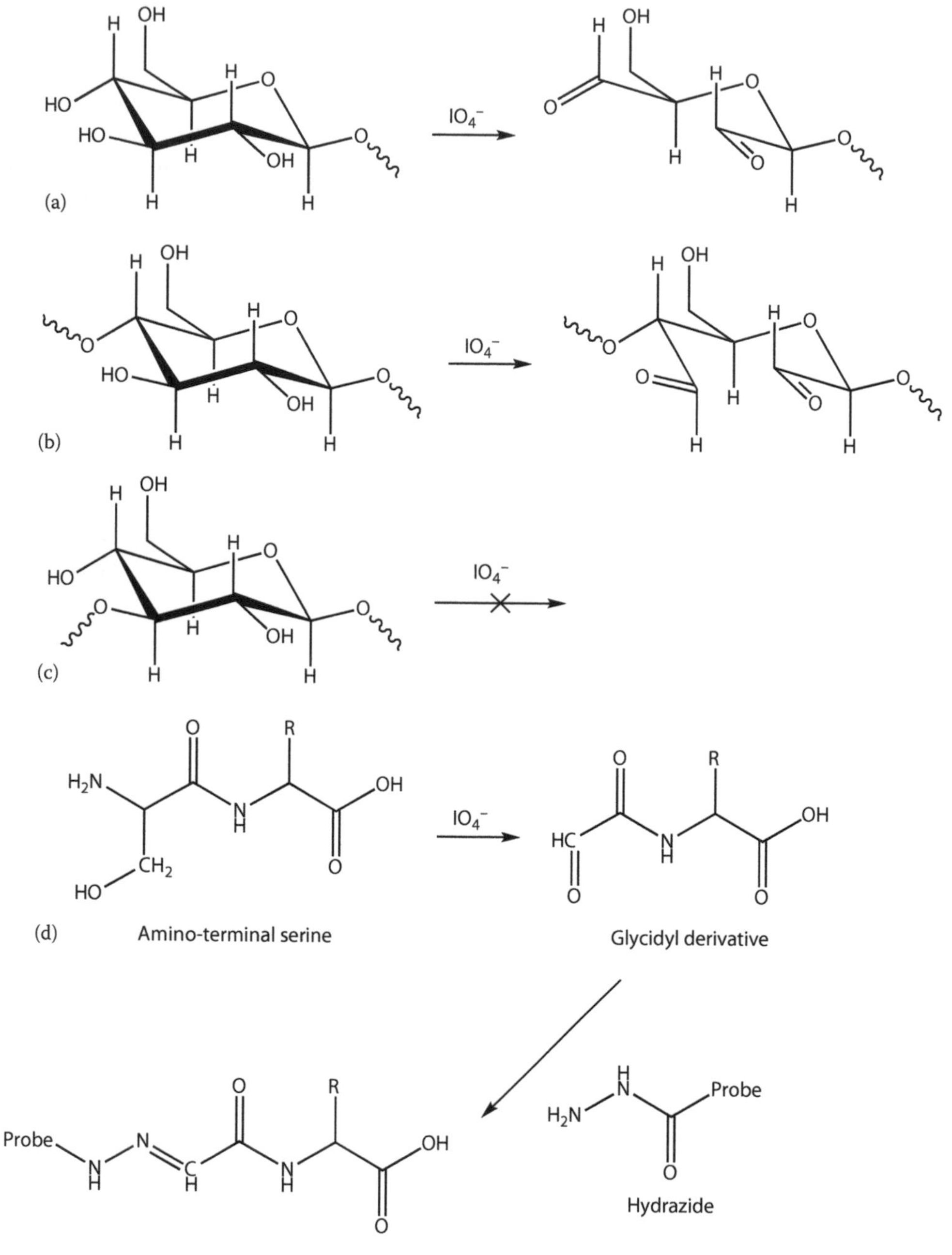

FIGURE 5.6 Reaction of periodate with proteins and glycoproteins. Shown at the top is the periodate oxidation of a polysaccharide. The top illustration (a) is the oxidation of a terminal nonreducing residue, the middle (b) a 1,4-linked residue, and the third (c) a 1,3-linked residue; the 1,3-linked residue lacks the vicinal diol necessary for periodate cleavage (Adapted from Kristiansen, K.A., Potthast, A., and Christensen, B.E., Periodate oxidation of polysaccharides for modification of chemical and physical properties, *Carbohydr. Res.* 345, 1264–1271, 2010). Shown at the bottom (d) is the periodate oxidation of an *N*-terminal serine to form a glycidyl derivative, which is in equilibrium with a glycoxylyl form (not shown) and subsequent reaction with a hydrazide. (See Filtz, T.M., Chumpradit, S., Kung, H.F., and Molinoff, P.B., Synthesis and applications of an aldehyde-containing analogue of SCH-23390, *Bioconjug. Chem.* 1, 394–399, 1990; Chelius, D. and Shaler, T.A., Capture of peptides with *N*-terminal serine and threonine: A sequence-specific chemical method for peptide mixture simplification, *Bioconjug. Chem.* 14, 205–211, 2003).

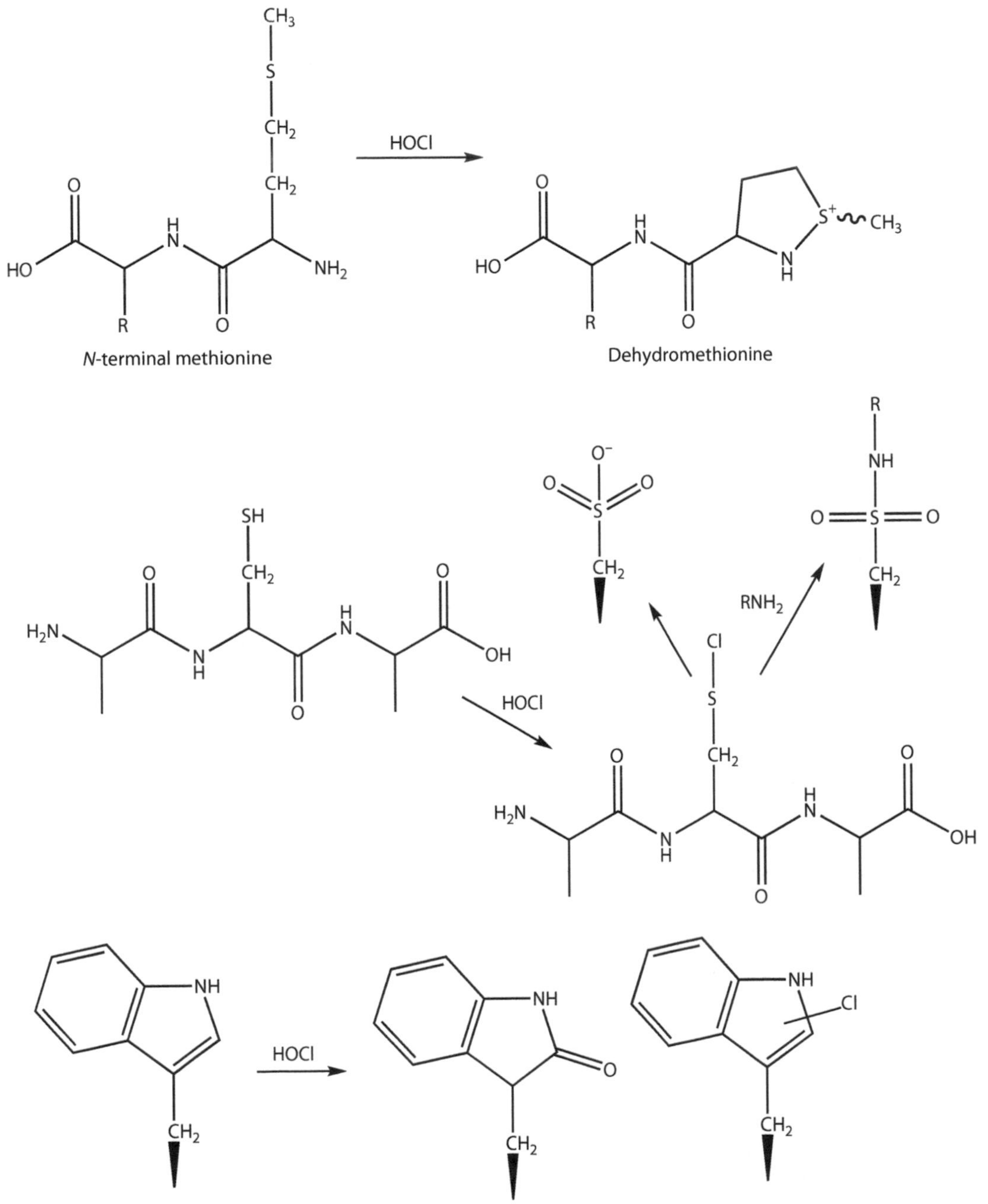

FIGURE 5.7 Some effects of hypochlorous acid on proteins. Shown at the top is the reaction of hypochlorous acid with an *N*-terminal methionine resulting in the formation of dehydromethionine (Adapted from Beal, J.L., Foster, S.B., and Ashby, M.T., Hypochlorous acid reacts with the *N*-terminal methionines of proteins to give dehydromethionine, a potential biomarker for neutrophil-induced oxidative stress, *Biochemistry* 48, 11142–11148, 2009). The following shows the various modifications of cysteine to form an *S*-chlorocysteine transient intermediate, which could hydrolyze to form cysteic acid or react with an amine to form a sulfonamide derivative. Also shown are putative reactions with tryptophan. (See Hawkins, C.L., Pattison, D.L., and Davies, M.J., Hypochlorite-induced oxidation of amino acids, peptides and proteins, *Amino Acids* 25, 259–274, 2003).

REFERENCES

1. Schoneich, C., Hagerman, M.J., and Borschardt, R.T., Stability of peptides and proteins, in *Controlled Drug Delivery*, ed. K. Park, pp. 205–228, American Chemical Society, Washington, DC, 1997.
2. Reis, A., Fonseca, C., Maciel, E. et al., Influence of amino acid relative position on the oxidative modification of histidine and glycine peptides, *Anal. Bioanal. Chem.* 399, 2779–2794, 2011.
3. Nguyen, T.H., Burnier, J., and Meng, W., The kinetics of relaxin oxidation by hydrogen peroxide, *Pharm. Res.* 10, 1563–1571, 1993.
4. Oztug Durer, Z.A., Kamal, J.K., Benchaar, S. et al., Myosin binding surface on actin probed by hydroxyl radical footprinting and site-directed labels, *J. Mol. Biol.* 414, 204–216, 2011.
5. Fujino, T., Kojima, M., Beppu, M. et al., Identification of the cleavage sites of oxidized protein that are susceptible to oxidized protein hydrolase (OPH) in the primary and tertiary structures of the protein, *J. Biochem.* 127, 1087–1093, 2000.
6. Houde, D., Peng, Y., Berkowitz, S.A., and Engen, J.R., Post-translational modifications differentially affect IgG1 conformation and receptor binding, *Mol. Cell Proteomics* 9, 1716–1728, 2010.
7. Hancock, J.T. (ed.), *Redox-Mediated Signal Transduction. Methods and Protocols*, Humana Press/Springer Science Media, New York, 2008.
8. Boschi-Muller, S., Gand, A., and Branlant, G., The methionine sulfoxide reductases: Catalysis and substrate specificities, *Arch. Biochem. Biophys.* 474, 266–273, 2008.
9. Moskovitz, J. and Oien, D.B., Protein carbonyl and the methionine sulfoxide reductase system, *Antioxid. Redox Signal.* 12, 405–415, 2010.
10. Murphy, M.P., Mitochondrial thiols in antioxidant protection and redox signaling: Distinct roles for glutathionylation and other thiol modifications, *Antioxid. Redox Signal.* 16, 476–495, 2012.
11. Lima, V.V., Spitler, K., Choi, H. et al., O-GlcNacylation and oxidation of proteins: Is signalling in the cardiovascular system becoming sweeter? *Clin. Sci. (Lond.)* 123, 473–486, 2012.
12. Izquierdo-Álvarez, A. and Martínez-Ruiz, A., Thiol redox proteomics seen with fluorescent eyes: The derivatization and 2-DE, *J. Proteomics* 75, 329–338, 2011.
13. Higdon, A., Diers, A.R., Oh, J.Y. et al., Cell signalling by reactive lipid species: New concepts and molecular mechanisms, *Biochem. J.* 442, 453–464, 2012.
14. Miki, H. and Funato, Y., Regulation of intracellular signalling through cysteine oxidation by reactive oxygen species, *J. Biochem.* 151, 255–261, 2012.
15. Jacob, C., Battaglia, E., Burkholz, T. et al., Control of oxidative posttranslational cysteine modifications: From intricate chemistry to widespread biological and medical applications, *Chem. Res. Toxicol.* 25, 588–604, 2012.
16. Liu, Q., Simpson, D.C., and Gronert, S., The reactivity of human serum albumin toward *trans*-4-hydroxy-2-nonenal, *J. Mass Spectrom.* 47, 411–424, 2012.
17. Lam, X.M. Lai, W.G., Chan, E.K. et al., Site-specific tryptophan oxidation induced by autocatalytic reaction of polysorbate 20 in protein formulation, *Pharm. Res.* 28, 2543–2555, 2011.
18. Singh, S.R., Zhang, J., O'Dell, C. et al., Effect of polysorbate 80 quality on photostability of a monoclonal antibody, *AAPS PharmSciTech* 13, 422–430, 2012.
19. Mozziconacci, O., Mirkowski, J., Rusconi, F. et al., Methionine residue acts as a prooxidant in the •OH-induced oxidation of enkephalins, *J. Phys. Chem. B* 116, 12460–12472, 2012.
20. Kishore, R.S.K., Kiese, S., Fischer, S. et al., The degradation of polysorbates 20 and 80 and its potential impact on the stability of biotherapeutics, *Pharm. Res.* 28, 1194–1210, 2011.
21. Hemenway, J.N., Carvalho, T.C., Rao, V.M. et al., Formation of reactive impurities in aqueous and neat polyethylene glycol 400 and effects of antioxidants and oxidation inducers, *J. Pharm. Sci.* 101, 3305–3318, 2012.
22. Li, Z., Jacobus, L.K., Wuelfing, W.P. et al., Detection and quantification of low-molecular-weight aldehydes in pharmaceutical excipients by headspace gas chromatography, *J. Chromatogr. A* 1104, 1–10, 2006.
23. Li, Z., Kozlowski, B.M., and Chang, E.F., Analysis of aldehydes in liquid/semi-solid formulations by gas chromatography-negative ionization mass spectrometry, *J. Chromatogr. A* 1160, 299–305, 2007.
24. Hambly, D.M. and Gross, M.L., Cold chemical oxidation of proteins, *Anal. Chem.* 81, 7245–7242, 2009.
25. Lam, X.M., Lani, W.G., Chan, E.K. et al., Site-specific, tryptophan oxidation induced by autocatalytic reaction of polysorbate 80 in protein formulation, *Pharm. Res.* 28, 2543–2555, 2011.
26. Froelich, J.M. and Reid, G.E., The origin and control of ex vivo oxidative peptide modifications prior to mass spectrometry analysis, *Proteomics* 8, 1334–1345, 2008.
27. Zang, L., Carlage, T., Murphy, D. et al., Residual metals cause variability in methionine oxidation measurements in protein pharmaceuticals using LC-UV/MS peptide mapping, *J. Chromatogr. B* 895–896, 2012.

28. Wang, W. and Kelner, D.N., Correlation of rFVIII inactivation with aggregation in solution, *Pharm. Res.* 20, 693–700, 2003.

29. Wang, W., Singh, S.K., Li, N. et al., Immunogenicity of protein aggregation—Concerns and realities, *Int. J. Pharm.* 431, 1–11, 2012.

30. Ripple, D.C. and Dimitrova, M.N., Protein particles: What do we know and what do we not know? *J. Pharm. Sci.* 101, 3568–3579, 2012.

31. Hermeling, S., Aranho, L., Dainen, J.M.A. et al., Structural characterization at immunogenicity in wild-type and immune tolerant mice of degraded recombinant human interferon alpha2B, *Pharm. Res.* 22, 1997–2006, 2005.

32. Tan, Z., Shang, S., and Danishefsky, S.J., Rational development of a strategy of modifying the aggregatability of proteins, *Proc. Natl. Acad. Sci. USA* 108, 4297–4302, 2011.

33. Thirumangalathu, R., Krishnan, S., Bondarenko, P. et al., Oxidation of methionine residues in recombinant human interleukin-1 receptor antagonist: Implications of conformational stability on protein oxidation kinetics, *Biochemistry* 46, 6213–6224, 2007.

34. De Filippis, V., Lancellotti, S., Maset, F. et al., Oxidation of Met1606 in von Willebrand facto is a risk factor for thrombotic and septic complications in chronic renal failure, *Biochem. J.* 442, 423–432, 2012.

35. Ravi, J., Hills, A.E., Cerasoli, E. et al., FTIR markers of methionine oxidation for early detection of oxidized protein therapeutics, *Eur. Biophys. J.* 40, 339–345, 2011.

36. Weber, D., Kneschke, N., Grimm, S. et al., Rapid and sensitive determination of protein-nitrotyrosine by ELISA: Application to human plasma, *Free Radic. Res.* 46, 276–285, 2012.

37. Oien, D.B., Canello, T., Gabizon, R. et al., Detection of oxidized methionine in selected proteins, cellular extracts and blood serums by novel anti-methionine sulfoxide antibodies, *Arch. Biochem. Biophys.* 485, 35–40, 2009.

38. Wehr, N.B. and Levine, R.L., Wanted and wanting: Antibody against methionine sulfoxide, *Free Radic. Biol. Med.* 53, 1222–1225, 2012.

39. Surh, Y.-J. and Packer, L. (eds.), *Oxidative Stress, Inflammation, and Health*, CRC/Taylor & Francis, Boca Raton, FL, 2005.

40. Armstrong, D. (ed.), *Advanced Protocols in Oxidative Stress*, Humana/Springer, New York, 2006.

41. Rahman, I. and Kinnula, V.L., Strategies to decrease ongoing oxidant burden in chronic obstructive pulmonary disease, *Expert Rev. Clin. Pharmacol.* 5, 293–309, 2012.

42. Auerbach, A. and Hernandez, M.L., The effect of environmental oxidative stress on airway inflammation, *Curr. Opin. Allergy Clin. Immunol.* 12, 133–139, 2012.

43. Cardosa, D.R., Libardi, S.H., and Skibsted, L.H., Riboflavin as a photosensitizer. Effects on human health and food quality, *Food Funct.* 3, 487–502, 2012.

44. Chen, J., Rempel, D.L., Gau, B.C., and Gross, M.L., Fast photochemical oxidation of proteins and mass spectrometry follow submillisecond protein folding at the amino-acid level, *J. Am. Chem. Soc.* 134, 18724–18731, 2012.

45. Tacal, O., Li, B., Lockridge, O., and Schopfer, L.M., Resistance of human butyrylcholinesterase to methylene blue catalyzed photoinactivation: Mass spectrometry analysis of oxidation products, *Photochem. Photobiol.* 89, 336–348, 2012.

46. Berrill, A., Ho, S.V., and Bracewell, D.G., Product and contaminant measurement in bioprocess development by SELDI-MS, *Biotechnol. Prog.* 26, 881–887, 2010.

47. Grosvenor, A.J., Morton, J.D., and Dyer, J.M., Profiling of residue-level photo-oxidative damage in peptides, *Amino Acids* 39, 285–296, 2010.

48. Pattison, D.I., Rahmanto, A.S., and Davies, M.J., Photo-oxidation of proteins, *Photochem. Photobiol. Sci.* 11, 38–53, 2012.

49. Solazzo, C., Dyer, J.M., Deb-Choudhury, S. et al., Proteomic profiling of the photo-oxidation of silk fibroin: Implications for historic tin-weighted silk, *Photochem. Photobiol.* 88, 1217–1226, 2012.

50. Kerwin, B.A. and Remmele, R.L., Jr., Protect from light: Photodegradation and protein biologics, *J. Pharm. Sci.* 96, 1468–1479, 2007.

51. Aldini, G., Yeum, K.-J., Niki, E., and Russell, R.M. (eds.), *Biomarkers for Antioxidant Defense and Oxidative Damage: Principles and Practical Applications*, Wiley-Blackwell, Ames, IA, 2010.

52. Bruschi, M., Candiano, G., Della Ciana, L. et al., Analysis of the oxido-redox status of plasma proteins. Technology advances for clinical applications, *J. Chromatogr. B Anal. Technol. Biomed. Life Sci.* 879, 1338–1344, 2011.

53. Garcia-Garcia, A., Rodriquez-Rocha, H., Madayiputhiya, N. et al., Biomarkers of protein oxidation in human disease, *Curr. Mol. Med.* 12, 681–697, 2012.

54. Khosroshahi, M.E., Nourbakhsh, M.S., Saremi, S. et al., Application of albumin protein and indocyanine green chromophore for tissue soldering by using an IR diode laser: Ex vivo and in vivo studies, *Photomed. Laser Surg.* 28, 723–733, 2010.

55. Westhead, E.W., Photooxidation with Rose Bengal of a critical histidine residue in yeast enolase, *Biochemistry* 4, 2135–2144, 1965.

56. Tudball, N. and Thomas, P., The inhibition of glutamate dehydrogenase by L-serine *O*-sulphate and related compounds and by photo-oxidation in the presence of Rose Bengal, *Biochem. J.* 123, 421–426, 1971.

57. Fahnestock, S.R., Evidence of the involvement of a 50S ribosomal protein in several active sites, *Biochemistry* 14, 5321–5327, 1975.

58. Auron, P.E., Erdelsky, K.J., and Fahnestock, S.R., Chemical modification studies of a protein at the peptidyltransferase site of the *Bacillus stearothermophilus* ribosome. The 50S ribosomal subunit is a highly integrated functional unit, *J. Biol. Chem.* 253, 6893–6900, 1978.

59. Dohme, F. and Fahnestock, S.R., Identification of proteins involved in the peptidyl transferase activity of ribosomes by chemical modification, *J. Mol. Biol.* 129, 63–81, 1979.

60. Cerna, J. and Rychlik, I., Photoinactivation of peptidyl transferase binding sites, *FEBS Lett.* 102, 277–281, 1979.

61. Nakamura, S. and Kaziro, Y., Selective photo-oxidation of histidine residues in polypeptide chain elongation factor Tu from *E. coli*, *J. Biochem.* 90, 1117–1124, 1981.

62. Funakoshi, T., Abe, M., Sakata, M., Shoji, S., and Kubota, Y., The functional site of placental anticoagulant protein: Essential histidine residue of placental anticoagulant protein, *Biochem. Biophys. Res. Commun.* 168, 125–134, 1990.

63. Kuno, S., Fukui, S., and Toraya, T., Essential histidine residues in coenzyme B12-dependent diol dehydrase: Dye-sensitized photo-oxidation and ethoxycarbonylation, *Arch. Biochem. Biophys.* 277, 211–217, 1990.

64. Medina, M.M. et al., Involvement of essential histidine residue(s) in the activity of Ehrlich cell plasma membrane NADH-ferricyanide oxidoreductase, *Biochim. Biophys. Acta* 1190, 20–24, 1994.

65. Chang, S.H. et al., Metal-catalyzed photo-oxidation of histidine in human growth hormone, *Anal. Biochem.* 244, 221–227, 1997.

66. Moor, A.C. et al., In vitro fluence rate effects in photodynamic reactions with AlPcS4 as sensitizer, *Photochem. Photobiol.* 66, 860–865, 1997.

67. Schey, K.L. et al. Photo-oxidation of lens alpha-crystallin by hypericin (active ingredient of St. John's Wort), *Photochem. Photobiol.* 72, 200–203, 2000.

68. Beghetto, C. et al., Implications of the generation of reactive oxygen species by photoactivated calcein for mitochondrial studies, *Eur. J. Biochem.* 267, 5585–5592, 2000.

69. Parish, C.R., Fluorescent dyes for lymphocyte migration and proliferation studies, *Immunol. Cell Biol.* 77, 499–508, 1999.

70. Kummrow, A., Frankowski, M., Bock, N. et al., Quantitative assessment of cell viability based on flow cytometry and microscopy, *Cytometry A* 83(2), 197–204, 2013.

71. Roberts, J.E. et al., Photo-oxidation of lens proteins with xanthurenic acid: A putative chromophore for cataractogenesis, *Photochem. Photobiol.* 74, 740–744, 2001.

72. Madian, A.G. and Regnier, F.E., Proteomic identification of carbonylated proteins and their oxidation states, *J. Proteome Res.* 9, 3766–3780, 2010.

73. Akagawa, M., Suyama, K., and Uchida, K., Fluorescent detection of α-aminoadipic and γ-glutamic semialdehyde in oxidized proteins, *Free Radic. Biol. Med.* 46, 701–706, 2009.

74. Cecil, R. and McPhee, J.R., The sulfur chemistry of proteins, *Adv. Protein Chem.* 14, 255–389, 1959.

75. Halliwell, B. and Gutterridge, J.M.C., *Free Radicals in Biology and Medicine*, 4th edn., Oxford University Press, Oxford, U.K., 2007.

76. Hrycay, E.G. and Bandiera, S.M., The monooxygenase, peroxidase, and peroxygenase properties of cytochrome P450, *Arch. Biochem. Biophys.* 522, 71–89, 2012.

77. Harpel, M.R., Serpersu, E.H., Lamerdin, J.A. et al., Oxygenation mechanism of ribulose-bisphosphate carboxylase/oxygenase. Structure and origin of 2-carboxytetritol 1,4-bisphosphate, a novel O_2-dependent side product generated by a site-directed mutant, *Biochemistry* 34, 11296–11306, 1995.

78. Webb, J.L., Oxidants, in *Enzyme and Metabolic Inhibitors*, Vol. II, Chapter 5, pp. 665–700, Academic Press, New York, 1966.

79. Webb, J.L., Iodosobenzoate, in *Enzyme and Metabolic Inhibitors*, Vol. II, Chapter 6, pp. 701–728, Academic Press, New York, 1966.

80. Permylakov, S.E., Yu Zernil, E., Knyazeva, E.L. et al., Oxidation mimicking substitution of conservative cysteine in recoverin suppresses its membrane association, *Amino Acids* 42, 1435–1442, 2012.

81. Miyata, Y., Rauch, J.N., Jinwal, U.K. et al., Cysteine reactivity distinguishes redox sensing by the heat-inducible and constitutive forms of heat shock protein 70, *Chem. Biol.* 19, 1391–1399, 2012.

82. Buettner, G.R., Superoxide dismutase in redox biology: The roles of superoxide and hydrogen peroxide, *Anticancer Med. Chem.* 11, 341–346, 2011.

83. Fukai, T. and Ushio-Fukae, M., Superoxide dismutases: Role in redox signaling, vascular function, and diseases, *Antioxid. Redox Signal.* 15, 1583–1606, 2011.

84. Bae, Y.S., Oh, H., Rhee, S.G. et al., Regulation of reactive oxygen species generation in cell signaling, *Mol. Cells* 32, 491–509, 2012.

85. Armogida, M., Nisticò, R., and Mercuri, N.B., Therapeutic potential of targeting hydrogen peroxide metabolism in the treatment of brain ischaemia, *Br. J. Pharmacol.* 166, 1211–1224, 2012.

86. Ha, E., Wang, W., and Wang, Y.J., Peroxide formation in polysorbate 80 and protein stability, *J. Pharm. Sci.* 91, 2252–2264, 2002.

87. Wasylaschuk, W.R., Harmon, P.A., Wagner, G. et al., Evaluation of hydroperoxides in common pharmaceutical excipients, *J. Pharm. Sci.* 96, 106–116, 2007.

88. Steinmann, D., Ji, J.A., Wang, Y.J., and Schöneich, C., Oxidation of human growth hormone by oxygen-centered radicals: Formation of Leu-101 hydroperoxide and Tyr-103 oxidation products, *Mol. Pharm.* 9, 803–814, 2012.

89. Kishore, R.S., Pappenberger, A., Dauphin, I.B. et al., Degradation of polysorbates 20 and 80: Studies on thermal autoxidation and hydrolysis, *J. Pharm. Sci.* 100, 721–731, 2011.

90. Burton, D., Tombs, M.P., and MacLagen, N.F., Perchloric acid-soluble proteins, *Biochem. J.* 84, 98P–99P, 1962.

91. Burton, D., Tombs, M.P., Apsey, M.E., and MacLagen, N.F., The perchloric acid soluble basic and acidic proteins of the cytoplasm: Variation in cancer, *Br. J. Cancer* 17, 162–178, 1963.

92. Srebreva, L., Ziatanova, J., Miloshev, G., and Tsanev, R., Immunological evidence for the existence of H1-like histone in yeast, *Eur. J. Biochem.* 165, 449–454, 1987.

93. Oestvold, A.C., Hullstein, I., and Sletten, K., A novel mammalian nucleic acid-binding protein with homology to the yeast ribosomal protein YL43, *FEBS Lett.* 298, 219–222, 1992.

94. Ghezzo, F., Berta, G.N., Bussolati, B. et al., Perchloric acid-soluble proteins from goat liver inhibit chemical carcinogenesis of Syrian hamster cheek-pouch carcinoma, *Br. J. Cancer* 79, 54–58, 1999.

95. Zougman, A. and Wisniewski, J.R., Beyond linker histones and high mobility proteins: Global profiling of perchloric acid soluble proteins, *J. Proteome Res.* 5, 925–934, 2006.

96. Williams, B.J., Russell, W.K., and Russell, D.H., High-throughput method for on-target performic acid oxidation of MALDI-deposited samples, *J. Mass Spectrom.* 45, 157–166, 2010.

97. Chang, Y.C., Huang, C.H., Lin, C.H. et al., Mapping protein cysteine sulfonic acid modifications with specific enrichment and mass spectrometry: An integrated approach to explore the cysteine oxidation, *Proteomics* 10, 2961–2971, 2010.

98. Chowdhury, S.K., Eshraghi, J., Wolfe, H. et al., Mass spectrometric identification of amino acid transformations during oxidation of peptides and proteins: Modification of methionine and tyrosine, *Anal. Chem.* 67, 390–398, 1995.

99. Dai, J., Zhang, Y., Wang, J. et al., Identification of degradation products formed during performic oxidation of peptides and proteins by high-performance liquid chromatography with matrix-assisted laser desorption/ionization and tandem mass spectrometry, *Rapid Commun. Mass Spectrom.* 19, 1130–1138, 2005.

100. Alvarez, B. and Radi, R., Peroxynitrite reactivity with amino acids and proteins, *Amino Acids* 25, 295–311, 2003.

101. Rubbo, H., Trostchansky, A., and O'Donnel, V.B., Peroxynitrite-mediated lipid oxidation and nitration: Mechanisms and consequences, *Arch. Biochem. Biophys.* 484, 167–172, 2009.

102. Liaudet, L., Vassalli, G., and Pacher, P., Role of peroxynitrite in the redox regulation of cell signal transduction pathways, *Front. Biosci.* 14, 4809–4814, 2009.

103. Ascenzi, P., di Masi, A., Sciorati, C., and Clementi, E., Peroxynitrite–An ugly biofactor? *Biofactors* 36, 264–273, 2010.

104. Mahoney, W.C. and Hermodson, M.A., High-yield cleavage of tryptophanyl peptide bonds by *o*-iodosobenzoic acid, *Biochemistry* 18, 3810–3814, 1979.

105. Sinha, S., Watorek, W., Karr, S. et al., Primary structure of human neutrophil elastase, *Proc. Natl. Acad. Sci. USA* 84, 2228–2232, 1987.

106. Mahoney, W.C., Smith, P.K., and Hermodson, M.A., Fragmentation of proteins with *o*-iodosobenzoic acid: Chemical mechanism and identification of *o*-iodoxybenzoic acid as a reactive contaminant that modifies tyrosyl residues, *Biochemistry* 20, 443–448, 1981.

107. Zhdankin, V.V., Koposov, A.Y., Litvinev, D.N. et al., Esters of 2-iodoxybenzoic acid: Hypervalent iodine oxidizing reagents with a pseudobenziodoxole structure, *J. Org. Chem.* 70, 6484–6491, 2005.

108. Arkin, A., The influence of certain oxidizing agents (sodium iodosobenzoate and sodium iodoxybenzoate) on phagocytosis, *J. Infect. Dis.* 11, 427–432, 1913.

109. Arkin, A., The influence of an oxidizing substance (sodium iodoxybenzoate) on immune reactions, *J. Infect. Dis.* 16, 349–360, 1915.

110. Kersley, G.D. and Simpson, N.R.W., Clinical trial of calcium ortho-iodoxybenzoate, *Ann. Rheumat. Dis.* 9, 174–175, 1950.

111. Trundle, D. and Cunningham, L.W., Iodine oxidation of the sulfhydryl groups of creatine kinase, *Biochemistry* 8, 1919–1925, 1969.

112. Cunningham, L.W., The reaction of α-lactoglobulin sulfenyl iodide with several antithyroid agents, *Biochemistry* 3, 1629–1634, 1964.

113. Reddy, K.M., Kumari, Y.B., Mallikharjunasaruma, D. et al., Large scale solid phase synthesis of peptide drugs: Use of commercial anion exchange resin as a quenching agent for removal of iodine during disulphide bond formation, *Int. J. Pept.* 2012, Article ID 323907, 2012.

114. Fromageot, C., Chatagner, F., and Bergeret, B., La formation d'alanine par désulfination enzymatique de l'acide l-cystéine-sulfinique, *Biochim. Biophys. Acta* 2, 294–301, 1948.

115. Montero, C., Histochemistry of protein-bound disulphide groups in the duct secretory granules of the rat submandibular gland, *Histochem. J.* 4, 259–266, 1972.

116. Sipponen, P., Histochemical reactions of gastrointestinal mucosubstances with orcein, high iron diamine and Alcian blue after prior oxidation of tissue sections, *Histochemistry* 59, 199–206, 1979.

117. Chung, S.S., Kwak, K.B., Lee, J.S. et al., Preferential degradation of the $KMnO_4$-oxidized or *N*-ethylmaleimide-modified form of sarcoplasmic reticulum ATPase by calpain from chick skeletal muscle, *Biochim. Biophys. Acta* 1041, 160–163, 1990.

118. Haas, W.J., Sizer, I.W., and Loofbourow, J.R., The effect of permanganate of the ultraviolet absorption spectra of aromatic amino acids and proteins, *Biochim. Biophys. Acta* 6, 601–605, 1951.

119. Benisek, W.F., Reaction of the catalytic subunit of *Escherichia coli* aspartate transcarbamylase with permanganate ion, a reactive structural analogue of phosphate ion, *J. Biol. Chem.* 246, 3151–3159, 1971.

120. Thomas, R.A. and Kirsch, J.F., Kinetics and mechanism of inhibition of *Escherichia coli* alkaline phosphatase by permanganate ion, *Biochemistry* 19, 5328–5334, 1980.

121. Bui, C.T., Rees, K., and Cotton, R.G.H., Permanganate oxidation reactions of DNA: Perspective in biological studies, *Nucleosides Nucleotides Nucleic Acids* 22, 1835–1855, 2003.

122. Gries, T.J., Kontur, W.S., Capp, M.W. et al., One-step DNA melting in the RNA polymerase cleft opens the initiation bubble to form an unstable open complex, *Proc. Natl. Acad. Sci. USA* 107, 10418–10423, 2010.

123. Wisdom, G.B., Conjugation of antibodies to horseradish peroxidase, *Methods Mol. Biol.* 295, 127–130, 2005.

124. Liu, B., Burdine, L., and Kodadek, T., Chemistry of periodate-mediated cross-linking of 3,4-dihydroxyphenylalanine-containing molecules to proteins, *J. Am. Chem. Soc.* 128, 15228–15235, 2006.

125. Wislocki, G.B., Rheingold, J.J., and Dempsey, E.W., The occurrence of the periodic acid-Schiff reaction in various normal cells of blood and connective tissue, *Blood* 4, 562–568, 1949.

126. Kilcoyne, M., Gerlach, J.Q., Farrell, M.P. et al., Periodic acid-Schiff's reagent for carbohydrates in a microtiter plate format, *Anal. Biochem.* 416, 18–26, 2011.

127. Gershoni, J.M., Bayer, E.A., and Wilchek, M., Blot analyses of glycoconjugates: Enzyme:hydrazide—A novel reagent for the detection of aldehydes, *Anal. Biochem.* 146, 59–63, 1985.

128. Bouchez-Mahiout, I., Doyen, C., and Laurière, M., Accurate detection of both glycoproteins and total proteins on blots: Control of side reactions occurring after periodate oxidation of proteins, *Electrophoresis* 20, 1412–1417, 1999.

129. Lewandrowski, U., Moebius, J., Walter, U., and Sickmann, A., Elucidation of *N*-glycosylation sites on human platelet proteins, *Mol. Cell. Proteomics* 5, 226–233, 2006.

130. Xuan, H. and Hage, D.S., Immobilization of α_1-acid glycoprotein for chromatographic studies of drug-protein binding, *Anal. Biochem.* 346, 300–310, 2005.

131. Kirkeby, S., Chemical modification of carbohydrates in tissue sections may unmask mucin antigens, *Biotech. Histochem.* 88, 19–26, 2013.

132. Kristiansen, K.A., Potthast, A., and Christensen, B.E., Periodate oxidation of polysaccharides for modification of chemical and physical properties, *Carbohydr. Res.* 345, 1264–1271, 2010.

133. Winterbourn, C.C. and Kettle, A.J., Biomarkers of myeloperoxidase-derived hypochlorous acid, *Free Radic. Biol. Med.* 29, 403–409, 2000.

134. Hawkins, C.L., Pattison, D.I., and Davies, M.J., Hypochlorite-induced oxidation of amino acids, peptides and proteins, *Amino Acids* 25, 259–274, 2003.
135. Mouls, L., Silajdzic, E., Horoune, N. et al., Development of novel mass spectrometric methods for identifying HOCl-induced modifications to proteins, *Proteomics* 9, 1617–1631, 2009.
136. Peskin, A.V., Turner, R., Maghzal, G.J. et al., Oxidation of methionine to dehydromethionine by reactive halogen species generated by neutrophils, *Biochemistry* 48, 101075–10182, 2009.
137. Petrônio, M.S. and Ximenes, V.F., Effects of oxidation of lysozyme by hypohalous acids and haloamines on enzymatic activity and aggregation, *Biochim. Biophys. Acta* 1824, 1090–1096, 2012.
138. Stacey, M.M., Cuddihy, S.L., Hampton, M.B., and Winterbourn, C.C., Protein thiol oxidation and formation of *S*-glutathionylated cyclophilin A in cells exposed to chloramines and hypochlorous acid, *Arch. Biochem. Biophys.* 527, 45–54, 2012.
139. Fu, X., Kassim, S.Y., Parks, W.C., and Heinecke, J.W., Hypochlorous acid oxygenates the cysteine switch domain of pro-matrilysin (MMP-7). A mechanism for matrix metalloproteinase activation of atherosclerotic plaque rupture by myeloperoxidase, *J. Biol. Chem.* 276, 41279–41287, 2001.
140. Szuchman-Sapir, A.J., Pattison, D.I., Davies, M.J., and Witting, P.K., Site-specific hypochlorous acid-induced oxidation of recombinant human myoglobin affects specific amino acid residues and the rate of cytochrome b5-mediated heme reduction, *Free Radic. Biol. Med.* 48, 35–46, 2010.
141. Beal, J.L., Foster, S.B., and Ashby, M.T., Hypochlorous acid reacts with the *N*-terminal methionines of protein to give dehydromethionine a potential biomarker for neutrophil-induced oxidative stress, *Biochemistry* 48, 11142–11148, 2009.
142. Nakamura, M., Shishido, N., Nunomura, A. et al., Specific reaction of Met 35 in amyloid β peptide with hypochlorous acid, *Free Radic. Res.* 44, 734–741, 2010.
143. Nagy, P. and Ashby, M.T., Reactive sulfur species: Kinetics and mechanisms of the oxidation of cysteine by hypohalous acid to give cysteine sulfenic acid, *J. Am. Chem. Soc.* 129, 14082–14091, 2007.
144. Pitt, A.R. and Spickett, C.M., Mass spectrometric analysis of HOCl- and free-radical-induced damage to lipids and proteins, *Biochem. Soc. Trans.* 36, 1077–1082, 2008.
145. Shao, B. and Heinecke, J.W., Using tandem mass spectrometry to quantify site-specific chlorination and nitration of proteins: Model systems with high-density lipoprotein oxidized by myeloperoxidase, *Methods Enzymol.* 440, 33–63, 2008.

6 Modification of Proteins with Reducing Agents

The interaction of reducing agents with proteins is more complicated than the reduction of disulfide bonds. While there is some mention of disulfide bond reduction in the current chapter, specific discussion of reduction of disulfide bonds is presented in Chapter 8. A reducing agent can be used for the reduction of other oxidized forms of cysteine such as cysteine sulfenic acid or S-nitrosocysteine, the reduction of imine bonds in Schiff bases, the reduction of metal ion centers in proteins such as those found in heme proteins, and the reduction of pertechnetate ($^{99m}TcO_4^-$) to product technetium (^{99m}Tc), which is then bound to reduced antibody or other ligand proteins for diagnostic use. In addition to the targets for reduction in protein chemistry, there are also a variety of reducing agents (see Table 6.1) whose function depends not just on intrinsic redox characteristics but the chemical structure being reduced. As an example, the selective reduction of disulfide bonds has been used for the study of the relationship of protein structure to function[1–4] as well as the importance of disulfide bonds in providing stability for native conformation.[5,6] Limited reduction of immunoglobulin proteins has proved useful for the determination of protein structure.[7–11] Early work on the selective reduction of disulfide bonds in immunoglobulin was part of the manufacturing process for a therapeutic intravenous immunoglobulin.[12–14] More recently, as described in the succeeding text, monoclonal antibodies have been reduced and then modified with maleimide-based drugs for therapeutic purposes.[15] Reducing agents are also used for the reduction of other cysteine oxidation products such as sulfenic acid.[16,17] Lindhoud and coworkers[17] observed that DTT could convert cysteine sulfenic acid or disulfides to cysteine but did not reduce either cysteine sulfinic acid or cystine sulfonic acid. This provides a useful method for distinguishing between cysteine oxidation products.

There is more limited use of some stronger reducing agents such as the borohydrides in protein chemistry for the reduction of Schiff bases formed between aldehydes (including reducing sugars) and amines (see Chapter 19) and the reduction of enol alcohols in the reaction of N-ethyl-5-phenylisoxazolium-3′-sulfonate with proteins.[18] There was, however, early use of sodium borohydride for the reduction of disulfide bonds in proteins as discussed in the succeeding text. Some of the reducing agents commonly used in organic chemistry such as lithium aluminum hydride are too potent to be used in protein chemistry (Table 6.1), while less potent hydrides such as sodium borohydride and sodium cyanoborohydride are used for the aforementioned applications in protein chemistry.

There is a tendency, not unreasonable, to think of reducing agents such as 2-mercaptoethanol and DTT as selective for the reduction of disulfide bonds; there are other interactions with proteins unrelated to disulfide bonds.[19,20] Alliegro[19] observed that DTT reduced the carbohydrate-binding activity of mutant of pigpen where cysteine residues had been removed; DTT also reduced the carbohydrate binding of the native protein. The effect of DTT on carbohydrate binding was shown not to be due to chelation of metal ions. Piedrafita and coworkers[21] showed that the product of the reaction between DTT and N-(2,3-dihydroxyphenyl) maleimide was a competitive inhibitor of catechol-O-methyltransferase; the maleimide did inactivate the enzyme by reaction with thiol groups on the protein. The reaction between DTT and N-(2,3-dihydroxyphenyl) maleimide is essentially instantaneous.[21] These observations combined with the variable susceptibility of disulfide bonds in proteins to reduction and the potential for the formation of mixed disulfides between reducing agents such as 2-mercaptoethanol and sulfhydryl groups in proteins.

TABLE 6.1

Properties of Some Reducing Agents for Proteins[a]

Compound	Comment[b]	E^o (V)[c]	References
2-Mercaptoethanol	Reduces disulfide, forms mixed disulfide with cysteine sulfenic acid.[d]	−0.25	6.1.1
2-Mercaptoethanol	Reduces disulfide, forms mixed disulfide with cysteine sulfenic acid.[d]	−0.24	6.1.2
2-Mercaptoethanol	Reduces disulfide, forms mixed disulfide with cysteine sulfenic acid.[d]	−0.20	6.1.3
2-Mercaptoethanol	Reduces disulfide, forms mixed disulfide with cysteine sulfenic acid.[d]	−0.32[e]	6.1.4
Thioglycolic acid (mercaptoacetic acid)	Reduces disulfide bonds; extensively used in cosmetology.	−0.30[e]	6.1.4
Thioglycolic acid (mercaptoacetic acid)	Reduces disulfide bonds; extensively used in cosmetology.	−0.14	6.1.5
Dithiothreitol	Reduces disulfide bonds, cysteine sulfenic acid, S-nitrosocysteine, and some metal ions.	−0.33	6.1.6
Dithiothreitol	Reduces disulfide bonds, cysteine sulfenic acid, S-nitrosocysteine, and some metal ions.	−0.39	6.1.7
Dithiothreitol	Reduces disulfide bonds, cysteine sulfenic acid, S-nitrosocysteine, and some metal ions.	−0.33	6.1.3
Dithiothreitol	Reduces disulfide bonds, cysteine sulfenic acid, S-nitrosocysteine, and some metal ions.	−0.33	6.1.8
Dithioerythritol	Reduces disulfide bonds.[f]		6.1.6
Dithiothreitol	Reduces disulfide bonds.[f]	−0.33[e]	6.1.4
Diselenothreitol	Reduces disulfide bonds.[g]	−0.464	6.1.9
Selenothiothreitol	Reduces disulfide bonds.[g]	−0.394	6.1.9
(2S)-2-Amino-1,4-dimercaptobutane (dithiobutylamine)	Reduces disulfide bonds.[h]	−0.32	6.1.3
Methylhydrazine	Reduces S-nitrosocysteine to cysteine.		6.1.8
Tris(2-carboxyethyl) phosphine	Reduces disulfide bonds. Unlike the thiol-based reagents, there is little or no reaction with sulfhydryl reagents such as maleimides or iodoacetamide.	−0.29	6.1.10
Tris(2-carboxyethyl) phosphine	Reduces disulfide bonds. Unlike the thiol-based reagents, there is little or no reaction with sulfhydryl reagents such as maleimides or iodoacetamide.	−0.29	6.1.11
Mercaptoacetic acid	Reduces mixed disulfide.	−0.30[e]	6.1.4
Reduced glutathione	Reduces disulfide bonds. Shown to be a nitric oxide donor (S-nitrosoglutathione).	−0.25	6.1.1
Reduced glutathione	Reduces disulfide bonds. Shown to be a nitric oxide donor (S-nitrosoglutathione).	−0.16	6.1.8
Reduced glutathione	Reduces disulfide bonds. Shown to be a nitric oxide donor (S-nitrosoglutathione).	−0.26	6.1.7
Reduced glutathione	Reduces disulfide bonds. Shown to be a nitric oxide donor (S-nitrosoglutathione).	−0.24[e]	6.1.4
Reduced lipoic acid	Reduces disulfide bonds. Serves as an in vivo oxidant and cofactor for enzyme-catalyzed reduction.	−0.35	6.1.7, 6.1.12
Ultraviolet light	There are several reports that disulfide bonds are reduced with UV-C light (200–280 nm). The effect may or may not reflect other factors such as the presence of metal ions or proximity of a tryptophanyl residue. In the case of tryptophan, it is suggested that there is a one-electron reduction resulting in a thiyl radical.	NA	6.1.13–6.1.16

TABLE 6.1 (continued)
Properties of Some Reducing Agents for Proteins[a]

Compound	Comment[b]	E^o (V)[c]	References
Sodium cyanoborohydride	Reduces Schiff base and collagen cross-links. Sodium cyanoborohydride is considered a milder reagent than sodium borohydride. Sodium cyanoborohydride is considered to be more hydrophobic than sodium borohydride. Does not reduce disulfide bonds.		6.1.17–6.1.22
Sodium borohydride	A stronger reducing agent than sodium cyanoborohydride and reduces disulfide bonds, Schiff bases, aldehydes, thioesters, protein carbonyl groups, and succinimides derived from aspartic/asparagine.		6.1.23–6.1.33
Ascorbic acid[e]	Reduces metal ions, disulfide bonds, and S-nitrosocysteine. Ascorbic acid is also known for generation of free radicals. Oxidized ascorbic acid can oxidize cysteinyl residues during protein folding.	−0.07[i]	6.1.34
Ascorbic acid[e]	Reduces metal ions, disulfide bonds, and S-nitrosocysteine. Ascorbic acid is also known for generation of free radicals. Oxidized ascorbic acid can oxidize cysteinyl residues during protein folding.	−0.13	6.1.8
Ascorbic acid	Reduces metal ions, disulfide bonds, and S-nitrosocysteine. Ascorbic acid is also known for generation of free radicals. Oxidized ascorbic acid can oxidize cysteinyl residues during protein folding.	−0.06[j]	6.1.35
Cysteine	Reduces disulfide bonds; forms disulfide bonds with cysteine sulfenic acid. Oxidation product is cystine, which, with cysteine, can form a redox pair for protein folding.	−0.21	6.1.6
Cysteamine (2-amino-ethanethiol, CAS 60-22-11)	Reduces disulfide bond. Can form a mixed disulfide. Can form a redox pair with cystamine 2,2′-dithiobis (ethylamine).	−0.20	6.1.3, 6.1.36
Homocysteine	Reduces disulfide bonds; forms mixed disulfides.	−0.23	6.1.2
Sodium dithionite (sodium hydrosulfite; NaS_2O_4)	Reduces disulfide bonds and heme iron and possible reduction of cysteine sulfenic acid and 3-nitrotyrosine.	−0.660[k]	6.1.37–6.1.40
Stannous chloride	Reduces disulfide bonds and pertechnetate (formation of ^{99m}Tc).	−0.151	6.1.41, 6.1.42
Dimethylborane	Reduces Schiff base.		6.1.43
Trimethylborane	Reduces Schiff base.		6.1.43
Pyridine borane	Reduces Schiff base.		6.1.44–6.1.46
2-Picoline borane	Reduces Schiff base.		6.1.46–6.1.48
2-Aminothiophenol	Reduce disulfide under acidic conditions.		6.1.49, 6.1.50
2,3-Dimercaptopropanol (BAL; British anti-Lewisite)	No use in protein chemistry; used as a therapeutic for mercury, arsenic poisoning.		6.1.51, 6.1.52

[a] It is recommended that the reader consider earlier work (Clark, W.M., *Oxidation-Reduction Potentials of Organic Systems*, Williams & Wilkins, Baltimore, MD, 1960).

[b] The susceptibility of disulfide bonds, cysteine sulfenic acid, and S-nitrosocysteine to reduction varies by individual protein. The mention of a specific chemical entity only indicates that there is at least one study demonstrating reduction of the specific target.

[c] At pH (pD) 7.0, 23°C–25°C unless otherwise indicated. There is some variability on the oxidation–reduction potentials and then numbers presented are those obtained in the cited work. Temperature and pH both affect measured oxidation/reduction potentials (Fischer, E.K., Oxidation-reduction potentials of certain sulfhydryl compounds, *J. Biol. Chem.* 89, 753–763, 1930). Values in this table are for pH 7.0, 25°C unless otherwise indicated.

(continued)

188 Chemical Reagents for Protein Modification

TABLE 6.1 (continued)
Properties of Some Reducing Agents for Proteins[a]

^d Dalle-Donne, I., Carini, M., Orioli, M. et al., Protein carbonylation: 2,4-dinitrophenylhydrazine reacts with both aldehydes/ketones and sulfenic acids, *Free Radic. Biol. Med.* 46, 1411–1419, 2009.

^e Values may vary as much as 10%.

^f While it is likely that DTE has the same activity as DTT, data are only available to support the reduction of disulfide bonds.

^g While it is likely that both selenothiothreitol and diselenothreitol reduce both cysteine sulfenic acid and *S*-nitrosocysteine, data are not available to support such action.

^h While it is likely that dithiobutylamine reduces both cysteine sulfenic acid and *S*-nitrosocysteine, data are not available to support such action.

ⁱ E^o (pH 2.04) = –0.283; measurement performed at 35.5°C; data for pH 7.0 obtained by extrapolation.

^j The oxidation–reduction potential for ascorbic acid is complicated by the presence of a free radical intermediate, monohydro-L-ascorbic acid.

^k Despite this value, there are few reports of the reduction of disulfide bonds in proteins with sodium dithionite. It is used for the reduction of nitrotyrosine in proteins without a report of reduction of disulfide bonds.

References to Table 6.1

6.1.1. Miles, K.K., Weaver, K.H., and Rabenstein, D.L., Oxidation/reduction potential of glutathione, *J. Org. Chem.* 58, 4144–4146, 1993.

6.1.2. Rabenstein, D.L., Redox potentials of cysteine residues in peptides and proteins, in *Oxidative Folding of Proteins and Peptides*, eds. J. Buchner and L. Moreder, RSC Publishing, Cambridge, U.K., 2009.

6.1.3. Lukesh, J.C., III, Palte, M.J., and Raines, R.T., A potent, versatile disulfide-reducing agent from aspartic acid, *J. Am. Chem. Soc.* 134, 4057–4059, 2012.

6.1.4. Rao, P.S., Evans, R.G., and Mueller, H.S., Experimental determination of activation potentials of CK-isoenzymes in human serum and their significance, *Biochem. Biophys. Res. Commun.* 78, 548–654, 1977.

6.1.5. Lundblad, R.L. and MacDonald, F.M., *Handbook of Biochemistry and Molecular Biology*, 4th edn., CRC Press, Boca Raton, FL, 2010.

6.1.6. Cleland, W.W., Dithiothreitol, a new protective reagent for SH groups, *Biochemistry* 4, 480–482, 1964.

6.1.7. Gilbert, H.F., Molecular and cellular aspects of thiol disulfide exchange, *Adv. Enzymol.* 63, 69–172, 1990.

6.1.8. Wiesweg, M., Berchner-Pfannschmidt, U., Fandrey, J. et al., Rocket fuel for the quantification of *S*-nitrosothiols. Highly specific reduction of *S*-nitrosothiols to thiols by methylhydrazine, *Free Radic. Res.* 47, 104–115, 2013.

6.1.9. Nauser, T., Steinmann, D., and Koppenol, W.H., Why do proteins use selenocysteine instead of cysteine? *Amino Acids* 42, 39–44, 2012.

6.1.10. Pullela, P.K., Chiku, T., Carvan, III, M.J., and Sem, D.S., Fluorescence-based detection of thiols in vitro and in vivo using dithiol probes, *Anal. Biochem.* 352, 265–273, 2006.

6.1.11. Peng, L., Xu, X., Guo, M. et al., Effects of metal ions and disulfide bonds on the activity of phosphodiesterase from *Trimeresurus stejnegeri* venom, *Metallomics* 5, 920–927, 2013.

6.1.12. Chen, H.J., Chen, Y.M., and Chang, C.M., Lipoyl dehydrogenase catalyzes reduction of nitrated DNA and protein adducts using dihydrolipoic acid or ubiquinol as the cofactor, *Chem. Biol. Interact.* 140, 199–213, 2002.

6.1.13. Permyakov, E.A. and Uversky, V.N., Ultraviolet illumination-induced reduction of α-lactalbumin disulfide bridges, *Proteins* 51, 498–503, 2003.

6.1.14. Verhaar, R., Dekkers, D.W., De Cuyper, I.M. et al., UV-C irradiation disrupts platelet surface disulfide bonds and activates the platelet integrin αIIbβ3, *Blood* 112, 4935–4939, 2008.

6.1.15. Roy, S., Mason, B.D., Schöneich, C.S. et al., Light-induced aggregation of type I soluble tumor necrosis factor receptor, *J. Pharm. Sci.* 98, 3182–3199, 2009.

6.1.16. Wang, L., Yan, X., Xu, C. et al., Photocatalytical reduction of disulphide bonds in peptide on Ag-loaded nano-TiO$_2$ for subsequent derivatization and determination, *Analyst* 136, 3602–3604, 2011.

6.1.17. Friedman, M., Williams, L.D., and Masri, M.S., Reductive alkylation of proteins with aromatic aldehydes and sodium borohydride, *Int. J. Pept. Protein Res.* 6, 183–185, 1974.

6.1.18. Robins, S.P. and Bailey, A.J., The chemistry of the collage cross-links. Characterization of the products of reduction of skin, tendon, and bone with sodium cyanoborohydride, *Biochem. J.* 163, 339–346, 1977.

6.1.19. Fager, R.S., Cyanoborohydride reduction of rhodopsin, *Methods Enzymol.* 81, 288–290, 1982.

6.1.20. Peng, L., Calton, G.J., and Burnett, J.W., Effect of borohydride reduction on antibodies, *Appl. Biochem. Biotechnol.* 14, 91–99, 1987.

TABLE 6.1 (continued)
Properties of Some Reducing Agents for Proteins[a]

6.1.21. Fishman, R.M., Moore, G.L., Zegna, A., and Marini, M.A., High-performance liquid chromatographic evaluation of pyridoxal 5′-phosphate hemoglobin derivatives produced by different reduction procedures, *J. Chromatogr.* 532, 55–64, 1990.

6.1.22. Madian, A.G. and Regnier, F.E., Profiling carbonylated proteins in human plasma, *J. Proteome Res.* 9, 1330–1343, 2010.

6.1.23. Fischer, E.H., Kent, A.B., Snyder, E.R., and Krebs, E.G., Reaction of sodium borohydride with muscle phosphorylase, *J. Am. Chem. Soc.* 80, 2906–2807, 1958.

6.1.24. Brown, W.D., Reduction of the disulfide bonds by sodium borohydride, *Biochim. Biophys. Acta* 44, 365–367, 1960.

6.1.25. Kahlenberg, A., Lack of stereospecificity of glucose binding to human erythrocyte membrane protein upon reduction with sodium borohydride, *Biochem. Biophys. Res. Commun.* 36, 690–695, 1969.

6.1.26. Bailey, A.J., Peach, C.M., and Fowler, L.J., Chemistry of the collagen cross-links. Isolation and characterization of two intermediate intermolecular cross-links in collagen, *Biochem. J.* 117, 819–831, 1970.

6.1.27. Zemel, E.S., Cartwright-Harwick, C.M., and Nilsson, U.R., Effect of reduction on the structural, biologic, and immunologic properties of the third component of human complement, *J. Immunol.* 125, 1099–1103, 1980.

6.1.28. Thomas, M.L., Davidson, F.F., and Tack, B.F., Reduction of the β-cys-γ-glu thiol ester bond of human C3 with sodium borohydride, *J. Biol. Chem.* 258, 13580–13586, 1983.

6.1.29. Dobryszycka, W. and Guszczynski, T., Reduction of disulphide bonds in human haptoglobin 2–1, *Biochim. Biophys. Acta* 829, 13–18, 1985.

6.1.30. Lenz, A.G., Costabel, U., Shaltiel, S., and Levine, R.L., Determination of carbonyl groups in oxidatively modified proteins by reduction with tritiated sodium borohydride, *Anal. Biochem.* 177, 419–425, 1989.

6.1.31. Carter, D.A. and McFadden, P.N., Trapping succinimides in aged polypeptides by chemical reduction, *J. Protein Chem.* 13, 89–96, 1994.

6.1.32. Li, Q., Ling, J., and Liu, W.Y., Partial restoration of inactivated ribosomes with sodium borohydride or amino acids, *FEBS Lett.* 370, 123–126, 1995.

6.1.33. Purich, D.L., Use of sodium borohydride to detect acyl-phosphate linkages in enzyme reactions, *Methods Enzymol.* 354, 168–177, 2002.

6.1.34. Nobile, S. and Woodhill, J.M., *Vitamin C: The Mysterious Redox System: A Trigger of Life?* Chapter 3, pp. 21–26, MTP Press, Lancaster, U.K., 1981.

6.1.35. Boorsook, H. and Keighly, G., Oxidation-reduction potential of ascorbic acid (vitamin C), *Proc. Natl. Acad. Sci. USA* 19, 875–878, 1933.

6.1.36. Aly, A.M., Arai, M., and Hoyer, L.W., Cysteamine enhances the procoagulant activity of Factor VIII-East Hartford, a dysfunctional protein due to a light chain thrombin cleavage site mutation (arginine-1689 to cysteine), *J. Clin. Invest.* 89, 1375–1381, 1992.

6.1.37. Senft, A.P., Dalton, T.P., and Shertzer, H.G., Determining glutathione and glutathione disulfide using the fluorescence probe o-phthalaldehyde, *Anal. Biochem.* 280, 80–86, 2000.

6.1.38. Knipp, M., Taing, J.J., and He, C., Reduction of the lipocalin type heme containing protein nitrophorin—Sensitivity of the fold-stabilizing cysteine disulfides toward heme-iron reduction, *J. Inorg. Biochem.* 105, 1405–1412, 2011.

6.1.39. Thakur, M.L., DeFulvio, J., Richard, M.D., and Park, C.H., Technetium-99m labeled monoclonal antibodies: Evaluation of reducing agents, *Nucl. Med. Biol.* 18, 227–233, 1991.

6.1.40. Sokolovsky, M., Riordan, J.F., and Vallee, B.L., Conversion of 3-nitrotyrosine to 3-aminotyrosine in peptides and proteins, *Biochem. Biophys. Res. Commun.* 27, 20–25, 1967.

6.1.41. Hainsworth, J.E., Harrison, P., and Mather, S.J., Novel preparation and characterization of a trastuzumab-streptavidin conjugate for pre-targeted radionuclide therapy, *Nucl. Med. Commun.* 27, 461–471, 2006.

6.1.42. Fernandes, C., Correla, J.D.G., Gano, L. et al., Dramatic effect of the tridentate ligand on the stability of ^{99m}Tc "3 + 1" oxo complexes bearing arylpiperazine derivatives, *Bioconjug. Chem.* 16, 660–669, 2005.

6.1.43. Geoghegan, K., Cabacungan, J.C., Dixon, H.B.F., and Feeney, R.E., Alternative reducing agents for reductive methylation of amino groups in proteins, *Int. J. Pept. Protein Res.* 17, 345–352, 1981.

6.1.44. Wong, W.S.D., Osuga, D.T., and Feeney, R.E., Pyridine borane as a reducing agent for proteins, *Anal. Biochem.* 139, 58–67, 1984.

6.1.45. Miron, T. and Wilchek, M., Fast mass spectrometry detection of tryptophan-containing peptides and proteins by reduction with pyridine-borane, *Anal. Biochem.* 440, 12–14, 2013.

(continued)

TABLE 6.1 (continued)

Properties of Some Reducing Agents for Proteins[a]

6.1.46. Ambrogelly, A., Cutler, C., and Paporello, B., Screening of reducing agents for the PEGylation of recombinant human IL-10, *Protein J.* 32, 337–342, 2013.

6.1.47. Ruhaak, L.R., Steenvoorden, E., Leoleman, C.A.M. et al., 2-Picoline borane: A non-toxic reducing agent for oligosaccharide labeling by reductive amination, *Proteomics* 10, 2330–2336, 2010.

6.1.48. Fiege, K., Luensdorf, H., and Mischnick, P., Aminoalkyl functionalization of dextran for coupling to bioactive molecules and nonstructure formation, *Carbohydr. Polym.* 95, 569–577, 2013.

6.1.49. Abe, Y., Ueda, T., and Imoto, T., Reduction of disulfide bonds in proteins by 2-aminothiophenol under weakly acidic conditions, *J. Biochem.* 115, 52–57, 1994.

6.1.50. Abe, Y., Ueda, T., and Imoto, T., An improved method for preparing lysozyme with chemically ^{13}C-enriched methionine residues using 2-aminothiophenol as a reagent of thiolysis, *J. Biochem.* 122, 1153–1159, 1997.

6.1.51. Peters, R.A., Stocken, L.A., and Thompson, R.H., British anti-lewisite (BAL), *Nature* 156, 616–619, 1945.

6.1.52. Beasley, D.M., Schep, L.J., and Slaughter, R.J., Full recovery from a potentially lethal dose of mercuric chloride, *J. Med. Toxicol.* 2013.

The inclusion of mild reducing agents such as 2-mercaptoethanol is a long-standing practice for the stabilization of enzymes containing cysteine residues.[22] Conversely, there are enzymes easily inactivated by 2-mercaptoethanol.[23] This latter study on nitrate reductase[23] also reported inactivation by sodium sulfite. Reducing agents such as 2-mercaptoethanol were also used early for the reactivation of enzymes with organic mercurials.[24] Another interesting observation is the different effects of various reducing agents in high-throughput screening (HTS) assays. Reducing agents such as 2-mercaptoethanol or DTT are frequently included in an HTS assay to prevent oxidation of target proteins. In several HTS assays, the choice of reducing agent is demonstrated to influence the observed interaction of proteins with probes thus affecting the avidity of the inhibitor.[25,26] As an example, Lee and coworkers[26] reported an IC_{50} of 48.4 μM for one candidate inhibitor for a protease (NS3) from hepatitis virus, which increased to 69.4 μM in the presence of reduced glutathione and decreased to 12.8 μM in the presence of tris(2-carboxyethyl)phosphine (TCEP); no inhibition ($IC_{50} > 200$ μM) was observed with either DTT or 2-mercaptoethanol. In another observation, Crowe and coworkers[27] observed that DTT participated in a redox cycle resulting in the generation of peroxides from certain pyrimidotriazines, benzofurans, porphyrins, and anthraquinone in an HTS assay for inhibitors of tau fibril formation. Other investigators[28] found that reducing DTT concentration improved signal stability in a luciferase-based HTS system.

In another study evaluating the difference between four reducing agents, DTT, GSH, ascorbic acid, and TCEP, Peng and coworkers[29] showed that while DTT, GSH, and TCEP could reduce disulfide bonds and cupric ion to cuprous ion in a snake venom diesterase, ascorbic acid could reduce cupric ion to cuprous ion but did not reduce disulfide bonds. Peng and coworkers[29] did show that ascorbic acid reduced free cupric ion and native phosphodiesterase to yield cuprous ion as detected with bicinchoninic acid; similar results were not obtained with the other reducing agents. It was suggested that the other reducing agents (DTT, GSH, and TCEP) bound cuprous ion more tightly than bicinchoninic acid. Peng and coworkers[29] did show EPR spectra suggesting that all four reducing agents could reduce cupric ion in the phosphodiesterase. Ascorbate was also not able to reduce the disulfide bonds in cyclic somatostatin analog in boiling water.[30] The disulfide bond in cyclic somatostatin was stable in the presence of stannous chloride under conditions adequate to reduce pertechnetate for complexation with peptides[30]; a small amount of reduction of disulfide bonds with stannous ions was observed with boiling water and a much greater reduction was observed with dithionite under the same conditions. These observations were part of a study to develop conditions for labeling somatostatin analogs with ^{99m}Tc. This technique is based on the reduction of pertechnetate to technetium and the chelation of the technetium with thiol groups of the antibody or antibody fragment; the thiol groups can be obtained by reduction of disulfide bonds in the antibody or the use of mutagenesis to insert cysteine residues into the protein. Qi and coworkers[31] used dithionite and/or ascorbate to reduce disulfide bonds in antibodies and dithionite to reduce the pertechnetate to technetium for the radiolabeling of antibody. These investigators also showed that ascorbate can stabilize the radiolabeled antibody to a cysteine challenge.[32] A cysteine challenge[32] is a method by which cysteine is incubated with the radiolabeled antibody to assess the integrity of the radiolabel; a stable radiolabel remains bound to the antibody in the face of a cysteine challenge (or on incubation with serum).

Malviya and coworkers[33] used 2-mercaptoethanol to reduce a therapeutic monoclonal antibody (rituximab, Rituxan®) for this application. More recent work has used engineered cysteine residues and stannous chloride reduction of pernechtate[34] as well as histidine tag for radiolabeling.[35] Zamora and Rhodes[36] earlier reported that imidazole binds technetium as efficiently as thiol groups. Greenland and Blower[37] have shown that sulfonated TPP could be used for the reduction of both the antibody and pertechnetate in the radiolabeling of salmon calcitonin. In another work, Landino and coworkers[38] reported the reduction of disulfide bonds in tubulin with ascorbate. A consideration of the literature suggested that there is some evidence to support reduction of metal ions by DTT.[39,40] DTT and GSH have been reported to reduce Co(III) to Co (II) in cobalamin.[39] Yu[40] has observed that ascorbate/Fe^{2+} activates protein phosphate 2A; DTT but not 2-mercaptoethanol could replace ascorbate. Previous work by Yu and coworkers[41] has shown that ascorbate/Fe^{2+} could activate an ATP/Mg-dependent type 1 protein phosphatase; DTT could not replace ascorbate in this system. The spectral determination of the complex of Co(II) and DTT can be used as a method for the determination of ischemia-modified albumin.[42,43] There is a limited literature on the reduction of metal ions by ascorbic acid[44–47] and somewhat larger literature on the generation of free radicals with cupric ions.[47–53] It has been suggested that for metal ions such as copper, reduction with ascorbate increases their oxidant activity.[52] Ascorbate has been used to reduce S-nitrosocysteine.[53] Derakhashan and coworkers[53] also suggested that the ascorbic acid reduction of disulfide is unlikely to occur during the biotin-switch reaction. The biotin-switch reaction is an analytical method developed for the measurement of S-nitrosocysteine in proteins that requires selective reduction of the S-nitrosocysteine and can be complicated by disulfide bond reduction.[38,54–58] Wiesweg and coworkers[59] have suggested that methylhydrazine is more specific for the reduction of S-nitrosothiols. The reader is referred to the recent work[60] for more information on the ability of hydrazine and derivatives to function as a reducing agent. Oxidized ascorbic acid is involved in the formation of disulfide bonds.[61]

Some of the various reagents that have been used for the reduction of disulfide bonds are shown in Table 6.1. The structures of some of these reagents are shown in Figure 6.1. Some solid-phase reagents[62–64] have been proposed for the reduction of disulfide bonds. It has been possible to use TCEP for the reduction of disulfide bonds of hexahistidine fusion protein bound to a Ni^{2+}-nitriloacetic acid agarose matrix.[65] Thiopropyl agarose has been used to selectively reduce immunoglobulin proteins[63] as well as the capture of modified cysteine residues in proteins.[66,67] In this latter application, it has been possible to identify S-acyl proteins[66] and S-nitroso proteins.[67]

Sodium borohydride and sodium cyanoborohydride are also listed in Table 6.1 but have seen somewhat limited use in solution protein chemistry compared to the other reducing agents. Sodium cyanoborohydride has been observed to reduce a mixed disulfide[68] but has been shown not to reduce protein disulfide bonds.[69] The most cited use for the use of sodium cyanoborohydride in protein chemistry is the reduction of the imine bond in the Schiff base formed between an aldehyde and an amine.[70–74] This chemistry is discussed in more detail in Chapter 19. Liu and Breslow[75] used sodium cyanoborohydride for reductive methylation with formaldehyde and sodium borohydride for reduction of disulfide bonds. Mosckovitz and Gershoni[76] observed that two disulfide bonds, Cys128Cys142 and Cys412Cys418, in the acetylcholine receptor could only be reduced with sodium borohydride. Sodium borohydride has been used to reduce sulfhydryl groups in peptides having the advantage here in being removed by acidification.[77] While there has been little recent use of sodium borohydride for the modification of protein, there is some early work that is of considerable interest. Light and Sinha[78] used sodium borohydride (pH 9.6) to selectively reduce two of six disulfide bonds in

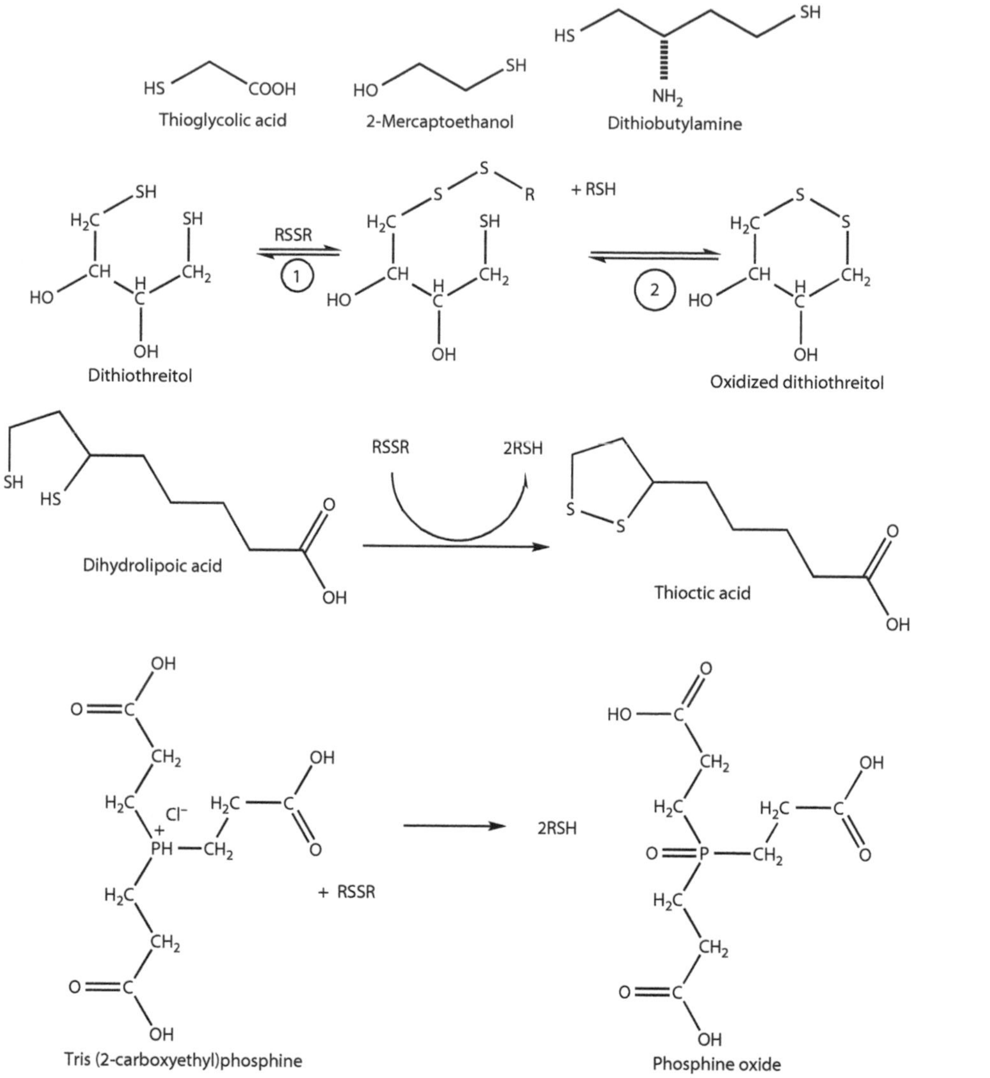

FIGURE 6.1 Some reducing agents for the modification of protein structure. Thioglycolic acid (mercapto-acetic acid) is a mild reducing agent similar to 2-mercaptoethanol. 2-Mercaptoethanol may be the most used reducing agent in biochemistry. (2S)-2-Amino-1,4-dimercaptobutane is a new reducing agent possessing an amino group that can provide coupling to a matrix (Lukesh, J.C., III, Palte, M.J., and Raines, R.T., A potent, versatile disulfide-reducing agent from aspartic acid, *J. Am. Chem. Soc.* 134, 4057–4059, 2012). It is similar to DTT shown in the succeeding text (Cleland, W.W., Dithiothreitol, a new protective reagent for SH groups, *Biochemistry* 4, 480–482, 1964). The effectiveness of this reagent is driven by the equilibrium constant of the second step that approaches 10^4. Shown in the succeeding text are dihydrolipoic acid and lipoic acid, which is a redox system similar to that described for DTT where the formation of the thiolane ring is associated with an increase in absorbance at 330 nm (Barltrop, J.A., Hayes, P.M., and Calvin, M., The chemistry of 1,2-dithiolane (trimethylene disulfide) as a model for the primary quantum conversion act in photosynthesis, *J. Am. Chem. Soc.* 76, 4346–4367, 1954; Jocelyn, P.C., The standard redox potential of cysteine-cystine from the thiol-disulphide exchange reaction with glutathione and lipoic acid, *Eur. J. Biochem.* 2, 327–331, 1967). At the bottom is the reaction of TCEP (Burns, J.A., Butler, J.C., Moran, J., and Whitesides, G.M., Selective reduction of disulfides by tris(2-carboxyethyl)phosphine, *J. Org. Chem.* 56, 2648–2650, 1991; Fischer, W.H., Rivier, J.E., and Craig, A.G., In situ reduction suitable for matrix-assisted laser desorption/ionization and liquid secondary ionization using tris (2-carboxyethyl) phosphine, *Rapid Commun. Mass Spectrom.* 7, 225–228, 1993). Note that the reaction of TCEP with disulfides is an irreversible reaction.

bovine trypsinogen without loss of activity; disulfide bonds in chymotrypsinogen are not reduced under these conditions. Results similar to trypsinogen were obtained with trypsin; all six disulfide bonds were reduced by sodium borohydride in the presence of 8 M urea. Kress and Laskowski[79] used the same approach to reduce a single disulfide bond (Cys14Cys38) in the basic trypsin inhibitor from bovine pancreas. The partially reduced inhibitor retained activity, while the carboxymethylated derivative was inactive. Air oxidation of the partially reduced inhibitor resulted in 38% loss of free sulfhydryl after 25 min; no oxidation was observed in a partially reduced inhibitor in complex with trypsin. Peptide bond cleavage with sodium borohydride as reported by Crestfield and coworkers[80] was not observed in these studies. Jering and Tschesche[81] used the work of Kress and Laskowski[79] to prepare the partially reduced bovine basic pancreatic trypsin inhibitor with disulfide bond. It was possible to cleave this derivative at the *active site* sequence of Lys15Ala16, which then could be reoxidized to form a modified inhibitor, which was suggested to form from the dissociation of the trypsin-inhibitor complex but had not been isolated. The modified inhibitor would have formed from the *forward* direction of the reaction. Jering and Tschesche[74] were able to form the native inhibitor by incubation of the reduced derivative with trypsin at pH 7.8 for 1.5 h followed by acidification (HCl) to pH 1.7 (kinetic control dissociation[82,83]). The synthesis of peptide bonds by proteases has been described by a number of investigators.[84–86] More recent work by Hansen and coworkers[87] has used sodium borohydride for the determination of disulfides in proteins. The assay is based on the reduction of disulfide bonds with sodium borohydride followed by acidification to destroy the residual borohydride. The liberated thiol groups are determined by reaction with 4,4'-dithiodipyridine yielding a product that can be determined by spectroscopy; 4,4'-dithiopyridine has the advantage of reacting with thiol groups at acid pH. This assay has the advantage of being performed in a single tube without any separation steps. It is noted that the reduction of cystine by sodium borohydride (or DTT) can be catalyzed by selenol.[88] Selenium and other chalcogens are discussed in the succeeding text in somewhat greater detail. Suffice to say, selenocysteine proves to be a more potent nucleophile than cysteine.[89] Sodium cyanoborohydride has been more recently used for the radiolabeling of interleukin-1 receptor antagonist with ^{18}fluoroacetaldehyde; in this work,[90] the Schiff base is converted to the fluoroethyl derivative with sodium cyanoborohydride. Sodium cyanoborohydride has also been used to stabilize 4-hydroxy-2-nonenal adducts in proteins[91] and for N-terminal protein conjugates.[92] More recently, Ambrogelly and coworkers[71] have noted (as has this author) that cyanide is a by-product of the use of sodium cyanoborohydride and presents issues for toxicity in pharmaceutical applications.[93] Ambrogelly and coworkers[71] were able to show that various borane derivatives such as those developed by Geoghegan and coworkers[69] could substitute for sodium cyanoborohydride. Pelter and coworkers[94] had previously noted the lack of toxicity with the use of dimethylborane and trimethylborane. The reader is also directed to a work by Brown,[95] which discussed early borane chemistry. Sodium borohydride and sodium cyanoborohydride have both been used to reduce the Schiff base formed between aldehydes formed from carbohydrates and amines in preparation affinity matrices containing covalently bound proteins. Peng and coworkers[96] compared sodium borohydride and sodium cyanoborohydride on the activity of several monoclonal antibodies. While there were differences in the loss of immunological activity of the several antibodies on reaction with two borohydrides, sodium cyanoborohydride was less destructive than sodium borohydride; the presence of Zn^{2+} or Al^{3+} increased the loss of immunological activity observed with sodium cyanoborohydride. In a more recent work, Weber and coworkers[97] compared sodium borohydride and sodium cyanoborohydride in the coupling of antibody to cellulose microparticles; increased activation of complement was observed with sodium borohydride. These investigators note that sodium borohydride reduces aldehydes and Schiff bases, while sodium cyanoborohydride reduces only Schiff bases.

Thioglycolic acid is used infrequently as a reducing agent but there are some interesting examples. It should be noted that thioglycolic acid has a long history of use in cosmetology.[98] In recent work, Suzuta and coworkers[99] used thioglycolic acid to reduce the disulfide bonds in swollen hair fibers in a model study of the permanent wave; thioglycolic acid has a long history of inclusion in acid hydrolysis used for the preparation of proteins for amino acid analysis.[100–104]

Hoogerheide and Campbell[105] used dithioglycolic acid as an additive in acid hydrolysis for the determination of cysteine and cystine.

Dithionite (sodium hydrosulfite), which was mentioned earlier in the discussion of radiolabeling immunoglobulins, is a reagent that reduces certain disulfide bonds and heme iron. The reaction with cystine (Figure 6.2) results in a free sulfhydryl and S-sulfocysteine.[106–111] S-sulfocysteine is also a naturally occurring amino acid.[112,113] Sodium sulfite and sodium tetrathionate also reacts with cystine to yield cysteine and S-sulfocysteine[114] Sodium sulfite has been included in acid hydrolysates of protein to protect cystine, methionine, and tyrosine.[115] Sodium dithionite is also used for the reduction of 3-nitrotyrosine to 3-aminotyrosine (see Chapter 17).[116,117]

The use of stannous chloride for the reduction of disulfide bonds in proteins, primarily antibodies or antibody fragments such as Fab or F(ab'), prior to radiolabeling with ^{99m}technetium, requires additional discussion. Ekelman and Richards[118] reported a rapid method for the synthesis of ^{99m}Tc-diethyleenetriaminepentaacetic acid from pertechnetate using SnCl$_2$ reduction in 1970. Subsequently Wong and coworkers[119] were able to label several plasma proteins with ^{99m}Tc in the presence of stannous chloride but the mechanism was not clear. Paik and coworkers[120] were the first to suggest that stannous chloride reduced cystine residues in proteins producing the thiol groups responsible for the binding of technetium. Mardirossian and coworkers[121] compared stannous ions (stannous tartrate–phthalate prepared from stannous chloride and sodium tartrate and potassium phthalate) with 2-mercaptoethanol for the reduction and subsequent labeling of antibody with ^{99m}Tc. Pertechnetate could be added directly to the antibody, which was reduced with stannous ions as stannous ions were not removed from the initial reduction of protein. In the case of antibody reduced with 2-mercaptoethanol, pertechnetate was added with a methylene diphosphonate bone imaging kit, which contained stannous fluoride. Both methods of reduction resulted in Fab' fragments. Zhang and coworkers[122] also used 2-mercaptoethanol and a bone imaging kit for technetium labeling of antibody. Jeong and coworkers[123] used 2-mercaptoethanol to reduce a lactosylated albumin that was subsequently labeled with pertechnetate in the presence of stannous ion for use in imaging the asialoglycoprotein receptor. There is other use of stannous ion for the labeling of albumin with technetium for diagnostic purposes.[124–126]

Glutathione and cysteine are considered to be in vitro reducing agents for proteins as is the thioredoxin system. However, these reducing agents are perhaps of more importance for in vivo redox regulation than for in vitro chemical modification studies. For example, the S-glutathionylation of proteins is considered to be an important posttranslational modification[127–129], and a monoclonal antibody has been developed to identify S-glutathionylation.[130] The immunogen used for the monoclonal antibody was a conjugate of reduced glutathione with an acrolein-modified lysine residue in keyhole limpet hemocyanin. The combination of reduced glutathione (GSH) and oxidized glutathione (GSSG) with components such as arginine[131] is used in the refolding of recombinant proteins.[132–138] There is somewhat less literature on the reaction of cysteine with proteins, although it is reported that a substantial amount of human serum albumin is present as mixed disulfide with cysteine.[139] Cysteine has been reported to react with the cysteine sulfenic acid derivative of albumin considerably faster than either reduced glutathione or homocysteine.[140] I could not find significant use of cysteine/cystine as a redox couple for protein folding. Boyle and coworkers[141] used catalytic amount of cystine with DTT for the refolding of a recombinant human cytokine protein expressed as an inclusion body in *Escherichia coli*.

British anti-Lewisite (BAL) (2,3-dimercaptopropanol) was developed during WWII as an antidote as arsenical-based vesicant.[142] There were several early papers that used BAL as a reducing agent for cystine,[143,144] but most use is as an antidote for arsenic and mercury poisoning.[145] After administration, BAL is found as a mixed disulfide with albumin and excreted as a polymer.[146]

For those of us who have considerable laboratory experience in protein chemistry, the smell of 2-mercaptoethanol was an indication of scientists at work. Others like the author also had purple hands from working with ninhydrin. The earliest citation [147] that I could find to 2-mercaptoethanol compares this thiol with various dithiols including 2,3-dimercaptopropanol (BAL) as antidotes for

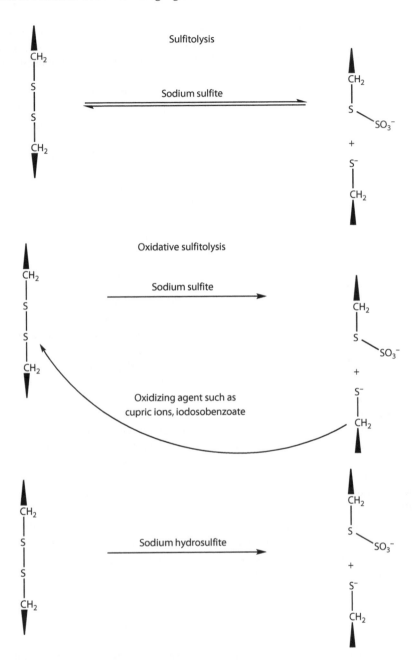

FIGURE 6.2 The reduction of cysteine to yield cysteine and cysteine-S-sulfonate. Shown at the top is the reaction of sodium sulfite with cysteine and oxidative sulfitolysis of cysteine (Jirgensons, B., Adams-Mayne, M.E., Gorguraki, V., and Migliore, P.J., Structure studies on human serum gamma-globulins and myeloma proteins. 3. Oxidative sulfitolysis of myeloma globulins and reconstitution of the macromolecules, *Arch. Biochem. Biophys.* 111, 283–295, 1965). Oxidative sulfitolysis is used for the quantitative conversion of cystine to *S*-sulfocysteine derivatives (Chapter 8). Shown at the bottom is the reaction of sodium hydrosulfite (sodium dithionite) with cysteine to yield cysteine and cysteine-S-sulfonate (Knipp, M., Taing, J.J., and He, C., Reduction of the lipocalin type heme containing protein nitrophorin—Sensitivity of the fold-stabilizing cysteine disulfides toward routine heme-iron reduction, *J. Inorg. Biochem.* 105, 1405–1412, 2011). Cysteine-S-sulfonate is also obtained from the reaction of sodium tetrathionate with cysteine (Church, J.S. and Evans, D.J., A spectroscopic investigation into the reaction of sodium tetrathionate with cysteine, *Spectrochim. Acta A Mol. Biomol. Spectrosc.* 69, 256–262, 2008).

arsenic; monothiols such as 2-mercaptoethanol were not as effective as dithiols as BAL. A later work[148] describes the synthesis of 2-chloroethyl 2-hydroxyethyl sulfide, a hydrolysis product of mustard gas (Chapter 2), from 2-mercaptoethanol and vinyl chloride. I have been unable to find the seminal work on 2-mercaptoethanol but suspect that this compound was developed during WWII and such information in a work[149] compiled on classified work done during that time. 2-Mercaptoethanol has a long history of use in the reduction of immunoglobulins.[150-164] The differential stability of IgG and IgM to reduction by 2-mercaptoethanol has been used to identify the immunoglobulin class responsible for a specific immunological action.[156-164] While most of the early work on the reduction of immunoglobulins used 2-mercaptoethanol, recent work has used DTT[165,166] or TCEP.[167]

Electrophoresis after reduction of the sample with 2-mercaptoethanol has been used for the characterization of disulfide bonds in proteins.[168-171] 2-Mercaptoethanol is also used for the extraction of proteins from biological substances[172-176] As with reduced glutathione, 2-mercaptoethanol has been demonstrated to form a mixed disulfide with cysteine in several proteins.[177-181] The mixed disulfide of 2-mercaptoethanol and cysteine is labile, and it is clear that mass spectrometry has been critical in the identification of this modification.[182,183]

Thioctic acid (lipoic acid, 1,2-dithiolane-3-pentanoic acid, CAS 62-46-6) (Figure 6.1) is better known as an antioxidant than a reducing agent,[184] and its action is complicated by the ability of the reduced dithiolane ring (thioctic acid) to bind metal ions.[185,186] The oxidation from the dithiol form (dihydrolipoic acid) to oxidized form (lipoic acid, thioctic acid) is associated with an increase in absorbance at 330 nm.[187] Thioctic acid was observed to inhibit caspace-3 activity as was 1,2-butanedithiol and cystine, while stimulation of activity was observed with thiols such as DTT and 2-mercaptoethanol.[25] It is noted that inactivation of papain by DTT was reported by Sluyterman and Wijdenes[188]; inactivation was not observed by 2-mercaptoethanol, cysteine, or 2,3-dimercaptopropanol. It was suggested that inactivation was due to autolysis when a single disulfide bond was reduced making the enzyme more susceptible to proteolysis by native papain. Stoyanovsky and coworkers[189] showed the thioctic acid reacted with S-nitrosopeptides and proteins to yield the parent cysteine and nitroxyl. The presence of a carboxyl group permitted the covalent binding of thioctic acid to an amine matrix via N-hydroxysuccinimide/carbodiimide chemistry.[190] The dithiolane ring could be reduced with sodium borohydride to yield the matrix-bound dihydrolipoic acid, which could serve as a reducing agent. Bienvenu and coworkers[191] used similar chemistry to link dihydrolipoic acid to a PEG aminomethyl-ChemMatrix®. These investigators also used sodium borohydride to reduce the matrix-bound thioctic acid. It is suggested that the matrix could be useful in reducing high-value proteins and organic disulfides. Roux and coworkers[192] used sodium borohydride to reduce thioctic acid to dihydrolipoic acid bound to gold nanoparticles. A urethane polymer containing the dithiolane structure was prepared in 1974 as a reducing agent.[193] These investigators also used sodium borohydride to reduce the matrix-bound thiolactic acid to the dihydrolipoyl derivative. The thiol-containing matrix could be used to reduce protein disulfide bonds.

Electrolytic/electrochemical reduction was used by Leach and coworkers in 1965[194] to reduce disulfide bonds in several proteins. These workers do cite earlier work on the electrolytic insulin and ribonuclease. The reaction was performed in the presence of 0.07 M mercaptoethanol, which served as current carrier. There was some reduction of disulfide bonds in the absence of electrolysis; electrolytic reduction resulted in extensive reduction of disulfide bonds in the absence of chaotropic agents with either bovine serum albumin or lysozyme. More recent work has used online electrolytic reduction of proteins prior to mass spectrometric analysis.[195,196] The electrolytic reduction of proteins during electrophoresis has been described in detail by Lee and Chang.[197]

I would be remiss if I did mention the use of selenol and some related chalcogen organics in the reduction of proteins. Sulfur and oxygen are members of group 16 in the periodic table[198]; selenium, tellurium, polonium, and ununhexium are also members of this group. The group is characterized by six valence electrons and form divalent anions (X^{-2}). Other than sulfur, the other chalcogens are very rare, and with the exception of selenium and tellurium,[199,200] there is little application of chalcogens in biochemistry. Selenocysteine is a naturally occurring amino acid in proteins[201] and demonstrates increased nucleophilicity.[202–204] Selenols such as selenocystamine are generated in situ and have been shown to enhance the reduction of antibodies.[88,205] Bramanti and coworkers[206] have shown that selenol enhanced the DTT reduction of other proteins such as albumin, lysozyme, and α-lactalbumin. It does appear that selenocystamine is a preferred source of *selenol* being generated from selenocystamine.[202] Nauser and coworkers[204] reported an $E°$ = −464V for diselenothreitol (Table 1). Günther[207] reported that hypophosphorous acid reduced diselenides but not disulfides. The hypophosphorous acid does assist in the reduction of disulfides by catalyzing the formation of selenols from diselenides. It was used by Clarembeau and coworkers[208] to provide methylselenol from dimethyldiselenide.

There are a variety of reducing agents listed in Table 6.1, but it is clear that there are two reagents that see the most use in protein chemistry as reducing agents: DTT and TCEP. The primary application of these two reagents is the reduction of disulfide bonds to cysteine and preservation of cysteine in the reduced state. As shown in Table 6.1, there is not much difference in oxidation potential between the two reagents. There are also few published studies that compare the two reagents; Getz and coworkers[209] compared some of the properties of DTT and TCEP as such properties pertain to use in protein chemistry. They observed that DTT and TCEP were equivalent in preserving the enzymatic activity of myosin, which has an oxidation-sensitive sulfhydryl at the active site. Both reagents reduced the extent of heavy meromyosin modification with tetramethylrhodamine-5-maleimide but had little effect on reaction with tetramethylrhodamine-5-iodoacetamide. TCEP was more stable in aqueous solution than DTT; there was an exception in the presence of EDTA. Shafer and coworkers[210] reported reaction of TCEP with either *N*-ethylmaleimide or iodoacetamide. Both of these studies reported end points, not rates. Subsequent workers have emphasized the ability to modify sulfhydryl groups in proteins in the presence of TCEP.[211] Shafer and coworkers[210] also reported anomalous behavior of TCEP on gel filtration, which presents issues for removal of the reagent prior to modification of thiol groups.

TCEP was developed by Burns and coworkers[212] as a water-soluble derivative for the reduction of disulfide bonds in proteins. There had been previous use of trialkylphosphines such as tributylphosphine (TBP) (CAS998-40-3) (Figure 6.4) for the cleavage of disulfide bonds and *S*-sulfocysteinyl residues in proteins.[213–215] While TBP is effective in reducing several oxidized forms of cysteine, it is an unpleasant material having a rather strong garlic-like odor and being insoluble in water. Röegg and Rudinger[214] included 1-propanol to an extent of 20%–50% to dissolve the TBP. Herbert and coworkers[216] reported that the use of TBP improved the solubility of proteins in 2D electrophoresis. TBP can be used in combination with a reagent such as 4-(aminosulfonyl)-7-fluoro-2,1,3-benzoxadiazole (ABD-F) for the combined determination of free cysteine and disulfide cysteine in proteins. Free cysteine can be determined first with ABD-F; ABD-F can subsequently be used for the determination of cysteine derived from cystine after reduction with TBP as TBP does not to be removed prior to reaction with ABD-F.[217–223] While some investigators[224] have had success with the use of TBP for reduction followed by alkylation with 2-vinylpyridine in the preparation of protein samples for two-dimensional electrophoresis, other investigators have found TBP to be less useful than DTT in the preparation of protein samples for electrophoretic analysis.[216,225] Herbert and coworkers[226] have replaced DTT with TBP in sample preparation and achieved greater solubility and combined reduction and alkylation in the same step. TBP has been useful in the use of organic solvents for the extraction of proteins for electrophoretic analysis.[227] Miller and coworkers[228] have provided a useful review of the various applications of TBP in proteomic analysis. TPP (CAS603-35-0) is used more infrequently than TBP. TPP is a solid material that is soluble in ether and ethanol but essentially insoluble in water. TPP has been used to reduce the iron–sulfur center in several enzymes with the formation of TPP sulfide.[229–231]

Li and coworkers[232] used TPP esters to reduce *S*-nitrosocysteinyl residues in proteins (Figure 6.5 later in the chapter). This is an example of the use of Staudinger ligation and is discussed by Bechtold and King.[233] Bechtold and King[233] also discuss the action of TPP on a nitrosothiol to yield *S*-substituted aza-ylides, which can undergo further reaction. Tris(hydroxymethyl)phosphine (Figure 6.3) has a long history of use as a flame retardant in the textile industry.[234–237] I have not

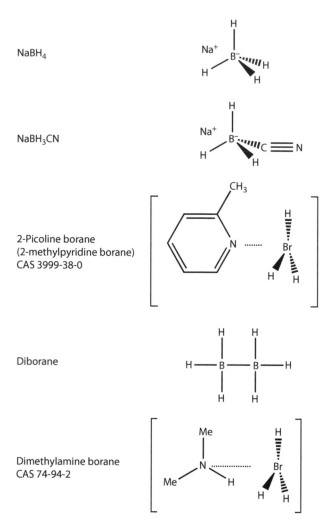

NaBH$_4$

NaBH$_3$CN

2-Picoline borane
(2-methylpyridine borane)
CAS 3999-38-0

Diborane

Dimethylamine borane
CAS 74-94-2

FIGURE 6.3 Some borohydrides and boranes used for the modification of proteins. Sodium borohydride and sodium cyanoborohydride are well-recognized reagents in organic chemistry. In addition, some borane derivatives have been proposed for substitution for sodium cyanoborohydride (Ambrogelly, A., Cutler, C., and Paporello, B., Screening of reducing agents for the PEGylation of recombinant human IL-10, *Protein J.* 32, 337–342, 2013).

found any use of tris(hydroxymethyl)phosphine as a reducing agent but there has been use for a cross-linking agent and a reagent for immobilizing proteins.[238–242] Tris(3-hydroxypropyl)phosphine has been used to reduce cysteine and homocysteine disulfides in plasma.[243] Tris(3-hydroxylpropyl) phosphine has also been used to measure the formation of superoxide anion by sulfhydryl oxidase.[244]

   Much of the use of TCEP is in the reduction of disulfide bonds and will be discussed in some detail in Chapter 8 (cystine). There is considerably less use for the maintenance of thiol groups in the reduced state. There are several studies that are not discussed in those chapters and need to be presented within the context of TCEP as a reducing agent. Chen and coworkers[245] have observed that TCEP promotes the reaction of cisplatin with Sp1 zinc finger protein via formation of a protein–platinum–TCEP complex associated with the release of the amine portion. TCEP has been shown to have an in vivo effect in protecting photoreceptors in a rodent model from photodegenera-tion.[246] Methyl ester derivatives of TCEP (Figure 6.4) have been developed to improve membrane permeability.[247]

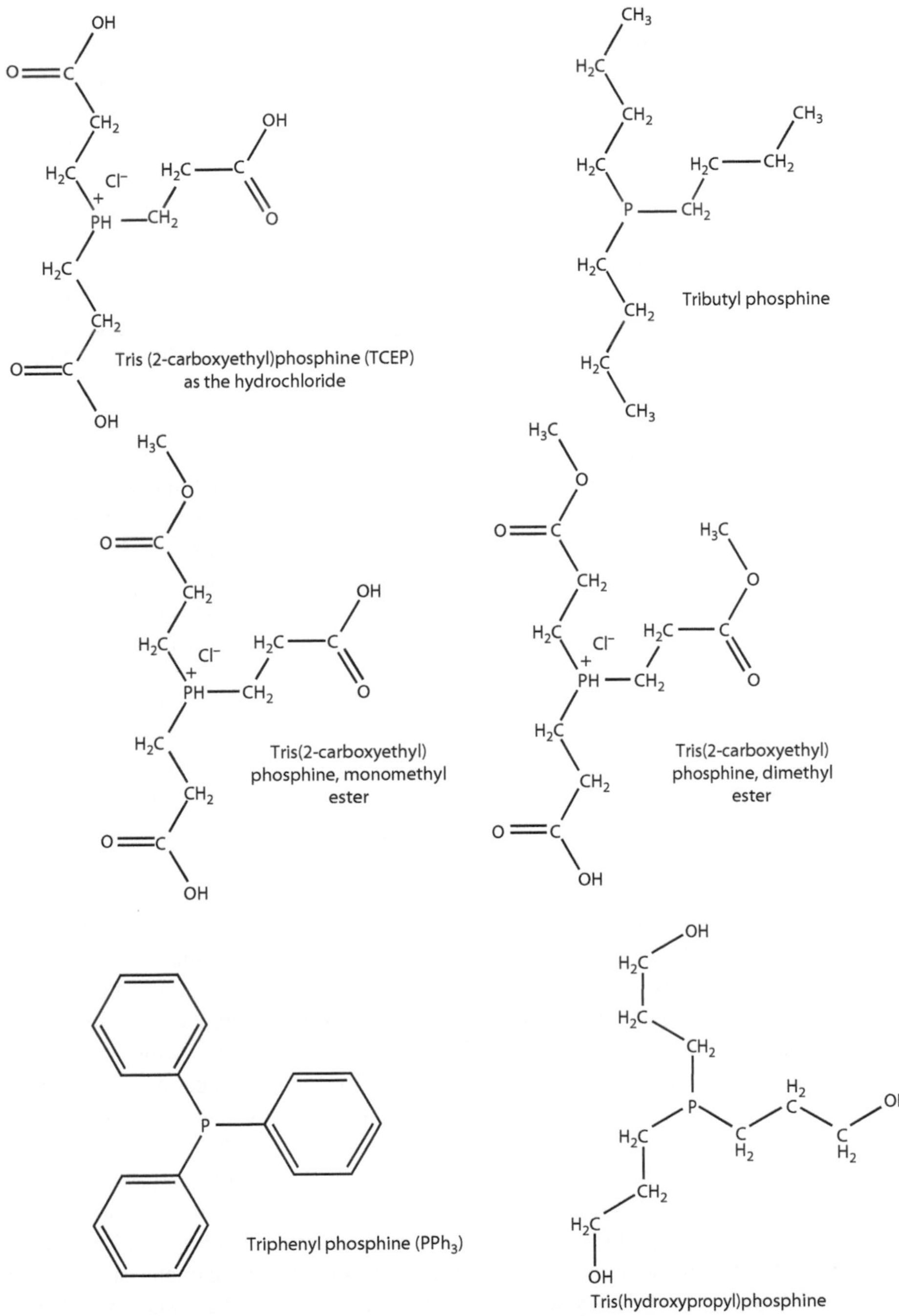

FIGURE 6.4 Structure of some phosphines used for the reduction of proteins. TCEP is the most frequently used phosphine for the reduction of proteins. The monomethyl, dimethyl, and trimethyl esters have been developed to improve membrane permeability (Cline, D.J., Redding, S.E., Brohawn, S.G. et al., New water-soluble phosphines as reductants of peptide and protein disulfide bonds: Reactivity and membrane permeability, *Biochemistry* 43, 15195–15203, 2004). TBP and TPP were earlier phosphines, which are far less soluble than TCEP. Tris(hydroxypropyl)phosphine has been used to reduce disulfide derivatives of cysteine and homocysteine in plasma (Švagera, Z., Hanzlíková, D., Šimek, P., and Hušek, P., Study of disulfide reduction and alkyl chloroformate derivatization of plasma sulfur amino acids using gas chromatography-mass spectrometry, *Anal. Bioanal. Chem.* 402, 2953–2963, 2012).

While it is generally accepted that TCEP does not react with maleimide, which is very selective for the modification of sulfhydryl groups,[211,248] there are several studies that show the reaction of TCEP with maleimide.[209,210,249] None of the studies on the interaction of TCEP and maleimides report the rate of reaction, but the study[249] on the effect of TCEP on the thiol–maleimide oligonucleotide cyclization reaction did demonstrate conversion of maleimides into succinimide derivatives. In this regard, there are several reports[64,250–253] of the use of insoluble TCEP (Figure 6.5), which obviates the potential of a reaction with a maleimide or other sulfhydryl reagents. There is considerable use of TCEP for the reduction of antibody prior to conjugation with maleimide derivatives of drugs.[254–256] Much of this work is either proprietary or covered by intellectual property. It should be noted that TCEP will reduce disulfide bonds at acid pH,[212,257,258] allowing the reduced cysteine residues to remain as thiols rather than more reactive thiolate ions, which would be present at higher pH. TCEP has been shown to weakly bind metals.[259,260] Kretzel and coworkers[260] also showed the phosphine oxide derived from TCEP weakly bound metal ions. In addition, these investigators noted that TCEP was a weak acid with a pKa of approximately 7.7; thus TCEP could function as a buffer. Finally, Kretzel[260] demonstrated that cupric ions could oxidize TCEP to phosphine oxide.[260] Finally, Wang and coworkers[261] have shown that TCEP is a better substrate for Erv2p, a sulfhydryl oxidase than DTT.

Despite the advantages of TCEP cited earlier, it is my sense that DTT is more widely used. DTT was developed by the late Mo Cleland at the University of Wisconsin in 1964.[262] It should be noted that while Cleland is likely best known for the discovery of DTT, he was responsible for major understanding of enzyme kinetics.[263–266] The reaction of DTT (Figure 6.1) is driven by a localized high concentration of sulfhydryl group to reduce the mixed disulfide formed in the initial reaction of DTT, and thus reaction 2 is much faster than reaction 1. It is also noted that, unlike TCEP, the oxidation of DTT is a reversible reaction. The oxidation of DTT is associated with a change in the UV spectra, which can be used to measure the extent of reduction of disulfide bonds[262,267,268] as is observed with thioctic acid.[184,185] DTT has been demonstrated to reduce disulfide bonds at neutral to alkaline pH[269–272] differing from TCEP which reduces disulfide at acid pH[212,257,258]. The selenium derivative has been synthesized by several groups[88,273] and reported as too air sensitive to be isolated by one group[273] but was used by the other group and shown to be a powerful reducing agent. DTT has been shown to bind arsenite[270,274,275] as well as metal ions. DTT can reduce cupric ions to cuprous ions[29,276] and has the property of binding cuprous ions. Smirnova and coworkers[276] demonstrated that DTT could remove bound cuprous ions from reduced human insulin–like growth factor 1. DTT can also reduce cysteine sulfenic acid to cysteine.[17] I could not find a reference to the use of TCEP for the reduction of sulfenic acid; there is at least one reference[16] for the use of ascorbate to reduce cysteine sulfenic acid. Dithioerythritol (DTE), the erythro isomer of DTT, has the same redox characteristics as the threo form but is rarely used in the reduction of disulfide bonds but does see a variety of other uses and is used to a lesser extent (less than 1000 citations). A brief consideration of search results did not reveal a unique use of the erythro isomer.[225,277–282] Zahler and Cleland[283] used both DTT and DTE for the determination of disulfides in proteins. pKa values were determined for the two isomers in this study: 8.3 and 9.5 for DTT and 9.0 and 9.9 for DTE. Singh and coworkers[284] developed several reagents related to DTT including bis(2-mercaptoethyl) sulfone and meso-2.5-N,N.N',N'-tetramethyladipamide. Despite useful characteristics, I could find no further use of these compounds beyond their description in 1995. Lukesh and coworkers[285] have recently described dithiobutylamine [(2S)-2-amino-1,4-dimercaptobutane] (Figure 6.1), which has characteristics similar to DTT (Table 6.1).

There are well over 10,000 citations to DTT in PubMed and it is not possible to cover all of the specific applications. Some of the applications to cysteine and cystine are discussed in Chapters 7 and 8, respectively. There is considerable use of DTT in commercial biotechnology for the processing of inclusion bodies derived from the expression of recombinant proteins in bacterial systems.[141,286–288] The general approach to the recovery of active protein from inclusion bodies involves reduction with DTT or other reducing agents in the presence of a chaotropic agent such as guanidine followed by removed of the chaotropic agents and reducing agents and slow

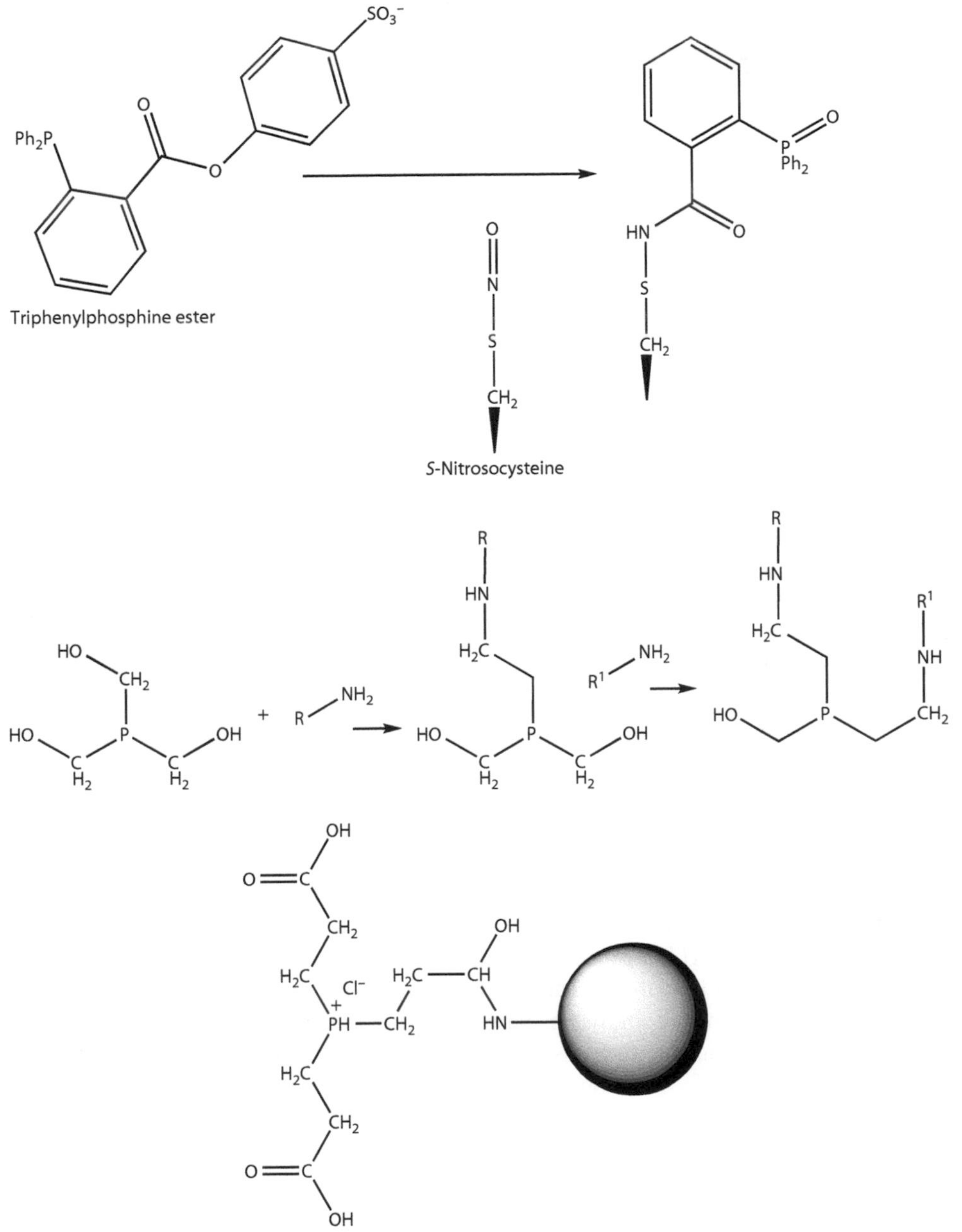

FIGURE 6.5 Some other reactions of phosphines with proteins. Shown at the top is the structure of a phenylsulfonate TPP ester, which can reduce nitrosocysteine residues in proteins to yield thiols (Li, S., Wang, H., Xian, M., and Whorton, A.R., Identification of protein nitrosothiols using phosphine-mediated selective reduction, *Nitric Oxide* 26, 20–26, 2012; Bechtold, E. and King, S.B., Chemical methods for the direct detection and labeling of *S*-nitrosothiols, *Antioxid. Redox Signal.* 17, 981–991, 2012). Shown in the middle is the use of tris(hydroxymethyl)phosphine as a crosslinking reagent (Renner, J.N., Cherry, K.M., Su, R.S.-C., and Liu, J.C., Characterization of resilin-based materials for tissue engineering applications, *Biomacromolecules* 13, 3678–3685, 2012) and for protein immobilization (Petach, H.H., Henderson, W., and Olsen, G.M., P(CH$_2$OH)$_3$—A new coupling reagent for the covalent immobilization of enzymes, *J. Chem. Soc. Chem. Commun.* (18), 2181–2182, 1994). At the bottom is shown an insoluble derivative of TCEP (Miralles, G., Verdié, P., Puget, K. et al., Microwave-mediated reduction of disulfide bridges with supported (tris(2-carboxyethyl)phosphine) as resin-bound reducing agent, *ACS Comb. Sci.* 15, 169–173, 2013).

oxidative refolding of the desired protein. This process is as much art as it is science and may be unique to each protein[289] and is also complicated by intellectual property issues.[290] A redox system is usually added during the refolding process including oxidized glutathione/reduced glutathione (GSH/GSSG).[136,291,292] The study by Vallejo and Rinas on the refolding and dimerization of bone morphogenetic protein-2 with the GSH/GSSG system is of considerable interest in that there is a careful study of refolding kinetics. The process of refolding/dimerization is very slow (second-order rate constants for intermediates range from 10^{-5} to 10^{-3} M^{-1} s^{-1}). The inclusion of a redox system such as GSH/GSSG allows the process of refolding to proceed through a series of disulfide exchange reactions.[293] In the absence of a redox system, it is likely that air oxidation proceeds through the formation of cysteine sulfenic acid.[294] Other reducing agents are used including 2-mercaptoethanol.[295–298] Refolding of these proteins is usually accomplished by air oxidation with participation of a redox pair. The use of 2-mercaptoethanol can markedly reduce specific process cost compared to DTT.[288] However, 2-mercaptoethanol may provide more issues with respect to toxicology than observed with DTT or GSH/GSSH.[299] Cystine has been added to the reduced protein to enhance the refolding process.[141,300,301]

There has been some interest in the use of DTT in the manufacture of hydrogels for protein drug delivery where the DTT is a structural component in a biodegradable matrix.[302–308] Some examples of these biomaterials are shown in Figure 6.6. DTT does provide a useful cross-linking reagent with functional groups such as acrylic acid and maleimides. Michael addition reactions proceed quickly at these sites and the choice of a terminal ester linker would provide a biodegradable matrix, while an amide link would be more resistant to degradation.[309] The use of dithiol derivatives for the manufacture of hydrogels for protein drug delivery dates to the work by Elbert and coworkers in 2001 (320). Elbert and coworkers[309] used PEG-dithiol form hydrogels by reaction with tri-, tetra-, and octaacylate derivatives of poly(ethylene glycol). The reaction was performed in the presence of a protein providing a protein drug delivery system. Most subsequent work using dithiol chemistry used DTT instead of PEG dithiol; PEG dithiol is considered a macro-cross-linker.[310]

The thrust of the work in this chapter and the entire book has been on in vitro reaction of synthetic organic compounds and metal ions with functional groups in proteins. Some organic compounds have both in vitro and in vivo action; one such organic compound is ascorbate (vitamin C), which is considered to be an in vivo antioxidant.[311,312] Ascorbate also is used occasionally as an in vitro reagent for selective reduction (Table 6.1). Ascorbate is used in a procedure called biotin-switch assay for S-nitrosocysteine.[54,313] It is assumed that ascorbate reduces S-nitrosocysteine to cysteine without reducing cystine, which would complicate the biotin-switch assay. However, data have appeared suggesting that ascorbate does reduce disulfide bonds.[314] Dehydroascorbate (oxidized ascorbate) is used for protein refolding.[315,316] It should be noted that ascorbate would appear to have activity beyond that which would be expected from its redox potential (Table 6.1). The reader is directed to a recent review by Du and coworkers[317] on complex chemistry of ascorbic acid. Another example of an in vivo reduction reaction is provided by sodium dithionite and 3-nitrotyrosine. Sokolovsky and coworkers[116] showed that 3-nitrotyrosine (see Chapter 11) is reduced to 3-aminotyrosine. The reduction of 3-nitrotyrosine in unfixed tissue can complicate the determination of 3-nitrotyrosine by immunoblotting.[318] Similar reduction did not occur in fixed tissues, suggesting the potential importance of heme groups in this process. Earlier work by Balabanli and coworkers[319] showed that while either a spleen homogenate or DTT alone was not effective in reducing nitrotyrosine, the combination of spleen homogenate and DTT was effective (100°C/10 min). Reaction at 37°C was not effective. The combination of hemoglobin and DTT was partially effective at 37°C with total reduction at 100°C; myoglobin and cytochrome C were also effective in combination with DTT. Some effect was also seen with pyrrole or vitamin B12 in combination with DTT. Cysteamine and ascorbic acid were somewhat effective in reducing 3-nitrotyrosine in combination with hemoglobin, while cysteine, glutathione, and 2-mercaptoethanol were not effective. Chen and coworkers[320] observed that hemoglobin and DTT, DTE, or dihydrolipoic acid could reduce nitro groups (8-nitroguanine) in DNA: ascorbate was much less effective but still showed activity.

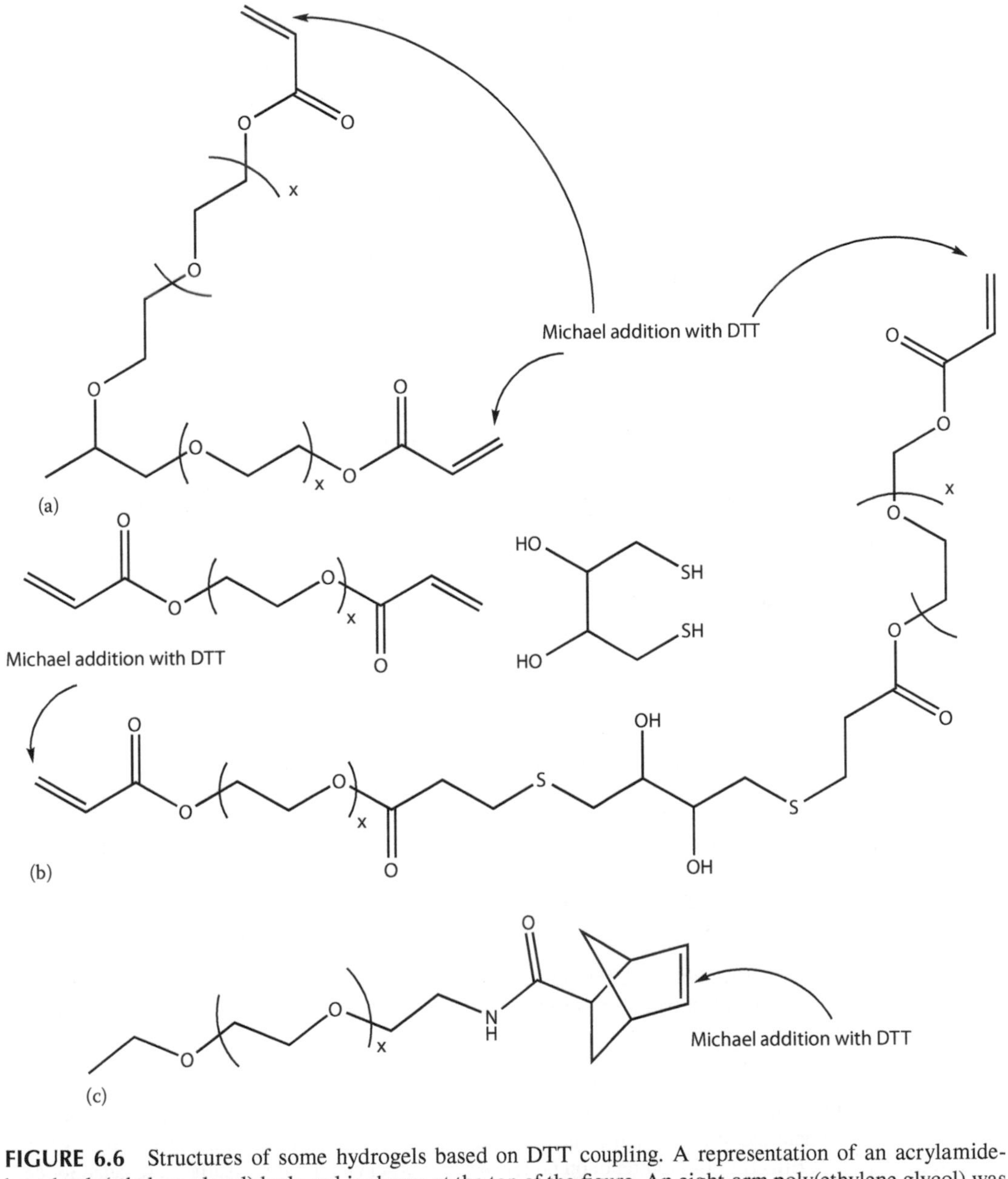

FIGURE 6.6 Structures of some hydrogels based on DTT coupling. A representation of an acrylamide-based poly(ethylene glycol) hydrogel is shown at the top of the figure. An eight-arm poly(ethylene glycol) was presented by van de Wetering and coworkers (van de Wetering, P., Metters, A.T., Schoenmakers, R.G., and Hubbell, J.A., Poly(ethylene glycol) hydrogels formed by conjugate addition with controllable swelling, degradation, and release of pharmaceutically active proteins, *J. Control. Release* 102, 619–627, 2005). Another poly(ethylene glycol) hydrogel based on an acrylamide for Michael addition with DTT is shown in the middle (Qiu, Y., Lim, J.J., Scott, L., Jr. et al., PEG-based hydrogels with tunable degradation characteristics to control delivery of marrow stromal cells for tendon overuse injuries, *Acta Biomater.* 7, 959–966, 2011). Other PEG-based hydrogels, which used DTT coupling to an amide norbornene, are shown at the bottom (Raza, A. and Lin, C.-C., The influence of matrix degradation and functionality on cell survival and morphogenesis in PEG-based hydrogels, *Macromol. Biosci.* 13, 1048–1058, 2013).

REFERENCES

1. Tomasi, M., Battistini, A., Araco, A. et al., The role of the reactive disulfide bond in the interaction of cholera toxin functional regions, *Eur. J. Biochem.* 93, 621–627, 1979.
2. Pace, C.N., Grimsley, G.R., Thomson, J.A., and Barnett, B.J., Conformational stability and activity of ribonucleased T_1 with zero, one, and two intact disulfide bonds, *J. Biol. Chem.* 265, 11820–11825, 1988.
3. Nnyepi, M.R., Peng, Y., and Broderick, J.B., Inactivation of *E. coli* pyruvate formate-lyase: Role of AdhE and small molecules, *Arch. Biochem. Biophys.* 459, 1–9, 2007.
4. Weber, W., Buck, F., Meyer, A., and Hilz, H., Prostate specific antigen: One out of five disulfide bridges determines inactivation by reduction, *Biochem. Biophys. Res. Commun.* 379, 1101–1106, 2009.
5. Hollecker, M., Vincent, M., Gallay, J. et al., Insight into the factors influencing the backbone dynamics of three homologous proteins, dendrotoxins I and K, and BPTI: FTIR and time-resolved fluorescence investigations, *Biochemistry* 41, 15267–15276, 2002.
6. David, C. and Enescu, M., Free energy calculations on disulfide bridges reduction in proteins by combining ab initio and molecular mechanics methods, *J. Phys. Chem. B* 114, 3020–3027, 2010.
7. Olins, D.E. and Edelman, G.M., Reconstitution of 7S molecules from L and H polypeptide chains of antibodies and gamma-globulins, *J. Exp. Med.* 119, 789–815, 1964.
8. Nisonoff, A., Wissler, F.C., Lipman, L.N., and Woernley, D.L., Separation of univalent fragments from the bivalent rabbit antibody molecule by reduction of disulfide bonds, *Arch. Biochem. Biophys.* 89, 230–244, 1960.
9. Milstein, C.P., Richardson, N.E., Daverson, E.V., and Feinstein, A., Interchain disulfide bridges of immunoglobulin M, *Biochem. J.* 151, 625–624, 1975.
10. Sears, D.W., Mohrer, J., and Beychok, S., A kinetic study in vitro of the reoxidation of interchain disulfide bonds in a human immunoglobulin $IgG1_\kappa$. Correlation between sulfhydryl disappearance and intermediates in covalent assembly of H_2L_2, *Proc. Natl. Acad. Sci. USA* 72, 353–357, 1975.
11. Sears, D.W., Mohrer, J., and Beychok, S., Relative susceptiblities of the interchain disulfides of an immunoglobulin G to reduction by dithiothreitol, *Biochemistry* 16, 2031–2035, 1977.
12. Schroeder, D.D., Tankersley, D.L., and Lundblad, J.L., A new preparation of modified immune serum globulin (human) suitable for intravenous administration. I. Standardization of the reduction and alkylation reaction, *Vox Sang.* 40, 373–382, 1981.
13. Schroeder, D.D. and Dumas, M.L, A preparation of modified immune serum globulin (human) for intravenous administration. Further characterization and comparison with pepsin-treated intravenous gamma globulin, *Am. J. Med.* 76, 33–39, 1984.
14. Rousell, R.H., Collins, M.S., Dobkin, M.S. et al., Antibody-levels in reduced alkylated intravenous immune globulin, *Am. J. Med.* 76, 40–45, 1984.
15. Jeffrey, S.C., Burke, P.J., Lyon, R.P. et al., A potent anti-CD70 antibody-drug conjugate combining a dimeric pyrrolobenzodiazepine drug with site-specific conjugation technology, *Bioconjug. Chem.* 24, 1256–1263, 2013.
16. Monteiro, G., Horta, B.B., Pimental, D.C. et al., Reduction of 1-cys peroxiredoxins by ascorbate changes the thiol-specific antioxidant paradigm, revealing another function of vitamin C, *Proc. Natl. Acad. Sci. USA* 104, 4886–4891, 2007.
17. Lindhoud, S., van den Berg, W.A., van den Heuvel, R.L. et al., Cofactor binding protects flavodoxin against oxidative stress, *PLoS One* 7(7), e41363, 2012.
18. Jennings, M.L. and Anderson, M.P., Chemical modification and labeling of glutamate residues at the stilbenedisulfonate site of human red blood cell band-3 protein, *J. Biol. Chem.* 262, 1691–1697, 1987.
19. Alliegro, M.C., Effects of dithiothreitol on protein activity unrelated to thiol-disulfide exchange for consideration in the analysis of protein function with Cleland's reagent, *Anal. Biochem.* 282, 102–106, 2000.
20. Kimber, M.S., Yu, A.Y.H., Borg, M. et al., Structural and theoretical studies indicate that the cylindrical protease ClpP samples extended and compact conformations, *Structure* 18, 798–808, 2010.
21. Piedrafita, F.J., Fernandez-Alvarez, E., Nieto, O., and Tipton, K.F., Kinetics and inhibition studies on catechol-*O*-methyltransferase affinity labelling by *N*-(3,4-dihydroxyphenyl)maleimide, *Biochem. J.* 286, 951–958, 1992.
22. Scrinivason, P.R. and Sprinson, D.B., 2-Keto-3-deoxy-D-arabo-heptonic acid 7-phosphate dehydrogenase, *J. Biol. Chem.* 234, 716–726, 1959.
23. Gómez-Moreno, C. and Palacián, E., Nitrate reductase from *Chlorella fusca*. Reversible inactivation by thiols and sulfite, *Arch. Biochem. Biophys.* 160, 269–273, 1974.
24. Webb, J.L., Reversal of enzyme inhibition, in *Enzyme and Metabolic Inhibitors*, Vol. 1, Chapter 13, Academic Press, New York, pp. 622–626, 1963.

25. Okun, I., Malarchuk, S., Dubrovskaya, E. et al., Screening for caspase-3 inhibitors: Effect of a reducing agent on identified hit chemotypes, *J. Biomol. Screen.* 11, 694–703, 2006.
26. Lee, H., Torres, J., Truong, L. et al., Reducing agents affect inhibitory activities of compounds: Results from multiple drug targets, *Anal. Biochem.* 423, 46–53, 2012.
27. Crowe, A., Ballatore, C., Hyde, S. et al., High throughput screening for small molecules inhibitors of heparin-induced tau fibril formation, *Biochem. Biophys. Res. Commun.* 358, 1–6, 2007.
28. Siebring-van Olst, E., Vermeulen, C., de Menezes, R.X. et al., Affordable luciferase reporter assay for cell-based high-throughput screening, *J. Biomol. Screen.* 18, 453–461, 2013.
29. Peng, L., Xu, X., Guo, M. et al., Effects of metal ions and disulfide bonds on the activity of phosphodi-esterase from *Trimeresurus stejnegeri* venom, *Metallomics* 5, 920–927, 2013.
30. Uehara, T., Arano, Y., One, M. et al., The integrity of the disulfide bond in a cyclic somatostatin analog during ^{99m}Tc complexation reactions, *Nucl. Med. Biol.* 26, 883–890, 1999.
31. Qi, P., Muddukrishna, S.N., Torok-Both, R. et al., Direct ^{99m}Tc-labeling of antibodies by sodium thionite reduction, and the role of ascorbate in cysteine challenge, *Nucl. Med. Biol.* 23, 827–835, 1996.
32. Hnatowich, D.J. Vivzi, F., Fogarasi, M. et al., Can a cysteine challenge assay predict the in vivo behavior of ^{99m}Tc-labeled antibodies?, *Nucl. Med. Biol.* 22, 1035–1044, 1994.
33. Malviya, G., Anzola, K.L., Podestà, E. et al., ^{99m}Tc-labeled rituximab for imaging B lymphocyte infiltra-tion in inflammatory autoimmune disease patients, *Mol. Imaging Biol.* 14, 637–646, 2012.
34. Tran, T., Engfeldt, T., Orlova, A. et al., In vivo evaluation of cysteine-base chelators for attachment of ^{ppm}Tc to tumor targeting affinity molecules, *Bioconjug. Chem.* 18, 549–558, 2007.
35. Xavier, C., Devoogdt, N., Hernot, S. et al., Site-specific labeling of his-tagged nanobodies with ^{99m}Tc: A practical guide, *Methods Mol. Biol.* 911, 485–490, 2012.
36. Zamora, P.O. and Rhodes, B.A., Imidazoles as well as thiolates in proteins bind technetium-99m, *Bioconjug. Chem.* 3, 493–498, 1992.
37. Greenland, W.E. and Blower, P.J., Water-soluble phosphines for direct labeling of peptides with tech-netium and rhenium: Insights from electrospray mass spectrometry, *Bioconjug. Chem.* 16, 939–946, 2005.
38. Landino, L.M., Koumas, M.T., Mason, C.E., and Alston, J.A., Ascorbic acid reduction of microtubule protein disulfides and its relevance to protein *S*-nitrosylation assays, *Biochem. Biophys. Res. Commun.* 340, 347–352, 2006.
39. Ramasamy, S., Kundu, T.K., Antholine, W. et al., Internal spin trapping of thiyl radical during the com-plexation and reduction of cobalamin with glutathione and dithiothreitol, *J. Porphyr. Phthalocyanines* 16, 25–36, 2012.
40. Yu, J.-S., Activation of protein phosphatase 2A by the Fe^{2+}/ascorbate system, *J. Biochem.* 124, 225–230, 1998.
41. Yu, J.S., Chan, W.H., and Yang, S.D., Activation of the ATP-Mg-dependent type 1 protein phosphatase by the Fe^{2+}/ascorbate system, *J. Protein Chem.* 15, 455–460, 1996.
42. Lu, J., Stewart, A.J., Sadler, P.J. et al., Allosteric inhibition of cobalt binding to albumin by fatty acid: Implications for the detections of myocardial ischemia, *J. Med. Chem.* 55, 4425–4430, 2012.
43. Gidenne, S., Ceppa, F., Fontana, E. et al., Analytical performance of the Albumin Cobalt Binding (ACB) test on the Cobas MIRA Plus analyzer, *Clin. Chem. Lab. Med.* 42, 455–461, 2004.
44. Bucci, R., Carunchio, V., Magri, A.D., and Magri, A.L., Spectroscopic and thermoanalytical studies on bicinchoninic acid and its copper (I) complexes, *Ann. Chim.* 84, 509–520, 1994.
45. Harel, S., Oxidation of ascorbic acid and metal ions as affected by NaCl, *J. Agric. Food Chem.* 42, 2402–2406, 1994.
46. Winkler, B.S., In vitro oxidation of ascorbic acid and its prevention by GSH, *Biochim. Biophys. Acta* 925, 258–263, 1987.
47. Kanazawa, H., Fujimoto, S., and Ohara, A., On the mechanism of inactivation of active papain by ascor-bic acid in the presence of cupric ions, *Biol. Pharm. Bull.* 17, 789–793, 1994.
48. Dikalov, S.I., Vitek, M.P., and Mason, R.P., Cupric-amyloid beta peptide complex stimulates oxidation of ascorbate and generation of hydroxyl radical, *Free Radic. Biol. Med.* 36, 340–347, 2004.
49. Zhu, B.Z., Antholine, W.E., and Frei, B., Thiourea protects against copper-induced oxidative damage by formation of a redox-inactive thiourea-copper complex, *Free Radic. Biol. Med.* 32, 1333–1338, 2002.
50. Shiraishi, N., Ohta, Y., and Nishikimi, M., The octapeptide repeat region of prion protein binds Cu(II) in the redox-inactive state, *Biochem. Biophys. Res. Commun.* 267, 398–402, 2000.
51. Biaglow, J.E., Manevich, Y., Uckun, F., and Held, K.D., Quantitation of hydroxyl radicals products by radiation and copper-linked oxidation of ascorbate by 2-deoxy-D-ribose method, *Free Radic. Biol. Med.* 22, 1129–1138, 1997.

52. Buettner, G.R. and Jurkiewicz, B.A., Catalytic metals, ascorbate and free radicals: Combinations to avoid, *Radiat. Res.* 145, 532–541, 1996.

53. Derakhashan, B., Wille, P.C., and Gross, S.S., Unbiased identification of cysteine *S*-nitrosylation sites on proteins, *Nat. Protoc.* 2, 1685–1691, 2007.

54. Forrester, M.T., Foster, M.W., and Stamler, J.S., Assessment and application of the biotin switch technique for examining protein *S*-nitrosylation under conditions of pharmacologically induced oxidative stress, *J. Biol. Chem.* 282, 13977–13983, 2007.

55. Huang, B. and Chen, C., An ascorbate-dependent artifact that interferes with the interpretation of the biotin switch assay, *Free Radic. Biol. Med.* 41, 562–567, 2006.

56. Kallakunta, V.M., Staruch, A., and Mutus, B., Sinapinic acid can replace ascorbate in the biotin switch assay, *Biochim. Biophys. Acta* 1800, 23–30, 2010.

57. Murray, C.I., Uhrigshardt, H., O'Meally, R.N. et al., Identification and quantification of *S*-nitrosylation by cysteine reactive tandem mass tag switch assay, *Mol. Cell. Proteomics* 11, M111.013441, 2011.

58. Zhang, Y., Keszler, A., Broniowska, K.A., and Hogg, N., Characterization and application of the biotin-switch assay for the identification of *S*-nitrosylated proteins, *Free Radic. Biol. Med.* 38, 874–881, 2005.

59. Wiesweg, M., Berchner-Pfannschmidt, U., Fandrey, J. et al., Rocket fuel for the quantification of *S*-nitrosothiols. Highly specific reduction of *S*-nitrosothiols by methylhydrazine, *Free Radic. Res.* 47, 104–115, 2013.

60. Schmidt, E.W., *Hydrazine and Its Derivatives: Preparation, Properties, and Applications*, 2nd edn., pp. 442–450, John Wiley & Sons, New York, 2001.

61. Ruddock, L.W., Low-molecular-weight oxidants involved in disulfide bond formation, *Antioxid. Redox Signal.* 16, 1129–1138, 2012.

62. Grazú, V., Ovsejevi, K., Cuadra, K. et al., Solid-phase reducing agents as alternative for reducing disulfide bonds in proteins, *Appl. Biochem. Biotechnol.* 110, 23–32, 2003.

63. Ferraz, N., Leverrier, J., Batista-Viera, F., and Manta, C., Thiopropyl-agarose as a solid phase reducing agents for chemical modification of IgG and G(ab')$_2$, *Biotechnol. Prog.* 24, 1154–1159, 2008.

64. Tzanavaras, P.D., Mitani, C., Anthemidis, A., and Themedis, D.G., On-line cleavage of disulfide bonds by soluble and immobilized tris-(2-carboxyethyl)phosphine using sequential injection analysis, *Talanta* 96, 21–25, 2012.

65. Bergendahl, V., Anthony, L.C., Heyduk, T., and Burgess, R.B., On-column tris(2-carboxyethyl)phosphine reduction and IC5-maleimide labeling during purification of a RpoC fragment on a nickel-nitrilotriacetic acid column, *Anal. Biochem.* 307, 368–374, 2002.

66. Forrester, M.T., Hess, D.T., Thompson, J.W. et al., Site-specific analysis of protein *S*-acylation by resin-assisted capture, *J. Lipid Res.* 52, 393–398, 2011.

67. Thompson, J.W., Forrester, M.T., Moseley, M.A., and Foster, M.W., Solid-phase capture for the detection and relative quantification of *S*-nitrosoproteins by mass spectrometry, *Methods*, 62, 130–137, 2013.

68. Zhang, W., Suzuki, M., Ito, Y., and Douglas, K.T., A chemically modified green fluorescent protein that responds to cleavage of an engineered disulfide bond by fluorescent energy transfer (FRET)-based changes, *Chem. Lett.* 34, 766–767, 2005.

69. Geoghegen, K.F., Cabacungen, J.C., Dixon, H.B., and Feeney, R.F., Alternative reducing agents for reductive methylation of amino groups in proteins, *Int. J. Protein Pept. Res.* 17, 345–352, 1981.

70. Chicooree, N., Connolly, Y., Tan, C.-T. et al., Enhanced detection of ubiquitin isopeptides using reductive methylation, *J. Am. Soc. Mass Spectrom.* 24, 421–430, 2013.

71. Ambrogelly, A., Cutler, C., and Paporello, B., Screening of reducing agents for the PEGylation of recombinant human IL-10, *Protein J.* 32, 337–342, 2013.

72. Yildrim, D. and Tukel, S.S., Immobilized *Pseudomonas* sp. lipase: A powerful biocatalyst for asymmetric acylation of (±)-2-amino-1-phenylethanols with vinyl acetate, *Process Biochem.* 48, 819–830, 2013.

73. Mohammed, N.A., Sabaa, M.W., El-Ghandour, A.H. et al., Quaternized *N*-substituted carboxymethyl chitosan derivatives as antimicrobial agents, *Int. J. Biol. Macromol.* 60, 156–164, 2013.

74. van der Madden, K., Koen, Y., Huxin, S. et al., Nanolayered chemical modification of silicon surfaces with ionizable surface groups for pH-triggered protein adsorption and release: Application to microneedles, *J. Mater. Chem. B: Mater. Biol. Med.* 1, 4466–4477, 2013.

75. Liu, L. and Breslow, R., A potent polymer/pyridoxamin enzyme mimic, *J. Am. Chem. Soc.* 124, 4978–4979, 2002.

76. Mosckovitz, R. and Gershoni, J.M., Three possible disulfides in the acetylcholine receptor α-subunit, *J. Biol. Chem.* 263, 1017–1022, 1988.

77. Gailit, J., Restoring free sulfhydryl groups in synthetic peptides, *Anal. Biochem.* 214, 334–335, 1993.

78. Light, A. and Sinha, N.K., Difference in the chemical reactivity of the disulfide bonds of trypsin and chymotrypsin, *J. Biol. Chem.* 242, 1358–1359, 1967.

79. Kress, L.F. and Laskowski, M., Sr., The basic trypsin inhibitor of bovine pancreas. VII. Reduction with borohydride of disulfide bond linking half-cystine residues 14 and 38, *J. Biol. Chem.* 242, 4925–4929, 1967.

80. Crestfield, A.M., Moore, S., and Stein, W.H., The preparation and enzymatic hydrolysis of reduced and *S*-carboxymethylated proteins, *J. Biol. Chem.* 238, 622–627, 1963.

81. Jering, H. and Tscheche, H., Preparation and characterization of the active derivative of bovine trypsin-kallikrein inhibitor (Kunitz) with the reactive site lysine 15-alanine-16 hydrolyzed, *Eur. J. Biochem.* 61, 443–452, 1976.

82. Kowalski, D. and Laskowski, M., Jr., Chemical-enzymatic replacement of Ile64 in the reactive site of soybean trypsin inhibitor (Kunitz), *Biochemistry* 15, 1300–1309, 1976.

83. Hixson, H.F., Jr. and Laskowski, M., Jr., Formation from trypsin and modified soybean trypsin inhibitor of a complex which upon kinetic control dissociation yields trypsin and virgin inhibitor, *J. Biol. Chem.* 245, 2027–2035, 1970.

84. Chu, S.C., Wang, C.C., and Brandenburg, D., Intramolecular enzymatic peptide synthesis: Trypsin-mediated coupling of the peptide bond between B22-arginine and B23-glycine in a split crosslinked insulin, *Hoppe Seyler's Z. Physiol. Chem.* 382, 647–654, 1981.

85. Kubiak, T. and Cowburn, D., Trypsin-catalyzed formation of pig des-(23–63)-proinsulin from desoctapeptide-(B23–30)-insulin, *Biochem. J.* 234, 665–670, 1986.

86. Morihara, K., Enzymatic semisynthesis of human insulin: An update, *J. Mol. Recognit.* 3, 181–186, 1990.

87. Hansen, R.E., Østergaard, H., Nørgaard, P., and Winther, J.R., Quantification of protein thiols and dithiols in the picomolar range using sodium borohydride and 4,4′-dithiodipyridine, *Anal. Biochem.* 363, 77–82, 2007.

88. Singh, R. and Katz, L., Catalysis of reduction of disulfide by selenol, *Anal. Biochem.* 232, 86–91, 1995.

89. Casi, G., Roelfes, G., and Hilvert, D., Selenoglutaredoxin as a glutathione peroxidase mimic, *ChemBioChem* 9, 1623–1631, 2008.

90. Prenant, C., Cawthorne, C., Fairclough, M. et al., Radiolabeling with fluorine-18 of a protein, interleukin-1 receptor antagonist, *Appl. Radiat. Isot.* 68, 1721–1727, 2010.

91. Wakita, C., Honda, K., Shibata, T. et al., A method for detection of 4-hydroxy-2-nonenal adducts n proteins, *Free Radic. Biol. Med.* 51, 2–4, 2011.

92. Kinstler, O., Molineux, G., Treuheit, M. et al., Mono-N-terminal poly(ethylene glycol)-protein conjugates, *Adv. Drug Deliv. Rev.* 54, 477–485, 2002.

93. Gidley, M.J. and Sanders, J.K. Reductive methylation of proteins with sodium cyanoborohydride. Identification, suppression and possible uses of *N*-cyanomethyl by-products, *Biochem. J.* 203, 331–334, 1982.

94. Pelter, A., Smith, K., and Brown, H.C., *Borane Reagents*, p. 46, Academic Press, New York, 1988.

95. Brown, H.C., *Boranes in Organic Chemistry*, Cornell University Press, Ithaca, NY, 1972.

96. Peng, L., Calton, G.J., and Burnett, J.W., Effect of borohydride reduction on antibodies, *Appl. Biochem. Biotechnol.* 14, 91–99, 1987.

97. Weber, V., Linsberger, I., Ettenauer, M. et al., Development of specific adsorbents for human tumor necrosis factor-α: Influence of antibody immobilization on performance and biocompatibility, *Biomacromolecules* 6, 1864–1870, 2005.

98. McCord, C.P., Toxicity of thioglycolic acid used in cold permanent wave process, *JAMA* 131, 776, 1946.

99. Suzuta, K., Ogawa, S., Takeda, Y. et al., Intermolecular disulfide cross-linked structural change induced by permanent wave treatment of human hair with thioglycolic acid, *J. Cosmet. Sci.* 63, 177–196, 2012.

100. Gardner, M.L.G., Cysteine: A potential source of error in amino acid analysis of mercaptoethane sulfonic or hydrochloric acid hydrolyzates of proteins and peptides, *Anal. Biochem.* 141, 429–431, 1984.

101. Yokote, Y., Arai, K.M., and Akahane, K., Recovery of tryptophan from 25-minute acid hydrolysates of protein, *Anal. Biochem.* 152, 245–249, 1986.

102. Hashimoto, Y., Yamagata, S., and Hayakawa, T., Amino acid analysis by high-performance liquid chromatography of a single stained protein band from a polyacrylamide gel, *Anal. Biochem.* 160, 362–367, 1987.

103. Sottrup-Jensen, L., Determination of half-cysteine in proteins as cysteine from reducing hydrolyzates, *Biochem. Mol. Biol. Int.* 30, 789–794, 1993.

104. Joergensen, L. and Thestrup, H.N., Determination of amino acids in biomass and protein samples by microwave hydrolysis and ion-exchange chromatography, *J. Chromatogr. A* 706, 421–428, 1995.

105. Hoogerheide, J.G. and Campbell, C.M., Determination of cysteine plus half-cystine in protein and peptide hydrolyzates: Use of dithiodiglycolic acid and phenylisothiocyanate derivatization, *Anal. Biochem.* 201(1):146–151.

106. Miller, S.M., Moore, M.J., Massey, V. et al., Evidence for the participation of Cys558 and Cys559 at the active site of mercuric reductase, *Biochemistry* 28, 1194–1205, 1989.

107. Thorpe, C. and Williams, C.H., Jr., Differential reactivity of the two active site cysteine residues generated on reduction of pig heart lipoamide dehydrogenase, *J. Biol. Chem.* 251, 3553–3557, 1976.

108. Dolin, M.I., Reduced diphosphopyridine nucleotide peroxidase. Intermediates formed on reduction of the enzyme with dithionite or reduced diphosphopyridine nucleotide, *J. Biol. Chem.* 250, 310–317, 1975.

109. Knipp, M., Taing, J.J., and He, C., Reduction of the lipocalin type heme containing protein nitrophorin— Sensitivity of fold-stabilizing cysteine disulfides toward routine heme-iron reduction, *J. Inorg. Biochem.* 105, 1405–1412, 2011.

110. Raje, S. and Thorpe, C., Inter-domain redox communication in flavoenzymes of the quiescin/sulfhydryl oxidase family: Role of a thioredoxin domain in disulfide bond formation, *Biochemistry* 42, 4560–4568, 2003.

111. Senft, A.P., Dalton, T.P., and Shertzer, H.G., Determining glutathione and glutathione disulfide using the fluorescence probe *o*-phthalaldehyde, *Anal. Biochem.* 280, 80–86, 2000.

112. Bermudez, M.A., Páez-Ochoa, M.A., Gotor, C., and Romero, L.C., Arabidopsis *S*-sulfocysteine synthase activity is essential for chloroplastic function and long-day light-dependent redox-control. *Plant Cell* 22, 403–416, 2010.

113. Abbas, A.K., Xia, W., Tranberg, M. et al., *S*-Sulfo-cysteine is an endogenous amino acid in neonatal rat brain but an unlikely mediator of cysteine neurotoxicity, *Neurochem. Res.* 33, 301–307, 2008.

114. Fukumoto, Y., Inai, S., Nagaki, K. et al., Interaction of *S*-sulfonated human IgG with human complement and its components, *J. Biochem.* 82, 955–960, 1977.

115. Swadesh, J.K., Thannhauser, T.W., and Scheraga, H.A., Sodium sulfite as an antioxidant in the acid hydrolysis of bovine pancreatic ribonuclease A, *Anal. Biochem.* 141, 397–401, 1984.

116. Sokolovsky, M., Riordan, J.F., and Vallee, B.L., Conversion of 3-nitrotyrosine to 3-aminotyrosine in peptides and proteins, *Biochem. Biophys. Res. Commun.* 27, 20–25, 1967.

117. Guo, J. and Prokai, L., Conversion of 3-nitrotyrosine to 3-aminotyrosine residues facilitates mapping of tyrosine nitration in proteins by electrospray ionization-tandem mass spectrometry using electron capture dissociation, *J. Mass Spectrom.* 47, 1601–1611, 2012.

118. Ekelman, W. and Richards, P., Instant ^{99m}Tc-DTPA, *J. Nucl. Med.* 11, 761, 1970.

119. Wong, D.W., Mishkin, F., and Lee, T., A rapid chemical method of labeling human plasma proteins with ^{99m}Tc-pertechnetate at pH 7.4, *Int. J. Appl. Radiat. Isot.* 29, 251–253, 1978.

120. Paik, C.H., Phan, L.N.B., Hong, J.J. et al., The labeling of high affinity sites of antibodies with ^{99m}Tc, *Int. J. Nucl. Med. Biol.* 12, 3–8, 1985.

121. Mardisossian, G., Wu, C., Rusckowski, M., and Hatowich, D.J., The stability of ^{99m}Tc directly labelled to an Fab′ antibody via stannous ion and mercaptoethanol reduction, *Nucl. Med. Commun.* 13, 503–512, 1992.

122. Zhang, Z.M., Ballinger, J.R., Sheldon, K., and Boxen, I., Evaluation of reduction-mediated labelling of antibodies with technetium-99m, *Int. J. Radiat. Appl. Instrum. B* 19, 607–609, 1992.

123. Jeong, J.M., Hong, M.K., Lee, J. et al., ^{99m}Tc-neolactosylated human serum albumin for imaging the hepatic asialoglycoprotein receptor, *Bioconjug. Chem.* 15, 850–855, 2004.

124. Hung, J.C., Gadient, K.R., Mahoney, D.W., and Murray, J.A., In-house preparation of technetium 99m-labeled human serum albumin for evaluation of protein-losing gastroenteropathy, *J. Am. Pharm. Assoc. (Wash)* 42, 57–62, 2002.

125. Hunt, A.P., Frier, M., Johnson, R.A. et al., Preparation of Tc-99m-macroaggregated albumin from recombinant human albumin for lung perfusion imaging, *Eur. J. Pharm. Biopharm.* 62, 26–31, 2006.

126. Wang, Y.F., Chuang, M.H., Chiu, J.S. et al., On-site preparation of technetium-99m-labeled human serum albumin for clinical application, *Tohoku J. Exp. Med.* 211, 379–385, 2007.

127. Sun, C., Shi, Z.-Z., Zhou, X. et al., Prediction of *S*-glutathionylation sites based on protein sequences, *PLoS One* 8(2), e5512, 2013.

128. Lillig, C.H. and Berndt, C., Glutaredoxins in thiol/disulfide exchange, *Antioxid. Redox Signal.* 18, 1654–1665, 2013.

129. Janssen-Heininger, Y.M.W., Nolin, J.D., Hoffman, S.M. et al., Emerging mechanisms of glutathione-dependent chemistry in biology and disease, *J. Cell. Biochem.* 114, 1962–1968, 2013.

130. Furuhata, A., Honda, K., Shibata, T. et al., Monoclonal antibody against protein-bound glutathione: Use of glutathione conjugate of acrolein-modified proteins as an immunogen, *Chem. Res. Toxicol.* 25, 1393–1401, 2012.

131. Rajan, R.S., Rsumoto, K., Tokunaga, M. et al., Chemical and pharmacological chaperones: Application for recombinant protein production and protein folding diseases, *Curr. Med. Chem.* 18, 1–15, 2011.

132. Zhang, Y. and Gray, R.D., Characterization of folded, intermediate, and unfolded states of recombinant human interstitial collagenase, *J. Biol. Chem.* 271, 8015–8021, 1996.

133. Frankel, A.E., Ramage, J., Latimer, A. et al., High-level expression and purification of the recombinant diphtheria fusion toxin DTGM for phase I clinical trials, *Protein Expr. Purif.* 16, 190–201, 1999.

134. Urieto, J.O., Liu, T., Black, J.H. et al., Expression and purification of the recombinant diphtheria fusion toxin DT388IL3 for phase I clinical trials, *Protein Expr. Purif.* 33, 123–133, 2004.

135. Gerami, S.J., Farajnia, S., Mahboudi, S. et al., Optimizing refolding condition for recombinant tissue plasminogen activator, *Iran. J. Biotechnol.* 9, 253–259, 2011.

136. Vallejo, L.F. and Rinas, U., Folding and dimerization kinetics of bone morphogenetic protein-2, a member of the transforming growth factor-β family, *FEBS J.* 280, 83–92, 2013.

137. Hill, H.E. and Pioszak, A.A., Bacterial expression and purification of a heterodimeric adrenomedullin receptor extracellular domain complex using Dsbs-assisted disulfide shuffling, *Protein Expr. Purif.* 88, 107–113, 2013.

138. Matsuda, T., Watanabe, S., and Kigawa, T., Cell-free synthesis system suitable for disulfide-containing proteins, *Biochem. Biophys. Res. Commun.* 431, 296–301, 2013.

139. Beck, J.L., Ambahera, S., Yong, S.R. et al., Direct observation of covalent adducts with Cys34 of human serum albumin using mass spectrometry, *Anal. Biochem.* 325, 326–336, 2004.

140. Turell, L., Botti, H., Carballal, S. et al., Reactivity of sulfenic acid in human serum albumin, *Biochemistry* 47, 358–367, 2008.

141. Boyle, D.M., Buckley, J.J., Johnson, G.V. et al., Use of the design-of-experiments approach for the development of a refolding technology for progenipoietin-1, a recombinant human cytokine fusion protein from *Escherichia coli inclusion bodies*, *Biotechnol. Appl. Biochem.* 54, 85–92, 2009.

142. Peters, R.A., Stocken, L.A., and Thompson, R.H.S., British anti-Lewisite (BAL), *Nature* 156, 616–619, 1945.

143. Cooperstein, S.J., The effect of disulfide bond reagents on cytochrome oxidase, *Biochim. Biophys. Acta* 73, 343–346, 1963.

144. Cooperstein, S.J., Reversible inactivation of cytochrome oxidase by disulfide bond reagents, *J. Biol. Chem.* 238, 3606–3610, 1963.

145. Flora, S.J. and Pachauri, V., Chelation in metal intoxication, *Int. J. Environ. Res. Public Health* 7, 2745–2788, 2010.

146. Maiorino, R.M., Xu, Z.F., and Aposhian, H.V., Determination and metabolism of dithiol chelating agents. XVII. In human, sodium 2,3-dimercapto-1-propanesulfonate is bound to plasma albumin via mixed disulfide formation and is found in the urine as cyclic polymeric disulfides, *J. Pharmacol. Exp. Ther.* 277, 375–384, 1996.

147. Stocken, L.A. and Thompson, R.H.S., British antilewisite. II. Dithiol compounds as antidotes for arsenic, *Biochem. J.* 40, 535–540, 1946.

148. Fuson, R.C. and Ziegler, J.B., Jr., 2-Chloroethyl 2-hydroxyethyl sulfide, *J. Org. Chem.* 11, 510–512, 1946.

149. U.S. Publication Board, *Bibliography of Scientific and Industrial Reports*, Department of Commerce, Washington, DC, 1946.

150. Payne, R.B., The reaction of rheumatoid factor with bovine IgG fragments obtained by papain digestion and by reduction, *Immunology* 9, 449–456, 1965.

151. Solheim, B.G. and Harboe, M., Reversible dissociation of reduced and alkylated IgM subunits to half subunits, *Immunochemistry* 9, 523–634, 1972.

152. Denk, H., Stemberger, H., Wiedermann, G. et al., The influence of 2-mercaptoethanol (2-ME) treatment of IgG antibody on its ability to induce cytotoxicity nonsensitized lymphocytes, *Cell. Immunol.* 13, 489–492, 1974.

153. Utzig, E. and Rialdi, G., Kinetics and thermodynamic study on the chemical reduction of macroimmunoglobulins, *Mol. Immunol.* 27, 343–350, 1990.

154. Ditzel, H., Erb, K., Leslie, G., and Jensenius, J.C., Preparation of antigen-binding monomeric and half-monomeric fragments from human monoclonal IgM antibodies against colorectal cancer-associated antigens, *Hum. Antibodies Hybridomas* 4, 86–93, 1993.

155. Kaul, M. and Loos, M., Dissection of C1q capability of interacting with IgG. Time-dependent formation of a tight and only partially reversible association, *J. Biol. Chem.* 272, 33234–33244, 1997.

156. Dias, C.R., Jeger, S., Osso, J.A., Jr. et al., Radiolabeling of rituximab with [188]Re and ^{99m}Tc using the tricarbonyl technology, *Nucl. Med. Biol.* 38, 19–28, 2011.

157. Deutsch, H.F. and Morton, J.I., Dissociation of human serum macroglobulin, *Science* 125, 600–601, 1957.

158. Grues, B. and Swahn, B., Destruction of some agglutinins but not of others by two sulfhydryl compounds, *Acta Pathol. Microb. Scand.* 43, 305–309, 1958.

159. Jerne, N.K., Nordin, A.A., and Henry, C., in *Cell-Bound Antibodies*, ed. H. Amos and H. Kiprowski, The Wistar Institute, Philadelphia, PA, pp. 109–116, 1963.

160. Plotz, P.H., Talal, N., and Asofsky, R., Assignment of direct and facilitated hemolytic plaques in mice to specific immunoglobulin classes, *J. Immunol.* 100, 744–751, 1968.

161. Hosono, M. and Muramatsu, S., Use of 2-mercaptothanol for distinguishing between IgM and IgG antibody-producing cells of mice immunized with bovine γ-globulin, *J. Immunol.* 109, 857–863, 1972.

162. Coe, J.E. and Hadlow, W.J., Studies on immunoglobulins of mink: Definition of IgG, IgA, and IgM, *J. Immunol.* 108, 530–537, 1972.

163. Aarden, L.A., Lakmaker, F., and De Groot, E., Immunology of DNA. VI. The effect of mercaptans on IgG and IgM anti-dsDNA, *J. Immunol. Methods* 16, 143–152, 1977.

164. Capel, P.J., Gerlag, P.G., Hageman, J.F., and Koene, R.A., The effect of 2-mercaptoethanol on IgM and IgG antibody activity, *J. Immunol. Methods* 36, 77–80, 1980.

165. Cherkaoui, S., Bettinger, T., Hauwel, M. et al., Tracking of antibody reduction fragments by capillary gel electrophoresis during the coupling to microparticles surface, *J. Pharm. Biomed. Anal.* 53, 172–178, 2010.

166. Gagnon, P., Rodriquez, G., and Zaidi, S., Dissociation and fractionation of heavy and light chains from IgG monoclonal antibodies, *J. Chromatogr. A* 1218, 2402–2404, 2011.

167. Cotham, V.C., Wine, Y., and Brodbelt, J.S., Selective 351 nm photodissociation of cysteine-containing peptides for discrimination of antigen-binding regions of IgG fragments in bottom-up liquid chromatography-tandem mass spectrometry workflows, *Anal. Chem.* 85, 5577–5585, 2013.

168. Marshall, T., Sodium dodecyl sulfate polyacrylamide gel electrophoresis of serum after protein denaturation in the presence or absence of 2-mercaptoethanol, *Clin. Chem.* 30, 475–479, 1984.

169. Marshall, T. and Williams, K.M., Artifacts associated with 2-mercaptoethanol upon high resolution two-dimensional electrophoresis, *Anal. Biochem.* 139, 502–505, 1984.

170. Jin, Y.J., Koyasu, S., Moingeon, P. et al., A fraction of CD3$_\varepsilon$ subunits exists as disulfide-linked dimers in both human and murine T lymphocytes, *J. Biol. Chem.* 265, 15850–15853, 1990.

171. Chevalier, F., Hirtz, C., Sommerer, N., and Kelly, A.L., Use of reducing/nonreducing two-dimensional electrophoresis for the study of disulfide-mediated interactions between proteins in raw and heated bovine milk, *J. Agric. Food Chem.* 57, 5948–5955, 2009.

172. De Graan, P.N., Moritz, A., de Wit, M., and Gispen, W.H., Purification of B-50 by 2-mercaptoethanol extraction from rat brain synaptosomal plasma membranes, *Neurochem. Res.* 18, 875–881, 1993.

173. King, B.J., Lee, L.S., Rackemann, R.G., and Scott, P.T., Preparation of extracts for electrophoresis from citrus leaves, *J. Biochem. Biophys. Methods* 29, 295–305, 1994.

174. Watanabe, Y., Aburatani, K., Mizumura, T. et al., Novel ELISA for the detection of raw and processed egg using extraction buffer containing a surfactant and a reducing agent, *J. Immunol. Methods* 300, 115–123, 2005.

175. Doña, V., Urrutia, M., Bayardo, M. et al., Single domain antibodies are specially suited for quantitative determination of gliadins under denaturing conditions, *J. Agric. Food Chem.* 58, 918–926, 2010.

176. Vinterováz, Z., Sanda, M., Dostál, J. et al., Evidence for the presence of proteolytically active secreted aspartic proteinase 1 of *Candida parapsilosis* in the cell wall, *Protein Sci.* 20, 2004–2012, 2011.

177. Schumacher, M., Glocker, M.O., Wunderlin, M., and Przybylski, M., Direct isolation of proteins from sodium dodecyl sulfate-polyacrylamide gel electrophoresis and analysis by electrospray-ionization mass spectrometry, *Electrophoresis* 17, 848–854, 1996.

178. Goch, G., Vdovenko, S., Kozlowska, H., and Bierzyñski, A., Affinity of S100A1 protein for calcium increases dramatically upon glutathionylation, *FEBS J.* 272, 2557–2565, 2005.

179. Voronova, A., Kazantseva, J., and Tuuling, M., Cox17, a copper chaperone for cytochrome c oxidase: Expression, purification, and formation of mixed disulfide adducts with thiol reagents, *Protein Expr. Purif.* 53, 138–144, 2007.

180. Zhukov, I., Ejchart, A., and Bierzyñski, A., Structural and motional changes induced in apo-S100A1 protein by the disulfide formation between its Cys85 residue and β-mercaptoethanol, *Biochemistry* 47, 640–650, 2008.

181. Stefanescu, R., Born, R., Moise, A. et al., Epitope structure of the carbohydrate recognition domain of asialoglycoprotein receptor to a monoclonal antibody revealed by high-resolution proteolytic excision mass spectrometry, *J. Am. Soc. Mass Spectrom.* 22, 148–157, 2011.

182. Noll, K.M., Rinehart, K.L., Jr., Tanner, R.S., and Wolfe, R.S., Structure of component B (7-mercaptoheptanoylthreonine phosphate) of the methylcoenzyme M methylreductase system of *Methanobacterium thermoautotrophicum*, *Proc. Natl. Acad. Sci. USA* 83, 4238–4242, 1986.

183. Sauer, F.D., Blackwell, B.A., and Kramer, J.K., Structure of purified cytoplasmic cofactor from *Methanobacterium thermoautotrophicum*, *Biochem. Biophys. Res. Commun.* 147, 1021–1026, 1987.

184. Deneke, S.M., Thiol-based antioxidants, *Curr. Top. Cell Regul.* 36, 151–180, 2000.

185. Lodge, J.K., Traber, M.G., and Packer, L., Thiol chelation of Cu^{2+} by dihydrolipoic acid prevents human low density lipoprotein peroxidation, *Free Radic. Biol. Med.* 25, 287–297, 1998.

186. Suh, J.H., Zhu, B.X., deSzoeke, E. et al., Dihydrolipoic acid lowers the redox activity of transition metal ions but does not remove them from the active site of enzymes, *Redox Rep.* 9, 57–61, 2004.

187. Jocelyn, P.C., The standard redox potential of cysteine-cystine from the thiol-disulphide exchange reaction with glutathione and lipoic acid, *Eur. J. Biochem.* 2, 327–331, 1967.

188. Sluyterman, L.A. and Wijdenes, J., The effect of thiol compounds on the autolysis of papain, *Eur. J. Biochem.* 113, 189–193, 1980.

189. Stoyanovsky, D.A., Tyurina, Y.Y., Tyurin, V.A. et al., Thioredoxin and lipoic acid catalyze the denitrosylation of low molecular weight and protein *S*-nitrosothiols, *J. Am. Chem. Soc.* 127, 15815–15823, 2005.

190. Gorecki, M. and Patchornik, A., Polymer-bound dihydrolipoic acid: A new insoluble reducing agent for disulfides, *Biochim. Biophys. Acta* 303, 36–43, 1973.

191. Bienvenu, C., Greiner, J., Vierling, P., and DiGiorgio, C., Convenient supported recyclable material based on dihydrolipoyl-residue for the reduction of disulfide derivatives, *Tetrahedron Lett.* 51, 3309–3311, 2010.

192. Roux, S., Garcia, B., Bridot, J.-L. et al., Synthesis, characterization of dihydrolipoic acid capped gold nanoparticles, and functionalization by the electroluminescent luminol, *Langmuir* 21, 2520–2536, 2005.

193. Fujimoto, A., Endo, T., and Okawara, M., Synthesis of a polymer containing the cyclic disulfide (1,2-dithiolane) structure, *Macromolekulare Chem.* 175, 3597–3602, 1974.

194. Leach, S.J., Meschers, A., and Swanepoel, O.A., The electrolytic reduction of proteins, *Biochemistry* 4, 23–27, 1965.

195. Li, J., Dewald, H.D., and Chen, H., Online coupling of electrochemical reactions with liquid sample desorption electrospray ionization-mass spectrometry, *Anal. Chem.* 81, 9716–9722, 2009.

196. Zhang, Y., Cui, W., Zhang, H. et al., Electrochemistry-assisted top-down characterization of disulfide-containing proteins, *Anal. Chem.* 84, 3838–3842, 2012.

197. Lee, D.-Y. and Chang, G.-D., Electrolytic reduction: Modification of proteins occurring in isoelectric focusing electrophoresis and in electrolytic reactions in the presence of high salts, *Anal. Chem.* 81, 3957–3964, 2009.

198. Cox, P.A., *Inorganic Chemistry*, Bios/Springer, Oxford, U.K., 2000.

199. Moroder, L., Isoteric replacement of sulfur with other chalcogens in peptides and proteins, *J. Pept. Sci.* 11, 187–214, 2005.

200. Chasteen, T.G., Fuentes, D.E., Tantaleán, J.C., and Vasquez, C.C., Tellurite: History, oxidative stress, and molecular mechanisms of resistance, *FEMS Microbiol. Rev.* 33, 820–832, 2009.

201. Wessjohann, L.A., Schneider, A., Abbas, M., and Brandt, W., Selenium in chemistry and biochemistry in comparison to sulfur, *Biol. Chem.* 388, 997–1006, 2007.

202. Pleasants, J.C., Guo, W., and Rabenstein, D.L., A comparative study of the kinetics of selenol/diselenide and thiol/disulfide exchange reactions, *J. Am. Chem. Soc.* 111, 6553–6558, 1989.

203. Jan, Y.-H., Heck, D.E., Gray, J.P. et al., Selective targeting of selenocysteine in thioredoxin reductase by the half-mustard 2-chloroethyl ethyl sulfide in lung epithelial cells, *Chem. Res. Toxicol.* 23, 1045–1053, 2010.

204. Nauser, T., Steinmann, D., and Koppenol, W.H., Why do proteins use selenocysteine instead of cysteine?, *Amino Acids* 42, 39–44, 2012.

205. Singh, R. and Maloney, E.K., Labeling of antibodies by in situ modification of thiol groups generated from selenol-catalyzed reduction of native disulfide bonds, *Anal. Biochem.* 304, 147–156, 2002.

206. Bramanti, E., Lomonte, C., Onor, M. et al., Study of the disulfide reduction of denatured proteins by liquid chromatography coupled with on-line cold-vapor-generation atomic-fluorescence spectrometry (LC-CVGAFS), *Anal. Bioanal. Chem.* 380, 310–318, 2004.

207. Günther, W.H.H., Hypophosphorous acid, a novel reagent for the reduction of diselenides and the selenol-catalyzed reduction of disulfides, *J. Org. Chem.* 31, 1202–1205, 1966.

208. Clarembeau, M., Cravador, A., Dumont, W. et al., Synthesis of selenoacetals, *Tetrahedron* 41, 4793–4812, 1985.

209. Getz, E.B., Xiao, M., Chakrabarty, T. et al., A comparison between the sulfhydryl reductants tris(2-carboxyethyl)phosphine and dithiothreitol for use in protein biochemistry, *Anal. Biochem.* 273, 73–80, 1999.

210. Shafer, D.E., Inman, J.K., and Lees, A., Reaction of tris(2-carboxyethyl)phosphine (TCEP) with maleimide and α-haloacyl groups: Anomalous elution of TCEP by gel filtration, *Anal. Biochem.* 282, 161–164, 2000.

211. Di Stefano, G., Lanza, M., Kratz, F. et al., A novel method for coupling doxorubicin to lactosaminated human albumin by an acid sensitive hydrazone bond: Synthesis, characterization and preliminary biological properties of the conjugate, *Eur. J. Pharm. Sci.* 23, 393–397, 2004.

212. Burns, J.A., Butler, J.C., Moran, J., and Whitesides, G.M., Selective reduction of disulfides by tris(2-carboxyethyl)phosphine, *J. Org. Chem.* 56, 2648–2650, 1991.

213. Rüegg, U.T. and Gattner, H.G., Reduction of S-sulpho groups by tributylphosphine: An improved method for the recombination of insulin chains, *Hoppe-Seyler's Z. Physiol. Chem.* 356, 1527–1533, 1975.

214. Rüegg, U.T. and Rudinger, J., Reductive cleavage of cystine disulfides with tributylphosphine, *Methods Enzymol.* 47, 111–116, 1977.

215. Rüegg, U.T., Reductive cleavage of S-sulfo groups with tributylphosphine, *Methods Enzymol.* 47, 123–126, 1977.

216. Herbert, B.R., Molloy, M.E., Gooley, A.A. et al., Improved protein solubility in two-dimensional electrophoresis using tributyl phosphine as reducing agent, *Electrophoresis* 19, 845–851, 1998.

217. Kirley, T.L., Determination of three disulfide bonds and one free sulfhydryl in the β subunit of (Na,K)-ATPase, *J. Biol. Chem.* 264, 7185–7192, 1989.

218. Chin, C.C.Q. and Wold, F., The use of tributylphosphine and 4-(aminosulfonyl)-7-fluoro-2,1,3-benzoxadiazole n the study of protein sulfhydryls and disulfides, *Anal. Biochem.* 214, 128–134, 1993.

219. Ivanenko, V.V., Murphey-Piedmonte, D.M., and Kirley, T.L., Bacterial expression, characterization, and disulfide bond determination of soluble human NTPDase6 (CD39L2) nucleotidase: Implications for structure and function, *Biochemistry* 42, 11726–11735, 2003.

220. Zinellu, A., Pinna, A., Zinellu, E. et al., High-throughput capillary electrophoresis method for plasma cysteinylglycine measurement: Evidence for a clinical application, *Amino Acids* 34, 69–74, 2008.

221. Kubalczyk, P. and Bald, E., Method for determination of total cystamine in human plasma by high performance capillary electrophoresis with acetonitrile stacking, *Electrophoresis* 29, 3636–3640, 2008.

222. Bald, E., Kaniowska, E., Chwatko, G., and Glowacki, R., Liquid chromatographic assessment of total and protein-bound homocysteine in human plasma, *Talanta* 50, 1233–1243, 2000.

223. Zinellu, A., Lepdda, A., Jr., Sotgia, S. et al., Albumin-bound low molecular weight thiols analysis in plasma and carotid plaques by CE, *J. Sep. Sci.* 33, 126–131, 2010.

224. Zhang, E., Chen, X., and Liang, X., Resolubilization of TCA-precipitated plant proteins for 2-D electrophoresis, *Electrophoresis* 32, 696–698, 2011.

225. Vâlcu, C.M. and Schlink, K., Reduction of proteins during sample preparation and two-dimensional gel electrophoresis of woody plant samples, *Proteomics* 6, 1599–1605, 2006.

226. Lee, K., Pi, K., and Lee, K., Buffer optimization for high resolution of human lung cancer tissue proteins by two-dimensional gel electrophoresis, *Biotechnol. Lett.* 31, 31–37, 2009.

227. Molloy, M.P., Herbert, B.R., Williams, K.L. and Gooley, A.A., Extraction of *Escherichia coli* proteins with organic solvents prior to two-dimensional electrophoresis, *Electrophoresis* 20, 701–704, 1999.

228. Miller, I., Eberini, I., and Gianazza, E., Other than IPG-DALT: 2-DE variants, *Proteomics* 10, 586–610, 2010.

229. Bayer, E. and Dieter, J., Untersuchungen zur Struktur des Planzenferredoxins, *Hoppe-Seyler's Z. Physiol. Chem.* 351, 537–543, 1970.

230. Kimura, T., Studies on adrenal steroid hydroxylases. The reactivity of triphenyl phosphine with labile sulfur in adrenal iron-sulfur protein (adrenodoxin), *J. Biol. Chem.* 246, 5140–5146, 1971.

231. Manabe, T., Goda, K., and Kimura, T., Chemical reactivity of labile sulfur of iron-sulfur proteins. The reaction of triphenyl phosphine, *Biochim. Biophys. Acta* 428, 312–320, 1976.

232. Li, S., Wang, H., Xian, M., and Whorton, A.R., Identification of protein nitrosothiols using phosphine-mediated selective reduction, *Nitric Oxide* 26, 20–26, 2012.

233. Bechtold, E. and King, S.B., Chemical methods for the direct detection and labeling of S-nitrosothiols, *Antioxid. Redox Signal.* 17, 981–991, 2012.

234. Daigle, D.J., Pepperman, A.B., Jr., Drake, G.L., Jr., and Reeves, W.A., Flame-retardant finish based on tris(hydroxymethyl)phosphine, *Textile Res. J.* 42, 347–353, 1972.

235. Hooper, G., Nakajima, W.N., and Herbes, W.F., *J. Coated Fabrics* 6, 105–120, 1976.

236. Pepperman, A.B., Jr., Vail, S.L., and Lyons, D.W., Microwave drying of FR cotton fabrics, *Textile Chem. Color.* 9, 137–141, 1977.

237. Nair, G.P., Flammability in textiles and routes to flame retardant textiles—XIII, *Colourage* 49, 59–60, 68, 2002.

238. Petach, H.H., Henderson, W., and Olsen, G.M., $P(CH_2OH)_3$—A new coupling reagent for covalent immobilization of enzymes, *J. Chem. Soc. Chem. Commun.* (18), 2181–2182, 1994.

239. Oswald, P.R., Evans, P.A., Henderson, W. et al., Properties of a thermostable β-glucosidase immobilized using tris(hydroxymethyl)phosphine as a highly effective coupling agent, *Enzyme Microb. Technol.* 23, 14–20, 1998.

240. Graham, J.S., Miron, Y., and Grandbois, M., Assembly of collagen fibril meshes using gold nanoparticles functionalized with tris(hydroxymethyl)phosphine-alanine as multivalent cross-linking agents, *J. Mol. Recognit.* 24, 477–482, 2011.

241. Renner, J.N., Cherry, K.M., Su, R.S.-C., and Liu, J.C., Characterization of resilin-based materials for tissue engineering applications, *Biomacromolecules* 13, 3678–3685, 2012.

242. Li, L., Tong, Z., Jia, X., and Kiick, K.L., Resilin-like polypeptide hydrogel engineered for versatile biological function, *Soft Matter* 9, 665–673, 2013.

243. Švagera, Z., Hanzlíková, D., Šimek, P., and Hušek, P., Study of disulfide reduction and alkyl chloroformate derivatization of plasma sulfur amino acids using gas chromatography-mass spectrometry, *Anal. Bioanal. Chem.* 402, 2953–2963, 2012.

244. Daithankar, V.N., Wang, W., Trujillo, J.R., and Thorpe, C., Flavin-linked Erv-family sulfhydryl oxidase release superoxide anion during catalytic turnover, *Biochemistry* 51, 265–272, 2012.

245. Chen, S., Jiang, H., Wei, K., and Liu, Y., Tris-(2-carboxyethyl) phosphine significantly promotes the reaction of cisplatin with Sp1 zinc finger protein, *Chem. Commun.* 49, 1226–1228, 2013.

246. Lieven, C.J., Ribich, J.D., Crowe, M.E., and Levin, L.A., Redox proteomic identification of visual arrestin dimerization in photoreceptor degeneration after photic injury, *Invest. Ophthalmol. Vis. Sci.* 53, 3990–3998, 2012.

247. Cline, D.J., Redding, S.E., Brohawn, S.G. et al., New water-soluble phosphines as reductants of peptide and protein disulfide bonds: Reactivity and membrane permeability, *Biochemistry* 43, 15195–15203, 2004.

248. Visser, C.C., Voorwinden, L.H., Harders, L.R. et al., Coupling of metal containing homing devices to liposomes via a maleimide linker: Use of TCEP to stabilize thiol-groups without scavenging metals, *J. Drug Target.* 12, 569–575, 2004.

249. Sánchez, A., Pedroso, E., and Grandas, A., Oligonucleotide cyclization: The thiol-maleimide reaction revisited, *Chem. Commun.* 49, 309–311, 2013.

250. Han, J., Clark, C., Han, G. et al., Preparation of 2-nitro-5-thiobenzoic acid using immobilized tris(2-carboxyethyl)phosphine, *Anal. Biochem.* 268, 404–407, 1999.

251. Christie, R.J., Tadiello, C.J., Camberlain, L.M., and Grainger, D.W., Optical properties and application of a reactive and bioreducible thiol-containing tetramethylrhodamine dimer, *Bioconjug. Chem.* 20, 476–480, 2009.

252. Da Pieve, C., WIlliams, P., Haddleton, D.M. et al., Modification of thiol functionalized aptamers by conjugation of synthetic polymers, *Bioconjug. Chem.* 21, 169–174, 2010.

253. Miralles, G., Verdié, P., Puget, K. et al., Microwave-mediated reduction of disulfide bridges with supported (tris(2-carboxyethyl)phosphine) as resin-bound reducing agent, *ACS Comb. Sci.* 15, 169–173, 2013.

254. Sun, M.M.C., Beam, K.S., Cerveny, C.G. et al., Reduction-alkylation strategies for the modification of specific monoclonal antibody disulfides, *Bioconjug. Chem.* 16, 1282–1290, 2005.

255. Sharma, H. and Mutharasan, R., Half antibody fragments improve biosensor sensitivity without loss of selectivity, *Anal. Chem.* 85, 2472–2477, 2013.

256. Stefano, J.E., Busch, M., Hou, L. et al., Micro-and mid-scale maleimide-based conjugation of cytotoxic drugs to antibody hinge region thiols for tumor targeting, *Methods Mol. Biol.* 1045, 145–171, 2013.

257. Gray, W.R., Echistatin disulfide bridges. Selective reduction and linkage assignment, *Protein Sci.* 2, 1749–1755, 1993.

258. Wechtersbach, L. and Cigic, B., Reduction of dehydroascorbic acid at low pH, *J. Biochem. Biophys. Methods* 70, 767–772, 2007.

259. Brobawn, S.G., Miksa, I.R., and Thorpe, C., Avian sulfhydryl oxidase is not a metalloenzyme: Adventitious binding of divalent metal ions to the enzyme, *Biochemistry* 42, 11074–11082, 2003.

260. Krezel, A., Latajka, R., Bujacz, G.D., and Bal, W., Coordination properties of tris(2-carboxyethyl)phosphine, a newly introduced thiol reductant, and its oxide, *Inorg. Chem.* 42, 1994–2003, 2003.

261. Wang, W., Winther, J.R., and Thorpe, C., Erv2p: Characterization of the Redox behavior of a yeast sulfhydryl oxidase, *Biochemistry* 46, 246–3254, 2007.

262. Cleland, W.W., Dithiothreitol, a new protective reagent for SH groups, *Biochemistry* 3, 480–482, 1964.

263. Cleland, W.W., Enzyme kinetics, *Annu. Rev. Biochem.* 36, 77–112, 1967.

264. Cleland, W.W., Low-barrier hydrogen bonds and enzymatic catalysis, *Arch. Biochem. Biophys.* 382, 1–5, 2000.

265. Cleland, W.W., The use of isotope effects to determine enzyme mechanisms, *Arch. Biochem. Biophys.* 433, 2–12, 2005.

266. Cleland, W.W. and Hengge, A.C., Enzymatic mechanisms of phosphate and sulfate transfer, *Chem. Rev.* 106, 3252–3278, 2006.

267. Morioka, Y. and Kobayahi, K., Colorimetric determination of cystine (disulfide bond) in hair using dithiothreitol, *Biol. Pharm. Bull.* 20, 825–827, 1997.

268. Seo, A., Jackson, J.L. Schuster, J.V., and Vardar-Ulu, D., Using UV-absorbance of intrinsic dithiothreitol (DTT) during RP-HPLC as a measure of experimental redox potential *in vitro*, *Anal. Bioanal. Chem.* 405, 6379–6384, 2013.

269. Rothwarf, D.M. and Scheraga, H.A., Equilibrium and kinetic constants for the thiol-disulfide interchange reaction between glutathione and dithiothreitol, *Proc. Natl. Acad. Sci. USA* 89, 7944–7948, 1992.

270. Le, M. and Means, G.E., A procedure for the determination of monothiols in the presence of dithiothreitol—An improved assay for the reduction of disulfides, *Anal. Biochem.* 229, 264–271, 1995.

271. Stokes, G.B. and Stumpf, P.K., Fat metabolism in higher plants. LXI. Nonenzymatic acylation of dithiothreitol by acyl coenzyme A, *Arch. Biochem. Biophys.* 162, 638–648, 1974.

272. Wilson, I., Wardman, P., Lin, T.S., and Sartorelli, A.C., Reactivity of thiols towards derivatives of 2- and 6-methyl-1,4-naphthoquinone bioreductive alkylating agents, *Chem. Biol. Interact.* 61, 229–240, 1987.

273. Iwaoka, M., Takahashi, ST., and Tomoda, S., Synthesis and structural characterization of water-soluble selenium reagents for the redox control of protein disulfide bonds, *Heteroatom Chem.* 12, 293–299, 2001.

274. Simons, S.S., Jr., Chakraborti, P.K., and Cavanaugh, A.H., Arsenite and cadmium (II) as probes of glucocorticoid receptor structure and function, *J. Biol. Chem.* 265, 1938–1945, 1990.

275. Kato, K., Ito, H., and Okamoto, K., Modulation of the arsenite-induced expression of stress proteins by reducing agents, *Cell Stress Chaperones* 2, 199–209, 1997.

276. Smirnova, J., Muhhina, J., Tõugu, V. et al., Redox and metal ion binding properties of human insulin-like growth factor 1 determined by electrospray ionization mass spectrometry, *Biochemistry* 1, 5851–5859, 2012.

277. Magnani, A., Barbucci, R., Lamponi, S. et al., Two-step elution of human serum proteins from different glass-modified bioactive surfaces: A comparative proteomic analysis of adsorption patterns, *Electrophoresis* 25, 2413–2424, 2004.

278. Pivetta, L.A., Dafre, A.L., Zeni, G. et al., Acetaldehyde does not inhibit glutathione peroxidase and glutathione reductase from mouse liver *in vitro*, *Chem. Biol. Interact.* 159, 196–204, 2006.

279. Giangregorio, N., Palmieri, F., and Indiveri, C., Glutathione controls the redox state of the mitochondrial carnitine/acylcarnitine carrier Cys residues by glutathionylation, *Biochim. Biophys. Acta* 1830, 5299–5304, 2013.

280. Milovanova, T.N., Bhopale, V.M., Sorokina, E.M. et al., Lactate stimulates vasculogenic stem cells via the thioredoxin system and engages an autocrine activation loop involving hypoxia-inducible factor 1, *Mol. Cell. Biol.* 28, 6248–6262, 2008.

281. Wagner, M., Morel, M.H., Bonicel, J., and Cuq, B., Mechanisms of heat-mediated aggregation of wheat gluten protein upon pasta processing, *J. Agric. Food Chem.* 59, 3146–3154, 2011.

282. Oppedisano, F., Catto, M., Koutentis, P.A. et al., Inactivation of the glutamine/amino acid transporter ASCT2 by 1,2,3-dithiazoles: Proteoliposomes as a tool to gain insights in the molecular mechanism of action and of antitumor activity, *Toxicol. Appl. Pharmacol.* 265, 93–102, 2012.

283. Zahler, W.L. and Cleland, W.W., A specific and sensitive assay for disulfides, *J. Biol. Chem.* 243, 716–719, 1968.

284. Singh, R., Lamoureux, G.V., Lees, W.J., and Whitesides, G.M., Reagents for rapid reduction of disulfide bonds, *Methods Enzymol.* 251, 167–173, 1995.

285. Lukesh, J.C., III, Palte, M.J., and Raines, R.T., A potent, versatile disulfide-reducing reagent from aspartic acid, *J. Am. Chem. Soc.* 134, 4057–4059, 2012.

286. Lee, S.H., Carpenter, J.F., Chang, B.S. et al., Effects of solutes on solubilization and refolding of proteins from inclusion bodies with high hydrostatic pressure, *Protein Sci.* 15, 304–313, 2006.

287. Akbari, N., Khajeh, K., Rezaie, S. et al., High-level expression of lipase in *Escherichia coli* and recovery of active recombinant enzyme through in vitro refolding, *Protein Expr. Purif.* 70, 75–80, 2010.

288. Freydell, E.J., van der Wielen, L.A., Eppink, M.H., and Ottens, M., Techno-economic evaluation of an inclusion body solubilization and recombinant protein refolding process, *Biotechnol. Prog.* 27, 1315–1328, 2011.

289. Yamaguchi, S., Yamamoto, E., Mannen, T. et al., Protein refolding using chemical refolding additives, *Biotechnol. J.* 8, 17–31, 2013.

290. Eiberle, M.K. and Junghauer, A., Technical refolding of proteins: Do we have freedom to operate?, *Biotechnol. J.* 5, 547–559, 2010.

291. Pham le, T.M., Kim, S.J., Song, B.K., and Kim, Y.H., Optimized refolding and characterization of *S*-peroxidase (CWPO_C of *Populus alba*) expressed in *E. coli*, *Protein Expr. Purif.* 80, 268–273, 2011.

292. Ito, L., Okumura, M., Tao, K. et al., Glutathione ethylester, a novel protein refolding reagent, enhances both the efficiency of refolding and correct disulfide formation, *Protein J.* 31, 499–503, 2012.

293. Okumura, M., Saiki, M., Yamaguchi, H., and Hidaka, Y., Acceleration of disulfide-coupled protein folding using glutathione derivatives, *FEBS J.* 278, 1137–1144, 2011.

294. Rehder, D.S. and Bores, C.R., Cysteine sulfenic acid as an intermediate in disulfide bond formation and nonenzymatic protein folding, *Biochemistry* 49, 7748–7755, 2010.

295. Esfandiar, S., Hashemi-Najafadi, S., Shojaosadati, S.A. et al., Purification and refolding of *Escherichia coli*-expressed recombinant human interleukin-2, *Biotechnol. Appl. Biochem.* 55, 209–214, 2010.

296. Min, C.K., Son, Y.J., Kim, C.K. et al., Increased expression, folding and enzyme reaction rate of recombinant human insulin by selecting appropriate leader peptide, *J. Biotechnol.* 151, 350–356, 2011.

297. Kumar, A., Tiwari, S., Thavaselvam, D. et al., Optimization and efficient purification of recombinant Omp28 protein of *Brucella melitensis* using Triton X-100 and β-mercaptoethanol, *Protein Expr. Purif.* 83, 226–232, 2012.

298. Meneses-Acosta, A., Vizcaino-Meza, L.R., Ayala-Castro, H.G. et al., Effect of controlled redox potential and dissolved oxygen on the in vitro refolding of an *E.coli* alkaline phosphatase and chicken lysozyme, *Enzyme Microb. Technol.* 52, 312–318, 2013.

299. White, K., Bruckner, J.V., and Guess, W.L., Toxicological studies of 2-mercaptoethanol, *J. Pharm. Sci.* 62, 237–241, 1973.

300. Tiwari, K., Shevannavar, S., Kattaavarapu, K. et al., Refolding of recombinant human granulocyte colony stimulating factor: Effect of cysteine/cystine redox system, *Indian J. Biochem. Biophys.* 49, 285–288, 2012.

301. Kawano, S., Iyaguchi, D., Okada, C. et al., Expression, purification, and refolding of an active recombinant human E-selectin lectin and EGF domains in *Escherichia coli*, *Protein J.* 32, 386–391, 2013.

302. Shu, X.Z., Liu, Y., Palumbo, F., and Prestwich, G.D., Disulfide-crosslinked hyaluronan-gelatin hydrogel films: A covalent mimic of the extracellular matrix for in vitro cell growth, *Biomaterials* 24, 3825–3334, 2003.

303. van de Wetering, P., Metters, A.T., Schoenmakers, R.G., and Hubbell, J.A., Poly(ethylene glycol) hydrogels formed by conjugate addition with controlled swelling, degradation, and release of pharmaceutically active proteins, *J. Control. Release* 102, 619–627, 2005.

304. Hudalla, G.A., Eng, T.S., and Murphy, W.L., An approach to modulate degradation and mesenchymal stem cell behavior in poly(ethylene glycol) networks, *Biomacromolecules* 9, 842–849, 2008.

305. Qiu, Y., Lim, J.J., Scott, L., Jr. et al., PEG-based hydrogels with tunable degradation characteristics to control delivery of marrow stromal cells for tendon overuse injuries, *Acta Biomater.* 7, 959–966, 2011.

306. Yu, Y. and Chau, Y., One-step "click" method for generating vinyl sulfone groups on hydroxyl-containing water-soluble polymers, *Biomacromolecules* 13, 937–942, 2012.

307. Peng, G., Wang, J., Yang, F. et al., In situ formation of biodegradable dextran-based hydrogel via Michael addition, *J. Appl. Polym. Sci.* 127, 577–584, 2013.

308. Raza, A. and Lin, C.C., The influence of matrix degradation and functionality on cell survival and morphogenesis in PEG-based hydrogels, *Macromol. Biosci.* 13, 1048–1058, 2013.

309. Elbert, D.L., Pratt, A.B., Lutoli, M.P. et al., Protein delivery from materials formed by self-selective conjugate addition reactions, *J. Control. Release* 76, 11–25, 2001.

310. Chohn S.Y., Cross, D., and Wang, C., Facile synthesis and characterization of disulfide-cross-linked hyaluronic acid hydrogels for protein delivery and cell encapsulation, *Biomacromolecules* 12, 1126–1136, 2011.

311. Birlouez-Aragon, I. and Tessier, F.J., Antioxidant vitamins and degenerative pathologies. A review of vitamin C, *J. Nutr. Health Aging* 7, 103–109, 2003.

312. Halliwell, B., Zhao, K., and Whiteman, M., The gastrointestinal tract: A major site of antioxidant action? *Free Radic. Res.* 33, 819–830, 2000.
313. Jaffey, S.R., Erdjument-Bromage, H., Ferris, C.D. et al., Protein *S*-nitrosylation: A physiological signal for neuronal nitric oxide, *Nat. Cell Biol.* 3, 193–197, 2001.
314. Giustarini, D., Dalle-Donne, I., Colombo, R. et al., Is ascorbate able to reduce disulfide bridges? A cautionary note, *Nitric Oxide* 19, 252–258, 2008.
315. Bánhegyi, G., Csala, M., Szarka, A. et al., Role of ascorbate in oxidative protein folding, *Biofactors* 17, 37–46, 2003.
316. Saaranen, M.J., Karala, A.R., Lappi, A.K., and Ruddock, L.W., The role of dehydroascorbate in disulfide bond formation, *Antioxid. Redox Signal.* 12, 15–25, 2010.
317. Du, J., Cullen, J.J., and Buettner, G.R., Ascorbic acid: Chemistry, biology and the treatment of cancer, *Biochim. Biophys. Acta* 1826, 443–457, 2012.
318. Söderling, A.S., Hultman, L., Pelbro, D. et al., Reduction of the nitro group during sample preparation may cause underestimation of the nitration level in 3-nitrotyrosine immunoblotting, *J. Chromatogr. B Analyt. Technol. Biomed. Life Sci.* 851, 277–286, 2007.
319. Balabanli, B., Kamsaki, Y., Martin, E., and Murad, F., Requirements for heme and thiols for nonenzymatic modification of nitrotyrosine, *Proc. Natl. Acad. Sci. USA* 96, 13136–13141, 1999.
320. Chen, H.J.C., Chang, C.M., and Chen, Y.M., Hemoprotein-mediated reduction of nitrated DNA bases in the presence of reducing agents, *Free Radic. Biol. Med.* 34, 254–268, 2003.

7 Chemical Modification of Cysteine

There are a variety of reagents available for the modification of cysteinyl residues in proteins (Table 7.1; Figure 7.1), which vary in their chemistry including reaction rate (Figure 7.2), solution behavior, and physical size. As such, it is important to determine the purpose of modification of cysteine prior to selection of a reagent and establishment of experimental conditions. The chemical modification of cysteine residues can proceed via a nucleophilic addition or displacement reaction with the thiolate anion as the nucleophile (Figure 7.2). The reaction with the α-ketohaloalkyl compounds such as iodoacetate is an example of a nucleophilic displacement reaction (S_N2 reaction), while the reaction of maleimide is a nucleophilic addition to an olefin (Michael reaction). Disulfide- and disulfide-like compounds such as the methylthiosulfonates react with cysteine to form mixed disulfides. The haloalkyl compounds, maleimides, and alkyl alkanethiosulfonates are also base structures for more complex molecules such as spectral probes and cross-linking agents, which are discussed elsewhere in the text. Metal ions such as mercuric can react with cysteine to form a presumably covalent derivative, while zinc ions form a less stable linkage; gold and silver are thought to form a covalent bond with cysteinyl residues. Many of the reagents used to modify cysteine residues are alkylating agents which are discussed in Chapter 2.

As an example, if the goal is to establish reactivity of a sulfhydryl group or groups in a protein, then it is most useful to consider iodoacetamide which reacts cysteine at a moderate rate which can usually be measured with some accuracy. However, it might be necessary to use a less reactive reagent such as chloroacetamide for highly reactive cysteinyl residues.[1] The acetamides are recommended instead of the acetate to minimize any microenvironmental effects based on the presence of charged groups around the target cysteine. Iodoacetate can function as an affinity label in reacting with active-site residues.[2,3] In both of these studies, iodoacetamide was ineffective and a sulfhydryl group was not modified; in the case of ribonuclease,[2] iodoacetate modified a histidine residue, while a methionine residue was modified in dehydroquinase.[3] However, the situation is complicated a bit by the hydrophobic nature of iodoacetamide (see Chapter 2).

Alkylation with N-alkylmaleimides[4–9] or haloalkyl derivatives[6,10] (see Chapter 2) is the best choice for cross-linking, bioconjugation, or attachment to a surface. The objective is a stable bond that forms in a reasonable period of time with a reagent not particularly susceptible to degradation (light should be avoided with the haloalkyl derivatives). Thioester chemistry has been useful in attaching peptides to microarray surfaces via native chemical ligation.[11,12] Alkylation is used for the attachment of PEG to an inserted cysteine residue in a therapeutic protein.[13,14] Alkyl methanethiosulfonates[15,16] and N-alkylmaleimides[17,18] are the preferred choice for the establishment of location of inserted cysteine residues in membrane proteins. Modifications obtained with the N-alkylmaleimide derivatives are more stable compared to the mixed disulfide derived obtained with the methanethiosulfonates. However, this is not critical as the modification studies with the methanethiosulfonate reagents are most often combined with mutagenesis studies that establish the site of modification.[15,16]

If the goal is to establish the presence of a functional cysteine residue in protein, then an organic mercurial (Table 7.2) may be the best choice based on speed of reaction and specificity.[17–21] The dipyridyl disulfides would also be a good choice given their reactivity over a broad pH range (Chapter 2) as would be the selenide reagents[22] described below (p. 301). Studies with the organic mercurials are rarely quantitative but are useful in a preliminary characterization of the catalytic

TABLE 7.1

Some Characteristics of Reagents Used to Modify Cysteine in Protein[a]

Reagent Type	Chemistry	Comment
Haloalkyl[b]	Alkylation; modification also reported for histidine, primary amines, carboxyl (γ-carboxyl group of glutamic acid)	Slow compared to other reagents. While there is the potential of reaction at other amino acid residues, the reaction at cysteine is fast. Reagents can be unstable at high pH
N-Alkyl maleimides	Lysine, tyrosine	More rapid than haloalkyl and somewhat more specific than haloalkyl derivatives.
DTNB (Ellman's reagent)	No other amino acids modified	Specific and fast; however, reaction is reversibly variable on modified residue characteristics.
Dithiodipyridine	No other amino acids modified	Specific and fast; can be performed at low pH (3–4).
p-Hydroxymercuribenzoate[c]	No other amino acids modified	Very fast, mostly reversible with excess thiol.
Alkylmethanethiosulfonates	No other amino acids modified	Fast and specific; reversible with excess thiol. Reagents are susceptible to hydrolysis.
Metal ions[d]	Various other amino acids.	Fast but not specific.

[a] The information in this table does not address the intrinsic reactivity of a specific cysteine residue in a protein and is a bit subjective as such is susceptible to the prejudices of the author. The reader is referred to the text and various references (Stark, G.R., Recent development in chemical modification and sequential degradation of proteins, *Adv. Protein Chem.* 24, 201–308, 1970; Brocklehurst, K., Specific covalent modification of thiols: Applications in the study of enzymes and other biomolecules, *Int. J. Biochem.* 10, 259–274, 1979; Means, G.E. and Feeney, R.E., Chemical modifications of proteins: History and applications, *Bioconjug. Chem.* 1, 2–12, 1990; Shalaby, W.S.W. and Park, K., Chemical modification of proteins and polysaccharides and its effect on enzyme-catalyzed degradation, in *Biomedical Polymers: Designed-to-Degrade Systems*, ed. S.W. Shaleby, Chapter 9, pp. 213–258, Hanson Publishers, Munich, Germany, 1994; Means, G.E., Zhang, H., and Le, M., The chemistry of protein functional groups, in *Protein: A Comprehensive Treatise*, Vol. 2, ed. G. Allen, Chapter 2, pp. 23–59, JAI Press, Stamford, CT, 1997; Stephanopoulos, N. and Francis, M.B., Choosing an effective protein bioconjugation strategy, *Nat. Chem. Biol.* 7, 876–884, 2011).

[b] This includes haloalkanes and α-haloketo compounds such as iodoacetic acid, iodoacetamide, and related derivatives.

[c] Included are related organic mercurial derivatives such as mercurinitrophenol derivatives and mercuriphenylsulfonate derivatives.

[d] Metal ions include, for example, silver, gold, mercury, and zinc. The binding to gold is important in the binding of thiols to surfaces for analytical purposes (Kick, A., Bönsch, M., Katzschner, B. et al., DNA microarrays for hybridization detection of surface plasmon resonance spectroscopy, *Bionsens. Bioelectron.* 26, 1543–1547, 2010). The binding of auric (Gold III) ions to cysteine is important for drug delivery (Glišić, B.D., Rychlewska, U., and Djuran, M.I., Reactions and structural characterization of gold (III) complexes with amino acids, peptides and proteins, *Dalton Trans.* 41, 6887–6901, 2012).

activity. The reader is directed to the early work of J. Leyden Webb[23] for a review of the earlier work on mercuric salts and organic mercurials. The reaction of organic mercurials with enzymes can be complex. The inactivation of phosphoglucose isomerase with PCMB[24] and a periplasmic dehydrogenase from *Desulfovibrio vulgaris*[25] consisted of two phases: an initial phase that could be reversed by thiols and a second phase for inactivation that was irreversible. In the latter study,[25] Fagan and Mayhew observed that the enzyme was resistant to inactivation with N-ethylmaleimide, iodoacetate, or 5,5′-dithiobis(2-nitrobenzoic acid) (DTNB). The data obtained with any single reagent have the potential to be somewhat misleading as shown by the work of Fagan and Mayhew.[25] Altamirano and colleagues[26] found that the reaction of either DTNB or methyl iodide with glucosamine-6-phosphate isomerase deaminase established the presence of two available cysteinyl residues and one cysteinyl that is buried and reacted only in the presence of denaturing agent. The enzyme modified with DTNB was inactive, while the enzyme with methyl iodide retained activity. A variety of reagents were used to define the role of cysteine in arginyl-tRNA synthase from *Escherichia coli*.[27]

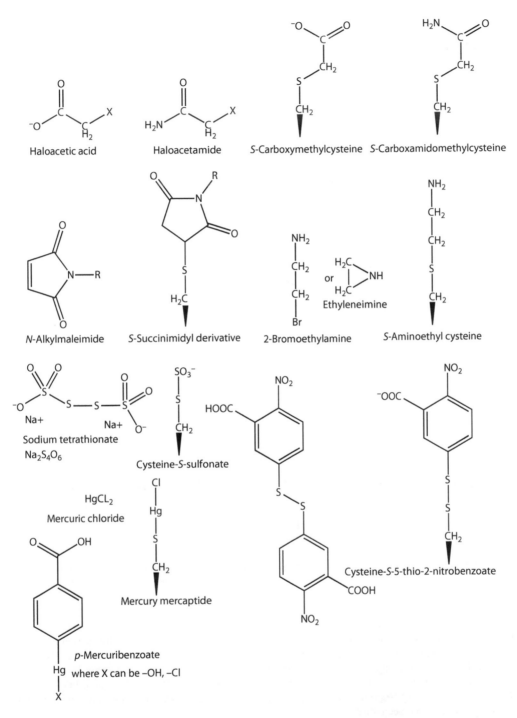

FIGURE 7.1 Examples of reagents used to modify cysteine residues in proteins. Shown are the haloalkanoic derivatives, an alkyl maleimide, and 2-bromoethylamine. These are examples of irreversible modifications. In the case of iodoacetate, a negative charge is added to the protein, and with 2-bromoethylamine, the addition of a positive charge. The product obtained with 2-bromoethylamine, S-aminoethyl cysteine, provides a site for tryptic cleavage. Modification with tetrathionate yields S-sulfocysteine, and the disulfide reagents, an example provided by DTNB can be reversed by mild thiol reagents such as dithiothreitol. Inorganic mercury compounds such as mercuric chloride tend to be irreversible, while organic mercurials such as p-mercuribenzoate can be reversed by mild thiols.

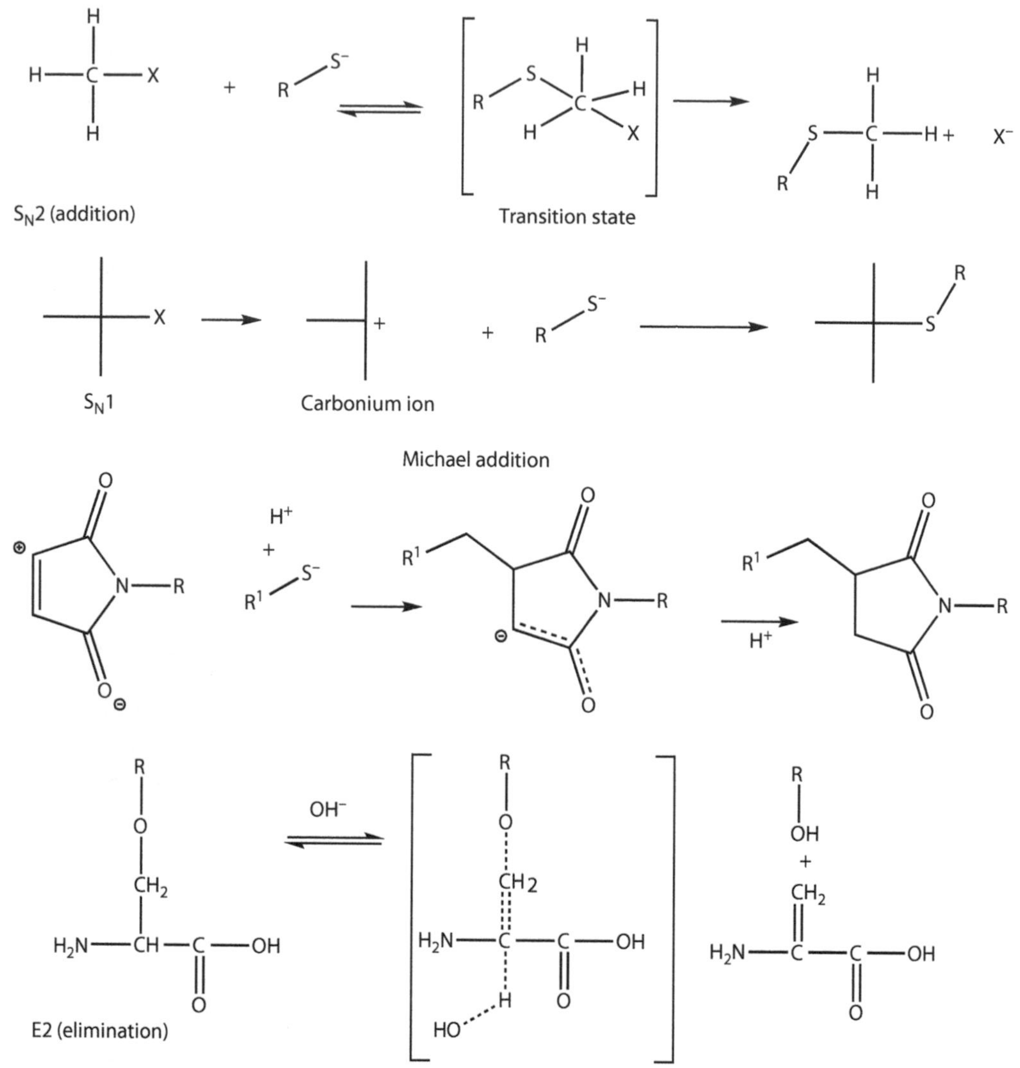

FIGURE 7.2 Organic reaction mechanisms for the modification of thiol groups in proteins. S_N1 reactions do not retain configuration, and there is racemization of product. S_N2 reaction occurs with an inversion of configuration. This is usually, but not always, not of significance with reaction of protein nucleophiles. The alkylation reactions using iodoacetate and iodoacetamide are examples of the S_N2 reaction. The Michael addition is a conjugate addition reaction where a nucleophile adds to α,β-unsaturated carbon; the modification of cysteine with N-ethylmaleimide is an example of a Michael addition. Also shown is the E2 elimination reaction that can result in the formation of β-alanine residues from serine or cysteine derivatives in proteins (Herbert, B., Hopwood, F., Oxley, D. et al., β-Elimination: An unexpected artifact in proteome analysis, *Proteomics* 3, 826–831, 2003). The reader is directed an organic chemistry text such as Roberts (Roberts, J.D. and Caserio, M.C., *Basic Principles of Organic Chemistry*, p.548, W.A. Benjamin, New York, 1964) for further information on organic reaction mechanisms.

This protein contains four cysteine residues. Analysis with DTNB suggested that two cysteine residues are located on the surface (reactive with DTNB) and two cysteine residues were buried (not reactive with DTNB). Iodoacetic acid reacted with one DTNB-sensitive residue with 50% loss of activity, while N-ethylmaleimide modified both DTNB-sensitive residues with a total loss of activity. However, replacement of the four cysteine residues with alanine via oligonucleotide-mediated mutagenesis had a minimal effect on activity. It was concluded that the loss of activity observed

TABLE 7.2

Some Rate Constants for the Reaction of Some Sulfhydryl Reagents with Model Thiol Compounds and Some Proteins[a]

Nucleophile	Reagent/Experimental Conditions	Rate (M^{-1} s^{-1})	Reference
Glutathione	PCMB in 0.02 M glycerophosphate HCl, pH 6.8; 30°C	6.2×10^6	7.2.1
2-Mercaptoethanol	PCMB in 0.02 M glycerophosphate HCl, pH 6.8; 30°C	6.5×10^6	7.2.1
2-Mercaptoethanol	PCMB in potassium phosphate-potassium chloride (ionic strength of 0.2), pH 7.63; 25°C	1.0×10^7	7.2.1
Phosphorylase B sulfhydryl groups (type 1)[b] (type 1)[b]	PCMB in 0.02 M glycerophosphate HCl, pH 6.8, 30°C	2.0×10^6	7.2.1
Glutathione	2-Mercuri-4-nitrophenol in 5 mM bis-tris (ionic strength 0.2 obtained with potassium sulfate), pH 7.0[c]	1.6×10^7 at 8°C[c] 3.0×10^7 at 25°C[c]	7.2.2
N-acetyl-L-cysteine	2-Mercuri-4-nitrophenol in 5 mM bis-tris (ionic strength 0.2 obtained with potassium sulfate), pH 6.9	1.0×10^7 at 8°C	7.2.2
2-Mercaptoethanol	Iodoacetate in 0.5 M, triethanolamine, pH 8.55	0.134	7.2.3
2-Mercaptoethanol	Iodoacetate in 0.5 M, triethanolamine, pH 7.55	0.026	7.2.3
2-Mercaptoethanol	Iodoacetate in 0.5 M, triethanolamine	2.5[d]	7.2.3
2-Mercaptoethanol	Iodoacetate in 0.5 M, triethanolamine, pH 7.40 with 1:1 zinc[d]	$11.8 \times 10^-$	7.2.3
Hamster NAT2[e]	Iodoacetamide; pH 7.0 (50 mM sodium pyrophosphate-80 mM NaCl-1% glycerol), 25°C	802.7	7.2.4
2-Mercaptoethanol	N-Ethylmaleimide (1.0 M KCl, 0.4 mM EDTA)	$1.8 \times 10^{5\,f}$	7.2.5
2-Mercaptoethanolamine	N-Ethylmaleimide (1.0 M KCl, 0.4 mM EDTA)	$6.6 \times 10^{4\,f}$	7.2.5
L-Cysteine	N-Ethylmaleimide (1.0 M KCl, 0.4 mM EDTA)	$1.8 \times 10^{5\,f}$	7.2.5
L-Cysteine	Chloroacetate	8.8×10^{-5}	7.2.6
GSH	Chloroacetamide, pH 9.0,[g] 25°	0.13[h]	7.2.7
Streptococcal proteinase	Chloroacetamide, pH 9.0,[g] 25°	13.3	7.2.7
GSH	Chloroacetic acid, pH 9.0,[g] 25°	$9 \times 10^{-3\,h}$	7.2.7
Streptococcal proteinase	Chloroacetic acid, pH 9.0,[g] 25°	0.25	7.2.7
2-Mercaptoethanol	2-Pyridylmercurial	3.5×10^3	7.2.8
L-Cysteine	Bromoacetate, 25°C, in phosphate–borate–acetate with NaCl, pH 8.0–8.4[i]	1.03	7.2.9
2-Mercaptoethanol	Bromoacetate, 25°C, in phosphate–borate–acetate with NaCl, pH 8.0–8.4[i]	1.97	7.2.9
L-Cysteine	Bromoacetamide, 25°C, in phosphate–borate–acetate with NaCl, pH 8.0–8.4[i]	5.71	7.2.9
2-Mercaptoethylamine	Bromoacetamide, 25°C, in phosphate–borate–acetate with NaCl, pH 8.0–8.4[i]	8.79	7.2.9
Hamster NAT2[e]	Bromoacetamide; pH 7.0 (50 mM sodium pyrophosphate, 80 mM NaCl, 1% glycerol), 25°C	427	7.2.4
Hamster NAT2[e]	Bromoacetanilide, pH 7.0 (50 mM sodium pyrophosphate, 1.0 mM DTT), 23°C	5.2	7.2.10
YADH	Bromoacetamide, pH 7.9 (50 mM sodium pyrophosphate), 25°C	0.37	7.2.11
YADH	Bromoacetamide, pH 7.9 (50 mM sodium pyrophosphate), 25°C, with 10 mM ADP	0.3	7.2.11
YADH	N^1-(2-bromoacetamidoethy)nicotinamide, pH 7.9 (50 mM sodium pyrophosphate), 25°C	0.1	7.2.11

(*continued*)

TABLE 7.2 (continued)

Some Rate Constants for the Reaction of Some Sulfhydryl Reagents with Model Thiol Compounds and Some Proteins[a]

Nucleophile	Reagent/Experimental Conditions	Rate ($M^{-1} s^{-1}$)	Reference
YADH	Bromoacetamide, pH 7.9 (50 mM sodium pyrophosphate), 25°C, with 10 mM ADP	0.45	7.2.11
Pyroglutamyl aminopeptidase	N^{α}-Carbobenzoxy-L-pyroglutamyl chloromethyl ketone, pH 7.8 (0.1 M Tris), 30°C	1.1×10^5	7.2.12
2-Mercaptoethanol	Methyl methanethiosulfonate	$3.0 \times 10^{6\,j}$	7.2.13
2-Mercaptoethylamine	5,5′-Dithiobis-(2-nitrobenzoate). pH 6.65[k]	4.6×10^3	7.2.14
2-Mercaptoethanol	5,5′-Dithiobis-(2-nitrobenzoate). pH 6.65[k]	2.20×10^2	7.2.14
Cysteine	5,5′-Dithiobis-(2-nitrobenzoate). pH 6.65[k]	9.40×10^2	7.2.14
Glutathione	5,5′-Dithiobis-(2-nitrobenzoate). pH 6.65[k]	2.70×10^2	7.2.14
2-Mercaptoethanol	Chlorodinitrobenzene (30°C, pH 8.0)	0.08	7.2.15
Cysteine	Chlorodinitrobenzene (30°C, pH 8.0)	0.2	7.2.15
Phosphorylase *b*	Chlorodinitrobenzene (30°C, pH 8.0)	1	7.2.15
2-Mercaptoethanol	DTNB, pH 6.0 (0.05 M phosphate, pH 6.0, 15°C)	30	7.2.16
2-Mercaptoethanol	5,5-Dithiobis-(2-nitro-N-trimethylbenzylammonium bromide), pH 6.0 (0.05 M phosphate, pH 6.0, 15°C)	7×10^2	7.2.16
2-Mercaptoethanol	5,5′-Dithiobis-(2-nitro-N-2′-hydroxyethyl benzamide), pH 6.0 (0.05 M phosphate, pH 6.0, 15°C)	3×10^2	7.2.16
Disulfiram[l]	Caspase-1, pH 7.5, assume 23°C	2.2×10^3	7.2.17
Disulfiram[l]	Caspase-3, pH 7.5, assume 23°C	4.5×10^2	7.2.17
Disulfiram[l]	Human arylamine NAT 1, 25 mM Tris, pH 7.5, 37°C[m]	6×10^3	7.2.18

[a] Also see Dahl, K.H. and McKinley-McKee, J.S., The reactivity of affinity labels: A kinetic study of the reactions of alkyl halides with thiolate anions—A model for protein alkylation, *Bioorg. Chem.* 10, 329–341, 1981; Schöneich, C., Kinetics of thiol reactions, *Methods Enzymol.* 251, 45–55, 1995.

[b] There are several classes of sulfhydryl groups in phosphorylase *b* that can be classified on the basis of reactivity with sulfhydryl reagents (Damjanovich, S. and Kleppe, K., The reactivity of SH groups in phosphorylase *b*, *Biochim. Biophys. Acta* 122, 145–147, 1966; Kleppe, K. and Damjanovich, S., Studies on the SH groups of phosphorylase *b*. Reaction with 5,5′-dithiobis-(2-nitrobenzoic acid), *Biochim. Biophys. Acta* 185, 88–102, 1969).

[c] At pH 9.0, the rate approaches 10^9 M^{-1} s^{-1}, which is considered diffusion-limited (Pinto, A.F., Rodriguez, J.V., and Teixeiro, M., Reductive elimination of superoxide: Structure and mechanism of superoxide reductases, *Biochim. Biophys. Acta* 1804, 285–297, 2010). The reaction was performed at 8°C in order to obtain useful data.

[c] The reactions were performed at 8°C as reaction at 25°C were to rapid to permit accurate measurement.

[d] The rate of reaction of iodoacetate with the thiolate form is some 200-fold greater than reaction with the zinc–thiol complex (Dahl, K.H. and McKinley-McKee, J.S., Enzymatic catalysis in the affinity labeling of liver alcohol dehydrogenase with haloacids, *Eur. J. Biochem.* 118, 507–513, 1981).

[e] Hamster NAT2; hamster NAT 2.

[f] pH-independent rate constant (see Jones, J.G., Otieno, S., Barnard, E.A., and Bhargava, A.K., Essential and nonessential thiols of yeast hexokinase reactions with iodoacetate and iodoacetamide, *Biochemistry* 14, 2376–2403, 1975).

[g] A wide range buffer consisting of 0.0001 M EDTA, 0.05 M boric acid, 0.05 M sodium phosphate, and 0.05 M sodium acid with NaCl to a final ionic strength of 0.19.

[h] Estimated from graphical data.

[i] The buffer is described as a wide range buffer, 16 mM each in phosphate, borate, and acetate with NaCl added to maintain ionic strength at 0.084 M.

[j] Data obtained at pH 4.0, 5.0, and 5.65 and corrected to reflect the amount of thiolate anion present. It has been observed that the rate of reaction of alkyl methanethiosulfonates (alkyl thiosulfonates) with sulfhydryl groups is rapid (Wynn, R. and Richards, F.M., Chemical modification of protein thiols: Formation of mixed disulfides, *Methods Enzymol.* 251, 351–356, 1995).

[k] Temperature not provided, assume 25°C.

TABLE 7.2 (continued)
Some Rate Constants for the Reaction of Some Sulfhydryl Reagents with Model Thiol Compounds and Some Proteins[a]

l Disulfiram, *bis*-(diethylthiocarbamoyl)disulfide.

m The inactivation of NAT1 was not reversed by excess thiols leading the authors to question whether the mechanism involved formation of a mixed disulfide. However, other data suggested that the inactivation did involve the reaction of the active-site sulfhydryl group. There are examples of the irreversible formation of mixed disulfides with peptide and proteins with glutathione (Thing, M., Zhang, J., Lawrence, J., and Topp, E.M., Thiol-disulfide interchange in the tocinoic acid/glutathione system during freezing and drying, *J. Pharm. Sci.* 99, 4949–4956, 2010; Cooper, A.J., Pinto, J.T., and Callery, P.S., Reversible and irreversible protein glutathionylation: Biological and clinical aspects, *Expert Opin. Drug. Metab.* 7, 891–910, 2011).

References to Table 7.2

7.2.1. Hasinoff, B.B., Madsen, N.B., and Avramovic-Zikic, O., Kinetics of the reaction of *p*-chloromercuribenzoate with the sulfhydryl groups of glutathione, 2-mercaptoethanol, and phosphorylase *b*, *Can. J. Biochem.* 49, 742–751, 1971.

7.2.2. Sanyal, G. and Khalifah, R.G., Kinetics of organomercurial reactions with model thiols: Sensitivity to exchange of the mercurial labile ligand, *Arch. Biochem. Biophys.* 196, 157–164, 1979.

7.2.3. Dahl, K.H. and McKinley-McKee, J.S., Enzymatic catalysis in the affinity labelling of liver alcohol dehydrogenase with haloacetate, *Eur. J. Biochem.* 118, 507–513, 1981.

7.2.4. Wang, H., Vath, G.M., Gleason, K.J. et al., Probing the mechanism of hamster arylamine *N*-acetyltransferase 2 Acetylation by active site modification, site-directed mutagenesis, and pre-steady state and steady-state kinetic studies, *Biochemistry* 43, 8234–8246, 2004.

7.2.5. Bednar, R.A., Reactivity and pH dependence of thiol conjugation to *N*-ethylmaleimide—Detection of a conformational changes in chalcone isomerase, *Biochemistry* 29, 3684–3690, 1990.

7.2.6. Sluyterman, L.A.Æ., The rate-limiting reaction in papain action as derived from the reaction of the enzyme with chloroacetic acid, *Biochim. Biophys. Acta* 151, 178–187, 1968.

7.2.7. Gerwin, B.L., Properties of a single sulfhydryl group of streptococcal proteinase. A comparison of the rate of alkylation by chloroacetic acid and chloroacetamide, *J. Biol. Chem.* 242, 451–456, 1967.

7.2.8. Baines, B.S. and Brocklehurst, K., A thiol-labelling reagent and reactivity probe containing electrophilic mercury and a chromophoric leaving group, *Biochem. J.* 179, 701–704, 1979.

7.2.9. Wandinger, A. and Creighton, D.J., Solvent isotope effects on the rates of alkylation of thiolamine models of papain, *FEBS Lett.* 116, 116–121, 1980.

7.2.10. Wang, H., Guo, Z., Vath, G.M. et al., Chemical modification of hamster arylamine *N*-acetyltransferase 2 with isozyme-selective and nonselective *N*-arylbromoacetamido reagents, *Protein J.* 23, 153–166, 2004.

7.2.11. Plapp, B.V., Woenckhaus, C., and Pfleiderer, G., Evaluation of N1-(ω-bromoacetamidoalkyl)nicotinamides as inhibitors of dehydrogenases, *Arch. Biochem. Biophys.* 128, 360–368, 1968.

7.2.12. Fujiwara, K., Matsumoto, E., Kitagawa, T., and Tsuru, D., Inactivation of pyroglutamyl aminopeptidase by N^α-carbobenzoxy-L-pyroglutamyl chloromethyl ketone, *J. Biochem.* 90, 433–437, 1981.

7.2.13. Lewis, S.D., Misra, D.C., and Shafer, J.A., Determination of interactive thiol ionizations in bovine serum albumin, glutathione, and other thiols by potentiometric difference titration, *Biochemistry* 19, 6129–6137, 1980.

7.2.14. Zhang, H., Le, M., and Means, G.E., A kinetic approach to characterize the electrostatic environments of thiol groups in proteins, *Bioorg. Chem.* 26, 356–364, 1998.

7.2.15. Gold, A.M., Sulfhydryl groups of rabbit muscle glycogen phosphorylase *b*. Reaction with dinitrophenylating agents, *Biochemistry* 7, 2106–2115, 1968.

7.2.16. Legler, G., 4,4′-Dinitrophenyldisulfides of different charge types as probes for the electrostatic environment of sulfhydryl groups, *Biochim. Biophys. Acta* 405, 136–145, 1976.

7.2.17. Nobel, C.S.I., Kimland, M., Nicholson, D.W. et al., Disulfiram is a potent inhibitor of proteases of the caspase family, *Chem. Res. Toxicol.* 10, 1319–1324, 1997.

7.2.18. Malka, F., Dairou, J., Ragunathan, N. et al., Mechanisms and kinetics of human arylamine *N*-acetyltransferase 1 inhibition by disulfiram, *FEBS J.* 276, 4900–4908, 2009.

with chemical modification was a result of steric hindrance rather than the loss of a specific chemical functional group.

It is useful to consider the merits of irreversible modification versus reversible modification of cysteine residues in the selection of the reagent. Reaction with the haloalkyl reagents or the N-alkylmaleimides is an irreversible modification. Modification with the alkyl methanethiosulfonates, DTNB, tetrathionate, and the dithiodipyridyl reagents can usually be reversible by a mild reducing agent such as 2-mercaptoethanol or dithiothreitol. Modification with organic mercurials can usually be reversed with reducing agents while modification with mercuric ions is more problematic, which may be a reflection of the interaction of mercuric ions with sites other than cysteine in proteins; the interaction of mercuric ions with two sulfhydryl groups is very stable and resistant to reversal.

Cysteine is the sulfur analog of serine where the hydroxyl group is replaced with a sulfhydryl group. The bond dissociation energy for sulfhydryl groups is substantially less than that of the corresponding alcohol function providing a basis for the increased acidity of sulfhydryl groups; the pKa for ethanethiol is approximately 10.6 while the pKa for ethanol is approximately 18.[28,29] Hederos and Baltzer[30] studied nucleophilic selectivity in an acyl transfer reaction (acyl transfer of S-glutathionyl benzoate catalyzed by an engineered human glutathione transferase A1-1); ethanethiol was far more effective than ethanol while ethylamine was ineffective in stimulating the acyl transfer reaction; it is noted that ethylamine would be fully protonated at the pH of the reaction (pH 7.0). Given this information, it is not surprising that while the reaction of the cysteine sulfhydryl group with haloacetates or haloacetamides such as chloroacetate or chloroacetamide is slow, reaction with the serine hydroxyl group is essentially nonexistent. However, as with serine, the presence of cysteine at the enzyme active site such as in papain greatly influences reactivity. The reaction of chloroacetic acid with a cysteine residue at the active site of papain is some 30,000 times faster (150 M^{-1} min^{-1}) than that of free cysteine (5.3×10^{-3} M^{-1} min^{-1}) at pH 6.0.[31] As a comparison, the rate of reaction of the cyanate with the cysteine residue at the active site of papain is 9.4×10^3 M^{-1} min^{-1} compared to the rate of reaction of cysteine (3.4 M^{-1} min^{-1});[32] thus, the rate of reaction of cyanate with the cysteine at the enzyme active site is some 3000-fold more rapid than reaction with free cysteine. Stark and coworkers[33] found a rate of reaction of 4.0 M^{-1} min^{-1} for the reaction of cyanate with free cysteine. Unlike the reaction of chloroacetic acid with cysteine, the reaction of cyanate with the active-site cysteine of papain is reversible.[31] The reader is referred to an excellent review by Janssen[34] that addresses the nucleophilicity of organic sulfur where he notes that thiolates are less basic than alkoxide anions but in general are better nucleophiles. A solvent does have an effect on the relative nucleophilicity of sulfur and oxygen compounds.[35] This is shown by changes in the reactivity of haloacids depending on the solvent as discussed below. The reader is directed to several discussions on the effect of solvent on organic chemical reactions;[36–38] there is also a discussion of solvent effects on reactions in Chapter 1 with emphasis on the effect of solvent on transition states; too often, there is tendency to forget that proteins (and other biological polymers) are complex organic polymers, albeit with unique patterns of reactivity. The chemistry of cysteine has been reviewed[39–42], and the reader is directed to these articles for more information.

There are two other issues that I wish to discuss before entering to a consideration of the various reagents used to modify cysteinyl residues in proteins. The first is the chemistry of thioesters. The second concerns two modification reactions, oxidation and nitrosylation, which are of considerable physiological importance but of lesser importance in solution protein chemistry.

Thioesters (Figure 7.3) are more reactive than oxyesters and, as is the case of biological thioesters such as acetyl CoA, are high-energy compounds.[43] The increased electrophilic nature of the carbonyl carbon in a thioester makes it far more susceptible to nucleophilic attack (see Chapter 3) such as that seen in the thiolactone structure at the metastable binding sites in α_2-macroglobulin and the C3 and C4 components of complement.[44,45] Thioester intermediates are also involved in the process of ubiquitin-like modifications.[46] Homocysteine thiolactone can modify several different functional groups in proteins including lysine[47] resulting in pathologies.[48,49] Betaine thioester can be used to modify proteins to improve solubility.[50] This later study also emphasizes the importance

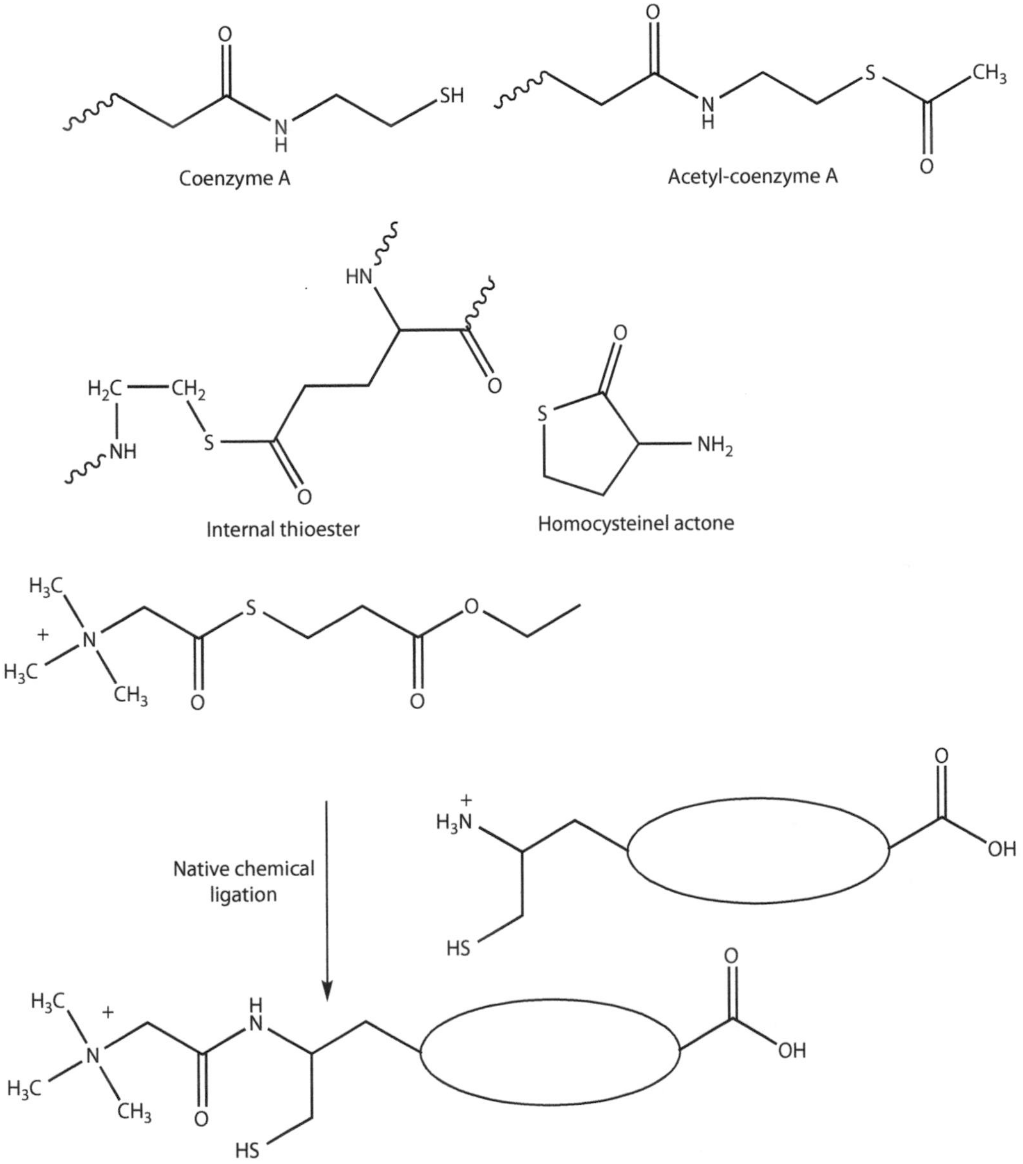

FIGURE 7.3 Examples of thioesters important in protein chemistry. CoA is the basic for a number of biologically active thioesters including acetyl CoA that is involved in a large number of biological processes. Not shown is palmitoyl CoA (Linder, M.E. and Deschenes, R.J., Palmitoylation: Policing protein stability and traffic, *Nat. Rev. Cell. Mol. Biol.* 8, 74–84, 2007), which is important in membrane protein function (Wei, Y., Di Vizio, D.K., Steeen, H., and Freeman, M.R., Peptide and protein thioester synthesis by *N,S* acyl transfer, *Mol. Cell. Proteomics* 9, 54–70, 2010). Homocysteine lactone is an internal thioester that has the potential of modifying different functional groups in proteins (Jop, E.C.A. and Bakhtiar, R., Homocysteine thiolactone and protein homocysteinylation: Mechanistic studies with model peptides and proteins, *Rapid Commun. Mass Spectrom.* 16, 1049–1053, 2002). Betaine thioester is used to increase the solubility of proteins via coupling to the amino terminal via native chemical ligation (Xiao, J., Burn, A., and Tolbert, T.J., Increasing solubility of proteins and peptides by site-specific modification with betaine, *Bioconj. Chem.* 19, 1113–1118, 2008).

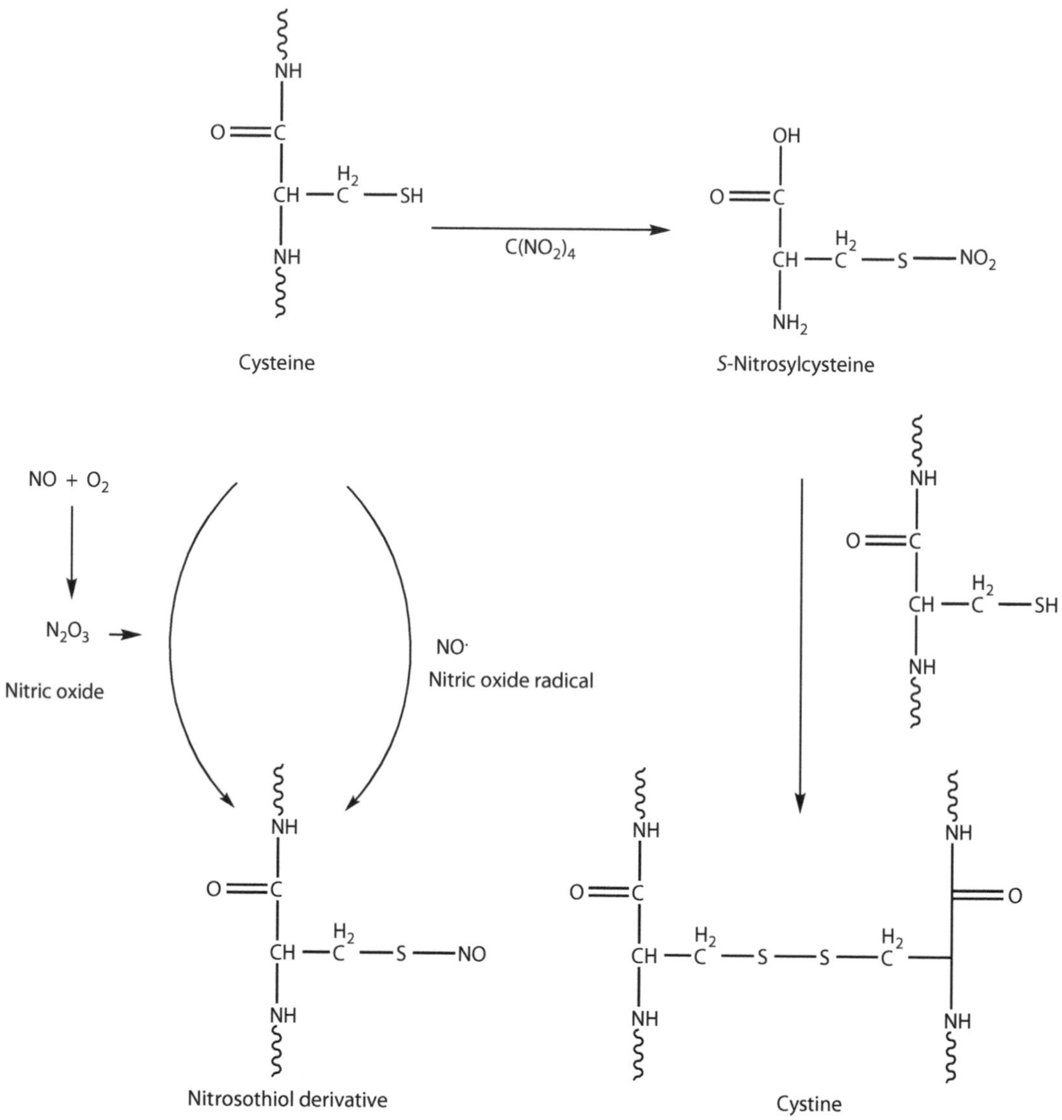

FIGURE 7.4 The oxidation or nitrosylation of thiols in biological polymers. Active nitrogen compounds such as nitric oxide and tetranitromethane. Cysteinyl residues in proteins can be modified with tetranitromethane as well as by reactive nitrogen species (Hess, D.T., Matsumoto, A., Kim, S.O. et al., Protein S-nitrosylation: Purview and parameters, *Nat. Rev. Mol. Cell. Biol.* 6, 150–166, 2005; Torta, F., Usuelli, V., Malgaroli, A., and Bachi, A., Proteomic analysis of protein S-nitrosylation, *Proteomics* 8, 4484–4494, 2008; Nagahara, N., Matsumura, T., Okamoto, R., and Kajihara, Y., Protein cysteine modifications: (1) Medical chemistry for proteomics, *Curr. Med. Chem.* 16, 4419–4444, 2009).

of thioesters in native chemical ligation.[51] There is evidence to show that the cysteinyl residue that is involved in the transesterification step in the formation of the thioester intermediate in protein splicing has a low pKa (5.8).[45,52] It is noted that a reactive histidine has a role in enhancing the nucleophilicity of the cysteine residue in protein splicing.[53]

Cysteine residues are also subject to oxidation (Figure 7.4) (modification by reactive oxygen species) and nitrosylation, while reversible modification of cysteine by oxidation or nitrosylation is considered an important process in redox regulation;[54–63] such reactions are rarely used for the chemical modification of cysteine. The review by Spadaro and coworkers[61] is of considerable interest as they argue that the specificity of modification by oxidation or nitrosylation is driven by the reactivity of the cysteine residues; cysteine residues with low pKa values are targets for modification by oxidation or nitrosylation with the reversible formation of nitrosocysteine and cysteine sulfenic acid providing the basis for specific regulation. Sulfinic acid may also have a regulatory role as there are enzymes for the reduction of this derivative. Chapter 8 will discusss the general chemistry of cystine and reaction of cystine with various chemical reagents.

The reaction of cysteine residues with chemical reagents depends on two major factors, physical accessibility and local electrostatic environment, and can be classified as freely reacting, sluggish, and masked.[64–70] Ueland and coworkers[64] characterized the reactivity of the 24 sulfhydryl groups in an adenosine 3′,5′-monophosphate-adenosine binding protein from mouse liver. It was demonstrated that four sulfhydryl groups reacted with DTNB rapidly (in s), four sulfhydryl groups reacted less rapidly, and 16 sulfhydryl groups reacted quite slowly and their reactivity might reflect protein denaturation. Floris and coworkers[68] studied the reactivity of sulfhydryl groups in diamine oxidase with 4,4-dithiopyridine. The two sulfhydryl groups per monomer unit are unreactive in the native enzyme but can be modified by 4,4-dithiopyridine after denaturation in 8 M urea or in the absence of oxygen and in the presence of substrate. Polyanovsky and colleagues[69] identified the exposed and buried cysteine residues (five total residues) in aspartate aminotransferase based on reaction with iodoacetate and isolation of peptides containing the modified cysteine residues. Two cysteine residues are exposed and react with iodoacetate without denaturation. The remaining three residues are partially buried (sluggish), buried, and the last buried to the extent that it is modified only with great difficulty in the presence of chaotropic agents. Fuchs and coworkers[70] studied the reactivity of cysteine in bovine cardiac troponin C using DTNB. There are two cysteine residues in bovine cardiac troponin C. Modification with DTNB occurs more rapidly (3.37 M^{-1} s^{-1}; $t_{1/2}$ 25 min) in the presence of calcium ions than in the absence of calcium ions (1.82 M^{-1} s^{-1}), while reaction is complete within 2 min in the presence of 6 M urea. Lee and Blaber[71] report that the solution stability of fibroblast growth factor is improved by mutagenesis of buried cysteine residues. Subsequent work[72] from this laboratory suggested that while this residue is conserved in the fibroblast growth factor family, the residue is vestigial. Cysteine can be considered to be a hydrophobic amino acid[73–78] and can be buried without major thermodynamic consequences.[74] Nagano and coworkers[73] examined the distribution of cysteine in a number of proteins and concluded that cysteine behaved as hydrophobic amino acid residue. Mima and coworkers[77] studied mutant forms for carboxypeptidase Y where Cys341 was replaced with a variety of amino acids and concluded that this residue that constitutes part of the S1 binding pocket had amphipathic properties that enhanced catalysis. While crystallography or one of the more advanced solution chemical techniques such as NMR, light scattering, and analytical ultracentrifugation has replaced cysteine exposure as a measure of protein conformational change, the accessibility of cysteinyl residues to chemical modification is of continuing interest.[79–81] Notwithstanding the lack of extensive crystallographic data supporting cysteine exposure, cysteine residues have been engineered into proteins to study conformation using chemical reactivity to measure accessiblity.[82–87] Akabas and coworkers[82] used the term water accessible to describe the substituted cysteine accessibility method (SCAM); this is described in more detail below.

The early work on the appearance of sulfhydryl groups on denaturation of proteins was summarized by Greenstein in 1939[88] who also noted the importance of the counterion in the use of guanidinium salts for protein denaturation. The exposure of cysteine occurring during protein denaturation was observed some 100 years ago but not associated with denaturation until some years later,[89] although it is clear that understanding of protein denaturation was an evolving art at that time.[90] What is clear is that early investigators considered native proteins devoid of sulfhydryl groups[91] and composition was usually reported as cystine.[92] These early studies used reaction with porphyrindin or nitroprusside (Figure 7.5) to detect sulfhydryl groups. Egg albumin, for example, did not react with porphyrindin or nitroprusside within the native state but did react after denaturation or digestion with pepsin. Anson in 1940[93] reported that cysteine in native egg albumin, while unreactive with nitroprusside or porphyrindin, could be modified with iodine or iodoacetamide.

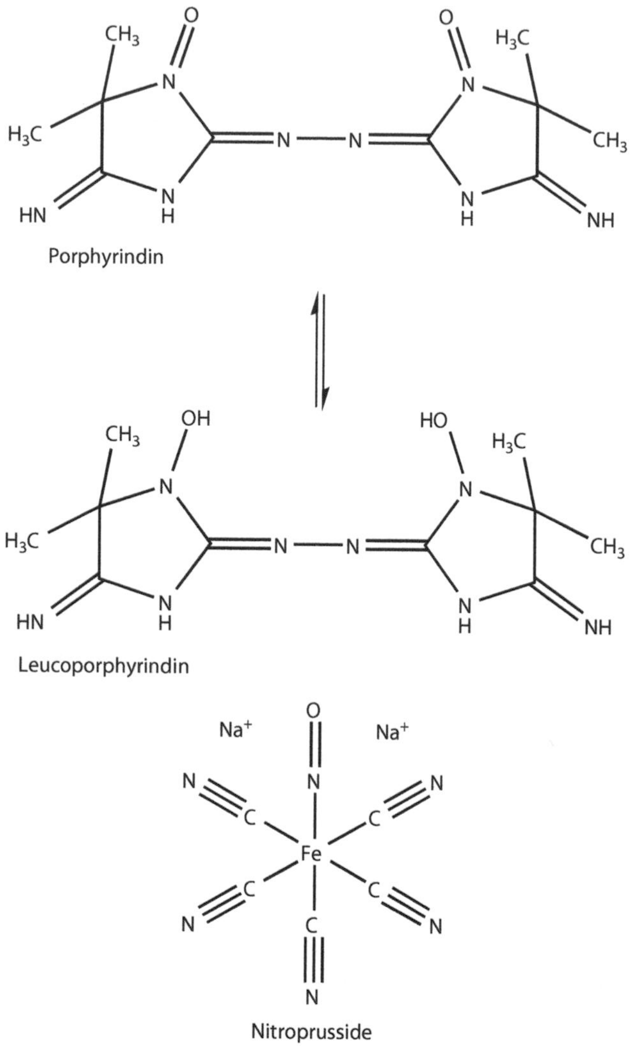

Porphyrindin

Leucoporphyrindin

Nitroprusside

FIGURE 7.5 The structures of porphyrindin and nitroprusside. These two reagents were used for the determination of sulfhydryl groups. See Greenstein, J.P. and Jenrette, W.V., Reactivity of porphyrindin in the presence of denatured proteins, *J. Biol. Chem* 142, 175–180, 1942; Christen, P. and Gasser, A., Visual detection of fructose-1,6-diphosphate aldolase after electrophoresis by its oxidative paracatalytic reaction, *Anal. Biochem.* 109, 270–272, 1980 for more information on porphyrindin and Johnson, M.D. and Wilkins, R.G., Kinetics of the primary interaction of pentacyanonitrosylferrate (2-) (nitroprusside) with aliphatic thiols, *Inorg. Chem.* 23, 231–235, 1984 for nitroprusside.

The reaction with iodine was performed as acid pH to prevent reaction with tyrosine. In a subsequent publication,[94] Anson demonstrated that *p*-mercuribenzoate reacted with denatured guanidine hydrochloride, but not native egg albumin. Bull[95] reviewed the concept of buried and exposed sulfhydryl groups in 1941. Nitroprusside was used early for the determination of cysteine and identification[96] and determination of glutathione.[97] More recently, nitroprusside has been used as donor for nitric oxide[98–101] and is used as a drug.[102–105]

While there are exceptions, the reactivity of cysteine in a native protein increases on denaturation (see above) and is lower in a native protein than with model sulfhydryl compounds. In addition to physical accessibility, local electrostatic environment (see Chapter 1) has a profound effect on the reactivity of cysteine residues in proteins. Rat brain tubulin dimer contains 20 cysteine residues: 12 residues in the α-subunit and eight in the β-subunit. Britto and colleagues[106] reported that 6–7 cysteines were designated as fast reacting and could not be distinguished with various reagents that included *syn*-monobromobimane, *N*-ethylmaleimide, iodoacetamide, and 5-((((2-iodoacetyl)amino) ethyl) amino) naphthalene-1-sulfonic acid AEDANS. Local electrostatic potential is suggested to be responsible for the reactivity of the cysteine residues; reactive cysteinyl residues are close to arginine and/or lysine residues while unreactive cysteinyl residues are associated with negatively charged local environments. There was some differentiation of reactivity based on the nature of the reagents; reaction is slower with iodoacetamide than with *N*-ethylmaleimide; a greater number of cysteine residues are modified with *N*-ethylmaleimide than with iodoacetamide, and the difference in the rates of reaction is ascribed to the differences in the chemistry of the reaction of the two compounds with the thiolate ion with the reaction with iodoacetamide being a nucleophilic displacement while the reaction with *N*-ethylmaleimide is an addition reaction. Britto and coworkers[106] were able to identify a single highly reactive cysteinyl residue using chloroacetamide. Parente and colleagues[107] suggested that the increased reactivity of cysteine residues in seminal ribonuclease at neutral pH was dependent on the proximity of a positively charged lysine residue. Heitmann[108] provided insight into the importance of local environmental factors on the reactivity of cysteine residues. The *N*-dodecanoyl derivative of cysteine was prepared and inserted into micelles and modified with either chloroacetamide, iodoacetamide, or *p*-nitrophenyl acetate. Incorporation into a cationic micelle (hexadecyltrimethylammonium bromide) increased the rate of reaction with any of the three reagents with an apparent decrease in the pKa of the cysteine derivative suggesting the importance of an adjacent positively charged residue on enhancing cysteine reactivity. The incorporation of dodecanoyl cysteine into micelles of *N*-dodecanoyl glycine resulted in the inhibition of the rate of reaction, while a neutral micelle (Brij 35) has little effect. This is not due to charge repulsion as neutral reagents, chloroacetamide, iodoacetamide, and *p*-nitrophenyl acetate, were used in this study. Gitler and coworkers[109] reviewed the enhancement of the reactivity of cysteine residues by cationic detergent such as cetyltrimethylammonium bromide. Lutolf and coworkers[110] studied the rate of reaction of pentapeptide thiols with PEG diacrylates. The pentapeptides contained arginine residues or aspartic acid residues or one aspartic residue and one arginine residues such that the peptides had net charges at pH 7.4 of −2, −1, 0, +1, or +2; there was variance in the placement of the charged amino acids. Net charge was the dominant factor with positively charged peptides reacting more rapidly than neutral or negatively charged peptides; the pKa of the cysteine was lowest in the doubly charged peptide (pKa = 8.12) and highest in the doubly negatively charged peptide (pKa = 8.93), while the pKa was 8.50 in the neutral peptide. Snyder and coworkers[111] studied the reaction of DTNB (Ellman's reagent, Table 7.12) with cysteine in peptides which contained either two positively charged neighboring amino acid residues, one positively charged residue, and a neutral residue. The rate observed with the cysteine having two positively charged residues was 132,000 $M^{-1} s^{-1}$, while the rate with only neutral residues was 367 $M^{-1} s^{-1}$. The presence of an attracting group may result in saturation kinetics for the reaction[2,3] where the reagent, in these studies iodoacetate, is behaving like an affinity reagent.

Organic mercurials (Figure 7.6) are highly specific for the modification of cysteine sulfhydryl groups in proteins (Table 7.1). Organic mercurials can be very useful for the specific modification of sulfhydryl groups in proteins as the reaction of organic mercurials with exposed sulfhydryl groups is rapid. While not as rapid as mercuric ion, the rate is still faster than most other reagents used to modify sulfhydryl groups in proteins (Table 7.2). The reaction product of organic mercurials with sulfhydryl groups is stable but may be reversed by the addition of reducing agents such as 2-mercaptoethanol or dithiothreitol; however, the reaction of organic mercurials with cysteinyl residues in proteins may be complex consisting of two phases, the second of which may be irreversible. In the absence of such a reducing agent, the binding is quite tight but not as tight as mercuric ions.[112] Methylmercury is produced in the environment from mercury,[113] and while a major factor in the toxicity of mercury,[114] methylmercury is used only occasionally for the modification of cysteine in proteins. Keller and coworkers[115] compared the effect of methylmercury chloride, PCMB, phenylmercury acetate, and methyl

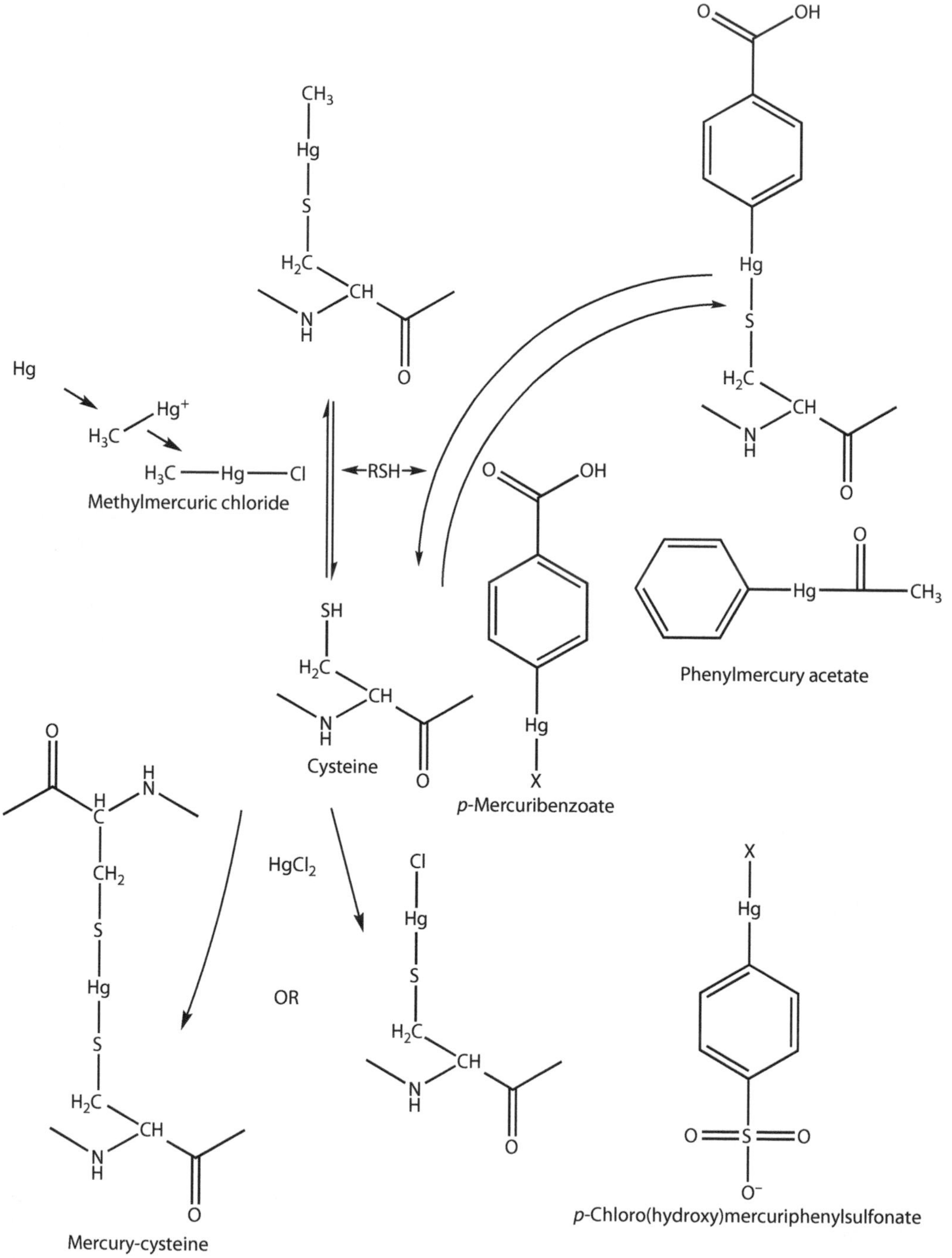

FIGURE 7.6 Organomercurial compounds. The structures and reactions of some organomercurial compounds. Shown are the structures and some reactions of organomercurial compounds. Shown also is the genesis of methylmercury. The reactions of most organic mercurial compounds with cysteine can be reversed by the addition of thiols such as 2-mercaptoethanol or dithiothreitol.

methanethiosulfonate on acetylcholine and vesamicol binding sies in the vesicular acetylcholine transporter of *Torpedo californica*. Modification with either PCMB or phenylmercury acetate reduced the binding affinity of vesamicol by 10^3, while only a modest sevenfold reduction was observed with either methylmercury chloride or methyl methanethiosulfonate. Modification with methylmercury chloride but not methyl methanethiosulfonate eliminated the binding of phenyl-mercury acetate. These investigators concluded that there are two classes of cysteine sulfhydryl groups within the vesicular acetylcholine receptor, one that reacts only with organic mercurials and second class that reacts with both mercurials and thiosulfonates. Subsequent work from this group[116] showed that modification with *p*-chloromercuriphenylsulfonate (Figure 7.7) was slower

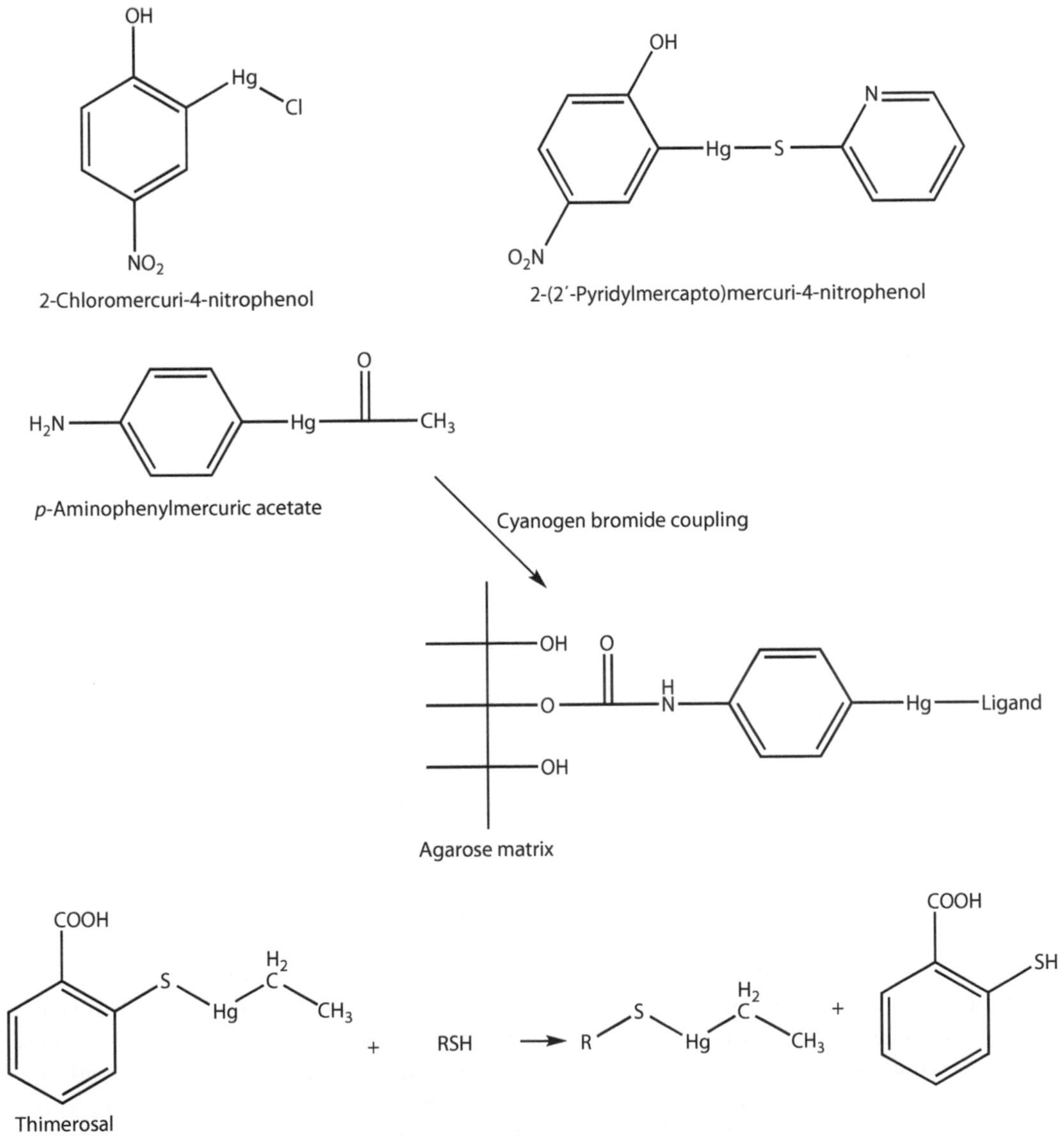

2-Chloromercuri-4-nitrophenol

2-(2′-Pyridylmercapto)mercuri-4-nitrophenol

p-Aminophenylmercuric acetate

Cyanogen bromide coupling

Agarose matrix

Thimerosal

FIGURE 7.7 Various mercury-containing reagents for the modification of cysteine sulfhydryl in protein. Shown are several chromogenic derivatives as well as the aminophenylmercurial derivative that can be coupled to an activated agarose matrix for use in affinity chromatography. Thimerosal is a compound used for the stabilization of vaccines (Wu, X., Liang, H., O'Hara, K.A. et al., Thiol-modulated mechanisms of the cytotoxicity of thimerosal and inhibition of DNA topoisomerase IIα, *Chem. Res. Toxicol.* 21, 483–493, 2008).

(0.11 min^{-1}) than with phenlymercury chloride (1.36 min^{-1}). p-Chloromercuriphenylsulfonate was used by this group earlier as a membrane-impermeant reagent.[117] p-Chloromercuriphenyl-sulfonate had been synthesized for Vellick[118] as a more soluble organic mercurial for his work on glyceraldehyde-3-phosphate dehydrogenase.

Chicken ovalbumin, which contains four sulfhydryl groups and one disulfide bond for a total of six cysteine residues, has been used as a model protein for the past century. The early use of cysteine exposure to measure protein denaturation with ovalbumin has been mentioned earlier, and cysteine exposure continues to be of use in this area.[119] These two studies used DTNB to measure cysteinyl sulfhydryl groups. The results obtained for the reaction of the sulfhydryl groups in ovalbumin with the various chemical reagents are instructive in considering the availability of cysteinyl residues in a protein for modification. Diez and coworkers[120] showed that native chicken ovalbumin had three masked sulfhydryl groups. The reaction of native ovalbumin with DTNB at pH 8.0 did not detect any sulfhydryl groups, while less than one sulfhydryl group was detected by reaction with N-ethylmaleimide (pH 7.0); both reagents detected approximately four (3.8) sulfhydryl groups in the denatured (0.45% SDS) proteins. Results by the investigators obtained with PCMB and iodine detected 2–3 sulfhydryl groups in the native protein and approximately 4 in the denatured protein. Reaction with PCMB required 30 min to obtain completion at pH 4.7, while almost 60 min is required for reaction completion at pH 7.0. The reaction of iodine with the native protein is almost complete after 3 min. Webster and Thompson[121] observed incomplete reaction of ovalbumin cysteinyl residues with iodoacetic acid in alkaline 8 M urea, but if denaturation was accomplished in acid with 8 M urea and then taken to alkaline pH, complete modification of cysteinyl residues could be accomplished. The reaction of DTNB with sulfhydryl groups in ovalbumin was more rapid in 6.0 M guanidine or 1% (w/v) SDS at pH 8.2 than in 8. M urea. Specific labeling of the cysteinyl residues participating in the disulfide bond of ovalbumin was accomplished with ^{14}C-labeled iodoacetate after alkylation of the four cysteine residues. Takahashi and Hirose[122] showed that there is a sharp transition in the concentration of urea required for reaction of sulfhydryl groups in ovalbumin as determined with reaction at DTNB. Approximately 0.3 cysteinyl residues were available for reaction in 6.0 M urea while 4.0 cysteinyl residues available for reaction at 9 M urea. Takatera and Watanabe[123] observed that 3 cysteinyl residues in chicken ovalbumin were available for reaction with several organic mercurial compounds under nondenaturing conditions (0.3 M sodium phosphate, pH 7). Reaction was most rapid with ethylmercuric chloride and slightly slower with mersalyl acid with PCMB at about half the rate; reaction with fluorescein mercuric acetate was the slowest with lower total reaction (ca. 1 mol/mol ovalbumin). It would appear that avidity of the organic mercurials is sufficient to drive the reaction of the three masked residues but the fourth residue remained unavailable. Takahashi and Hirose[122] reported that 9.0 M urea was required for the reaction of the four sulfhydryl groups in ovalbumin with iodoacetate with no reaction detected below 5.5 M urea. Taketara and Watanabe[123] observed that three of the four sulfhydryl groups in chicken ovalbumin could be modified by the various organic mercurials under nondenaturing conditions. The most rapid rate of modification (0.3 M sodium phosphate, pH 7.0) was observed with ethylmercuric chloride ($t_{1/2}$, 1 min) with a lower rate with methylmercuric chloride ($t_{1/2}$, 3 min), PCMB ($t_{1/2}$, 8 min), and p-chloromercuribenzene-sulfonate ($t_{1/2}$, 50 min). Bramanati and coworkers[124] used mercury cold vapor atomic fluorescence spectrometry to measure the binding of mercury and an organic mercurial to ovalbumin at neutral pH. Approximately three sulfhydryl groups are available for modification; reaction with mercuric ions is accomplished in 1 min while 24 h is required for reaction with PCMB. Protein conformation has a major role in the reactivity of the sulfhydryl groups in ovalbumin with organic mercurials such as PCMB, which is consistent with the very early observations of Anson.[94] In work cited above, Webster and Thompson[121] observed that ovalbumin unfolded more rapidly in acid urea than alkaline urea as previously reported by McKenzie and coworkers.[125] These investigators also reported that PCMB increased the rate of unfolding for ovalbumin in urea. This is consistent with the observations of Batra and coworkers[126] who showed that modification of the sulfhydryl groups of reduced albumin (iodoacetamide, DTNB) caused unfolding of the protein. It is possible that modification of a conformer of ovalbumin occurs with PCMB, which in turns drives further conformational change. Conformational change driven by chaotropic agents is important for the reaction of ovalbumin sulfhydryl groups by studies cited above and by the results of Takeda and coworkers with DTNB.[127] Yang and coworkers[128] reported that the inactivation of mitochondrial succinate-ubiquinone reductase by DTNB required a conformational change prior to reaction. Sluyterman and coworkers[129] studied the reaction of mercuric chloride or *meso*-1,4-bis(acetatomercuri)-2,3-diethoxybutane with papain. A conformational change was required for reaction of either mercuric ions or *meso*-1,4-bis(acetatomercuri)-2,3-diethoxybutane with the active-site cysteine (Cys25). A dimer of papain is formed with either reagent resulting from the reaction of mercuric ion with two cysteine residues. There are several other studies on changes in protein conformation resulting from interaction with mercuric ions[130–132] and such conformational change may represent irreversible inactivation.

Aromatic mercurial derivatives were synthesized and characterized by Whitmore in 1921.[133] Boyer in 1954[134] described the use of organic mercurials to the study of sulfhydryl groups in proteins with emphasis on the use of PCMB (Figure 7.7). Boyer notes that the designation of PCMB is generally used in the description of the use of this reagent; however, dependent on the solution, the mercuribenzoate may be combined with hydroxyl or other monovalent anion. It is probably best to refer the compound as *p*-mercuribenzoate.[135] While both mercuric ions and *p*-mercuribenzoate react with cysteinyl residues, the product obtained with *p*-mercuribenzoate is almost always reversed by thiols, while the modification seen with mercuric ions may be irreversible. Mercuric ions are more promiscuous than *p*-mercuribenzoate and irreversibility of activity loss with mercuric ions may represent interaction at sites other than sulfhydryl groups.

Hasinoff and coworkers[136] have studied the reaction rate of *p*-mercuribenzoate with glutathione, 2-mercaptoethanol, and phosphorylase *b*. The rates of reactions with the model compounds with *p*-mercuribenzoate are shown in Table 7.2. The rates are quite rapid comparable to those with *N*-ethylmaleimide and much faster than those with haloalkyl derivatives. These investigators also studied the reaction of the sulfhydryl groups in phosphorylase *b*; one group of cysteine residues reacted as rapidly as the model compounds ($3 \times 10^6 \, M^{-1} \, s^{-1}$), while the other cysteines could be divided into two rather slowly reacting groups ($4 \times 10^2 \, M^{-1} \, s^{-1}$ and $3.4 \times 10^2 \, M^{-1} \, s^{-1}$). *p*-Mercuribenzoate does have the advantage of providing a spectral change that can be useful to quantitate modification (and measure sulfhydryl groups). The absorbance change at 255 nm upon modification is $6200 \, M^{-1} \, cm^{-1}$ at pH 4.6 and $7600 \, M^{-1} \, cm^{-1}$ at pH 7.0. Bai and Hayashi[137] have examined the reaction of organic mercurials with yeast carboxypeptidase (carboxypeptidase Y); the treatment of the modified enzyme with millimolar cysteine resulted in virtually complete recovery of catalytic activity. The inactivation of chalcone isomerase by PCMB and mercuric chloride has been studied by Bednar and coworkers,[138] and as with carboxypeptidase Y, the modified protein could be readily reactivated by treatment with either thiols or KCN. The reactivation by KCN is based on the formation of a tight complex between cyanide and either organic mercurial. There a number of other studies[139–145] on the use of *p*-mercuribenzoate in protein chemistry, but interest seems to be moving toward the use of this chemical in analytical methods. Kutscher and coworkers[146,147] used *p*-hydroxymercuribenzoic acid (pHMB) to modify sulfhydryl groups in ovalbumin in developing a method for the quantitation of proteins by mass spectrometry. Lu and coworkers[148] have proposed the use of that pHMB for the determination of exposed cysteinyl residues in proteins. This approach is based on the premise pHMB will not react with *buried* sulfhydryl groups based on the size of the reagent. It is not clear that this premise is justified considering the results obtained with ovalbumin as described above. Lu and coworkers[148] used a 10-min period of reaction for the labeling processes, while the modification of the buried sulfhydryl groups in ovalbumin is accomplished in 20–24 h.

There are several chemical reagents that have become *signature reagents* in defining a group of functionally related proteins. Examples include the *N*-ethylmaleimide sensitive factor that is involved in membrane fusion and exocytosis,[149,150] *p*-bromophenacyl bromide inactivation of phospholipase A2,[151,152] and diisopropylphosphorofluoridate (DFP) for serine proteases.[153,154] A consideration of the literature suggests that *p*-aminophenylmercuric acetate is a signature reagent.

p-Aminophenylmercuric acetate (Figure 7.7) is described as an activator of various MMPs.[155–159] Uitto and coworkers[160] observed that either *N*-ethylmaleimide or *p*-aminophenylmercuric acetate stimulated human leukocyte collagenase. Subsequent work from a number of laboratories demonstrated that the activation of matrix MMPs by sulfhydryl reagents such as *p*-aminophenylmercuric acetate was due to reaction with a cysteine residue in the amino terminal domain that bind a zinc atom at the enzyme active site.[161] The situation may be more complex than just blockage of a cysteinyl residues in the propeptide domain.[162] 4-Aminophenylmercuric acetate is included in the assay for matrix MMPs for the activation of the latent enzyme.[163] The organic mercurial may be a problem for the assay of MMP activity after activation. Huse and coworkers[164] developed a clever method for MMP activation and purification of the activated enzyme with removal of the 4-aminophenylmercuric acetate in a single step using a thiophilic matrix. As an aside, matrix MMPs are of importance in oncology and wound healing and the cysteinyl residue in the propeptide domain is a therapeutic target.[165]

2-Chloromercuri-4-nitrophenol (Figure 7.7) is a compound related to the organic mercurial described previously. This reagent was developed by McMurray and Trentham[166] in 1969 together with other mercurinitrophenol derivatives and their reaction with glyceraldehyde 3-phosphate dehydrogenase. 2-Chloromercuri-4-nitrophenol has proved useful as a reporter group where changes in A_{410} can be used to measure polarity of the solvent surrounding the nitrophenol group. The ligand anion for organic mercurials was mentioned earlier in the solution substitution of hydroxyl for chloride in PCMB. Sanyal and Khalifah[167] studied the effect of various monovalent anions on the reaction of 2-mercuri-4-nitrophenol with glutathione. Maximum stimulation was observed with azide, while inhibition was observed with adenosine monophosphate (AMP) and adenosine diphosphate (ADP); the stimulation of the reaction was also observed with Tris buffer. This work was initiated following the observation that the presence of cyanide resulted in a 75-fold rate enhancement of the reaction of 2-chloromercuri-4-nitrophenol with a *buried* cysteine (Cys212) in carbonic anhydrase.[168] McMurray and Trentham used the mercuri-4-nitrophenol group on glyceraldehyde 3-phosphate dehydrogenase to measure conformation change on binding of phosphate and divalent cations. Marshall and Cohen[169] used mercuri-4-nitrophenol as a reporter group in ornithine transcarbamylase. They also reported useful pKa values and spectral data for various derivatives of 2-chloromercuri-4-nitrophenol including 2-mercaptoethanol and glutathione. The pKa of 2-chloromercuri-4-nitrophenol is 6.59, 7.17 for the glutathione derivative, and as high as 8.81 when bound to protein consistent with a more hydrophobic environment. Scawen and coworkers[170] modified the sulfhydryl groups in plastocyanin from *Cucurbita pepo* with 2-chloromercuri-4-nitrophenol. Previous work by Katoh and workers[171] showed the presence of two sulfhydryl groups in spinach plastocyanin on titration with *p*-mercuribenzoate in the presence of 6 M urea and suggested that the sulfhydryl groups are important for the binding of copper. The reaction of plastocyanin with 2-chloromercuri-4-nitrophenol was more rapid in the denatured protein than with the native protein. The modification of the native protein with 2-chloromercuri-4-nitrophenol was also associated with the loss of copper. These investigators observed that the reaction of plastocyanin with mercuric ion was more rapid than with 2-chloromercuri-4-nitrophenol. The protein modified with mercuric ion was obtained by gel filtration and treated with reduced glutathione to obtain the apoprotein. Stinson[172] modified the single sulfhydryl group in yeast phosphoglycerate kinase with 2-chloromercuri-4-nitrophenol without loss of activity. The pKa of the nitrophenol group in the modified enzyme was 8.30 compared to 6.8 in the reagent suggesting a hydrophobic environment in the vicinity of the reactive sulfhydryl group. Subsequent work by Stinson[173] used the mercuri-4-nitrophenol group to follow the conformational changes on urea denaturation (the pKa of the nitrophenol group changed from 8.20 in the native enzyme to 7.36 in the presence of 5.0 M urea consistent with exposure to solvent; there was also a hyperchromic shift). Stinson also observed a major increase in the rate of reaction of the single sulfhydryl group in yeast phosphoglycerate kinase in the presence of urea; the rate increased from 2.6×10^3 M^{-1} s^{-1} (140 mM triethanolamine, pH. 7.5, 22°C) in the absence of urea to 7×10^5 M^{-1} s^{-1} in the presence of 5.0 M urea.

Creatine kinase can be inactivated by iodoacetic acid[174] and iodoacetamide[175] demonstrating the importance of a cysteinyl residue(s). It was then surprising when Quiocho and Thomson[176] reported that the reaction of creatine kinase with 2-chloromercuri-4-nitrophenol resulted in the modification of cysteine without loss of activity; the modified enzyme could be inactivated by iodoacetate. The mercuri-4-nitrophenol bound to the enzyme showed a shift in the pKa of the nitrophenol from 6.50 in reagent to 7.9 when bound to the protein. This is consistent with what has been reported for other enzymes and supports a hydrophobic environment for the modified cysteinyl residue; there is also a modest hyperchromic response as well as a red shift (405 nm to 416 nm). Subsequent studies from this laboratory[177] reported the modification of creatine kinase with 2-chloromercuri-4-nitrophenol in the absence of substrates (creatine, MgATP, nitrate) was rapid ($k = 4.5 \times 10^6$ M^{-1} s^{-1}) at pH 8.0 (20°C); the addition of various components of the substrate complex resulted in only a modest decrease in the rate. Somerville and Quiocho[178] showed that mercuri-4-nitrophenol bound to cysteine residues could be used to monitor conformational change in creatine kinase on binding the competitive inhibitor, 2′,4′,5′,66-tetraiodofluorescein. Binding of the dye resulted in a hyperchromic response with a further red shift in the spectrum. While it is clear that the mercuri-4-nitrophenol group was useful as a reporter group in the modified enzyme, the issue of the inactivation obtained with iodoacetate acid and reaction with the organic mercurial without loss of activity was not resolved. While not providing complete resolution of the issue, Wu and coworkers[179] reported that the reaction of creatine kinase with 2-chloromercuri-4-nitrophenol did result in total inactivation and the conditions of the coupled assay system used in the previous studies (bovine serum albumin and glyceraldehyde-3-phosphate dehydrogenase) resulted in reactivation of the modified enzyme. Wu and coworkers[179] also discussed issues with interpretation of the reporter group studies. Wang and coworkers[180] demonstrated that the active thiol in creatine kinase has a low pKa so facile reactivation of the enzyme modified with 2-chloromercuri-4-nitrophenol is reasonable. There are other studies[181,182] that suggest that the modification of cysteine in creatine kinase is somewhat complicated. This does emphasize the importance of appreciating the potential lability of the modification of a particular amino acid residue in a protein; a good example is provided by the acetylation of histidine at the active site of trypsin[183] and the reaction of thrombin with N-butyrylimidazole.[184]

Yang and coworkers[185] modified lactose repressor protein with 2-chloromercuri-4-nitrophenol or 2-bromoacetamido-4-nitrophenol; the modification with organomercurial could be reversed with 2-mercaptoethanol or dithiothreitol, while the alkylation with the bromoacetamido derivative was stable. The two reagents demonstrated differences in specificity of modification of the several cysteine residues in lactose repressor protein. 2-Chloromercuri-4-nitrophenol reacted with Cys281$_a$ (90%), Cys107 (70%), and Cys140 (40%) while 2-bromoacetamido-4-nitrophenol reacted with Cys140 (85%) and Cys107 (35%). Denaturation of the protein exposed all three cysteine residues for modification. Earlier work from this laboratory[186] showed that the reaction of lactose repressor protein with 2-chloromercuri-4-nitrophenol was quite slow. Friedman and coworkers[186] also observed that the reaction of 2-chloromercuri-4-nitrophenol with the lactose repressor was strictly first-order. These investigators estimated a $t_{1/2}$ of 5 s for the reaction of 2-chloromercuri-4-nitrophenol with lactose repressor compared to a $t_{1/2}$ of 5 ms for the reaction of this reagent with creatine kinase.[177] Friedman and colleagues[186] also noted that the cysteine residues in lactose repressor are refractory to reaction with either iodoacetate or iodoacetamide. Sams and coworkers[187] reported on the spectral characteristics of lactose repressor modified either with 2-chloromercuri-4-nitrophenol or 2-bromo-acetamido-4-nitrophenol. As with other proteins, the pKa of the organic mercurial increased from 6.75 to 8.3; the pKa for the alkyl nitrophenol could be determined as it was below pH 7.0 where the repressor protein is unstable. The interaction of modified repressor protein with binding of inducer (isopropyl-β-D-thiogalactoside) caused different responses from the two chromophoric groups. I could find no published work on 2-chloromercuri-4-nitrophenol since 1996.[188]

Baines and Brocklehurst[189] described the synthesis and characterization of 2-(2′-pyridylmercapto)-mercuri-4-nitrophenol (Figure 7.7). The pyridine-2-thione as the leaving group provides a method for measurement for the reaction based on absorbance at 343 nm ($\varepsilon_{343} = 7300$ M^{-1} cm^{-1}). The rate of reaction of 2-(2′-pyridylmercapto)-mercuri-4-nitrophenol with sulfhydryl groups (2-mercaptoethanol; 3.5×10^3 M^{-1} s^{-1}) is somewhat slower than that for other organomercurials

(Table 7.2) which might be useful in characterizing the reactivity of different cysteinyl residues. Scott-Ennis and Noltmann[190] compared seven reagents for the modification of cysteinyl residues in pig muscle phosphoglucoisomerase. Each subunit of the enzyme contains three cysteine residues. Scott-Ennis and Noltman[190] also found that two of three cysteinyl residues reacted slowly with PHMB and one was available for reaction only after denaturation of the protein. Several other sulfhydryl reagents including 2-(2′-pyridylmercapto)mercuri-4-nitrophenol, iodoacetamide, 2,2′-dithiopyridine, and DTNB react with only one sulfhydryl; iodoacetate did not react with any of the three sulfhydryl groups in the native protein. These investigators suggest that reagent charge rather than physical size is the determining factor in reactivity.

Haloalkyl compounds (see Chapter 2) such as methyl iodide, iodoacetate, iodoacetamide, and more complex derivatives have been used to modify cysteine residues in proteins. These various reagents react with cysteine via an S_N2 reaction mechanism to give the corresponding carboxymethyl or carboxamidomethyl derivatives (Figure 7.8). Specificity is an issue as α-ketoalkyl halides can also modify histidine, lysine, tyrosine, and glutamic acid. In general, the rate of reaction at sulfhydryl groups is much more rapid than that of other functional groups. Dahl and McKinley-McKee[191] have made a rather detailed study of the reaction of alkyl halides with thiols. It is emphasized that reactivity of alkyl halides depends not only on the halogen but also on the nature of the alkyl groups. These investigators emphasized that the reactivity of an alkyl halide such as iodoacetate depends

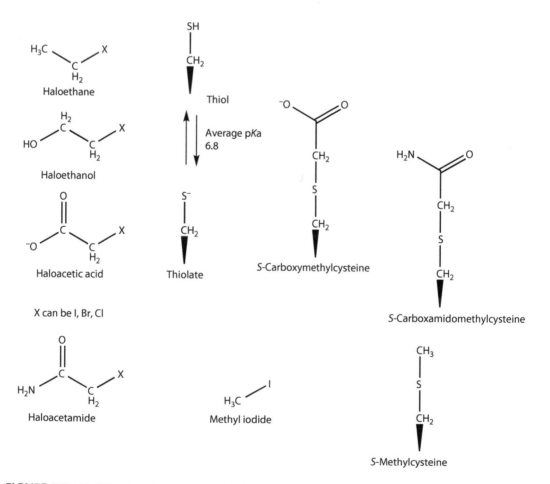

FIGURE 7.8 Modification of cysteine with haloacetic and haloacetamide. Also shown is the structure of methyl iodide that modifies cysteine to yield *S*-methylcysteine (Refsvik, T., The mechanism of biliary excretion of methyl mercury: Studies with methylthiols, *Acta Pharmacol. Toxicol.*, 53, 153–158, 1983).

not only on the leaving potential of the halide substituent (I > Br>>> CI; 130:90:1) but also on the nature of the alkyl group (Table 7.2). The rate of reaction of 2-bromoethanol with the sulfhydryl group of l-cysteine (pH 9.0) is approximately 1000 times less than that observed with bromoacetic acid. The reactions are extremely pH dependent, emphasizing the importance of the thiolate anion in the reaction. As a general rule, when a rapid reaction is desired, the iodine-containing compounds are generally used, but there can be issues with respect to specificity of modification.

Iodoacetic acid and iodoacetamide are used more often than bromo- or chloroderivatives. Fluoroacetic acid or fluoroacetamide derivatives are rarely used. Sodium fluoroacetamide is used as a rodenticide on the basis of metabolic formation of fluorocitrate, an inhibitor of aconitase in the tricarboxylic acid cycle.[192,193] Chloroacetate is also toxic both as a strong acid and also by reaction with sulfhydryl groups.[194,195] It is of historical interest that the first use of α-haloacids was with bromoacetate in 1874[196] with iodoacetate introduced in 1930.[197] Both of these studies used whole animals, and the reader is directed to Webb for more detail.[198] Some differences in reaction characteristics of acid and amide derivatives have been discussed previously (see also Table 2.3). There are studies[3,199] where iodoacetate is an affinity label based on the demonstration of saturation kinetics (Chapter 1). Iodoacetate and, to a somewhat lesser extent, iodoacetamide are used to probe sulfhydryl group reactivity in proteins. Chloro- and bromoderivatives are less reactive and are more useful in affinity labeling. Chloro- and bromo-derivatives are less reactive than the iodo-derivatives and are more useful in affinity labeling where reaction is driven by reagent binding rather than residue reactivity. Affinity labeling is a very useful technique for identifying binding sites on proteins, and the reader is directed to an early review[200] by Bryce Plapp for more details on this technique. With nitrogen as the nucleophile, the order of reaction of α-haloacetic acids with 4-(p-nitrobenzyl) pyridine was I > Br >> Cl consistent with the order of reaction with thiols as described by Dahl and McKinley-McKee;[191] the order of reactivity with the histidine residues in RNAse was Br > I > Cl and Br > I > Cl with DNAse.

Reduced thioredoxin, which functions as a disulfide reductase/disulfide isomerase, contains two cysteinyl residues at the active center, Cys32 and Cys35. The oxidation of reduced thioredoxin results in the formation of an internal disulfide between Cys32 and Cys35. The work by Kallis and Holmgren[201] showed that only one of the cysteinyl residues, Cys32, is alkylated by iodoacetate below pH 8.0. Above pH 8.0 and at pH 7.0 in the presence of 4.5 M guanidine, both cysteinyl residues in reduced thioredoxin are alkylated by iodoacetate. The reactivity of Cys 32 is suggested to be due to a lower pKa value (6.7) reflecting the presence of a lysine residue (Lsy36). The presence of basic amino acid residues increases the reactivity of cysteinyl residues in proteins (Chapter 2). The rate of reaction with iodoacetamide at pH 2 (107 M^{-1} s^{-1} is close to 20-fold higher than that seen with iodoacetate at pH 7.5 (6 M^{-1} s^{-1}). This is somewhat more rapid than observed for model compounds (see Table 7.2). This work was followed with similar observations on the human protein disulfide isomerase by Hawkins and Freedman,[202] leading to the development of a mechanism for thioredoxin involving the nucleophilic attack of a thioredoxin sulfhydryl (Cys32) on the substrate forming a transient mixed disulfide following the reaction of Cys35 to form a disulfide ring. While there have been some adjustments over the years, the basic mechanism remains a useful guide to the action of thioredoxin and related proteins.[203]

Soybean β-amylase contains five sulfhydryl groups. Mikami and coworkers[204] showed that the two of these sulfhydryl groups could be modified by DTNB, iodoacetamide, or iodoacteate at high ionic strength (0.8 M KCl, pH 8.2) with inactivation of the enzyme. At low ionic strength, only one sulfhydryl group is modified by DTNB or iodoacetate, and activity is not lost. The modified enzyme could be isolated and subsequently modified with DTNB or iodoacetamide. All five sulfhydryl groups are available for modification with PHMB in 0.1 M phosphate, pH 7.0; 3 M guanidine is required for reaction with DTNB at pH 8.0. The two sulfhydryl groups that could be modified under nondenaturing conditions by iodoacetamide were designated *most reactive* and *essential*. The reaction of the *essential* sulfhydryl group with iodoacetamide (8.3 × 10^{-3} M^{-1} s^{-1}) is some 60 more rapid than reaction with iodoacetate (1.3 × 10^{-4} M^{-1} s^{-1}) while the rate of reaction of iodoacetamide with the "most reactive" sulfhydryl group (1.9 × 10^{-1} M^{-1} s^{-1}) is approximately eight-fold faster than the rate of reaction with iodoacetate (2.4 × 10^{-2} M^{-1} s^{-1}). The rate of reaction of DTNB at high ionic strength (0.8 M KCl) is 35 $M^{-1} s^{-1}$ for *most reactive* sulfhydryl group and 1.9 $M^{-1} s^{-1}$ for the *essential* sulfhydryl group. Approximately one sulfhydryl group is available for reaction with DTNB at low ionic strength (0.05 M Tris, pH 7.4) with a rate of 14 $M^{-1} s^{-1}$ without loss of activity. The message from this study is that the most reactive functional group is not necessarily the one important for biological activity.

The chemical modification of the cysteinyl sulfhydryl groups important in the function of aldehyde dehydrogenases is a useful study of functional group reactivity. There are several families of aldehyde dehydrogenases[205,206] with two families, aldehyde dehyrogenase-1 (cytoplasmic) and aldehyde dehydrogenase-2 (mitochondrial), of interest in the study of the reactivity of cysteine residues in proteins. These enzymes are tetramers with two active sites and demonstrate 70% homology.[207] Aldehyde dehydrogenase-1 and aldehyde dehydrogenase-2 have dehydrogenase activity toward various aldehyde substrates,[208] acetaldehyde being the signature substrate, and esterase activity toward generic substrates such as *p*-nitrophenyl acetate.[209,210] Hempel and Pietruszko[211] have shown that human liver aldehyde dehydrogenase is inactivated by iodoacetamide, but not by iodoacetic acid. Saturation kinetics were observed for the reaction of iodoacetamide with either aldehyde dehydrogenase isozyme suggesting that there is a noncovalent binding of the reagent prior to reaction with the active-site cysteine, establishing that iodoacetamide is an affinity label for human liver aldehyde dehydrogenase, as described for iodoacetate in studies on pancreatic nucleases[2] and dehydroquinonase[3] as cited previously (p. 217).[2,3] The analysis of peptides derived from either isozyme showed modification at a single site. Subsequent work from this laboratory[212] identified the residue in the cytoplasmic enzyme (E_1) reacting with iodoacetamide and confirmed that 5%–10% activity was remaining after total modification. The cysteine residue at the active site reacting with iodoacetamide was identified at Cys302; there is a second cysteine residue at position 455 that is involved with coenzyme binding. These residues are conserved in the horse cytoplasmic enzyme.[213] Bromoacetophenone (phenacyl bromide) and chloroacetophenone were identified as affinity labels for both cytoplasmic and mitochondrial aldehyde dehydrogenases;[214] substantial inactivation was observed after the incorporation of two moles of bromoacetophenone/mole of proteins (78% for cytoplasmic, E1; 58% for mitochondrial, E2) with complete inactivation after 4 mol of reagent/mole protein. The first phase of the reaction with bromoacetophenone was subsequently shown to involve modification at Glu268.[215] Alkylation of Cys302 with bromoacetophenone also occurs during the reaction but is not associated with the loss of activity and accounts for the second 2 mol of reagent incorporated in the reaction. Total loss of activity does not occur until after modification of Glu268* is complete.[216] The inactivation of aldehyde dehydrogenase by *N*-tosyl-L-phenylalanine chloromethyl ketone was associated with the modification of Glu398.[217] Takahashi and coworkers[218] did demonstrate the modification of a glutamic acid residue by iodoacetate in ribonuclease T_1; iodoacetamide was ineffective. Subsequent work by Takahashi[219] extended this observation to a variety of other haloalkanoic acids such as β-bromopyruvate. Erlanger and coworkers[220] reported the inactivation of pepsin with bromoacetophenone; bromoacetophenone was 10% as effective as *p*-bromophenacyl bromide. Subsequent work from this group[221] showed that the inactivation of pepsin with *p*-bromophenacyl bromide was associated with the modification of aspartic acid. Thus, there is a history of bromoacetophenone and chloroacetophenone reacting with carboxyl groups in proteins. The reaction of a peptide chloromethyl ketone with a carboxyl group appears to be a unique observation, although here is some literature on the reaction of iodoacetate with glutamic acid ribonuclease T1. Johansson and coworkers[222] reported the modification of Cys301 and Cys302 in cytoplasmic aldehyde dehydrogenase with iodoacetamide with 20% residual activity. Aldehyde dehydrogenase is also inactivated by haloenol lactones in a mechanism-based process presumably involving Cys302.[223] Further support for a role for Cys302 was obtained from modification with a vinyl ketone[224] and from isolation of covalent intermediate from the action of aldehyde dehydrogenase with 4-*trans*-(*N*,*N*'-dimethylamino)cinnamaldehyde and

---

* This is residue 268 in this work but was changed to 281 in later work (Matthews, K.S., Tryptic core protein of lactose repressor binds operator DNA, *J. Biol. Chem.* 254, 3348–3353, 1979).

4-*trans*-(*N,N'*-dimethylamino)cinnamoylimidazole.[225] Other work with *N*-ethylmaleimide resulted in the modification of Cys49 and Cys162 with *N*-ethylmaleimide.[226] Cys49 was implicated in dehydrogenase activity and esterase activity, while Cys162 is only involved in esterase activity. The mitochondrial enzyme has been shown to be involved in the metabolism of nitroglycerin that generates nitric oxide.[227] Wenzl and coworkers[228] mutated Cys302 to a serine in mitochondrial aldehyde dehydrogenase resulting greater than 90% loss of both esterase and dehydrogenase activities as well as the denitration (reduction) of nitroglycerin (propanetriol trinitrate) to 1,2-glyceryl dinitrate and nitrite. These investigators also mutated Glu268 to Gln, which reduced esterase and dehydrogenase activities but did not reduce the rate of denitration of nitroglycerin. Subsequent work by Wenzl and coworkers[229] described the mutation of the two cysteine residues adjacent to C302 (C301 and C303) to serine. This mutation did not affect dehydrogenase activity and increased nitric oxide production. The role of aldehyde dehydrogenases in the production of vasoactive nitric oxide is the subject of considerable investigation.[230–232] The previously presented results combining chemical modification and site-specific mutagenesis suggest that there are amino residues that are important in various catalytic activities but not essential. The issue is made both more complicated and of greater importance as additional activities are described for the aldehydrogenase.[233–237]

It would be remiss to leave the subject of aldehyde dehydrogenase without mention of disulfiram [antabuse; *bis*(diethylthiocarbamoyl)disulfide] (Figure 7.9). This material likely is more suited for inclusion in the discussion of disulfide exchange; however, considering the previous discussion on cysteine reactivity in aldehyde dehydrogenase, it is useful to place consideration of disulfiram here. This drug was discovered by a Danish group in 1948[237,238] who were quite observant on their own work and very knowledgable of the literature. The paper is well worth reading as it is likely an example of clever research and clinical observation. The observation was that exposure to alcohol after ingestion of disulfiram caused an uncomfortable physical response. These investigators were astute enough to recognize the commercial possibility of this material and coined the term antabuse. Subsequent work by this group[239] showed elevation of acetaldehyde in subject ingesting alcohol after antabuse administration. Shortly after this work, Eldjarn[240] suggested that diethyldithiocarbamate (Figure 7.9) was the active species in disulfiram therapy. While disulfiram can modify purified aldehydrogenase and the chemistry of modification has been described, the in vivo action of disulfiram appears to be mediated through a metabolite, *S*-methyl-*N,N*-diethylthiolcarbamate sulfoxide.[241,242] This is a result of the rapid absorption of metabolism of disulfiram to diethyldithiocarbamate.[243] Earlier work of the reaction of disulfiram with purified cytoplasmic aldehyde dehydrogenase (E1, LADH1) by Vallari and Pietruszko[244] had suggested that the inactivation was due to the formation of an internal disulfide resulting from the initial formation of a mixed disulfide product (Figure 7.9); this has been reported after the reaction of thiol groups in proteins with Ellman's reagent.[245,246] Purified mitochondrial aldehyde dehydrogenase (E2, LADH2) was not inhibited by disulfiram but was inhibited by diethyldithiocarbamic acid methanethiol mixed disulfide[247,248]; it was suggested that the size of disulfiram prevented reaction with the mitochondrial enzyme. Sanny and Weiner[249] were able to inhibit horse liver mitochondrial aldehyde dehydrogenase with the tetramethyl derivative, tetramethylthiuram disulfide. A conjugate of *S*-methyl-*N,N*-diethylcarbamate sulfoxide and glutathione (Figure 7.9) is thought to be responsible for the efficacy of disulfiram in treating cocaine dependency.[250] Disulfiram has been reported to inactivate other enzymes including caspase-1 and caspase-3[251] where inactivation is shown to involve the formation of a mixed disulfide and is prevented by the presence of dithiothreitol. It was a bit surprising to find that disulfiram (cited as tetraethylthiuram disulfide) and the immediate product of in vivo reduction, diethyldithiocarbamate (as zinc–thiol complex), are accelerators in the vulcanization of rubber and latex products and are suggested to act as haptens in the development of allergic responses.[252]

The use of iodoacetate and iodoacetamide appeared to peak in the 1970s (Table 7.3) and seemed to parallel the number of publications on the chemistry of enzyme active sites. However, there is still use of these reagents with more use of iodoacetamide than iodoacetate. In what is *back to the future*, iodoacetate is used to inhibit glycolysis in intact animals,[253] isolated tissue such as muscle,[254] and ex

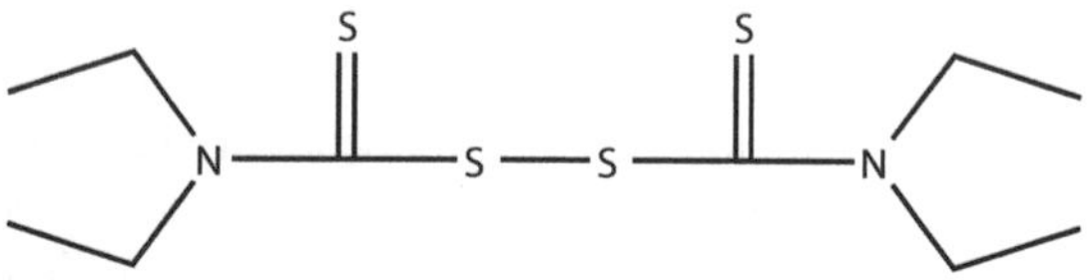

Disulfiram

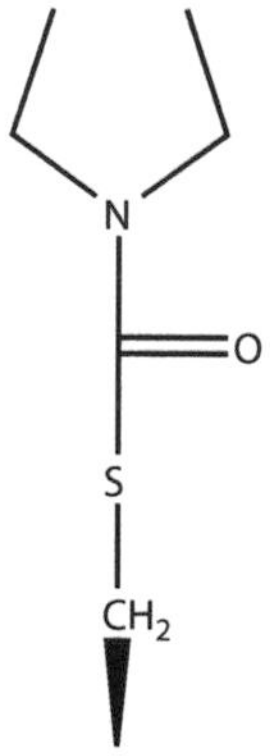

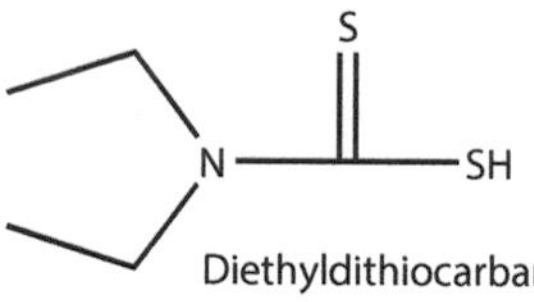

Diethyldithiocarbamate

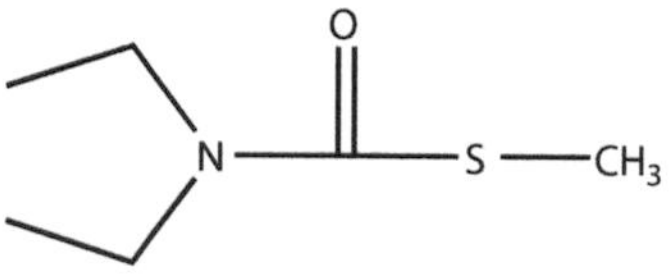

S-Methyl-N,N-diethylthiolcarbamate

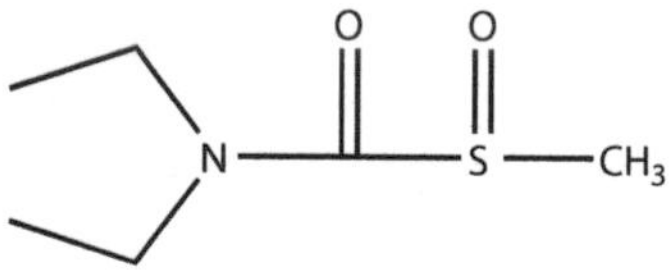

S-Methyl-N,N-diethylthiolcarbamate sulfoxide

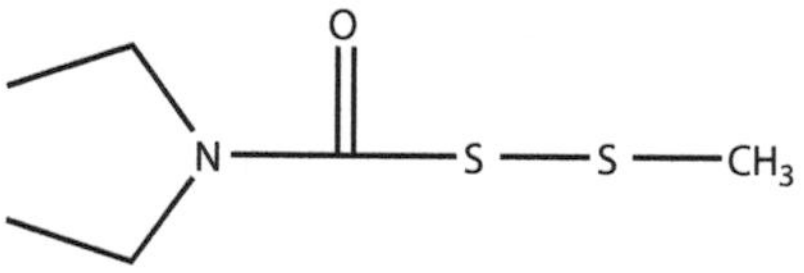

Diethyldithiocarbamic acid methanethiol mixed disulfide

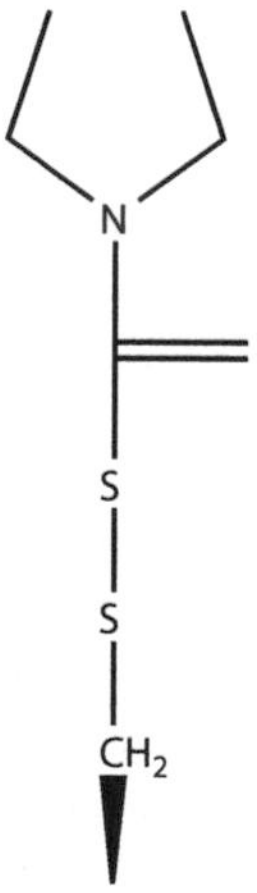

Carbamoyl derivative
of Cys302

Mixed disulfide derivative

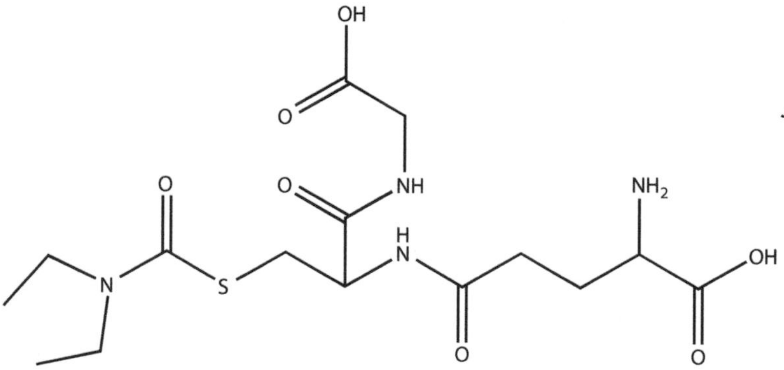

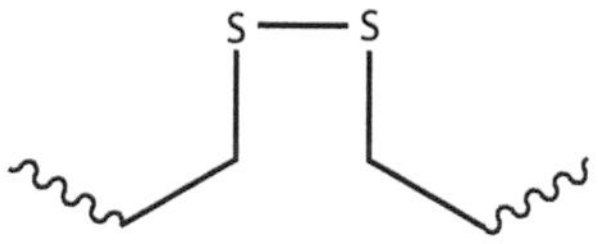

Disulfide Product

Carbamathione

FIGURE 7.9 (continued)

TABLE 7.3
Chronological Record of Iodoacetate Use[a]

Time Period	Number of Citations	
	PubMed[b]	SciFinder[©c]
1920–1930	(4)[d]	—
1930–1940	(224)[d]	—
1940–1950	(145)[d]	—
1950–1960	161 (764)[d]	782
1960–1970	1337	1403
1970–1980	2292	1664
1980–1990	1023	1625
1990–2000	771	1433
2000–2010	343	800[e]

[a] The citations have not been selected for application. Thus, for example, this number would include the use of iodoacetate as an enzyme inhibitor, its use in the carboxymethylation of proteins for structural analysis, or whole cell metabolism studies.

[b] PubMed is a service through the National Center for Biotechnology Information in the National Library of Medicine within the National Institute of Health (Bethesday, MD). PubMed contains material from the biomedical literature and, more recently, online books.

[c] SciFinder is a literature service provided by the American Chemical Society (Washington, DC), which has evolved from Chemical Abstracts Service (American Chemical Society, Columbus, OH). This database contains material from the chemistry journals (including biochemical journals), selected books, abstracts from American Chemical Society Meetings, and patents from various geographies. It is usual to obtain a larger number of citations from SciFinder than from PubMed.

[d] Numbers in parenthesis are obtained from Webb, J.L., *Enzyme and Metabolic Inhibitors*, Vol. 1, Preface (page x), Academic Press, New York, 1963.

[e] 1367 references for iodoacetamide.

FIGURE 7.9 (continued) The structure of disulfiram (tetraethylthiuram disulfide) and metabolites and derivative products. Diethyldithiocarbamate is the immediate in vivo product derived from disulfiram that is then converted via cytochrome P450 to a sulfoxide and subsequently a sulfone; the sulfoxide and sulfone are considered products that are responsible for the in vivo inactivation of aldehyde dehydrogenase (Mays, D.C., Nelson, A.N., Lam-Holt, J. et al., S-Methyl-N,N-diethylthiocarbamate sulfoxide and S-methyl-N,N-diethylthiocarbamate sulfone, two candidates for the active metabolite of disulfiram, *Alcohol Clin. Exp. Res.* 20, 595–600, 1996; Pike, M.G., Martin, Y.N., Mays, D.C. et al., Roles of FMO and CYP450 in the metabolism in human liver microsomes of S-methyl-N,N-diethyldithiocarbamate, a disulfiram metabolite, *Alcohol Clin. Exp. Res.* 23, 1173–1179, 1999). Also shown are various products that have been suggested to be derived from the action of disulfiram and disulfiram metabolites on various enzymes. Also shown in the product of the reaction of reduced glutathione and S-methyl-N,N-diethylthiocarbamate sulfoxide (Hochreiter, J., McCance-Katz, E.F., Lapham, J. et al., Disulfiram metabolite S-methyl-N,N-diethylthiocarbamate quantitation in human plasma with reverse phase ultra performance liquid chromatography and mass spectrometry, *J. Chromatogr. B* 897, 80–84, 2012).

vivo cell culture.[255-257] As would be expected, iodoacetamide has been observed to be more effective than iodoacetate.[257] Iodoacetamide was used by Rodriguez and coworkers[258] used reaction with iodoacetamide to identifiy Cys351 an essential residue in catalysis by *Porphyromonas gingivalis* deiminase; this was subsequently confirmed by site-specific mutagenesis. Iodoacetic acid has been labeled with O^{18} by exchange with H_2O^{18} and used for protein quantitation in mass spectrometry.[259] Iodoacetate is used to induce osteoarthritis in experimental animals,[260-269] while iodoacetamide is used to induce experimental ulcerative colitis.[270,271]

In addition to the use as reagents for the modification of amino acid residues in proteins. The haloacetyl moiety is used as reactive function to couple probes such as chromophores, fluorophores, or spin labels (Figure 7.10) to cysteine residues in proteins. Some probes for protein structure were mentioned previously in the discussion of organomercurials and other will be mentioned in the following in association with maleimide chemistry and mixed disulfides. Examples of haloalkanoic acid/amide-based probes include 5-iodoacetamido fluorescein,[272-276] 5-[2-((iodoacetyl)amino)-ethyl]naphthalene-1-sulfonic acid (1,5-IAEDANS)[277-289] and 4-(2-iodoacetamido)-TEMPO,[290-293] 3-(2-iodoacetamido)-PROXYL,[294-296] 3-[2-(2-iodoacetamido)acetamido]-PROXYL,[294] and cyanine dyes.[297,298]

The covalent attachment of probes to proteins via modification of sulfhydryl groups is a very useful technique. First, there are several chemistries available including haloalkyl and maleimide functions that are very specific for sulfhydryl groups forming a stable covalent bond. Second, the number of sulfhydryl groups available for reaction with a probe is usually small,[299] and three sulfhydryl groups can be introduced into a protein either by site-specific mutagenesis[300,301] or by chemical modification.[302]

The covalent attachment of fluorescent probes dates least back to the work of Creech and Jones on fluorescein isocyanate derivates and their reaction with albumin in 1941,[303,304] the development and application of dansyl chloride by Weber in 1952,[305] and the work of Emmert on fluorescein, fluorescein derivatives, and conjugates in 1958.[306] These were useful materials, but the chemistry of attachment can be somewhat nonspecific. A major advance in the specificity of modification was the development of two isomers of *N*-(iodoacetylaminoethyl)-naphthalene sulfonic acid (1,5-I-AEDANS and 1,8-I-AEDANS) by Hudson and Weber[307] in 1973. Fluorescent labels, including 1,5-IAEDANS and 1,8-IAEDANS, based on the iodoacetyl group were used by Wu and Stryer[308] as a spectroscopic ruler based on energy transfer in rhodopsin. Zukin and coworkers[309] extended these observations in using 5-iodoacetamido-fluorescein to measure conformational change in bacterial membrane sensor. There has been extensive use of 1,5-IAEDANS for the study of protein conformation over the past 40 years. One of the more interesting areas of the application of fluorescent labels the study of myosin. Takashi and coworkers[310] compared the reaction of 1,5-IAEDANS and iodoacetamide with sulfhydryl groups in myosin. The *fast-reacting* thiol in myosin (subfragment-1) is modified more rapidly by 1,5-IAEDANS than iodoacetamide. Subsequent work by Duke and coworkers[311] showed the complex formation of myosin subfragment with actin decreased the reactivity of the *fast reacting* thiol in myosin while increased the reaction of Cys10 in actin. In later work, Baba and coworkers[312] observed that the reaction of iodoacetamide with sulfhydryl groups in rabbit muscle sarcoplasmic reticulum was slower than that observed with either 1,5-IAEDANS, 5-iodoacetamido fluorescein, or 5-iodoacetamido eosin. These investigators also reported that the sulfhydryl group(s) modified by iodoacetate was different from those modified with *N*-ethylmaleimide. Later work by Yamashita and Kawakita[313] showed that iodoacetamide and 1,5-IAEDANS modified the same cysteinyl residue in sarcoplasmic reticulum. A lysine residue in myosin subfragment-1 has been shown to be important for the reactivity of the *fast-reacting* thiol group (SH1).[314] Modification of SH1 with 1,5-IAEDANS decreased the rate of trinitrophenylation the lysine residues by TNBS by 52%, while modification of SH1 with iodoacetamide decreased the rate of modification with TNBS by 24%; TNBS modification of lysine had a modest effect on the rate of reaction of subfragment 1 with 1,5-IAEDANS. There have been a number of studies in the intervening years on the use of fluorescent probes and the modification of the *fast-reacting* cysteinyl residue in relation to the myosin ATPase activity and muscle contraction.[290,315-318] There is one study by Hiratsuka and coworkers[319] that demonstrates the alteration of reactivity of the cysteine residues in myosin on the binding ATP analogs. These authors suggest that the reactivity of the cysteine residues (SH1 and SH2) in the C-terminal region of the motor domain of myosin is a marker of conformation change around the converter. This work emphasizes the use of sulfhydryl modification as an approach to understanding the function of myosin and the role of ATP in muscle contraction. However, a discussion of the physiology of muscle contraction is beyond the scope of this work. Suffice to say, the 30+ year period of the study

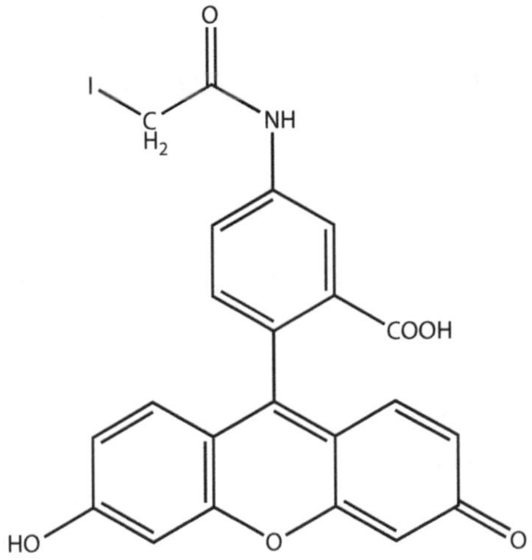

5-(Iodoncetoamido)fluorescein

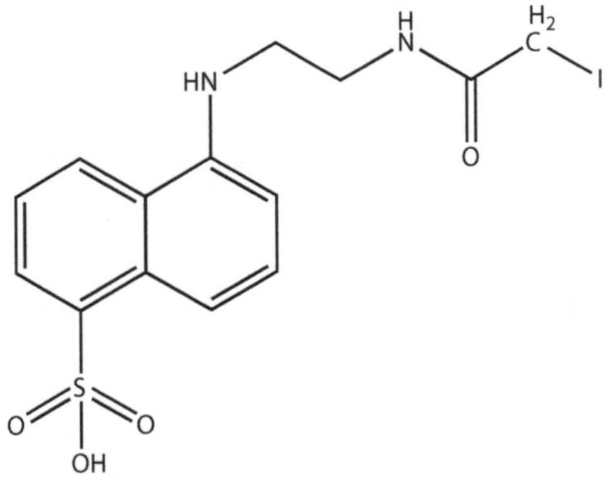

1,5 IAEDANS
(N-Iodoacetaminoethyl)-1-naphthylamine-5-sulfonic acid

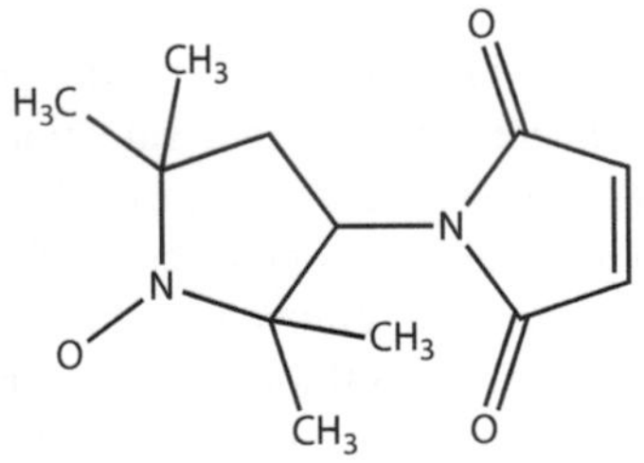

3-Maleimido-2,2,5,5-tetramethyl-1-pyrrolidinyloxyl

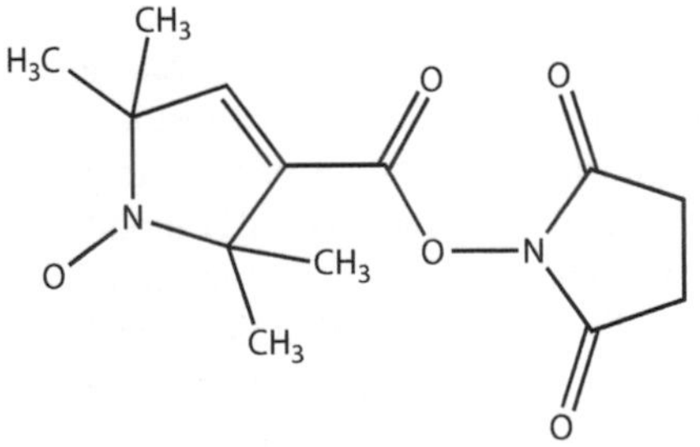

2,2,5,5-Tetramethyl-3-pyrrolin-1-oxyl-
3-carboxylic acid N-hydroxysuccinimide

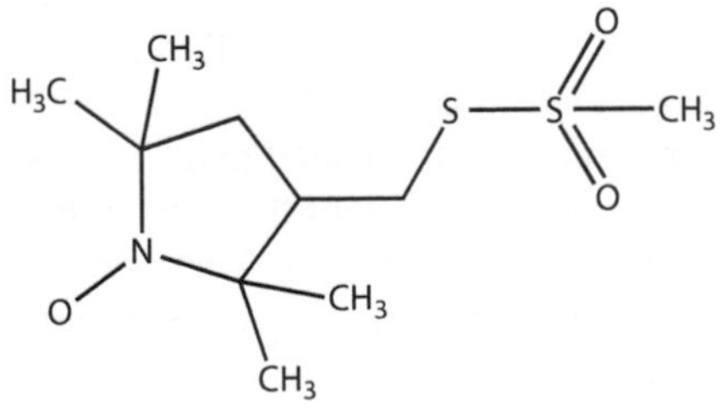

1-Oxyl-2,2,5,5-tetramethylpyrolidin-3-yl methyl methanethiosulfonate

FIGURE 7.10

(*continued*)

FIGURE 7.10 (continued)   Some structural probes based on haloacetyl functional groups. Shown are the structures of 5-(iodoacetamido)fluorescein (Zukin, R.S., Hartig, P.R., and Koshland, D.E., Jr., Use of a distant reporter group as evidence for a conformational change in a sensory receptor, *Proc. Natl. Acad. Sci. USA* 74, 1932–1936, 1977; Wang, Y.L., and Taylor, D.L., Preparation and characterization of a new molecular cyto-chemical probe: 5-iodoacetamidofluorescein-labeled actin, *J. Histochem. Cytochem.* 28, 1198–1206, 1980), 1,5-IAEDANS [*N*-(iodoacetaminoethyl)-1-naphthylamine-5-sulfonic acid] (Hudson, E.N. and Weber, G., Synthesis and characterization of two fluorescent sulfhydryl reagents, *Biochemistry* 12, 4154–4161, 1973; Takahashi, R., Duke, J., Ue, K., and Morales, M.F., Defining the "fast-reacting" thiols of myosin by reaction with 1,5,-IAEDANS, *Arch. Biochem. Biophys.* 175, 279–283, 1976), 3-(2-acetamido)PROXYL (PROXYL is 2,2,5,5,-tetramethylpyrrolidone-1-oxyl) (Keana, J.F., Lee, T.D., and Bernard, E.M., Side-chain substituted 2,2,5,5-tetramethylpyrrolidone-*N*-oxyl (proxyl) nitroxides. A new series of lipid spin labels showing improved properties of the study of biological membranes, *J. Am. Chem. Soc.* 98, 3052–3053, 1976; Gettins, P., Beth, A.H., Cunningham, L.W., Proximity of thiol esters and bait region in human α_2-macroglobulin: Paramagnetic mapping, *Biochemistry* 27, 2905–2911, 1988) and 4-(2-iodoacetamido)TEMPO (IASL; 4-(2-acetamido)-2,2,6,6-tetramethyl-1-piperidine-1-oxyl) (Ho, C., Baldassare, J.J. and Charache, S., Electron paramagnetic resonance studies of spin-labeled hemoglobins and their implications to the nature of cooperative oxygen binding to hemoglobin, *Proc. Natl. Acad. Sci. USA* 66, 722–729, 1970; Xie, L., Li, W.X., Barnett, V.A., and Schoenberg, M., Graphical evaluation of alkylation of myosin's SH1 and SH2: The *N*-phenylmaleimide reaction, *Biophys. J.* 72, 858–865, 1997).

of the reaction of myosin with haloalkyl derivatives is instructive even if it seems, in my opinion, to be incomplete. The difference in reactivity of iodoacetate and the larger iodoacetyl fluorescent derivatives, most likely reflects a specific binding of the fluorescent derivatives, but I could not find any evidence of studies in this area. The other observation on the difference between iodoacetamide and *N*-ethylmaleimide is important when assessing whether the lack of reactivity with a *sulfhydryl-specific* reagent is a definitive indication of the lack of a cysteinyl residue. A difference with iodoac-etate would be easier to explain. In light of these comments, it is worth considering the work of Botts and coworkers[320] who reported that 12 cysteinyl residues were modified in myosin subfragment 1 with PHMB in the presence of 8 M urea. Methyl methanethiosulfonate modified seven residues in myosin subfragment 1 previously modified by 1,5-IAEDANS. While this work is interesting, there has been no further work on the reaction of MMTS with either myosin or myosin derivatives. The reaction of myosin with DTNB resulted in the modification of SH1 permitting modification of SH2; the removal of the thionitrobenzoate group with reducing agent allowed the evaluation of the SH2 modification independent of SH1 modification.[321] Reisler and coworkers[322] performed a similar experiment where myosin was first modified at SH1 with 1-fluoro-2,4-dinitrobenzene (FDNB) and then modified with *N*-ethylmaleimide; the dinitrophenyl group was then removed by thiolysis with 2-mercaptoethanol[323]. Shaltiel[324] had previously demonstrated the ability of thiolysis with 2-mercap-toethanol to remove the dinitrobenzyl group from histidine, tyrosine, and cysteine with 1-thio-2,4-dinitrobenzene as the product with the unmodified amino acid.

It is most likely that the various studies on the chemical modification of sulfhydryl groups in myosin will be instructive when applying such technology to the study of complex biological sys-tems. The studies with myosin do show that the use of covalently bound fluorescent probes are very useful for exploring conformational change during physiological function.[325–329] To reiterate, the chemical modification of sulfhydryl groups in myosin is complicated and depends on solvent conditions, temperature, and reagent, and consideration of the review of the area by Reisler[330] is of great importance for the design and interpretation of the various studies. There are some 40 plus cysteine residues in myosin of which 12 or so are in the head, which is a globular domain separate from the rod domain obtained by proteolytic cleavage of myosin.[331] There are two subfragment 1[332] in each globular domain containing the enzymatic (ATPase) activities that are considered to exhibit the motor properties of intact myosin.[333,334]

Cyanine dyes (Figure 7.11) have been primarily developed to modify proteins for the purpose of identification of proteins after electrophoretic separation. Cyanine dyes have a long history as

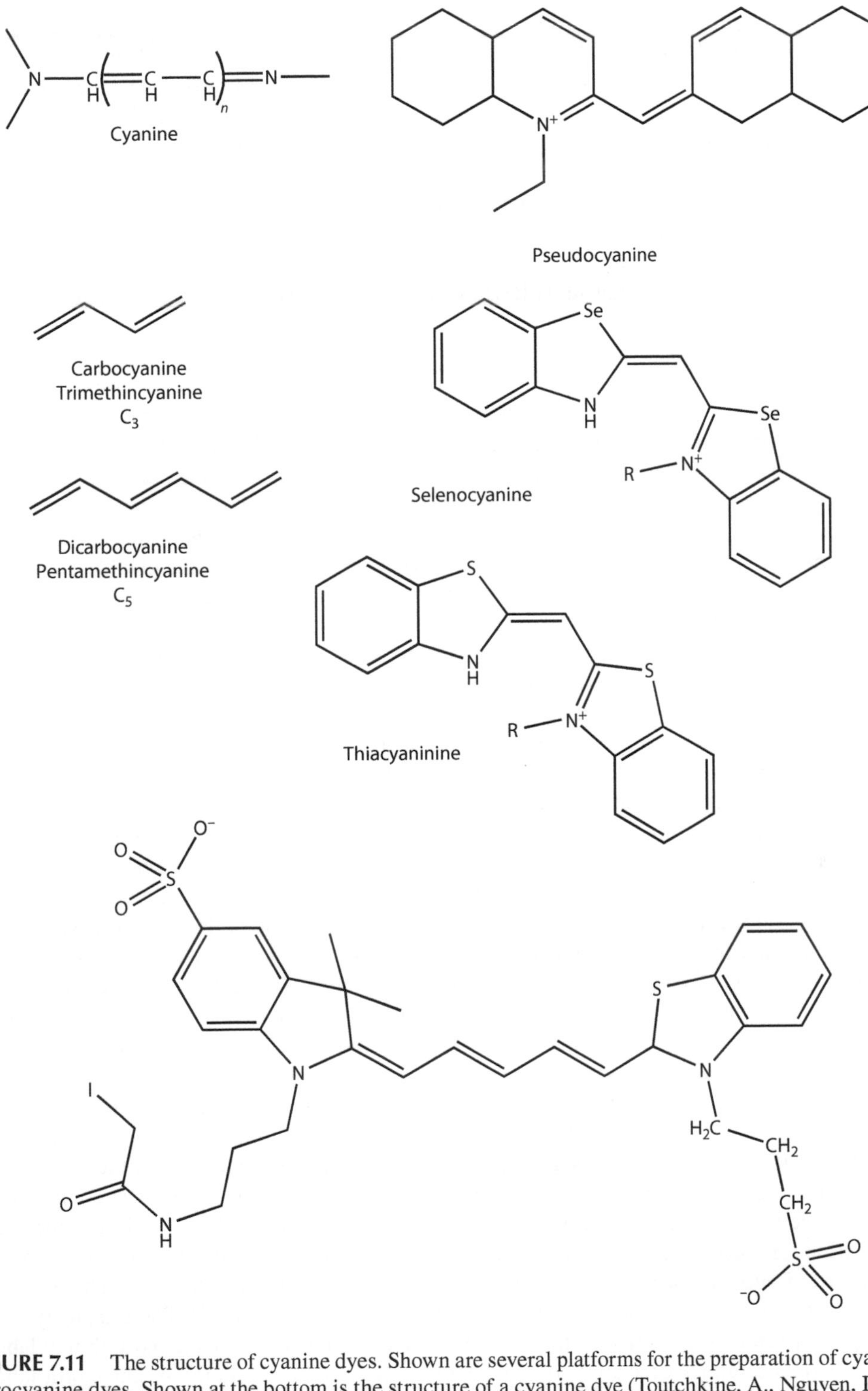

FIGURE 7.11 The structure of cyanine dyes. Shown are several platforms for the preparation of cyanine and merocyanine dyes. Shown at the bottom is the structure of a cyanine dye (Toutchkine, A., Nguyen, D.-V., and Hahn, K.M., Simple one-pot preparation of water-soluble, cysteine-reactive cyanine and merocyanine dyes for biological imaging, *Bioconjug. Chem.* 18, 1344–1348, 2007). The iodoacetamide derivative is shown; the rate of reaction of the chloroderivative was approximately 300-fold less than the iodo derivative.

photographic sensitizers[335–337] and have more recent uses as probes for biological systems.[338–340] Cyanine dyes are one of the series of dyes consisting of two heterocyclic groups (usually a quinoline nucleus) connected by a chain of conjugated double bonds containing an odd number of carbon atoms. An iodoacetamide derivative of a cyanine dye has been developed[341], but there has been limited application of this probe in biological systems.[342–344] Lin and coworkers[342] developed several near-infrared fluorescent probes for imaging use. Chan and coworkers[343] used iodoacetamide derivatives of cyanine dyes to measure redox changes in proteomic research. Human mammary epithelial cells were treated with hydrogen peroxide in the model of oxidative stress. Experimental and control samples were then examined by 2D differential gel electrophoresis (2D-DIGE).[345] Kim and coworkers[344] used 5-iodoacetamido fluorescein as one of the three receptor fluorophore in a fluorescence resonance energy transfer (FRET) system to study estrogen receptor ligand exchange.

Solvatochromic dyes (Figure 7.12) have optical properties (e.g., quantum yield), which are sensitive to changes in solvent polarity[346–348] and are of considerable current interest[349–351] including application to the study of cell migration.[352,353] Loving and coworkers[352] also present an excellent and concise description of the various solvatochromic dyes and conjugation to proteins. Toutchkine and coworkers[354] introduced several solvatochromic dyes with superior solubility and (Figure 7.9). An iodoacetamide function can be attached to these dyes permitting covalent attachment to sulfhydryl groups in proteins. These investigators attached a benzothiophen-3-one-1,1-dioxide iodoacetamide derivative to an inserted cysteine residue in Wiskott–Aldrich syndrome protein (WASP) fragment. This dye–protein conjugate was used as a biosensor for the activated GTP-bound form of Cdc42. Binding of the WASP fragment dye conjugate to the activated Cdc42 was associated with a threefold increase in fluorescence-indicated formation of a hydrophobic pocket and change in hydrogen bonding with solvent; a similar change was not seen with Cy3 or Cy5 dyes, which are sensitive to change in a hydrophobic environment but not to change in hydrogen bonding. Subsequent work from this laboratory[1] has reported an improved method for the synthesis of iodoacetamide derivatives. Garrett and coworkers[355] labeled S100A4 with the same dye using iodoacetamide chemistry. The labeled S100A4 undergoes a threefold increase in fluorescence on activation with calcium ions. Loving and Imperiali[356] have reported the synthesis of a solvatochromic dyes (Figure 7.12) based on 4-*N*,*N*-dimethylamino-1,8-napthalimide (4-DMN) for the study of conformational change in proteins. The dyes were coupled with an iodoacetamide function or a maleimide function, which was coupled to single cysteinyl residues inserted at various points in the calmodulin sequence. One dye–calmodulin conjugate (4-DMN with a 6 Å spacer and maleimide functional group) with the S38C mutant (amino-terminal domain) was observed to undergo a 100-fold increase in intensity following binding to calcium ions. A maleimide derivative with a 9 Å spacer coupled at the same site demonstrated an 80-fold enhancement of fluorescence intensity on binding of calcium ions. Derivatives with bromoacetamido functional groups showed a lesser increased in fluorescence intensity on binding calcium ions. Conjugates with other fluorophores such as 5-IAEDANS did not demonstrate fluorescence intensity increases of comparable magnitude; a maleimide derivative [7-diethylamino-3-((((2-maleimidyl)ethyl)amino)carbonyl)-coumarin] did show approximately 1/3 the response of the 4-DMN probe. Denaturation of the protein conjugates muted the fluorescence response. These derivatives have been used for extracellular studies, but there are complications with the intracellular use of these derivatives[353,357] Goguen and Imperiali[353] do discuss the use of microinjection for placing a dye–protein conjugate inside a cell. Microinjection and electroporation have a record of success for placing dyes and modified proteins inside cells.[358–362] There are also reports of cell penetrating peptides for the delivery of modified proteins.[363] The iodoacetyl function group has been used for introducing a biotin probe in proteins.[364–367] Chu and Fukui[364] labeled myosin with iodoacetamide derivatives of either biotin or tetramethylrhodamine. The biotin-labeled myosin was used to determine the effect of microinjection on the amoeba and ability of the biotin-labeled myosin to associate with the intracellular actin cytoskeleton. Biotin-myosin was observed to associate with specific actin populations. The tetramethylrhodamine derivative was then used to study intracellular association of myosin with actin over a long period of time. The iodoacetamide

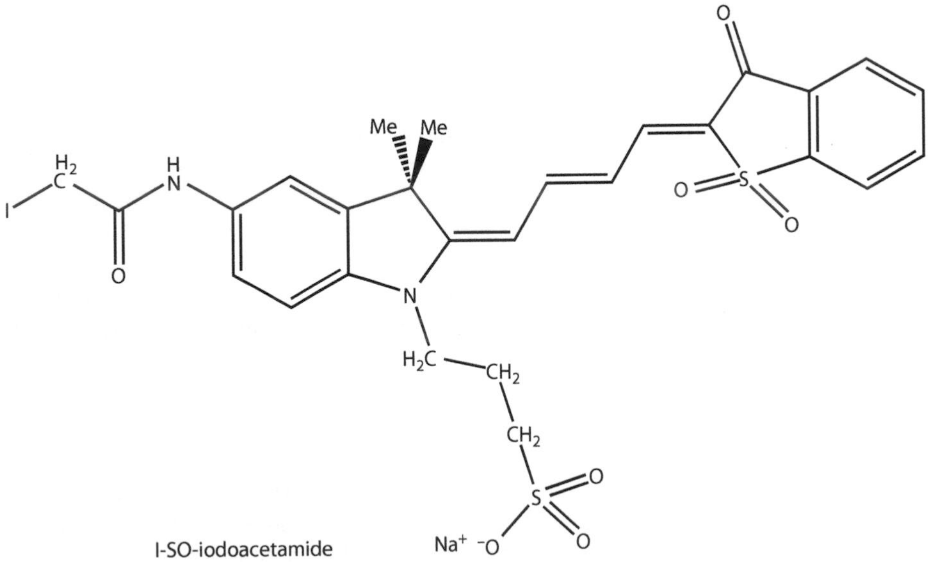

I-SO-iodoacetamide

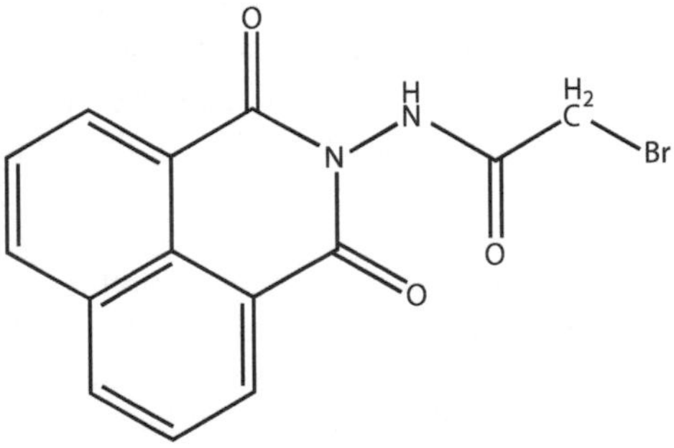

4-*N,N*-Dimethylamino-*N*-[2-bromoacetamido]-1,8-naphthalimide

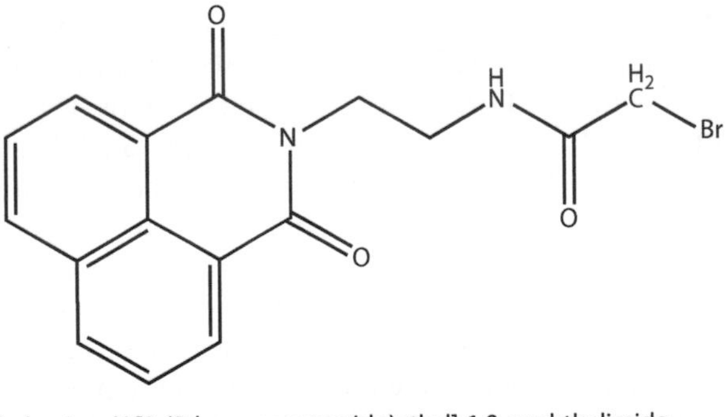

4-*N,N*-Dimethylamino-*N*-[2-(2-bromoacetamido)ethyl]-1,8-naphthalimide

FIGURE 7.12 The structures of several iodoacetamide derivatives of solvatochromic dyes. Shown is the structure of I-SO-iodoacetamide [sodium-3-{(2Z)-5-[(iodoacetyl)amino]2-[2*E*,4*Z*04-(1,1-dioxido-3-oxo-1benzothien-2(3*H*)-ylidene)but-2-enylidene]-3–3-dimethyl2,3-dihydro-1*H*-indol-1-yl}propane-1-sulfonate] (Toutchkine, A., Kraynov, V., and Hahn, K., Solvent-sensitive dyes to report protein conformational changes in living cells, *J. Am. Chem. Soc.* 125, 4132–4245, 2003; Garrett, S.C., Hodgson, L., Rybin, A. et al., A biosensor of S100A4 metastasis factor activation: Inhibitor screening and cellular activation dynamics, *Biochemistry* 47, 986–996, 2008). Also shown are the structures for two bromoacetamido derivatives of a naphthalamide-based fluorophore (Loving, G. and Imperiali, B., Thiol-reactive derivatives of the solvatochromic 4-*N,N*-dimethylamino-1,8-naphthalamide fluorophore: A highly sensitive toolset for the detection of biomolecular interactions, *Bioconjug. Chem.* 20, 2133–2141, 2009).

derivative of biotin has been used to identify cysteine residues in proteins[365,366] including definition of the thiol proteome.[367]

Arylamine N-acetyltransferases (NATs) are intracellular enzymes that catalyze the acetylation of arylamines, arylhydroxylamines, and arylhydrazines in an acetyl CoA-dependent reaction.[368] There are two isozymes, NAT1 and NAT2, in humans and rodents that have different specificities and different tissue distributions.[369] There are also polymorphisms that may have functional consequences resulting in isoforms that are *slow acetylators* or *fast acetylators*.[370] The association of NAT1 expression with breast cancer and response to tamoxifein[371] has increased interest in NAT1 as a therapeutic target.[372] Some of the work that is described below was performed some years before the evidence was obtained on the role of polymorphisms in defining the catalytic activities of the NAT and before there was an understanding of the differences in the specificities of NAT1 and NAT2. Both NAT1 and NAT2 catalyze the acetylation of xenobiotic compounds, and NAT1 is also suggested to have a role in the catabolism of folate through the acetylation of p-aminobenzoylglutamate.

Bartsch and coworkers[373] showed that the transacetylase activity in the cytosol of rat liver was inhibited by either iodoacetamide, PHMB, or N-ethylmaleimide. This work is likely the first to suggest a role for cysteine in the activity of arylamine N-acetyltransferase. PHMB resulted in 62% inhibition of transacetylase activity while mM concentrations of either iodoacetamide or N-ethylmaleimide resulted in total inhibition of transacetylase activity; 30 mM cysteine did not influence the inhibition by either iodoacetate or N-ethylmaleimide but did block inhibition with PHMB. There will be more comment on this observation below. Mangold and Hanna[374] reported that the partially purified N,O-acyltransferase activity from rodent liver was inhibited by either iodoacetamide, PHMB, or N-ethylmaleimide. In the work with the NAT from pigeon liver, Andres and coworkers demonstrated the presence of an enzyme-bound acetyl intermediate[375] in the transfer of acetyl function from acetyl CoA to substrates such as 4-aminobenzoic acid by isolation of a cysteinyl thioester. These investigators showed the reversible (with dithiothreitol) inactivation of the pigeon liver enzyme with DTNB or hydrogen peroxide; inactivation with PHMB was only partially reversible. The results obtained with hydrogen peroxide suggest the formation of sulfenic acid (see below). These workers also reported inactivation of the NAT by iodoacetic acid was irreversible. Several years later, Saito and coworkers[376] showed that NAT activity isolated from hamster liver was inhibited by iodoacetamide consistent with the several previous suggestions on the participation of cysteine in this enzyme. Additional evidence for the participation of a cysteine residue in catalysis by NAT was presented by Cheon and Hanna based on reaction with N-ethylmaleimide[377] demonstrating effective inactivation of either NAT1 or NAT2 in the presence of 0.25 mM dithiothreitol (dithiothreitol is required for the stabilization of enzyme activity; it is stated that the concentration of dithiothreitol does not interfere with N-ethylmaleimide reaction. These investigators also demonstrated inactivation of the NAT by diethylpyrocarbonate (histidine) and phenylglyoxal (arginine). The combined results from the previous studies support the importance of a cysteine residue in the activity of arylamine N-transacetylases and suggest the formation of a thioester intermediate.

The work of Bartsch and coworkers[373] demonstrated that the inactivation of their cytosol enzyme preparations by either iodoacetamide or N-ethylmaleimide occurs in the presence of a substantial molar excess of cysteine (30 mM) implying a hyperactive residue. In looking at the literature, I was surprised by the lack of studies on the effect of low molecular weight thiols on the inactivation of sulfhydryl enzymes. The Bartsch study[373] was one of the few studies that examined this question. Cheon and Hanna[377] did observe that the presence of 0.25 mM dithiothreitol did not affect inactivation of the NAT by N-ethylmaleimide. Andreas and coworkers[378] stated that at pH 6.9 (100 mM MOPS with 1 mM EDTA, 25°C), 500 μM iodoacetamide reacted with no more that 10% (10 μM) of the 100 μM dithiothreitol present in the reaction mixture. The thiol groups on dithiothreitol are not particularly nucleophilic with the reducing power being driven by the formation of the cyclic disulfide.[379] Smythe[380] reported that iodoacetamide reacted with thioglycol more slowly ($t_{1/2} = 2.0$ min) than with cysteine ($t_{1/2} = 0.81$ min) at pH 7.1; no reaction was observed with ethyl mercaptan. Andreas and coworkers[378] also reported the lack of reaction between bromoacetanilide (200 μM) and dithiothreitol (100–200 µM at 25°C in 100 mM MOPS, pH 7.5). I was unable to find information on the reaction of bromoacetanilide with model sulfhydryl compounds. There was early interest in the reaction of low molecular thiols such as glutathione in the early work on the use of iodoacetate in whole animals and tissue extracts.[198] Dickens[381] observed that the iodoacetate inactivation of glyoxalase was due to the reaction with glutathione, now known to be a cofactor in the glyoxalase system. Schroeder and coworkers[382] studied the inhibitory effect of iodoacetic acid on yeast fermentation observing that the observed effect was not due entirely to glutathione depletion. In a more recent study, Zeng and Davies[383] reported that low molecular weight thiols protected proteins from glycation. The point is that reaction of sulfhydryl reagents with low molecular weight thiols needs to be documented rather than assumed. Such information would certainly be required for process validation in a GMP manufacturing situation.

There were a number of subsequent studies on the modification of the active site cysteine residue of NAT (NAT1 and NAT2) derived from hamster. Guo and coworkers[384] reported on the modification of hamster NAT1 with either 2-(bromoacetylamino)fluorene or bromoacetanilide (Figure 7.13). 2-(bromoacetylamino)fluorene was developed as a result of the earlier work[385] on rat hepatic NAT isoforms. Wick and coworkers[385] used 1-(fluoren-2-yl)-2-propen-1-one (Figure 7.10) as an affinity label for NAT2 based on the structure of N-hydroxy-2-acetylaminofluorene, a substrate for the acetylation of 4-aminobenzoic acid by NAT2 (Figure 7.10). The rapid rate of NAT inactivation by 1-(fluoren-2-yl)-2-propen-1-one precluded determination of rate constants for the reaction; attempts to work at a lower temperature were frustrated by low solubility of 1-(fluoren-2-yl)-2-propen-1-one. Guo and workerers[384] reported that the inactivation of NAT1 by 2-(bromoacetylamino)fluorene demonstrated saturation kinetics (second-order rate constant, 1587 $M^{-1} s^{-1}$) consistent with affinity labeling[200] while bromoacetanilide is shown to behave kinetically as a simple alkylating agent reacting by an S_N2 reaction with a second-order rate constant of 7.3 $M^{-1} s^{-1}$ (25°C, pH 7.0). Wang and coworkers[386] reported a second-order rate constant of 5.2 $M^{-1} s^{-1}$ for the inactivation of NAT2 by bromoacetanilide. These investigators also observed saturation kinetics for the reaction of NAT2 with 2-(bromoacetylamino)fluorene consistent with affinity labeling; however, the second-order rate constant of 35 $M^{-1} s^{-1}$ obtained for the NAT2 reaction is much less than the value reported for the same reaction with NAT1.[384] It is noted that a previous study[378] suggested that bromoacetanilide demonstrated saturation kinetics with the liver enzyme. A value of 24.9 $M^{-1} s^{-1}$ for the inactivation was calculated by this author (RLL) from data in this study,[378] which are close to the value obtained for 2-(bromoacetylamino)fluorene for NAT2 from hamster liver but much less than the value obtained for reaction with NAT1. Andreas and coworkers[378] also reported other findings of interest with the reactivity of cysteine in the NAT derived from rapid acetylator rabbit liver. Chloroacetanilide was much less effective that the bromoacetanilide (at 50 µM reagent concentration, 0°C, pH 7.5); the $t_{1/2}$ was 5.5 min for the bromine derivative and 10 h for the chlorine derivative. At 25°C (100 µM reagent concentration), the $t_{1/2}$ was 20 min for chloroacetanilide and 47 min for iodoacetic acid. In a later study, Wang and coworkers[387] reported on the reaction of iodoacetamide or bromoacetamide with hamster NAT2 as measured by the loss of enzyme activity; no significant loss of enzyme activity was reported for reaction with either iodoacetic acid or bromoacetic acid (data not shown). At pH 7.0, a second-order rate constants of 802.7 $M^{-1} s^{-1}$ was obtained for reaction with iodoacetamide, while a value of 426.9 $M^{-1} s^{-1}$ was obtained for bromoacetamide. These rates are comparable to rates obtained for reaction with active-site cysteine in thiol proteases such as streptococcal proteinase. Brenda Gerwin who worked with Stanford Moore and William Stein on the 5th floor of Flexner Hall at the Rockefeller University demonstrated the exceptional reactivity of the cysteine residue at the active site of streptococcal proteinase.[388] Chloroacetic acid and chloroacetamide demonstrated vastly different properties. Chloroacetic acid show a bell-shaped pH dependence with an optimum at approximately pH 7, while pH dependence for chloroacetamide was a sigmoidal curve similar to that shown by glutathione reaction with a maximum value above pH 10. The rate constant at pH 9.0 for reaction of streptococcal proteinase with chloroacetic acid is 15 M^{-1} min^{-1} (~20 M^{-1} min^{-1} at pH 7), while it is 800 M^{-1} min^{-1} for chloroacetamide. Chaiken

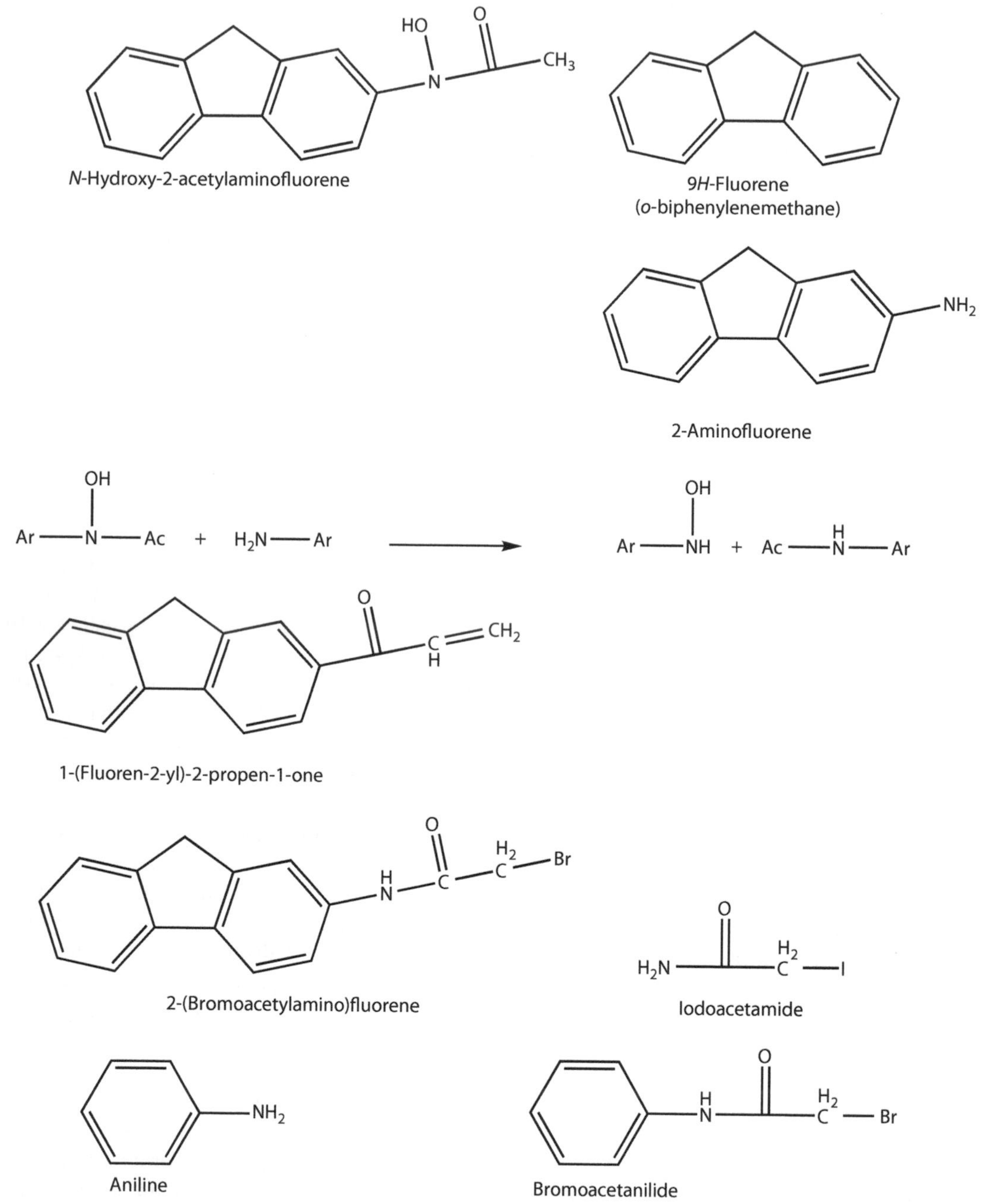

FIGURE 7.13 Structure of some reagents for modification of cysteine in NATs. Shown are the structures of *N*-hydroxy-2-acetylaminofluorene, a substrate for NAT2 and 1-(fluoren-2-yl)-2-propen-1-one, an affinity label (Wick, M.J., Yeh, H.-M., and Hanna, P.E., An isozyme-selective affinity label for rat hepatic acetyltransferases, *Biochem. Pharmacol.* 40, 1389–1398, 1990). Shown below are the structures for 2-(bromoacetylamino) fluorene, an affinity label for NAT1 and NAT2 and bromoacetanilide (Cheon, H.G., Boteju, L.W., and Hanna, P.E., Affinity alkylation of hamster hepatic arylamine *N*-acetyltransferases: Isolation of a modified cysteine residue, *Mol. Pharmacol.* 42, 82–93, 1992; Guo, Z., Vath, G.M., Wagner, C.R., and Hanna, P.E., Arylamine *N*-acetyltransferases: Covalent modification and inactivation of hamster NAT1 by bromoacetamido derivatives of aniline and 2-aminofluorene, *J. Protein Chem.* 22, 631–642, 2003). Also shown for reference is the structure of 9*H*-fluorene, 2-aminofluorene, and aniline for reference.

and Smith later showed similar results for the reaction of chloroacetic acid or chloroacetamide with papain: a bell-shaped curve was obtained for reaction with chloroacetic acid,[389] while a sigmoidal curve was obtained for chloroacetamide.[390] The pH dependence for the inactivation of hamster NAT2 by iodoacetamide[387] shows a rapid increase from pH 4.5 to pH 5.5 (apparent $pKa = 5.23$) and then slightly decreases to a plateau extending beyond pH 9.5. This profile is dramatically different from data obtained with papain[389,390] or streptocooccal proteinase.[388] The low cysteine pKa value is consistent with an exceptionally reactive thiol nucleophile[391–393] although other factors may be involved.[394–397] The fact is that NATs have an exceptionally reactive thiol residue at the enzyme active site.[398] As an aside, I did find that NAT1 is inhibited by disulfiram[399] with a second-order rate constant of $6 \times 10^4\,M^{-1}\,min^{-1}$. As a comparison, Nobel and coworkers[253] reported a second-order rate constant of $2.2 \times 10^3\,M^{-1}\,s^{-1}$ for the inactivation of caspase-1 (Table 7.2.).

The development of proteomic technologies such as mass spectrometry enabled the quantitative analysis of protein expression (Table 7.4) in cells in response to challenges such as pharmaceuticals. One approach, SILAC, involves the use of stable-isotope (i.e., ^{13}C) labeling in yeast and bacteria.[400–403] Here, the proteins are chemically identical differing in only mass that can be identified by mass spectrometry. Isotope-coded tags provide a way of differentially labeling control and experimental protein mixtures by chemical modification with chemically identical tags differing in mass (usually by incorporation of ^{13}C or ^2H [deuterium]); the differentiating mass label is applied after synthesis rather than being incorporated in the synthesis process. There are several approaches to the use of this technology. Isotope-coded protein labeling (ICPL) is a technique that uses nonisobaric reagents (Figure 7.14) based on nicotinic acid to modify lysine residues in protein for subsequent proteomic analysis.[404–407] Coombs[406] compares ICPL with other labeling technologies such as tandem mass tags and isotope-coded affinity tagging while Lottspeich and Kellerman[407] provide a concise description of the laboratory technology. Tandem mass tags are complex, isobaric molecules consisting of a reactive functional group and a tandem mass tag fragment that serves to differentiate the tag (Figure 7.14). Tandem mass tags are used to modify cysteine residues in reduced proteins in cell lysates (proteins are denatured and reduced prior to modification). iTRAQ are also isobaric reagents for multiplex analysis of proteins; iTRAQ reagents have been used to modify amino groups in proteins. The process differs from tandem mass tagging (TMT) in that the protein fraction from the various fractions is obtained, denatured, and reduced and coupled with the iTRAQ reagents. The digested sample is then mixed and subjected to MS/MS analysis.

The use of isotope-coded affinity tags[408,409] for studying differential protein expression[410–412] has fostered the development of novel reagents based on the iodoacetyl function. Examples include N-(13-iodoacetamido-4,7,10-trioxatridecanyl) biotinamide,[413] N-ethyliodoacetamide,[414] and N-t-butyliodoacetamide and iodoacetanilide.[415] These reagents are prepared as the *light* form (hydrogen) or ^{12}C-labeled and the *heavy* form (deuterium) or ^{13}C.[416] The biotin derivatives are referred to as ICAT reagents (Figure 7.15).[416,417] The biotin *tag*, the affinity portion of the ICAT reagent, enables the rapid purification of peptides/proteins from mixtures after modification of cysteine residues with this reagent. Metal-coded affinity tags (Figure 7.15) have been developed for quantitative proteomics; the first generation of reagent used maleimide functional groups,[418] while more recent generation of reagents has used the iodoacetamide function.[419]

Another approach to cysteine modification for proteomics is the use of ethyleneimine (Chapter 2) for an in-gel derivatization of proteins after electrophoretic separation[420] introducing a new site for cleavage by trypsin. Finally, vinyl pyridine (Chapter 2) is an alternative for the alkylation of cysteine residues in proteins[421] and has been used in proteomic analysis.[422–425] Deuterated derivatives have been prepared for use in the study of differential protein expression.[421,426] The use of iodoacetate and iodoacetamide for protein structural studies was briefly mentioned in Chapter 2. As most proteins are resistant to the proteolysis required to obtain peptide for structural analysis, reduction and carboxymethylation has a long history of use in structural analysis.[427] There has been continued use of iodoacetate for a variety of purposes[428,429] including structural analysis of proteins.[430] Iodoacetamide has been used with increasing frequency[430–433] as it is less likely that iodoacetamide will influenced

TABLE 7.4
Methods for Quantitative Proteomics[a]

Technique	Description	Reference
SILAC	Stable-isotope labeling in culture[b]	7.4.1–7.4.3
SILAM	Stable-isotope labeling of mammals[c]	7.4.4–7.4.7
CDIT	Culture-derived isotope tags[d]	7.4.8–7.4.10
AQUA	Absolute quantitation[e]	7.4.11–7.4.13
MRM-MS	Multiple reaction monitoring mass spectrometry[f]	7.4.14–7.4.16
ICPL	Isotope-coded protein labeling[g]	7.4.17–7.4.21
TMT	Tandem mass tagging[h]	7.4.22–7.4.25
iTRAQ	Isotope-tagging agents for absolute quantitation[i]	7.4.26–7.4.28
ICAT	Isotope-coded affinity tagging[j]	7.4.29–7.4.31

[a] Quantitative proteomics describes the various methods used to quantitatively measure protein expression in the cell either in relative term or absolute term. The ultimate goal would be to quantitatively measure the proteome at any time. Relative methods provide information on relationships in the expression of specific proteins in complex mixtures in response to various stimuli; relative methods do not provide absolute values.

[b] SILAC is a method to evaluate the effect of a stimulus on cells based on the incorporation of stable-isotope -labeled amino acids (e.g., ^{13}C, ^{15}N) into proteins during cell growth. Most SILAC experiments used radiolabeled lysine and arginine with trypsin used to generate peptides for analysis (since trypsin cleaves at arginine are lysine, all derived peptides will contain radiolabel). A test culture is mixed with a control culture and the relative distribution of isotope in proteins evaluated by LC/MS.

[c] SILAM is the application of stable isotope incorporation into proteins in intact animals with analysis as with SILAC with MS. In essence, it is SILAC applied to an intact animal. SILAM should not be confused with System for Integrated modeLling of Atmospheric coMposition (SILAM) (Sofiev, M., Slijamo, P., Ranta, H., and Rantio-Lehtimäke, A., Towards numerical forecasting of long-range air transport of birch pollen: Theoretical considerations and a feasibility study, *Int. J. Biometeorol.* 50, 392–402, 2006).

[d] Culture-derived isotope tags (CDIT) (also cytochrome P450 drug interaction table; Polasek, T.M., Lin, F.P., Miners, J.O., and Doogue, M.P., Perpetrators of pharmacokinetic drug–drug interactions arising from altered cytochrome P450 activity: A criteria-based assessment, *Br. J. Clin. Pharmacol.* 71, 727–736, 2011) is a method of obtaining a stable-isotope labeled standard. A consideration of the literature shows limited application with use restricted to nervous system tissue. Cells are grown in culture with stable isotope labeled amino acids and added to the standard tissue sample and the control tissue sample to provide an internal standard for LC/MS analysis following extraction of protein and proteolytic digestion. Synthetic unlabeled peptides can be used to quantify the isotopically labeled protein standards.

[e] Absolute quantitation (AQUA) is based on isotope dilution theory and the use of synthetic peptides containing stable isotope-labeled amino acids. The synthetic isotope-labeled peptide(s) must be identical to the peptide(s) derived from the experimental proteins. A variant approach is used in CDIT above.[d]

[f] Multiple reaction monitoring mass spectrometry (MRM-MS) (also known as selected reaction monitoring). The application of selected reaction monitoring to multiple product ions derived from one or more precursor ions (Roepstorff, P. and Fohlman, J., Proposal for a common nomenclature for sequence ions in mass spectra of peptides, *Biomed. Mass Spectrom.* 11, 601, 1984). This term should not be confused with consecutive reaction monitoring, which involves the serial application of three or more stages of selected reaction monitoring (*J. Chromatogr.*, glossary, 2010). The technology is based on the targeted separation of selected ions in tandem through three quadrupole mass spectrometers. Quantitation is based on isotope dilution using ^{13}C-labeled peptide standards similar to those used in AQUA.

[g] ICPL labels proteins with stable-isotope-labeled reagents. The majority of studies use *N*-hydroxysuccinimide for the modification of protein amino groups. The experimental and control samples are processed separately, fractionated, and digested with a protease such as endoproteinase Glu-C (modification at lysine precludes the use of trypsin). Analysis is performed using LC/MS. While the majority of studies use modification at amino groups, an isotope-coded dimedone procedure has been used for sulfenic acid. A process described as dual stable isotope coding uses ^{13}C-labeled acrylamide for labeling cysteine residues and ^{13}C-labeled succinic anhydride for labeling lysine residues in a variation of ICPL.

TABLE 7.4 (continued)
Methods for Quantitative Proteomics[a]

h   TMT is a technique based on the use of MS/MS (McLafferty, F.W., Tandem mass spectrometry, *Science* 214, 280–287, 1981). The reagent (Figure 12.11) consists of a mass reporter group, a mass normalizer group, and a functional group. Reagents are intended to react with sulfhydryl groups and lysine residues. It is possible to eightplex samples with reagents that are isobaric and not subject to mass-based variation in separation technologies. Hydrazide and aminoxy functionalized reagents have been developed for reaction with glycans in proteins. These are isobaric reagents so there is no issue of different molecular mass influencing separation on high-performance liquid chromatography (HPLC) (Berg, T. and Strand, C.H., ^{13}C labelled internal standards—A solution to minimize ion suppression effects in liquid chromatography-tandem mass spectrometry analyses of drugs in biological samples? *J. Chromatogr.* 1218, 9366–9374, 2011).

i   Isotope tag for relative and absolute quantitation (iTRAQ). ITRAQ® reagents are stable-isotope labeled reagents that react with amino groups in proteins. It is possible multiplex samples for relative and absolute quantitation of proteins. The reagents are similar to those used in MS/MS including a mass reporter group, a linker/mass normalization group, and a reactive function, frequently *N*-hydroxysuccinimide. These are isobaric reagents so there is no issue of different molecular mass influencing separation on HPLC (Berg, T. and Strand, C.H., ^{13}C labelled internal standards—A solution to minimize ion suppression effects in liquid chromatography-tandem mass spectrometry analyses of drugs in biological samples? *J. Chromatogr.* 1218, 9366–9374, 2011).

j   Isotope-coded affinity tags (ICAT) (Figure 12.15) are reagents that can be heavy (containing a stable isotope such as deuterium or ^{13}C). Most ICAT reagents contain iodoacetamide as tag for the cysteine residues in the target protein or peptide; there are other novel tags such as dimedone for sulfenic acid (Truong, T.H., Garcia, F.J., Ho Seo, Y., and Carroll, K.S., Isotope-coded chemical reporter and acid-cleavable affinity reagents for monitoring protein sulfenic acids, *Bioorg. Med. Chem.* 21, 5015–5020, 2011). A linker region contains the heavy or light atoms that are terminated by an affinity label such as biotin. The linker region may contain an acid-labile bond that can be cleaved to release the peptide/protein from the matrix, which, in the case of biotin, contains avidin.

References to Table 7.4

7.4.1.   Schulze, W.X. and Mann, M., A novel proteomic screen for peptide–protein interactions, *J. Biol. Chem.* 279, 10756–10764, 2004.

7.4.2.   Ong, S.-E. and Mann, M., Stable isotope labeling by amino acids in cell culture for quantitative proteomics, in *Quantitative Proteomics by Mass Spectrometry*, ed. S. Sechi, Chapter 3, pp. 37–52, Humana Press, Totowa, NJ, 2007.

7.4.3.   Munday, D.C., Surtees, R., Emmett, E. et al., Using SILAC and quantitative proteomics to investigate the interactions between viral and host proteomes, *Proteomics* 12, 66–672, 2012.

7.4.4.   Liao, L., McClatchy, D.B., Park, S.K. et al., Quantitative analysis of brain nuclear phosphoproteins identifies developmentally regulated phosphorylation event, *J. Proteome Res.* 7, 4743–4755, 2008.

7.4.5.   McClatchy, D.B., Liao, L., Park, S.K. et al., Differential proteomic analysis of mammalian tissues using SILAM, *PLoS One*, 6(1), e16039, 2011.

7.4.6.   Lu, X.M., Tompkins, R.G., and Fischman, A.J., SILAM for quantitative proteomics of liver Akt1/PKBα after burn injury, *Int. J. Mol. Med.* 29, 461–471, 2012.

7.4.7.   McClatchy, D.B., Liao, L, Lee, J.H. et al., Dynamics of subcellular proteomes during brain development, *J. Proteome Res.* 11, 2467–2479, 2012.

7.4.8.   Ishihama, Y., Sato, T., Tabata, T. et al., Quantitative mouse brain proteomics using culture-derived isotope tags as internal standards, *Nat. Biotechnol.* 23, 617–621, 2005.

7.4.9.   Sato, T., Ishihama, Y., and Oda, Y., Quantitative proteomics of mouse brain and specific protein-interaction studies using stable isotope labeling, in *Quantitative Proteomics by Mass Spectrometry*, ed. S. Sechi, Chapter 4, pp. 53–70, Humana Press, Totowa, NJ, 2007.

7.4.10.   Li, C., Ruan, H.Q., Liu, Y.S. et al., Quantitative proteomics reveal upregulated protein expression of the SET complex associated with hepatocellular carcinoma, *J. Proteome Res.* 11, 871–885, 2012.

7.4.11.   Kirkpatrick, D.S., Gerber, S.A., and Gygi, S.P. The absolute quantification strategy: A general procedure for the quantification of proteins and post-translational modifications, *Methods* 35, 265–273, 2005.

(continued)

TABLE 7.4 (continued)

Methods for Quantitative Proteomics[a]

7.4.12. Gerber, S.A., Kettenbach, A.N., Rush, J., and Gygi, S.P., The absolute quantification strategy. Application to phosphorylation profiling of human Separase Serine 1126, in *Quantitative Proteomics by Mass Spectrometry*, ed. S. Sechi, Chapter 5, pp. 71–86, Humana Press, Totowa, NJ, 2007.

7.4.13. Phillips, C.M., Iavarone, A.T., and Marletta, M.A., Quantitative proteomic approach for cellulose degradation by *Neurospora crassa*, *J. Proteome Res.* 10, 4177–4185, 2011.

7.4.14. Zakett, D., Flynn, R.G.A., and Cooks, R.G., Chlorine isotope effects in mass spectrometry by multiple reaction monitoring, *J. Phys. Chem.* 82, 2359–2362, 1978.

7.4.15. Kuhn, E., Wu, J., Karl, J. et al., Quantification of C-reactive protein in the serum of patients with rheumatoid arthritis using multiple reaction monitoring mass spectrometry and 13C-labeled peptide standards, *Proteomics* 4, 1175–1186, 2004.

7.4.16. Chen, Y.T., Chen, H.W., Domanski, D. et al., Multiplexed quantification of 63 proteins in human urine by multiple reaction monitoring-based spectrometry for discovery of potential bladder cancer biomarkers, *J. Proteomics* 75, 3529–3545, 2012.

7.4.17. Bisle, B., Schmidt, A., Scheibe, B. et al., Quantitative profiling of the membrane proteome in a halophilic archaeon, *Mol. Cell. Proteomics* 5, 1543–1558, 2006.

7.4.18. Paradela, A., Marcilla, M., Navajas, R. et al., Evaluation of isotope-coded protein labeling (ICPL) in the quantitative analysis of complex proteomes, *Talenta* 80, 1496–1502, 2010.

7.4.19. Kellermann, J. and Lottspeich, F., Isotope-coded protein label, in *Quantitative Methods in Proteomics*, ed. K.Marcus, Chapter 11, pp. 143–153, Springer Science, New York, 2012.

7.4.20. Wang, H., Wong, C.-H., Chin, A. et al., Quantitative serum proteomics using dual stable isotope coding and nano LC-MS/MSMS, *J. Proteome Res.* 8, 5412–5422, 2009.

7.4.21. Truong, T.H., Garcia, F.J., Seo, Y.H., and Carroll, K.S., Isotope-coded chemical reporter and acid-cleavable affinity reagents for monitoring protein sulfenic acids, *Bioorg. Med. Chem. Lett.* 21, 5015–5020, 2011.

7.4.22. Thompson, A., Schäfer, J., Kuhn, K. et al., Tandem mass tags: A novel quantification strategy for comparative analysis of complex protein mixtures by MS/MS, *Anal. Chem.* 75, 1895–1904, 2003.

7.4.23. Hahne, H., Neubert, P., Kuhn, K. et al., Carbonyl-reactive tandem mass tags for the proteome-wide quantification of N-linked glycans, *Anal. Chem.* 84, 3716–3724, 2012.

7.4.24. Werner, T., Becher, I., Sweetman, G. et al., High-resolution enabled TMT 8-plexing, *Anal. Chem.* 84, 7188–7194, 2012.

7.4.25. Tandem Mass Tags-Cysteine Reagents. Thermo Scientific, January, 2014, http://www.piercenet.com/browse.cfm?fldID=CB4C02C4-5056-8A76-4EF8-D72 D59A2EAB9.

7.4.26. Zieske, L.R., A perspective on the use of iTRAQ™ reagent technology for protein complex and profiling studies, *J. Exp. Bot.* 57, 1501–1508, 2006.

7.4.27. Unwin, R.D., Quantification of proteins by iTRAQ, in *LC-MS/MS in Proteomics*, eds. P.R. Cutillas and J.F. TImms, Chapter 12, pp. 205–215, Springer Science-Business Media, Berlin, Germany, 2010.

7.4.28. McDonagh, B., Martinez-Acedo, P., Vásquez, J. et al., Application of iTRAQ reagents to relatively quantify the reversible redox state of cysteine residues, *Int. J. Proteomics* 2012, 514847, 2012.

7.4.29. Gygi, S.P., Rist, B., Gerber, S.A. et al., Quantitative analysis of complex protein mixtures using isotope-coded affinity tags, *Nat. Biotechnol.* 17, 994–999, 1999.

7.4.30. Zhou, H., Ranish, J.A., Watts, J.D. et al., Quantitative proteome analysis by solid-phase isotope tagging and mass spectrometry, *Nat. Biotechnol.* 19, 512–515, 2002.

7.4.31. Esteban-Fernández, D., Scheler, C., and Linscheid, M.W., Absolute protein quantification by LC-ICP-MS using MeCAT peptide labeling, *Anal. Bioanal. Chem.* 401, 657–666, 2011.

by local environmental factors and carboxamidomethylation does not result in the addition of a charged group. Carboxymethylation or carboxamidomethylation of bovine serum albumin results in a protein product with different physical characteristics.[434] Batra and coworkers[435] found that the reaction of bovine serum albumin with iodoacetate (or DTNB) formed substantial β-structure as well as random coil in the presence of guanidine (6 M) or urea (8 M), while modification with

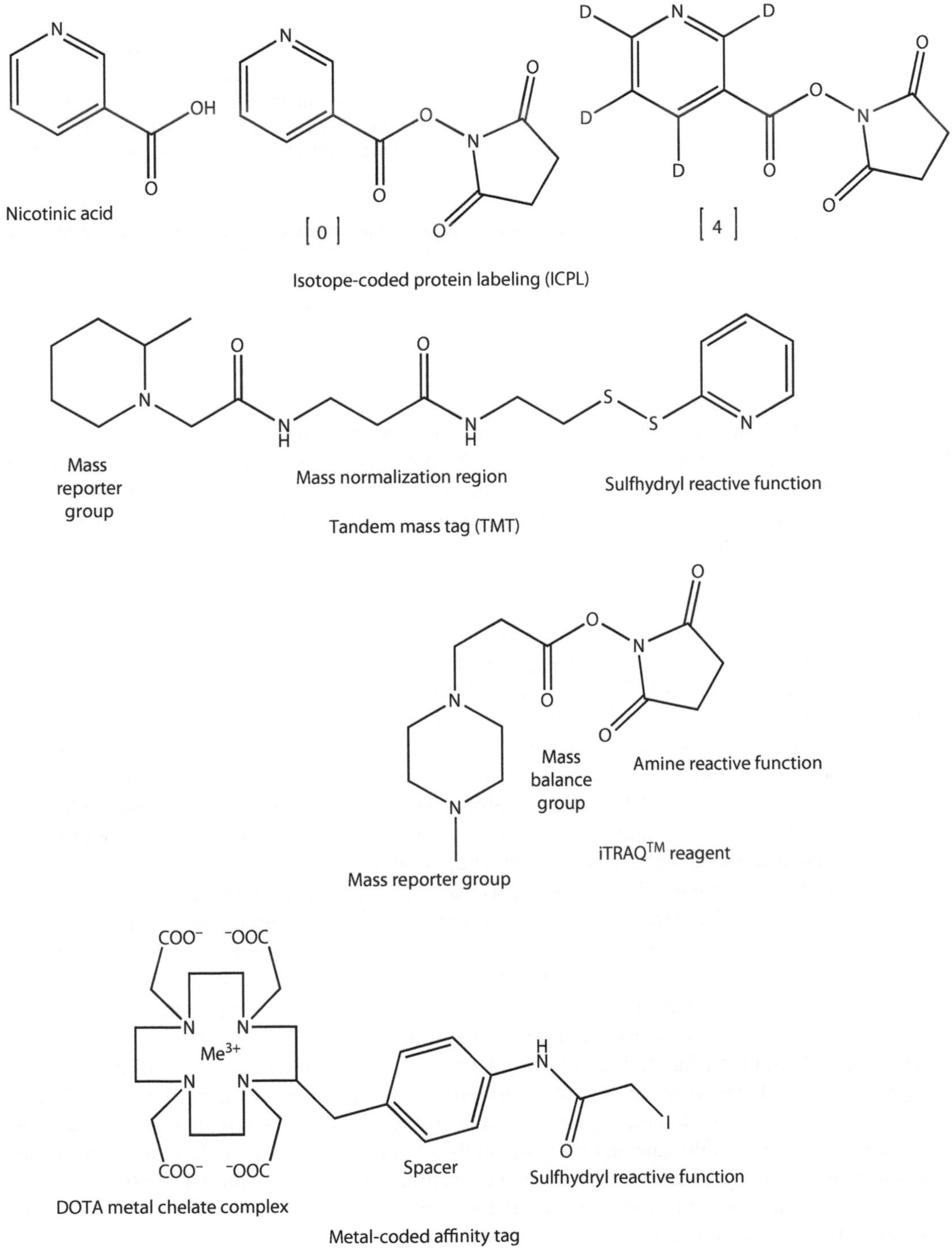

Nicotinic acid

[0]

[4]

Isotope-coded protein labeling (ICPL)

Mass reporter group

Mass normalization region

Sulfhydryl reactive function

Tandem mass tag (TMT)

Mass balance group

Amine reactive function

iTRAQ™ reagent

Mass reporter group

DOTA metal chelate complex

Spacer

Sulfhydryl reactive function

Metal-coded affinity tag

FIGURE 7.14

(*continued*)

FIGURE 7.14 (continued) Chemical modification reagents for use in quantitative proteomics. Shown at the top are derivatives of nicotinic acid that are used in ICPL. The derivatives shown are used for labeling amino groups and variation of deuterium labeling and ^{13}C permits four reagents that allow a fourplex study (Schmidt, A., Kellerman, J., and Lottspeich, F., A novel strategy for quantitative proteomics using isotope-coded protein labels, *Proteomics* 5, 4–15, 2005). Below the ICPL reagent is a tandem mass tag consisting of a mass reporter group, a linker/mass normalized region, and a sulfhydryl reactive function (Murray, C.I., Uhrigshardt, H., O'Meally, R.N. et al., Identification and quantification of S-nitrosylation by cysteine reactive tandem mass tag switch assay, *Molec. Cell. Proteomics* 11:10.1074/mcp.M111.013441, 1–12, 2012). These reagents are available from ThermoScientific (CysTmT®; http://www.piercenet.com); the pyridyl disulfide derivatives shown are being replaced by haloacetyl derivatives. Shown at the bottom is the amine-reactive isobaric tagging reagent that is now described as an iTRAQ™ reagent (Ross, P.L., Huang, Y.N., Marchese, J.N. et al., Multiplexed protein quantitation in *Saccharomyces cerevisiae* using amine-reactive isobaric tagging reagents, *Mol. Cell. Proteomics* 3, 1154–1168, 2004; Zieske, L.R., A perspective on the use of iTRAQ™ reagent technology for protein complex and profiling studies, *J. Exp. Bot.* 57, 1501–1508, 2006). A metal-coded affinity tag is shown at the bottom (Schwarz, G., Beck, S., Weller, M.G., and Linscheid, M.W., MeCAT—New iodoacetamide reagents for metal labeling of proteins and peptides, *Anal. Bioanal. Chem.* 401, 1203–1209, 2011).

iodoacetamide provided a product with substantially more random coil conformation and less β-structure than observed with carboxylation in the presence of chaotropic agents. Huggins and Jensen[436] reported that treatment of serum with iodoacetate prevented heat aggregation while iodoacetamide did not; iodoacetate prevented aggregation of albumin on heating while iodoacetamide promoted aggregation. It is presumed that there is a single site of modification (Cys34) in albumin resulting in substantial change(s) in physical properties. Trevidi and coworkers[437] modified the single sulfhydryl group of the EC1 domain of E-cadherin with iodoacetate, iodoacetamide, or a maleimide-PEG derivative. The native EC1 domain protein expressed in *E. coli* undergoes dimerization in solution followed by aggregation. These investigators modified the native protein as expressed in *Esherichia coli*. The carboxymethylated protein did not fold properly, while the carboxyamidomethylated protein folded in a manner similar to the native protein. I would be remiss if I did not note that a number of investigators refer to the product of iodoacetamide reaction with a nucleophile as carbamidomethylation rather than carboxamidomethylation dating to work by C.H. Li in San Francisco in 1966.[438] I have been unable to ascertain if there is nomenclature ruling on this matter, but the term carboxamidomethylation will be used in the current study to refer to the reaction of a haloacetamide with a protein-bound nucleophile.

Reduction and carboxymethylation of proteins was used for the preparation of proteins for classical structure analysis[427] and more recently for the preparation of samples for structural analysis by mass spectrometry.[432] An important point raised by Galvani and coworkers[432] is that the alkylation process can take a significant period of time (hours as opposed to minutes). Alkylation can also occur at sites other than sulfhydryl groups in peptides.[439] Oxidation is also proving to be a useful approach to modifying cysteine residues for the structural analysis of proteins.[440,441]

The Michael addition (Figure 7.16) was originally used to describe the addition of an activated methylene (Michael donor) to an activated unsaturated system such as an α,β-ketone (Michael acceptor) but has a broader definition today in biochemistry where it is used to describe the addition of a nucleophile, such as thiolate, across a double bond conjugated with an electron-withdrawing group such as a carbonyl function.[442–444] Examples include the reaction of proteins with maleimide derivatives, quinones, as well as acrylamides and 4-hydroxy-2-nonenal (4-HNE) as Michael acceptors and sulfhydryl and amine functions as Michael donors.

While most applications of the Michael addition involve a Michael acceptor reacting such as a maleimide reacting with a protein-bound Michael donor such as the thiolate anion, Michael acceptors have been formed in proteins, mostly by β-elimination of serine or threonine residues in forming dehydroalanine or dehydrobutyric acid that then reacts with a Michael donor in the solution phase (Figure 7.17) or undergoes an intramolecular reaction forming cross-links. Nisin

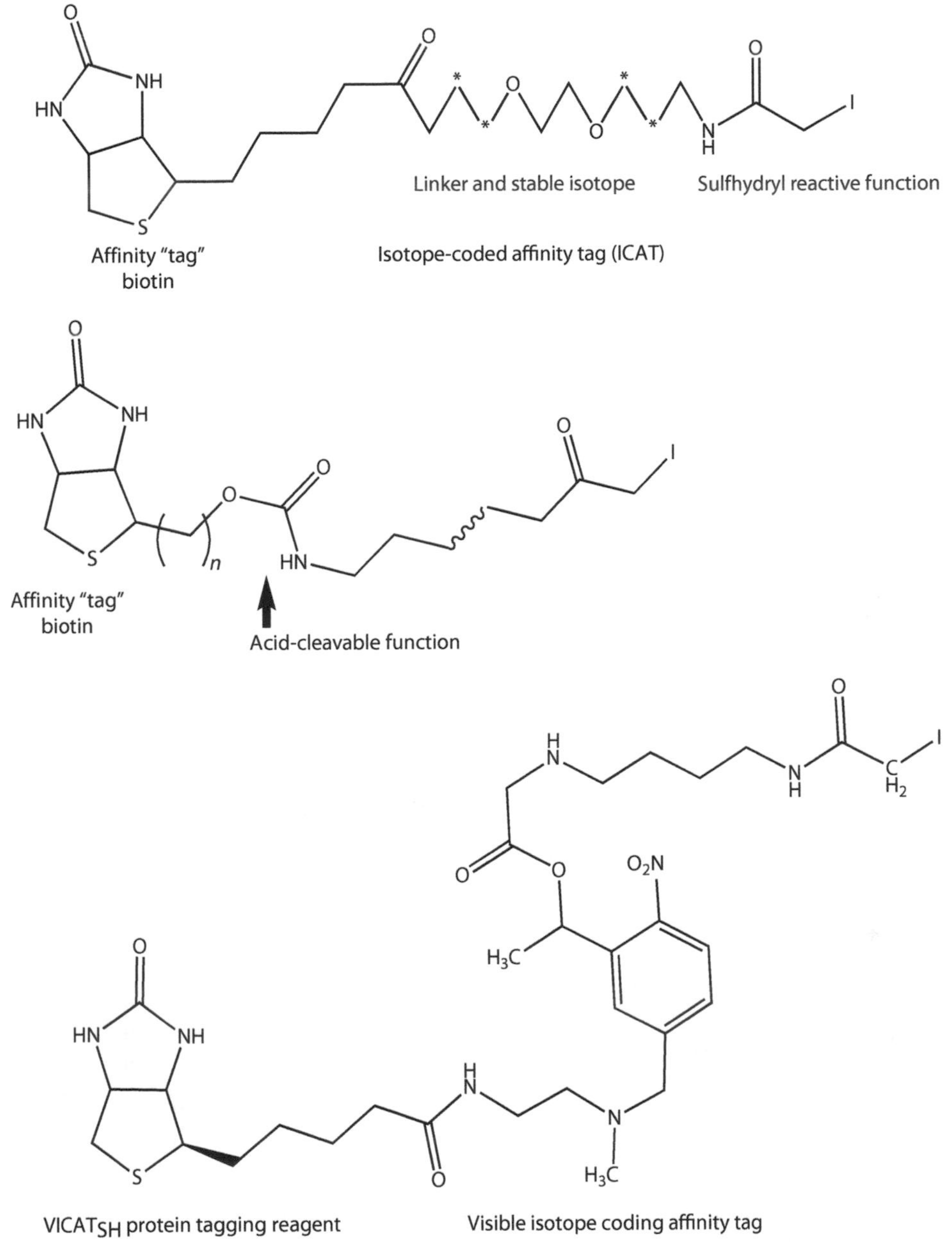

FIGURE 7.15 The variety of reagents used in isotope-coded affinity tags. ICAT, as with ICPL tags, are synthesized with and without isotope. Other than the isotope labeling, the structures are identical containing the affinity tag moiety, in this case, a biotin that can be captured with streptavidin; a middle mass differentiating portion/linker with stable isotope position(s) indicated with asterisks; and reactive function, in this case, an iodoacetyl function that reacts with sulfhydryl groups (Smolka, M.B., Zhou, H., Purkayastha, S., and Aebersold, R., Optimization of the isotope-coded affinity tag-labeling procedure for quantitative proteome analysis, *Anal. Biochem.* 297, 25–31, 2001). An acid-cleavable derivative is shown in part below (Fauq, A.H., Karche, R., Khan, M.A., and Vega, I.E., Synthesis of acid-cleavable light isotope-coded affinity tags (ICAT-*L*) for potential use in proteomic expression profiling analysis, *Bioconjug. Chem.* 17, 248–254, 2006). Shown below the acid-cleavable isotope-coded affinity tag is a visible-coded affinity tag (Lu, Y., Bottari, P., Turecek, F. et al., Absolute quantification of specific proteins in complex mixtures using visible isotope-coded affinity tags, *Anal. Chem.*76, 4104–4011, 2004).

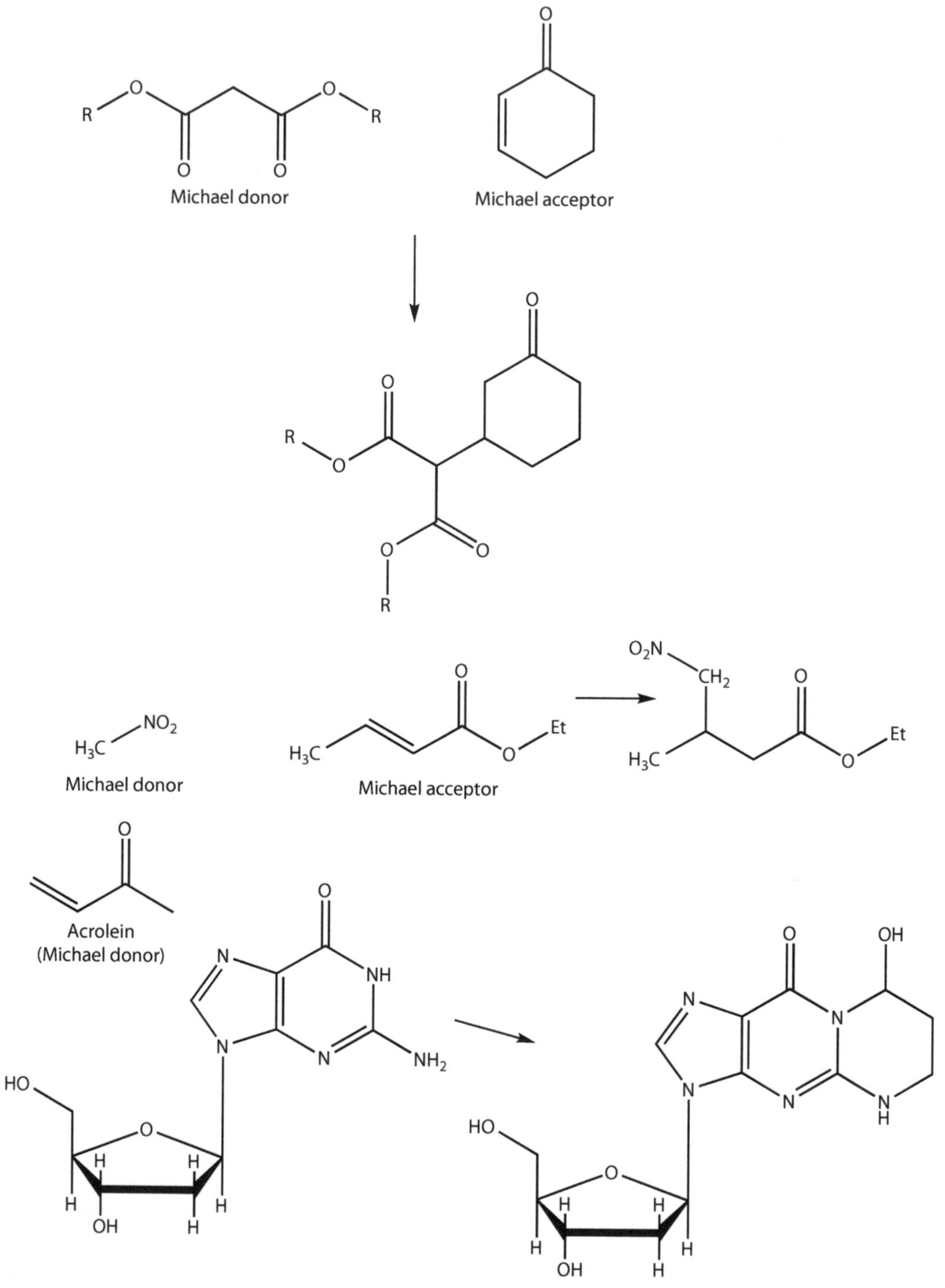

FIGURE 7.16 The Michael addition. The material has been adapted from several sources including Orchin, M., Kaplan, F., MaComber, R.A., Wilson, R.M., and Zimmer, H., *The Vocabulary of Organic Chemistry*, p. 301 (10.480), p. 385 (11.890), John Wiley & Sons, New York, 1980; Li, J.J., *Name Reactions. A Collection of Detailed Reaction Mechanisms*, p. 232, Springer, Berlin, Germany, 2005. Shown at the bottom is a Michael addition of acrolein with deoxyguanosine (Minko, I.G., Kozekov, I.D., Harris, T.M. et al., Chemistry and biology of DNA containing 1,N^2-deoxyguanosine adducts of the α,β-unsaturated aldehydes acrolein, crotonaldehyde, and 4-hydroxynonenal, *Chem. Res. Toxicol.* 22, 759–778, 2009).

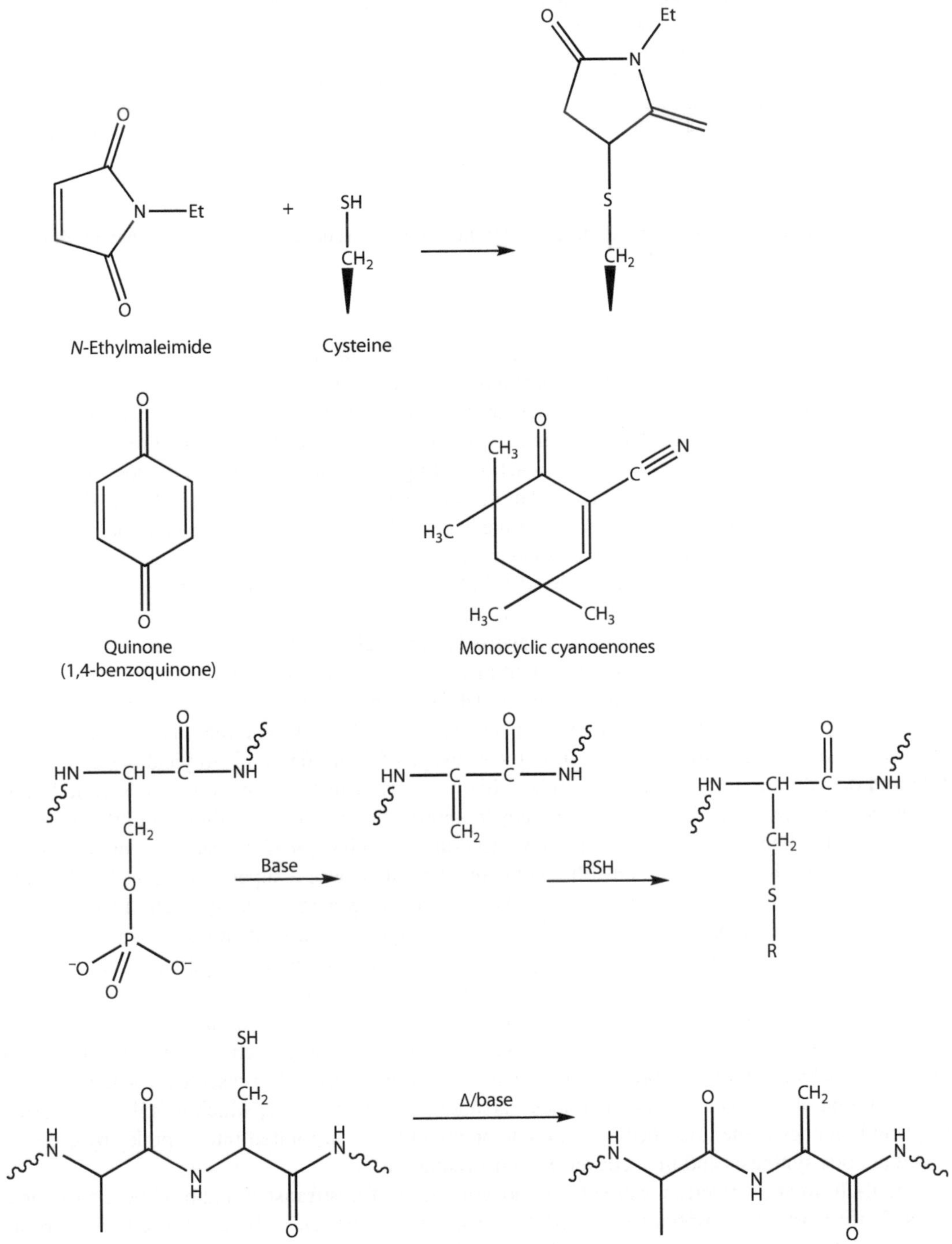

FIGURE 7.17

(*continued*)

FIGURE 7.17 (continued) Michael acceptors of biological importance. Shown is the reaction of *N*-ethylmaleimide with cysteine to form *N*-ethylsuccinimide cysteine. Also shown are the structures of quinine (benzoquinone) and a monocyclic cyanoenone that are Michael acceptors with somewhat more complex reaction characteristics (Hanzlik, R.P., Harriman, S.P., and Frauenhoff, M.M., Covalent binding of benzoquinone to reduced ribonuclease. Adduct structures and stoichiometry, *Chem. Res. Toxicol.* 7, 177–184, 1994; Zheng, S., Laxmi, Y.R.S., David, E. et al., Synthesis, chemical reactivity as Michael acceptors, and biological potency of monocyclic cyanoenones, novel and highly potent anti-inflammatory and cytoprotective agents, *J. Med. Chem.* 55, 4837–4846, 2012). Shown below is the formation of a Michael acceptor in a protein by β-elimination to form dehydroalanine (Tseng, H.-C., Ovaa, H., Wei, N.J.C. et al., Phosphoproteomic analysis with a solid-phase capture-release-tag approach, *Chem. Biol.* 12, 769–777, 2005). The formation of dehydroalanine from cysteine with heat in the presence of base is shown at the bottom of the figure (Bar-Or, R., Rael, L.T., and Bar-Or, D., Dehydroalanine derived from cysteine is a common post-translational modification in human serum albumin, *Rapid Commun. Mass Spectrom.* 22, 711–716, 2008).

is a low molecular weight antibiotic peptide of in the food industry.[445] Nisin undergoes a number of posttranslational including the formation of dehydroalanine and dehydrobutyrine.[446] Some of the dehydroalanine residues undergo intramolecular reaction with cysteine to form lanthionine (β-amino-β-carboxyethyl sulfide), while dehydrobutyrine residues can form β-methyllanthionine, both of which are intramolecular cross-links, while the free dehydro amino acid residues appear to be important for antibiotic function as reaction with glutathione[447] results in the loss of activity. Lysinoalanine was formed from dehydroalanine and lysine in tubulin at high pH (0.1 M glycine, pH 10) and elevated temperature (37°C) resulting in cross-link formation.[448] Lysoalanine and lanthionine have also been found in processed food products resulting from the formation of dehydroalanine.[449,450] Lanthionine is also formed from disulfide in the presence of base;[451] the base degradation of disulfide bridges is discussed in Chapter 8. The formation of lanthionine from cystine in proteins proceeds through the formation of dehydroalanine. Dehydroalanine has been reported to be a common derivative in human serum albumin obtained from normal individuals and in commercial human serum albumin therapeutics[452]; dehydroalanine can also be derived from cysteine in human serum albumin in the presence of heat and base. Dehydroalanine has also been found in IgG molecules[453] possibly resulting from the cleavage of disulfide bonds to yield cysteine disulfide and dehydroalanine; cysteine disulfide is likely involved in the formation of trisulfide bonds (Chapter 8). Cysteine can be converted to dehydroalanine by oxidative elimination in the presence of *O*-mesitylsulfonylhydroxylamine at pH 8.0 (sodium phosphate, 4°C).[454] Subsequent work from this group[455] established that a 1,4-dibromobutane derivative was a superior reagent for the conversion of cysteine to dehydroalanine and was used for the generation of synthetic histones with multiple posttranslational modifications.[456] These investigators[456] also demonstrated the coupling of the thiol of *N*-acetylglucosamine(*N*-acetyl-1-thio-β-D-glucosamine) to dehydroalanine to yield a mimic of serine *O*-*N*-acetylglucosamine in histones.[457] There is a useful review from this group on the use of dehydroalanine for protein modification,[458] and it is noted that the earlier work by Chalker and coworkers[456] also contains a thorough discussion of various approaches for the conversion of cysteine to dehydroalanine in proteins. Levengood and van der Donk[459] presented a scheme for the incorporation of hydroalanine into peptides starting with *O*-tosylserine, which was then converted to phenylselenocysteine; the phenylselenocysteine could be incorporated into peptides by conventional peptide synthesis and then converted to hydroalanine.

The β-elimination reaction is of value for locating sites of posttranslational modification in proteins. Early work on phosphoserine in proteins used β-elimination in base followed by coupling with ethanethiol to yield *S*-ethylcysteine[460,461] or with methylamine to yield β-methylaminoalaine.[462] Subsequent workers have used 1,2-ethanedithiol (Figure 7.18) for reaction with the dehydro amino acid derivatives. Adamczyk and coworkers[463] conjugated 1,2-ethanedithiol to dehydro amino acids obtained from β-elimination of phosphoserine or phosphothreonine in proteins and coupled the free thiol of the resulting 2-(mercaptoethyl)cysteine with a pyridyldithio derivative of biotin; the biotin permits isolation of peptides with the modified amino acid residue to be captured by affinity chromatography on a streptavidin matrix. The reduction of the pyridyldithio linkage permits elution of the phospho peptides. McLachlin and Chait[464] coupled the 2-(mercaptoethyl)cysteine peptides to an activated thiol resin (pyridyldithio matrix). The peptides could then be obtained by reduction and taken for analysis by mass spectrometry. Ahn and coworkers[465] subjected phosphopeptides to β-elimination (barium in base) in the presence of mercaptoethylguanidine (guanidinoethanethiol) to obtain an arginine mimic, S-(2-guanidinoethyl)cysteine improving the sensitivity of mass spectral analysis. A more recent application of β-elimination for identification of O-phosphorylated

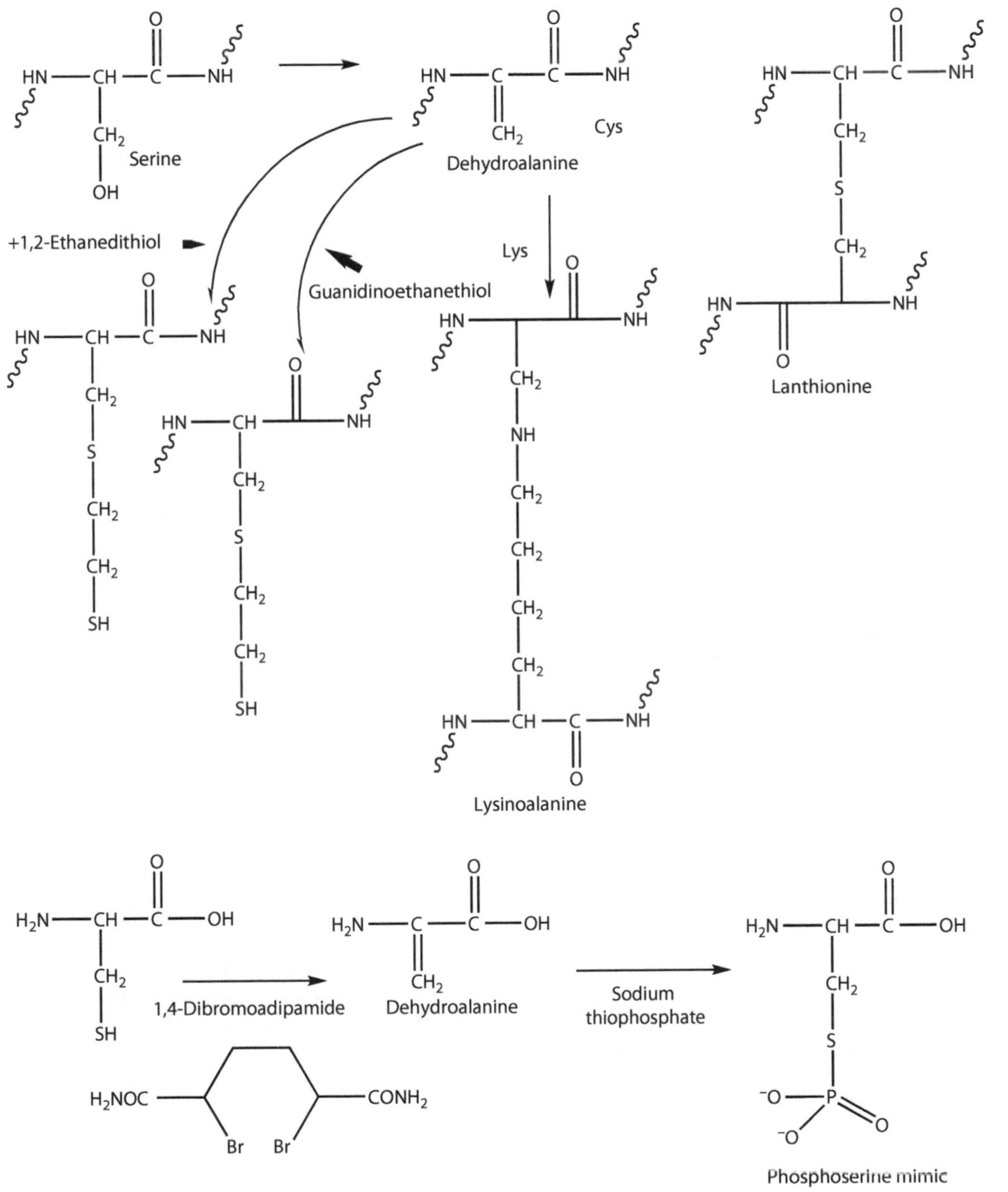

FIGURE 7.18

(continued)

FIGURE 7.18 (continued)   The chemical formation and reactions of dehydroalanine in proteins. Shown at the top is the conversion of serine to dehydroalanine with the subsequent formation of lanthionine via conjugation with cysteine (Lagrain, B., De Vleeschouwer, K., Rombouts, I. et al., The kinetics of β-elimination of cystine and the formation of lanthionine in gliadin, *J. Agric. Food Chem.* 58, 10761–10767, 2010) and the formation of lysinoalanine via conjugation of dehydroalanine and lysine (Correla, J.J., Lipscomb, L.D., and Lobert, S., Nondisulfide crosslinking and chemical cleavage of tubulin subunits: pH and temperature dependence, *Arch. Biochem. Biophys.* 300, 105–114, 1993). Also shown is the reaction of 1,2-ethanedithiol with dehydroalanine with the formation of *S*-(2-mercaptoethyl)cysteine (Adamczyk, M., Gebler, J.C., and Wu, J., Selective analysis of phosphopeptides within a protein mixture by chemical modification, reversible biotinylation and mass spectrometry, *Rapid Commun. Mass Spectrom.* 15, 1481–1488, 2001) or guanidinethanethiol to form *S*-ethylguanidinocysteine (Ahn, Y.H., Ji, E.S., Lee, J.Y. et al., Arginine-mimic labeling with guanidinoethanethiol to increase mass sensitivity of lysine-terminated phosphopeptides by matrix-assisted laser desorption/ionization time-of-flight mass spectrometry, *Rapid Commun. Mass Spectrom.* 21, 2204–2210, 2007). Shown below is the conversion of cysteine to dehydroalanine as procedure for placing a site in a protein for site-specific modification (Bernardes, G.J.L., Chalker, J.M., Errey, J.C., and Davis, B.G., Facile conversion of cysteine and alkyl cysteines to dehydroalanine on protein surfaces: Versatile and switchable access to functionalized proteins, *J. Am. Chem. Soc.* 130, 5052–5053, 2008; Chalker, J.M., Cunnoo, S.B., Boutureira, O. et al., Methods for converting cysteine to dehydroalanine on peptides and proteins, *Chem. Sci.* 2, 1666–1676, 2011). The example shown is the reaction of the dehydroalanine residues with sodium thiophosphate to form a mimic of phosphoserine in a histone (Chalker, J.M., Lercher, L., Rose, N.R. et al., Conversion of cysteine into dehydroalanine enables access to synthetic histones bearing diverse post-translational modifications, *Angew. Chem. Int. Edn.* 51, 1835–1839, 2012).

amino acid residues in proteins is provided in the work of Tseng and coworkers using a solid-phase capture method.[466] Peptides containing phosphoserine and/or phosphothreonine exposed to base (50 mM barium hydroxide, pH 10.5) at 37°C for 16 h and resulting dehydro amino acids (dehydroalanine or dehydrobutyrine) are captured with 2-mercaptoacetyl-hexanoyl beads.

β-Elimination can also be used to identify sites of *O*-glycosylation at serine and threonine residues in proteins (Figure 7.19). The base-catalyzed release of oligosaccharides from serine by the process of β-elimination has been known for some time[467,468] and was subsequently recognized as the method for identifying the site of *O*-*N*-acetylglucosamine (O-GlcNAc) in proteins.[469] Modification with *O*-GlncNAc is thought to be regulatory modification as phosphorylation.[470,471] Well and coworkers[472] initially used dithiothreitol as the Michael donor for modification of the dehydro amino acids resulting from the β-elimination of serine and threonine, which are sites of *O*-GlcNac modification. These investigators also used biotin pentylamine as a tag. Overath and coworkers[473] used biotin cystamine as a tag for anhydro amino acids derived from sites of *O*-GlnNAc attachment in the subunits of 20 S proteasomes. These investigators used performic acid oxidation followed by tryptic digestion. The resulting peptides are separated into two parts, one treated with phosphatase and glycosidase and the other only with phosphatase. The peptides are labeled with *light* and *heavy* derivatives of biotin cystamine and analyzed by mass spectrometry. Furukawa and coworkers[474] used β-elimination of residues containing *O*-linked oligosaccharides in the presence of pyrazolone to identify sites of glycosylation.

There are several reagents (Figure 7.20) for the modification of sulfhydryl groups via conjugate nucleophilic addition/Michael addition in proteins including acrylamide and *N*-ethylmaleimide; 4-hydroxy-2-nonenal and quinones also can modify sulfhydryl groups via a Michael addition. Originally considered to be an unwanted complication of the electrophoresis of reduced proteins in acrylamide gel systems,[475] modification with acrylamide is now considered to be a useful site-specific modification of cysteine in proteins.[476–482] Some of these studies[479] focus on the molecular mechanism for acrylamide toxicity. Acrolein is unsaturated aldehyde that can form Michael addition products with a variety of protein nucleophiles such as lysine and cysteine in proteins. It is of interest that the presence of Tris buffer prevented effective immunodetection of acrolein-modified proteins,[483] likely as a result of reaction of the Tris buffer with a carbonyl product as triethanolamine did not interfere with immunodetection.

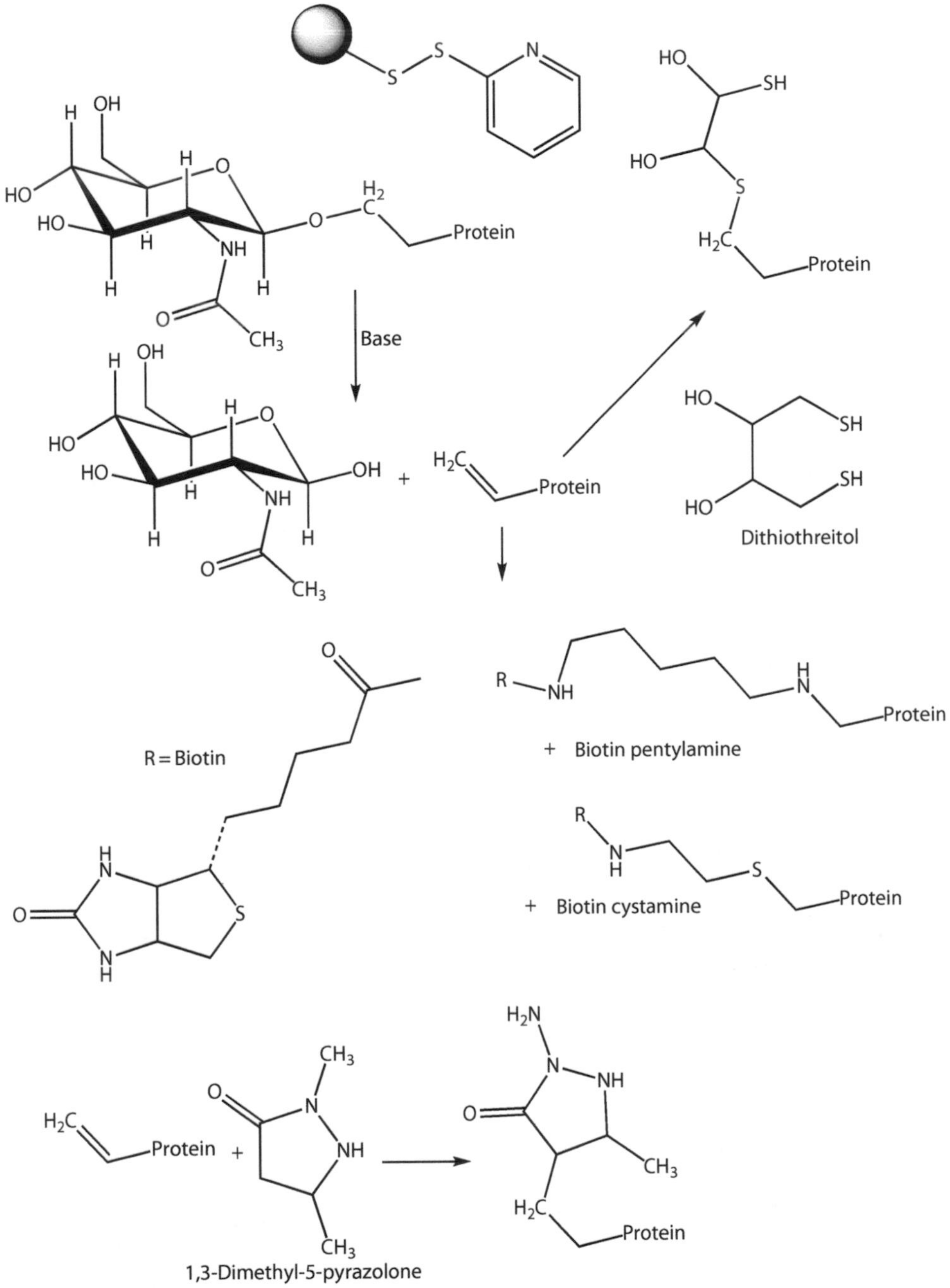

FIGURE 7.19 The conversion of *O*-GlnNAc to anhydroamino acids and subsequent modification for analysis. Shown is the conversion of an *O*-GlncNAc amino acid (serine or threonine) to the corresponding anhydroderivative and the reaction with either dithiothreitol or 5-(biotinamido)pentylamine (Wells, L., Vosseller, K., Cole, R.N. et al., Mapping sites of *O*-GlcNAc modification using affinity tags for serine or threonine post-translational modifications, *Mol. Cell. Proteomics* 1, 791–804, 2002). Biotin cystamine can also be used to couple with the anhydro amino acid (Overath, T., Kuckelkorn, U., Henklein, P. et al., Mapping of *O*-GlnNAc sites of 20 S proteasome subunits and Hsp90 by a novel biotin-cystamine tag, *Mol. Cell. Proteomics* 11, 467–477, 2012). Shown at the bottom is a Michael addition with 1,3-dimethyl-5-pyrazolone (Furukawa, J.-I., Fujitani, N., Araki, K. et al., A versatile method for analysis of serine/threonine posttranslational modifications by β-elimination in the presence of pyrazolone analogues, *Anal. Chem.* 83, 9060–9067, 2011).

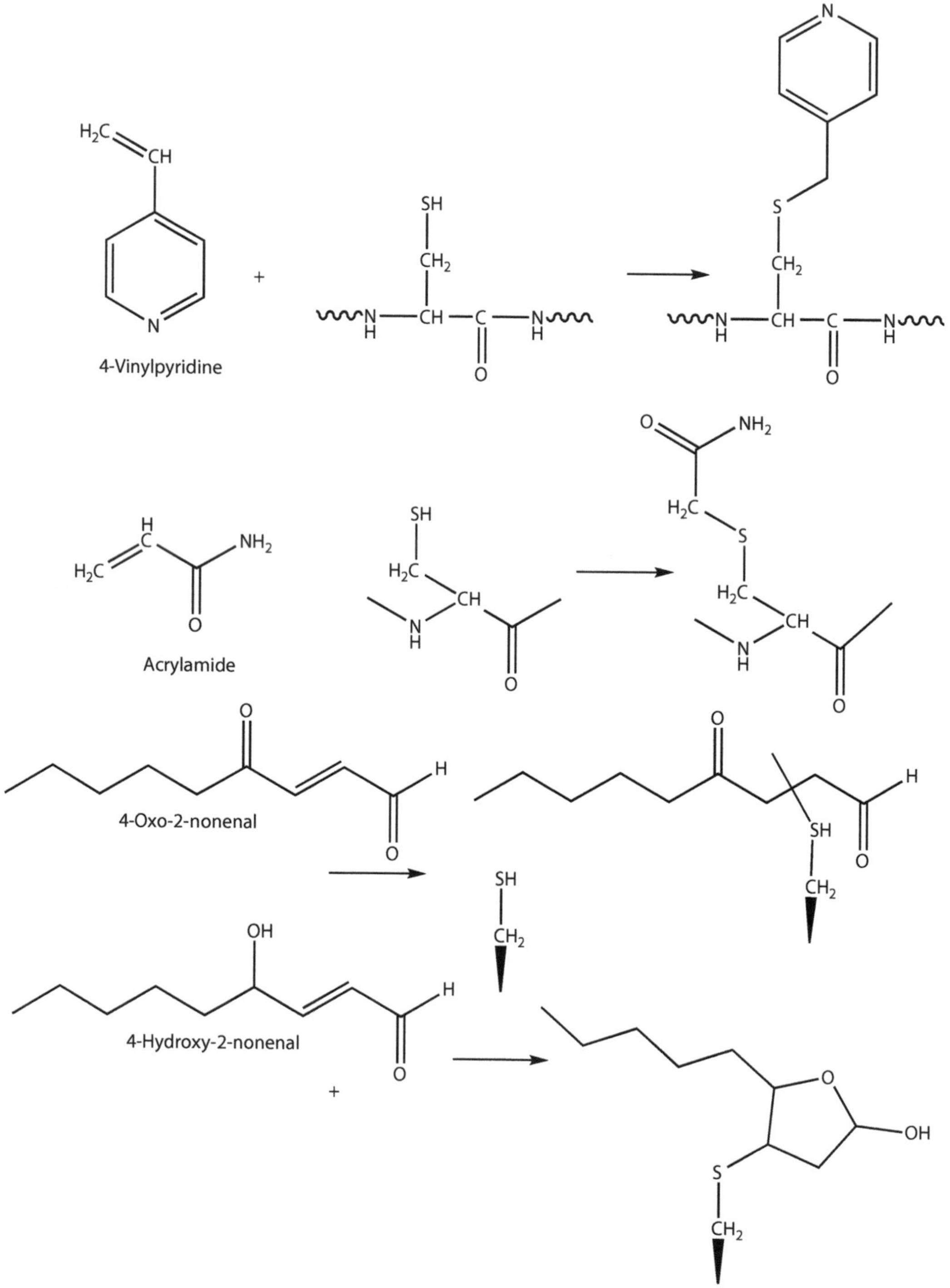

FIGURE 7.20 (continued)

4-Hydroxy-2-nonenal (4-HNE) is a product produced from the oxidation of lipids that modifies sulfhydryl groups (Figure 7.20). 4-Hydroxy-2-nonenal is somewhat nonspecific in that it reacts with sulfhydryl groups, imidazole rings, and amino groups in proteins.[484–488] Doorn and Peterson[489] have reported on the reaction of 4-HNE and 4-oxo-2-nonenal (4-ONE) with various nucleophilic amino residues. They reported that 4-HNE reacted with cysteine, histidine, and amino groups via a Michael addition; reaction was not observed with arginine or methionine. 4-ONE was more reactive than 4-HNE and reacted with cysteine, histidine, amino groups, and the guanidino group arginine; reaction with methionine was not observed. Some data on the reaction of various Michael acceptors with protein-bound sulfhydryl and model compounds are given in Table 7.5. Some data on the reaction of N-ethylmaleimide with cysteine and papain are also shown. While methods are available for the preparation of 4-HNE and 4-ONE,[490] these materials are available from at least one commercial source.[491–493] 4-HNE and 4-ONE are of more interest for studies in pharmacology and physiology[494] than for protein function studies.

N-Ethylmaleimide reacts with sulfhydryl groups in proteins via a Michael addition (Figure 7.16) with considerable specificity yielding the S-succinimidyl derivative.[495–497] The specificity of modification is not complete as a reaction is said to occur at amino groups and histidine; however, there is little information available to support reaction at nucleophiles other than cysteine. Some evidence for the reaction of other α,β-unsaturated carbonyl compounds with amino groups is presented in Table 7.5. LoPachin and coworkers[498] have an excellent presentation on the application of hard and soft acids and bases (HSAB) theory to the reaction of these compounds with protein nucleophiles. Sulfur is considered a soft nucleophile and nitrogen a hard nucleophile; the α,β-unsaturated carbonyl compounds are considered as a class of soft electrophiles and would be expected to have a low rate of reaction with a hard nucleophile such as nitrogen or oxygen. Although the reagent is reasonably specific for cysteine, reaction with other nucleophiles such as histidine and lysine must be considered.[499–503] Smyth and coworkers[499] presented evidence supporting the reaction of NEM with histidine (glycylhistidine amide), the α-amino group (leucinamide, Leu-Leu, Tyr-Tyr), and cysteine. Later work from this group[501] described the reaction of NEM with several peptides with $t_{1/2}$ values of 1–3 h at pH 7.4. Subsequent work by Sharpless and Flavin[504] provided second-order rate constants for the reaction of NEM with some amino acids, amino acid derivatives, and glycylglycine dipeptide. Some of these data are given in Table 7.6 and compared with data from the reaction of NEM with L-cysteine. Brewer and Riehm[506] studied the reaction of NEM with two proteins, lysozyme and bovine pancreatic ribonuclease, which do not contain sulfhydryl groups. They also studied the reaction of NEM with tosyl-L-lysine and N-acetyl-L-histidine. The reaction of 0.1 M NEM with 0.01 M tosyl-L-lysine at pH 8.0 resulted in substantial alkylation of the ε-amino groups of lysine; the reaction of NEM with N-acetyl-L-histidine under the same conditions resulted in the formation of material consistent with alkylation of the imidazole ring. Reaction of lysozyme and bovine pancreatic ribonuclease with 0.1 M NEM at pH 8.0 resulted in substantial modification of lysine

FIGURE 7.20 (continued)   A variety of reagents that add to sulfhydryl groups via Michael addition. Shown is the reaction of vinyl pyridine with cysteine (Friedman, M., Application of the S-pyridylethylation reaction to the elucidation of the structures and functions of proteins, *J. Protein Chem.* 20, 431–453, 2001; Kleinova, M., Belgacem, O., Pock, K. et al., Characterization of cysteinylation of pharmaceutical-grade human serum albumin by electrospray ionization mass spectrometry and low-energy collision-induced dissociation tandem mass spectrometry, *Rapid Commun. Mass Spectrom.* 19, 2965–2973, 2005), the reaction of acrylamide with thiols (LoPachin, R.M., Gavin, T., Geohagen, B.C., and Das, S., Neurotoxic mechanisms of electrophilic type-2 alkenes: soft interactions described by quantum mechanical parameters, *Toxicol. Sci.* 98, 561–570, 2007), and 4-hydroxy-2-nonenal (4-HNE) (Wakita, C., Maeshima, T., Yamazaki, A. et al., Stereochemical configuration of 4-hydroxy-2-nonenal -cysteine adducts and their stereoselective formation in a redox-regulated protein, *J. Biol. Chem.* 284, 28810–28827, 2009) with cysteine. Also shown at the bottom is the reaction of cysteine with 4-oxo-2-nonenal where reaction may occur at the 2carbon or 3carbon in the 4-oxo-2-nonenal (Doorn, J.A. and Petersen, D.R., Covalent modification of amino acid nucleophiles by the lipid peroxidation products 4-hydroxy-2-nonenal and 4-oxo-2-nonenal, *Chem. Res. Toxicol.* 15, 1445–1450, 2002).

TABLE 7.5

Reaction of Some α,β-Unsaturated Carbonyl Compounds (Michael Acceptors) with Model Nucleophiles

Michael Acceptor	Michael Donor/Reaction Conditions	Reaction Rate($M^{-1} s^{-1}$)	Reference
Acrylamide	GAPDH[a]/5 mM Tris-acetate, 13.6 mM sodium arsenate, pH 7.4/30°C	0.053[b]	7.5.1
Acrylamide	GAPDH/5 mM Tris-acetate, 13.6 mM sodium arsenate, pH 8.5/30°C	0.267[b]	7.5.1
Methyl vinyl ketone	GAPDH[a]/5 mM Tris-acetate, 13.6 mM sodium arsenate, pH 7.4/30°C	128[b]	7.5.1
Acrolein	GAPDH[a]/5 mM Tris-acetate, 13.6 mM sodium arsenate, pH 7.4/30°C	297[b]	7.5.1
4-HNE[c]	Ac-Lysine amide/50 mM sodium phosphate, pH 7.4/23°C	1.33×10^{-3}	7.5.2
4-ONE[d]	Ac-Lysine amide/50 mM sodium phosphate, pH 7.4/23°C	7.46×10^{-3}	7.5.2
4-HNE	Ac-Histamine/50 mM sodium phosphate, pH 7.4/23°C	2.14×10^{-3}	7.5.2
4-ONE	Ac-Histamine/50 mM sodium phosphate, pH 7.4/23°C	2.21×10^{-2}	7.5.2
4-HNE	Ac-Cystamine/50 mM sodium phosphate, pH 7.4/23°C	1.21	7.5.2
4-ONE	Ac-Cystamine/50 mM sodium phosphate, pH 7.4/23°C	186	7.5.2
4-HNE	Glutathione/50 mM sodium phosphate, pH 7.4/23°C	1.33	7.5.2
4-ONE	Glutathione/50 mM sodium phosphate, pH 7.4/23°C	145	7.5.2
N-Ethylmaleimide	Cysteine/pH 7.0/25°C	1530	7.5.3
N-Ethylmaleimide	Cysteine/pH 6.0/25°C	153	7.5.3
N-Ethylmaleimide	Papain/pH 6.0/25°C	0.661	7.5.4

[a] GAPDH, glyceraldehyde-3-phosphate dehydrogenase.

[b] The rate of inactivation of GAPDH is measured and assumed to reflect reaction at Cys152. The pKa of this residue is reported to be 6.03.

[c] 4-HNE, 4-hydroxy-2-nonenal.

[d] 4-ONE, 4-oxo-2-nonenal.

References to Table 7.5

7.5.1. Martyniuk, C.J., Fang, B., Koomen, J.M. et al., Molecular mechanism of glyceraldehyde-3-phosphate dehydrogenase inactivation by α,β-unsaturated carbonyl derivatives, *Chem. Res. Toxicol.* 24, 2302–2311, 2011.

7.5.2. Doorn, J.A. and Petersen, D.R., Covalent modification of amino acid nucleophiles by the lipid peroxidation products 4-hydroxy-2-nonenal and 4-oxo-2-nonenal, *Chem. Res. Toxicol.* 15, 1445–1450, 2002.

7.5.3. Gorin, G., Martic, P.A., and Doughty, G., Kinetics of the reaction of N-ethylmaleimide with cysteine and some congeners, *Arch. Biochem. Biophys.* 115, 593–597, 1966.

7.5.4. Brubacher, L.J. and Glick, B.R., Inhibition of papain by N-ethylmaleimide, *Biochemistry* 13, 915–920, 1974.

and histidine; reaction of these proteins with 0.001 M (1 mM) NEM at pH 6.0 (*normal* reaction conditions for modification of sulfhydryl groups in proteins) resulted in only slight or no detectable reaction with proteins. In particular, the alkylation of histidine with NEM is only observed under the most rigorous experimental conditions. Smith did use maleimide-functionalized spin labels to modify a lysine residue, presumably Lys41, in bovine pancreatic ribonuclease.[505] Data were not provided regarding the rate of modification, but the time of modification is referred in hours to days at pH 8.5. Papini and coworkers[503] modified several peptides at the amino-terminal amino acid with N-hydroxysuccinimido maleoyl-β-alanate (3-maleimidopropionic acid), which then reacted with an internal histidine residues to form a cyclic peptide. While quantitative data were not provided, the cyclization reaction was more rapid than the modification of NEM in a solution.

There is considerable variability in the rate of modification of cysteinyl residues by NEM in proteins (Table 7.7). Also, while there are several exceptions, the rate of modification of

TABLE 7.6
Reaction of N-Ethylmaleimide with Amines and
Model Sulfhydryl Compounds

Michael Donor/Conditions	Rate ($M^{-1}\,s^{-1}$)	Reference
Benzylamine, pH 8.5, 25°C	0.025	7.6.1
Glycylglycine, pH 8.5, 25°C	0.022	7.6.1
Piperidine, pH 8.5, 25°C	0.10	7.6.1
L-Proline, pH 8.5, 25°C	0.059	7.6.1
Glycine, pH 7.0/25°C	0.000062	7.6.2
L-Cysteine, pH 3.0, 25°C	0.346	7.6.3
L-Cysteine, pH 4.95, 25°C	14.1[a]	7.6.3

[a] A value of 1530 $M^{-1}\,s^{-1}$ was estimated for the reaction of
N-ethylmaleimide and L-cysteine at pH 7.0/25°C.

References to Table 7.6

7.6.1. Sharpless, N.E. and Flavin, M., The reactions of amines and amino acids with maleimides. Structure of the reaction products deduced from infrared and nuclear magnetic resonance spectroscopy, *Biochemistry* 5, 2963–2971, 1966.

7.6.2. Leslie, J., Kinetics of the reaction of glycine with N-ethylmaleimide, *Can. J. Chem.* 48, 507–508, 1970.

7.6.3. Gorin, G., Martic, P.A., and Doughty, G., Kinetics of the reaction of N-ethylmaleimide with cysteine and some congeners, *Arch. Biochem. Biophys.* 115, 593–597, 1966.

cysteinyl residues in proteins is less than in model compounds such as cysteine or glutathione. Also, as with most reagents, there is a considerable variability with the rate of thiol modification in different proteins. The studies by Britto and coworkers on the modification of tubulin by N-ethylmaleimide[106] have been discussed above. These investigators noted that the reaction of N-ethylmaleimide was more rapid than iodoacetamide with the cysteine residues in tubulin; a similar difference in reactivity was observed with model sulfhydryl groups (see Table 7.2). Bednar has examined the chemistry of the reaction of N-ethylmaleimide with cysteine and other thiols in some detail.[506] The pH-independent second-order rate constant for the reaction of NEM with the thiolate anion of 2-mercaptoethanol is in the range of 10^7 M^{-1} min^{-1}, which is only slightly less than the rate of aromatic mercuric compounds with thiol groups; it is suggested that this value is at least 5×10^{10} greater that the reaction with the thiol (2×10^{-4} M^{-1} min^{-1}). This study also reports data for the decomposition of NEM in several buffers and should be considered for the determination of truly accurate kinetic data. This study[506] was focused on the reaction of N-ethylmaleimide with chalcone isomerase. The pH dependence of the results suggests that the reaction is dependent on protein conformation; at high pH (pH ≥ 8), the cysteinyl residues are much less reactive than a model thiol at the same pH (pH ≥ 8) (see Table 7.7), while below pH 8, the rate of reaction of NEM with the cysteinyl residue in chalcone isomerase approaches that of a model thiol. Jack bean urease is inactivated by NEM, and the inactivation is inhibited by cysteine, glutathione, or dithiothreitol.[507] This effect differs from the effect of thiol compounds on the inactivation of arylamine transferase by iodoacetate described above.[373]

The reaction of NEM with a cysteinyl residue can be followed by the decrease in absorbance at 300 nm, the absorbance maximum of N-ethylmaleimide. The extinction coefficient of

TABLE 7.7

Rate of Reaction of N-Ethylmaleimide with Thiol Groups in Proteins and Model Thiol Compounds[a]

Protein or Model Compound/Conditions	Rate ($M^{-1}s^{-1}$)	Reference
Papain/pH 6.0 (0.1 phosphate)/25°C	0.661[a]	7.7.1
Papain/pH 6.8/0.13 M potassium phosphate/25°C	2.77[a]	7.7.1[a]
N-Acetyl-L-cysteine, pH 6.0/25°C	26.7[b]	7.7.2
Jack bean urease/20 mM phosphate, pH 7.4/25°C	0.92[a]	7.7.3
Jack bean urease/20 mM phosphate, pH 8.3/25°C	3.17[a]	7.7.3
Apo-CadC/5 mM MES-0.40 M NaCl, pH 7.0/25°C	32.2[c]	7.7.4
Chalcone isomerase, 100 mM PIPES ($I=0.15$ with KCl), pH 6.8, 25°C	0.14[a]	7.7.5
Chalcone isomerase, 50 mM TAPS ($I=0.10$ with KCl), pH 8.5, 25°C	0.95[a]	7.7.5
L-Cysteine hydrochloride,[d] pH 3.0/25°C	0.196[b]	7.7.6
Cysteine/pH 3.0/25°C	0.346[e]	7.7.7
Cysteine/pH 4.95/25°C	14.1[b]	7.7.7
Glutathione/pH 4.95/25°C	10.6[b]	7.7.7
Cysteine/pH 6.0/25°C	153[f]	7.7.7
Cysteine/pH 7.0/25°C	1530[f]	7.7.7
Cysteine, pH 4.75/25°C	8.1[b]	7.7.8
Cysteine, pH 4.75/25°C/1% SDS	6.7[b]	7.7.8
B-Lactoglobulin/pH 6.0/25°C/1% SDS	1.4[b,g]	7.7.8
YADH/50 mM sodium pyrophosphate, pH 7.0, 20°C	0.22[a]	7.7.9
Cysteine/25 mM sodium acetate, pH 5.0,[h] 25°C	20.2[b]	7.7.9
Glutathione/25 mM sodium acetate, pH 5.0,[h] 25°C	25.8[b]	7.7.9
Rabbit muscle L-α-glycerophosphate dehydrogenase/50 mM Tris, pH 7.0/10°C	0.043	7.7.10
L-serine-activating enzyme from E. coli/100 mM phosphate, pH 6.6, 0°C	77	7.7.11
Opiate membrane receptor (cell membrane preparations)/pH 7.4(50 mM Tris)[i]/37°C	0.18	7.7.12
Plasma membrane H+-ATPase (Neurospora crassa)/50 mM Tris, pH 8.7 with 5% dimethyl sulfoxide[j]/0°C	205 (fast site[k]) 0.078 (slow site)	7.7.13
Bacterial luciferase (luminous strain MAV)/20 mM phosphate, pH 7.0/25°C	27.8	7.7.14

[a] Reaction rate determined by the loss of enzyme activity.

[b] Reaction rates determined by changes in the spectra of NEM (Friedmann, E., Spectrophotometric investigation of the interaction of glutathione with maleimide and N-ethylmaleimide, *Biochim. Biophys. Acta* 9, 65–75, 1952; Gregory, J.D., The stability of N-ethylmaleimide and its reaction with sulfhydryl groups, *J. Am. Chem. Soc.* 77, 3922–3923, 1955; Roberts, E. and Rouser, G., Spectrophotometric assay for reaction of N-ethylmaleimide with sulfhydryl groups, *Anal. Chem.* 30, 1291–1292, 1958). It should be noted that this method measures the disappearance of reactant rather than the formation of product. Thus, it is possible, but not likely, that reaction with amino groups (Sharpless, N.E. and Flavin, M., The reactions of amines and amino acids with maleimides. Structure of the reaction products deduced from infrared and nuclear magnetic resonance spectroscopy, *Biochemistry* 5, 2963–2971, 1966) might be a complication. It would be useful to validate the spectrophotometric method by measurement of product(s). It is also necessary to correct for hydrolysis of the N-ethylmaleimide (Sharpless, N.E. and Flavin, M., The reaction of amines and amino acids with maleimides. Structure of the reaction products deduced from infrared and nuclear magnetic resonance spectroscopy, *Biochemistry* 5, 2963–2971, 1966; Bednar, R.A., Reactivity and pH dependence of thiol conjugation to N-ethylmaleimide: Detection of a conformational change in chalcone isomerase, *Biochemistry* 29, 3684–3690, 1990). The oxidation of the Michael donor should also be considered (at pH 9.0/25°C: 2-mercaptoethanol, 0.054 $M^{-1}s^{-1}$; cysteine, 0.113 $M^{-1}s^{-1}$. Schelté, P., Boeckler, C., Frisch, B., and Schuber, F., Differential reactivity of maleimide of bromoacetyl functions with thiols: Application to the preparation of liposomal diepitope constructs, *Bioconjug. Chem.* 11, 118–123, 2000).

[c] For reaction at Cys7 (determined by mass spectrometry).

TABLE 7.7 (continued)
Rate of Reaction of *N*-Ethylmaleimide with Thiol Groups in Proteins and Model Thiol Compounds[a]

[d] The authors comment that it was not feasible to use cysteine instead of cysteine hydrochloride in aqueous solution at the rate was too fast measure at the higher pH of an unbuffered cysteine solution.

[e] The authors note that this value is of the same order of magnitude as the value obtained in Reference 7.7.6 in an unbuffered solution.

[f] Estimated from data obtained at lower pH values.

[g] Native β-lactoglobulin is a dimer where each monomer has 2 disulfide bonds and a single sulfhydryl group; the sulfhydryl group is not available for modification with denaturation (Croguennec, T., Bouhallab, S., Mollé, D. et al., Stable monomeric intermediate with exposed Cys-119 is formed during heat denaturation of β-lactoglobulin, *Biochem. Biophys. Res. Commun.* 301, 465–471, 2003).

[h] The rate of reaction of NEM with cysteine or glutathione at pH 7.0 was too rapid to measure with accuracy.

[i] Experimental detail is missing concerning the solvent conditions of the modification reaction; the conditions cited are those used for the bremazocine binding assay.

[j] The sulfhydryl reagents were introduced into the reaction mixture with dimethylsulfoxide and stated that the dimethylsulfoxide concentration was maintained at 5% (assume V/V). It is stated that the presence of 5% dimethylsulfoxide reduced the rate of NEM inhibition by 50%.

[k] The reaction of NEM with the *fast site* had to be studied in an indirect manner (by protection against modification with *N*-pyrenemaleimide).

References to Table 7.7

7.7.1. Brubacher, L.J. and Glick, B.R., Inhibition of papain by *N*-ethylmaleimide, *Biochemistry* 13, 915–920, 1974.

7.7.1a. Anderson, B.M. and Vasini, E.C., Nonpolar effects in reaction of sulfhydryl groups of papain, *Biochemistry* 9, 3348–3352, 1970.

7.7.2. Glick, B.R. and Brubacher, L.J., The reaction between *N*-ethylmaleimide and ribosomes, *J. Mol. Biol.* 93, 319–321, 1975.

7.7.3. Kot, M. and Bicz, A., Inactivation of jack bean urease by *N*-ethylmaleimide: pH dependence, reversibility and thiols influence, *J. Enzyme Inhib. Med. Chem.* 23, 514–520, 2008.

7.7.4. Apuy, J.L., Busenlehner, L.S., Russell, D.H., and Giedroc, D.P., Ratiometric pulsed alkylation mass spectrometry as a probe of thiolate reactivity in different metalloderivatives of *Staphylococcus aureus* p1258 CadC, *Biochemistry* 43, 3824–3834, 2004.

7.7.5. Bednar, R.A., Reactivity and pH dependence of thiol conjugation to *N*-ethylmaleimide: Detection of a conformational change in chalcone isomerase, *Biochemistry* 29, 3684–3690, 1990.

7.7.6. Lee, C.C. and Samuels, E.R., The kinetics of reaction between L-cysteine hydrochloride and some maleimides, *Can. J. Chem.* 42, 168–170, 1964.

7.7.7. Gorin, G., Martic, P.A., and Doughty, G., Kinetics of the reaction of *N*-ethylmaleimide with cysteine and some congeners, *Arch. Biochem. Biophys.* 115, 593–597, 1966.

7.7.8. Franklin, J.G. and Leslie, J., The kinetics of the reaction of *N*-ethylmaleimide with denatured β-lactoglobulin and ovalbumin, *Biochim. Biophys. Acta* 160, 333–339, 1968.

7.7.9. Heitz, J.R., Anderson, C.D., and Anderson, B.M., Inactivation of yeast alcohol dehydrogenase by *N*-alkylmaleimides, *Arch. Biochem. Biophys.* 127, 627–636, 1968.

7.7.10. Anderson, B.M., Kim, S.J., and Wang, C.-N., Inactivation of rabbit muscle L-α-glycerophosphate dehydrogenase by *N*-alkylmaleimides, *Arch. Biochem. Biophys.* 138, 66–72, 1970.

7.7.11. Bryce, G.F., Enzymes involved in the biosynthesis of cyclic tris-(*N*-2,3-dihydroxybenzoyl-L-seryl)in *Escherichia coli*: Kinetic properties of the L-serine-activating enzyme, *J. Bacteriol.* 116, 790–796, 1973.

7.7.12. Shahrestanifar, M.S. and Howells, R.D., Sensitivity of opioid receptor binding to N-substituted maleimides and methanethiosulfonate derivatives, *Neurochem. Res.* 21, 1295–1299, 1996.

7.7.13. Davenport, J.W. and Slayman, C.W., The plasma membrane H+-ATPase of *Neurospora crassa*. Properties of two reactive sulfhydryl groups, *J. Biol. Chem.* 263, 16007–16013, 1988.

7.7.14. Nicoli, M.Z. and Hastings, J.W., Bacterial luciferase. The hydrophobic environment of the reactive sulfhydryl, *J. Biol. Chem.* 249, 2393–2396, 1974.

N-ethylmaleimide is 620 M^{-1} cm^{-1} at 302 nm.[495] This technique is used most frequently for the measurement of rate of reaction of NEM with model sulfhydryl compounds (Table 7.7). It is difficult to use this technique with proteins, and the loss of biological activity is most often used for the measurement of the rate of reaction. The product resulting from the reaction of cysteine and NEM, N-ethylsuccinimide cysteine, can be determined by amino acid analysis following acid hydrolysis (6 N HCL, 22 h, 110°C) as ethylamine and S-(2-succinyl)cysteine (Figure 7.21).[508,509] S-(2-succinyl) cysteine coelutes with carboxymethylcysteine while ethylamine elutes with arginine on the ion-exchange columns used for amino acid analysis. S-(2-succinyl)cysteine has also been described as occurring in plasma proteins and other tissue proteins arising from the reaction of fumaric acid with cysteine (Figure 7.21).[510] The presence of S-(2-succinyl)cysteine has been shown to be a biomarker for mitochondrial stress in obesity and/or type 2 diabetes.[511–513] The standard for the determination of S-(2-succinyl)cysteine was obtained from the acid hydrolysis of the product obtained from the reaction of NEM with cysteine.[510] MS has largely replaced amino acid analysis as a method for determining the covalent modification of proteins with NEM.[514–516] The availability of a polyclonal antibody to S-(2-succinyl)cystine[517] has proved useful in studies on the distribution of this posttranslational modification.[512,517] It is noted that these latter studies[512,517] refer to the reaction of fumarase with cysteine as succination and the product as S-(2-succino)cysteine. It is noted that the reaction of NEM with proteins can be monitored by determination of unreacted sulfhydryl groups with DTNB or incorporation of radiolabeled reagent.[518,519] Martinez-Carrion and Sneden[519] modified a cytoplasmic aspartate transaminase in crystal form with NEM; modification in solution or crystalline form occurs only in the presence of a substrate. The extent of modification was measured either with radiolabeled reagent or by loss of reactivity toward DTNB.

A *diagonal* procedure for the isolation of cysteine-containing peptides modified with N-ethylmaleimide has been reported.[520,521] This procedure is based on the base hydrolysis (ammonia vapor) of the reaction product obtained from reaction with N-ethylmaleimide to cysteine-S-N-ethyl succinamic acid (Figure 7.21) generating a new negative charge on the peptide/protein. While conditions of acid hydrolysis used for the preparation of proteins for amino acid analysis convert the N-ethylmaleimide cysteine to S-(2-succinyl)cysteine and ethyl amine, the mild acid conditions used for peptide separation and storage (10%formic acid, 50% acetic acid) did affect the maleimide-cysteine derivative. It is noted that Gehring and Christen did observe chymotryptic cleavage at the N-ethylmaleimide cysteine residue in proteins.[508] Allesandro and coworkers[522] suggest that cleavage at N-ethylmaleimide-modified cysteine residues could be responsible for increased heterogeneity in NEM-treated red blood cell concentrates.

There are observations that suggest that the reaction of NEM with cysteinyl residues is reversible.[523–527] Granted that these studies were performed in relatively crude systems and reversibility is based on the recovery of biological activity rather than chemical measurement, it is possible that the observed phenomena are based on noncovalent interaction of NEM and protein.[527] It is possible to develop a maleimide reagent that reacts with cysteinyl residues in a reversible manner.[528] Tedaldi and coworkers[529] introduced bromomaleimides (Figure 7.22) as reagents for the reversible modification of cysteine residues in proteins. The reaction differs from the Michael addition chemistry of maleimides in that there is a nucleophilic substitution reaction preserving the double bond in the maleimide ring. The resulting thiomaleimide may be converted to dehydroalanine by base (potassium carbonate) or with reducing agent such as phosphine to yield the original cysteine. These authors noted that the conversion of the thiomaleimide to dehydroalanine does not proceed effectively in water. It is likely that the hydrolysis of the maleimide to maleamic acid (Figure 7.22) is a factor in the conversion to dehydroalanine in aqueous solvent. It is also noted that a phosphine reducing agent, tris-(2-carboxyethyl)phosphine, does not interfere with the maleimide reaction with thiols.[530] Following the initial observations of Tedaldi and coworkers,[529] Smith and coworkers[528] showed that the reaction of the Grb2 SH2 domain with N-methylbromomaleimide could subsequently be modified by glutathione in a Michael addition to yield a vicinal bis-thioether. This group subsequently developed N-phenylbromomaleimide (Figure 7.22).[531] The reaction of N-phenylbromomaleimide with a protein thiol is reversible; the hydrolysis of the maleimide ring to yield the maleamic acid derivatives makes thiol adduct formation irreversible (Figure 7.22). N-arylmaleimides are more susceptible to hydrolysis to maleamic acid derivatives than N-alkylmaleimides.[532,533] Machida and coworkers[532] observed that N-phenylmaleimide is at least an order of magnitude less stable than N-ethylmaleimide at pH 7.0. Moody and coworkers[534] have recently demonstrated that conjugates obtained by reaction of maleimide or bromomaleimide dyes with thiol groups on proteins could

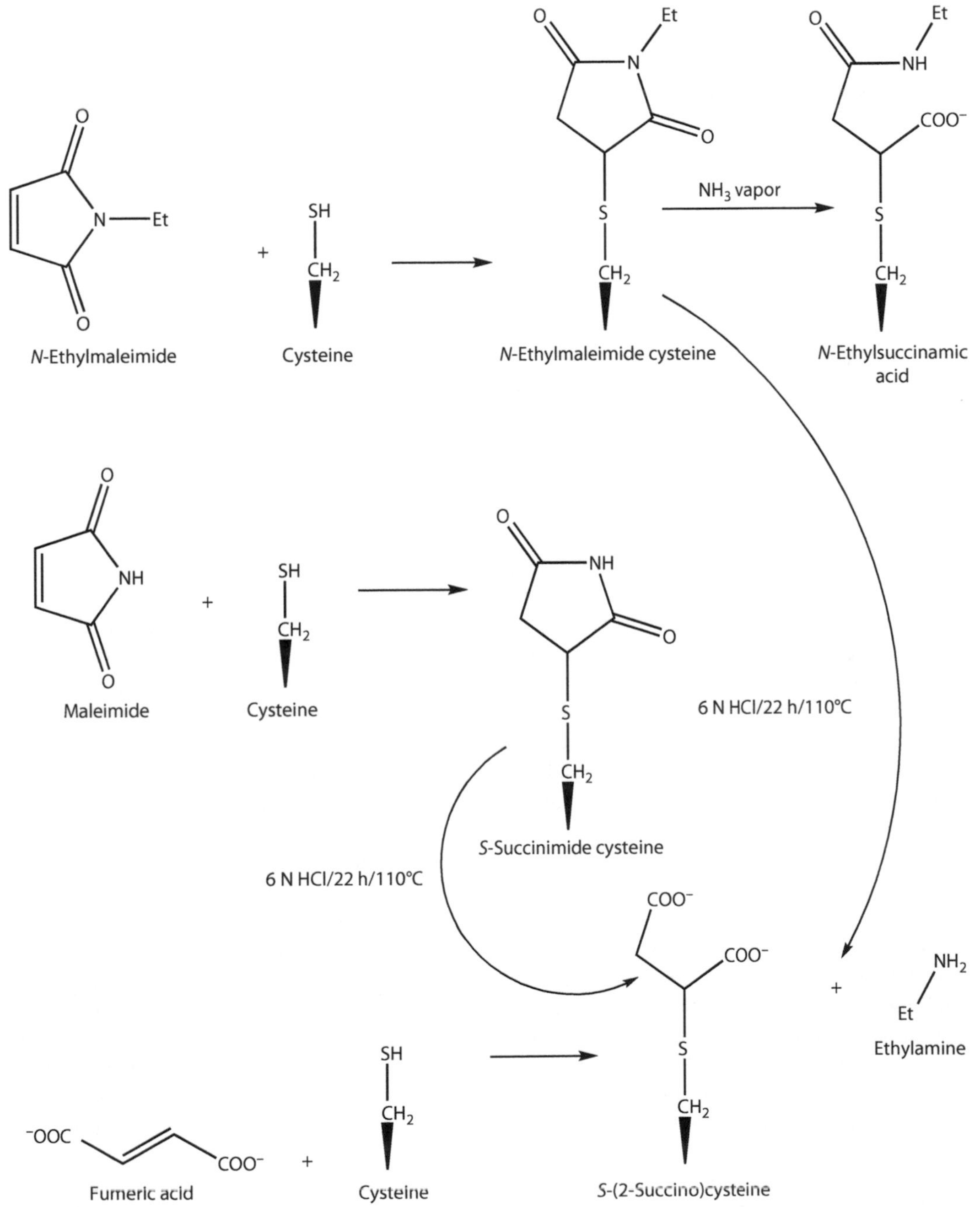

FIGURE 7.21

(continued)

FIGURE 7.21 (continued)   *N*-Ethylmaleimide cysteine, maleimide cysteine. Succinimide cysteine and succinyl cysteine. The reaction of *N*-ethylmaleimide with cysteine yields *N*-ethylmaleimide cysteine that yields *S*-succinylcysteine and ethyl amine on acid hydrolysis (Gehring, H. and Christen, P., Syncatalytic conformational changes in mitochondrial aspartate aminotransferases, Evidence from modification and demodification of Cys 166 in the enzyme from chicken and pig, *J. Biol. Chem.* 253, 3158–3163, 1978). Exposure of *N*-ethylmaleimide cysteine to ammonia vapor results in the formation of *N*-ethylsuccinamic acid (Gehring, H. and Christen, P., Isolation of sulfhydryl peptides alkylated with *N*-ethylmaleimide by diagonal electrophoresis, *Methods Enzymol.* 91, 392–396, 1983). Shown below is the reaction of maleimide with cystine yielding *S*-succinimido cysteine that also yields *S*-succinylcysteine on acid hydrolysis (Abe, Y., Ueda, T., and Imoto, T., Reduction of disulfide bonds in proteins by 2-aminothiophenol under weakly acidic conditions, *J. Biochem.* 115, 52–57, 1994). While it is not shown, the reaction of maleimide with cysteine yields a diastereomeric pair. Shown at the bottom is the reaction of fumarate with cysteine to yield *S*-(2-succinyl)cysteine (Alderson, N.L., Wang, Y., Blatnik, M. et al., *S*-(2-succinyl)cysteine: A novel chemical modification of tissue proteins by a Krebs cycle intermediate, *Arch. Biochem. Biophys.* 450, 1–8, 2006). A later work from this group suggested preference for *S*-(2-succino)cysteine to describe the product of the reaction between fumarate and cysteine (Thomas, S.A., Storey, K.B., Baynes, J.W., and Frizzell, N., Tissue distribution of *S*-(2-succino)cysteine (2SC), a biomarker of mitochondrial stress in obesity and diabetes, *Obesity* 20, 263–269 2012).

be cleaved *in vitro* with physiological concentrations of glutathione; intracellular cleavage of conjugate of *N*-rhodaminemaleimide with green fluorescent protein was also demonstrated following injection. Another approach for developing a reversible Michael acceptor has been advanced by Serafimova and coworkers[535] who showed that modification of acrylamide derivatives (*activated olefins*) to increase electrophilicity resulted in both increasing reactivity and increased ability to reverse formation of the thiol adduct (Figure 7.23). Baldwin and Kiick[536] showed that a thiomaleimide adduct formed with 4-mercaptophenylacetic acid could undergo a retroreaction in the presence of another Michael donor such as reduced glutathione; maleimide ring opening is a competing reaction with the formation of the substituted maleamic acid. These basic studies have great importance in the design of prodrugs and related biopharmaceuticals.[533] While not of direct relevance, it is useful to note that there are *N*-ethylmaleimide reductases in bacteria.[537–539]

The maleimide functional group is used to introduce probes into proteins. The specificity of the maleimide for reaction at cysteine combined with the small amount of cysteine in protein provides a major advantage. The ability to use cysteine mutagenesis for designing probe placement is also useful.[18,540–543] The chemistry of the various probes based on maleimide chemistry will be discussed below (Figure 24.26). *N*-Alkyl and *N*-aryl maleimide derivatives other than *N*-ethyl, while not structural probes such as the solvatochromic derivatives, are used to study the microenvironment of cysteine residues in proteins. Work on papain, a sulfhydryl protease, provides an interesting story. Data from the reaction of various sulfhydryl reagents with papain are presented in Table 7.8. Increasing the hydrophobicity of the alkyl function of the maleimide derivative results in a marked increased in the reaction of reaction at Cys25 as evaluated by the loss of enzymatic activity. The nature of the alkyl function does not appear to influence the rate of maleimide reaction with cysteine. Heitz and coworkers[544] showed the rates of reaction of *N*-ethylmaleimide and *N*-heptylmaleimide with cysteine or glutathione at pH 5.0 (acetate)/25°C; the rate of reaction of two maleimides was the same with cysteine (20.2 $M^{-1} s^{-1}$) or glutathione (25.8 $M^{-1} s^{-1}$). There was also a suggestion of saturation kinetics in the reaction of higher-chain alkyl derivatives with papain[545] consistent with affinity binding. Papain is also inactivated by TLCK or TPCK presumably by reaction at Cys25 as opposed to alkylation of a histidine residue.[546–548] The rate of reaction of either TLCK or TPCK with papain is greater than that observed with either trypsin or chymotrypsin respectively.[547] In fact, the rate of reaction of TLCK with papain was too rapid to measure even at pH 5.2 at 0°C.[547] Whitaker and Perez-Villaseñor[547] also studied the reaction of TLCK with free cysteine in 0.2 M phosphate, pH 6.57 at 25°C; a second-order rate constant of 6.76×10^{-4} $M^{-1} s^{-1}$ was obtained that is less than that obtained (2.39×10^{-3} $M^{-1} s^{-1}$) for the reaction of chloroacetamide and cysteine at pH 6.5 (pH stat, 0.1 M KCl) at 30°C.[549] Wolthers[548] was able to measure the rate of reaction of

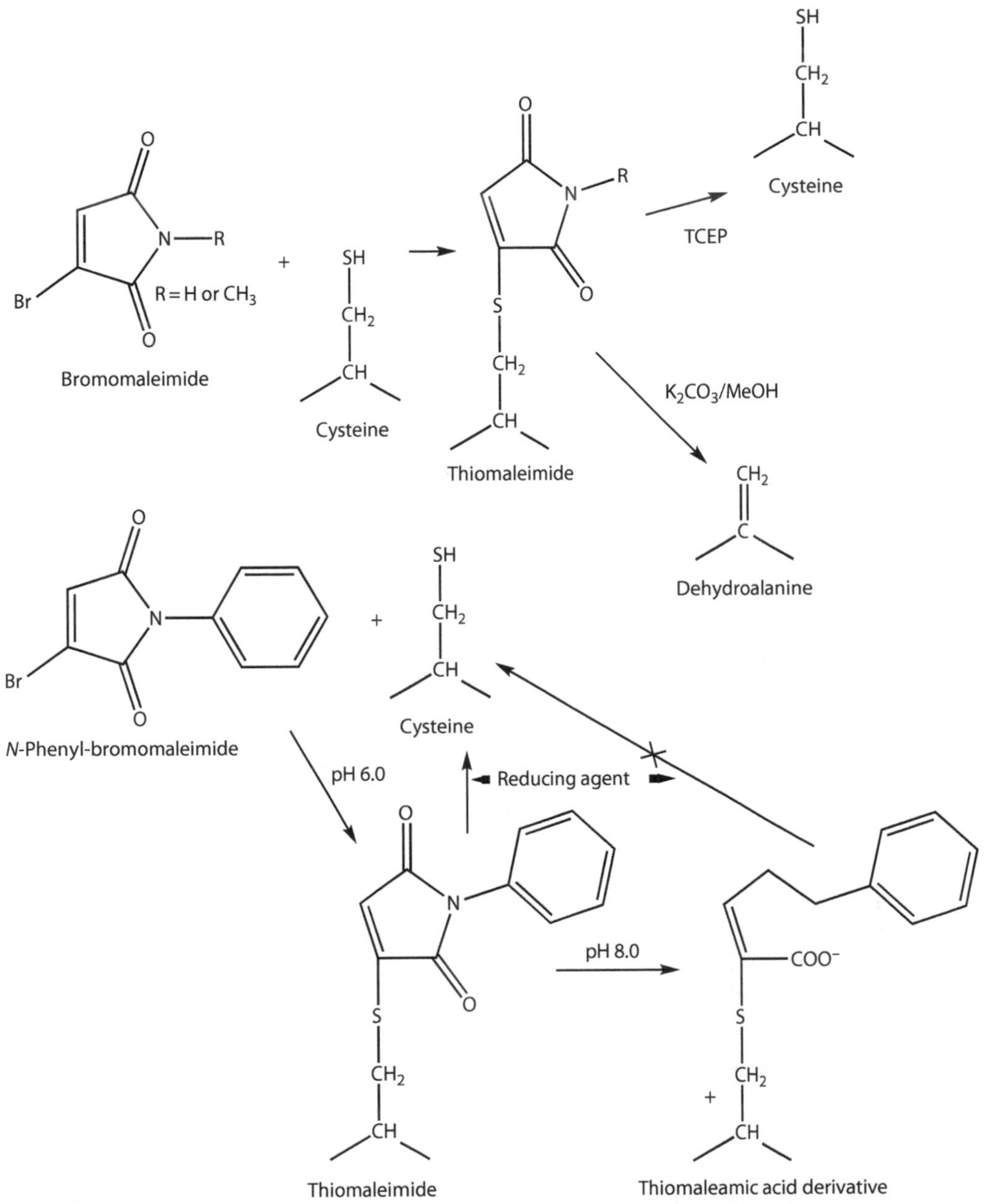

FIGURE 7.22 The reversible reaction of bromomaleimide and other Michael acceptors with cysteine and cysteinyl residues. Shown at the top is the reaction of bromomaleimide or *N*-methylbromomaleimide with cysteine to form the thiomaleimide, which can then be treated with TCEP [tris-(2-carboxyethyl)phosphine] (or dithiothreitol) to yield cysteine or with potassium carbonate in methanol to yield dehydroalanine. The reaction does not proceed in water most likely due to hydrolysis of the maleimide (Tedaldi, L.M., Smith, M.E.B., Nathani, R.I., and Baker, J.R., Bromomaleimides: New reagents for the selective and reversible modification of cysteine, *Chem. Commun.* (43), 6583–6585, 2009). Shown below is the reaction of *N*-phenylbromomaleimide with cysteine at pH 6.0 to yield the thiomaleimide derivative that can be reduced to obtain cysteine. The treatment of the thiomaleimide at pH 8.0 hydrolyzes the maleimide to a maleamic acid that which cannot be reduced to obtain cysteine (Ryan, C.P., Smith, M.E.B., Schumacher, F.F. et al., Tunable reagents for multifunctional bioconjugation: Reversible or permanent chemical modification of proteins and peptides by control of maleimide hydrolysis, *Chem. Commun.* 47, 5452–5454, 2011).

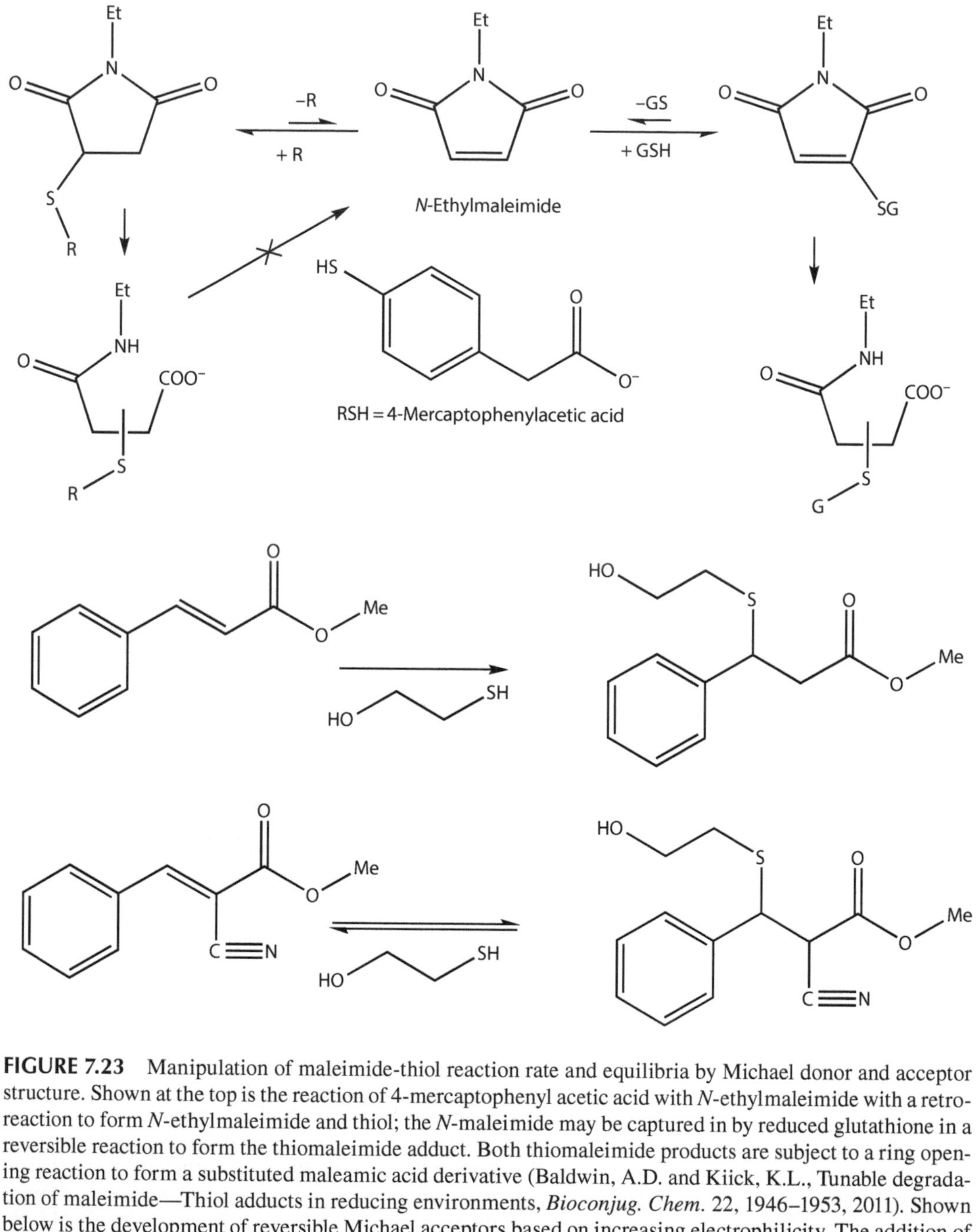

FIGURE 7.23 Manipulation of maleimide-thiol reaction rate and equilibria by Michael donor and acceptor structure. Shown at the top is the reaction of 4-mercaptophenyl acetic acid with *N*-ethylmaleimide with a retro-reaction to form *N*-ethylmaleimide and thiol; the *N*-maleimide may be captured in by reduced glutathione in a reversible reaction to form the thiomaleimide adduct. Both thiomaleimide products are subject to a ring opening reaction to form a substituted maleamic acid derivative (Baldwin, A.D. and Kiick, K.L., Tunable degradation of maleimide—Thiol adducts in reducing environments, *Bioconjug. Chem.* 22, 1946–1953, 2011). Shown below is the development of reversible Michael acceptors based on increasing electrophilicity. The addition of a nitrile group both increases both the rate of reaction with 2-mercaptoethanol and facilitates the rapid elimination of the thiol adducts to form the parent cyanoacrylamide (Serafimova, I.M., Pufall, M.A., Krishnan, S. et al., Reversible targeting of noncatalytic cysteines with chemically tuned electrophiles, *Nat. Chem. Biol.* 8, 471–476, 2012).

TABLE 7.8
Reaction of Various Sulfhydryl Reagents with Papain

Reagent	Reaction Conditions	Rate (M^{-1} s^{-1})	Reference
N-Ethylmaleimide	pH 6.0 (0.1 phosphate)/25°C	0.661[a]	7.8.1
N-Ethylmaleimide	pH 6.8/0.13 M potassium phosphate/25°C	2.77[a]	7.8.2
N-Ethylmaleimide	pH 6.5/0.105 M sodium phosphate/26°C	2.55[a,b]	7.8.3
N-Butylmaleimide[c]	pH 6.8/0.13 M potassium phosphate/25°C	4.33[a]	7.8.2
N-Heptylmaleimide	pH 6.8/0.13 M potassium phosphate/25°C	50.83[a]	7.8.2
N-Nonylmaleimide	pH 6.8/0.13 M potassium phosphate/25°C	135[a]	7.8.2
N-Decylmaleimide	pH 6.8/0.13 M potassium phosphate/25°C	358[a]	7.8.2
N-Phenylmaleimide	pH 6.8/0.13 M potassium phosphate/25°C	8.33[a]	7.8.2
N-Ethylmaleimide	Hydroxynitrobenzyl-papain, pH 6.5/0.105 M sodium phosphate/26°C	3.24	7.8.3
Chloroacetamide	pH 6.5/0.105 M sodium phosphate/26°C	0.144	7.8.3
Chloroacetate	pH 6.5/0.105 M sodium phosphate/26°C	2.75	7.8.3
Chloroacetamide	Hydroxynitrobenzyl-papain, pH 6.5/0.105 M sodium phosphate/26°C	0.179	7.8.3
Chloroacetate	Hydroxynitrobenzyl-papain, pH 6.5/0.105 M sodium phosphate/26°C	1.53	7.8.3
MMTS[d]	pH 6.0/0.025 M sodium potassium phosphate, 25°C	2.0×10^5 [e]	7.8.4
2PDS[f]	pH 3.80/formate/25°C	1.71×10^4	7.8.5
2PDS[f]	pH 6.28/phosphate/25°C	9×10^2	7.8.5
DTNB[g]	pH 6.0/25°C	63[e]	7.8.6
DTNB[g]	pH 8.0/25°C	79[e]	7.8.6
DTNB[g]	pH 9.0/25°C	316	7.8.6
Cyanate	pH 6.0/25°C	157	7.8.7
TPCK[h]	pH 6.75/0°	1.43	7.8.8
TPCK[h]	Limit value (pKa = 8.9) at 0°C	200	7.8.8
TPCK[h]	Limit value (pKa = 8.28) at 25°C	1100	7.8.9
TPCK[h]	pH 7.0 (pH stat) with 0.3 M KCl and 1 mM EDTA/25°C with 9.7 mM BAEE[i]	57	7.8.10
TLCK[i]	pH 7.0 (pH stat) with 0.3 M KCl and 1 mM EDTA/25°C with 7.8 mM BAEE[j,k]	910	7.8.10

[a] Reaction rate determined by the loss of enzyme activity.

[b] Alkylation of the active site cysteine (Cys25) in papain could also be monitored by change in intrinsic fluorescence; activity loss on alkylation correlated with change in fluorescence.

[c] Other alkylmaleimides were studied by these investigators but have not been included in this table.

[d] Methyl methanethiosulfonate.

[e] Estimated from graph.

[f] 2PDS, 2,2'-dipyridyl disulfide (dithiodipyridine; 2-di-pyridyl disulfide).

[g] DTNB, 5,5'-dithiobis(2-nitrobenzoic acid); Ellman's reagent.

[h] α-N-tosyl-L-phenylalanine chloromethylketone.

[i] α-N-tosyl-L-lysine chloromethylketone.

[j] BAEE, α-N-benzoyl-L-arginine ethyl ester (competitive inhibitor of papain).

[k] Work from other workers (Whitaker, J.R. and Perez-Villaseñor, J., Chemical modification of papain. I. Reaction with the chloromethyl ketones of phenylalanine and lysine and with phenylmethylsulfonyl fluoride, *Arch. Biochem. Biophys.* 124, 70–78, 1968) suggested that the rate of papain with TLCK was too rapid to allow determination of rate even at 0°C, pH 5.2.

(*continued*)

TABLE 7.8 (continued)

Reaction of Various Sulfhydryl Reagents with Papain

References to Table 7.8

7.8.1. Brubacher, L.J. and Glick, B.R., Inhibition of papain by *N*-ethylmaleimide, *Biochemistry* 13, 915–920, 1974.

7.8.2. Anderson, B.M. and Vasini, E.C., Nonpolar effects in reaction of sulfhydryl groups of papain, *Biochemistry* 9, 3348–3352, 1970.

7.8.3. Evans, B.L.B., Knopp, J.A., and Horton, H.R., Effect of hydroxynitrobenzylation of tryptophan-177 on reactivity of active-site cysteine -25 in papain, *Arch. Biochem. Biophys.* 206, 362–371, 1981.

7.8.4. Roberts, D.D., Lewis, S.D., Ballou, D.P. et al., Reactivity of small thiolate anions and cysteine-25 in papain toward methyl methanethiosulfonate, *Biochemistry* 25, 5595–5601, 1986.

7.8.5. Brocklehurst, K. and Little, G., A novel reactivity of papain and a convenient active site titration in the presence of other thiols, *FEBS Lett.* 9, 113–116, 1970.

7.8.6. Brocklehurst, K., Mushiri, S.M., Patel, G., and Willenbrock, F., A marked gradation in active-centre properties in the cysteine proteinases revealed by neutral and anion reactivity probes. Reactivity characteristics of the thiol groups of actinidin, ficin, papain and papaya peptidase A towards 4,4′dipyridyl disulphide and 5,5′-dithionbis-(2-nitrobenzoate) dianion, *Biochem. J.* 209, 873–879, 1983.

7.8.7. Sluyterman, L.A.Æ., Reversible inactivation of papain by cyanate, *Biochim. Biophys. Acta* 135, 439–449, 1967.

7.8.8. Whitaker, J.R. and Perez-Villaseñor, J., Chemical modification of papain. I. Reaction with the chloromethyl ketones of phenylalanine and lysine and with phenylmethylsulfonyl fluoride, *Arch. Biochem. Biophys.* 124, 70–78, 1968.

7.8.9. Bender, M.L. and Brubacher, L.J., The kinetics and mechanism of papain-catalyzed hydrolysis, *J. Am. Chem. Soc.* 88, 5880–5889, 1966.

7.8.10. Wolthers, B.C., Kinetics of inhibition of papain by TLCK and TPCK in the presence of BAEE as substrate, *FEBS Lett.* 2, 143–145, 1969.

TLCK with papain in the presence of substrate, α-*N*-benzoyl arginine ethyl ester (BAEE). Papain is also inhibited by phenylmethylsulfonyl fluoride via formation of a *S*-sulfonyl derivative.[547] The influence of increased hydrophobicity on maleimide reactivity is consistent with the presence of a tryptophanyl residue at the papain active site.[550] The observations of Evans and coworkers[551] on the modification of trp177 in papain with hydroxynitrobenzyl bromide (see Chapter 12) resulted in an increased rate of modification by NEM and chloroacetamide (Table 7.8) but a decrease in the rate of modification with a polar reagent, chloroacetic acid. The data obtained with TLCK are remarkable, but it is recognized that papain has a specificity for cleavage of peptide bonds containing arginine or lysine. Nevertheless, the rate of inactivation of papain by TLCK by modification of cysteine is much larger than that reported by Shaw and coworkers[552] (5.61 M^{-1} s^{-1}) for the inactivation of trypsin by TLCK by alkylation of histidine at pH 6.0.

The ease of preparing *N*-alkyl derivatives of maleimide[544,553,554] and subsequent commercial availability has provided the opportunity to use such materials as probes for cysteine reactivity in a variety of proteins (Table 7.9). Heitz and coworkers[544] were the first group to explore the reaction of series of *N*-alkylmaleimides of increasing hydrophobicity (Figure 7.24) with cysteinyl residues in a protein. In studies with YADH, these investigators showed an increasing rate of reaction with increasing chain length of the *N*-alkyl group. Solubility considerations precluded evaluation of *N*-nonylmaleimide and *N*-decylmaleimide. There was no evidence to support a saturation effect on the reaction of the *N*-alkylmaleimides with YADH; however, sufficient data to support this suggestion were obtained only for NEM. Fonda and Anderson[555] studied the reaction of several alkyl and aryl maleimides with D-amino acid oxidase. While 5–6 cysteine residues per dimer react with the maleimides, the rate of inactivation involved one cysteine per subunit. The data obtained at pH 7.0 are shown; the rate of inactivation is at least doubled at pH 7.5. There are data supporting saturation

TABLE 7.9
Rate Constants for Reaction of Alkyl Maleimides with Model Compounds and Cysteinyl Residues in Proteins

Reagent	Rate (M^{-1} s^{-1}) YADH[a]	Rate (M^{-1} s^{-1}) DAAO[b]	Rate (M^{-1} s^{-1}) GPD[c]
N-Ethylmaleimide	0.22	0.45	0.04
N-Butylmaleimide	0.56	1.21	0.19
N-Pentylmaleimide	0.84	1.83	0.62
N-hexylmaleimide	1.13	2.85	2.37
N-Heptylmaleimide	1.88	4.70	8.67
N-Octylmaleimide	2.16	6.72	28.8
N-Nonylmaleimide	—	—	91.7
N-Benzylmaleimide	1.34	—	—
N-Phenylmaleimide	—	5.33	—

[a] YADH; yeast alcohol dehydrogenase/0.1 M sodium pyrophosphate, pH 7.0 with 2% EtOH/20°C (Heitz, J.R., Anderson, C.D., and Anderson, B.M., Inactivation of yeast alcohol dehydrogenase by N-alkylmaleimides, *Arch Biochem. Biophys.* 127, 627–636, 1968).

[b] DAAO; D-amino acid oxidase/50 mM sodium pyrophosphate, pH 7.0, 25°C (Fonda, M.L. and Anderson, B.M., D-Amino acid oxidase IV. Inactivation by maleimides, *J. Biol. Chem.* 244, 666–674, 1969).

[c] GPD, rabbit muscle L-α-glycerophosphate dehydrogenase/50 mM Tris–HCl, pH 7.0 with 2% EtOOH/10°C (Anderson, B.M., Kim, S.J., and Wang, C.-N., Inactivation of rabbit muscle L-glycerophosphate dehydrogenase by N-alkylmaleimides, *Arch. Biochem. Biophys.* 138, 66–72, 1970).

suggesting binding of reagent prior to the modification of sulfhydryl groups. These experiments are difficult to perform with the higher-chain-length alkyl derivatives reflecting solubility issues. Anderson and coworkers[556] also studied the effect of increasing chain length on the rate of inactivation of rabbit muscle L-α-glycerophosphate dehydrogenase by N-alkylmaleimides. Again, increasing alkyl chain length results in an increased rate of inactivation. This suggests the likely importance of reagent binding by the enzyme prior to alkylation of a cysteinyl residue at the active site of this enzyme. The concept of reagent partitioning is discussed in Chapter 1. Octanol–water partitioning has been correlated with the reactivity of alkyl (N-ethyl, N-butyl) and aryl (N-benzyl) maleimides with a cysteine residue in succinic dehydrogenase.[557]

The versatility of reagent structure provided by N-substituted maleimides has been of major interest in the study of membrane proteins. This versatility is shared with the methanethiosulfonate reagents described below (Figure 7.30). The localization of sulfhydryl groups within membranes has been achieved through the comparison of the reaction with membrane-permeant and membrane-impermeant maleimide derivatives.[558–560] The use of these reagents can be traced back to the original observations of Abbott and Schachter in 1976.[561] The basic concept is to provide either a polar derivative or a derivative with steric considerations that preclude passage through or into the membranes. Maleimide derivatives of glucosamine have been synthesized as affinity labels for the human erythrocyte hexose transport protein.[562] Le-Quoc and colleagues have examined the effect of the nature of the N-substituent groups (Figure 7.24) on the rate of sulfhydryl group modification in membrane-bound succinate dehydrogenase.[557] The derivatives used were N-ethylmaleimide, N-butylmaleimide, and N-benzylmaleimide. The rate of modification (inactivation) of membrane-bound succinate dehydrogenase correlated with the octanol–water partitioning coefficient. The partition coefficient for NEM is 3, and the rate of modification was 3.67 mM^{-1} min^{-1}; the partition coefficient for N-benzylmaleimide was infinite (could not be measured), and the rate of

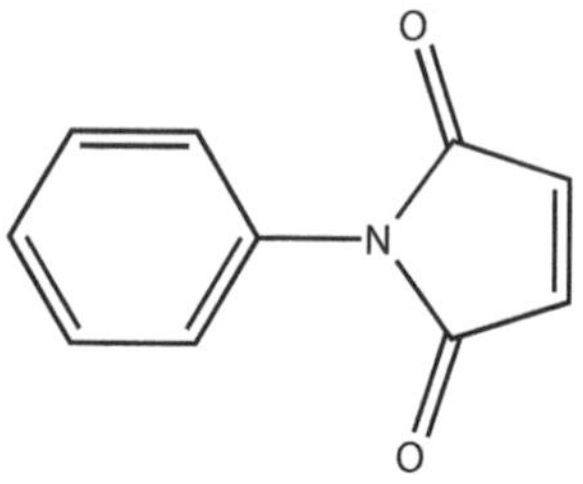

N-Phenylmaleimide

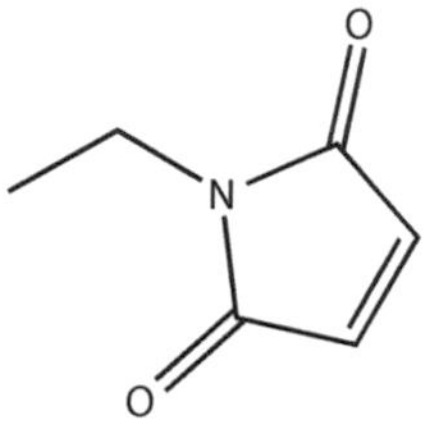

N-Ethylmaleimide

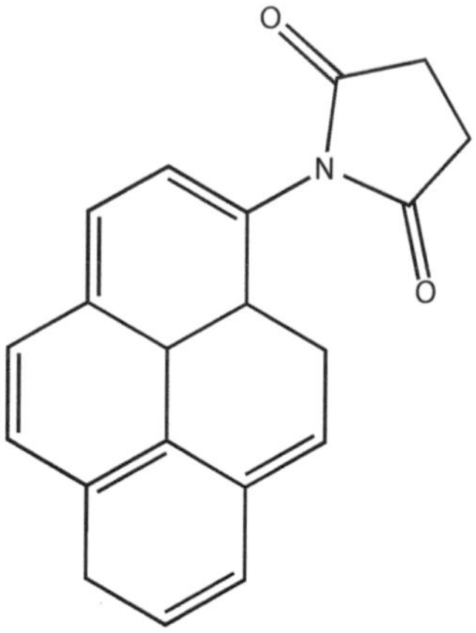

N-(3-pyrene)-maleimide

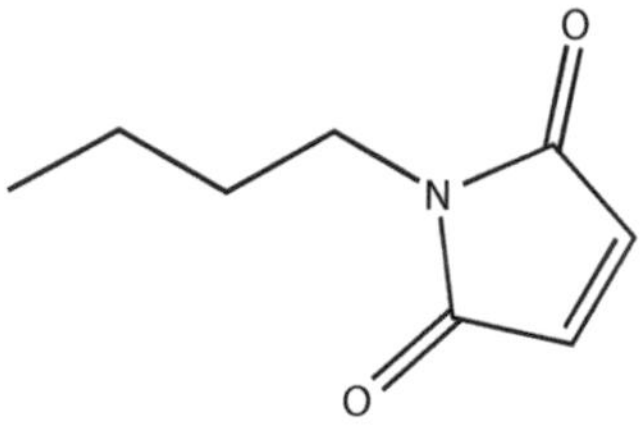

N-Butylmaleimide

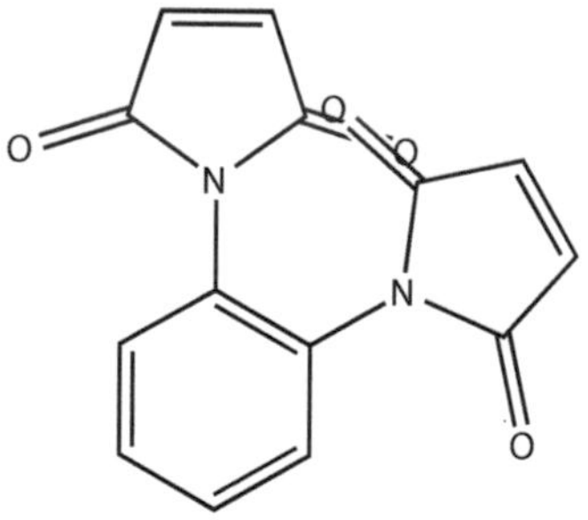

N,N'-o-Phenylene dimaleimide

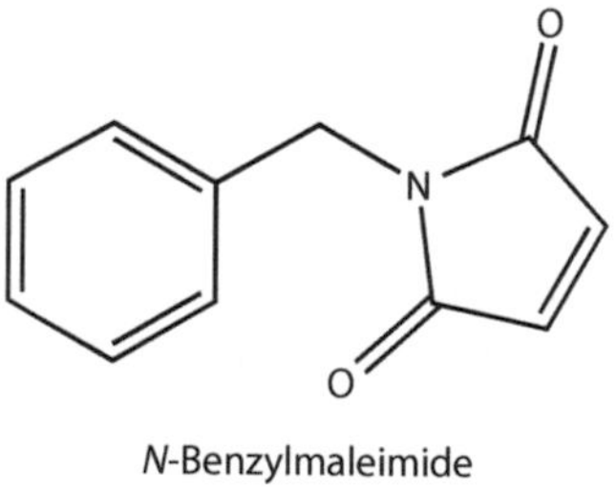

N-Benzylmaleimide

FIGURE 7.24 Hydrophobic maleimide probes. Some hydrophobic maleimide chemistry for membrane protein modification. Shown is *N*-phenyl maleimide that has been used to modify cysteine residues in a hydrophobic environment (Indiveri, C., Giangregorio, N., Iacobazzi, V., and Palmeri, F., Site-directed mutagenesis and chemical modification of the six native cysteine residues of the rat mitochondrial carnitine carrier: Implications for the role of Cysteine-136, *Biochemistry* 41, 8649–8656, 2002) and three maleimide derivatives that were used to differentiate sulfhydryl reactivity with a membrane-bound succinate dehydrogenase (LéQuôc, K., LéQuôc, D., and Gauderner, Y., Evidence for the existence of two classes of sulfhydryl groups essential for membrane-bound succinate dehydrogenase activity, *Biochemistry* 20, 1705–1710, 1981). *N*-pyrene maleimide is a hydrophobic fluorescent probe for membrane structure (Brown, R.D. and Matthews, K.S., Chemical modification of lactose repressor protein using *N*-substituted maleimides, *J. Biol. Chem.* 254, 5128–5134, 1979; Mancek-Keber, M., Gradisar, N., Iñigo Pestaña, M. et al., *J. Biol. Chem.* 284, 19493–19500, 2009). Also shown is the structure of *N,N'-o*-phenylenedimaleimide (Rial, E., Aréchaga, I., Sainz-de-la-Maza, E., and Nicholls, D.G., Effect of hydrophobic sulphydryl reagents on the uncoupling protein and inner-membrane anion channel of brown-adipose-tissue mitochondria, *Eur. J. Biochem.* 182, 187–193, 1989); this reagent is also used as a site-specific cross-linking reagent (Rimon, A., Tzubery, T., Galili, L., and Padan, E., Proximity of cytoplasmic and periplasmic loops in NhaA Na+/H antiporter of *Escherichia coli* as determined by site-directed thiol cross-linking, *Biochemistry* 41, 14897–14905, 2002).

modification was 15.4 mM^{-1}min^{-1}. N-Butylmaleimide was intermediate. Later studies by various investigators[563-568] extended information on the use of hydrophobic maleimide derivatives that can also be environmentally sensitive (Figure 7.25) including N-polymethylenecarboxymaleimide (Figure 7.25),[563-565] N-tetramethylrhodamine maleimide (Figure 7.25),[566] aryl-indole maleimide,[567] and N-arachidonylmaleimide.[568] N-(Pyrene) maleimide (Figure 7.24) is also a fluorescent probe.[569] Ridge and coworkers[564] compared the reaction of some membrane-impermeant polymethylenecarboxy maleimides, such as maleimido-N-hexanoic acid and maleimido-N-propionic acid, with reactivity with N-ethylmaleimide to map cysteine substitution mutants in rhodopsin. Of interest was the observation that the mutant cysteine residues while reacting with N-ethylmaleimide and the polymethylenecarboxymaleimides did not react with iodoacetate or iodoacetamide. Wache and coworkers[570] have reported the development of a diaminoterephthalate maleimide derivative, which is a *turn-on* fluorescent probe; the maleimide probe, NiWa Blue (Figure 7.25), develops fluorescence on reaction with a cysteine residue in a protein.

Modification with N-ethylmaleimide has found considerable use in the modification of inserted cysteine residues in the study of transmembrane proteins.[571] The synthesis of deuterated N-alkylmaleimides has been reported[572], and these reagents have been used for quantitative peptide analysis in the study of differential protein expression. An electroactive probe has been developed, which uses maleimide coupled to sulfhydryl groups in proteins.[573] A particularly novel approach to this problem has been used by Falke and Koshland to analyze aspartate receptor structure.[574] In this study, site-specific mutagenesis was used to place cysteinyl residues at six positions in the peptide chain. A novel membrane-impermeant maleimide derivative [N-(6-phosphonyl-n-hexyl)-maleimide] was used to study the reactivity of the individual sulfhydryl residues. From these studies, it was possible to *map* the domain structure of the receptor protein. Cysteinyl residues placed in the surface area could be modified by aqueous reagents, while transmembrane areas could be excluded by the lack of reaction with membrane-impermeant reagents.

Hydrophobic derivatives of N-ethylmaleimide (Figure 7.24) can be used as probes of the environment surrounding a sulfhydryl group in membrane anion channels.[575] Reaction with N-ethylmaleimide, N-benzylmaleimide, and N,N'-1,2-phenylenedimaleimide was evaluated and the reaction rate increased with increasing hydrophobicity. N-Phenylmaleimide has been used for the modification of a sulfhydryl group in the acetylcholine receptor.[576,577] Detergent was required for the modification reaction (10 mM MOPS to 100 mM NaCl to 0.1 mM EDTA with 0.02% sodium azide and 1% sodium cholate). These studies identified cysteine residues potentially important in membrane function. Subsequent studies using site-specific mutagenesis have supported the importance of these cysteinyl residues.[577] Site-directed modification of cysteine has been demonstrated to be useful in the study of the relationship between structure and function in proteins. An equal number of proteins where endogenous cysteine residues have been modified are included in this tabulation for comparison of reaction conditions.

Maleimide PEG derivatives (Figure 7.26) have been used to modify cysβ93 in human adult hemoglobin.[578] Subject maleimides include maleimide phenyl carbamyl (O''-methyl PEG 5000) and O'-(2-maleimidoethyl)–O'–methyl-PEG5000. Modification with the various N-alkylmaleimides was accomplished in phosphate-buffered saline, pH. 7.4 at various temperatures. Other applications include recombinant engineered human granulocyte-macrophage colony-stimulating factor (GM-CSF),[579,580] an erythropoietin mutant with an engineered cysteine,[581] and recombinant human interleukin-1 receptor.[582] A related study[583] describes the synthesis of maleimide derivatives of carbohydrates as *chemoselective* tags for site-specific glycosylation of proteins (Figure 7.26). Amino derivatives of carbohydrates were first coupled to 6-maleimidohexanoic acid N-hydroxysuccinimide ester to form the 6'-maleimdohexanamidoethyl-glycosides, which can then be coupled to a cysteinyl residue in the protein. This is an interesting approach within the area of glycoengineering.[584] Maleimide was used for the preparation of carbohydrate microarrays.[585,586] A related application is the coupling of doxorubicin to lactosamined human albumin for drug delivery to liver cells via the asialoglycoprotein receptor.[587]

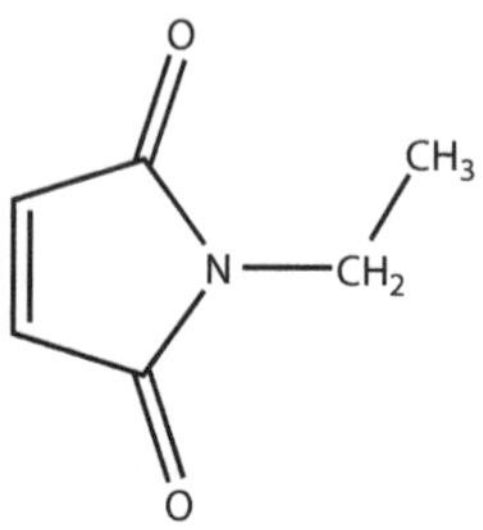

N-Ethylmaleimide

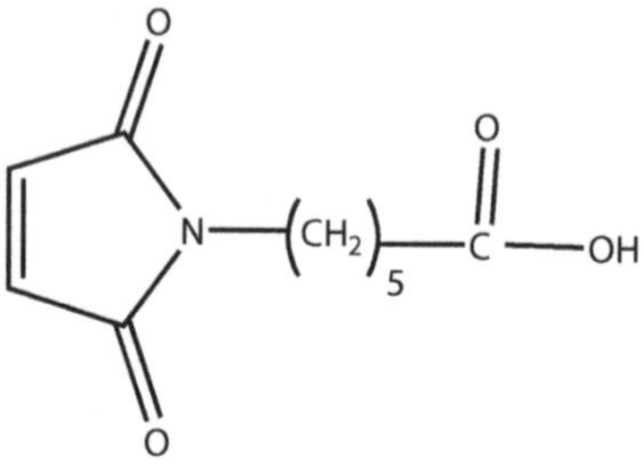

N-Hexanoic acid maleimide
(maleimido-N-hexanoic acid)

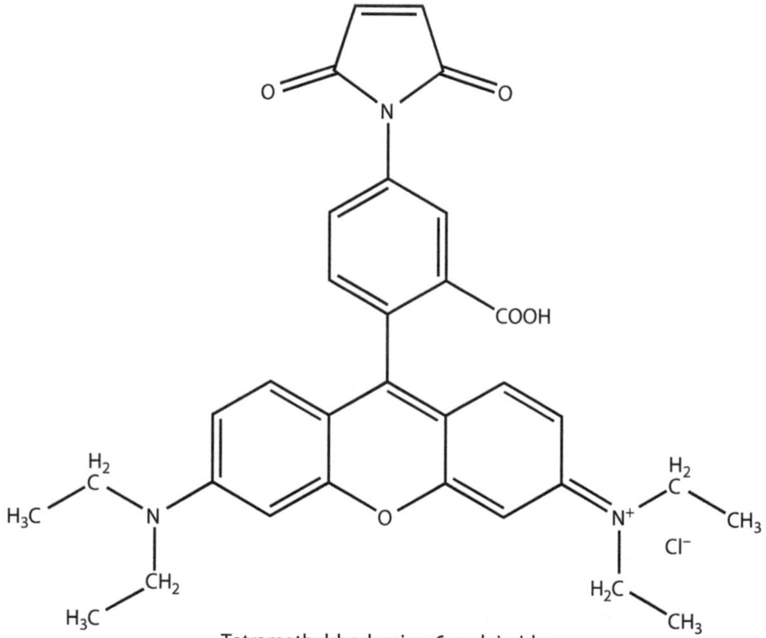

Tetramethylrhodamine-6-maleimide

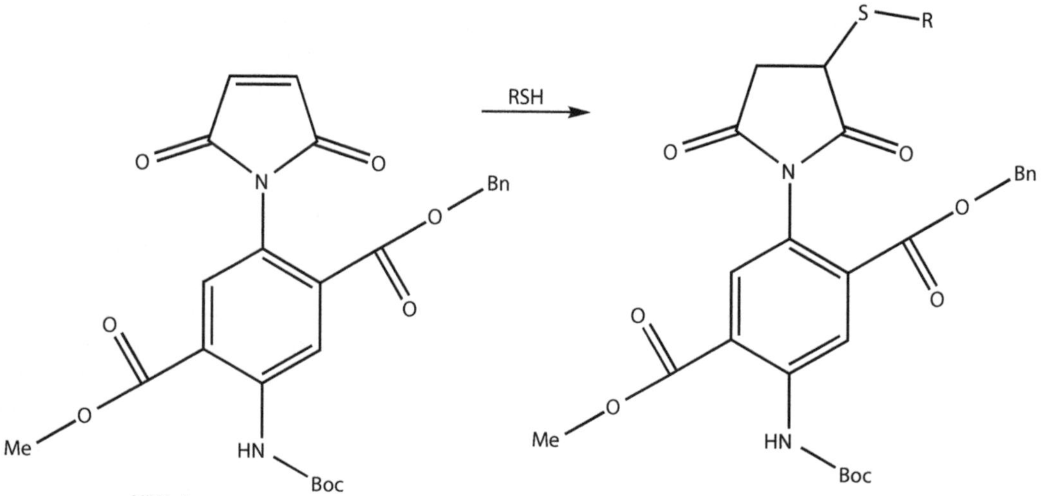

RSH

NiWa I

Fluorescent product

FIGURE 7.25 (continued)

Brown and Matthews[516,588] have studied the reaction of lactose repressor protein with N-ethylmaleimide, two spin-labeled derivatives of N-ethylmaleimides, and a fluorophore derivative. The spin-labeled compounds showed the same pattern of reaction with the three cysteinyl residues as seen with N-ethylmaleimide. The fluorophore derivative (N-(3-pyrene)maleimide) shows a slightly different reaction pattern. Other probes include 4-maleimido-2,2,6,6-tetramethylpiperidine-1-oxyl,[589] maleimidotetra-methylrhodamine,[589] N-(1-pyrenyl)maleimide,[590] 2-(4-maleimidoanilino) naphthalene-6-sulfonic acid,[591] 2,5-dimethoxy-4-stil-benzylmaleimide,[591] rhodamine maleimide,[591] and eosin-5-maleimide.[592]

Pulse-chase labeling using deuterated N-ethylmaleimide has been used to determine the relative reactivity of cysteine residues in the zinc finger regions of the MRE (metal response element)-binding domain of MRE-binding transcription factor-1 (MTF-1).[593] The results are used as a surrogate measurement of the zinc chelate stability of individual zinc fingers in MTF-1. As with the reaction of metallothionein with aromatic mercurials or N-ethylmaleimide. Thiol reactivity with maleimide in MRE-binding domain is more rapid in the absence of metal ion. In these experiments, the protein is first reacted with a 10-fold molar excess (to cysteine) of d_5-N-ethylmaleimide and at various time points, portions are removed and added to a solution containing H_5-N-ethylmaleimde (at a 100-fold molar excess to cysteine) and trypsin or chymotrypsin. Mass spectrometric analysis is used to determine the ratio of deuterium to hydrogen label at each cysteine residue, which is, in turn, a measure of reactivity and can be used to derive second-order rate constants for the alkylation reaction at each individual cysteine. Cysteine reactivity was also influenced by the binding of oligonucleotide. Apuy and coworkers[594] applied this technology to the study of thiolate reactivity in a metal-regulated transcriptional repressor. It is noted that Apuy and coworker originally used the radiometric pulsed alkylation technique for the study of protein folding.[595]

Acid-labile isotope-coded extractants (ALICE) are N-alkyl maleimide derivatives (Figure 7.27) where there is an acid-labile function between the alkylmaleimide and a polymer matrix.[596] Cysteine-containing peptides are *captured* by reaction with the maleimide function bound via linker that can contain either hydrogen or deuterium to a matrix via an acid-labile bond. The bound polymer peptide can be removed by filtration and the peptides released from the matrix with 5% trifluoroacetic acid in dichloromethane. Another group[597] converted phosphopeptides to thiol-containing peptides and subsequent capture of the peptide with a maleimide function linked to a matrix by an acid-labile bond. This approach was used by other investigators[598] with an iodoacetyl matrix for the capture of thiol peptide derived from phosphopeptides. Fauq and coworkers[599] used an acid-labile biotin reagent that can contain either maleimide or haloalkyl function. A clever application of the use of Michael addition is the selective modification of a disulfide bond for the insertion of a PEG chain (Figure 7.28).[600–602]

FIGURE 7.25 (continued) Hydrophobic maleimide structural probes. Shown are some reagents that have been used in the environment surrounding reactive cysteine residues in membranes and other hydrophobic environments. Some of these probes are also solvatochromic probes. Shown is the structure of a membrane-impermeant polymethylenecarboxylmaleimide, N-hexanoic acid maleimide (maleimido-N-hexanoic acid) (Ridge, K.D., Zhang, C., and Khorana, H.G., Mapping of the amino acids in the cytoplasmic loop connecting helices C and D in rhodopsin. Chemical reactivity in the dark state following single cysteine replacements, *Biochemistry* 34, 8804–8811, 1995). Shown below is the structure of tetramethylrhodamine-6-maleimide that is a membrane-permeant(hydrophobic) fluorescent probe (Duerr, K.L., Tavraz, N.N., Zimmerman, D. et al., Characterization of Na,K-ATPase and H,K-ATPase enzymes with glycosylation -deficient β-subunit variants by voltage-clamp fluorometry in *Xenopus oocytes*, *Biochemistry* 47, 4288–4297, 2008; Omoto, J.J., Maestas, M.J., Rahnama-Vaghef, A. et al., Function consequences of sulfhydryl modification of the γ-aminobutyric acid transporter 1 at a single solvent exposed cysteine residue, *J. Membr. Biol.* 245, 841–857, 2012). Shown at the bottom is the structure of a diaminoterephthalate compound that is a fluorogenic reagent for sulfhydryl groups in proteins (Wache, N., Schröder, C., Koch, K.W., and Christoffers, J., Diaminoterephthalate turn-on fluorescence probes for thiols—Tagging of recoverin and tracking of its conformational change, *ChemBioChem* 13, 993–998, 2012).

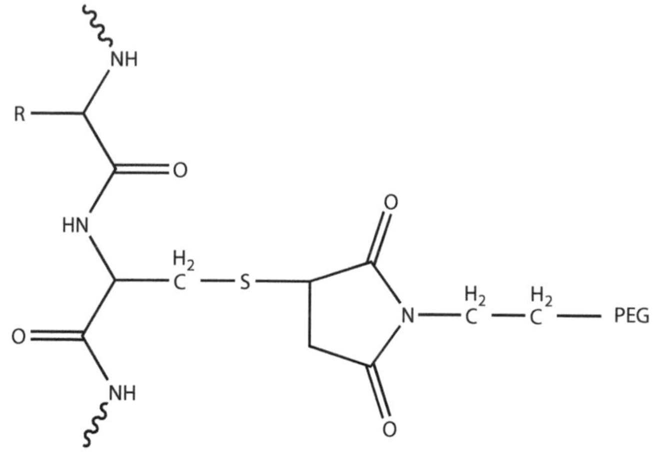

PEG-modified protein

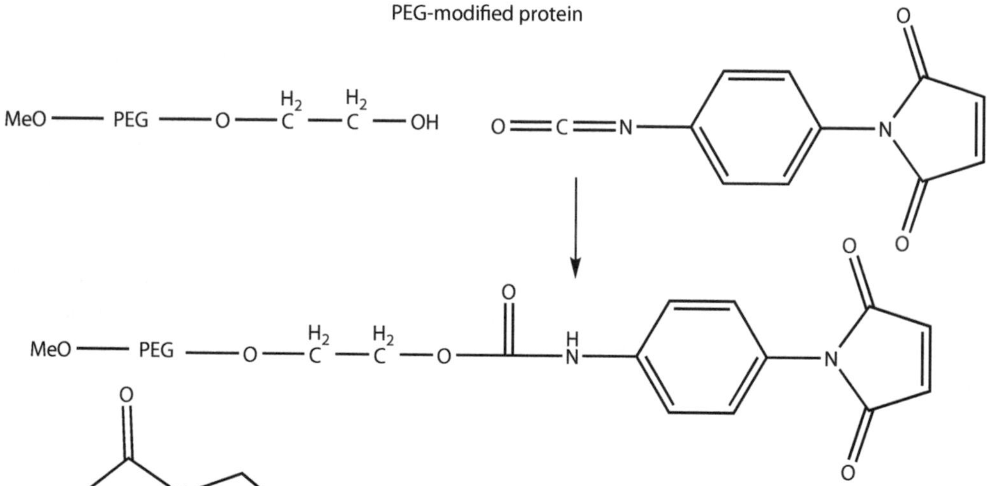

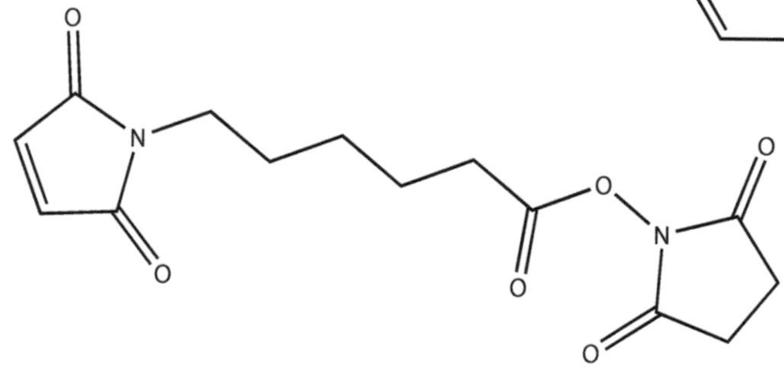

6-Maleimidohexanoic acid N-hydroxysuccinimide ester

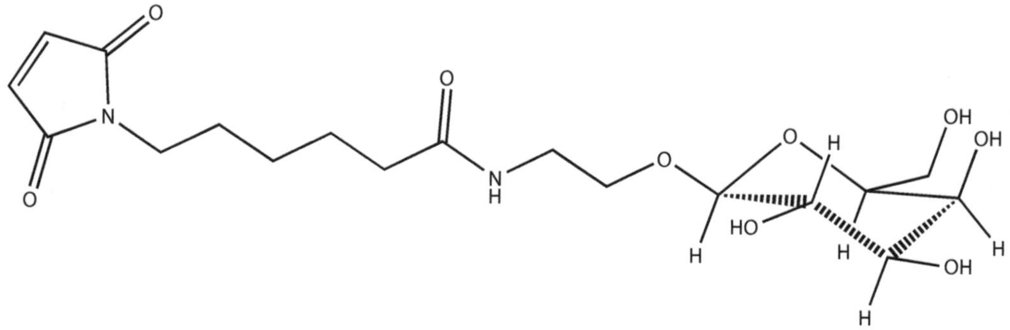

FIGURE 7.26 (continued)

A consideration of Table 7.2 shows that the reaction of model thiols with *N*-ethylmaleimide is much faster than with haloalkyl compounds. A direct comparison of the reactivities of *N*-ethylmaleimide and iodoacetic acid/iodoacetamide with proteins in tissues was performed by Rogers and coworkers.[603] These workers observed that the reaction of NEM with protein sulfhydryl groups in tissue (murine myofibril tissue extracts) (1) was much faster than either iodoacetate or iodoacetamide, (2) could proceed at acidic pH, and (3) required less reagent. Earlier work by Crawhall and Segal[604] used *N*-ethylmaleimide to determine the intracellular ratio of cysteine and cystine in several rat tissues including kidney cortex, liver, diaphragm, jejunum, and brain. Cysteine was present mostly in the reduced form ranging from 54% in brain to 97% in jejunum; correction for extracellular fluid increased the values to 87% and 100%, respectively. Hill and coworkers[371] have reviewed the application of biotin- and fluorophore derivatives of maleimide for measurement of intracellular thiols (thiol proteome).

ALKYL ALKANETHIOSULFONATES

Alkyl alkanethiosulfonates (e.g., methyl methanethiosulfonate, MMTS) (Figure 7.29) have been extensively used in the past decade for the modification of cysteine residues in proteins. These reagents have the advantage of specificity, speed of reaction, and reversibility. Methyl, ethyl, and trichloromethyl derivatives of methanethiosulfonate and propylpropanethiosulfonate were described by Smith and colleagues in 1975.[605] These alkyl alkanethiosulfonates form mixed disulfides with sulfhydryl group of cysteine with the release of methylsulfinic acid. The mixed disulfides are going to be of variable stability, more stable than thioesters but less stable than thioethers. It is likely that the stability of the mixed disulfide is influence by the chemistry of the alkyl function. The modification is highly specific for sulfhydryl groups and is usually reversed by mild reduced agents such as 2-mercaptoethanol or dithiothreitol. Reaction could occur at sites other than cysteine in proteins.[606] Such reactions would most likely occur at amino groups, and these modifications would be irreversible not being reduced by reducing agents such as dithiothreitol or 2-mercaptoethanol. Kluger and Tsui[606] observed that the inactivation of D-3-hydroxybutyrate dehydrogenase by MMTS could not be reversed with hydroxylamine, 2-mercaptoethanol, or tributylphosphine suggesting that reaction occurred at residue other than cysteine. The modification of the enzyme with 2,3-dimethylmaleic anhydride, a reversible modifier of amino groups in proteins, protected the enzyme from inactivation. NMR studies on the reaction of MMTS with Gly-Gly suggested the formation of a sulfenamide derivative. While, to the best of my knowledge, there are no reports of the modification of amino groups in proteins with alkylthiosulfonates, there are modifications with alkylthiosulfonates that are only partially reversed on reduction.[607] Pathak and coworkers modified a cysteine residue in the Wbp1p subunit of *Saccharomyces cerevisiae* oligosaccharyl transferase MMTS. Catalytic activity was lost on reaction with MMTS but not by reaction with iodoacetate, iodoacetamide, or *N*-ethylmaleimide. Activity in the MMTS-modified enzyme could be partially recovered (50%) with dithiothreitol; the use of tributylphosphine resulted in the loss of activity in the modified and unmodified enzyme. Nishimura and coworkers[608] reported one of the early studies

FIGURE 7.26 (continued) Some maleimide derivatives. Shown at the top is the product obtained with the reaction of maleimide derivatives of PEG (Juszczak, L.J., Manjula, B., Bonaventura, C. et al., UV resonance Raman study of β93-modified hemoglobin A: Chemical modifier-specific effects and added influences of attached poly(ethylene glycol) chains, *Biochemistry* 41, 376–285, 2002). Shown below is the illustration of a method for preparing a maleimide derivative of methoxy PEG (Ananda, K., Nacharaju, P., Smith, P.K. et al., Analysis of functionalization of methoxy-PEG as maleimide-PEG, *Anal. Biochem.* 374, 231–242, 2008). Shown at the bottom is 6-maleimidohexanoic acid *N*-hydroxysuccinimide ester that is used to prepare maleimide derivatives of carbohydrates for coupling to proteins (Ni, J., Singh, S., and Wang, L.-X., Synthesis of maleimide-activated carbohydrates as chemoselective tags for site-specific glycosylation of peptides and proteins, *Bioconjug. Chem.* 14, 232–238, 2003).

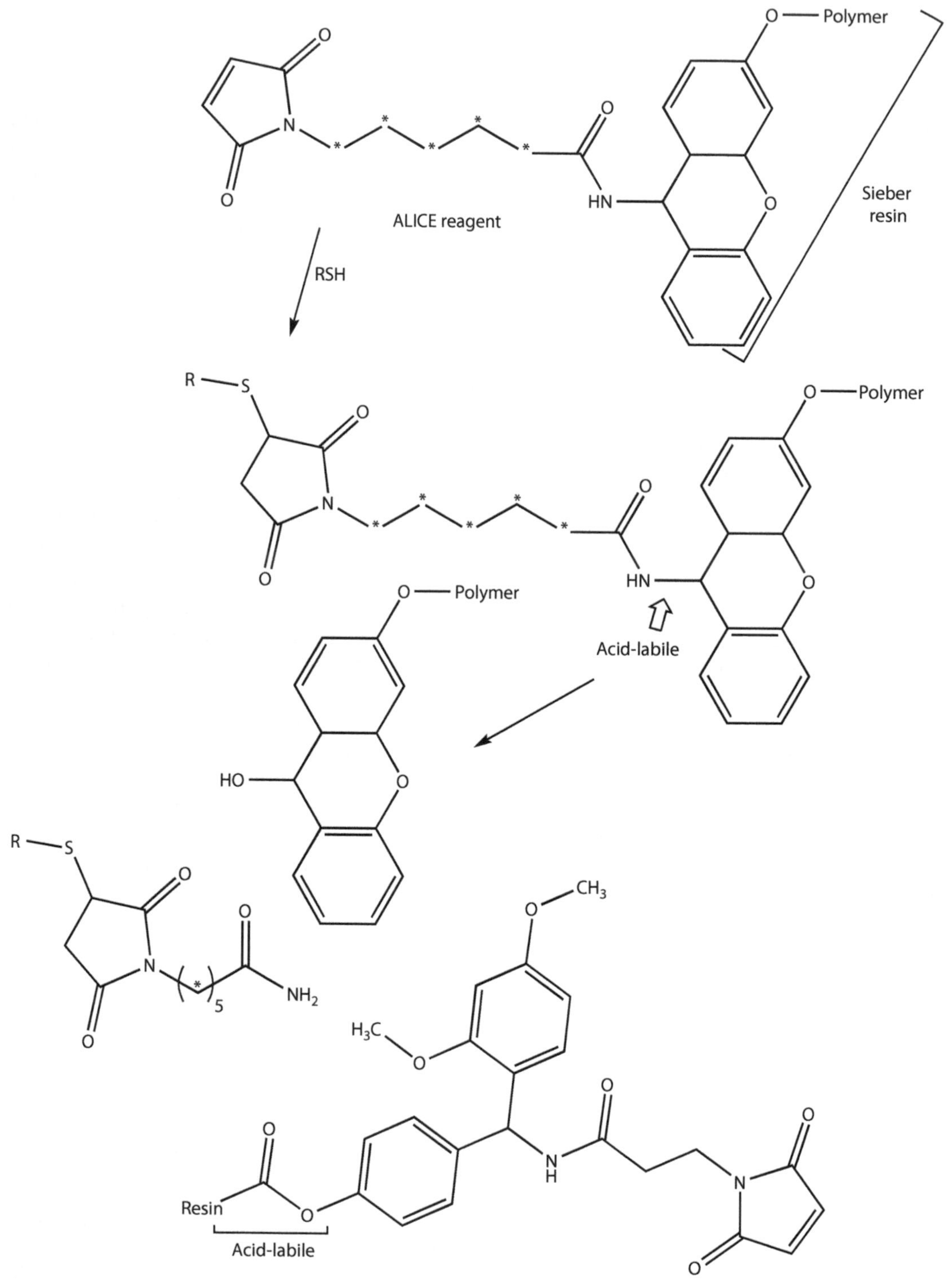

FIGURE 7.27 (continued)

on the reactivation of MMTS-modified proteins. These investigators reported the inactivation of *Escherichia coli* succinic thiokinase with either MMTS or methoxycarbonylmethyl disulfide and reactivation with tributylphosphine but not with dithiothreitol. Other studies do suggest that the reaction of MMTS-inactivated proteins is not a trivial matter. The failure of dithiothreitol to recover activity while recovery was observed with tributylphosphine is similar to the pattern observed with cysteine sulfenic acid (Chapter 8). Roberts[609] compared the reactivity of MMTS with several small thiols and the active-site cysteine in papain. The reaction of MMTS with the thiolate is at least 5×10^9 times more rapid than the corresponding thiol, and the rate of reaction depends on thiol basicity. The reader is also directed to a study by Stauffer and Karlin[610] on the effect of ionic strength on the reaction of alkylthiosulfonates with simple and protein-bound thiols. The reaction of 2-aminoethyl methanethiosulfonate, [2-(trimethylammonium)ethyl] methanesulfonate, (2-sulfonatoethyl)meth-ane thiosulfonate, and methyl methanethiosulfonate (Figure 7.30) with the acetylcholine binding site on the acetylcholine receptor was studied as a function of ionic strength. The goal of the study was to evaluate long-range electrostatic interactions in the acetylcholine binding site. The positively charged (2-aminoethyl methanethiosulfonate and [2-(trimethylamino)ethyl]methanethiosulfonate reacted more rapidly with simple thiols such as 2-mercaptoethanol, 2-mercaptoethylamine, and 2-mercaptoethylsulfonic acid than did the negatively charged (2-sulfonatoethyl)methane-thiosulfo-nate. Both negatively and positively charged derivatives were more active than the neutral methyl methanethiosulfonate. Both negatively charged and positively charged derivatives were more active than the neutral MMTS. Both positively charged derivatives reacted much more rapidly with the sulfhydryl group in the acetylcholine binding site than did the neutral MMTS that was, in turn, far more potent than the negatively charged reagent. There was a significant effect of ionic strength on the reaction of positively charged reagents consistent with the presence of two or three negative charges at the binding site. In later work,[611,612] Kenyon and Bruice extended the understanding of the use of this class of reagents as part of a more general review of the modification of cysteine residues. Most work has used alkyl derivatives of methanethiosulfonate (MTS) such these reagents are frequently referred to as MTS reagents.[613]

The alkyl methylthiosulfonates (MTS derivatives) have received much more use as reagents for the determination of cysteine accessibility[614] (SCAM,[615] scanning cysteine accessibility mutations; substituted cysteine accessibility method) in proteins than as reagents for the modification of func-tionally important residues. Karlin and Akabas[614] reviewed the use of MTS reagents for SCAM in 1998. In the most simple terms, MTS reagents can be considered large, small, positively charged, negatively charged, and neutral (Figure 7.30). The stability of reagents must be considered as the rate of hydrolysis (Table 7.10) does vary with the alkyl function. The rate of hydrolysis of (2-ami-noethyl)methane thiosulfonate is more rapid ($t_{1/2} \sim 12$ min) at pH 7.0 (20°C) than at pH 6.0 (20°C) ($t_{1/2} \sim 92$ min); at lower temperature (4°C), the $t_{1/2}$ at pH 7.0 is ~ 116 min. (2-Sulfonatoethyl)meth-ane thiosulfonate is considerably more stable ($t_{1/2} \sim 370$ min) at pH 7.0 than the positively charged

FIGURE 7.27 (continued) Some maleimide reagents for purification of peptides and proteins. Shown at the top is an ALICE for the purification of cysteine-containing amino acids. The asterisks indicate methylene groups that can either be light (hydrogen) or heavy (deuterium) based on the quality of the 6-aminocaproic acid used for the synthesis of the linker. The cysteine peptide is captured with maleimide function and recov-ered from the Sieber peptide resin (Sieber, P., A new acid-labile anchor group for the solid-phase synthe-sis of C-terminal peptide amides by the FMOC method, *Tetrahedron Lett.* 28, 2107–2110, 1987) by mild acid cleavage (Qiu, Y., Sousa, E.A., Hewick, R.M., and Wang, J.H., Acid-labile isotope-coded extractants: A class of reagents for quantitative mass spectrometric analysis of complex peptide mixtures, *Anal. Chem.* 74, 4969–4979, 2002). Shown below is a maleimide reagent for isolating cysteine-containing peptides gener-ated from phosphorylated peptides. Dehydroalanine residues are obtained by β-elimination of phosphoserine residues that are in turn derivatized by 1,2-dithioethane in a Michael addition providing a free sulfhydryl group available for reaction with a maleimide-based reagent (Chowdhury, S.M., Munske, G.R., Siems, W.F., and Bruce, J.E., A new maleimide-bound acid-cleavable solid-support reagent for profiling phosphorylation, *Rapid Commun. Mass Spectrom.* 19, 899–909, 2005).

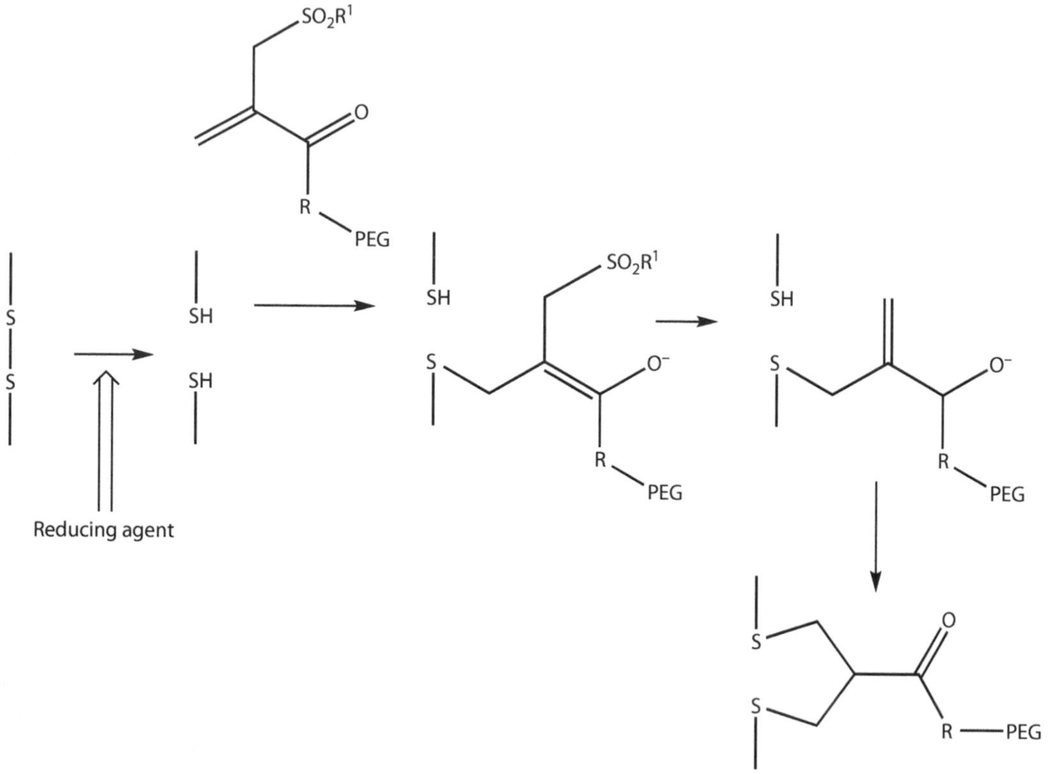

FIGURE 7.28 Michael addition for addition of PEG to disulfide bond. The insertion of a PEG moiety into a protein via a disulfide bond using successive Michael additions. The disulfide bond is reduced with a suitable reducing agent [dithiothreitol or tris-(carboxyethyl)-phosphine]. The addition of an α,β-unsaturated-β′-sulfone functionalized PEG results in the covalent linkage of the reagent via successive Michael additions. See Brocchini, S., Godwin, A., Balan, S., Disulfide bridge based PEGylation of proteins, *Adv. Drug Deliv. Rev.* 60, 3–12, 2008.

(2-aminoethyl)methanethiosulfonate. As with other studies,[610] these investigators also show that the positively charged reagents react more rapidly with 2-mercaptoethanol than either negatively charged or neutral MTS reagents (Table 7.10). A number of studies (Table 7.11) have used MTS reagents to probe cysteine reactivity in membranes with particular emphasis on ion channels.[616–628] Membrane permeability is dependent on charge and hydrophobicity. Generally, a charged reagent is not membrane permeable; with the primary amino reagent, the charge would be pH dependent. Jha and Udgaonkar[629] used MMTS to study the folding of a small protein, barstar. As they needed a *baseline* value for cysteine reaction, a value of 4.8×10^5 M^{-1} s^{-1} was obtained for a cysteine (Cys 82) residue in urea-denatured proteins (pH 9.2, 25°C). This is less than the value obtained for reaction of MMTS with 2-mercaptoethanol (Table 7.2).

The production of recombinant proteins in bacterial expression systems such as *Escherichia coli* frequently involves reduction under denaturing conditions, renaturation, and purification.[630–632] It is useful to utilize the purification step prior to renaturation/refolding but is difficult to do with free sulfhydryl groups. Inoue and coworkers[633] blocked the sulfhydryl groups in recombinant BDNF with trimethylammoniopropyl methanethiosulfonate. The derivatized protein was partially purified by chromatography on SP-Sepharose® in the presence of urea. The blocking groups are removed by reduction and allowed to refold in the presence of oxidized glutathione, dithiothreitol, and a lower urea concentration.[634] The use of trimethylammoniopropyl methanethiosulfonate is an example of protein catonization that is useful for internalizing proteins into cells.[635,636]

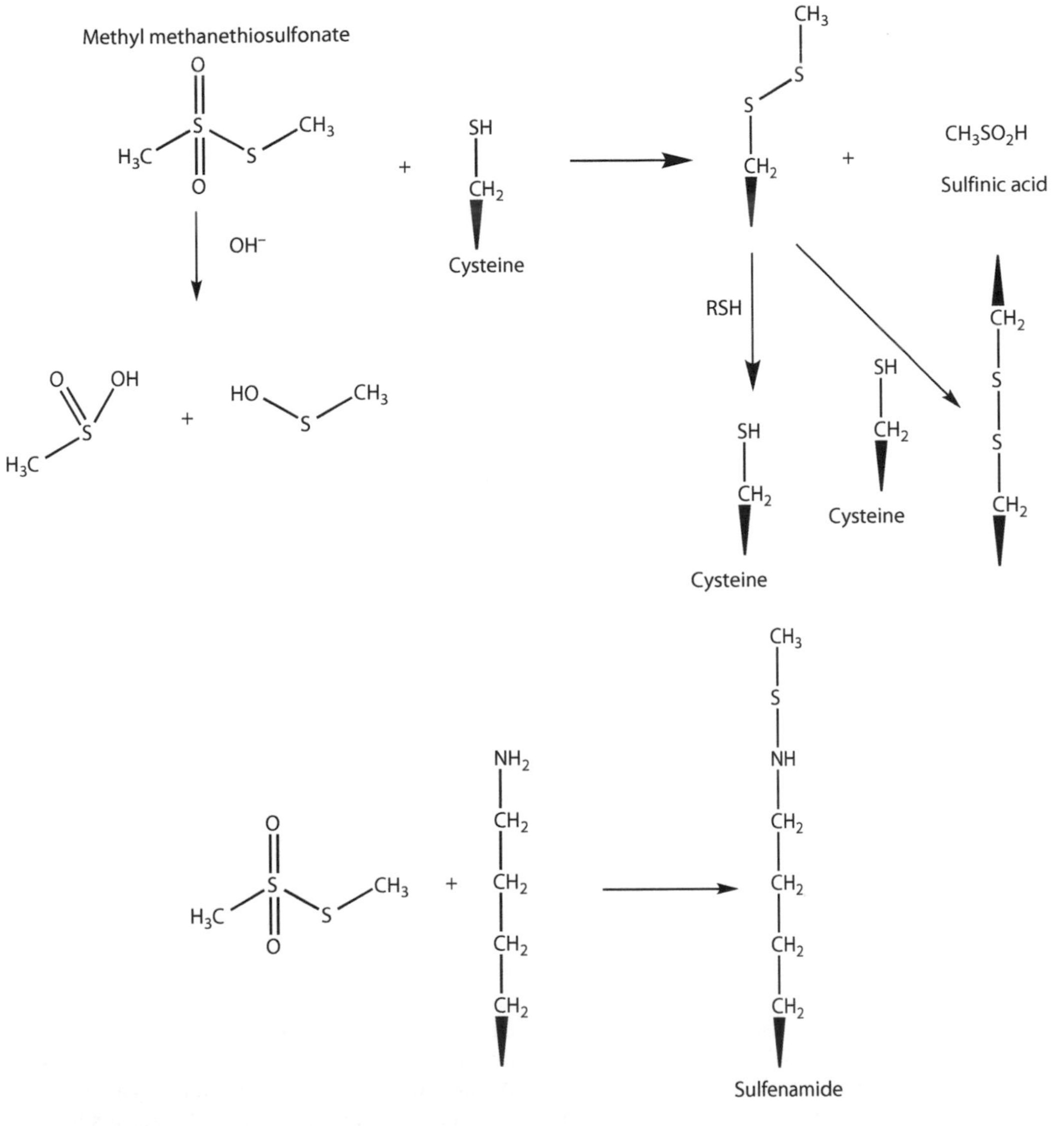

FIGURE 7.29 Reactions of alkyl alkanethiosulfonates with proteins. Shown is the reaction of methyl methanethiosulfonate with a cysteine sulfhydryl groups in proteins. The formation of the mixed disulfide is, in principle, reversible in the presence of added sulfhydryl reagents such as dithiothreitol, 2-mercaptoethanol, or trialkylphosphines. Also shown at the top is the hydrolysis of MMTS that is a significant issue (Karlin, A. and Akabas, M.H., Substituted cysteine accessibility method, *Methods Enzymol.* 293, 123–145, 1998). The potential formation of a new disulfide from the mixed disulfide formed with methyl methanethiosulfonate is shown on the right (Karala, A.R. and Ruddock, L.W., Does *S*-methyl methanethiosulfonate trap the thiol-disulfide state of proteins? *Antioxid. Redox Signal.* 9, 527–531, 2007). Shown at the bottom is the reaction of methyl methanethiosulfonate with an amine to form a sulfenamide derivative (Kluger, R. and Tsui, W.-C., Amino group reactions of the sulfhydryl reagent methyl methanethiosulfonate. Inactivation of D-3-hydroxybutyrate dehydrogenase and reaction with amines in water, *Can. J. Biochem.* 58, 639–632, 1980). While this reaction is suggested to occur under more drastic conditions (Pathak, R., Hendrickson, T.L., and Imperiali, B., Sulfhydryl modification of the yeast Wbp1p inhibits oligosaccharyl transferase activity, *Biochemistry* 34, 4179–4185, 1995), The formation of a sulfenamide between cysteine and lysine has been demonstrated in a crystal structure (Rodkey, E.A., Drawz, S.M., Sampson, J.M. et al., Crystal structure of a preacylation complex of the β-lactamase inhibitor sulbactam bound to a sulfenamide bond-containing thiol-β-lactamase, *J. Am. Chem. Soc.* 134, 16798–16804, 2012).

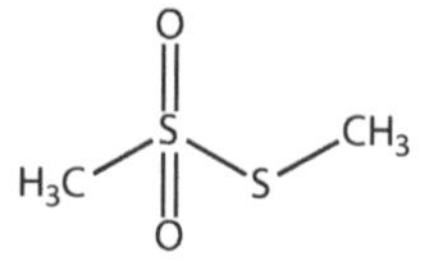

Methyl methanethiosulfonate (MMTS)   1.3

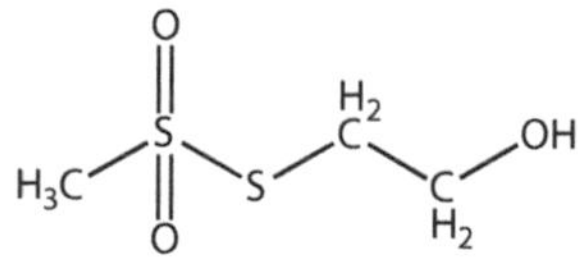

2-Hydroxyethyl methanethiosulfonate (MMSEH)   3.2

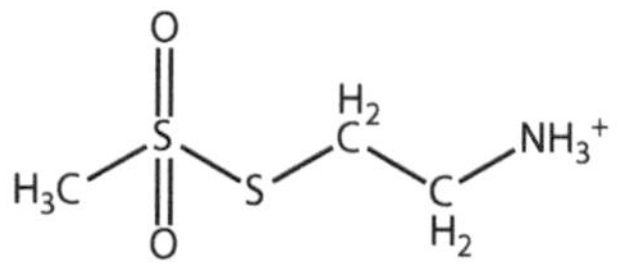

2-Aminoethyl methanethiosulfonate (MTSEA)   400

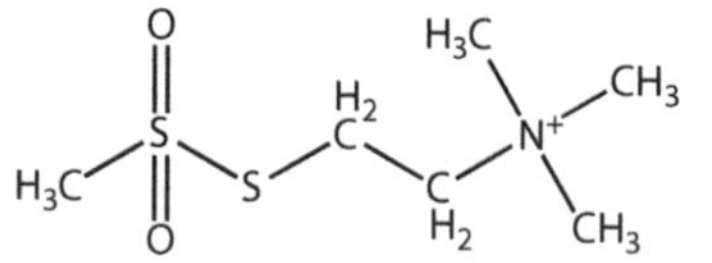

2-Trimethyl ammoniumethyl methanesulfonate (MTSET)   600

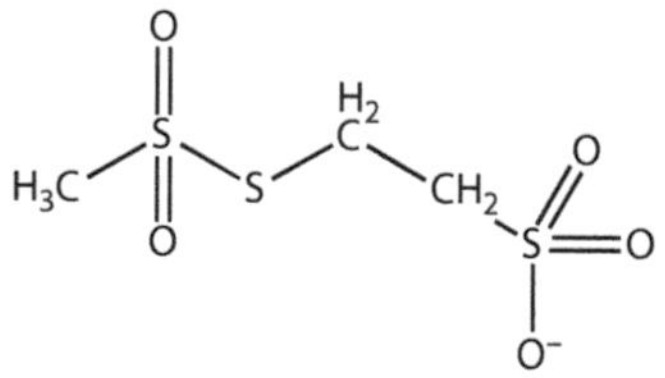

2-Sulfonatoethyl methanethiosulfonate (MTSES)   2500

FIGURE 7.30 Anionic and cationic alkyl methanethiosulfonates. Shown are various alkyl methanethiosulfonate derivatives that are charged and are membrane-impermeant. These reagents are used for substituted cysteine accessibility studies (Karlin, A. and Akabas, M.H., Substituted cysteine accessibility method, *Methods Enzymol.* 293, 123–145, 1998; Sobczak, I. and Lolkema, J.S., Accessibility of cysteine residues in a cytoplasmic loop of CitS and *Klebsiella pneumoniae* is controlled by the catalytic state of the transporter, *Biochemistry* 42, 9789–9790, 2003; Giangregorio, N., Tonazzi, A., Indiveri, C., and Pamieri, F., Conformation-dependent accessibility of Cys-136 and Cys-155 of the mitochondrial rat carnitine/acylcarnitine carrier to membrane impermeable SH reagents, *Biochim. Biophys. Acta* 1767, 1331–1339, 2007). Also shown is the octanol/water partition coefficient, which is an approximation of hydrophobicity/membrane permeability (Kah, M. and Brown, C.D., Log D: Lipophilicity for ionizable compounds, *Chemosphere* 72, 1401–1408, 2008).

   The reaction between the alkyl alkanethiosulfonates and cysteine residues in proteins is an example of the formation of a mixed disulfide (Figure 7.31). There are numerous other examples of the use of this reaction for the site-specific modification of cysteine residues in proteins. Cystine or cystamine have proved effective in the inactivation of guanylate cyclase.[637] This reaction involves the formation of a mixed disulfide and is easily reversed by the addition of dithiothreitol. The formation of a mixed disulfide can be regarded as an oxidative reaction when two thiols are involved and is referred to as *S*-thiolation and is observed with a variety of proteins.[638–642] The chemistry of these reactions is poorly understood. Guanylate cyclase is activated by oxidizing agents such as nitric

TABLE 7.10
Some Properties of MTS Reagents[a]

MTS Reagent	Rate Constant[b]	Partition Coefficient[c]	Hydrolysis Rate[d]
(2-Aminoethyl)	7.6	400	12
(2-Hydroxyethyl)	0.95	1.3	—
(2-Sulfonatoethyl)	1.7	2500	370
[2-(trimethyl-ammonium)ethyl]	21.2	690	11.2

[a] Data adapted from Karlin, A. and Akabas, M.H., Substituted-cysteine accessibility method, *Methods Enzymol.* 293, 123–145, 1998.

[b] Reaction with 2-mercaptoethanol in 58 mM sodium phosphate—0.1 mM EDTA, pH. 7.0 at 20°C; ($\times 10^{-4}$) $M^{-1}\,s^{-1}$.

[c] Partition in water–*n*-octanol.

[d] $t_{1/2}$, 20°C, 190 mM sodium phosphate, pH 7.0, for hydrolysis to sulfenic acid and sulfinic acid derivatives.

oxide[643,644] under conditions consistent with the formation of cysteine sulfenic acid. Cunningham and Nuenke[645,646] demonstrated that the controlled oxidation of a protein sulfhydryl group with iodine resulted in the formation of cysteine sulfenyl iodide (Figure 7.31), which can be used to reversibly modify proteins via the formation of a mixed disulfide.[646] There has been limited application of the this technology as (1) it is technically challenging and (2) similar derivatives can be formed via MTS technology. In addition, there is the possibility of hydrolysis of the sulfenyl iodide derivative to sulfenic acid with the potential of the irreversible conversion to cysteine sulfinic acid/ cysteine sulfonic acid. The formation of the sulfenic acid and sulfonic acid derivatives is essentially irreversible. Cupa and Pace[647] used iodine oxidation to modify the single sulfhydryl group in β-lactoglobulin B with aminoethyl (2-mercaptoethylamine), propyl(1-propanethiol), carboxyethyl (3-mercaptopropionic acid), and hydroxyethyl(2-mercaptoethanol). Modification with the 2-mercaptoethanolamine resulted in a large change in stability to urea denaturation and demonstrated conformation change at lower urea concentrations than the other derivatives; the other derivatives were somewhat less stable than the native protein. A related approach was recently developed by Mecinović and coworkers to measure the nucleophilicity of cysteine residues in prolyl hydroxylase domain 2.[648] These investigators used controlled oxidation of cysteine with hydrogen peroxide that could then be coupled with aromatic thiols to form mixed disulfides that were measured with mass spectrometry. A single cysteine residue (Cys20) was identified as being highly susceptible to oxidation; the same residue was preferentially modified with *N*-ethylmaleimide or a spin-labeled methanethiosulfonate. This approach is based on the assumption that the nucleophilicity (pKa) of cysteine is a major determinant of susceptibility to oxidation. Sanchez and coworkers[649] have identified cysteine pKa as a determinant of susceptibility to oxidation; other determinants include distance to another cysteine and solvent accessibility. The reader is also directed to other studies on the relationship of cysteine oxidation and thiol pKa.[650–653] Wynn and Richards provided a review of the various approaches in using mixed disulfide for the chemical modification of proteins in 1995.[623]

The reaction of sodium tetrathionate (Figure 7.32) with cysteine residues yields the *S*-sulfonate derivative,[634–657] which can be used for the *protection* of cysteinyl residues[658–660] and promoting protein solubilization.[661] Darrel (Teh-Yung) Liu working in the laboratories of Stanford Moor and William Stein at the Rockefeller University in New York City blocked the active-site cysteine in streptococcal proteinase[662] with tetrathionate permitting modification of the active-site histidine with *N*α-bromoacetylarginine methyl ester. The active-site histidine was not modified with iodoacetamide or iodoacetate suggesting the importance of the negative charge introduced by the sulfonyl group at the active site. Liu used the procedure of Pihl and Lange[664] for the use of sodium

TABLE 7.11
Reaction of Methanethiosulfonate Reagents with Proteins

Protein	MTS Reagent	Conditions	Reference
Nicotinic receptor[a]	(2-aminoethyl)-]MTS [2-(trimethylammonium) ethyl]-MTS (2-sulfonatoethyl)-MTS	Sodium phosphate, pH 7.0[b]	7.11.1
Yeast oligosaccharyl transferase Wbplp	Methyl-MTS (N-biotinylamino)ethyl-MTS	50 mM HEPES—140 mM sucrose, pH 7.5 with 0.6% nonIdet P-40	7.11.2
Leukotriene A(4) hydrolase/aminopeptidase	Methyl-MTS	20 mM MOPS, pH 8.0 at 2°C	7.11.3
Lamb uterine estrogen receptor alpha (ER alpha)	Methyl-MTS	20 mM Tris–HCl, pH 8.5, 4 h, 0°C	7.11.4
Human sodium/bile acid Cotransporter	(2-sulfonatoethyl)-MTS [2-(trimethylammonium) ethyl-MTS	Modified Hank's balanced salt solution	7.11.5
Bovine heart mitochondrial ADP/ATP carrier	Methyl-MTS	250 mM sucrose—0.2 mM EDTA-10 mM PIPES, pH 7.2, 25°C	7.11.6
Nicotinic acetylcholine receptor	(2-aminoethyl)-MTS [2-(trimethylammonium) ethyl]-MTS [3-(trimethylammonium) propyl]-MTS	50 mM HEPES—96 mM NaCl—2 mM KCl-0.3 mM CaCl$_2$—1 mM MgCl$_2$, pH 7.6	7.11.7
Oxalate-formate transporter of *Oxalobacter formigenes*	(2-carboxyethyl)-MTS (2-sulfonatoethyl)-MTS [2-(trimethylammonium)ethyl]-MTS	20 mM potassium phosphate, pH 8.0, 23°C	7.11.8
Mitochondrial ornithine citrulline carrier	(2-aminoethyl)-]MTS [2-(trimethylammonium) ethyl]-MTS (2-sulfonatoethyl)-MTS	10 mM HEPES—60 mM sucrose, pH 8.0	7.11.9

[a] This study contains a comparison of the rates of reaction of MTS compounds with simple thiols and protein-bound thiols.
[b] A variety of buffers were used in this study with varying concentrations of NaCl and detergent.

References to Table 7.11

7.11.1. Stauffer, D.A. aqnd Karin, A., Electrostatic potential of the acetylcholine binding sites in the nicotinic receptor probed by reactions of binding-site cysteines with charged methanethiosulfonates. *Biochemistry* 33, 6840–6849, 1994.

7.11.2. Pathak, R., Hendrickson, T.L., and Imperiali, B., Sulfhydryl modifications of the yeast Wbplp inhibits oligosaccharide transferase activity. *Biochemistry* 34, 419–4185, 1995.

7.11.3. Orning, L. and Fitzpatrick, F.A., Modification of leukotriene A(4) hydrolase/aminopeptidase by sulfhydryl-blocking reagents: Differential effects on dual enzyme activities by methyl methanethiosulfonate. *Arch. Biochem. Biophys.* 368, 131–138, 1999.

7.11.4. Aliau, S. et al., Steroidal affinity labels of the estrogen receptor alpha. 4. Electrophilic 11beta-aryl derivatives of estradiol, *J. Med. Chem.* 43, 613–628, 2000.

7.11.5. Hallén, S., Frykland, J., and Sachs, G., Inhibition of the human sodium/bile acid cotransporter by site-specific methanethiosulfonate sulfhydryl reagents. *Biochemistry* 39, 6743–6750, 2000.

7.11.6. Hashimoto, M. et al., Irreversible extrusion of the first loop facing the matrix of the bovine heart mitochondrial ADP/ATP carrier by labeling the Cys (56) residue with the SH-reagent methyl methanethiosulfonate, *J. Biochem.* 127, 443–449, 2000.

7.11.7. Sullivan, D.A. and Cohen, J.B., Mapping the agonist binding site of the nicotinic acetylcholine receptor. Orientation requirements for activation by covalent agonists, *J. Biol. Chem.* 275, 12651–12660, 2000.

7.11.8. Fu, D. et al., Structure/function relationships in OXIT, the oxalate-formate transporter of *Oxalobacter formigenes*. Assignment of transmembrane helix II to the translocation pathway, *J. Biol. Chem.* 276, 8753–8760, 2001.

7.11.9. Tonazzi, A. and Indiveri, C., Chemical modification of the mitochondrial ornithine/citrulline carrier by SH reagents: Effects on the transport activity and transition from carrier to pore-like function, *Biochim. Biophys. Acta* 1611, 123–130, 2003.

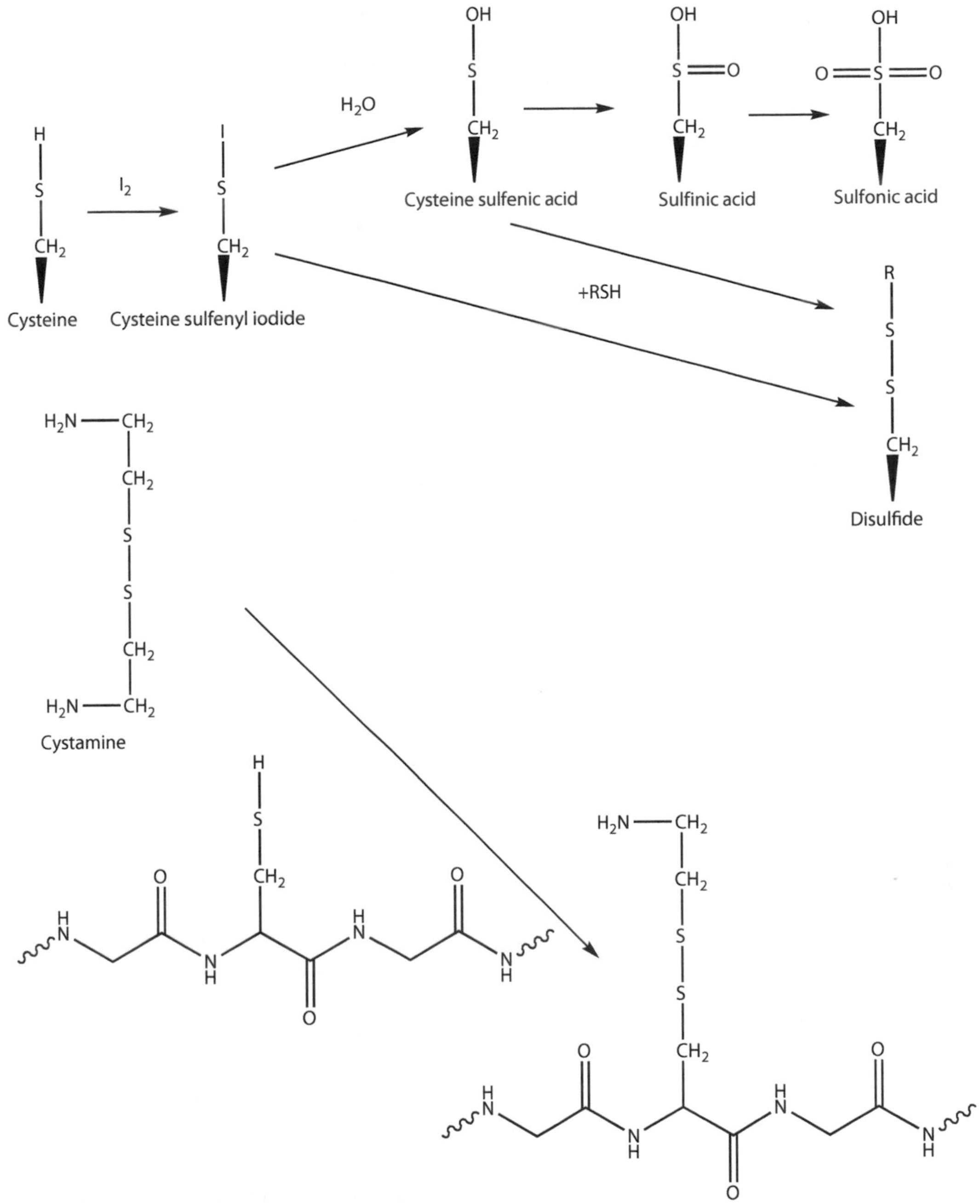

FIGURE 7.31 The formation of mixed disulfides in proteins. Shown is the formation of a mixed disulfide between cysteine sulfenic acid and a thiol such as 2-mercaptoethanol (Peshenko, I.V. and Shichi, H., Oxidation of active center cysteine of bovine 1-Cys peroxiredoxin to the cysteine sulfenic acid form by peroxide and peroxynitrite, *Free Radic. Biol. Med.* 31, 292–303, 2001). Cystamine has been demonstrated to form mixed disulfides with a variety of peptides and proteins (Tamai, K., Shen, H.X., Tsuchida, S. et al., Role of cysteine residues in the activity of rat glutathione transferase P (7-7): Elucidation by oligonucleotide site-directed mutagenesis, *Biochem. Biophys. Res. Commun.* 179, 790–797, 1991; Cappiello, M., Del Corso, A., Camici, M., and Mura, U., Thiol and disulfide determination by free zone capillary electrophoresis, *J. Biochem. Biophys. Methods* 26, 335–341, 1993; Maret, M., Metallothionein/disulfide interactions, oxidative stress, and the mobilization of cellular zinc, *Neurochem. Int.* 27, 111–117, 1995).

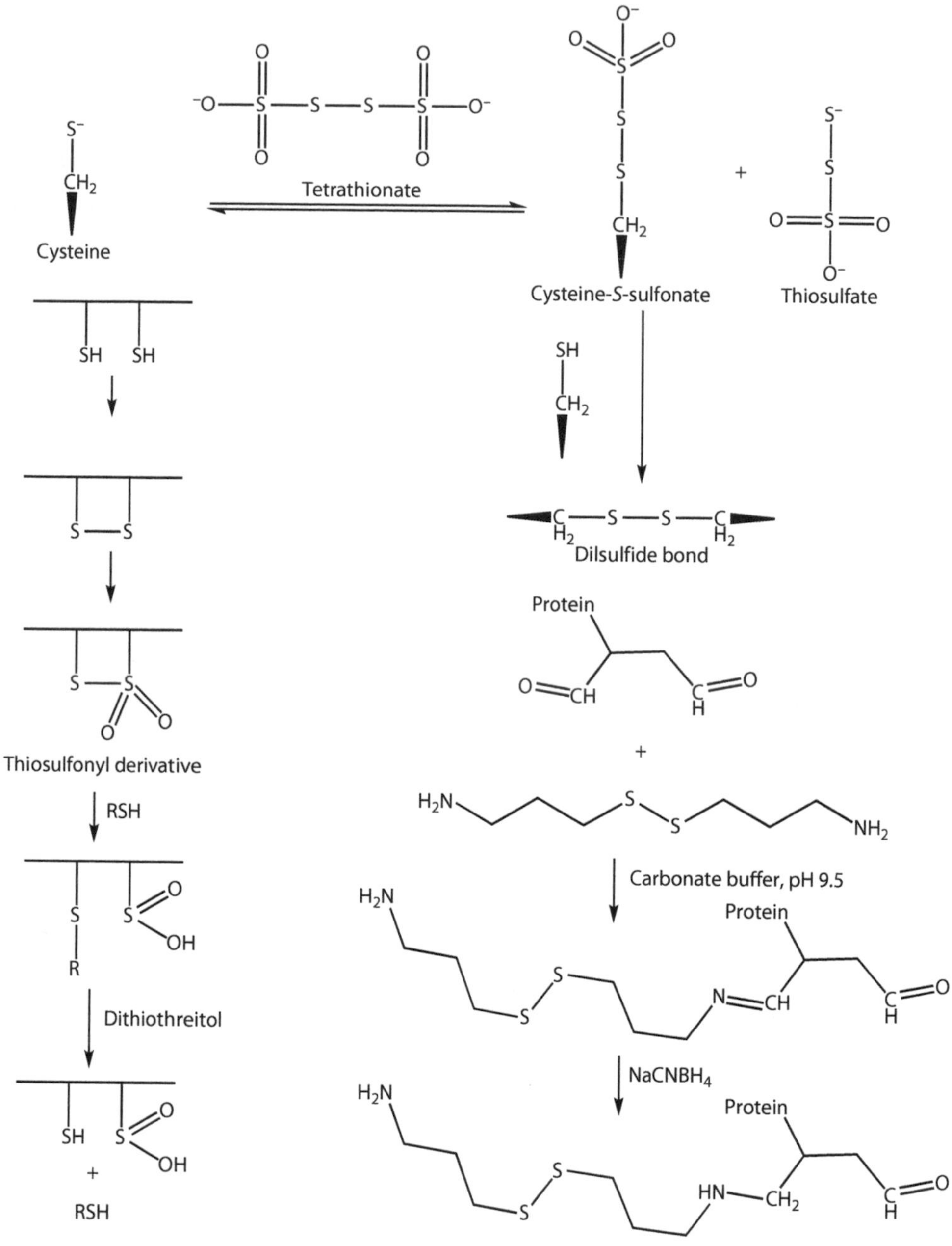

FIGURE 7.32 The reaction of sodium tetrathionate with proteins and some methods of reversible protein immobilization via disulfide bond formation. Shown is the reaction of sodium tetrathionate with proteins to form *S*-sulfonyl derivatives and subsequent formation of disulfide bonds. Also shown is the electrochemical formation of solid-phase thiosulfonates for the formation of matrix-bound mixed disulfides (Batista-Viera, F., Barbieri, M., Ovesejevi, K. et al., A new method for reversible immobilization of thiol biomolecules based on solid-phase bound thiosulfonate groups, *Appl. Biochem. Biotechnol.* 31, 175–195, 1991; Pavlovic, E., Quist, A.P., Gelius, U. et al., Generation of thiolsulfinates/thiosulfonates by electrooxidation of thiols on silicon surfaces for reversible immobilization of molecules, *Langmuir* 19, 4217–4221, 2003). Also shown is a method for glycoprotein immobilization using periodate oxidation, coupling with cystamine and reduction with sodium cyanoborohydride to stabilize the Schiff base; the disulfide may be then bound to a gold surface (Suárez, G., Jackson, R.J., Spoors, J.A., and McNeil, C.J., Chemical introduction of disulfide groups on glycoproteins: A direct protein anchoring scenario, *Anal. Chem.* 79, 1961–1969, 2007).

tetrathionate. Pihl and Lange[664] had shown the selective modification of three sulfhydryl groups in rabbit muscle D-glyceraldehyde 3-phosphate dehydrogenase with sodium tetrathionate. Kingsbury and coworkers[663,664] have shown that the S-sulfonation of cysteinyl residues in transthyretin stabilizes tetramer formation, while S-cysteinylation promotes dissociation of the native tetramer. Previous work by Zhang and Kelly[665] suggested that the formation of a mixed disulfide of Cys10 with cysteine, GlyCys, or glutathione increased the amyloidogenecity rate of transthyretin above pH 4.6. Transthyretin modified with sodium tetrathionate (S-sulfonation) was less amyloidogenic than native transthyretin. As shown in Figure 7.32, sodium tetrathionate will also *catalyze* the crosslinking of proteins via sulfhydryl groups.[666–671] Chung and Folk[657] observed that the reaction of guinea pig liver transglutaminase with sodium tetrathionate resulted in total inactivation of the enzyme with far less than stoichiometric incorporation (0.1–0.15 mol/mol) of ^{35}S-labeled tetrathionate. Further work showed that two intramolecular disulfide bonds had been formed as a result of reaction with sodium tetrathionate. Neither of the two new disulfide bonds formed on reaction of the protein with tetrathionate contained the cysteine residue involved in transglutaminase activity. The treatment of the modified enzyme with dithiothreitol resulted in the initial cleavage of one disulfide bond without recovery of activity; the cleavage of the second disulfide bond resulted in the total recovery of transglutaminase activity. A consideration of the literature does not clearly indicate the conditions favorable for the formation of a disulfide bond versus the formation of the S-sulfonate. It is likely that pH < 7 favors the S-sulfonate formation while pH > 7 may favor disulfide formation.

Mixed disulfide formation can be used for the reversible immobilization of thiol proteins. Solid-phase-bound thiosulfonate groups (Figure 7.32) were proposed by Batista-Viera and coworkers[672] almost 20 years ago for the reversible immobilization of thiol proteins. The solid-phase thiol/disulfides were converted to the reactive thiosulfonate groups with hydrogen peroxide. Pavlovic and coworkers[673] have used electroxidation of vicinal thiols on a silicon surface to produce thiosulfinate and thiosulfonates. Suárez and coworkers[674] introduced a disulfide into glycoproteins, which was subsequently used to anchor the modified protein to a gold surface. The glycoprotein was oxidized with periodate and coupled to cystamine. The imine bond was reduced to a stable amine with sodium cyanoborohydride (which does not reduce disulfide bonds) (Figure 7.32). The resulting derivative can be bound to a gold surface.

The reaction of Ellman's reagent (DTNB) and related disulfide-based reagents has been quite useful as these reagents form colorimetric derivatives on reaction with cysteinyl residues (Figure 7.33). George Ellman developed bis (p-nitrophenyl)disulfide (PNPD) for the measurement of thiols in 1958.[675] This reagent was developed in response to a need for a more quantitative method for the measurement of mercaptans in biological samples. This reagent was effective and sensitive. Reaction with a thiol-containing compound (p-nitrobenzylthiol, 2-mercaptoethanol, glutathione) yielded a mixed disulfide and p-nitrobenzylthiol. At pH values above 8, the nitrobenzylthiol anion was yellow with a molar extinction coefficient of $13,600$ M^{-1} cm^{-1}. While PNPD was useful, solubility problems precluded extensive use. In 1959, Ellman developed DTNB (Ellman's reagent),[676] which today is one of the more popular reagents for the modification and determination of the sulfhydryl group (Figure 7.33). Reaction with sulfhydryl groups in proteins results in the release of 2-nitro-5-mercaptobenzoic acid which was a molar extinction coefficient of $13,600$ M^{-1} cm^{-1} at 410 nm. The chemistry of DTNB was been the subject of continuing investigation after the original description with most of the concern focusing on the extinction coefficient.[677–681] Eyer and coworkers[681] suggest a value for the extinction coefficient of 2-nitro-5-mercaptobenzoic acid of 14.15×10^3 M^{-1} cm^{-1} at 25°C and 13.8×10^3 M^{-1} cm^{-1} at 37°C (412 nm, 0.1 M sodium phosphate, pH 7.4). While it is important to have accurate values for extinction coefficients, the difference between the original value determined by Ellman in 1959 and the most recent value obtained in 2003 is on the order of 4%. Riddles and coworkers[679] studied the chemistry of DTNB in alkaline solutions as well as the reaction of DTNB with thiols. They concluded that the rate of reaction of DTNB was dependent on the pH and the pKa of the thiol residue.

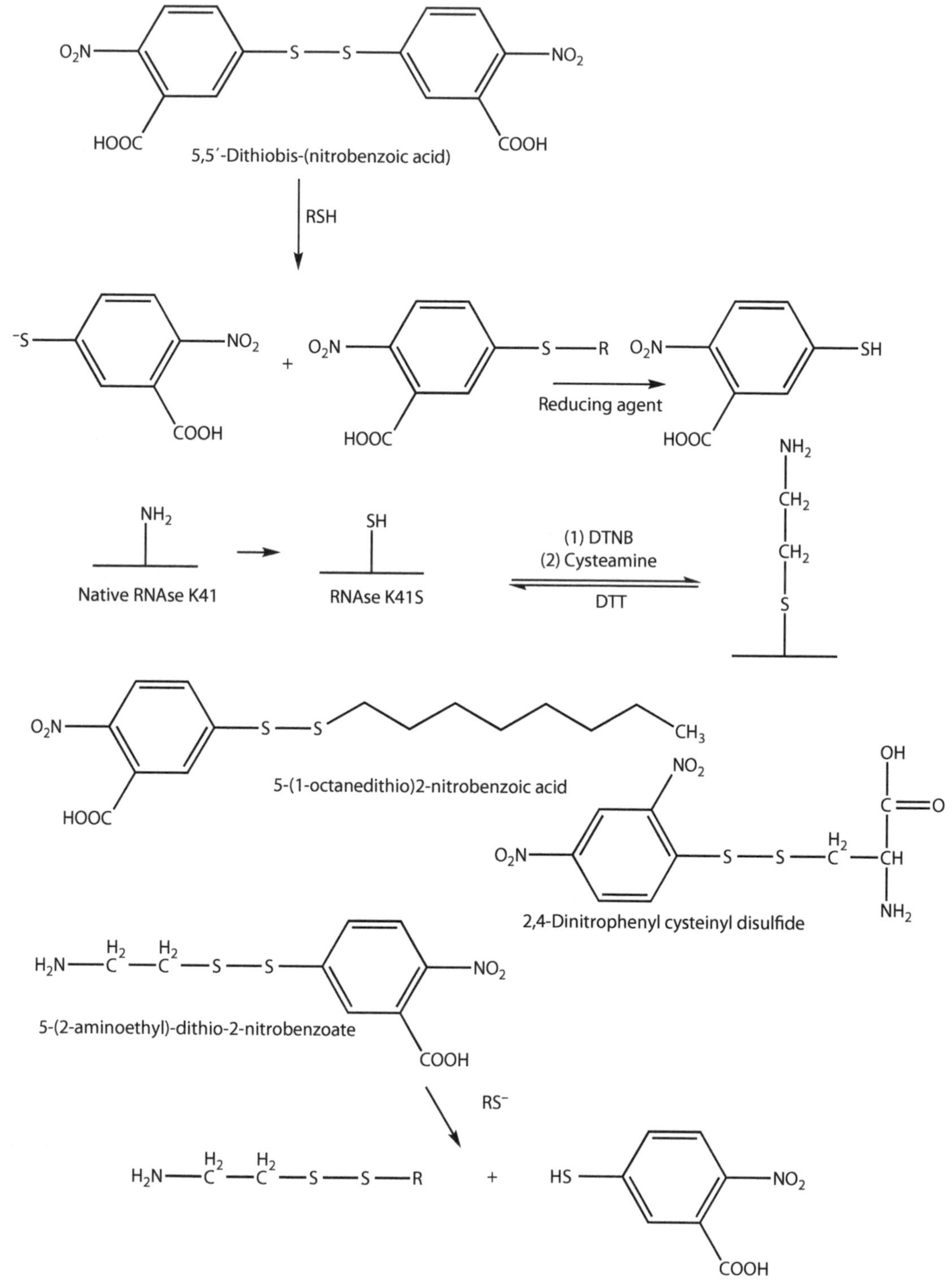

FIGURE 7.33 (continued)

The early use of DTNB focused on the modification of cysteine residues in proteins. Zhang and coworkers[682] have prepared a positive-charged derivative of DTNB, 5-(2-aminoethyl)-dithio-2-nitrobenzoate (ADNB) (Figure 7.33). Depending on the model thiol, the reaction with ADNB was either faster (glutathione, 3-mercaptopropionic acid, thioglycolic acid) or slower (2-mercaptoethyl-amine, 2-mercaptothanol, cysteine, homocysteine) than the corresponding reaction with DTNB at pH 6.65. Increasing ionic strength markedly enhanced the rate of reaction of DTNB with mercaptoacetic acid with a lesser effect on the rate of DTNB reaction with 2-mercaptoethanol; inhibition was observed with the reaction of DTNB with 2-mercaptoethylamine. The reaction of ADNB was far less sensitive to ionic strength and suggested to be of value in modifying sulfhydryl groups in anionic environments This suggestion was tested by a study of the modification of the single sulfhydryl group (Cys34) in serum albumin. ADNB reacted far more rapidly (208 $M^{-1} s^{-1}$) than did DTNB (1.1 $M^{-1} s^{-1}$) with the sulfhydryl group in bovine serum albumin(10 mM MES, pH 6.65) where there is glutamic acid residue (Glu82) close to Cys34; the rate of reaction of DTNB with the sulfhydryl (70.1 $M^{-1} s^{-1}$) equine albumin where the Glu82 is replaced with an alanine is similar to that seen with ADNB 69.7 $M^{-1} s^{-1}$. Stewart and colleagues[683] argued for a role of Tyr84 in the reactivity of Cys34 in human albumin where there is also a glutamic residue at position 82 in the sequence. Replacing Tyr84 with phenylalanine resulted in a fourfold increase in the rate of reaction of Cys34 with DTNB. Later studies by Zhu and coworkers[684] demonstrated that ADNB is more stable than DTNB under basic conditions (pH $\geq$ 10). Legler[685] has reported on the synthesis and characterization of a positively charged (5,5'-dithiobis-(2-nitro-N-trimethylbenzyl ammonium iodide) and neutral (5,5'-dithiobis-(2-nitro-N-2'-hydroxyethyl benzamide) analog of Ellman's reagent (Table 7.12). Legler[685] stressed the necessity of performing experiments for the determination of reaction rates with model thiols at pH6.0 (0.05 M phosphate [counterion not stated]) and 15°C; rates were too rapid at higher pH for accurate measurement. The reader is also directed to a study by Whitesides and coworkers[686] on the rate of reaction of Ellman's reagent with monothiols and dithiols. The rate of reaction correlated with pKa of the thiol. As with the reaction of cysteine with other reagents as discussed above, the presence of a positive charge close to the thiol function results in enhanced reactivity. It is clear from these studies that DTNB and related derivatives react rapidly with model thiols and many sulfhydryl groups in proteins (Table 7.12). Faulstich and Heintz[687] discuss the use of DTNB for the reversible introduction of thiols into protein for the formation of mixed disulfides; the dithiopyridyl disulfides can also be used for this purpose as discussed below. Messmore and coworkers[688] used this chemistry for the preparation of some novel derivatives of bovine pancreatic ribonuclease A (RNAse A). These investigators prepared a mutant where Lys41 is replaced by cysteine thereby reducing catalytic efficiency (kcat/Km) by 10^5. The mutant protein is modified with DTNB (there are no other free cysteinyl residues in RNAse A other than the cysteine introduced at Lys41). It is possible to introduce cysteamine, which has the net effect of introducing an amino

FIGURE 7.33 (continued)   The structure of Ellman's reagent (DTNB) and some related reagents. shown is the structure of DTNB and the reaction with a sulfhydryl group. The modified thiol groups is an activated disulfide (Faulstich, H. and Heintz, D., Reversible introduction of thiol compounds into proteins by use of activated mixed disulfides, *Methods Enzymol.* 251, 357–366, 1995) and can be reduced to restore the thiol. Shown below is the application the DTNB technology for "rescue" of a mutant RNAse (Messmore, J.M., Holmgren, S.K., Grilley, J.E., and Raines, R.T., Sulfur shuffle: Modulating enzymatic activity by thiol disulfide interchange, *Bioconjug. Chem.* 11, 401–413, 2000). Shown at the bottom is 5-(1-octanedithio)-2-nitrobenzoic acid that can be used for the study of thiol groups in biological membranes (Czerski, L. and Sanders, C.R., Thiol modification of diacylglycerol kinase: Dependence upon site membrane disposition and reagent hydrophobicity, *FEBS Lett.* 472, 225–229, 2000). ADNB was developed to probe the electrostatic environment around cysteine residues in proteins (Zhang, H., Le, M., and Means, G.E., A kinetic approach to characterize the electrostatic environments of thiol groups in proteins, *Bioorg. Chem.* 26, 356–364, 1998). 2,4-dinitrophenyl cysteine disulfide (Drewes, G. and Faulstich, H., 2,4-Dintrophenyl [^{14}C]cysteinyl disulfide allows selective radioactive labeling of protein thiols under spectrophotometric control, *Anal. Biochem.* 188, 109–113, 1990) was developed as reagent to radiolabel proteins.

TABLE 7.12

Reaction of 5,5′-Dithiobis(2-Nitrobenzoic Acid) (DTNB) and Related Reagents with Some Model Thiols and Proteins[a]

Reagent	Thiol/Conditions	Rate ($M^{-1} s^{-1}$)	Reference
DTNB[b]	Glutathione/100 mM Tes, pH 7.0 with 6% glycerol and 0.16 mM EDTA/25°C	1.6×10^3	7.12.1
DTNB	Phosphoenolpyruvate carboxykinase (GTP)/100 mM Tes, pH 7.0 with 6% glycerol and 0.16 mM EDTA/25°C	2×10^4	7.12.1
DTNB	Bovine serum albumin/10 mM MES with 1 mM EDTA, pH 6.65[c]	1.1	7.12.2
DTNB	Canine serum albumin/10 mM MES with 1 mM EDTA, pH 6.65[b]	22.9	7.12.2
DTNB	Bovine serum albumin/10 mM MES with 1 mM EDTA, pH 6.65[b]	70.1	7.12.2
DTNB	Horse muscle phosphoglycerate kinase[d]/100 mM Tris, pH 7.0/20°C	640	7.12.3
DTNB	Pig muscle phosphoglycerate kinase[d]/100 mM Tris, pH 7.0/20°C	1.1×10^3	7.12.3
DTNB	Guinea pig hemoglobin/100 mM sodium/potassium phosphate, pH 7.0/4°C	1.46×10^4	7.12.4
DTNB	2-Mercaptoethanol/0.05 M phosphate, pH 6.0/15°C	30	7.12.5
DTNHEB[b]	2-Mercaptoethanol/0.05 M phosphate, pH 6.0/15°C	300	7.12.5
DTNTMB[b]	2-Mercaptoethanol/0.05 M phosphate, pH 6.0/15°C	700	7.12.5
DTNB	Glutathione/0.05 M phosphate, pH 6.5/15°C	160	7.12.5
DTNHEB[b]	Glutathione/0.05 M phosphate, pH 6.5/15°C	1350	7.12.5
DTNTMB[b]	Glutathione/0.05 M phosphate, pH 6.5/15°C	4570	7.12.5
2PDS[e]	Papain/formate ($I=0.1$), pH 3.80/25°C	1.71×10^4	7.12.6
2PDS[e]	L-Cysteine/formate ($I=0.1$), pH 3.90/25°C	130	7.12.6
2PDS[e]	2-Mercaptoethanol/acetate ($I=0.1$), pH 4.58/25°C	14	7.12.6
2PDS	Papain/phosphate ($I=0.1$), pH 6.28/25°C	9×10^2	7.12.6
2PDS	L-Cysteine/phosphate ($I=0.1$), pH 6.25/25°C	1.3×10^3	7.12.6
2PDS	2-Mercaptoethanol/phosphate($I=0.1$), pH 6.60/25°C	105	7.12.6
2PDS	Papain/Tris ($I=0.1$), pH 8.05/25°C	1.14×10^3	7.12.6
2PDS	L-Cysteine/Tris ($I=0.1$), pH 8.00/25°C	9.3×10^3	7.12.6
2PDS	2-Mercaptoethanol/Tris($I=0.1$), pH 8.10/25°C	4.3×10^3	7.12.6
2PDS	Ficin/formate ($I=0.1$), pH 3.2[f]/25°C	8×10^4	7.12.7
4PDS[g]	Ficin/pH ≈ 5[h]/25°C	400[h]	7.12.8
4PDS[g]	Ficin/pH ≈ 8[h]/25°C	6×10^{3} [h]	7.12.8
2PDS	Human albumin/0.1 M formate, pH 3.0/23°C[i]	87	7.12.9
2PDS	Human albumin/0.1 M acetate, pH 5.0/23°C[i]	4.2	7.12.9
2PDS	Human albumin/0.1 M phosphate, pH 7.0/23°C[i]	58	7.12.9
2PDS	Human albumin/0.1 M borate, pH 9.0/23°C[i]	500	7.12.9
2PDS	Glutathione/0.1 M formate, pH 3.4[f]/23°C[i]	50[f]	7.12.9
2PDS	Glutathione/0.1 M formate, pH 5.0[f]/23°C[i]	83[f]	7.12.9
2PDS	Glutathione/0.1 M Tris, pH 8.0[f]/23°C[i]	>1600[f]	7.12.9
2PDS	Acetylcholinesterase (*Torpedo nobiliana*)/0.18 M NaCl, pH 7.8 (pH stat)/25°C	13[j]	7.12.10
2PDS	Urease (*Klebsiella aerogenes*)/80 mM HEPES with 8 mM EDTA, pH 7.75/37°C	22[k]	7.12.11
DTNB	Urease (*Klebsiella aerogenes*)/80 mM HEPES with 8 mM EDTA, pH 7.75/37°C	7.2[l,m]	7.12.11
4PDS	Diamine oxidase (*Euphorbia characias* latex)/0.1 M phosphate, pH 7.0/38°C	0.33	7.12.12
DTNB	Thiolase 1 (porcine heart)/0.2 M potassium phosphate, pH 7.5/25°C	4	7.12.13
2PDS	Thiolase 1 (porcine heart)/0.2 M potassium phosphate, pH 7.5/25°C	42	7.12.13
4PDS	Thiolase 1 (porcine heart)/0.2 M potassium phosphate, pH 7.5/25°C	9.7	7.12.13
4PDS	L-Galactonolactone oxidase/50 mM Tris, pH 8.6 with 0.1% Tween and 1 mM EDTA/25°C	1.1×10^4	7.12.14

TABLE 7.12 (continued)
Reaction of 5,5'-Dithiobis(2-Nitrobenzoic Acid) (DTNB) and Related Reagents with Some Model Thiols and Proteins[a]

[a] Some data for the reaction of DTNB with model thiols are contained in Table 12.2.

[b] DTNB, 5,5'-dithiobis(2-nitrobenzoic acid) (Nbs$_2$); DTNHEB, 5,5'-dithiobis-(2-nitro-N-2'-hydroxyethyl benzamide) (a neutral analog of DTNB); DTNTMB, 5,5'-dithiobis-(2-nitro-N-trimethylbenzyl ammonium iodide) (a positively charged analog of DTNB).

[c] Temperature not provided.

[d] There are seven cysteine residues, two of which are *fast-reacting* and are measured in the cited rate constants.

[e] 2PDS, 2,2'-dipyridyl disulfide, 2,2'-dithiopyridine.

[f] pH estimated from graph.

[g] 4PDA, 4,4'-dipyridyl disulfide, 4,4'-dithiopyridine.

[h] pH and rate constant estimated from graph (Figure 2 in Ref. [7.12.7]).

[i] Reactions performed at room temperature, stabilized by circulating water bath; ambient temperature assumed to be 23°C.

[j] 2PDS also is a competitive inhibitor of the *Torpedo* enzyme (K_i = 0.042 mM) and the *Electrophorus electricus* enzyme (K_i = 0.37 mM). The second-order rate constant reported in Table 12.12 for the reaction of 2PDS with acetylcholinesterase from *Torpedo nobiliana* is an equilibrium constant.

[k] The 2PDS-inactivated enzyme is reactivated with dithiothreitol with an observed second-order rate constant of 2.7 $M^{-1}s^{-1}$.

[l] DTNB shows saturation kinetics.

[m] The DTNB-inactivated enzyme is reactivated with dithiothreitol with an observed second-order rate constant of 3.1 $M^{-1}s^{-1}$.

References to Table 7.12

7.12.1. Carlson, G.M., Colombo, G., and Lardy, H.A., A vicinal dithiol containing an essential cysteine in phosphoenol-pyruvate carboxykinase (Guanosine triphosphate) from cytosol of rat liver, *Biochemistry* 17, 5329–5338, 1978.

7.12.2. Zhang, H., Le, M., and Means, G.E., A kinetic approach to characterize the electrostatic environments of thiol groups in proteins, *Bioorg. Chem.* 26, 356–364, 1998.

7.12.3. Minard, P., Desmadril, M., Ballery, N. et al., Study of the fast-reacting cysteines in phosphoglycerate kinase using chemical modification and site-directed mutagenesis, *Eur. J. Biochem.* 185, 419–423, 1989.

7.12.4. Miranda, J.J., Highly reactive cysteine residues in rodent hemoglobins, *Biochem. Biophys. Res. Commun.* 275, 517–523, 2000.

7.12.5. Legler, G., 4,4'-dinitrophenyldisulfides of different charge type as probes for the electrostatic environment of sulf-hydryl groups, *Biochim. Biophys. Acta* 405, 136–143, 1975.

7.12.6. Brocklehurst, K. and LIttle, G., A novel reactivity of papain and a convenient active site titration in the presence of other thiols, *FEBS Lett.* 9, 113–116, 1970.

7.12.7. Malthouse, J.P.G. and Brocklehurst, K., Preparation of fully active ficin from *Ficus glabrata* by covalent chromatography and characterization of its active centre by using 2,2'dipyridyl disulphide as a reactivity probe, *Biochem. J.* 159, 221–234, 1976.

7.12.8. Malthous, J.P.G. and Brocklehurst, K., A kinetic method for the study of solvent environments of thiol groups in proteins involving the use of a pair of isomeric reactivity probes and a differential solvent effect, *Biochem. J.* 185, 217–222, 1980.

7.12.9. Pedersen, A.O. and Jacobsen, J., Reactivity of the thiol group in human and bovine albumin at pH 3–9, as measured by exchange with 2,2'-dithiopyridine, *Eur. J. Biochem.* 106, 291–295, 1980.

7.12.10. Salih, E., Howard, S., Chishti, S.B. et al., Labeling of cysteine 231 in acetylcholinesterase from *Torpedo nobiliana* by the active-site directed reagent, 1-bromo-2-[14C]pinacolone. Effects of 2,2'-dipyridyl disulfide and other sulfhydryl reagents, *J. Biol. Chem.* 268, 245–251, 1993.

7.12.11. Todd, M.J. and Hausinger, R.P., Reactivity of the essential thiol of *Klebsiella aerogenes* urease. Effect of pH and ligands on thiol modification, *J. Biol. Chem.* 266, 10260–10267, 1991.

7.12.12. Floris, G., Giartosio, A., and Rinaldi, A., Essential sulfhydryl groups in diamine oxidase from *Euphorbia characias* latex, *Arch. Biochem. Biophys.* 220, 623–627, 1983.

7.12.13. Izbicka-Dimitrijevic, E. and Gilbert, H.F., Multiple oxidation products of sulfhydryl groups near the active site of thiolase I from porcine heart, *Biochemistry* 23, 4318–4324, 1984.

7.12.14. Noguchi, E., Nishikimi, M., and Yagi, K., Studies on the sulfhydryl groups of L-galactonolactone oxidase, *J. Biochem.* 90, 33–38, 1981.

function at position 41 resulting in a 10^3 increase in catalytic activity; the enzyme is inactivated by treatment with dithiothreitol that removes the cysteamine (Figure 7.33). If mercaptopropylamine is introduced by the same chemistry, the activity is further reduced below that is found in the mutant (K41C) enzyme. Drewes and Faulstich[689] prepared 2,4-dinitrophenyl-^{14}C-cysteinyl disulfide (Figure 7.34) via a facile synthetic method as a means for introducing radiolabeled cysteine into proteins via disulfide exchange with free thiols. The reaction can be monitored by following the release of 2,4-dinitrophenol at 408 nm ($\varepsilon = 12,700$ M^{-1} cm^{-1}). The specificity of this reagent corresponded to that obtained with DTNB. Reaction with the sulfhydryl groups of papain was more rapid than that observed with DTNB. The resulting derivative can be easily reversed with thiols but is stable to cyanogen bromide degradation and peptide purification. The activating ability of the thionitrobenzoyl anion as a activating group[687] permits the development of reagents such as 5-(1-octanedithio)-2-nitrobenzoic acid (Figure 7.34), which was used for the modification of cysteine residues in proteins.[690]

It would appear that most recent studies with DTNB have used this reagent to measure cysteine residues in proteins rather than for the site-specific chemical modification of proteins. A number of these studies have been cited above as have studies where analysis with DTNB has been used to determine the extent of cysteine modification by other reagents. Accessibility of cysteinyl residues to reaction has also been determined by reaction with DTNB although as cited above,[682] local environment has a major influence. As with other reagents, the reaction of protein sulfhydryl groups with DTNB is dependent on protein conformation. Takeda and coworkers[128] studied the effect of two chaotropic agents, urea and guanidine, on the reaction of DTNB with bovine serum albumin and ovalbumin. While DTNB readily reacts with bovine serum albumin in the absence of chaotropic agents, ovalbumin requires the presence of a chaotropic agent such as urea or guanidine for DTNB reaction with the four cysteinyl residues that are buried. Fernandez-Diaz and coworkers[691] used reaction with DTNB to measure denaturation of ovalbumin in pulsed electric fields. The reaction of DTNB with fibronectin provides a useful example of the application of DTNB to measure conformational change in proteins. Narasimhan and coworkers[692] showed the one sulfhydryl group in fibronectin, which is not available for reaction with DTNB in the absence of chaotropic agents, became available for reaction with DTNB on binding to a polystyrene surface. Subsequent studies with monoclonal antibodies also showed a conformational change in fibronectin upon binding to a polystyrene surface.[693]

2,2'-dithiopyridine (2,2-dipyridyl disulfide, [2PDS]) and 4,4'-dithiopyridine (4,4'-dipyridyl disulfide, [4PDS]) (Figure 7.34) are similar to DTNB in that a mixed disulfide is formed between a cysteinyl residue in the protein and the reagent with the concomitant release of pyridiine-2-thione (2-thiopyridone)($\varepsilon_{343} = 7.06 \times 10^3$ or pyridine-4-thione (4-thiopyridone)($\varepsilon_{324} = 1.98 \times 10^4$).[694,695] Nonspecific binding of the thiopyridone product may frustrate accurate determination of reaction stoichiometry as Harris and Hodgins[696] observed that binding of 2-thiopyridone to protein resulted in an ipsochromic shift from 343 nM to 302 nM. Data on the reaction of 2PDS and 4PDS with some model thiols and protein sulfhydryl groups are shown in Table 7.12. 2PDS and 4PDS were developed by Grassetti and Murray[694] as reagents for the measurement of thiols. 2PDS is still extensively used for the measurement of thiols[695–700] and there is continued use of 4PDS.[701–703] Erwim and Gruber[704] reported that 4PDS was superior to DTNB for the determination of various alkyl thiol compounds. It is also noted that the reaction of 4PDS or 2PDS with a thiol is essentially irreversible and the equilibrium of the product is toward the thiopyridone (pyridylthione); the thione/thiol ratio is estimated to be 4.9×10^4 for the 2PDS product and 3.5×10^4 for the 4-PDS product.[705] The mixed disulfide formed with protein cysteinyl residues and 2PDS or 4PDS is usually reversed with treatment with dithiothreitol.[706] However, Tatsumi and Hirose[707] observed the modification of a single sulfhydryl group in ovalbumin at pH 2.2 (0.1 M potassium phosphate-HCl, pH 2.2; modification required 2 h at 25°C). Unlike most mixed disulfides formed with 2PDS, reversal with cysteine occurred only in the presence of a chaotropic agent (urea).

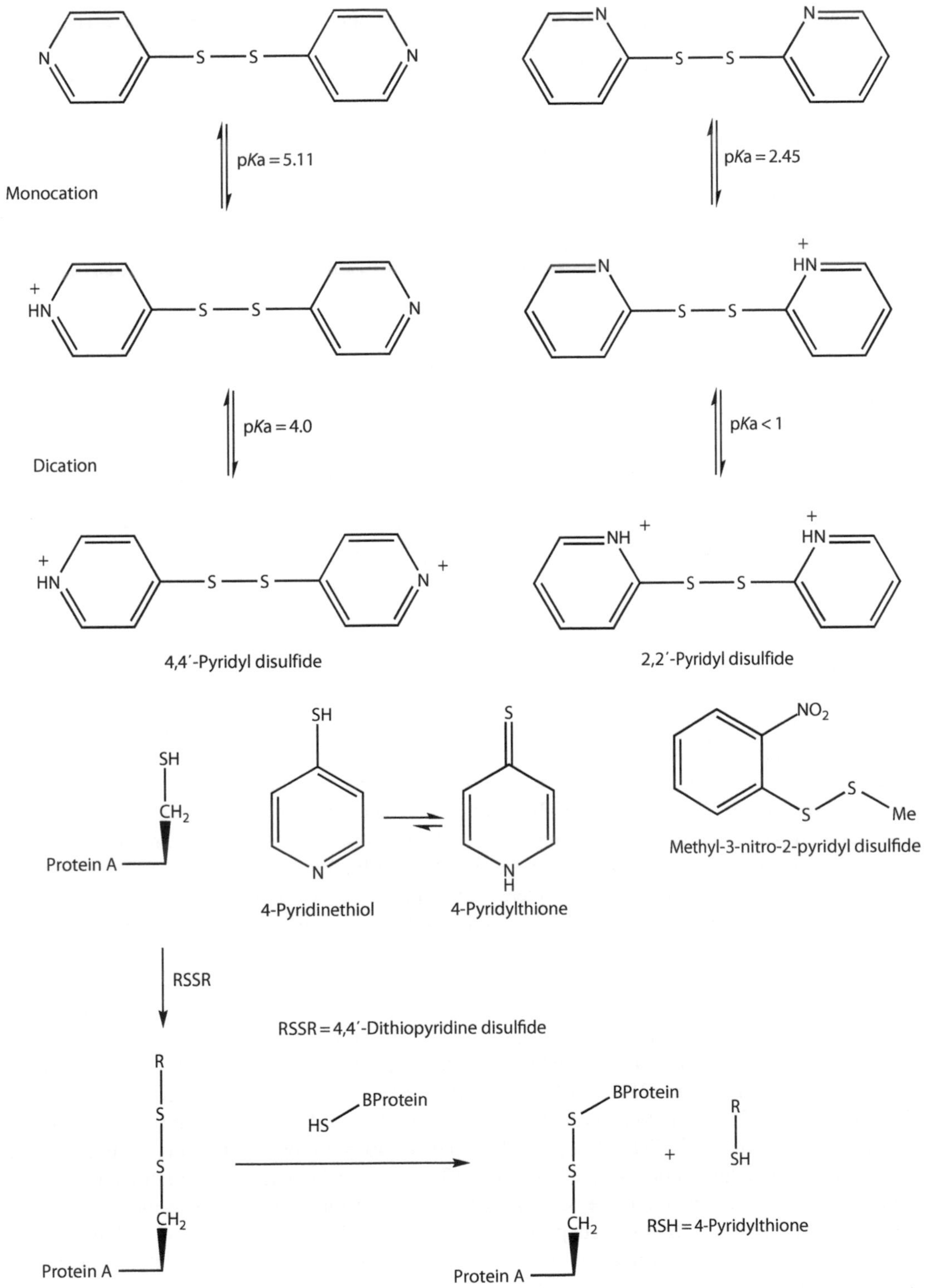
pKa = 5.11
Monocation
pKa = 4.0
Dication
4,4′-Pyridyl disulfide
pKa = 2.45
pKa < 1
2,2′-Pyridyl disulfide
Protein A
4-Pyridinethiol
4-Pyridylthione
Methyl-3-nitro-2-pyridyl disulfide
RSSR
RSSR = 4,4′-Dithiopyridine disulfide
BProtein
RSH = 4-Pyridylthione

FIGURE 7.34

(continued)

FIGURE 7.34 (continued) Dithiopyridyl disulfide and derivative reagents. Shown are the structures of 2PDS and 4PDS that were developed reagents for the analysis of thiols (Grassetti, D.R. and Murray, J.F., Jr., Determination of sulfhydryl groups with 2,2'- or 4,4'-dithiodipyridine, *Arch. Biochem. Biophys.* 119, 41–49, 1967) and subsequently used for the modification of sulfhydryl groups in proteins. The enhanced reactivity at low pH is due to the protonation of the reagent (Brocklehurst, K. and Little, G., Reactions of papain and of low-molecular-weight thiols with some aromatic disulphides. 2PDS as a convenient active-site titration for papain even in the presence of other thiols, *Biochem. J.* 133, 67–80, 1973). The monoprotonated states are thought to be the species responsible for enhanced reactivity at low pH. The protein mixed disulfide product is an activated disulfide and can be coupled to another sulfhydryl group for the preparation of a bioconjugate (King, T.P., Li, Y., and Koehoumian, L., Preparation of protein conjugates via intermolecular disulfide bond formation, *Biochemistry* 17, 1499–1506, 1978). The reaction of the dithiopyridyl disulfides with a thiol is essentially irreversible as the pyridylthione form is much more favorable than the pyridinethiol; the ratio of 4-pyridylthione to 4-pyridinethiol is estimated at 3.5×10^4 (Grimshaw, C.E., Whistler, R.L., and Cleland, W.W., Ring opening and closing rates for thiosugars, *J. Am. Chem. Soc.* 101, 1521–1532, 1979). Also shown is the structure of 3-nitro-2-pyridyl disulfide that result in thiomethylation of a protein (Kimura, T., Matsueda, R., Nakagawa, Y., and Kaiser, E.T., New reagents for the introduction of the thiomethyl group at sulfhydryl residues of proteins with concomitant spectrophotometric titration of sulfhydryls: Methyl-3-nitro-2-pyridyl disulfide and methyl-2-pyridyl disulfide, *Anal. Biochem.* 122, 274–282, 1982).

Brocklehurst and Little[708] reported that 2PDS was an effective inhibitor of papain demonstrating remarkable activity (Table 7.12) at low pH (pH 3.8) with a second pH optima at alkaline pH. A comparison of the effect of pH on the reaction of 2PDS with papain, L-cysteine, and 2-mercaptoethanol is shown in Table 7.12. While the rate of reaction of 2PDS with the model thiols is rapid at low pH, the rate of reaction with papain is several orders of magnitude larger; however, at higher pH (8.0), the rate of reaction with the model thiols is larger than that observed for papain. Brocklehurst and Little were able to take advantage of the rapid reaction of 2PDS with papain at acid pH (3.5–4.5) for active-site titration. It was possible to perform the active-site titration of papain in the presence of low molecular weight thiols.[707,708] Brocklehurst and Little[709] showed that the enhanced reactivity of the cysteine residue (Cys25) in papain was unique to the native protein as denaturation by acid or heat eliminated the enhanced reactivity. The active site of papain could be *titrated* in the presence of a 10-fold excess of cysteine or 100-fold excess of 2-mercaptoethanol. Brocklehurst and Little[709] have compared the reaction of 2PDS and 4PDS with papain and L-cysteine; the reaction of 2PDS with papain has a sharp optima at pH 4, while there is small increase in the reaction of 4PDS with papain at pH 5.0. 2PDS has a faster reaction with L-cysteine than 4PDS. These investigators argue that 2PDS is unique in its interaction with the active site of papain.[708] The enhanced rate of reaction of 2PDS at acidic pH is observed with ficin[710] as well as human albumin or bovine albumin.[711] Pedersen and Jacobsen[711] also showed that the rate of 2PDS modification of glutathione was comparable to that observed for either human albumin or bovine albumin at acid pH 3; unlike either human or bovine albumin, there was no decrease in reactivity at pH 5.0. 2PDS and 4PDS have the advantage of being active at lower pH than DTNB for reaction with thiols.[713] The enhanced reactivity of 2PDS and 4PDS at lower pH is suggested to be due to the enhanced electrophilicity of sulfur in protonated reagent (2PDS, $pK_a = 2.74$; 4PDS, $pK_a = 5.1$).[709] The enhanced electrophilicity of the monocation form of PDS (Figure 7.26) does permit the modification of the buried active-site cysteine in ficin at pH < 4.[723] DTNB does not have a mechanism for protonation that would increase nucleophilicity and thus is not active at lower pH.[712] The increased reactivity of 2PDS with papain as compared to 4PDS reflects the increased nucleophilicity of the unionized thiol at the active site.[723] The unique reactivity of 2PDS is supported by the work of Malthouse and Brocklehurst[713] who reported that while the rate of reaction of 2,2'-dipyridyldisulfide with 2-mercaptoethanol decreases in the presence of either dimethylformamide or dioxane, there was no effect of solvent on the rate of reaction of either 4,4'-dipyridyldisulfide or DTNB with 2-mercaptoethanol. Modification with either 2PDS or 4PDS is used reversibly by the addition of a reducing agent such as dithiothreitol. Disulfide exchange with a pyridyl disulfide matrix has been developed.

The activity of disulfide-based reagents described above, DTNB, 2PDS, and 4PDS, is based, in part, on the *activated* nature of these compounds.[687] The modification of a thiol group in protein yields an *activated* mixed disulfide that can then couple to another thiol as shown below for factor VIII. Kimura and coworkers[714] introduced methyl 3-nitro-2-pyridyl disulfide and methyl 2-pyridyl disulfide. Both of these reagents modify sulfhydryl groups forming the thiomethyl derivative. The spectrum of 3-nitro-2-pyridone is pH dependent. There is an isosbestic point at 310.4 nm which can be used to determine the extent of the reaction of methyl-3-nitro-2-pyridyl disulfide with sulfhydryl groups. The difference in spectrum obtained does not show the pH dependence of the nitropyridyl derivative. At 343 nm, the change in extinction coefficient is 7060 M^{-1} cm^{-1}. The extinction coefficient (7600 M^{-1} cm^{-1}) of the 2-thiopyridinone at 343 nm is relatively stable from pH 3 to 8.0.[714] There is a marked decrease in absorbance above pH 8.0 reflecting the loss of a proton. Reaction with the sulfhydryl group in the protein clearly proceeds more rapidly at alkaline pH. The thiomethyl group is also the product of methyl methanethiosulfonate,[605] which is described above. S-(2-thiopyridyl)-L-cysteine hydrazide (Figure 7.35) has been developed as a reagent for the preparation of immunoconjugates based on coupling to periodate-oxidized immunoglobulins.[715] S-(2-thiopyridyl)-L-cysteine was the material used for the preparation of the hydrazide; this disulfide has also been used for reversible coupling of fatty acids to protein (lipidization). The N-hydroxysuccinimide ester of palmitic acid is coupled to the amino group of cysteine and the resulting disulfide conjugate couple to protein by disulfide exchange.[716] Other derivatives (Figure 7.35) include S-[2-(4-azidosalicylamido)ethylthio)-2-thiopyridine[717] and 2-S-(2'-thiopyridyl)-6-hydroxynaphthyldisulfide.[718] S-[2-(4-azidosalicylamido)ethylthio)-2-thiopyridine[717] was developed as a cross-linking agent, while 2-S-(2'-thiopyridyl)-6-hydroxynaphthyldisulfide[718] was used for the purification of a low molecular weight thiol, mycothiol. The thiol was coupled to the reagent, isolated by hydrophobic affinity chromatography, and recovered by the reduction of the isolated product. The modification of a membrane protein, lambda holin (S105), with 2PDS permitted purification with protection of the thiol group that was regenerated by reduction following IMAC chromatography.[719] This approach has been described above in the use of S-sulfonation with sodium tetrathionate for the protection of sulfhydryl groups during protein purification. 2PDS has been used for the preparation of an *activated disulfide* matrix (agarose-glutathione-2-pydridyl disulfide) (Figure 7.35)[720] for the purification of thiol proteins[721] and thiol peptides.[722] Another approach was used by Harris and coworkers for the purification of human blood coagulation factor VIII.[723] Crude plasma factor VIII was obtained from plasma containing dithiothreitol and modified with 2PDS. After purification on a polyelectrolyte column, the modified protein was taken to a thiopropyl-agarose column and active protein eluted with dithiothreitol. Proteins modified with dithiopyridyl disulfide have been used to prepare various conjugates.[724–727] These products used modification of amino groups with 2-iminothiolane[714,726] or with N-acetyl-DL-homocysteinethiolactone[725] to generate a sulfhydryl function for reaction with dithiopyridyl disulfide. Russell-Jones and coworkers[728] prepared a long alkyl chain dithiopyridyl disulfide derivative of vitamin B_{12} that was coupled to a buried sulfhydryl group in granulocyte-colony-stimulating factor for oral drug delivery. The synthesis of a selenium analog of this class of reagents, 6,6-diselenobis-(3-nitrobenzoic acid), has been reported.[729] The selenium-containing reagent has similar reaction characteristics as the sulfur-containing compound in terms of specificity of reaction with cysteinyl residues in proteins. The reaction is monitored by spectroscopy following the release of 6-seleno-3-nitrobenzoate, which has a maximum at 432 nm. The extinction coefficient for the 6-seleno-3-nitrobenzoate anion varies slightly from 9,532 (with excess reagent) to 10,200 M^{-1} cm^{-1} (with either excess cysteine or excess 2-mercaptoethanol). Other than commenting that the reaction of 6,6'-diselenobis(3-nitrobenzoic acid) with sulfhydryl groups was rapid, there is no mention of reaction rate. There has been limited use of the ^{77}Se derivative as an NMR probe in proteins.[730–732] Pleasants and coworkers[733] have made a comparative study of the rates of selenol/diselenide (selenocysteine/selenocystamine) and thiol/disulfide (cysteamine/cystamine) exchange using NMR spectroscopy and reported that the rate observed with selenocysteamine/selenocystamine was 1.2×10^7 faster than cysteamine/cystamine.

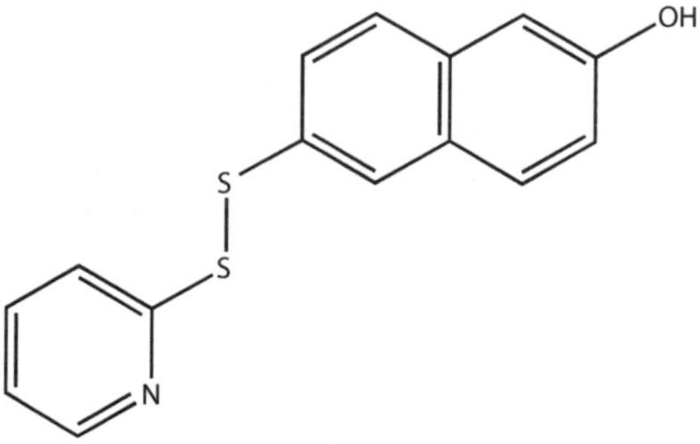

2-S-(2′-thiopyridyl)-6-hydroxynapthyldisulfide

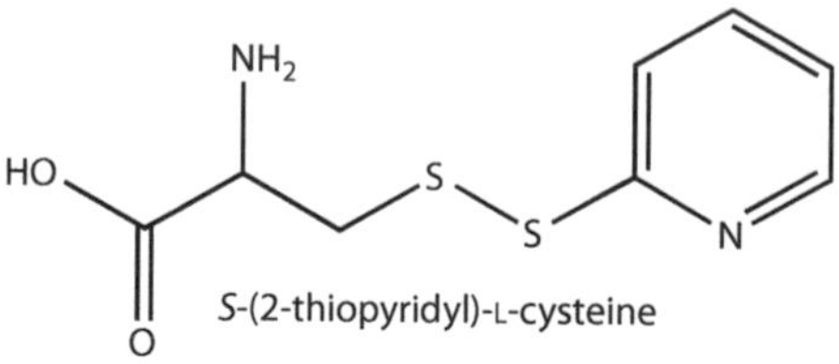

S-(2-thiopyridyl)-L-cysteine

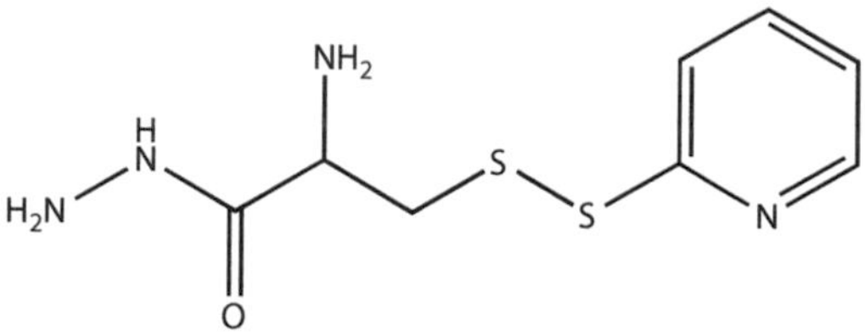

S-(2-thiopyridyl)-L-cysteine hydrazide

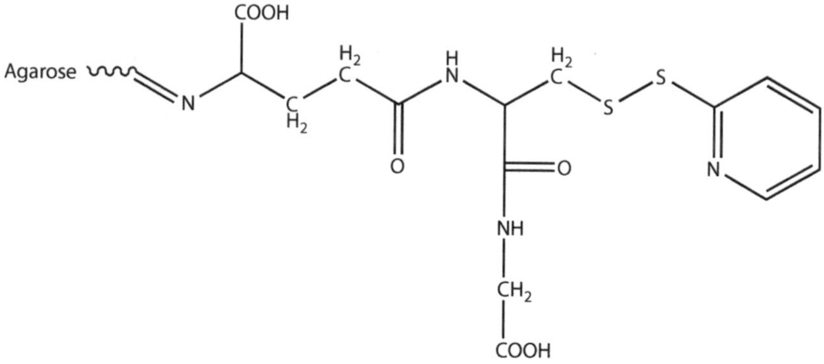

Agarose-(glutathione-2-pyridyl disulfide)

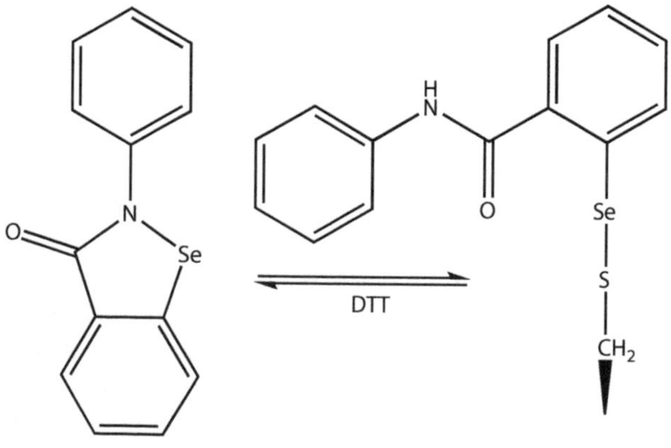

2-Phenyl-1,2-benzisoselenazol-3(2H)-one

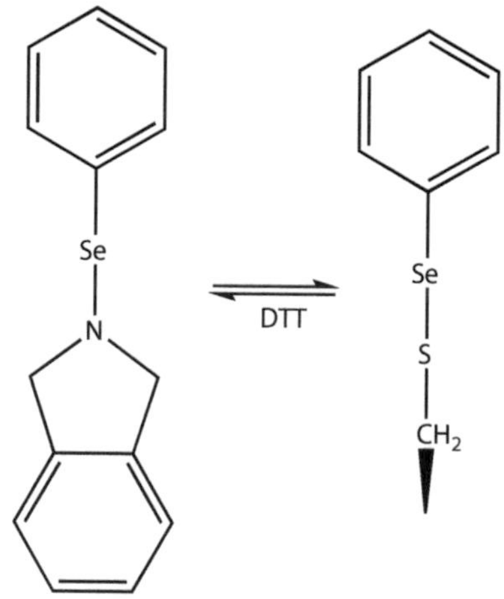

N-(Phenylseleno)phthalimide

FIGURE 7.35 (continued)

The difference in these rates reflect both the higher nucleophilicity of the alkyl selenoates and the ability of alkyl selenoates to serve as better leaving groups in the S_N2 reaction when compared to the corresponding thiolates. Alkyl selenium compounds are considered to be considerably more nucleophilic than the corresponding sulfur compounds.[734,735] Xu and coworkers[736] have reported the modification of thiols with Se-N compounds such as N-(phenylseleno)phthalimide (Figure 7.35) forming a Se–S bond. The modification is selective and very rapid and reversed with dithiothreitol. These authors note that Se-S derivative obtained with 2-phenyl-1,2-benzisoselenazol-3-(2H)-one was cleaved during collision-induced-dissociation MS while the derivative with N-(phenylseleno) phthalimide was stable presenting analytical opportunities. Later work[737] with N-phenylseleno) phthalimide with a modified protocol where the reagent is dissolved in neat acetonitrile showed that the quantitative modification of sulfhydryl groups was achieved in seconds with a modest molar excess. The modification of cysteinyl thiol groups with N-(phenylseleno)phthalimide is sensitive to the environment; the proximity of basic amino acid residues enhances reaction. Selenol has been reported to catalyze the thiol–disulfide interchange reaction.[738]

There are several other methods for the modification of sulfhydryl groups in proteins which have not been used extensively but which have considerable historical value. O-Methylisourea reacts with cysteinyl residues to form the S-methyl derivative.[739] S-methylation of proteins does occur in vivo.[740] O-Methylisourea is used far more frequently for the modification of lysine.[741–743] Dimethylsulfonate can also methylate cysteine thiol groups.[744] Cyanate also can modify sulf-hydryl groups resulting in the formation of S-carbamoyl cysteine (Figure 7.34).[745,746] While this reaction is more rapid for sulfhydryl groups than for other nucleophiles such as amino groups, reaction product stability is an issue; the carbamoyl derivative of cysteine is stable at acid pH, but rapidly decomposes at alkaline pH.[747] Stark[746] estimated that the half-life of S-carbamoylcysteine (decomposition to cysteine and cyanate) was 1–2 h at pH 6.8/30°C. There are isolated studies on the reaction of cyanate with cysteine in proteins. Hu and coworkers[748] reported that the car-bamoylation of the sulfhydryl groups of glutathione increased with a decrease in pH from 7.4 to 6.6. While carbamylation of proteins has been a historic problem,[10] current work suggests that carbamylation of cysteine is not an issue in structural analysis.[749] Concern in past years came from problems with urea purity that could result in the dismutation of urea to form cyanate.[750–752] Barrett and coworkers[753] reported the reaction of hypothiocyanous acid with active site cysteine of several intracellular enzymes including creatine kinase forming cysteinesulfenyl thiocyanate, which is suggested to hydrolyze to cysteinesulfenic acid.

2-Nitro-5-thiocyanobenzoic acid, structurally related to DTNB, reacts with cysteine in pro-tein resulting in the formation of S-cyanocysteine and the formation of 2-mercapto-5-nitrobenzoic acid (Figure 7.36), which could be used for the quantitative determination of sulfhydryl groups.

FIGURE 7.35 (continued) Activated mixed disulfides based on pyridyl disulfides and seleno derivatives. Shown at the top is 2-S-(2'-thiopyridyl)-6-hydroxynapthyldisulfide that is used to purify low molecular weight thiols by formation of a mixed disulfide (Steenkamp, D.J. and Vogt, R.N., Preparation and utili-zation of a reagent for the isolation and purification of low-molecular thiols, *Anal. Biochem.* 325, 21–27, 2004). Also shown at the top is the structure of S-(2-thiopyridyl)-L-cysteine hydrazide that can be used to form a reversible cross-link between the aldehyde formed from periodate oxidized carbohydrate and a cysteine residue in a protein; the hydrazide is derived from the mixed disulfide between 2-thiopyridine and cysteine (Zara, J.J., Wood, R.D., Boon, P. et al., A carbohydrate-directed heterobifunctional cross-linking reagent for the synthesis of immunoconjugates, *Anal. Biochem.* 194, 156–162, 1991). Shown below is the structure of an agarose-glutathione-2-pyridyl disulfide used for thiol–disulfide exchange chromatography (Brocklehurst, K., Carlsson, J., Kierstan, M.P.J., and Crook, E.M., Covalent chromatography. Preparation of fully active papain from dried papaya latex, *Biochem. J.* 133, 573–584, 1973). Shown at the bottom is the reaction of 2-phenyl-1,2-benzisoselenazol-3(2H)-one or N-(phenylseleno)phthalimide with a thiol. Reversal of the modification is accomplished with dithiothreitol (DTT). See Xu, K., Zhang, Y., Tang, B., et al., Study of highly selective and efficient thiol derivatization using selenium reagents by mass spectrometry, *Anal. Chem.* 82, 6926–6932, 2010.

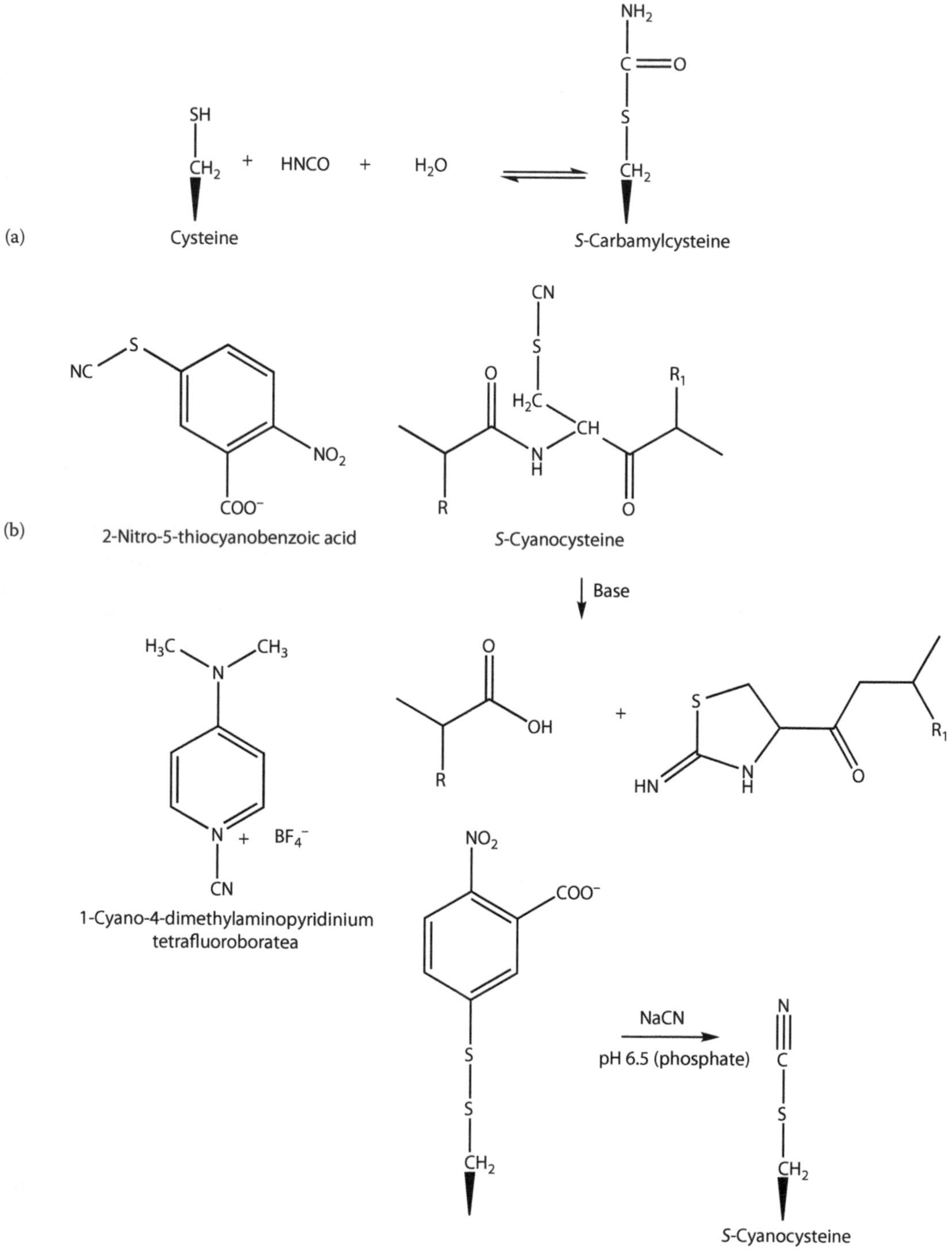
SH
CH2
+
HNCO
+
H2O
NH2
C
O
S
CH2
(a)
Cysteine
S-Carbamylcysteine
NC
S
NO2
COO−
(b)
2-Nitro-5-thiocyanobenzoic acid
CN
S
O
H2C
CH
R1
N
H
R
O
S-Cyanocysteine
Base
H3C
CH3
N
N
+
BF4−
CN
1-Cyano-4-dimethylaminopyridinium
tetrafluoroboratea
O
OH
R
+
S
HN
N
H
O
R1
NO2
COO−
S
S
CH2
NaCN
pH 6.5 (phosphate)
N
C
S
CH2
S-Cyanocysteine

FIGURE 7.36 (continued)

2-Mercapto-5-nitrobenzoic acid has an absorbance maximum at 412 nm with a molar extinction coefficient of 13,600 M^{-1} cm^{-1}.[676] Pecci and coworkers[754] have characterized the reaction of rhodanese with 2-nitro-5-thiocyanobenzoic acid. These investigators used a 1.3 molar excess of reagent in 0.050 M phosphate buffer, pH 8.0 at 18°C. The reaction was followed spectrophotometrically by the release of 2-mercapto-5-nitrobenzoic acid and was complete after 6 h. There has been more use for S-cyanocysteine can be used for structural analysis where S-cyanocysteine (β-thiocyanatoalanine) can be used as an infrared chromophore for measuring protein conformation.[755–757]

Peptide bond cleavage can be accomplished on the amino-terminal side of S-cyanocysteine. Cleavage at S-cyanocysteinyl residues was first studied by Vanaman and Stark.[758] These investigators modified the catalytic subunit of E. coli aspartate transcarbamylase with DTNB and modified the 5-thio-2-nitrobenzoyl derivative with cyanide that resulted in the S-thiocyano derivative (Figure 7.36) (it should be obvious that this approach is closely related to the use of 2-nitro-5-thiocyanobenzoic acid described above). Witkowska and coworkers[759] used this approach for the modification of a cysteinyl residue in thioesterase II and subsequent cleavage at the modified cysteine. More recent work has used 2-nitro-5-thiocyanobenzoic acid for the modification of cysteine for peptide bond cleavage.[760–762] Wu and Watson[763] used 1-cyano-4-dimethylaminopyridinium tetrafluoroborate (Figure 7.36) for the modification of the cysteine residues prior to peptide bond cleavage.

4-Chloro-7-nitrobenzo-2-oxa-1,3-diazole (4-chloro-7-nitrobenzofurazan; Nbd-Cl) (Figure 7.37) is a reagent developed for the modification of amino groups[764] but is found to be more applicable in the modification of sulfhydryl groups and is useful in that it introduces a fluorescent probe.[765–769] Nbd-Cl is useful for the assay of cysteine in proteins[770] and in drugs.[771] Nitta and coworkers[768] have noted that there are several possible reaction products of Nbd-Cl with protein nucleophiles including sulfhydryl groups. The reaction of Nbf-Cl with sulfhydryl groups in glutathione reductase and lipoamide dehydrogenase has also been reported.[772] Nbd-Cl is also useful for the determination of cysteine sulfenic acid[773–776] but does not distinguish between cysteine and cysteine sulfenic acid.

FIGURE 7.36 (continued) Various mechanisms for the reaction of cyanate and cyanide with cysteine. (a) The reversible reaction of cyanate with cysteine to form the carbamoyl derivative (see Stark, G.R., Reversible reaction of cyanate with sulfhydryl groups and the determination of NH_2-terminal cysteine and cystine in proteins, *J. Biol. Chem.* 239, 1411–1414, 1964). (b) The reaction of 2-nitro-5-thiocyanobenzoate with cysteine to yield S-cyanocysteine, a derivative that is cleaved based on an iminothiazolidine derivative; not shown is the β-elimination reaction resulting in dehydroalanine (Iwasaki, M., Masuda, T., Tomita, M., and Ishihama, Y., Chemical cleavage-assisted tryptic digestion for membrane proteome analysis, *J. Proteome Res.* 8, 3169–3175, 2008). Also shown is the structure of 1-cyano-4-dimethylaminopyridinium tetrafluoroborate (Wu, J. and Watson, J.T., Optimization of the cleavage reaction for cyanylated cysteinyl proteins for efficient and simplified mass mapping, *Anal. Biochem.* 258, 268–276, 1998) and the generation of cyanylated cysteine (β-thiocyanatoalanine) by reaction of S-(5-thio-5-nitrobenzoyl)cysteine with cyanate (Alfieri, K.N., Vienneau, A.R., and Londergan, C.H., Using infrared spectroscopy of cyanylated cysteine to map the membrane binding structure and orientation of the hybrid antimicrobial peptide CM15, *Biochemistry* 50, 11097–11108, 2011).

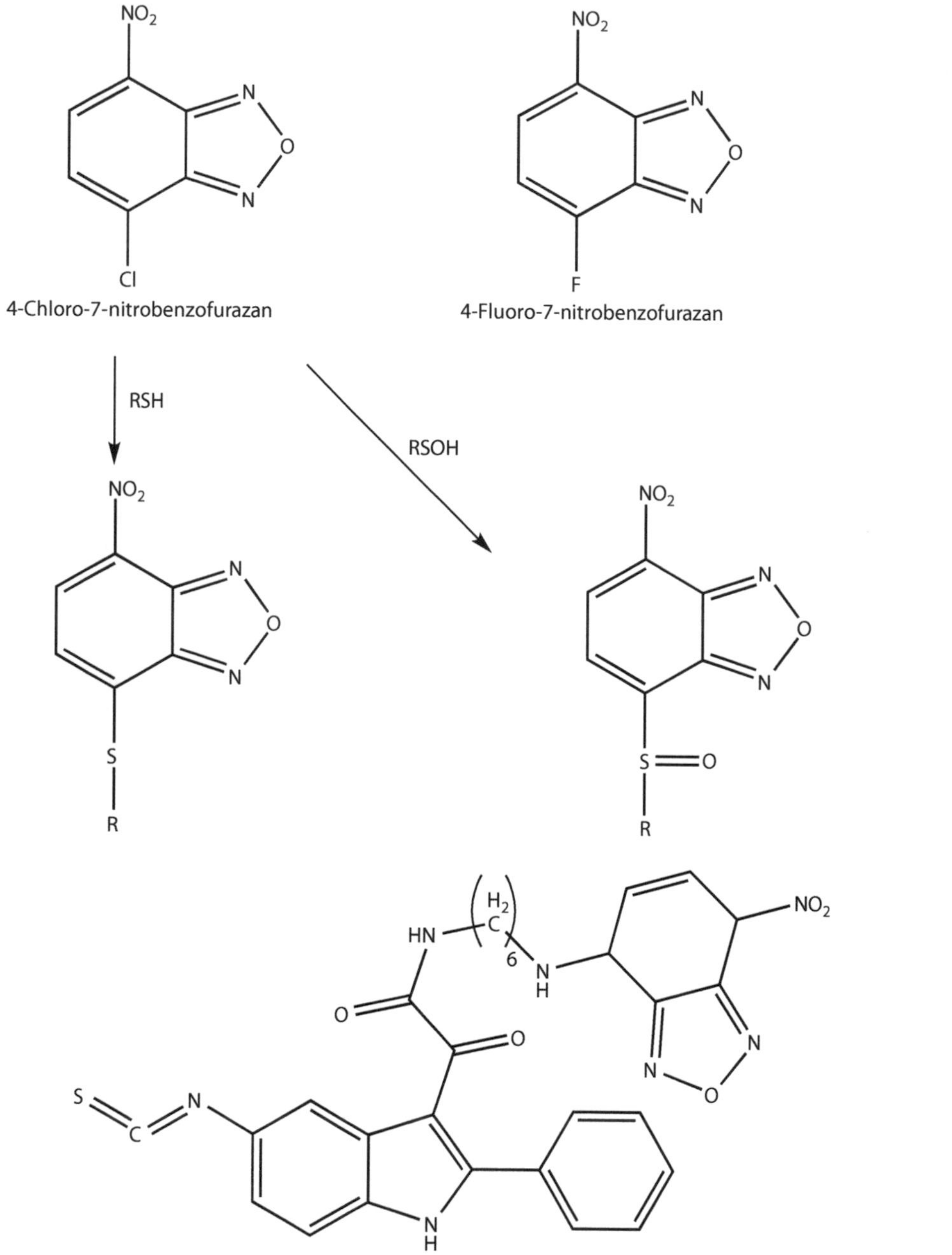

FIGURE 7.37 The structure of 4-chloro-7-nitrobenzofurazan (Nbf-Cl) and 4-fluoro-7-nitrobenzofurazan and the reaction of Nbf-Cl with a thiol. Also shown is the reaction of Nbf-Cl with a sulfenic acid (Poole, L.B. and Ellis, H.R., Identification of cysteine sulfenic acid in AhpC of alkyl hydroperoxide reductase, *Methods Enzymol.* 348, 297–305, 2002). 4-Fluoro-7-nitrobenzofurazan has been used to modify lysine and tyrosine in proteins (Luo, J., Fukuda, E., Takase, H. et al., Identification of the lysine residue responsible for coenzyme A binding in the heterodimeric 2-oxoacid:ferredoxin oxidoreductase from *Sulfolobus tokodaii*, a thermophilic archaeon, using 4-fluoro-7-nitrobenzofurazan as an affinity label, 345, 1963).

REFERENCES

1. Toutchkine, A., Nguyen, D.-V., and Hahn, K.M., Simple one-pot preparation of water-soluble, cysteine-reactive cyanine and merocyanine dyes for biological imaging, *Bioconjug. Chem.* 18, 1344–1348, 2007.

2. Plapp, B.V., Mechanisms of carboxymethylation of bovine pancreatic nucleases by haloacetates and tosylglycolate, *J. Biol. Chem.* 248, 4896–4900, 1973.

3. Kleanthous, C., Campbell, D.G., and Coggins, J.R., Active site labeling of the shikimate pathway enzyme, dehydroquinonase and dehydroquinate synthase, *J. Biol. Chem.* 265, 10929–10934, 1990.

4. Lutter, L.C. and Kurland, C.G., Chemical determination of protein neighbourhoods in a cellular organelle, *Mol. Cell. Biochem.* 7, 105–116, 1975.

5. Mello, R.N. and Thomas, D.D., Three distinct actin-attached structural states of myosin in muscle fibers, *Biophys. J.* 102, 1088–1096, 2012.

6. Tinianow, J.N., Gill, H.S., Ogasawara, A. et al. Site-specifically ^{89}Zr-labeled monoclonal antibodies for ImmunoPET, *Nucl. Med. Biol.* 37, 289–297, 2010.

7. Wang, H., Gao, H., Guo, N. et al., Site-specific labeling of scVEGF with fluorine-18 for positron emission tomography imaging, *Theranostics* 2, 607–617, 2012.

8. Ravi, S., Krishnamurthy, V.R., Caves, J.M. et al., Maleimide-thiol coupling of a bioactive peptide to an elastin-like protein polymer, *Acta Biomater.* 8, 627–635, 2012.

9. Lyon, R.P., Meyer, D.L., Setter, J.R., and Senter, P.D., Conjugation of anticancer drugs through endogenous monoclonal antibody cysteine residues, *Methods Enzymol.* 502, 123–138, 2012.

10. Giron, P., Dayon, L., Mihala, N. et al., Cysteine-reactive covalent capture tags for enrichment of cysteine-containing peptides, *Rapid Commun. Mass Spectrom.* 23, 3377–3386, 2009.

11. Lesaicherre, M.L., Uttamchandani, M., Chen, G.Y., and Yao, S.Q., Developing site-specific immobilization strategies of peptides in a microarray, *Bioorg. Med. Chem. Lett.* 12, 2079–2083, 2002.

12. Chattopadhaya, S., Abu Bakar, F.B., and Yao, S.Q., Use of intein-mediated protein ligation strategies for the fabrication of functional protein arrays, *Methods Enzymol.* 462, 195–223, 2009.

13. Mei, B., Pan, C., Jiang, H. et al., Rational design of a fully active, long-acting PEGylated factor VIII for hemophilia A treatment, *Blood* 116, 270–279, 2010.

14. Li, L., Crow, D., Turatti, F. et al., Site-specific conjugation of monodispersed DOTA-PEGn to a thiolated diabody reveals the effect of increasing peg size on kidney clearance and tumor uptake with improve 64-copper PET imaging, *Bioconjug. Chem.* 22, 709–716, 2011.

15. Park, J.S., Hughes, S.J., Cunningham, F.K., and Hammond, J.R., Identification of cysteines involved in the effects of methanethiosulfonate reagents on human equilibrative nucleoside transporter 1, *Mol. Pharmacol.* 80, 735–746, 2011.

16. Roberts, J.A., Asllsopp, R.C., El Ajous, S. et al., Agonist binding evokes extensive conformational changes in the extracellular domain of the ATP-gated human P2X1 receptor ion channel, *Proc. Natl. Acad. Sci. USA* 109, 4663–4667, 2012.

17. Koch, S., Fritsch, M.J., Buchanan, G., and Palmer, T., *Escherichia coli* TatA and TatB proteins have N-out, C-in topology in intact cells, *J. Biol. Chem.* 287, 14420–14431, 2012.

18. Kim, H.S. and Nikaido, H., Different functions of MdtB and MdtC subunits in the heterotrimeric efflux transporter MdtB$_2$C complex of *Escherichia coli*, *Biochemistry* 51, 4188–4197, 2012.

19. Rashid, M., Arumugam, T.V., and Karamyan, V.T., Association of the novel non-AT1, non-AT2 angiotensin binding site with neuronal cell death, *J. Pharmacol. Exp. Ther.* 335, 754–761, 2010.

20. Yadav, S., Saxena, J.K., and Dwivedi, U.N., Purification and characterization of *Plasmodium yoelii* adenosine deaminase, *Exp. Parasitol.* 129, 368–374, 2011.

21. Pleszczynska, M., Wiater, A., Skowronek, M., and Szczodrak, J., Purification and characterization of mutanase produced by *Paenibacillus curdlanolyticus* MP-1, *Prep. Biochem. Biotechnol.* 42, 335–347, 2012.

22. Chen, X., Zhou, Y., Peng, X., and Yoon, J., Fluorescent and colorimetric probes for detection of thiols, *Chem. Soc. Rev.* 39, 2120–2135, 2010.

23. Webb, J.L., Organic mercurials, in *Enzyme and Metabolic Inhibitors*, Vol. II, Chapter 7, pp. 729–985, Academic Press, New York, 1966.

24. Chatterjee, G.C. and Noltmann, E.A., Reaction of sulfhydryl groups in phosphoglucose mutase with organic mercurials, *J. Biol. Chem.* 242, 3440–3448, 1967.

25. Fagan, T.F. and Mayhew, S.G., Effects of thiols and mercurials on the periplasmic hydrogenase from *Desulfovibrio vulgaris* (Hildenborough), *Biochem. J.* 293, 237–241, 1993.

26. Altamirano, M.M., Mulliert, G., and Calcagno, M., Sulfhydryl groups of glucosamine-6-phosphate isomerase deaminase from *Escherichia coli*, *Arch. Biochem. Biophys.* 258, 95–106, 1987.

27. Liu, M., Huang, Y., Wu, J. et al., Effect of cysteine residues on the activity of arginyl-tRNA synthetase from *Escherichia coli.*, *Biochemistry* 38, 11006–11011, 1999.

28. Ohno, A. and Oae, S., Thiols, in *Organic Chemistry of Sulfur*, ed. S. Oae, Chapter 4, pp. 119–187, Plenum Press, New York, 1977.

29. Silva, C.O., da Silva, E.C., and Nascimento, M.A.C., Ab initio calculations of absolute pKa values in aqueous solutions II. Aliphatic alcohols, thiols, and halogenated carboxylic acids, *J. Phys. Chem. A* 104, 2402–2409, 2000.

30. Hederos, S. and Baltzer, L., Nucleophile selectivity in the acyl transfer reaction of a designed enzyme, *Biopolymers* 79, 292–299, 2005.

31. Sluyterman, L.A.A., The rate-limiting reaction in papain action as derived from the reaction of the enzyme with chloroacetic acid, *Biochim. Biophys. Acta* 151, 178–187, 1968.

32. Sluyterman, L.A.A., Reversible inactivation of papain by cyanate, *Biochim. Biophys. Acta* 139, 439–449, 1967.

33. Stark, G.R., Stein, W.H., and Moore, S., Reactions of cyanate present in aqueous urea with amino acids and proteins, *J. Biol. Chem.* 235, 3177–3181, 1960.

34. Janssen, M.J., Nucleophilicity of organic sulfur compounds, in *Sulfur in Organic and Inorganic Chemistry*, ed. A. Senning, Chapter 3, pp. 355–377, Dekker, New York, 1972.

35. Pienta, N.J. and Kessler, R.J., Pentaenyl cations from the photolysis of retinyl acetate. Solvent effects on the leaving group ability and relative nucleophilicities: An unequivocal and quantitative determination of the importance of hydrogen bonding, *J. Am. Chem. Soc.* 114, 2419–2428, 1992.

36. Reichart, C., *Solvent and Solvent Effects in Organic Chemistry* 3rd edn., Wiley-VCH, Weinheim, Germany, 2003.

37. Modena, G., Paradisi, C., and Scorrano, G., Solvation effects on basicity and nucleophilicity. in *Organic Sulfur Chemistry. Theoretic and Experimental Advances*, eds. F. Bernardi, I.G. Csizmadia, and A. Mongini, Chapter 10, pp. 569–597, Elsevier, Amersterdam, the Netherlands, 1985.

38. Davies, R.E., The oxibase scale, in *Organosulfur Chemistry*, ed. M.J. Janssen, Chapter 18, pp. 311–328, Wiley Interscience, New York, 1967.

39. Cecil, R. and McPhee, J.R., The sulfur chemistry of proteins, *Adv. Protein. Chem.* 14, 255–389, 1959.

40. Liu, T.-Y., The role of sulfur in proteins, in *The Proteins*, Vol. 3, 3rd edn., eds. H. Neurath and R.L. Hill, Academic Press, New York, 1977.

41. Torchinsky, Y.M., *Sulfur in Proteins*, Pergamon Press, Oxford, U.K., 1981.

42. Kooyman, E.C. Some characteristics of organosulfur compound and intermediates, in *Organosulfur Chemistry*, ed. M.J. Janssen, Chapter 1, pp. 1–10, Wiley Interscience, New York, 1967.

43. Ogino, K. and Fujihara, H., Biochemical reactions involving thioesters, in *Organic Sulfur Chemistry*, eds. S. Oae and T. Okuyama, Chapter 3, pp. 71–136, CRC Press, Boca Raton, FL, 1992.

44. Khan, S.A., Sekulski, J.M., and Erickson, B.W., Peptide models of protein metastable binding sites: Competitive kinetics of isomerization and hydrolysis, *Biochemistry* 25, 5165–5171, 1986.

45. Isenman, D.E., The role of the thioester bond in C3 and C4 in the determination of the conformational and functional states of the molecule, *Ann. N. Y. Acad. Sci.* 421, 277–290, 1983.

46. Song, J., Wang, J., Jazwiak, A.A. et al., Stability of thioester intermediates in ubiquitin-like modifications, *Protein Sci.* 18, 2482–2499, 2009.

47. Hop, C.E. and Bakhtier, R., Homocysteine thiolactone and protein homocysteinylation: Mechanistic studies with model peptides and proteins, *Rapid Commun. Mass Spectrom.* 16, 1049–1059, 2002.

48. Jakubowski, H., Homocysteine thiolactone: Metabolic origin and protein homocysteinylation in human, *J. Nutr.* 130, 377S–381S, 2000.

49. Paoli, P., Sbrana, E., Tiribilli, B. et al., Protein N-homocysteinylation induces formation of toxic amyloid-like protofibrils, *J. Mol. Biol.* 400, 889–897, 2010.

50. Xiao, J., Burn, A., and Tolbert, T.J., Increasing solubility of proteins and peptides by site-specific modifications with betaine, *Bioconjug. Chem.* 19, 1113–1118, 2008.

51. Rohde, H. and Seitz, O., Ligation-desulfurization: A powerful combination in the synthesis of peptides and glycopeptides, *Biopolymers* 94, 551–559, 2010.

52. Shingledecker, K., Jiang, S.-Q., and Paulus, H., Reactivity of the cysteine residues in the protein splicing active center of the *Mycobacterium tuberculosis* RecA intein, *Arch. Biochem. Biophys.* 375, 138–145, 2000.

53. Du, Z., Shemella, P.T., Liu, Y. et al., Highly conserved histidine plays a dual catalytic role in protein splicing: A pKa shift mechanism, *J. Am. Chem. Soc.* 131, 11581–11589, 2009.

54. Furuta, S., Ortriz, F., Zhu Sun, X. et al., Copper uptake is required for pyrrolidone dithiocarbamate-mediated oxidation and protein level increase of p53 in cells, *Biochem. J.* 365, 639–648, 2002.

55. Sharp, J.S., Becker, J.M., and Hettich, R.L., Protein surface mapping by chemical oxidation: Structural analysis by mass spectrometry, *Anal. Biochem.* 313, 216–225, 2003.

56. Mannick, J.B. and Schonhoff, C.M., Nitrosylation: The next phosphorylation, *Archs. Biochem. Biophys.* 408, 1–6, 2002.

57. Tao, L. and English, A.M., Mechanism of S-nitrosylation of recombinant human brain calbindin D$_{28K}$, *Biochemistry* 42, 3326–3334, 2003.

58. Nogueira, L., Figueiredo-Freitas, C., Casimiro-Lopes, G. et al., Myosin is reversibly inhibited by S-nitrosylation, *Biochem. J.* 424, 221–231, 2009.

59. Ralat, L.A., Ren, M., Schilling, A.B, adn Tang, W.J., Protective role of Cys-178 against the inactivation and oligomerization of human insulin-degrading enzyme by oxidation and nitrosylation, *J. Biol. Chem.* 284, 34005–24018, 2009.

60. Marino, S.M. and Gladyshev, V.N., Structural analysis of cysteine S-nitrosylation: A modified acid-based motif and the emerging role of *trans*-nitrosylation, *J. Mol. Biol.* 395, 844–859, 2010.

61. Spadaro, D., Yun, B.W., Spoel, S.H. et al., The redox switch: Dynamic regulation of protein function by cysteine modifications, *Physiol. Plant.* 138, 360–371, 2010.

62. Spickett, C.M. and Pitt, A.R., Protein oxidation: Role in signalling and detection by mass spectrometry, *Amino Acids*, 42, 5–21, 2010.

63. Aesif, S.W., Janssen-Heininger, Y.N., and Reynaert, N.L., Protocols for the detection of S-glutathionylated and S-nitrosylated proteins *in situ*, *Methods Enzymol.* 474, 289–296, 2010.

64. Ueland, P.M., Skoolard, T., Doskeland, S.O., and Flatmark, T., An adenosine 3′,5′-monophosphate-adensine binding protein from mouse liver: Some physicochemical properties, *Biochim. Biophys. Acta* 533, 57–65, 1978.

65. Arai, K., Arai, T., Kawakita, M., and Kaziro, Y., Conformational transitions of polypeptide elongation factor Tu. I. Studies with hydrophobic probes, *J. Biochem.* 77, 1096–1108, 1975.

66. Nakamura, S., Ohta, S., Arai, K., and Arai, N., Studies on polypeptide chain-elongation factors from an extreme thermophile, *Thermus thermophilus* HB8. 3. Molecular properties, *Eur. J. Biochem.* 92, 533–543, 1978.

67. Whitfield, C.E. and Schworer, M.E., Locus of N-ethylmaleimide action on sugar transport in nucleated erythrocytes, *Am. J. Physiol.* 241, C33–C41, 1981.

68. Floris, G., Giartosio, A., and Rinaldi, A., Essential sulfhydryl groups in diamine oxidase from *Euphorbia characias* latex, *Arch. Biochem. Biophys.* 220, 623–627, 1983.

69. Polyanovsky, O.L., Novikov, V.V., Deyev, S.M. et al., Location of exposed and buried cysteine residues in the polypeptide chain of aspartate aminotransferase, *FEBS Lett.* 35, 322–326, 1973.

70. Fuchs, F., Liou, Y.-M., and Grabarek, Z., The reactivity of sulfhydryl groups of bovine cardiac troponin C, *J. Biol. Chem.* 264, 20344–20349, 1989.

71. Lee, J. and Blaber, M., The interaction between thermodynamic stability and buried free cysteines in regulating the functional half-life of fibroblast growth factor-1, *J. Mol. Biol.* 393, 113–127, 2009.

72. Lee, J. and Blaber, M., Structural basis of conserved cysteine in the fibroblast growth factor family: Evidence for a vestigial half-cysteine, *J. Mol. Biol.* 393, 129–139, 2009.

73. Nagano, N., Ota, M, and Nishikawa, K., Strong hydrophobic nature of cysteine residues in proteins, *FEBS Lett.* 458, 69–71, 1999.

74. Zhou, H. and Zhou, Y., Quantifying the effect of burial of amino acid residues on protein stability, *Proteins*, 54, 315–322, 2003.

75. You, C., Huang, Q., Xue, H. et al., Potential hydrophobic interaction between two cysteines in interior hydrophobic region improves thermostability of a family 11 xylanase from *Neocallimastix patriciarum*, *Biotechnol. Bioeng.* 105, 861–870, 2010.

76. Carter, J.R., Jr., Role of sulfhydryl groups in erythrocyte membrane structure, *Biochemistry* 12, 171–176, 1973.

77. Mima, J., Jung, G., and Onizuki, T. et al., Amphipathic property of free thiol group contributes to an increase in the catalytic efficiency of carboxypeptidase Y, *Eur. J. Biochem.* 269, 3220–3225, 2002.

78. Kaplan, H., Long, B.G., and Young, N.M., Chemical properties of functional groups of immunoglobulin IgA, IgG2, and IgM mouse myeloma proteins, *Biochemistry* 19, 2821–2827, 1980.

79. Shaw, C.F., 3rd, Coffer, M.T., Klingbell, J., and Mirabelli, C.K., Application of phosphorous-31 NMR chemical shift: Gold affinity correlation to hemoglobin-gold binding and the first inter-protein gold transfer reaction, *J. Am. Chem. Soc.* 110, 729–734, 1988.

80. Paul, I., Cul, J., and Maynard, E.L., Zinc binding to the HCCH motif of HIV-1 virion infectivity factor induces a conformational change that mediates protein-protein interactions, *Proc. Natl. Acad. Sci. USA* 103, 18475–18480, 2006.

81. Ingraham, R.H. and Hodges, R.S., Effects of calcium and subunit interactions on surface accessibility of cysteine residues in cardiac troponin, *Biochemistry* 27, 5891–5898, 1988.

82. Akabas, M.H., Kaufmann, C., Archdeacon, P., and Karlin, A., Identification of acetylcholine receptor channel-lining residues in the entire M2 segment of the alpha subunit, *Neuron* 13, 919–927, 1994.

83. Brown, L.S., Needleman, R., and Lanyl, J.K., Conformational changes of the E-F interhelical loops in the M photointermediate of bacteriorhodopsin, *J. Mol. Biol.* 317, 471–478, 2002.

84. Visudtiphole, V., Chalton, D.A., Hong, Q., and Lakey, J.H., Determining OMP topology by computation, surface plasmon resonance and cysteine labelling: The test case of OMPG, *Biochem. Biophys. Res. Commun.* 351, 113–117, 2006.

85. Nagler, C., Nagler, G., and Kuhn, A., Cysteine residues in the transmembrane regions of M13 procoat protein suggest that oligomeric coat proteins assemble onto phage progeny, *J. Bacteriol.* 189, 2897–2905, 2007.

86. Nair, M.S. and Dean, D.H., All domains of Cry1A toxins insert into insect brush border membranes, *J. Biol. Chem.* 283, 26324–26331, 2008.

87. Girard, F., Vachon, V., Labele, G. et al., Chemical modification of *Bacillus thuringiensis* Cry1Aa toxin single-cysteine mutants reveals the importance of domain I structural elements in the mechanism of pore formation, *Biochim. Biophys. Acta* 1788, 575–580, 2008.

88. Greeenstein, J.P., Sulfhydryl groups in proteins. III. Effect on egg albumin of various salts of guanidine, *J. Biol. Chem.* 130, 519–526, 1939.

89. Hopkins, F.G., Denaturation of proteins by urea and related substances, *Nature* 126, 328–330, 1930.

90. Putnam, F.W., Protein denaturation, in *The Proteins*, Vol. 1., Pt.B., eds. H. Neurath and K. Bailey, Chapter 9, pp. 807–892, Academic Press, New York, 1953.

91. Hopkins, F.G., CXXI. Glutathione. Its influence in the oxidation of fats and proteins, *Biochem. J.* 19, 787–819, 1925.

92. Osborne, T.B., *The Proteins of the Whole Wheat Kernel*, Carnegie Institute of Washington, Washington, DC, 1902.

93. Anson, M.L., The reactions of iodine and iodoacetamide with native egg albumin, *J. Gen. Physiol.* 23, 321–331, 1940.

94. Anson, M.L., The sulfhydryl groups of egg albumin, *J. Gen. Physiol.* 24, 399–421, 1941.

95. Bull, H.B., Protein structure, *Adv. Enzymol.* 1, 1–42, 1941.

96. Hopkins, F.G., XXXII. On an autoxidisable constituent of the cell, *Biochem. J.* 15, 286–305, 1921.

97. Callow, A.B. and Robinson, M.E., The nitroprusside reaction of bacteria, *Biochem. J.* 19, 19–24, 1925.

98. Katsuki, S., Arnold, W., Mittal, C., and Murad, F., Stimulation of guanylate cyclase by sodium nitroprusside, nitroglycerin and nitric oxide in various tissue preparations and comparison to the effects of sodium azide and hydroxylamine, *J. Cyclic Nucleotide Res.* 3, 23–35, 1977.

99. Bates, J.N., Baker, M.T., Guerra, R., Jr., and Harrison, D.G., Nitric oxide generation from nitroprusside by vascular tissue. Evidence that reduction of the nitroprusside anion and cyanide loss are required, *Biochem. Phamacol.* 43(suppl), S157–S165, 1991.

100. Foresti, R., Clark, J.E., Green, C.J., and Motterlini, R., Thiol compounds interact with nitric oxide in regulating heme oxygenase-1 induction in endothelial cells. Involvement of superoxide and peroxynitrite anions, *J. Biol. Chem.* 272, 18411–18417, 1997.

101. Vuppugalla, R. and Mehvar, R., Short-term inhibitory effects of nitric oxide on cytochrome P450-mediated drug metabolism: Time dependency and reversibility profiles in isolated perfuse rate livers, *Drug Metab. Dispos.* 32, 1446–1454, 2004.

102. Griffiths, D.P., Cummings, B.G., Greenbaum, R. et al., Cerebral blood flow and metabolism during hypotension induced with sodium nitroprusside, *Br. J. Anaesth.* 46, 671–679, 1974.

103. Chantler, P.D., Nussbacher, A., Gerstenblith, M.G. et al., Abnormalities in arterial-ventricular coupling in older healthy persons are attenuated by sodium nitroprusside, *Am. J. Physiol. Heart. Circ. Physiol.* 300, H1914–1922, 2011.

104. Fumagalli, F. and Ristagno, G., The patient is in cardiac arrest!, Let's be snappy: Prepare a bolus of sodium nitroprusside while I compress the chest. It's not a joke!, *Crit. Care. Med.* 39, 1548–1549, 2012.

105. Clark, D., Tesseneer, S., and Tribble, C.G., Nitroglycercin and sodium nitroprusside: Potential contributors to postoperative bleeding?, *Heart Surg. Forum* 15, E92–96, 2012.

106. Britto, P.J., Knipling, L., and Wolff, J., The local electrostatic environment determines cysteine reactivity of tubulin, *J. Biol. Chem.* 277, 29018–29027, 2002.

107. Parente, A., Merrifield, B., Geraci, G., and D'Allessio, G., Molecular basis of superreactivity of cysteine residues 31 and 32 of seminal ribonuclease, *Biochemistry* 24, 1098–1104, 1985.

108. Heitmann, P., A model for sulfhydryl groups in protein. Hydrophobic interactions of the side chain in micelles, *Eur. J. Biochem.* 3, 346–350, 1968.

109. Gitler, C., Zarmi, B., and Kalef, E., Use of cationic detergents to enhance reactivity of protein sulfhydryls, *Methods Enzymol.* 251, 366–375, 1995.

110. Lutolf, M.P., Tirelli, N., Cerritelli, S. et al., Systematic modulation of Michael Type reactivity of thiols through the use of charged amino acids, *Bioconjug. Chem.* 12, 1051–1056, 2001.

111. Snyder, G.H., Cennerazzo, M.J., Karalis, A.J., and Field, D., Electrostatic influence of local cysteine environments on disulfide exchange kinetics, *Biochemistry* 20, 6509–6519, 1981.

112. Simpson, R.B., Association constants of methylmercury with sulfhydryl and other bases, *J. Am. Chem. Soc.* 83, 4711–4717, 1961.

113. Crespo-López, M.E., Macêdo, G.L., Pereira, S.I.D. et al., Mercury and human genotoxicity: Critical considerations and possible molecular mechanisms, *Pharmacol. Res.* 60, 212–220, 2009.

114. *Toxicological Effects of Methylmercury*, National Research Council, National Academies of Science, Washington, DC, 2000.

115. Keller, J.R., Bravo, D.T., and Parsons, S.M., Modification of cysteines reveals linkage to acetylcholine and vesamicol binding sites in the vesicular acetylcholine transporter of *Torpedo californica*, *J. Neurochem.* 74, 1739–1748, 2000.

116. Keller, J.E. and Parsons, S.M., Diffusion pathways to critical cysteines in the vesicular acetylcholine transporter of *Torpedo*, *Neurochem. Res.* 26, 477–482, 2003.

117. Kornreich, W.D. and Parsons, S.M., Sidedness and chemical and kinetic properties of the vesamicol receptor of cholinergic synaptic vesicles, *Biochemistry* 27, 5262–5267, 1988.

118. Velick, S.F., Coenzyme binding and the thiol groups of glyceraldehyde-3-phosphate dehydrogenase, *J. Biol. Chem.* 203, 563–573, 1953.

119. Li, C.-P., Hayashi, Y., Shinohara, H. et al., Phosphorylation of ovalbumin by dry-heating in the presence of pyrophosphate—Effects of protein structure and some properties, *J. Agric. Food Chem.* 53, 4962–4967, 2005.

120. Diez, M.J.F., Osuga, D.T., and Feeney, R.E., The sulfhydryls of avian ovalbumins, bovine β-lactoglobulin, and bovine serum albumin, *Arch. Biochem. Biophys.* 107, 449–458, 1964.

121. Webster, D.M. and Thompson, E.O., Carboxymethylation of thiol groups in ovalbumin: Implications for proteins that contain both thiol and disulfide groups, *Aust. J. Biol. Sci.* 35, 125–135, 1982.

122. Takahashi, N. and Hirose, M., Determination of sulfhydryl groups and disulfide bonds in a protein by polyacrylamide gel electrophoresis, *Anal. Biochem.* 188, 359–365, 1990.

123. Takatera, K. and Watanabe, T., Determination of sulfhydryl groups in ovalbumin by high-performance liquid chromatography with inductively coupled plasma mass spectrometric detection, *Anal. Chem.* 65, 3644–3646, 1993.

124. Bramanti, E., D'Ulivo, A., Lampugnani, L. et al., Application of mercury cold vapor atomic fluorescence spectrometry to the characterization of mercury-accessible -SH groups in native proteins, *Anal. Biochem.* 274, 163–173, 1999.

125. McKenzie, H.A., Smith, M.B., and Wake, R.G., The denaturation of proteins. I. Sedimentation, diffusion optical rotation, viscosity, and gelation in urea solutions of ovalbumin and bovine serum albumin, *Biochim. Biophys. Acta* 69, 222–239, 1963.

126. Batra, P.P., Sasa, K., Ueki, T., and Takeda, K., Circular dichroism study of conformational changes in ovalbumin induced by modification of sulfhydryl groups and disulfide reduction, *J. Protein. Chem.* 8, 609–617, 1989.

127. Takeda, K., Shigemura, A., Hamada, S. et al., Dependence of reaction rate of 5,5′-dithiobis-2-(nitrobenzoic acid) to free sulffydryl groups of bovine serum albumin and ovalbumin on the protein conformation, *J. Protein. Chem.* 11, 187–192, 1992.

128. Yang, Y., Wang, H.-R., Xu, J.-X., and Zhou, H.-M., Kinetics of modification of the mitochondrial succinate-ubiquinone reductase by 5,5′-dithiobis-(2-nitrobenzoic acid), *J. Protein Chem.* 15, 169–176, 1996.

129. Sluyterman, L.A.Æ, Wijdnes, J. and Voorn, G., Dimerization of papain induced by mercuric chloride and a bifunctional organic mercurial, *Eur. J. Biochem.* 77, 107–111, 1977.

130. Barone, L.M., Shih, C., and Wasserman, B.P., Mercury-induced conformational changes and identification of conserved surface loops in plasma membrane aquaporins from higher plants. Topology of PMIP31 from *Beta vulgaris* L, *J. Biol. Chem.* 272, 30672–30677, 1997.

131. Lin, J.C., Xie, X.L., Gong, M. et al., Effects of mercuric ion on the conformation and activity of *Penaeus vannamei* β-N-acetyl-d-glucosaminidase, *Int. J. Biol. Macromol.* 36, 327–330, 2005.

132. Frasco, M.F., Colletier, J.P., Weik, M. et al., Mechanisms of cholinesterase inhibition by inorganic mercury, *FEBS J.* 274, 1849–1861, 2007.

133. Whitmore, F.C., *Organic Compounds of Mercury,* American Chemical Society, The Chemical Catalogue Company, New York, 1921.
134. Boyer, P.D., Spectrophotometric study of the reaction of protein sulfhydryl groups with organic mercurials, *J. Am. Chem. Soc.* 76, 4331–4437, 1954.
135. Riordan, J.F. and Vallee, B.L., Reactions with *N*-ethylmaleimide and *p*-chlorobenzoate, *Methods Enzymol.* 11, 541–548, 1967.
136. Hasinoff, B.B., Madsen, N.B., and Avramovic-Zikic, O., Kinetics of the reaction of *p*-chloromercuribenzoate with the sulfhydryl groups of glutathione, 2-mercaptoethanol, and phosphorylase *b*, *Can. J. Biochem.* 49, 742–751, 1971.
137. Bai, Y. and Hayashi, R., Properties of the single sulfhydryl group of carboxypeptidase Y. Effects of alkyl and aromatic mercurials on activities toward various synthetic substrates, *J. Biol. Chem.* 254, 8473, 1979.
138. Bednar, R.A., Fried, W.B., Lock, Y.W., and Pramanik, B., Chemical modification of chalcone isomerase by mercurials and tetrathionate. Evidence for a single cysteine residue in the active site, *J. Biol. Chem.* 264, 14272, 1989.
139. Ojcius, D.M. and Solomon, A.K., Sites of p-chloromercuribenzenesulfonate inhibition of red cell urea and water transport, *Biochim. Biophys. Acta* 942, 73, 1988.
140. Clark, S.J. and Ralston, G.B., The dissociation of peripheral proteins from erythrocyte membranes brought about by p-mercuribenzenesulfonate, *Biochim. Biophys. Acta* 1021, 141, 1990.
141. Pode, L.B., and Claiborne, A., The non-flavin redox center of the Streptococcal NADH peroxidase. I. Thiol-reactivity and redox balance in the presence of urea, *J. Biol. Chem.* 264, 12322, 1989.
142. Yao, S.Y.M. et al., Identification of cys[140] in helix 4 as an exofacial cysteine residue within the substrate-translocation channel of rat equilibrative nitrobenzylthioinosine (NBMPR)-insensitive nucleoside transporter rENT2, *Biochem. J.* 353, 387–393, 2001.
143. Fann, M.C., Busch, A., and Maloney, P.C., Functional characterization of cysteine residues in G1pT, the glycerol 3-phosphate transporter of *Escherichia coli*, *J. Bacteriol.* 185, 3863–3870, 2003.
144. Ding, Z. et al., Inactivation of the human P2Y$_{12}$ receptor by thiol reagents requires interaction with both extracellular cysteine residues, Cys17 and Cys 270, *Blood* 101, 3908–2914, 2003.
145. Jagtap, S. and Rao, M., Conformation and microenvironment of the active site of a low molecular weight 1,4-β-D-glucan glucanohydrolase from an alkalothermophilic *Thermomonospora* sp.: Involvement of lysine and cysteine residues, *Biochem. Biophys. Res. Commun.* 347, 428–432, 2006.
146. Kutscher, D.J., Estele del Castillo Busto, M., Zinn, N. et al., Protein labelling with mercury tags: Fundamental studies on ovalbumin derivatised with *p*-hydroxymercuribenzoic acid (pHMB), *J. Anal. At. Spectrom.* 23, 1359–1364, 2008.
147. Kutshcer, D.J. and Bettmer, J., Absolute and relative protein quantification with the use of isotopically labeled *p*-hydroxymercuribenzoic acid and complementary MALDI-MS and ICPMS detection, *Anal. Chem.* 81, 9172–9177, 2009.
148. Lu, M., Li, X.-F., Le, X.C. et al., Identification and characterization of cysteinyl exposure in proteins by selective mercury labeling and nano-electrospray ionization quadrupole time-of-flight mass spectrometry, *Rapid Commun. Mass Spectrom.* 24, 1523–1532, 2010.
149. Block, M.R., and Rothman, J.E., Purification of *N*-ethylmaleimide-sensitive fusion protein, *Methods Enzymol.* 219, 300–309, 1992.
150. Whiteheart, S.W., Rossnagel, K., Buhrow, S.A. et al., *N*-Ethylmaleimide-sensitive fusion protein: A trimeric ATPase whose hydrolysis of ATP is required for membrane fusion, *J. Cell Biol.* 126, 946–954, 1994.
151. da Silva Cunha, K.C., Fuly, A.L., and de Araujo, E.G., A phospholipase A$_2$ isolated from *Lachesis muta* snake venom increases the survival of retinal ganglion cells in vitro, *Toxicon* 57, 580–585, 2011.
152. Dames, J.E. and Cake, M.H., An albumin-associated PLA$_2$-like activity inactivates surfactant phosphatidylcholine secreted from fetal type II pneumocytes, *Am. J. Physiol. Lung Cell Mol. Physiol.* 301, L966–L974, 2011.
153. Gorman, M.J., Wang, Y., Jiang, H., and Kanost, M.R., *Manduca sexta* hemolymph proteinase 21 activates prophenoloxidase-activating proteinase 3 in an insect innate immune response proteinase cascade, *J. Biol. Chem.* 282, 11742–11749, 2007.
154. Tsang, J.L., Parodo, J.C., and Marshall, J.C., Regulation of apoptosis and priming of neutrophil oxidative burst by diisopropyl fluorophosphate, *J. Inflamm.* 7(7), 32, 2010.
155. Azzo, W. Woessner, J.F., Jr., Purification and characterization of an acid metalloproteinase from human articular cartilage, *J. Biol. Chem.* 261, 5434–5441, 1986.
156. Murphy, G., Ward, R., Hambry, R.M. et al., Characterization of gelatinase from pig polymorphonuclear leukocytes. A metalloproteinase resembling tumor type IV collagenase, *Biochem. J.* 258, 463–472, 1989.

157. Freimark, B.D., Fenser, W.S., and Rosenfeld, S.A., Multiple sites of the propeptide region of human stromelysin-1 are required for maintaining a latent form of the enzyme, *J. Biol. Chem.* 269, 26962–26988, 1994.

158. Sanderson, M.P., Keller, S., Alonso, A. et al., Generation of novel, secreted epidermal growth factor receptor (EGFR/ErbB1) isoforms via metalloprotease-dependent ectodomain shedding and exosome secretion, *J. Cell. Biochem.* 103, 1783–1797, 2008.

159. Pavlaki, N, Giannopoulou, E., Niariakis, A. et al., Walker256 (W256) cancer cells secrete tissue inhibitor of metalloproteinase-free metalloproteinase-9, *Mol. Cell. Biochem.* 328, 189–199, 2009.

160. Uitto, V.-J., Turto, H., Huttunen, A. et al., Activation of human leukocyte collagenase by compounds reacting with sulfhydryl groups, *Biochim. Biophys. Acta* 613, 168–177, 1980.

161. Shapiro, S.D., Fliszar, C.J., Broekelmann, T.J. et al., Activation of the 93-kDa gelatinase by stromelysin and 4-aminophenylmercuric acetate. Differential processing and stabilization of the carboxyl-terminal domain by tissue inhibitors of metalloproteinases (TIMP), *J. Biol. Chem.* 270, 6351–6356, 1995.

162. Galazka, G., Windsor, L.J., Birkedal-Hansen, H., and Engler, J.A., APMA (4-aminophenylmercuric acetate) activation of stromelysin-1 involves protein interactions in addition to those with cysteine-75 in the propeptide, *Biochemistry* 35, 1121–11227, 1996.

163. Grierson, C., Miller, D., LaPan, P. et al., Utility of combining MMP-9 enzyme-linked immunosorbent assay and MMP-9 activity assay data to monitor enzyme specific activity, *Anal. Biochem.* 404, 232–234, 2010.

164. Huse, K., Wippich, P., Gutknecht, D. et al., Purification of aminophenyl mercuryacetate-activated human matrix metalloproteinase 1 and removal of the organomercurial in a single-step chromatography, *Bioseparation* 7, 281–286, 1999.

165. Pei, P., Horan, M.P., Hille, R. et al., Reduced nonprotein thiols inhibit activation and function of MMP-9: Implications for chemoprevention, *Free Radic. Biol. Med.* 41, 1315–1324, 2006.

166. McMurray, C.H. and Trentham, D.R., A new class of chromophoric organomercurials and their reactions with D-glyceraldehyde 3-phosphate dehydrogenase, *Biochem. J.* 115, 913–921, 1969.

167. Sanyal, G. and Khalifah, R.G., Kinetics of organomercurial reactions with model thiols: Sensitivity to exchange of the labile ligand, *Arch. Biochem. Biophys.* 196, 157–164, 1979.

168. Khalifah, R.G., Sanyal, G., Strader, D.J., and Sutherland, W.M., Facile ligand exchange of organomercurials alters their reactivity toward protein and other thiols, *J. Biol. Chem.* 254, 602–604, 1979.

169. Marshall, M. and Cohen, P.P., The essential sulfhydryl group of ornithine transcarbamylases—pH dependence of the spectra of its 2-mercuri-4-nitrophenol derivative, *J. Biol. Chem.* 255, 7296–7300, 1980.

170. Scawen, M.D., Hewitt, E.J., and James, D.M., Preparation, crystallization and properties of *Cucurbita pepo* plastocyanin and ferredoxin, *Phytochemistry* 14, 1225–1233, 1975.

171. Katoh, S., Shiratori, I., and Takamiya, A., Purification and some properties of spinach plastocyanin, *J. Biochem.* 51, 32–40, 1962.

172. Stinson, R.A., Chromophoric labeling of yeast 3-phosphoglycerate kinase with an organomercurial, *Biochemistry* 13, 4523–4529, 1974.

173. Simpson, R.A., Use of isoelectric focusing and a chromophoric organomercurial to monitor urea-induced conformational changes of yeast phosphoglycerate kinase, *Biochem. J.* 167, 65–70, 1977.

174. Watts, D.C., Rabin, B.R., and Crook, E.M., The number of catalytic sites in creatine phosphokinase as determined by a study of its reactive sulfhydryl groups, *Biochem. J.* 82, 417–417, 1962.

175. Milner-White, E.J. and Watts, D.C., Inhibition of adenosine 5′-triphosphate-creatine phosphotransferase by substrate-anion complexes, *Biochem. J.* 122, 727–740, 1971.

176. Quiocho, F.E. and Thomson, J.W., Substrate binding to an active creatine kinase with a thiol-bound mercurinitrophenol chromophoric probe, *Proc. Natl. Acad. Sci. USA* 70, 2858–2862, 1973.

177. Quiocho, F.A. and Olson, J.S., The reaction of creatine kinase with 2-chloromercuri-4-nitrophenol, *J. Biol. Chem.* 249, 5885–5888, 1974.

178. Somerville, L.L. and Quiocho, F.A., The interaction of tetraiodofluorescein with creatine kinase, *Biochim. Biophys. Acta* 481, 493–499, 1977.

179. Wu, H., Yao, Q.-Z., and Tsou, C.-L., Creatine kinase is modified by 2-chloromercuri-4-nitrophenol at the active site thiols with complete inactivation, *Biochim. Biophys. Acta* 997, 78–82, 1989.

180. Wang, P.-F., McLeish, M.J., Kneen, M.M. et al., An usually low pK_a for Cys282 in the active site of human muscle creatine kinase, *Biochemistry* 40, 11628–11703, 2001.

181. Hou, L.X. and Zhou, J.X., Creatine kinase: The characteristics of the enzyme regenerated from the thiomethylated creatine kinase reflect a differentiation in function between the two reactive thiols, *Biochemie* 78, 219–226, 1996.

182. Yang, Y. and Zhou, H.-M., Reactivation kinetics of 5,5′-dithiobis-(2-nitrobenzoic acid)-modified creatine kinase reactivated by dithiothreitol, *Biochim. Biophys. Acta* 1388, 190–198, 1998.

183. Houston, L.L. and Walsh, K.A., The transient inactivation of trypsin by mild acetylation with *N*-acetylimidazole, *Biochemistry* 9, 156–166, 1970.
184. Lundblad, R.L., The reaction of bovine thrombin with *N*-butyrylimidazole. Two different reactions resulting in the inhibition of catalytic activity, *Biochemistry* 14, 1033–1037, 1975.
185. Yang, D.S., Burgum, A.A., and Matthews, K.S., Modification of the cysteine residues of the lactose receptor protein using chromophoric probes, *Biochim. Biophys. Acta* 493, 24–36, 1977.
186. Friedman, B.E., Olson, J.S., and Matthews, K.S., Kinetics studies of inducer binding to lactose repressor protein, *J. Biol. Chem.* 251, 1171–1174, 1976.
187. Samsm, C.F., Friedman, B.E., Burgum, A.A. et al., Spectral studies of lactose repressor protein modified with nitrophenol reporter groups, *J. Biol. Chem.* 252, 3153–3159, 1977.
188. Wang, H.-R., Bai, J.-H., Zheng, S.-Y. et al., Ascertaining the number of essential thiol groups for the folding of creatine kinase, *Biochem. Biophys. Res. Commun.* 221, 174–180, 1996.
189. Baines, B.S. and Brocklehurst, K., A thiol-labelling reagent and reactivity probe containing electrophilic mercury and a chromophoric leaving group, *Biochem. J.* 179, 701–704, 1979.
190. Scott-Ennis, R.J. and Noltmann, E.A., Differential response of cysteine residues in pig muscle phospho-glucose isomerase to seven sulfhydryl modifying reagents, *Arch. Biochem. Biophys.* 239, 1–11, 1985.
191. Dahl, K.S. and McKinley-McKee, J.S., The reactivity of affinity labels: A kinetic study of the reaction of alkyl halides with thiolate anions—A model reaction for protein alkylation, *Bioorg. Chem.* 10, 329–341, 1981.
192. Proudfoot, A.T., Bradberry, S.M., and Vale, J.A., Sodium fluoroacetate poisoning, *Toxicol. Rev.* 25, 213–217, 2006.
193. Tsuji, H., Shimizu, H., Dote, T. et al., Effects of sodium monofluoroacetate on glucose, amino-acid, and fatty-acid metabolism and risk assessment of glucose supplementation, *Drug. Chem. Toxicol.* 32, 353–361, 2009.
194. Kulling, P., Andersson, H., Boström, K. et al., Fatal systemic poisoning after skin exposure to monochloroacetic acid, *J. Toxicol. Clin. Toxicol.* 30, 643–652, 1992.
195. Pirson, J., Toussaint, P., Sejers, N., An unusual cause of burn injury: Skin exposure to monochloroacetic acid, *J. Burn Care Rehabil.* 24, 407–409, 2003.
196. Steinauer, E., *Arch. Pathol. Anat. Physiol.* 59, 65, 1874, cited in Webb, J.L., *Enzyme and Metabolic Inhibitors*, Vol. III, Chapter 1, pp. 1–2, Academic Press, New York, 1966. (Format from *The Chicago Manual Of Style*)
197. Lundsgaard, E. [Further studies on muscle contraction without lactic acid formation] *Biochem. Z.* 217, 162–177, 1930. (brackets indicate translation from original copy)
198. Webb, J.L., Iodoacetate and iodoacetamide, in *Enzyme and Metabolic Inhibitors*, Vol. III, Chapter 1, pp. 1–283, Academic Press, New York, 1966.
199. Syvertsen, C. and McKinly-McKee, J.S., Binding of ligands to the catalytic zinc in the horse liver alcohol dehydrogenase, *Arch. Biochem. Biophys.* 299, 159–169, 1984.
200. Plapp, B.V., Application of affinity labeling for studying structure and function of enzymes, *Methods Enzymol.* 87, 469–499, 1982.
201. Kallis, G.-B. and Holmgren, A., Differential reactivity of the functional sulfhydryl groups of cysteine-32 and cysteine-35 present in the reduced form of thioredoxin from *Escherichia coli*, *J. Biol. Chem.* 255, 10261–10265, 1980.
202. Hawkins, H.C. and Freedman, R.B., The reactivities and ionization properties of the active-site dithiol groups of mammalian protein disulphide-isomerase, *Biochem. J.* 275, 335–339, 1991.
203. Chang, Z.Y., Zhang, J.E., Ballou, D.P., and Williams, C.H., Reactivity of thioredoxin as a protein disulfide oxidoreductase, *Chem. Rev.* 111, 5768–5783, 2011.
204. Mikami, B., Aibara, S., and Morita, Y., Chemical modification of sulfhydryl groups in soybean α-amylase, *J. Biochem.* 88, 103–111, 1980.
205. Marchitti, S.A., Brocker, C., Stagos, D. et al., Non-P450 aldehyde oxidizing enzymes: The aldehyde dehydrogenase superfamily, *Expert. Opin. Drug. Metab. Toxicol.* 4, 697–720, 2008.
206. Song, B.J., Abdelmegeed, M.A., Yoo, S.H. et al., Post-translational modifications of mitochondrial aldehyde dehydrogenase and biomedical implications, *J. Proteomics* 74, 2691–2702, 2011.
207. Hempel, J., Kaiser, R., and Jörnvall, H., Mitochondrial aldehyde dehydrogenase from human liver. Primary structure, differences in relation to the cytosolic enzyme, and functional correlations, *Eur. J. Biochem.* 153, 13–28, 1985.
208. Kitson, T.M., Effect of some thiocarbamate compounds on aldehyde dehydrogenase and implications for the disulfiram ethanol reaction, *Biochem. J.* 278, 189–192, 1991.

209. Kitson, T.M. and Kitson, K.E., Studies of the esterase activity of cytosolic aldehyde dehydrogenase with resorufin acetate as substrate, *Biochem. J.* 322, 701–708, 1997.

210. Strickland, K.D., Krupenko, N.I., Dubard, M.E. et al., Enzymatic properties of ALDH1L2, mitochondrial 10-formyltetrahydrofolate dehydrogenase, *Chem. Biol. Interact* 191, 129–136, 2011.

211. Hempel, J.D. and Pietruszko, R., Selective chemical modification of human liver aldehyde and dehydrogenases E_1 and E_2 by iodoacetamide, *J. Biol. Chem.* 256, 10889–10996, 1981.

212. Hempel, J., Pietruszko, R., Fietzek, P., and Jörnvall, H., Identification of a segment containing a reactive cysteine residue in human liver cytoplasmic aldehyde dehydrogenase (Isoenzyme E_1), *Biochemistry* 21, 6834–6838, 1982.

213. von Bahr-Lindström, H., Hempel, J., and Jörnvall, H., The cytoplasmic isoenzyme of horse aldehyde dehydrogenase. Relationship to the corresponding human isoenzyme, *Eur. J. Biochem.* 141, 37–42, 1984.

214. Mackerell, A.D., MacWright, R.S., and Pietruszko, R., Bromoacetophenone as an affinity reagent for human-liver aldehyde dehydrogenase, *Biochemistry* 25, 5182–5189, 1986.

215. Abriola, D.P., Fields, R., Stein, S. et al., Active site of human liver aldehyde dehydrogenase, *Biochemistry* 26, 5679–5684, 1987.

216. Abriola, D.P., MacKerrell, A.D., Jr., and Pietruszko, R., Correlation of loss of activity of human aldehyde dehydrogenase with reaction of bromoacetophenone with Glutamic acid-268 and cysteine-302 residues, *Biochem. J.* 266, 179–187, 1990.

217. Dryjanski, M., Kosley, L.L., and Pietruszko, R., *N*-Tosyl-L-phenylalanine chloromethyl ketone: A serine protease inhibitor, identifies glutamate 398 as the coenzyme-binding site of human aldehyde dehydrogenase. Evidence for a second "naked anion" at the active site, *Biochemistry* 37, 14151–14156, 1998.

218. Takahashi, K., Stein, W.H., and Moore, S., The identification of a glutamic acid residue as part of the active site of ribonuclease T-1, *J. Biol. Chem.* 242, 4682–4690, 1967.

219. Takahashi, K., Structure and function of ribonuclease T1. X. Reactions of iodoacetate, iodoacetamide, and related alkylating agents with ribonuclease T1, *J. Biochem.* 68, 5170527, 1970.

220. Erlanger, B.F., Vratsanos, S.M., Wasserman, N., and Cooper, A.G., Specific and reversible inactivation of pepsin, *J. Biol. Chem.* 240, PC3447–PC3448, 1965.

221. Erlanger, B.F., Vratsanos, S.M., Wasserman, N., and Cooper, A.G., A chemical investigation of the active center of pepsin, *Biochem. Biophys. Res. Commun.* 23, 243–245, 1966.

222. Johansson, J., Fleetwood, L., and Jörnvall, H., Cysteine reactivity in sorbitol and aldehyde dehydrogenases. Differences towards the pattern in alcohol dehydrogenases, *FEBS Lett.* 202, 1–3, 1992.

223. Mukerjee, N., Dryjanski, M., Dai, W. et al., Haloenol lactones as inactivators and substrates of aldehyde dehydrogenase, *J. Protein Chem.* 15, 639–648, 1996.

224. Blatter, E.E., Tasayco, M.L., Prestwich, G. and Pietruszko, R., Chemical modification of aldehyde dehydrogenase by a vinyl ketone analogue of an insect pheromone, *Biochem. J* 272, 351–358, 1990.

225. Blatter, E.E., Abriola, D.P., and Pietruszko, R., Aldehyde dehydrogenase: Covalent intermediate in aldehyde dehyrogenase and ester hydrolysis, *Biochem. J.* 282, 353–360, 1992.

226. Tu, G.C. and Weiner, H., Evidence for two distinct active sites on aldehyde dehydrogenase, *J. Biol. Chem.* 263, 1218–1222, 1988.

227. Kollau, A., Hofer, A., Russwurm, M. et al., Contribution of aldehyde dehydrogenase to mitochondrial activation of nitroglycerin: Evidence for the activation of purified soluble guanylate cyclase through direct formation of nitric oxide, *Biochem. J.* 385, 769–777, 2005.

228. Wenzl, M.V., Beretta, M., Gorren, A.C. et al., Role of the general base Glu-268 in nitroglycerin bioactivation and superoxide formation by aldehyde dehydrogenase-2, *J. Biol. Chem.* 284, 19878–19886, 2009.

229. Wenzl, M.V., Beretta, M., Griesberger, M. et al., Site-directed mutagenesis of aldehyde dehydrogenase-2 suggests three distinct pathways of nitroglycerin biotransformation, *Mol. Pharmacol.* 80, 258–266, 2011.

230. Tsou, P.S., Page, N.A., Lee, S.G. et al., Differential metabolism of organic nitrates by aldehyde dehydrogenase 1a1 and 2: Substrate selectivity, enzyme inactivation, and active cysteine sites, *AAPS J.* 13, 548–555, 2011.

231. Beretta, M., Wölkart, G., Schernthaner, M. et al., Vascular bioactivation of nitroglycerin is catalyzed by cytosolic aldehyde dehydrogenase -2, *Circ. Res.* 110, 385–393, 2012.

232. Garcia-Bou, R., Rocha, M., Apostolova, N. et al., Evidence for a relationship between mitochondrial Complex I activity and mitochondrial aldehyde dehydrogenase during nitroglycerin tolerance, *Biochim. Biophys. Acta* 1817, 828–837, 2012.

233. Marcato, P., Dean, C.A., Giacomantonio, C.A., and Lee, P.W., Aldehyde dehydrogenase: Its role as a cancer stem cell marker comes down to the specific isoform, *Cell Cycle* 10, 1378–1384, 2011.

234. Marchitti, S.A., Chen, Y., Thompson, D.C., and Vasiliou, V., Ultraviolet radiation: Cellular antioxidant response and role of ocular aldehyde dehydrogenase enzymes, *Eye Contact Lens* 37, 206–213, 2011.

235. Muzio, G., Maggiora, M., Paiuzzi, E. et al., Aldehyde dehydrogenases and cell proliferation, *Free Radic. Biol. Med.* 52, 735–746, 2012.

236. Burgos-Ojeda, D., Rueda, B.R., and Buckanovich, R.J., Ovarian cancer stem cell markers: Prognostic and therapeutic implications, *Cancer Lett.* 322, 1–7, 2012.

237. Schmitt, F., Ricardo, S., Vieira, A.F. et al., Cancer stem cell markers in breast neoplasias: Their relevance and distribution in distinct molecular subtypes, *Virchows Arch.* 460, 545–553, 2012.

238. Hald, J., Jacobsen, E., and Larsen, V., The sensitizing effect of tetraethylthiuramdisulphide (Antabuse) to ethyl alcohol, *Acta Pharmacol.* 4, 285–296, 1948.

239. Hald, J., Jacobsen, E., and Larsen, V., Formation of acetaldehyde in the organism in relation to dosage of antabuse(tetraethylthiuramdisulphide) and to alcohol-concentration in blood, *Acta Pharmacol.* 5, 179–188, 1979.

240. Eldjarn, L., The metabolism of tetraethylthiuramdisulphide (antabuse; aversan) in man, investigated by means of radioactive sulphide, *Scand. J. Clin. Lab. Invest.* 2, 202–208, 1950.

241. Hart, B.W. and Faiman, M.D., In vitro and in vivo inhibition of rat liver aldehyde dehydrogenase by *S*-methyl-*N,N*-diethylthiolcarbamate, *Biochem. Pharmacol.* 43, 403–406, 1992.

242. Mays, D.C., Tomlinson, A.J., Johnson, K.L. et al., Inhibition of human mitochondrial aldehyde dehydrogenase by metabolites of disulfiram and structural characterization of the enzyme adduct by HPLC-tandem mass spectrometry, *Adv. Exp. Med. Biol.* 463, 61–70, 1999.

243. Peachey, J.E., Brien, J.F., Roach, C.A., and Loomis, C.W., A comparative review of the pharmacological and toxicological properties of disulfiram and calcium carbimide, *J. Clin. Psychopharmacol.* 1, 21–26, 1981.

244. Vallari, R.C. and Pietruszko, R., Human aldehyde dehydrogenase: Mechanism of inhibition by disulfiram, *Science* 216, 637–639, 1982.

245. Holtzer, A., Phenomenological analysis of the kinetics of the production of interchain disulfide cross-links in two-chain, coiled-coil proteins by reaction with 5,5′-dithiobis (2-nitrobenzoate), *Biochemistry* 25, 3008–3012, 1986.

246. Rajasekharan, K.N. and Burke, M., Trypsinolysis promotes disulfide formation between 21- and 50-kilodalton segments of myosin subfragment 1 during reaction with 5,5′-dithiobis(2-nitrobenzoic acid), *Biochemistry* 28, 6473–6477, 1989.

247. Huber, P.J., Brunner, U.T., and Schaub, M.C., Disulfide formation within the regulatory light chain of skeletal muscle myosin, *Biochemistry* 28, 9116–9123, 1989.

248. Mackerell, A.D., Vallari, R.C., and Pietruszko, R., Human mitochondrial aldehyde dehydrogenase inhibition by diethyldithiocarbamic acid methanethiol disulfide: A derivative of disulfiram, *FEBS Lett.* 179, 71–81, 1985.

249. Sanny, C.G. and Weiner, H., Inactivation of horse liver mitochondrial aldehyde dehydrogenase by disulfiram, *Biochem. J.* 242, 499–503, 1987.

250. Hochreiter, J., McCance-Katz, E.F., Lapham, J. et al., Disulfiram metabolite S-methyl-N,N-diethyldithiocarbamate quantitation in human plasma with reverse phase ultra performance liquid chromatography and mass spectrometry, *J Chromatogr B.* 897, 80–84, 2012.

251. Nobel, C.S.I., Kimland, M., Nicholson, D.W. et al., Disulfiram is a potent inhibitor of proteases of the caspase family, *Chem. Res. Toxicol.* 10, 1319–1324, 1997.

252. Chipinda, I., Hettick, J.M., Simoyi, R.H., and Siegel, P.D., Zinc diethyldithiocarbamate allergenicity: Potential haptenation mechanism, *Contact Dermatitis* 59, 79–89, 2008.

253. Estrada-Sanchez, A.M., Montiel, T., and Massieu, L., Glycolysis inhibition decreases the glutamate transporters and enhances glutamate neurotoxicity in the R6/2 Huntington's disease mice, *Neurochem. Res.* 35, 1156–1163, 2010.

254. Kikuchi, K., Yamada, T., and Sugi, H., Effects of adrenaline on glycogenolysis in resting anaerobic frog muscles studied by ^{31}P-NMR, *J. Physiol. Sci.* 59, 439–446, 2009.

255. Schmidt, M.M. and Dringen, R., Differential effects of iodoacetamide and iodoacetate on glycolysis and glutathione metabolism of cultured astrocytes,. *Front. Neuroenergetics* 1, 1, 2009.

256. Wartenberg, M., Richter, M., Datchev, A. et al., Glycolytic pyruvate regulates P-glycoprotein expression in multicellular tumor spheroids via modulation of the intracellular redox state, *J. Cell. Biochem.* 109, 434–446, 2010.

257. Bhardwaj, V., Vikas, N., Lai, M.B. et al., Glycolytic enzyme inhibitors affect pancreatic cancer survival by modulating its signaling and energetics, *Anticancer Res.* 30, 743–750, 2010.

258. Rodriguez, S.B., Stitt, B.L., and Ash, D.E., Cysteine 351 is an essential nucleophile in catalysis by *Porphyromonas gingivalis* peptidylarginine deiminase, *Arch. Biochem. Biophys.* 504, 190–196, 2010.

259. Wang, S. and Kallashov, I.A., A new strategy of using o^{18}-labeled iodoacetic acid for mass spectrometry-based protein quantitation, *J. Am. Soc. Mass Spectrom.* 23, 1293–1297, 2012.

260. Bar-Yehuda, S., Rath-Wolfson, L., Del Valle, L. et al., Induction of an antiinflammatory effect and prevention of cartilage damage in rat knee osteoarthritis by CF101 treatment, *Arthritis Rheum.* 60, 3061–3071, 2009.

261. Al-Saffar, R.J., Ganabadi, S., Fakurazi, S. et al., Chondroprotective effect of zerumbone on monosodium iodoacetate induced osteoarthritis in rats, *J. Appl. Sci.* 10, 248–260, 2010.

262. Sniekers, Y.H., van Osch, G.J.V.M., Jahr, H. et al., Estrogen modulates iodoacetate-induced gene expression in bovine cartilage explants, *J. Orthop. Res.* 28, 607–615, 2010.

263. Kalff, K.-M., El Mouedden, M., van Egmond, J. et al., Pre-treatment with capsaicin in rat osteoarthritis model reduces the symptoms of pain and bone damage induced by monosodium acetate, *Eur. J. Pharmacol.* 64, 106–113, 2010.

264. Sniekers, Y.H., Weinans, H., van Osch, J.V.M. et al., Oestrogen is important for maintenance of cartilage and subchondrial bone in a murine model of knee osteoarthritis, *Arthritis Res. Ther.* 12, R182, 2010.

265. Sagar, D.R., Staniaszek, L.E., Okine, B.N. et al., Tonic modulation of spinal hyperexcitability by the endocannabinoid receptor system in a rat model of osteoarthritis pain, *Arthritis Rheum.* 6, 3666–3676, 2010.

266. Ahmed, A.D., Li, J., Erlandsson-Harris, H. et al., Suppression of pain and joint destruction by inhibition of the proteasome system in experimental osteoarthritis, *Pain* 153, 18–26, 20102.

267. Nagase, H., Kumakura, S., and Shimada, K., Establishment of a novel objective and quantitative method to assess pain-related behavior in monosodium iodoacetate-induced osteoarthritis in rat knee, *J. Pharmacol. Toxicol. Methods* 65, 29–36, 2012.

268. Xie, L.Q., Lin, A.S., Kundu, K. et al., Quantitative imaging of cartilage and bone morphology: Reactive oxygen species, and vascularization in a rodent model of osteoarthritis, *Arthritis Rheum.* 64, 1899–1908, 2012.

269. Lin, X., Shanmugasundaram, S., Liu, Y. et al., B2A peptide induces chronogenic differentiation in vitro and enhances cartilage repair in rats, *J. Orthop. Res.* 30, 1221–1228, 2012.

270. Paunovic, B., Deng, X., Khomenko, T. et al., Molecular mechanisms of basic fibroblast growth factor effect on healing of ulcerative colitis in rats, *J. Pharmacol. Exp. Ther.* 339, 430–437, 2011.

271. Tolstanova, G., Deng, X., French, S.W. et al., Early endothelial damage and increased colonic vascular permeability in the development of experimental ulcerative colitis in rats and mice, *Lab. Invest.* 92, 9–21, 2012.

272. Seifried, S.E., Wang, Y., and Von Hippel, P.H., Fluorescent modification of the cysteine 202 residue of *Escherichia coli* transcription termination factor rho, *J. Biol. Chem.* 263, 13511–13514, 1988.

273. Chumsae, C., Gaza-Bulseco, G., and Liu, H., Identification and localization of unpaired cysteine residues in monoclonal antibodies by fluorescence labeling and mass spectrometry, *Anal. Chem.* 81, 6449–647, 2009.

274. Zinellu, A., Lepedda, A., Jr., Sorgia, S. et al., Albumin-bound low molecular weight thiols analysis in plasma and carotid plaques by CE, *J. Sep. Sci.* 33, 136–131, 2010.

275. Smith, C., Mehta, R., Gibson, D.E. et al., Characterization of a recombinant form of annexin VI for detection of apoptosis, *Bioconjug. Chem.* 212, 1554–1558, 2010.

276. Landino, L.M., Hagedorn, T.D., Kim, S.E., and Hogan, K.M., Inhibition of tubulin polymerization by hypochlorous acid and chloramines, *Free Radic. Biol. Med.* 50, 1000–1008, 2011.

277. Pardo, J.P. and Slayman, C.W., Cysteine 532 and cysteine 545 are the N-ethylmaleimide-reactive residues of the *Neurospora* plasma membrane H$^+$-ATPase, *J. Biol. Chem.* 264, 9373–9379, 1989.

278. Miyanishi, T. and Borejdo, J., Differential behavior of two cysteine residues on the myosin head in muscle fibers, *Biochemistry* 28, 1287–1294, 1989.

279. Bishop, J.E., Squier, T.C., Bigelow, D.J., and Inesi, G., (Iodoacetamido) fluorescein labels a pair of proximal cysteines on the Ca^{2+}-ATPase of sarcoplasmic reticulum, *Biochemistry* 27, 5233–5240, 1988.

280. First, E.A. and Taylor, S.S., Selective modification of the catalytic subunit of cAMP-dependent protein kinase with sulfhydryl-specific fluorescent probes, *Biochemistry* 28, 3598–3605, 1989.

281. Ramalingam, T.S., Das, P.K., and Podder, S.K., Ricin-membrane interaction: Membrane penetration depth by fluorescence quenching and resonance energy transfer, *Biochemistry* 33, 12247–12254, 1994.

282. Hamman, B.D., Oleinikov, A.V., Jokhadze, G.G. et al., Rotational and conformational dynamics of *Escherichia coli* ribosomal protein L7/L12, *Biochemistry* 35, 16672–16679, 1996.

283. Karlstrom, A. and Nygren, P.A., Dual labeling of a binding protein allows for specific fluorescence detection of native protein, *Anal. Biochem.* 295, 22–30, 2001.

284. Teilum, K. et al., Early kinetic intermediate in the folding of acyl-CoA binding protein detected by fluorescence labeling and ultrarapid mixing, *Proc. Natl. Acad. Sci. USA* 99, 9807–9812, 2002.

285. Tripet, B., De Crescenzo, G., Grothe, S. et al., Kinetic analysis of the interactions between troponin C and the C-terminal troponin I regulatory region and validation of a new peptide delivery/capture system used for surface plasmon resonance, *J. Mol. Biol.* 323, 345–362, 2002.

286. Ujfalusi, Z., Barkó, S., Hild, G., and Nyitrai, M., The effects of formins on the conformation of subdomain 1 in actin filaments, *J. Photochem. Photobiol. B.* 98, 7–11, 2010.

287. Raha, P., Chattopadhyay, S., Mukerjee, S. et al., Alternative sigma factors in the free state are equilibrium mixtures of open and compact conformations, *Biochemistry* 49, 9809–9819, 2010.

288. Rysev, N.A., Karpicheva, O.E., Redwood, C.S. et al., The effect of Asp 175Asn and Glu180Gly TPM1 mutations on actin-myosin interaction during the ATPase cycle, *Biochim. Biophys. Acta* 1824, 366–373, 2012.

289. Santos, A., Duarte, A.G., Fedorov, A. et al., Rubredoxin mutant A51C unfolding dynamics: A Förster resonance energy transfer study, *Biophys. Chem.* 148, 131–137, 2010.

290. Xie, L., Li, W.X., Barnett, V.A., and Schoenberg, M., Graphical evaluation of alkylation of myosin's SH1 and SH2: The N-phenylmaleimide reaction, *Biophys. J.* 72, 858–865, 1997.

291. Holyoak, T. and Nowak, T., Structural investigation of the binding of nucleotide to phosphoenolpyruvate carboxykinase by NMR, *Biochemistry* 40, 11037–11047, 2001.

292. Gwozdzinski, K., Pieniazek, A., Sudak, B., and Kaca, W., Alterations in human red blood cell membrane properties induced by the lipopolysaccharide from *Proteus mirabilis* S1959, *Chem. Biol. Interact.* 146, 73–80, 2003.

293. Hurth, K.M., Nilges, M.J., Carlson, K.E. et al., Ligand-induced changes in estrogen receptor conformation as measured by site-directed spin labeling, *Biochemistry* 43, 1891–1907, 2004.

294. Gettins, P., Beth, A.H., and Cunningham, L.W., Proximity of thiol esters and bait region in human α_2-macroglobulin: Paramagnetic mapping, *Biochemistry* 27, 2905–2911, 1988.

295. Kirby, T.L., Karim, C.B., and Thomas, D.D., Electron paramagnetic resonance reveals a large-scale conformational change in the cytoplasmic domain of phospholamban upon binding to the sarcoplasmic reticulum Ca-ATPase, *Biochemistry* 43, 5842–5852, 2004.

296. Claxton, D.P., Zou, P., and Mchaourab, H.S., Structure and orientation of T4 lysozyme bound to the small heat shock protein α-crystallin, *J. Mol. Biol.* 375, 1026–1039, 2008.

297. Toutchkine, A., Nalbant, P., and Hahn, K.M., Facile synthesis of thiol-reactive Cy3 and Cy5 Derivatives with enhanced water solubility, *Bioconjug. Chem.* 13, 387–383–391, 2002.

298. Bruschi, M., Grilli, S., Candiano, G. et al., New iodo-acetamido cyanines for labeling cysteine thiol residues. A strategy for evaluating plasma proteins and their oxido-redox status, *Proteomics* 9, 460–469, 2009.

299. Marino, S.M. and Gladyshev, V.N., Cysteine function governs its conservation and degeneration and restricts its utilization on protein surfaces, *J. Mol. Biol.* 404, 902–916, 2010.

300. Johnson, S., Evan, D., Laurenson, S. et al., Surface-immobilized peptide aptamers as probe molecules for protein detection, *Anal. Chem.* 80, 978–983, 2008.

301. Kostiainen, M.A., Kotimaa, J., Laukkanen, M.-L., and Pavan, G.M., Optically degradable dendrons for temporary adhesion of proteins to DNA, *Chemistry* 16, 6912–6918, 2010.

302. Ji, T., Muenker, M.C., Papineni, R.V.L. et al., Increased sensitivity in antigen detection with fluorescent latex nanosphere-IgG antibody conjugates, *Bioconjug. Chem.* 21, 427–435, 2010.

303. Creech, H.J. and Jones, R.N., The conjugation of horse serum albumin with isocyanates of certain polynuclear aromatic hydrocarbons, *J. Am. Chem. Soc.* 63, 1661–1669, 1941.

304. Creech, H.J. and Jones, R.N., Conjugates synthesized from various proteins and the isocyanates of certain aromatic polynuclear hydrocarbons, *J. Am. Chem. Soc.* 63, 1670–1673, 1941.

305. Weber, G., Polarization of the fluorescence of macromolecules. 2. Fluorescent conjugates of ovalbumin and bovine serum albumin, *Biochem. J.* 51, 155–167, 1952.

306. Emmert, E.W., Observations on the absorption spectra of fluorescein, fluorescein derivatives and conjugates, *Arch. Biochem. Biophys.* 73, 1–8, 1958.

307. Hudson, E.N. and Weber, G., Synthesis and characterization of two fluorescent sulfhydryl reagents, *Biochemistry* 12, 4154–4161, 1973.

308. Wu, C.-W. and Stryer, L., Proximity relationships in rhodopsin, *Proc. Natl. Acad. Sci. USA* 69, 1104–1108, 1972.

309. Zukin, R.S., Hartig, P.R., and Koshland, D.E., Jr., Use of a distant reporter group as evidence for a conformational change in a sensory receptor, *Proc. Natl. Acad. Sci. USA* 74, 1932–1936, 1977.

310. Takashi, R., Duke, K., Ue, K., and Morales, M.F., Defining the "fast-reacting" thiols of myosin by reaction with 1,5,-IAEDANS, *Arch. Biochem. Biophys.* 175, 279–283, 1976.
311. Duke, J., Takashi, K., Ue, K., and Morales, M.F., Reciprocal reactivities of specific thiols when actin binds to myosin, *Proc. Natl. Acad. Sci. USA* 73, 302–306, 1976.
312. Baba, A., Nakamura, T., and Kawakita, M., Chemical modification and fluorescence labeling study of Ca^{2+}, Mg^{2+}-adenosine triphosphatase of sarcoplasmic reticulum using iodoacetamide and its N-substituted derivatives, *J. Biochem.* 100, 1137–1147, 1986.
313. Yamashita, T. and Kawakita, M., Reactive sulfhydryl groups of sarcoplasmic reticulum ATPase. II. Site of labeling with iodoacetate and its fluorescent derivative, *J. Biochem.* 101, 377–385, 1987.
314. Takashi, R., Muhlrad, A., and Botts, J., Spatial relationship between a fast-reacting thiol and a reactive lysine residue of myosin subfragment-1, *Biochemistry* 21, 5661–5668, 1982.
315. Ajtai, I. and Burghard, T.P., Selective fluorescent labeling of the 50-kilodalton, 26-kilodalton, and 20-kilodalton heavy-chain segments of myosin ATPase, *J. Biochem.* 101, 1457–1462, 1987.
316. Phan, B.C., Peyser, Y.M., Reisler, E., and Muhlrad, A., Effect of complexes of ADP and phosphate analogs on the conformation of the Cys707-Cys697 region of myosin subfragment 1, *Eur. J. Biochem.* 243, 636–642, 1997.
317. Smyczynski, C. and Kasprzak, A.A., Effect of nucleotides and actin on the orientation of the light chain-binding domain in myosin subfragment 1, *Biochemistry* 36, 13201–13207, 1997.
318. Tiago, T., Aureliano, M., and Gutiérrez-Merino, C., and Decavanadate binding to a high affinity site near the myosin catalytic centre inhibits F-actin-stimulated myosin ATPase activity, *Biochemistry* 43, 551–5561, 2004.
319. Hiratsuka, Y., Eto, M., Yazawa, M., and Morita, F., Reactivities of Cys707 SH1) in intermediate states of myosin subfragment -1 ATPase, *J. Biochem.* 124, 609–614, 1998.
320. Botts, J., Ue, K., Hozumi, T., and Samet, J., Consequences of reacting the thiols of myosin subfragment 1, *Biochemistry* 18, 5157–5163, 1979.
321. Seidel, J.C., Similar effects on enzymic activity due to chemical modification of either of two sulfhydryl groups of myosin, *Biochim. Biophys. Acta* 180, 216–219, 1969.
322. Reisler, E., Burke, M., and Harrington, W.F., Cooperative role of two sulfhydryl groups in myosin triphosphatase, *Biochemistry* 13, 2014–2022, 1974.
323. Balain, G. and Bárány, M., Thiolysis of dinitrophenylated myosin with restoration of adenosine triphosphatase activity, *J. Biol. Chem.* 247, 7815–7821, 1972.
324. Shaltiel, S., Thiolysis of some dinitrophenyl derivatives of amino acids, *Biochem. Biophys. Res. Commun.* 29, 178–183, 1967.
325. Squire, T.C., Bigelow, D.J., Garcia de Ancos, J., and Inesi, G., Localization of site-specific probes on the Ca-ATPase of sarcoplasmic reticulum using fluorescence energy transfer, *J. Biol. Chem.* 262, 4748–4754, 1987.
326. Kolega, J., Fluorescent analogues of myosin II for tracking the behavior of different myosin isoforms in living cells, *J. Cell. Biochem.* 68, 389–401, 1998.
327. Pronina, O.E., Wrzosek, A., Dabrowska, R., and Borovikov, Y.S., Effect of nucleotides on the orientation and mobility of myosin subfragment -1 in ghost muscle fiber, *Biochemistry* 70, 1140–1144, 2005.
328. Dong, W.J., An, J., Xing, J., and Cheung, H.C., Structural transition of the inhibitory region of troponin I within the regulated cardiac thin filament, *Arch. Biochem. Biophys.* 456, 135–142, 2006.
329. Rysev, N.A., Karpicheva, O.E., Redwood, C.S., and Borikov, Y.S., The effect of the Asp175Asn and Gly180Glye TPM1 mutations on actin-myosin interaction during the ATPase cycle, *Biochim. Biophys. Acta* 1824, 366–373, 2012.
330. Reisler, E., Sulfhydryl modification and labeling of myosin, *Methods Enzymol.* 85, 84–93, 1982.
331. Lowey, S., Slayter, H.S., Weeds, A.G., and Baker, H., Substructure of the myosin molecule. I. Subfragments of myosine by enzymatic degradation, *J. Mol. Biol.* 42, 1–29, 1969.
332. Hayashi, Y. and Tonomura, Y., On the active site of myosine A-adenosine triphosphatase. X. Functions of the two subfragment s, S-1, of the myosin molecule, *J. Biochem.* 68, 665–680, 1970.
333. Spudich, J.A. and Watt, S., The regulation of rabbit skeletal muscle contraction. I. Biochemical studies of the interaction of the tropomyosin-troponin complex with actin and the proteolytic fragments of myosin, *J. Biol. Chem.* 246, 4866–4871, 1971.
334. Holmes, K.C., Myosin structure, in *Myosins. A Super Family of Molecular Motors* (Vol. 7 in *Proteins and Cell Regulation*), ed. L.M. Colucci, Chapter 2, pp. 35–54, Springer, Dordrecht, the Netherlands, 2008.
335. Venkataramen, K., *The Chemistry of Synthetic Dyes*, Vol. 1, Academic Press, New York, 1952.
336. Gómez-Hens, A. and Aguilar-Caballos, M.P., Long-wavelength fluorophores: New trends in their analytical use, *Trends Anal. Chem.* 23, 127–136, 2004.

337. Hamer, F.A., *The Cyanine Dyes and Related Compounds*, *Chemistry of Heterocyclic Compounds*, Vol. 19, Interscience (John Wiley), New York, 1964.

338. Wang, M. and Armitage, B.A., Colorimetric detection of PNA-DNA hybridization using cyanine dyes, in *Peptide Nucleic Acids; Methods and Protocols*, ed. P.E. Nielsen, Chapter 9, pp. 131–142, Humana Press, Towata, NJ, 2002.

339. Kane, M.D., Jatkoe, T.A., Stumpf, C.R., Lu, J., Thomas, J.D., and Madre, S.J., Assessment of the sensitivity and specificity oligonucleotide (50 mer) microarrays, *Nucleic Acid Res.* 28, 4552–4557, 2000.

340. Albers, A.E., Chan, E.M., McBride, P.M. et al., Dual-emitting quantum dot/quantum rod-based nano-thermometers with enhanced response and sensitivity in live cells, *J. Am. Chem. Soc.* 134, 9565–9568, 2012.

341. Ernst, L.A., Gupta, R.K., Mujumdar, R.B., and Waggoner, A.S., Cyanine dye labeling reagents for sulf-hydryl groups, *Cytometry* 10, 3–10, 1989.

342. Lin, Y., Weissleder, R., and Tung, C.H., Synthesis and properties of sulfhydryl-reactive near-infrared cyanine fluorochromes for fluorescence imaging, *Mol. Imaging* 2, 87–92., 2003.

343. Chan, H.L., Sinclair, J., and Timms, J.F., Proteomic analysis of redox-dependent changes using cysteine-labeling 2D DIGE, *Methods Mol. Biol.* 854, 113–128, 2012.

344. Kim, S.H., Gunther, J.R., and Katzenellenbogen, J.A., Monitoring a coordinated exchange process in a four-component biological interaction system: Development of a time-resolved terbium-based one-donor three-acceptor multicolor FRET system, *J. Am. Chem. Soc.* 132, 4685–4692, 2010.

345. Viswanathan, S., Unlü, M., and Minden, J.S., Two-dimensional difference gel electrophoresis, *Nat. Protocol.* 1, 1351–1358, 2006.

346. Kolling, O.W. and Goodnight, J.L., Phenol blue as a solvent indicator for binary aprotic solvents, *Anal. Chem.* 45, 160–164, 1973.

347. Gaines, G.L., Jr., Solvatochromic compound as an acid indicator in nonaqueous media, *Anal. Chem.* 48, 450–451, 1976.

348. Deye, J.F., Berger, T.A., and Anderson, A.G., Nile Red as a solvatochromic dye for measuring solvent strength in normal liquids and mixtures of normal liquids with supercritical and near critical fluids, *Anal. Chem.* 62, 615–622, 1990.

349. Cano-Sarabia, M., Angelova, A., Ventosa, N. et al., Cholesterol induced CTAB micelle-to-vesicle phase transitions, *J. Colloid Interface Sci.* 350, 10–15, 2010.

350. Breton, M., Prével, G., Audibert, J.F. et al., Solvatochromic dissociation of non-covalent fluorescent organic nanoparticles upon cell internalization, *Phys. Chem. Chem. Phys.* 134, 13268–13276, 2011.

351. Kudo, K., Momotake, A., Tanaka, J.K. et al., Environmental polarity estimation in living cells by use of quinoxaline-based full-colored solvatochromic fluorophore PQX and its derivatives, *Photochem. Photobiol. Sci.* 11, 674–678, 2012.

352. Loving, G.S., Sainlos, M., and Imperiali, B., Monitoring protein interactions and dynamics with solvato-chromic fluorophores, *Trends Biotechnol.* 28, 73–83, 2010.

353. Goguen, B.N. and Imperiali, B., Chemical tools for studying directed cell migration, *ACS Chem. Biol.* 6, 1164–1174, 2011.

354. Toutchkine, A., Kraynov, V., and Hahn, K., Solvent-sensitive dyes to report protein conformational changes in living cells, *J. Am. Chem. Soc.* 125, 4132–4145, 2003.

355. Garrett, S.C., Hodgson, L., Rybin, A. et al., A biosensor of S100A4 metastasis factor activation: Inhibitor screening and cellular activation dynamics, *Biochemistry* 47, 986–996, 2008.

356. Loving, G. and Imperiali, B., Thiol-reactive derivatives of the solvatochromic 4-*N,N*-dimethylamino-1,8-naphthalimide fluorophore: A highly sensitive toolset for the detection of biomolecular interactions, *Bioconjug. Chem.* 20, 2133–2141, 2009.

357. Sharma, V. and Lawrence, D.S., Über-responsive peptide-based sensors of signaling proteins, *Angew. Chem. Int. Ed. Engl.* 48, 7290–7292, 2009.

358. Gu, L. and Mohanty, S.K., Targeted microinjection into cells and retina using optoporation, *J. Biomed. Opt.* 15, 128003, 2011.

359. Muro, E., Fragola, A., Pons, T. et al., Comparing intracellular stability and targeting of sulfobetaine quantum dots with other surface chemistries in live cells, *Small* 8, 1029–1037, 2012.

360. Choi, S.O., Kim, Y.C., Lee, J.W. et al., Intracellular protein delivery and gene transfection by electropora-tion using a microneedle electrode array, *Small* 8, 1081–1091, 2012.

361. Le Gac, S. and van den Berg, A., Single cell electroporation using microfluidic devices, *Methods Mol. Biol.* 853, 65–82, 2012.

362. Steinmeyer, J.D. and Yanik, M.F., High-throughput single-cell manipulation in brain tissue, *PLoS One* 7, e35603, 2012.

363. Medintz, I.L., Pons, T., Delehanty, J.B. et al., Intracellular delivery of quantum dot-protein cargos mediated by cell penetrating peptides, *Bioconjug. Chem.* 19, 1785–1795, 2008.

364. Chu, Q. and Fukui, Y., In vivo dynamics of myosin II in *Dictyostelium* by fluorescent analogue cytochemistry, *Cell. Motil. Cytoskeleton* 35, 254–268, 1996.

365. Kim, J.R., Yoon, H.W., Kwon, K.S. et al., Identification of proteins containing cysteine residues that are sensitive to oxidation by hydrogen peroxide at neutral pH, *Anal. Biochem.* 283, 214–221, 2000.

366. Requejo, R., Chouchani, E.T., James, A.M. et al., Quantification and identification of mitochondrial proteins contains vicinal dithiols, *Arch. Biochem. Biophys.* 504, 228–235, 2010.

367. Hill, B.G., Reilly, C., Oh, J.Y. et al., Methods for the determination and quantification of the reactive thiol proteome, *Free Radic. Biol. Med.* 47, 675–683, 2009.

368. Sun, E., Lack, N., Wang, C.-J. et al., Arylamine *N*-acetyltransferases: Structural and functional implications of polymorphisms, *Toxicology* 254, 170–183, 2009.

369. Butcher, N.J., and Minchin, R.F., Arylamine *N*-acetyltransferase 1: A novel drug target in cancer development, *Pharmacol. Rev.* 64, 147–165, 2012.

370. Walker, K., Ginsberg, G., Hattis, D. et al., Genetic polymorphism in *N*-acetyltransferase (NAT): Population distribution of NAT1 and NAT2 activity, *J. Toxicol. Environ. Health B Crit. Rev.* 12, 440–472, 2009.

371. Sim, E., Walters, K., and Boukouvala, S., Arylamine *N*-acetyltransferases: From structure to function, *Drug Metab. Rev.* 40, 479–510, 2008.

372. Rodrigues-Lima, F., Dairou, J., Busi, F., and Dupret, J.M., Human arylamine *N*-acetyltransferase 1: A drug-metabolizing enzyme and a drug target?, *Curr. Drug Targets* 11, 759–766, 2010.

373. Bartsch, H., Dworkin, M., Miller, J.A., and Miller, E.C., Electrophilic *N*-acetoxyaminoarenes derived from carcinogenic *N*-hydroxy-*N*-acetylaminoarenes by enzymatic deacetylation and transacetylation in liver, *Biochim. Biophys. Acta* 286, 272–298, 1972.

374. Mangold, B.L.K. and Hanna, P.E., Arylhydroxamic acid *N,O*,-acyltransferase substrates. Acetyl transfer and electrophilic generating activity of *N*-hydroxy-*N*-(4-alkyl, 4-alkenyl, and 4-cyclohexylphenyl)acetamidates, *J. Med. Chem.* 25, 630–638, 1982.

375. Andres, H.H., Kolb, H.J., Schreiber, R.J., and Weiss, L., Characterization of the active site, substrate specificity and kinetic properties of acetyl-CoA:arylamine *N*-acetyltransferase from pigeon liver, *Biochim. Biophys. Acta.* 746, 193–201, 1983.

376. Saito, K., Shinohara, A., Kamataki, T., and Kato, R., *N*-Hydroxyarylamine *O*-acetyltransferase in hamster liver: Identity with arylhydroxamic acid *N,O*,-acetyl transferase and arylamine *N*-acetyltransferase, *J. Biochem.* 99, 1689–1697, 1986.

377. Cheon, H.G. and Hanna, P.E., Effect of group-selective modification reagents on arylamine *N*-acetyltransferase activities, *Biochem. Pharmacol.* 43, 2255–2268, 1992.

378. Andres, H.H., Klem, A.J., Schofer, L.M. et al., On the active site of liver acetyl-CoA arylamine *N*-acetyltransferase from rapid acetylator rabbits (III/J), *J. Biol. Chem.* 263, 7521–7527, 1988.

379. Cleland, W.W., Dithiothreitol, a new protective reagent for SH groups, *Biochemistry* 4, 480–482, 1984.

380. Smythe, C.V., The reaction of iodoacetate and iodoacetamide with various sulfhydryl groups, with urease, and with yeast preparations, *J. Biol. Chem.* 114, 601–612, 1936.

381. Dickens, F., CLII. Interaction of halogenacetates and SH compounds. The reaction of halogenacetic acids with glutathione and cysteine. The mechanism of iodoacetate poisoning of glycoxalase, *Biochem. J.* 27, 1141–1151, 1933.

382. Schroeder, E.F., Woodward, G.E., and Platt, M.E., The relation of sulfhydryl to inhibition of yeast fermentation by iodoacetic acid, *J. Biol. Chem.* 101, 133–144, 1933.

383. Zeng, J. and Davies, M.J., Protein and low molecular mass thiols as targets and inhibitors of glycation reactions, *Chem. Res. Toxicol.* 19, 1668–1676, 2006.

384. Guo, Z., Vath, G.M., Wagner, C.R. et al. Arylamine *N*-acetyltransferases: Covalent modification and inactivation of hamster NAT1 by bromoacetamido derivatives of aniline and 2-aminofluorene, *J. Protein. Chem.* 22, 631–642, 2003.

385. Wick, M.J., Yeh, H.-M., and Hanna, P.E., An isozyme-selective affinity label for rat hepatic acetyltransferases, *Biochem. Pharmacol.* 40, 1389–1398, 1990.

386. Wang, H., Guo, Z., Vath, G.M. et al., Chemical modification of hamster arylamine *N*-acetyltransferase 2 with isozyme-selective and nonselective *N*-arylbromoacetamido reagents, *Protein J.* 23, 153–166, 2004.

387. Wang, H., Vath, G.M., Gleason, K.J. et al., Probing the mechanism of hamster arylamine *N*-acetyltransferase 2 acetylation by active site modification, site-directed mutagenesis, and pre-steady state and steady state kinetic studies, *Biochemistry* 43, 8234–8246, 2004.

388. Gerwin, B.I., Properties of the single sulfhydryl group of streptococcal proteinase. A comparison of the rates of alkylation of chloroacetic acid and chloroacetamide, *J. Biol. Chem.* 242, 451–456, 1967.

389. Chaiken, I.M. and Smith, E.L., Reaction of the sulfhydryl group of papain with chloroacetic acid, *J. Biol. Chem.* 244, 5095–5099, 1969.

390. Chaiken, I.M. and Smith, E.L., Reaction of chloroacetamide with the sulfhydryl group of papain, *J. Biol. Chem.* 244, 5087–5094, 1969.

391. Baker, L.M., Baker, P.R., Golin-Bisello, F. et al., Nitro-fatty acid reaction with glutathione and cysteine: Kinetic analysis of thiol alkylation by a Michael addition reaction, *J. Biol. Chem.* 282, 31085–31093, 2007.

392. Dourado, D.F.A.R., Fernandes, P.A., Mannervik, B., and Ramos, M.J., Glutathione transferase: New model for glutathione activation, *Chemistry* 14, 9591–9598, 2008.

393. Martyniuk, C.J., Fang, B., Koomen, J.M. et al., Molecular mechanism of glyceraldehye-3-phosphate dehydrogenase inactivation by α,β-unsaturated carbonyl derivatives, *Chem. Res. Toxicol.* 24, 2302–2311, 2011.

394. Ferrer-Sueta, G., Manta, B., Botti, H. et al., Factors affecting protein thiol reactivity and specificity in peroxide reduction, *Chem. Res. Toxicol.* 24, 434–450, 2011.

395. McGrath, N.A. and Raines, R.T., Chemoselectivity in chemical biology: Acyl transfer reactions with sulfur and selenium, *Acc. Chem. Res.* 44, 752–761, 2011.

396. Sengupta, R.N., Herschlag, D., and Piccirilli, J.A., Thermodynamic evidence for negative charge stabilization by a catalytic metal ion within an RNA active site, *ACS Chem. Biol.* 7, 294–299, 2012.

397. Rinaldo-Mathis, A., Ahmad, S., Welterholm, A. et al., Pre-steady-state kinetic characterization of thiolate anion formation in human leukotriene C4 synthase, *Biochemistry* 51, 848–856, 2012.

398. Zhou, X., Zhang, N.-X., Liu, L. et al., Probing the catalytic potential of the hamster arylamine *N*-acetyltransferase 2 catalytic triad by site-directed mutagenesis of the proximal conserved residue, Tyr190, *FEBS J.* 276, 6928–6941, 2009.

399. Malka, F., Dairou, J., Ragunathan, N. et al., Mechanisms and kinetics of human arylamine *N*-acetyltransferase 1 inhibition by disulfiram, *FEBS J.* 278, 4900–4908, 2009.

400. Ong, S.E., Blagoev, B., Kratchmarova, I. et al., Stable isotope labeling by amino acids in cell culture, SILAC, as a simple and accurate approach to expression proteomics, *Mol. Cell. Proteomics* 1, 376–386, 2002.

401. Berger, S.J., Lee, S.W., Anderson, G.A. et al., High-throughput global peptide proteomic analysis by combining stable isotope amino acid labeling and data-dependent multiplexed-MS/MS. *Anal. Chem.* 74, 4994–5000, 2002.

402. Everley, P.A., Krijgsveld, J., Zetter, B.R., and Gygi, S.P., Quantitative cancer proteomics: Stable isotope labeling with amino acids in cell culture (SILAC) as a tool for prostate cancer research, *Mol. Cell. Proteomics* 3, 729–735, 2004.

403. Frohn, A., Eberl, H.C., Stöhr, J. et al., Dicer-dependent and -independent Argonaute2 protein interaction networks in mammalian cells, *Mol. Cell. Proteomics*, 11, 1442–1456, 2012.

404. Tebbe, A., Schmidt, A., Konstantinidis, K. et al., Life-style changes of a halophilic archaeon by quantitative proteomics, *Proteomics* 9, 3843–3855, 2009.

405. Baptiste, L., Rosier, C., Erculisse, V. et al., Differential proteomic analysis using isotope-coded protein-labeling strategies: Comparison, improvements and application to simulated microgravity effect on *Cupriavidus metallidurans* CH34, *Proteomics* 10, 2281–2291, 2010.

406. Coombs, K., Quantitative proteomics of complex mixtures, *Expert Rev. Proteomics* 8, 659–677, 2011.

407. Lottspeich, F. and Kellerman, J., ICPL labeling strategies for proteome research, *Methods Mol. Biol.* 753 (Gel-Free Proteomics), 55–64, 2011.

408. Jeffrey, D.A. and Bogyo, M., Chemical proteomics and its application to drug discovery, *Curr. Opin. Biotechnol.* 14, 87–95, 2003.

409. Mann, M., Quantitative proteomics? *Nat. Biotechnol.* 17, 954–955, 1999.

410. Smolka, M., Zhou, H., and Aebersold, R., Quantitative protein profiling using two-dimensional gel electrophoresis, isotope-coded affinity tag labeling, and mass spectrometry, *Mol. Cell. Proteomics* 1, 19–29, 2002.

411. Gygi, P., Rist, B., Gerber, S.A. et al., Quantitative analysis of complex protein mixtures using isotope-coded affinity tags, *Nat. Biotechnol.* 17, 994–999, 1999.

412. Tao, W.A. and Aebersold, R., Advances in quantitative proteomics via stable isotope tagging and mass spectrometry, *Curr. Opin. Biotechnol.* 14, 100–118, 2003.

413. Zhang, Z., Edwards, P.J., Roeske, P.W., and Guo, L., Synthesis and self-alkylation of isotope-coded affinity tag reagents, *Bioconjug. Chem.* 16, 458–464, 2005.

414. Shen, M., Guo, L., Wallace, A. et al., Isolation and isotope labeling of cysteine- and methionine-containing tryptic peptides: Application to the study of cell surface proteolysis, *Mol. Cell. Proteomics* 2, 315–324, 2003.

415. Pasquarello, C., Sanchez, J.C., Hochstrasser, D.F., and Corthalls, G.L., *N-t*-butyliodoacetamide and iodoacetanilide: Two new cysteine alkylating reagents for relative quantitation of proteins, *Rapid Commun. Mass Spectrom.* 18, 117–127, 2004.

416. Sebastiano, R., Citterio, A., Lapadula, M., and Righetti, P.G. A new deuterated alkylating agent for quantitative proteomics, *Rapid Commun. Mass Spectrom.* 17, 2380–2386, 2003.

417. Hansen, K.C., Schmitt-Ulms, G., Chalkley, R.J. et al., Mass spectrometric analysis of protein mixtures at low levels using cleavable ^{13}C-isotope-coded affinity tag and multidimensional chromatography, *Mol. Cell. Proteomics* 2, 299–314, 2003.

418. Ahrends, R., Pieper, S., Neumann, B. et al., Metal-coded affinity tag labeling: A demonstration of analytical robustness and suitability for biological applications, *Anal. Chem.* 81, 2176–2184, 2009.

419. Schwarz, G., Beck, S, Weller, M.G., and Linscheid, M.W., MeCAT—New iodoacetamide reagents for metal labeling of proteins and peptides, *Anal. Bioanal. Chem.* 401, 1203–1209, 2011.

420. Thevis, M., Loo, R.R.O., and Loo, J.A., In-gel derivatization of proteins for cysteine-specific cleavages and their analysis by mass spectrometry, *J. Proteome Res.* 2, 163–172, 2003.

421. Friedman, M., Application of the S-pyridylethylation reaction to the elucidation of the structure and function of proteins, *J. Protein Chem.* 20, 431–453, 2001.

422. Righetti, P.G., Castagna, A., Antonucci, F. et al., Critical survey of quantitative proteomics in two-dimensional electrophoretic approaches, *J. Chromatogr. A,* 1051, 3–17, 2004.

423. Bai, F., Liu, S., and Witzmann, F.A., A "de-streaking" method for two-dimensional electrophoresis using tris(2-carboxyethyl)-phosphine hydrochloride and alkylating agent vinylpyridine, *Proteomics* 5, 2043–2047, 2005.

424. Maeda, K., Finnie, C., and Svensson, B., Identification of thioredoxin h-reducible disulphides in proteomes by differential labelling of cysteines: Insights into recognition and regulation of proteins in barley seeds by thioredoxin h, *Proteomics* 5, 1634–1644, 2005.

425. Chowdhury, S.M., Munske, G.R., Ronald, R.C., and Bruce, J.E., Evaluation of low energy CID and ECD fragmentation behavior of mono-oxidized thio-ether bonds in peptides, *J. Am. Soc. Mass Spectrom.* 18, 493–501, 2007.

426. Righetti, P.G., Sebastiano, R., and Citterio, A., Isotope-coded two-dimensional maps: Tagging with deuterated acrylamide and 2-vinylpyridine, *Methods Mol. Biol.* 424, 87–99, 2008.

427. Crestfield, A.M., Moore, S., and Stein, W.H., The preparation and enzymatic hydrolysis of reduced and *S*-carboxymethylated proteins, *J. Biol. Chem.* 238, 622–627, 1963.

428. Borromeo, V., Gaggioli, D., Berrini., A. et al., Monoclonal antibodies as a probe for the unfolding of porcine growth hormone, *J. Immunol. Methods* 272, 107–115, 2003.

429. Brock, J.W., Hinton, D.J., Cotham, W.E. et al., Proteomic analysis of the site specificity of glycation and carboxymethylation of ribonuclease, *J. Proteome Res.* 2, 506–513, 2003.

430. Simpson, R.J., Reduction and a carboxymethylation of proteins: Large-scale method, *CSH Protocol.* 2007:pdb.prot4569, 2007.

431. Sutton, M.R. and Brew, K., Purification and characterization of the seven cyanogen bromide fragments of human serum transferrin, *Biochem. J.* 139, 163–168, 1974.

432. Galvani, M., Hamden, M., Herbert, B., and Righetti, P.G., Alkylation kinetics of proteins in preparation for two-dimensional maps: A matrix assisted laser desorption/ionization-mass spectrometry investigation, *Electrophoresis* 22, 2058–2065, 2001.

433. Kim, J.S., Fillmore, T.L., Liu, T. et al., ^{18}O-labeled proteome reference as global internal standards for targeted quantification by selected reaction monitoring mass-spectrometry, *Mol. Cell. Proteomics* 10, M110.007302, 2011.

434. Jagtap, D.D., Narahari, A., Swamy, M.J., and Mahale, S.D., Disulphide bond reduction and *S*-carboxamidomethylation of PSP94 affects its conformation but not the ability to bind immunoglobulin, *Biochim. Biophys. Acta* 1774, 723–731, 2007.

435. Batra, P.P., Sasa, K., Ueki, T., and Takeda, K., Circular dichroic study of the conformational stability of sulfhydryl-blocked bovine albumin, *Int. J. Biochem.* 21, 857–862, 1989.

436. Huggins, C. and Jensen, E.V., Thermal coagulation of serum proteins; the effects of iodoacetate, iodoacetamide, and thiol compounds on coagulation, *J. Biol. Chem.* 179, 645–654, 1949.

437. Trivedi, M., Lawrence, J.S., Williams, T.D. et al., Improving the stability of the EC1 domain of E-cadherin by thiol alkylation of the cysteine residues, *Int. J. Pharm.* 43, 16–25, 2012.

438. Li, C.H., Retention of the biological potency of human pituitary growth hormone after reduction and carbamidomethylation, *Science* 154, 785–786, 1996.

439. Woods, A.G., Sokolowska, L., and Darie, C.C., Identification of consistent alkylation of cysteine-less peptides in a proteomics experiment, *Biochem. Biophys. Res. Commun.* 419, 305–308, 2012.

440. Matthiesen, R., Bauw, G., and Welinder, K.G., Use of performic acid oxidation to expand the mass distribution of tryptic peptides, *Anal. Chem.* 76, 6848–6852, 2004.
441. Samgina, T.Y., Artemenko, K.A., Gorshkov, V.A. et al., Oxidation versus carboxamidomethylation of S-S bond in ranid frog peptides: Pros and contra for de novo MALDI-MS sequencing, *J. Am. Soc. Mass Spectrom.* 19, 479–487, 2008.
442. Sharp, D.W.A.. ed., *Penguin Dictionary of Chemistry*, 2nd edn., p. 261, Penguin, London, U.K., 1990.
443. Orchin, M., Kaplan, F., MaComber, R.A., Wilson, R.M., and Zimmer, H., *The Vocabulary of Organic Chemistry*, p. 301 (10.480), p. 385, (11.890), John Wiley & Sons, New York, 1980.
444. Li, J.J., *Name Reactions. A Collection of Detailed Reaction Mechanisms*, p. 232, Springer, Berlin, Germany, 2005.
445. Mills, S., Stanton, C., Hill, C. et al., New developments and applications of bacteriocins and peptides in foods, *Annu. Rev. Food Sci. Technol.* 2, 299–329, 2011.
446. Klaenhammer, T.R., Genetics of bacteriocins produced by lactic acid bacteria, *FEMS Microbiol. Lett.* 12, 39–86, 1993.
447. Ross, N.L., Sporns, P., Dodd, H.M. et al., Involvement of dehydroalanine and dehydrobutyrine in the addition of glutathione to nisin, *J. Agric. Food Chem.* 51, 3174–3178, 2003.
448. Correla, J.J., Lipscomb, L.D., and Lobert, S., Nondisulfide crosslinking and chemical cleavage of tubulin subunits: pH and temperature dependence, *Arch. Biochem. Biophys.* 300, 105–114, 1993.
449. Gilani, G.S., Cockell, K.A., and Sepehr, E., Effects of antinutritional factors on protein digestibility and amino acid availability in foods, *J. AOCA Int.* 88, 967–987, 2005.
450. Rombouts, I., Lagrain, B., Brijs, K., and Delcour, J.A., Cross-linking of wheat gluten proteins during productions of hard pretzels, *Amino Acids* 42, 2429–2438, 2012.
451. Galande, A.K., Trent, J.O, and Spatola, A.F., Understanding base-assisted desulfurization using a variety of disulfide-bridged peptides, *Biopolymers* 71, 534–551, 2003.
452. Bar-Or, R., Rael, L.T., and Bar-Or, D., Dehydroalanine derived from cysteine is a common post-translational modification in human serum albumin, *Rapid Commun. Mass Spectrom.* 22, 711–716, 2008.
453. Liu, H. and May, K., Disulfide bond structures of IgG molecules: Structural variations, chemical modifications and possible impacts to stability and biological function, *MAbs* 4, 17–23, 2012.
454. Bernardes, G.J.L., Chalker, J.M., Errey, J.C., and Davis, B.G., Facile conversion of cysteine and alkyl cysteines to dehydroalanine on protein surfaces: Versatile and switchable access to functionalized proteins, *J. Am. Chem. Soc.* 130, 5052–5053, 2008.
455. Chalker, J.M., Gummoo, S.B., Boutureira, O. et al., Methods for converting cysteine to dehydroalanine on peptides and proteins, *Chem. Sci.* 2, 1666–1676, 2011.
456. Chalker, J.M., Conversion of cysteine into dehydroalanine enables access to synthetic histones bearing diverse post-translational modifications, *Angew. Chem. Int. Ed. Engl* 51, 1835–1839, 2012.
457. Sakabe, K.,Wang, Z., and Hart, G.W., β-*N*-acetylglucosamine (*O*-GlCNac) is part of the histone code, *Proc. Natl. Acad. Sci. USA* 107, 19915–19920, 2010.
458. Chalker, J.M., Bernardes, G.C.L., and Davis, B.G., A "tag-and-modify" approach to site-selective protein modification, *Acc. Chem. Res.* 44, 730–741, 2011.
459. Levengood, M.R. and van der Donk, W.A., Dehydroalanine-containing peptides: Preparation from phenylselenocysteine and utility in convergent ligation strategies, *Nat. Protoc.* 1, 3001–3010, 2006.
460. Marshak, D.R. and Carroll, D., Synthetic peptide substrates for casein kinase II, *Methods Enzymol.* 200, 134–156, 1991.
461. Jaffe, H., Veeranna, and Pant, H.C., Characterization of serine and threonine phosphorylation sites in β-elimination/ethanethiol addition-modified proteins, *Biochemistry* 37, 16211–16224, 1998.
462. Byford, M.R., Rapid and selective modification of phosphoserine resides catalyzed by Ba^{2+} ions for their detection during peptide microsequencing, *Biochem. J.* 280, 261–265, 1991.
463. Adamczyk, M., Gebler, J.C., and Wu, J., Selective analysis of phosphopeptides within a protein mixture by chemical modification, reversible biotinylation and mass spectrometry, *Rapid Commun. Mass Spectrom.* 15, 1481–1488, 2001.
464. McLachlin, D.T. and Chait, B.T., Improved β-elimination-based affinity purification strategy for enrichment of phosphopeptides, *Anal. Chem.* 75, 6826–6836, 2003.
465. Ahn, Y.H., Ji, E.S., Lee, J.Y. et al., Arginine-mimic labeling with guanidinoethanethiol to increase mass sensitivity of lysine-terminated phosphopeptides by matrix-assisted laser desorption/ionization time-of-flight mass spectrometry, *Rapid Commun. Mass Spectrom.* 21, 2204–2210, 2007.
466. Tseng, H.-C., Ovaa, H., Wei, N.J.C. et al., Phosphoproteomic analysis with a solid-phase capture-release-tag approach, *Chem. Biol.* 12, 769–777, 2005.

467. Meyer, K. and Anderson, B., The chemical specificity of the mucopolysaccharides of the cornea, *Exp. Eye Res.* 4, 346–348, 1965.

468. Bella, A., Jr. and Danishefsky, I., Dermatan sulfate-protein linkage region, *J. Biol. Chem.* 243, 2660–2664, 1968.

469. Greis, K.D., Hayes, B.K., Comer, F.I. et al., Selective detection and site-analysis of *O*-GlcNAc-modified glycopeptides by β-elimination and tandem electrospray mass spectrometry, *Anal. Biochem.* 234, 38–49, 1996.

470. Lima, V.V., Giachini, F.R., Hardy, D.M. et al., *O*-GlcNAcylation: A novel pathway contributing to the effects of endothelin in the vasculature, *Am. J. Physiol. Regul. Integr. Comp. Physiol.* 300, R236–R250, 2011.

471. Darley-Usmar, V.M., Ball, L.E., and Chatham, J.C., Protein *O*-linked β-acetylglucosamine: A novel effector of cardiomyocyte metabolism and function, *J. Mol. Cell Cardiol.* 52, 538–549, 2012.

472. Wells, L., Vosseller, K., Cole, R.N. et al., Mapping sites of *O*-GlcNAc modification using affinity tags for serine and threonine post-translational modification, *Mol. Cell. Proteomics* 1, 791–804, 2002.

473. Overath, T., Kuckelkorn, U., Henklein, P. et al., Mapping of *O*-GlnNAc sites of 20 S proteaosome subunits and Hsp90 by a novel biotin-cystamine tag, *Mol. Cell. Proteomics* 11, 467–477, 2012.

474. Furukawa, J.-I., Fujitani, N., Araki, K. et al., A versatile method for analysis of serine/threonine post-translational modifications by β-elimination in the presence of pyrazolone analogues, *Anal. Chem.* 83, 9060–6067, 2011.

475. Yan, J.X., Keet, W.C., Herbert, B.R. et al., Identification and quantitation of cysteine in proteins separated by gel electrophoresis, *J. Chromatogr.* 813, 187–200, 1998.

476. Bordini, E., Hamdan, M., and Righetti, P.G., Probing acrylamide alkylation sites in cysteine-free proteins by matrix-assisted laser desorption/ionization time-of-flight. *Rapid Commun. Mass Spectrom.* 14, 840–848, 2000.

477. Mineki, R. et al., In situ alkylation with acrylamide for identification of cysteinyl residues in proteins during one- and two-dimensional sodium dodecyl sulphate-polyacrylamide gel electrophoresis, *Proteomics* 2, 1672–1681, 2002.

478. Cahill, M.A. et al., Analysis of relative isopologue abundances for quantitative profiling of complex protein mixtures labelled with acrylamide/D-3-acrylamide alkylation tag system, *Rapid Commun. Mass Spectrom.* 17, 1283–1290, 2003.

479. LoPachin, R.M., Molecular mechanisms of the conjugated α,β-unsaturated carbonyl derivatives: Relevance to neurotoxicity and neurodegenerative diseases, *Toxicol. Sci.* 104, 235–249, 2008.

480. Lü, Z.R., Zou, H.C., Park, S.J. et al., The effects of acrylamide on brain creatine kinase: Inhibition kinetics and computational docking simulation, *Int. J. Biol. Macromol.* 44, 128–132, 2009.

481. Ghahghaei, A., Rekas, A., Carver, J.A., and Augusteyn, R.C., Structure/function studies on dogfish α-crystallin, *Mol. Vis.* 15, 2411–2420, 2009.

482. Sciandrelio, G., Mauro, M., Cardonna, F. et al., Acrylamide catalytically inhibits topoisomerase II in V79 cells, *Toxicol. In Vitro* 24, 830–834, 2010.

483. Burcham, P.C., Fontaine, F.R., Petersen, D.R., and Pyke, S.M., Reactivity with Tris(hydroxymethyl)aminomethane confounds immunodetection of acrolein-adducted proteins, *Chem. Res. Toxicol.* 16, 1196–1201, 2003.

484. Stevens, S.M., Jr., Rauniyar, N., and Prokai, L., Rapid characterization of covalent modifications to rat brain mitochondrial proteins after ex vivo exposure to 4-hydroxy-2-nonenal by liquid chromatography-tandem mass spectrometry using data-dependent and neutral-driven MS3 acquisition, *J. Mass Spectrom.* 42, 1599–1605, 2007.

485. Zhu, X., Gallogly, M.M., Mieyal, J.J. et al., Covalent cross-linking of glutathione and carnosine to proteins by 4-oxy-2-nonenal, *Chem. Res. Toxicol.* 22, 1050–1059, 2009.

486. Wakita, C., Maeshima, T., Yamazaki, A. et al., Stereochemical configuration of 4-hydroxy-2-nonenal-cysteine adducts and their stereoselective formation in a redox-regulated protein, *J. Biol. Chem.* 284, 28810–28822, 2009.

487. Rauniyar, N. and Prokai, L., Detection and identification of 4-hydroxy-2-nonenal Schiff-base adducts along with products of Michael addition using data-dependent neutral loss-driven MS3 acquisition: Method evaluation through an in vitro study on cytochrome c oxidase modifications, *Proteomics* 9, 5188–5193, 2009.

488. Chavez, J., Chung, W.G., Miranda, C.L. et al., Site-specific protein adducts of 4-hydroxy-2(E)-nonenal in human THP-1 monocytic cells: Protein carbonylation is diminished by ascorbic acid, *Chem. Res. Toxicol.* 23, 37–47, 2010.

489. Doorn, J.A. and Petersen, D.R., Covalent modification of amino acid nucleophiles by the lipid peroxidation products 4-hydroxy-2-nonenal and 4-oxo-2-nonenal, *Chem. Res. Toxicol.* 15, 1445–1450, 2002.

490. Esterbauer, H., Zollner, H., and Lang, J., Metabolism of the lipid peroxidation product 4-hydroxynonenal by isolated hepatocytes and liver cytosolic fractions, *Biochem. J.* 228, 363–368, 1985.

491. Lee, S.H., Goto, T., and Oe, T., A novel 4-oxo-2(E)-nonenal-derived modification to angiotensin II: Oxidative decarboxylation of N-terminal aspartic acid, *Chem. Res. Toxicol.* 21, 2237–2237, 2008.

492. Nässtrom, T., Fagerqvist, T., Barbu, M. et al., The lipid peroxidation products 4-oxo-2-nonenal and 4-hydroxy-2-nonenal promote the formation of α-synuclein oligomers with distinct biochemical, morphological, and functional properties, *Free Radic. Biol. Med.* 50, 428–437, 2011.

493. https://www.caymanchem.com.

494. Armstrong, D., ed., *Oxidative Stress Biomarkers and Antioxidant Protocols*, Springer Protocols, New York, 2002.

495. Gregory, J.D., The stability of N-ethylmaleimide and its reaction with sulfhydryl groups, *J. Am. Chem. Soc.*, 77, 3922–3923, 1955.

496. Leslie, J., Spectral shifts in the reaction of N-ethylmaleimide with proteins, *Anal. Biochem.* 10, 162–167, 1965.

497. Gorin, G., Martic, P.A., and Doughty, G., Kinetics of the reaction of N-ethylmaleimide with cysteine and some congeners, *Arch. Biochem. Biophys.*, 115, 593–597, 1966.

498. LoPachin, R.M., Gavin, T., Petersen, D.R., and Barber, D.S., Molecular mechanisms of 4-hydroxy-2-nonenal and acrolein toxicity: Nucleophilic targets an adduct formation, *Chem. Res. Toxicol.* 22, 1499–1508, 2009.

499. Smyth, D.G., Nagamatsu, A., and Fruton, J.S., Reactions of N-ethylmaleimide, *J. Am. Chem. Soc.* 82, 4600–4604, 1960.

500. Brewer, C.F. and Riehm, J.P., Evidence for possible nonspecific reactions between N-ethylmaleimide and proteins, *Anal. Biochem.* 18, 248–255, 1967.

501. Smyth, D.G., Blumenfeld, O.O., and Konigsberg, W., Reaction of N-ethylmaleimide with peptides and amino acids, *Biochem. J.* 91, 589–595, 1964.

502. Morodor, L., Musiel, H.J., and Scharf, R., Aziridine-2-carboxylic acid: A reactive amino acid unit for a new class of cysteine proteinase inhibitors, *FEBS Lett.* 299, 51–53, 1992.

503. Papini, A., Rudolph, S., Siglmueller, G. et al., Alkylation of histidine with maleimido-compounds, *Int. J. Pept. Protein Res.* 39, 345–355, 1992.

504. Sharpless, N.E. and Flavin, M., The reactions of amines and amino acids with maleimides. Structure of the reaction products deduced from infrared and nuclear magnetic resonance spectroscopy, *Biochemistry* 5, 2963–2971, 1966.

505. Smith, C.P., A study of the conformational properties of bovine pancreatic ribonuclease A by electron paramagnetic resonance, *Biochemistry* 7, 745–757, 1968.

506. Bednar, R.A., Reactivity and pH dependence of thiol conjugation to N-ethylmaleimide: Detection of a conformational change in chalcone isomerase, *Biochemistry* 29, 3684–3690, 1990.

507. Kot, M. and Bicz, A., Inactivation of jack bean urease by N-ethylmaleimide: pH dependence, reversibility and thiols influence, *J. Enzyme Inhib. Med. Chem.* 23, 514–524, 2008.

508. Gehring, H. and Christen, P., Syncatalytic conformational changes in mitochondrial aspartate aminotransferases Evidence from modification and demodification of cys 166 in the enzyme from chick and pig, *J. Biol. Chem.* 253, 3158–3163, 1978.

509. Brown, R.D. and Matthews, K.S., Chemical modification of lactose repressor protein using N-substituted maleimides, *J. Biol. Chem.* 254, 5128–5134, 1979.

510. Alderson, N.L., Wang, Y., Blatnik, M. et al., S-(2-succinyl)cysteine: A novel chemical modification of tissue proteins by a Krebs cycle intermediate, *Arch. Biochem. Biophys.* 450, 1–8, 2006.

511. Blatnik, M., Frizzell, N., Thorpe, S.R., and Baynes, J.W., Inactivation of glyceraldehyde-3-phosphate dehydrogenase by fumarate in diabetes—Formation of S-(2-succinyl)cysteine, a novel chemical modification of protein and possible biomarker of mitochondrial stress, *Diabetes* 57, 41–49, 2008.

512. Thomas, S.A., Storey, K.B., Baynes, J.W., and Frizzell, N., Tissue distribution of S-(2-succino)cysteine (S2C), a biomarker of mitochondrial stress in obesity and diabetes, *Obesity* 20, 263–269, 2012.

513. Lin, H.N., Su, X.Y., and He, B., Protein lysine acylation and cysteine succination by intermediates of energy metabolism, *ACS Chem. Biol.* 7, 947–960, 2012.

514. Knee, K.M., Roden, C.K., Flory, M.R., and Mukerji, I., The role of β93 Cys in the inhibition of Hb S fiber formation, *Biophys. Chem.* 127, 181–193, 2007.

515. Purvis, A.R., Gross, J., Dang, L.T. et al., Two cys residues essential for von Willebrand factor multimer assembly in the Golgi, *Proc. Natl. Acad. Sci. USA* 104, 15647–15652, 2007.

516. Jones, L.N., Baldwin, S.A., Henderson, P.J., and Ashcroft, A.E., Defining topological features of membrane proteins by nanoelectrospray ionization mass spectrometry, *Rapid Commun. Mass Spectrom.* 24, 276–284, 2010.

517. Nagai, R., Brock, J.W., Blatnik, M. et al., Succination of protein thiols during adipocyte maturation A biomarker of mitochondrial stress, *J. Biol. Chem.* 282, 34219–34228, 2007.

518. Boettcher, B. and Martinez-Carrion, M., Glutamine aspartate transaminase modified at cysteine 390 with enriched carbon-13 cyanide, *Biochem. Biophys. Res. Commun.* 64, 28–33, 1975.

519. Martinez-Carrion, M. and Sneden, D., Substrate-mediated increased reactivity of a critical sulfhydryl group crystals of cytoplasmic aspartate transaminase, *Arch. Biochem. Biophys.* 202, 624–628, 1980.

520. Gehring, H. and Christen, P., A diagonal procedure for isolating sulfhydryl peptides alkylated with N-ethylmaleimide, *Anal. Biochem.* 107, 358–341, 1980.

521. Gehring, H. and Christen, P., Isolation of sulfhydryl peptides alkylated with *N*-ethylmaleimide by diagonal electrophoresis, *Methods Enzymol.* 91, 392–396, 1983.

522. D'Alessandro, A., D'Amici, G.M., Vaglio, S., and Zolla, L., Time-course investigation of SAGM-stored leukocyte-filtered red blood cell concentrates: From metabolism to proteomics, *Haematologica* 97, 107–115, 2012.

523. Mickelson, M.N., Glucose transport in *Streptococcus agalactiae* and its inhibition by lactoperoxidase-thiocyanate-hydrogen peroxide, *J. Bacteriol.* 132, 541–548, 1977.

524. Lin, C., Sarath, G., Frank, J.A., and Krueger, R.J., Bivalent ACTH antagonists: Influence of peptide and spacer components on potency enhancement, *Biochem. Pharmacol.* 41, 789–795, 1991.

525. Tsutsui, I. and Ohkawa, T., *N*-Ethylmaleimide blocks the hydrogen ion pump in the plasma membrane of *Chara corallina* internodal cells, *Plant Cell Physiol.* 34, 1159–1162, 1993.

526. Naganuma, T., Murayama, T., and Nomura, Y., Modifications of Ca^{2+} mobilization and noradrenaline release by *S*-nitroso-cysteine in PC12 cells, *Arch. Biochem. Biophys.* 364, 133–142, 1999.

527. Menshikova, E.V., Cheong, E., and Salama, G., Low *N*-ethylmaleimide concentrations activate ryanodine by a reversible interaction, not an alkylation of critical thiols, *J. Biol. Chem.* 275, 36775–36780, 2000.

528. Smith, M.E.B., Schumacher, F.F., Ryan, C.P. et al., Protein modification, bioconjugation, and disulfide bridging using bromomaleimides, *J. Am. Chem. Soc.* 132, 1960–1965, 2010.

529. Tedaldi, L.M., Smith, M.E.B., Nathani, R.I., and Baker, J.R., Bromomaleimides: New reagents for the selective and reversible modification of cysteine, *Chem. Commun.* (43), 6583–6585, 2009.

530. Scales, C.W., Convertine, A.J., and McCormick, C.L., Fluorescent labeling of RAFT-generated poly(*N*-isopropylacrylamide) via a facile maleimide-thiol coupling reaction, *Biomacromolecules* 7, 1389–1392, 2006.

531. Ryan, C.P., Smith, M.E.B., Schumacher, F.F. et al., Tunable reagents for multi-functional bioconjugation: Reversible or permanent modulation of proteins and peptides by control maleimide hydrolysis, *Chem. Commun.* 47, 5452–5454, 2011.

532. Machida, M., Machida, M.L., and Kanaoka, Y., Hydrolysis of *N*-substituted maleimides: Stability of fluorescence thiol reagents in aqueous media, *Chem. Pharm. Bull.* 25, 2739–2743, 1977.

533. Schlesinger, J., Fischer, C., Koezle, I. et al., Radiosynthesis of new [90Y]-DOTA-based maleimide reagents suitable for the prelabeling of thiol-bearing L-oligonucleotides and peptides, *Bioconjug. Chem.* 20, 1340–1348, 2009.

534. Moody, P., Smith, M.E.B., Ryan, C.P. et al., Bromomaleimide-linked bioconjugates are cleavable in mammalian cells, *Chembiochem* 13, 39–41, 2012.

535. Serafimova, I.M., Pufall, M.A., Krishnan, S. et al., Reversible targeting of noncatalytic cysteines with chemically tuned electrophiles, *Nat. Chem. Biol.* 8, 471–476, 2012.

536. Baldwin, A.D. and Kiick, K.L., Tunable degradation of maleimide-thiol adducts in reducing environments, *Bioconjug. Chem.* 22, 1946–1953, 2011.

537. Miura, K., Tomioka, Y., Hoshi, Y. et al., The effects of unsaturated fatty acids, oxidizing agents and Michael acceptors on the induction of *N*-ethylmaleimide reductase in *Escherichia coli*: Possible application for drug design of chemoprotectors, *Methods Find. Exp. Clin. Pharmacol.* 19, 147–151, 1997.

538. Mizugaki, M., Miura, K., Yamamoto, H. et al., Pro-R hydrogen of NADPH was abstracted for enzymatic hybrids transfer by *N*-ethylmaleimide-reductase of *Yarrowia lipolytica*, *Tohoku J. Exp. Med.* 181, 447–457, 1997.

539. Umezawa, Y., Shimada, T., Kori, A. et al., The uncharacterized transcription factor YdhM is the regulator of the nemA gene, encoding *N*-ethylmaleimide reductase, *J. Bacteriol.* 190, 5890–5897, 2008.

540. Eder, M., Krivoshein, A.V., Backer, M. et al., ScVEGF-PEG-HBED-CC and scVEGF-PEG-NOTA conjugates: Comparison of easy-to-label recombinant proteins for [68Ga]PET imaging of VEGF receptors in angiogenic vasculature, *Nucl. Med. Biol.* 37, 405–412, 2010.

541. Tolmachev, V., Altai, M., Sandström, M. et al., Evaluation of a maleimido derivative of NOTA for site-specific labeling of affibody molecules, *Bioconjug. Chem.* 22, 894–902, 2001.

542. Li, W., Niu, G., Lang, L. et al., PET imaging of EGF receptors using [^{18}F]FBEM-EGF in a head and neck squamous cell carcinoma model, *Eur. J. Nucl. Med. Mol. Imaging* 39, 300–308, 2012.

543. Kurtz, L., Kao, L., Newman, D. et al., Integrin αIIbβ3 inside-out activation: An in situ conformational analysis reveals a new mechanism, *J. Biol. Chem.* 287, 23255–23565, 2012.

544. Heitz, J.R., Anderson, C.D., and Anderson, B.M., Inactivation of yeast alcohol dehydrogenase by N-alkylmaleimides, *Arch. Biochem. Biophys.* 127, 627–636, 1968.

545. Anderson, B.M. and Vasini, E.C., Nonpolar effects in reaction of the sulfhydryl group of papain, *Biochemistry* 9, 3348–3352, 1970.

546. Bender, M.L. and Brubacher, L.J., The kinetics and mechanism of papain-catalyzed hydrolyses, *J. Am. Chem. Soc.* 88, 5580–5589, 1966.

547. Whitaker, J.R. and Perez-Villaseñor, J., Chemical modification of papain. I. Reaction with the chloromethyl ketones of phenylalanine and lysine and with phenylmethylsulfonyl fluoride, *Arch. Biochem. Biophys.* 124, 70–78, 1968.

548. Wolthers, B.C., Kinetics of inhibition of papain by TLCK and TPCK in the presence of BAEE as substrate, *FEBS Lett.* 2, 143–145, 1969.

549. Lindley, H., A study of the kinetics of the reaction between thiol compounds and chloroacetamide, *Biochem. J.* 74, 577–584, 1960.

550. Gul, S., Hussain, S., Thomas, M.P. et al., Generation of nucleophilic character n the cys25/his159 ion pair of papain involve trp177 and not asp158, *Biochemistry* 47, 2025–2035, 2008.

551. Evans, B.L., Knopp, J.A., and Horton, H.R., Effect of hydroxynitrobenzylation of tryptophan-177 on reactivity of active site cysteine-25 in papain, *Arch. Biochem. Biophys.* 206, 362–371, 1981.

552. Shaw, E., Mares-Guia, M., and Cohen, W., Evidence for an active-center histidine in trypsin through use of a specific reagent, 1-chloro-3-tosylamido-7-amino-2-heptanone, the chloromethyl ketone derived from N$^\alpha$-L-lysine, *Biochemistry* 4, 2219–2224, 1965.

553. Coleman, L.E., Jr., Bork, J.F., and Dunn, H., Jr., Reaction of primary aliphatic amines with maleic anhydride, *J. Org. Chem.* 29, 135–136, 1959.

554. Song, H.Y., Ngai, M.H., Song, Z.Y. et al., Practical synthesis of maleimides and coumarin-linked probes for protein and antibody labelling *via* reduction of native disulfides, *Org. Biomol. Chem.* 7, 3400–3406, 2009.

555. Fonda, M.L. and Anderson, B.M., D-Amino acid oxidase IV. Inactivation by maleimides, *J. Biol. Chem.* 244, 666–674, 1969.

556. Anderson, B.M., Kim, S.J., and Wang, C.-N., Inactivation of rabbit muscle L-α-glycerophosphate dehydrogenase by N-alkylmaleimides, *Arch. Biochem. Biophys.* 138, 66–72, 1970.

557. Le-Quoc, K., Le-Quoc, D., and Gaudemer, Y., Evidence for the existence of two classes of sulfhydryl groups essential for membrane-bound succinate dehydrogenase activity, *Biochemistry* 20, 1705–1710, 1981.

558. Abbott, R.E. and Schachter, D., Topography and functions of sulfhydryl groups of the human erythrocyte glucose transport mechanism, *Mol. Cell. Biochem.* 82, 85–90, 1988.

559. May, J.M., Reaction of an exofacial sulfhydryl group on the erythrocyte hexose carrier with an impermeant maleimide. Relevance to the mechanism of hexose transport, *J. Biol. Chem.* 263, 13635–13640, 1988.

560. May, J.M., Interaction of a permeant maleimide derivative of cysteine with the erythrocyte glucose carrier. Differential labelling of an exofacial carrier thiol group and its role in the transport mechanism, *Biochem. J.*, 263, 875–881, 1989.

561. Abbott, R.E. and Schachter, D., Impermeant maleimides. Oriented probes of erythrocyte membrane proteins, *J. Biol. Chem.*, 251, 7176–7183, 1976.

562. May, J.M., Selective labeling of the erythrocyte hexose carrier with a maleimide derivative of glucosamine: Relationship of an exofacial sulfhydryl to carrier conformation and structure, *Biochemistry* 28, 1718–1725, 1989.

563. Griffiths, D.G., Partis, M.D., Sharp, R.N., and Beechey, R.B., N-Polymethylenecarboxy-maleimides—A new class of probes for membrane sulfhydryl groups, *FEBS Lett.* 134, 261–263, 1981.

564. Griffiths, D.G., Partis, M.D., Churchill, P. et al., The use of amphipathic maleimides to study membrane-associated proteins, *J. Bioenerg. Biomembr.* 22, 691–707, 1990.

565. Ridge, K.D., Zhang, C., and Khorana, H.G., Mapping of the amino acids in the cytoplasmic loop connecting helices C and C in rhodopsin. Chemical reactivity in the dark state following single cysteine replacements, *Biochemistry* 34, 8804–8811, 1995.

566. Hirano, M., Takeuchi, Y., Aoki, T. et al., Rearrangements in the cytoplasmic domain underlie its gatings, *J. Biol. Chem.* 385, 3777–3783, 2010.

567. Lu, Q., Chen, Z., Perumattam, J. et al., Aryl-indole maleimides as inhibitors of CAMKIIδ Part 3: Importance of the indole orientation, *Bioorg. Med. Chem. Lett.* 18, 2399–2403, 2008.

568. Labar, G., Bauvois, C., Bore, F. et al., Crystal structure of the human monoacylglycerol lipase, a key factor in endocannabinoid signalling, *ChemBioChem.* 11, 218–227, 2010.

569. Mann, R.S., Usova, E.V., Cass, C.E., and Eriksson, S., Fluorescence energy transfer studies of human deoxycytidine kinase: Role of cysteine 185 in the conformational changes that occur upon substrate binding, *Biochemistry* 45, 3534–3541, 2006.

570. Wache, N., Schröder, C., Koch, K.W., and Cristoffers, J., Diaminoterephthalate turn-on fluorescence probe for thiol—Tagging of recoverin and tracking of its conformational change, *Chembiochem.* 13, 993–998, 2012.

571. Niwayama, S., Kurano, S., and Matsumoto, N., Synthesis of *d*-labeled N-alkylmaleimides and application to quantitative peptide analysis by isotope differential mass spectrometry, *Bioorg. Med. Chem. Lett.* 11, 2257–2261, 2001.

572. Jones, P.C., Sivaprasadarao, A., Wray, D., and Findlay, J.B., A method for determining transmembrane protein structure. *Mol. Membr. Biol.* 13, 53–60, 1996.

573. Di Gleria, K. et al., N-(2-Ferrocene-ethyl)maleimide: A new electroactive sulfhydryl-specific reagent for cysteine-containing peptides and proteins, *FEBS Lett.* 390, 142–144, 1996.

574. Falke, J.J., Dernburg, A.F., Sternberg, D.A. et al., Structure of a bacterial sensory receptor. A site-directed sulfhydryl study, *J. Biol. Chem.*, 263, 14850–14858, 1988.

575. Rial, E., Aréchaga, I., Sainz-de-la-Maza, E., and Nicholls, D.G., Effect of hydrophobic sulphydryl reagents on the uncoupling protein and inner-membrane anion channel of brown-adipose-tissue mitochondria, *Eur. J. Biochem.*, 182, 187–193, 1989.

576. Yee, A.S., Corley, D.E., and McNamee, M.G., Thiol-group modification of *Torpedo californica* acetylcholine receptor: Subunit localization and effects on function, *Biochemistry* 25, 2110–2119, 1986.

577. Pradier, L., Yee, A.S., and McNamee, M.G., Use of chemical modifications and site-directed mutagenesis to probe the functional role of thiol groups on the gamma subunit of *Torpedo californica* acetylcholine receptor, *Biochemistry* 28, 6562–6571, 1989.

578. Juszczak, L.J., Manjula, B., Bonaventura, C. et al., UV resonance Raman study of β93-modified hemoglobin A: Chemical modification-specific effects and added influences of attached poly(ethylene glycol) chains, *Biochemistry* 41, 376–385, 2002.

579. Doherty, D.H., Rosendahl, M.S., Smith, D.J. et al., Site-specific PEGylation of engineered cysteine analogues of recombinant human granulocyte-macrophage colony-stimulated factor, *Bioconjug. Chem.* 16, 1291–1298, 2005.

580. Salmaso, S. Bersani, S., Scomparin, A. et al., Tailored PEG for rh-G-CSF analogue site-specific conjugation, *Bioconjug. Chem.* 29, 1179–1189, 2009.

581. Long, D.L., Doherty, D.H. Eisenberg, S.P. et al., Design of homogeneous, nonpegylated erythropoietin analogues with preserved in vitro bioactivity, *Exp. Hematol.* 34, 697–704, 2006.

582. Yu, P., Zheng, C., Chen, J. et al., Investigations on PEGylation strategy of recombinant human interleukin-1 receptor antagonist, *Bioorg. Med. Chem.* 15, 5396–5405, 2007.

583. Ni, J., Singh, S., and Wang, L.-X., Synthesis of maleimide-activated carbohydrates as chemoselective tags for site-specific glycosylation of peptides and proteins, *Bioconjug. Chem.* 14, 232–238, 2003.

584. Gamblin, D.P., Scanlan, E.M., and Davis, B.G., Glycoprotein synthesis: An update, *Chem. Rev.* 109, 131–163, 2009.

585. Disney, M.D. and Seeberger, P.H., The use of carbohydrate microarrays to study carbohydrate-cell interactions to detect pathogens, *Chem. Biol.* 11, 1701–1707, 2004.

586. Yatawara, A.K., Achani, K., Tiruchinapally, G. et al., Carbohydrate surface attachment characterized by sum frequency generation spectroscopy, *Langmuir* 25, 1901–1904, 2009.

587. Di Stefano, G., Lanza, M., Kratz, F. et al., A novel method for coupling doxorubicin to lactosaminated human albumin by an acid sensitive hydrazone bond: Synthesis, characterization and preliminary biological properties of the conjugate, *Eur. J. Pharm. Sci.* 23, 393–397, 2004.

588. Brown, R.D. and Matthews, K.S., Spectral studies on Lac repressor modified with N-substituted maleimide probes, *J. Biol. Chem.*, 254, 5135–5143, 1979.

599. Perussi, J.R., Tinto, M.H., Nascimento, O.R., and Tabak, M., Characterization of protein spin labeling by maleimide: Evidence for nitroxide reduction, *Anal. Biochem.* 173, 289–295, 1988.

590. Marquez, J., Iriarte, A., and Martinez-Carrion, M., Covalent modification of a critical sulfhydryl group in the acetylcholine receptor: Cysteine-222 of the α-subunit, *Biochemistry*, 28, 7433–7439, 1989.

591. Mills, J.S., Walsh, M.P., Nemcek, K., and Johnson, J.D., Biologically active fluorescent derivatives of spinach calmodulin that report calmodulin target protein binding, *Biochemistry* 27, 991–996, 1988.

592. Jezek, P. and Drahota, Z., Sulfhydryl groups of the uncoupling protein of brown adipose tissue mitochondria—Distinction between sulfhydryl groups of the H^+-channel and the nucleotide binding site, *Eur. J. Biochem.* 183, 89–95, 1989.

593. Apuy, J.L., Chen, X., Russell, D.H. et al., Radiometric pulsed alkylation/mass spectrometry of the cysteine pairs in individual zinc fingers of MRE-binding transcription factor-1 (MTF-1) as a probe of zinc chelate stability, *Biochemistry* 40, 15164–15175, 2001.

594. Apuy, J.L., Busenlehner, L.A., Russell, D.H., and Giedroc, D.P., Radiometric pulsed alklylation mass spectrometry as a probe of thiolate reactivity in different metalloderivatives of *Staphylococcus aureus* p1258, CadC, *Biochemistry* 43, 3824–3834, 2004.

595. Apuy, J.L., Park, Z.Y., Swartz, P.D. et al., Pulsed-alkylation mass spectrometry for the study of protein folding and dynamics: Development and application to the study of a folding/unfolding intermediate of bacterial luciferase, *Biochemistry* 40, 15153–15163, 2001.

596. Qiu, Y., Sousa, E.A., Hewick, R., and Wang, J.H., Acid-labile isotope-coded extractants: A class of reagents for quantitative mass spectrometric analysis of complex protein mixtures, *Anal.. Chem.* 74, 4969–4979, 2002.

597. Chowdhury, S.M., Munske, G.R., Siems, W.F., and Bruce, J.E., A new maleimide-bound acid-cleavable solid-support reagent for profiling phosphorylation, *Rapid Commun. Mass Spectrom.* 19, 899–909, 2005.

598. Qian, W.J., Gosche, M.B., Camp, D.G. et al., Phosphoprotein isotope-coded solid-phase tag approach for enrichment and quantitative analysis of phosphoproteins from complex mixtures, *Anal. Chem.* 75, 5441–5450, 2003.

599. Fauq, A.H., Kache, R., Khan, M.A., and Vega, J.E., Synthesis of acid-cleavable light isotope-coded affinity tags (ICAT-L) for potential use in proteomic expression profiling analysis, *Bioconjug. Chem.* 17, 248–254, 2006.

600. Shaunak, S., Godwin, A., Choi, J.W. et al., Site-specific PEGylation of native disulfide bonds in therapeutic proteins, *Nat. Chem. Biol.* 2, 312–313, 2006.

601. Brocchini, S., Godwin, A., Balan, S. et al., Disulfide bridge based PEGylation of proteins, *Adv. Drug Deliv. Rev.* 60, 3–12, 2008.

602. Balan, S., Choi, J.W., Godwin, A. et al., Site-specific PEGylation of protein disulfide bonds using a three-carbon bridge, *Bioconjug. Chem.* 18, 61–76, 2007.

603. Rogers, L.K., Leinweber, B.L., and Smith, C.V., Detection of reversible protein thiol modifications in proteins, *Anal. Biochem.* 358, 171–184, 2006.

604. Crawhall, J.C. and Segal, S., The intracellular ratio of cysteine and cystine in various tissues, *Biochem. J.* 105, 891–896, 1967.

605. Smith, D.J., Maggio, E.T., and Kenyon, G.L., Simple alkane thiol groups for temporary blocking of sulfhydryl groups of enzymes, *Biochemistry* 14, 766, 1975.

606. Kluger, R. and Tsue, W.-C., Amino group reactions of the sulfhydryl reagent methyl methanethiolsulfonate. Inactivation of D-3-hydroxybutyrate and reaction with amines in water. *Can. J. Biochem.* 58, 629–632, 1980.

607. Pathak, R., Hendrickson, T.L., and Imperiali, B., Sulfhydryl modification of the yeast Wbp1p inhibits oligosaccharide transferase activity, *Biochemistry* 34, 4179–4181, 1995.

608. Nishimura, J.S., Kenyon, G.L., and Smith, G.L., Reversible modification of the sulfhydryl groups of *Escherichia coli* succinic thiokinase with methane thiolating reagents, 5,5′-dithio-bis(2-nitrobenzoic acid), *p*-hydroxymercuribenzoate, and ethylmercuri thiosalicylate, *Arch. Biochem. Biophys.* 170, 461–467, 1975.

609. Roberts, D.D., Lewis, S.D., Ballou, D.D. et al., Reactivity of small thiolate anions and cysteine-25 in papain toward methyl methanethiolsulfonate, *Biochemistry* 25, 5595–5601, 1986.

610. Stauffers, D.A. and Karlin, A. Electrostatic potential of the acetylcholine binding site in the nicotinic receptor probed by reaction of binding-site cysteines with charged methanethiolsulfonates, *Biochemistry* 33, 6840–6849, 1994.

611. Kenyon, G.L. and Bruice, T.W., Novel sulfhydryl reagents, *Methods Enzymol.* 47, 407–430, 1977.

612. Bruice, T.W. and Kenyon, G.L., Novel alkyl alkanethiolsulfonate sulfhydryl reagents. Modification of derivatives of l-cysteine, *J. Protein Chem.* 1, 47–58, 1982.

613. Toronto Research Chemicals, Inc., Methanethiolsulfonate reagents: Application to the study of protein topology and ion channels, http://www.trc-canada.com/white_papers.lasso.

614. Karlin, A. and Akabas, M.H., Substituted-cysteine accessibility method, *Methods Enzymol.* 293, 123–145, 1998.
615. Karlin, A., Scam feels the pinch, *J. Gen. Physiol.* 117, 235–238, 2001.
616. Park, J.S. and Hammond, J.P., Cysteine residues in the transmembrane (TM) 9 to 11 domains of the human equilibrative nucleoside transporter subtype 1 play an important role in inhibitor binding and transmembrane function, *Mol. Pharmacol.* 82, 784–794, 2012.
617. Wang, Y., Zhang, M., Xu, Y. et al., Probing the structural basis for differential KCNQ1 modulation by KCNE1 and KCNE2, *J. Gen. Physiol.* 140, 653–669, 2012.
618. Endeward, B., Butterwick, J.A., MacKinnon, R., and Prisner, T.F., Pulsed electron-electron double-resonance determination of spin-label distances and orientations on the tetrameric potassium ion channel KcsA, *J. Am. Chem. Soc.* 131, 15246–15250, 2009.
619. Passero, C.J., Okumura, S., and Carattino, M.D., Conformational changes associated with proton-dependent gating of ASIC1a, *J. Biol. Chem.* 284, 36473–36481, 2009.
620. Matsuda, J.J., Filali, M.S., Collins, M.M. et al., The ClC-3 Cl$^-$/H$^+$ antiporter becomes uncoupled at low extracellular pH, *J. Biol. Chem.* 285, 2569–2579, 2010.
621. McNally, B.A., Yamashita, M., Engh, A., and Prakriya, M., Structural determinants of ion permeation in CBAC channels, *Proc. Natl. Acad. Sci. USA* 106, 22516–22521, 2009.
622. Nichols, A.S. and Luetje, C.W., Transmembrane segment 3 of *Drosophila melanogaster* odorant receptor subunit 85b contributes to ligand-receptor interactions, *J. Biol. Chem.* 285, 11854–11862, 2010.
623. Zhang, X.D., Yu, W.P., and Chen, T.Y., Accessibility of the CLC-0 pore to charged methanethiosulfonate reagents, *Biophys. J.* 98, 377–385, 2010.
624. Belvy, V., Anishkin, A., Kamaraju, K. et al., The tension-transmitting 'clutch' in the mechanosensitive channel MscS, *Nat. Struct. Mol. Biol.* 17, 451–458, 2010.
625. Bruhova, L. and Zhorov, B.S., A homology model for the pore domain of a voltage-gated calcium channel is consistent with available SCAM data, *J. Gen. Physiol.* 135, 261–274, 2010.
626. Bargeton, B. and Kellenberger, S., The contact region between the three domains of the extracellular loop of ASIC1a is critical for channel function, *J. Biol. Chem.* 285, 13816–13826, 2010.
627. Kurata, H.T., Zhu, E.A., and Nichols, C.S., Locale and chemistry of spermine binding in the archetypal inward rectifier Kir2.1, *J. Gen. Physiol.* 135, 495–508, 2010.
628. Wu, D., Delaloye, K., Zaydman, M.A. et al., State-dependent electrostatic interactions of S4 arginines with E1 in S2 during Kv7.1 activation, *J. Gen. Physiol.* 135, 595–606, 2010.
629. Jha, S.K. and Udagaonkar, J.B., Exploring the cooperativity of the fast folding reaction of a small protein using pulsed thiol labeling and mass spectrometry, *J. Biol. Chem.* 282, 37479–37391, 2007.
630. Patra, A.K., Mukhopadhyay, R., Mukhija, R. et al., Optimization of inclusion body solubilization and renaturation of recombinant human growth hormone from *Escherichia coli*, *Protein Expr. Purif.* 18, 182–192, 2000.
631. Razeghifard, M.R., On-column refolding of recombinant human interleukin-4 from inclusion bodies, *Protein Expr. Purif.* 37, 180–186, 2004.
632. Cindrić, M., Cepo, T., Marinc, S. et al., Determination of dithiothreitol in complex protein mixtures by HPLC-MS, *J. Sep. Sci.* 31, 3489–3496, 2008.
633. Inoue, M. Akimaru, J., Nishikawa, T. et al., A new derivatizing agent, trimethylammoniopropyl methanethiosulfonate, is efficient for preparation of recombinant brain-derived neurotrophic factor from inclusion bodies, *Biotechnol. Appl. Biochem.* 28, 207–213, 1998.
634. Gazaryan, I.G., Doseeva, V.V., Galkin, A.G., and Tishkov, V.I., Effect of single-point mutations Phe[41]→His and Phe[143]→Glu on folding and catalytic properties of recombinant horseradish peroxidase expressed in *E. coli*, *FEBS Lett.* 354, 248–250, 1994.
635. Murata, H., Sakaguchi, M., Futami, J. et al., Denatured and reversibly cationized p53 readily enters cells and simultaneously folds to the functional protein in the cells, *Biochemistry* 45, 6124–6132, 2006.
636. Futami, J., Kitazoe, M., Murata, H., and Yamada, H., Exploiting protein catonization techniques in future drug development, *Expert Opin.* 2, 261–269, 2007.
637. Brandwein, H.J., Lewicki, J.A., and Murad, F., Reversible inactivation of guanylate cyclase by mixed disulfide formation, *J. Biol. Chem.* 256, 2958–2962, 1981.
638. Riddles, P.W., Andrews, R.K., Blakeley, R.L., and Zerner, B., Jack bean urease 6. Determination of thiol and disulfide content—Reversible inactivation of the sulfhydryl-group oxidation and reduction, *Biochem. Pharmacol.* 32, 811–818, 1983.
639. Ziegler, D.M., Role of reversible oxidation-reduction of enzyme thiols-disulfides in metabolic regulation, *Annu. Rev. Biochem.* 54, 305–329, 1985.

640. Thomas, J.A., Chai, Y.-C., and Jung, C.-H., Protein S-thiolation and dethiolation, *Methods Enzymol.* 233, 385–394, 1994.
641. Lim, A. et al., Identification of *S*-sulfonation and *S*-thiolation of a novel transthyretin Phe33Cys variant from a patient diagnosed with familial transthyretin amyloidosis, *Protein Sci.* 12, 1775–1785, 2003.
642. Shenton, D. and Grant, C.M., Protein S-thiolation targets glycolysis and protein synthesis in response to oxidative stress in the yeast *Saccharomyces cerevisiae*, *Biochem. J.* 374, 513–519, 2003.
643. Braughler, J.M., Soluble guanylate-cyclase activation by nitric oxide and its reversal—Involvement of sulfhydryl group oxidation and reduction, *Biochem. Pharmacol.* 32, 811–818, 1983.
644. Wu, X.B., Brune, B., Vonappen, F., and Ullrich, V., Reversible activation of soluble guanylate cyclase by oxidizing agents, *Arch. Biochem. Biophys.* 294, 75–82, 1992.
645. Cunningham, L.W. and Nuenke, B.J., Physical and chemical studies of a limited reaction of iodine with proteins, *J. Biol. Chem.* 234, 1447–1451, 1959.
646. Cunningham, L.W. and Nuenke, B.J., Analysis of modified β-lactoglobulins and ovalbumins prepared from sulfenyl iodide intermediates, *J. Biol. Chem.* 235, 1711–1715, 1960.
647. Cupo, J.F. and Pace, C.N., Conformational stability of mixed disulfide derivatives of β-lactoglobulin B, *Biochemistry* 22, 2654–2658, 1983.
648. Mecinović, J., Chowdhury, R., Flashman, E., and Schofield, C.J., Use of mass spectrometry to probe the nucleophilicity of cysteinyl residues of prolyl hydroxylase domain 2, *Anal. Biochem.* 393, 215–221, 2009.
649. Sanchez, R., Riddle, M., Woo, J., and Momand, J., Prediction of reversibly oxidized protein cysteine thiols using protein structure properties, *Protein Sci.* 17, 473–481, 2008.
650. Peskin, A.V. and Winterbourn, C.C, Taurine chloramine is more selective than hypochlorous acid at targeting critical cysteines and inactivating creatine kinase and glyceraldehyde-3-phosphate dehydrogenase, *Free Radic. Biol. Med.* 40, 45–53, 2006.
651. Tew, K., Redox in redux: Emergent roles of glutathione *S*-transferase P in regulation of cell signalling and *S*-glutathionylation, *Biochem. Pharmacol.* 73, 1257–1269, 2007.
652. Ochoa, G., Gutierrez, C., Ponce, I. et al., Reactivity trends of surface-confined Co-tetraphenyl porphyrins and vitamin B12 for the oxidation of 2-aminoethanethiol: Comparison with Co-phthalocyanines and oxidation of other thiols, *J. Electroanal. Chem.* 639, 88–94, 2010.
653. Wynn, R. and Richards, F.M., Chemical modification of protein thiols: Formation of mixed disulfides, *Methods Enzymol.* 251, 351–356, 1995.
654. Pihl, A. and Lange, R., The interaction of oxidized glutathione, cystamine, monosulfoxide, and tetrathionate with the -SH groups of rabbit muscle, D-glyceraldehyde 3-phosphate dehydrogenase, *J. Biol. Chem.* 237, 1356–1362, 1962.
655. Parker, D.J. and Allison, W.S., The mechanism of inactivation of glyceraldehyde 3-phosphate dehydrogenase by tetrathionate, *o*-iodosobenzoate, and iodine monochloride, *J. Biol. Chem.* 244, 180–189, 1969.
656. Barbehenn, E.K. and Kaufman, B.T., Activation of chicken liver dihydrofolate reductase by tetrathionate, *Biochem. Biophys. Res. Commun.* 85, 402–407, 1978.
657. Church, J.S. and Evans, D.J., A spectroscopic investigation into the reaction of sodium tetrathionate with cysteine, *Spectrochim. Acta A Mol. Biomol. Spectrosc.* 69, 256–262, 2008.
658. Mukhopadhyay, A., Reversible protection of disulfide bonds followed by oxidative folding render recombinant hCGbeta highly immunogenic, *Vaccine* 18, 1802–1810, 2000.
659. Tikhonov, R.V. et al., Recombinant human insulin. VIII. Isolation of fusion proteins –*S*– sulfonate, biotechnological precursor of human insulin, from the biomass of transformed *Escherichia coli* cells, *Protein Expr. Purif.* 21, 176–182, 2001.
660. Llerena-Suster, C.R., Priolo, N.S., and Morcelle, S.R., Sodium tetrathionate effect on papain purification from different Carica papaya latex crude extracts, *Prep. Biochem. Biotechnol.* 41, 107–121, 2011.
661. Esipov, R.S., Stepanenko, V.N., Chupova, L.A., and Miroshnikov, A.I., Production of recombinant oxytocin through sulfitolysis of intein containing fusion proteins, *Protein Pept. Lett.* 19, 479–484, 2012.
662. Liu, T.-Y., Demonstration of the presence of a histidine residue at the active site of streptococcal proteinase, *J. Biol. Chem.* 242, 4029–4032, 1967.
663. Kingsbury, J.S., Laue, T.M., Klimtchuk, E.S. et al., The modulation of transthyretin tetramer stability by cysteine 10 adducts and the drug diflunisal. Direct analysis by fluorescence-detected analytical ultracentrifugation, *J. Biol. Chem.* 283, 11887–11895, 2008.
664. Kingsbury, J.S., Klimtchuk, E.S., Théberge, R. et al., Expression, purification, and in vitro cysteine-10 modification of native sequence recombinant human transthyretin, *Protein Expr. Purif.* 53, 370–377, 2007.

665. Zhang, Q. and Kelly, J.W., Cys10 mixed disulfide make transthyretin more amyloidogenic under mildly acidic conditions, *Biochemistry* 42, 8756–8761, 2003.

666. Chung, S.I. and Folk, J.E., Mechanism of the inactivation of guinea pig liver transglutaminase by tetrathionate, *J. Biol. Chem.* 245, 681–689, 1970.

667. Kaufmann, S.H. and Shaper, J.H., A subset of non-histone nuclear proteins reversibly stabilized by the sulfhydryl cross-linking reagent tetrathionate—Polypeptides of the internal nuclear matrix, *Exp. Cell Res.* 155, 477–495, 1984.

668. Zhang, P., Davis, A.T., and Ahmed, K., Mechanism of protein kinase CK2 association with nuclear matrix: Role of disulfide bond formation, *J. Cell. Biochem.* 69, 211–220, 1998.

669. Hahn, S.K., Park, J.K., Tomimatsu, T. et al., Synthesis and degradation test of hyaluronic acid hydrogels, *Int. J. Biol. Macromol.* 40, 374–380, 2007.

670. Hann, S.K., Kim, M.S., and Shimobouji, T., Injectable hyaluronic acid microhydrogels for controlled release formulation of erythropoietin, *J. Biomed. Mater. Res. A* 80, 916–924, 2007.

671. Haratake, M., Hongoh, M., Ono, M., and Nakayama, M., Thiol-dependent membrane transport of selenium through an integral protein of the red blood cell membrane, *Inorg. Chem.* 48, 7805–7811, 2009.

672. Batista-Viera, F., Barbieri, M., Ovsejevi, K. et al., A new method for reversible immobilization of thiol biomolecules based on solid-phase bound thiolsulfonate groups, *Appl. Biochem. Biotechnol.* 31, 175–195, 1991.

673. Pavlovic, E., Quist, A.P., Gelius, U. et al., Generation of thiolsulfinates and thiolsulfonates by electrooxidation of thiols on silicon surfaces for reversible immobilization of molecules, *Langmuir* 19, 4217–4221, 2003.

674. Suárez, G., Jackson, R.L., Spoors, J.A., and McNeil, C.J., Chemical introduction of disulfide groups on glycoproteins: A direct protein anchoring scenario, *Anal. Chem.* 79, 1961–1969, 2007.

675. Ellman, G.L., A colorimetric method for determining low concentrations of mercaptans, *Arch. Biochem. Biophys.* 74, 443–450, 1958.

676. Ellman, G.L., Tissue sulfhydryl groups, *Arch. Biochem. Biophys.* 82, 70–77, 1959.

677. Habeeb, A.F.S.A., Reaction of protein sulfhydryl groups with Ellman's reagent, *Methods Enzymol.* 25, 457–464, 1972.

678. Collier, H.B., A note on the molar absorptivity of reduced Ellman's reagent, 3-carboxylato-4-nitrothiophenolate, *Anal. Biochem.* 56, 310–311, 1973.

679. Riddles, P.W., Blakeley, R.L., and Zerner, B., Ellman's reagent: 5,5′-dithiobis(2-nitrobenzoic acid)—A reexamination, *Anal. Biochem.* 94, 75–81, 1979.

680. Riddles, P.W., Blakeley, R.L., and Zerner, B., Reassessment of Ellman's reagent, *Methods Enzymol.* 91, 49–60, 1983.

681. Eyer, P., Worek, F., Kiderlen, D., et al., Molar absorption coefficients for the reduced Ellman reagent: Reassessment, *Anal. Biochem.* 312, 224–227, 2003.

682. Zhang, H., Le, M., and Means, G.E., A kinetic approach to characterize the electrostatic environments of thiol groups in proteins, *Bioorg. Chem.* 26, 356–364, 1998.

683. Stewart, A.J., Blindauer, C.A., Berezenko, S. et al., Role of Tyr84 in controlling the reactivity of Cys34 of human albumin, *FEBS J.* 272, 353–362, 2004.

684. Zhu, J., Dhimitruka, I., and Pei, D., 5-(2-Aminoethyl)dithio-2-nitrobenzoate as a more base-stable alternative to Ellman's reagent, *Org. Lett.* 6, 3809–3812, 2004.

685. Legler, G., 4,4′-Dintriophenyldisulfides of different charge types as probes for the electrostatic environment of sulfhydryl groups, *Biochim. Biophys. Acta* 405, 136–143, 1975.

686. Whitesides, G.M., Lilburn, J.E., and Szajewski, R.P., Rates of thiol-disulfide interchange reactions between mono- and dithiols and Ellman's reagent, *J. Org. Chem* 42, 332–338, 1977.

687. Faulstich, H. and Heintz, D., Reversible introduction of thiol compounds into proteins by use of activated mixed disulfides, *Methods Enzymol.* 251, 357–366, 1995.

688. Messmore, J.M., Holmgren, S.K., Grilley, J.E., and Raines, R.T., Sulfur shuttle: Modulating enzymatic activity by thiol disulfide interchange, *Bioconjug. Chem.* 11, 408–413, 2000.

689. Drewes, G. and Faulstich, H., 2,4-Dinitrophenyl [^{14}C]cysteinyl disulfide allows selective radiolabeling of protein thiols under spectrophotometric control, *Anal. Biochem.* 188, 109–113, 1990.

690. Czerski, L. and Sanders, C.R., Thiol modification of diacylglycerol kinase: Dependence upon site membrane disposition and reagent hydrophobicity, *FEBS Lett.* 472, 225–229, 2000.

691. Fernandez-Diaz, M., Barsotti, L., Dumay, E., and Cheftel, J.C., Effects of pulsed electric fields on ovalbumin solutions and dialyzed egg white, *J. Agric. Food Chem.* 48, 2332–2339, 2000.

692. Narasimhan, C., Lai, C.-S., Haas, A., and McCarthy, J., One free sulfhydryl group of plasma fibronectin becomes titratable upon binding of the protein to solid substrates, *Biochemistry* 27, 4970–4973, 1988.

693. Underwood, P.A., Steele, J.G., and Dalton, B.A., Effects of polystyrene surface chemistry on the biological activity of solid phase fibronectin, analyzed with monoclonal antibodies, *J. Cell Sci.* 104, 793–803, 1993.

694. Grassetti, D.R. and Murray, J.F., Jr., Determination of sulfhydryl groups with 2,2'- or 4,4'-dithiodipyridine, *Arch. Biochem. Biophys.* 119, 41–49, 1967.

695. Brocklehurst, K. and Little, G., Reactions of papain and low-molecular-weight thiols with some aromatic disulphides. 2,2'-Dipyridyl disulphide as a convenient active-site titrant for papain even in the presence of other thiols, *Biochem. J.* 133, 67–80, 1973.

696. Harris, R.B. and Hodgins, L.T., Anomalous interaction of 2-thiopyridione with proteins during thiol-disulfide interchange reactions, *Anal. Biochem.* 109, 247–249, 1980.

697. Glatz, Z. and Masianova, H., Specific thiol determination by micellar chromatography and on-column detection reaction with 2,2'-dipyridyl disulfide, *J. Chromatogr. A* 895, 179–187, 2000.

698. Sevcikova, P., Glatz, Z., and Tomandi, J., Determination of homocysteine in human plasma by micellar electrokinetic chromatography and in-capillary detection reaction with 2,2'-dipyridyl disulfide, *J. Chromatogr. A* 990, 197–204, 2003.

699. Nogueira, R., Laemmerhofer, M., Maier, N.M., Lindner, W., Spectrophotometric determination of sulfhydryl concentration on the surface of thiol-modified chromatographic silica particles using 2,2'-dipyridyl disulfide reagent, *Anal. Chim. Acta* 533, 179–183, 2005.

700. Nguyen, T.H., Giri, A., and Oshima, T., A rapid HPLC post-column reaction analysis for the quantification of ergothioneine in edible mushrooms and in animals fed a diet supplemented with extracts from the processing waste of cultivated mushrooms, *Food Chem.* 133, 585–591, 2012.

701. Le, M. and Means, G.E., A procedure for the determination of monothiols in the presence of dithiothreitol—An improved assay for the reduction of disulfides, *Anal. Biochem.* 229, 264–271, 1995.

702. Hansen, R.E., Østergaard, H., Nørgaard, P. et al., Quantification of protein thiols and dithiols in the picomolar range using sodium borohydride and 4,4'-dithiodipyridine, *Anal. Biochem.* 363, 77–82, 2007.

703. Ou, W.B., Yi, T., Kim, J.M., and Khorana, H.G., The roles of transmembrane domain helix-III during rhodopsin photoactivation, *PLoS One* 6, e17398, 2011.

704. Egwin, I.O. and Gruber, H.J., Spectrophotometric measurement of mercaptans with 4,4'-dithiopyridine, *Anal. Biochem.* 288, 188–194, 2001.

705. Grimshaw, C.E., Whistler, R.L., and Cleland, W.W., Ring opening and closing rates for thiosugars, *J. Am. Chem. Soc.* 101, 1521–1532, 2001.

706. Islam, M.S., Kindmark, H., Larsson, O., and Berggren, P.O., Thiol oxidation by 2,2'-dithiodipyridine causes a reversible increase in cytoplasmic free Ca^{2+} concentration in pancreatic β-cells. Role for inositol 1,4,5-trisphosphate-senstive Ca^{2+} stores, *Biochem. J.* 321, 347–354, 1997.

707. Tasumi, E. an Hirose, M., Highly ordered molten globule-like state of ovalbumin at acidic pH: Native-like fragmentation by protease and selective modification of Cys367 with dithiopyridine, *J. Biochem.* 122, 300–308, 1997.

708. Brocklehurst, K. and Little, G., A novel reactivity of papain and a convenient active site titration in the presence of other thiols, *FEBS Lett.* 9, 113–116, 1970.

709. Brocklehurst, K. and Little, G., Reactivities of the various protonic states in the reaction of papain and L-cysteine with 2,2'- and with 4,4'-dipyridyl disulphide: Evidence for nucleophilic reactivity in the un-ionized thiol group of cysteine-25 residue occasioned by its interaction with the histidine-159-asparagine-175 hydrogen-bonded system, *Biochem. J.* 128, 471–474, 1972.

710. Malthouse, J.P.G. and Brocklehurst, K., Preparation of fully active ficin from *Ficus glabrata* by covalent chromatography and characterization of its active centre by using 2,2'-dipyridyl disulphide as a reactivity probe, *Biochem. J.* 159, 221–234, 1976.

711. Pedersen, A. and Jacobsen, J., Reactivity of the thiol group in human and bovine albumin at pH 3–9, as measured by exchange with 2,2'-dithiopyridine, *Eur. J. Biochem.* 106, 291–295, 1980.

712. Little, G. and Brocklehurst, K., Kinetics of the reversible reaction of papain with 5,5'-dithiobis-(2-nitrobenzoate)dianion: Evidence for nucleophilic reaction in the un-ionized thiol group of cysteine-25 and for general acid catalysis by histidine-159 of the reaction of the 5-mercapto-2-nitrobenzoate dianion with the papain-5-mercapto-2-nitrobenzoate mixed disulphide, *Biochem. J.* 128, 475–477, 1972.

713. Malthouse, J.P.G. and Brocklehurst, K., A kinetic method for the study of solvent environments of thiol groups in proteins involving the use of a pair of isomeric reactivity probes and a differential solvent effect, *Biochem. J.* 185, 217–222, 1980.

714. Kimura, T., Matsueda, R., Nakagawa, Y., and Kaiser, E.T., New reagents of the introduction of the thiomethyl group at sulfhydryl residues of proteins with concomitant spectrophotometric titration of the sulfhydryl: Methyl 3-nitro-2-pyridyl disulfide and methyl 2-pyridyl disulfide, *Anal. Biochem.* 122, 274–282, 1982.

715. Zara, J.J., Wood, R.D., Boon, P. et al., A carbohydrate-directed heterobifunctional cross-linking reagent for the synthesis of immunoconjugates, *Anal. Biochem.* 194, 156–162, 1991.

716. Ekrami, H.M., Kennedy, A.R., and Chen, W.-C., Water-soluble fatty acid derivatives as acylating agents for reversible lipidization of polypeptides, *FEBS Lett.* 371, 281–286, 1995.

717. Ebright, Y.W.,Chen, Y., Kim, Y., and Ebright, R.H., *S*-[2-(4-azidosalicylamido)ethylthio]-2-thiopyridine: Radioiodinatable, cleavable, photoactivatable cross-linking agent, *Bioconjug. Chem.* 7, 380–384, 1996.

718. Steenkamp, D.J. and Vogt, R.N., Preparation and utilization of a reagent for the isolation and purification of low-molecular-mass thiols, *Anal. Biochem.* 325, 21–27, 2004.

719. Dewey, J.S., Struck, D.K., and Young, R., Thiol protection in membrane protein purifications: A study with phage holins, *Anal. Biochem.* 390, 221–223, 2009.

720. Brocklehurst, K., Carlsson, J., Kierstan, M.P., and Crook, E.M., Covalent chromatography. Preparation of fully active papain from dried papaya latex, *Biochem. J.* 133, 573–584, 1973.

721. Brocklehurst, K., Carlsson, J., Kierstan, M.P., and Crook, E.M., Covalent chromatography by thiol-disulfide interchange, *Methods Enzymol.* 34, 531–544, 1974.

722. Egorov, T.A., Svenson, A., Ryden, L., and Carlsson, J., A rapid and specific methods for isolation of thiol-containing peptides from large proteins by thiol-disulfide exchange on a solid support, *Proc. Natl. Acad. Sci. USA* 72, 3029–3033, 1975.

723. Harris, R.B., Johnson, A.J., and Hodgins, L.T., Partial purification of biologically active, low molecular weight, human antihemophilic factor free of Von Willebrand factor. II. Further purification with thiol-disulfide interchange chromatography and additional evidence for disulfide bonds susceptible to limited reduction, *Biochim. Biophys. Acta* 668, 471–480, 1981.

724. King, T.P., Li, Y., and Kochoumian, L., Preparation of protein conjugates via intermolecular disulfide bond formation, *Biochemistry* 17, 1499–1506, 1978.

725. Wang, D., Preparation of the bifunctional enzyme ribonuclease-deoxyribonuclease by cross-linkage, *Biochemistry* 18, 4449–4452, 1979.

726. Alagon, A.C. and King, T.P., Activation of polysaccharides with 2-iminothiolane and its uses, *Biochemistry* 19, 4341–4345, 1980.

727. Lambert, A.C., Senter, P.D., Yau-Young, A. et al., Purified immunotoxins that are reactive with human lymphoid cells. Monoclonal antibodies conjugated to the ribosome-inactivating proteins gelonin and the pokeweed antiviral proteins, *J. Biol. Chem.* 260, 12035–12041, 1985.

728. Russell-Jones, G.I., Westwood, S.W., and Habberfield, A.D., Vitamin B$_{12}$ mediated oral delivery systems for granulocyte-colony stimulating factor and erythropoietin, *Bioconjug. Chem.* 6, 459–465, 1995.

729. Luthra, M.P., Dunlap, R.B., and Odom, J.D., Characterization of a new sulfhydryl group reagent: 6,6′-diselenobis-(3-nitrobenzoic acid), a selenium analog of Ellman's reagent, *Anal. Biochem.* 117, 94–102, 1981.

730. Luthra, N.P., Costello, R.C., Odom, J.D., and Dunlap, R.B., Demonstration of the feasibility of observing nuclear magnetic resonance signals of ^{77}Se covalently attached to proteins, *J. Biol. Chem.* 257, 1142–1144, 1982.

731. Gettins, P. and Wardlaw, S.A., NMR relaxation properties of ^{77}Se-labeled proteins, *J. Biol. Chem.* 266, 3422–3426, 1991.

732. Gettins, P. and Crews, B.C., ^{77}Se NMR characterization of ^{77}Se-labeled ovine erythrocyte glutathione peroxidase, *J. Biol. Chem.* 266, 4804–4809, 1991.

733. Pleasants, J.C., Guo, W., and Rabenstein, D.L., A comparative study of the kinetics of selenol/diselenide and thiol/disulfide exchange reactions, *J. Am. Chem. Soc.* 111, 6553–6558, 1989.

734. Kang, S.I. and Spears, C.P., Structure-activity studies on organoselenium alkylating agents, *J. Pharm. Sci.* 79, 57–62, 1990.

735. Arnér, E.S., Selenoproteins—What unique properties can arise with selenocysteine in place of cysteine? *Exp. Cell Res.* 316, 1296–1303, 2010.

736. Xu, K., Zhang, Y., Tang, B. et al., Study of highly selective and efficient thiol derivatization using selenium reagents by mass spectrometry, *Anal. Chem.* 82, 6926–6932, 2010.

737. Wang, Z., Zhang, Y., Zhang, H. et al., Fast and selective modification of thiol proteins/peptides by *N*-(phenylseleno)phthalimide, *J. Am. Soc. Mass Spectrom.* 23, 520–529, 2012.

738. Singh, R. and Whitesides, G.M., Selenols catalyze the interchange reactions of dithiols and disulfides in water, *J. Org. Chem.* 56, 6931–6933, 1991.

739. Banks, T.E. and Shafer, J.A., Inactivation of papain by S-methylation of its cysteinyl residue with O-methylisourea, *Biochemistry*, 11, 110–114, 1972.

740. Lapko, V.N., Smith, D.L., and Smith, J.B., Methylation and carbamylation of human γ-crystallins, *Protein Sci.* 12, 1762–1774, 2003.

741. Beardsley, R.L., Karty, J.A, and Reilly, J.P., Enhancing the intensities of lysine-terminated tryptic peptide ions in matrix-assisted laser desorption/ionization mass spectrometry, *Rapid. Commun. Mass Spectrom.* 14, 2147–2153, 2000.

742. Brancia, F.L., Montgomery, H., Tanaka, K., and Kumashiro, S., Guanidino labeling derivatization strategy for global characterization of peptide mixtures by liquid chromatography matrix-assisted laser desorption/ionization mass spectrometry, *Anal. Chem.* 76, 2748–2755, 2004.

743. Carbetta, V.J., Lit, T., Shakya, A. et al., Integrating Lys-N-proteolysis and N-terminal guanidination for improved fragmentation and relative quantification of singly-charged ions, *J. Am. Soc. Mass Spectrom.* 21, 1050–1060, 2010.

744. Eyem, J., Sjödahl, J., and Sjöquist, J., S-Methylation of cysteine residues in peptides and proteins with dimethylsulfate, *Anal. Biochem.* 74, 359–368, 1976.

745. Stark, G.R., On the reversible reaction of cyanate with sulfhydryl groups and the determination of NH_2-terminal cysteine and cystine in proteins, *J. Biol. Chem.* 239, 1411–1414, 1964.

746. Stark, G., Modification of proteins with cyanate, *Methods Enzymol.* 11, 590–594, 1967.

747. Lippincot, J. and Apostol, I., Carbamylation of cysteine: A potential artifact in peptide mapping of hemoglobins in the presence of urea, *Anal. Biochem.* 267, 57–64, 1999.

748. Hu, J.J., Dimaira, M.J., Zirvi, K.A. et al., Influence of pH on the modification of thiols by carbamoylation agents and effects on glutathione levels in normal and neoplastic cells, *Cancer Chemother. Pharmacol.* 24, 95–101, 1989.

749. Hebert, B. Hopwood, F., Oxley, D. et al., β-Elimination: An unexpected artefact in proteome analysis, *Proteomics* 3, 826–831, 2003.

750. Cole, E.G. and Mecham, D.K., Cyanate formation and electrophoretic behavior of proteins in gels containing urea, *Anal. Biochem.* 14, 215–222, 1966.

751. Hagel, P., Gerding, J.J.T., Fieggen, W., and Bloemendal, H., Cyanate formation in solutions of urea. I. Calculation of cyanate concentrations at different temperatures and pH, *Biochim. Biophys. Acta* 243, 366–373, 1971.

752. Wrigley, C.W., Sensitive procedure for determining cyanate in urea solutions, *J. Chromatogr.* 66, 189–190, 1972.

753. Barrett, T.J., Pattison, D.I., Leonard, S.E. et al., Inactivation of thiol-dependent enzymes by hypothiocyanous acid: Role of sulfenyl thiocyanate and sulfenic acid intermediates, *Free Radic. Biol. Med.* 52, 1075–1085, 2012.

754. Pecci, L., Cannella, C., Pensa, B., Costa, M., and Cavallini, D., Cyanylation of rhodanese by 2-nitro-5-thiocyanobenzoic acid, *Biochim. Biophys. Acta*, 623, 348–353, 1980.

755. Fafarman, A.T., Webb, L.J., Chuang, J.I., and Boxer, S.G., Site-specific conversion of cysteine thiols into thiocyanate creates at IR probe for electric fields in proteins, *J. Am. Chem. Soc.* 128, 13356–13357, 2006.

756. Edelstein, L., Stetz, M.A., McMahon, H.A., and Londergan, C.H., The effects of α-helical structure and cyanylated cysteine on each other, *J. Phys. Chem. B* 114, 4931–4936, 2010.

757. Alfieri, K.N., Vienneau, A.R., and Londergan, C.H., Using infrared spectroscopy of cyanylated cysteine to map the membrane binding structure and orientation of the hybrid antimicrobial peptide CM15, *Biochemistry* 50, 11097–11108, 2011.

758. Vanaman, T.C. and Stark, G.C., A study of the sulfhydryl groups of the catalytic subunit of Escherichia coli aspartate transcarbamylase. The use of enzyme-5-thio-2-nitrobenzoate mixed disulfides as intermediates in modifying enzyme sulfhydryl groups, *J. Biol. Chem.* 245, 3565–3573, 1970.

759. Witkowska, H.E., Green, B.N., and Smith, S., The carboxyl-terminal region of thioesterase II participates in the interaction with fatty acid synthetase. Use of electrospray ionization mass spectrometry to identify a carboxyl-terminally truncated form of the enzyme, *J. Biol. Chem.* 265, 5662–5665, 1990.

760. Tang, H.Y. and Speicher, D.W., Identification of alternative products and optimization of 2-nitro-5-thiocyanatobenzoic acid cyanylation and cleavage at cysteine residues, *Anal. Biochem.* 334, 48–61, 2004.

761. Belghazi, M., Klett, D., Caoreau, C., and Combarnous, Y., Nitro-thiocyanobenzoic acid (NTCB) reactivity of cysteines β100 and β110 in porcine luteinizing hormone: Metastability and hypothetical isomerization of the two disulfide bridges of its β-subunit seatbelt, *Mol. Cell. Endocrinol.* 247, 175–182, 2006.

762. Koehn, H., Clerens, S. Deb-Choudhury, S. et al., Higher sequence coverage and improved confidence in the identification of cysteine-rich proteins from the wool cuticle using combined chemical and enzymatic digestion, *J. Proteomics* 73, 323–330, 2009.

763. Wu, J. and Watson, J.T., Optimization of the cleavage reaction for cyanylated cysteinyl peptides for efficient and simplified mass mapping, *Anal. Biochem.* 258, 268–276, 1998.

764. Ghosh, P.B. and Whitehouse, M.W., 7-Chloro-4-nitrobenzo-2-oxa-1,3-diazole: A new fluorigenic reagent for amino acids and other amines, *Biochem. J.* 108, 155–156, 1968.

765. Birkett, D.J., Price, N.D., Radda, G.K., and Salmon, A.G., The reactivity of SH groups with a fluorogenic reagent, *FEBS Lett.* 6, 346–348, 1970.

766. Birkett, D.J., Dwek, R.A., Radda, G.K. et al., Probes for the conformational transitions of phosphorylase *b*. Effect of ligands studied by proton relaxation enhancement, fluorescence and chemical reactivities, *Eur. J. Biochem.* 20, 494–508, 1971.

767. Lad, P.M., Wolfman, N.M., and Hammes, G.G., Properties of rabbit muscle phosphofructokinase modified with 7-chloro-4-nitrobenzo-2-oxa-1,3-diazole, *Biochemistry* 16, 4802–4806, 1977.

768. Nitta, K., Bratcher, S.C., and Kronman, M.J., Anomalous reaction of 4-chloro-7-nitrobenzofurazan with thiol compounds, *Biochem. J.* 177, 385–392, 1979.

769. Dwek, R.A., Radda, G.A., Richards, R.E., and Salmon, A.G., Probes for the conformational transitions of phosphorylase a. Effect of ligands studied by proton-relaxation enhancement, and chemical reactivities, *Eur. J. Biochem.* 29, 494–508, 1972.

770. Aruna, B., Ghosh, S., Singh, A.K. et al., Human recombinant resistin protein displays tendency to aggregate by forming internal disulfide linkages, *Biochemistry* 42, 10554–10559, 2003.

771. Haggag, R., Belal, S., and Shaalan, R., Derivatization with 4-chloro-7-nitro-2,1,3-benzoxadiazole for the spectrophotometric and differential pulse polarographic determination of acetylcysteine and captopril, *Sci. Pharm.* 76, 33–48, 2008.

772. Carlberg, I., Sahlman, L., and Mannervik, B., The effect of 2,4,6-trinitrobenzenesulfonate on mercuric reductase, glutathione reductase, and lipoamide dehydrogenase, *FEBS Lett.* 180, 102–106, 1985.

773. Demasi, M., Silva, G.M, and Netto, L.E.S., 20 S proteasome from *Saccharomyces cerevisiae* is responsible to redox modifications and is *S*-glutathionylated, *J. Biol. Chem.* 278, 679–685, 2003.

774. Carballai, S., Radi, R., Kirk, M.C. et al., Sulfenic acid formation in human serum albumin by hydrogen peroxide and peroxynitrite, *Biochemistry* 42, 9906–9914, 2003.

775. Silva, G.M., Netto, L.E.S., Discola, K.E. et al., Role of glutaredoxin 2 and cytosolic thioredoxins is cysteinyl-based redox modification of the 20S proteasome, *FEBS J.* 275, 2942–2955, 2008.

776. Godat, E., Herve-Grepinet, V., and Veillard, F., Regulation of cathepsin K activity by hydrogen peroxide, *Biol. Chem.* 389, 1123–1126, 2008.

8 Chemical Modification of Cystine

Cystine is one of the several oxidized forms of cysteine (Figure 8.1). Cystine is a disulfide; there are trisulfide and tetrasulfide forms of cysteine that are discussed at the end of this chapter. Early literature recognized cystine as a symmetrical amino acid distinct from cysteine.[1] However, it was also recognized that cystine was formed from cysteine. While amino acid composition of a protein determined by amino acid analysis as a characteristic has been largely replaced by MS, early literature found such useful. Cystine and cysteine were expressed as 1/2 cystine (1/2Cys). Cysteine content of a protein was usually determined by a separate colorimetric procedure such as the nitroprusside reaction.[2] The reduction of a protein was recognized early as a mild method for the characterization of a protein and determination of the importance of disulfide bonds in biological activity.[3] These early studies did report the difficulty in the restoration of activity on oxidation of the reduced protein. These early reviews by Desnuelle,[1] Tristram,[2] and Putnam[3] contain information on the early work on cystine. Cecil[4,5] provides more recent consideration of the early work on sulfur in proteins. The reader is also directed to a more recent volume of *Methods in Enzymology* for a more recent discussion of redox chemistry and thiol/disulfides in proteins.[6] There has been a long-standing interest in the analysis and modification of cystine in wool.[7–19] There has also been interest in the modification of cystine in hair during permanent waving.[20–22] The oxidation of wool is an interesting application of protein chemistry. Oxidation of fabric was a process used to reduce shrinkage of the final product. Early work used chlorine/hypochloride solutions to *shrink-proof* finish fabric.[23] Since the shrinkproofing process was based on oxidation, there was a search for other more environmentally friendly oxidizing agents. Denning and coworkers[24] reviewed the development of new oxidizing agents for wool resulting in the development of permonosulfate (Oxone®; potassium peroxymonosulfate; $KHSO_5$) for the oxidation of wool followed by the addition of sodium sulfite/bisulfite yielding the *S*-sulfonate (Bunte salt; thiosulfates), cysteic acid, and cystine monoxide (Figure 8.2). Diz and coworkers[25] were able to couple a cationic surfactant (Figure 8.2) with oxidized wool to yield a fabric with antibacterial properties. These investigators used the process of sulfitolysis (Figure 8.6) to generate cysteine-*S*-sulfonate in wool. Coupling the thiol compound could be accomplished either by reaction with the *S*-sulfonate derivate or by disulfide exchange. Gao and Cranston[26] showed that peroxymonosulfate/sulfite treatment enabled productive binding of polyhexamethylene biguanide to wool fabric yielding a product with antimicrobial properties. Although not discussed, coupling of oxidized wool to a thiol-containing compound could also be accomplished via the cystine monoxide[27] or cystine dioxide.[28]

Early work on cysteine and cystine in proteins suggested that cysteine was found only in intracellular proteins and disulfide bonds found in extracellular proteins.[29] While there is a definite difference in the relative amounts of cysteine and cystine in intracellular proteins and extracellular proteins,[30] the difference is not absolute in that while intracellular proteins do not contain disulfide bonds, some 10% of extracellular proteins contain sulfhydryl groups that, if exposed, would be susceptible to oxidation. Bessette and coworkers[31] noted that the cytoplasm (intracellular) has a reducing environment, while the extracellular environment is characterized by an oxidizing environment; it is possible to make the cytoplasm of *Escherichia coli* sufficiently oxidizing to promote disulfide formation without compromising cell viability; the ability to manipulate the cytoplasmic

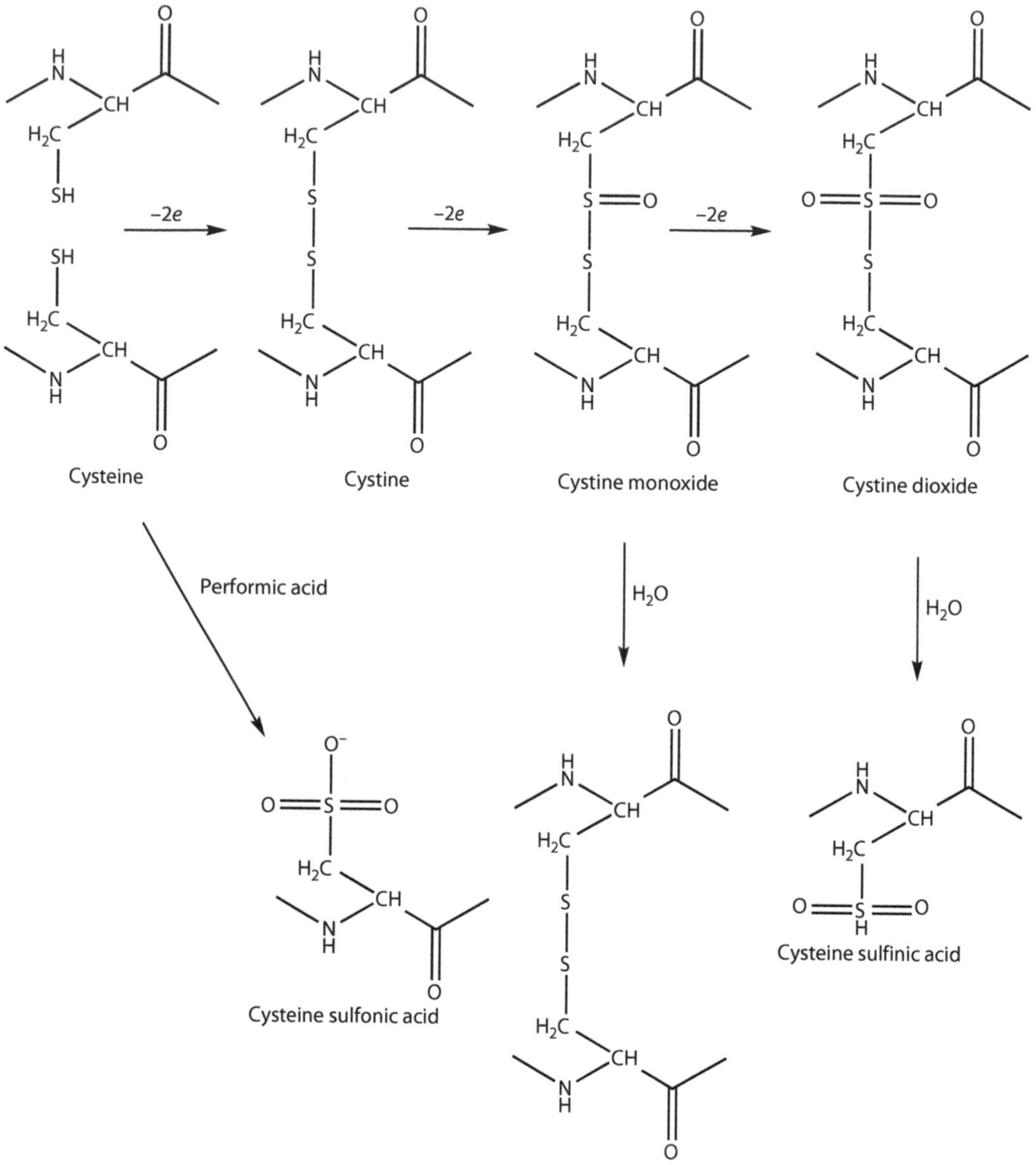

FIGURE 8.1 Various oxidation products of cystine.

redox environment is of great potential value for the use of microorganisms to produce properly folded therapeutic proteins.[32–34]

Cystine is the oxidation product of cysteine, and if present in a protein, the disulfide cross-link is assumed to be critical for maintaining the native structure of an extracellular protein. However, there are examples where there are nonessential disulfide bonds. Staley and Kim[35] reported that bovine pancreatic trypsin inhibitor could fold to a native conformation with only one of the two disulfide bonds; the cysteine residues that would have contributed to the second disulfide bond were replaced by alanine. A more recent study[35a] demonstrated that either one of the disulfide bonds in human coagulation factor VIII could be removed without an effect on secretion and likely activity. So it is likely that a protein could fold effectively to a native conformation without disulfide bonds, but there would likely be a substantial effect on stability. On the other hand, cleavage of

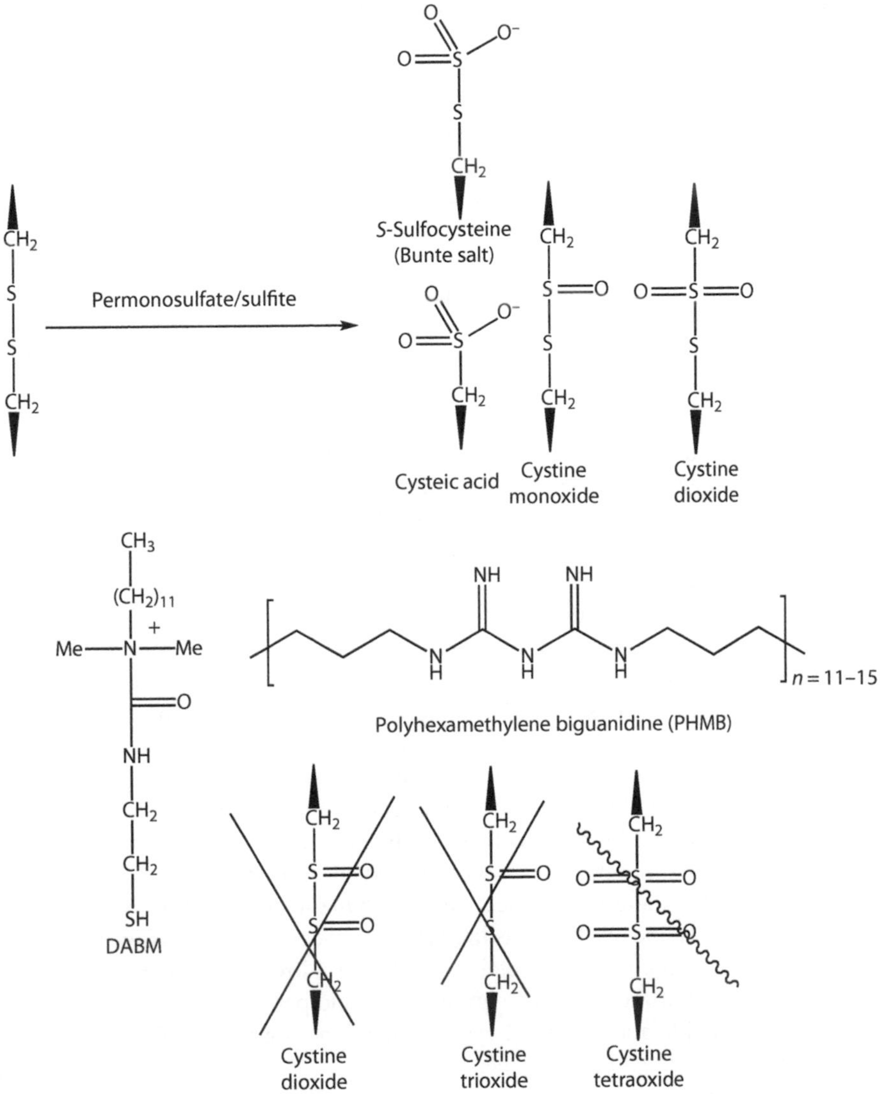

FIGURE 8.2 Oxidation of cystine in wool. Shown at the top are some of the products from the oxidation of wood with permonosulfate (Oxone®; potassium peroxymonosulfate) (Denning, R.J., Freeland, G.N., Guise, G.B., and Hudson, A.H., Reaction of wool with permonosulfate and related oxidants, *Textile Res. J.* 64, 413–422, 1994). The *S*-sulfonate derivate of cysteine can also be obtained by sulfitolysis of the disulfide bond (Cecil, R. and McPhee, J.R., A kinetic study of the reactions on some disulphides with sodium sulphite, *Biochem. J.* 60, 496–506, 1955), which was used for the coupling of DABM (*N*-dodecyl-*N*-dimethyl glycine cystamine) to wool for preparation of an antimicrobial fabric (Diz, M., Infante, M.R., and Erra, P., Antimicrobial activity of wool treated by a new thiol cationic surfactant, *Textile Res. J.* 71, 695–700, 2001). Also shown is the structure of a poly-hexamethylene biguanidine that is noncovalently bound to oxidized wool (Gao, Y. and Cranston, R., An effective antimicrobial treatment for wool using polyhexamethylene biguanide as the biocide, Part 1: Biocide uptake and antimicrobial activity, *J. Appl. Polym. Sci.* 117, 3075–3082, 2010). Shown at the bottom is an early structure for cystine dioxide (thiolsulfonate) that was discarded (Axelson, G., Hamkin, K., Fahlman, A. et al., Electron spectroscopic evidence of the thiolsulphonate structure of cystine *S*-dioxide, *Spectrochim. Acta* 23A, 2015–2020, 1067). The trioxide has never been described. There was a single report describing the tetraoxide; however, attempts to synthesize the tetraoxide have been unsuccessful (Setiawan, L.D., Baumann, H., and Gribbin, D., Surface studies of keratin fibres and related model compounds using ESCA. I-Intermediate oxidation products of the model compound L-cystine and their hydrolytic behaviour, *Surf. Interface Anal.* 7, 188–195, 1985).

disulfide bonds in a protein by reduction or oxidation results in loss of biological activity. There are several examples where cleavage of a single disulfide bond in a protein results in loss of activity. For example, reduction of the single disulfide bond in the ABA-1 allergen of the nematode *Ascaris* results in the loss of structural integrity and function.[36] Mallon and Rossman[37] reported the cleavage of the single disulfide bond in DNA polymerase I with bisulfite resulting in decreased activity and accuracy of DNA synthesis. Deitcher and coworkers reduced one of the four disulfide bonds in α-lactalbumin as part of a study on the effect of conformation on protein adsorption in hydrophobic exchange chromatoagaphy.[38] Reduction of the native protein with DTT followed by either carboxymethylation or carboxamidomethylation had no effect on chromatography in the absence of calcium ions, while there was enhanced absorption in the presence of 12 mM $CaCl_2$; the authors suggest that the increased absorption is a reflection of decreased stability of the modified protein. Pace and workers[39] prepared derivatives of ribonuclease T_1 with one or both disulfide bonds reduced and carboxamidomethylated. Ribonuclease T_1 could fold and function from a denatured state but did have reduced stability (urea denaturation, thermal stability). A more recent work on the insertion or deletion of disulfide bonds in proteins has used protein engineering rather than chemical modification. One example is provided by the work on leech carboxypeptidase inhibitor[40] where elimination of single disulfide bonds results in markedly decreased stability with modest effect on functional activity. Reznik and coworkers[41] expressed plasma retinol-binding protein where the six cysteine residues are replaced with serine residues. The engineered protein has reduced stability and reduced, but not absent, activity in binding retinol. Support for the importance of disulfide bonds in conformational stability can be obtained from studies where the insertion of an additional disulfide into a protein increases stability.[42–46] Bull and Breese[47] studied the denaturation of various proteins in guanidine hydrochloride and observed that protein expand on denaturation with an increase in viscosity; the presence of disulfide bonds limited the expansion of the protein molecules during denaturation. Hen egg white lysozyme (HEWL) is one of several proteins that have a sweet taste.[48] Both sweetness and enzymatic activity are lost on reduction and alkylation, while it was possible to inhibit enzyme activity without losing sweetness.[49] The sweetness was also lost by maintaining at 95°C for 18 h. This suggests that the sweetness is a product of the tertiary structure of the protein stabilized by disulfide bonds and not a reflection of enzymatic activity. Subsequent work showed that the positive charge of lysine residues was important in sweetness.[50] Extensive modification with pyridoxal-5-phosphate (>3 mol per mol protein) resulted in a loss of sweetness activity that could be restored by the removal of the phosphate groups with acid phosphatase. Additional work[51] with mutants and chemical modification with 1,2-cyclohexanedione suggests a role for arginine residues as well as lysine in the *sweetness* of lysozyme.

The author would be remiss, for both personal and scientific reasons, if lima bean trypsin inhibitor was not mentioned at this juncture. Lima bean trypsin inhibitor is a group of six or more closely related proteins of low molecular weight (ca. 10 kDa) and some 20% cysteine/cystine by weight. The cysteine is present as cystine so the protein is highly cross-linked (one-half cystine every six residues) and very stable; the inhibitory activity is stable to boiling. Jones and coworkers[52] purified several of the variants and noted that while inhibitory activity was lost on reduction and carboxymethylation by classical methods, the protein was still resistant to trypsin. As told to the author of the current text (RLL), one of the senior authors on the Jones paper (SM) was quite bothered by that such that he encouraged a subsequent fellow, William Ferdinand, to take a another look at the problem. Ferdinand and coworkers[53] did indeed solve the problem. Reduction and carboxymethylation were determined to be incomplete under the conditions used in the Jones study[52]; Ferdinand and coworkers found it necessary to use 5 M guanidine hydrochloride in place of 8 M urea, and the concentrations of 2-mercaptoethanol and iodoacetamide were increased 10-fold. Even under these conditions, reduction and carboxymethylation were incomplete, and an increase in temperature was suggested. It is clear that the stability of lima bean trypsin inhibitor is a product of the high disulfide bond content. This study would imply that proteins and peptides with high disulfide content are very stable; however, there are exceptions. Cyclotides are a group of cyclic peptides containing six conserved cysteine residues present as three disulfide bonds that are found in plants.[54] These are also referred to as cystine knot peptides with a variety of biological functions.[55–57] Despite the highly disulfide-cross-linked structure, the analysis of these proteins[58,59] has not required the drastic conditions required for other highly cross-linked proteins.

Thiol–disulfide exchange refers to a reversible reaction between a thiol and a protein disulfide process resulting in an equilibrium mixture of thiol, disulfide, and mixed disulfide consisting of the attacking thiol and one of the two cysteine partners in the protein disulfide (Figure 8.3). A mixed disulfide may also result from the reaction of an activated disulfide compound such as DTNB or MMTS with a sulfhydryl group (see Chapter 7); the reaction of S-sulfocysteine and cystine oxides

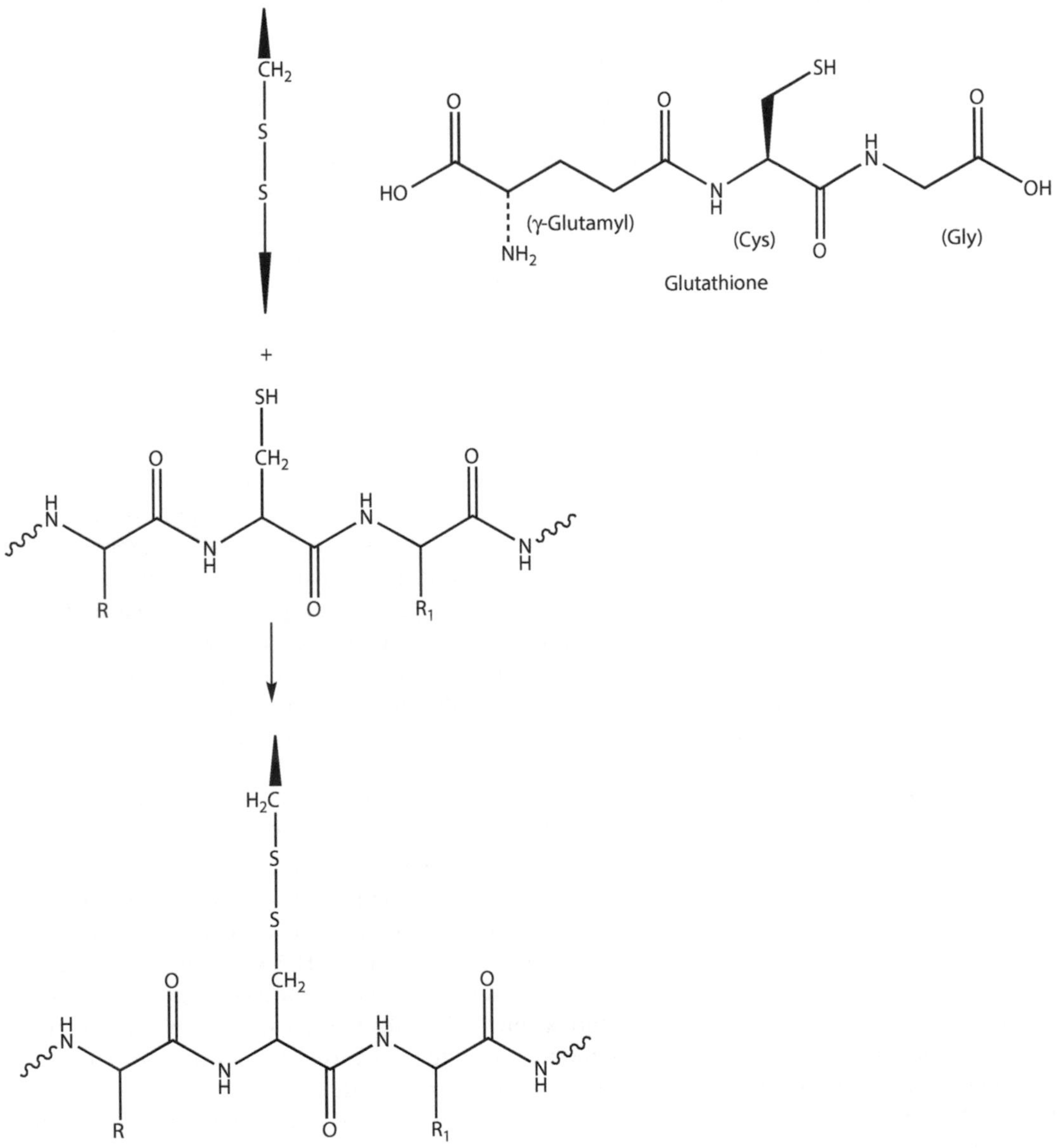

FIGURE 8.3 Formation of mixed disulfides in proteins. The reaction of a suitably reactive thiol such as glutathione with a cystine residue in a protein can result in the formation of a mixed disulfide. As shown in Chapter 7, mixed disulfides may also form upon reaction of a disulfide, such as oxidized glutathione, with a cysteine residue. The formation of a mixed disulfide in a protein is an active process.

with cysteine can also result in mixed disulfides. In 1935, Mirsky and Anson[60] at the then Rockefeller Institute used reaction with cystine for the determination of cysteine in proteins. Somewhat later, Glazer and Smith[61] used *N,N'*-dinitrophenyl-L-cystine in 9.6 M HCl for the determination of cystine plus cysteine in proteins. Cupric sulfate has been reported to catalyze the reaction of cystine with β-globulin.[62] Glutamic acid was observed to inhibit the reaction based on its ability to bind cupric ions; there is evidence to support the formation of a stable complex between cupric ions and glutamic acid.[63] Other investigators have also reported the catalysis of mixed disulfide formation by cupric ions.[64] Recent work by Rigo and coworkers[65] demonstrated that cupric ions perform a catalytic function in the reduction of cysteine to cystine in an anaerobic reaction. A mixture of penicillamine disulfide and cupric ions inhibits the denaturation of human IgG[66]; it is suggested that this effect involved the formation of a mixed disulfide with protein since inhibition of denaturation is also observed with NEM, PHMB, or iodoacetamide. Membrane fouling by denatured protein can be a problem in bioprocessing. Such fouling by bovine serum albumin is blocked by the use of reagents such as EDTA and citrate, which would bind cupric ions.[67] It is suggested that protein aggregation responsible for membrane fouling involves the cupric ion–catalyzed formation of intermolecular disulfide bonds that would be blocked by the chelating agents. The use of glutamic acid in the formulation buffer[68] is suggested to be responsible for the successful preparation of a freeze-dried somatostatin analog; the metal chelating ability of glutamate is mentioned as a quality factor in its selection as an excipient. Thiol–disulfide exchange has an important role in redox regulation and protein folding with glutathione having a central role.[69–75]

The thiol–disulfide reaction is an S_N2 reaction with nucleophilic addition by a thiolate function occurring along the *S–S* axis of the disulfide bond. Eldgarn and Phil[76] measured equilibrium constants for several thiol–disulfide pairs such as *N*-acetylcystamine/cystine, *N*-acetylcystamine/oxidized glutathione, *N*-dimethylcystamine/cystine, *N*-dimethylcystamine/oxidized glutathione, *N*-piperidylcystamine/cystine, and *N*-piperidylcystamine/oxidized glutathione and observed that none were far from unity, implying that there will be a large amount of mixed disulfide present in thiol/disulfide mixture in addition to the starting materials. This study was published in 1957 and was an early use of affinity (paper) chromatography. The progress of disulfide exchange was measured by paper chromatography on mercuric ion–impregnated paper; the free thiol would be selectively retarded during the development of the chromatogram. An organomercurial affinity column was later used for the purification of a thiol protease[77]; a more recent work by Raftery[78] used an organomercurial agarose column for purification of cysteine-containing peptides from yeast cell lysates. Wu and coworkers[79] studied the effect of neighboring charged groups on the process of disulfide exchange in a group of peptides in work on use of disulfide bonds in drug delivery systems. Homodimers linked by cystine were prepared from a series of hexapeptides (N-terminal acetylated and C-terminal aminidated) containing an internal cysteine residue. The disulfide exchange of the various peptides with glutathione (10 mM) was measured at pH 4.9 (acetate). The $t_{1/2}$ values ranged from 48.05 h (WEECEE) to 0.12 h (WRRCRR). These were the extreme examples. The data obtained from these and other peptides supported the suggestion that the presence of arginine increases the rate of exchange, while the presence of glutamic acid decreased exchange rate. These investigators note that it was necessary to work at acid pH; at pH 7.4 (phosphate), the rate of exchange of the basic and neutral peptides was too rapid for accurate measurement. The inclusion of 40% (v/v) ethanol into the reaction mixture as method for decreasing solvent polarity to mimic some intracellular compartments increased the reaction rate for the five peptides studied. The $t_{1/2}$ for the neutral peptide, WGGCGG decreased from 1.54 h in acetate only to 0.94 h in 40% (v/v) ethanol, from 0.49 to 0.23 h in 40% (v/v) ethanol for WGGCRG, and from 8.72 to 5.70 h 40% (v/v) ethanol for WGGCEE.

The structural analysis of proteins usually required the cleavage of disulfide bonds prior to analysis. It is noted that advances in MS are moving toward the elimination of the necessity for this step in the structural analysis of proteins[80–82]; however, challenges remain.[83] One approach is to cleave by oxidation and the other by reduction. The second approach usually takes two steps, one to reduce the disulfide bonds and second to block the liberated sulfhydryl groups by a process such as carboxymethylation. Both oxidation and reduction of disulfide bonds have significant importance in biological function.

Cystine is an oxidation product of cysteine and in turn is susceptible to oxidation to disulfide-S-monoxide and disulfide S-dioxide (Figure 8.4), which in turn can be hydrolyzed to other products.[84] The structure of cystine dioxide was originally suggested to be a disulfoxide[85] but was later established as a thiosulfonate.[86] Cystine thiosulfonate (disulfide dioxide) was characterized before the monoxide derivative. Cystine monoxide was synthesized and characterized by Savige and coworkers.[87] The synthesis of either monoxide (e.g., 2 N sulfuric acid with performic acid for 2 h at 0°C)[87] or dioxide (performic acid in concentrated hydrochloric acid at 20°C)[88] requires conditions that would be considered drastic for a protein. The disulfide oxides are relatively unstable in aqueous solution,[88,89] which has presented challenges to study. Cystine monoxide is more stable than cystine dioxide; stability is maximal at pH 3–5 and unstable above pH 7.0. Cysteine dioxide is hydrolyzed to cystine monoxide and cysteine sulfinic acid; cystine monoxide in turn is hydrolyzed to cysteine sulfinic acid and cystine. At pH < 1, cystine monoxide disproportionates to cystine dioxide and cystine. Thiosulfinite derivatives of cystine can be converted to unsymmetrical cystines and lanthionines with tris(dialkylamino)phosphines.[90] This reaction is thought to involve the formation of a sulfenic acid and a thiol as intermediates in this reaction. Alkyl thiolsulfinates were identified more than 60 years ago as an antibacterial component derived from garlic.[91] Hunter and coworkers[92] have reviewed the antibacterial function of the thiolsulfinate allicin from garlic as a starting point for the development of new antimicrobial agents. Baumann and coworkers[93] did identify cystine monoxide, cystine sulfinic acid, and cystine sulfonic acid in wool treated with performic acid. Disulfide oxides do react with mercuric chloride or organic mercurials such as p-hydroxymercuribenzoate giving a falsely high value for thiols in peracetic-oxidized wool or bovine serum albumin.[94] Increased reactivity was not observed with iodoacetate. It is reported that the monoxide derivative can react with N-ethylmaleimide.[89] Cysteine sulfenic acid (alanine-3-sulfenic acid) in AhpC of alkyl hydroperoxide reductase has been observed to react with both N-ethylmaleimide and iodoacetate.[95] There has been limited additional work on cystine dioxide in the wool literature.[96]

Given the relatively harsh conditions required for the in vitro formation of cystine oxides as described earlier, it was therefore surprising to see the formation of these derivatives under physiological conditions. Disulfide oxides are products of the in vivo oxidation of biological thiols such as glutathione[97] that formed under conditions of oxidative stress.[98] Disulfide oxides are reactive species[98,99] that can modify proteins.[100–102] Huang and Huang[97] reported that thiosulfinate (monoxide) and thiosulfonate(dioxide) are metabolites of S-nitrosoglutathione. Glutathione thiosulfinate (monoxide) is very reactive toward any thiol forming a mixed disulfide. These investigators reported that the reaction of 5-thio-2-nitrobenzoate [derived from 5,5′-dithiobis(2-nitrobenzoic acid)] with glutathione thiosulfinate was 20-fold more rapid than the similar reaction with S-nitrosoglutathione. This group subsequently reported on the reaction of glutathione oxides and captopril oxides with protein kinase C.[100] There is difference in the reaction with protein kinase C with monoxide and dioxide forms of the two thiols. However, there has been little subsequent work in this area. More recent work on the formation of mixed protein disulfide with glutathione has focused on the role of glutathione sulfenic acid and S-nitrosoglutathione as the active species.[103] The reader is directed to an excellent review by Jacob and coworkers[84] on the role of oxidation state in protein function.

The various disulfide oxides are likely intermediates in the oxidation of cystine by a variety of agent resulting in the formation of cysteine sulfenic acid, cysteine sulfinic acid, and cysteine sulfonic acid. The conversion of cystine to cysteic acid by iodate in acid has been reported.[104] This reaction has been applied to insulin with millimolar concentrations of iodate in 0.5 M HCl. The reaction product was not completely characterized, but the consumption of iodate and the nature of the product are consistent with the oxidation of cystine to cysteic acid. The reaction of iodate with proteins may be of some industrial interest[105] but has not seen any active use in solution protein chemistry. Nowduri and coworkers[106] have studied the oxidation of cystine with

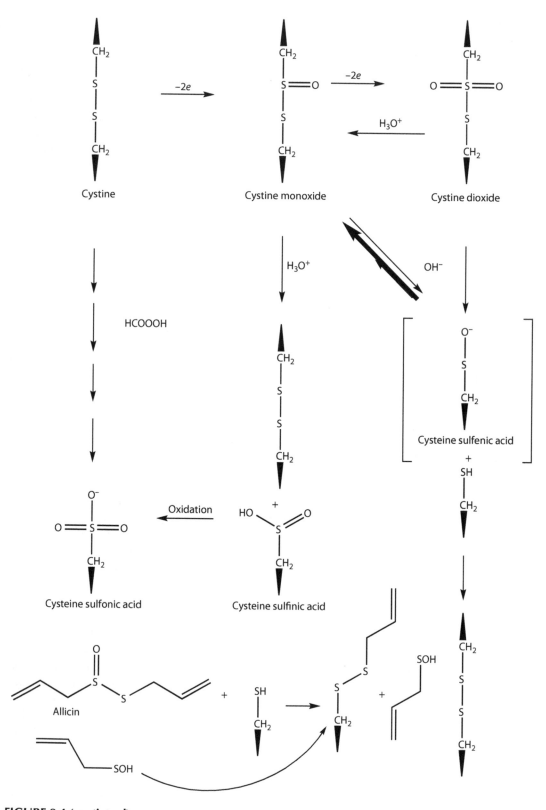

FIGURE 8.4 (continued)

hexacyanoferrate(III) in alkaline medium suggesting a mechanism proceeding through cysteine sulfenic acid to cysteic acid (3-sulfoalanine) as the terminal product. Oxidation of cystine in proteins for structural analysis is usually accomplished under more vigorous conditions with reagents such as performic acid (Figure 8.1).[107] This is a procedure used for the cleavage of disulfide bonds in proteins for structural analysis[108,109] and for accurate determination of cysteine/cystine in proteins using amino acid analysis.[110,111] More recently, performic acid oxidation has proved of value in preparation of proteins for mass spectrometric analysis.[112,113] Chang and coworkers[114] have used a polyarginine-coated probe for the enrichment of peptides containing cysteine sulfonic acid. Cao and coworkers[115] observed that performic acid oxidation improved the solubility of integral membrane proteins for structural analysis.

The cleavage of disulfide bonds by base (Figure 8.5) was reported by Challenger and Rawlings in 1937[116] who suggested the products were a mercaptan (thiol) and a sulfenic/sulfinic acid. Some years later, Donovan[117] observed a time-dependent increase in absorbance of phenolic groups in ovomucoid at high pH (≥12.5). Subsequently, Donovan[118] reported that protein (ovomucoid) disulfide bonds were cleaved under those conditions resulting in an increase in absorbance at 240 and 290 nm. There was production of material that reacted with Elman's reagent at a rate similar to the rate of change in absorbance. The results were consistent with the mechanism proposed by Challenger and Rawlings, although the possibility of the formation of dehydroalanine and a persulfide was mentioned. Cavallini and coworkers[119] established that persulfide was formed in insulin under alkaline conditions; these investigators also reviewed the various products from the alkaline treatment of disulfide bonds. Helmerhorst and Stokes[120] presented more data to support the heterolytic cleavage of the disulfide bonds with the formation of a protein-bound persulfide group. Some 10 years later, Florence[121] extended these observations substantiating the heterolytic nature of disulfide bond cleavage in base with the formation of persulfide and dehydroalanine; dehydroalanine can subsequently form lanthionine by reaction with cysteine and lysinoalanine by reaction with lysine. Both of these modifications are examples of a Michael addition. Florence[121] observed that the susceptibility to alkaline degradation of disulfide bonds in proteins was sensitive to conformation and degradation of the disulfide bonds in bases and was enhanced by the presence of a chaotropic agent. The degradation of the disulfide bonds to form the persulfide is associated with an increase in absorbance at 335 nm.[120,121] Lu and Chang[123] recently reported that the reaction of protein disulfide with DTT in a base (pH 8.5) resulted in the quantitative formation of 2 mol of dehydroalanine that could then be coupled, as previously discussed, with lysine or cysteine forming lysinoalanine or lanthionine, respectively. These investigators propose that there is a symmetrical reduction of the disulfide followed by base-catalyzed β-elimination of the cysteine residues forming the dehydroalanine residues. Angeli and coworkers[124] observed that oxidized glutathione reacted with PHMB in moderate

FIGURE 8.4 (continued) Reactions of the oxidation products of cystine. Cystine is oxidized to cystine thiosulfinate (cystine monoxide) and cystine thiosulfonate (cystine dioxide). Cystine thiosulfonate is less stable than cystine thiosulfinate. A variety of products are formed from the hydrolysis of both the monoxide and dioxide, and the distribution is dependent on pH. At or close to physiological pH, the products are cystine and cysteine sulfinic acid. Also shown is the oxidation of cystine to cysteic acid (cystine sulfonic acid) by performic acid oxidation as used in the structural analysis of proteins. The cysteine sulfenic acid shown is considered transient, but it is likely the active species in forming a new disulfide bond. Oxidation products identified in proteins (and glutathione) include cystine thiosulfonate (cystine dioxide), cystine thiosulfinite (cystine monoxide), cystine sulfinic acid, and cystine sulfonic acid (see Baumann, H., Setiawan, L.D., and Gribbin, D., Surface studies of keratin fibers and related model compounds using ESCA. 2-Intermediate oxidation products of cystyl residues on keratin fiber surfaces and their hydrolytic stability, *Surf. Interface Anal.* 8, 219–225, 1986; Huang, K.-P. and Huang, F.L., Glutathionylation of proteins by glutathione disulfide S-oxide, *Biochem. Pharmacol.* 64, 1049–1056, 2002). At the bottom is shown the structure of allicin, a thiosulfinate from garlic, and the reaction with thiols to form mixed disulfides (Hunter, R., Caira, M., and Stellenboom, N., Thiolsulfinate allicin from garlic. Inspiration for a new antimicrobial agent, *Ann. N. Y. Acad. Sci.* 1056, 234–241, 2005).

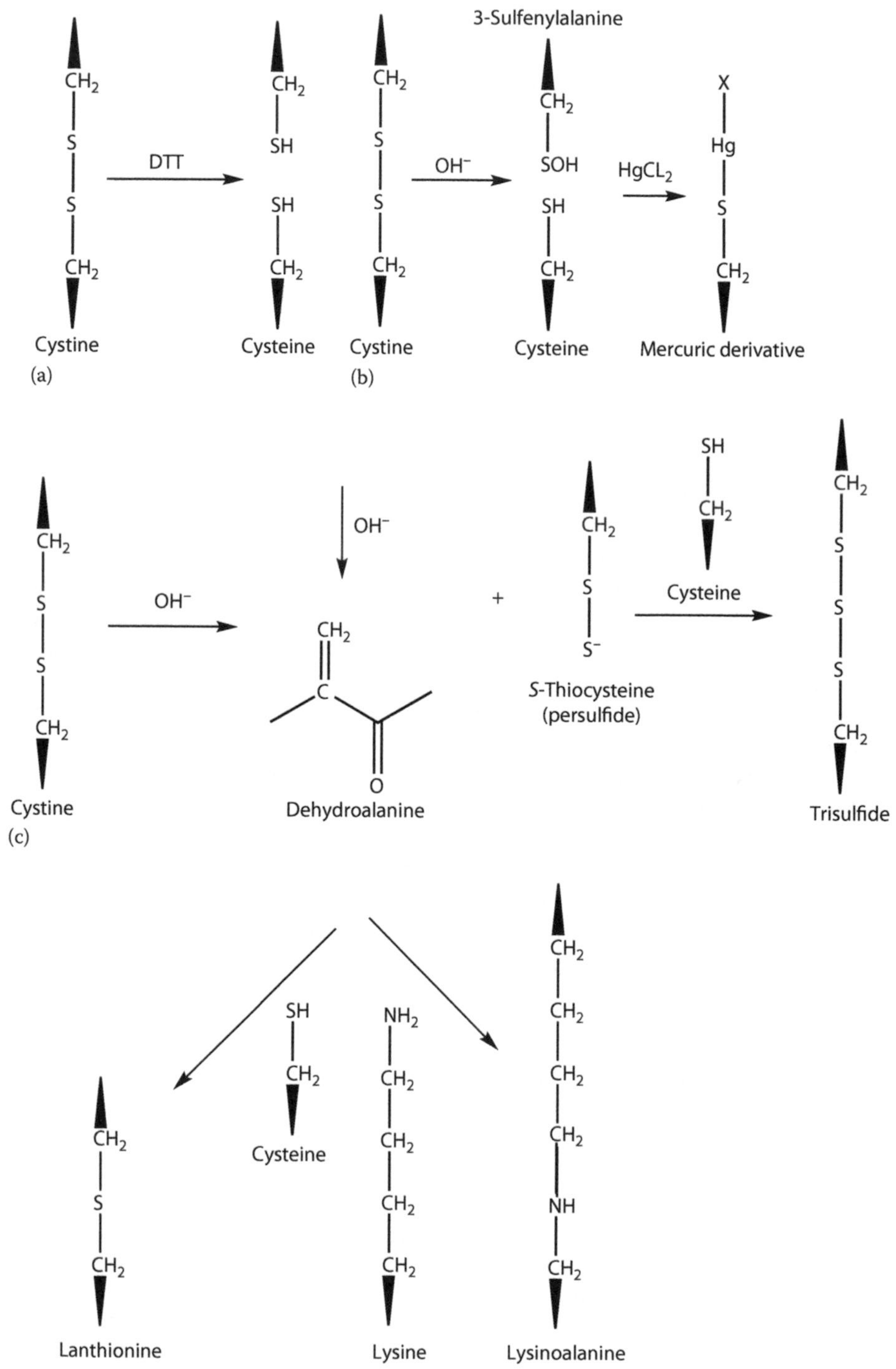

FIGURE 8.5 (continued)

base (0.1 M NaOH), while cystine did not. This would be consistent with a cleavage reaction that forms cystine sulfenic acid and a free thiol.[125] Stricks and Kolthoff[125] had observed that oxidized glutathione did react with mercuric chloride in 0.026 M NaOH, while there was no reaction with cystine under these reaction conditions. A consideration of all of the data suggesting the effect of a base on cystine depends on the environment of the cystine residue and solvent conditions including the strength of the base. In this sense, base cleavage of disulfides is similar to the effect of oxidizing agents on disulfides where there is both homolytic and heterolytic cleavage of disulfide bonds.[89] A summary of the various observations on base cleavage of cystine is shown in Figure 8.5.

The reaction of sulfite with cystine yields the S-sulfo derivative of cysteine and cysteine (Figure 8.6). The reaction of cystine with sulfite to form cysteine and S-sulfocysteine was described by Clarke[126] and Lugg[127] in 1932. Stricks and Kolthoff[128] report that the reaction between sulfite and cystine to yield the thiolate and S-sulfonyl derivative is reversible in alkaline solution. Carter[129] used sulfite to measure disulfide bonds in proteins in 8 M urea. Cecil and McPhee[130] studied the kinetics of the reaction and reported that the reaction of cystine and sulfite is a simple reversible reaction above pH 9.0 (below pH 9.0, the reaction is more complex reflecting the ionization of HSO_3^{-1}; disulfide bonds react with SO_3^{-2}, not with HSO_3^{-1}); the reaction can be taken to completion by recycling the thiolate anion by the use of an oxidizing agent. These investigators noted that the presence of a negative charge (ionized carboxyl group) markedly inhibited the reaction. The earlier literature and chemistry of this reaction has been reviewed by Cole.[131] As noted earlier, the reaction of sulfite and cystine is readily reversible but can be taken to completion by recycling the thiolate anion with an oxidizing agent such as cupric ions or o-iodosobenzoate (oxidative sulfitolysis).[132] As with the analogous process of disulfide bond reduction, complete reaction of disulfide bonds to sulfite usually requires protein denaturation.[129] Würfel and colleagues studied the reaction of sodium sulfite with a number of bacterial and plant thioredoxins.[133] The sulfitolysis reaction was performed in the presence or absence of guanidine hydrochloride. The process of sulfitolysis was measured by the reaction of the liberated thiol group with radiolabeled iodoacetate. As with the reduction of disulfide bonds (Figure 8.7), the extent of sulfitolysis was, in general, more marked in the presence of guanidine, but the extent of sulfitolysis appeared to depend on the primary structure of the protein. The extent of sulfitolysis in wild-type $E.\ coli$ thioredoxin increased from 2% to 25% in the presence of 6 M guanidine hydrochloride (0.1 M Tris, pH 8.0 with 2 mM EDTA), from 6% to 100% with the D26A mutant, but only from 16% to 19% with the P34H mutant. A subsequent study from this group suggests that the low results obtained with carboxymethylation reflected accessibility issues for the cysteine residue(s) formed from sulfotolysis.[134] There is an increase in fluorescence at 345 nm (excitation of tryptophan at 280 nm) reflecting the cleavage of disulfide bonds and the concomitant loss of the quenching effect on tryptophan fluorescence such that changes in fluorescence can

FIGURE 8.5 (continued) Alkaline degradation of cystine residues in proteins. (a) Shown is the cleavage of cystine by dithreitol at pH 13.5 with subsequent conversion to dehydroalanine. The dehydroalanine can then react with cysteine to form lanthionine or with lysine to form lysinoalanine. (b) Shown is the cleavage of cystine in a base to form cysteine sulfenic acid (3-sulfenylalanine) and cysteine. It has been suggested that the presence of mercuric chloride *drives* the reaction to completion by reacting with cysteine (Stricks, W. and Kolthoff, I.M., Amperometric titration of oxidized glutathione in presence of cystine, *Anal. Chem.* 25, 1050–1057, 1953; Angeli, V., Chen, H., Mester, Z. et al., Derivatization of GSSG by pHMB in alkaline media. Determination of oxidized glutathione in blood, *Talanta* 82, 815–820, 2010). (c) Shown is the heterolytic cleavage of cystine in base yielding thiocysteine (a persulfide) and dehydroalanine; dehydroalanine could then react with cysteine to form lanthionine or with lysine to form lysinoalanine. See Cavallini, D., Federici, G., Barboni, E., and Marcucci, M., Formation of persulphide groups in alkaline treated insulin, *FEBS Lett.* 10, 125–128, 1970; Florence, T.M., Degradation of protein disulphide bonds in dilute alkali, *Biochem. J.* 189, 507–520, 1980; Helmerhorst, E. and Stokes, G.B., Generation of an acid-stable and protein-bound persulfide-like residue in alkali- or sulfhydryl-treated insulin by a mechanism consonant with the β-elimination hypothesis of disulfide bond lysis, *Biochemistry* 1983, 69–75, 1983.

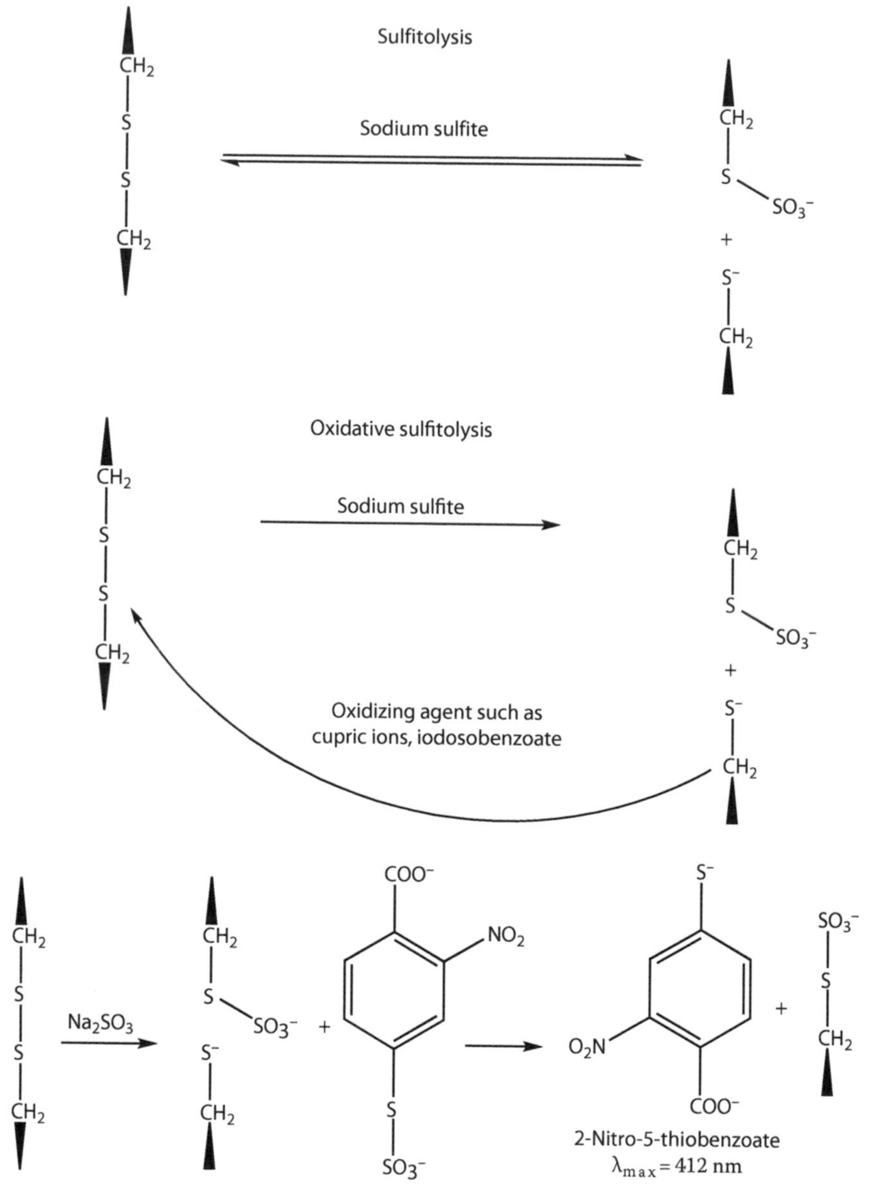

FIGURE 8.6 Reaction of sulfite with cystine to yield cysteine and *S*-sulfocysteine. The reaction of sulfite with cystine to form cysteine and *S*-sulfocysteine is an equilibrium reaction referred to as sulfitolysis; inclusion of an oxidizing agent is referred to as oxidative sulfitolysis. See Raftery, M.J., Selective detection of thiosulfate-containing peptides using tandem mass spectrometry, *Rapid Commun. Mass Spectrom.* 19, 674–682, 2005; Patrick, J.S. and Lagu, A.L., Determination of recombinant proinsulin fusion protein produced in *E. coli* using oxidative sulfitolysis and two-dimensional HPLC, *Anal. Chem.* 64, 507–511, 1992. Shown at the bottom is the reaction of the free thiol formed in sulfitolysis with iodoacetate providing the basis for quantitation of disulfide bonds (Würfel, M.W., Haberlein, I., and Follmann, H., Facile sulfitolysis of the disulfide bonds in oxidized thioredoxin and glutaredoxin, *Eur. J. Biochem.* 211, 609–614, 1999). Shown at the bottom is the sulfitolysis of a disulfide bond yielding the *S*-sulfocysteine derivative and cysteine thiolate anion. 2-Nitro-5-thiosulfobenzoate, in the presence of sulfite, reacts with the thiolate anion yielding a second *S*-sulfocysteine and 2-nitro-5-thiobenzoate, which is a strong chromophore ($\varepsilon = 13,900\ M^{-1}\ cm^{-1}$ at 412 nm). See Thannhauser, T.W., Konishi, Y., and Scheraga, H.A., Analysis for disulfide bonds in peptides and proteins, *Methods Enzymol.* 143, 115–119, 1987.

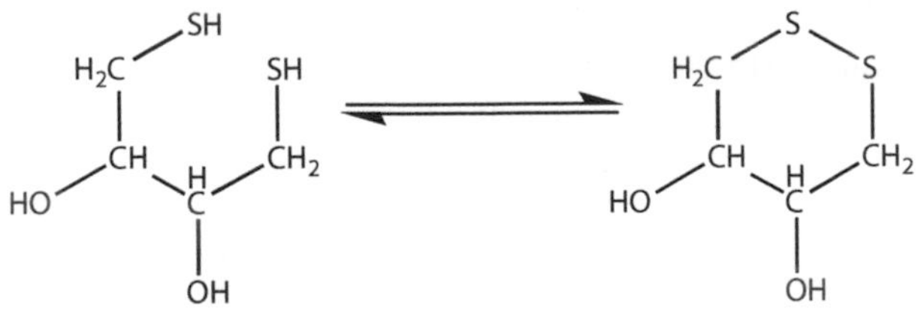

Dithiothreitol/dithioerythritol

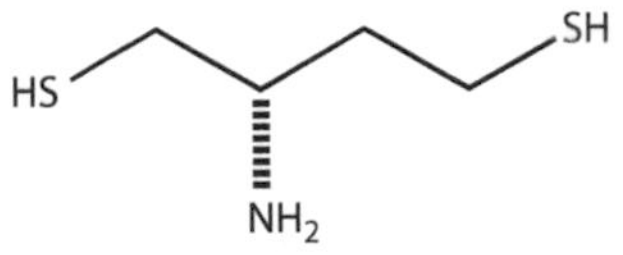

(2S)-2-Amino-1,4-dimercaptobutane

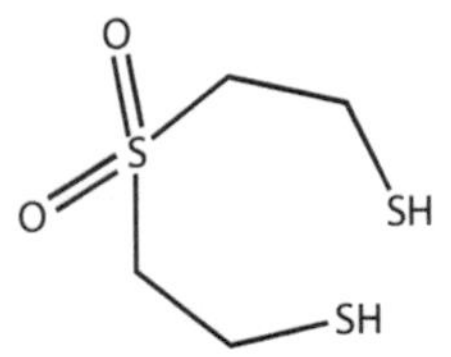

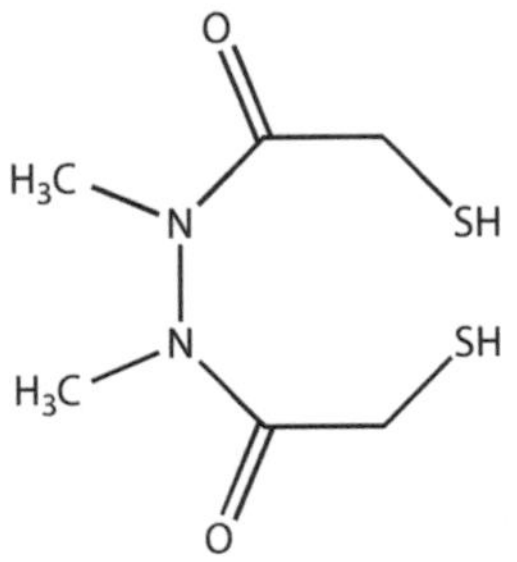

bis(2-Mercaptoethyl)sulfone   N,N'-Dimethyl-N,N'-bis(mercaptoacetyl)hydrazine

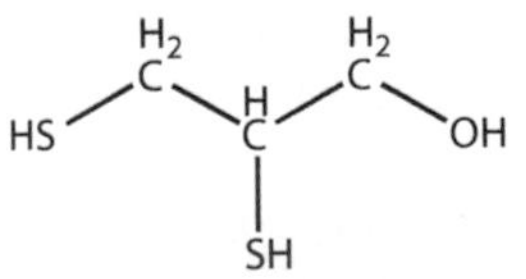

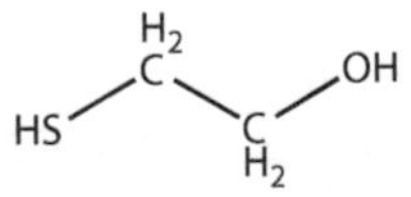

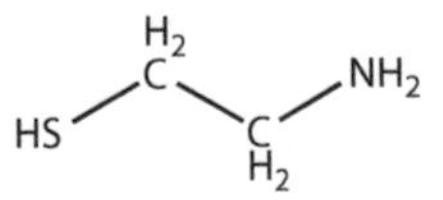

2,3-Dimercaptopropanol   2-Mercaptoethanol   2-Mercaptoethylamine

FIGURE 8.7 Structure and mechanism of reducing agents used for disulfide bonds in proteins. Shown are the structures of various reagents used for the reduction of disulfide bonds in proteins. Shown is DTT (and its stereoisomer dithioerythritol) (Cleland, W.W., Dithiothreitol, a new protective reagents for SH groups, *Biochemistry* 3, 480–482, 1964), (2S)-2-amino-1,4-dimercaptobutane (Lukesh, J.C., III, Palte, M.J., and Raines, R.T., A potent, versatile disulfide-reducing agent from aspartic acid, *J. Am. Chem. Soc.* 134, 4057–4059, 2012), bis-(2-mercaptoethyl)sulfone and N,N'-dimethyl-N,N'-bis(mercaptoacetyl)hydrazine (Singh, R., Lamoureux, G.V., Lees, W.J., and Whitesides, G.M., Reagents for rapid reduction of disulfide bonds, *Methods Enzymol.* 251, 167–173, 1885), and 2,3-dimercaptopropanol (Simpson, S.D. and Young, L., Biochemical studies of toxic agents; experiments with radioactive 2:3-dimercaptopropanol (British anti-lewisite), *Biochem. J.* 46, 634–640, 1950). Also shown at the bottom is 2-mercaptoethanol and 2-mercaptoethylamine. The various dimercapto compounds are the more efficient reducing agents based on their ability to cyclize as shown at the top for DTT.

be used to measure the sulfitolysis of the disulfide bond. In studies performed in the absence of a denaturing agent, the change of Asp26 to an alanine residue (D26A) had a small effect on sulfitolysis yield, while the change of Lys36 to glutamic acid (K36E) leads to a decrease in sulfitolysis yield (60%–27%) and requires a higher concentration of sulfite for the reaction. The enhancement of sulfitolysis observed when a positive charge is close to the cystine is consistent with the enhancement of cysteine nucleophilicity by adjacent positive charges as discussed earlier. Furthermore, the replacement of lysine by glutamic acid places a negative charge adjacent to the cysteine residue, which, as noted by Cecil and McPhee,[132] would decrease the rate of modification of cysteine. The influence of primary structure on the susceptibility of cystine to sulfitolysis is provided by a study by Kristjánsson and coworkers comparing a subtilisin-like proteinase from a psychrotrophic *Vibrio* species, proteinase K, and aqualysin.[135] Psychrotrophic proteases are obtained from cold-adapted organisms and tend to be more flexible (less rigid) than corresponding proteases from mesophilic or thermophilic organisms.[136–138] It is suggested that there are less intramolecular contacts in psychro-trophic proteins. The disulfide bonds of the *Vibrio* protease were cleaved by sulfitolysis (0.2 M Tris-HCl, 20 mM EDTA, 0.1 M Na$_2$SO$_3$, pH 9.5) in the presence of 2-nitro-5-sultothiobenzoate where the disulfide bonds of either proteinase K or aqualysin were resistant to sulfitolysis. The *Vibrio* protease contains four disulfide bonds as determined with 2-nitro-5-thiosulfobenzoate[135] in the presence of 3.0 M guanidine hydrochloride. DTT (10 mM) caused a small loss of activity in the *Vibrio* protease that was partially reversed on oxidation. Incubation of *Vibrio* protease in the presence of sodium sulfite and absence of guanidine caused approximately 90% loss of activity with almost total modi-fication of cystine; approximately 3.5 of the 4 cystine groups are available for modification within 15 min. In comparison, 2.5 M guanidine hydrochloride was required for complete sulfitolysis of the disulfide bonds of proteinase K and 5.5 M guanidine hydrochloride for aqualysin I.

Thannhauser and colleagues[139] used the reaction with 2-nitro-5-sulfothiobenzoate with the free thiol to measure disulfide bond content in proteins after sulfitolysis (Figure 8.6). This concept is similar to the use of iodoacetate to measure the liberated thiol as described by Würfel and cowork-ers[133] in their work on thioredoxins. The process of sulfitolysis yields *S*-sulfocysteine and cyste-ine thiolate; 2-nitro-5-thiosulfobenzoate reacts with the thiolate anions resulting in a second *S*-sulfocysteine, and 2-nitro-5-thiobenzoate can be measured at 412 nm and is proportional to the cystine residues in the protein. The reaction is reversible to form cysteine upon treatment with a thiol such as 2-mercaptoethanol or DTT. Guanidine hydrochloride (3 M) was required for determination of the disulfide bonds in RNAse A. Urea was unsatisfactory for use in this procedure as were higher concentrations of guanidine hydrochloride. Damodaran[140] reported that the 2-nitro-5-thiobenzoate anion that is measured in the previous reaction is sensitive to light in the presence of sulfite so that it is important to perform this assay in the dark.

It is possible to obtain complete sulfitolysis of disulfide bonds in proteins in the absence of denaturing agent by using oxidative sulfitolysis (Figure 8.6). Kella and Kinsella[141] used sodium sulfite in the presence of cupric ions for the progressive cleavage of disulfide bonds in bovine serum albumin and several other proteins. The time required varied from 2.66 h for chymotrypsin to 6 h for trypsin; approximately 3 h were required for the complete sulfitolysis of bovine serum albu-min. The *S*-sulfonated bovine serum albumin has decreased solubility, a lower isoelectric point, and changes in conformation as assessed by viscosity measurements, UV difference spectroscopy, intrinsic fluorescence, and binding of 1-amino-8-naphthlene sulfonate (ANS).[142] Modification of proteins with sodium sulfite is of considerable interest for food processing[143–145] and the prepara-tion of protein films.[146,147] 2-Nitro-5-sulfothiobenzoate with sodium sulfite (pH 9.0, 200-fold molar excess of sodium sulfite) was also used for the processing of a fusion protein containing the extracel-lular domain of the P$_2$X$_2$ ion channel.[148] Sulfitolysis in the presence of either cupric ions or sodium tetrathionate has proved useful for the quantitative conversion of cystine to *S*-sulfocysteine in the processing of biotherapeutics.[147,149–154] This is mostly used when there is the need to reduce and reoxidize a protein expression in a bacterial expression system. Given the history of use of sulfite in food processing[143–145], there is somewhat less concern from a product safety point of view. Sato and Aimoto[155] used *S*-sulfonation for protection of thiol groups in peptide ligation using thioesters.

The reduction and alkylation of cysteine residues in proteins for structural analysis have been discussed earlier. It is possible to selectively reduce cystine disulfide bonds in proteins as a chemical modification. Reduction of cystine in proteins is usually accomplished with a mild reducing agent such as β-mercaptoethanol, DTT, tris-(2-carboxyethyl)phosphine (TCEP), or cysteine and may or may not involve chaotropic agents such as urea or guanidine hydrochloride. These studies are tech-nically difficult but can provide useful information.

Disulfide groups were originally considered to be masked in the sense of being unavailable for chemical reaction.[3,156,157] The reagents commonly used for the reduction of disulfide bonds (Figure 8.7) will also react with cysteine sulfenic acid[158,159] and cystine thiosulfinate[91,92] but do not react with other functional groups in proteins (see Chapter 6).[3,160] Olcott and Fraenkel-Conrat[160] have reviewed the early work on the reduction of disulfide bonds in proteins, reporting that the rate of reduction of disulfide bonds varied for the various proteins under the same reaction conditions and was greatly influenced by what was described as the degree of denaturation. The data at that time suggested that proteins that have a high content of cystine were sensitive to and adversely affected (solubility) by a low concentration of reducing agent. It must be emphasized that the assays were somewhat less sophisticated than those currently available. As an example, Walker[161] added cyanide to the nitroprusside assay for cysteine. Cyanide had been described as a reagent that would *reduce* disulfide bonds.[162] Walker also reported that the cyanide/nitroprusside reaction was negative for egg albumin or serum unless the sample had been heated. This early observation is consistent with later studies showing that many proteins contain disulfide bonds that are not available unless the protein is denatured. The reaction of cyanide with disulfide bonds deserves additional comment. The formation of S-cyanocysteine has been described earlier; S-cyanocysteine is a product of the reaction of cyanide with cystine and is used for peptide bond cleavage[163,164] (Figure 8.8). Catsimpoolos and Wood[164] also used 2-nitro-5-thiocyanobenzoate in their studies. Catsimpoolos and Wood[165] had previously studied the reaction of cyanide with bovine serum albumin (approximately 3000-fold molar excess; 50 mg albumin, 20 mg sodium cyanide, 18 h at 37°C) and observed the formation of thiocyanate only above pH 8.0. At pH 7.0, no thiocyanate is formed and is consistent with the formation of iminothiazolidine instead of thiocyanoalanine.

DTT was one of the several polyhydric alcohols synthesized by Evans and coworkers in 1949.[166] Some years later, William Cleland, best known for his excellent work on kinetics, observed that DTT was useful for the reduction of disulfide bonds in proteins.[167] DTT and the isomeric form dithioerythritol are each capable of the quantitative reduction of disulfide bonds in proteins (Figure 8.7). Furthermore, the oxidized form of DTT has an absorbance maximum at 283 nm ($\Delta\varepsilon = 273$), which can be used to determine the extent of disulfide bond cleavage.[168] DTT has been extensively used for the reduction of disulfide bonds in proteins under native conditions; 2-mercaptoethanol continues to be used more in the structural analysis of proteins as described earlier. The previous discussions support the concept that the susceptibility of a disulfide bond to reduction is a function of disulfide bond location (primary, secondary, and tertiary structure) and reaction conditions (pH, temperature, and solvent). The below studies are provided as examples of work of the selective reduction of disulfide bonds over the past 30 years. Lukesh and coworkers[169] have introduced a new reducing agent, (2S)-2-amino-1,4-dimercaptobutane (DTBA), which is said to have an advantage over DTT in that it will function under more acidic conditions; the pKa for the first thiol in DTBA is 8.2 compared to 9.2 in DTT.

The use of trivalent phosphorus nucleophiles (Figure 8.9) to reduce organic disulfides has been known for some time.[170] Tri-n-butylphosphine will reduce disulfide bonds in proteins[171] and will also convert S-sulfocysteine to cysteine.[172] The reaction is performed under alkaline conditions (pH 8.0, 0.1 M Tris, or 0.5 M bicarbonate) with n-propanol added (50/50, v/v) to dissolve tri-n-butylphosphine that is insoluble in strictly aqueous solutions. The reduction of disulfide bonds with thiols such as 2-mercaptoethanol or DTT is a reversible reaction; while it could be *driven* by concomitant modification of the newly formed sulfhydryl groups, this is not feasible in the presence of large quantities of competing thiols. The cleavage of disulfide bonds with alkylphosphines with the formation of phosphines is an irreversible reaction, and the presence of the alkylphosphines does not compete with alkylation of the newly formed sulfhydryl groups. Kirly[173] and Chin and Wold[174] used the simultaneous reaction of tri-n-butylphosphine to reduce disulfide bonds and alkylation with 4-(aminosulfonyl)-7-fluoro-2,1,3-benzoxadiazole for the fluorescent labeling of the newly formed sulfhydryl groups. The lack of reaction of 4-(aminosulfonyl)-7-fluoro-2,1,3-benzoxadiazole with tri-n-butylphosphine permits the simultaneous use of the two reagents to modify proteins.

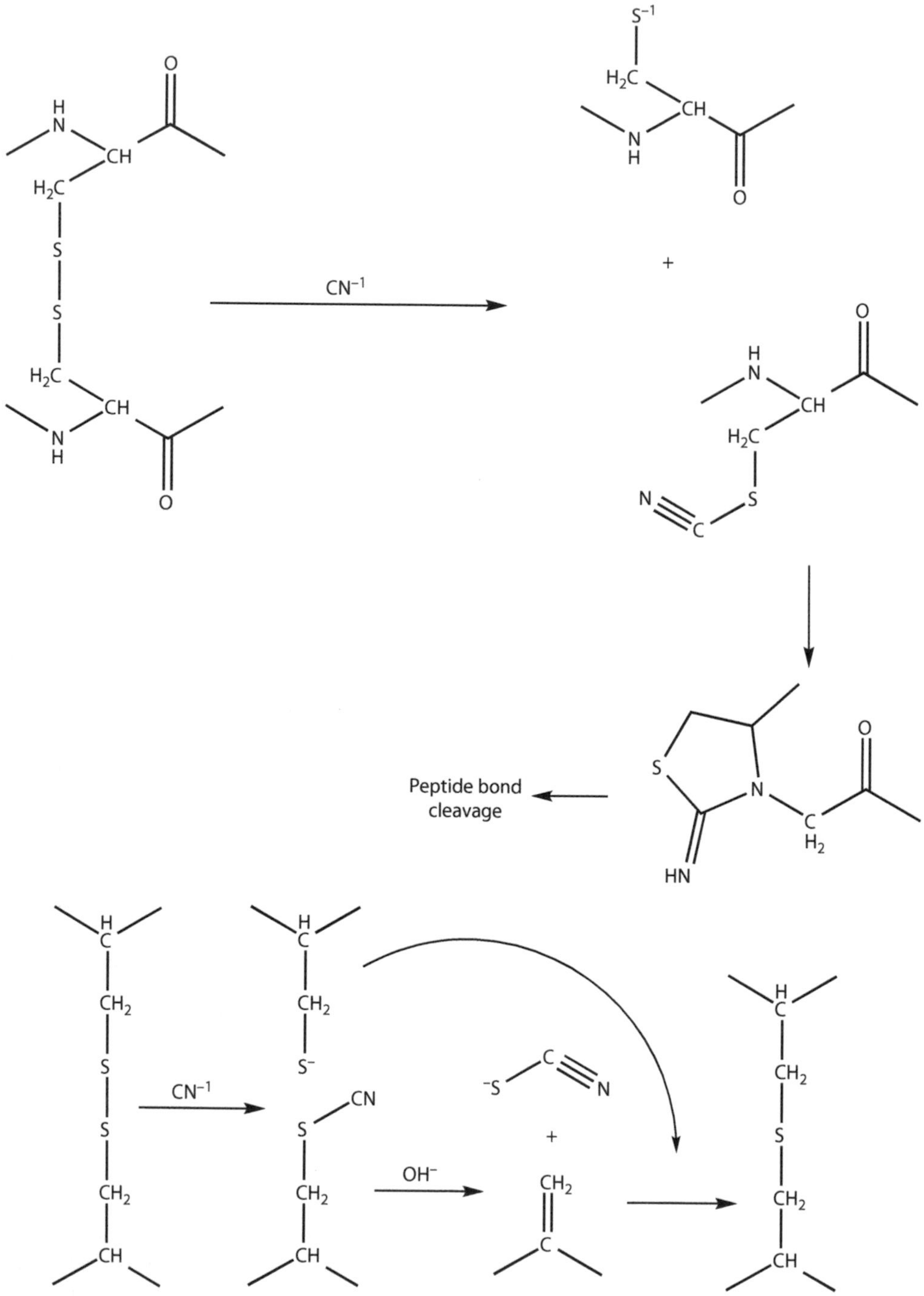

FIGURE 8.8 Reaction of cystine peptides with cyanide. Cyanide can react with a disulfide bond. The reaction is asymmetric and depends on the anionic stability of the leaving thiolate anion (Hiskey, R.G. and Harpp, D.N., Chemistry of aliphatic disulfides. VII. Cyanide cleavage in the presence of thiocyanates, *J. Am. Chem. Soc.* 86, 2014–2018, 1964). The scheme for peptide bond cleavage is adapted from Catsimpoolas, N. and Wood, J.L., The cleavage of cystine peptides by cyanide, *J. Biol. Chem.* 241, 1790–1796, 1966. The bottom scheme shows a suggestion for the reaction of cyanide with proteins at alkaline pH (Catsimpoolas, N. and Wood, J.L., The reaction of cyanide with bovine serum albumin, *J. Biol. Chem.* 239, 4132–4137, 1964).

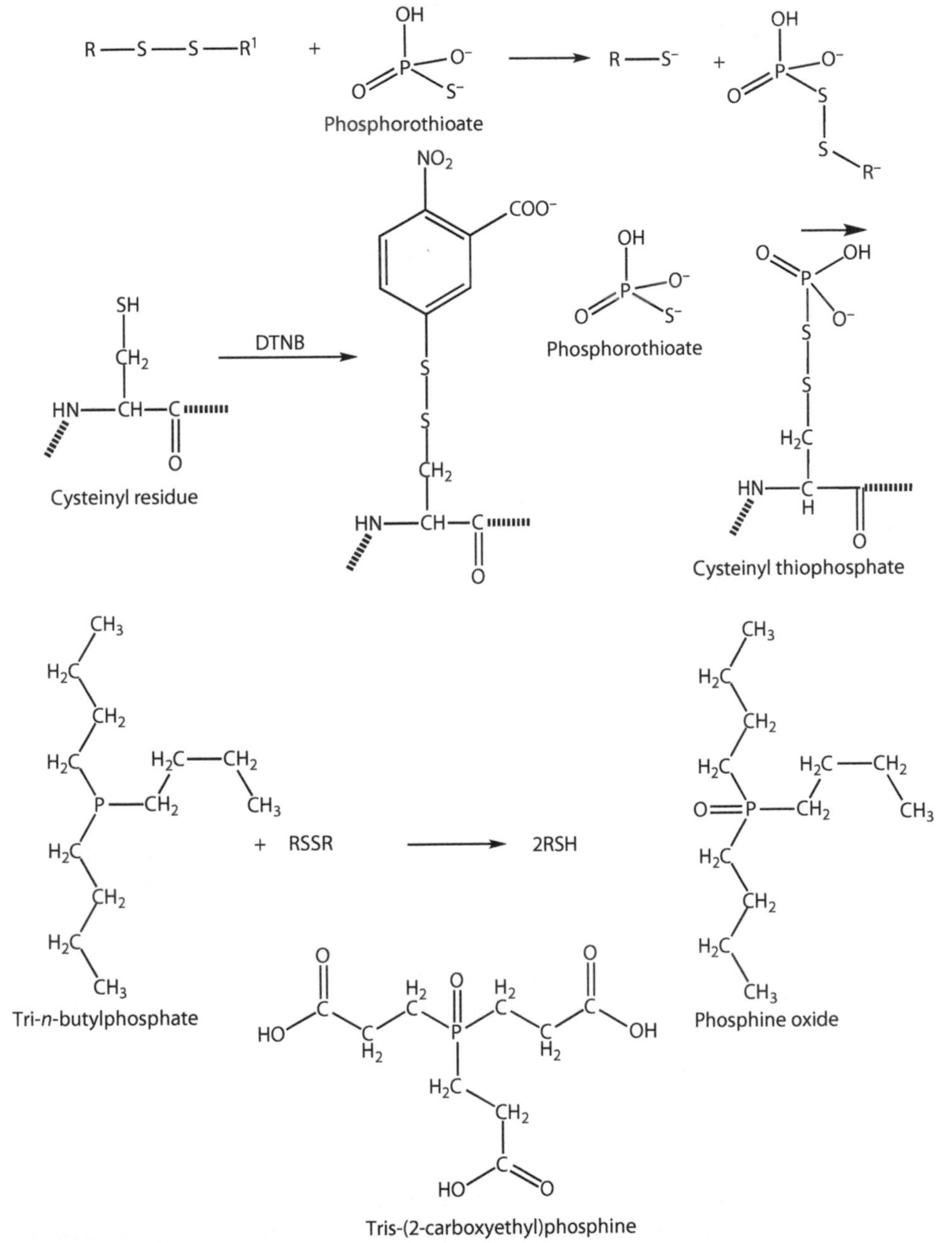

FIGURE 8.9 Use of phosphorothioate and alkyl phosphines for the reduction of disulfide bonds in proteins. Shown is the reduction of a disulfide bond by phosphorothioate (Neumann, H. and Smith, R.A., Cleavage of the disulfide bonds of cystine and oxidized glutathione by phosphorothioate, *Arch. Biochem. Biophys.* 122, 354–361, 1967). Also shown is the formation of cysteine thiophosphate by the reaction of phosphorothioate with DTNB-modified protein (Saxl, R.L., Anand, G.S., and Stock, A.M., Synthesis and biochemical characterization of a phosphorylated analogue of the response regulator CheB, *Biochemistry* 40, 12896–12903, 2001). Shown below is tri-*n*-butylphosphine and the reduction of a disulfide with the formation of phosphine oxide (Rüegg, U.T. and Rudinger, J., Reductive cleavage of cystine disulfide with tributylphosphine, *Methods Enzymol.* 47, 111–116, 1977); also shown is the structure of tris-(2-carboxyethyl)phosphine (Burns, J.A., Butler, J.C., Moran, J., and Whitesides, G.M., Selective reduction of disulfides by tris-(2-carboxyethyl)phosphine, *J. Org. Chem.* 56, 2648–2650, 1991).

Smejkal and coworkers[175] reported the simultaneous reduction and alkylation of proteins with tri-*n*-butylphosphine and acrylamide. The reader is recommended to a review by Overman and coworkers[176] for additional references to early work on the trialkylphosphines and triarylphosphines. Burns and coworkers[176] reported a synthetic process for TCEP and the selective reduction of disulfides by this reagent. The extensive application of trialkylphosphines/triarylphosphines for the modification of cystine residues in proteins was hampered by the insolubility of reagents such as tri-*n*-butylphosphine, and there are no current citations to the use of this reagent in protein chemistry. Current activity on phosphines in protein chemistry focuses on tris-(2-carboxyethyl)phosphine; there are studies comparing tri-*n*-butylphosphine and tris-(2-carboxyethyl)phosphine in the preparation of samples for proteomics.[177]

Tris-(2-carboxyethyl)phosphine has effectively replaced tri-*n*-butylphosphine in the preparation of samples for analysis of homocysteine in plasma.[178–180] There has been some comparison of TCEP with DTT for the reduction of proteins in tissue extracts.[181] Rogers and coworkers[181] found that DTT and TCEP were equivalent in reducing proteins in myofibrillar tissue extracts but that the resolution of fluorescently labeled proteins (MBB) on electrophoresis was superior with the DTT-reduced samples. Humphreys and coworkers[182] showed that TCEP was far more potent than 2-mercaptoethanol, 2-mercaptoethylamine, reduced glutathione, or DTT in reducing the disulfide bonds of a Fab' fragment.

The synthesis of a water-soluble phosphine, TCEP, was a significant advance.[176] The early development of this reagent[177] described the properties of the reagent. It is quite soluble in water (310 g L^{-1}), and dilute solutions (5 mM) are reasonably stable at acid pH values; at pH values above 7, the rate of conversion of the reagent to the oxide is significant and should be considered in experimental design. Unlike thiols such as DTT, the reduction of disulfides with TCEP proceeds very rapidly at pH 4.5 and below that makes alkylation of the liberated thiol optional for further process. Gray[183] extended his early work in developing the use of TCEP reduction to establish the position of disulfide bonds in proteins. Since the reduction is performed at low pH, the reduction was performed in 0.1% trifluoroacetic acid with 1–10 μM TCEP, making it possible to directly obtain partially reduced peptides by HPLC; alkylation of the free thiols in the isolated peptides with 4-vinylpyridine permitted subsequent structural analysis of the peptide and disulfide bond assignment. In a subsequent study,[184] Gray applied the use of TCEP to the assignment of disulfide bonds in echistatin. Other studies have used trialkylphosphines for the reduction of disulfide bonds prior to electrophoretic analysis in proteomic research.[185–187] White and coworkers[188] were able to use partial reduction with TECP as described by Gray[183] to map the disulfide bonds in the fourth and fifth EGF-like domains of thrombomodulin. Wu and Watkins[189] have used TCEP reduction followed by modification with cyanate[190] and subsequent peptide bond cleavage in ammonia[191,192] to assign disulfide bonds in proteins. While TCEP has been demonstrated to be unstable at pH 7.0 and above,[177] useful studies with this reagent have been performed at pH 7.0–8.0.[193–198] While reference can be found for the reduction of aromatic sulfenic acids to sulfhydryl groups,[199,200] citation could not be found for the reduction of cysteine sulfenic acids by alkylphosphines.

The bulk of the use of TCEP has been for the preparation of samples for proteomic analysis. There are, however, several studies that have used TCEP for the chemical modification of proteins. Tetenbaum and Miller[193] used TCEP to reduce disulfide bonds in soybean trypsin inhibitor (SBTI). The use of TCEP permitted the use of sulfur x-ray absorption spectroscopy to follow the reduction of the disulfide bonds of the protein in real time. The disulfide bonds in SBTI were reduced in a noncooperative manner within 5 min. Circular dichroism (CD) and FTIR-IR spectroscopy established that the protein structure collapsed after the reduction of the disulfide bonds. Maciel and coworkers[194] used x-ray photoelectron spectroscopy to measure sulfur oxidation in self-assembled monolayers emphasizing the value of advanced spectral techniques for real-time measurements of oxidation of thiol groups. Zhang and coworkers[195] reported the use of TCEP for reducing of disulfide bonds with simultaneous proteolysis to measure H/D exchange (HDX). sRAGE proteins were taken into a deuterium solvent, and exchange was allowed to proceed. At various times, a portion of the HDX reaction is taken into a mixture of TCEP, urea, and protease (pepsin or *Aspergillus saitoi* protease type XIII) at pH 2.3–2.5. The rate and extent of HDX is determined by MS. This work is intended to extend the use of HDX to proteins with disulfide bonds. Side reactions with TCEP have been observed including cleavage at cysteine residues.[196] These investigators suggest that a *few* percent of the total protein population is degraded with TCEP (1–10 mM, 37°C, 16 h); the pH optimum for degradation is 8. Degradation can occur at lower temperatures over longer periods of time (40% of a protein cleaved after 2 weeks at 4°C).

DTT, tris-(2-carboxyethyl)phosphine, and 2-mercaptoethanol are the most frequently used reagents for the reduction of disulfide bonds. However, there are other reagents that are used for the reduction of disulfide bonds in proteins. Phosphorothioate (Figure 8.9) had been demonstrated to effectively cleave disulfide bonds in proteins forming the S-phosphorothioate derivative.[201,202] Related reducing agents including bis(2-mercaptoethyl) sulfone and N,N'-dimethyl-N,N'-bis(mercaptoacetyl)-hydrazine have been described (Figure 8.8)[203] but have not been used despite being somewhat more effective than DTT. Sodium borohydride can be used for the reduction of disulfide bonds in proteins[204] but has seen little use; it is somewhat difficult to use, and while it has an advantage in terms of being able to be destroyed rather than removed before subsequent reactions, it has a disadvantage in stability in aqueous solvent (in addition, there may be foaming). Light and coworkers have examined the susceptibility of disulfide bonds in trypsinogen to reduction.[205] At pH 9.0 (0.1 M sodium borate), a single disulfide bond (Cys179–Cys203) is cleaved in trypsinogen by 0.1 M NaBH$_4$. The resulting sulfhydryl groups are *blocked* by alkylation. The characterization of the modified protein has been performed by the same group.[206] The disulfide bond that is modified under these conditions was identified as critical for the primary specificity site in trypsin.

It is usually necessary to remove the reducing reagents from the reaction mixture before further processing (an exception is the TCEP as this reagent can be used at low pH, conditions where disulfide bond formation and other oxidation are avoided). Efficiency would be improved by avoiding this step by the use of a reagent that could be readily removed from the reaction mixture. Insolubilized dihydrolipoic acid (Figure 8.10) has also been proposed for use in the quantitative reduction of disulfide bonds.[207] Scouten and coworkers[208] developed lipoic acid coupled to a glass matrix for the affinity purification of lipoamide dehydrogenase and later suggested use as an insoluble reducing agent.[209] Bienvenu and coworkers[210] developed a recyclable matrix (Aminomethyl-ChemMatrix®) containing dihydrolipoamide for the reduction of disulfides. Al-Dubai and coworkers[211] used insoluble dihydrolipoamide to reduce the disulfide bonds in antibody proteins. Thiopropyl-agarose (Figure 8.10) is another insoluble reagent for the reduction of disulfide bonds.[212]

The selective or limited reduction of protein disulfide bonds can be used to study structure and function as well as providing interesting derivatives. Mise and Bahl[213] used limited reduction with DTT to assign disulfide bonds in the α-subunit of human chorionic gonadotropin. Since disulfide bonds in a protein will demonstrate different susceptibility to reduction, distribution of radiolabeled iodoacetate will permit the assignment of disulfide bonds. However, the results are not as consistent as one would like as shown by the various reports on the reduction of HEWL. Gorin and coworkers[214] have examined the rate of reaction of HEWL with various thiols. At pH 10.0 (0.025 M borate), the relative rates of reaction were 1.0 for DTT, 0.4 for 3-mercapto-proptionate, and 0.2 for 2-mercaptoethanol. The rate of disulfide bond reduction with 2-mercaptoethanol greatly decreased from pH 10.0 to 9.0, while there was little change in the rate of disulfide bond reduction with DTT until the pH was decreased from 9.0 to 7.8. These experiments were performed in the absence of chaotropic agents. Bewley and coworkers[215] demonstrated the quantitative reduction of disulfide bonds in human pituitary growth hormone, bovine serum albumin, and HEWL with DTT. Warren and Gordon[216] reported that HEWL in water was resistant to reduction with 002 M DTT or 0.09 M 2-mercaptoethanol. One group[217] states that the disulfide bonds in HEWL are resistant to reduction without the addition of chaotropic agents such as dimethylforamide.[218] Radford and coworkers[219] used DTT at pH 7.8 to reduce one of the four disulfide bonds of HEWL and blocked the two cysteinyl residues (Cys6, Cys127) with carboxymethylation. It is noted that the partially reduced HEWL is in a nonnative conformation based on the reaction with monoclonal antibody.[220]

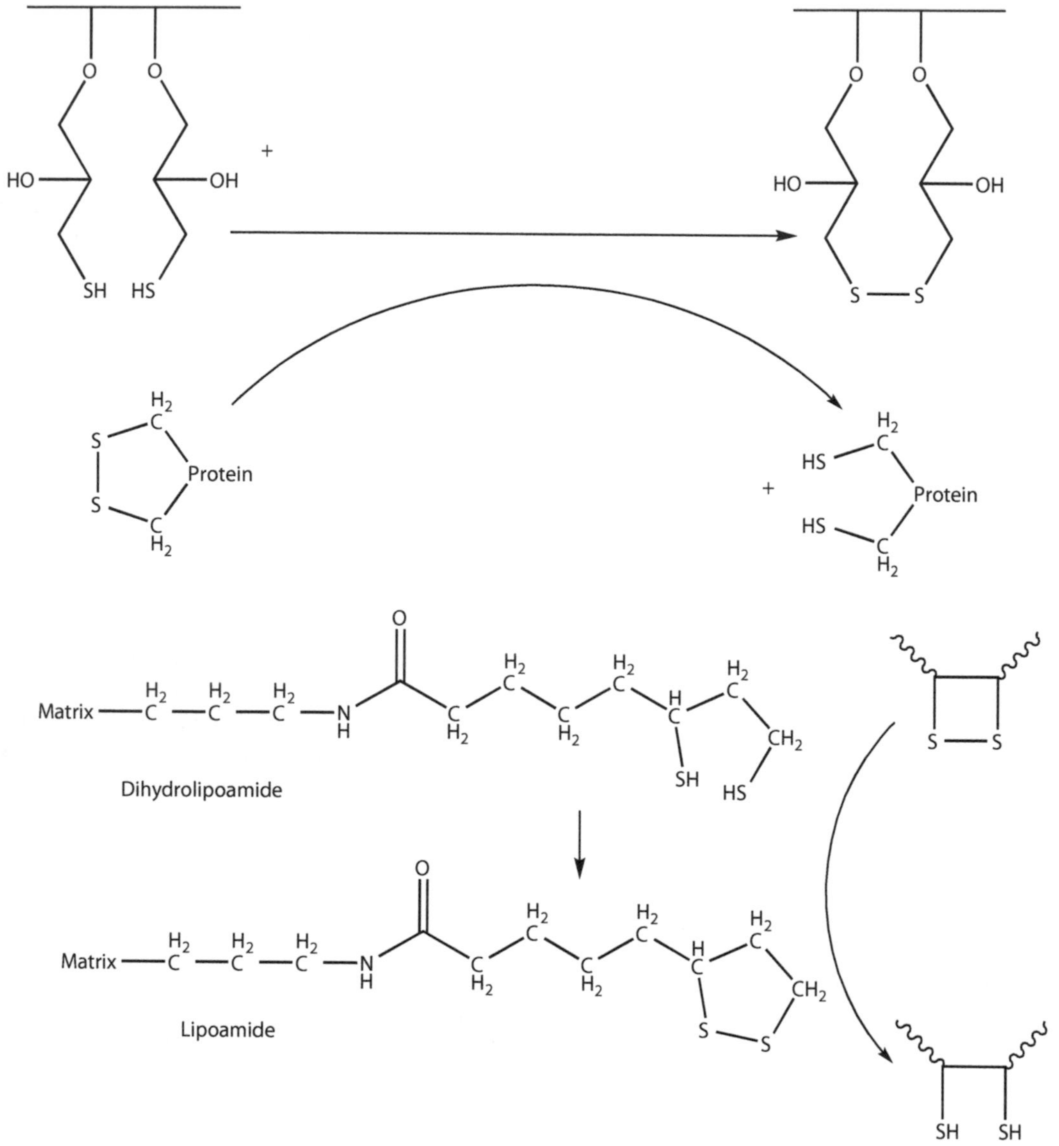

FIGURE 8.10 Structures and mechanism of insoluble reagents for protein disulfide reduction. At the top is thiopropyl-agarose (Ferraz, N., Leverrier, J., Batista-Viera, F., and Mantu, C., Thiopropyl-agarose as a solid phase reducing agent for chemical modification of IgG and F(ab')$_2$, *Biotechnol. Prog.* 24, 1154–1159, 2008), while the bottom of the figure shows the mechanism for insoluble lipoic acid (Gorecki, M. and Patchornik, A., Polymer-bound dihydrolipoic acid: A new insoluble reducing agent for disulfides, *Biochim. Biophys. Acta* 303, 36–43, 1973; Scouten, W.H. and Firestone, G.L., *N*-Propylhydrolipoamide glass beads an immobilized reducing agent, *Biochim. Biophys. Acta* 453, 277–284, 1976).

While the Cys6–Cys127 disulfside bond in HEWL is not considered hyperactive, the corresponding disulfide bonds in equine lysozyme and bovine α-lactalbumin are considered superactive.[221] Wu and Watson[191] presented an approach using cyanylation of cysteine sulfhydryl groups obtained from partial reduction of proteins for the assignment of disulfide linkages in proteins of known sequence. These investigators used limited reduction with TCEP in pH 3.0 citrate buffer with 6 M guanidine hydrochloride (TCEP added equivalent to cystine content of protein [in the case of pancreatic ribonuclease A that has four cystine residues, 40 nmol TCEP added to 10 nmol protein] at

23°C for 10–15 min) followed by reaction with 1-cyano-4-dimethylamino-pyridinium tetrafluo-roborate (CDAP). Cleavage at the cyanylated cysteine residues yielded a mixture of peptides linked by cystine residues that are isolated by HPLC with one portion analyzed by MS and the other portion, cleaved by a base (NH_4OH), reduced by TCEP and subjected to analysis by MS. A more recent work by Peng and coworkers[222] used mass spectrometric analysis of partially reduced human salivary α-amylase (5 mM TCEP in 0.1 M citric buffer, pH 3.0 with shaking for 10 min at 23°C) compared to analysis of fully reduced protein (25 mM TCEP in 0.1 M citrate, pH 3.0 at 65°C for 10 min) to identify disulfide bonds. These investigators emphasize the ability to use TCEP at acidic pH that obviates the necessity to alkylate the cysteine residues.

Bewley and Li[223] used a 20-fold molar excess of DTT to reduce disulfide bonds in various proteins under native conditions (25°C, 0.1 M KCl). Quantitative reduction was observed with lysozyme, prolactin, and insulin, while 14 of 18 disulfide bonds were reduced in bovine serum albumin; bovine pancreatic RNAse was not reduced to any extent in the absence of chaotropic agents. White[224] used dithreitol to obtain full reduction of HEWL, while 2-mercaptoethanol is used for partial reduction of HEWL. It is noted that Bradshaw and coworkers[225] observed that the mixed disulfide of cysteine with HEWL can be easily reactivated in the presence of 2-mercaptoethanol, cysteine, or 2-mercaptoethylamine. Homandberg and Wai[226] demonstrated that the reduction of urokinase by DTT in the presence of arginine allows the selective reduction of a disulfide bond joining the catalytically active chain to a nonessential 13 amino acid peptide. A synthetic peptide may then be coupled to the free sulfydryl group. This work was based on the ability to more selectively reduce interchain disulfide bond between the catalytic B-chain of urokinase and the A-chain. This was accomplished by including arginine with the DTT, which permitted some specificity in the reduction of the interchain disulfide bonds; intrachain disulfide bonds were still reduced in the A-chain, but it was possible to refold the protein. Synthetic peptide added to the refolded protein was incorporated at the free sulfhydryl group with a yield of approximately 40%. The success of this work was based on the ability of arginine to slow the rate of disulfide bond resulting in loss of activity (presumably interchain disulfide bonds in the catalytic B-chain) compared to the lesser effect on the disulfide bond between the B-chain and the A-chain. Arginine was required for the correct refolding of the protein. The difficulty in obtained selective reduction of the interchain disulfide bond complicated the success of this approach. Singh and Chang[227] used limited reduction to evaluate the structural stability of human α-thrombin. Human α-thrombin is derived from pre-thrombin II by cleavage of a single peptide bond resulting in this two-chain structure consisting of two peptide chains: a 259 residue heavy chain (B-chain) that contains all of the various functional sites and three internal disulfide bonds and a 36 residue chain (A-chain) joined by a single disulfide bond. Reduction of native human α-thrombin with DTT in the absence of chaotropic agents resulted in the initial reduction of an internal B-chain disulfide bond followed by the reduction of the intrachain disulfide between the catalytic B-chain and the A-chain; reduction of the intrachain disulfide occurred with 5 mM DTT, while 50 mM DTT was required for the interchain disulfide.

Disulfide scrambling is the cleavage and reformation of disulfide bonds in a protein potential resulting in a protein with an alignment of disulfide bonds different from the native protein. While it is accepted that primary structure dictates the tertiary structure of a protein, the folding patterns of protein can be markedly different[228], and reduction and reoxidation of a disulfide bond may or may not result in the native conformation and could provide a problem in the processing of samples for MS.[229] A fully processed protein may not have all of the primary structure information necessary for folding. An extreme example of this is provided by insulin. Insulin is a two-chain protein (or large peptide) that is synthesized as a precursor.[230] The two-chain structure frustrated many good peptide chemists who attempted the synthesis of this material for use in the treatment of diabetes. Tang and Tsou[231] were able to obtain 20%–30% yields of insulin from DTT-reduced insulin or from S-sulfo derivatives on refolding in the presence of DTT. The yield is higher from the S-sulfo chains. The addition of the C-chain, the chain joining the A and B chains in proinsulin, did not influence the yield. These studies were performed in the presence of DTT and protein disulfide isomerase. Ryle and Sanger described disulfide interchange reactions in their early structural work on insulin.[232] Limited hydrolysis of insulin in dilute acid (0.1–5 M HCl) provided many more cystine peptides than could be accounted for by a unique structure of insulin. Disulfide exchange was also observed under neutral (pH 7.2 phosphate) conditions. The presence of a thiol (cysteine) promoted exchange at pH 7.2 but inhibited exchange under acid solvent conditions. Exchange at neutral pH was also inhibited by N-ethylmaleimide. Disulfide bridges have been observed to stay intact during the conversion of native insulin into amyloid fibrils (pH 1.9, 65°C, 16 h).[233] In the studies with thrombin described previously,[227] disulfide scrambling was accomplished with several chaotropic agents (guanidine hydrochloride, urea, guanidine thiocyanate) in the presence of catalytic amounts of 2-mercaptoethanol. The unfolding of thrombin under these conditions was assessed by reverse-phase HPLC and demonstrated low conformational stability (extensive scrambling of disulfide bonds). While there is some continued interest in the use of disulfide scrambling for measurement of conformational stability in proteins,[234] there are a variety of other approaches.[235]

The correct formation of disulfide bonds is an issue in the manufacture and storage of therapeutic proteins produced by recombinant DNA technology. For the purpose of this discussion, there are two problems. One is the correct folding of a protein after synthesis; this is mostly a problem with protein expressed in bacterial systems.[236] The second is disulfide scrambling in the processing and storage of therapeutic proteins.[237–243] While the presence of a reducing agent such as an intrinsic cysteine or exogenous thiol in catalytic amounts definitely enhances disulfide scrambling, disulfide scrambling can occur in the absence of such influence. Kaneko and Kiabatake[243] observed disulfide scrambling in thaumatin, a protein that does not contain cysteine, with thermal stress. Analysis of the products of the thermal stress supported a hypothesis where β-elimination of a cystine yielding dehydroalanine and thiocysteine/cysteine would then have provided the initial reducing power for the disulfide scrambling. A subsequent report[245] on the thermal cleavage of cystine supported this mechanism. This heterolytic cleavage of cysteine is that observed with degradation of disulfide bonds in base, which is discussed earlier. Disulfide scrambling can also confound the electrophoretic analysis of proteins[229,246–248] as well as MS.[249–252] Hapuarachchi and coworkers[248] used capillary electrophoresis to evaluate an Fc fusion protein expressed in E. coli. An extra peak was observed on nonreduced electrophoresis, which was a disulfide-scrambled form and isobaric with the correctly folded material. The order of the migration of the disulfide-scrambled form is reversed on SDS-PAGE analysis where the disulfide-scrambled form elutes before the main peak. Other studies on selective reduction include the study of bovine seminal ribonuclease, which is a dimer with two interchain disulfide bonds[253–255] and dimeric α-cobratoxin.[256]

The limited reduction of immunoglobulin proteins has been used to develop highly nucleophilic sites for the attachment of probes and drugs. Dolder[257] used 2-thiopyridine to dissociate IgM. The rate of reaction was much slower than that observed with DTT. The rate and extent of reduction of IgM was much higher at an alkaline pH (9.0–12.0). I was unable to find any further work on the use of 2-thiopyridine as a reducing agent but do note that it is formed from the reaction of 2,2′-dipyridyl disulfide and exists in equilibrium with the pyridone form. 2-Thiopyridone has been reported to form a complex with reduced and nonreduced proteins, which have been suggested to interfere with the use of 2,2′-dipyridyl disulfide for the measurement of cysteine residues in proteins.[258] The lack of reaction of iodoacetamide with the reduced IgM was attributed to disulfide interchange converting intersubunit bonds to intrasubunit bonds. Hong and Nisonff[259] evaluated the stability of disulfide bonds in IgG. This includes disulfide bonds between the heavy chains and between heavy chains and light chains. There may be two to four disulfide bonds between the heavy chains of IgG depending on the IgG class and species and a variable number of intrachain disulfides in light-chain and heavy-chain domains.[260,261] Bunting and Cathou[262] prepared a fluorescent derivative of an antidansyl (DNS) immunoglobulin by a disulfide interchange reaction. An IgG or F(ab′) fragment was reduced with 2-mercaptoethanol (0.1 M 2-mercaptoethanol at pH 8.2)[259] cleaving the disulfide bond between the heavy chain or heavy-chain fragment (in rabbit IgG, there is a single disulfide bond between the two heavy chains). The heavy–light chain derivative is isolated by gel filtration in acetic acid and coupled with bis(fluoresceinthiocarbamyl)-DL-cysteine. Measurement of fluorescence transfer suggested that the epitope binding site (binding site for DNS) is at the tip of the Fab fragment and the disulfide is at the edge of the C_LC_H1 domain of the immunoglobulin. Sun and coworkers[263] developed methods for the reduction of some interchain disulfides in an anti-CD30 monoclonal antibody producing thiols that could be alklylated with monomethyl auristatin E. The disulfide bonds in monoclonal antibodies can have properties somewhat different than a native antibody population. Ferraz and coworkers[212] used thiopropyl-agarose as a solid-phase reducing agent for IgG and F(ab)$_2$. The use of thiopropyl-agarose produced six thiol groups in the F(ab)$_2$ fragment and three thiol groups in IgG. The use of DTT also produced 6 thiol groups in the F(ab)$_2$ but 11 thiol groups in IgG. The IgG reduced by thiopropyl-agarose retained the same level of biological activity as the native protein thus providing a method for the development of probe and drug conjugates. The extent of modification of IgG with the thiopropyl matrix did depend on temperature with more extensive reduction (from three thiol groups to six thiol groups) when the temperature was increased from 22°C to 37°C. Aono and coworkers[264] used reduction of a recombinant IgG1 bound to protein A using cysteine to convert a trisulfide linkage to a disulfide linkage. Early work by Mozen and Schroeder[265] used DTT reduction of IgG followed by alkylation to reduce/eliminate anticomplement activity; the derivative immunoglobulin protein retained full potency as well as native pharmacokinetic activity.

The studies discussed have used the reaction of a reducing agent (soluble or insoluble) in a solution with a protein. Since reduction in its most basic form increases the electronegativity of an atom (e.g., $Cu^{2+} \rightarrow Cu^{1+}$), the process can be considered as the addition of an electron with the hydrogen coming along for the ride. Thus, any process that would permit the electron transfer would be a reduction. A unique study of the reduction of a disulfide bond is provided by the study of the Rieske [2Fe-2S] center of *Thermus thermophilus*.[266] It is possible to reduce this disulfide bond by protein-film voltammetry. Reversible oxidation and reduction of the soluble Rieske domain of *T. thermophilus* were observed with direct absorption of the protein (the soluble Rieske protein fragment obtained by heterologous expression in *E. coli*) on a pyrolytic graphite-edge electrode. It has been demonstrated that disulfide bridges in α-lactalbumin are reduced by electron transfer from an excited tryptophan residue.[267,268] In studies with human α-lactalbumin,[268] UV light (270–290 nm, 1mW cm^{-2}, 2–4 h, 50 mM HEPES-150 mM KCl, pH 7.8, 20°C) irradiation resulted in a 10 nm *red shift* of its tryptophan fluorescence emission spectrum (324–334 nm). Reaction of the irradiated protein with 5,5'-dithiobis(2-nitrobenzoic acid) demonstrated the presence of free sulfhydryl groups. It is possible that a similar phenomenon is involved in the changes in a cyclic-nucleotide-gated ion channel by UV irradiation.[269] The cleavage of a disulfide bond mediated by UV light has also been reported for bovine somatropin.[270] Lyophilized recombinant bovine somatotropin was photolyzed by UV light (305–410 nm, λ_{max} = 350 nm). The protein had been lyophilized from carbonate buffer and the cake contained 6% moisture. Unlike the other examples of disulfide bond reduction mediated via tryptophan, the cysteine residues in photolyzed somatotropin appear to donate electrons back to tryptophan leaving a pair of thiyl radicals that subsequently add oxygen to form the sulfonate. Disulfide bonds can also be cleaved by x-radiation (synchrotron radiation).[271] This study reported the reduction of a redox-active disulfide in a tryporedoxin. The radiation dose was less than that required to break a structural disulfide in lysozyme.[272] Leach and coworkers[273] recognized that while small peptides containing disulfide bonds could be reduced using cathodic reduction, there would likely be problems with proteins because of size and tertiary structure considerations. Therefore, a small thiol (2-mercaptoethanol) was used as a catalyst for the reduction. Electrolysis markedly increased the extent of disulfide bond reduction in the presence of 70 mM 2-mercaptoethanol. The extent of disulfide bond reduction did increase with increasing pH; as with other reducing agents, the susceptibility of disulfide bonds to reduction is dependent on their location within the protein. A more vigorous electrolytic reduction of proteins has been recently reported by Lee and Chang[274] who reported the reduction of acids to aldehydes or alcohols with an increase in the isoelectric point during 2D electrophoresis. The extent of reduction depended on the salts present in the electrophoresis system and resulted in artifacts in 2D electrophoresis. Li and coworkers[275] reported that the combination of a thin-layer electrochemical flow cell (EC) and liquid sample desorption electrospray ionization mass spectrometry (DESI-MS) resulted in disulfide bond cleavage of insulin. There is limited use of electrolytic reduction of proteins prior to analysis by MS.[275–277] Huang and coworkers[278] reported the reduction of disulfide bonds in SBTI with ultrasound; there were also conformational changes observed with CD. It is clear that the possible cleavage of disulfide bonds when subjected to electrical fields has the potential of resulting in disulfide cleavage/interchange.

Ascorbic acid (vitamin C) is considered to be an antioxidant that can control free radicals.[279] It also has been used on occasion as a potential reducing agent for disulfide bonds in proteins,[280–282] although it is recognized that it is extremely weak ($E° \approx 0.06$) when compared to cysteine ($E° \approx -0.33$).[283] Ascorbic acid has been used to *selectively* reduce S-nitroso groups in the biotin-switch assay, but the ability of ascorbate to reduce disulfide bonds has created problems.[284] Recent work by Thompson and colleagues has provided an alternative approach using capture of the S-nitrosoprotein by thiol agarose.[285] Digestion of the captured protein bound to the matrix has provided a facile method for providing useful samples for analysis by MS.

In conclusion, it is useful to consider the chemistry of trisulfide bonds (Figure 8.11). Low molecular trisulfide compounds such as diallyl trisulfide (DATS) are found in garlic and related vegetables and are of interest as nutriceutical products.[286–289] Trisulfide in proteins can be traced to studies showing the presence of such derivatives of cysteine in the acid hydrolysates of wool.[290] Yanagi and coworkers[291] have detected the presence of trisulfide in human hair by electron spin resonance (ESR) spectrometry. Breton and coworkers[292] detected traces of cysteine trisulfide in a mutein of interleukin-6; an anionic form of the mutein where the N-terminal 22 residues were deleted and Cys23 and Cys29 were replaced with serine residues was isolated by isoelectric focusing. This anionic (pI 6.7 compared to native pI of 6.56) form contained a trisulfide linkage (MS) that could be converted to the native form by mild reduction (10 mM DTT). Pristasky and coworkers[293] found trisulfide bonds in a recombinant variant of a human IgG2 monoclonal antibody that, as with the interleukin-6 mutein, could be eliminated by mild reduction. Although there is not an extensive literature, trisulfide linkages may occur more frequently than assumed[294,295] and may, in particular, be present in recombinant antibodies.[296] The various observations suggest that conversion of the trisulfide links to disulfide occurs readily in the presence of mild reducing agents.

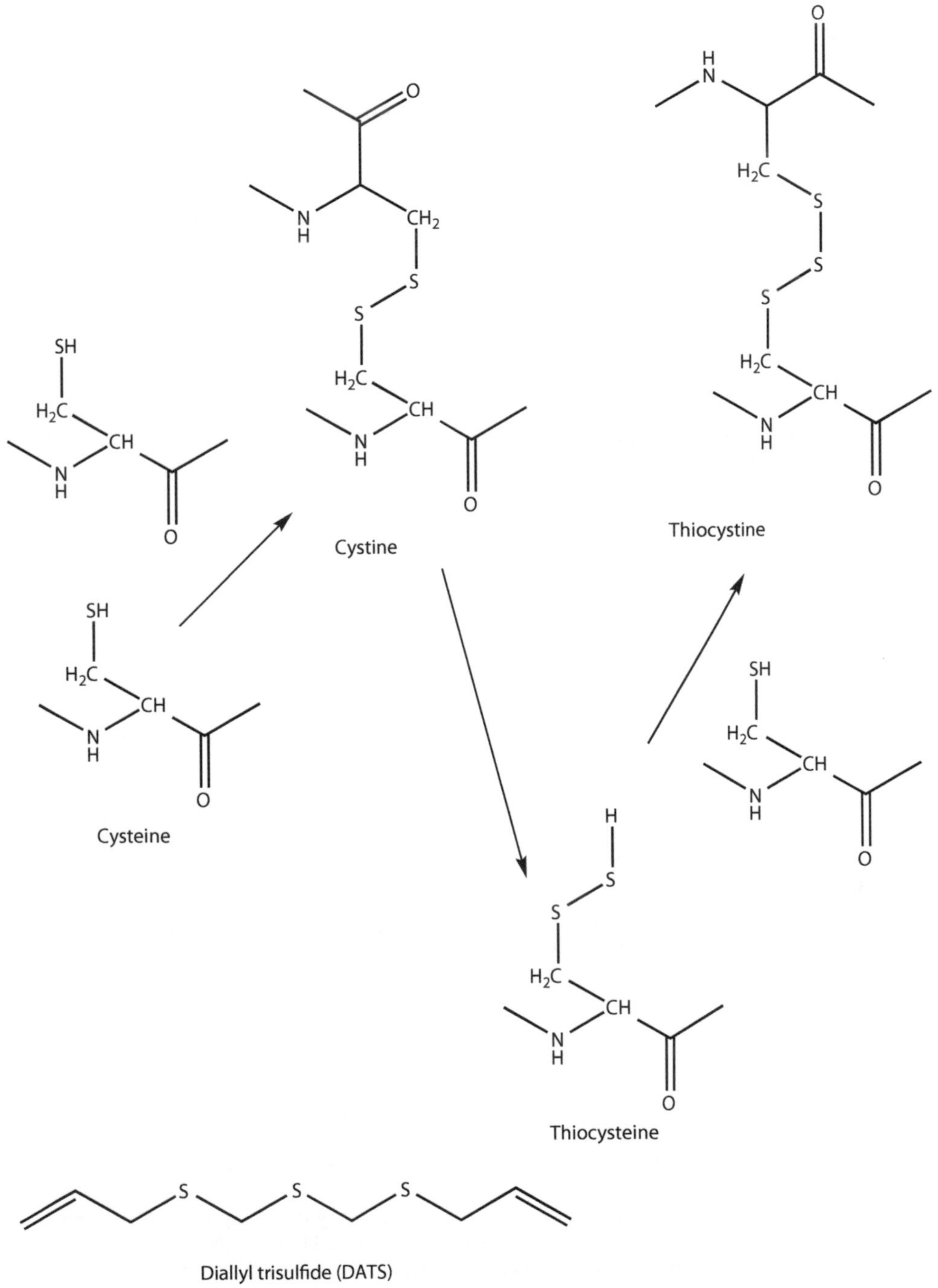

FIGURE 8.11 Structure of cysteine trisulfide and DATS. Shown is the structure of cysteine trisulfide with a putative mechanism for its formation. (See Breton, J., Avanzi, N., Valsasina, B. et al., Detection of traces of a trisulphide derivative in the preparation of a recombinant truncated interleukin-6 mutein, *J. Chromatogr. A* 709, 135–146, 1995; Gu, S., Wen, D., Weinreb, P.H. et al., Characterization of trisulfide modification in antibodies, *Anal. Biochem.* 400, 89–98, 2010). Also shown is the structure of DATS that is considered to be a nutriceutical derived from garlic (see Koizumi, K., Iwasaki, Y., and Narukawa, M., Diallyl sulfides in garlic activate both TRPA1 and TRPV1, *Biochem. Biophys. Res. Commun.* 382, 545–548, 2009).

REFERENCES

1. Desnuelle, P., The general chemistry of amino acids and peptides, in *The Proteins: Chemistry, Biological Activity, and Methods*, Vol. 1, Part A, Chapter 2, ed. H. Neurath and K. Bailey, pp. 87–180, Academic Press, New York, 1953.
2. Tristram, G.R., The amino acid composition of proteins, in *The Proteins: Chemistry, Biological Activity, and Methods*, Vol. 1, Part A, Chapter 3, ed. H. Neurath and K. Bailey, pp. 181–233, Academic Press, New York, 1953.
3. Putnam, F.W., The chemical modification of proteins, in *The Proteins: Chemistry, Biological Activity, and Methods*, Vol. 1, Part B, Chapter 10, ed. H. Neurath and K. Bailey, pp. 893–972, Academic Press, New York, 1953.
4. Cecil, R. and McPhee, J.R., The sulfur chemistry of proteins, *Adv. Protein Chem.* 14, 255–389, 1959.
5. Cecil, R., Intramolecular bonds in proteins, in *The Proteins: Chemistry, Biological Activity, and Methods*, 2nd edn., Vol. 1, Chapter 5, ed. H. Neurath, pp. 379–476, Academic Press, New York, 1963.
6. Cadenas, E. and Packer, L., ed., *Thiol Redox Transitions in Cell Signaling. Part A, Chemistry and Biochemistry of Low Molecular Weight and Protein Thiols (Methods in Enzymology)*, Vol. 473, Academic Press, San Diego, CA, 2010.
7. Lindley, H. and Phillips, H., The action of sulphites on the cystine disulphide linkages of wool: 5. The effect of some chemical modifications of the wool on the magnitude of the bisulphite-reactive fraction (A + B) and the relative magnitudes of subfractions A and B, *Biochem. J.* 41, 34–38, 1947.
8. Farnworth, A.J., The reactivity of the cystine linkages in wool towards reducing agents, *Biochem. J.* 60, 626–635, 1955.
9. Earland, C. and Wiseman, A., The relationship between the chemical heterogeneity of wool keratin and the mode of incorporation of the cystine residues, *Biochim. Biophys. Acta* 36, 273–275, 1959.
10. Lindley, H. and Ellerman, T.C., The preparation and properties of a group of proteins from the high-sulphur fraction of wool, *Biochem. J.* 128, 859–867, 1972.
11. Lindley, H. and Cranston, R.W., The reactivity of the disulphide bonds of wool, *Biochem. J.* 138, 515–523, 1974.
12. Swart, L.S. and Parris, D., Periodicity in high sulphur proteins from wool, *Nature* 249, 580–581, 1974.
13. Williams, A.J., Metabolism of cysteine by Merino sheep genetically different in wool production IV. Rates of entry of cystine into plasma, measured with a single intravenous injection of L-^{35}S-cystine, and the subsequent incorporation of ^{35}S into wool fibres, *Aust. J. Biol. Sci.* 29, 513–524, 1976.
14. Friedman, M., Finley, J.W., and Yeh, L.S., Reactions of proteins with dehydroalanines, *Adv. Exp. Med. Biol.* 86B, 213–224, 1977.
15. Ahmadi, B. and Speakman, P.T., Suberimidate crosslinking shows that a rod-shaped, low cystine, high helix protein prepared by limited proteolysis of reduced wool has four protein chains, *FEBS Lett.* 94, 365–367, 1978.
16. Douthwaite, E.J. and Lewis, D.M., The formation of cysteine-S-sulphonate groups in wool and the effect on shrink resistance, *J. Soc. Dyers Color.* 110, 304–307, 1994.
17. Hill, R.R. and Ghadimi, M., Alkali-promoted yellowing of wool. Yellow degradation products from a model for protein-bound cystine, *J. Soc. Dyers Color.* 112, 148–152, 1996.
18. Diz, M., Infante, M.R., Ezra, P., and Manresa, A., Antimicrobial activity of wood treated with a new thiol cationic surfactant, *Textile Res. J.* 71, 795–800, 2001.
19. Barba, C., Méndez, S., Roddick-Lanzilotta, A. et al., Wood peptide derivatives for hand care, *J. Cosmet. Sci.* 58, 99–107, 2007.
20. Hilterhaus-Bong, S. and Zahn, H., Contributions to the chemistry of human hair: III. Protein chemical aspects of permanent waving treatments, *Int. J. Cosmet. Sci.* 11, 221–231, 1989.
21. Kijima, K., Analysis of permanent wave products by HPLC, *Fragrances* 21, 49–52, 1993.
22. Inoue, T., Takehara, K., Simizu, N. et al., Application of XANES profiles to x-ray spectromicroscopy for biomedical specimens: Part II. Mapping oxidation state of cysteine in human hair, *J. Xray Sci. Technol.* 19, 313–320, 2011.
23. Makinson, K.R., *Shrinkproofing of Wool*, Marcel Dekker, New York, 1979.
24. Denning, R.J., Freeland, G.N., Guise, G.B., and Hudson, A.H., Reaction of wool with permonosulfate and related oxidants, *Textile Res. J.* 64, 413–422, 1994.
25. Diz, M., Infante, M.R., and Erra, P., Antimicrobial activity of wool treated with a new thiol cationic surfactant, *Textile Res. J.* 71, 695–700, 2001.

26. Gao, Y. and Cranston, R., An effective antimicrobial treatment for wool using polyhexamethylene biguanide as the biocide, Part 1: Biocide uptake and antimicrobial activity, *J. Appl. Polym. Sci.* 117, 3075–3082, 2010.

27. Nagy, P., Lemma, K., and Ashby, M.T., Reactive sulfur species: Kinetics and mechanisms of the reaction of cysteine thiosulfinate ester with cysteine to give cysteine sulfinic acid, *J. Org. Chem.* 72, 8838–8846, 2007.

28. Giles, G.I., Tasker, K.M., and Jacob, C., Oxidation of biological fluids by highly reactive disulfide-*S*-oxides, *Gen. Physiol. Biophys.* 21, 65–72, 2002.

29. Thornton, J.M., Disulphide bridges in globular proteins, *J. Mol. Biol.* 151, 261–287, 1981.

30. Fiser, A. and Simon, I., Predicting the redox states of cysteine in proteins, *Methods Enzymol.* 353, 10–44, 2002.

31. Bessette, P.H., Aslund, F., Beckwith, J., and Georgiou, G., Efficient folding of proteins with multiple disulfide bonds in the *Escherichia coli* cytoplasm, *Proc. Natl. Acad. Sci. USA* 96, 13703–13708, 1999.

32. Woycechowsky, K.J. and Raines, R.T., Native disulfide bond function in proteins, *Curr. Opin. Chem. Biol.* 4, 533–539, 2000.

33. Baneyx, F. and Mujacic, M., Recombinant protein folding and misfolding in *Escherichia coli*, *Nat. Biotechnol.* 22, 1399–1408, 2004.

34. Kadokura, H. and Beckwith, J., Mechanism of oxidative protein folding in the bacterial cell envelope, *Antioxid. Redox Signal.* 13, 1231–1246, 2010.

35. Staley, J.P. and Kim, P.S., Complete folding of bovine pancreatic trypsin inhibitor with only a single disulfide bond, *Proc. Natl. Acad. Sci. USA* 89, 1519–1523, 1992.

35a. Selveraj, S.R., Scheller, A.N., Miao, H.Z. et al., Bioengineering of coagulation factor VIII for efficient expression through elimination of a dispensable disulfide loop, *J. Thromb. Haemost.* 10, 107–115, 2012.

36. McDermott, L., Moore, J., Brass, A. et al., Mutagenic and chemical modification of the ABA-1 allergen of the nematode *Ascaris*: Consequences for structure and lipid binding properties, *Biochemistry* 40, 9918–9926, 2001.

37. Mallon, R.G. and Rossman, T.G., Effects of bisulfite (sulfur dioxide) on DNA synthesis and fidelity by DNA polymerase I, *Chem. Biol. Interact.* 4, 101–108, 1983.

38. Deitcher, R.W., Xiao, Y., O'Connell, J.P., and Fernandez, E.J., Protein instability during HIC: Evidence of unfolding reversibility, and apparent adsorption strength of disulfide bond-reduced α-lactalbumin variants, *Biotechnol. Bioeng.* 102, 1416–1427, 2009.

39. Pace, C.N., Grimsley, G.R., Thomas, J.A., and Barnett, B.J., Conformational stability and activity of ribonuclease T_1 with zero, one, and two intact disulfide bonds, *J. Biol. Chem.* 263, 11820–11825, 1995.

40. Arolas, J.L., Castillo, V., Bronsoms, S. et al., Designing out disulfide bonds of leech carboxypeptidase inhibitor: Implications for its folding, stability and function, *J. Mol. Biol.* 392, 529–546, 2009.

41. Reznik, G.O., Yu, Y., Tarr, G.E., and Cantor, C.R., Native disulfide bonds in plasma retinol-binding protein are not essential for all-*trans*-retinol binding activity, *J. Proteome Res.* 2, 243–248, 2003.

42. Ikegaya, K. et al., Kinetic analysis of enhanced thermal stability of an alkaline protease with engineered twin disulfide bridges and calcium-dependent stability, *Biotechnol. Bioeng.* 81, 187–192, 2003.

43. Hagihara, Y., Mine, S., and Uegaki, K., Stabilization of an immunoglobulin fold domain by an engineered disulfide bond at the buried hydrophobic region, *J. Biol. Chem.* 282, 36489–36495, 2007.

44. Kolmar, H., Natural and engineered cystine knot miniproteins for diagnostic and therapeutic applications, *Curr. Pharm. Des.* 17, 4329–4336, 2011.

45. Radtke, K.P., Griffith, J.H., Riceberg, J., and Gale, A.J., Disulfide bond-stabilized factor VIII has prolonged factor VIIIa activity and improved potency in whole clotting assays, *J. Thromb. Haemost.* 5, 102–108, 2007.

45a. Wakabayashi, H., Griffiths, A.E., and Fay, P.J., Increasing hydrophobicity or disulfide bridging at the factor VIII A1 and C2 domain interface enhances procofactor stability, *J. Biol. Chem.* 286, 25748–25755, 2011.

46. Zhu, F., Liu, Z., Miao, J. et al., Enhanced plasma factor VIII activity in mice via cysteine mutation using dual vectors, *Sci. China Life Sci.* 55, 521–526, 2012.

47. Bull, H.B. and Breece, K., Stability of proteins in guanidine HCl solutions, *Biopolymers* 14, 2197–2209, 1975.

48. Maehashi, K. and Udaka, S., Sweetness of lysozymes, *Biosci. Biotechnol. Biochem.* 62, 605–606, 1998.

49. Masuda, T., Ueno, Y., and Kitabatake, N., Sweetness and enzymatic activity of lysozyme, *J. Agric. Food Chem.* 49, 4937–4947, 2001.

50. Masuda, T., Ide, N., and Kitabatake, N., Effect of chemical modification of lysine residues on the sweetness of lysozyme, *Chem. Senses* 30, 253–264, 2005.

51. Masuda, T., Ide, N., and Kitabatake, N., Structure-sweetness relationship in egg white lysozyme: Role of lysine and arginine residues on the elicitation of lysozyme sweetness, *Chem. Senses* 30, 667–681, 2005.

52. Jones, G., Moore, S., and Stein, W.H., Properties of chromatographically purified trypsin inhibitors from lima beans, *Biochemistry* 2, 66–71, 1963.

53. Ferdinand, W., Moore, S., and Stein, W.H., Susceptibility of reduced, alkylated trypsin inhibitors from lima beans to tryptic action, *Biochim. Biophys. Acta* 96, 524–527, 1965.

54. Craik, D.J., Henriques, S.T., Mylne, J.S., and Wang, C.K., Cyclotide isolation and characterization, *Methods Enzymol.* 516, 37–62, 2012.

55. Pi, L., Shenoy, A.K., Liu, J. et al., CCN2/CTGF regulates neovessel formation via targeting structurally conserved cystine knot motifs in multiple angiogenic regulators, *FASEB J.* 26, 3365–3379, 2012.

56. Ayroza, G., Ferreira, I.L., Sayegh, R.S. et al., Juruin: An antifungal peptide from the venom of the Amazonian Pink Toe spider, *Avicularia juruensis*, which contains the inhibitory cystine knot, *Front. Microbiol.* 3, 324, 2012.

57. Eliasen, R., Andresen, T.L., and Conde-Frieboes, K.W., Handling a tricycle: Orthogonal versus random oxidation of the tricyclic inhibitor cystine knotted peptide gumarin, *Peptides* 37, 144–149, 2012.

58. Haniu, M., Hui, J., Young, Y. et al., Glial cell line-derived neurotrophic factor: Selective reduction of the intermolecular disulfide linkage and characterization of its disulfide structure, *Biochemistry* 35, 16799–16805, 1996.

59. He, W., Chan, L.Y., Zeng, G. et al., Isolation and characterization of cytotoxic cyclotides from *Viola philippica*, *Peptides* 32, 1719–1723, 2011.

60. Mirsky, A.E. and Anson, M.L., Sulfhydryl and disulfide groups of proteins. I. Methods of estimation, *J. Gen. Physiol.* 18, 307–323, 1935.

61. Glazer, A.N. and Smith, E.L., Estimation of cystine plus cysteine in proteins by the disulfide interchange reaction, *J. Biol. Chem.* 236, 416–421, 1961.

62. Williamson, M.B., Catalysis of formation of mixed disulfides between cystine and β-globulins, *Biochim. Biophys. Acta* 39, 379–383, 1970.

63. Tommel, D.K., Vliegenthart, J.F., Penders, T.J., and Arens, J.F., A method for the separation of peptides and α-amino acids, *Biochem. J.* 107, 335–340, 1968.

64. Sakuri, H., Yokoyama, A., and Tanaka, H., Sulfur-containing chelating agents. 31 Catalytic effect of copper(II) ion to formation of mixed disulfide, *Chem. Pharm. Bull.* 19, 1416–1423, 1971.

65. Rigo, A., Corazza, A., di Paolo, M.L. et al., Interaction of copper with cysteine: Stability of cuprous ion complexes and catalytic role of cupric ions in anaerobic thiol oxidation, *J. Inorg. Biochem.* 98, 1495–1501, 2004.

66. Gerber, D.A., Inhibition of denaturation of human gamma-globulin by a mixture of D-penicillamine disulfide and copper-possible mechanism of action of D-penicillamine in rheumatoid arthritis, *Biochem. Pharmacol.* 27, 469–472, 1978.

67. Kelly, S.T. and Zydney, A.L., Effects of intermolecular thiol-disulfide interchange reactions on BSA fouling during microfiltration, *Biotechnol. Bioeng.* 44, 972–982, 1994.

68. Pourrat, H., Barthomeuf, C., Pourrat, A. et al., Stabilization of Octostatin®, a somatostatin analog. Preparation of freeze-dried products for parenteral injection, *Biol. Pharm. Bull.* 18, 766–771, 1995.

69. Ghezzi, P., Regulation of protein function by glutathionylation, *Free Radic. Res.* 39, 573–580, 2005.

70. Toledano, M.B., Kumar, C., Le Moan, N. et al., The system biology of thiol redox system in *Escherichia coli* and yeast: Differential functions in oxidative stress, iron metabolism and DNA synthesis, *FEBS Lett.* 581, 3598–3607, 2007.

71. Dalle-Donne, I., Milzani, A., Gagliano, N. et al., Molecular mechanisms and potential clinical significance of S-glutathionylation, *Antioxid. Redox Signal.* 10, 445–473, 2008.

72. Gallogly, M.M., Starke, D.W., and Mieyal, J.J., Mechanistic and kinetics details of catalysis of thiol-disulfide by glutaredoxins and potential mechanisms of regulation, *Antioxid. Redox Signal.* 11, 1059–1081, 2009.

73. Regazzoni, L., Panusa, A., and Yeum, K.J., Hemoglobin glutathionylation can occur through cysteine sulfenic acid intermediate: Electrospray ionization LTQ-Orbitrap hybrid mass spectrometry studies, *J. Chromatogr. B* 877, 3456–3461, 2009.

74. Dalle-Donne, I., Rossi, R., Colombo, G. et al., Protein S-glutathionylation: A regulatory device from bacteria to humans, *Trends Biochem.* 34, 85–96, 2009.

75. Iversen, R., Andersen, P.A., Jensen, K.S. et al., Thiol-disulfide exchange between glutaredoxin and glutathione, *Biochemistry* 49, 810–820, 2010.

76. Eldjarn, L. and Pihl, A., The equilibrium constants and oxidation-reduction potentials of some thiol-disulfide systems, *J. Am. Chem. Soc.* 79, 4584–4593, 1957.

77. Shenolikar, S. and Stevenson, K.J., Purification and partial characterization of a thiol proteinase from the thermophilic fungus *Humicola lanuginosa*, *Biochem. J.* 205, 147–152, 1982.
78. Raftery, M.J., Enrichment by organomercurial agarose and identification of cys-containing peptides from yeast cell lysates, *Anal. Chem.* 80, 3334–3341, 2008.
79. Wu, C., Belenda, C., Leroux, J.-C., and Gauthier, M.A., Interplay of chemical microenvironment and redox environment on thiol-disulfide exchange kinetics, *Chem. Eur. J.* 17, 10064–10079, 2011.
80. Everley, R.A., Mott, T.M., Toney, D.M., and Croley, T.R., Characterization of *Clostridium* species utilizing liquid chromatography/mass spectrometry of intact proteins, *J. Microbiol. Methods* 77, 152–158, 2009.
81. Haselberg, R., Ratnayake, C.K., de Jong, G.J., and Somsen, G.W., Performance of a sheathless porous tip sprayer for capillary electrophoresis-electrospray ionization-mass spectrometry of intact proteins, *J. Chromatogr. A* 1217, 7605–7611, 2010.
82. Sikanen, T., Aura, S., Franssila, S. et al., Microchip capillary electrophoresis-electrospray ionization-mass spectrometry of intact proteins using uncoated Ormocomp microchips, *Anal. Chim. Acta* 711, 69–79, 2012.
83. Tsybin, Y.O., Fornelli, L., Stoermer, C. et al., Structural analysis of intact monoclonal antibodies by electron transfer dissociation mass spectrometry, *Anal. Chem.* 83, 8919–8927, 2001.
84. Jacob, C., Giles, G.I., Giles, N.M., and Sies, H., Sulfur and selenium: The role of oxidation state in protein structure and function, *Angew. Chem. Int. Ed.* 42, 4742–4758, 2003.
85. Toennies, G. and Lavine, T.F., Oxidation of cystine in non-aqueous media. V. Isolation of a disulfoxide of l-cystine, *J. Biol. Chem.* 113, 571–582, 1936.
86. Axelson, G., Hamrin, K., Fahlman, A., and Nordling, C., Electron spectroscopic evidence of the thiosulphonate structure of cystine *S*-dioxide, *Spectrochim. Acta* 23A, 2015–2020, 1967.
87. Savige, W.E., Eager, J., Maclaren, J.A., and Roxburgh, C.M., The *S*-monoxides of cystine, cystamine, and homocystine, *Tetrahedron Lett.* 5(44), 3289–3293, 1964.
88. Setiawan, L.D., Baumann, H., and Gribbin, D., Surface studies of keratin fibers and related model compounds using ESCA. I-Intermediate oxidation products of the model compounds l-cystine and their hydrolytic behavior, *Surf. Interface Anal.* 7, 188–195, 1985.
89. Savige, W.E. and Maclaren, J.A., Oxidation of disulfides with special reference to cysteine, in *The Chemistry of Organic Sulfur Compounds*, Vol. 1, ed. N. Khaarasch and C.Y. Meyers, Pergamon Press, Oxford, U.K., 1966.
90. Olsen, R.K., Kini, G.D., and Hennen, W.J., Conversion of thiosulfinate derivatives of cystine to unsymmetrical cystines and lanthinines by reaction with tris(dialkylamino)phosphines, *J. Org. Chem.* 50, 4332–4326, 1985.
91. Small, L.D., Bailey, J.H., and Cavallito, C.J., Alkyl thiolsulfinates, *J. Am. Chem. Soc.* 69, 1710–1713, 1947.
92. Hunter, R., Caira, M., and Stellenboom, N., Thiolsulfinate allicin from garlic: Inspiration for a new antimicrobial agent, *Ann. N. Y. Acad. Sci.* 1056, 234–241, 2005.
93. Baumann, H., Seitawan, L.D., and Gribbin, D., Surface studies of keratin fibres and related model compounds using ESCA. 2-Intermediate oxidation products of cystyl residues on keratin fibre surfaces and their hydrolytic stability, *Surf. Interface Anal.* 8, 219–225, 1986.
94. Maclaren, J.A., Savige, W.E., and Sweetman, B.J., Disulphide monoxide groups in oxidized proteins, *Aust. J. Chem.* 18, 1655–1665, 1965.
95. Poole, L.B. and Ellis, H.R., Identification of cysteine sulfenic acid in AhpC of alkyl hydroperoxide reductase, *Methods Enzymol.* 348, 122–136, 2002.
96. Yang, M., Yan, K., and Zhou, A., A study on Xe excilamp treatment of wool fibers, *J. Appl. Poly. Sci.* 118, 241–246, 2010.
97. Huang, K.-P. and Huang, F.L., Glutathionylation of proteins by glutathione disulfide *S*-oxide, *Biochem. Pharmacol.* 64, 1049–1056, 2002.
98. Giles, G.I. and Jacob, C., Reactive sulfur species: An emerging concept in oxidative stress, *Biol. Chem.* 383, 375–383, 2002.
99. Brannan, R.G., Reactive sulfur species act as prooxidants in liposomal and skeletal model systems, *J. Agric. Food Chem.* 58, 3767–3771, 2010.
100. Huang, K.-P., Huang, F.L., Shetty, P.K., and Yergey, A.L., Modification of protein by disulfide *S*-monoxide and disulfide *S*-oxide: Distinctive effects on PKC, *Biochemistry* 46, 1961–1971, 2007.
101. Sakuma, S., Fujita, J., Nakanishi, M. et al., Disulfide *S*-monoxides convert xanthine dehydrogenase into oxidase in rat liver cytosol more potently than their respective disulfides, *Biol. Pharm. Bull.* 31, 1013–1015, 2008.

102. Bowles, E., Alves de Sousa, R., Galardon, E. et al., Direct synthesis of a thiolate-*S* and thiolate-*S* Co[III] complex related to the active site of nitrile hydratase: A pathway to the post-translational oxidation of the protein, *Angew. Chem. Int. Ed.* 44, 6162–6165, 2005.

103. Hill, B.G. and Bhatnagar, A., Protein *S*-glutathionylation: Redox-sensitive regulation of protein function, *J. Mol. Cell. Cardiol.* 52, 559–567, 2012.

104. Gorin, G. and Godwin, W.E., The reaction of iodate with cystine and with insulin, *Biochem. Biophys. Res. Commun.* 25, 227–232, 1966.

105. Mamani, F.P. and Pomeranz, Y., Isolation and characterization of wheat flour proteins. IV. Effects on wheat flour proteins of dough mixing and of oxidizing agents, *J. Sci. Food Agric.* 17, 339–343, 1966.

106. Nowduri, A., Adari, K.K., Gollapalli, N.R., and Parvataneni, V., Kinetics and mechanism of oxidation of L-cystine by hexacyanoferrate (III) in alkaline medium, *E-J. Chem.* 6, 93–98, 2009.

107. Moore, S., On the determination of cystine as cysteic acid, *J. Biol. Chem.* 239, 235–237, 1963.

108. Kuromizo, K., Abe, O., and Maeda, H., Location of the disulfide bonds in the antitumor neocarzinostatin, *Arch. Biochem. Biophys.* 286, 569–573, 1991.

109. Cobb, K.A. and Novolny, M.V., Peptide mapping of complex proteins at the low-picomole level with capillary electrophoresis separations, *Anal. Chem.* 64, 879–886, 1992.

110. Gehrke, C.W., Wall, L., Absheer, J.S. et al., Sample preparation for chromatography of amino acids: Acid hydrolysis of proteins, *J. Assoc. Anal. Chem.* 68, 811–821, 1991.

111. Varga-Visi, E., Terlaky-Balla, E., Pohn, G. et al., RPHPLC determination of L- and C-cystine and cysteine as cysteic acid, *Chromatographia* 51(Suppl 1), S325–S327, 2000.

112. Pesavento, J.J., Garcia, B.A., Streeky, J.A. et al., Mild performic acid oxidation enhances chromatographic and top down mass spectrometric analysis of histones, *Mol. Cell. Proteomics* 6, 1510–1526, 2007.

113. Williams, B.J., Russell, W.K., and Russell, D.H., High-throughput method for on-target performic acid oxidation of MADLI-deposited samples, *J. Mass Spectrom.* 45, 157–166, 2010.

114. Chang, Y.C., Huang, C.N., Lin, C.H. et al., Mapping protein cysteine sulfonic acid modifications with specific enrichment and mass spectrometry: An integrated approach to explore the cysteine oxidation, *Proteomics* 10, 2961–2971, 2010.

115. Cao, R., Liu,Y., Chen, P. et al., Improvement of hydrophobic integral membrane protein identification by mild performic acid oxidation-assisted digestion, *Anal. Biochem.* 407, 196–204, 2010.

116. Challenger, F. and Rawllings, A.A., The formation of organo-metalloidal and similar compounds by micro-organisms. Part V. Methylated alkyl sulphides. The fission of the disulphide link, *J. Chem. Soc.* 1937, 868–875, 1937.

117. Donovan, J.W., A spectrophotometric and spectrofluorometric study of intramolecular interactions of phenolic groups in ovomucoid, *Biochemistry* 6, 3918–3927, 1967.

118. Donovan, J.W., Spectrophotometric observation of the alkaline hydrolysis of protein disulfide bonds, *Biochem. Biophys. Res. Commun.* 29, 734–740, 1967.

119. Cavallini, D., Federici, G., Barboni, E., and Marcucci, M., Formation of persulfide groups in alkaline treated insulin, *FEBS Lett.* 10, 125–128, 1970.

120. Helmerhorst, E. and Stokes, G.B., Generation of an acid-stable and protein-bound persulfide-like residue in alkali- or sulfhydryl-treated insulin by a mechanism consistent with the β-elimination hypothesis of disulfide bond lysis, *Biochemistry* 22, 69–75, 1983.

121. Florence, T.M., Degradation of protein disulphide bonds in dilute alkali, *Biochem. J.* 189, 507–520, 1980.

122. Abdolrasulnia, R. and Wood, J.L., Persulfide properties of thiocystine and related trisulfides, *Bioorg. Chem.* 9, 253–260, 1980.

123. Lu, B.-Y. and Chang, J.-Y., Rapid and irreversible reduction of protein disulfide bonds, *Anal. Biochem.* 405, 67–72, 2010.

124. Angeli, V., Chen, H., Mester, Z. et al., Derivatization of GSSG by pHMB in alkaline media. Determination of oxidized glutathione in blood, *Talanta* 82, 815–820, 2010.

125. Stricks, W. and Kolthoff, I.M., Amperometric titration of oxidized glutathione in presence of cystine, *Anal. Chem.* 25, 1050–1057, 1953.

126. Clarke, H.T., The action of sulfite upon cystine, *J. Biol. Chem.* 97, 235–248, 1932.

127. Lugg, J.W.H., CCLIII. The application of phospho-18-tungstic acid (Folin's reagent) to the colorimetric determination of cysteine, cystine, and related substances, *Biochem. J.* 26, 2144–2159, 1932.

128. Stricks, W. and Kolthoff, I.M., Equilibrium constants of the reactions of sulfite with cystine and with dithioglycolic acid, *J. Am. Chem. Soc.* 72, 4569–4574, 1951.

129. Carter, J.R., Amperometric titration of disulfide and sulfhydryl in proteins in 8 M urea, *J. Biol. Chem.* 234, 1705–1709, 1959.

130. Cecil, R. and McPhee, J.R., A kinetic study of the reaction on some disulphides with sodium sulphite, *Biochem. J.* 60, 496–506, 1955.

131. Cole, R.D., Sulfitolysis, *Methods Enzymol.* 11, 206–208, 1967.

132. Tikhonov, R.V., Pechenov, S.E., and Belacheu, I.A., Recombinant human insulin, *Protein Exp. Purif.* 21, 176–182, 2001.

133. Würfel, M., Häberlein, I., and Follman, H., Facile sulfitolysis of the disulfide bonds in oxidized thioredoxin and glutaredoxin, *Eur. J. Biochem.* 211, 609–614, 1993.

134. Häberlein, I., Structure requirements for disulfide bridge sulfitolysis of oxidized *Escherichia coli* thioredoxin studied by fluorescence spectroscopy, *Eur. J. Biochem.* 223, 473–479, 1994.

135. Kristjánsson, M.M., Magnússon, O.Th., Gudmundsson, H.M. et al., Properties of a subtilisin-like proteinase from a psychrotrophic *Vibrio* species. Comparison with proteinase K and aqualysin I, *Eur. J. Biochem.* 260, 752–760, 1999.

136. Tadokoro, T., You, D.J., Abe, Y. et al., Structural, thermodynamic, and mutational analyses of a psychrotrophic RNase H1, *Biochemistry* 46, 7460–7468, 2007.

137. Gráczer, E., Varga, A., Hajdú, I. et al., Rates of unfolding, rather than refolding, determine thermal stabilities of thermophilic, mesophilic, and psychrotrophic 3-isopropylmalate dehydrogenases, *Biochemistry* 46, 11536–11549, 2007.

138. Kasana, R.C., Proteases from psychrotrophs: An overview, *Crit. Rev. Microbiol.* 36, 134–145, 2010.

139. Thannhauser, T.W., Konishi, Y., and Scheraga, H.A., Sensitive quantitative analysis of disulfide bonds in polypeptides and proteins, *Anal. Biochem.* 138, 181–188, 1984.

140. Damodaran, S., Estimation of disulfide bonds using 3-nitro-5-thiosulfobenzoic acid: Limitations, *Anal. Biochem.* 145, 200–205, 1985.

141. Kella, N.K.D. and Kinsella, J.E., A method for the controlled cleavage of disulfide bonds in proteins in the absence of denaturants, *J. Biochem. Biophys. Methods* 11, 251–263, 1985.

142. Kella, N.K.D., Kang, Y.J., and Kinsella, J.E., Effect of oxidative sulfitolysis of disulfide bonds of bovine serum albumin on its structural properties: A physicochemical study, *J. Protein Chem.* 7, 535–548, 1988.

143. Kella, N.K.D., Yang, S.Y., and Kinsella, J.E., Effect of disulfide bond cleavage on structural and interfacial properties of whey proteins, *J. Agric. Food Chem.* 37, 1203–1210, 1989.

144. Petruccelli, S. and Añón, M.C., Partial reduction of soy protein isolate disulfide bonds, *J. Agric. Food Chem.* 43, 2001–2006, 1995.

145. Wang, H., Faris, R.J., Wang, T. et al., Increased in vitro and in vivo digestibility of soy proteins by chemical modification of disulfide bonds, *J. Am. Oil Chem. Soc.* 86, 1093–1099, 2009.

146. Morel, M.H., Bonicel, J., Micard, V., and Guilbert, S., Protein insolubilization and thiol oxidation in sulfite-treated wheat gluten films during aging at various temperatures and relative humidities, *J. Agric. Food Chem.* 48, 186–192, 2000.

147. Taylor, D.P., Carpenter, C.E., and Walsh, M.K., Influence of sulfonation on the properties of expanded extrudates containing 32% whey protein, *J. Food Sci.* 71, E17–E21, 2006.

148. Kim, M., Yoo, O.J., and Choe, S., Molecular assembly of the extracellular domain of P2X$_2$, an ATP-gated ion channel, *Biochem. Biophys. Res. Commun.* 240, 618–622, 1997.

149. Patrick, J.S. and Lagu, A.L., Determination of recombinant human proinsulin fusion protein produced in *Escherichia coli* using oxidative sulfitolysis and two-dimensional HPLC, *Anal. Chem.* 64, 507–511, 1992.

150. Nilsson, J. et al., Integrated production of human insulin and its C-peptide. *J. Biotechnol.* 48, 241–259, 1996.

151. Mukhopadhyay, A., Reversible protection of disulfide bonds followed by oxidative folding render recombinant hCGβ highly immunogenic, *Vaccine* 18, 1802–1810, 2000.

152. Tikhonov, R.V., Pechenov, S.E., Belacheu, L.A.I. et al., Recombinant human insulin. VII. Isolation of fusion protein-*S*-sulfonate, biotechnological precursor of human insulin from the biomass of transformed *Escherichia coli* cells, *Protein Exp. Purif.* 21, 176–182, 2001.

153. Tikhonov, R.V., Pechenov, S.E., Belacheu, L.A.I. et al., Recombinant human insulin. IX. Investigation of factors, influencing the folding of fusion protein-*S*-sulfonates, biotechnological precursors of human insulin, *Protein Exp. Purif.* 26, 187–193, 2002.

154. Esipov, R.S., Stepanenko, V.N., Chupova, L.A., and Miroshnikov, A.I., Production of recombinant oxytocin through sulfitolysis of intein containing fusion protein, *Protein Pept. Lett.* 19, 479–484, 2012.

155. Sato, T. and Aimoto, S., Use of thiosulfonate for the protection of thiol groups in peptide ligation by the thioester method, *Tetrahedron Lett.* 44, 8085–8087, 2003.
156. Harris, L.J., On the existence of an unidentified sulphur grouping in the protein molecule. Part I. On the denaturation of proteins, *Proc. R. Soc. Lond. B* 94, 426–441, 1923.
157. Harris, L.J., On the existence of an unidentified sulphur grouping in the protein molecule. Part II. On the estimation of cystine in certain protein, *Proc. R. Soc. Lond. B* 94, 441–450, 1923.
158. Miller, H. and Claiborne, A., Peroxide modification of monoalkylated glutathione reductase. Stabilization of an active-site cysteine-sulfenic acid, *J. Biol. Chem.* 266, 19342–19350, 1991.
159. Nagagara, N. and Katayama, A., Post-translational regulation of mercaptopyruvate sulfurtransferase via a low redox potential cysteine-sulfenate in the maintenance of redox homeostasis, *J. Biol. Chem.* 280, 34569–34576, 2005.
160. Olcott, H.S. and Fraenkel-Conrat, H., Specific group reagents for proteins, *Chem. Rev.* 41, 151–197, 1947.
161. Walker, E., CLIV. A colour reaction for disulphides, *Biochem. J.* 19, 1082–1084, 1925.
162. Mauthner, C., Z. *Cystin, Physiol. Chem.* 78, 28–36, 1912.
163. Lotter, H. and Timpl, R., Disulfide-linked cyanogen bromide peptides of bovine fibrinogen. I. Isolation of peptide F-CB3 and characterization of its single disulfide bond by cleavage with cyanide, *Biochim. Biophys. Acta* 427, 558–568, 1976.
164. Catsimpooles, N. and Wood, J.L., Specific cleavage of cystine peptides by cyanide, *J. Biol. Chem.* 241, 1790–1796, 1966.
165. Catsimpoolas, N. and Wood, J.L., The reaction of cyanide with bovine serum albumin, *J. Biol. Chem.* 239, 4132–4136, 1964.
166. Evans, R.M., Fraser, J.B., and Owen, L.N., Dithiols. III. Derivatives of polyhydric alcohols, *J. Chem. Soc.* 1949, 248–255, 1949.
167. Cleland, W.W., Dithiothreitol, a new protective reagent for SH groups, *Biochemistry* 3, 480–483, 1964.
168. Iyer, K.S. and Klee, W.A., Direct spectrophotometric measurement of the rate of reduction of disulfide bonds. The reactivity of the disulfide bonds of bovine α-lactalbumin, *J. Biol. Chem.* 248, 707–710, 1973.
167. Lukesh, J.C., III, Palte, M.J., and Raines, R.T., A potent, versatile disulfide-reducing agent from aspartic acid, *J. Am. Chem. Soc.* 134, 4057–4059, 2012.
170. Overman, L.E., Matzinger, D., O'Connor, E.M., and Overman, J.D., Nucleophilic cleavage of the sulfur-sulfur bond by phosphorus nucleophiles. Kinetic study of the reduction of aryl disulfides with triphenylphosphine and water, *J. Am. Chem. Soc.* 96, 6081–6089, 1975.
171. Rüegg, U.T. and Rudinger, J., Reductive cleavage of cystine disulfides with tributylphosphine, *Methods Enzymol.* 47, 111–116, 1977.
172. Rüegg, U.T., Reductive cleavage of *S*-sulfo groups with tributylphosphine, *Methods Enzymol.* 47, 123–126, 1977.
173. Kirley, T.L., Reduction and fluorescent labeling of cyst(e)ine-containing proteins for subsequent structural analysis, *Anal. Biochem.* 180, 231–236, 1989.
174. Chin, C.C.Q. and Wold, F., The use of tributylphosphine and 4-(aminosulfonyl)-7-fluoro-2,1,3-benzoxadiazole in the study of protein sulfhydryls and disulfides, *Anal. Biochem.* 214, 128–134, 1993.
175. Smejkal, G.B., Li, C., Robinson, M.H. et al., Simultaneous reduction and alkylation of protein disulfides in a centrifugal ultrafiltration device prior to two-dimensional gel electrophoresis, *J. Proteome Res.* 5, 963–967, 2006.
176. Burns, J.A., Butler, J.C., Moran, J., and Whitesides, G.M., Selective reduction of disulfides by tris-(2-carboethoxyethyl)-phosphine, *J. Org. Chem.* 56, 2648–2650, 1991.
177. Valcu, C.-M. and Schlink, K., Reduction of proteins during sample preparation and two-dimensional gel electrophoresis of woody plant samples, *Proteomics* 6, 1599–1605, 2006.
178. Gilfix, B.M., Evans, D.W., and Rosenblatt, D.S., Novel reductants for determination of total plasma homocysteine, *Clin. Chem.* 43, 687–688, 1997.
179. Pfeiffer, C.M., Huff, D.L., and Gunter, E.W., Rapid and accurate HPLC assay for plasma total homocysteine and cysteine in a clinical laboratory setting, *Clin. Chem.* 45, 290–292, 1999.
180. Krijt, J., Vackova, M., and Kozich, V., Measurement of homocysteine and other aminothiols in plasma: Advantages of using tris-(2-carboxyethyl)phosphine as reductant compared with tri-*n*-butylphosphine, *Clin. Chem.* 47, 1821–1828, 2003.
181. Rogers, L.K., Leinweber, B.L., and Smith, C.V., Detection of reversible protein thiol modifications in proteins, *Anal. Biochem.* 358, 171–184, 2006.
182. Humphreys, D.P., Heywood, S.P., Henry, A. et al., Alternative antibody Fab' fragment PEGylation strategies: Combination of strong reducing agents, disruption of the interchain disulphide bond and disulphide engineering, *Protein Eng. Des. Sel.* 20, 227–234, 2007.

183. Gray, W.R., Disulfide structures of highly bridged peptides: A new strategy for analysis, *Protein Sci.* 2, 1732–1748, 1993.
184. Gray, W.R., Echistatin disulfide bridges: Selective reduction and linkage assignment, *Protein Sci.* 2, 1749–1755, 1993.
185. Vuong, G.L. et al., Improved sensitivity proteomics by postharvest alkylation and radioactive labeling of proteins, *Electrophoresis* 21, 2594–2605, 2000.
186. Shaw, J., Rowlinson, R., Nickson, J. et al., Evaluation of saturation labelling two-dimensional difference gel electrophoresis fluorescent dyes, *Proteomics* 3, 1181–1195, 2003.
187. Shaw, M.M. and Riederer, B.M., Sample preparation for two-dimensional electrophoresis, *Proteomics* 3, 1408–1417, 2003.
188. White, C.E., Hunter, M.J., Meininger, D.P. et al., The fifth epidermal growth factor-like domain of thrombomodulin does not have an epidermal growth factor-like disulfide bonding pattern, *Proc. Natl. Acad. Sci. USA* 93, 10177–10182, 1996.
189. Wu, J. and Watson, J.T., A novel methodology for assignment of disulfide bond pairings in proteins, *Protein Sci.* 6, 391–398, 1997.
190. Qi, J., Wu, J., Somkuti, G.A., and Watson, J.T., Determination of the disulfide structure of sillucin, a highly knotted cysteine-rich peptide by cyanylation/cleavage mass trapping, *Biochemistry* 40, 4531–4538, 2001.
191. Stark, G.R., Cleavage at cysteine after cyanylation, *Methods Enzymol.* 47, 129–132, 1977.
192. Daniel, R., Caminade, E., Martel, A. et al., Mass spectrometric determination of the cleavage sites in *Escherichia coli* dihydroorotase induced by a cysteine-specific agent, *J. Biol. Chem.* 272, 26934–26939, 1997.
193. Tetenbaum, J. and Miller, L.M., A new spectroscopic approach to examining the role of disulfide bonds in the structure and unfolding of soybean trypsin inhibitor, *Biochemistry* 40, 12215–12219, 2001.
194. Maciel, J., Martins, M.C.L., and Barbosa, M.A., The stability of self-assembled monolayers with time and under biological conditions, *J. Biomed. Mater. Res. A* 94A, 833–843, 2010.
195. Zhang, H.M., McLoughlin, S.M., Frausto, S.D. et al., Simultaneous reduction and digestion of proteins with disulfide bonds for hydrogen/deuterium exchange monitored by mass spectrometry, *Anal. Chem.* 82, 1450–1454, 2010.
196. Liu, P., O'Mara, B.W., Warrack, R.M. et al., A tris(2-carboxyethyl) phosphine (TCEP) related cleavage on cysteine-containing proteins, *J. Am. Soc. Mass Spectrom.* 21, 837–844, 2010.
197. Singh, R. and Maloney, E.K. Labeling of antibodies by in situ modification of thiol groups generated from selenol-catalyzed reduction of native disulfide bonds, *Anal. Biochem.* 304, 147–156, 2002.
198. Carl, P., Kwok, C.H., Manderson, G. et al., Forced unfolding modulated by disulfide bonds in the Ig domain of a cell adhesion molecule, *Proc. Natl. Acad. Sci. USA* 98, 1565–1570, 2001.
199. Goto, K., Shimada, K., Nagabama, M. et al., Reaction of stable sulfenic and selenic acids containing a bowl-type steric protection group with a phosphine. Elucidation of the mechanism of reduction of sulfenic and selenic acids, *Chem. Lett.* 32, 108–1081, 2003.
200. Dansette, P.M., Thibault, S., Bertho, G., and Mansuy, D., Formation and fate of a sulfenic acid intermediate in the metabolic activation of the antithrombotic prodrug Prasugrel, *Chem. Res. Toxicol.* 12, 1268–1274, 2010.
201. Neumann, H. and Smith, R.L., Cleavage of the disulfide bonds of cystine and oxidized glutathione by phosphorothioate, *Arch. Biochem. Biophys.* 122, 354–361, 1967.
202. Borman, C.D., Wright, C., Twitchett, M.B. et al., Pulse radiolysis studies on galactose oxidase, *Inorg. Chem.* 41, 2158–2163, 2002.
203. Singh, R., Lamoureux, G.V., Lees, W.J., and Whitesides, G.M., Reagents for rapid reduction of disulfide bonds, *Methods Enzymol.* 251, 167–178, 1995.
204. Hansen, R.L., Østergaard, H., Nørgaard, P., and Winther, J.R., Quantitation of protein thiols and dithiols in the picomolar range using sodium borohydride and 4,4'-dithiodipyridine, *Anal. Biochem.* 363, 77–82, 2007.
205. Light, A., Hardwick, B.C., Hatfield, L.M., and Sondack, D.L., Modification of a single disulfide bond in trypsinogen and the activation of the carboxymethyl derivative, *J. Biol. Chem.* 244, 6289–6296, 1969.
206. Knights, R.J. and Light, A., Disulfide bond-modified trypsinogen. Role of disulfide 179–203 on the specificity characteristics of bovine trypsin toward synthetic substrates, *J. Biol. Chem.* 251, 222–228, 1976.
207. Gorecki, M. and Patchornik, A., Polymer-bound dihydrolipoic acid: A new insoluble reducing agent for disulfides, *Biochim. Biophys. Acta* 303, 36–43, 1973.
208. Scouten, W.H., Torok, F., and Gitmer, W., Purification of lipoamide dehydrogenase by affinity chromatography on propyllipoamide glass columns, *Biochim. Biophys. Acta* 309, 521–524, 1973.

209. Scouten, W.H. and Firestone, G.L., N-Propylhydrolipoamide glass beads. An immobilized reducing agent, *Biochim. Biophys. Acta* 453, 227–283, 1976.
210. Bienvenu, C., Greiner, J., Vierling, P., and Di Giorgio, C., Convenient supported recyclable material based on dihydrolipoyl-residue for the reduction of disulfide derivatives, *Tetrahedron Lett.* 51, 3309–3311, 2010.
211. Al-Dubai, H., Oberhofer, G., Kerleta, V. et al., Cleavage of antibodies using dihydrolipoamide and anchoring of antibody fragments on a to biocompatible coated carriers, *Monatsch. Chem.* 141, 485–490, 2010.
212. Ferraz, N., Leverrier, J., Batista-Viera, F., and Manta, C., Thiopropyl-agarose as a solid phase reducing agent for chemical modification of IgG and F(ab')₂, *Biotechnol. Prog.* 24, 1154–1159, 2008.
213. Mise, T. and Bahl, O. P., Assignment of disulfide bonds in the α-subunit of human chorionic gonadotropin, *J. Biol. Chem.* 255, 8516, 1980.
214. Gorin, G., Fulford, R., and Deonier, R.C., Reaction of lysozyme with dithiothreitol and with other mercaptans, *Experientia* 24, 26–27, 1968.
215. Bewley, T.A., Dixon, J.S., and Li, C.H., Human pituitary growth hormone. XVI. Reduction with dithiothreitol in the absence of urea, *Biochim. Biophys. Acta* 154, 420–421, 1968.
216. Warren, J.R. and Gordon, J.A., Denaturation of globular proteins. II. The interaction of urea with lysozyme, *J. Biol. Chem.* 245, 4097–4104, 1979.
217. Yamanaka, K., Nakajima, H., and Wada, Y., Kinetic study of denaturation and subsequent reduction of disulfide bonds of lysozyme by the rapid ultrasonic absorption measurement, *Biopolymers* 17, 2159–2169, 1978.
218. Hamaguchi, K., Structure of muramidase (lysozyme). V. Effect of N,N-dimethylformamide and the role of disulfide bonds in the stability of muramidase, *J. Biochem.* 55, 333–339, 1964.
219. Radford, S.E., Woolfson, D.N., Martin, S.R. et al., A three-disulphide derivative of hen lysozyme, *Biochem. J.* 273, 211–217, 1991.
220. Oda, M., Kitai, A., Murakami, A. et al., Evaluation of the conformational equilibrium of reduced hen egg white lysozyme by antibodies to the native form, *Arch. Biochem. Biophys.* 494, 145–150, 2010.
221. Golda, S., Shimizu, A., Ikeguchi, M., and Sugai, S., The superreactive disulfide bonds in α-lactalbumin and lysozyme, *J. Protein Chem.* 14, 731–737, 1995.
222. Peng, Y., Chen, X., Sato, T. et al., Purification and top-down mass spectrometric characterization of human salivary α-amylase, *Anal. Chem.* 84, 3329–3346, 2012.
223. Bewley, T.A. and Li, C.H., The reduction of protein disulfide bonds in the absence of denaturants, *Int. J. Protein Res.* 1, 117–124, 1969.
224. White, F.H., Jr., Studies on the relationship of disulfide bonds to the formation and maintenance of secondary structure in chicken egg white lysozyme, *Biochemistry* 21, 967–977, 1982.
225. Bradshaw, R.A., Kanarek, L., and Hill, R.L., The preparation, properties, and reactivation of the mixed disulfide derivative of egg white lysozyme and L-cystine, *J. Biol. Chem.* 242, 3789–3798, 1967.
226. Homandberg, G.A. and Wai, T., Reduction of disulfides in urokinase and insertion of a synthetic peptide, *Biochim. Biophys. Acta* 1038, 209–215, 1990.
227. Rajesh, S.R. and Chang, J.Y., Structural stability of human α-thrombin studied by disulfide reduction and scrambling, *Biochim. Biophys. Acta* 1651, 85–92, 2003.
228. Chang, J.Y., Distinct folding pathways of two homologous disulfide proteins: Bovine pancreatic trypsin inhibitor and tick anticoagulant peptide, *Antioxid. Redox Signal.* 14, 127–135, 2011.
229. Pompach, P., Man, P., Kavan, D. et al., Modified electrophoretic and digestion conditions allow a simplified mass spectrometric evaluation of disulfide bonds, *J. Mass Spectrom.* 44, 1571–1578, 2009.
230. Steiner, D.F., Adventures with insulin in the islets of Langerhans, *J. Biol. Chem.* 286, 17399–17421, 2011.
231. Tang, J.G. and Tsou, C.L., The insulin A and B chains contain structural information for the formation of the native molecule. Studies with protein disulfide isomerase, *Biochem. J.* 268, 429–435, 1990.
232. Ryle, A.P. and Sanger, F., Disulphide interchange reactions, *Biochem. J.* 60, 535–560, 1955.
233. Kurouski, D., Washington, J., Ozbil, M. et al., Disulfide bridges remain intact while native insulin converts into amyloid fibrils, *PLoS One* 7, e36989, 2012.
234. Allison, J.R., Moll, G.P., and van Gunseren, W.F., Investigation of stability and disulfide bond shuffling of lipid transfer proteins by molecular dynamics simulation, *Biochemistry* 49, 6916–6927, 2010.
235. Lundblad, R.L., *Approaches to the Conformational Analysis of Biopharmaceuticals*, CRC Press, Boca Raton, FL, 2010.
236. Zhang, L., Chou, C.P., and Moo-Young, M., Disulfide bond formation and its impact on the biological activity and stability of recombinant therapeutic proteins produced by *Escherichia coli* expression system, *Biotechnol. Adv.* 29, 923–929, 2011.

237. Costantino, H.R., Langer, R., and Klibanov, A.M., Solid-phase aggregation of proteins under pharmaceutically relevant conditions, *J. Pharm. Sci.* 83, 1662–1669, 1994.

238. Fransson, J.R., Oxidation of human insulin-like growth factor I in formulation studies. 3. Factorial experiments of the effects of ferric ions, EDTA, and visible light on methionine oxidation and covalent aggregation in aqueous solution, *J. Pharm. Sci.* 86, 1046–1050, 1997.

239. Costantino, H.R., Schwendeman, S.P., Langer, R., and Klibanov, A.M., Deterioration of lyophilized pharmaceutical proteins, *Biochemistry (Mosc.)* 63, 357–363, 1998.

240. Bartkowski, R., Kitchel, R., Peckham, N., and Margulis, L., Aggregation of recombinant bovine granulocyte colony stimulating factor in solution, *J. Protein Chem.* 21, 137–143, 2002.

241. Mousavi, S.H., Bordbar, A.K. and Haertlé, T., Changes in structure and in interactions of heat-treated bovine β-lactoglobulin, *Protein Pept. Lett.* 15, 818–825, 2008.

242. Thing, M., Zhang, J., Lawrence, J., and Topp, E.M., Thiol-disulfide interchange in the tocinoic acid/glutathione system during freezing and drying, *J. Pharm. Sci.* 99(12), 4849–4856, 2010.

243. Wang, W. and Roberts, C.J., ed., *Aggregation of Therapeutic Proteins*, John Wiley & Sons, Hoboken, NJ, 2010.

244. Kaneko, R. and Kitabatake, N., Heat-induced formation of intermolecular disulfide linkages between thaumatin molecules that do not contain cysteine residues, *J. Agric. Food Chem.* 47, 4950–4955, 1999.

245. Kim, J.-S. and Kim, H.-J., Matrix-assisted laser desorption/ionization time-of-flight mass spectrometry observation of a peptide triplet induced by thermal cleavage of cystine, *Rapid Commun. Mass Spectrom.* 15, 2296–2300, 2001.

246. Taylor, F.R., Prentice, H.L., Garber, E.A. et al., Suppression of sodium dodecyl sulfate-polyacrylamide gel electrophoresis sample preparation artifacts for analysis of IgG4 half-antibody, *Anal. Biochem.* 353, 204–208, 2006.

247. Liu, H., Gaza-Bulseco, G., Chumsae, C., and Newby-Kew, A., Characterization of lower molecular weight artifact bands of recombinant monoclonal IgG1 antibodies on non-reducing SDS-PAGE, *Biotechnol. Lett.* 29, 1611–1622, 2007.

248. Hapuarachchi, S., Fodor, S., Apostol, I., and Huang, G., Use of capillary electrophoresis-sodium dodecyl sulfate to monitor disulfide scrambled forms of an Fc fusion protein during purification process, *Anal. Biochem.* 414, 187–195, 2011.

249. Wu, S.L., Jiang, H., Lu, Q. et al., Mass spectrometric determination of disulfide linkages in recombinant therapeutic proteins using online LC-MS with electron-transfer dissociation, *Anal. Chem.* 91, 112–122, 2009.

250. Zhao, L., Almaraz, R.T., Ziang, F. et al., Gas-phase scrambling of disulfide bonds during matrix-assisted laser desorption/ionization mass spectrometry analysis, *J. Am. Soc. Mass Spectrom.* 20, 1603–1616, 2009.

251. Xia, Y. and Cooks, R.G., Plasma induced oxidative cleavage of disulfide bonds in polypeptides during nanospray ionization, *Anal. Chem.* 82, 2856–2864, 2010.

252. Wang, Y., Lu, Q., Wu, S.L. et al., Characterization and comparison of disulfide linkages and scrambling patterns in therapeutic monoclonal antibodies: Using LC-MS with electron transfer dissociation, *Anal. Chem.* 83, 3133–3140, 2011.

253. D'Alessio, G., Malorni, M.C., and Parente, A., Dissociation of bovine seminal ribonuclease into catalytically active monomers by selective reduction and alkylation of the intersubunit disulfide bridges, *Biochemistry* 14, 1116–1122, 1975.

254. Piccoli, R., Tamburrini, M., Piccialli, G. et al., The dual-mode quaternary structure of seminal RNase, *Proc. Natl. Acad. Sci. USA* 89, 1870–1874, 1992.

255. Picone, D., Di Fiore, A., Ercole, C. et al., The role of the hinge loop in domain swapping: The special case of bovine seminal ribonuclease, *J. Biol. Chem.* 280, 13771–13778, 2005.

256. Osipov, A.V., Rucktooa, P., Kasheverov, I.E. et al., Dimeric α-cobratoxin x-ray structure: Localization of intermolecular disulfides and possible mode of binding to nicotinic acetylcholine receptors, *J. Biol. Chem.* 287, 6725–6734, 2012.

257. Dolder, F., pH-Dependent selective reduction of disulfide bonds in human immunoglobulin M using heterocyclic compounds, *Biochim. Biophys. Acta* 207, 286–289, 1970.

258. Harris, R.B. and Hodgins, L.T., Anomalous interaction of 2-thiopyridone with proteins during thiol-disulfide interchange reactions, *Anal. Biochem.* 109, 247–249, 1980.

259. Hong, R. and Nisonoff, A., Relative labilities of the two types of interchain disulfide bond of rabbit γG-immunoglobulin, *J. Biol. Chem.* 240, 3883–3891, 1965.

260. Liu, H., Gaza-Bulseco, G., Faldu, D. et al., Heterogeneity of monoclonal antibodies, *J. Pharm. Sci.* 97, 2426–2447, 2008.

261. Liu, H. and May, K., Disulfide bond structures of IgG molecules: Structural variations, chemical modifications and possible impacts to stability and biological function, *MAbs* 4, 17–23, 2012.

262. Bunting, J.R. and Cathou, R.E., Energy transfer distance measurements in immunoglobulins. II. Localization of the hapten binding sites and the interchain disulfide bond in rabbit antibody, *J. Mol. Biol.* 77, 223–235, 1973.

263. Sun, M.M.C., Beam, K.S., Cerveny, C.G. et al., Reduction-alkylation strategies for the modification of specific monoclonal antibody disulfides, *Bioconjug. Chem.* 16, 1282–1290, 2005.

264. Aono, H., Wen, D., Zang, L. et al., Efficient on-column conversion of IgG1 trisulfide linkages to native disulfides in tandem with protein A affinity chromatography, *J. Chromatogr. A* 1217, 5225–5232, 2010.

265. Mozen, M.M., Schroeder, D.D., and Cabasso, V.J., Ein neues Immunserumglobulinpräparat zur intravenösen Verabreichung, *Arzneimittal Forsch.* 30, 1484–1486, 1980.

266. Zu, Y., Fee, J.A., and Hirst, J., Breaking and re-forming the disulfide bond at the high potential respiratory-type Rieske [2Fe-2S] center of *Thermos thermophilus*: Characterization of the sulfhydryl states by protein-film voltammetry, *Biochemistry* 41, 14054–14065, 2002.

267. Vanhooren, A., Devreese, B., Vanhee, K. et al., Photoexcitation of tryptophan groups induces reduction of two disulfide bonds in goat α-lactalbumin, *Biochemistry* 41, 11035–11043, 2002.

268. Permyakov, E.A., Permyakov, S.E., Deikus, G.Y. et al., Ultraviolet illumination-induced reduction of α-lactalbumin disulfide bridges, *Proteins* 51, 498–503, 2003.

269. Middendoft, T.R., Aldrich, R.W., and Baylor, P.A., Modification of cyclic nucleotide-gated ion channels by ultraviolet light, *J. Gen. Physiol.* 116, 227–252, 2000.

270. Miller, B.L. et al., Solid-state photodegradation of bovine somatotropin (bovine growth hormone): Evidence for tryptophan-mediated photooxidation of disulfide bonds, *J. Pharma. Sci.* 92, 1698–1709, 2003.

271. Alphey, M.S., Gabrielsen, M., Micossi, E. et al., Tryparedoxins from *Crithidia fasciculata* and *Trypanosoma brucei*: Photoreduction of the redox disulfide using synchrotron radiation and evidence for a conformational switch implicated in function, *J. Biol. Chem.* 278, 25919–25925, 2003.

272. Ravelli, R.B.G. and McSweeney, S.M., The "fingerprint" that x-rays can leave on structures, *Structure* 8, 315–328, 2000.

273. Leach, S.J., Meschers, A., and Swanepoel, O.A., The electrolytic reduction of proteins, *Biochemistry* 4, 23–27, 1965.

274. Lee, D.-Y. and Chang, G.-D., Electrolytic reduction: Modification of proteins occurring in isoelectric focusing electrophoresis and in electrolytic reactions in the presence of high salts, *Anal. Chem.* 81, 3957–3964, 2009.

275. Li, J., Dewald, H.D., and Chen, H., Online coupling of electrochemical reactions with liquid sample desorption electrospray ionization-mass spectrometry, *Anal. Chem.* 81, 9716–9722, 2009.

276. Zhang, Y., Dewald, H.D., and Chen, H., Online mass spectrometric analysis of proteins/peptides following electrolytic cleavage of disulfide bonds, *J. Proteome Res.* 10, 1293–1304, 2011.

277. Zhang, Y., Cui, W., Zhang, H. et al., Electrochemistry-assisted top-down characterization of disulfide-containing proteins, *Anal. Chem.* 84, 3838–3842, 2012.

278. Huang, H., Kwok, K.-C., and Liang, H.-H., Inhibitory activity and conformation changes of soybean trypsin inhibitors induced by ultrasound, *Ultrason. Sonochem.* 15, 724–730, 2008.

279. Foti, M.C. and Amorati, R., Non-phenolic radical-trapping antioxidants, *J. Pharm. Pharmacol.* 61, 1435–1448, 2009.

280. Elliott, S.P., Proteolytic enzymes produced by group A Streptococci with special reference to its effect on the type-specific M antigen, *J. Exp. Med.* 81, 573–592, 1945.

281. Caldwell, M.J. and Seegers, W.H., Inhibition of prothrombin, thrombin, and autoprothrombin C with enzyme inhibitors, *Thromb. Diath. Haemorrh.* 13, 373–386, 1965.

282. Iznaga Escobar, N., Morales, A., and Nuñez, G., A computer program for quantification of SH groups generated after reduction of monoclonal antibodies, *Nucl. Med. Biol.* 23, 635–639, 1996.

283. Grob, D., Proteolytic enzymes: I. The control of their activity, *J. Gen. Physiol.* 29, 219–247, 1946.

284. Giustarini, D., Dalle-Donne, I., Colombo, R. et al., Is ascorbate able to reduce disulfide bridges? A cautionary note, *Nitric Oxide* 19, 252–258, 2008.

285. Thompson, J.W., Forrester, M.T., Moseley, M.A., and Foster, M.W., Solid-phase capture for the detection and relative quantification of *S*-nitrosoproteins by mass spectrometry, *Methods*, 62(2), 130–137, 2012.

286. Yeh, Y.Y. and Liu, L., Cholesterol-lowering effect of garlic extracts and organosulfur compounds: Human and animal studies, *J. Nutr.* 131(3s), 989S–993S, 2001.

287. Powolny, A.A. and Singh, S.V., Multitargeted prevention and therapy of cancer by diallyl trisulfide and related *Allium* vegetable-derived organosulfur compounds, *Cancer Lett.* 269, 305–314, 2008.

288. Seki, T., Hosono, T., Hosono-Fukao, T. et al., Anticancer effects of diallyl trisulfide derived from garlic, *Asia Pac. J. Clin. Nutr.* 17(Suppl 1), 249–252, 2008.

289. Wang, H.C., Yang, J.H., Hsieh, S.C., and Sheen, L.Y., Allyl sulfides inhibit cell growth of skin cancer cells through induction of DNA damage mediated G2/M arrest and apoptosis, *J. Agric. Food Chem.* 58, 7096–7103, 2010.

290. Fletcher, J.C. and Robson, A., The occurrence of Bis-(2-amino-2-carboxyethyl)trisulphide in hydrolysates of wool and other proteins, *Biochem. J.* 87, 553–557, 1963.

291. Yanagi, N., Niwa, M., Sakurai, Y. et al., ESR spectrum attributed to trisulfide neutral radical [RSS(R)SR] of protein observed for α-keratin present in white human hair, *Chem. Lett.* 39, 756–757, 2010.

292. Breton, J., Avanzi, N., Valsasina, B. et al., Detection of traces of a trisulphide derivative in the preparation of a recombinant truncated interleukin-6 mutein, *J. Chromatogr. A* 709, 135–146, 1995.

293. Pristasky, P., Cohen, S.L., Krantz, D. et al., Evidence for trisulfide bonds in a recombinant variant of a human IgG$_2$ monoclonal antibody, *Anal. Chem.* 81, 6148–6155, 2009.

294. Haselberg, R., Brinks, V., Hawe, A. et al., Capillary electrophoresis -mass spectrometry using noncovalently coated capillaries for the analysis of biopharmaceuticals, *Anal. Bioanal. Chem.* 400, 295–300, 2011.

295. Nielsen, R.W., Tachibana, C., Hansen, N.E., and Winther, J.R., Trisulfides in proteins, *Antioxid. Redox Signal.* 15, 67–75, 2011.

296. Gu, S., Dingyi, W., Weinreb, P. et al., Characterization of trisulfide modifications in antibodies, *Anal. Biochem.* 400, 89–98, 2010.

9 Modification of Methionine

Methionine (2-amino-4-thiomethylbutanoic acid; Figure 9.1) is one of the less commonly occurring amino acids in proteins with a prevalence of 2.6%[1] but is suggested to have a high frequency of occurrence at protein interfaces.[2,3] As such, methionine is an attractive target for modification but is generally accomplished with considerable difficulty. The small amount of methionine residues in proteins has been useful in obtaining polypeptide fragments for structural analysis via cyanogen bromide cleavage.[4] The ability of methionine to function as a nucleophile at low pH allows for selective (and reversible) modification with reagents such as iodoacetate under acidic conditions.[5] While other residues such as cysteine and histidine are susceptible to alkylation, these residues are protonated and resist modification under acid conditions. The susceptibility of methionine to oxidation by peroxides occurring in biopharmaceutical manufacturing is an important issue.[6] The in vivo oxidation of methionine has been of greater interest as part of redox system.[7] Table 9.1 lists some reagents used for the modification of methionine.

Current interest in the oxidation of methionine (Figure 9.1) stems in part from issues in association with the manufacture of recombinant proteins[8–21] and part from the increase in the understanding of the role of methionine in biological redox systems.[22–26] The reader is directed to several recent review articles[27–30] for a discussion of the chemical and biological oxidation of methionine. The oxidation of methionine proceeds initially to methionine sulfoxide, which is a reversible process.[31–35] Saunders and Stites[36] have developed an electrophoretic assay for the presence of methionine sulfoxide in proteins. The assay is based on the modification of methionine with methyl methanesulfonate at acid pH (Figure 9.1), which is described by the authors as a more potent alkylating agent than iodoacetate or iodoacetamide. The alkylation of the sulfur of methionine results in a positively charged sulfonium derivative. A peptide containing the sulfonium salt derived from an unmodified methionine with a positive charge would be separated by electrophoresis at acid pH from a peptide containing methionine sulfoxide that would be neutral; methionine sulfoxide would not be modified by alkylation. The addition of a negative charge by, for example, the formation of cysteic acid in a peptide would neutralize the added positive charge and complicate the application of this approach; phosphorylation would also create a problem. It should be noted that Savige and Fontana[31] had reported the use of alkylation of oxidized proteins as an indirect method for determination of methionine sulfoxide. Houghten and coworkers[37] used incorporation of radiolabeled iodoacetic acid (pH 3.5, 8 M urea) to measure modified methionine residues and thus, indirectly, the extent of methionine oxidation. Conversion of methionine sulfoxide to methionine sulfone is essentially irreversible under common solvent conditions and usually requires more vigorous reagents such as performic acid.[38] Homocysteic acid (Figure 9.1) can be formed from methionine under fairly drastic conditions.[39–41] Homocysteic acid may also be formed from the oxidation of homocysteine.[42]

Reaction of proteins with hydrogen peroxide at acid pH results in the preferential modification of methionine residues to the sulfoxide although formation of the sulfone is not unknown.[43] Ichiba and coworkers[44] reported that the oxidation of the C-terminal octapeptide of cholecystokinin with hydrogen peroxide resulted in the oxidation of Met28 and Met31 to methionine sulfoxide, while a Fenton system ($FeSO_4 + H_2O_2$) produced methionine sulfone and oxidation at the single tryptophanyl residue in this peptide. In the example cited by Neumann,[43] the modification of pancreatic ribonuclease A was performed at pH 2.5 at 30°C producing the sulfoxide derivative. Neumann also noted that the formation of the sulfoxide derivative in proteins is usually but not always reversed with DTT or other reducing agents. Modification of methionine can occur at alkaline pH (8.5) in the absence of cysteine.[37,45] Luo and workers[46] have shown that the oxidation of cysteine to cysteine

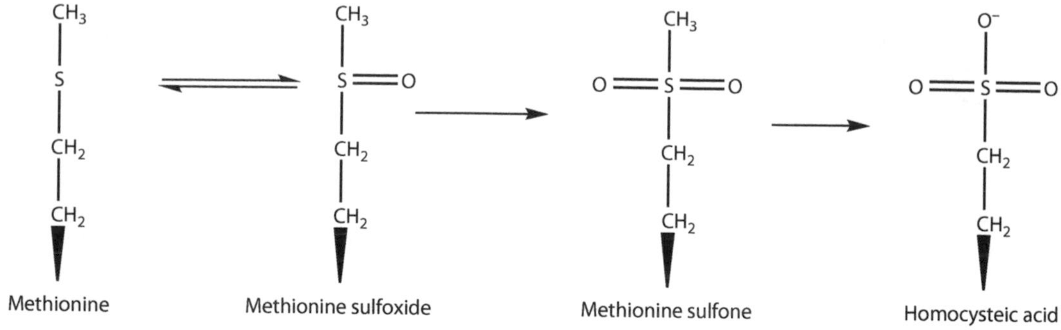

FIGURE 9.1 The structure and oxidative modification of methionine. The reversible oxidation of methionine to methionine sulfoxide is shown as well as the subsequent further irreversible oxidation of methionine sulfoxide to methionine sulfone.

TABLE 9.1
Reagents Used for the Modification of Methionine in Proteins

Reagent	Other Amino Acids Modified	References
Chloramine-T	Cysteine, cystine, histidine, tyrosine, tryptophan	9.1.1–9.1.3
Hydrogen peroxide	Cysteine, cystine, histidine, tryptophan	9.1.4
Iodoacetate	Cysteine, histidine, lysine	9.1.5
N-Chlorosuccinimide	Tryptophan, cysteine	9.1.6
t-Butyl hydroperoxide	Tryptophan, cysteine	9.1.7

References to Table 9.1

9.1.1. Oda, T. and Tokushige, M., Chemical modification of tryptophanase by chloramine T: A possible involvement of the methionine residue in enzyme activity, *J. Biochem.* 104, 178, 1988.

9.1.2. Cutruzzola, F., Ascenzi, P., Barra, D., Bolognesi, M., Menegatti, E., Sarti, P., Schenebli, H.-P., Tomova, S., and Amiconi, G., Selective oxidation of Met-192 in bovine alpha-chymotrypsin. Effect on catalytic and inhibitor binding properties, *Biochim. Biophys. Acta* 1161, 201, 1993.

9.1.3. Hussain, A.A., Awad, R., Crooks, P.A., and Dittert, L.W., Chloramine T in radiolabeling techniques. I. Kinetics and mechanism of the reaction between chloramine-T and amino acids, *Anal. Biochem.* 214, 495, 1993.

9.1.4. Drozdz, R., Naskalski, J.W., and Sznajd, J., Oxidation of amino acids and peptides in reaction with myeloperoxidase, chloride and hydrogen peroxide, *Biochim. Biophys. Acta* 957, 47, 1988.

9.1.5. Kleanthous, C., Campbell, D.G., and Coggins, J.R., Active site labeling of the shikimate pathway enzyme, dehydroquinase. Evidence for a common substrate binding site within dehydroquinase and dehydroquinate synthase, *J. Biol. Chem.* 265, 10929, 1990.

9.1.6. Padrines, M., Rabaud, M., and Bieth, J.G., Oxidized alpha-1-proteinase inhibitor: A fast-acting inhibitor of human pancreatic elastase, *Biochim. Biophys. Acta* 1118, 174, 1992.

9.1.7. Keck, R.G., The use of t-butyl hydroperoxide as a probe for methionine oxidation in proteins, *Anal. Biochem.* 236, 56–62, 1996.

9.1.8. Spiess, P.C., Morin, D., Jewell, W.T., and Buckpitt, A.R., Measurement of protein sulfhydryls in response to cellular oxidative stress using gel electrophoresis and multiplexed fluorescent imaging analysis, *Chem. Res. Toxicol.* 21, 1074–1085, 2008.

9.1.9. Hensel, M., Steurer, R., Fichtl, J. et al., Identification of potential sites for tryptophan oxidation in recombinant antibodies using *tert*-butylhydroperoxide and quantitative LC-MS, *PLoS One* 6, e17708, 2011.

sulfenic acid occurs very slowly at pH 4.0 with rapid reaction above pH 8.0 consistent with oxidation of the thiolate anion to cysteine sulfenic acid followed by formation of cystine as product. Subsequent work by these investigators[47,48] demonstrated the reaction of cysteine with hydrogen peroxide in an amorphous polyvinylpyrrolidone[47] or trehalose[48] lyophile yielded cysteine sulfinic acid and cysteine sulfonic acid in addition to cystine. It is of interest that while the oxidation of cysteine in solution shows pH dependence consistent with the involvement of the thiolate anion, such pH dependence is not observed within either of the lyophiles. Thus, while I know of no demonstration of methionine sulfone in a lyophilized biopharmaceutical product, such a possibility cannot be excluded. Hydrogen peroxide together with other oxidizing agents such as superoxide anion is responsible for the in vivo oxidation of methionine residues in proteins in activated neutrophils.[49] These investigators also reported that methionine sulfoxide bonds are not cleaved by cyanogen bromide (Figure 9.3) providing another assay for methionine oxidation in proteins.[31]

Methionine sulfoxide can be reduced to methionine; the extent to which reduction occurs appears to be quiet variable. A systematic study[33,50] of the reduction of methionine sulfoxide has shown that of four reducing agents tested, mercaptoacetic acid, 2-mercaptoethanol, DTT, and *N*-methylmercaptoacetamide, the last was the most effective. The reactions demonstrated little pH dependence, but did not proceed well at concentrations of acetic acid above 50% (v/v). Complete regeneration of methionine could be accomplished with 0.7–2.8 M reagent at 37°C for 21 h. There are also methionine sulfoxide reductases[51–57] that can catalyze the conversion of methionine sulfoxide to methionine. Recent work has shown that methionine sulfoxide reductase may have a role beyond the methionine–methionine sulfoxide redox system.[58,59]

The stability of methionine sulfoxide on acid hydrolysis prior to amino acid analysis is not clearly understood.[61] However, MS has largely replaced amino acid analysis as a method for determining methionine sulfoxide or methionine sulfone. Regardless, understanding of the chemistry of a derivative(s) that has potential as biomarker is useful. Methionine sulfoxide is converted back to methionine on acid hydrolysis (6 N HCl) in the presence of reducing agents such as DTT.[61] Keutmann and Potts[60] had previously observed that the presence of 2-mercaptoethanol in 6 N HCl hydrolysis of proteins improved the recovery of methionine. Joergensen and Thestrup[62] reported that thorough degassing and the addition of thioglycolic acid improved the recovery of methionine (and tryptophan) after 6 N HCl hydrolysis. Most current work on amino acid analysis for the measurement of methionine sulfoxide uses *p*-toluenesulfonic acid[63] or methanesulfonic acid[64,65] for the hydrolysis of proteins under conditions where methionine sulfoxide is stable. Other approaches for maintaining methionine sulfoxide during hydrolysis of proteins for amino acid analysis used alkaline hydrolysis[66] or microwave[67] or flash[68] hydrolysis. Antibodies against methionine sulfoxide have been used to detect this modification in proteins.[69,70] There has been interest in the formation of methionine sulfoxide in Alzheimer disease and for use as a biomarker.[70–78] Methionine oxidation in prion protein has been suggested to be involved in neurodegenerative disease.[79] Wolschner and coworkers[80] used periodate to oxidize methionine residues in recombinant cellular prion protein PrP(C) as part of an effort to understand oxidative stress in neurodegenerative diseases. The extent of methionine oxidation was determined by MS and was correlated with the tendency to aggregate. It is suggested that methionine coupled with methionine sulfoxide reductase serves as an antioxidant system.[81–83] Recent work has established that methionine sulfoxide reductase A is a stereospecific enzyme that catalyzes both the oxidation and reduction of methionine sulfoxide.[84,85]

While not of direct interest for the in vitro chemical modification of methionine, several of the studies have sufficient importance to the chemistry of methionine oxidation to merit comment. Simon and coworkers[72] used the oxidation of a methionine-containing peptide derived from the ApoE4 protein as a proteotypic peptide in absolute quantitation of this protein in plasma. A proteotypic peptide is a peptide derived from a specific protein that can be determined with absolute certainty and is a characteristic of the specific protein on mass spectrometric analysis.[73,74] In the work by Simon and others,[73] used performic acid[39] to oxidize the putative proteotypic peptide from the ApoE4 protein converting methionine to methionine sulfone (Figure 9.1) which was determined by mass spectrometry. Conversion to the sulfone obviated the intrinsic heterogeneity in methionine sulfoxide content (the content of monoxized peptide in their clinical cohort varied from 0% to 10%) as well as changes in the oxidation state of methionine during the storage and processing of sample.[18,86,87]

Chemical oxidation (footprinting) is a method for mapping surface residues in proteins greatly enabled by the use of MS for identification of modified residues.[88] Some current approaches to oxidative surface mapping use hydroxyl radicals ($\cdot$OH)[89] generated from water by radiolysis or by a Fenton reagent.[90] As an example, Sharp and coworkers[91] used a Fenton reagent to catalyze the formation of hydroxyl radicals from hydrogen peroxide. These investigators showed that Met131 in apomyoglobin was modified with hydroxyl radicals despite a lack of surface accessibility suggesting either a different mechanism (radical transfer from an exposed residues) or such oxidation occurred in the processing of the sample. Subsequent work[92] from this group supported the concept that methionine could be oxidized via hydroxyl radical without solvent exposure. Methionine can also act as mediator in hydroxyl radical action on protein. Mozziconacci and coworkers[93] studied the oxidation of enkephalins with hydroxyl radicals (pulse radiolysis) and reported that methionine served a prooxidant with the initial formation of a hydroxyl sulfuryl radical with subsequent transfer of an electron to a tyrosine with formation of a tyrosine radical and regeneration of methionine. Bern and coworkers[94] have shown the presence of homocysteic acid (Figure 9.1) in highly oxidized samples for proteomic analysis. As mentioned earlier, homocysteic acid is formed from methionine under what can be considered rigorous conditions. Methionine-containing peptides and protein samples for MS have been oxidized by H_2O_2 directly on MALDI-MS target plates.[95] This approach permits the more facile identification of proteins from database searches; the oxidation of methionine by hydrogen peroxide increases the mass by 16 Da per methionine residue. The technique allows spectra to be obtained before and after methionine oxidation. Pan and coworkers[96] used insertional mutagenesis to place methionine residues in bacteriorhodopsin; hydroxyl radicals were then used to oxidize the inserted methionine residues in a study of folding. Methionine sulfoxide residues resulting from hydroxyl radical were identified by MS. Watson and Sharp[97] have applied hydroxyl radical footprinting to the conformational analysis of several therapeutic proteins.

Methionine oxidation is of great interest to the biopharmaceutical industry as the oxidation of therapeutic proteins can result in defective product with reduced stability and conformational change.[98] Examples of where methionine oxidation may result in conformational change include recombinant human interferon α-2b,[99] recombinant human interleukin-1 receptor antagonist,[100] immunoglobulin light chain LEN,[101] and oxidatively stressed monoclonal antibodies.[102] It should be noted that considering that methionine residues are buried, oxidation may be well following conformational change. There are examples where methionine oxidation has occurred without measured conformational change.[103,104] Kim and coworkers[105] have compared the effect of methionine oxidation with H_2O_2 (25 mM sodium phosphate, 100 mM NaCl, pH 7.0, 1.1% H_2O_2, room temperature, 30 min) and oligonucleotide-directed mutagenesis on the stability of staphylococcal nuclease. Oxidation of two methionine residues (M65, M95) resulted in a considerable loss of protein stability, while modification of the other two residues (M26, M32) had little effect. Substitution of a leucine residue at position 95 (M95L) resulted in a loss of stability similar to that seen with oxidation, while substitution at position 65 (M65L) was less deleterious. A similar experimental approach was taken by Chien and coworkers[106] to a N-carbamoyl D-amino acid aminohydrolase from *Agrobacterium radiobacter*. This protein contains nine methionine residues that were individually replaced with leucine. Two of the mutants had activity similar to wild type, while the other seven had reduced activity. The three mutants with solvent-accessible residues retained activity on oxidation with H_2O_2 (0.1–1.0 mM H_2O_2 in 200 mM sodium phosphate, pH 7.0, 25°C, 15 min). The other mutants were more resistant to loss of activity on oxidation than wild type. Caldwell and coworkers[109] used hydrogen peroxide to oxidize the methionine residues in *Escherichia coli* ribosomal protein L12; the extent of methionine oxidation was measured by reaction with radiolabeled iodoacetate at pH 3.0 (Figure 9.2). The loss of activity on the oxidation of L12 by H_2O_2 can be reversed by 2-mercaptoethanol; oxidation of the

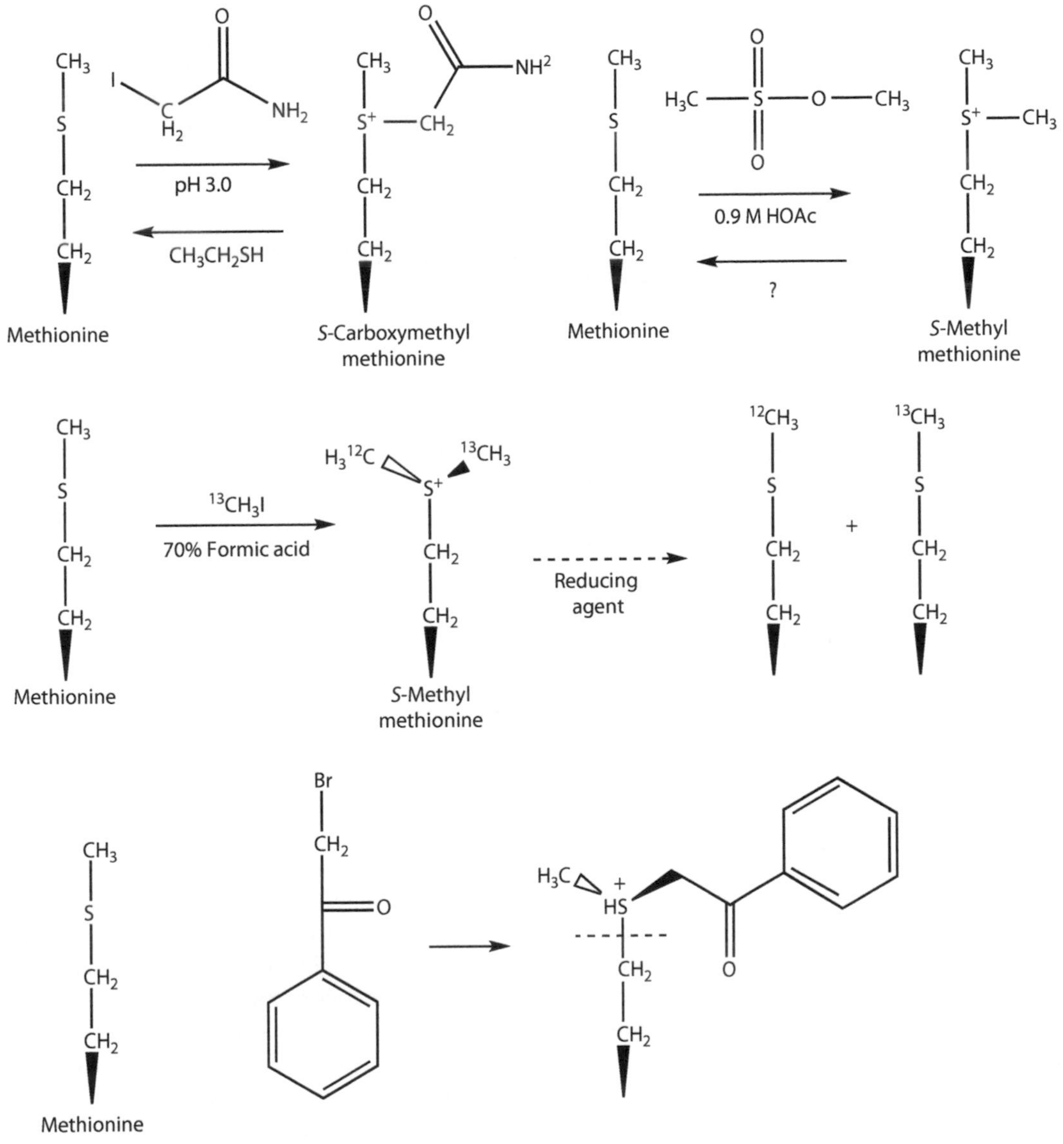

FIGURE 9.2 The alkylation of methionine. Shown is the alkylation of methionine with iodoacetamide and the *S*-methylation of methionine with methyl methanesulfonate. The alkylation by iodoacetamide is reversed by the process of thiolysis (Degen, J. and Kyte, J., The purification of peptides which contain methionine residues, *Anal. Biochem.* 89, 529–539, 1978). Also shown is the formation of *S*-methylmethionine with methyl iodide in 70% formic acid; reducing agents have a variable record of success in converting the sulfonium salt back to the parent methionine residue (Jones, W.C., Jr., Rothgeb, T.M., and Gurd, F.R.N., Specific enrichment with ¹³C of the methionine methyl groups of sperm whale myoglobin, *J. Am. Chem. Soc.* 97, 3875–3877, 1975; Matta, M.S., Henderson, P.A., and Patrick, T.R., Preparation and ¹³C NMR characterization of [[ε-¹³C] methionine-192]-α-chymotrypsin, *J. Biol. Chem.* 256, 4172–4174, 1981). Also shown is the reaction of phenacyl bromide (ω-bromoacetophenone) with methionine and the cleavage to yield neutral product ion for MS/MS analysis (Reid, G.E., Roberts, K.D., and Simpson, R.J., Selective identification and quantitative analysis of methionine containing peptides by charge derivatization and tandem mass spectrometry, *J. Am. Soc. Mass Spectrom.* 16, 1131–1150, 2005).

methionine residues in L12 did convert the dimer to monomer form. Kachurin and coworkers[107] used H_2O_2 to oxidize a methionine residue in an α-galactosidase from *Trichoderma reesei*. The modified enzyme demonstrated an increase in activity with *p*-nitrophenyl-α-D-galactoside; a similar effect was observed with periodate and permanganate. One of the five methionine residues was modified; there was no modification of tryptophan, histidine, tyrosine, and cysteine. Further oxidation of the methionine residues could be accomplished with a longer reaction time or denaturation of the protein. Kornfelt and coworkers[108] used H_2O_2 to oxidize two of the four methionine residues in blood coagulation factor VIIa. Three oxidation products were obtained from the reaction of H_2O_2 with factor VIIa that could be separated by RP-HPLC using a C_4 column. Nomura and coworkers[109] have observed that methionine oxidation decreases the stability of the p53 tetramer; the oxidized protein is more susceptible to tryptic digestion. In the work previously cited, Watson and Sharp[97] used hydroxyl radical footprinting to evaluate the conformation of several therapeutic proteins.

The development of *t*-butyl hydroperoxide (Figure 9.1) by Keck[110] as a selective oxidizing agent for methionine in proteins represented a significant advance. Results obtained with native recombinant interferon and recombinant tissue-type plasminogen activator showed that this reagent was selective for the oxidation of exposed methionine residues in proteins. The reaction were performed in 5 mM succinate-0.1% Tween 20, pH 5.0 recombinant interferon, or in 0.2 M arginine-0.1 M sodium phosphate-0.1% Tween 80, pH 7.0 for recombinant tissue-type plasminogen activator. *t*-Butyl hydroperoxide (0–73 mM) was added and the reaction allowed to proceed overnight at room temperature. Tryptic peptides were separated by HPLC and analyzed by MS. Two methionine residues were oxidized in recombinant interferon with *t*-butyl hydroperoxide, while all five residues were oxidized to a varying extent by H_2O_2 under the same reaction conditions. Three methionine residues were oxidized in native tissue-type plasminogen activator; all five residues were oxidized to a varying degree in the presence of 8.0 M urea. *t*-Butyl hydroperoxide has been successfully used for recombinant human leptin[111] and recombinant human granulocyte colony-stimulating factor.[112] Chumsae and coworkers[113] used *t*-butyl hydroperoxide for the oxidation of methionine in a fully human monoclonal antibody as a probe for conformational change during thermal stressing. The rationale for these studies is based on the susceptibility of methionine to oxidation with exposure of methionine considered an index of conformational change. Subsequent work by this group[114] established that this modification decreased binding affinity to protein A or protein G affinity columns. This latter observation is of significance given the importance of protein A and/or protein G affinity columns in antibody purification. Further work from this group[115] compared the photooxidation and *t*-butylperoxide oxidation of the antibody. Another group[116] confirmed that the oxidation of antibody with *t*-butylperoxide decreased affinity to protein A; it was also observed that affinity for the FcRn receptor was reduced suggesting that protein A binding is a surrogate measure for FcRn binding of antibody. The reduced affinity of oxidized monoclonal antibody to protein A or protein G has provided the basis of an analytical method for determination of methionine oxidation in monoclonal antibodies. Loew and coworkers[117] used gradient elution (Dulbecco's phosphate-buffered saline to 100 mM acetic acid—150 mM NaCl, pH 2.8); oxidized protein was eluted earlier than unmodified material.

Periodate is another relatively mild oxidizing agent that can modify methionine and cysteine, with lesser modification of tryptophan and tyrosine. However, periodate is most often used to oxidize vicinal diols in carbohydrates yielding aldehydes, which can, for example, be coupled to amines in proteins as used for the preparation of a conjugate polysaccharide vaccine.[118] Periodate is rarely, if ever, used for the modification of amino acids in proteins; oxidation of amino acids is usually an unwanted side reaction that should be considered when periodate is used for the modification of proteins or peptides.[119] Antibodies can be coupled to matrices via periodate oxidation of the carbohydrate in the Fc domain.[120] The periodic acid Schiff reagent is used for the staining of glycoproteins on gel electrophoresis.[121] Yamasaki and coworkers[122] established that periodate under mild conditions modified a variable number of methionine residues in various proteins under various condition. With pancreatic ribonuclease, 2 out of 4 methionine residues were oxidized at pH 4.0 (5 mM sodium periodate at 25°C), while all 16 methionine residues in chicken ovalbumin were oxidized under the same conditions; one of the two methionine residues in chymotrypsin was modified by 5 mM periodate at pH 5.0 (5°C). Yamasaki and coworkers[122] demonstrated some specificity in the oxidation of methionine residues in proteins with periodate. Careful considerations of the concentration of peroxide permit the selective formation of the sulfoxide and the sulfone. The authors note that when periodate is used for the oxidation of vicinal diols (carbohydrates) in glycoproteins, possible oxidation of methionine should be considered. Knowles[123] showed that the modification of a single methionine in chymotrypsin resulted in a partially active enzyme with an increased Km for aromatic substrates such as N-acetyl-L-tryptophan ethyl ester, while the Km is unchanged for nonaromatic substrates such as N-acetyl-L-valine ethyl ester. de la Llosa and coworkers[124] used periodate or chloramine-T (see below) to oxidize methionine residues in bovine luteinizing hormone. Periodate oxidized seven methionine residues, while chloramine-T oxidized approximately six methionine residues; periodate oxidized most of the fucose association with the peptide hormone, while a small amount was lost with chloramine-T. Most of the activity is lost with the oxidation of three methionine residues (presumably to the sulfoxide); the modification of methionine residues with cyanogen bromide that forms homoserine with associated peptide bond cleavage resulted in smaller loss of activity. Periodate was used by Gleisner and Liener[125] to oxidize a single methionine residues in ficin after protection of the active-site sulfhydryl group with tetrathionate permitting the modification of the active-site histidine residue with bromoacetone in 2.0 M urea. Chloramine-T has been used for the selective oxidation of methionine in proteins.

Shechter and coworkers[61] demonstrated that chloramine-T could oxidize methionine residues in proteins at neutral or slightly alkaline pH (pH 8.5); cysteine was also oxidized to cysteine under these conditions, but the modification of other amino acid was not observed. The modification of tryptophan with chloramine-T was observed at pH 2.2. Similar results were obtained with N-chlorosuccinimide, while N-bromosuccinimide (NBS) was more promiscuous with tyrosine and histidine was also modified at pH 2.2; tyrosine was also modified with NBS at pH 8.5 with histidine modification occurring only with higher excesses of reagent. Methionine sulfoxide was converted to methionine sulfone by NBS at pH 8.5 only after complete modification of histidine. This study also reported that oxidized methionine residues were not cleaved by cyanogen bromide providing for an assay for oxidized methionine residues in proteins. Chloramine-T has been used for the selective oxidation of methionine in human recombinant secretory leukocyte proteinase inhibitor,[126] large conductance calcium-activated potassium channels,[127] actin,[128] and kininogens.[129] The oxidation of methionine residues occurring when chloramine-T is used for protein iodination can pose a problem for subsequent analysis.[130,131] The oxidation of methionine in recombinant human interleukin-2 by potassium peroxodisulfate has been reported.[132]

Methionine can be modified with various alkylation agents such as iodoacetamide, methyl iodide, and phenacyl bromide (Figure 9.2). The reaction of iodoacetate with methionine has been examined in some detail by Gundlach and coworkers.[133] The reaction of iodoacetate with methionine was not pH dependent and proceeds much slower than the reaction with cysteine under the mildly alkaline conditions used for reduction and carboxymethylation. The resulting sulfonium salt yields homoserine and homoserine lactone when heated at 100°C at pH 6.5. On acid hydrolysis (6 N HCl, 110°C, 22 h), a mixture of methionine and S-carboxymethyl homocysteine together with a small amount of homoserine lactone was obtained. In general, methionine residues only react with the α-halo acids after the disruption of the secondary and tertiary structure of a protein.[5] Selectivity in the modification of methionine in proteins by α-halo acids can be achieved by performing the reaction at acid pH (pH 3.0 or less), but there are examples of modification under less rigorous conditions. The modification of methionine in porcine kidney acyl-CoA dehydrogenase occurs with iodoacetate (0.030 M) in 0.1 M phosphate, pH 6.6 at ambient temperature.[134] The identification of methionine as the residue modified by iodoacetate in this protein was supported by the comparison of the chromatogram of the acid hydrolyzate of the modified protein (reacted with ^{14}C-iodacetate) with that of the acid hydrolyzate of synthetic S-([1-^{14}C]carboxymethyl)-methionine.[133] This is necessary since the S-carboxymethyl derivative yielded several different compounds on acid hydrolysis.[133,135] Cummings and coworkers[136] reported that the single methionine residue in pig kidney medium-chain acyl-CoA dehydrogenase[134] modified by iodoacetate was not directly involved in catalysis, but the modification had a major effect on K_m. The modification of methionine by ethyleneimine has been reported in a reaction producing a sulfonium salt derivative.[137] The modification of methionine in azurin with bromoacetate has been reported.[138] In the protein, four of six methionine residues were modified at pH 4.0, while all methionine residues were reactive at pH 3.2. These modification reactions were performed in 0.1 M sodium formate at ambient temperatures for 24 h with 0.16 M bromoacetate.

Naider and Bohak[139] have reported that the sulfonium salt derivatives of methionine (e.g., S-carboxymethyl methionine, the reaction product of methionine and iodoacetic acid) can be converted to methionine by reaction with a suitable nucleophile. For example, the reaction of S-carboxamidomethyl methionine (in the peptide Gly-Met-Gly) with a sixfold molar excess of mercaptoethanol at pH 8.9 at a temperature of 30°C resulted in the complete regeneration of methionine after 24 h of reaction. The S-phenacyl derivative of methionine (in the peptide Gly-Met-Gly) was converted to methionine in 1 h under the same reaction conditions. Naider and Bohak[139] showed that chymotrypsin was inactivated with phenacyl bromide as a result of alkylation at Met 198; activity was recovered with 2-mercaptoethanol in sodium phosphate, pH 7.5. It is of interest that the S-phenacyl methionine in chymotrypsin is converted to methionine at a substantially faster rate than the tripeptide derivative. The authors speculate that the increased reactivity of the chymotrypsin derivative is a reflection of interaction of the phenacyl moiety with the substrate-binding site. Reid and coworkers[141] have used phenacyl bromide modification of methionine as a method for the analysis of methionine-containing peptides by charge derivatization and MS. Alkylation of methionyl residues in pituitary thyrotropin and lutropin with iodoacetic acid has been reported.[142] Differential reactivity of various methionyl residues was reported on reaction with iodoacetate in 0.2 M formate, pH 3.0 for 18 h at 37°C. The reversible alkylation of methionine by iodoacetate in dehydroquinase has been reported by Kleanthous and coworkers.[143] In this reaction, iodoacetate behaves kinetically as an affinity label with a K_i of 30 μM and a k_{inact} of 0.014 min^{-1}, pH 7.0 (50 mM potassium phosphate). There is no reaction with iodoacetamide. Two methionine residues are modified during the reaction of dehydroquinase with iodoacetate. In a companion study, Kleanthous and Coggins[144] demonstrated that 2-mercaptoethanol treatment under alkaline conditions (0.5% ammonium bicarbonate, 37°C) could reverse modification at one of the two residues. If the modified protein is denatured, there is no reversal of modification at either residue. The results are interpreted in terms of the proximity of a positive charge (i.e., lysine) in close proximity to one of the two methionine residues that (1) provides the basis for the affinity labeling and (2) provides the basis for the 2-mercaptoethanol-mediated reversal of modification.

Methylation of the methionine residues with radiolabeled (^{13}C) methyl iodide yields the sulfonium salt (Figure 9.2), which is diastereoisomer. This reaction can usually be reversed by a reducing agent such as DTT resulting in the insertion of a ^{13}C (methyl) probe to a level of 50%. Jones and coworkers[145] modified sperm whale myoglobin with ^{13}CH$_3$I at pH 4.0 (18 h, 23°C, dark; $t_{1/2}$ = 2.7 h for the methylation reaction) to produce the sulfonium salt. The sulfonium salt (S-methylmethionine) could be placed with DTT (18 h, 37°C) producing an equimolar mixture of ^{12}C-methionine and ^{13}C-methionine. Somewhat later, Harino and coworkers[146,147] modified bovine pancreatic trypsin inhibitor with ^{13}CH$_3$I in 70% formic acid (18 h, 23°C). These investigators did not find it necessary to remove the methyl group from the sulfonium salt prior to NMR analysis: the two methyl groups that are diastereoisomeric and are nonequivalent in the NMR signal reflecting local environment factors including protein conformation.[147] Matta and coworkers[148] modified α-chymotrypsin with methyl iodide (pH 4.0, 16 h, 23°C, dark). The modified enzyme preparation was fractionated on solid phase (Sepharose) lima bean trypsin inhibitor resulting in three fractions. The first peak that was not retained by the column (equilibrated with 0.05 M Tris, pH 8.0) contained denatured protein, the second peak (eluted with the addition of 0.12 M CaCl$_2$ and 0.1 M KCl) was more active than the native protein in the hydrolysis p-nitrophenyl-3-phenylpropionate, and amino acid analysis showed equimolar amounts of *S*-methyl methionine and methionine (chymotrypsin contains two methionine residues). A third fraction containing unmodified protein was eluted with pH 2.0 HCl. This latter material was found to be [*S*-[^{13}C] methylmethionine-192-α-chymotrypsin and has equal or enhanced activity in the hydrolysis of *p*-nitrophenyl-3-phenylpropioniate. The sulfonium salt could not be converted back to the native protein by reduction without denaturation of the protein. Demethylation of [*S*-[^{13}C] methylmethionine-192-α-chymotrypsin was accomplished with a reagent directed toward the enzyme active site, 2-mercaptoacetyl-4'-methoxyanilide.[149] ^{13}C-methionine can be incorporated into proteins during synthesis,[2,150–152] and it has been shown that NMR analysis can distinguish ^{13}C-methionine sulfoxide from ^{13}C-methionine.[150] More recent work has used protein engineering to insert ^{13}C-labeled methionine residues into protein.[153]

The ability to reverse the alkylation of methionine under relatively mild conditions as described earlier has resulted in the development of several approaches to the purification of methionine-containing peptides. In 1967, Tang and Hartley[154] used a diagonal electrophoretic method previously developed for identification of disulfide bridges[155] for the purification of methionine peptides. These procedures involve the paper electrophoresis of a peptide mixture, in situ alkylation of the peptides with iodoacetamide, and a second electrophoresis at right angles to the original electrophoresis. A *diagonal* would be formed by the peptides with the positively charged methionine peptides off of the diagonal. A decade later, Degen and Kyte[156] described a method for the purification of methionine peptides from proteins conceptually related to the diagonal concept where modification of one amino acid in a peptide, such as methionine, changes the behavior of that peptide in a given separation system, while all other amino acids are not modified such that the behavior of peptides not containing the target amino acid is unchanged in the separation system. As an example, Cruikshank and coworkers[157] used the selective modification of tyrosine to identify tyrosine-containing peptides in a paper chromatography system. In a work by Degen and Kyte,[156] a proteolytic digest of a protein was modified with radiolabeled iodoacetamide under conditions specific for methionine (30% acetic acid, 57°C, 2–3 h) and then fractionated on a cation-exchange resin (Aminex A-5). The methionine-containing peptides are identified by radioactivity, subjected to thiolysis with 2-mercaptoethanol, and again fractionated on an Aminex-5 column to obtain highly purified peptides. More recent work is on the use of diagonal chromatography to separate methionine peptides using oxidation with H_2O_2 (methionine sulfoxide) and performic acid (methionine sulfone) to change chromatographic behavior.[158,159] Here, the oxidation of methionine in a peptide forms the sulfoxide, which is somewhat more polar and is eluted earlier from a reverse-phase column. The general use of diagonal reverse-phase chromatography combined fractional diagonal chromatography (COFRADIC®)[160,161] has been reviewed by Gevaert and coworkers.[162]

The solid-phase alkylation of methionine in peptides has provided a clever affinity approach to the purification of methionine peptides. Several groups[163–165] have reported the isolation of methionine peptide by reaction with bead containing a bromoacetyl function under acidic conditions (e.g., 25% acetic acid) and subsequent thiolysis with a reducing agent under alkaline conditions to release the bound peptide (Figure 9.2).

Cleavage of methionine-containing peptide bonds with cyanogen bromide (CNBr)[4,166,167] is a useful method for specific chemical cleavage of peptide bonds (Figure 9.3). Maximum use of this technology was in the 1980s with 241 PubMed citations in 1988; there were 19 citations in 2010. The CNBr reaction cleaves a peptide bond (Figure 9.3) in which methionine contributes the carboxyl moiety. Methionine is converted into homoserine lactone and homoserine during this process with the loss of methyl thiocyanate. This means, for example, if one has an ^{35}S-labeled methionine in one's protein, the radiolabel will be lost during the cyanogen bromide cleavage. For the skeptical, there are several examples of this being brought to the author's attention. The reaction is reasonably quantitative although, as indicated in the following, variable amounts of CNBr might be required. The methionine content of most proteins is low[1] enough that a reasonably small number of fragments are obtained from an average protein providing a distinct advantage in primary structure analysis. Early work on the *top-down* strategy for protein structure analysis by MS utilized CNBr

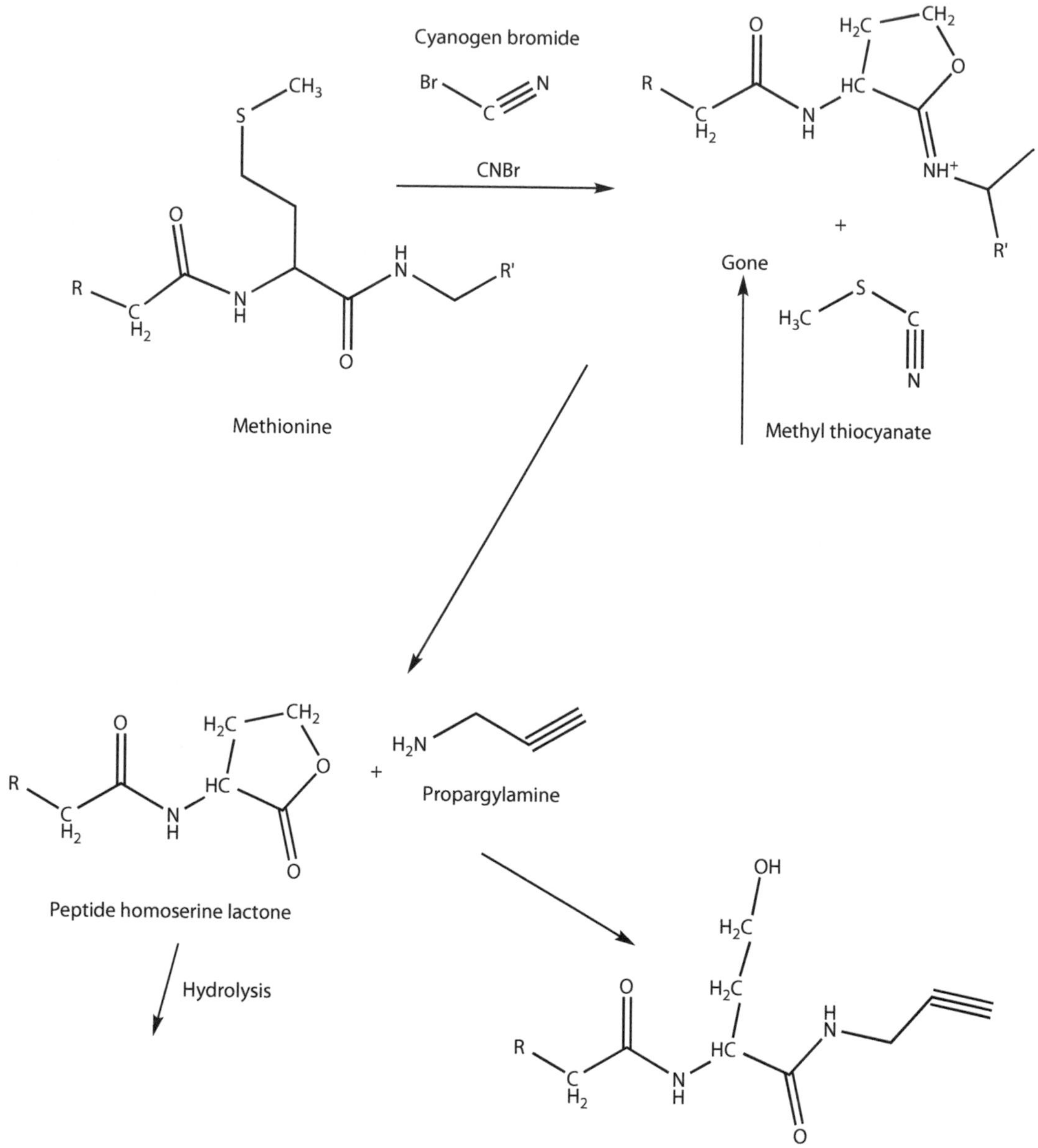

FIGURE 9.3 The cleavage of methionine-containing peptide bonds and reaction of homoserine lactone with nucleophiles. Note the *S*-methyl group is lost as methyl thiocyanate in the process of homoserine lactone (Gross, E., The cyanogen bromide reaction, *Methods Enzymol.*, 11, 238, 1967). Shown, for example, is the reaction of propargylamine with a C-terminal homoserine lactone to form a derivative (Fricke, T., Mart, R.J., Watkins, C.L. et al., Chemical synthesis of cell-permeable apoptotic peptides from in vivo produced proteins, *Bioconjug. Chem.* 22, 1763–1767, 2011).

fragmentation.[168,169] However, advances in MS are reducing the necessity for a preliminary protein fragmentation prior to MS analysis[170]; however, CNBr is still useful for the study of protein structure[171,172] and for preparing fragments for the study of biological function.[173,174] Cyanogen bromide cleavage can also be used to identify site of methionine oxidation in proteins.[175–177] Cyanogen bromide cleavage is used for the preparation of protein fragments for *semisynthesis* of proteins.[178,179] The homoserine lactone arising from the peptide bond cleavage is a reactive function subject to nucleophilic attack; with an amine, there is the formation of an amide bond (Figure 9.3). The concept of using the lactone formed from cyanogen bromide cleavage at methionine for the *semisynthesis* of proteins was introduced by Offord in 1972.[180] Offord[180] reported the rate of hydrolysis at pH 10 in a aqueous solution of triethylamine (20°) and determined that opening of the lactone ring was complete in 20 min; the lactone ring could be closed with trifluoroacetic acid in 60 min at 20°C. Offord also reported experiments showing the reaction of homoserine lactone hydrazine in dimethylformamide (DMF) to yield the hydrazide derivative. Wright and coworkers[181] reported CNBr cleavage of glucagon resulting in a 27-amino acid fragment with a C-terminal homoserine lactone. Reaction with butylamine, hydrazine, and 6-aminohexyl biotinamide was achieved in aqueous solution at alkaline pH (2 days at 23°). Kullopulos and Walsh[182] were able to prepare a variety of derivatives from peptides with C-terminal homoserine lactone. Reaction occurs with primary amine derivatives of biotin and fluorescein. Reaction was accomplished in DMF in the presence of triethylamine. The homoserine lactone-derivatized peptide was stable at –20° for 6 months; there was limited hydrolysis of the lactone that was easily reversed with trifluoroacetic acid. Shi and coworkers[183] also mentioned the use of trifluoroacetic acid to assure maximal amounts of the lactone prior to coupling. These investigators couple homoserine lactone-containing peptides with primary amine-containing affinity tags, tetrahistidine or biotin, in anhydrous DMSO. Fricke and coworkers[184] coupled C-terminal homoserine lactone peptides derived from CNBr cleaved with propargylamine (Figure 9.3) providing an approach for the addition of a variety of compounds via click chemistry. Fricke and coworkers[184] coupled C-terminal homoserine lactone peptides derived from CNBr-cleaved proteins with propargylamine (Figure 9.3) yielding a terminal homoserine propylargylamide. The peptidylpropylargylamide could then be coupled azide derivatives of the compounds of interest to giving triazole-coupled compounds in 80% and better yields. Success on reaction with the homoserine lactone appears to depend in part on the nucleophilicity of the coupling partner. It is recommended that coupling be performed at neutral pH with reaction preferred in anhydrous organic solvents such as DMF or DMSO[183]; hydrolysis of the lactone ring is a competing reaction at alkaline pH. For reference, the pKa of propargylamine used with success by Fricke and coworkers[184] is 7.87[185]; the reader is also directed to the recent work by Garel and Tawfik[186] on the hydrolysis/aminolysis of homocysteine thiolactone.

The chemistry of the cyanogen bromide cleavage of peptide bonds (Figure 9.3) is straightforward, involving the nucleophilic attack of the thioether sulfur on the carbon in CNBr followed by cyclization to form the iminolactone, which is hydrolyzed by water resulting in cleavage of the peptide bond. At acid pH, this reaction does not generally, in and by itself, affect any other amino acid with the exception of cysteine, which is converted to cysteic acid. In this regard, it is noted that one would rarely be working with a protein or peptide containing free sulfhydryl groups. The yield of cleavage can be measured either by the loss of methionine or by the sum of homoserine and homoserine lactone after acid hydrolysis. This value is probably best determined by allowing complete conversion to homoserine with base at room temperature. Cleavage of peptide chains at methionine with CNBr proceeds best with a fully denatured protein in mild acid. Early work with this reaction used 0.1 M HCl as the solvent or 0.1 M HCl in 6 M guanidine hydrochloride. Recent studies have used 70%[187] or 80% formic acid,[188,189] trifluoroacetic acid,[190,191] or an equal mixture of formic acid and trifluoroacetic acid.[192] The use of formic acid was shown to result in the blocking of amino-terminal residues via reaction with formaldehyde (present as a contaminant in the formic acid).[193] More recent work has confirmed this observation as well as the formylation of serine and threonine residues in proteins.[194–196] Grozav and coworkers[187] reported the bromination of a tyrosine residue during the CNBr cleavage reaction in 70% formic acid. The reaction proceeds effectively with a 20- to 100-fold molar excess of CNBr (added as either a solid to the protein or a peptide dissolved in the solvent of choice). Solutions of acetic acid could also be used as solvent for the CNBr reaction. The molar ratio of CNBr to methionine residues needs to be established for each peptide and protein under study. In the work on the structure of the pancreatic deoxyribonuclease, it was necessary to use a 3000-fold molar excess to cleave a particular methionine–serine peptide bond.[197] Kaiser and Metzka[198] have developed conditions for improving the efficiency of the CNBr cleavage at Met–Ser and Met–Thr peptide bonds. Decreasing the acid concentration and increasing acid concentration improved cleavage yields in model peptides. For example, cleavage of PFAMSL was 60% in 0.1 N HCl, 0% in 0.1 N HCl in 100% acetonitrile, and 36% in 30% formic acid. In this regard, it is of interest that Met123 in human serum albumin is converted to homoserine lactone by treatment with CNBr without concomitant peptide bond cleavage.[199] The conversion of methionine to methionine sulfoxide under conditions used for the CNBr cleavage has also been reported.[200] With a 10-fold molar excess of CNBr for 22 h at ambient temperature, 1% conversion to methionine sulfoxide was observed in 70% formic acid; 8% conversion in 0.1 M HCl; 64% conversion in 0.1 M citrate, pH 3.5; and 97% conversion in 0.1 M phosphate, pH 6.5. As noted earlier, cyanogen bromide does not cleave methionine sulfoxide-containing peptide bonds. In this regard, Hwang and coworkers[201] show that complete reduction methionine is required for cyanogen bromide cleavage. A recent report by Andreev and coworkers[202] suggested that desalting of the protein is not necessary prior to CNBr cleavage; acidification to 0.5 M HCl suffices for cleavage efficiency. Compagnini and colleagues[203] have shown that Tris buffer reacts with homoserine lactone provided a product with a mass difference of 103 Da compared to the homoserine peptide. The early work on the CNBr cleavage of proteins[166] showed that the homoserine lactone product was in equilibrium with homoserine, which yields a poorly resolved doublet with mass spectrometric analysis. The reaction with Tris provided easy resolution between homoserine and homoserine lactone.

There was considerable application of the CNBr cleavage to proteins on membranes and/or in electrophoretic gel matrices. I could find no direct work in this area since 2002.[204] The following material is included in this edition to assure that there is some record of this technology.

Simpson and Nice[205] developed a procedure for the in situ CNBr cleavage of proteins absorbed to the glass-fiber membranes developed for the gas-phase protein–peptide sequenator. The procedure was originally developed to determine internal sequence information from N-terminal blocked proteins. Xu and Shively[206] have described improvements on the electroblotting of proteins. These investigators reported higher degrees of success with polyvinyldifluoride (PVDF) membranes. Transfer yields were markedly improved upon pretreatment of the membranes with Polybrene. Scott and coworkers[207] described an additional approach to the CNBr cleavage of proteins on PDVF membranes followed by elution of the reaction products from the membranes with 2% sodium dodecyl (SDS) sulfate per 1% Triton X-100 in 50 mM Tris, pH 9.2. This solvent had previously been evaluated by Szewczyk and Summers[208] as useful for elution of proteins and protein fragments from PDVF membranes. These investigators used bovine serum albumin as a model protein and determined that the efficiency of elution with this solvent was 100% with 90% recovery of protein; with SDS alone, efficiency of elution was 20% with 10% recovery of eluted protein. Sokolov and coworkers[209] describe a modified method for direct CNBr cleavage directly within the polyacrylamide gel. After identification of the proteins by either staining or autoradiography, the gel is sliced and dried. An alternative approach has been developed by Jahnen and coworkers.[210] These investigators isolated the fragments from the CNBr cleavage of proteins on polyacrylamide gel (the gel slices were dried by lyophilization prior to the CNBr cleavage step) either by a second electrophoretic step or by HPLC after elution. The electrophoretic step is recommended over the HPLC step. Robillard and coworkers[211] have provided a method for the in-gel cleavage of integral membranes proteins for subsequent analysis by MALDI-TOF MS. The CNBr cleavage reaction was performed in 70% TFA (trifluoroacetic acid). One small CNBr crystal was dissolved in 200–300 μL; 70% TFA and added to the gel slice for 14 h in the dark at room temperature. The digested gel piece was sonicated for 5 min and then extracted twice with sonication with 30 μL 60% acetonitrile and 1% TFA and concentrated in vacuo prior to analysis.

Cyanogen bromide has proved useful for the cleavage of a variety of fusion proteins.[212–224] The caveat is that the protein/peptide of interest must be stable to acidic conditions. This has made it more applicable to small peptides and proteins than to large proteins that might be more sensitive to denaturation in acid. There are some interesting observations from these studies that are important to the use of cyanogen bromide. Rodriguez and coworkers[219] reported that the use of 70% trifluoroacetic acid for CNBr cleavage of hydrophobic proteins resulted in both incomplete cleavage at expected sites and the production of products from degradation of the substrate protein; the use of 70% formic acid or 0.1 M HCl in 6 M guanidine hydrochloride resulted in the expected cleavage products. Wong and coworkers[220] observed deamidation (0.018 h^{-1}) of glutamine during CNBr cleavage (70% formic acid, 23°C). Hwang and coworkers[201] have reported the necessity of reducing the methionine residue in the expressed fusion protein before cyanogen bromide cleavage. The reduction of the fusion protein was accomplished after elution from a nickel-affinity column. The eluted protein was acidified, potassium iodide added, and allowed to stand overnight at 23°C. Pretreatment with potassium iodide was more effective than 2-mercaptoethanol.

REFERENCES

1. White, S.H., Global statistics of protein sequence: Implications for the origin, evolution, and prediction of structures, *Ann. Rev. Biophys. Biomol. Struct.* 23, 407–439, 1994.
2. Lo Conte, L., Chothia, C., and Janin, J., The atomic structure of protein-protein recognition sites, *J. Mol. Biol.* 285, 2177–2198, 1999.
3. Weininger, U., Liu, Z., McIntyre, D.D. et al., Specific $^{12}C^{\beta}D_2{}^{12}C^{\gamma}D_2S^{13}C^{\epsilon}HD_2$ Isotopomer labeling of methionine to characterize protein dynamics by ^{1}H and ^{13}C NMR relaxation, *J. Am. Chem. Soc.* 134, 18562–18565, 2012.
4. Moerman, P.P., Sergeant, K., Debyser, G. et al., A new chemical approach to differentiate carboxy terminal peptide fragments in cyanogen bromide digests of proteins, *J. Proteomics* 73, 1454–1460, 2010.
5. Gurd, F.R.N., Carboxymethylation, *Methods Enzymol.* 25, 424–438, 1972.
6. Pan, H., Chen, K., Cho, L. et al., Methionine oxidation in human IgG2 Fc decreases binding affinities for Protein A and FcRn, *Protein Sci.* 18, 424–433, 2009.
7. Stadtman, E.R., Protein oxidation and aging, *Free Radic. Res.* 40, 1250–1258, 2006.
8. Jensen, J.L., Kolvenbach, C., Roy, S. et al., Metal-catalyzed oxidation of brain-derived neurotrophic factor (BDNF): Analytical challenge for the identification of modified sites, *Pharm. Res.* 17, 190–196, 2000.
9. Duenas, E.T., Keck, R., De Vos, A. et al., Comparison between light induced and chemically induced oxidation of rhVEGF, *Pharm. Res.* 18, 1455–1460, 2001.
10. Shapiro, R.I., Wen, D., Levesque, M. et al., Expression of sonic hedgehog-Fc fusion protein in *Pichia pastoris*. Identification and control of post-translational, chemical, and proteolytic modifications, *Protein Expr. Purif.* 29, 272–283, 2003.
11. Pan, B., Abel, J., Ricci, M.S. et al., Comparative oxidation studies of methionine residues reflect a structural effect on chemical kinetics, *Biochemistry* 45, 15430–15443, 2006.
12. Liu, H., Gaza-Bulseco, G., Xiang, T., and Chumsae, C., Structural effect of deglycosylation and methionine oxidation on a recombinant monoclonal antibody, *Mol. Immunol.* 45, 701–708, 2008.
13. Pipes, G.D., Campbell, P., Bondarenko, P.V. et al., Middle-down fragmentation for the identification and quantitation of site-specific methionine oxidation in an IgG1 molecule, *J. Pharm. Sci.* 99, 4469–4476, 2010.
14. Time, V., Gruber, P., Wasilu, M. et al., Identification and characterization of oxidation and deamidation sites in monoclonal rat/mouse hybrid antibodies, *J. Chromatogr. B* 878, 777–784, 2010.
15. Schneiderheinze, J., Walden, Z., Dufield, R., and Demarest, C., Rapid online proteolytic mapping of PEGylated rhGH for identity confirmation, quantitation of methionine oxidation and quantitation of UnPEGylated N-terminus using HPLC with UV detection, *J. Chromatogr. B* 877, 4065–4070, 2009.
16. Ohkubo, T., Inagaki, S., Min, J.Z. et al., Rapid determination of oxidized methionine residues in recombinant human basic fibroblast growth factor by ultra-performance liquid chromatography and electrospray ionization quadrupole time-of-flight mass spectrometry with in-source collision-induced dissociation, *Rapid Commun. Mass Spectrom.* 23, 2053–2060, 2009.
17. Jiang, H., Wu, S.L., Karger, B.L., and Hancock, W.S., Mass spectrometric analysis of innovator, counterfeit, and follow-on recombinant human growth hormone, *Biotechnol. Prog.* 25, 207–218, 2009.
18. Ren, D., Ratnaswarmy, G., Beierle, J. et al., Degradation products analysis of an Fc fusion protein using LC/MS methods, *Int. J. Biol. Macromol.* 44, 81–85, 2009.
19. Silva, M.M., Lamarre, B., Cerasoli, E. et al., Physicochemical and biological assays for quality control of biopharmaceuticals: Interferon α-2 case study, *Biologicals* 36, 383–392, 2008.

20. Zamani, L., Andersson, F.O., Edebrink, P. et al., Conformational studies of a monoclonal antibody, IgG1, by chemical oxidation: Structural analysis by ultrahigh-pressure LC-electrospray ionization time-of-flight MS and multivariate data analysis, *Anal. Biochem.* 380, 155–163, 2008.

21. Zang, L., Carlage, T., Murphy, D. et al., Residual metals cause variability in methionine oxidation measurements in protein pharmaceuticals using LC-UV/MS peptide mapping, *J. Chromatogr. B* 895–896, 71–76, 2012.

22. Tien, M., Berlett, B.S., Levine, R.L. et al., Peroxynitrite-mediated modification of protein at physiological carbon dioxide concentration: pH dependence of carbonyl formation, tyrosine nitration, and methionine oxidation, *Proc. Natl. Acad. Sci. USA* 96, 7809–7814, 1999.

23. Hawkins, C.L. and Davies, M.J., Hypochlorite-induced oxidation of proteins in plasma: Formation of chloramines and nitrogen-centered radicals and their role in protein fragmentation, *Biochem. J.* 340, 539–545, 1999.

24. Davies, M.J., Singlet oxygen-mediated damage to proteins and its consequences, *Biochem. Biophys. Res. Commun.* 305, 761–770, 2003.

25. Imlay, J.A., Pathways of oxidative damage, *Annu. Rev. Microbiol.* 57, 395–418, 2003.

26. Droge, W., Oxidative stress and aging, *Adv. Exp. Med. Biol.* 543, 191–200, 2003.

27. Vogt, W., Oxidation of methionyl residues in proteins: Tools, targets, and reversal, *Free Radic. Biol. Med.* 18, 93–105, 1995.

28. Emes, M.J., Oxidation of methionine residues: The missing link between stress and signalling responses in plants, *Biochem. J.* 422, e1–e2, 2011.

29. Lee, B.C. and Gladyshev, V.N., The biological significance of methionine sulfoxide stereochemistry, *Free Radic. Biol. Med.* 50, 221–227, 2011.

30. Cui, Z.J., Han, Z.Q., and Li, Z.Y., Modulating protein activity and cellular function by methionine residue oxidation, *Amino Acids* 43, 505–517, 2012.

31. Savige, W.E. and Fontana, A., Interconversion of methionine and methionine sulfoxide, *Methods Enzymol.* 47, 453–459, 1977.

32. Caldwell, P., Luk, D.C., Weissbach, H., and Brot, N., Oxidation of the methionine residues of *Escherichia coli* ribosomal protein L12 decreases the protein's biological activity, *Proc. Natl. Acad. Sci. USA* 75, 5349–5352, 1978.

33. Houghten, R.A. and Li, C.H., Reduction of sulfoxides in peptides and proteins, *Methods Enzymol.* 91, 549–559, 1983.

34. Sánchez, J., Nikolau, B.J., and Stumpf, P.K., Reduction of N-acetyl methionine sulfoxide in plants, *Plant Physiol.* 73, 619–623, 1983.

35. Brot, N., Fliss, H., Coleman, T., and Weissbach, H., Enzymatic reduction of methionine sulfoxide residues in proteins and peptides, *Methods Enzymol.* 107, 352–360, 1984.

36. Saunders, C.C. and Stites, W.E., An electrophoretic mobility shift assay for methionine sulfoxide in proteins, *Anal. Biochem.* 421, 767–769, 2012.

37. Houghten, R.A., Glaser, C.G., and Li, C.H., Human somatotropin. Reaction with hydrogen peroxide, *Arch. Biochem. Biophys.* 178, 350–355, 1977.

38. Hirs, C.H.W., Performic acid oxidation, *Methods Enzymol.* 11, 197–199, 1967.

39. Butz, L. and du Vigneaud, V., The formation of a homolog of cystine by decomposition of methionine with sulfuric acid, *J. Biol. Chem.* 99, 135–142, 1932.

40. Kopoldova, J., Kolvesek, J., Babicky, A., and Liebster, J., Degradation of DL-methionine by radiation, *Nature* 182, 1074–1076, 1958.

41. Floyd, N.F., Cammaroti, M.S., and Lavine, T.F., The decomposition of DL-methionine in 6 N hydrochloric acid, *Arch. Biochem. Biophys.* 102, 343–345, 1963.

42. Hayward, M.A., Campbell, E.B., and Griffith, O.W., Sulfonic acids: L-Homocysteine sulfonic acid, *Methods Enzymol.* 143, 279–281, 1987.

43. Neumann, N.P., Oxidation with hydrogen peroxide, *Methods Enzymol.* 25, 393–400, 1972.

44. Ichiba, H., Nakamoto, M., Yajima, T. et al., Analysis of oxidation products of cholecystokinin octapeptide with reactive oxygen species by high-performance liquid chromatography and subsequent electrospray ionization mass spectrometry, *Biomed. Chromatogr.* 24, 140–147, 2010.

45. Houghten, R.A. and Li, C.H., Reaction of human choriomammatropin with hydrogen peroxide, *Eur. J. Biochem.* 77, 119–123, 1977.

46. Luo, D., Smith, S.W., and Anderson, B.D., Kinetics and mechanism of the reaction of cysteine and hydrogen peroxide in aqueous solution, *J. Pharm. Sci.* 94, 304–316, 2005.

47. Luo, D. and Anderson, B.D., Kinetics and mechanism for the reaction of cysteine with hydrogen peroxide in amorphous polyvinylpyrrolidone lyophiles, *Pharm. Res.* 23, 2239–2253, 2006.

48. Luo, D. and Anderson, B.D., Application of a two-stage kinetic model to the heterogeneous kinetics of reaction between cysteine and hydrogen peroxide in amorphous lyophiles, *J. Pharm. Sci.* 97, 3907–3926, 2008.

49. Fliss, H., Weissbach, H., and Brot, N., Oxidation of methionine residues in proteins of activated human neutrophils, *Proc. Natl. Acad. Sci. USA* 90, 7160–7164, 1983.

50. Houghten, R.A. and Li, C.H., Reduction of sulfoxides in peptides and proteins, *Anal. Biochem.* 98, 36–46, 1979.

51. Brot, N., Weissbach, L., Werth, J., and Weissbach, H., Enzymatic reduction of protein-bound methionine sulfoxide, *Proc. Natl. Acad. Sci. USA* 78, 2155–2158, 1981.

52. Brot, N. and Weissbach, H., Peptide methionine reductase: Biochemistry and physiological role, *Biopolymers* 55, 288–296, 2000.

53. Hoshi, T. and Heinemann, S., Regulation of cell function by methionine oxidation and reduction, *J. Physiol.* 531, 1–11, 2001.

54. Weissbach, H., Etienne, F., Hoshi, T. et al., Peptide methionine sulfoxide reductase: Structure, mechanism of action, and biological function, *Arch. Biochem. Biophys.* 397, 172–178, 2002.

55. Stadtman, E.R., Moskovitz, J., Berlett, B.S., and Levine, R.L., Cyclic oxidation and reduction of protein methionine residues is an important antioxidant mechanism, *Mol. Cell. Biochem.* 234–235, 3–9, 2002.

56. Antoine, M., Boschi-Muller, S., and Branlant, G., Kinetic characterization of the chemical steps involved in the catalytic mechanism of methionine sulfoxide reductase A from *Neisseria meningitidis*, *J. Biol. Chem.* 278, 45352–45357, 2003.

57. Kwak, G.H., Hwang, K.Y., and Kim, H.Y., Analyses of methionine sulfoxide reductase activities toward free and peptidyl methionine sulfoxides, *Arch. Biochem. Biophys.* 527, 1–5, 2012.

58. Kim, J.Y., Choi, S.H., Lee, E. et al., Methionine sulfoxide reductase A attenuates heme oxygenase-1 induction through inhibition of Nrf2 activation, *Arch. Biochem. Biophys.* 528, 134–140. 2012.

59. Stryskal, J., Nwagwu, F.A., and Watkins, Y.N., Methionine sulfoxide reductase A affects resistance by protecting insulin receptor function, *Free Radic. Biol. Med.* 56, 123–132, 2012.

60. Keutmann, H.T. and Potts, J.T., Jr., Improved recovery of methionine after acid hydrolysis using mercaptoethanol, *Anal. Biochem.* 29, 175–183, 1969.

61. Shechter, Y., Burstein, Y., and Patchornik, A., Selective oxidation of methionine residues in proteins, *Biochemistry* 14, 4497–4503, 1975.

62. Joergensen, L. and Thestrup, H.N., Determination of amino acids in biomass and protein samples by microwave hydrolysis and ion-exchange chromatography, *J. Chromatogr. A* 706, 412–428, 1995.

63. Hayashi, R. and Suzuki, F., Determination of methionine sulfoxide in proteins in protein and food by hydrolysis with *p*-toluenesulfonic acid, *Anal. Biochem.* 149, 521–528, 1985.

64. Weiss, M., Manneberg, M., Juranville, J.-F. et al., Effect of hydrolysis of method on the determination of the amino acid composition of proteins, *J. Chromatogr. A* 795, 263–275, 1998.

65. Sochaski, M.A., Jenkins, A.J., Lyons, T.J. et al., Isotope dilution gas chromatography/mass spectrometry method for the determination of methionine sulfoxide in protein, *Anal. Chem.* 73, 4662–4667, 2001.

66. Todd, J.M., Marable, N.L., and Kehrberg, N.L., Methionine sulfoxide determination after alkaline hydrolysis of amino acid mixtures, model protein systems, soy products, and infant formulas, *J. Food Sci.* 49, 1547–1551, 1984.

67. Molnar-Perl, I., HPLC of amino acids as phenylthiocarbamyl derivatives, *J. Chromatogr. Library* 70, 137–162, 2005.

68. Fujii, K., Yahashi, Y., Nakano, T. et al., Simultaneous detection and determination of the absolute configuration of thiazole-containing amino acids in a peptide, *Tetrahedron* 58, 6873–6879, 2002,

69. Wang, X.S., Shao, B., Oda, M.N. et al., A sensitive and specific ELISA detects methionine sulfoxide-containing apolipoprotein A-1 in HDL, *J. Lipid Res.* 50, 586–594, 2009.

70. Schöneich, C., Methionine oxidation by reactive oxygen species: Reaction mechanisms and relevance to Alzheimer's disease, *Biochim. Biophys. Acta* 1703, 111–119, 2005.

71. Oien, D.B., Canello, T., Gabizon, R. et al., Detection of oxidized methionine in selected proteins, cellular extracts and blood serums by novel anti-methionine sulfoxide antibodies, *Arch. Biochem. Biophys.* 485, 35–40, 2009.

72. Simon, R., Girud, M., Fonbonne, P. et al., Total ApoE and ApoE5 isoform assays in an Alzheimer's disease case control study by targeted mass spectrometry (n = 669): A pilot assay for methionine-containing proteotypic peptides, *Mol. Cell. Proteomics* 11, 1389–1403, 2012.

73. Craig, R., Cortens, J.P., and Beavis, R.C., The use of proteotypic peptide libraries for protein identification, *Rapid Commun. Mass Spectrom.* 19, 1844–1859, 2005.

74. Kuster, B., Schirle, M., Mallick, P., and Aebersold, R., Scoring proteomes with proteotypic peptide probes, *Nat. Rev. Mol. Cell Biol.* 6, 577–583, 2005.

75. Long, L.H., Wu, P.F., Guan, X.L. et al., Determination of protein-bound methionine oxidation in the hippocampus of adult and old rats by LC-ESI-ITMS method after microwave-assisted proteolysis, *Anal. Bioanal. Chem.* 399, 2267–2274, 2011.

76. Moskovitz, J., Maiti, P., Lopes, D.H. et al., Induction of methionine-sulfoxide reductases protects neurons from amyloid β-protein insults in vitro and in vivo, *Biochemistry* 50, 10687–10697, 2011.

77. Bhatia, S., Knock, B., Wong, J. et al., Selective reduction of hydroperoxyeicosatetraenoic acids to their hydroxyl derivatives by apolipoprotein D: Implications for lipid antioxidant activity and Alzheimer's disease, *Biochem. J.* 442, 713–721, 2012.

78. Ringman, J.M., Fithian, A.T., Gylys, K. et al., Plasma methionine sulfoxide levels with familial Alzheimer's disease mutations, *Dement. Geriatr. Cogn. Disord.* 33, 219–225, 2012.

79. Silva, C.J., Onisko, B.C., Dynin, I. et al., Assessing the role of oxidized methionine at position 213 in the formation of prions in hamsters, *Biochemistry* 9, 1854–1861, 2010.

80. Wolschner, C., Giese, A., Kretzchmar, H.A. et al., Design of anti- and pro-aggregation variants to assess the effects of methionine oxidation in human prion protein, *Proc. Natl. Acad. Sci. USA* 106, 7756–7761, 2009.

81. Stadtman, E.R., Moskovitz, J., and Levine, R.L., Oxidation of methionine residues of proteins: Biological consequences, *Antioxid. Redox Signal.* 5, 577–582, 2003.

82. Brennan, L.A. and Kantorow, M., Mitochondrial function and redox control in the aging eye: Role of MsrA and other repair systems in cataract and macular degenerations, *Exp. Eye Res.* 88, 195–203, 2009.

83. Pal, R., Oien, D.B., Ersen, F.Y., and Moskovitz, J., Elevated levels of brain-pathologies associated with neurodegenerative diseases in the methionine sulfoxide reductase A knockout mouse, *Exp. Brain Res.* 180, 765–774, 2007.

84. Lim, J.C., You, Z., Kim, G., and Levine, R.L., Methionine sulfoxide reductase A is a stereospecific methionine reductase, *Proc. Natl. Acad. Sci. USA* 108, 10472–10477, 2011.

85. Lim, J.C., Gruschus, J.M., Kim, G. et al., A low pKa cysteine at the active of mouse methionine sulfoxide reductase A, *J. Biol. Chem.* 287, 25596–25601, 2012.

86. Yokota, H., Saito, H., Masuoka, K. et al., Reversed phase HPLC of Met58 oxidized rhIL-11: Oxidation enhanced by plastic tubes, *J. Pharm. Biomed. Anal.* 24, 317–324, 2000.

87. Takenawa, T., Yokota, A., Oda, M. et al., Protein oxidation during long storage: Identification of the oxidation sites in dihydrofolate reductase from *Escherichia coli* through LC-MS and fragment studies, *J. Biochem.* 145, 517–523, 2009.

88. Pan, Y., Ruan, X., Valvano, M.A., and Konnermann, L., Validation of membrane protein topology models by oxidative labeling and mass spectrometry, *J. Am. Soc. Mass Spectrom.* 23, 889–898, 2012.

89. Xu, G. and Chance, M.R., Radiolytic modification of acidic amino acid residues in peptides: Probes for examining protein-protein interaction, *Anal. Chem.* 76, 1213–1221, 2004.

90. Chance, M.R., Unfolding of apomyoglobin examined by synchrotron footprinting, *Biochem. Biophys. Res. Commun.* 287, 614–621, 2001.

91. Sharp, J.S., Becker, J.M., and Hettich, R.L., Protein surface mapping by chemical oxidation: Structural analysis by mass spectrometry, *Anal. Biochem.* 313, 216–225, 2003.

92. Sharp, J.S., Becker, J.M., and Hettich, R.L., Analysis of protein surface accessible residues by photochemical oxidation and mass spectrometry, *Anal. Chem.* 76, 672–683, 2004.

93. Mozziconacci, O., Mirkowski, J., Rusconi, F. et al., Methionine residues acts as a prooxidant with ·OH-induced oxidation of enkephalins, *J. Phys. Chem. B* 116, 12460–12472, 2012.

94. Bern, M., Saladino, J., and Sharp, J.S., Conversion of methionine into homocysteic acid in heavily oxidized proteomics samples, *Rapid Commun. Mass Spectrom.* 24, 768–772, 2010.

95. Corless, S. and Cramer, R., On-target oxidation of methionine residues using hydrogen peroxide for composition-restricted matrix-assisted laser desorption/ionization peptide mass-mapping, *Rapid Commun. Mass Spectrom.* 17, 1212–1215, 2003.

96. Pan, Y., Brown, L., and Konnermann, L., Site-directed mutagenesis combined with oxidative methionine labeling for probing structural transitions of a membrane protein by mass spectrometry, *J. Am. Soc. Mass Spectrom.* 21, 1947–1956, 2010.

97. Watson, C. and Sharp, J.S., Conformational analysis of therapeutic proteins by hydroxyl radical protein footprinting, *AAPS* 14, 206–217, 2012.

98. Jenkins, N., Modifications of therapeutic proteins: Challenges and prospects, *Cytotechnology* 53, 121–125, 2007.

99. Cindrić, M., Galić, N., Vuletić, M. et al., Evaluation of recombinant human interferon α-2b structure and stability by in-gel tryptic digestion, H/D exchange and mass spectrometry, *J. Pharm. Biomed. Anal.* 40, 781–787, 2006.

100. Thirumangalathu, R., Krishnan, S., Bondarenko, P. et al., Oxidation of methionine residues in recombinant human interleukin-1 receptor antagonist: Implications of conformational stability on protein oxidation kinetics, *Biochemistry* 46, 6213–6224, 2007.

101. Hu, D., Qin, Z., Zue, B. et al., Effects of methionine oxidation on the structural properties, conformational stability, and aggregation of immunoglobulin light chain LEN, *Biochemistry* 47, 8665–8677, 2008.

102. Burkitt, W., Domann, P., and O'Connor, G., Conformational changes in oxidatively stressed monoclonal antibodies studied by hydrogen exchange mass spectrometry, *Protein Sci.* 19, 826–835, 2010.

103. Labrenz, S.R., Calmann, M.A., Heavner, G.A., and Tolman, G., The oxidation of methionine-54 of epotinum α does not affect molecular structure or stability, but does decrease activity, *PDA J. Pharm. Sci.* 62, 211–223, 2008.

104. Mulinacci, F., Bell, S.E., Capelle, M.A. et al., Oxidized recombinant human growth hormone that maintains conformational integrity, *J. Pharm. Sci.* 100, 110–122, 2010.

105. Kim, Y.H., Berry, A.H., Spencer, D.S., and Stiles, W.E., Comparing the effect on protein stability of methionine oxidation versus mutagenesis: Steps toward engineering oxidation resistance in proteins, *Protein Eng.* 14, 343–347, 2001.

106. Chien, H.-C. et al., Enhancing oxidative resistance of *Agrobacterium radiobacter* N-carbamoyl D-amino acid aminohydrolase by engineering solvent-accessible methionine residues. *Biochem. Biophys. Res. Commun.* 297, 282–287, 2002.

107. Kachurin, A.M., Golubev, A.M., Geisow, M.M. et al., Role of methionine in the active site of α-galactosidase from *Trichoderma reesei, Biochem. J.* 308, 955–964, 1995.

108. Kornfelt, T., Persson, E., and Palm, L., Oxidation of methionine residues in coagulation factor VIIa, *Arch. Biochem. Biophys.* 363, 43–54, 1999.

109. Nomura, T., Kamada, R., Ito, I. et al., Oxidation of methionine residue at hydrophobic core destabilizes p53 tetramer, *Biopolymers* 91, 78–84, 2009.

110. Keck, R.G., The use of t-butyl hydroperoxide as a probe for methionine oxidation in proteins, *Anal. Biochem.* 236, 56–62, 1996.

111. Liu, J.L., Lu, K.V., Eris, T. et al., In vitro methionine oxidation of recombinant human leptin, *Pharm. Res.* 15, 632–640, 1998.

112. Lu, H.S., Fausset, P.R., Narhi, L.O. et al., Chemical modification and site-directed mutagenesis of methionine residues in recombinant human granulocyte colony-stimulating factor: Effect on stability and biological activity, *Arch. Biochem. Biophys.* 362, 1–11, 1999.

113. Chumsae, C., Gaza-Bulseco, G., Sun, J., and Liu, H., Comparison of methionine oxidation in thermal stability and chemically stressed samples of a fully human monoclonal antibody, *J. Chromatogr. B* 850, 285–294, 2007.

114. Gaza-Bulseco, G., Faldu, S., Hurkmans, K. et al., Effect of methionine oxidation of a recombinant monoclonal antibody on the binding affinity to protein A and protein G., *J. Chromatogr. B* 870, 55–62, 2008.

115. Liu, H., Gaza-Bulseco, G., and Zhou, L., Mass spectrometry analysis of photo-induced methionine oxidation of a recombinant human monoclonal antibody, *J. Am. Soc. Mass Spectrom.* 20, 525–528, 2009.

116. Fan, H., Chen, K., Chu, L. et al., Methionine oxidation in human IgG2 Fc decreases binding affinities to protein A and FcRn, *Protein Sci.* 18, 424–433, 2009.

117. Loew, C., Knoblich, C., Fichtl, J. et al., Analytical Protein A chromatography as a quantitative tool for the screening of methionine oxidation in monoclonal antibodies, *J. Pharm. Sci.* 101, 4248–4257, 2012.

118. Lee, C.J., Quality control of polyvalent pneumococcal polysaccharide-protein conjugate vaccine by nephelometry, *Biologicals* 30, 97–103, 2002.

119. Yamasaki, R.B., Osuga, D.T., and Feeney, R.E., Periodate oxidation of methionine in proteins, *Anal. Biochem.* 126, 183–189, 1982.

120. Hage, D.S., Wolfe, C.A., and Oates, M.R., Development of a kinetic model to describe the effective rate of antibody oxidation by periodate, *Bioconjug. Chem.* 8, 914–920, 1997.

121. Riebe, D. and Thorn, W., Influence of carbohydrate moieties of human serum transferrin on the determination of its molecular mass by polyacrylamide gradient gel electrophoresis and staining with periodic acid-Schiff reagent, *Electrophoresis* 12, 287–293, 1991.

122. Yamasaki, R.B., Osuga, D.T., and Feeney, R.E., Periodate oxidation of methionine in proteins, *Anal. Biochem.* 126, 183–189, 1982.

123. Knowles, J.R., The role of methionine in α-chymotrypsin-catalyzed reactions, *Biochem. J.* 95, 180–190, 1965.
124. de la Llosa, P., El Abed, A., and Roy, M., Oxidation of methionine residues in lutropin, *Can. J. Biochem.* 58, 745–748, 1980.
125. Gleisner, J.M. and Liener, I.E., Chemical modification of the histidine residue located at the active site of ficin, *Biochim. Biophys. Acta* 317, 482–491, 1973.
126. Tomova, S., Cutruzzolá, F., Barra, D. et al., Selective oxidation of methionyl residues in the recombinant human secretory leukocyte proteinase inhibitor. Effect on inhibitor binding properties, *J. Mol. Recogn.* 7, 31–37, 1994.
127. Tang, X.D., Daggett, H., Hanner, M. et al., Oxidative regulation of large conductance calcium-activated potassium channels, *J. Gen. Physiol.* 117, 253–274, 2001.
128. Daile-Donne, I., Rossi, R., Giustarini, D. et al., Methionine oxidation as a major cause of the functional impairment of oxidized actin, *Free Radic. Biol. Med.* 32, 927–937, 2002.
129. Nieziolek, M., Kot, M., Pyka, K. et al., Properties of chemically oxidized kininogens, *Acta Biochim. Pol.* 50, 753–763, 2003.
130. Bauer, R.J., Leigh, S.D., Birr, C.A. et al., Alteration of the pharmacokinetics of small proteins by iodination, *Biopharm. Drug Dispos.* 17, 761–774, 1996.
131. Kumar, C.C., Nie, H., Armstrong, L. et al., Chloramine T-induced structural and biochemical changes in echistatin, *FEBS Lett.* 429, 239–248, 1998.
132. Cadée, J.A., van Steenbergen, M.J., Versluis, C. et al., Oxidation of recombinant human interleukin-2 by potassium peroxodisulfate, *Pharm. Res.* 18, 1461–1467, 2001.
133. Gundlach, H.G., Moore, S., and Stein, W.H., The reaction of iodoacetate with methionine, *J. Biol. Chem.*, 234, 1761–1764, 1959.
134. Mizzer, J.P. and Thorpe, C., An essential methionine in pig kidney general acyl-CoA dehydrogenase, *Biochemistry* 19, 5500–5504, 1980.
135. Goren, H.J., Glick, D.M., and Barnard, E.A., Analysis of carboxymethylated residues in proteins by an isotopic method and its application to the bromoacetate-ribonuclease reaction, *Arch. Biochem. Biophys.* 126, 607–623, 1968.
136. Cummings, J.G., Lau, S.-M., Powell, P.J., and Thorpe, C., Reductive half-reaction in medium-chain acyl-CoA dehydrogenase: Modulation of internal equilibrium by carboxymethylation of a specific methionine residue, *Biochemistry* 31, 8523–8529, 1992.
137. Schroeder, W.A., Shelton, J.R., and Robberson, B., Modification of methionyl residues during aminoethylation, *Biochim. Biophys. Acta* 147, 590–592, 1967.
138. Marks, R.H.L. and Miller, R.D., Chemical modification of methionine residues in azurin, *Biochem. Biophys. Res. Commun.* 88, 661–667, 1979.
139. Naider, F. and Bohak, Z., Regeneration of methionyl residues from their sulfonium salts in peptides and proteins, *Biochemistry* 11, 3208–3211, 1972.
140. Schramm, H.J. and Lawson, W.B., Über das activ Zentrum von Chymotrypsin. II. Modifizierung eines Methioninrestes in Chymotrypsin durch einfache Benzolderivate, *Hoppe-Seyler's Z. Physiol. Chem.* 332, 97–100, 1963.
141. Reid, G.E., Roberts, K.D., and Simpson, R.J., Selective identification and quantitative analysis of methionine containing peptides by charge derivatization and tandem mass spectrometry, *J. Am. Soc. Mass Spectrom.* 16, 1131–1150, 2005.
142. Goverman, J.M. and Pierce, J.G., Differential effects of alkylation of methionine residues on the activities of pituitary thyrotropin and lutropin, *J. Biol. Chem.* 256, 9431–9435, 1981.
143. Kleanthous, C., Campbell, D.G., and Coggins, J.R., Active site labeling of the shikimate pathway enzyme, dehydroquinase. Evidence for a common substrate binding site within dehydroquinase and dehydroquinate synthase, *J. Biol. Chem.* 265, 10929–10934, 1990.
144. Kleanthous, C. and Coggins, J.R., Reversible alkylation of an active site methionine residue in dehydroquinase, *J. Biol. Chem.* 265, 10935–10939, 1990.
145. Jones, W.C., Jr., Rothgeb, T.M., and Gurd, F.R.N., Specific enrichment with ^{13}C of the methionine methyl groups of sperm whale myoglobin, *J. Am. Chem. Soc.* 97, 3875–3877, 1975.
146. Harino, B.M., Dyckes, D.F., Willcott, M.R., III., and Jones, W.C., Jr., ^{13}C-enriched S-methylmethionyl residues as a probe of protein conformation, *J. Am. Chem. Soc.* 100, 4897–4899, 1978.
147. Harino, B.M., Dyckes, D.F., Willcott, M.R., III., and Jones, W.C., Jr., Denaturation studies by ^{13}C nuclear magnetic resonance on modified basic pancreatic trypsin inhibitor using the novel S-[^{13}C] methyl methionyl probe, *J. Am. Chem. Soc.* 102, 1120–1124, 1980.

148. Matta, M.S., Landis, M.E, Patrick, T.B. et al., ^{13}C-enriched *S*-methyl probe at the active site of an enzyme: [*S*-[^{13}C]methyl methionine-192]-α-chymotrypsin. *J. Am. Chem. Soc.* 102, 7151–7152, 1980.
149. Matta, M.S., Henderson, P.A., and Patrick, T.P., Preparation and ^{13}C NMR characterization of [[ε-^{13}C] methionine-192]-α-chymotrypsin. The demethylation of [*S*-[^{13}C]methyl methionine-192]-α-chymotrypsin by an active-site-directed thiol. *J. Biol. Chem.* 256, 4172–4174, 1981.
150. Cohen J.S., Yariv, J., Kalb (Gilboa), A.J. et al., ^{13}C NMR analysis of methionine sulfoxide in protein, *J. Biochem. Biophys. Methods* 1, 145–151, 1979.
151. Gifford, J.L., Ishida, H., and Vogel, H.J., Fast methionine-based solution structure determination of calcium-calmodulin complexes, *J. Biomol. NMR* 50, 71–81, 2011.
152. O'Grady, C., Rempel, B.L., Sokaribo, A. et al., One-step amino acid selective isotope labeling of proteins in prototrophic *Escherichia coli* strains, *Anal. Biochem.* 426, 126–128, 2012.
153. Stoffregen, M.C., Schwer, M.M., Renschler, F.A. et al., Methionine scanning as an NMR tool for detecting and analyzing biomolecular interaction surfaces, *Structure* 20, 573–581, 2012.
154. Tang, J. and Hartley, B.S., A diagonal electrophoretic method for selective purification of methionine peptide, *Biochem. J.* 102, 593–599, 1967.
155. Brown, J.R. and Hartley, B.S., Location of disulphide bridges by diagonal paper electrophoresis. The disulphide bridges of bovine chymotrypsinogen A, *Biochem. J.* 101, 214–228, 1966.
156. Degen, J. and Kyte, J., The purification of peptides which contain methionine residues, *Anal. Biochem.* 89, 529–539, 1978.
157. Cruikshank, W.H., Malchy, B.L., and Kaplan, H., Diagonal chromatography for the selective purification of tyrosyl peptides, *Can. J. Biochem.* 52, 1013–1017, 1974.
158. Gevaert, K., Van Damme, J., Goethals, M. et al., Chromatographic isolation of methionine-containing peptides for gel-free proteome analysis—Identification of more than 800 *Escherichia coli* proteins, *Mol. Cell. Proteomics* 1, 896–903, 2002.
159. Gevaert, K., Pinxteren, J., Demol, H. et al., Four stage chromatographic selection of methionyl peptides for peptide-centric proteome analysis: The proteome of human multipotent adult progenitor cells, *J. Proteome Res.* 5, 1415–1428, 2006.
160. Larsen, T.F., Bache, N., Gramsbergen, J.B., and Roepstorff, P., Identification of nitrotyrosine containing peptides using combined fractional diagonal chromatography (COFRADIC) and off-line nano-LC-MALDI, *J. Am. Soc. Mass Spectrom.* 22, 989–996, 2011.
161. Staes, A., Impens, F., Van Damme, P. et al., Selecting protein *N*-terminal peptides by combined fractional diagonal chromatography, *Nat. Protoc.* 6, 1130–1041, 2011.
162. Gevaert, K., Van Damme, P., Martens, L., and Vandekerckhove, J., Diagonal reverse-phase chromatography applications in peptide-centric proteomics: Ahead of catalogue-omics?, *Anal. Biochem.* 345, 18–29, 2005.
163. Weinberger, S.R., Viner, R.J., and Ho, P., Tagless extraction-retentate chromatography: A new global protein digestion strategy for monitoring differential protein expression, *Electrophoresis* 23, 3182–3192, 2002.
164. Grunert, T., Pock, K., Buchacher, A., and Allmaier, G., Selective solid-phase isolation of methionine-containing peptides and subsequent matrix-assisted laser desorption mass spectrometric detection of methionine- and methionine-sulfoxide-containing tryptic peptides, *Rapid Commun. Mass Spectrom.* 17, 1815–1824, 2003.
165. Shen, M., Guo, L., Wallace, A. et al., Isolation and isotope labeling of cysteine- and methionine-containing tryptic peptides, *Mol. Cell. Proteomics* 2, 315–324, 2003.
166. Gross, E., The cyanogen bromide reaction, *Methods Enzymol.* 11, 238, 1967.
167. Smith, B.J., Chemical cleavage of proteins, *Methods Mol. Biol.* 32, 197–309, 1994.
168. Kelleher, N.L. et al., Top down versus bottom up protein characterization by tandem high-resolution mass spectrometry. *J. Am. Chem. Soc.* 121, 806–812, 1999.
169. McLoughlin, S.M., Mazur, M.T., Miller, L.M. et al., Chemoenzymatic approaches for streamlined detection of active site modifications on thiotemplate assemble lines using mass spectrometry, *Biochemistry* 44, 14159–14169, 2005.
170. Fornelli, L., Damoc, E., and Thomas, P.M., Top-down analysis of monoclonal antibody IgG1 by electron transfer dissociation Orbitrap FTMS, *Mol. Cell. Proteomics* 11, 1758–1767, 2012.
171. Morla, A., Poirier, F., Pons, S. et al., Analysis of high molecular mass proteins larger than 150 kDa using cyanogen bromide cleavage and conventional 2-DE, *Electrophoresis* 29, 4158–4168, 2008.
172. Lee, J.E., Kwon, J., and Baek, M.C., A combination method of chemical with enzyme reactions for identification of membrane proteins, *Biochim. Biophys. Acta* 1814, 397–404, 2011.

173. Sánchez, D., Moussaoui, M., Carreras, E. et al., Mapping the eosinophil cationic protein antimicrobial activity by chemical and enzymatic cleavage, *Biochimie* 93, 331–338, 2011.

174. Adel-Patient, K., Nutten, S., Bernard, H. et al., Immunomodulatory potential of partially hydrolyzed β-lactoglobulin and large synthetic peptides, *J. Agric. Food Chem.* 60, 10858–10866, 2012.

175. Hollemeyer, K., Heinzle, E., and Tholey, A., Identification of oxidized methionine residues in peptides containing two methionine residues by derivatization and matrix-assisted laser desorption-ionization mass spectrometry, *Proteomics* 2, 1524–1531, 2002.

176. Anraku, M., Kragh-Hansen, U., Kawai, K. et al., Validation of the chloramine-T induced oxidation of human serum albumin as a model for oxidative damage *in vivo*, *Pharm. Res.* 20, 684–692, 2003.

177. Kothari, R., Kumar, V., Jena, R. et al., Modes of degradation and impurity characterization in rhPTH (1–34) during stability studies, *PDS J. Pharm. Sci. Technol.* 65, 348–362, 2011.

178. Pál, G., Santamaria, F., Kossiakoff, A.A., and Lu, W., The first semi-synthetic serine protease made by native chemical ligation, *Protein Expr. Purif.* 29, 185–192, 2003.

179. Richardson, J.P. and Macmillan, D., Optimisation of chemical protein cleavage for erythropoietin semi-synthesis using native chemical ligation, *Org. Biomol. Chem.* 6, 3977–3982, 2008.

180. Offord, R.E., The possible use of cyanogen bromide fragments in the semisynthesis of proteins and polypeptides, *Biochem. J.* 129, 499–501, 1972.

181. Wright, D.E., Hruby, V.J., and Rodbell, M., Preparation and properties of glucagon analogs prepared by semi-synthesis from CNBr-glucagon, *Biochim. Biophys. Acta* 631, 49–55, 1980.

182. Kullopulos, A. and Walsh, C.T., Production, purification, and cleavage of tandem repeats of recombinant peptides, *J. Am. Chem. Soc.* 116, 4599–4607, 1994.

183. Shi, T., Weerasekera, R., Yan, C. et al., Method for the affinity purification of covalently linked peptides following cyanogen bromide cleavage of proteins, *Anal. Chem.* 81, 9885–9895, 2009.

184. Fricke, T., Mart, R.J., Watkins, C.L. et al., Chemical synthesis of cell-permeable apoptotic peptides from in vitro produced proteins, *Bioconjug. Chem.* 22, 1763–1767, 2011.

185. Hine, J., Yeh, C.Y., and Schmalstieg, F.C., Polar effects on the formation of imines from isobutyraldehyde and primary aliphatic amines, *J. Org. Chem.* 35, 340–344, 1970.

186. Garel, J. and Tawfik, D.S., Mechanism of hydrolysis and aminolysis of homocysteine thiolactone, *Chemistry* 12, 4144–4152, 2006.

187. Grozav, A.G., Willard, B.B., Kozuki, T. et al., Tyrosine 656 in topoisomerase IIβ is important for the catalytic activity of the enzyme: Identification based on artifactual +80-Da modification at this site, *Proteomics* 11, 829–842, 2011.

188. Parthasarathy, K., Lu, H., Surya, W. et al., Expression and purification of coronavirus envelope proteins using a modified β-barrel construct, *Protein Expr. Purif.* 85, 133–141, 2012.

189. Mauldin, K., Lee, B.L., Loeszczuk, M. et al., The carboxyl-terminal segment of apolipoprotein A-V undergoes a lipid-induced conformation change, *Biochemistry* 49, 4821–4826, 2010.

190. Morrison, J.R., Fidge, N.N., and Grego, H., Studies on the formation, separation, and characterization of CNBr fragments of human A1 apolipoprotein, *Anal. Biochem.* 186, 145, 1990.

191. Arsenault, J., Cabana, J., Fillion, D. et al., Temperature dependent photolabeling of the human angiotensin II type 1 receptor reveals insights into its conformational landscape and its activation mechanism, *Biochem. Pharmacol.* 80, 990–999, 2010.

192. Shively, J.E., Reverse-phase HPLC isolation and microsequence analysis, in *Methods of Protein Microcharacterization*, ed. J.E. Shively, p. 65, Humana Press, Clifton, NJ, 1986.

193. Shively, J.E., Hawke, D., and Jones, B.N., Microsequence analysis of peptides and proteins. III. Artifacts and the effects of impurities on analysis, *Anal. Biochem.* 120, 312, 1982.

194. Goodlett, D.R., Armstrong, F.B., Creech, R.J., and VanBremmen, R.B., Formylated peptides from cyanogen-bromide digests identified by fast atom bombardment mass-spectrometry, *Anal. Biochem.* 186, 116–120, 1990.

195. Duewell, H.S. and Honek, J.F., CNBr/formic acid reactions of methionine- and trifluoromethionine-containing lambda lysozyme: Probing chemical and positional reactivity and formylation side reactions by mass spectrometry, *J. Protein Chem.* 17, 337–350, 1998.

196. Loo, R.R.O. and Loo, J.A., Matrix-assisted laser desorption/ionization-mass spectrometry of hydrophobic proteins in mixtures using formic acid, perfluorooctanoic acid, and sorbitol, *Anal. Chem.* 79, 1115–1125, 2007.

197. Liao, T.-H., Salnikow, J., Moore, S., and Stein, W.H., Bovine pancreatic deoxyribonuclease A. Isolation of CNBr peptides; complete covalent structure of the polypeptide chain, *J. Biol. Chem.* 248, 1489–1495, 1973.

198. Kaiser, R. and Metzka, L., Enhancement of cyanogen bromide cleavage yields for methionyl-serine and methionyl-threonine peptide bonds, *Anal. Biochem.* 266, 1–8, 1999.

199. Doyen, N. and LaPresle, C., Partial non-cleavage by CNBr of a methionine-cystine bond from human serum albumin and bovine α-lactalbumin, *Biochem. J.* 177, 251–254, 1979.

200. Joppich-Kuhn, R., Corkill, J.A., and Giese, R.W., Oxidation of methionine to methionine sulfoxide as a side reaction of CNBr cleavage, *Anal. Biochem.* 119, 73–77, 1982.

201. Hwang, P.M., Pan, J.S., and Sykes, B.D., A PagP fusion protein system for the expression of intrinsically disordered proteins in *Escherichia coli*, *Protein Expr. Purif.* 85, 145–151, 2012.

202. Andreev, Y.A., Kozlov, S.A., Vassilevski, A.A., and Grishin, E.V., Cyanogen bromide cleavage of proteins in salt and buffer solutions, *Anal. Biochem.* 407, 144–146, 2010.

203. Compagnini, A., Cunsolo, V., Foti, S., and Saletti, R., Improved accuracy in the matrix-assisted laser desorption/ionization-mass spectrometry determinations of the molecular mass of cyanogen bromide fragments of proteins by post-cleavage reaction with tris(hydroxymethyl)aminomethane, *Proteomics* 1, 967–974, 2001.

204. Takeuchi, K., Nakamura, K., Fujimoto, M. et al., Heat stress-induced loss of eukaryotic initiation factor 5A (eIF-5A) in a human pancreatic cancer cell line, MIA PaCa-2, analyzed by two-dimensional gel electrophoresis, *Electrophoresis* 23, 662–669, 2002.

205. Simpson, R.J. and Nice, E.C., In situ CNBr cleavage of N-terminally blocked proteins in a gas-phase sequencer, *Biochem. Int.* 8, 787–791, 1984.

206. Xu, Q.-Y. and Shively, J.E., Microsequence analysis of peptides and proteins. VIII. Improved electroblotting of proteins onto membranes and derivatized glass-fiber sheets, *Anal. Biochem.* 170, 19–30, 1988.

207. Scott, M.G., Crimmins, D.L., McCourt, D.W. et al., A simple in situ CNBr cleavage method to obtain internal amino acid sequence of proteins electroblotted to polyvinyldifluoride membranes, *Biochem. Biophys. Res. Commun.* 155, 1353–1359, 1988.

208. Szewczyk, B. and Summers, D.F., Preparative elution of proteins blotted to immobilon membranes, *Anal. Biochem.* 168, 48–53, 1988.

209. Sokolov, B.P., Sher, B.M., and Kalinin, V.N., Modified method for peptide mapping of collagen chains using CNBr-cleavage of protein within polyacrylamide gels, *Anal. Biochem.* 176, 365–367, 1989.

210. Jahnen, W., Ward, L.D., Reid, G.E., Moritz, R.L., and Simpson, R.J., Internal amino acid sequencing of proteins by in situ CNBr cleavage in polyacrylamide gels, *Biochem. Biophys. Res. Commun.* 166, 139–145, 1990.

211. van Montfort, B.A. et al., Improved in-gel approaches to generate peptide maps of integral membrane proteins with matrix-assisted laser desorption/ionization time-of-flight mass spectrometry, *J. Mass. Spectrom.* 37, 322–330, 2002.

212. Olson, H. et al., Production of a biologically active variant form of recombinant human secretin, *Peptides* 9, 301–307, 1988.

213. Forsberg, G., Baastrup, B., Brobjer, M., Lake, M., Jörnvall, H., and Hartmanis, M., Comparison of two chemical cleavage methods for preparation of a truncated form of recombinant human insulin-like growth factor I from a secreted fusion protein, *BioFactors* 2, 105–112, 1989.

214. Husken, D., Beckers, T., and Engels, J.W., Overexpression in *Escherichia coli* of a methionine-free designed interleukin-2 receptor (Tac protein) based on a chemically cleavable fusion proteins, *Eur. J. Biochem.* 193, 387–394, 1990.

215. Myers, J.A. et al., Expression and purification of active recombinant platelet factor 4 from a cleavable fusion protein, *Protein Expr. Purif.* 2, 136–143, 1991.

216. Dobeli, H., Andres, H., Breyer, N. et al., Recombinant fusion proteins for the industrial production of disulfide bridge containing peptides: Purification, oxidation without concatemer formation, and selective cleavage, *Protein Expr. Purif.* 12, 404–414, 1998.

217. Rais-Beghdadi, C., Roggero, M.S., Fasel, N., and Reymond, C.D., Purification of recombinant proteins by chemical removal of the affinity tag, *Appl. Biochem. Biotechnol.* 74, 95–103, 1998.

218. Fairlie, W.D., Uboldi, A.D., De Souza, D.P. et al., A fusion protein system for the recombinant production of short disulfide-containing peptides, *Protein Expr. Purif.* 26, 171–178, 2002.

219. Rodriguez, J.C., Wong, L., and Jennings, P.A., The solvent in CNBr cleavage reactions determines the fragmentation efficiency of ketosteroid isomerase fusion proteins used in the production of recombinant peptides, *Protein Expr. Purif.* 28, 224–231, 2003.

220. Wong, J.-I., Meagher, R.J., and Barron, A.E., Characterization of glutamine deamidation in a long repetitive protein polymer via bioconjugate capillary electrophoresis, *Biomacromolecules* 5, 618–627, 2004.

221. Englander, J., Cohen, L., Arshava, B. et al., Selective labeling of a membrane peptide with ^{15}N-amino acids using cells grown in rich medium, *Biopolymers* 84, 508–518, 2006.
222. Morin, K.M., Arcidiacono, S., Beckwitt, R., and Mello, C.M., Recombinant expression of indolicidin in *Escherichia coli*, *Appl. Microbiol. Biotechnol.* 70, 698–704, 2006.
223. Wang, Y.-Q. and Cai, J.-Y., High-level expression of acidic partner-mediated antimicrobial peptide from tandem genes in *Escherichia coli*, *Appl. Biochem. Biotechnol.* 141, 203–213, 2007.
224. Lee, E.F., Yao, S., Sao, J.K. et al., Peptide inhibitors of the malaria surface protein, apical membrane antigen 1: Identification of key binding residues, *Biopolymers* 95, 354–364, 2011.

10 Modification of Carboxyl Groups in Proteins

Aspartic acid and glutamic acid are dicarboxylic amino acids (Figure 10.1), which are among the 20 amino acids considered normal monomer units of proteins. Other amino acids that contain a *hydroxyl* function are serine, threonine, and tyrosine, which are not considered to be ionizable groups of proteins while the carboxyl groups of aspartic acid and glutamic acid are ionizable groups.[1] γ-Carboxyglutamic acid (Figure 10.1) is a product of the posttranslational modification of glutamic acid. Carboxyl groups are unique in that there are two carbon–oxygen bonds with different bond lengths.[2] Early work on the amino acid composition of proteins did not discriminate between the free acid and amide forms of aspartic acid and glutamic acid, and reported composition in terms of Asx and Glx[3] as only aspartic acid and glutamic acid would be measured after the acid hydrolysis of proteins (asparagine and glutamine would be converted to the free acid forms). More recent work that is based on DNA sequence analysis permits the reporting of both free acid and amide forms.[4] An average content in proteins of 6% has been reported for aspartic acid, while a value of 6.3% has been reported for glutamic acid.[4] The selective hydrolysis of the carboxyl peptide bond of aspartic acid in dilute acid has been known for some time[5–8] and continues to see active use for structural analysis of proteins.[9–11] This reaction has been the subject of considerable study, and a mechanism involving a succinimide intermediate (Figure 10.1) was proposed by Piszkiewicz and coworkers[12] noting the particular sensitivity of Asp–Pro peptide bond. The succinimide intermediate, which also results from asparagine, can be *trapped* by reduction with borohydride to yield homoserine and isohomoserine.[13]

Aspartic acid and glutamic acid are defined as acidic amino acids in distinction to lysine, arginine, and histidine, which are defined as basic amino acids. The various acidic and basic amino acids are jointly considered hydrophilic amino acids as differentiated from hydrophobic amino acids.[14] The average pKa for the β-carboxyl group of aspartic acid has been reported as 3.5 ± 1.2, while a value of 4.2 ± 0.9 was reported for the γ-carboxyl group of glutamic acid.[15] The pKa for the α-carboxyl groups is approximately 2 for both amino acids (1.95 for aspartic acid and 2.18 for glutamic acid),[16] while the average pKa for the C-terminal carboxyl group in a protein is 3.3 ± 0.8 (range 2.4–5.9).[15] There is a wide range of reported values, 0.5–9.0 for aspartic and 2.1–8.8 for glutamic acid.[17] The dicarboxylic acids are less prevalent at protein interfaces than on the surface of a protein.[18] However, carboxyl groups may be *buried*, which may result in a decreased pKa value[19,20] or increased pKa value.[21] A number of factors can influence the observed pKa of a carboxyl group including hydrogen bonding.[20] The reader is directed to a useful report by Li and coworkers[22] on the various factors, which can influence the ionization of carboxyl groups in proteins. Dicarboxylic acids can be directly involved in enzyme action as in pepsin[23] or indirectly as in the classical charge relay system in the serine proteases[24] or in the S_1 binding site of tryptic-like serine proteases.[25–27] The sensitivity of aspartyl peptide bond to cleavage in mild acid or thermal stress has been discussed above. A conserved aspartic acid residue (D422) in the *Mycobacterium tuberculosis* RecA intein is suggested to have an important role in the process of protein spicing.[28] This aspartic acid residue has an elevated pKa (6.1), which is coupled to the cysteine residue involved in the initial $N \rightarrow S$ shift. This role is different from that of the asparagine residue, which forms the succinimide intermediate prior to peptide bond cleavage.[29] A similar cleavage of an autotransporter protein is obtained from a gram-negative bacterium.[30] In the case of Hbp protein, replacement of one of the

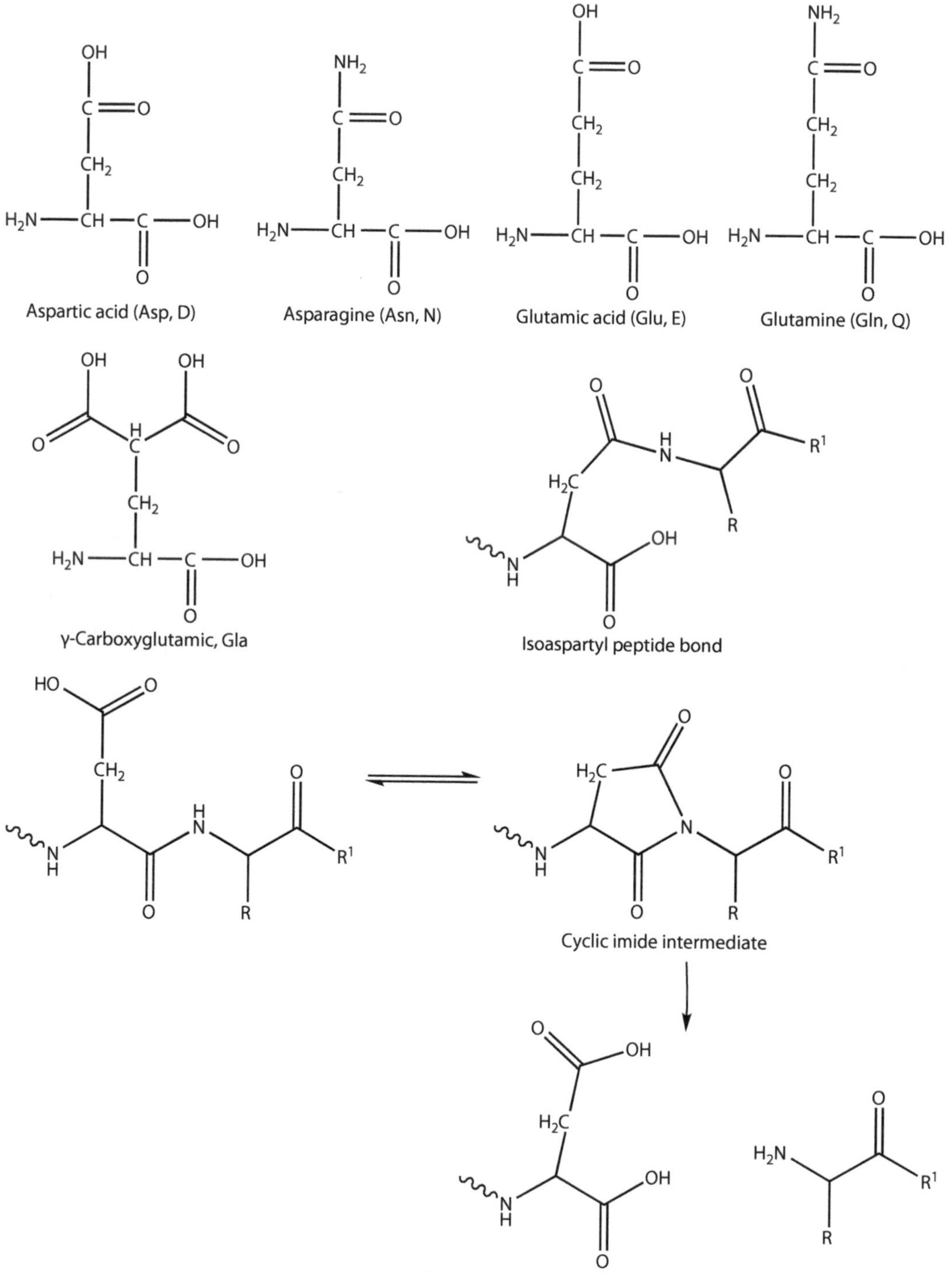

FIGURE 10.1 (continued)

two conserved asparagine residues with an aspartic acid prevents proteolytic cleavage with the release of the passenger domain. The reader is directed to several useful articles on autotransporter proteins.[31,32] Notwithstanding the importance of asparagine residues in the intein cleavage process, there is a report of the presence of glutamine (Chilo iridescent virus [CIV]) and aspartic acid (*Carboxydothermus hydrogenoformans* [Chy]) in the C-terminal position in the place of asparagine.[33] It was observed that replacement of the C-terminal aspartic acid in Chy intein abolished cleavage. These investigators also noted the absence of evidence for an anhydride intermediate in the deamination of glutamine residues in proteins for the generation of a glutaramide intermediate similar to the succinimide intermediate reported for aspartic acid and asparagine.

The carboxyl groups of aspartic acid and glutamic acid are primarily responsible for the negative charges in proteins, and there is evidence for cation-binding exosites on proteins, which are important for specific binding events.[34–39] Zuh and Karlin[40,41] have identified statistically significant charge clusters in the 3D structures of proteins and found roughly equivalent amounts in mammalian and bacterial (*Escherichia coli*) sequences. These acidic clusters are suggested to be important in the binding of metal ions, and other functions have been suggested.[42] The presence of a sequence (Gla domain) with a large number of γ-carboxyglutamic acid residues in the amino-terminal region is unique to vitamin-K-dependent proteins and critical for the function of several of the blood coagulation factors.[43,44] Kyte[45] discussed the importance of charged amino acids such as aspartate and glutamate in heterologous interfaces between proteins where such interactions are transient in nature. Ionic strength also has an effect on charge–charge interactions,[46] and protonation of residues can be considered as a posttranslational modification.[47]

The presence of dicarboxylic amino acids on the surface of proteins and their importance in determining the overall charge of a protein have generated interest in the modification of carboxyl groups resulting either in charge neutralization[48–50] or charge reversal (Figure 10.2).[51–53] Such a change in surface properties can have interesting consequences. Bokhari and coworkers[54] reported a marked increase in the optimum temperature of a carboxymethylcellulase from *Scopulariopsis* sp. on modification of carboxyl groups with aniline. Of somewhat indirect interest is the generation of carboxyl groups on carbon nanotubes with strong oxidizing agents such as mixture of sulfuric acid and nitric acid.[55] The carboxyl groups thus generated can be used for coupling proteins via carbodiimide technology.[56] There are other methods of inserting carboxyl groups on carbon matrices such as the electrochemical deposition of aromatic diazonium salts.[57] This approach was used by Lerner and coworkers[58] for insertion of a 4-carboxyphenyl group on a carbon matrix via electrochemical deposition of 4-carboxybenzene diazonium tetrafluoroborate (Figure 10.2).

The site-specific modification of carboxyl groups in proteins is somewhat difficult to achieve as is the differentiation between aspartyl residues and glutamyl residues. The modification of carboxyl functions in proteins is accomplished by the formation of esters through several different types

FIGURE 10.1 (continued) Structures of dicarboxylic amino acids and related compounds. Shown are the structures of aspartic acid and glutamic acid and their corresponding amide derivatives, asparagine and glutamine. Also shown is the structure of γ-carboxyglutamic acid (Stenflo, J., Fernlund, P., Egan, W., and Roepstorff, P., Vitamin K dependent modifications of glutamic acid residues in prothrombin, *Proc. Natl. Acad. Sci. USA* 71, 2730–2733, 1974; Nelsestuen, G.L., Zytkovicz, T.H., and Howard, J.B., The mode of action of vitamin K. Identification of γ-carboxyglutamic acid as a component of prothrombin, *J. Biol. Chem.* 249, 6347–6350, 1974). Shown at the bottom is a mechanism for the cleavage of aspartic carboxyl peptide bonds in proteins (Piszkiewicz, D., Landon, M., and Smith, E.L., Anomalous cleavage of aspartyl-proline peptide bonds during amino acid sequence determinations, *Biochem. Biophys. Res. Commun.* 40, 1173–1178, 1970; Herrmann, K.A., Wysocki, V.H., and Vorpagel, E.R., Computational investigation and hydrogen/deuterium exchange of the fixed charge derivative tris(2,4,6-trimethoxyphenyl) phosphonium: Implications for the aspartic acid cleavage mechanism, *J. Am. Soc. Mass Spectrom.* 16, 1067–1080, 2005). Also shown is the formation of an isoaspartyl peptide bond from the cyclic imide intermediate (Aswad, D.W., Pananandi, M.V., and Schurter, B.T., Isoaspartate in peptides and proteins: Formation, significance, and analysis, *J. Pharm. Biomed. Anal.* 21, 1129–1136, 2000).

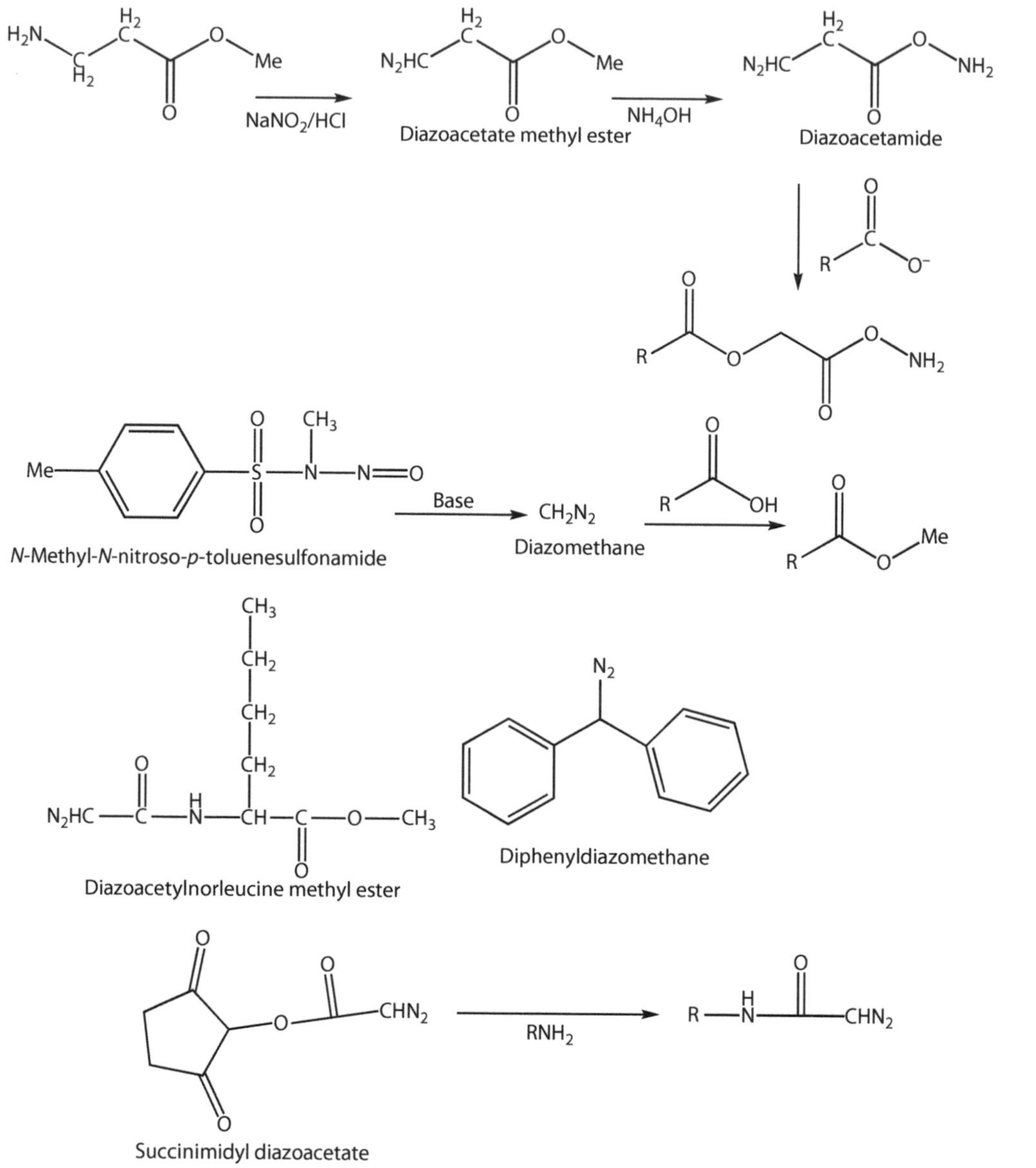

FIGURE 10.2 Diazo compounds used for the modification of carboxyl groups in proteins. Shown at the top is the synthesis of diazoacetamide and subsequent reaction of the diazo aliphatic compound with a carboxyl group (Wilcox, P.E., Esterification, *Methods Enzymol.* 11, 605–617, 1967; Doscher, M.S. and Wilcox, P.E., Chemical derivatives of α-chymotrypsinogen IV. A comparison of the reactions of α-chymotrypsinogen and simple carboxylic acids with diazoacetamide, *J. Biol. Chem.* 236, 1326–1337, 1961). Shown below is the structure of Diazald® (*N*-methyl-*N*-nitroso-*p*-toluenesulfonamide), which is used to generate diazomethane for the modification of carboxyl groups. At the lower left is the structure of diazoacetylnorleucine methyl ester developed as an inhibitor for pepsin (Rajagopalan, T.G., Stein, W.H., and Moore, S., The inactivation of pepsin by diazoacetylnorleucine methyl ester, *J. Biol. Chem.* 241, 4295–4297, 1966). Diphenyldiazomethane was another compound used to modify carboxyl groups in proteins (Delpierre, G.R. and Fruton, J.S., Inactivation of pepsin by diphenyldiazomethane, *Proc. Natl. Acad. Sci. USA* 54, 1161–1167, 1965). Shown below is succinimidyl diazoacetate, which can serve as precursor to diazoacetyl derivatives via coupling to an amino group via NHS technology (Ohiha, A., René, J., Pascard, C., and Badet, B., A new diazoacetylating reagent: Preparation, structure, and use of succinimidyl diazoacetate, *J. Org. Chem.* 58, 1641–1642, 1993). Shown as an example is the coupling with an amine to provide the *N*-diazoacetyl derivative.

of chemistries. Early work on the modification of carboxyl groups in proteins used esterification under somewhat protein-unfriendly conditions. As an example, Carr and coworkers[59] modified the carboxyl groups in insulin with acidic methanol (0.75 M HCl in ethanol: 3 volumes of ethyl alcohol, 1 volume of 3 N HCl) for 30 h at 20°C; reaction was considered complete when maximum inactivation had been obtained. The rate of inactivation could be increased by increasing temperature, acid concentration, or ethyl alcohol concentration; the activity could be recovered by reaction in 0.08 N NaOH at 0°C for 17 h. Subsequent work by Charles and Scott[60] demonstrated inactivation by acid alcohol with partial reactivation; the formation of a methyl–nitrogen derivative was suggested to rationalize the partial reactivation in base. The early work on the chemical modification of insulin was reviewed by Jensen and Evans in 1934.[61] There were several applications of the acid alcohol approach to the modification of proteins following this early work. Fraenkel-Conrat and Olcott[62] reported on the reaction of polyglutamic with various alcohols in the presence of HCl. The extent of esterification with methanol was greater than that observed with ethanol; n-propanol was less reactive, while isopropanol and benzyl alcohol were essentially unreactive. Propylene glycol was as reactive as ethanol in the presence of hydrogen chloride. Acetyl chloride was as effective as hydrogen chloride in promoting the esterification of polyglutamic acid with methanol. These investigators also reported on the methylation of various proteins in the presence of either hydrogen chloride or acetyl chloride; greater than 90% modification was reported for most proteins. Mommaerts and Neurath[63] showed that the careful control of the conditions developed by Fraenkel-Conrat and Olcott[62] could yield the monomethyl ester derivative of insulin. Alexander and coworkers[64] studied the esterification of carboxyl groups in wool with acid/alcohol. The rate of esterification with ethyl alcohol increased with increasing acid concentration. They also observed that the rate and extent of esterification in acid/alcohol decreased as alkyl size increased; approximately 70% of the carboxyl groups were modified by methanol (0.1 M HCl/100°C), while 15% modification was achieved with isopropyl alcohol and 12% with benzyl alcohol. As would be expected, the ester groups were more stable in acid than in base, and some suggested that esters formed with higher molecular weight alcohols such as isopropyl were hydrolyzed in base at a slower rate than those formed with a low molecular weight alcohol such as methanol, but the difference is not striking. In other work, Pénasse and coworkers[65] determined the carboxyl terminal phenylalanine in ovomucoid by analysis of the protein after reduction with lithium aluminum hydride (lithium aluminum hydride; AlH_4Li) and hydrolysis. The C-terminal amino acid is determined as the corresponding amino alcohol following on the earlier work of Fromageot and coworkers.[66] The use of acid/methanol and diazoacetate for the esterification of carboxyl groups in proteins was reviewed by Wilcox in 1967.[67] Diazo compounds are discussed below in greater detail. Acid/alcohol modification has continued to see sporadic use over the last 40+ years. Parsons and coworkers[68] modified the carboxyl groups in lysozyme with ethyl alcohol in the presence of HCl and observed the modification of 5-carboxyl groups and suggested that there is one very labile ester than undergoes rapid hydrolysis. These investigators also evaluated triethyloxonium fluoroborate (Figure 10.3), which is somewhat more specific. Acharya and Vithayalthil[69] modified the carboxyl groups of bovine pancreatic ribonuclease A in anhydrous methanol/HCl as developed by Vithayathil and Richards[70] for ribonuclease S peptide. A dry preparation of protein is stirred into anhydrous methanol yielding a suspension, and HCl in methanol is added to provide a final concentration of 0.1 M HCl. The addition of the acid provided a transient clearing, which subsequently became turbid. Loss of activity was associated with the modification of 5 of the 11 carboxyl groups that were modified; all 11 carboxyl groups were modified after 24 h of reaction.[71] Some immunological reactivity was lost after modification of four carboxyl groups, and the fully methylated product retained 30% immunological reactivity. Subsequent work showed that the initial product of reaction was dimethyl derivatives with esterification at Glu49 and Asp53, which retained approximately 60% of native activity.[72] The reader is also directed toward the paper of Vithayathil and Richards[70] for a consideration of methods for the measurement of methoxy groups. Vakos and coworkers[73] reported on the en vacuo modification of ribonuclease with gaseous methanol/HCl and observed approximately that same extent of modification as observed by Acharya and Vithayalthil[37] for modification in solution. It is also suggested that the en vacuo reaction proceeds more rapidly than the solution phase reaction. It is not known as to whether the same residues, which are modified en vacuo, are those identified in solution. It has been observed that methylation of proteins can occur during the staining of proteins following electrophoresis.[74-77] Haebel and workers[74] observed that under their conditions for Coomassie blue staining, trichloroacetic acid and methanol, there was some specificity for the modification of the γ-carboxyl group of glutamic acid. These in vitro methylation reactions are not to be confused with the in vivo enzyme-catalyzed methylation of carboxylic acids, which is considered to be a useful posttranslational modification.[78-83] The reader is directed to an excellent review[84] on S-adenosylmethionine and the methyltransferases for more information of this process. As an aside, the methylation of proteins during processing for proteomic analysis has resulted in the development of alternatives, which avoid this artifact.[85,86] There has been limited use of esterification in the presence of acid (Fischer–Speier reaction) for the modification of carboxyl groups in proteins in recent years. Two groups have used acid alcohol esterification for determination of C-terminal amino acid residues.[87,88] Another group modified peptides with deuterated methanol or nonisotopically labeled methanol to (1) block carboxyl groups permitting the more efficient isolation of peptides by immobilized metal ion chromatography (IMAC) and (2) to differentially label peptides permitting comparison of samples.[89] This is a variation on ICPL.[90] There is likely more use of esterification for protein modification than can be determined from the journal-based literature; however, I have not used the patent literature for support of the current work. There are two examples of commercial application of Fischer–Speier esterification to proteins. The first is the modification of soy protein to improve processing characteristics in the food industry,[91] while the second is the modification of zein protein for potential use as a biopolymer to replace petroleum-based polymers.[92]

Diazoacetate and related compounds (Figure 10.2) are also used for the formation of ester bonds on protein carboxyl groups.[67] Doscher and Wilcox[93] reported on the reaction of diazoacetamide with α-chymotrypsinogen. There were an average of 3.0 mol of reagent incorporated per mole of protein after reaction at pH 5.5 (pH-stat with perchloric acid). The reaction was allowed to proceed for 24–48 h at which point 95% of the reagent had decomposed. The chromatography of the reaction product on carboxymethyl cellulose demonstrated a heterogeneous product. These investigators also reported on the rates of reaction of diazoacetamide with a variety of carboxylic acids. They noted that the differences in the rate of reaction of the various monocarboxylic acids is not sufficient to explain the difference in the rates of the reaction of the various carboxyl groups in α-chymotrypsinogen. The studies with the various model monocarboxylic acids suggested that the rate of reaction of diazoacetamide was more rapid with undissociated acid that the dissociated acid by perhaps a factor of 5. Modification of amino acid residues other than carboxylic acids was not observed. Riehm and Scheraga[94] modified bovine pancreatic ribonuclease A with diazoacetoglycinamide (Figure 10.2). The reaction was performed at pH 4.5 with perchloric acid (pH-stat) at 10°C. The reaction was allowed to proceed 24 h with addition of diazoacetoglycinamide at 0, 6, and 12 h; the hydrolysis of the diazo compounds is a significant factor.[93] The product of the reaction was heterogeneous and could be separated into several components by ion-exchange chromatography (IRC-50). The major component contained a single modified residue (Asp-53), which was fully active; this residue is also preferentially modified with acid/methanol.[72] Diazoacetylglycinamide had a brief history of use as an alkylating agent for DNA.[95-97] Nguyen and coworkers[98] used diazoacetamide to modify carboxyl groups in a DNA-specific autoantibody, which had activity in the hydrolysis of phosphodiester and glycosidic bonds. Modification of the carboxyl groups in this antibody reduced the activity in the hydrolysis of p-nitrophenyl-β-D-N-acetylglucosamide. Diazomethane is one of the older reagents[99] used for the modification of proteins but is a dangerous chemical (a toxic gas, mp −23°C), which requires drastic conditions (soluble in ether, dioxane)[100] and requires in situ generation of reagent.[101] There is current use of diazomethane in organic synthesis.[102,103] Considering the drastic reaction conditions, there has been limited use of diazomethane for the modification of proteins. In addition to the work on hair mentioned earlier[100] where the reaction was performed in ether with Diazald® (N-methyl-N-nitroso-p-toluene sulfonamide) (Figure 10.2), Osbahr[104] used diazomethane to modify carboxyl groups in bovine fibrinogen. The reaction was performed in 0.3 M sodium phosphate, pH 7.4 at 2°C. This study[104] also described the use of methanol/thionyl chloride for the modification of fibrinogen. Thionyl chloride is commonly used in the acid-catalyzed esterification of carboxyl groups.[105–108] There is one early example of the use of thionyl chloride by Bello for the modification of carboxyl groups in gelatin.[109] I could not find any other application of thionyl chloride/methanol for the direct modification of proteins; thionyl chloride has, however, been used for the preparation of matrices for immobilizing proteins[110–112] and in manufacture of hydrogels.[113] Osbahr also used DMS for the modification of carboxyl groups in fibrinogen. DMS is used for the modification of carboxyl groups in organic chemistry[114,115] and occasionally for the modification of proteins.[116,117] Diazomethane, thionyl chloride/methanol, and DMS have also been observed to modify lysine and tyrosine residues in some of the aforementioned studies. The modification of carboxyl groups in bovine fibrinogen was observed to produce aggregation.[104,118] Osbahr[119] subsequently showed that trimethyloxonium fluoroborate (see in the succeeding text) was a more selective reagent for the modification of carboxyl groups in fibrinogen again resulting in the polymerization of the fibrinogen.

There have been some diazo derivatives that have been shown to be useful for the specific modification of carboxyl groups in proteins. Early work at Yale University had shown that diphenyl-diazomethane (Figure 10.2) could inactivate pepsin[120] and α-chymotrypsin.[121] These results were interpreted as inactivation resulting from the modification of carboxyl groups. The lack of solubility of diphenyldiazomethane in aqueous solvents presented a challenge in the aforementioned studies. Some 60 miles south (more or less) at 66th and York Avenue on the fifth floor of Flexner Hall, T.G.(Raj) Rajagopalan and coworkers[122] found that diazoacetyl-DL-norleucine methyl ester (DANLeu) was a potent inhibitor of pepsin in the presence of cupric ions resulting in the incorporation of 1 mol of norleucine per mole of pepsin (norleucine had been chosen to allow determination of stoichiometry by amino acid analysis after acid hydrolysis of the modified protein); the inactivation reaction proceed much slower (hours instead of minutes) with the incorporation of 3.5 mol of norleucine per mole of protein.

I arrived at the Rockefeller University, and Raj was headed back to India and continued the work on DANLeu under the direction of William H. Stein. This was an experience, which has increased in value over the years. I would be remiss if I did not mention that he took the first draft of the manuscript describing these studies and tossed it into the waste basket; such is described today as a learning moment. I should also mention that he congratulated me once on the synthesis of sodium chloride in one my early attempts at the synthesis of glycyl-norleucine methyl ester. I did eventually meet the Ted Bella standard for C, H, and N—this was long before NMR became the standard. Lundblad and Stein[123] did suggest that the effect of cupric ions was to form a copper-complex carbene intermediate, which then reacted with a carboxyl group at pepsin active site. They also observed that Ag(I) would also stimulate the inactivation reaction and that dimethylsulfonium phenacylide will also form a copper-complexed carbene-inactivated pepsin. These investigators also reported that diazoacetic acid methyl ester rapidly inactivated pepsin in a reaction dependent on the presence of cupric ions, but more than 1 mol of reagent was incorporated in the reaction. These studies were performed with porcine pepsin; Meitner[124] has reported similar results with bovine pepsin for reaction with DANLeu including isolation of a peptide containing a modified aspartic acid residue. DANLeu has been used to characterize other acid proteases[125–139] such that it is considered a *classic* reagent for the modification of aspartic acid proteases. There are several instances where unexpected results were obtained. Niishi and coworkers[140] showed that physaro-lisin, an acid protease from *Physarum polycephalum*, was a serine-carboxyl protease,[141] which is inhibited by both DFP and DANLeu; the DFP reacts with a serine residue while DANLeu reacts with Asp529. In another study where unusual results were obtained, Ackerman and coworkers[142] observed that DANLeu in the presence of cupric ions inhibited neutrophil chemotaxis in response to either pepstatin or formylmethionine peptides.[143] There was no effect on the chemotactic response to zymosan-activated serum, and it was found that DANLeu competed with formyl-Met-Leu-Phe (fMLP) for binding to neutrophils. It is noted that DANLeu was dissolved in DMSO in the presence or absence of anhydrous cupric sulfate; the cupric sulfate was poorly soluble in DMSO and appeared to be solubilized by DANLeu. The stock reagents were diluted 1:1000 into the cell suspensions. The addition of the DANLeu and cupric sulfate separately did not have an effect, and it was postulated that the inactivation of chemotaxis required the formation of a reactive intermediate such as described by Lundblad and Stein.[123] Subsequent work by Ackerman and coworkers[144] showed that *p*-bromophenacyl bromide (see in the succeeding text) or epoxy-*p*-nitrophenoxypropane also inhibited chemotaxis but did not compete for the binding of fMLP. It is noted that proteolysis of fMLP and internalization of the liberated phenylalanine by alveolar macrophages has been reported by Basilan and coworkers.[144] Diazoacetyl-L-phenylalanine methyl ester has also been shown to inactivate pepsin in a reaction stimulated by cupric ions.[145–147] Bayliss and coworkers[145] also observed some incorporation of reagent into pepsinogen at pH 6.0, which was greatly reduced when the reaction was performed at pH 7.0. Structural analysis[148,149] of pepsinogen has provided evidence for a preformed active site in pepsinogen that is exposed during activation to pepsin. Bayliss and coworkers[145] commented on the challenges of isolating the peptide containing the modified aspartic acid residue. The derivative is a glycolic acid ester, which is labile at pH 7 and above resulting in the formation of *N*-glycolyl-L-phenylalanine methyl ester. Thus, the digestion of the modified protein and subsequent peptide fractionation were performed at an acidic pH. The lability of the modified peptide was used by these investigators in *diagonal electrophoresis*[150,151] for final isolation of the modified peptide (the second dimension was performed after treatment with triethylamine). Ouihia and coworkers have developed a novel approach to the synthesis of diazoacetyl derivatives using succinimidyl diazoacetate (Figure 10.2).[152]

Frank Putnam reviewed the chemical modification of proteins in 1953.[153] While this work is dated with respect to some content, other content such as that on factors influencing functional group reactivity is as applicable today as it was 60 years ago. Reagents listed by Putnam[153] for the modification of carboxyl groups included epoxides, mustard gas, acid/alcohol, and diazoacetamide. Acid/alcohol and diazoacetamide have been discussed earlier. There is limited information on the reaction of epoxide with carboxylic acid groups in proteins to form an ester derivative. The reaction of a simple epoxide with a protein would be expected to be slow based on the results of Wu and Soucek[154] on the reaction of cyclohexene oxide with acetic acid in the presence of triethylamine in dichloromethane to form *trans*-2-acetoxy cyclohexenol ($6.7\ \mathrm{M^{-1}\ s^{-1}}$ at 25°C). Given the poor reactivity with carboxyl groups, the several observations on the reaction of epoxides with protein carboxyl groups have involved compounds with affinity for the protein.[155–159] The majority interest in the reaction of epoxides with proteins is directed toward matrix-bound epoxy groups with sulfhydryl groups[160] and amino groups.[161,162] Mateo and coworkers[163] have a recent review of the development of various epoxide-based supports for the immobilization of enzymes. Heath and coworkers[164] have reviewed the epoxide reaction with collagen in the process of leather tanning where it may be a more useful alternative from environmental considerations.

The mustards (nitrogen and sulfur) are powerful alkylating agents (Chapter 2), which can react with carboxyl groups. Moore and coworkers[165] described the reaction between sulfur mustard and carboxyl groups in 1946. This work was performed several years earlier and was not published until the end of WWII. Stanford Moore and William H. Stein stayed at the Rockefeller to form the Moore–Stein laboratory, while Joseph Fruton went north to Yale University. These three investigators with other investigators including Hans Neurath and John Edsall created the field of protein chemistry after their work during WWII. These early observations were extended by Goodlad[166] who observed that the ester product formed with the nitrogen mustard was more labile than that formed with the sulfur mustard. Davis and Ross[167] also reported that esters derived from nitrogen mustards, sulfur mustards, and some epoxides were more labile than the more simple alkyl esters. I could only find two later references on the reaction of nitrogen mustards with proteins.[168,169] There is a recent review[170] on the chemistry of the several mustards, which is quite useful.

The majority of current work on the chemistry of carboxyl groups in proteins has used carbodiimide chemistry. The use of carbodiimide chemistry for protein modification can be said to have developed from the work of Khorana on the synthesis of oligonucleotides[171] and saw early application to the study of carboxyl group function in the study of mitochondrial proteins.[172–176] These latter studies used N,N'-dicyclohexylcarbodiimide (DCCD) (Figure 10.4). The use of the various carbodiimides is discussed in greater detail in the succeeding text. Other reagents, which have been used to study carboxyl group function, include N-ethylphenylisoxazolium tetrafluoroborate (Figure 10.3)[177,178] and Woodward's reagent K (Figure 10.3).[179–181] Woodward's reagent K was developed for peptide synthesis[180] where it provided good yields and retention of configuration. Application to the modification of proteins came later[182] and in that study to the modification of bacterial cell surface carboxyl groups. These investigators also reported that diazomethane provided a greater extent of modification of surface carboxyl groups than observed with Woodward's reagent K. The common characteristic of carbodiimides and the isoxazolium salts is the formation of *active* ester intermediate such as an enol ester on the protein carboxyl groups, which could either rearrange into a stable derivative or react with a nucleophile such as an amino acid derivative such as glycine ethyl ester.

Another reagent used for the modification of carboxyl groups, triethyloxonium tetrafluoroborate (Meerwein's salt) (Figure 10.3), is an alkylating reagent, which can react with a variety of functional groups in proteins including carboxyl groups.[183–186] Triethyloxonium fluoroborate (Meerwein's salt) (Figure 10.3) is used for the modification of carboxyl groups in proteins.[183–186] Triethyloxonium fluoroborate is not specific for the modification of carboxyl groups in proteins; the reagent does preferentially react with carboxylate groups as opposed to carboxyl groups. Chen and Benoitin[187] showed that the reaction of trimethylisozolium tetrafluoroborate with an N-acetylamino acid yielded an imino ether that could be reduced to the free N-ethylamino acid of reduction with sodium borohydride. Hydrolysis of the imino ether resulted in deacylation of the amino acid derivative. Triethyloxonium tetrafluoroborate does appear to be somewhat more specific than other reagents, which modify carboxyl groups.[119] Parsons and coworkers used triethyloxonium fluoroborate to modify the β-carboxyl group of an aspartic residue essential for the enzymatic activity of lysozyme.[68,188] Parsons and coworkers, as with Osbahr,[119] observed that triethyloxonium tetrafluoroborate was more specific than acid/alcohol in the modification of carboxyl groups in lysozyme as it reacts with ionized carboxyl groups; oxonium ions react with anions more rapidly than with neutral (protonated) carboxyl groups. The reaction of triethyloxonium tetrafluoroborate with lysozyme does provide a heterogeneous product (heterogeneity depended somewhat on pH of reaction), which can be resolved into separate species by ion-exchange chromatography (Bio-Rex®-70; cation-exchange resin). One carboxyl group was modified at pH 4.0 resulting in reduced catalytic activity (57% if assayed immediately on elution from the column as the ethyl ester derivative is not stable). Reaction at pH 4.5 yielded three modified derivatives, one apparently identical to the derivative obtained at pH 4.0 and two other derivatives (II, IV), which were essentially inactive; one of the two derivatives reverted to other inactive derivative (II) on standing. This derivative (IV) contained two modified carboxyl groups, which reverted to II (one modified carboxyl group) suggesting an ester group of unusual stability in II. Subsequent work showed that the modified residue is Asp52.[188] There are two carboxyl groups at the active site, one of which is ionized and is considered to be subject to modification with triethyloxonium tetrafluoroborate. The reader is directed to an article by Rebek and coworkers[189] for additional consideration of an active site with multiple carboxyl groups. Paterson and Knowles[23] used trimethyloxonium fluoroborate to determine the number of carboxyl groups in pepsin, which are essential for catalytic activity. This article discusses in some depth the rigorous precautions necessary for the preparation of this reagent. This reagent is highly reactive, and considerable care is required for its introduction into the reaction mixture containing protein. The reaction is performed at pH 5.0 (0.020 M sodium citrate, pH maintained at 5.0 with 2.5 M NaOH). These investigators also report the preparation of the ^{14}C-labeled reagent from sodium methoxide and [^{14}C]-methyl iodide. Waley[190] reported on the inactivation of β-lactamase I with triethyloxonium tetrafluoroborate, N-ethyl-5-phenylisoxazolium-3'-sulfonate, or 1-cyclohexyl-3-(2-morphilinoethyl)-carbodiimide. Triethyloxonium tetrafluoroborate

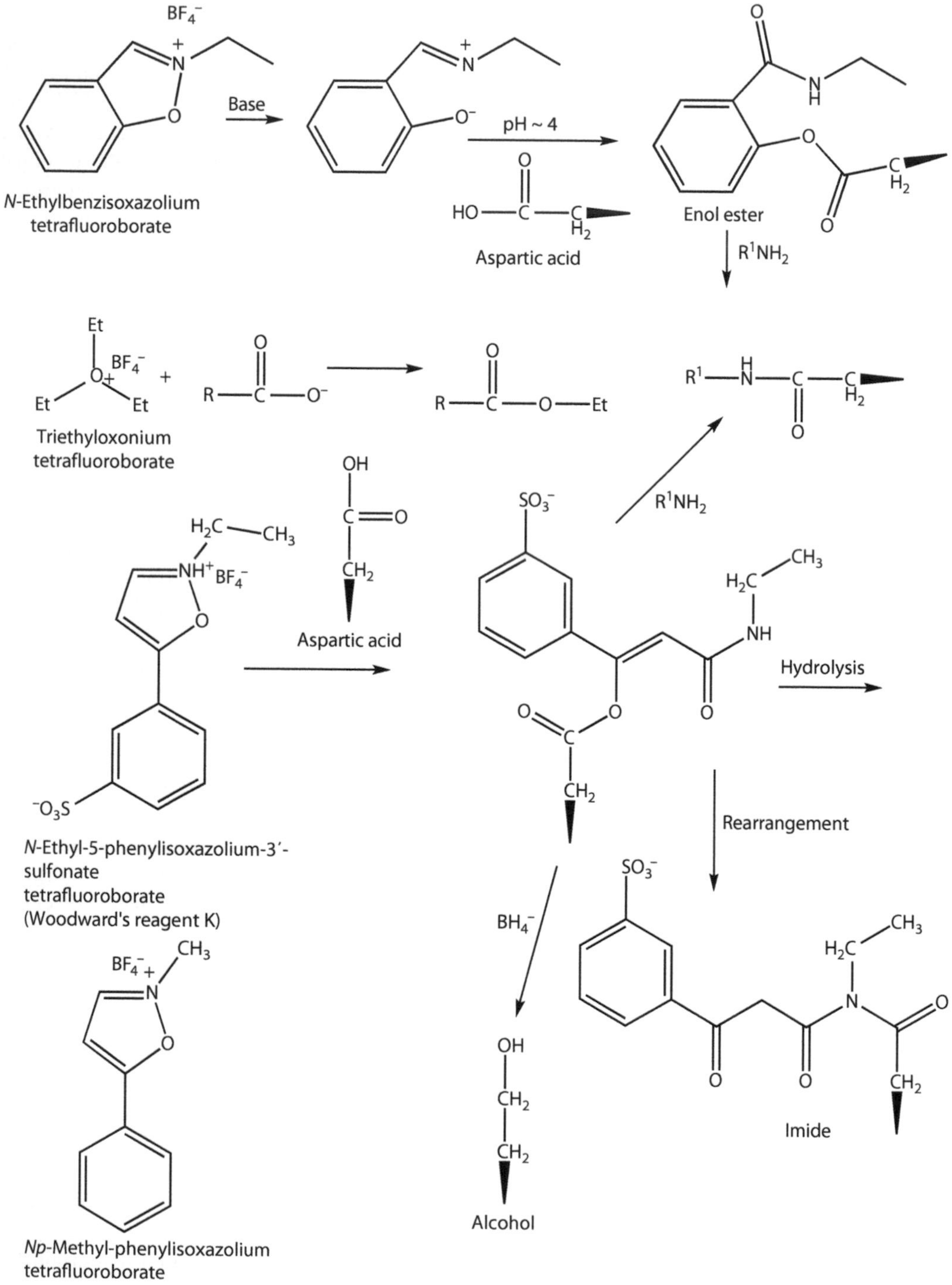

FIGURE 10.3 (continued)

was difficult to work with considering the instability in aqueous media, but the reagent is smaller than other reagents. Llewellyn and Moczydlowski[191] have modified saxiphilin with trimethyloxonium tetrafluoroborate. In consideration of the instability of trimethyloxonium tetrafluoroborate in water,[192] the reaction was performed under stringent conditions. Trimethyloxonium tetrafluoroborate was placed in a tube, which was flushed with nitrogen, and then sealed. A buffered solution (100 mM Tris HCl–100 mM NaCl, pH 8.6) of saxiphilin and bovine serum albumin was introduced, and the reaction was allowed to proceed for 10 min. In the studies on the reaction of lysozyme with triethyloxonium tetrafluoroborate[68] described earlier, Parsons and coworkers either used a single addition of triethyloxonium tetrafluoroborate in acetonitrile to yield a final concentration of 0.2 M while maintaining pH with NaOH (initial pH of 4.0 or 4.5 established with perchloric acid), sequential additions of reagent with pH maintained with NaOH, or reaction with 1 M triethyloxonium tetrafluoroborate with pH maintained with 8 M NaOH. Cherbavez[193] observed that modification of batrachotoxin-activated Na channels with trimethyloxonium tetrafluoroborate altered function. Gómez and coworkers[194] used triethyloxonium tetrafluoroborate to modify carboxyl groups at the active site of a β-xylosidase from *Trichoderma reesei* QM9414. The reagent in dichloromethane was added to the protein in 20 mM MES, pH 5.1. More recently, Kim and coworkers[195,196] used triethyloxonium tetrafluoroborate to synthesize ethyl picolinimidate, which was then used to modified peptide amino groups to improve signal in MALDI MS. Triethyloxonium tetrafluoroborate is reasonably specific for ionized carboxyl groups, but other amino acid functional group may also be modified.[197]

N-ethylbenzisoxazolium tetrafluoroborate was introduced by Kemp and Chien in 1967,[177] but while useful in peptide synthesis, this reagent has seen only limited potential application with proteins.[178] As seen is Figure 10.3, the activation of the carboxyl group with *N*-ethylbenzisoxazolium tetrafluoroborate, as with Woodward's reagent K, is mediated through an enol ester intermediate. Bodlaender and coworkers[198] compared the reaction of *N*-methylbenzisoxazolium tetrafluoroborate with trypsin to that of other isoxazolium salts such as Woodward's reagent K. The reaction of *N*-methylbenzisoxazolium tetrafluoroborate with trypsin was more rapid than that observed with the other isoxazolium salts, but there was more modification of lysine than observed with the other reagents. *N*-methylbenzisoxazolium tetrafluoroborate was also less stable than the other isoxazolium salts undergoing rapid hydrolysis at pH 4; the other isoxazolium salts were stable at pH < 5.

Woodward and coworkers[179–181] developed *N*-ethyl-5-phenylisoxazolium-3′-sulfonate (Woodward's reagent K) (Figure 10.3) and various other *N*-alkyl-5-phenylisoxazolium fluoroborates as reagents for the *activation* of carboxyl groups for synthetic purposes. Dunn and coworkers[199] have studied the kinetics of the aqueous hydrolysis of this reagent and reaction with staphylococcal nuclease. The results in this study suggested that Woodward's reagent K is very unstable in aqueous

FIGURE 10.3 (continued) Structure and reactions of *N*-ethylbenzisoxazolium tetrafluoroborate, *N*-triethyloxonium tetrafluoroborate, and *N*-ethyl-5-phenylisoxazolium-3′-sulfonate (Woodward's reagent K). Shown at the top is the reaction of a carboxylic acid with *N*-ethylbenzisoxazolium tetrafluoroborate to form an enol ester, which then can react with a nucleophile (an alkyl amine is shown) to form a peptide bond (Kemp, D.S. and Chien, S.W., A new peptide coupling reagent, *J. Am. Chem. Soc.* 89, 22743–2744, 1967; Fieser, L.F. and Fieser, M., *Reagents for Organic Synthesis*, Vol. 1, p. 364, John Wiley & Sons, New York, 1967). Shown below is the reaction of triethyloxonium tetrafluoroborate with a carboxyl group to yield an ethyl ester (Parsons, S.M., Jao, L., Dahlquist, F.W. et al., The nature of amino acid side chains which are critical for the activity of lysozyme, *Biochemistry* 8, 700–712, 1969). Shown below is the reaction of *N*-ethyl-5-phenylisoxazolium-3′-sulfonate with aspartic acid. The resulting enol ester can undergo rearrangement to the imide, undergo hydrolysis to the enol ester with the release of reagent, undergo substitution with a nucleophile (an amine is shown) to yield a stable derivative, or be reduced with borohydride to yield the derivative alcohol (Bodlaender, P., Feinstein, G., and Shaw, E., The use of isoxazolium salts for carboxyl group modification in proteins. Trypsin, *Biochemistry* 8, 4941–4948, 1969; Jennings, M.L. and Anderson, M.P., Chemical modification and labeling of glutamate residues at the stilbenedisulfonate site of human red blood cell band 3 protein, *J. Biol. Chem.* 262, 1691–1697, 1987).

solution above pH 3.0. Studies on the rate of enzyme inactivation by this reagent should be corrected for reagent hydrolysis to obtain accurate second-order rate constants. Bodlaender and coworkers[198] used N-ethyl-5-phenylisoxazolium-3-sulfonate, N-methyl and N-ethyl derivatives of 5-phenylisoxazolium fluoroborate, or N-methylbenzisoxazolium fluoroborate to modify carboxyl groups in trypsin. The reaction is more specific for carboxyl groups when performed at pH < 4.75; the derivative enol esters (Figure 10.3) can subsequently react with nucleophiles such as glycine ethyl ester to form stable derivatives. The enol esters can also rearrange to form a stable imide derivative. The N-alkyl substituent is retained with the formation of the imide derivative but lost on reaction with a nucleophile. The amount of enol ester or imide was determined by amino acid analysis for either methyl amine or ethylamine depending on the nature of the N-alkyl group. The extent of modification obtained ranged from approximately three residues modified with (N-metyl-5-phenylisoxazolium fluoroborate or N-ethyl-5-phenylisoxazolium fluoroborate, pH 3.80, 20°C, 80 min) to approximately 11 residues modified with (N-methyl-5-phenylisoxazolium fluoroborate, pH 6.0, 20°C, 10 min). The modification appears fairly selective for carboxyl groups, although some modification of lysine was observed under conditions where extensive modification was obtained (250-fold molar excess of N-methyl-5-phenylisoxazolium fluoroborate, pH 4.75, 20°C, 72 min); greater modification of lysine was obtained with the use of N-ethylbenzisoxazolium tetrafluoroborate. Bodlaender and coworkers[198] also showed that Tris[tris(hydroxymethylaminomethyl)methane] had a marked influence on the stability of the enol ester derivative catalyzing the rearrangement to the imide and some hydrolysis of the enol ester. Subsequently, Feinstein and coworkers[200] showed that solvent conditions (pH) and reagent structure influenced the specificity of modification. The use of either N-methyl-5-phenylisoxazolium tetrafluoroborate (Figure 10.3) or N-ethyl-5-phenylisoxazolium tetrafluoroborate at pH 3.8 resulted in the modification of 2–3 carboxyl groups in trypsin with complete loss of catalytic activity. The enol esters could be converted to an amide by reaction with glycine ethyl ester or O-methylhydroxylamine; reaction with N-methylhydroxylamine removed the reagent with restoration of catalytic activity. Saini and Van Etten[201] reported on the reaction of N-ethyl-5-phenylisoxazolium-3′-sulfonate with human prostatic acid phosphatase. The modification was performed with a 4,000–10,000-fold molar excess of reagent in 0.020 M pyridinesulfonic acid, pH 3.6 at 25°C. Ethylamine was utilized as the attacking nucleophile to determine the extent of modification. A substantial number of carboxyl groups in the protein were modified under these experimental conditions. Arana and Vallejos[202] have compared the reaction of chloroplast coupling factor with Woodward's reagent K and DCCD. Reaction with Woodward's reagent K was accomplished at 25°C in 0.040 M Tricine, pH 7.9, while reaction with DCCD was accomplished at 30°C in 0.040 M MOPS, pH 7.4. ATP and derivatives such as ADP and inorganic phosphate protect against the loss of activity occurring upon reaction with Woodward's reagent K, but they do not have any effect on inactivation by DCCD. The reverse was seen with divalent cations such as Ca^{2+}. The modification of an essential carboxyl group in pancreatic phospholipase A_2 by Woodward's reagent K has been reported.[203] The reaction was performed in 0.01 M sodium phosphate, pH 4.75 (pH-stat) at 25°C. A second-order rate constant of 25.5 M^{-1} min^{-1} was obtained for the loss of catalytic activity. This rate inactivation increased more than twofold in the presence of 30 mM $CaCl_2$ (69.3 M^{-1} min^{-1}). Quantitative information on the extent of modification is obtained with [^{14}C]-glycine ethyl ester. It is of interest that treatment with a water-soluble carbodiimide, 1-(3′-dimethylaminopropyl)-3-ethylcarbodiimide, results in the loss of catalytic activity in a reaction with characteristics different from those seen with Woodward's reagent K. Kooistra and Sluyterman[204] modified guanidated mercuripapain with N-ethylbenziosoxazolium tetrafluoroborate at pH 4.2, 0°C, to yield a protein modified with N-ethylsalicylamide esters. When the active ester groups on the protein were allowed to undergo aminolysis (2.0 M ammonium acetate, pH 9.2) and the esters were converted to the corresponding amides (an isosteric modification), the conversion to the amides resulted in the loss of a bulky substituent group with an increase in activity. Johnson and Dekker[205] have reported the modification of histidine and cysteine residues in L-threonine dehydrogenase. Woodward's reagent K has been used by several groups[206–209] to modify Glu681 in human erythrocyte band 3 protein. Jennings and Anderson[206] showed that the enol ester derivative could be reduced with borotriitide to yield a radiolabeled alcohol of the modified amino acid (Figure 10.3).

The use of carbodiimide-mediated modification of carboxyl functional groups in proteins[210,211] (Table 10.1; Figure 10.4) may be used more than other methods for the study of such functional groups; carbodiimide chemistry is used more for the attachment of proteins to matrices and for the preparation of conjugates. Carbodiimides most likely react with protonated carboxyl groups to yield an activated intermediate, an *O*-acylisourea, which can then react with a nucleophile such as an amine (Figure 10.4).[212] The *O*-acylisourea may also rearrange to form the more stable *N*-acylurea (Figure 10.4).[213] Weare and Reichert[213] also observed that the *N*-acylurea would bear a positive charge at neutral pH such that the modification would represent charge reversal as opposed to the derivative obtained from the reaction of a nucleophile such as glycine methyl ester which would yield a neutral derivative. Chan and coworkers[214] reported that the presence of a nucleophile such as glycinamide decreased the rate of inactivation of thrombin by EDC (75 mM HEPES–25 mM MES, pH 6.0), while the presence of hydroxylamine essentially prevented the inactivation of thrombin by EDC. Faijes and colleagues[215] studied the modification of carboxyl groups in *Bacillus licheniformis* 1,3–1,4-β-glucanase. Modification of the native enzyme with EDC (20 mM MES–20 mM HEPES–40 mM 4-hydroxy-1-methylpiperidine, pH 5.5, 25°C) resulted in the rapid loss of activity to a residual level of 23%. This loss of activity was irreversible most likely reflecting the formation of the *N*-acylurea product. However, if glycine ethyl ester was included, the amount of residual activity was reduced to 8%. It was concluded that the loss of activity reflected the modification of two carboxyl groups at the active site where modification of one carboxyl resulted in a total loss of activity, while modification of the other carboxyl group resulted in a partial loss of activity; the modification of either of the carboxyl groups precluded modification of the other carboxyl group.

Carbodiimides are also used for zero-length cross-linking of proteins between proximate lysine residues and carboxyl groups.[216–221] As will be referenced in greater detail in the succeeding text, carbodiimides are also used for forming peptide bonds between carboxyl groups on matrices and amino groups on ligands[222] and for conjugates.[223] There is also use of carbodiimide chemistry in the manufacture of functional hydrogels.[224–227] The study by Hwang and Lyubovitsky[227] used Raman spectroscopy to characterize the modification of collagen in the hydrogel. There can be stability problems with the *O*-acylurea intermediate for use in bioconjugation studies,[228] which can be addressed by the formation of an NHS intermediate.[229–232] In one of these studies,[230] the carboxyl groups modified were on the ligand, a tridentate chelator, which reacts with a lysine residue on a f(ab) or f(ab′)$_2$ fragment.

Water-insoluble carbodiimides such as DCCD (Figure 10.5) have a long history of use for the modification of carboxyl groups,[233–235] but my sense is that there is a current preference for the use of water-soluble carbodiimides such as 1-ethyl-3-(3-dimethylaminopropyl) carbodiimide (*N*-ethyl-*N*′-(dimethylaminopropyl) carbodiimide; EDC) (Figure 10.5).[236–239] There is, however, sustained use of DCCD for the construction of hydrogels and the study of membrane proteins. One remarkable application of DCCD to the study of membrane proteins was described by Kluge and Dimroth in their studies on the F_1F_0 ATPase.[240,241] The pH optimum for the reaction was found to be 7.0, which decreased to pH 6.2 in the presence of 0.5 mM Na$^+$. The initial work[240] with DCCD indicated a very rapid reaction. Subsequent study of the reaction provided a second-order rate constant of 1.2×10^5 M^{-1} min^{-1} at pH 5.6/0°C. This is a very rapid reaction as the reaction of carbodiimide with carboxyl groups can be a slow reaction occurring for a period of hours.[242] In this study, Tohri and coworkers[242] modified the carboxyl groups in a 23 kDa protein derived from the spinach photosystem II in a reaction with EDC and glycine methyl ester at pH 6.2, which proceeded for 12 h at 20°C. In order to put the results obtained by Kluge and Dimroth[241] in some perspective, Volkova and coworkers[243] reported a rate of reaction of DCCD with human acetyl cholinesterase of 2.2 M^{-1} min^{-1} (pH 7.5, 25°C) (the estimated rate from the F_1F_0 ATPase is approximately 1.0×10^5 M^{-1} min^{-1} at pH 7.5, 25°C[240]). Huynh[244] reported a rate of 2.2 M^{-1} min^{-1} (pH 5.5, 25°C) for EDC in the presence of glycine methyl ester for the inactivation of 5-enolpyruvylshikimate-3-phosphate synthase from

TABLE 10.1
Reaction of Carbodiimides with Some Proteins

Protein	Reaction Conditions	Modification	Reference
Lysozyme	pH 4.75 at 25°C, pH-stat,[a] 0.1 M BDC[b] with 1.0 M glycine methyl ester	2.1 (5 min)[c] 4.7 (1 h)[c] 8.1 (5–6 h)[c]	10.1.1
Lysozyme	pH 4.75 at 25°C, pH-stat, 0.1 M BDC with 0.1 M nitrotyrosine ethyl ester	1.4[c,d]	10.1.1
Trypsin	pH 4.75 at 25°C, pH-stat, 0.1 M BDC with1.0 M glycine methyl ester	4.6 (5 min)[c] 8.8 (1 h)[c] 12.5 (5–6 h)[c]	10.1.1
Chymotrypsin	pH 4.75 at 25°C, pH-stat,[a] 0.1 M BDC[b] with 1.0 M glycine methyl ester	6.2 (5 min)[c] 11.8 (1 h)[c] 15.5 (5–6 h)[c]	10.1.1
Chymotrypsinogen	pH 4.0 at 25°C (pH-stat), EDC[e] with 1.0 M glycine ethyl ester pH 5.0 at 25°C (pH-stat), EDC with 1.0 M glycine ethyl ester pH 6.0 at 25°C (pH-stat), EDC with 1.0 M glycine ethyl ester pH 8.0 at 25°C (pH-stat), EDC with 1.0 M glycine ethyl ester	13[f] 10[f] 3[f] 1[f]	10.1.2
Chymotrypsin	pH 4.75 at 25°C (pH-stat) EDC, with 1.0 M glycine methyl ester	12.7 15.6[g]	10.1.3
Chymotrypsinogen	pH 4.75 at 25°C (pH-stat) EDC, with 1.0 M glycine methyl ester	10.6 13.5[g]	10.1.3
Lysozyme	pH 4.75 at 25°C (pH-stat) with BDC and 0.25 M aminomethanesulfonic acid	8.5–9.5[h]	10.1.4
Lysozyme	pH 4.75 at 25°C (pH-stat) with BDC and 1.0 M glycine methyl ester	6.5–7.5[h]	10.1.4
Trypsin	pH 4.75 at 25°C (pH-stat), EDC with 1.0 M glycinamide, 3 h reaction	7.9 7.1[i] 3–5 Tyr[j]	10.1.5
Chymotrypsin	pH 4.0 (pH-stat)at 25°C, EDC with glycine ethyl ester	15[k]	10.1.6
Phosphorylase *b*	pH 5.1 (1 mM glycerophosphate with 0.5 mM EDTA) at 25°C, CMC with glycine ethyl ester[l]	3[m]	10.1.7

[a] It is generally necessary to add dilute HCl (0.2 M) during the course of the reaction. This was best accomplished by the automatic addition of acid by a pH-stat. This allows the maintenance of constant pH in a reaction without the need for a buffer.

[b] BCD, N-benzoyl-N'-(3-ethylaminopropyl) carbodiimide.

[c] Portions of the reaction mixtures were taken into 10 volumes of 1.0 M sodium acetate, pH 4.75, to terminate the reaction at indicated times.

[d] Time of reaction was not given. Based on results from other investigators (Pho, D.B., Routan, C., Toi, A.N., and Pradel, L.A., Evidence for an essential glutamyl residue in yeast hexokinase, *Biochemistry* 16, 4533–4537, 1977), the use of nitrotyrosine ethyl ester as a nucleophile for reaction with the *O*-acylurea intermediate can affect reaction rate but not stoichiometry.

[e] EDC, N-ethyl-N'-(3-dimethylaminopropyl) carbodiimide (1-ethyl-3-(3-dimethylaminopropyl) carbodiimide).

[f] Determined by incorporation of ^{14}C-glycine ethyl ester. There is marked decrease in incorporation between pH 4 and pH 5 and then a second marked decrease at pH 6.5 dropping to a value of approximately 1 at pH 8.0.

[g] This value was obtained from reaction in 7.5 urea.

[h] Note the difference in the extent of modification depending on nucleophile.

[i] This reaction was performed in the presence of benzamidine, a competitive inhibitor of trypsin. The reaction of this modified enzyme with ^{14}C-labeled glycinamide resulted in the modification of ASP177.

[j] The tyrosine residues were regenerated in 0.5 M hydroxylamine, pH 7.1, without effect on the inactivation of the enzyme from reaction with EDC/glycinamide.

TABLE 10.1 (continued)
Reaction of Carbodiimides with Some Proteins

Protein	Reaction Conditions	Modification	Reference

k The modification of all available carboxyl groups was accomplished by reaction at pH 4.0 on the assumption, and such groups would be protonated and available for reaction with the carbodiimide.

l The presence of the nucleophile, either glycine ethyl ester or N-(2,4-dinitrophenyl)-ethylenediamine, accelerated the rate of reaction. This has been observed with the reaction of nitrotyrosine ethyl ester and CMC with other enzymes (c.f. Lacombe, G., Van Thiem, N., and Swynghedsauw, B., Modification of myosin subfragment 1 by carbodiimide in the presence of a nucleophile. Effect on adenosinetriphosphatase activities, *Biochemistry* 20, 3648–3655, 1981).

m The use of ^{14}C-labeled CMC showed the incorporation of 3 mol of reagent per mole of monomer (phosphorylase *b* is a dimer) at complete inactivation. In the presence of N-(2,4-dinitrophenyl)-ethylenediamine, 2 mol of CMC and 1 mol of N-(2,4-dinitrophenyl)-ethylenediamine are incorporated at complete inactivation.

References to Table 10.1

10.1.1. Hoare, D.G. and Koshland, D.E., Jr., A procedure for the selective modification of carboxyl groups in proteins, *J. Am. Chem. Soc.* 88, 2057–2058, 1966.

10.1.2. Abita, J.P., Maroux, S., Delaage, M., and Lazdunski, M., The reactivity of carboxyl groups in chymotrypsinogen, *FEBS Lett.* 4, 203–206, 1969.

10.1.3. Carraway, K.L., Spoerl, P., and Koshland, D.E., Jr., Carboxyl group modification in chymotrypsin and chymotrypsinogen, *J. Mol. Biol.* 42, 133–137, 1969.

10.1.4. Lin, T.-Y. and Koshland, D.E., Jr., Carboxyl group modification and the activity of lysozyme, *J. Biol. Chem.* 244, 505–508, 1969.

10.1.5. Eyl, A.W., Jr. and Inagami, T., Identification of essential carboxyl groups in the specific binding site of bovine trypsin by chemical modification, *J. Biol. Chem.* 246, 738–746, 1971.

10.1.6. Johnson, P.E., Stewart, J.A., and Allen, K.G.D., Specificity of α-chymotrypsin with exposed carboxyl groups blocked, *J. Biol. Chem.* 251, 2353–2362, 1976.

10.1.7. Ariki, M. and Fukui, T., Modification of rabbit muscle phosphorylase b by a water-soluble carbodiimide, *J. Biochem.* 83, 183–190, 1978.

Escherichia coli. It is suggested that the inactivation is associated with the modification of Glu418. Chen and coworkers[245] observed that thymidylate synthase was inactivated by EDC at a rate of 2.1 M^{-1} min^{-1} (50 mM MES, pH 6.0, 30°C), while the rate with 1-phenyl-3-(3-trimethylaminopropyl) carbodiimide was 200 M^{-1} min^{-1}. Bulone and coworkers[246] reported a rate of 0.29 M^{-1} s^{-1} (EDC, pH 4.5–5.5) for the inactivation of a $(1 \rightarrow 3)$-β-glucan synthase. These investigators demonstrated a rapid decrease in reaction rate above pH 5.5. The extent of inactivation of the $(1 \rightarrow 3)$-β-glucan synthase was increased by the inclusion of glycine methyl ester in the reaction with the carbodiimide; there was little inactivation with Woodward's reagent K. I could find one report[247] on the rate of reaction of a carbodiimide with several model compounds containing a carboxylic acid function where values of 1.3–3.02 M^{-1} min^{-1} were obtained for some derivatives of phthalamic acid.

It is clear from a consideration of the studies that the rate of reaction of DCCD with the F_1F_0-ATPase reported by Kluge and Dimroth[241] is at least three orders of magnitude faster than that the carbodiimide modification of other protein carboxyl groups. In subsequent work,[248] Kluge and Dimroth reported on the reaction of DCCD with isolated subunit c monomer obtained from the F_1F_2-ATPase. There are several observations of interest from this study. First, DCCD was observed to react with the isolated subunit c monomer in chloroform/methanol solvent. The protein was transferred into an aqueous solvent in the presence of dodecyl-β-D-maltoside. The isolated subunit did not have catalytic activity, but a rate for reaction of 70 M^{-1} min^{-1} was obtained for the reaction with DCCD through the use of radiolabel. Subsequent work by Meier and coworkers[249] studied the reaction of DCCD with the undecameric c subunit obtained from the F_1F_0-ATPase by detergent dissociation. The reactive glutamic acid (Glu65) can be modified in the isolated subunit (reconstituted

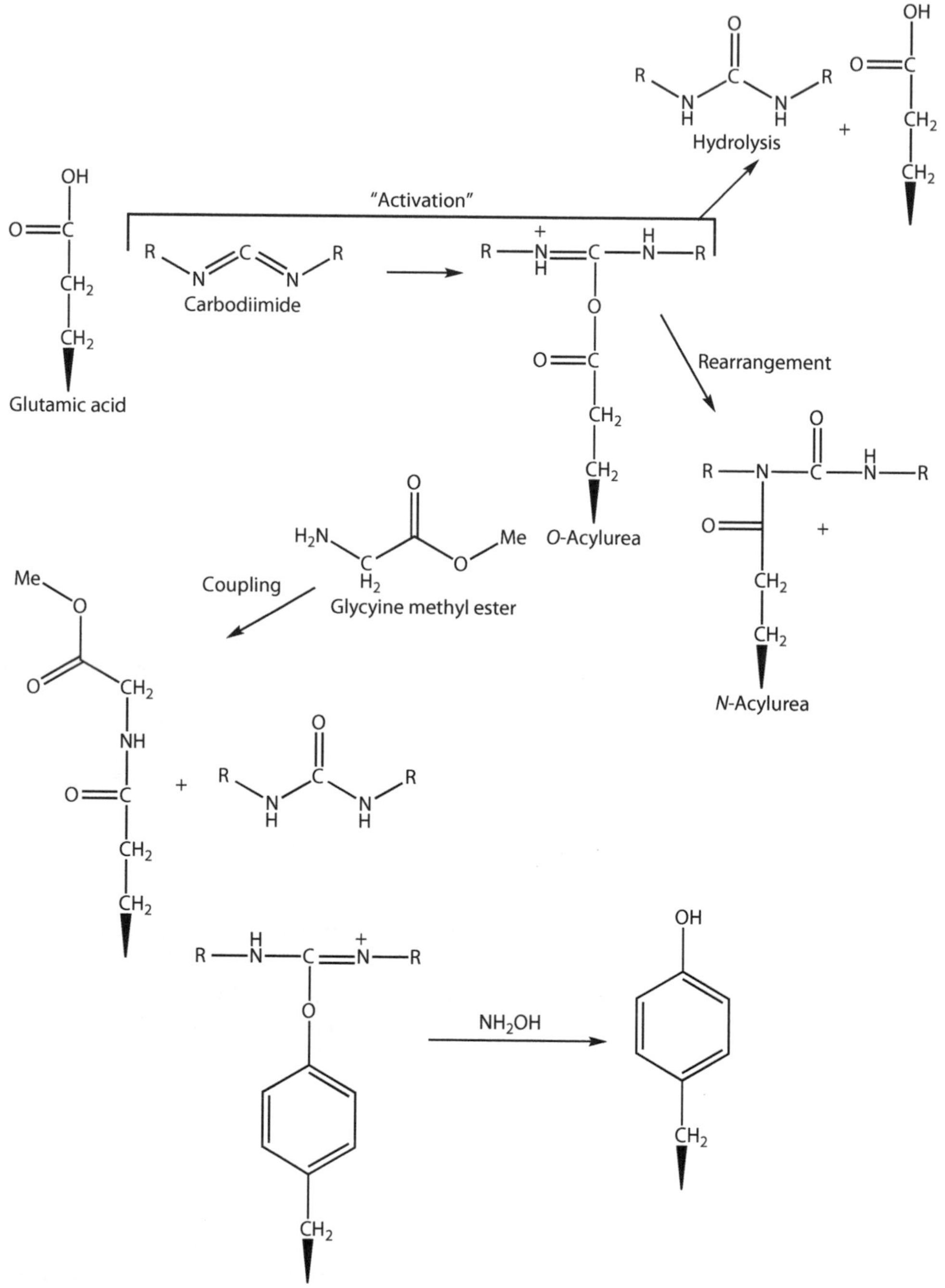

FIGURE 10.4 The modification of a dicarboxylic acid with a carbodiimide. Shown is the formation of the O-acyl derivative that is an active ester intermediate, which may undergo hydrolysis to regenerate the original carboxylic acid and a urea, rearrange to form an N-acyl derivative, or react with a suitable nucleophile (glycine methyl ester is shown as an example) (Hoare, D.G. and Koshland, D.E., Jr., A method for the quantitative modification and estimation of carboxylic acid groups in proteins, *J. Biol. Chem.* 242, 2447–2453, 1967; Carrway, K.L. and Koshland, D.E., Jr., Carbodiimide modification of proteins, *Methods Enzymol.* 25, 616–623, 1972). At the bottom is a structure suggested for the reaction of a carbodiimide with tyrosine (Carraway, K.L. and Koshland, D.E., Jr., Reaction of tyrosine residues in proteins with carbodiimide reagents, *Biochim. Biophys. Acta* 160, 272–274, 1968). This derivative can be converted back to tyrosine by neutral hydroxylamine.

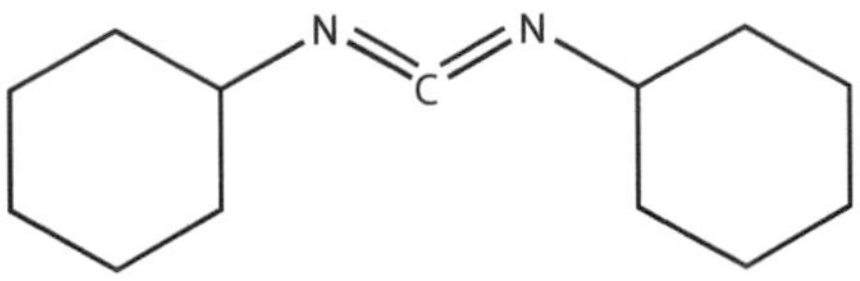

1,3-Dicyclohexylcarbodiimide (DCCD)

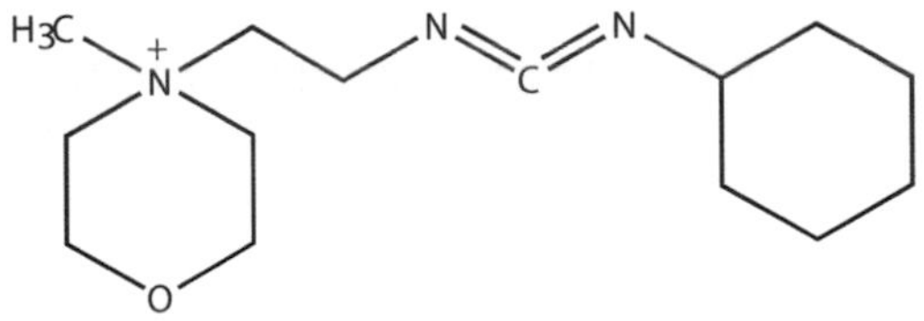

1-Cyclohexyl-2-(2-morpholinethyl)-carbodiimide

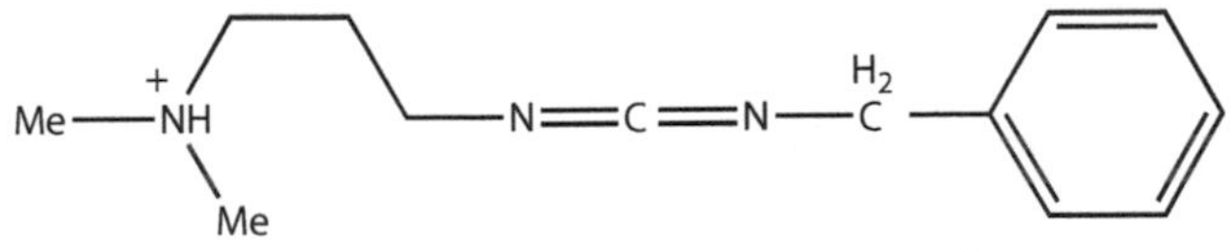

N-Benzyl-N′-(3-dimethylpropylcarbodiimide)

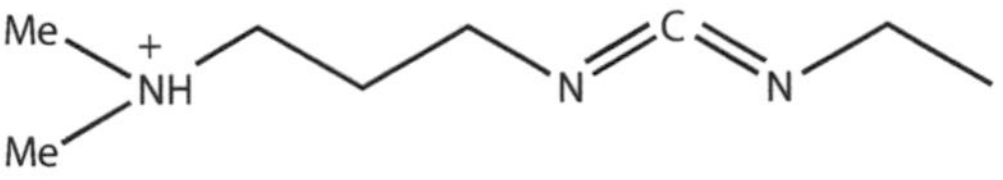

1-Ethyl-3-(3-dimethylaminopropyl)carbodiimide (EDC)
N-Ethyl-N′-(dimethylaminopropyl)carbodiimide

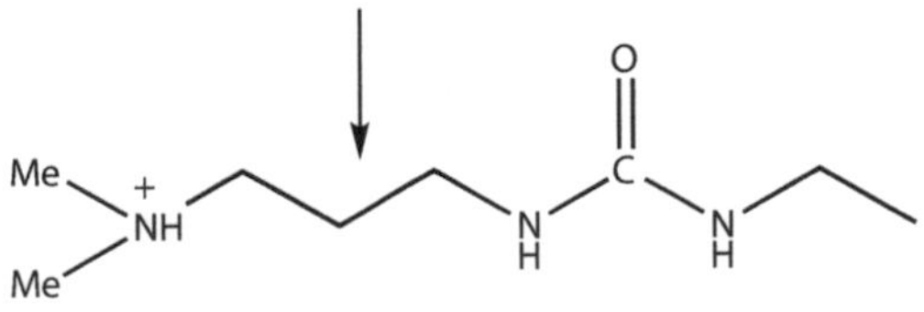

N-Ethyl-N′-(dimethylaminopropyl)urea

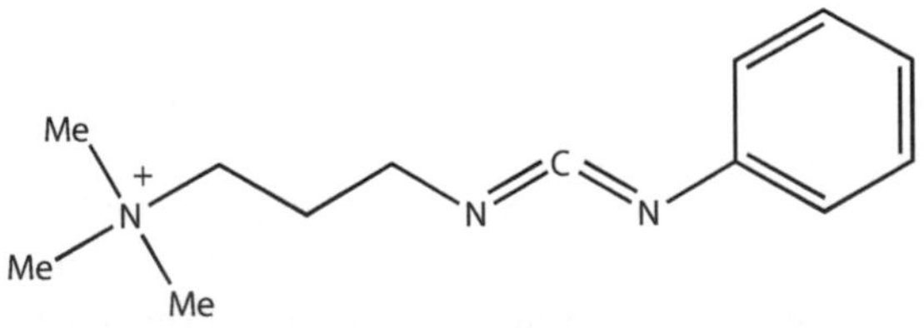

N-Phenyl-N′-(3-trimethylaminopropyl)carbodiimide

FIGURE 10.5 Structures of some commonly used carbodiimides and product of reactions with some amino acids. Shown is the structure of DCCD in a hydrophobic reagent (Rea, P.A., Griffith, C.J., and Sanders, D., Purification of the N,N′-dicyclohexylcarbodiimide-binding proteolipid of a higher plant tonoplast proton ATPase, *J. Biol. Chem.* 262, 14745–14752, 1987). Solubility issues that prompted the synthesis of N-cyclohexyl-N′-(2-morpholinoethyl) carbodiimide metho-*p*-toluenesulfonate (Sheehan, J.C. and Hlavka, J.J., The use of water-soluble and basic carbodiimides in peptide synthesis, *J. Org. Chem.* 21, 439–441, 1956). N-benzyl-N′-(3-dimethylaminopropyl)carbodiimide (Hoare, D.G. and Koshland, D.E., Jr., A method for the selective modification of carboxyl groups in proteins, *J. Am. Chem. Soc.* 88, 2057–2058, 1966) and N-ethyl-N′-(3-dimethylaminopropyl) carbodiimide (EDC) (Wilchek, M., Frensdorff, A., and Sela, M., Modification of the carboxyl groups of ribonuclease by attachment of glycine or alanylglycine, *Biochemistry* 6, 247–252, 1967) were later developed. EDC is like the most frequently used water-soluble carbodiimide. Also shown is the structure of the substituted urea derived from the hydrolysis of the parent carbodiimide (Lei, Q.P., Lamb, D.H., Heller, R.K. et al., Kinetic studies on the rate of hydrolysis of N-ethyl-N′-(dimethylaminopropyl) carbodiimide in aqueous solutions using mass spectrometry and capillary electrophoresis, *Anal. Biochem.* 310, 122–124, 2002). N-Phenyl-N′-(3-trimethylaminopropyl) carbodiimide (Chen, D.H., Daron, H.H., and Aull, J.L., 1-Phenyl-3-trimethylaminopropyl carbodiimide: A new inhibitor of thymidylate synthase, *J. Enzyme Inhib.* 5, 259–268, 1992) is shown at the bottom.

into proteoliposomes) and shows reaction characteristic to the native enzyme. It is noted that the isolated undecameric c subunit does not have catalytic activity, but the modified enzyme can be separated from the unmodified protein by HPLC on a weak anion-exchange matrix (WAX300 column, SynChrom). It is of interest that the undecameric c subunit retains the reactivity of the native protein with DCCD, which is absent in the monomer.[248]

Water-soluble carbodiimides were developed by Sheehan and Hlavka.[250] Previous work from this laboratory had demonstrated the value of carbodiimide chemistry in the synthesis of peptides using DCCD.[251] Difficulty in separating peptide product from the acylurea derived from the reagent prompted the development of water-soluble carbodiimides by the insertion of a tertiary or quaternary amine yielding 1-ethyl-3-(2-morpholinyl-(4)-ethyl)carbodiimide metho-p-toluenesulfonate. It is of interest that the first application of this reagent to proteins was the zero-length cross-linking of collagen.[252] Riehm and Scheraga[253] advanced the use of water-soluble carbodiimides for proteins in a study on the modification of ribonuclease with 1-cyclohexyl-3-(2-morpholinylethyl)carbodiimide at pH 4.5 with a pH-stat. A number of different products were obtained, which could be separated by ion-exchange chromatography (BioRex® 70). These investigators also suggested that the mechanism of reaction between a carbodiimide and a carboxyl group resulted in the formation of an unstable acylisourea, which would either decompose to an acylurea derivative or react with a nucleophile. Hoare and Koshland[254] used N-benzyl-N'-3-dimethylaminopropylcarbodiimide. Subsequent studies used EDC[255] and resulted in the development of a quantitative method for the measurement of carboxyl groups in proteins. These initial studies also introduced the concept of using a unique nucleophile such as norleucine methyl ester, aminomethanesulfonic acid, and norvaline. These initial studies used 0.1 M carbodiimide, pH 4.75 with 1.0 M glycine methyl ester in a pH-stat at 25°C. The possibility of a side reaction was discussed with reference to the possible modification of the phenolic hydroxyl of tyrosine to form the O-arylisourea (see the following text). Few, if any, investigators perform the reaction in a pH-stat although, again as noted in the succeeding text, there are issues with respect to the effect of buffer components. Many investigators use the *Good* buffers such as MES [2-(N-morpholino)ethanesulfonic acid][256,257] developed by Norman Good and colleagues.[258,259] The studies by Moffett and coworkers[256] used EDC as a tissue fixative. There is significant use of carbodiimides as tissue fixatives for histochemistry likely based on their ability to form zero-length cross-links.[260,261]

The effect of buffer and pH on carbodiimide stability was reported by Metz and Brown in 1969[262] as part of a study on the use of carbodiimides for the study of the conformation of nucleic acids. Metz and Brown investigated the effect of solvent conditions on the stability of N-cyclohexyl-N'-β-(4-methylmorpholino)ethylcarbodiimide (p-toluenesulfonate salt). The $t_{1/2}$ was 240 h in H_2O; 110 h in 0.1 M sodium borate, pH 8.0; 32 h in 0.1 M Tris, pH 7.6; 0.92 h in 0.1 M sodium phosphate, pH 7.0; 3.4 h in 0.1 M sodium cacodylate, pH 7.0; 1.0 h in 0.1 M sodium cacodylate, pH 6.0; and 0.19 h in 0.1 M sodium acetate, pH 5.0 (all at 30°C). Instability in the presence of phosphate is somewhat obvious as the reaction of carbodiimide with phosphate is well documented as DCCD has a long history in the use of synthesis of phosphodiester bonds in oligonucleotide synthesis.[263] More recent studies have seen the use of DCCD for the synthesis of polyphosphates via N-phosphoryl intermediate.[264] Popisov and coworkers[265] measured the disappearance of EDC by spectrophotometry as a measure of reactivity (either hydrolysis or formation of O-acylurea). These investigators showed that the rate of reaction of EDC with acetic acid or glutaric acid had a pH optima of 4; while the decrease in reaction above pH 4.0 is understood, the reason for the decrease in reaction rate below 4 is not clear. The rate of reaction of EDC with a high molecular weight (100–150 kDa) copolymer of acrylic acid and acrylamide showed a pH optimum of approximately 5, while reaction with chondroitin sulfate had a pH optimum of 4.0. Other investigators also observed the rapid loss of EDC in the presence of phosphate with a pH optimum of approximately 2.0. The reaction of carbodiimide with carboxyl groups can be separated into several phases, the first being the activation phase that is the formation of the O-acylurea (Figure 10.4). Following activation, the O-acylurea may undergo hydrolysis to regenerate the carboxyl group, rearrange to form the N-acylurea, or, as shown in Figure 10.4, react with an amino group to form an amide bond. As noted earlier, the O-acylurea may also react with a phosphate group forming phophodiesters.[266] Bettendorf and coworkers[266] used DCCD for the chemical synthesis of labeled or unlabeled nucleotide triphosphates. Murphy[267] reported the synthesis of an adduct between ATP and EDC (Figure 10.4), which was used to inactivate a calcium ATPase of sarcoplasmic reticulum possibly by cross-linking the protein. Earlier work by Chu and coworkers[268] demonstrated the formation of a phosphoroimidazole derivative in the presence of EDC, which could then form a phosphoroamidate with an amine. Rao and coworkers[269] used this chemistry for the immobilization of oligonucleotides onto silica particles.

Gilles and colleagues[270] have evaluated the stability of EDC in aqueous solution. EDC has a $t_{1/2}$ of 37 h (pH 7.0), 20 h (pH 6.0), and 3.9 h (pH 5.0) in 50 mM 2-(N-morpholino)ethanesulfonic acid at 25°C; in the presence of 100 mM glycine, the $t_{1/2}$ values were 15.8 h (pH 7.0), 6.7 h (pH 6.0), and 0.73 h (pH 5.0). The authors suggest that this supports the use of EDC at pH 6.0 or 7.0, but at pH 5.0, stability would be an issue. This study also reported a major decrease in the stability of EDC under the aforementioned solvent conditions in the presence of 10 mM phosphate ($t_{1/2}$ 1.43 h at pH 5.0) or 10 mM ATP/Mg^{2+} ($t_{1/2}$ 0.19 h at pH 5.0) consistent with the earlier observations of Metz and Brown.[262] Lei and coworkers[271] have reported kinetic studies on the hydrolysis of EDC in aqueous solution under acidic conditions. The conversion of EDC to N-ethyl-N'-(dimethylaminopropyl) urea (Figure 10.5) was measured by MS and capillary electrophoresis. Consistent with the results of Gilles and cowokers,[270] the rate of decomposition increased with increasing acidity (decreased pH). The observed rate of hydrolysis of EDC was 0.00243 min^{-1} at pH 6.5, 0.045 min^{-1} at pH 5.1, and 0.354 min^{-1} at pH 4.00. These rates are comparable to the results obtained from Popisov and coworkers[265] for the loss of EDC in acetate buffer at pH 4.0, which result from a combination of reaction of EDC with acetate to form the O-acylurea and hydrolysis to the substituted urea. Janolino and Swaisgood[272] compared a sequential method with a simultaneous method for the immobilization of a protein (sulfhydryl oxidase) onto a matrix (porous glass). Silinization was used to attach an aminopropyl group to the porous glass matrix, which was then reacted with succinic anhydride to provide a succinylaminopropyl function. The two-step process used activation with EDC at pH 4.75 followed by a wash with phosphate buffer at pH 7.0 (0°C–4°C). Comparable results were obtained with the sequential and simultaneous coupling methods. These investigators did use phosphate buffer but did not comment on stability issues. Using glycine ethyl ester as nucleophile, they observed that the rate of reaction of the O-acylurea with the nitrogen nucleophile (9.4×10^{-5} M^{-1} s^{-1}) was approximately eightfold higher than the rate of hydrolysis (1.2×10^{-5} M^{-1} s^{-1}). Wrobel and colleagues[273] used the decrease in absorbance at 214 nm ($\varepsilon = 6.3 \times 10^{-3}$ M^{-1} cm^{-1}) to measure the stability of EDC. These investigators also report decreased stability with decreasing pH. The presence of citrate, acetate, or phosphate increased the rate of EDC decomposition. Shegal and Vijay[274] have optimized the conditions for EDC-mediated coupling of a carboxyl-containing compound to an amine matrix (Affi-Gel® 102). These investigators noted that the presence of NHS greatly improved the coupling of butyric acid to the matrix.

Carbodiimides have been shown to react with sulfhydryl groups[275] and the phenolic hydroxyl group of tyrosine.[276] Carraway and Triplett[275] showed that EDC reacted with the sulfhydryl group of 2-mercaptoethanol at pH 5.0, 25°C, at a rate of 2.3×10^{-2} s^{-1}; this rate is comparable to the rate of reaction of EDC with carboxyl groups at the same pH. These investigators also demonstrated the modification of sulfhydryl groups in bovine serum albumin and β-lactoglobulin. There was negligible regeneration of sulfhydryl reactivity with reduction, hydroxylamine, or 2.0 M ammonia (pH 12.0). The modification of the active-site cysteinyl group in papain has been reported[277] as occurring under conditions (pH 4.75, 25°C) where 6 of 14 carboxyl groups are modified together with 9/19 tyrosyl residues. Modification of the tyrosyl residues is reversed by 0.5 M hydroxylamine, pH 7.0 (5 h at 25°C) as first demonstrated by Carraway and Koshland.[276] Perfetti and coworkers[277] observed that reaction of papain with EDC resulted in the irreversible loss of activity; reaction of mercuripapain also resulted in the irreversible loss of activity again with the modification of 9 of the 19 tyrosine residues. Treatment of the modified enzyme with hydroxylamine reversed the modification of the tyrosine residues without restoration of activity. There are other studies[278,279] where a functional sulfhydryl group was protected, as with mercuripapain, prior to modification with carbodiimide. There is an early review by Carraway and Koshland,[280] which reviews the early work on the use of carbodiimides. There are several studies which, while not of direct importance to the reaction of carbodiimides with proteins, merit mention. Millotti and coworkers[281] attached 6-mercaptonicotinic acid to chitosan using carbodiimide chemistry. The chitosan (mw 400 kDa) was dissolved in water at pH 3 and the 6-mercaptonicotinic acid added and the pH adjusted to 5. EDC was added, the pH adjusted to 6 and allowed to proceed at 23°C with vigorous stirring for 7 h at which point, tris(2-carboxyethyl)phosphine (TCEP) was added. The viscosity of this reaction markedly increased forming a gel which could be converted to a liquid by introduction of a reducing agent. It is suggested that the product has value as mucoadhesive formulation. I could not find any other studies, but the aforementioned work would suggest that TCEP as the more potent reducing agent would regenerate the product from the reaction of a carbodiimide with a sulfhydryl group. Another approach used the conversion of the O-acylurea to the NHS ester prior to coupling as Kafedjiiski and coworkers[282] did in the preparation of a chitosan–glutathione conjugate. The other approach is to protect the sulfhydryl group with a reversible modification prior to coupling.[283,284]

There is the potential to use the carbodiimide reaction to insert a different functionality on the carboxyl group. Ammonium ions can be used as the attacking nucleophile to generate asparaginyl and glutaminyl residues from *exposed* carboxyl groups.[285] The modification was accomplished in 5.5 M NH_4Cl, pH 4.75 for 3 h at 25°C. Under these conditions, approximately 11 of the 15 free carboxyl groups in chymotrypsinogen were converted to the corresponding amide. 1,2-Diaminoethane or diaminomethane can be coupled to aspartic acid residues to produce a trypsin-sensitive bond.[286] Another example of using this chemistry to introduce a new functional group into a protein is provided by the work of Lin and coworkers.[287] The first step involves the water-soluble carbodiimide-mediated coupling of cystamine to protein carboxyl groups. Reduction of the coupled cystamine with dithiothreitol results in 2-aminothiol functional groups bound to protein carboxyl groups. Montes and coworkers[288] modified the carboxyl groups in penicillin G acylase with EDC and ethylenediamine to change the net positive charge on the protein permitting binding of the enzyme to a carboxymethyl or dextran sulfate support. It is suggested that this approach permits immobilization of the enzyme for an industrial application; succinylation allowed binding to an anion-exchange matrix (e.g., DEAE).

There are several studies that have used nitrotyrosine ethyl ester as the modifying nucleophile with a carbodiimide.[289–293] This permits the formation of an adduct, which can be measured by UV absorbance or by reaction with a specific antibody, and a derivative protein/peptide, which can be isolated by immunoaffinity chromatography (see Chapter 11 for a discussion of the properties of nitrotyrosine). Pho and coworkers[289] reported the inactivation of yeast hexokinase with 1-cyclo-hexyl-3-(2-morpholinoethyl)carbodiimide metho-p-sulfonate (CMC). The reaction was performed in 0.1 M phosphate, pH 6.0, and was accelerated by the addition of nitrotyrosine ethyl ester. The effect of nitrotyrosine on the rate of inactivation of hexokinase is substantial with an eightfold increase in rate, while glycine ethyl ester had no effect; tyrosine ethyl ester that would be expected to be a weaker nucleophile provided a threefold increase in rate. The nitrotyrosine nucleophile provides a chromophoric label for the modified carboxyl group ($\varepsilon = 4.6 \times 10^3$ M^{-1} cm^{-1}, 0.1 M NaOH). Desvages and coworkers[290] showed that nitrotyrosine ethyl ester also markedly increased the rate of inactivation of yeast 3-phosphoglycerate kinase by CMC. The modified glutamic acid residues were identified after isolation of modified peptide by immunoaffinity chromatography using an immobilized antibody directed against nitrotyrosine.[290] Lacombe and coworkers[278] showed that the adenosine triphosphatase activity of rabbit skeletal myosin subfragment 1 was inactivated by CMC, and, as with the aforementioned studies, the rate was markedly enhanced by the inclusion of nitrotyrosine ethyl ester. These investigators did observe that the combination of CMC and nitrotyrosine ethyl ester was far more potent ($t_{1/2} = 0.5$ min) than a combination of EDC and nitrotyrosine ethyl ester ($t_{1/2}$ = 8 min); the $t_{1/2}$ for CMC alone was approximately 15 min. Imidazole also accelerated the rate of inactivation with CMC ($t_{1/2}$ = 6 min). Körner and coworkers[291] obtained similar results with the reaction of CMC and nitrotyrosine ethyl ester with canine cardiac myosin. The cysteine residues in the myosin subfragment 1 were blocked by cyanylation (see Chapter 7) prior to reaction with carbodiimide. The studies with the canine cardiac myosin protein showed that the incorporation of nitrotyrosine ethyl ester was the same with native or cyanylated subfragment 1. In a subsequent study, Körner and coworkers[292] identified the essential carboxyl group that was modified by CMC/ nitrotyrosine ethyl ester. These investigators also showed that immunoblotting with antinitrotyrosine antibody was effective in identifying fragments containing the modified carboxyl groups. Finally, despite the relatively slow rate of reaction, carbodiimide chemistry is suggested for use in *footprinting* of protein.[294–296]

A number of studies have used several different reagents for the site-specific modification of carboxyl groups in the same research study. There are several that compare a water-insoluble carbodiimide, *N,N'*-dicyclohexylcarbodiimide (DCCD) and EDC. Sigrist-Nelson and Azzi[297] studied the effect of carboxyl-modifying reagents on proton translocation in a liposomal chloroplast ATPase complex. Proton transport was inhibited by DCCD but not by EDC. Sussman and Slayman[298] reported similar differences with *Neurospora crassa* plasma membrane ATPase. This difference most likely reflects the hydrophobic nature of DCCD permitting it to pass through membranes (membrane-permeant reagent). There are subsequent studies that demonstrate that EDC can inhibit the purified F_1 ATPase,[299] but the reaction is complex and multiple products, including cross-linked derivatives, are obtained. Safir and coworkers[300] showed that voltage-dependent anion channel activity in skeletal muscle sarcoplasmic reticulum was inhibited by DCCD (20 mM MES–0.2 M NaCl, pH 6.4) but not by EDC or Woodward's reagent K. Yang and colleagues[301] modified purified vacuolar H+-pyrophosphatase (*Vigna radiata* L.) with several carboxyl reagents (50 mM MOPS-20% [v/v] glycerol–1 mM EGTA–1 mM DTT, pH 7.2, 33°C). Rapid inactivation was observed with DCCD and much slower inactivation by either EDC or Woodward's reagent K. Bjerrum and colleagues[302] used an impermeant, water-soluble carbodiimide, 1-ethyl-3-(1-azonia-4,4-dimethylpentyl) carbodiimide, to inhibit anion exchange in human red blood cell membranes. The rate and extent of inactivation was increased by the addition of tyrosine ethyl ester.

There are examples of carboxyl group modification with reagents expected to react far more effectively with other nucleophiles (Figure 10.6). An example of this is the reaction of iodoacetamide with ribonuclease T_1 to form the glycolic acid derivative of the glutamic acid residue as elegantly shown by Takahashi and coworkers.[303] While this modification resulted in the loss of enzymatic activity, the ability to bind substrate was retained. Subsequent work by Kojima and coworkers[304] using ^{1}H NMR demonstrated that this modification was associated with increased thermal stabilization of the modified enzyme compared to the native protein. A related modification of human fibroblast collagenase by chloroacetyl *N*-hydroxypeptide derivatives was reported by Lin and Kuo.[305] Chloroacetyl-*N*-hydroxy-Leu-Ala-Gly (Figure 10.6) inhibited human fibroblast collagenase. The inactive enzyme was modified with EDC/glycinamide in the presence of 4 M guanidine to amidate the other carboxyl groups. The modified protein was treated with imidazole and modified with radiolabeled glycinamide/EDC; approximately 1 mol of glycinamide was incorporated per mole of protein. The enzyme inactivated with the *N*-hydroxypeptide could be reactivated with either imidazole or hydroxylamine. It is noted that an *N*-hydroxypeptide showed an enhanced rate of hydrolysis by chymotrypsin.[306] Another example is the modification of a specific carboxyl group in pepsin by *p*-bromophenacyl bromide[307] (the use of *p*-bromophenacyl bromide in the specific modification of proteins is not uncommon, but is generally associated with the modification of cysteine, histidine, or methionine). In the study of pepsin, optimal inactivation (approximately 12-fold molar excess of reagent, 3 h, 25°C) was obtained in the pH range of 1.5–4.0 with a rapid decrease in the extent of inactivation at pH 4.5 and above (the effect of pH greater than 5.5–6.0 on the modification of pepsin cannot be studied because of irreversible denaturation of pepsin at pH 6.0 and above). In studies with a 10% M excess of *p*-bromophenacyl

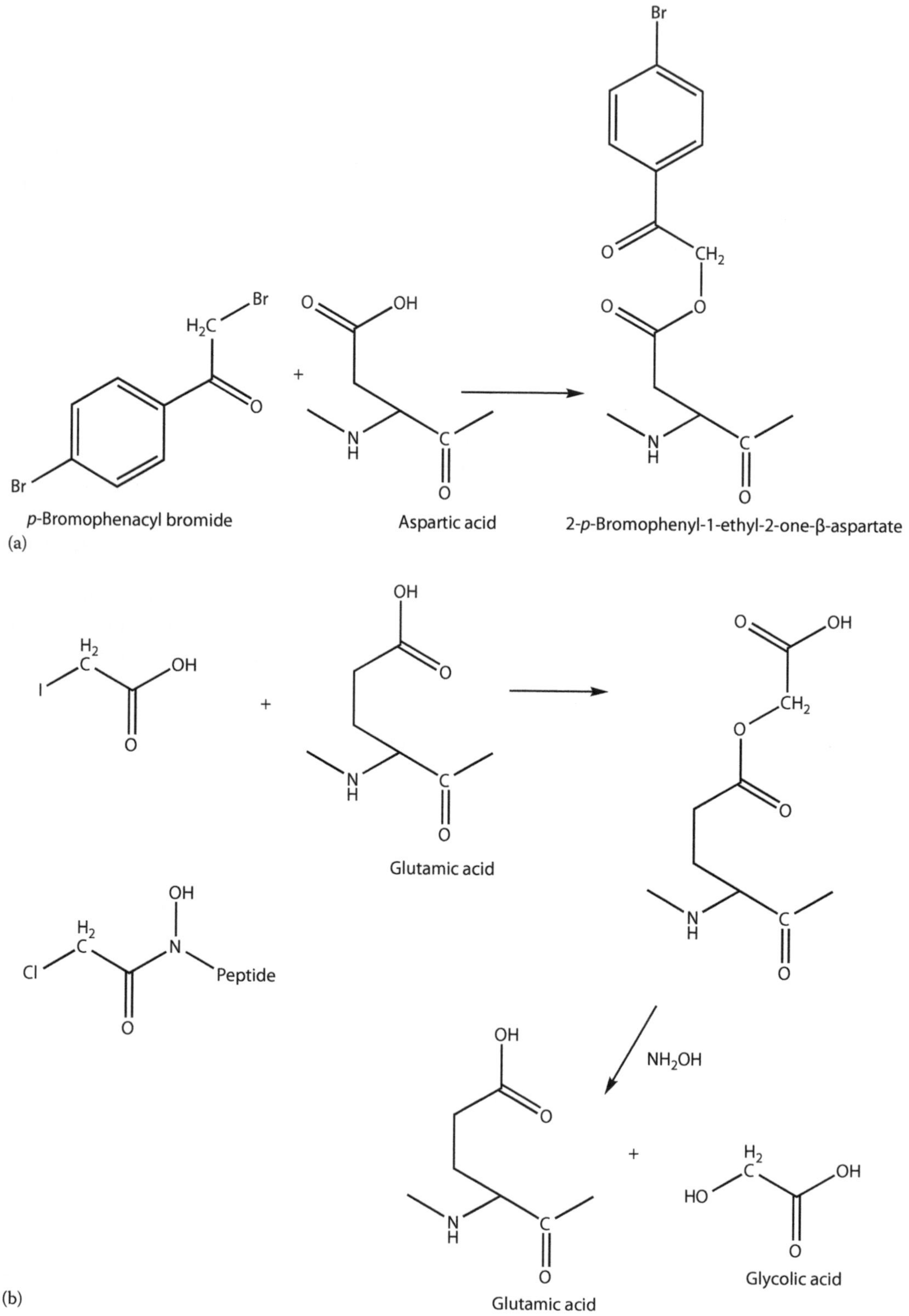

p-Bromophenacyl bromide
Aspartic acid
2-p-Bromophenyl-1-ethyl-2-one-β-aspartate
(a)
Glutamic acid
Peptide
NH2OH
Glutamic acid
Glycolic acid
(b)

FIGURE 10.6 (continued)

FIGURE 10.6 (continued)   Reaction of *p*-bromophenacyl bromide or iodoacetate with carboxyl groups in proteins. Reaction a shows the reaction of *p*-bromophenacyl bromide with the β-carboxyl group of pepsin. Although this reaction forms an ester, the reaction is not reversed by hydroxylamine but is reversed by 2-mercaptoethanol (see Gross, E. and Morell, J.L., Evidence for an active carboxyl group in pepsin, *J. Biol. Chem.* 241, 3638–3639, 1966). Reaction b shows the reaction of iodoacetic acid with a glutamic acid in ribonuclease T₁. This reaction is reversed by hydroxylamine (see Takahashi, K., Stein, W.H., and Moore, S., The identification of a glutamic acid residue as part of the active site of ribonuclease T₁, *J. Biol. Chem.* 242, 4682–4690, 1967). Shown at the bottom is a chloroacetyl-*N*-hydroxy peptide, which has been shown to react with a carboxyl group in a human fibroblast collagenase (Lin, T.-Y. and Kuo, D.W., Inactivation of human fibroblast collagenase by chloroacetyl *N*-hydroxypeptide derivatives, *J. Enzyme Inhib.* 5, 33–40, 1991).

bromide, pH 2.8 at 37°C for 3 h, complete inactivation of the enzyme was obtained concomitant with the incorporation of 0.93 mol of reagent per mole of pepsin (assessed by bromide analysis). Attempts to reactivate the modified enzyme with a potent nucleophile such as hydroxylamine were unsuccessful, but reactivation could be obtained with sulfhydryl-containing reagents (i.e., β-mercaptoethanol, 2,3-dimercaptopropanol, thiophenol). It has been subsequently established that reaction occurs at the β-carboxyl group of an aspartic acid residue (formation of 2-*p*-bromophenyl-1-ethyl-2-one-β-aspartate).[308] These investigators noted that the reduction of enzyme under somewhat harsh conditions (LiBH₄ in tetrahydrofuran) resulted in the formation of homoserine. Chacur and colleagues[309] have modified Asp49 in the phospholipase A from *Bothrops asper* snake venom. This modification eliminates enzymatic activity and also abolishes the hyperalgesia effect. More recently, Pickering and Davies[310] have modified a carboxyl group in a protein with 7-amino-4-methylcoumarin in the presence of sodium cyanoborohydride.

REFERENCES

1. Simms, H.S., The nature of the ionizable groups in proteins, *J. Gen. Physiol.* 11, 629–640, 1928.
2. Simonetta, M. and Carrà, S., General and theoretical aspects of the COOH and COOR groups, in *The Chemistry of Carboxyl Acids and Esters*, ed. S. Patai, Chapter 1, pp. 1–52, Wiley Interscience, London, U.K., 1969.
3. Cornish-Bowden, A., The amino acid composition of proteins are correlated with their molecular sizes, *Biochem. J.* 213, 271–274, 1983.
4. Gromiha, M.M. and Suwa, M., A simple statistical method for discriminating outer membrane proteins with better accuracy, *Comput. Biol. Chem.* 29, 136–147, 2005.
5. Schultz, J., Allison, H., and Grice, M., Specificity of the cleavage of proteins by dilute acid. I. Release of aspartic acid from insulin, ribonuclease, and glucagon, *Biochemistry* 1, 694–698, 1962.
6. Tsung, C.M. and Fraenkel-Conrat, H., Preferential release of aspartic acid by dilute acid treatment of tryptic peptides, *Biochemistry* 4, 793–801, 1965.
7. Schultz, J., Cleavage at aspartic acid, *Methods Enzymol.* 11, 255–263, 1967.
8. Miekke, S.T., Heat-induced fragmentation of human plasma fibronectin, *Biochim. Biophys. Acta* 748, 374–380, 1983.
9. Fenselau, C., Laine, O., and Swatkoski, S., Microwave assisted cid cleavage for denaturation and proteolysis of intact human adenovirus, *Int. J. Mass Spectrom.* 301, 7–11, 2011.
10. Osula, O., Swatkoski, S., and Cotter, R.J., Identification of protein SUMOylation sites by mass spectrometry using combined microwave-assisted aspartic acid cleavage and tryptic digestion, *J. Mass Spectrom.* 47, 644–654, 2012.
11. Cannon, J.R., Edwards, N.J.U., and Fenselau, C., Mass-biased partitioning to enhance middle down proteomics analysis, *J. Mass Spectrom.* 48, 340–343, 2013.
12. Piszkiewicz, D., Landon, M., and Smith, E.L., Anomalous cleavage of aspartyl-proline peptide bonds during amino acid sequence determinations, *Biochem. Biophys. Res. Commun.* 40, 1173–1178, 1970.
13. Carter, D.A. and McFadden, P.N., Trapping succinimides in aged polypeptides by chemical reduction, *J. Protein Chem.* 13, 89–96, 1994.
14. Barrett, G.C. and Elmore, D.T., *Amino Acids and Peptides*, p. 4, Cambridge University Press, Cambridge, U.K., 1998.

15. Grimsley, G.R., Scholtz, J.M., and Pace, C.N., A summary of the measured pK values of the ionizable groups in folded proteins, *Protein Sci.* 18, 247–251, 2009.

16. Lundblad, R.L. and MacDonald, F.M. (eds.), *Handbook of Biochemistry and Molecular Biology*, 4th edn., CRC Press/Taylor & Francis, Boca Raton, FL, 2010.

17. Pace, C.N., Grimsley, G.R., and Scholtz, J.M., Protein ionizable groups: pK values and their contribution to protein stability and solubility, *J. Biol. Chem.* 284, 13285–13289, 2009.

18. Bahadur, R.P., Chakrabarti, P., Rodier, F., and Janin, J., Dissecting subunit interfaces in homodimeric proteins, *Protein Struct. Funct. Genet.* 53, 708–719, 2003.

19. Laurents, D.V., Huyghues-Despointes, B.M., Bruix, M. et al., Charge-charge interactions are key determinants of the pK values of ionizable groups in ribonuclease Sa (pI = 3.5) and a basic variant (pI = 10.2), *J. Mol. Biol.* 325, 1077–1092, 2003.

20. Thurkill, R.L., Grimsely, G.R., Scholtz, J.M., and Pace, C.N., Hydrogen bonding markedly reduces the pK of buried carboxyl groups in proteins, *J. Mol. Biol.* 362, 594–604, 2006.

21. Harms, M.J., Castañeda, C.A., Schlessman, J.L. et al., The pKa values of acidic and basic residues buried at the same internal location in a protein are governed by different factors, *J. Mol. Biol.* 389, 34–47, 2009.

22. Li, H., Robertson, A.D., and Jensen, J.H., Very fast empirical prediction and rationalization of protein pKa values, *Proteins* 61, 704–721, 2005.

23. Paterson, A.K. and Knowles, J.R., The number of catalytically essential carboxyl groups in pepsin. Modification of the enzyme by trimethyloxonium fluoroborate, *Eur. J. Biochem.* 31, 510–517, 1972.

24. Blow, D.M., Birktoft, J.J., and Hartley, B.S., Role of a buried acid group in the mechanism of action of chymotrypsin, *Nature* 221, 337–340, 1969.

25. Craik, C.S., Roczniak, S., Sprang, S. et al., Redesigning trypsin via genetic engineering, *J. Cell Biochem.* 33, 199–211, 1987.

26. Gershenfeld, H.K., Hershberger, R.J., Mueller, C., and Weissman, I.L., A T cell- and natural killer cell-specific, trypsin-like serine protease. Implications of a cytolytic cascade, *Ann. N. Y. Acad. Sci.* 532, 367–379, 1988.

27. Mangel, W.F., Singer, P.T., Cyr, D.M. et al., Structure of an acyl-enzyme intermediate during catalysis: (Guanidinobenzoyl) trypsin, *Biochemistry* 29, 8351–8357, 1990.

28. Du, Z., Zheng, Y., Patterson, M. et al., pKa coupling at the intein active site: Implications for the coordination mechanism of protein splicing with a conserved aspartate, *J. Am. Chem. Soc.* 133, 10275–10282, 2011.

29. Majika, J.I., Lopez, X., and Mulholland, A.J., Modeling protein splicing: Reaction pathways for C-terminal splice and intein scission, *J. Phys. Chem. B* 113, 5607–5616, 2009.

30. Tajima, N., Kawai, F., and Park, S.-Y., A novel intein-like autoproteolytic mechanism in autotransporter proteins, *J. Mol. Biol.* 402, 645–656, 2010.

31. Nishimura, K., Tajima, N., Yoo, Y.-H. et al., Autotransporter passenger proteins: Virulence factors with common structural themes, *J. Mol. Med.* 88, 451–458, 2010.

32. Barnard, T.J., Gumbart, J., Peterson, J.H. et al., Molecular basis for the activation of a catalytic asparagine residue in a self-cleaving bacterial autotransporter, *J. Mol. Biol.* 415, 128–142, 2012.

33. Amitai, G., Dassa, B., and Pietrokovski, S., Protein splicing of inteins with atypical glutamine and aspartate C-terminal residue, *J. Biol. Chem.* 279, 3121–3131, 2004.

34. Merenmies, J., Pihlaskari, R., Laitinen, J. et al., 30-kDa heparin-binding protein of brain (amphoterin) involved in neurite outgrowth. Amino acid sequence and localization in the filopodia of the advancing plasma membrane, *J. Biol. Chem.* 266, 16722–16729, 1991.

35. Dhe-Paganon, S., Sirano, S., and Ron, C., Crystal structure of human frataxin, *J. Biol. Chem.* 275, 30753–30756, 2000.

36. Fu, J.R., Balan, S., Potty, A. et al., Enhanced protein affinity and selectivity of clustered-charge anion-exchange adsorbents, *Anal. Chem.* 79, 9060–9065, 2007.

37. Chen, W.-H., Fu, J.Y., Kourentzi, K., and Wilson, R.C., Nucleic acid affinity of clustered-charge anion exchange adsorbents. Effects of ionic strength and ligand density, *J. Chromatogr. A* 1218, 258–262, 2011.

38. Daniels, C.R., Kisley, L., Kim, H. et al., Fluorescence correlation spectroscopy study of protein transport and dynamic interactions with clustered-charge peptide adsorbents, *J. Mol. Recogn.* 25, 435–442, 2012.

39. Wang, X., Schroder, H.C., Schlossmacher, U. et al., Biosilica aging: From enzyme-driven gelation via syneresis to chemical/biochemical hardening, *Biochim. Biophys. Acta* 1830, 3437–3446, 2013.

40. Karlin, S. and Zhu, Z.-Y., Characterization of diverse residue clusters in protein three-dimensional structures, *Proc. Natl. Acad. Sci. USA* 93, 8344–8349, 1996.

41. Zhu, Z.-Y. and Karlin, S., Clusters of charged residues in protein three-dimensional structures, *Proc. Natl. Acad. Sci. USA* 93, 8350–8355, 1996.

42. Choura, M. and Rebai, A., Exploring charge biased regions in the human proteome, *Gene* 515, 277–280, 2013.
43. Nelsestuen, G.L., Zytkovicz, T.H., and Howard, J.B., The mode of action of vitamin K. Identification of γ-carboxyglutamic acid as a component of prothrombin, *J. Biol. Chem.* 249, 6347–6350, 1974.
44. Thørgersen, H.C., Petersen, T.E., Sottrup-Jensen, L. et al., The N-terminal sequences of blood coagulation factor X1 and X2 light chains. Mass-spectrometric identification of twelve residues of γ-carboxyglutamic acid in their vitamin K-dependent domains, *Biochem. J.* 175, 613–627, 1978.
45. Kyte, J., *Structure in Protein Chemistry*, 2nd edn., p. 513, Garland Science/Taylor & Francis, New York, 2007.
46. Lindman, S., Linse, S., Mulder, F.A., and André, I., Electrostatic contributions to residue-specific protonation equilibria and proton binding capacitance for a small protein, *Biochemistry* 45, 13993–14002, 2006.
47. Schönichen, A., Webb, B.A., Jacobson, M.P., and Barber, D.L., Considering protonation as a posttranslational modification regulated protein structure and function, *Annu. Rev. Biophys.* 42, 289–314, 2013.
48. Stroffekova, K., Kupert, E.Y., Malinowska, D.H., and Cuppoletti, J., Identification of the pH sensor and activation by chemical modification of the CLC-2G Cl⁻ channel, *Am. J. Physiol.* 275, C1113–C1123, 1998.
49. Frankel, L.K., Cruz, J.A., and Bricker, T.M., Carboxylate groups on the manganese-stabilizing protein are required for its efficient binding to photosystem II, *Biochemistry* 38, 14271–14278, 1999.
50. Siddiqui, K.S., Najmus Squibb, A.A., Rashid, M.H., and Rajoka, M.I., Carboxyl group modification significantly altered the kinetic constants of purified carboxymethylcellulase from *Aspergillus niger*, *Enzyme Microb. Technol.* 27, 467–474, 2000.
51. Hopman, A.H., Wiegard, J., and van Duijn, P., A new hybridocytochemical method based on mercurated nucleic acid probes and sulfhydryl-hapten ligands. II. Effects of variations in ligand structure on the in situ detection of mercurated probes, *Histochemistry* 84, 179–185, 1986.
52. Rashid, M.H. and Siddiqui, K.S., Thermodynamic and kinetic stability of the native and chemically modified β-glucosidases from *Aspergillus niger*, *Process Biochem.* 33, 109–115, 1998.
53. Zhang, Z. and Grelet, E., Tuning chirality in the self-assembly of the rod-like viruses by chemical surface modifications, *Soft Matter* 9, 1015–1024, 2013.
54. Bokhari, S.A., Afzai, A.J., Rashid, M.H. et al., Coupling of surface carboxyls of carboxymethylcellulase with aniline via chemical modification: Extreme thermostabilization in aqueous and water-miscible organic mixtures, *Biotechnol. Prog.* 18, 278–281, 2002.
55. Park, S.J. and Khang, D., Conformational changes of fibrinogen in dispersed carbon nanotubes, *Int. J. Nanomed.* 7, 4325–4333, 2012.
56. Gao, Y. and Kyratzis, I., Covalent immobilization of proteins on carbon nanotubes using the cross-linker 1-ethyl-3-(3-dimethylaminopropyl) carbodiimide—A critical assessment, *Bioconjug. Chem.* 19, 1945–1950, 2008.
57. Saby, C., Ortiz, B., Champagne, G.Y., and Bélanger, D., Electrochemical modification of glassy carbon electrode using aromatic diazonium salts. 1. Blocking effect of 4-nitrophenyl and 4-carboxyphenyl groups, *Langmuir* 13, 6805–6813, 1997.
58. Lerner, M.B., D'Souza, J., Pazina, T. et al., Hybrids of a genetically engineered antibody and a carbon nanotube transistor for detection of prostate cancer biomarkers, *ACS Nano* 6, 5143–5149, 2012.
59. Carr, F.H., Culhane, K., Fuller, A.T., and Underhill, S.W.F., CIX. A reversible inactivation of insulin, *Biochem. J.* 23, 1010–1021, 1929.
60. Charles, A.F. and Scott, D.A., Action of acid alcohol on insulin, *J. Biol. Chem.* 92, 289–302, 1931.
61. Jensen, H. and Evans, E.A., Jr., The chemistry of insulin, *Physiol. Rev.* 14, 188–209, 1934.
62. Fraenkel-Conrat, H. and Olcott, H.S., Esterification of proteins with alcohols of low molecular weight, *J. Biol. Chem.* 161, 259–268, 1945.
63. Mommaerts, W.F. and Neurath, H., Insulin methyl ester. I. Preparation and properties, *J. Biol. Chem.* 185, 909–917, 1950.
64. Alexander, P., Carter, D., Earland, C., and Ford, O.E., Esterification of the carboxyl groups in wool, *Biochem. J.* 48, 629–637, 1951.
65. Pénasse, L., Jutisz, M., Fromageot, C., and Fraenkel-Conrat, H., La détermination des groupes carboxyliques des protéines. II. Le groupe carboxylique terminal de l'ovomucoide, *Biochim. Biophys. Acta* 9, 511–557, 1952.
66. Fromageot, C., Jutisz, M., Meyer, D., and Pénasse, L., Methode pour la characterisation groupes carboxyliques terminaux les protéines. Application à l'insuline, *Biochim. Biophys. Acta* 6, 283–289, 1950.
67. Wilcox, P.E., Esterification, *Methods Enzymol.* 11, 605–617, 1967.
68. Parsons, S.M., Jao, L., Dahlquist, F.W. et al., The nature of amino acid side chains which are critical for the activity of lysozyme, *Biochemistry* 8, 700–712, 1969.

69. Acharya, A.S. and Vithayathil, P.J., On the reactivity of carboxyl groups of ribonuclease-A in anhydrous methanol, *Int. J. Pept. Protein Res.* 7, 207–219, 1975.

70. Vithayathil, P.J. and Richards, F.M., The carboxyl and amide groups of the peptide component of ribonuclease-S, *J. Biol. Chem.* 236, 1380–1385, 1961.

71. Acharya, A.S., Manula, B.N., Murthy, G.S., and Vithayathil, P.J., The influence of esterification of carboxyl groups of ribonuclease-A on its structure and immunological activity, *Int. J. Pept. Protein Res.* 9, 213–219, 1987.

72. Acharya, A.S., Manjula, B.N., and Vithayathil, P.J., Structure and enzymic activity of ribonuclease-A esterified at glutamic acid-49 and aspartic acid-53, *Biochem. J.* 173, 821–830, 1978.

73. Vakos, H.T., Black, B., Dawson, B. et al., In vacuo esterification of carboxyl groups in lyophilized proteins, *J. Protein Chem.* 20, 521–531, 2001.

74. Haebel, S., Albrecht, T., Sparkler, K. et al., Electrophoresis-related protein modification: Alkylation of carboxy residues revealed by mass spectrometry, *Electrophoresis* 19, 679–686, 1988.

75. Jung, S.Y., Li, Y., Wang, Y. et al., Complications in the assignment of 14 and 28 Da mass shifts detected by mass spectrometry as in vivo methylation from endogenous proteins, *Anal. Chem.* 80, 1721–1729, 2008.

76. Xing, G., Zhang, J., Chen, Y., and Zhao, Y., Identification of four novel types of in vitro protein modification, *J. Proteome Res.* 7, 4603–4608, 2008.

77. Chen, G., Liu, H., Wang, X., and Li, Z., In vitro methylation by methanol: Proteomic screening and prevalence investigation, *Anal. Chim. Acta* 661, 67–75, 2010.

78. Diliberto, E.J., Jr., Viveros, O.H., and Alexrod, J., Subcellular distribution of protein carboxymethylase and its endogenous substrates in the adrenal medulla: Possible role in excitation-secretion coupling, *Proc. Natl. Acad. Sci. USA* 73, 4050–4054, 1976.

79. Kim, S., Lew, B., and Chang, F.N., Enzymatic methyl esterification of *Escherichia coli* ribosomal proteins, *J. Bacteriol.* 130, 839–845, 1977.

80. Galletti, P., Ingrosso, D., Nappi, A. et al., Increased methyl esterification of membrane proteins in aged red blood cells. Preferential esterification of ankyrin and band-4.1 cytoskeletal proteins, *Eur. J. Biochem.* 135, 25–31, 1983.

81. Ingrosso, D., D'Angelo, S., Di Carlo, E. et al., Increased methyl esterification of altered aspartyl residues in erythrocyte membrane proteins in response to oxidative stress, *Eur. J. Biochem.* 267, 4397–4405, 2000.

82. Cimmino, A., Capasso, R., Muller, F. et al., Protein isoaspartate methyltransferase prevents apoptosis induced by oxidative stress in endothelial cells: Role of Bcl-xl deamidation and methylation, *PLoS One* 3, e3258, 2008.

83. Sprung, R., Chen, Y., Zhang, K. et al., Identification and validation of eukaryotic aspartate and glutamate methylation in proteins, *J. Proteome Res.* 7, 1001–1006, 2008.

84. Grillo, M.A. and Colombatto, S., *S*-Adenosylmethionine and protein methylation, *Amino Acids* 28, 357–362, 2005.

85. Yasumitsu, H., Ozeki, Y., Kawsar, S.M.A. et al., RAMA stain: A fast, sensitive and less protein-modifying CBB R250 stain, *Electrophoresis* 31, 1913–1917, 2010.

86. Yasumistu, H., Pzeki, Y., Kawsar, S.M.A. et al., CGP stain: An inexpensive, odorless, rapid, sensitive, and, in principle, in vitro methylation-free Coomassie Brilliant BLue stain, *Anal. Biochem.* 406, 86–88, 2010.

87. Julka, S., Dielman, D., and Young, S.A., Detection of C-terminal peptide of proteins using isotope coding strategies, *J. Chromatogr. B* 874, 101–110, 2008.

88. Xu, G., Sung, B.Y.S., and Jaffrey, S.R. Chemoenzymatic labeling of protein C-termini for positive selection of C-terminal peptides, *ACS Chem. Biol.* 6, 1015–1020, 2011.

89. Platt, M.D., Salicioni, A.M., Hunt, D.F., and Visconti, P.E., Use of differential isotopic labeling and mass spectrometry to analyze capacitation-associated change in the phosphorylation status of mouse sperm proteins, *J. Proteome Res.* 8, 1431–1440, 2009.

90. Paradela, A., Marcila, M., Navajas, R. et al., Evaluation of isotope-coded protein labeling (ICPL) in the quantitative analysis of complex proteomes, *Talanta* 80, 1496–1502, 2010.

91. Wang, G. and Wang, T., Improving foaming properties of yolk-contaminated egg albumen by basic soy protein, *J. Food Sci.* 74, C581–C587, 2009.

92. Wheelwright, W.V.K., Easteal, A.J., Ray, S., and Niewoudt, M.K., A one-step approach for esterification of zein with methanol, *J. Appl. Polym. Sci.* 127, 3500–3505, 2013.

93. Doscher, M.S. and Wilcox, P.E., Chemical derivatives of α-chymotrypsinogen. IV. A comparison of the reaction of α-chymotrypsinogen and of simple carboxylic acids with diazoacetamide, *J. Biol. Chem.* 236, 1328–1337, 1961.

94. Riehm, J.P. and Scheraga, H.A., Structural studies of ribonuclease. XVII. A reactive carboxyl group in ribonuclease, *Biochemistry* 4, 772–782, 1965.

95. Giraldi, T. and Baldini, L., Mechanism of inhibition of DNA synthesis in Ehrlich ascites tumour cells by diazoacetyl glycine-amide, *Biochem. Pharmacol.* 22, 1793–1799, 1973.

96. Giraldi, T., Guanino, A.M., Nisi, C., and Baldini, L., Selective antimetastatic effects of *N*-diazoacetylglycine derivatives in mice, *Eur. J. Cancer* 15, 603–607, 1979.

97. Giraldi, T., Sava, G., and Nisi, C., Mechanism of the antimetastatic action of *N*-diazoacetylglycinamide in mice bearing Lewis lung carcinoma, *Eur. J. Cancer* 16, 87–92, 1980.

98. Nguyen, H.T.T., Jan, Y.-J., Jeong, S., and Yu, J., DNA-specific autoantibody cleaves DNA by hydrolysis of phosphodiester and glycosidic bond, *Biochem. Biophys. Res. Commun.* 311, 767–773, 2003.

99. Landsteiner, K., Ueber die Antigeneigenschaften vn methyliertem Eiwisz. VII. Miteillungen über antigen, *Zeitschrift für Immunitätsforschung und Experimentale Therapie* 26, 122–133, 1917.

100. Stout, P.R., Clafferty, D.J., and Ruth, J.A., Chemical factors involved in accumulation and retention of fentanyl in hair after external exposure or in vivo deposition, *Drug Metab. Dispos.* 26, 689–700, 1998.

101. http://www.sigmaaldrich.com/chemistry/chemical-synthesis/technology-spotlights/diazald.html.

102. Siciliano, C., De Marco, R., Guidi, L.E. et al., A one-pot procedure for the preparation of *N*-9-fluorenylmethyloxycarbonyl-α-amino diazoketones from α-amino acids, *J. Org. Chem.* 77, 10575–10582, 2012.

103. Franssen, N.M., Finger, M., Reek, J.N., and de Bruin, B., Propagation and termination steps in Rh-mediated carbene polymerisation using diazomethane, *Dalton Trans.* 42, 4139–4152, 2013.

104. Osbahr, A.J., Chromatographic isolation and identification of methylated derivatives obtained from modified fibrinogen, *Biomaterials* 1, 83–88, 1980.

105. Brenner, M. and Huber, W., Herstellung von α-aminosäureestern durch alkoholyse der methylester, *Helv. Chim. Acta* 36, 1109–1115, 1953.

106. Korhone, L.K., Specific methylation of carboxyl groups by thionyl chloride in methanol, *Acta Histochem.* 26, 80–86, 1967.

107. Campana, S.P. and Goissis, G., Kinetics and yield of the esterification of amino acids with thionyl chloride in *n*-propanol, *J. Chromatogr.* 236, 197–200, 1982.

108. Pavlik, J.W. and Laohhasurayotin, S., Synthesis and spectroscopic properties of some deuterated cyanopyridines, *J. Heterocycl. Chem.* 42, 73–76, 2005.

109. Bello, J., The specific esterification of the carboxyl groups of gelatin with methanol and thionyl chloride, *Biochim. Biophys. Acta* 20, 426–427, 1956.

110. Horton, H.R. and Swaisgood, H.E., Covalent modification of proteins by techniques which permit subsequent release, *Methods Enzymol.* 135, 130–141, 1987.

111. Stabel, T.J., Casele, E.S., Swaisgood, H.E., and Horton, H.R., Anti-IgG immobilized controlled-pore glass. Thionyl chloride-activated succinamidopropyl-glass as a covalent immobilization matrix, *Appl. Biochem. Biotechnol.* 36, 87–96, 1992.

112. Aksoy, S., Tumturk, H., and Hasirci, N., Stability of α-amylase immobilized on poly(methyl methacrylate-acrylic acid) microspheres, *J. Biotechnol.* 60, 37–46, 1998.

113. Porjazoska, A., Yilmaz, O.K., Baysal, K. et al., Synthesis and characterization of poly(ethylene glycol)-poly (D,L-lactide-co-glycolide) poly(ethylene glycol) tri-block co-polymers modified with collagen: A model surface suitable for cell interaction, *J. Biomater. Sci. Polym. Ed.* 17, 323–340, 2006.

114. Ogawa, H., Chibara, T., and Taya, K., Selective monomethyl esterification of dicarboxylic acids by use of monocarboxylate chemisorption on alumina, *J. Am. Chem. Soc.* 107, 1365–1369, 1985.

115. Manetsch, R., Zheng, L., Reymond, M.T. et al., A catalytic antibody against a tocopherol cyclic inhibitor, *Chem. Eur. J.* 10, 2481–2506, 2004.

116. Bowen, W.J. and Field, V.L.S., Chemistry, microscopy, and performance of methylated glycerol treated muscle fibers, *Biochemistry* 6, 1127–1136, 1967.

117. Staab, H.J. and Anderer, F.A., Structure and immunogenic behavior of methylated tobacco mosaic virus, *Biochim. Biophys. Acta* 427, 453–464, 1976.

118. Osbahr, A.J., Fibrinogen-induced polymerization via the process of methylation, *Thromb. Haemost.* 42, 1396–1410, 1980.

119. Osbahr, A.J., Esterification of the carboxyl groups in fibrinogen by the application of highly specific methylating agent, *Thromb. Haemost.* 48, 226–231, 1982.

120. Delpierre, G.R. and Fruton, J.S., Inactivation of pepsin by diphenyldiazomethane, *Proc. Natl. Acad. Sci. USA* 54, 1161–1167, 1965.

121. Aboderin, A.A. and Fruton, J.S., Inactivation of chymotrypsin by diphenyldiazomethane, *Proc. Natl. Acad. Sci. USA* 56, 1252–1259, 1966.

122. Rajagopalan, T.G., Stein, W.H., and Moore, S., The inactivation of pepsin by diazoacetylnorleucine methyl ester, *J. Biol. Chem.* 241, 4295–4297, 1966.

123. Lundblad, R.L. and Stein, W.H., On the reaction of diazoacetyl compounds with pepsin, *J. Biol. Chem.* 244, 154–160, 1969.
124. Meitner, P.A., Bovine pepsinogens and pepsins. Sequence around a reactive aspartyl residue, *Biochem. J.* 124, 673–676, 1971.
125. Keilová, H., Inhibition of cathepsin D by diazoacetylnorleucine methyl ester, *FEBS Lett.* 6, 312–314, 1970.
126. Sodek, J. and Hoffmann, T., Amino acid sequence around the active site aspartic acid in penicillopepsin, *Can. J. Biochem.* 48, 1014–1016, 1970.
127. Mizobe, F., Takahashi, K., and Ando, T., Structure and function of acid proteases. I. Specific inactivation of an acid protease from *Rhizopus chinensis* by diazoacetyl-DL-norleucine methyl ester, *J. Biochem.* 73, 61–68, 1973.
129. Liu, C.L., Ohtsuki, K., and Hatano, H., Inactivation of *Rhodotorula glutinis* acid proteases by diazoacetyl compounds, *J. Biochem.* 73, 671–673, 1973.
130. Chang, W.-J. and Takahashi, K., Structure and function of acid proteases. III. Isolation and characterization of the active-site peptides from bovine rennin, *J. Biochem.* 76, 467–474, 1974.
131. Rickert, W.S. and McBride-Warren, P.A., Structural and functional determinants of *Mucor miehei* protease VI. Inactivation of the enzyme by diazoacetyl norleucine methyl ester, pepstatin and 1,2-epoxy-3-(*p*-nitrophenoxy)propane, *Biochim. Biophys. Acta* 480, 262–274, 1977.
132. Dionyssious-Asteriou, A. and Rakitzis, E., Inactivation of cathepsin D from human gastric mucosa and from stomach carcinoma by diazoacetyl-DL-norleucine methyl ester, *Biochem. Pharmacol.* 27, 827–829, 1978.
133. Frith, G.J.T., Peoples, M.B., and Dalling, M.J., Proteolytic enzymes in green wheat-leaves. III. Inactivation of acid proteinase II by diazoacetyl-DL-norleucine methyl ester by 1,2-epoxy-3-(*p*-nitrophenoxy)-propane, *Plant Cell Physiol.* 19, 819–824, 1978.
134. Kageyama, T. and Takahashi, K., A cathepsin D-like acid proteinase from human gastric mucosa. Purification and characterization, *J. Biochem.* 87, 725–735, 1980.
135. Hislop, E.C., Pavel, J.L., and Keon, J.P.R., An acid protease produced by *Monilinia fructigena* in vitro and in infected apple fruits, and its possible role in pathogenesis, *J. Gen. Microbiol.* 128, 799–807, 1982.
136. Portillo, F. and Gancedo, C., Purification and properties of three intracellular proteinases from *Candida albicans*, *Biochim. Biophys. Acta* 881, 299–235, 1986.
137. Toogood, H.S., Prescott, M., and Daniel, R.M., A pepstatin-insensitive aspartic proteinase from a thermophilic *Bacillus* sp., *Biochem. J.* 307, 783–789, 1995.
138. Valdivieso, E., Dagger, F., and Rascón, A., *Leishmania mexicana*: Identification and characterization of an aspartyl proteinase activity, *Exp. Paristol.* 116, 77–82, 2007.
139. Jeong, Y., Cheong, H., Choi, G. et al., An HrpB-dependent but type III-independent extracellular aspartic protease is a virulence factor of *Ralstonia solanacearum*, *Mol. Plant Pathol.* 12, 373–380, 2011.
140. Nishii, W., Ueki, T., Miyashita, R. et al., Structural and enzymatic characterization of physarolisin (formerly physaropepsin) proves that it is a unique serine-carboxyl proteinase, *Biochem. Biophys. Res. Commun.* 301, 1023–1029, 2003.
141. Wlodawer, A., Li, M., Dauter, Z. et al., Carboxyl proteinase from *Pseudomonas* defines a novel family of subtilisin-like enzymes, *Nat. Struct. Biol.* 8, 442–446, 2001.
142. Ackerman, S.K., Matter, L., and Douglas, S.D., Neutrophil chemotaxis and enzyme release. Competitive inhibition by a diazoacetamide pepsin inhibitor, *Biochim. Biophys. Acta* 629, 470–481, 1980.
143. Harvath, L. and Leonard, E.J., Two neutrophil populations in human blood with different chemotactic activities: Separation and chemoattractant binding, *Infect. Immun.* 36, 443–449, 1982.
144. Basilon, J.P., Stickle, D.E., and Holian, A., Extracellular hydrolysis of formyl peptides and subsequent uptake of liberated amino acids by alveolar macrophages, *Biochim. Biophys. Acta* 886, 255–256, 1986.
145. Bayliss, R.S., Knowles, J.R., and Wybrandt, G.B., Aspartic acid residue at the active site of pepsin. The isolation and sequence of the heptapeptide, *Biochem. J.* 113, 377–386, 1969.
146. Hamilton, G.A., Fry, K.T., Kim, O.-K., and Spona, J., Site of reaction of a specific diazo inactivator of pepsin, *Biochemistry* 9, 4624–4632, 1970.
147. Iliadis, G., Zundel, G., and Brzezinski, B., Aspartic proteinases-Fourier transform IR studies of the aspartic carboxylic groups in the active site of pepsin, *FEBS Lett.* 352, 315–317, 1994.
148. Marciniszyn, J., Jr., Huang, J.S., Hartsuck, J., and Tang, J., Mechanism of intramolecular activation of pepsinogen. Evidence for an intermediate δ and the involvement of the active site of pepsin in the intramolecular activation of pepsinogen, *J. Biol. Chem.* 251, 7095–7102, 1976.
149. Hartsuck, J.A., Koelsch, G., and Remington, S.J., The high resolution crystal structure of porcine pepsinogen, *Protein Struct. Funct. Genet.* 13, 1–25, 1992.
150. Brown, J.R. and Hartley, B.S., Location of disulphide bridges by diagonal paper electrophoresis. The disulphide bridges of bovine chymotrypsinogen A, *Biochem. J.* 101, 214–228, 1966.

151. Tang, J. and Hartley, B.S., A diagonal electrophoretic method for selective purification of methionine peptides, *Biochem. J.* 102, 593–599, 1967.
152. Ohihia, A., René, L., Guilhem, J. et al., A new diazoacylating reagent: Preparation and use succinimidyl diazoacetate, *J. Org. Chem.* 58, 1641–1642, 1993.
153. Putnam, F.W., The chemical modification of proteins, in *The Proteins: Chemistry, Biological Activity and Methods*, Vol. 1, Part B, Chapter 10, eds. H. Neurath and K. Bailey, pp. 893–972, Academic Press, New York, 1953.
154. Wu, S. and Soucek, M.D., Kinetic modeling of crosslinking reaction for cycloaliphatic epoxides with hydroxyl- and carboxyl-functionalized acrylic copolymers: 1. pH and temperature effects, *Polymer* 39, 5747–5769, 1998.
155. Skipper, P.L., Naylor, S., Gan, L.-S. et al., Origin of tetrahydrotetrols derived from human hemoglobin adducts of benzo[α]pyrene, *Chem. Res. Toxicol.* 2, 280–281, 1989.
156. Naylor, S., Gan, L.-S., Day, B.W. et al., Benzo[α]pyrene diol epoxide adduct formation in murine and human hemoglobin: Physiochemical basis for dosimetry, *Chem. Res. Toxicol.* 3, 111–117, 1990.
157. Day, B.W., Skipper, P.L., Rich, R.H. et al., Conversion of hemoglobin α-chain aspartate(47) ester to *N*-(2,3-dihydroxypropyl) asparagine as a method for identification of the principal binding site for benzo[α]pyrene anti-diol epoxide, *Chem. Res. Toxicol.* 4, 359–363, 1991.
158. Day, B.W., Skipper, P.L., Zain, J. et al., Benzo[α] pyrene anti-diol epoxide covalently modifies human serum albumin carboxylate side chains and imidazole side chains of histidine (146), *J. Am. Chem. Soc.* 113, 8505–8509, 1991.
159. Kimura, A., Takata, M., Fukushi, Y. et al., A catalytic amino acid and primary structure of active site of *Aspergillus niger* α-glucosidase, *Biosci. Biotech. Biochem.* 61, 1091–1098, 1997.
160. Grazu, V., Abian, O., Mateo, C. et al., Novel bifunctional epoxy/thiol-reactive support to immobilize thiol containing proteins by the epoxy chemistry, *Biomacromolecules* 4, 1495–1501, 2003.
161. Szili, E.J., Kumar, S., DeNichio, M. et al., Development of surface modification techniques for the covalent of insulin-like growth factor-1 (IGF-1) on PECVD silica-coated titanium, *Surf. Coat. Technol.* 205, 1630–1635, 2010.
162. Coad, B.R., Jasieniak, M., Griesser, S.S., and Griesser, H.J., Controlled covalent surface immobilization of proteins and peptides using plasma methods, *Surf. Coat. Technol.* 233, 169–177, 2013.
163. Mateo, C., Grazú, V., Pessela, B.C.C. et al., Advances in the design of new epoxy supports for enzyme immobilization-stabilization, *Biochem. Soc. Trans.* 35, 1593–1601, 2007.
164. Heath, R.J., Di, Y., Clara, S. et al., Epoxide tannage: A way forward, *J. Soc. Leath. Technol. Chem.* 89, 186–193, 2005.
165. Moore, S., Stein, W.H., and Fruton, J.S., Chemical reactions of mustard gas and related compounds. II. The reaction of mustard gas with carboxyl groups and the amino groups of amino acids and proteins, *J. Org. Chem.* 11, 675–680, 1946.
166. Goodlad, G.A.J., Esterification of protein and amino acid carboxyl groups by mustard gas and related compounds, *Biochim. Biophys. Acta* 24, 645–646, 1951.
167. Davis, W. and Rss, W.C.J., Aryl-2-halogenalkylamines. Part VIII. A comparison of the stability of esters derived from some aryl-2-halogenoalkylamines with those obtained from other radiomimetic compounds, *J. Chem. Soc.* 1950, 3056–2062, 1950.
168. Roth, E.F., Jr., Amone, A., Bookchin, R.M., and Nagel, R.L., Chemical modification of human hemoglobin by antisickling concentrations of nitrogen mustard, *Blood* 58, 300–308, 1981.
169. Harris, R.B. and Wilson, I.B., Irreversible inhibition of bovine lung angiotensin I-converting enzyme with *p*[*N,N*-bis(chloroethyl)amino]phenylbutyric acid (chlorambucil) and chlorambucyl L-proline and with evidence that an active site carboxyl group is labeled, *J. Biol. Chem.* 257, 811–815, 1982.
170. Wang, Q.Q., Begum, R.A., Day, V.W. et al., Sulfur, oxygen, and nitrogen mustards: Stability and reactivity, *Org. Biomol. Chem.* 10, 8786–8793, 2012.
171. Khorana, H.G., Synthesis in the study of nucleic acids, *Biochem. J.* 109, 709–725, 1968.
172. Fillingame, R.H., Identification of the dicyclohexylcarbodiimide-reactive protein component of the adenosine-5-triphosphate energy transducing system of *Escherichia coli*, *J. Bacteriol.* 124, 879–883, 1975.
173. Altendorf, K. and Zitzman, W., Identification of a DCCD (*N,N'*-dicyclohexylcarbodiimide)-reactive protein of the energy transducing adenosinetriphosphatase complex, *FEBS Lett.* 59, 268–272, 1975.
174. Azzi, A., Casey, R.P., and Nalecz, M.J., The effect of *N,N'*-dicyclohexylcarbodiimide on enzymes of bioenergetics relevance, *Biochim. Biophys. Acta* 768, 209–226, 1984.
175. Parsonage, D., Wilke-Mounts, S., and Senior, A.E., Directed mutagenesis of the dicyclohexylcarbodiimide-reactive carboxyl residues in β-subunit of F1-ATPase of *Escherichia coli*, *Arch. Biochem. Biophys.* 261, 222–225, 1988.

176. Hassinen, I.E. and Vuokila, P.T., Reaction of dicyclohexylcarbodiimide with mitochondrial proteins, *Biochim. Biophys. Acta* 1144, 107–124, 1993.
177. Kemp, D.S. and Chien, S.W., A new peptide coupling reagent, *J. Am. Chem. Soc.* 89, 2743–2744, 1967.
178. Koolstra, C. and Sluyterman, L.A.A., Isosteric conversion of protein carboxyl groups into carboxamide groups. I. Examination of alkyl and aryl esters as model compounds, *Int. J. Pept. Protein Res.* 29, 347–356, 1987.
179. Woodward, R.B., Olofson, R.A., and Mayer, H., A new synthesis of peptides, *J. Am. Chem. Soc.* 83, 1010–1012, 1961.
180. Woodward, R.B., Olofson, R.A., and Mayer, H., A useful synthesis of peptides, *Tetrahedron* 22(Suppl. 8), 321–326, 1966.
181. Woodward, R.B. and Olofson, R.A., Peptide synthesis using *N*-ethyl-5-phenylisoxazolium-3-sulfonate, carbobenzoxy-L-asparaginyl-L-leucine methyl ester and *N*-carbobenzoxy-3-hydroxy-L-prolylglycylglycine ethyl ester, *Org. Synth.* 56, 88–985, 1977.
182. Gittens, G.J. and James, A.M., Some physical investigations on the behavior of bacterial surfaces VI. Chemical modification of surface components, *Biochim. Biophys. Acta* 66, 237–249, 1963.
183. Meerwein, H., Hinz, G., Hofmann, G. et al., Über tertiäre oxoniumsalze I, *J. Prakt. Chem.* 147, 257–285, 1937.
184. Meerwein, H., Borner, P., Fuchs, O. et al., Reaktionen mit alkylkationen, *Chem. Ber.* 89, 2060–2079, 1957.
185. Meerwein, H., Triethyloxonium tetrafluoroborate, *Org. Synth.* 46, 113, 1966.
186. Pichlmair, S., $R_3O^+BF_4^-$: Meerwein's salt, *Synlett* 2004(1), 195–196, 2004.
187. Chen, F.M.F. and Benoiton, N.L., *N*-Ethylamino acid synthesis and *N*-acylamino acid cleavage using Meerwein's reagent, *Can. J. Chem.* 55, 1433–1435, 1977.
188. Parsons, S.M. and Raftery, M.A., The identification of aspartic acid residue 52 as being critical to lysozyme activity, *Biochemistry* 8, 4199–4205, 1969.
189. Rebek, J., Jr., Marshall, L., Wolak, R. et al., Convergent functional groups: Synthetic and structural studies, *J. Am. Chem. Soc.* 107, 7476–7481, 1985.
190. Waley, S.G., The pH-dependence and group modifications of β-lactamase I, *Biochem. J.* 149, 547–551, 1975.
191. Llewellyn, L.E. and Moczydlowski, E.G., Characterization of saxitoxin binding to saxiphilin, a relative of the transferrin family that displays pH-dependent ligand binding, *Biochemistry* 33, 12312–12322, 1994.
192. Mackinannon, R. and Miller, C., Functional modification of a Ca^{2+}-activated K^+ channel by trimethyloxonium, *Biochemistry* 28, 8087–8092, 1989.
193. Cherbavez, D.B., Trimethyoxonium modification of batrachotoxin-activated Na channels alters functionally important pore residues, *Biophys. J.* 68, 1337–1346, 1995.
194. Gómez, M., Isarna, P., Rojo, M., and Estrada, P., Chemical modification of β-xylosidase from *Trichoderma reesei* QM 9414: pH-dependence of kinetic parameters, *Biochimie* 83, 961–967, 2001.
195. Kim, J.-S., Kim, J.-H., and Kim, H.-J., Matrix-assisted laser desorption/ionization signal enhancement of peptides of picolinimidation of amino groups, *Rapid Commun. Mass Spectrom.* 22, 495–502, 2008.
196. Kim, J.-S., Song, J.-S., Yongju, P. et al., De novo analysis of protein N-terminal sequence utilizing MADLI signal enhancing derivatives with Br, signaling, *Anal. Bioanal. Chem.* 402, 1911–1919, 2012.
197. Yonemitsu, O., Hamada, T., and Kanaoka, H., Esterification of peptides in aqueous solution, *Tetrahedron Lett.* 10(23), 1819–1820, 1969.
198. Bodlaender, P., Feinstein, G., and Shaw, E., The use of isoxazolium salts for carboxyl group modification in proteins. Trypsin, *Biochemistry* 8, 4941–4949, 1969.
199. Dunn, B.M., Anfinsen, C.B., and Shrager, R.I., Kinetics of Woodward's Reagent K hydrolysis and reaction with staphylococcal nuclease, *J. Biol. Chem.* 249, 3717–3723, 1974.
200. Feinstein, G., Bodlaender, P., and Shaw, E., The modification of essential carboxylic acid side chains of trypsin, *Biochemistry* 8, 4949–4955, 1969.
201. Saini, M.S. and Van Etten, R.L., An essential carboxylic acid group in human prostate acid phosphatase, *Biochim. Biophys. Acta* 568, 370–376, 1979.
202. Arana, J.L. and Vallejos, R.H., Two different types of essential carboxyl groups in chloroplast coupling factor, *FEBS Lett.* 123, 103–106, 1981.
203. Dinur, D., Kantrowitz, E.R., and Hajdu, J., Reaction of Woodward's Reagent K with pancreatic porcine phospholipase A2: Modification of an essential carboxylate residue, *Biochem. Biophys. Res. Commun.* 100, 785–792, 1981.
204. Kooistra, C. and Sluyterman, L.A.A., Isosteric acid and non-isosteric modification of carboxyl groups of papain, *Biochim. Biophys. Acta* 997, 115–120, 1989.

205. Johnson, A.R. and Dekker, E.E., Woodward's Reagent K inactivation of *Escherichia coli* L-threonine dehydrogenase: Increased absorbance at 340–350 nm is due to modification of cysteine and histidine residues, not aspartate or glutamate carboxyl groups, *Protein Sci.* 5, 382–390, 1996.

206. Jennings, M.L. and Anderson, M.P., Chemical modification and labeling of glutamate residues at the stilbenedisulfonate site of human blood cell band 3 protein, *J. Biol. Chem.* 262, 1691–1697, 1987.

207. Jennings, M.L. and Al-Rhaiyel, S., Modification of a carboxyl group that appears to cross the permeability barrier in the red blood cell anion transporter, *J. Gen. Physiol.* 92, 162–178, 1988.

208. Bahar, S., Gunter, C.T., Wu, C. et al., Persistence of external chloride and DIDS binding after chemical modification of Glu681 in human band 3, *Am. J. Physiol.* 277, C791–C799, 1999.

209. Salhany, J.M., Sloan, R.L., and Cordes, K.S., The carboxyl side chain of glutamate 681 interacts with a chloride binding modifier site that allosterically modulates the dimeric conformational state of band 3 (AEI). Implications of the mechanism of anion/proton cotransport, *Biochemistry* 42, 1589–1602, 2003.

210. Hoare, D.G. and Koshland, D.E., Jr., A method for the quantitative modification and estimation of carboxyl groups in proteins, *J. Biol. Chem.* 242, 2447–2453, 1967.

211. George, A.L., Jr. and Border, C.L., Jr., Essential carboxyl groups in yeast enolase, *Biochem. Biophys. Res. Commun.* 87, 59–65, 1979.

212. Khorana, H.G., The chemistry of carbodiimides, *Chem. Rev.* 53, 145–166, 1953.

213. Weare, J.A. and Reichert, L.K., Jr., Studies with carbodiimide-cross-linked derivatives of bovine lutropin. I. Effects of specific group modifications on receptor site binding in testes, *J. Biol. Chem.* 254, 6964–6971, 1979.

214. Chan, V.W.F., Jorgensen, A.M., and Border, C.L., Jr., Inactivation of bovine thrombin by water-soluble carbodiimide: The essential carboxyl group has a pKa of 5.51, *Biochem. Biophys. Res. Commun.* 151, 709–716, 1988.

215. Faijes, M., Pérez, X., Pérez, O., and Plantas, A., Glycosynthase activity of *Bacillus licheniformis* 1,3–1,4-β-glucanase mutants: Specificity, kinetics, and mechanism, *Biochemistry* 42, 13304–13318, 2003.

216. Kunkel, G.R., Mehrabian, M., and Martinson, H.G., Contact-site cross-linking agents, *Mol. Cell Biochem.* 34, 2–13, 1981.

217. Minami, Y., Kawasaki, H., Suzuki, K., and Yahara, I., The calmodulin-binding domain of the mouse 90-kDa heat shock protein, *J. Biol. Chem.* 268, 9604–9610, 1993.

218. Okamura-Ikeda, K., Fujiwara, K., and Motokawa, Y., The amino-terminal region of the *Escherichia coli* T-protein of the glycine cleavage system is essential for proper association with H-protein, *Eur. J. Biochem.* 264, 446–452, 1999.

219. Iwamoto, H., Oiwa, K., Suzuki, T., and Fujisawa, T., States of thin filament regulatory proteins as revealed by combining cross-linking/x-ray diffraction techniques, *J. Mol. Biol.* 317, 707–720, 2002.

220. Gu, L.-S., Kim, Y.K., Yan, T. et al., Immobilization of a phosphonated analog of matrix phosphoproteins within cross-linked collagen as a templating mechanism for biomimetic mineralization, *Acta Biomater.* 7, 268–277, 2011.

221. Hwang, Y.-J., Granelli, J., and Lyubovitsky, J., Effects of zero-length and non-zero length cross-linking reagents on the optical spectral properties and structure of collagen hydrogels, *ACS Appl. Mater. Interfaces* 4, 261–267, 2012.

222. Tian, F., Wu, W., Broderick, M. et al., Novel microbiosensors prepared utilizing biomimetic silicification method, *Biosens. Bioelectron.* 25, 2406–2413, 2010.

223. Tamburro, D., Fredolini, C., Espina, V. et al. Multifunctional core-shell nanoparticles: Discovery of previously invisible biomarkers, *J. Am. Chem. Soc.* 133, 19178–19188, 2011.

224. Fischer, S.E., Mi, L., Mao, H.-Q. et al., Biofunctional coatings via targeted covalent cross-linking of association triblock proteins, *Biomacromolecules* 10, 2408–2417, 2009.

225. Ma, D., Tu, K., and Zhang, L.-M., Bioactive supramolecular hydrogel with controlled dual drug release characteristics, *Biomacromolecules* 11, 2204–2212, 2010.

226. Raks, F., Mihic, A., Reis, L. et al., Hydrogels modified with QHREDGS peptide support cardiomyocyte survival in vitro and after sub-cutaneous implantation, *Soft Matter* 6, 5089–5099, 2010.

227. Hwang, Y.J. and Lyubovitsky, J.G., The structural analysis of three-dimensional fibrous collagen hydrogels by Raman microspectrometry, *Biopolymers* 99, 349–356, 2013.

228. Zambaux, M.F., Bonneaux, F., and Deliacherie, E., Covalent fixation of soluble derivatized dextrans to model proteins in low-concentration medium: Application to factor IX and protein C, *J. Protein Chem.* 17, 279–284, 1998.

229. Vinayaka, A.C. and Thakur, M.S., Photoabsorption and resonance energy transfer phenomenon in CdTe-protein bioconjugates: An insight into QD-biomolecular interactions, *Bioconjug. Chem.* 22, 968–975, 2011.

230. Misri, R., Saatchi, K., and Häfali, U.O., Radiolabeling of fab and f(ab')$_2$ antibody fragments with ^{99m}Tc(I) tricarbonyl core using a new bifunctional tridentate ligand, *Nuclear Med. Commun.* 32, 324–329, 2011.

231. Hsu, J.-C., Huang, C.-C., Ou, K.-L. et al., Silica nanohybrids integrated with CuInS$_2$/ZnS quantum dots and magnetite nanocrystals: Multifunctional agents for dual-mobility imaging and drug delivery, *J. Mater. Chem.* 21, 19257–19266, 2011.

232. Kar, P., Talard, F., Lamblin, G. et al., Silver nanoparticles to improve electron transfer at interfaces of gold electrodes modified by biotin or avidin, *J. Electroanal. Chem.* 69, 17–25, 2013.

233. Kiss, T., Erdei, A., and Kiss, L., Investigation of the active site of the extracellular β-D-xylosidase from *Aspergillus carbonarius*, *Arch. Biochem. Biophys.* 399, 188–194, 2002.

234. Cook, G.M., Keis, S., Morgan, H.W. et al., Purification and biochemical characterization of the F$_1$F$_0$ ATP synthase from thermophilic *Bacillus* sp. strain TA2.A1, *J. Bacteriol.* 185, 4442–4449, 2003.

235. Das, A. and Ljungdahl, L.G., *Clostridium pasteurianum* F$_1$F$_0$ ATP synthase: Operon, composition and some properties, *J. Bacteriol.* 185, 5527–5535, 2003.

236. Peacock, E.E., Jr., Inter- and intramolecular bonding in collagen of healing wounds by insertion of methylene and amide cross-links into scar tissue: Tensile strength and thermal shrinkage in rats, *Ann. Surg.* 163(1), 1–9, 1966.

237. Johnson, H.M., Brenner, K., and Hall, H.E., Use of a water-soluble carbodiimide as a coupling reagent in the passive hemagglutination test, *J. Immunol.* 97, 791–796, 1966.

238. Stason, W.B., Vallotton, M.B., and Haber, E., Synthesis of an antigenic copolymer of angiotensin and succinylated poly-L-lysine, *Biochim. Biophys. Acta* 133, 582–584, 1967.

239. Carpenter, R.H. and Barsales, P.B., Uptake by mononuclear phagocytes of protein-coated bentonite particles stabilized with a carbodiimide, *J. Immunol.* 98, 844–853, 1967.

240. Kluge, C. and Dimroth, P., Specific protection by Na$^+$ or Li$^+$ of the F$_1$F$_0$-ATPase of *Propionigenium modestum* from the reaction with dicyclohexylcarbodiimide, *J. Biol. Chem.* 268, 14557–14460, 1993.

241. Kluge, C. and Dimroth, P., Kinetics of inactivation of the F$_1$F$_0$ ATPase of *Propionigenium modestum* by dicyclohexylcarbodiimide in relationship to H$^+$ and Na$^+$ concentration: Probing the binding site for the coupling ions, *Biochemistry* 32, 10378–10386, 1993.

242. Tohri, A., Dohmae, H., Suzuki, T. et al., Identification of the domains on the extrinsic 23 kD protein possibly involved in electrostatic interaction with the extrinsic 33 kDa protein in spinach photosystem II, *Eur. J. Biochem.* 271, 962–971, 2004.

243. Volkova, R.I. and Kochetova, L.M., Effect of carbodiimide on the catalytic properties of cholinesterases, *Biokhimiya* 46, 1823–1831, 1981.

244. Huynh, Q.K., Evidence for a reactive γ-carboxyl group (Glu-418) at the herbicide glyphosate binding site of 5-enolpyruvylshikimate-3-phosphate synthase from *Escherichia coli*, *J. Biol. Chem.* 263, 11631–11635, 1988.

245. Chen, D.H., Daron, H.H., and Aull, J.L., 1-Phenyl-3-trimethylaminopropyl carbodiimide: A new inhibitor of thymidylate synthase, *J. Enzyme Inhib.* 5, 259–268, 1992.

246. Bulone, V., Lam, B.-I., and Stone, B.A., The effect of amino acid modifying reagents on the activity of a (1 → 3)-β-glucan synthase from Italian ryegrass (*Lolium multiflorum*) endosperm, *Phytochemistry* 50, 9–15, 1999.

247. Skuratovskaya, T.N., Mironova, D.F., and Dvorko, G.F., Kinetics and mechanism of condensations with the participation of carbodiimides. III. Dehydration of phthalamic acid and its derivatives under the influence of dicyclohexylcarbodiimide, *Ukr. Khim. Zh.* 35, 947–952, 1969.

248. Kluge, C. and Dimroth, P., Modification of isolated subunit c of the F$_1$F$_0$-ATPase from *Propionigenium modestum* by dicyclohexylcarbodiimide, *FEBS Lett.* 349, 245–248, 1994.

249. Meier, T., Matthey, U., von Ballmoos, C. et al., Evidence for structural integrity in the undecameric c-rings isolated from sodium ATP synthases. *J. Mol. Biol.* 325, 389–397, 2003.

250. Sheehan, J.C. and Hlavka, J.J., The use of water-soluble and basic carbodiimides in peptide synthesis, *J. Org. Chem.* 21, 439–440, 1956.

251. Sheehan, J.C. and Hess, G.P., A new method of forming peptide bonds, *J. Am. Chem. Soc.* 77, 1067–1068, 1955.

252. Sheehan, J.C. and Hlavka, J.J., The cross-linking of gelatin using a water-soluble carbodiimide, *J. Am. Chem. Soc.* 79, 4528–4529, 1957.

253. Riehm, J.P. and Scheraga, H.A., Structural studies on ribonuclease. XXI. The reaction between ribonuclease and a water-soluble carbodiimide, *Biochemistry* 5, 99–115, 1966.

254. Hoare, D.G. and Koshland, D.E., Jr., A procedure for the selective modification of carboxyl groups in proteins, *J. Am. Chem. Soc.* 88, 2057, 1966.
255. Lin, T.-Y. and Koshland, D.E., Jr., Carboxyl group modification and the activity of lysozyme. *J. Biol. Chem.* 244, 505–508, 1969.
256. Moffett, J.B., Namboodiri, M.A.A., and Neale, J.H., Enhanced carbodiimide fixation for immunohistochemistry: Application to the comparative distributions of *N*-acetylaspartylglutamate and *N*-acetylaspartate immunoreactivities in rat brain, *J. Histochem. Cytochem.* 41, 559–570, 1995.
257. Hsieh, S.-C., Tang, C.-M., Huang, W.-T. et al., Comparison between two different methods of immobilizing NGF in poly (DL-lactic acid-co-glycolic acid) conduit for peripheral nerve regeneration by EDC/NHS/MES and genipin, *J. Biomed. Mater. Res.* 99A, 576–585, 2011.
258. Good, N.E., Winget, G.D., Winter, W. et al., Hydrogen ion buffers for biological research, *Biochemistry* 5, 467–477, 1966.
259. Good, N.E. and Izawa, S., Hydrogen ion buffers, *Methods Enzymol.* 24, 53–68, 1972.
260. Pena, J.T., Sohn-Lee, C., Rouhanifard, S.H. et al., miRNA in situ hybridization in formaldehyde and EDC-fixed tissues, *Nat. Methods* 5, 139–141, 2009.
261. Takei, S., Hasegawa-Ishii, S., Uekawa, A. et al., Immunohistochemical demonstration of increased prostaglandin $F_2\alpha$ levels in the rat hippocampus following kainic acid-induced seizures, *Neuroscience* 218, 295–304, 2012.
262. Metz, D.H. and Brown, G.L., The investigation of nucleic acid secondary structure by means of chemical modification with a carbodiimide reagent. I. The reaction between *N*-cyclohexyl-*N′*-β-(4-methylmorphilium) ethyl carbodiimide and model nucleotides, *Biochemistry* 8, 2312–2328, 1969.
263. Gilham, P.T. and Khorana, H.G., Studies on polynucleotides. I. A new and general method for the chemical synthesis of the C_5'-C_3' internucleotidic linkage. Syntheses of deoxyribo-dinucleotides, *J. Am. Chem. Soc.* 80, 6212–6222, 1958.
264. Glonek, T., Kleps, R.A., Van Wazer, J.R., and Myers, T.C., Carbodiimide-intermediated esterification of the inorganic phosphates and the effect of tertiary amine base, *Bioinorg. Chem.* 5, 283–310, 1976.
265. Popisov, M.I., Maksimenko, A.H., and Torchilin, V.P., Optimization of reaction conditions during enzyme immobilization on soluble carboxyl-containing carriers, *Enzyme Microb. Technol.* 7, 11–16, 1981.
266. Bettendorff, L., Nghlem, H.O., Wins, P., and Lakaye, B., A general method for the chemical synthesis of γ-^{32}P-labeled or unlabeled nucleoside 5′-triphosphates and thiamine triphosphate, *Anal. Biochem.* 322, 190–197, 2003.
267. Murphy, A.J., Reaction of a carbodiimide adduct of ATP at the active site of sarcoplasmic reticulum calcium, *Biochemistry* 29, 11236–11242, 1990.
268. Chu, B.C.F., Wahl, G.M., and Orgel, L.E., Derivatization of unprotected polynucleotides, *Nucleic Acids Res.* 11, 6513–6529, 1983.
269. Rao, K.S., Rani, S.U., Kamakshaiah, D. et al., A novel route for immobilization of oligonucleotides onto modified silica nanoparticles, *Anal. Chim. Acta* 576, 177–183, 2006.
270. Gilles, M.A., Hudson, A.Q., and Borders, C.L., Jr., Stability of water-soluble carbodiimides in aqueous solution, *Anal. Biochem.* 184, 244–248, 1990.
271. Lei, P.Q., Lamb, D.H., Heller, R.K. et al., Kinetic studies on the rate of hydrolysis of *N*-ethyl-*N′*-(dimethylaminopropyl) carbodiimide I aqueous solution using mass spectrometry and capillary electrophoresis, *Anal. Biochem.* 310, 122–124, 2002.
272. Janolino, V.G. and Swaisgood, H.E., Analysis and optimization of methods using water-soluble carbodiimide for immobilization of biochemicals to porous glass, *Biotechnol. Bioeng.* 24, 1069–1080, 1982.
273. Wrobel, N., Schinkinger, M., and Mirsky, V.M., A novel ultraviolet assay for testing side reactions of carbodiimide, *Anal. Biochem.* 305, 135–138, 2003.
274. Sehgal, D. and Vijay, I.K., A method for the high efficiency of water-soluble carbodiimide-mediated amidation, *Anal. Biochem.* 218, 87–91, 1994.
275. Carraway, K.L. and Triplett, R.B., Reaction of carbodiimides with protein sulfhydryl groups, *Biochim. Biophys. Acta* 200, 564–566, 1970.
276. Carraway, K.L. and Koshland, D.E., Jr., Reaction of tyrosine residues in proteins with carbodiimide reagents, *Biochim. Biophys. Acta* 160, 272–274, 1968.
277. Perfetti, R.B., Anderson, C.D., and Hall, P.L., The chemical modification of papain with 1-ethyl-3(3-dimethylaminopropyl) carbodiimide, *Biochemistry* 15, 1735–1743, 1976.
278. Lacombe, G., Van Thiem, N., and Swyngehedauw, B., Modification of myosin subfragment 1 by carbodiimide in the presence of a nucleophile. Effect on adenosinetriphosphatase activities, *Biochemistry* 20, 3648–3653, 1981.

279. Inano, H. and Tamaoki, B., The presence of essential carboxyl group for binding of cytochrome c in rat hepatic NADPH-cytochrome P-450 reductase by the reaction with 1-ethyl-3-(3-dimethylaminopropyl) carbodiimide, *J. Enzyme Inhib.* 1, 47–59, 1985.

280. Carraway, K.L. and Koshland, D.E., Jr., Carbodiimide modification of proteins, *Methods Enzymol.* 25, 616–623, 1972.

281. Millotti, G., Samberger, C., Fröhlich, E., and Bernkop-Schnürch, A., Chitosan-*graft*-6-mercaptonicotinic acid: Synthesis, characterization, and biocompatibility, *Biomacromolecules* 10, 3023–3027, 2009.

282. Kafedjiiski, K., Foeger, F., Werle, M., and Bernkop-Schnürch, A., Synthesis and in vitro evaluation of a novel chitosan-glutathione conjugate, *Pharm. Res.* 22, 1480–1488, 2005.

283. Gianolio, D.A., Philbrook, M., Avila, L.B. et al., Synthesis and evaluation of hydrolyzable hyaluronan-tethered bupivacaine delivery systems, *Bioconjug. Chem.* 16, 1512–1518, 2005.

284. van der Wlies, A.J., O'Neil, C.P., Hasegawa, U. et al., Synthesis of pyridyl disulfide-functionalized nanoparticles for conjugating thiol-containing small molecules, *Bioconjug. Chem.* 21, 653–662, 2010.

285. Lewis, S.D. and Shafer, J.A., Conversion of exposed aspartyl and glutamyl residues in proteins to asparaginyl and glutaminyl residues, *Biochim. Biophys. Acta* 303, 284–291, 1973.

286. Wang, T.-T. and Young, N.M., Modification of aspartic acid residues to induce trypsin cleavage, *Anal. Biochem.* 91, 696–699, 1978.

287. Lin, C.M., Mihal, K.A., and Krueger, R.J., Introduction of sulfhydryl groups into proteins at carboxyl sites, *Biochim. Biophys. Acta* 1038, 382–385, 1990.

288. Montes, T., Grazu, V., Lopez-Gallego, F. et al., Chemical modification of protein surfaces to improve their reversible enzyme immobilization on ionic exchangers, *Biomacromolecules* 7, 3052–3058, 2006.

289. Pho, D.B., Roustan, C., Toi, A.N., and Pradel, L.-A., Evidence for an essential glutamyl residue in yeast hexokinase, *Biochemistry* 16, 4533–4537, 1977.

290. Desvages, G., Roustan, C., Fattoum, A., and Pradel, L.A., Structural studies on yeast 3-phosphoglycerate kinase. Identification by immuno-affinity chromatography of one glutamyl residue essential for yeast 3-phosphoglycerate kinase activity. Its location in the primary structure. *Eur. J. Biochem.* 105, 259–266, 1980.

291. Körner, M., Van Thiem, N., Lacombe, G., and Swynghedauw, B., Cardiac myosin subfragment 1 modification by carbodiimide in the presence of nucleophile, *Biochem. Biophys. Res. Commun.* 105, 1198–1207, 1982.

292. Körner, M., VanThiem, N., Cardinaud, R., and Lacombe, G., Location of an essential carboxyl group along the heavy chain of cardiac and skeletal myosin subfragments 1, *Biochemistry* 22, 5843–5847, 1983.

293. Hegde, S.S. and Blanchard, J.S., Kinetic and mechanistic characterization of recombinant *Lactobacillus viridescens* FemX (UDP-*N*-acetylmuramoyl pentapeptide-lysine N^6-alanine transferase, *J. Biol. Chem.* 278, 22861–22867, 2003.

294. Zhang, H., Shen, W., Rempel, D. et al., Carboxyl-group footprinting maps the dimerization interface and phosphorylation-induced conformational changes of a membrane-associated tyrosine kinase, *Mol. Cell Proteomics* 10(6), M110.005678, 2011.

295. Zhang, H., Wen, J., Huang, R.Y.-C. et al., Mass spectrometry-based carboxyl footprinting of proteins: Method evaluation, *Int. J. Mass Spectrom.* 312, 78–86, 2012.

296. Collier, T.S., Diraviyam, K., Monsey, J. et al., Carboxyl group footprinting mass spectrometry and molecular dynamics identify key interactions in the HER2-HER3 receptor tyrosine kinase interface, *J. Biol. Chem.* 288(35), 25254–25264, 2013.

297. Sigrist-Nelson, K. and Azzi, A., The proteolipid subunit of the chloroplast adenosine triphosphatase complex. Reconstitution and demonstration of proton-conductive properties, *J. Biol. Chem.* 255, 10638–10643, 1980.

298. Sussman, M.R. and Slayman, C.W., Modification of the *Neurospora crassa* plasma membrane (H+)-ATPase with *N,N'*-dicyclohexylcarbodiimide, *J. Biol. Chem.* 258, 1839–1843, 1983.

299. Lotscher, H.R., deJay, C., and Capaldi, R.A., Inhibition of the adenosinetriphosphatase activity of *Escherichia coli* F_1 by the water-soluble carbodiimide, 1-ethyl-3-(3-dimethylaminopropyl) carbodiimide is due to modification of several carboxyls in the beta subunit, *Biochemistry* 23, 4134–4140, 1984.

300. Shafir, I., Feng, W., and Shoshan-Barmatz, V., Dicyclohexylcarbodiimide interaction with the voltage-dependent anion channel from sarcoplasmic reticulum, *Eur. J. Biochem.* 253, 627–636, 1998.

301. Yang, S.J., Jiang, S.S., Kuo, S.Y. et al., Localization of a carboxylic residue possibly involved in the inhibition of vacuolar H+-pyrophosphatase by *N,N'*-dicyclohexylcarbodiimide, *Biochem. J.* 342, 641–646, 1999.

302. Bjerrum, P.J., Anderson, O.S., Borders, C.L., Jr., and Wieth, J.O., Functional carboxyl groups in the red blood cell anion exchange protein, *J. Gen. Physiol.* 93, 813–839, 1989.

303. Takahashi, K., Stein, W.H., and Moore, S., The identification of a glutamic acid residue as part of the active site of ribonuclease T1, *J. Biol. Chem.* 242, 4682–4690, 1967.
304. Kojima, M., Mizukoshi, T., Miyano, H. et al., Thermal stabilization of ribonuclease T1 by carboxy-methylation at Glu-58 as revealed by ^{1}H nuclear magnetic resonance spectrometry, *FEBS Lett.* 351, 389–392, 1994.
305. Lin, T.-Y. and Kuo, D.W., Inactivation of human fibroblast collagenase by chloroacetyl *N*-hydroxypeptide derivatives, *J. Enzyme Inhib.* 5, 33–40, 1991.
306. Bianco, A., Kaiser, D., and Jung, G., *N*-Hydroxy peptides as substrates for α-chymotrypsin, *J. Pept. Res.* 54, 544–548, 1999.
307. Erlanger, B.F., Vratsanos, S.M., Wassermann, M., and Cooper, A.G., Specific and reversible inactivation of pepsin, *J. Biol. Chem.* 240, PC3447, 1965.
308. Gross, E. and Morell, J.L., Evidence for an active carboxyl group in pepsin, *J. Biol. Chem.* 241, 3638–3639, 1966.
309. Chacur, M., Longo, I., Picolo, G. et al., Hyperalgesia induced by Asp49 and Lys49 phospholipases A_2 from *Bothrops asper* snake venom: Pharmacological mediation and molecular determinants, *Toxicon* 41, 667–678, 2003.
310. Pickering, A.M. and Davies, K.J.A., A simple fluorescence labeling method for studies of protein oxidation, protein modification, and proteolysis, *Free Radic. Biol. Med.* 52, 239–246, 2012.

11 Modification of Tyrosine Residues

Tyrosine represents approximately 3%–4% of the amino composition of an average protein.[1-3] There are differences in composition relating to protein size and cellular location; however, these are not significant for tyrosine.[4] Some amino acids such as cysteine/cystine, arginine, and lysine are markedly increased in small proteins, while aspartic acid and glutamic acid are decreased in small proteins.[5] Pace and coworkers[6] have summarized the pKa values for the various ionizable groups in proteins. They reported that 67% of the tyrosine residues were buried with an average pKa value of 10.3 with a low value of 6.1 and a high value of 12.1. The pKa for the tyrosyl residue in the pentapeptide, AlaAlaTryAlaAla, was found to be 9.8. Tyrosine is considered to be a hydrophobic amino acid[7,8] and is found at protein interfaces.[8-15] As one example of this function, Ploug and coworkers[16] used TNM to identify tyrosine residues at the interface between urokinase-type plasminogen activator and a glycolipid-anchored receptor.

Tyrosyl residues in proteins are subject to posttranslational modification including phosphorylation,[17-21] sulfation,[22-24] and nitration.[25-28] There will be only limited discussion of the physiological consequences of such in vivo modifications in the current work. It is not without a certain amount of bitterness that I recall that an NIH study section stated that our work[29] on nitrotyrosine would be of little value and that I should discontinue that line of research.

It is possible to modify tyrosyl residues in proteins under relatively mild conditions with reasonably high specificity with a variety of reagents (Table 11.1; Figure 11.1) obtaining, in turn, a variety of useful derivatives. For example, NAI acetylated the phenolic hydroxyl group in a reversible reaction. TNM nitrates tyrosyl residues to yield the 3-nitro derivative that markedly lowers the pKa of the phenolic hydroxyl group (see above). The 3-nitro function can be reduced under mild conditions to give the 3-amino derivatives that can be subsequently modified with a variety of useful compounds such as dansyl chloride 5-(dimethylamino)-1-napthalenesulfonyl chloride or biotinamidohexanoic acid sulfo-N-hydroxysuccinimide ester, sodium salt (sulfosuccinimidyl-6-biotinamidohexanoate). Reaction with TNM can also result in zero-length cross-linkage in proteins via the formation of dityrosine. Peroxynitrite also modifies tyrosine residues in proteins yielding products similar to those observed with TNM, the processes for the production of peroxynitrite and the subsequent modification of tyrosine (and tryptophan and cysteine).

Factors that influence the reactivity of tyrosyl residues in proteins are, at best, poorly understood. It would seem that the reactivity is influenced most by the ionization state of tyrosine that is a function of the microenvironment around the residue. Tyrosine residues were considered *exposed* or buried on the basis of UV–VIS spectrophotometric measurement as a function of pH and perturbation.[30] The advent of chemical modification created a considerable debate over *exposed* versus *buried* tyrosine residues, which were not entirely solvent with crystallographic analysis. Glazer provides an excellent summary of the issue in a 1976 review chapter.[31] Exposed and buried tyrosine residues continue to be a subject of spectral analysis albeit with more sophisticated NMR techniques.[32,33]

It is extremely useful for investigators to review early literature on the factors influencing tyrosine ionization in proteins.[34-39] It is also useful to appreciate that iodination and nitration are examples of electrophilic substitution of aromatic rings,[40] while reactions such as acetylation at the phenolic hydroxyl groups are acylation reactions dependent in part on the nucleophilicity of the said

TABLE 11.1
Techniques and Reagents Used to Modify Tyrosyl Residues in Proteins

Reagent	Advantages and Disadvantages	References
Iodination	The reaction results in the formation of 3(5) monoiodotyrosine or 3,5-diiodotyrosine. This reaction is rarely used today for the modification of tyrosyl residues in proteins but rather for the preparation of radiolabeled proteins for use in research or therapeutic applications. This is a mature technology with a limited new work on the protein chemistry. Discussion of specific techniques is presented in Table 11.2.	11.1.1–11.1.5
Acetic anhydride	Formation of O-acetyl derivative of the phenolic hydroxyl group of tyrosine. As an ester bond, this is not a stable modification and can be reversed by base or hydroxylamine. It can be a complicating reaction when acetic anhydride is used for modification of lysine residues in proteins but has been used for surface residue modification. The presence of acetic anhydride in cigarette smoke has been shown to produce modification of tyrosine residues. There is additional information in Table 11.3	11.1.6–11.1.9
NAI	Relatively specific for the modification of tyrosine residues in proteins by forming the O-acetyl derivative. Transient reaction has been reported at histidine residues in serine proteases; modification of primary amino groups is rare. Reaction occurs under mild conditions using phosphate or borate buffers in the pH range of 6–8. The hydrolysis of reagent is a complication that needs to be considered in kinetic studies. See Table 11.3 for more experimental detail.	11.1.10–11.1.14
Peroxynitrite [oxidoperoxynitrate(1-)]	Peroxynitrite can be formed in vitro by a variety of reactions including the action of hydrogen peroxide on nitrous acid and more recently the action of hydrogen peroxide on isoamyl nitrite; in vivo formation from nitric oxide and superoxide anion. Peroxynitrite is unstable at acid pH but is stable at alkaline pH and is commercially available as the sodium salt in sodium hydroxide. Peroxynitrite reacts with tyrosine to form 3-nitrotyrosine residues in proteins and can also oxidize cysteine and methionine. There is only limited use of peroxynitrite for the *in vitro* modification of tyrosyl residues in proteins but far more interest in the use of 3-nitrotyrosine as a biomarker for the *in vivo* production of peroxynitrite in oxidative stress. Peroxynitrite has been shown to be present in cigarette smoke and can react with proteins. There is additional information in Table 11.4 and Chapter 4.	11.1.15–11.1.23
TNM	A classic method for the in vitro nitration of tyrosyl residues in proteins. The reaction occurs under mild conditions with most studies performed at pH 8.0. Oxidation of sulfhydryl groups can occur as well as cross-linking of tyrosine residues. There is also potential for the nitration of tryptophan. There is a listing of the application of TNM for the modification of selected proteins in Table 11.5.	11.1.24–11.1.30
Diazonium salts	These reagents are of considerable historical interest as early dyes for textiles, then later as reagents for the modification of proteins. After a period of some inactivity, there has been a renaissance of activity focused on the specificity for modification of tyrosine residues and the ability of the aromatic ring structure as platform for other functional groups and for the attachment of PEG. The recent activity has been stimulated by the development of more stable reagents and a careful study of reaction conditions. Diazonium-1-H-tetrazole was the reagent used in early studies on protein chemistry and was generated in situ by the action of nitrous acid on 5-amino-1-H-tetrazole; the dry reagent is explosive and poses a danger.	11.1.31–11.1.35
NBSF	A reagent with considerable specificity for tyrosine but with little use in the last 20 years.	11.1.36–11.1.39

TABLE 11.1 (continued)

Techniques and Reagents Used to Modify Tyrosyl Residues in Proteins

References to Table 11.1

11.1.1. Espuna, G., Andreu, D., Barluenga, J. et al., Iodination of proteins by IPy$_2$BF$_4$, a new tool in protein chemistry, *Biochemistry* 45, 5957–5963, 2006.

11.1.2. Sohoel, A., Plum, A., Frokjaer, S., and Thygesen, P., ^{125}I used for labeling of proteins in an absorption model changes the absorption rate of insulin aspart, *Int. J. Pharm.* 330, 114–120, 2007.

11.1.3. Vaidyanathan, G. and Zalutsky, M.R., Synthesis of *N*-succinimidyl-4-guanidinomethyl-3[*I]iodobenzoate, a radio-iodination agent for labeling internalizing proteins and peptides, *Nat. Protoc.* 2, 282–286, 2007.

11.1.4. Inlesta, J., Cooper, H.J., Marshall, A.G. et al., Specific electrochemical iodination of horse heart myoglobin at tyrosine 103 as determined by Fourier transform ion cyclotron resonance mass spectrometry, *Arch. Biochem. Biophys.* 474, 1–7, 2008.

11.1.5. Waentig, L., Jakubowski, N., Hayen, H., and Roos, P.H., Iodination of proteins, proteomes and antibodies with potassium triodide for LA-ICP-MS based proteomic analyses, *J. Anal. At. Spectrom.* 26, 1610–1618, 2011.

11.1.6. Stratilová, E., Dzúrová, M., Markovic, O., and Jörnvall, H., An essential tyrosine residue of Aspergillus polygalacturonase, *FEBS Lett.* 382, 164–166, 1996.

11.1.7. King, B., Normant, L., King, D., and Storey, D., Acetylation labeling mass spectrometry: A method for studying protein conformations and interactions, *Proc. Indiana Acad. Sci.* 118, 107–113, 2009.

11.1.8. Liao, R., Wu, H., Deng, H. et al., Specific and efficient *N*-propionylation of histones with propionic acid *N*-hydroxysuccinimide ester for histone marks characterization by LC-MS, *Anal. Chem.* 85, 2253–2259, 2013.

11.1.9. Takahashi, Y., Horiyama, S., Honda, C. et al., A chemical approach to searching for bioactive ingredients in cigarette smoke, *Chem. Pharm. Bull.* 61, 85–89, 2013.

11.1.10. Cymes, G.D., Iglesias, M.M., and Wolfenstein-Todal, C., Chemical modification of ovine prolactin with *N*-acetylimidazole, *Int. J. Pept. Protein Res.* 42, 33–38, 1993.

11.1.11. Zhang, F., Gao, J., Weng, J. et al., Structural and functional differentiation of three groups of tyrosine residues by acetylation of *N*-acetylimidazole in manganese stabilizing protein, *Biochemistry* 44, 719–725, 2006.

11.1.12. Gao, J., Zhang, F., Weng, J. et al., Tyrosine residues of the extrinsic 23 kDa protein are important for its interaction with spinach PSII membranes, *Protein Pept. Lett.* 13, 539–544, 2006.

11.1.13. Khan, F., Ahmad, A., and Khan, M.I., Steady state and time resolved fluorescence quenching and chemical modification studies of a lectin from endophytic fungus, *Fusarium solani*, *J. Fluoresc.* 20, 305–313, 2010.

11.1.14. Kumar, U., Ranjan, A.K., Sharan, C. et al., Green approaches to size controlled synthesis of biocompatible antibacterial metal nanoparticles in aqueous phase using lysozyme, *Curr. Nanosci.* 8, 130–140, 2012.

11.1.15. Alvarez, B. and Radi, R., Peroxynitrite reactivity with amino acids and proteins, *Amino Acids* 25, 295–311, 2003.

11.1.16. Robinson, K.M. and Beckman, J.S., Synthesis of peroxynitrite from nitrite and hydrogen peroxide, *Methods Enzymol.* 396, 207–214, 2005.

11.1.17. Uppu, R.M., Synthesis of peroxynitrite using isoamyl nitrite and hydrogen peroxide in a homogeneous solvent system, *Anal. Biochem.* 354, 165–168, 2006.

11.1.18. Hughes, M.N., Chemistry of nitric oxide and related species, *Methods Enzymol.* 436. 3–19, 2008.

11.1.19. Goldstein, S. and Merényi, G., The chemistry of peroxynitrite: Implications for biological activity, *Methods Enzymol.* 436, 49–61, 2008.

11.1.20. Rebrin, I., Bregere, C., Gallaher, T.K., and Sohal, R.S., Detection and characterization of peroxynitrite-induced modifications of tyrosine, tryptophan, and methionine residues by tandem mass spectrometry, *Methods Enzymol.* 441, 283–294, 2008.

11.1.21. Alvarez, B., Caballal, S., Turell, L., and Radi, R., Formation and reactions of sulfenic acid in human serum albumin, *Methods Enzymol.* 473, 117–136, 2010.

11.1.22. Hof, D., Cooksley-Decasper, S., Moergli, S., and von Eckardstein, A., Generation of novel recombinant antibodies against nitrotyrosine by antibody phage display, *Hum. Antibodies* 20, 15–27, 2011.

11.1.23. Chen, H.J. and Chen, Y.C., Reactive nitrogen oxide species-induced post-translational modifications in human hemoglobin and the association with cigarette smoking, *Anal. Chem.* 84, 7881–7890, 2012.

11.1.24. Riordan, J.F. and Valle, B.L., Nitration with tetranitromethane, *Methods Enzymol.* 25, 515–521, 1972.

11.1.25. Liu, H., Gaza-Bulseco, G., Chumsae, C., and Radziejewski, C.H., Mass spectrometry analysis of in vitro nitration of a recombinant human IgG1 monoclonal antibody, *Rapid Commun. Mass Spectrom.* 22, 1–10, 2008.

(continued)

TABLE 11.1 (continued)

Techniques and Reagents Used to Modify Tyrosyl Residues in Proteins

11.1.26. Bolivar, J.G., Soper, S.A., and McCarley, R.L., Nitroavidin as a ligand for the surface capture and release of biotinylated proteins, *Anal. Chem.* 80, 9336–9342, 2008.

11.1.27. Fujisawa, Y., Kato, K., and Giulivi, C., Nitration of tyrosine residues 368 and 345 in the β-subunit elicits FoF1-ATPase activity loss, *Biochem. J.* 423, 219–231, 2009.

11.1.28. Ghesquière, B., Colaert, M., Libert, C. et al., In vitro and in vivo protein-bound tyrosine nitration characterized by diagonal chromatography, *Mol. Cell. Proteomics* 8, 2642–2652, 2009.

11.1.29. Zhang, Y., Yang, H., and Pöschl, U., Analysis of nitrated proteins and tryptic peptides by HPLC-chip-MS/MS: Site-specific quantification, nitration degree and reactivity of tyrosine residues, *Anal. Bioanal. Chem.* 399, 459–471, 2011.

11.1.30. Gong, B., Ramos, A., Vásquez-Fernández, E. et al., Probing structural difference between PrP(C) and PrP(Sc) by surface nitration and acetylation: Evidence of conformation change in the C-terminus, *Biochemistry* 50, 4963–4972, 2011.

11.1.31. Pauly, H. and Binz, A., Silk and wool as dye producers, *Zeitschrift fuer Farben- und Textil-Chemie* 3, 373–374, 1904.

11.1.32. Horinishi, H., Hachimori, Y., Kurihara, K., and Shibata, K., States of amino acid residues in proteins. III Histidine residues in insulin, lysozyme, albumin and proteinases as determined with a new reagent of diazo-1-H-tetrazole, *Biochim. Biophys. Acta* 86, 477–489, 1964.

11.1.33. Riordan, J.F. and Vallee, B.L., Diazonium salts as specific reagents and probes of protein conformation, *Methods Enzymol.* 25, 521–531, 1972.

11.1.34. Gavvrilyuk, J., Ban, H., Nagano, M. et al., Formylbenzene diazonium hexafluorophosphate reagent for tyrosine-selective modification of proteins and the introduction of a bioorthogonal aldehyde, *Bioconjug. Chem.* 23, 2321–2328, 2012.

11.1.35. Jones, M.W., Mantovani, G., Blindauer, C.A. et al., Direct peptide bioconjugation/PEGylation at tyrosine with linear and branched polymeric diazonium salts, *J. Am. Chem. Soc.* 134, 7406–7413, 2012.

11.1.36. Liao, T.H., Ting, R.S., and Yeung, J.E., Reactivity of tyrosine in bovine pancreatic deoxyribonuclease with *p*-nitrobenzenesulfonyl fluoride, *J. Biol. Chem.* 257, 5637–5644, 1982.

11.1.37. Blanchardie, P., Denis, M., Orsonneau, J.L., and Lustenberger, P., Reaction of tyrosyl-modifying reagents with the ligand- and DNA-binding domains of the rabbit liver glucocorticoid receptor, *J. Steroid Biochem.* 36, 15–23, 1990.

11.1.38. Zweidler, A., Role of individual histone tyrosines in the formation of the nucleosome complex, *Biochemistry* 31, 9205–9211, 1992.

11.1.39. Chang, L. and Lin, S., Modification of tyrosine −3(63) and lysine −6 of Taiwan cobra phospholipase A$_2$ affects its ability to enhance 8-anilinonaphthalene-1-sulfonate fluorescence, *Biochem. Mol. Biol. Int.* 40, 235–241, 1996.

functional group. The ability to measure the ionization of the tyrosine hydroxyl with a UV–VIS spectrometer made such measurement an attractive tool in determining conformational change.[34,41] The nitration of tyrosine is possibly by earliest, well-described chemical modification of tyrosine.[42] The work by Johnson and Kohmann[42] not only provided a structure for nitrotyrosine but also a history of the nitration of proteins with nitric acid. The term xanthoprotein was used to describe the yellow product arising from the action of nitric acid on proteins. The xanthoproteic reaction was used as a method for the determination of protein concentration.[43–45] Wornall also showed that the nitration of casein did increase antigenicity most likely via the generation of nitrotyrosine epitopes (neoantigenicity).[46] Nitrated proteins (xanthoproteins) were used as nonspecific substrates for the assay of proteolytic enyzmes.[47,48]

Iodination has a long history of use for the modification of tyrosyl residues in proteins.[49–51] The process of iodination of proteins[52,53] is still of considerable value in the pharmacokinetic studies; intact murine lymphocytes have been labeled with radioactive iodine for the purpose of studying migration into peripheral tissue and brain.[54] It is, of course, of critical importance to appreciate the strength of the elemental halides as oxidizing agents modifying residues other than tyrosine during iodination procedures. The chemical methods for iodination are examples of electrophilic aromatic

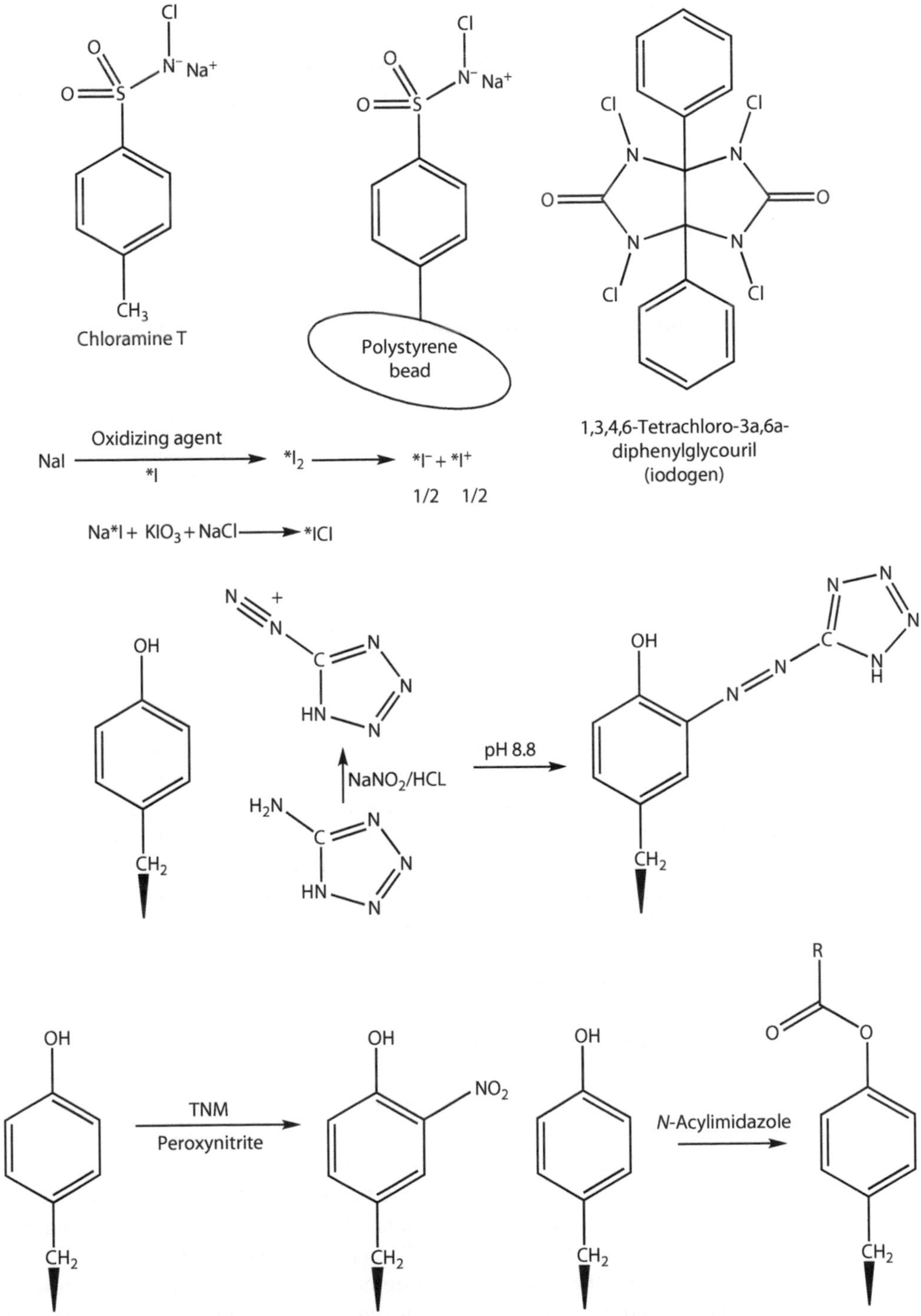

FIGURE 11.1 Reagents used to modify tyrosyl residues in proteins. Shown at the top are various chemical reagents used to iodinate tyrosyl residues in proteins including chloramine T. Shown below is the use of diazonium-1-H-tetrazole to modify a tyrosyl residue in a protein (Horinishi, H., Hachimori, Y., Kurihara, K., and Shibata, K., States of amino acid residues in proteins. III. Histidine residues in insulin, lysozyme, albumin, and proteinases as determined with a new reagent of diazo-1-H-tetrazole, *Biochem. Biophys. Acta* 86, 477–489, 1964). Shown at the bottom is the nitration and acetylation of tyrosine by TNM and NAI, respectively.

substitution with iodonium ion (I⁺)/hypoiodous acid (HOI) as the attacking species.[54] The processes described below describe methods for the generation of I⁺ from I₂. Various methods for the iodination of proteins are shown in Table 11.2. A recent work has used the photolysis of iodobenzoic acid to generate a putative iodine radical for the surface labeling of proteins.[55] Chen and coworkers[55] observed that only histidyl and tyrosyl residues are modified suggesting that the iodine radical may be more specific than hydroxyl radicals. Iodination has been demonstrated to occur as a side reaction in the sodium periodate oxidation step in the purification of glycopeptides by coupling to a hydrazide matrix (solid-phase extraction).[56]

Iodination has been utilized to study the reactivity of tyrosyl residues in cytochrome b_5.[57] Iodination is accomplished with a 10-fold molar excess of I_2 (15 mM I_2 in 30 mM KI) in 0.025 M sodium borate, pH 9.8. Iodination with limiting amounts of iodine is accomplished with a two- to six-fold molar excess of iodine in 0.020 M potassium phosphate, pH 7.5 at 0°C. Monoiodination and diiodination of tyrosyl residues are observed. Iodination with a tenfold molar excess of I_2 results in the formation of 3 mol of diiodotyrosine per mole of cytochrome c.[58] The fourth tyrosyl residue is modified only in the presence of 4.0 M urea. Iodination of tyrosine results in a decrease in the pKa of the phenolic hydroxyl groups similar to that observed with nitration. Iodination with a limiting amount of iodine as described earlier results first in the formation of 2 mol of monoiodotyrosine, then 1 mol of diiodotyrosine, and 1 mol of monoiodotyrosine. Tyrosyl residues that can be iodinated are also available for O-acetylation with acetic anhydride. The modification of tyrosyl residues in phosphoglucomutase by iodination has been reported.[59] Modification is achieved by reaction in 0.1 M borate, pH 9.5 with 1 mM I_2 (obtained by an appropriate dilution of a stock iodine/iodide solution, 0.05 M I_2 in 0.24 M KI) at 0°C for 10 min. Complete loss of enzymatic activity was observed with these reaction conditions, but the stoichiometry of modification was not established. Nitration of 7 of 20 tyrosyl residues also resulted in an 83% loss of catalytic activity. These investigators also studied the reaction of phosphoglucomutase with diazotized sulfanilic acid and NAI.

The previously mentioned studies used iodine/iodide in alkaline solution for the modification of tyrosyl residues in proteins. Iodination of tyrosyl residues may also be accomplished with iodine monochloride (ICl) at mildly alkaline pH as described for the modification of galactosyltransferase.[60] The reaction is initiated by the addition of reagent from a stock solution of ICl to the protein in 0.2 M sodium borate, pH 8.0. The reaction is initiated by a desired amount of a stock solution of ICl.[61] Silvia and Ebner[60] observed that five tyrosine residues were modified in galactosyltransferase with a total loss of enzyme activity. Tyrosine was the only amino acid modified in the reaction; the major product was 3,5-diiodotyrosine with small amounts of 3-iodotyrosine. The degree of inactivation of galactosyltransferase was dependent on the extent of tyrosine modification. α-Lactalbumin provided limited protection to galactosyltransferase from inactivation as did substrates; three tyrosyl residues could be modified in α-lactalbumin under these reaction conditions. Galactosyltransferase was protected from iodination by α-lactalbumin, while α-lactalbumin had no effect on the iodination of lysozyme suggesting a specific interaction between galactosyltransferase and α-lactalbumin in the absence of substrate. The interaction between galactosyltransferase and α-lactalbumin is of considerable importance in the regulation of lactose synthesis.[62] Parker and Allison[63] reported that the oxidation of cysteine to cysteine sulfenyl iodide with ICl at 0°C resulted in loss of enzyme activity; partial recovery of activity was possible with DTT. When the reaction with ICl was performed at 26°C, the loss of activity could be reversed with DTT. Parker and Allison[63] suggest that the reversible inactivation is associated with the formation of cysteine sulfenyl iodide while the irreversible inactivation is due to the formation of higher oxidation states of cysteine such as sulfinic acid or sulfonic acid. DTT would reduce the sulfenyl iodide to the sulfhydryl group but would have no effect on the higher oxidation products of cysteine. Allen and Harris[64] later showed that tyrosine residues could be modified with KI_3 in glyceraldehyde-3-phosphate dehydrogenase from *Bacillus stearothermophilus* after protection of the active-site sulfhydryl group with tetrathionate.

Iodination can also be accomplished using lactoperoxidase, hydrogen peroxide, and iodide, such as sodium iodide.[65] This work by the late Jack Marchalonis is a landmark contribution that has been

TABLE 11.2
Modification of Tyrosine Residues in Proteins by Iodination

Reagent[a]	Reaction Conditions	Other Residues Modified[b]	References
Chloramine T (N-chloro-4-methyl-benzenesulfonamide); available as a solid-phase reagent (Pierce Iodination Beads, previously Iodo-Beads®)	pH 7.5, phosphate, terminate with excess of tyrosine or sodium metabisulfite ($Na_2S_2O_5$). Mechanism involves the oxidation of iodide to produce I_2 followed by electrophilic aromatic substitution with I^+. Alternatively, radiolabeled I_2 can be added that in turn results in I^+. As with other reactions, HOI may be an intermediate species.	Methionine, cysteine, cystine histidine, peptide bond cleavage, and protein aggregation	11.2.1–11.2.8
ICl	$^{131}I_2$ with ICl in 0.2 M borate, pH 8.0 to produce ^{131}ICl. The ^{131}ICl is then added to the peptide/protein followed in turn by albumin. Phosphate buffer (pH 7.5) and acetate buffer (pH 4.0) can also be used. The use of a lower pH can increase specificity of modification for tyrosine residues. Reaction at 30°C for 2 min or 0°C for 30 min. The absence of an oxidizing agent during the reaction of ICl with the protein is thought to minimize reaction with other amino acid residues.	Histidine, disulfide bonds, cysteine	11.2.4, 11.2.7, 11.2.9–11.2.18
Iodogen (1,3,4,6-tetra-chloro-3a,6a-diphenyl-glycouril)	An oxidizing agent for the production of iodine from iodide. It is a poorly soluble material than can be *absorbed* to the surface of a reaction vessel minimizing contact with the protein. Oxidation of methionine residues to methionine sulfoxide has been reported; DTT has been shown to reverse oxidation. Suggested to be less rapid than chloramine T or ICl.	Histidine, methionine	11.2.4, 11.2.5, 11.2.7, 11.2.18–11.2.25
Lactoperoxidase	A mild technique for the iodination of proteins that have high specificity. However, it is difficult to obtain products of high specific radioactivity.		11.2.4, 11.2.5, 11.2.26–11.2.30

[a] The specific iodination technique used can influence the quality of the radiolabeled product (Mayers, G.L. and Klostergaard, J., The use of protein A in solid-phase binding assays: A comparison of four radiolabeling techniques, *J. Immunol. Methods* 57, 235–246, 1983; Thean, E.T., Comparison of specific radioactivities of human α-lactalbumin iodinated by three different methods, *Anal. Biochem.* 188, 330–334, 1990; Kienhuis, C.B., Heuvel, J.J., Ross, H.A. et al., Six methods for direct radioiodination of mouse epidermal growth factor compared: Effect of nonequivalence in binding behavior between labeled and unlabeled ligand, *Clin. Chem.* 37, 1749–1755, 1991).

[b] The extent of modification of amino acid residues other than tyrosine is dependent on protein substrate and reaction conditions.

References for Table 11.2

11.2.1. Hunter, W.M. and Greenwood, F.C., Preparation of Iodine-131 labelled human growth hormone of high specific activity, *Nature* 194, 495–496, 1962.

11.2.2. McConahey, P.J. and Dixon, F.J., Radioiodination of proteins by the use of the chloramine-T method, *Methods Enzymol.* 70, 210–213, 1980.

11.2.3. Siekierka, J.J. and DeGuidicibus, S., Radioiodination of interleukin 2 to high specific activities by the vapor-phase chloramine T method, *Anal. Biochem.* 172, 514–517, 1988.

(*continued*)

TABLE 11.2 (continued)
Modification of Tyrosine Residues in Proteins by Iodination

11.2.4. Kienhuis, C.B., Heuvel, J.J., Ross, H.A. et al., Six methods for direct radioiodination of mouse epidermal growth factor compared: Effect of nonequivalence in binding behavior between labeled and unlabeled ligand, *Clin. Chem.* 37, 1749–1755, 1991.

11.2.5. Bennett, G.L. and Horuk, R., Iodination of chemokines for use in receptor binding analysis, *Methods Enzymol.* 288, 134–148, 1997.

11.2.6. Ong, G.L., Elsamura, S.E., Goldenberg, D.M., and Mattes, M.J., Single-cell cytotoxicity with radiolabeled antibodies, *Clin. Cancer Res.* 7, 192–201, 2001.

11.2.7. Sobal, G., Resch, U., and Sinzinger, H., Modification of low-density lipoprotein by different radioiodination methods, *Nucl. Med. Biol.* 31, 381–388, 2004.

11.2.8. Holmberg, M., Stibius, K.B., Ndoni, S. et al., Protein aggregation and degradation during iodine labeling and its consequences for protein adsorption to biomaterials, *Anal. Biochem.* 361, 120–125, 2007.

11.2.9. McFarlane, A.S., Efficient trace-labelling of proteins with iodine, *Nature* 183, 53, 1953.

11.2.10. McFarlane, A.S., In vivo behavior of I-Fibrinogen, *J. Clin. Invest.* 42, 346–361, 1963.

11.2.11. Izzo, J.L., Bale, W.F., Izzo, M.J., and Roncone, A., High specific activity of insulin with ^{131}I, *J. Biol. Chem.* 239, 3743–3748, 1964.

11.2.12. Helmkamp, R.W., Contreras, M.A., and Bale, W.F., I^{131}-labeling of protein by the iodine monochloride method, *Int. J. Appl. Radiat. Isot.* 18, 737–746, 1967.

11.2.13. Helmkamp, R.M., Conreras, M.A., and Izzo, M.J., I^{131}-labeling of proteins at high activity with I131Cl produced by oxidation of total iodine in NaI^{131} preparations, *Int. J. Appl. Radiat. Isot.* 18, 747–754, 1967.

11.2.14. Parker, D.J. and Allison, W.S., Mechanism of inactivation of glyceraldehyde-3-phosphate dehydrogenase by tetrathionate, *o*-iodosobenzoate and iodine monochloride, *J. Biol. Chem.* 244, 180–189, 1969.

11.2.15. Hung, L.T., Fermandian, S., Morgal, J.L., and Fromageot, P., Peptide and protein labeling with iodine. Iodine monochloride reaction with aqueous solution of L-tyrosine, L-histidine, L-histidine-peptide, and his-effect on some simple disulfide bridges, *J. Labelled Comp.* 10, 3–21, 1974.

11.2.16. Contreras, M.A., Bale, W.F., and Spar, I.L., Iodine monochloride (ICl) iodination techniques, *Methods Enzymol.* 92, 277–292, 1983.

11.2.17. Siebenlist, K.R., Meh, D.A., and Mosesson, M.W., Position of γ-chain carboxy-terminal regions in fibrinogen/fibrin cross-linking mixtures, *Biochemistry* 39, 14171–14175, 2000.

11.2.18. Sobal, G., Resch, U., and Sinzinger, H., Modification of low-density lipoprotein by different radioiodination methods, *Nucl. Med. Biol.* 31, 381–388, 2004.

11.2.19. Wood, W.G., Wachter, C., and Scriba, P.C., Experiences using chloramine-T and 1,3,4,6-tetrachloro-3α, 6α-diphenylglycoluril (Iodogen) for radioiodination of materials for radioimmunoassay, *J. Clin. Chem. Clin. Biochem.* 19, 1051–1056, 1981.

11.2.20. Seet, L., Fabri, L., Nice, E.C., and Baldwin, G.S., Comparison of iodinated [Nle^{15}]—and [Met^{15}]-gastrin$_{17}$ prepared by reversed-phase HPLC, *Biomed. Chromatogr.* 2, 159–163, 1987.

11.2.21. Vigna, S.R., Giraud, A.S., Reeve, J.R., Jr., and Walsh, J.H., Biological activity of oxidized and reduced iodinated bombesins, *Peptides* 9, 923–926, 1988.

11.2.22. Li, J., Shi, L., Wang, C. et al., Preliminary biological evaluation of ^{125}I-labeled anti-carbonic anhydrase IX monoclonal antibody in the mice bearing HT-29 tumors, *Nucl. Med. Commun.* 32, 1190–1193, 2011.

11.2.23. Zalutsky, M.R., Boskovitz, A., Kuan, C.T. et al., Radioimmunotargeting of malignant glioma by monoclonal antibody D2C7 reactive against both wild-type and variant III mutant epidermal growth factor receptors, *Nucl. Med. Biol.* 39, 23–34, 2012.

11.2.24. Pruszynski, M., Koumarianou, E., Vaidyanathan, G. et al., Targeting breast carcinoma with radioiodinated anti-HER2 nanobody, *Nucl. Med. Biol.* 40, 52–59, 2013.

11.2.25. Cona, M.M., Li, J., Chen, F. et al., A safety study on single intravenous dose of tetrachloro-diphenyl glycoluril [iodogen] dissolved in dimethyl sulphoxide (DMSO), *Xenobiotica* 43(8), 730–737, 2013.

11.2.26. Marchalonis, J.J., An enzymic method for the trace iodination of immunoglobulins and other proteins, *Biochem. J.* 113, 299–305, 1969.

11.2.27. Kuo, B.S., Nordblom, G.D., and Wright, D.S., Perturbation of epidermal growth factor clearance after radioiodination and its implications, *J. Pharm. Sci.* 86, 290–296, 1997.

TABLE 11.2 (continued)

Modification of Tyrosine Residues in Proteins by Iodination

11.2.28. Bhuyan, B.J. and Mugesh, G., Heme peroxidase-catalyzed iodination of human angiotensins and the effect of iodination on angiotensin converting enzyme activity, *Inorg. Chem.* 47, 6569–6571, 2008.

11.2.29. Vergote, V., Baert, B., Vandermeulen, E. et al., LC-UV/MS characterization and DOE optimization of the iodinated peptide obestatin, *J. Pharm. Biomed. Anal.* 46, 127–136, 2008.

11.2.30. Fu, Y., Létourneau, M., Chatenet, D. et al., Characterization of iodinated adrenomedullin derivatives suitable for lung nuclear medicine, *Nucl. Med. Biol.* 38, 867–874, 2011.

cited well over 1000 times since its publication in 1969. Citation is given to some very selected works citing this study[66–70] including a recent critical review of peptide iodination technology.[71]

Iodination of tyrosyl residues in peptides and proteins can also be accomplished with chloramine T.[72–74] The solution structure of insulin-like growth factor was investigated by iodination of tyrosyl residues mediated by either chloramine T or lactoperoxidase.[75] Chloramine T was more effective than lactoperoxidase. There have been improvements in the chemistry of chloramine T iodination.[76,77] Tashtoush and coworkers[77] developed penta-*O*-acetyl-*N*-chloro-*N*-methylglucamine as a milder oxidizing agent to replace chloramine T. While the use of chloramine T can result in unwanted side reactions that compromise biological activity, it is recognized that reaction conditions for iodination with chloramine T can be selected to maximize radiolabel-specific activity and preservation of biological activity.

Radioisotope labeling of proteins with isotopes of iodine has been used extensively to study protein turnover (catabolism) and in vivo distribution.[78,79] Caution should be used in the interpretation of such results as most iodination techniques result in heterogeneous products that are trace labeled with iodine isotopes,[80–82] and there is the potential of asymmetric distribution of tyrosyl residues.[81] Site-specific modification of tyrosine with iodine has been used to obtain protein derivatives useful in crystallographic analysis.[83–86]

The phenolic hydroxyl of tyrosine can be modified by acid anhydrides and acyl chlorides.[87–89] In general, these reactions are not as specific as the other procedures such as NAI or TNM. The preparation of *O*-acyl derivatives via the action of carboxylic acid anhydrides (i.e., acetic anhydride) has been used for some time, but it is very difficult to obtain selective modification of tyrosine as these reagents readily react with primary amines to form stable *N*-acyl derivatives.[88,90] It was possible to obtain the selective modification of tyrosine in an α-amylase with acetic anhydride (Figure 11.2) by reaction at mildly acidic pH (1.0 M acetate, pH 5.8 at 25°C), approximately 20,000-fold molar excess of acetic anhydride (5.1×10^{-2} M acetic anhydride, 2.9×10^{-6} M enzyme).[91] Bernad and colleagues[92] described an extensive study comparing the modification of lysyl and tyrosyl residues in lysozyme with dicarboxylic acid anhydrides. In 50 mM HEPES, 1.25 M NaCl, pH 8.2, amino groups (primarily lysine residues) were far more reactive than hydroxyl groups (including tyrosine, serine, and threonine). A consideration of the recent literature suggests that the use of organic acid anhydrides for the modification of hydroxyl functions in biological polymers is not common, but modification of tyrosine residues must be considered with acetic anhydride used for mapping.[93] As indicated by the results obtained with α-amylase,[91] reaction under acid conditions would be expected to improve specificity as acid-catalyzed esterification would be favored over acetylation of amino groups at lower pH, although modification of lysine residues at mildly acid pH has been reported[94]; the acetylation of amino groups with acetic anhydride is enhanced by deprotonation.[95] Tyrosine-specific alkylation has recently been reported using π-allylpalladium complexes (Figure 11.2).[96] The preference of acetic anhydride for primary amines such as the ε-amino groups of lysine (Chapter 19), there has been very little use of acetic anhydride for the modification of tyrosine in recent years. While there is very little work on the use of acetic anhydride to modify tyrosine residues in proteins, there is one interesting observation

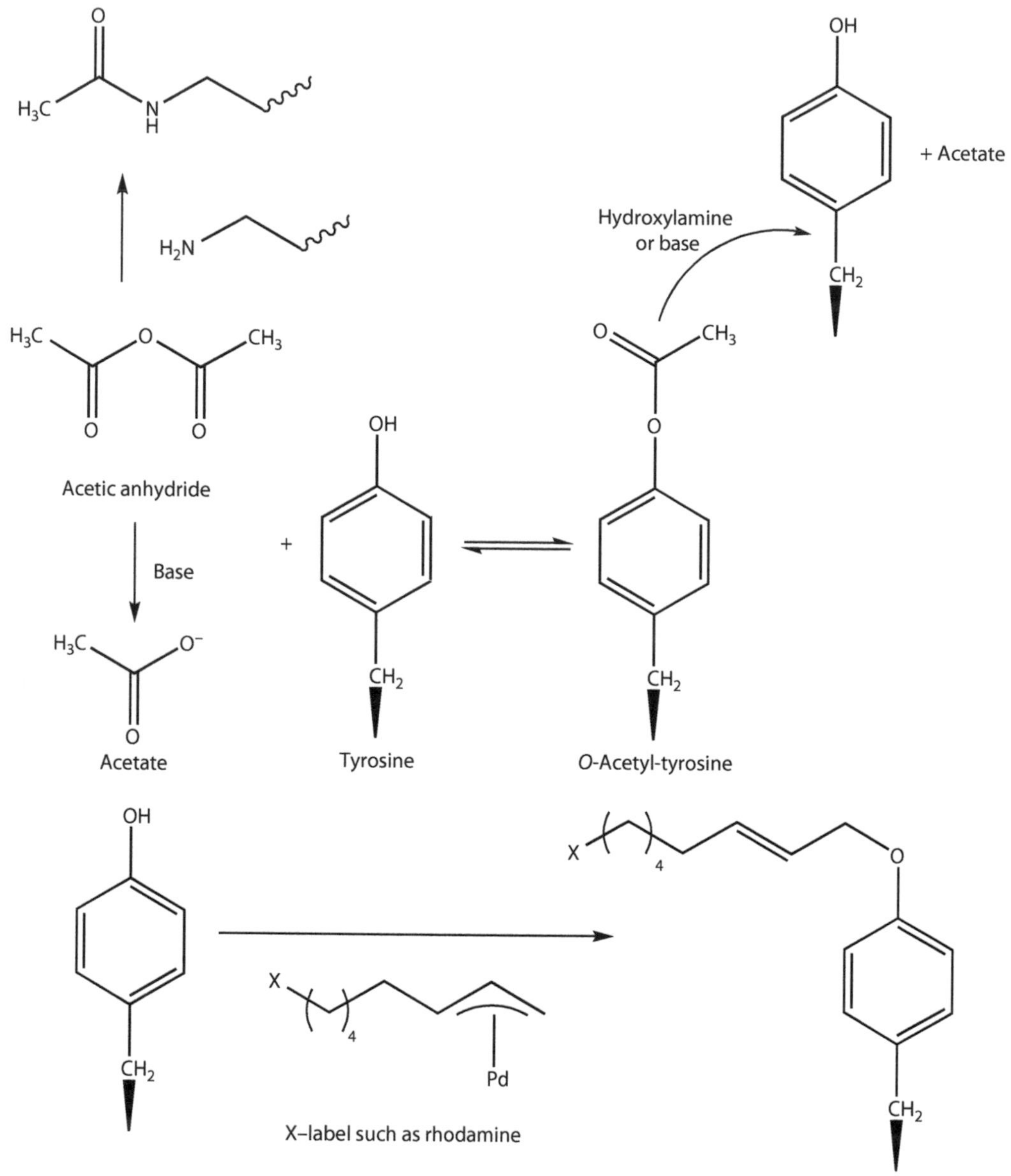

FIGURE 11.2 Modification of tyrosine with acetic anhydride. The modification of tyrosine by acetic anhydride is a readily reversible reaction. It is usually performed at pH 7.5–8.0 where the reaction rate and product stability are optimal. It is rarely used for the modification of tyrosine and more frequently for the modification of amino groups; the modification of amino groups is not reversible. See Riordan, J.F. and Vallee, B.L., *O*-Acetylation, *Methods Enzymol.* 25, 500–506, 1972. Shown at the bottom is the modification of tyrosine by π-allylpalladium complexes. (Adapted from Tilley, S.D. and Francis, M.B., Tyrosine-selective protein alkylation using π-allylpalladium complexes, *J. Am. Chem. Soc.* 128, 1080–1081, 2006; see also Lin, Y.Y.A., Chalker, J.M., and Davis, B.G., Olefin metathesis for site-selective protein modification, *ChemBioChem* 10, 959–969, 2009; Basel, E., Joubert, N., and Pucheault, M., Protein chemical modification on endogenous amino acids, *Chem. Biol.* 17, 213–227, 2010).

concerning the presence of acetic anhydride and methyl vinyl ketone in cigarette smoke, which react with tyrosine to form *N*-(3-oxobutyl)tyrosine, *N*-acetyltyrosine, and *O*-acetyltyrosine.[97]

The development of NAI as a reagent for the selective modification of tyrosyl residues (Figure 11.3) can, in part, be traced to the early observations[98–102] that NAI is, in fact, an energy-rich compound. Early work also focused on the study of NAI as a model for the study of the role of histidine in catalysis.[103,104] Atkinson and Green[103] in a work published in 1957 reported on the hydrolysis of NAI as a model for catalysis by cholinesterase. It should be noted that it was not clear as to the separate functions of the serine residue and histidine residue at the enzyme active site and there was interest in the possibility of an acetyl group on the imidazole ring of histidine as an intermediate in the hydrolysis of, for example, acetylcholine. It was observed that the rate of hydrolysis of NAI was more rapid in phosphate buffer than bicarbonate buffer (both at pH 7.6) and that the rate was dependent on phosphate concentration. These investigators also reported on the hydrolysis of *N*-di-isopropylphosphorylimidazole. A year later, Koltun and coworkers[104] published studies on the hydrolysis of NAI showing optimal stability between pH 6.0 and pH 7.5 with the rate of hydrolysis rapidly increasing at acid and basic pH.

NAI was first used as a reagent for the modification of tyrosyl residues in bovine pancreatic carboxypeptidase.[105] This same group of investigators subsequently reported on the use of NAI for the determination of *free* tyrosyl residues in proteins[106] as opposed to *buried* residues (Chapter 1); this has not necessarily proved to be the case.[107] *N*-acyl derivatives other than NAI have been prepared.[102,108] *N*-Butyrylimidazole was demonstrated to a more potent inhibitor of thrombin than *N*-acetylimidazole with modification occurring at an active-site histidine as well as at a single tyrosine residue.[109] This is in contrast with the modification of thrombin with *N*-acetylimidazole[110] where four tyrosine residues are modified. There was no evidence of histidine modification with *N*-acetylimidazole. It was suggested[109] that specificity of modification was a reflection of the more hydrophobic character of *N*-butyrylimidazole. El Kabbaj and Latruffe[111] examined the membrane penetration of NAI and observed that NAI was similar to TNM in the ability to preferentially inactivate D-3-hydroxybutyrate dehydrogenase in *inside-out* membranes when compared to intact mitochondria. These investigators reported that NAI showed a percent partition of 31/69 in H_2O/1-octanol while the percent partition was 11/89 for TNM; the percent partition of 7-chloro-4-nitrobenzo-2-oxa-1,3 diazole (NBD-Cl) is 5/95. Zhang and coworkers[112] have used reaction with NAI to differentiate between classes of tyrosine residues in green plant manganese-stabilizing protein; 5 tyrosine residues were easily modified, while 1–2 residues were modified only at higher concentrations of NAI, and 1–2 residues were only modified in urea-denatured protein. Martin and coworkers have demonstrated that NAI will not acetylate 3-nitrotyrosine or 3,5-dinitrotyrosine.[113] The p*K*a value for the phenolic hydroxyl group is markedly decreased on nitration that would decrease the stability of the ester bond formed on acetylation. This study also examined the rates of acetylation of tyrosine and 3-fluorotyrosine as a function of pH between 7.5 and pH 9.5. The formation of the *O*-acetylated derivative was evaluated by HPLC analysis. As NAI is unstable at increasing pH, it was not possible to obtain reliable second-order rate constants; thus, relative rates of reaction were reported. Since most of the reagents used for the modification of biological polymers are subject to nucleophilic attack (see Chapter 1) and water is a modest nucleophile,[114–116] a competing reaction in an aqueous system is to be expected. 3-Fluorotyrosine was acetylated more rapidly than tyrosine at pH 7.5, while the opposite was true at pH 9.5; the increased reactivity of 3-fluorotyrosine at pH 7.5 reflects the lower p*K*a of 3-fluorotyrosine; at pH 9.5, both tyrosine and 3-fluorotyrosine are deprotonated, and the intrinsic rate of tyrosine modification by NAI is 17 times greater than that of 3-fluorotyrosine.

N-Acetylimidazole, tetranitromethane, peroxynitrite, and iodine modify tyrosine residues by different chemistry yielding products with different characteristics, and there are several studies that compare the proteins modified with different reagents. Ji and Bennett[117] showed the peroxynitrite could activate (4–5 fold) reduced rat liver microsomal glutathione *S*-transferase; activity was also increased on reaction with TNM and NAI. Peroxynitrite and TNM also caused oxidation of a cysteine residue, but this oxidation was not considered to be responsible for the increase in activity. These investigators also stated that oxidation of cysteine did occur with NAI. I could not find

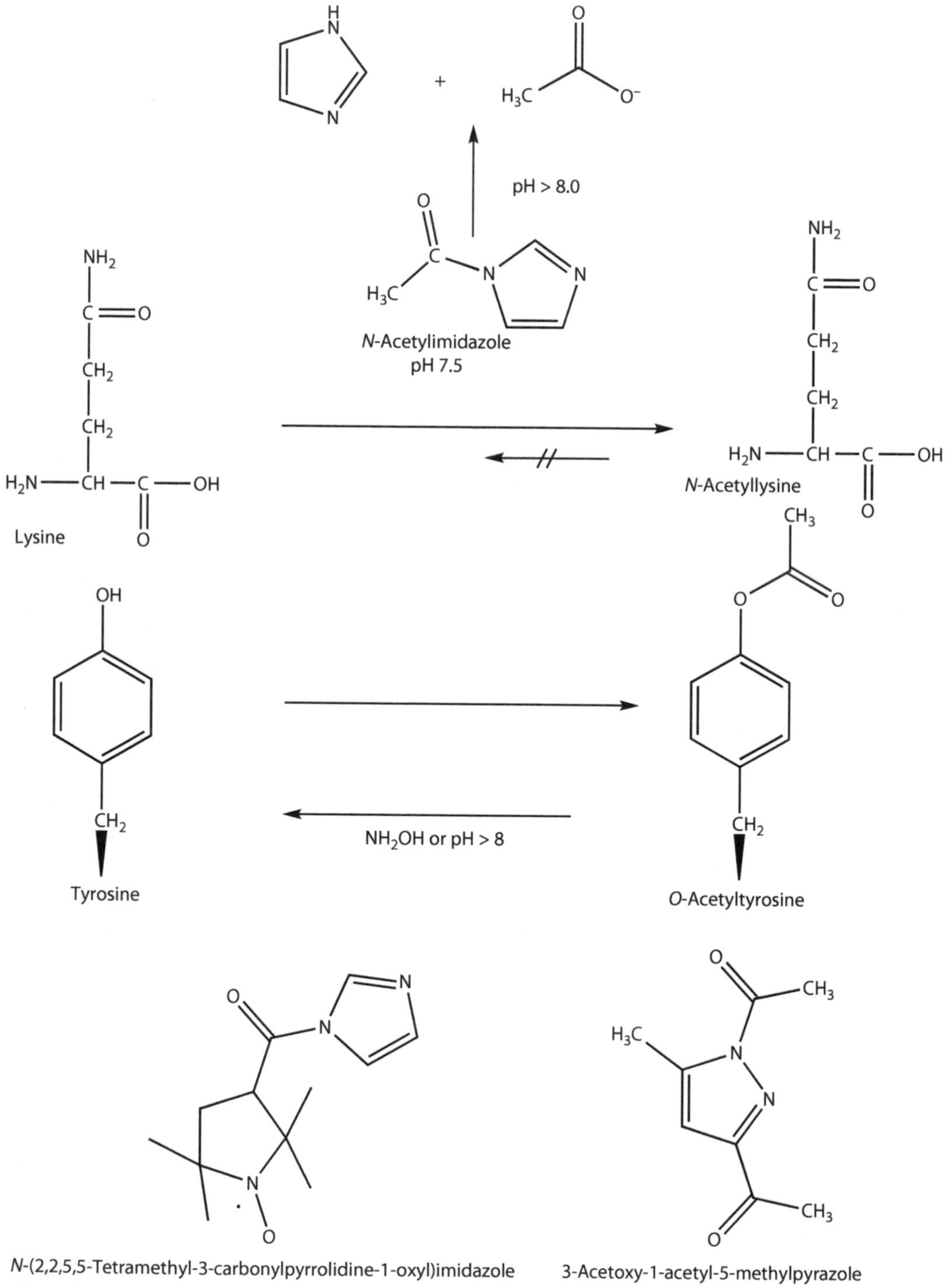

H
N
N
+
O
H_3C
O^-
pH > 8.0
O
C
H_3C
N
N
N-Acetylimidazole
pH 7.5
NH_2
C=O
CH_2
CH_2
H_2N—CH—C—OH
O
Lysine
NH_2
C=O
CH_2
CH_2
H_2N—CH—C—OH
O
N-Acetyllysine
OH
CH_2
Tyrosine
CH_3
O
O
CH_2
O-Acetyltyrosine
NH_2OH or pH > 8
O
N
N
N
O
O
CH_3
H_3C
N
N
CH_3
O
N-(2,2,5,5-Tetramethyl-3-carbonylpyrrolidine-1-oxyl)imidazole
3-Acetoxy-1-acetyl-5-methylpyrazole

FIGURE 11.3 (continued)

another example of cysteine oxidation by NAI in the literature. Elce[118] observed that the reaction of NAI with γ-glutamyltransferase resulted in loss of enzyme activity with modification of an amino group (loss of activity not reversed by hydroxylamine suggesting that inactivation did not involve tyrosine, histidine, or cysteine). Modification of γ-glutamyltransferase did result in the exposure of a cysteine residue for reaction with iodoacetamide. The cysteine residue in microsomal gluta-thione transferase is considered to be a sensor[119] and has unusual chemical properties.[120,121] Kuhn and Geddes[122] did observe that peroxynitrite did inactivate tryptophan hydroxylase by oxidation of cysteine; the concomitant nitration of tyrosine had little effect on enzyme activity. Subsequent work from JI and coworkers[123] demonstrated the increase in activity was due to modification of Tyr92. The sum of the work on the modification of tyrosine in microsomal glutathione transferase I is interesting because (1) similar results were obtained with both TNM and NAI and (2) modi-fication increased enzyme activity. The modification of the sensor cysteine residue (Cys69) in microsomal glutathione transferase 1 with NEM[124] also resulted in increased activity, and it is this cysteine residue that is thought to be the indicator of oxidative stress; modification with NEM does result in altered conformation as measured with hydrogen–deuterium exchange.

While the chemical modification of proteins is frequently used to increase thermal stability[125] or, in the case of poly(ethylene) glycol, to improve therapeutic half-life,[126,127] it is rare that chemical modification increases in catalytic activity. It is noted that the reaction of peroxynitrite with cathep-sin D is associated with an increase in activity.[128] Basu and Kirley[129] showed that the modification of Try252 in an ectonucleoside triphosphate diphosphohydrolase 3 with NAI resulted in an increase in enzyme activity. It is noted that the earlier cited paper on increase in thermal stability of a cold-adapted lipase with an oxidized polysaccharide[125] also resulted in increased activity but, as noted by the authors, this observation defies the general principle about modification increasing stability at the expense of activity. Selected studies on the reaction of proteins with NAI are presented in Table 11.3.

Peroxynitrite (Figure 11.4) has come to dominate the literature on the nitration of tyrosine resi-dues in proteins. There were limited studies[130,131] on peroxynitrite prior to the 1992 publication by Ischiropoulos and coworkers on the nitration of tyrosine catalyzed by superoxide dismutase.[132] This study may not have been the seminal paper on the reaction of with tyrosyl residues in proteins but it is certainly one of the earliest studies in this area. The reaction of peroxynitrite is of great interest in the physiology of nitric oxide and biological oxidations[133–135]; peroxynitrite is of less interest to protein chemist as peroxynitrite is not as a convenient reagent as TNM. There are, however, facile methods for the synthesis and storage of the reagent.[136–138] A search using SciFinder® yielded some-what more than 12,000 citations for peroxynitrite and approximately 3,000 for TNM.

The substantial amount of the in vitro work on peroxynitrite has involved the nitration of proteins in membranes.[139–142] Zhang and coworkers[139] used a 23-residue transmembrane peptide as a probe for nitration of tyrosine residues by peroxynitrite in membranes. A series of 23-mer peptides were prepared containing a single tyrosine at position 4, position 8, or position 12. The peptides were

FIGURE 11.3 (continued) *O*-Acetylation of tyrosine with NAI and reversal with hydroxylamine. The mod-ification of tyrosine by NAI proceeds optimally at pH 7.0–8.0. At value of pH greater than 8.0, base-catalyzed hydrolysis of the reagent becomes a significant problem; free imidazole also catalyzes the hydrolysis of the reagent. See Riordan, J.F. and Vallee, B.L., *O*-Acetylation, *Methods Enzymol.* 25, 500–506, 1972; Fife, T.H., Natarajan, R., and Werner, M.H., Effect of the leaving group in the hydrolysis of *N*-acylimidazoles. The hydroxide ion, water, and general base-catalyzed hydrolysis of *N*-acyl-4(5)-nitroimidazoles, *J. Org. Chem.* 52, 740–746, 1987. Also shown is the reaction of NAI to yield *N*ᵉ-acetyllysine; this reaction is irreversible under normal solvent conditions; the amide bond formed between the acetyl group and the ε-amino group of lysine is susceptible to hydrolysis in 6 N HCl/22 h/110°C. Shown at the bottom are two nitroxide spin label deriva-tives based on NAI (Adackaparayll, M. and Smith, J.H., Preparation and reactivity of a new spin label reagent, *J. Org. Chem.* 42, 1655–1656, 1977). Also shown is 3-acetoxy-1-acetyl-5-methylpyrazole that is less specific than NAI in the acetylation of phenolic hydroxyl groups and aliphatic hydroxyl groups (Irie, M., Miyaska, T., and Arakawa, K., 3-Acetoxy-1-acetyl-5-methylpyrazole, a novel acetylating reagent in proteins, *J. Biochem.* 72, 65–72, 1972).

TABLE 11.3
Modification of Proteins with NAI

Protein	Reaction Conditions	NAI/Protein[a]	O-AcTyr/Tyr[b]	Reference
Carboxypeptidase[c]	0.02 M sodium barbital, 2.0 M NaCl, pH 7.5 at 23°C	60	4.3/19[d]	11.3.1
Pepsinogen	0.02 M sodium Veronal, 2.0 M NaCl, pH 7.5 at 25°C	60	7/16[e]	11.3.2
Pepsin	2.0 M NaCl, pH 5.8 at 25°C[f]	60	9/16[g]	11.3.2
Trypsin	0.01 M sodium borate, 0.01 M CaCl$_2$, pH 7.6 at 0°C	30	1.7/10[h]	11.3.3
Trypsin	0.01 M sodium borate,[i] 0.01 M CaCl$_2$, pH 7.6 at 0°C	465	3.0/10[j]	11.3.3
α-Amylase[k]	0.02 M Tris-Cl, pH 7.5 at 25°C	500	3.5/12[l]	11.3.4
Subtilisin novo	0.016 M barbital, pH 7.5	100	7/10[m]	11.3.5
Subtilisin Carlsberg	0.016 M barbital, pH 7.5	130	8.4/13[m]	11.3.5
Hemerythrin	0.05 M sodium borate, 0.05 M Tris, pH 7.5 at 0°C	800	—[n]	11.3.6
Thrombin	0.02 M Tris, 0.02 *M* imidazole, 0.02 M acetate, pH 7.5 at 23°C	300[o]	4.4/12	11.3.7
Bowman–Birk soybean proteinase inhibitor	50 mM sodium borate, pH 7.5, 24°C	1500	1/2[p]	11.3.8
Bowman–Birk soybean proteinase inhibitor	50 mM sodium borate, pH 7.5, 24°C, 8 M urea	1500	1/2[p]	11.3.8
Bowman–Birk soybean proteinase inhibitor	50 mM sodium borate, pH 7.5, 24°C, 6 M guanidine	1500	2/2[p]	11.3.8
Pancreatic α-amylase	0.01 M phosphate, pH 7.5, 0.1 mM CaCl at 25°C	120[q]	5.9/18	11.3.9
Sweet potato α-amylase	0.01 M acetate, pH 7.5 at 25°C[r]	120[q]	5.3/17	11.3.9
Aspergillus niger glucamylase	0.01 M acetate, pH 7.5 at 25°C	120[q]	11.3/33	11.3.9
α$_1$-Proteinase inhibitor	50 mM sodium borate, pH 7.5, 25°C, 1 h	60[r]	3	11.3.10
Emulsin β-D-glucosidase	0.01 M phosphate, pH 6.5/25°C	300	9[s]	11.3.11
Rat uterus estrogen receptor	50 mM HEPES, pH 7.5, 15 min at 37°C or 16 h at 0°C–4°C[t]	—	—	11.3.12
Renal Na, K-ATPase	50 mM sodium borate (pH 7.5 at 20°C)–2 mM EDTA, 0°C[u]	—	—	11.3.13
Prostaglandin endoperoxide synthase	50 m M sodium phosphate, pH 7.2 with 0.01% octyl glycoside, 5 min at 23°C[w]	100	2.5(1)[v]	11.3.14
Leukotriene A$_4$ hydrolase	50 mM HEPES, pH 7.5, 25°C for 1 h[x]	—	—	11.3.15
D-Galactose-binding lectin from *Erythrina speciosa*	50 mM Tris, pH 8.0, 25°C	—	6[y]	11.3.16

[a] Moles NAI per mole of protein unless otherwise indicated.

[b] Moles *O*-acetyltyrosine per moles of tyrosine in modified protein.

[c] Bovine pancreatic carboxypeptidase A-Anson (Anson, M.L., Carboxypeptidase: I. The preparation of crystalline carboxypeptidase, *J. Gen. Physiol.* 20, 663–669, 1937; Pétra, P.H. and Neurath, H., The heterogeneity of bovine carboxypeptidase A. I. The chromatographic purification of carboxypepidase A (Anson), *Biochemistry* 8, 2566–2475, 1969).

[d] Primary amino groups were not acetylated under those reaction conditions. *O*-Acetylation of tyrosyl residues in carboxypeptidase A increases esterase activity but inhibits peptidase activity. Changes in catalytic activity reversed by treatment with 0.01 M hydroxylamine, pH 7.5 at 23°C; the altered spectra of the modified protein are also returned to that of the modified protein by hydroxylamine. Two tyrosyl residues are protected from modification by β-phenylpropionate, a competitive inhibitor; these two tyrosyl residues are also rapidly deacylated in the presence of hydroxylamine.

TABLE 11.3 (continued)
Modification of Proteins with NAI

^{e} Five out of ten lysine residues modified are also modified in pepsinogen; the modification of lysyl residues in pepsin was not demonstrated. It was not possible to demonstrate the modification of lysyl residues in peptide on reaction of NAI with pepsin. The reaction of NAI with pepsin was performed at pH 5.8 where the modification of the ε-amino group of lysine is unfavorable; the modification of the N-terminal residue was not excluded. The extent of tyrosine modification by NAI as determined by difference spectroscopy was consistent with that determined by hydroxamate formation (Balls, A.K. and Wood, H.N., Acetyl chymotrypsin and its reaction with ethanol, *J. Biol. Chem.* 219, 245–256, 1956).

^{f} pH maintained by NaOH from pH stat.

^{g} Lysine not acetylated under these conditions. Reaction with 1.0 M hydroxylamine, pH 5.8 (60 min; 37°C), reversed changes in catalytic activity produced on reaction with NAI and presumably deacetylated O-acetyl tyrosyl residues.

^{h} Also 1.0 serine and 0.3 lysine.

^{i} Also used Tris, TES, HEPES, and barbital buffers without any significant difference in nature of the reaction.

^{j} Also 1.7 (probably serine and histidine) and 2.5 lysine residues modified.

^{k} From *Bacillus subtilis*.

^{l} Approximately two lysine residues modified under these conditions. Only a single tyrosine residue is modified with TNM. Either reagent (TNM or NAI) led to a 70%–80% loss of catalytic activity.

^{m} The reaction with NAI was performed with subtilisin preparation previously treated with phenylmethanesulfonyl fluoride. The active enzyme catalyzes the rapid hydrolysis of NAI under reaction conditions.

^{n} Reaction performed on protein where lysine residue had been previously blocked by reaction with ethyl acetimidate. NAI was added in four 200-fold molar excess portion at 2 h intervals.

^{o} NAI was taken to dryness from a stock solution in benzene prior to addition to the components of the reaction mixture. Later studies used 50 mM sodium phosphate, pH 7.5, because of reagent stability issues. The extent of tyrosine was determined in difference spectroscopy and confirmed by the hydroxamate assay (Tildon, J.T. and Ogilvie, J.W., The esterase activity of bovine mercaptalbumin. The reaction of the protein with *p*-nitrophenyl acetate, *J. Biol. Chem.* 247, 1265–1271, 1972).[11.3.20] Later work with *N*-butylimidazole established the importance of a single tyrosine residue as well as butylation at the enzyme active site (Lundblad, R.L., The reaction of bovine thrombin with *N*-butyrylimidazole. Two different reactions resulting in the inhibition of catalytic activity, *Biochemistry* 14, 1033–1037, 1975).

^{p} The Bowman–Birk soybean trypsin inhibitor and lima bean trypsin inhibitor are *double-headed* inhibitors in that trypsin and chymotrypsin are inhibited at separate, independent sites. The acetylation of Tyr69 in the Bowman–Birk inhibitor that occurs in the absence of chaotropic agents results in partial loss of trypsin inhibition, while the acetylation of Try55 occurs in the presence of guanidine and not urea, which results in the loss of activity against chymotrypsin. The total loss of activity against trypsin that occurs in the presence of either urea or guanidine is thought to reflect acetylation of a lysine residue.

^{q} NAI added as a solid; pH maintained at 7.5 with pH stat.

^{r} Sixty-fold molar excess with respect to tyrosine. Difference spectroscopy showed the modification of 3 mol of tyrosine per mole of protein. Analysis with TNBS showed no modification of amino groups. There was a complete loss of ability to inhibit elastase with little effect on the ability to inhibit either trypsin or chymotrypsin. Inhibitory activity toward elastase was recovered (90%) on treatment of the acetylated protein with hydroxylamine. Similar results were obtained with TNM with 3 mol of tyrosine nitrated per mole of protein; one tyrosine residue was protected from modification when α_1-proteinase inhibitor was in complex with elastase.

^{s} Nine tyrosyl residues modified after 40 min of reaction with approximately 60% loss of β-glucosidase and approximately 80% loss of β-galactosidase activity. Both activities are recovered with hydroxylamine.

^{t} Loss of binding occurred with NAI in 15 min at 37°C; little loss of binding on reaction for 16 h at 0°C–4°C (it is likely that there was considerable hydrolysis of reagent during this time period). Hydroxylamine at 37°C reversed inhibition. Modification with TNM at pH 8 resulted in the loss of binding; reaction at pH 6 had no effect.

(*continued*)

TABLE 11.3 (continued)
Modification of Proteins with NAI

^u The pH of the borate buffer was determined at 20°C but the modification reaction was performed at 0°C. Stoichiometry of reaction was not determined but inactivation reaction consisted of two phases. Activity was recovered with hydroxylamine; full activity recovery was not possible as the control enzyme lost some activity (approximately 15%) in the presence of hydroxylamine. The existence of a fast phase and a slow phase for inactivation was not due to reagent instability; the $t_{1/2}$ for hydrolysis of NAI under these reaction conditions was 2.8 h ($k=4.1\times10^{-3}$ min^{-1}). The presence of ATP protected the enzyme from inactivation, and the acetylated enzyme demonstrated reduced binding of ADP.

^v Spontaneous reactivation occurred when the acetylated apoenzyme was taken into 0.1 M Tris, pH 8.1 at 22°C; there was very little recovery at pH 7 with more rapid recovery at pH 9.0. Recovery of activity also occurred with the addition of hydroxylamine.

^w The extent of tyrosine modification was determined by the formation of a hydroxymate (Tildon, J.T. and Ogilvie, J.W., The esterase activity of bovine mercaptablumin. The reaction of the protein with *p*-nitrophenyl acetate, *J. Biol. Chem.* 247, 1265–1271, 1972). The reaction of NAI with holoenzyme (prostaglandin endoperoxide synthase contains a prosthetic heme group) results in the incorporation of one acetyl group without loss of enzyme activity; the modification of apoenzyme results in 90% loss of activity with the incorporation of 2.5 acetyl groups; if the reaction with hydroxylamine is performed at pH 10 instead of pH 8.2, the presence of an acetylated aliphatic alcohol (either serine or threonine) is detected (see Balls, A.K. and Wood, H.N., Acetyl chymotrypsin and its reaction with ethanol, *J. Biol. Chem.* 219, 245–256, 1956; Riordan, J.F. and Vallee, B.L., Acetylation, *Methods Enzymol.* 25, 494–499, 1972).

^x NAI in dry toluene was taken to dryness with a stream of nitrogen prior to addition of enzyme solution. Both epoxide hydrolase and peptidase activities are inhibited by NAI; the activity is recovered by the addition of hydroxylamine (250 mM). Stoichiometry was not established for NAI; however, similar degree of inactivation was obtained with TNM. 2.5 tyrosine residues were modified by TNM in the absence of a competitive inhibitor, while approximately 0.5 tyrosine residues are modified in the presence of bestatin.

^y Six tyrosyl residues were modified at 50% loss of activity; the addition of more NAI resulted in the eventual modification of 20 tyrosyl residues with no further loss of activity.

References for Table 11.3

11.3.1. Simpson, R.T., Riordan, J.F., and Vallee, B.L., Functional tyrosyl residues in the active center of bovine pancreatic carboxypeptidase A, *Biochemistry* 2, 616–622, 1963.

11.3.2. Perlmann, G.E., Acetylation of pepsin and pepsinogen, *J. Biol. Chem.* 241, 153–157, 1966.

11.3.3. Houston, L.L. and Walsh, K.A., The transient inactivation of trypsin by mild acetylation with *N*-acetylimidazole, *Biochemistry* 9, 156–166, 1970.

11.3.4. Connellan, J.M. and Shaw, D.C., The inactivation of *Bacillus subtilis* α-amylase by *N*-acetylimidazole and tetranitromethane. Reaction of tyrosyl residues, *J. Biol. Chem.* 245, 2845–2851, 1970.

11.3.5. Myers, B., II and Glazer, A.N., Spectroscopic studies of the exposure of tyrosine residues in proteins with special reference to the subtilisins, *J. Biol. Chem.* 246, 412–419, 1971.

11.3.6. Fan, C.C. and York, J.L., The role of tyrosine in the hemerythrin active site, *Biochem. Biophys. Res. Commun.* 47, 472–476, 1972.

11.3.7. Lundblad, R.L., Harrison, J.H., and Mann, K.G., On the reaction of purified bovine thrombin with *N*-acetylimidazole, *Biochemistry* 12, 409–413, 1973.

11.3.8. Kay, E., Structure-function relationships of proteinase inhibitors from soybean (Bowman-Birk) and lima bean, *J. Biol. Chem.* 254, 7648–7650, 1979.

11.3.9. Hoschke, A., Laszlo, E., and Hollo, J., A study of the role of tyrosine groups at the active centre of amylolytic enzymes, *Carbohydr. Res.* 81, 157–166, 1981.

11.3.10. Feste, A. and Gan, J.C., Selective loss of elastase inhibitory activity of α1-proteinase inhibitor upon chemical modification of its tyrosyl residues, *J. Biol. Chem.* 256, 6372–6380, 1981.

11.3.11. Kiss, L., Korodi, I., and Nanasi, P., Study on the role of tyrosine side-chains at the active center of emulsin β-D-glucosidase, *Biochem. Biophys. Acta* 662, 308–311, 1981.

11.3.12. Koffman, B., Modarress, K.J., Beckerman, T., and Bashirelahi, N., Evidence for involvement of tyrosine in estradiol binding by rat uterus estrogen receptor, *J. Steroid Biochem. Mol. Biol.* 38, 135–139, 1991.

11.3.13. Arguello, J.M. and Kaplan, J.H., *N*-acetylimidazole inactivates renal Na, K-ATPase by disrupting ATP binding at the catalytic site, *Biochemistry* 29, 5775–5782, 1990.

TABLE 11.3 (continued)

Modification of Proteins with NAI

11.3.14. Scherer, H.-J., Karthein, R., Strieder, S., and Ruf, H.H., Chemical modification of prostaglandin endoperoxide synthase by *N*-acetylimidazole. Effect on enzymic activities and EPR spectroscopic properties, *Eur. J. Biochem.* 205, 751–757, 1992.

11.3.15. Mueller, M.J., Samuelsson, B., and Haeggstrom, J.Z., Chemical modification of leukotriene A4 hydrolase. Indications for essential tyrosyl and arginyl residues at the active site, *Biochemistry* 34, 3546–3542, 1995.

11.3.16. Konozy, E.H.E., Bernades, E.S., Rosa, C. et al., Isolation, purification, and physicochemical characterization of a D-galactose-binding protein lectin from seeds of *Erythrina speciosa*, *Arch. Biochem. Biophys.* 410, 222–229, 2003.

inserted into a multilamellar liposome composed of 1,2-dilauroyl-*sn*-glycero-3-phosphatidyl choline. When peroxynitrite is generated in situ, nitration of the peptide in the liposome is greater than that of peptide in aqueous solution. Furthermore, nitration of the tyrosine residues increased with depth of penetration of the peptide into the liposome in that nitration of the Tyr12 derivatives is greater than that of the Tyr8 derivative that is turn greater than that of the Tyr4 peptide. It is suggested that peroxynitrous acid diffuses into the membrane where it undergoes decomposition to form nitric oxide radical that then reacts with tyrosine to form nitrotyrosine. Hydrophobic substrates such as *N-t*-butyloxycarbonyl-tyrosine *t*-butyl ester have been developed for study of nitration by peroxynitrate in membranes.[140–142] Some examples of the use of peroxynitrite for the modification of tyrosyl residues in proteins are presented in Table 11.4. A consideration of the information in Table 11.4 allows several conclusions. First, the reaction of peroxynitrite with functional groups in proteins is very rapid.[143] Second, the stability of peroxynitrite in the various reaction mixtures is problematic with a half-life of <1 s.[143,144] Third, it is possible to obtain a stable peroxynitrite reagent by several synthetic procedures or commercial sources that can be stored at −80°C; there are several sources that use −20°C but I would not recommend that temperature. Fourth, rapid mixing of the reaction mixture after addition of the peroxynitrite reagent is recommended, and the reaction is likely complete within seconds; many investigators use a control where the peroxynitrite is allowed to decompose to nitrite and nitrate before addition to the reaction mixture.[145,146] As with TNM, peroxynitrite reaction of tyrosine can result in dimer formation. Finally, the major side reaction with the reaction of peroxynitrite with protein is the oxidation of cysteine to cysteine sulfenic acid; reaction can also occur with methionine and tryptophan.

The importance of peroxynitrite and in vivo tyrosine nitration has resulted in the development of a number of approaches to the analysis of 3-nitrotyrosine in proteins. As with other chemical modification reactions, MS is the method of choice for the analysis of nitrotyrosine in proteins.[147–158] Antibodies to 3-nitrotyrosine in proteins have been developed and are useful not only for the analysis of purified proteins[159–160] but also for the enrichment of nitrotyrosine-containing proteins,[160] for in situ localization in cells,[161] and for identification on 2D gel electrophoretograms.[162–166] Sharov and coworkers[167] developed a fluorescence-based approach as an alternative to immunohistochemistry for the visualization of nitrated proteins in tissue. These investigators reduced the 3-nitrotyrosine derivative into a 3-aminotyrosine with dithionite. The resulting 3-aminotyrosine was used as a site for reaction with 4-(aminomethyl)benzene sulfonate (ANS) yielding a fluorescent product. The pKa of the 3-aminotyrosine is considerably lower than the pKa of the epsilon amino group of lysine permitting specificity of modification.

The development of neoantigens on protein nitration was an early observation[46,168] that has been shown to have physiological consequences.[169,170] It should be noted that the iodination of proteins also results in the formation of neoantigens.[48,168,171] Benyamin and coworkers[172] reported on the effect of chemical modification on the antigenicity of lobster arginine kinase. Modification of the essential histidine residues with diethylpyrocarbonate and modification of the essential lysine

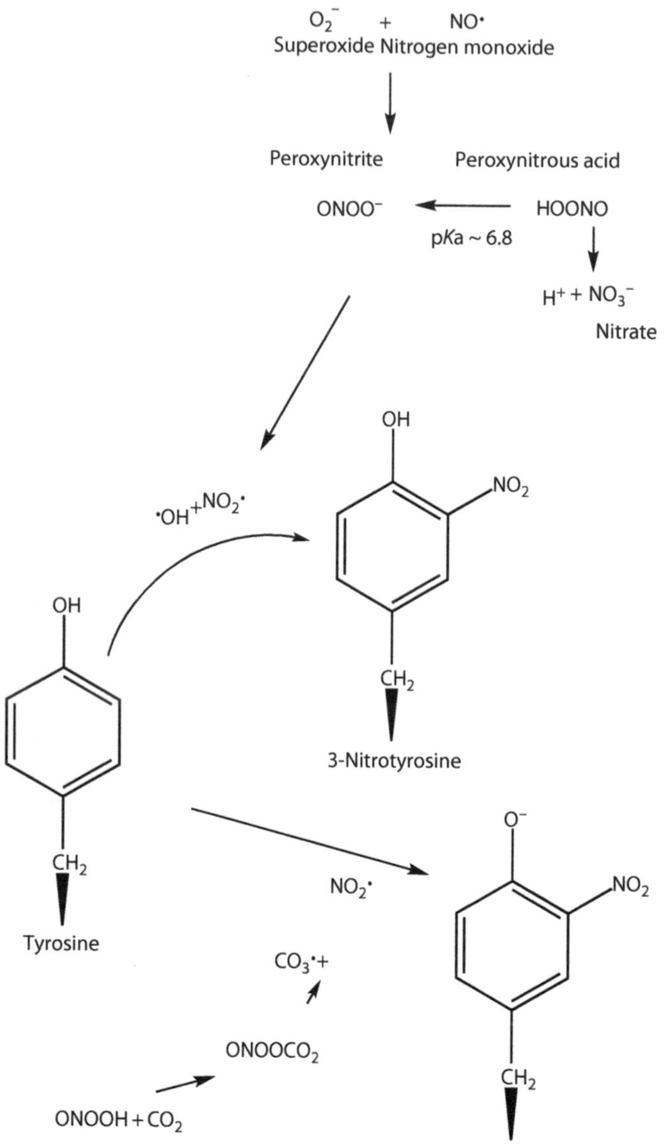

FIGURE 11.4   Reaction of peroxynitrite with tyrosine residues in proteins. Peroxynitrous acid/peroxynitrite is formed from superoxide and nitrogen monoxide. Peroxynitrous acid (pKa=6.8) rapidly decomposed to yield nitrate; also, the reaction is more complex than shown (Kirsh, M., Korth, H.-G., Wensing, A. et al., Product formation and kinetic simulations in the pH range 1–14 account for a free-radical mechanism of peroxynitrite decomposition, *Arch. Biochem. Biophys.* 418, 133–150, 2003; Goldstein, S. and Merényi, G., The chemistry of peroxynitrite: Implications for biological activity, *Methods Enzymol.* 436, 49–61, 2008). A variety of mechanisms have been proposed for the nitration of tyrosine by peroxynitrite including a concerted mechanism and a radical mechanism (Gunaydin, H. and Houk, K.N., Mechanisms of peroxynitrite-mediated nitration of tyrosine, *Chem. Res. Toxicol.* 22, 894–898, 2009). The radical mechanism is shown.

residue with dansyl chloride had no effect on antigenicity. The modification of the free sulfhydryl groups with potassium tetrathionite also had no effect on activity; the modification of the tyrosine residues in the *S*-sulfenylsulfo derivative with TNM did result in the loss of 20% of antigenic reactivity against a rabbit antisera raised against the native lobster arginine kinase; antigenicity was recovered on the reduction of the 3-nitroderivative to the 3-amino derivative sodium hydrosulfite

TABLE 11.4
Modification of Proteins, Peptides, and Related Compounds with Peroxynitrite[a]

Protein	Reaction Conditions	Reference
α_1-Antiproteinase inhibitor	Peroxynitrite forms from hydrogen peroxide and potassium nitrite (in 0.6 M HCl) at 0°C for one second followed by the addition of 1.2 M NaOH at 0°C. This was frozen overnight (−20°C) and the top layer taken for reagent.[b] The reaction with the protein was performed in 500 mM potassium phosphate, pH 7.4 (temperature not provided).[c]	11.4.1
Human serum albumin	Peroxynitrite was prepared by passing ozone through alkaline sodium azide at 0°C.[d] The concentration of peroxynitrite was determined by spectroscopy ($\varepsilon = 1670$ M^{-1} cm^{-1} at 302 nm). The reactions were performed in 150 mM potassium phosphate–25 mM sodium bicarbonate, pH 7.2, at 23°C. The conditions varied from a 30-fold molar excess of albumin to a 140-fold molar excess of peroxynitrite. The reaction was vortexed for a minute after the addition of the peroxynitrite and the reaction mixture transferred into deionized water by dialysis. It is estimated that the half-life of the peroxynitrite was less than one second under these reaction conditions. MS was used to determine sites of tyrosine nitration.[e]	11.4.2
Annexin II	Peroxynitrite was synthesized by a procedure described in the literature[f] and stored at −80°C (peroxynitrite is stable for several months under these conditions). Nitration reaction was performed in 100 mM phosphate, pH 7.4, with 0.1 mM diethylenetriaminepentaacetic acid at 23°C. In consideration of the instability of peroxynitrite under the reaction conditions, a 1 μL peroxynitrite reagent is placed on the reaction tube just above the level of the annexin solution; the solutions are mixed by vortexing. The reaction is continued for 30 min.[g]	11.4.3
Calmodulin	Peroxynitrite was prepared by passing ozone through alkaline sodium azide at 0°C.[d] The concentration of peroxynitrite was determined by spectroscopy ($\varepsilon = 1670$ M^{-1} cm^{-1} at 302 nm). Reaction with peroxynitrite with calmodulin was performed in 200 mM potassium phosphate, pH 7.0 (temperature not provided but assumed to be 23°C).[h] A *drop* of the peroxynitrite solution was placed above the protein solution and the reaction initiating by vortex mixing.	11.4.4
Thr-Tyr-Ser	Peroxynitrite obtained from commercial source (33–48 mM in 0.3 M NaOH); said to have >90% purity with remainder as nitrate. Peroxynitrite added to protein solution in 20 mM NH$_4$HCO$_3$, pH 7.4 at 23°C and reaction continued for 1 h.[i]	11.4.5
Prostaglandin H$_2$ synthase	Peroxynitrite obtained from commercial source (33–48 mM in 0.3 M NaOH); said to have >90% purity with remainder as nitrate. Peroxynitrite added to peptide solution in 20 mM NH$_4$HCO$_3$, pH 7.4 at 23°C and reaction continued for 1 h. The reaction mixture contained 5 mM Tris derived from the protein purification process.[j]	11.4.5
Microsomal glutathione *S*-transferase	Peroxynitrite forms from hydrogen peroxide and potassium nitrite (in 0.6 M HCl) at 0°C for 1 s followed by the addition of 1.2 M NaOH at 0°C.[b] This material was stored at −70°C prior to use. The concentration of peroxynitrite was determined by spectroscopy ($\varepsilon = 1670$ M^{-1} cm^{-1} at 302 nm). The nitration reaction was performed in 100 mM potassium phosphate, pH 7.0, containing 0.1 mM diethylenetriaminepentaacetic acid at 23°C.[k]	11.4.6

(continued)

TABLE 11.4 (continued)

Modification of Proteins, Peptides, and Related Compounds with Peroxynitrite[a]

Protein	Reaction Conditions	Reference
H2A histone	Peroxynitrite is prepared from the reaction of sodium nitrite and hydrogen peroxide in HCl (for 1 s) followed by rapid quenching with NaOH to stabilize the peroxynitrite.[l] The peroxynitrite was stored at −20°C and the concentration established by spectrophotometry prior to use. Peroxynitrite was added to the protein in 10 mM sodium phosphate, pH 7.4, containing 100 mM NaCl and 0.1 mM diethylenetriaminepentaacetic acid. The reaction was continued for 30 min at 37°C. The final pH of the reaction mixture was 10–11.[m]	11.4.7
Mitochondrial carbamoyl phosphate synthetase 1	Commercial peroxynitrite was added to mitochondrial fractions in 50 mM glycylglycine (pH not provided) at 37°C for 10 min.[n]	11.4.8

[a] Peroxynitrite is used infrequently for the in vitro specific modification of tyrosyl residues in proteins. The use of peroxynitrite in vitro for the modification of proteins is complicated by reagent instability and the potent oxidizing activity of this reagent (Radi, R., Beckman, J.S., Bush, K.M., and Freeman, B.A., Peroxynitrite oxidation of sulfhydryls. The cytotoxic potential of superoxide and nitric oxide, *J. Biol. Chem.* 266, 4244–4250, 1991). As of this writing (May 2013), sodium peroxynitrite is available from several commercial sources, but many investigators prepare the reagent using established protocols (Uppu, R.M. and Pryor, W.A., Synthesis of peroxynitrite in a two-phase system using isoamyl nitrite and hydrogen peroxide, *Anal. Biochem.* 236, 242–249, 1996; Uppu, R.M., Squadrito, G.L., Cueto, R., and Pryor, W.A., Selecting the most appropriate synthesis of peroxynitrite, *Methods Enzymol.* 269, 285–296, 1996. Koppenol, W.H., Kissner, R., and Beckman, J.S., Syntheses of peroxynitrite: To go with the flow or solid grounds?, *Methods Enzymol.* 269, 296–302, 1996; White, R., Crow, J., Spear, N. et al., Making and working with peroxynitrite, *Methods Mol. Biol.* 100, 215–230, 1998). Peroxynitrite is relatively stable in base but undergoes rapid decomposition in acid. The concentration of peroxynitrite may be determined by spectroscopy ($\varepsilon_{302\ nm} = 1670$); Hughes, M.N. and Nicklin, H.G., The chemistry of pernitrites. Part 1. Kinetics of decomposition of pernitrous acid, *J. Chem. Soc. (A)*, 450–452, 1968). The rate of reaction of peroxynitrite with proteins for oxidation and nitration is in the range of 10^4–10^8 M^{-1} s^{-1} (Kissler, R., Beckman, J.S., and Koppenol, W.H., Peroxynitrite studied by stopped-flow spectroscopy, *Methods Enzymol.* 301, 342–352, 1999; Goldstein, S. and Merenyl, G., The chemistry of peroxynitrite. Implications for biological activity, *Methods Enzymol.* 436, 49–61, 2008). It should be noted that the use of peroxynitrite for the modification of proteins is usually associated with rapid mixing of the various components of the reaction, although longer periods of reaction have been used.

[b] Beckman, J.S., Chen, J., Ischiropoulos, H., and Crow, J.P., Oxidative chemistry of peroxynitrite, *Methods Enzymol.* 233, 229–240, 1994.

[c] The reaction was allowed to proceed for 5 min and 70% inactivation of inhibitory activity was observed. Incubation of the peroxynitrite in solvent for 5 min at 37°C prior to addition to protein obviates inactivation. Partial protection is provided by reduced glutathione but not oxidized glutathione; protection is also provided by lipoic acid, methionine, and dihydrolipoic acid. Subsequent work showed protection by high concentrations (250–1000 μM) of mercaptoethylguanidine, while inactivation of α_1-antiproteinase inhibitor by peroxynitrite is enhanced at lower concentrations (1–60 μM) of mercaptoethylguanidine (Whiteman, M., Szabo, C., and Halliwell, B., Modulation of peroxynitrite- and hypochlorous-induced inactivation of α_1-antiproteinase by mercaptoethylguanidine, *Br. J. Pharmacol.* 120, 1646–1652, 1999).

[d] Pryor, W.A., Cueto, R., Jin, X. et al., A practical method for preparing peroxynitrite solutions of low ionic strength and free of hydrogen peroxide, *Free Radic. Biol. Med.* 18, 75–83, 1995.

[e] There was nitration of many of the 15 tyrosyl residues in human serum albumin; Tyr138 and Tyr411 were preferentially nitrated. Tyr411 is also nitrated by TNM.

[f] White, R., Crow, J., Spear, N. et al., Making and working with peroxynitrite, *Methods Mol. Biol.* 100, 215–230, 1998.

[g] The formation of nitrotyrosine in annexin II treated with peroxynitrite was established by the use of an antibody against nitrotyrosine on western blot analysis. Cross-linkage was also observed.

[h] The modification of Tyr99 was observed as well as oxidation of methionine residues. Replacement of the methionine residues with leucine residues did not affect nitration of the tyrosine residue suggesting that methionine does not act as scavenger for peroxynitrite. It is also suggested that nitration of calmodulin can serve as a biomarker.

[i] Reaction with tripeptide was used to compare the efficiencies of several nitration reagents. TNM was the most effective (11.5% nitrated product), while there was 4.5% product with peroxynitrite. Nitrogen dioxide or 1-hydroxy-2-oxo-3-(*N*-methyl-aminopropyl)-3-methyl-1-triazene both yielded less than 1% nitrated peptide product.

TABLE 11.4 (continued)

Modification of Proteins, Peptides, and Related Compounds with Peroxynitrite[a]

j   There was 2 mol of nitrotyrosine per protein subunit. Western blot analysis with antinitrotyrosine antibodies demonstrated the presence of nitrated prostaglandin H_2 synthase in atherosclerotic plaque.

k   Reaction of microsomal glutathione *S*-transferase with peroxynitrite results in oxidation of cysteine (reactivity with DTNB) and nitration of tyrosine residues (spectral analysis at 430 nm and antibody to nitrotyrosine) concomitant with an increase in enzyme activity. The increase in enzyme activity on reaction with peroxynitrite occurs in a matter of seconds. There was also formation of dimer and trimer forms of the enzyme. Reaction with TNM or NAI also increased enzyme activity. Subsequent work (Ji, Y., Neverova, I., Van Eyk, J.E. et al., Nitration of tyrosine 92 mediates the activation of rat microsomal glutathione *S*-transferase by peroxynitrite, *J. Biol. Chem.* 281, 1986–1991, 2006) showed that two tyrosyl residues, Tyr92 and Tyr153, of the seven tyrosine residues in rat microsomal glutathione *S*-transferase are nitrated by peroxynitrite. Site-specific mutagenesis of Tyr92 to Phe (Y92F) no longer showed activation with peroxynitrite.

l   Koppenol, W.H., Kissner, R., and Beckman, J.S., Syntheses of peroxynitrite: To go with the flow or on solid grounds?, *Methods Enzymol.* 269, 296–302, 1996.

m   Reaction of H2A histone with peroxynitrite resulted in the formation of carbonyl groups, nitrotyrosine, and dityrosine. The peroxynitrite-modified protein was a potent immunogen, and specificity was shown to the immunogen and to the nitrotyrosine sites in other proteins.

n   The extent of nitration was determined by western blot with antinitrotyrosine antibodies.

References to Table 11.4

11.4.1. Whiteman, M., Tritschler, H., and Halliwell, B., Protection against peroxynitrite-dependent tyrosine nitration and α_1-antiproteinase inactivation by oxidized and reduced lipoic acid, *FEBS Lett.* 379, 74–76, 1996.

11.4.2. Jiao, K., Mandapati, S., Skipper, P.L. et al., Site-selective nitration of tyrosine in human serum albumin by peroxynitrite, *Anal. Biochem.* 293, 43–52, 2001.

11.4.3. Rowan, W.H., III, Sun, P., and Liu, L., Nitration of annexin II tetramer, *Biochemistry* 41, 1409–1420, 2002.

11.4.4. Smallwood, H.S., Galeva, N.A., Bartlett, R.K. et al., Selective nitration of Tyr99 in calmodulin as a marker of cellular conditions of oxidative stress, *Chem. Res. Toxicol.* 16, 95–102, 2003.

11.4.5. Deeb, R.S., Resnick, M.J., Mittar, D. et al., Tyrosine nitration in prostaglandin H_2 synthase, *J. Lipid Res.* 45, 1718–1726, 2002.

11.4.6. Ji, Y. and Bennett, B.M., Activation of microsomal glutathione *S*-transferase by peroxynitrite, *Mol. Pharmacol.* 63, 136–146, 2003.

11.4.7. Khan, M.A., Dixit, K., Jabeen, S. et al., Impact of peroxynitrite modification of structure and immunogenicity of H2A histone, *Scand. J. Immunol.* 69, 99–109, 2009

11.4.8. Takahusa, H., Mohar, I., Kavanagh, T.J. et al., Protein tyrosine nitration of mitochondrial carbamoyl phosphate synthetase 1 and its functional consequences, *Biochem. Biophys. Res. Commun.* 420, 54–60, 2012.

(sodium dithionite).[173] Sodium tetrathionate modification of cysteine had been previously used by Darrell Liu to demonstrate the importance of a histidyl residue at the active site of streptococcal proteinase.[174] Conversion of nitrotyrosine to aminotyrosine via reduction with sodium dithionite[173] also improves the specificity for the detection of nitrotyrosine-containing proteins on western blots; nitrotyrosine-positive bands are eliminated on reduction leaving the false-positive spots.[163,164] As noted earlier,[167] aminotyrosine may be modified to yield a fluorescent product for enhanced detection. Nitrated proteins are being developed as biomarkers using antibodies and/or mass spectrometric analysis.[175–188] There have been issues with the use of commercial immunoassays for nitrotyrosine in plasma[189,190] but progress is being made in this area.[191]

The reaction of TNM for the modification of proteins in place of the mixture of nitric acid and sulfuric acid appears to date in the work by Wormall in 1930[46] that was, in turn, based on an earlier work in 1920 by Schmidt and Fischer[192] and Baillie and coworkers.[193] The concept of the use of TNM for protein modification was extended in a review by Roger Herriott in 1947.[194] However, it was not until some two decades later that studies of Vallee, Riordan, Sokolovsky, and Harell established the specificity and characteristics of the reaction of TNM (Figure 11.5) with proteins.[195,196] The modification proceeds optimally at mildly alkaline pH. The rate of modification of

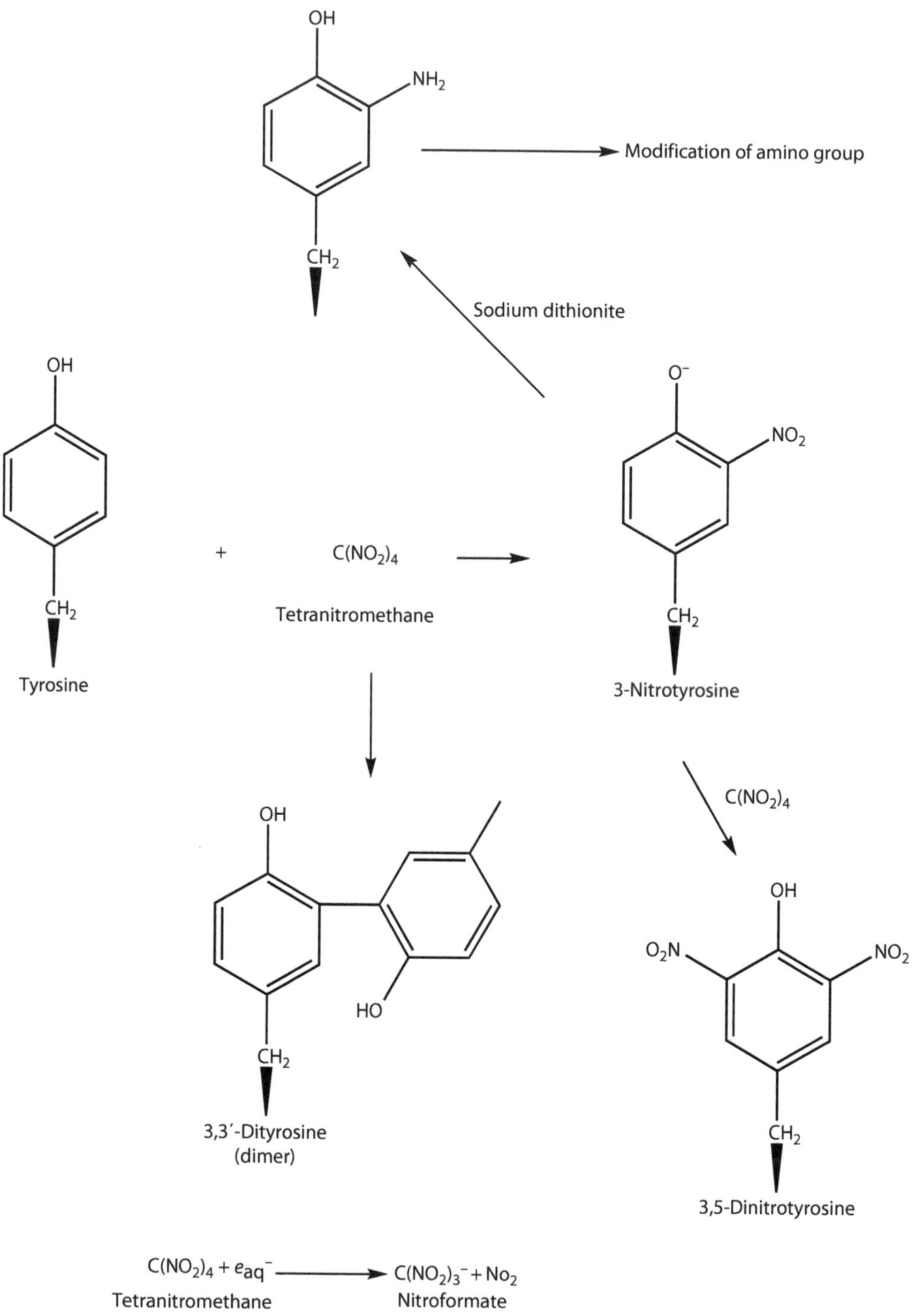

FIGURE 11.5 Reaction of TNM with tyrosine. The major product of the reaction is 3-nitrotyrosine; there is lesser formation of 3,5-dinitrotyrosine and cross-linkage to form 3,3′-dityrosine dimer. The nitrotyrosine can be reduced to 3-aminotyrosine with dithionite. The nitration reaction is an example of an electrophilic aromatic substitution reaction with the formation of nitroform (nitroformate). See Riordan, J.F. and Vallee, B.L., Nitration with tetranitromethane, *Methods Enzymol.* 25, 515–521, 1972; Capellos, C., Iyer, S., Liang, Y., and Garmms, L.A., Transient species and product formation from electronically excited tetranitromethane, *J. Chem. Soc. Faraday Trans.* 2 82, 2195–2206, 1986.

N-acetyltyrosine is twice as rapid at pH 8.0 as at pH 7.0; it is approximately ten times as rapid at pH 9.5 as at pH 7.0. The reaction of TNM with tyrosine produces 3-nitrotyrosine, nitroformate (trinitromethane anion), and two protons. The spectral properties of nitroformate (ε at 350 nm = 14,000)[197] suggested that monitoring the formation of this species would be a sensitive method for monitoring the time course of the reaction of TNM with tyrosyl residues.[195–197] Although determining the rate of nitroformate production appears to be effective in studying the reaction of TNM with model compounds such as *N*-acetyltyrosine, it has not proved useful with proteins.[196,198] Although the reaction of TNM with proteins is reasonably specific for tyrosine, the oxidation of sulfhydryl groups has also been reported[196,197] as has reaction with histidine,[196] methionine,[197] and tryptophan.[196,199] The oxidation of cysteine with TNM or peroxynitrite can confound the use of such reactions with proteins containing sulfhydryl groups. Cysteine residues in proteins can be oxidized by TNM.[200] Palamalai and Miyagi[200] were able to modify tyrosine residues in glyceraldehyde-3-phosphate dehydrogenase and establish their role in enzyme function but were successful only after reversibly blocking the active-site sulfhydryl group by formation of a mixed disulfide with DTNB (see Chapter 6).

Reaction of proteins with TNM can also result in the covalent cross-linkage of tyrosyl residues resulting in inter- and intramolecular association of peptide chains.[201,202] The cross-linking of tyrosyl residues in proteins via reaction with TNM is an example of zero-length or contact-site cross-linking.[203,204] In studies of the mechanism of TNM nitration of phenols, Bruice and coworkers[205] observed more *cross-linkage* (formation of Plummer's ketone) than nitration with the reaction of TNM with *p*-cresol at neutral pH. The magnitude of this problem is dependent on variables such as protein, protein concentration, and solvent conditions (i.e., pH, presence of organic solvents).[206] For example, reaction of pancreatic deoxyribonuclease with TNM results in extensive formation of dimer.[207] The dimerization of insulin on reaction with TNM[206,208,209] has also been reported by several groups. Both DNAse and insulin are known to form noncovalent dimers in solutions.[210–212] Chen and coworkers[55] used photochemical iodination to establish the reactivity of tyrosyl residues in insulin oligomers. Mädler and coworkers[212] reported a cross-link with tyrosine in insulin with NHS esters.

A related reaction is the free radical–induced cross-linking between tyrosyl residues and thymine providing a basis for the formation of nucleic acid–protein conjugates occurring as a result of ionizing radiation.[213] Treatment of apoovotransferrin with periodate (50 mM HEPES, pH 7.4 with 5 mM sodium periodate) resulted in protein cross-linking via 3,3′-dityrosine.[214] The reader is also directed to the earlier cited work by Bartesaghi and coworkers[142] on the dimerization of tyrosine mediated by lipid peroxyl radicals in membranes. It is recognized that peroxynitrite does modify DNA by both oxidation and nitration.[215,216]

The use of immunochemistry either in solid-phase assays, western blotting, and immunohistocytochemistry for the identification and measurement of nitrotyrosine in proteins has been discussed earlier, and such technologies are more related to the study of biological polymer modification in biological fluids and tissues. While these analytical techniques are quite useful for the physiologists and cell biologists, such techniques are of less value to those interested in solution chemistry of biological polymers where solution analysis is a critical part of the process of characterization. The extent of modification of tyrosyl residues by peroxynitrite or TNM in proteins can be assessed either by spectrophotometric means or by amino acid analysis.[198,217] At alkaline pH (pH $\geq$ 8), 3-nitrotyrosine has an absorption maximum at 428 nm with ε = 4100 M^{-1} cm^{-1}; the absorption maximum of tyrosine at 275 nm increases from ε = 1360 to 4000 M^{-1} cm^{-1}. At acid pH (pH $\leq$ 6), the absorption maximum is shifted from 428 to 360 nm, with an isosbestic point at 381 nm (ε = 2200 M^{-1} cm^{-1}). Amino acid analysis after acid hydrolysis has also proved to be a convenient method of assessing the extent of 3-nitrotyrosine formation. 3-Nitrotyrosine is stable to acid hydrolysis (6 N HCl, 105°C, 24 h). This approach has the added advantage that other modifications of tyrosine such as free radical–mediated cross-linkage can be either excluded or quantitatively determined. If nitration to form 3-nitrotyrosine is the only modification of tyrosyl residues in a protein occurring on reaction with TNM, the sum of 3-nitrotyrosine and tyrosine should be equivalent to the amount of tyrosine in the unmodified protein. The material on amino acid analysis is provided more for historical interest as MS is the dominant analytical technique.[218–220]

There are several consequences of the nitration of a tyrosyl residue by nitric acid, TNM, or peroxynitrite. First, there is the bulk of the nitro group addition *ortho* to the phenolic hydroxyl function. The other consequence is the lowering of the pKa of the phenolic hydroxyl groups, lowering the pKa of the phenolic hydroxyl from approximately 10.3 to 7.3. The lowering of the pKa is due to a combination of electrostatic effects of the nitro group with resonance stabilization as well as an inductive effect.[221–223] This of course means that the phenolic hydroxyl of the nitrated tyrosyl residue can be in a partially ionized state at physiological pH. It is suggested that the decrease in the activity of hen egg white lysozyme seen on modification with TNM[224] is due to the ionization of the nitrotyrosyl residue(s) that would repel the negatively charge substrate; the removal of the cell wall anionic polymer, teichuronic acid, from the *Micrococcus luteus* cell wall fragment eliminated this effect. The tight binding of biotin to avidin is reduced by modification of a tyrosine residue in avidin by TNM.[225] The modified avidin can then be used for the affinity chromatography of biotinylated proteins under relatively mild conditions.[226,227] Tawfik and coworkers[228] showed that nitrated antibodies demonstrate pH-dependent binding near physiological pH. The modification of monoclonal antibodies with TNM resulted in the loss of antigen binding at pH 8.0 that is regained at pH 6.0. This change is related to the decrease in the pKa of the phenolic hydroxyl group (which would be ionized at pH 8.0 but protonated at pH 6.0).

3-Nitrotyrosine can be reduced to the corresponding amine under relatively mild conditions (Na$_2$S$_2$O$_4$, 0.05 M Tris, pH 8.0).[173] The conversion of 3-nitrotyrosine to 3-aminotyrosine is associated with the loss of the absorption maximum at 428 nm and the change in the pKa of the phenolic hydroxyl group from approximately 7.0 to 10.0. On occasion, reduction of the nitro function in this manner reverses the modification of function observed on nitration. The resultant amine function can be subsequently modified.[167,229] Haas and colleagues[230] modified the single tyrosine residue in *Escherichia coli* acyl carrier protein with TNM, subsequently reduced the 3-nitrotyrosyl residue to 3-aminotyrosine with sodium dithionite, and modified the 3-aminotyrosine with dansyl chloride at pH 5.0 (50 mM sodium acetate) to obtain dansyl acyl carrier protein. The dansyl acyl carrier protein was subsequently used for fluorescence anisotropy studies of enzyme–substrate complex formation in stearoyl-ACP desaturase.[231] The 3-amino tyrosine derivative has been used as a *handle* for the purification of peptides containing nitrotyrosine. Nikov and coworkers[232] coupled biotin to aminotyrosine-containing peptides using a cleavable derivative [sulfosuccinimidyl-2-(biotinamido)ethyl-1,3-dithioproprionate]; peptides were purified by affinity chromatography using streptavidin columns with elution by reductive cleavage of the label. This approach has been used by other investigators for the identification of nitrotyrosine in proteins.[233–235] It should be noted that the low pKa for 3-aminotyrosine (ca. 4.7) provides the potential for specific modification of this functional group in a protein.[199]

The modification of protein-bound tyrosine with TNM also introduces a spectral probe (3-nitrotyrosine) that can be used to detect conformational change in the protein. 3-Nitrotyrosine has an absorption maximum at 428 nm at alkaline pH. This spectral property was first used by Riordan and coworkers with studies on nitrated carboxypeptidase A[236] to study changes in the microenvironment around the modified residue. The addition of β-phenylpropionate, a competitive inhibitor of carboxypeptidase and nitrated carboxypeptidase, decreased the absorbance of mononitrocarboxypeptidase at 428 nm. This change is consistent with an increase in the hydrophobic quality of the microenvironment surrounding the modified tyrosyl residue. The spectral characteristics of nitrotyrosine are thus useful as a reporter group. Reporter groups are probes that have absorbance or fluorescence properties that can be used to monitor conformational change in proteins.[237–241] There has only been limited use of the UV spectral characteristics of 3-nitrotyrosine as a reporter group[242,243]; there has been a use of the fluorescence quenching properties of nitrotyrosine as a conformational probe.[244–247] There are several unique applications that deserve specific mention. Herz and coworkers[248] coupled nitrotyrosine methyl ester to a carboxyl group in bacteriorhodopsin for use as reporter groups.

TNM is not considered to be water soluble[111,249] but is soluble in polar solvents such as ethanol and apolar solvents such as ethyl ether. The presence of impurities is a potential problem and may present a safety hazard. It is possible to *wash* TNM with water to remove impurities.[198] High concentrations of reagent in reaction mixtures should be avoided to obviate the possibility of phase separation. Rabani and coworkers also describe an aqueous extraction process to remove impurities from TNM.[197] In his description of the synthesis of TNM from acetic anhydride and nitric acid in 1955, Liang[250] reported that TNM separates from the aqueous phase during the process. However, studies on the chemistry of TNM at concentrations of 50 μM in aqueous solutions have been performed without reported difficulty.[251] It should be noted that most experiments using TNM involve the introduction of the reagent as a solution in ethanol or less frequently with methanol. The final concentration of ethanol is frequently 5%–10% of the final reaction volume that likely enhances the solubility of TNM in the reaction mixture. The author is not aware of any studies that specifically address the effect of ethanol (or other organic water-miscible solvents) on the solubility of TNM in an aqueous solution.

The reaction mechanism as suggested by Walters and Bruice[252] postulates the formation of an intermediate phenoxide–TNM charge transfer complex involving electron transfer from the aromatic ring. This would imply that the reaction of TNM with tyrosine residues occurs with the ionized species. These studies were performed with TNM and *p*-methylphenol (4-methylphenol; *p*-cresol), and the reaction was followed by the disappearance of TNM as measured by the formation of trinitromethane anion (nitroformate) at 350 nm. The products of the reaction were trinitromethane anion, nitrite, and 4-methyl, 3-nitrophenol. The rate of reaction ($k_2 = min^{-1}$ 5.1×10^4 M^{-1}) was slower with phenol ($k_2 = 2.0 \times 10^3$ M^{-1} min^{-1}) and much slower with either 4-chlorophenol (5.7×10^2 M^{-1} min^{-1}) or 4-cyanophenol (0.26 M^{-1} min^{-1}). There is no obvious correlation between these rates and the dissociation constant for the phenolic hydroxyl group; there is a correlation with the reaction rate and the Hammett constant σ.[253] The analysis of the products of the reaction of TNM and 4-methylphenol demonstrated the presence of 4-methyl, 3-nitrophenol in 23% yield and a 30% yield of 1,2,10,11-tetrahydro-6,11-dimethyl-2-oxodibenzfuran (Pummerer's ketone), a product derived from the free radical–mediated cross-linking of the parent 4-methylphenol. Given the average pKa value for tyrosine of 10.13 (see Table 1.1), this would imply that the average tyrosine residues would be mostly unreactive within the pH range of 6.0–8.0, which is suggested as the optimal range for the modification of tyrosine in proteins. This is supported by Bruice and coworkers[205] who demonstrated that the rate of reaction of TNM with the phenolic form must be at least four orders of magnitude smaller than those observed with the phenolate form. At pH values above 8, the rate of reaction of TNM with tyrosine would appear to increase consistent with the importance of the ionization of the phenolic function; however, the side reactions with methionine, tryptophan, and histidine increase. The studies on the reaction with 4-methylphenol and TNM yielded an approximate 20% yield of 3-nitro-4-methylphenol, which could be only modestly increased by excess TNM, while reaction in proteins could yield 100% of the modification of a given tyrosyl residue.[254] Early studies[195,196] suggested that the formation of trinitromethane anion (nitroformate) could be used to stoichiometrically follow the reaction of TNM with tyrosyl residues in proteins. This has not proved to be the situation as, in general, the release of trinitromethane anion greatly exceeds the extent of modification of tyrosine and continues beyond the nitration reaction in proteins. While it is not clear as to why this occurs, Fendler and Liecht[251] have shown that the certain detergent micelles catalyze the reaction of TNM with hydroxide ion. Micellar hexadecyltrimethylammonium bromide and polyoxyethylene (15) nonylphenol (Igepal CO-730) enhance the rate constant for the reaction, while micellar SDS has no effect. Nitroformate is derived from nitrotyrosine by reduction (addition of an electron)[197,255] and this reaction is used for the measurement of superoxide radical.[256,257] The reaction of hydroxyl radical with TNM also yields nitroformate.[258]

Photolysis is used to generate the nitrogen dioxide radical and the nitroformate anion from TNM.[259] This process can result in the nitration of suitable targets and the formation of trinitromethyl derivatives.[260] There is at least one study[261] on the photochemical reactions of aromatic compounds with TNM that may be applicable to the interpretation of the reaction of TNM with proteins. This author is also aware of at least one study[262] that examined the effect of light on the reaction of TNM with proteins; however, the effect of light is on protein, not TNM. That study reported the effect of light on the reaction of TNM with bacteriorhodopsin with light where light influenced the intrinsic properties of the protein. Bacteriorhodopsin is a bacterial membrane protein that is activated by light to generate an electrochemical gradient. Absorption of light by the protein initiates a cycle of reactions that results in proton translocation across the membrane.[263] Tyr26 was preferentially nitrated in the dark, while Tyr64 was preferentially modified in the presence of light.[262] In a subsequent study,[264] selective ionization of Tyr64 (pKa approximately 9.0) was observed to occur with illumination. Photochemically induced dynamic nuclear polarization NMR spectroscopy of bacteriorhodopsin suggested that only one tyrosine residue is exposed to a solvent.[265] Light also influenced the modification of bacteriorhodopsin with a carbodiimide.[266]

Other studies on the chemistry of TNM that might be instructive include a study by Capellos and others[267] that showed that, upon excitation with an excimer laser, TNM formed the NO_2 radical and the trinitromethane anion in either acrated or deaerated polar solvents (e.g., methanol), which the trinitromethyl radical was suggested to form in a nonpolar solvent (e.g., hexane). TNM also reacts with unsaturated compounds[268] and has been used as a colorimetric method[269] for the determination of unsaturation. The study on the difference of reactions in the polar and nonpolar solvents might be useful in interpreting environmental effects on the reaction of TNM with individual residues in proteins. Given the complexity of the interpretation of the previously mentioned studies and extension to the reaction of tyrosine in proteins, it is clear that the reaction of TNM with biological polymers can be complex and care should be taken in the interpretation of experimental results.[270]

The next section will discuss several studies on the reaction of TNM with tyrosine in proteins that I consider useful in that the studies use well-characterized proteins. Strosberg[271] and coworkers reported on the reaction of TNM with hen egg white lysozyme. A derivative containing 1 mol of nitrotyrosine per mole protein is obtained by reaction with a 12-fold molar excess of TNM in 50 mM Tris–1.0 M NaCl, pH 8.0 (20°C/1 h); a derivative containing 2 mol of nitrotyrosine per mole protein is obtained with 47-fold molar excess, while a derivative with 3 mol nitrotyrosine per mole protein is obtained with a 100-fold molar excess in 50 mM Tris-HCl–1 M NaCl–8.0 M urea, pH 8.0 at 50°C for 1 h. The analysis of the various products showed that with a 12-fold molar excess of reagent, two mononitrated derivatives are obtained; the major one (69%) is modified at Tyr23 and the minor product (21%) at Tyr20. With a 47-fold molar excess of TNM, two dinitrated products are obtained. The major product (71%) is modified at both Tyr23 and Tyr20, while there is a minor product (15%) modified at Tyr23 and Tyr53. The derivative modified with 100-fold molar excess of reagent in the presence of 8.0 M urea was homogeneous and was modified at Tyr23, Tyr20, and Tyr53. Nitration at Tyr20 or Try23 reduced immunoreactivity by 20% with an antilysozyme antibody preparation; a higher antibody concentration is also required. The modification of Tyr23 and Tyr50 decreased immunoreactivity by 40%, and a higher antibody concentration was required for maximum precipitation. A later study[272] examined the characteristics of hen egg while lysozyme modified at Tyr23 or at both Try23 and Tyr20. Both derivatives were obtained by reaction in 50 mM Tris-HCl, pH 8.0 at 37°C for 1 h. The Tyr23/Tyr20 derivative was obtained at a 15-fold molar excess of TNM, while the Tyr23 was selectively obtained at twofold molar excess of reagent. The Tyr23 derivative could also be obtained by electrochemical nitration.[273] Skawinski and coworkers[274] have used [^{15}N]TNM to modify tyrosine residues in lysozyme. Two or three ^{15}N resonances were obtained dependent on the extent of modification; the pH dependence of the detected resonances depicted the apparent microscopic pKa values. These investigators also studied the modification of ribonuclease with [^{15}N]TNM. Santrůcek and coworkers[275] showed only small differences in the quality and quantity of modification of lysozyme with TNM and iodine using MALDI-TOF MS. The order of reactivity of tyrosine residues with TNM and iodine (iodine + potassium iodide at pH 7.0) is similar in hen egg white lysozyme. There was no clear correlation between surface accessibility of residues and modification with TNM or iodine.

Crambin is a small water-insoluble protein that contains two tyrosine residues. Reaction with TNM was performed in 50% EtOH/H_2O at pH 9.0 (0.0005 M Tris) and was monitored by NMR spectroscopy.[276] Tyr29 is modified with a 50-fold molar excess of reagent in 10 min at 298°K (25°C). Addition of a second 50-fold molar excess of TNM and continuation of the reaction resulted in the additional modification of Tyr44. A reverse susceptibility is observed with iodination (I_2, pH 8.5 with 0.0005 M Tris-HCl in 50% EtOH/H_2O). The reaction of tyrosine residues with TNM is consistent with solvent-exposed residues being more susceptible to reaction, while the iodination pattern is not. It was concluded that the susceptibility to modification by either TNM or I_2 is controlled by the microenvironment of the residue independent of solvent exposure; it was concluded that the preferential iodination of Tyr44 was facilitated by an electrostatic effect from Asp43. The authors also suggest that iodination or nitration should not be taken as a single criterion for the exposure of tyrosyl residues. El Kabbaj and Latruffe[111] studied the ability of TNM to penetrate the inner mitochondrial membrane by evaluating the inactivation of D-3-hydroxybutyrate dehydrogenase by TNM in its normal location and in *inside-out* membranes. TNM readily inactivated the enzyme on *inside-out* membranes but was much less effective with intact mitochondrial membrane. On the basis of these studies, the authors suggest that the behavior of TNM is consistent with both partial amphiphilic and hydrophobic properties. The modification of tyrosine residues in thermolysin with TNM was performed in 40 mM Tris, pH 8.0 containing 10 mM $CaCl_2$ at 25°C 1 h.[277] Thermolysin contains 28 tyrosine residues and approximately 9 residues were modified under these conditions as determined by absorbance at 381 nm. When the reaction was allowed to proceed for 15 h under the same conditions, approximately 16 residues were modified. The analysis of the reaction rate allowed the identification of three classes of reactive tyrosyl residues that reacted at different rates with TNM. The apparent second-order rate constants for the three classes are 3.32, 0.52, and 0.18 M^{-1} min^{-1} compared to a second-order rate constant for the reaction of TNM with *N*-acetyltyrosine ethyl ester of 1.99 M^{-1} min^{-1} under the same reaction conditions. In the same study, spectrophotometric (295 nm) titration of thermolysin by pH jump demonstrated that 16 tyrosine residues were readily ionized, while 12 additional tyrosine residues required an apparent conformational change in the protein for ionization after adjustment to pH $\geq$ 12. These investigators divided the readily ionized tyrosine residues into three classes with pKa values of 10.2, 11.4, and 11.8. These investigators concluded that the microenvironment of the more slowly reacting tyrosine residues is either negatively charged or hydrophobic. The effect of nitration on the catalytic activity of thermolysin was examined in a subsequent communication from the same research group.[278] Activity was reduced to 10% of that of the native enzyme with nitration of 16 tyrosine residues (the effect was on k_{cat}) with partial recovery on reduction of the nitro groups to amino groups; activation by NaCl was also decreased on nitration. In another study, Muta and coworkers[279] showed that alkaline pKa for human matrilysin-7 shifted from 9.8 to 10.3–10.6 with nitration; the pKa returned to 9.8 on reduction of the nitro groups to amino groups.

The protection of an enzyme from loss of activity and lack of modification of an amino acid residue(s) is taken to infer the importance of the said residue(s) in the catalytic activity of the enzyme. However, Christen and Riordan[280] observed that aspartate aminotransferase was readily inactivated by TNM only in the presence of both substrates, glutamate and α-ketoglutarate (*syncatalytic* modification). This inactivation was associated with the modification of an additional tyrosyl residue. The syncatalytic modification of other enzymes has been subsequently reported including the inactivation of chicken liver xanthine oxidase by hydrogen peroxide generated during oxidation of the substrate by oxygen,[281] the syncatalytic modification of adrenocortical cytochrome P-450$_{scc}$ with TNM,[282] and the formation of a disulfide in aldolase.[283] Another anomaly is a situation when modifying agents such as TNM are substrates for enzymes such as isocitrate lyase from germinating castor seed endosperm where the enzyme catalyzes the reaction of TNM with isocitrate to form nitroform.[284] Medda and coworkers[285] demonstrated the formation of nitroform in the oxidation of putrescine by lentil seedling amine oxidase. In an earlier study,[286] Mekhanik and Torchinskii showed that TNM reacted with a pyridoxal phosphate–substrate carbanion during enzymatic decarboxylation with α-methylglutamate or aspartate as substrate for glutamate dehydrogenase.

This limited discussion suggests that it is not possible to accurately predict the environment of a tyrosyl residue modified by TNM in a protein. Care should be exercised in the interpretation of results when such observations are used to describe the topography of a protein.[287–290] The reader is again directed to the review by Glazer[31] for a consideration of the concept of exposed and buried residues. Selected examples on the application of TNM for the site-specific modification of protein are presented in Table 11.5.

Tyrosyl residues in proteins can also be modified by reaction with cyanuric fluoride.[291–293] The reaction proceeds at alkaline pH (9.1) via modification of the phenolic hydroxyl group with a change in the spectral properties of tyrosine. The phenolic hydroxyl groups must be ionized (phenoxide ion) for reaction with cyanuric fluoride. The modification of tyrosyl residues in elastase[294] and yeast hexokinase[295] with cyanuric fluoride has been reported. Coffe and Pudles[295] compared the effect of cyanuric fluoride, TNM, and NAI in the modification of tyrosine residues in yeast hexokinase; two tyrosyl residues are modified with either cyanuric fluoride or TNM, while only one is modified with NAI. While it is suggested that cyanuric fluoride is relatively specific for the modification of tyrosine in proteins,[291] there are no direct data available with a direct comparison to either cyanuric chloride or cyanuric bromide. Cyanuric chloride has been demonstrated to react with both tyrosine and amino groups in proteins and has been used for the coupling of dextran to proteins[296]; in this latter study, the inactivation associated with lysine residues was obviated by prior modification of the protein with citraconic anhydride. The author was not able to find information on the reaction of cyanuric bromide with any biological polymer. It is noted that the various cyanuric halides are precursors in the synthesis of various dyes. Cyanuric bromide can be an impurity in preparations of cyanogen bromide.[297,298] Modification of tyrosyl residues can occur with other residue-specific reagents not considered to be directed at tyrosine. One example is 7-chloro-4-nitrobenzo-2-oxa-1,3-diazole (7-chloro-4-nitro-benzofurazan; NBD-Cl; Nbf-Cl).[299,300] NBD-Cl reacts primarily with amino groups and sulfhydryl groups in proteins.[301–303] That said, there has been some work using NBD-Cl for proteins[304] and peptides.[305] The reaction product obtained from the reaction of NBD-Cl with tyrosine, unlike that obtained with either amino groups or sulfhydryl groups, is not fluorescent and has an absorption maximum at 385 nm compared to 475 nm for amino derivatives and 425 nm for sulfhydryl derivatives.[299] The reaction of NBD-Cl with sulfenic acid has been reported[302] and is used for the detection of sulfenic acid in oxidized proteins,[306,307] although dimedone appears to be more specific.[308,309]

Diazonium salts (Figure 11.6) readily couple with tyrosine, lysine, and histidine residues in proteins at alkaline pH to form colored derivatives with interesting spectral properties.[310–315] The reaction of chymotrypsinogen A with diazotized arsanilic acid has been investigated.[316] The extent of the formation of monoazotyrosyl and monoazohistidyl derivatives is determined by spectral analysis.[314,315] The arsaniloazo functional group provides a spectral probe that can be used to study conformational change in proteins. In this particular study, there was a substantial change in the circular dichroism spectrum (extrinsic Cotton effect) during the activation of the modified chymotrypsinogen preparation by trypsin.[316] The reaction of α-chymotrypsin with three diazonium salt derivatives of N-acetyl-D-phenylalanine methyl ester[317] (Figure 11.7) has also been studied. As noted in Figure 11.7, these diazonium salts are analogs of allosteric activators of chymotrypsin. Subsequent analysis showed that Tyr146 is modified by each of the three reagents. It was observed that the peptide with the modified tyrosine residue (possessing a yellow color) absorbs to the gel filtration matrix (G-10 equilibrated with 0.001 M HCl) and was eluted with 50% acetic acid. This phenomenon is somewhat similar to that observed with tryptophan-containing peptides, which have been modified with HNB.[318] A single tyrosine residue in bovine pancreatic ribonuclease has been modified by a diazonium salt derivative of uridine 2′(3′)5′-diphosphate [5′-(4-diazophenyl phosphoryl)-uridine-2′(3′)-phosphate].[319] These investigators also examined the reaction of ribonuclease with p-diazophenylphosphate under the same conditions of solvent and temperature. Reaction with this reagent was far less specific, with losses of lysine, histidine, and tyrosine (3 mol per mol ribonuclease).[319] Vallee and coworkers[320,321] used diazotized p-arsanilic acid to obtain specific modification of Tyr248. The reaction of bovine carboxypeptidase A with diazotized 5-amino-1H-tetrazole has also

TABLE 11.5
Modification of Tyrosyl Residues in Proteins with TNM[a]

Protein	Reaction Condition	Molar Excess[b]	NO$_2$Tyr	Reference
Carboxypeptidase A	50 mM Tris, 2 M NaCl/20°C[c]	4	1.2/18[d]	11.5.1
Staphylococcal nuclease	50 mM Tris, pH 8.1/23°C	2	1.1/7[e]	11.5.2
Horse heart cytochrome c	50 mM Tris, pH 8.0/23°C	16	2/4	11.5.3
Aspartate aminotransferase	50 mM Tris, pH 7.5/22°C	30	1.7/24[f]	11.5.4
Thrombin	30 mM sodium phosphate, pH 8.0/24°C	1000	4.9/12[g]	11.5.5
Porcine carboxypeptidase-B	0.05 M Tris, pH 8.0/23°C	8	1.2/21[h]	11.5.6
Bovine pituitary growth hormone	0.05 M Tris, pH 8.0/0°C	30	2.7/6[i]	11.5.7
α$_1$-Antiprotease inhibitor	0.05 M Tris, pH 8.0/25°C	120	3/7[j]	11.5.8
Carboxypeptidase A crystals	0.05 M Tris, pH 8.0/20°C	~4	1/9[k]	11.5.9
Bovine growth hormone	0.03 M ringer phosphate/25°C	12	3/7[l]	11.5.10
Lactose repressor protein	0.1 M Tris, pH 7.8 with 0.1 M mannose[m]/23°C	800	2.4/8[n]	11.5.11
Human serum albumin	0.1 M Tris—0.1 M KCl, pH 8.0/23°C	80	9/18[o]	11.5.12
Porcine pancreatic phospholipase	0.05 M Tris, 0.1 M NaCl, 0.01 M CaCl$_2$, pH 8.0/30°C; reaction also performed with the inclusion of egg yolk lysolecithin	10	1–2[p]	11.5.13
Bacillus subtilis neutral Protease	50 mM Tris-Cl, pH 8.0 with 5 mM CaCl$_2$/25°C[q]	60	2	11.5.14
Fructose-1,6-bisphosphatase	50 mM Tris-Cl, pH 8.0/23°C[r]	50	4[s]	11.5.15
Lipase/acyltransferase from *Aeromonas hydrophila*	50 mM Tris—100 mM NaCl, pH 8.0/23°C	53	2/13[t]	11.5.16
HlyC (acyl-ACP-proHlyA acyl-transferase)	50 mM Tris, pH 8.0 or 100 mM sodium phosphate, pH 6.0[u]	—	—	11.5.17
Pseudomonas cepacia lipase	50 mM Tris, pH 8.0 with 5% hexane[v]/23°C in the dark	300	11/14[w]	11.5.18
Human angiotensin II	50 mM Tris, pH 8.0/20°C	50	1[x]	11.5.19
Bovine serum albumin	50 mM Tris, pH 8.0/20°C	50	4[x]	11.5.19
Glycine receptor in phospholipid vesicles	25 mM sodium phosphate, pH 7.4/23°C	10–100	6/16[y]	11.5.20
Human matrilysin	50 mM HEPES, pH 7.5–10 mM CaCl$_2$ with 0.05% Brig-35/25°C	2000–7900	2.8–4.8/8[z]	11.5.21
Recombinant hamster prion protein (90–232)	50 mM sodium acetate, pH 5.0/23°C	100–1000	6[aa]	11.5.22
Recombinant IgG1 monoclonal antibody	10 mM Tris, pH 8.3/37°C/20 min	10–600	—[bb]	11.5.23

[a] The modification is performed in aqueous media with TNM added from a stock solution in absolute ethanol. TNM is a potent oxidizing agent and is considered to be a hazardous material with potential explosive characteristics and some toxicity. I personally never had a problem with working with TNM; however, caution is warranted. Reaction is usually terminated by dialysis for quenching with 2-mercaptoethanol.

[b] Reagent to protein unless otherwise indicated. In general, a lower molar ration of TNM is required compared to NAI— most likely reflecting the susceptibility of NAI to hydrolysis under aqueous reaction conditions (Kohtun, W.L., Dexter, R.N., Clark, R.E., and Gurd, F.R.N., Coordination complexes and catalytic properties of proteins and related substances. I. Effect of cupric and zinc ions on the hydrolysis of *p*-nitrophenyl acetate by imidazole, *J. Am. Chem. Soc.* 80, 1188–1194, 1958).

(continued)

TABLE 11.5 (continued)
Modification of Tyrosyl Residues in Proteins with TNM[a]

[c] Buffers were extracted with 0.1% dithizone in carbon tetrachloride to remove any adventitious metal ions. The reaction shows marked pH dependence with no effect on catalytic activity with reaction at pH 6.0. The choice of buffer is important with lower rates in phosphate or borate buffers. It is stated that phosphate and borate are inhibitors of carboxypeptidase A but data or citation is provided. High concentrations of phosphate were reported to interfere with the reaction of *P*-mercuribenzoate with the cysteinyl residue in apocarboxypeptidase A (Coombs, T.L., Omote, Y., and Vallee, B.L., The zinc-binding groups of carboxypeptiase A, *Biochemistry* 3, 653–662, 1964) and several organic phosphates were reported as inhibitors of carboxypeptidase A (Adelman, R.C. and Lacko, A.G., The inhibition of bovine pancreatic carboxypeptidase A by dihydroxyacetone phosphate, *Biochem. Biophys. Res. Commun.* 33, 596–601, 1968). Modification was performed in 0.05 M Tris–2.0 M NaCl with a similar rate reported in Veronal buffer. The solubility of TNM in the 0.05 M Tris–2.0 M NaCl, pH 8.0 was 5×10^{-4} M.

[d] Amino acid analysis showed only modification of tyrosine. A 0.9 mol of 3-nitrotyrosine was found with amino acid analysis after acid hydrolysis, while 1.2 mol of nitrotyrosine was found with spectrophotometry; 0.3 mol of nitrotyrosine was found when modification was performed in the presence of β-phenylpropionate. Nitration of carboxypeptidase A was associated with an increase in esterase activity and a decrease in peptidase activity. While analytical data was not provided, it appeared that the TNM did not react with *O*-acetyltyrosine (the product of reaction of tyrosine with *N*-acteylimidazole) but did react with iodinated protein giving results consistent with the displacement of the iodo group on the phenolic ring of tyrosine with the nitro group resulting in the formation of 3-nitrotyrosine. It was possible to modify more tyrosyl residues with more TNM; 6.7 tyrosine residues were modified with a 64-fold molar excess.

[e] Increasing the ration of TNM to protein increased the extent of modification with approximately five of the seven total tyrosine residues modified at a 60-fold molar excess; the presence of 4.0 M guanidine at this concentration was necessary to nitrate all seven tyrosine residues. If the reaction is performed in the presence of a competitive inhibitor and calcium ions, activity is retained and only 2–3 tyrosine residues are nitrated. Performing the nitration reaction at a lower molar excess of reagent (fourfold) results in the nitration of Tyr85; with the reaction in the presence of the competitive inhibitor and calcium ions, nitration at Tyr85 is greatly reduced with nitration occurring instead at Tyr115. It is suggested that Tyr115 is inaccessible for modification in the native protein but is exposed on the binding of the competitive inhibitor.

[f] Inactivation of the enzyme by TNM occurred in the presence of substrates, glutamate and α-ketoglutarate, resulting in the use of the term syncatalytic to describe an event occurring simultaneously with catalysis. In the presence of substrate together with 97% inactivation, 1.7 mol of 3-nitrotyrosine was found; 0.7 mol of 3-nitrotyrosine was found in the absence of substrate together with 5% inactivation. There was slightly more oxidation of sulfhydryl groups in the presence of substrates (1.7 sulfhydryl groups) than in the absence of substrates (1.4 sulfhydryl groups). It is considered unlikely that the oxidation of sulfhydryl groups was responsible for the loss of activity.

[g] Subsequent work (Lundblad, R.L., Noyes, C.M., Featherstone, G.L. et al., The reaction of bovine α-thrombin with tetranitromethane. Characterization of the modified protein, *J. Biol. Chem.* 263, 3729–3734, 1988) demonstrated that the reaction proceeded more rapidly in 50 mM Tris–100 mM NaCl, pH 8.0, than the phosphate buffer. There is considerable evidence to suggest an interaction of phosphate with thrombin. Analysis of the kinetics of inactivation (Levy, H.M., Leber, P.D., and Ryan, E.M., Inactivation of myosin by 2,4-dinitrophenol and protection by adenosine triphosphate and other phosphate compounds, *J. Biol. Chem.* 238, 3654–3659, 1963) suggested that the loss of activity was associated with the modification of one mole tyrosine per mole of thrombin. Structural analysis showed the modification of Tyr71 and Tyr85 but it was not possible to separate the derivative forms.

[h] Amino analysis showed that only tyrosine was modified in porcine carboxypeptidase B by TNM. The analysis of the rate of inactivation as function of tyrosine modification suggested that the change in catalytic activity is due to the modification of a single tyrosine residue.

[i] Data are shown for the modification of bovine pituitary growth hormone; similar results were obtained from ovine pituitary growth hormone at 0°C. Both proteins had two completely modified tyrosyl residues, two partially modified tyrosyl residues, and two unmodified tyrosyl residues. All six tyrosine residues in the bovine protein are modified with TNM at 25°C. The nitrotyrosyl residues in bovine protein could be reduced to aminotyrosyl residues. The nitrated proteins and the reduced product retained growth-promoting activity. Attempts to obtain a high-degree modification resulted in the formation of polymeric products.

TABLE 11.5 (continued)
Modification of Tyrosyl Residues in Proteins with TNM[a]

j   There was some polymerization of α_1-antitrypsin with TNM; however, there was a 65% yield of monomer. The modification of α_1-antitrypsin with a 120-fold molar excess of TNM resulted in the formation of 3 mol of nitrotyrosine per mole of protein; reaction with a 60-fold molar excess in the presence of 5 M guanidine hydrochloride resulted in the modification of all seven tyrosyl residues. The rate of nitration was determined by increase in absorbance at 428 nm; the extent of modification was determined by spectral analysis of the modified product and by amino acid analysis. Subsequent work (Mierzwa, S. and Chen, S.K., Chemical modification of human α_1-antiproteinase inhibitor by tetranitromethane. Structure-function relationship, *Biochem. J.* 246, 37–42, 1987) also showed substantial polymerization of the protein on modification with TNM. Of interest was the observation that nitration abolished 95% of inhibitory activity against elastase with retention of 95% of the inhibitory activity against trypsin. Structure analysis showed the varying (40%–80%) modification of four tyrosine residues with two unmodified tyrosine residues. An earlier work (Feste, A. and Gan, J.C., Selective loss of elastase inhibitory activity of α-1-antiproteinase inhibitor upon chemical modification of its tyrosyl residues, *J. Biol. Chem.* 256, 6374–6380, 1981) had also shown the selective loss of inhibitory activity against elastase; there was one less nitrotyrosine reported on nitration of the elastase-α_1-antiproteinase inhibitor complexes. There was no decrease in the nitration of the complexes of α_1-antiproteinase inhibitor with either trypsin or chymotrypsin. There is no evidence to support the modification of an amino acid residue other than tyrosine in the reaction of TNM and α_1-antitrypsin.

k   The modification of the crystalline protein resulted in the modification of the same amino acid residue, Tyr248, as observed in the solution reaction. The reaction of TNM with carboxypeptidase in solution is said to result in polymerization as a side reaction; polymerization is not observed with the reaction of TNM with the crystalline protein. While not mentioned in earlier work, the early studies on the reaction of TNM with carboxypeptidase were associated with the formation of considerable polymeric product; carboxypeptidase A has been reported to be subject to aggregation under various solvent conditions (Katzav-Gonzansky, T., Hanan, E., and Solomon, B., Effect of monoclonal antibodies in preventing carboxypeptidase A aggregation, *Biotechnol. Appl. Biochem.* 23, 227–230, 1996).

l   Approximately 3 mol of nitrotyrosine was formed with 140-fold molar excess of TNM with either bovine pituitary growth hormone (3 mol of NO_2Tyr) or equine pituitary growth hormone (3.1 mol of NO_2Tyr). Two tyrosine residues in the bovine or equine proteins were unreactive, while the remaining four residues were modified to varying degrees. The pattern of modification was similar but not identical with the bovine and equine proteins. There was some polymerization and 60% nitration of the single tryptophanyl residue in either protein. Reaction in 50 mM Tris, pH 8.0, with 8 M urea resulted in the modification of all tyrosine residues and full modification of the single tryptophanyl residue.

m   Mannose was included in some of the reactions as a control for the potential effect of a sugar on the nitration reaction. Galactose, which is an inducer and thus expected to interact with the lactose repressor protein, did result in a decrease in the modification of the protein, while mannose did not affect the reaction.

n   Two tyrosine residues (Tyr7 and Tyr17) were modified by a 5–10 M excess of TNM. Increasing TNM concentration to a 200-fold molar excess resulted in modification of four more residues while increasing the molar ratio to a 800-fold molar excess without further modification of tyrosyl residues. Polymerization was observed at the higher concentrations of TNM. There was oxidation of the three cysteine residues with TNM. Subsequent work (Hsieh, W.-T. and Matthews, K.S., Tetranitromethane modification of the tyrosine residues of the lactose repressor, *J. Biol. Chem.* 256, 4856–4862, 1961) used NEM to block the cysteine residues from modification with TNM as this modification does not affect the binding activities of the lactose repressor protein. Reaction of TNM with the NEM-modified protein resulted in the loss of both specific and nonspecific DNA-binding activities. The reduction of the nitrotyrosyl residues to aminotyrosyl residues partially restored specific DNA-binding activity (operator activity) and fully restored the nonspecific binding of DNA. This work establishes the importance of tyrosyl residues for operator function but leaves open the question of whether oxidation of cysteine to cysteic acid inhibits lactose repressor protein function.

(continued)

TABLE 11.5 (continued)

Modification of Tyrosyl Residues in Proteins with TNM[a]

[o]   A 99-fold excess of TNM resulted in the modification of approximately 9 of the 18 tyrosine residues in human serum albumin. The inclusion of 8 M urea did not result in the formation of additional nitrotyrosine but did decrease the concentration required for maximum modification. As a note, the albumin concentration had to be decreased in the urea experiments because of irreversible gel formation. The work is to be commended for the excellent UV–VIS spectroscopy studies on nitrated proteins and 3-nitrotyrosine. Subsequent work on the reaction of TNM with albumin demonstrated the presence of one highly reactive tyrosine residue (Fehske, K.J., Möller, W.E., and Wollert, U., Direct demonstration of the highly reactive tyrosine residue of human serum albumin located in fragment 299–585, *Arch. Biochem. Biophys.* 205, 217–221, 1982). More recent work showed the modification of 10 residues of human albumin with TNM, while only three residues are modified by peroxynitrite (Zhang, Y., Yang, H., and Pöschl, U., Analysis of nitrated proteins and tryptic peptides by HLPC-chip-MS/MS: Site-specific quantification, nitration degree, and reactivity of tyrosine residues *Anal. Biochem. Chem.* 394, 459–471, 2011).

[p]   The formation of 1 mol of nitrotyrosine per mole of protein is observed with porcine phospholipase A_2 that is associated with the loss of enzyme activity; the product of reaction can be separated into two mononitrated derivatives (Tyr69 and Tyr123). Reduction of the nitrotyrosine proteins to the aminotyrosine derivatives is associated with an increase in activity for the NO_2Tyr69 derivative but a further decrease in activity for the NO_2Tyr123 derivative. Nitration performed in the presence of egg yolk lysolecithin results in an increased rate of inactivation with predominant formation of a dinitrotyrosine derivative where nitration has occurred at both Tyr69 and Tyr123; the formation of protein dimer is also eliminated with TNM reaction in the presence of egg yolk lysolecithin. The Tyr69 is an invariant residue in the various phospholipase A_2 proteins. Only modification at Tyr69 is observed with the bovine protein, while modification at Tyr19 and Tyr63 is observed in the equine protein; Tyr19 in the equine protein is only nitrated in the presence of egg yolk lysolecithin. The reduction of the nitrotyrosine to aminotyrosine in the bovine protein yields a modest increase in activity, while in the equine protein, the reduction of the Tyr19 to the aminotyrosyl derivative results in a further loss of enzyme activity, while the reduction of the Tyr69 derivative, as with porcine and bovine proteins, results in an increase in enzyme activity. The accelerating effect of egg yolk lysolecithin on the TNM reaction is observed with all three proteins, and it is thought to serve a method for concentrating the TNM reagent and delivering it to the protein. Subsequent work used dansyl chloride modification of the aminotyrosine residues for use as conformational probes (Meyer, H., Puijk, W.C., Dijkman, R. et al., Comparative studies of tyrosine modification in pancreatic phospholipases. 2. Properties of the nitrotyrosyl, aminotyrosyl, and dansylaminotyrosyl derivatives of pig, horse, and ox phospholipase A_2 and their zymogens, *Biochemistry* 18, 3589–3597, 1979). It is of interest that two tyrosyl residues considered critical for catalytic activity were not available for reaction with TNM (Dupureur, C.M., Yu, B.Z., Jain, M.K. et al., Phospholipase A_2 engineering. Structural and functional roles of highly conserved active site residues tyrosine-52 and tyrosine-73, *Biochemistry* 31, 6402–6413, 1992).

[q]   The reaction of TNM with the protein was terminated by addition of 2-mercaptoethanol. Nitration of the enzyme results in the loss of approximately 80% of the activity as assayed by digestion of casein, while the effect on hydrolysis of low-molecular-weight substrates is less marked. Analysis suggested that nitration of Tyr158 is responsible for the observed changes in catalytic activity; nitration also occurs at Tyr21.

[r]   In the absence of information in the published manuscript, it is assumed that the experiments were performed at room temperature that is assumed to be 23°C (approximately 73°F). It is recognized that this is likely a high value; there is one study where room temperature was cited as ±16°C (Hornishi, H., Hachimore, Y., Kurihara, K., and Shibata, K., States of amino acid residues in proteins. III. Histidine residues in insulin, lysozyme, albumin and proteinases as determined with a new reagent of diazo-1-н-tetrazole, *Biochim. Biophys. Acta* 86, 477–489, 1964). Room temperature is defined at 22°C–25°C for regulatory purposes (Cohen, V., Jellinek, S.P., Teperikidis, L. et al., Room-temperature storage of medications labeled for refrigeration, *Am. J. Health Syst. Pharm.* 64, 1711–1715, 2007).

[s]   Samples from time course reactions of TNM with fructose-1,6-bisphosphatase were obtained by taken portions from the reaction into a 100-fold molar excess (to TNM). Reagent removed from reactions intended to prepare nitrated protein for subsequent studies was taken to an Amicon Centricon® 10 microconcentrator for solvent exchange. One tyrosine residue is protected from nitration by AMP. This work contains NMR studies of nitrated tyrosyl residues.

[t]   The modification of two tyrosine residues resulted in 80% loss of enzyme activity. Three possible sites of modifications were identified, Tyr30, Tyr226, and Try230. Replacement of Tyr230 with phenylalanine by site-specific mutagenesis resulted in an enzyme with catalytic properties similar to the nitrated protein.

TABLE 11.5 (continued)
Modification of Tyrosyl Residues in Proteins with TNM[a]

[u] The chemical modification studies used a protein with an *S*-tag (Raines, R.T., McCormick, M., Van Osbress, T.R., and Mierendorf, R.C., The S.Tag fusion for protein purification, *Methods Enzymol.* 326, 362–376, 2000). Reaction with TNM at pH 8 resulted in the loss of enzyme activity, while reaction at pH 6.0 had only a modest effect on activity (18% inactivation compared to a control enzyme preparation; control preparations had greatly reduced activity at this pH). Modification with NAI did not result in inhibition of activity. Site-specific mutagenesis was employed to identify a critical tyrosine residue. This study, while lacking quantitative data, does show that NAI and TNM can have quite different effects on protein function.

[v] The hexane is included in the reaction mixture to promote opening of the *lid* on the lipase active-site region (Jaeger, K.E., Ransac, S., Dikjstra, B.W. et al., Bacterial lipases, *FEMS Microbiol. Rev.* 15, 29–63, 1994).

[w] It is suggested that while 11 tyrosyl residues are modified by TNM, it is the modification of Tyr29 that influenced the enantioselectivity of the enzyme by hydrogen bonding to a transition state intermediate. Nitration of *P. cepacia* lipase increased the enantioselectivity in the hydrolysis of 2-phenoxy-1-propyl acetate by increasing the strength of the hydrogen bond. Acetylation with NAI decreased the enantioselectivity of the reaction.

[x] This is an early study describing the application of MS for the analysis of protein nitration. The study compared MALDI-TOF and ESI MS.

[y] Six tyrosine residues were modified in recombinant glycine receptor, which was reconstituted in phospholipid vesicles (1 h reaction at 23°C and terminated with 2-mercaptoethanol; reagent removed by centrifugation). The modified residues were identified by HPLC/mass spectrometric analysis of tryptic (or endo-Glu-C) peptides.

[z] Approximately five of the eight tyrosyl residues in matrilysin were modified at a ratio of TNM/protein of 2000/1 with no additional modification at a ratio to 7900/1. There was no effect of activity at neutral pH, but there was a decrease in activity at more acidic pH that is reflected in asymmetry in the plot of activity versus pH.

[aa] Nitration of prion protein with TNM resulted in a lower level of modification than peroxynitrite. Two tyrosyl residues were modified at 100-fold molar excess of TNM with six tyrosyl residues modified at a 1000-fold molar excess. The β-oligomer is prepared by denaturation in 6 M guanidine–10 mM potassium phosphate, pH 8.0, followed by dilution into conversion buffer (60 mM sodium acetate–160 mM NaCl–3.6 M urea, pH 3.7) and incubated overnight at 37°C for 16 h. Two residues, Tyr225 and Tyr225, are less reactive to nitration in the β-oligomer, while two tyrosine residues, Tyr149 and Tyr150, which are not nitrated in the native protein, are susceptible to nitration.

[bb] Nitration at molar ratio (TNM/protein) of greater than 60/1 resulted in precipitation. Modification at ratios less that 20/1 resulted in a satisfactory product. In general, there was greater nitration in the variable domains than in the constant domains.

References to Table 11.5

11.5.1. Riordan, J.F., Sokolovsky, M., and Vallee, B.L., The functional tyrosyl residues of carboxypeptidase A. Nitration with tetranitromethane, *Biochemistry* 6, 3609–3617, 1967.

11.5.2. Cuatrecasas, P., Fuchs, S., and Anfinsen, C.B., The tyrosyl residues at the active site of staphylococcal nuclease. Modifications by tetranitromethane, *J. Biol. Chem.* 243, 4787–4798, 1968.

11.5.3. Skov, K., Hofmann, T., and Williams, G.R., The nitration of cytochrome c, *Can. J. Biochem.* 47, 750–752, 1969.

11.5.4. Christen, P. and Riordan, J.F., Syncatalytic modification of a functional tyrosyl residue in aspartate aminotransferase, *Biochemistry* 9, 3025–3034, 1970.

11.5.5. Lundblad, R.L. and Harrison, J.H., The differential effect of tetranitromethane on the proteinase and esterase activity of bovine thrombin, *Biochem. Biophys. Res. Commun.* 45, 1344–1349, 1971.

11.5.6. Sokolovsky, M., Porcine carboxypeptidase B. Nitration of the functional tyrosyl residue with tetranitromethane, *Eur. J. Biochem.* 25, 267–273, 1972.

11.5.7. Glaser, C.B., Bewley, T.A., and Li, C.H., Reaction of bovine and ovine pituitary growth hormones with tetranitromethane, *Biochemistry* 12, 3379–3387, 1973.

11.5.8. Busby, T.F. and Gan, J.C., The reaction of tetranitromethane with human plasma α$_1$-antitrypsin, *Int. J. Biochem.* 6, 835–841, 1975.

11.5.9. Muszynska, G. and Riordan, J.F., Chemical modification of carboxypeptidase A crystals. Nitration of tyrosine 248, *Biochemistry* 15, 46–51, 1976.

(*continued*)

TABLE 11.5 (continued)
Modification of Tyrosyl Residues in Proteins with TNM[a]

11.5.10. Daurat-Larroque, S.T., Portuguez, M.E.M., and Santome, J.A., Reaction of bovine and equine growth hormones with tetranitromethane, *Int. J. Pept. Protein Res.* 9, 119–128, 1977.

11.5.11. Alexander, M.E., Burgum, A.A., Noall, R.A., Shaw, M.D., and Matthews, K.S., Modification of tyrosine residues of the lactose repressor protein, *Biochem. Biophys. Acta* 493, 367–379, 1977.

11.5.12. Malan, P.G. and Edelhoch, H., Nitration of human serum albumin and bovine and human goiter thyroglobulins with tetranitromethane, *Biochemistry* 9, 3205–3214, 1970.

11.5.13. Meyer, H., Verhoef, H., Hendriks, F.F.A. et al., Comparative studies of tyrosine modification in pancreatic phospholipases. I. Reaction of tetranitromethane with pig, horse, and ox phospholipases A₂ and their zymogens, *Biochemistry* 18, 3582–3588, 1979.

11.5.14. Kobayashi, R., Kanatani, A., Yoshimoto, T., and Tsuru, D., Chemical modification of neutral protease from *Bacillus subtilis* var. *amylosacchariticus* with tetranitromethane: Assignment of tyrosyl residues nitrated, *J. Biochem. (Tokyo)* 106, 1110–1113, 1989.

11.5.15. Liu, F. and Fromm, H.J., Investigation of the relationship between tyrosyl residues and the adenosine 5′-monophosphate binding site of rabbit liver fructose-1,6-bisphosphatase as studied by chemical modification and nuclear magnetic resonance spectroscopy, *J. Biol. Chem.* 264, 18320–18325, 1989.

11.5.16. Robertson, D.L., Hilton, S., Wong, K.R. et al., Influence of active site and tyrosine modification on the secretion and activity of the *Aeromonas hydrophila* lipase/acyltransferase, *J. Biol. Chem.* 269, 2146–2150, 1994.

11.5.17. Trent, S.M., Warsham, L.M.S., and Ernst-Fonberg, M.L., HlyC, the internal protein acyltransferase that activates hemolysin toxin: The role of conserved tyrosine and arginine residues in enzymatic activity as probed by chemical modification and site-directed mutagenesis, *Biochemistry* 38, 8831–8838, 1999.

11.5.18. Tuomi, W.V. and Kazlauskas, R.J., Molecular basis for enantioselectivity of lipase from *Pseudomonas cepacia* toward primary alcohols. Modeling, kinetics, and chemical modification of Tyr29 to increase or decrease enantioselectivity, *J. Org. Chem.* 64, 2638–2647, 1999.

11.5.19. Petersson, A.-S., Steen, H., Kalume, D.E. et al., Investigation of tyrosine nitration in proteins by mass spectrometry, *J. Mass. Spectrom.* 36, 616–625, 2001.

11.5.20. Leite, J.F. and Cascio, M., Probing the topology of the glycine receptor by chemical modification coupled to mass spectrometry, *Biochemistry* 41, 6140–6148, 2002.

11.5.21. Muta, Y., Oneda, H., and Inouye, K., Anomalous pH-dependence of the activity of human matrilysin (matrix metalloproteinase-7) as revealed by nitration and amination of its tyrosine residues, *Biochem. J.* 386, 363–270, 2005.

11.5.22. Lennon, C.W., Cox, H.D., Hennelly, S.P. et al., Probing structural differences in prion protein isoforms by tyrosine nitration, *Biochemistry* 46, 4850–4860, 2007.

11.5.23. Liu, H., Gaza-Bulseco, G., Chumsae, C., and Radziejewski, C.H., Mass spectrometry analysis of in vitro nitration of a recombinant human IgG1 monoclonal antibody, *Rapid Commun. Mass Spectrom.* 22, 1–10, 2008.

been reported.[322] Diazotized 5-amino-1H-tetrazole also specifically reacts with Tyr248 in bovine carboxypeptidase A; the modification of this tyrosine residue permits the subsequent modification of Tyr198 by TNM suggesting that the modification of Tyr248 with the diazonium compound results in a conformational change in protein.

After a period of inactivity, there has been a renaissance of interest in diazonium chemistry. A good part of this interest is the use of coupling diazo derivatives to carbon surface[323–325] that is beyond the scope of the current work. Advances in the chemistry of diazonium salts have been used for the preparation of conjugates with proteins that have been used for functionalization (Figure 11.7). Schlick and coworkers[326] used a variety of diazonium salts based on substituted anilines, such as *p*-aminoacetophenone, to attach probes to the surface of tobacco mosaic virus. Bruckman and coworkers[327] used the diazonium salts derived from 3-ethynylaniline to provide a site for the application of *click* chemistry on the surface of tobacco mosaic virus. In more recent work, Gavrilyuk and coworkers[328] used formylbenzene diazonium hexafluorophosphate to introduce an aldehyde function onto a protein surface. The use of hexafluorophosphate provided enhanced

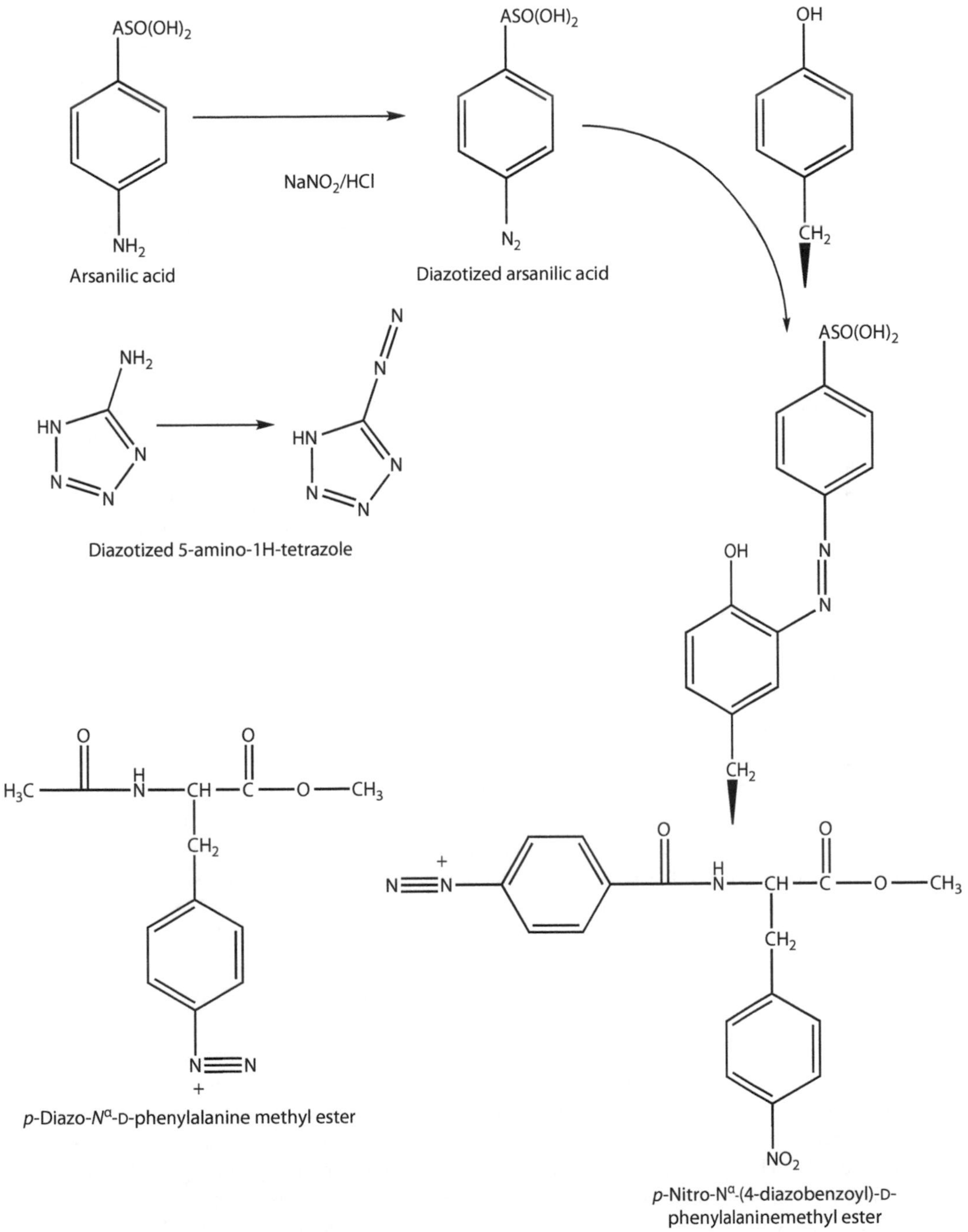

FIGURE 11.6 (continued)

FIGURE 11.6 (continued)   Various diazonium salts. Shown are some diazonium salts that have been used for the modification of proteins. See Riordan, J.F. and Vallee, B.L., Diazonium salts as specific reagents and probes of protein conformation, *Methods Enzymol.* 25, 521–531, 1972. See also Pielak, G.J., Urdea, M.S., and Legg, J.I., Preparation and characterization of sulfanilazo and arsaniloazo proteins, *Biochemistry* 23, 596–603, 1984; Pielak, G.J., Gurusiddaiah, S., and Legg, J.I., Quantification of azo-coupled lysine in azo proteins by amino-acid-analysis, *Anal. Biochem.* 156, 403–405, 1986. Shown on the left is the diazotization of 5-amino-1H-tetrazole that has been used to modify tyrosyl residues in proteins; reaction also occurs at histidyl residues (Sokolovsky, M. and Vallee, B.L., The reaction of diazonium-1H-tetrazole with proteins. Determination of tyrosine and histidine content, *Biochemistry* 5, 3574–3581, 1966; Cueni, L. and Riordan, J.F., Functional tyrosyl residues of carboxypeptidase A. The effect of protein structure on the reactivity of tyrosine-198, *Biochemistry* 17, 1834–1842, 1978). Also shown are two diazo derivatives of substrates used for the modification of chymotrypsin (Gorecki, M., Wilchek, M., and Blumberg, S., Modulation of the catalytic properties of α-chymotrypsin by chemical modification at Tyr 146, *Biochim. Biophys. Acta* 535, 90–99, 1978).

stability for the diazonium derivative. Jones and coworkers[329] used a PEG ester of *p*-diazobenzoic acid for PEGylation of peptides such as an enkephalin. These investigators showed that reaction at pH 4.5 enhanced specificity for modification at tyrosyl residues by decreases in reaction at nitrogen nucleophiles.

NBSF is another reagent that can be used for the modification of tyrosyl residues in proteins (Figure 11.8). This reagent was developed by Liao and coworkers for the selective modification of tyrosyl residue pancreatic DNase.[330] The modification reaction with NBSF can be performed in solvents (i.e., 0.1 M Tris-Cl, pH 8.0; 0.1 M *N*-ethylmorpholine acetate, pH 8.0) typically used for the modification of tyrosine residues by other reagents such as TNM or NAI. The rate of reagent hydrolysis is substantial and increases with increasing pH. Calcium ions are important for DNAse structure[331] and influence reaction with NBSF; in the presence of calcium ions, a single tyrosyl residue is modified, while in the absence of calcium ions, approximately three tyrosyl residues are modified. Liao and coworkers[330] also evaluated related compounds for the inactivation of DNAse; *p*-nitrobenzenesulfonyl chloride (NBSC) also inactivates DNAse, but both tyrosine and lysine residues are modified. NBSF reacts only with tyrosine in amino acid mixtures, while NBSC reacts with both tyrosine and lysine; NBSF did oxidize glutathione. The data obtained in this study suggest that NBSF reacts specifically with tyrosyl residues in proteins. There has been limited use of NBSF for the modification of tyrosyl residues in proteins since the work of Liao and coworkers.[330] NBSF has been used to characterize tyrosyl residues in NAD(P)H/quinone reductase.[332] The analysis of the product of the reaction showed that NBSF-modified tyrosyl residues were located in hydrophobic regions of the protein. In another study,[333] the reaction of NBSF with tyrosyl residues in human placental taurine transporter was compared with modification observed with TNM, NAI, and NBD-Cl. TNM and NBD-Cl were the most effective reagents. NBSF was an order of magnitude less effective, while NAI was 500 times less potent. Gitlin and coworkers[334] used NBSF to modify the tyrosyl residue in the biotin-binding site of avidin and streptavidin, while Suzuki and coworkers used NBSF to modify the tyrosyl residue at the active site of a catalytic antibody.[335] NBSF has also been used to modify the tyrosyl residue important for the activity of snake venom phosphatases.[336–338]

In addition to NBD-Cl that is discussed earlier, there are other examples of reagents that are generally considered to be *specific* for other functional groups in proteins. While I could not find the reference, I do think that Efraim Racker[339] once said "no reagent is as specific as when it is discovered." That said, the modification of the phenolic hydroxyl group of tyrosine with 2,4-dinitrofluorobenzene, a reagent considered specific for lysine, has been reported.[340] A novel reaction of PMSF with a tyrosyl residue in an archaeon superoxide dismutase has been reported.[341] Murachi and coworkers[342] reported on the reaction of diisopropylphosphorofluoridate with a tyrosyl residue (Figure 11.8) in stem bromelain, and it is noted that Means and Wu reported the modification of a tyrosine residue in human serum albumin with diisopropylphosphorofluoridate.[343] The modification of albumin by related organophosphorous compounds[344,345] is suggested to be useful as a biomarker of exposure to organophosphorous pesticides.[346,347]

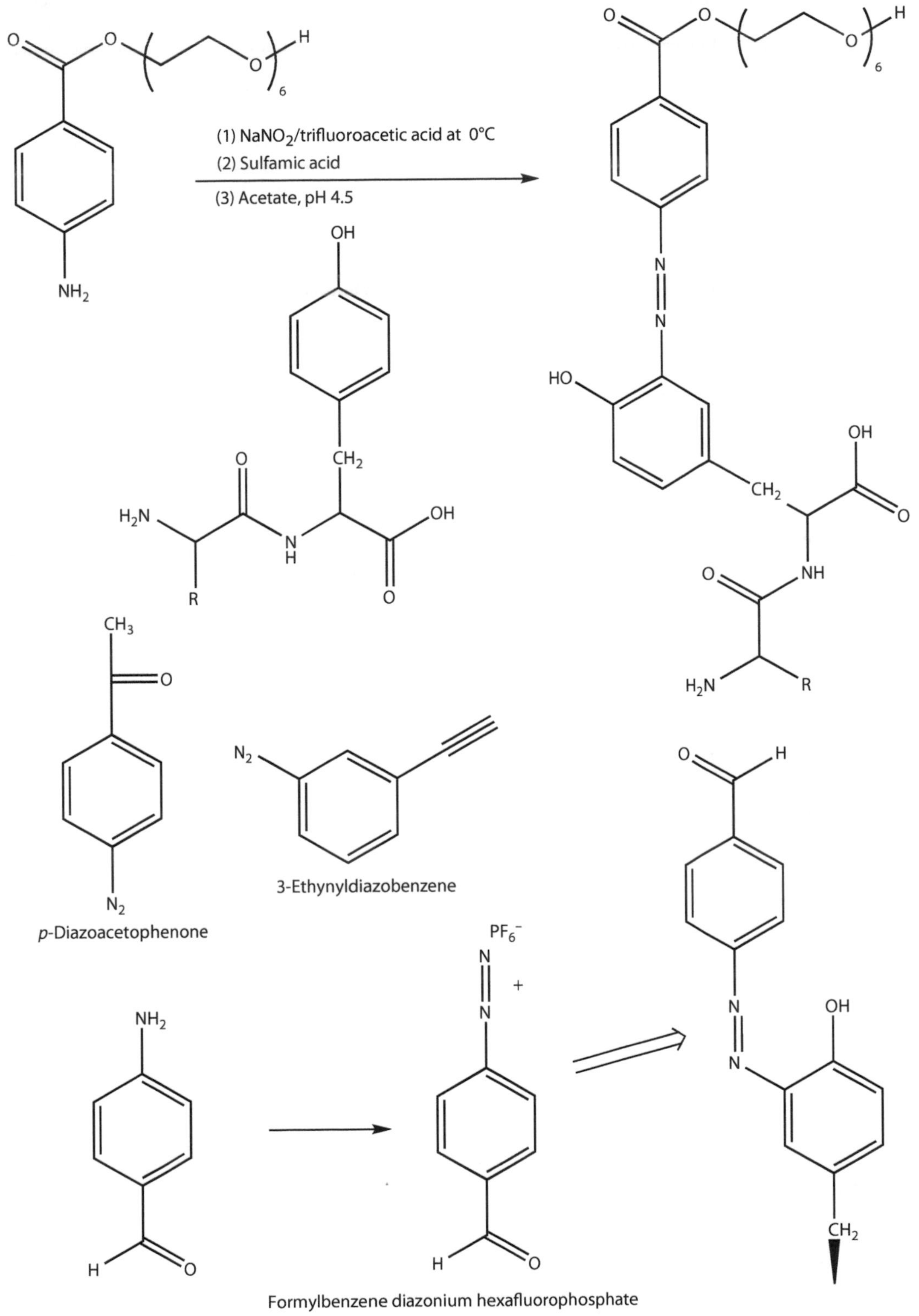

(1) NaNO$_2$/trifluoroacetic acid at 0°C
(2) Sulfamic acid
(3) Acetate, pH 4.5

3-Ethynyldiazobenzene

p-Diazoacetophenone

Formylbenzene diazonium hexafluorophosphate

FIGURE 11.7 (continued)

FIGURE 11.7 (continued) Some recent advances in diazonium chemistry. Shown at the top is the reaction of the diazo form of a PEG derivative of *p*-aminobenzoic acid with a peptide to form a PEGylated product. The *p*-aminobenzoyl PEG derivative was converted to the diazo form by sodium nitrite in the presence of trifluoroacetic acid. The diazotization reaction was terminated by the addition of sulfamic acid and added to an enkephalin peptide in acetate buffer at pH 4.5 to minimize reaction with nitrogen nucleophiles in the protein (Jones, M.W., Mantovani, G., Blindauer, C.A. et al., Direct peptide bioconjugation/PEGylation at tyrosine with linear and branched polymeric diazonium salts, *J. Am. Chem. Soc.* 134, 7406–7413, 2012). Shown at the left is *p*-diazoacetophenone, one of several diazonium salts based on aminobenzene derivatives (Schlick, T.L., Ding, Z., Kovacs, E.W., and Francis, M.B., Dual-surface modification of the tobacco mosaic virus, *J. Am. Chem. Soc.* 127, 3718–3723, 2005) and the diazonium salt derived from 3-ethynylaniline that, after coupling to tyrosyl residues, serves a site for the application of *click* chemistry (Bruckman, M.A., Kaur, G., Lee, L.A. et al., Surface modification of tobacco mosaic virus with *click* chemistry, *ChemBioChem* 9, 519–523, 2008). Shown at the bottom is the formation of formylbenzene diazonium hexafluorophosphate that can form a diazo derivative of tyrosine. The reagent contains an aldehyde function that can be used, for example, for coupling to a hydroxamate or hydrazine (Gavrilyuk, J., Ban, H., Nagano, M. et al., Formylbenzene diazonium hexafluorophosphate reagent for tyrosine-selective modification of proteins and the introduction of a bioorthogonal aldehyde, *Bioconjug. Chem.* 23, 2321–2328, 2012).

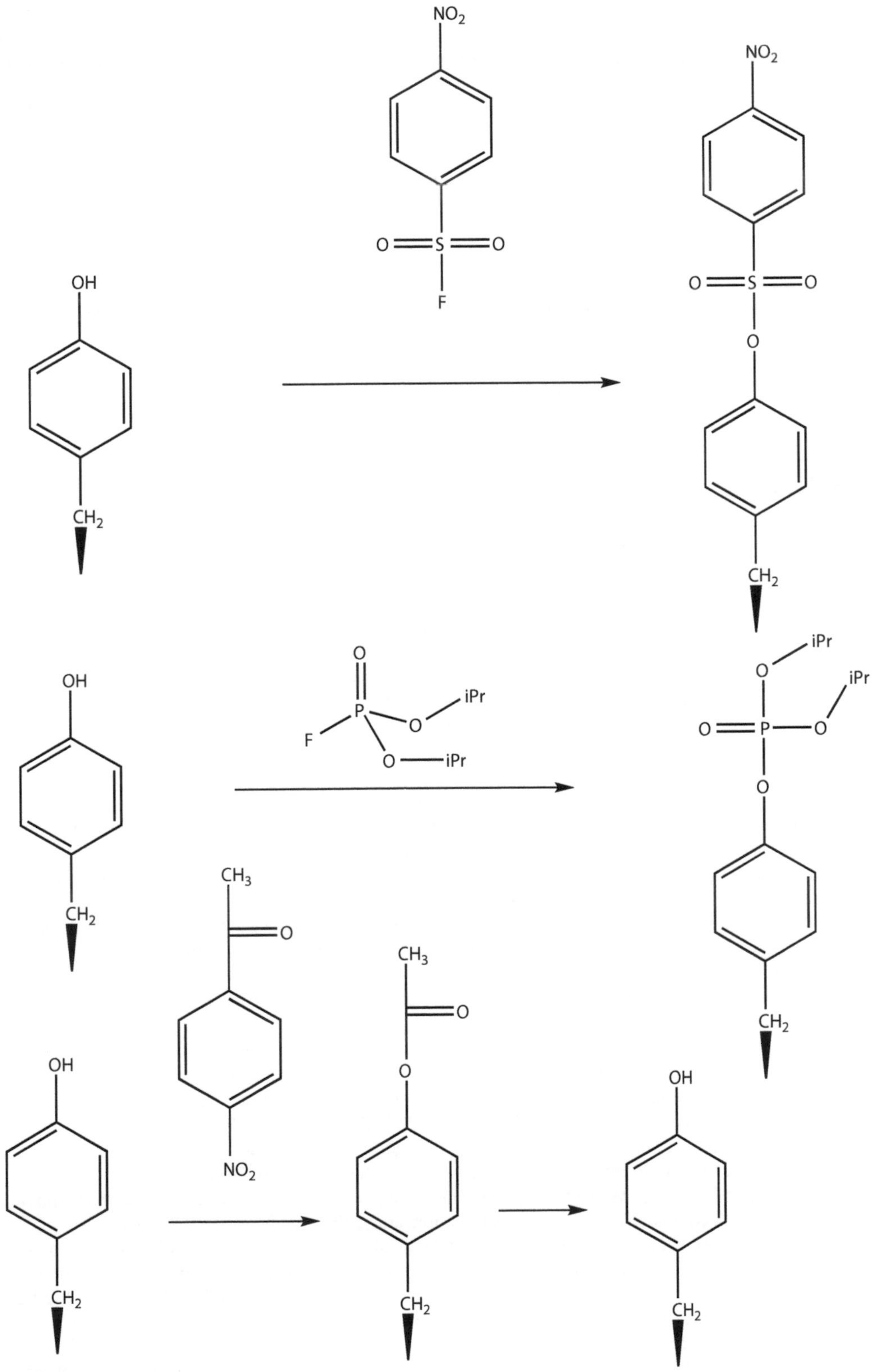

FIGURE 11.8 (continued)

FIGURE 11.8 (continued) Modification of tyrosine with NBSF, diisopropylphosphorofluoridate, and *p*-nitrophenyl acetate. Shown at the top is the reaction of NBSF with tyrosine. There is no evidence to suggest that the reaction is reversible and the product is stable to denaturation, digestion with trypsin at pH 8.0, and acid hydrolysis for amino acid analysis (Liao, T.-H., Ting, R.S., and Yeung, J.E., Reactivity of tyrosine in bovine pancreatic deoxyribonuclease with *p*-nitrobenzenesulfonyl fluoride, *J. Biol. Chem.* 257, 5637–5644, 1982; Gitlin, G., Bayer, E.A., and Wilchek, M., Studies on the biotin-binding sites of avidin and streptavidin. Tyrosine residues are involved in the binding site, *Biochem. J.* 269, 527–530, 1990). Shown below is the reaction of diisopropylphosphorofluoridate with tyrosine. The phosphorylated tyrosine is stable to acid hydrolysis and to tryptic hydrolysis (Murachi, T., Inagami, T., and Yasui, M., Evidence for alkylphosphorylation of tyrosyl residues of stem bromelain by diisopropylphosphorofluoridate, *Biochemistry* 4, 2815–2825, 1965; Means, G.E. and Wu, H.L., The reactive tyrosine residue of human serum albumin: Characterization of the its reaction with diisopropylfluorophosphate, *Arch. Biochem. Biophys.* 194, 526–530, 1979). At the bottom is the reaction of *p*-nitrophenyl acetate with tyrosine. The product undergoes hydrolysis and stability depends on the particular tyrosyl residue (Awad-Elkarim, A. and Means, G.E., The reactivity of *p*-nitrophenyl acetate with serum albumin, *Comp. Biochem. Physiol. B* 91, 267–272, 1988; Ascenzi, P., Gioia, M., Fanali, G. et al., Pseudo-enzymatic hydrolysis of 4-nitrophenyl acetate by human serum albumin: pH-dependence of rates of individual steps, *Biochem. Biophys. Res. Commun.* 424, 451–455, 2012).

REFERENCES

1. Tristram, G.R., The amino acid composition of protein, in *The Proteins*, Vol. I, Part A, Chapter 3, ed. H. Neurath, pp. 181–233, Academic Press, New York, 1953.
2. Cornish-Bowden, A., The amino acid composition of proteins are correlated with their molecular sizes, *Biochem. J.* 213, 271–274, 1983.
3. Nishikawa, K., Kubota, Y., and Ooi, T., Classification of proteins into groups based on amino acid composition and other characteristics II. Grouping into four groups, *J. Biochem.* 94, 997–1007, 1983.
4. Gromiha, M.M. and Suwa, M., A simple statistical method for discriminating outer membrane proteins with better accuracy, *Bioinformatics* 21, 961–968, 2005.
5. White, S.H., Amino acid preferences of small proteins. Implications for protein stability and evolution, *J. Mol. Biol.* 227, 991–995, 1992.
6. Pace, C.N., Grimsley, G.R., and Scholtz, J.M., Protein ionizable groups: pK values and their contribution to protein stability and solubility, *J. Biol. Chem.* 284, 13285–13289, 2009.
7. Whitney, P.L. and Tanford, C., Solubility of amino acids in aqueous urea solutions and its implications for the denaturation of proteins by urea, *J. Biol. Chem.* 237, PC1737–PC1739, 1962.
8. Rall, S.C. and Cole, R.D., Amino acid sequence and sequence variability of the amino-terminal regions of lysine-rich histones, *J. Biol. Chem.* 246, 7175–7190, 1971.
9. Bogan, A.A. and Thorn, K.S., Anatomy of hot spots in protein interfaces, *J. Mol. Biol.* 280, 1–9, 1998.
10. KIllian, J.A. and von Heijne, G., How proteins adapt to a membrane-water interface, *Trends Biochem. Sci.* 25, 429–434, 2000.
11. Dey, S., Pal, A., Chakrabarti, P., and Janin, J., The subunit interfaces of weakly associated homodimeric proteins, *J. Mol. Biol.* 398, 146–160, 2010.
12. Garnham, C.P., Campbell, R.L., Walker, V.K., and Davies, P.L., Novel dimeric β-helical model of an ice nucleation protein with bridged active sites, *BMC Struct. Biol.* 11, 36, 2011.
13. Dicontanzo, A.C., Thompson, J.R., Peterson, F.C. et al., Tyrosine residues mediate fibril formation in a dynamic light chain dimer interface, *J. Biol. Chem.* 287, 27997–28006, 2012.
14. DiConstanzo, A.C., Thompson, J.R., Peterson, F.C. et al., Tyrosine residues mediate fibrin formation in a dynamic light chain dimer interface, *J. Biol. Chem.* 287, 27997–28006, 2012.
15. Luitz, M.P. and Zacharias, M., Role of tyrosine hot-spot residues at the interface of colicin E9 and immunity protein 9: A comparative free energy simulation study, *Proteins* 81, 461–468, 2013.
16. Ploug, M., Rahbek-Nielsen, H., Ellis, V. et al., Chemical modification of the Urokinase-type plasminogen activator and its receptor using tetranitromethane. Evidence for the involvement of specific tyrosine residues in both molecules during receptor-ligand interaction, *Biochemistry* 34, 12524–12534, 1995.
17. Machida, K., Mayer, B.J., and Nollau, P., Profiling the global tyrosine phosphorylation state, *Mol. Cell. Proteomics* 2, 215–223, 2003.
18. Leitner, A., Sturm, M., and Lindner, W., Tools for analyzing the phosphoproteome and other phosphorylated biomolecules: A review, *Anal. Chim. Acta* 703, 19–30, 2011.

19. Dushek, O., Goyette, J., and van der Merwe, P.A., Non-catalytic tyrosine-phosphorylated receptors, *Immunol. Rev.* 250, 258–276, 2012.
20. Liu, B.A. and Nash, P.D., Evolution of SH2 domains and phosphotyrosine signalling networks, *Philos. Trans. R. Soc. Lond. B Biol. Sci.* 367, 2556–2573, 2012.
21. Hunter, T., Why nature chose phosphate to modify proteins, *Philos. Trans. R. Soc. Lond. B Biol. Sci.* 367, 2513–2516, 2012.
22. Monigatti, F., Hekking, B., and Steen, H., Protein sulfation analysis—A primer, *Biochim. Biophys. Acta* 1764, 1904–1913, 2006.
23. Seibert, C. and Sakmar, T.P., Toward a framework for sulfoproteomics: Synthesis and characterization of sulfotyrosine-containing peptides, *Biopolymers* 90, 459–477, 2008.
24. Stone, M.J., Chuang, S., Hou, X. et al., Tyrosine sulfation: An increasingly recognized post-translational modification of secreted proteins, *Nat. Biotechnol.* 25, 299–317, 2009.
25. Abello, N., Kerstjens, H.A., Postma, D.S., and Bischoff, R., Protein tyrosine nitration: Selectivity, physicochemical and biological consequences, dinitration, and proteomic methods for the identification of tyrosine-nitrated proteins, *J. Proteome Res.* 8, 3222–3238, 2009.
26. Ytterberg, A.J. and Jensen, O.N., Modification-specific proteomics in plant biology, *J. Proteomics* 73, 2249–2266, 2010.
27. Castro, L., Demicheli, V., Tórtora, V., and Radi, R., Mitochondrial protein tyrosine nitration, *Free Radic. Res.* 45, 37–52, 2011.
28. Aslan, M. and Dogan, S., Proteomic detection of nitroproteins as potential biomarkers for cardiovascular disease, *J. Proteomics* 74, 2274–2288, 2011.
29. Lundblad, R.L., Noyes, C.M., Featherstone, G.L. et al., The reaction of bovine α-thrombin with tetranitromethane. Characterization of the modified protein, *J. Biol. Chem.* 263, 3729–3734, 1988.
30. Nicholson, B.H., Location of aromatic amino acids and helix content in *Escherichia coli* ribonucleic acid polymerase, *Biochem. J.* 123, 117–122, 1971.
31. Glazer, A.N., The chemical modification of proteins by group-specific and site-specific reagents, in *The Proteins*, Vol. II, Chapter 1, ed. H. Neurath and R.L. Hill, pp. 1–103, Academic Press, New York, 1976.
32. Barurin, S.J., Olson, M., and McIntosh, L.P., Structure, dynamics, and ionization equilibria of the tyrosine in *Bacillus xylanase*, *J. Biomol. NMR* 51, 279–284, 2011.
33. Oktaviani, N.A., Pool, T.J., Kamikubo, H. et al., Comprehensive determination of protein tyrosine pKa values for photoactive yellow protein using indirect ^{13}C NMR spectroscopy, *Biophys. J.* 102, 579–586, 2012.
34. Donovan, J.W., Changes in ultraviolet absorption produced by alteration of protein conformation, *J. Biol. Chem.* 244, 1961–1967, 1969.
35. Markland, F.S., Phenolic hydroxyl ionization in two subtilisins, *J. Biol. Chem.* 244, 694–700, 1969.
36. Laws, W.R. and Shore, J.D., Spectral evidence for tyrosine ionization linked to a conformational change in liver alcohol dehydrogenase ternary complex, *J. Biol. Chem.* 254, 2582–2584, 1979.
37. Kuramitso, S. et al., Ionization of the catalytic groups and tyrosyl residues in human lysozyme, *J. Biochem.* 87, 771–778, 1980.
38. Kobayashi, J., Hagashijima, T., and Miyazawa, T., Nuclear magnetic resonance analyses of side chain conformations of histidine and aromatic acid derivatives, *Int. J. Pept. Protein Res.* 24, 40–47, 1984.
39. Poklar, N., Vesnaver, G., and Laponje, S., Studies by UV spectroscopy of thermal denaturation of beta-lactoglobulin in urea and alkylurea solutions. *Biophys. Chem.* 47, 143–151, 1993.
40. Taylor, R., *Electrophilic Aromatic Substitution*, John Wiley & Sons, Ltd., Chichester, U.K., 1990.
41. Atkins, W.M., Dietze, E.C., and Ibarra, C., Pressure-dependent ionization of Tyr 9 in glutathione *S*-transferase A1-1: Contribution of the C-terminal helix to a "soft" active site, *Protein Sci.* 6, 873–881, 1997.
42. Johnson, T.B. and Kohmann, E.F., Studies on nitrated proteins: I. The determination of the structure of nitrotyrosine, *J. Am. Chem. Soc.* 37, 1863–1884, 1915.
43. Zinsser, H. and Parker, J.T., Further studies on bacterial hypersusceptibility. II, *J. Exp. Med.* 37, 275–302, 1923.
44. Lille, R.D. and Donaldson, P.T., Histochemical azo coupling of protein histidine. Brunswick's nitration method, *J. Histochem. Cytochem.* 20, 929–937, 1972.
45. Larsson, L., Sörbo, B., Tiselius, H.G., and Öhman, S., A method for quantitative wet chemical analysis of urinary calculi, *Clin. Chim. Acta* 140, 9–20, 1984.
46. Wormall, A., The immunological specificity of chemically altered proteins, *J. Exp. Med.* 51, 295–317, 1930.

47. von Pechmann, E., The enzymatic hydrolysis of xanthoproteins, and its use in colorimetric determination of proteolytic ferments, *Biochem. Z.* 321, 248–260, 1950.

48. Noack, R., Koldavský, O., Friedrich, M. et al., Proteolytic and peptidase activities of the jejunum and ileum of the rat during postnatal development, *Biochem. J.* 100, 775–778, 1966.

49. Roholt, O.A. and Pressman, D., Iodination-isolation of peptides from the active site, *Method Enzymol.* 25, 438–449, 1972.

50. Tsomides, T.J. and Eisen, H.N., Stoichiometric labeling of peptides by iodination on tyrosyl or histidyl residues, *Anal. Biochem.* 210, 129–135, 1993.

51. Rosenfeld, R. et al., Sites of iodination in recombinant human brain-derived neurotrophic factor and its effect on neurotrophic activity, *Protein Sci.* 2, 1664–1674, 1993.

52. León-Tamariz, F., Vergaeys, I., and Van Boven, M., Biodistribution and pharmacokinetics of PEG-10 kDa-choleocystokinin-10 in rats after different routes of administration, *Curr. Drug Deliv.* 7, 137–143, 2010.

53. Kennel, S.J., Huang, Y., Zeng, W.B. et al., Kinetics of vascular targeted monoclonal antibody, *Curr. Drug Deliv.* 7, 428–435, 2010.

54. Banks, W.A., Niehoff, M.L., Ponzio, N.M. et al., Pharmacokinetics and modeling of immune cell trafficking: Quantifying differential influences of target tissues versus lymphocytes in SJL and lipopolysaccharide-treated mice, *J. Neuroinflammation* 9, 231, 2012.

55. Chen, J., Cui, W., Giblin, D., and Gross, M.L., New protein footprinting: Fast photochemical iodination combined with top-down and bottom-up mass spectrometry, *J. Am. Soc. Mass Spectrom.* 23, 1306–1318, 2012.

56. Cohen, A.M., Kostyleva, R., Chisholm, K.A., and Pinto, D.M., Iodination on tyrosine residues during oxidation with sodium periodate in solid phase extraction of N-linked glycopeptides, *J. Am. Soc. Mass Spectrom.* 23, 68–75, 2012.

57. Huntley, T.E. and Strittmatter, P., The reactivity of the tyrosyl residues of cytochrome b$_5$, *J. Biol. Chem.* 247, 4648–4653, 1972.

58. McGowan, E.B. and Stellwagen, E., Reactivity of individual tyrosyl residues of horse heart ferricyto-chrome c toward iodination, *Biochemistry* 9, 3047–3053, 1970.

59. Layne, P.P. and Najjar, V.A., Evidence for a tyrosine residue at the active site of phosphoglucomutase and its interaction with vanadate, *Proc. Natl. Acad. Sci. USA* 76, 5010–5013, 1979.

60. Silva, J.S. and Ebner, K.E., Protection by substrates and -lactalbumin against inactivation of galactosyl-transferase by iodine monochloride, *J. Biol. Chem.* 255, 11262–11267, 1980.

61. Izzo, J.L., Bale, W.F., Izzo, M.J., and Roncone, A., High specific activity labeling of insulin with ^{131}I, *J. Biol. Chem.* 239, 3743–3748, 1964.

62. Mercer, N., Ramakrishnan, B., Boeggeman, E., and Qasba, P.K., Applications of site-specific labeling to study HAMLET, a tumoricidal complex of α-lactalbumin and oleic acid, *PLoS One* 6, e26093, 2011.

63. Parker, D.J. and Allison, W.S., Mechanism of inactivation of glyceraldehyde-3-phosphate dehydrogenase by tetrathionate, *J. Biol. Chem.* 244, 180–189, 1969.

64. Allen, G. and Harris, J.L., Iodination of glyceraldehyde-3-phosphate dehydrogenase from *Bacillus stearothermophilus*, *Biochem. J.* 155, 523–544, 1976.

65. Marchalonis, J.J., An enzymic method for the trace iodination of immunoglobulins and other proteins, *Biochem. J.* 113, 299–305, 1969.

66. Morrison, M. and Schonbaum, G.R., Peroxidase-catalyzed halogenation, *Annu. Rev. Biochem.* 45, 861–888, 1976.

67. Karonen, S.L., Development in techniques for radioiodination of peptide hormones and other proteins, *Scand. J. Clin. Lab. Invest. Suppl.* 201, 135–138, 1990.

68. LaRue, B., Hogg, E., Sagare, A. et al., Method for measurement of the blood-brain barrier permeability in the perfused mouse brain: Application to amyloid-β peptide in wild type and Alzheimer's Tg2576 mice, *J. Neurosci. Method* 138, 233–242, 2004.

69. Mushunje, A., Evens, G., Brennan, S.O. et al., Latent antithrombin and its detection, formation and turn-over in the circulation, *J. Thromb. Haemost.* 2, 2170–2177, 2004.

70. Turatti, F., Mezzanzanica, D., Nardini, E. et al., Production and validation of the pharmacokinetics of a single-chain Fv fragment of the MGR6 antibody for targeting of tumors expressing HER-2, *Cancer Immunol. Immunother.* 49, 679–686, 2001.

71. de Blois, E., Chan, H.S., and Breeman, W.A.P., Iodination and stability of somatostatin analogues: Comparison of iodination techniques. A practical overview, *Curr. Top. Med. Chem.* 23, 2668–2676, 2012.

72. Hunter, W.M. and Greenwood, F.C., Preparation of iodine-131 labelled human growth hormone of high specific activity, *Nature* 194, 495–496, 1962.

73. Heber, D., Odell, W.D., Schedewie, H., and Wolfsen, A.F., Improved iodination of peptides for radioimmunoassay and membrane radioreceptor assay, *Clin. Chem.* 24, 796–799, 1978.

74. Vergote, V., Bodé, S., Peremans, K. et al., Analysis of iodinated peptides by LC-DAD/ESI ion trap mass spectrometry, *J. Chromatogr. B* 850, 213–230, 2007.

75. Maly, P. and Lüthi, C., The binding sites of insulin-like growth factor I (IGF I) to type I IGF receptor and to a monoclonal antibody. Mapping by chemical modification of tyrosine residues, *J. Biol. Chem.* 263, 7068–7072, 1988.

76. Hussain, A.A., Jona, J.A., Yamada, A., and Dittert, L.W., Chloramine T in radiolabeling techniques. II. A nondestructive method for radiolabeling biomolecules by halogenation, *Anal. Biochem.* 224, 221–226, 1995.

77. Tashtoush, B.M., Traboulsi, A.A., Dittert, L., and Hussain, A.A., Chloramine T in radiolabeling techniques. IV. Penta-*O*-acetyl-*N*-chloro-*N*-methylglucamine as an oxidizing agent, *Anal. Biochem.* 288, 16–21, 2001.

78. Bauer, R.J., Leigh, S.D., Birr, C.A. et al., Alteration of the pharmacokinetics of small proteins by iodination, *Biopharm. Drug Dispos.* 17, 761–774, 1996.

79. Linde, S., Hansen, B., and Lemmark, A., Preparation of stable radioiodinated polypeptide hormones and proteins using polyacrylamide gel electrophoresis, *Method Enzymol.* 92, 309–335, 1983.

80. Kamatso, Y. and Hayashi, H., Revaluating the effects of tyrosine iodination of recombinant hirudin on its thrombin inhibitor kinetics, *Thromb. Res.* 87, 343–352, 1997.

81. Nikula, T.K., Bocchia, M., Curcio, M.J. et al., Impact of the high tyrosine fraction in complementarity determining regions: Measured and predicted effects of radioiodination on IgG immunoreactivity, *Mol. Immunol.* 32, 865–872, 1995.

82. Braschi, S., Neville, T.A., Maugeais, C. et al., Role of the kidney in regulating the metabolism of HDL in rabbits: Evidence that iodination alters the catabolism of apolipoprotein A-1 by the kidney, *Biochemistry* 39, 5441–5449, 2000.

83. Ghosh, D., Erman, M., Sawicki, M. et al., Determination of a protein structure by iodination: The structure of iodinated acetylxylan esterase, *Acta Crystallogr. D Biol. Crystallogr.* 55, 779–784, 1999.

84. Leinala, E.K., Davies, P.L., and Jia, Z., Elevated temperature and tyrosine iodination aid in the crystallization and structure determination of an antifreeze protein, *Acta Crystallogr. D Biol. Crystallogr.* 58, 1081–1083, 2002.

85. Miyatak, H., Hasegawa, T., and Yamano, A., New methods to prepare iodinated derivatives by vaporizing iodine labeling (VIL) and hydrogen peroxide VIL (HYPER-VIL), *Acta Crystallogr. D Biol. Crystalogr.* 62, 280–289, 2006.

86. Schneggenburger, P.E., Beerlink, A., Worbs, B. et al., A novel heavy-atom label for site-specific peptide iodination: Synthesis, membrane incorporation and x-ray reflectivity, *Chemphyschem* 10, 1567–1576, 2009.

87. Fraenkel-Conrat, H. and Colloms, M.D., Reactivity of tobacco mosaic virus and its protein toward acetic anhydride, *Biochemistry* 6, 2740–2745, 1967.

88. Riordan, J.F. and Vallee, B.L., Acetylation, *Methods Enzymol.* 11, 565–576, 1967.

89. Riordan, J.F. and Vallee, B.L., *O*-acetylation, *Methods Enzymol.* 25, 570–576, 1972.

90. Karibian, D., Jones, C., Gertler, A., Dorrington, K.J., and Hofmann, T., On the reaction of acetic and maleic anhydrides with elastase. Evidence for a role of the NH$_2$-terminal valine, *Biochemistry* 13, 2891–2897, 1974.

91. Ohnishi, M., Suganuma, T., and Hiromi, K., The role of a tyrosine residue of bacterial liquefying α-amylase in the enzymatic hydrolysis of linear substrates as studied by chemical modification with acetic anhydride, *J. Biochem. (Tokyo)* 76, 7–13, 1974.

92. Bernad, A., Nieto, M.A., Vioque, A., and Palaciáan, E., Modification of the amino groups and hydroxyl groups of lysozyme with carboxylic acid anhydrides: A comparative study, *Biochim. Biophys. Acta* 873, 350–355, 1986.

93. Sanchez, A., Ramon, Y., Solano, Y. et al., Double acylation for identification of amino-terminal peptides of proteins isolated by polyacrylamide gel electrophoresis, *Rapid Commun. Mass Spectrom.* 21, 2237–2244, 2007.

94. Olsen, D.B., Hepburn, T.W., Lee, S.L. et al., Investigation of the substrate binding and catalytic groups of the P-C bond cleaving enzyme, phosphonoacetaldehyde hydrolase, *Arch. Biochem. Biophys.* 296, 144–151, 1992.

95. Kaplan, H., Stevenson, K.J., and Hartley, B.S., Competitive labelling, a method for determining the reactivity of individual groups in proteins. The amino groups of porcine elastase, *Biochem. J.* 124, 289–299, 1971.

96. Tilley, S.D. and Francis, M.B., Tyrosine-selective protein alkylation using π-allylpalladium compounds, *J. Am. Chem. Soc.* 128, 1080–1081, 2006.

97. Takahashi, Y., Horiyama, S., Honda, C. et al., A chemical approach to searching for bioactive ingredients in cigarette smoke, *Chem. Pharm. Bull.* 61, 85–89, 2013.

98. Wieland, T. and Schneider, G., *N*-acylimidazoles as acyl derivatives of high energy, *Ann. Chem. Justus Liebigs* 580, 159–168, 1953.

99. Stadtman, E.R. and White, F.H., Jr., The enzymic synthesis of *N*-acetylimidazole, *J. Am. Chem. Soc.* 75, 2022, 1953.

100. Stadtman, E.R., On the energy-rich nature of acetyl imidazole, an enzymatically active compound, in *A Symposium on the Mechanism of Enzyme Action*, ed. W.D. McElroy and B. Glass, Johns Hopkins Press, Baltimore, MD, 1954.

101. Fife, T.H., Steric effects in the hydrolysis of *N*-acylimidazoles and ester of *p*-nitrophenol, *J. Am. Chem. Soc.* 87, 4597–4600, 1965.

102. Lee, J.P., Bembi, R., and Fife, T.H., Steric effects in the hydrolysis reactions of *N*-acylimidazoles. Effect of aryl substitution in the leaving group, *J. Org. Chem.* 62, 2872–2876, 1997.

103. Atkinson, B. and Green, A.L., The hydrolysis of *N*-acetyl and *N*-di-isopropylphosphorylimidazoles, *Trans. Faraday Soc.* 53, 1334–1340, 1957.

104. Koltun, W.L., Dexter, R.N., Clark, R.E., and Gurd, F.R.N., Coordination complexes and catalytic properties of proteins and related substances. I. Effect of cupric and zinc ions on the hydrolysis of *p*-nitrophenyl acetate by imidazole, *J. Am. Chem. Soc.* 80, 1188–1194, 1958.

105. Simpson, R.T., Riordan, J.F., and Vallee, B.L., Functional tyrosyl residues in the active center of bovine pancreatic carboxypeptidase A, *Biochemistry* 2, 616–622, 1963.

106. Riordan, J.F., Wacker, W.E.C., and Vallee, B.L., *N*-Acetylimidazole: A reagent for determination of "free" tyrosyl residues of proteins, *Biochemistry* 4, 1758–1765, 1965.

107. Myers, B., II and Glazer, A.N., Spectroscopic studies of the exposure of tyrosine residues in proteins with special reference to the subtilisins, *J. Biol. Chem.* 246, 412–419, 1971.

108. Cronan, J.E., Jr. and Klages, A.L., Chemical synthesis of acyl thioesters of acyl carrier protein with native structure, *Proc. Natl. Acad. Sci. USA* 78, 5440–5444, 1981.

109. Lundblad, R.L., The reaction of bovine thrombin with *N*-butyrylimidazole. Two different reactions resulting in the inhibition of catalytic activity, *Biochemistry* 14, 1033–1037, 1975.

110. Lundblad, R.L., Harrison, J.H., and Mann, K.G., On the reaction of purified bovine thrombin with *N*-acetylimidazole, *Biochemistry* 12, 409–413, 1973.

111. El Kabbaj, M.S. and Latruffe, N., Chemical reagents of polypeptide side chains. Relationship between solubility properties and ability to cross the inner mitochondrial membranes, *Cell. Mol. Biol.* 40, 781–786, 1994.

112. Zhang, F., Gao, J., Weng, J. et al., Structural and functional differentiation of three groups of tyrosine residues by acetylation of *N*-acetylimidazole in manganese stabilizing protein, *Biochemistry* 44, 719–725, 2005.

113. Martin, B.L., Wu, D., Jakes, S., and Graves, D.J. Chemical influences on the specificity of tyrosine phosphorylation, *J. Biol. Chem.* 265, 7108–7111, 1990.

114. Wehtje, E. and Adlercreutz, P., Water activity and substrate concentration effects on lipase activity, *Biotechnol. Bioeng.* 55, 796–806, 1997.

115. Kurzawa, J. and Suszka, A., Kinetics and mechanism of the nucleophilic cleavage of disulfide bond in 2,2′-dithio-diimidazoles with hydroxide ions, *Polish J. Chem.* 8, 1487–1494, 2007.

116. Lawlor, D.A., More O'Farrall, R.A., and Rao, S.N., Stabilities and partitioning of arenonium ions in aqueous media, *J. Am. Chem. Soc.* 130, 17997–18007, 2008.

117. Ji, Y. and Bennett, B.M., Activation of microsomal glutathione S-transferase by peroxynitrite, *Mol. Pharmacol.* 63, 136–146, 2003.

118. Elce, J.S., Active-site amino acid residues in γ-glutamyltransferase and the nature of the γ-glutamyl-enzyme bond, *Biochem. J.* 185, 473–481, 1980.

119. Schaffert, C.S., Role of MGST1 in reactive intermediate-induced injury, *World J. Gastroenterol.* 17, 2552–2557, 2011.

120. Aniya, Y. and Daido, A., Organic hydroperoxide-induced activation of liver microsomal glutathione-*S*-transferase of rates in vitro, *Jpn. J. Pharmacol.* 62, 9–14, 1993.

121. Rinaldi, R., Aniya, Y., Svensson, R. et al., NADPH dependent activation of microsomal glutathione transferase 1, *Chem. Biol. Interact.* 147, 163–172, 2004.

122. Kuhn, D.M. and Geddes, T.J., Peroxynitrite inactivates tryptophan hydroxylase via sulfhydryl oxidation. Coincident nitration of enzyme tyrosyl residues has minimal impact on catalytic activity, *J. Biol. Chem.* 274, 29726–29732, 1999.
123. Ji, Y., Neverova, I., Van Eck, J.E., and Bennett, B.M., Nitration of tyrosine 92 mediates the activation of rat microsomal glutathione S-transferase by peroxynitrite, *J. Biol. Chem.* 281, 1986–1991, 2006.
124. Busenlehner, L.S., Codreanu, S.G., Holm, P.J. et al., Stress sensor triggers conformational response of the integral membrane protein microsomal glutathione transferase 1, *Biochemistry* 43, 11145–11152, 2004.
125. Sidddiqui, K.S. and Cavicchioli, R., Improved thermal stability and activity in the cold-adapted lipase B from *Candida antarctica* following chemical modification with oxidized polysaccharides, *Extremophiles* 9, 471–476, 2005.
126. Ryan, S.M., Mantovani, G., Wang, X. et al., Advances in PEGylation of important biotech molecules: Delivery aspects, *Expert Opin. Drug Deliv.* 5, 371–383, 2008.
127. Jevsevar, S., Kunstelj, M., and Porekar, V.G., PEGylation of therapeutic proteins, *Biotechnol. J.* 5, 113–128, 2010.
128. Zaragoza, R., Torres, L., Garcia, C. et al., Nitration of cathepsin D enhances its proteolytic activity during mammary gland remodeling after lactation, *Biochem. J.* 419, 279–288, 2009.
129. Basu, S. and Kirely, T.L., Identification of a tyrosine residue responsible for *N*-acetylimidazole -induced increase of activity of ecto-nucleoside triphosphate diphosphohydrolase 3, *Purinergic Signal.* 1, 271–280, 2005.
130. Yagil, G. and Anbar, M., Formation of peroxynitrite by oxidation of chloramine, hydroxylamine, and nitrohydroxamate, *J. Inorg. Nuclear Chem.* 26, 453–460, 1964.
131. Petriconi, G.L. and Papee, H.M., Aqueous solutions of sodium "pernitrite," *Can. J. Chem.* 44, 977–980, 1966.
132. Ishiropoulos, H., Zhu, L., Chen, J. et al., Peroxynitrite-mediated tyrosine nitration catalyzed by superoxide dismutase, *Arch. Biochem. Biophys.* 298, 431–437, 1992.
133. Greenacre, S.A.B. and Ischeriopoulos, H., Tyrosine nitration: Localisation, quantification, consequences for protein function and signal transduction, *Free Radic. Res.* 34, 541–481, 2001.
134. Ischiropoulos, H., Biological tyrosine nitration: A pathophysiological function of nitric oxide and reactive oxygen species, *Arch. Biochem. Biophys.* 356, 1–11, 1998.
135. Ischiropoulos, H., Biological selectivity and functional aspects of protein tyrosine nitration, *Biochem. Biophys. Res. Commun.* 305, 776–783, 2003.
136. Beckman, J.S., Chen, J., Ischiropoulos H., and Crow, J.P., Oxidative chemistry of peroxynitrite, *Methods Enzymol.* 233, 229–240, 1994.
137. Pryor, W.A., Cueto, R., Jin, X. et al., A practical method for preparing peroxynitrite solutions of low ionic strength and free of hydrogen peroxide, *Free Radic. Biol. Med.* 18, 75–83, 1995.
138. Uppu, R.M., Squadrito, G.L., Cueto, R., and Pryor, W.R., Selecting the most appropriate synthesis of peroxynitrite, *Methods Enzymol.* 269, 285–295, 1996.
139. Zhang, H., Bhargava, K., Kessler, A. et al., Transmigration nitration of hydrophobic tyrosyl peptides. Localization, characterization, mechanisms of nitration, and biological implications, *J. Biol. Chem.* 278, 8969–8978, 2003.
140. Bartesaghi, S., Valez, V., Trujillo, M. et al., Mechanistic studies of peroxynitrite-mediated tyrosine nitration in membranes using the hydrophobic probe *N-t*-BOC-L-tyrosine -*tert*-butyl ester, *Biochemistry* 45, 6813–6825, 2006.
141. Romero, N., Peluffo, G., Bartesaghi, S. et al., Incorporation of the hydrophobic probe *N-t*-BOC-L-tyrosine-*tert*-butyl ester to red cell membranes to study peroxynitrite-dependent reactions, *Chem. Res. Toxicol.* 20, 1638–1648, 2007.
142. Bartesaghi, S., Peluffo, G., Zhang, H. et al., Tyrosine nitration, dimerization, and hydroxylation by peroxynitrite in membrane studied by the hydrophobic probe *N-t*-BOC-L-tyrosine *tert*-butyl ester, *Methods Enzymol.* 441, 217–236, 2008.
143. Radi, R., Denicola, A., Alvarez, B. et al., The biological chemistry of peroxynitrite, in *Nitric Oxide: Biology and Pathobiology*, ed. L.J. Ignarro, pp. 57–82, Academic Press, San Diego, CA, 2000.
144. Sharov, V.S., Driomina, E.S., Briviva, K., and Sies, H., Sensitization of peroxynitrite chemiluminescence by the triplet carbonyl sensitizer coumarin-525. Effect of CO_2, *Photochem. Photobiol.* 68, 797–801, 1998.
145. Wardman, P., Methods to measure the reactivity of peroxynitrite-derived oxidants toward reduced fluoresceins and rhodamines, *Methods Enzymol.* 411, 261–282, 2008.

146. Nelson, K.J., Klomsiri, C., Codreanu, S.G. et al., Use of dimedone-based chemical probes for sulfenic acid detection methods to visualize and identify labeled proteins, *Methods Enzymol.* 473, 95–115, 2010.

147. Schmidt, P. et al., Specific nitration at tyrosine 430 revealed by high resolution mass spectrometry as basis for redox regulation of bovine prostacyclin synthase, *J. Biol. Chem.* 278, 12813–12819, 2003.

148. Petersson, A.-S., Steen, H., Kalume, D.E. et al., Investigation of tyrosine nitration in proteins by mass spectrometry, *J. Mass Spectrom.* 36, 616–625, 2001.

149. Willard, B.B., Ruse, C.I., Keightley, J.A. et al., Site-specific quantitation of protein nitration using liquid chromatography/tandem mass spectrometry, *Anal. Chem.* 75, 2370–2376, 2003.

150. Liu, B., Tewari, A.K., Zhang, L. et al., Proteomic analysis of protein tyrosine nitration after ischemia reperfusion injury: Mitochondria as a major target, *Biochim. Biophys. Acta* 1794, 476–485, 2009.

151. Chiappetta, G., Corbo, C., Palmese, A. et al., Quantitative identification of protein nitration sites, *Proteomics* 9, 1524–1537, 2009.

152. Sultana, R., Reed, T., and Butterfield, D.A., Detection of 4-hydroxy-2-nonenal- and 3-nitrotyrosine-modified proteins using a proteomics approach, *Methods Mol. Biol.* 519, 351–361, 2009.

153. Lee, J.R., Lee, S.J., Kim, T.W. et al., Chemical approach for specific enrichment and mass analysis of nitrated peptides, *Anal. Chem.* 81, 6620–6626, 2009.

154. Jones, A.W., Mikhailov, V.A., Iniesta, J., and Cooper, H.J., Electron capture dissociation mass spectrometry of tyrosine nitrated peptides, *J. Am. Soc. Mass Spectrom.* 21, 268–277, 2010.

155. Zhan, X. and Desiderio, D.M., Mass spectrometric identification of in vivo nitrotyrosine sites in the human pituitary tumor proteome, *Methods Mol. Biol.* 566, 137–163, 2009.

156. Abello, N., Barroso, B., Kerstjens, H.A. et al., Chemical labeling and enrichment of nitrotyrosine-containing peptides, *Talanta* 80, 1503–1512, 2010.

157. Tsumoto, H., Taguchi, R., and Kohda, K., Efficient identification and quantification of peptides containing nitrotyrosine by matrix-assisted laser desorption/ionization time-of-flight mass spectrometry after derivatization, *Chem. Pharm. Bull. (Tokyo)* 58, 488–494, 2010.

158. Daiber, A., Bachschmid, M., Kawaklik, C. et al., A new pitfall in detecting biological end products of nitric oxide—Nitration, nitros(yl)ation and nitrite/nitrate artifacts during freezing, *Nitric Oxide* 9, 44–52, 2003.

159. Irie, Y., Saekii, M., Kamisaki, Y. et al., Histone H1.2 is a substrate for dinitrase, an activity that reduces nitrotyrosine immunoreactivity in proteins, *Proc. Natl. Acad. Sci. USA* 100, 5634–5639, 2003.

160. Nikov, G., Bhat, Y., Wishak, J.S., and Tannenbaum, S.R., Analysis of nitrated proteins by nitrotyrosine-specific affinity probes and mass spectrometry, *Anal. Biochem.* 320, 214–222, 2003.

161. Ogino, K., Nakajima, M., Kodama, N. et al., Immunohistochemical artifact for nitrotyrosine in eosinophils or eosinophil containing tissue, *Free Radic. Res.* 36, 1163–1170, 2002.

162. Miyagi, M., Sakajushi, H., Darrow, R.M. et al., Evidence that light modulates protein nitration in rat retina, *Mol. Cell. Proteomics* 1, 293–303, 2003.

163. Viera, L., Ye, Y.Z., Estévez, A.G., and Beckman, J.S., Immunohistochemical methods to detect nitrotyrosine, *Methods Enzymol.* 301, 373–381, 1999.

164. Ogino, K., Nakajima, M., Kodama, N. et al., Immunohistochemical artifact for nitrotyrosine in eosinophils or eosinophil containing tissue, *Free Radic. Res.* 36, 1163–1170, 2002.

165. Aulak, K.S., Koeck, T., Crabb, J.W., and Stuehr, D.J., Proteomic method for identification of tyrosine-nitrated proteins, *Methods Mol. Biol.* 279, 151–165, 2004.

166. Koeck, T., Willard, B., Crabb, J.W. et al., Glucose-mediated tyrosine nitration in adipocytes: Targets and consequences, *Free Radic. Biol. Med.* 46, 884–892, 2009.

167. Sharov, V.S., Pal, R., Dremina, E.S. et al., Fluorogenic tagging of protein 3-nitrotyrosine with 4-(aminomethyl)benzene sulfonate in tissues: A useful alternative to immunohistochemistry for fluorescence microscopy imaging of protein nitration, *Free Radic. Biol. Med.* 53, 1877–1885, 2012.

168. Johnson, L.R. and Wornall, A., Immunological properties of alkali-treated proteins, *Biochem. J.* 26, 1202–1213, 1932.

169. Dixit, K., Khan, K.M., Sharma, Y.D. et al., Peroxynitrite-induced modification of H2A histone presents epitopes which are strongly bound by human anti-DNA antibodies. Role of peroxynitrite-modified-H2A in SLE induction and progression, *Hum. Immunol.* 72, 219–225, 2011.

170. Karle, A.C., Oostingh, G.J., Mutschlechner, S. et al., Nitration of the pollen allergen Bet v1.0101 enhances the presentation of Bet v 1-derived peptides by HLA-DR on human dendritic cells, *PLoS One* 7, e31483, 2012.

171. Saboori, A.M., Rose, N.R., Bresler, H.S. et al., Iodination of human thyroglobulin (Tg) alters its immunoreactivity. I. Iodination alters multiple epitopes of human Tg, *Clin. Exp. Immunol.* 113, 297–302, 1998.

172. Benyamin, Y., Robin, Y., and van Thoai, N., Immunochemistry of lobster arginine kinase. Effect of chemical modifications of the essential amino-acid residues on the antigenic reactivity, *Eur. J. Biochem.* 37, 459–466, 1973.

173. Sokolovsky, M., Riordan, J.F., and Vallee, B.L., Conversion of 3-nitrotyrosine to 3-aminotyrosine in peptides and proteins, *Biochem. Biophys. Res. Commun.* 27, 20–25, 1967.

174. Liu, T.-Y., Demonstration of the presence of a histidine residue at the active site of streptococcal proteinase, *J. Biol. Chem.* 242, 4029–4032, 1967.

175. Sun, H., He, J., Ru, Y. et al., Monitoring succinyl-CoA: 3-oxoacid CoA transferase nitration in mitochondria using monoclonal antibodies, *Biochem. Biophys. Res. Commun.* 415, 239–244, 2011.

176. Aslan, M. and Dogan, S., Proteomic detection of nitroproteins as potential biomarkers for cardiovascular disease, *J. Proteomics* 74, 2274–2288, 2011.

177. Zhan, X. and Desiderio, D.M., Nitroproteins identified in human ex-smoker bronchoalveolar lavage fluid, *Aging Dis.* 2, 100–115, 2011.

178. Ercolesi, E., Tedeschi, G., Fiore, G. et al., Protein nitration as footprint of oxidative stress-related nitric oxide signaling pathways in developing *Ciona intestinalis*, *Nitric Oxide* 27, 18–24, 2012.

179. Martinez, M., Cuker, A., Mills, A. et al., Nitrated fibrinogen is a biomarker of oxidative stress in venous thromboembolism, *Free Radic. Biol. Med.* 53, 230–236, 2012.

180. Ishii, N., Carmines, P.K., Yokoba, M. et al., Angiotensin-converting enzyme inhibition curbs tyrosine nitration of mitochondrial proteins in the renal cortex during the early stage of diabetes mellitus in rats, *Clin. Sci.* 124, 543–552, 2013.

181. Halliwell, B., What nitrates tyrosine? Is nitrotyrosine specific as a biomarker of peroxynitrite formation in vivo?, *FEBS Lett.* 411, 157–160, 1997.

182. van derVliet, A., Eiserich, J.P., Kaur, H. et al., Nitrotyrosine as biomarker for reactive nitrogen species, *Methods Enzymol.* 269, 175–184, 1996.

182. Drel, V.R., Lupachyk, S., Shevalye, H. et al., New therapeutic and biomarker discovery for peripheral diabetic neuropathy: PARP inhibitor, nitrotyrosine, and tumor necrosis factor-α, *Endocrinology* 151(6), 2547–2555, 2010.

183. Reddy, S. and Bradley, J., Immunohistochemical demonstration of nitrotyrosine, a biomarker of oxidative stress, in islets cells of the NOD mouse, *Ann. N. Y. Acad. Sci.* 1037, 199–202, 2004.

184. Souza, J.M., Peluffo, G., and Radi, R., Protein tyrosine nitration—Functional alteration or just a biomarker?, *Free Radic. Biol. Med.* 45, 357–366, 2008.

185. Radabaugh, M.R., Nemirovskiy, O.V., Misko, T.P. et al., Immunoaffinity liquid chromatography—Tandem mass spectrometry detection of nitrotyrosine in biological fluids: Development of a clinical translatable biomarker, *Anal. Biochem.* 380, 68–76, 2008.

186. Khan, M.A., Dixit, K., Uddin, M. et al., Role of peroxynitrite-modified H2A histone in the induction and progression of rheumatoid arthritis, *Scand. J. Rheumatol.* 41, 426–433, 2012.

187. Buizza, L., Cenini, G., Lanni, C. et al., Conformational altered p53 as an early marker of oxidative stress in Alzheimer's disease, *PLoS One* 7, e29789, 2012.

188. Sabuncuoğlu, S., Öztaş, Y., Çetinkaya, D.U. et al., Oxidative protein damage with carbonyl levels and nitrotyrosine expression after chemotherapy in bone marrow transportation patients, *Pharmacology* 89, 283–286, 2012.

189. Safinowski, M., Wilhelm, B., Reimer, T. et al., Determination of nitrotyrosine concentrations in plasma samples of diabetes mellitus patients by four different immunoassays leads to contradictive results and disqualifies the majority of the tests, *Clin. Chem. Lab. Med.* 47, 483–488, 2009.

190. Tsikas, D., Measurement of nitrotyrosine in plasma by immunoassays is fraught with danger: Commercial availability is no guarantee of analytical reliability, *Clin. Chem. Lab. Med.* 48, 141–143, 2010.

191. Weber, D., Kneschke, N., Grimm, S. et al., Rapid and sensitive determination of protein-nitrotyrosine by ELISA: Application to human plasma, *Free Radic. Res.* 46, 276–285, 2012.

192. Schmidt, E. and Fischer, H., Tetranitromethane. II. Tetranitromethane as nitrating agent, *Ber. Deut. Chem. Gesellschaft B. Abhandlungen* 53B, 1529–1537, 1920.

193. Baillie, A., Macbeth, A.K., and Maxwell, N.I., Effect of reducing agents on tetranitromethane, and a rapid method of estimation, *J. Chem. Soc. (Trans.)* 117, 880–884, 1920.

194. Herriott, R.M., Reactions of native proteins with chemical reagents, *Adv. Protein Chem.* 3, 169–225, 1947.

195. Riordan, J.F., Sokolovsky, M., and Vallee, B.L., Tetranitromethane. A reagent for the nitration of tyrosine and tyrosyl residues in proteins, *J. Am. Chem. Soc.* 88, 4104–4105, 1966.

196. Sokolovsky, M., Harell, D., and Riordan, J.F., Reaction of tetranitromethane with sulfhydryl groups in proteins, *Biochemistry* 8, 4740–4745, 1969.

197. Rabani, J., Mulae, W.A., and Matheson, M.S., The pulse radiolysis of aqueous tetranitromethane. I. Rate constants and the extinction coefficient of e_{aq}^-. II. Oxygenated solutions, *J. Phys. Chem.* 69, 53–70, 1965.

198. Riordan, J.F. and Vallee, B.L., Nitration with tetranitromethane, *Methods Enzymol.* 25, 515–521, 1972.

199. Cuatrecasas, P., Fuchs, S., and Anfinsen, C.B., The tyrosyl residues at the active site of staphylococcal nuclease. Modifications by tetranitromethane, *J. Biol. Chem.* 243, 4787–4798, 1968.

200. Palamalai, V. and Miyagi, M., Mechanism of glyceraldehyde-3-phosphate dehydrogenase inactivation by tyrosine nitration, *Protein Sci.* 19, 255–262, 2010.

201. Doyle, R.J., Bello, J., and Roholt, O.A., Probable protein crosslinking with tetranitromethane, *Biochim. Biophys. Acta* 160, 274–276, 1970.

202. Boesel, R.W. and Carpenter, F.H., Crosslinking during the nitration of bovine insulin with tetranitromethane, *Biochem. Biophys. Res. Commun.* 38, 678–682, 1970.

203. Kunkel, G.R., Mehrabian, M., and Martinson, H.G., Contact-site cross-linking agents, *Mol. Cell. Biochem.* 34, 3–13, 1981.

204. Nadeau, O.W., Traxler, K.W., and Carlson, G.M., Zero-length crosslinking of the beta subunit of phosphorylase kinase to the N-terminal half of its regulatory alpha subunit. *Biochem. Biophys. Res. Commun.* 251, 637–641, 1998.

205. Bruice, T.C., Gregory, M.J., and Walters, S.L., Reactions of tetranitromethane. I. Kinetics and mechanism of nitration of phenols by tetranitromethane. *J. Am. Chem. Soc.* 90, 1612, 1968.

206. Hass, G.M. and Gentry, L., Nitration of polypeptides using ethanol in reaction buffers minimizes crosslinking, *J. Biochem. Biophys. Methods* 1, 257–261, 1979.

207. Hugli, T.E. and Stein, W.H., Involvement of a tyrosine residue in the activity of bovine pancreatic deoxyribonuclease A, *J. Biol. Chem.* 246, 7191–7200, 1971.

208. Boesel, R.W. and Carpenter, F.H., Crosslinking during the nitration of bovine insulin with tetranitromethane, *Biochem. Biophys. Res. Commun.* 38, 678–682, 1970.

209. Cutfield, S.M., Dodson, G.G., Ronco, N., and Cutfield, J.F., Preparation and activity of nitrated insulin dimer, *Int. J. Pept. Protein Res.* 27, 335–343, 1986.

210. Sorrentino, S., Yakovlev, G.I., and Libonati, M., Dimerization of deoxyribonuclease I, lysozyme, and papain. Effects of ionic strength on enzymic activity, *Eur. J. Biochem.* 124, 183–189, 1982.

211. Strazza, S., Hunter, R., Walker, E., and Darnall, D.W., The thermodynamics of bovine and porcine insulin and proinsulin association determined by concentration difference spectroscopy, *Arch. Biochem. Biophys.* 238, 30–42, 1985.

212. Mädler, S., Bich, C., Touboul, D. et al., Chemical cross-linking with NHS esters: A systematic study on amino acid reactivities, *J. Mass Spectrom.* 44, 694–706, 2009.

213. Margolis, S.A., Coxon, B., Gajewski, E., and Dizdaroglu, M., Structure of a hydroxyl radical induced cross-link of thymine and tyrosine, *Biochemistry* 27, 6353–6359, 1988.

214. Hsuan, J.J., The cross-linking of tyrosine residues in apo-ovotransferrin by treatment with periodate anions, *Biochem. J.* 247, 467–473, 1987.

215. Yermilov, V., Rubio, J., and Ohshima, H., Formation of 8-nitroguanine in DNA treated with peroxynitrite in vitro and its rapid removal from DNA by depurination, *FEBS Lett.* 376, 207–210, 1995.

216. Niles, J.C., Wishnok, J.S., and Tannenbaum, S.R., Peroxynitrite-induced oxidation and nitration products of guanine and 8-oxoguanine: Structure and mechanisms of product formation, *Nitric Oxide* 14, 109–121, 2006.

217. Crow, J.P. and Ishiropoulos, H., Detection and quantitation of nitrotyrosine residues in proteins: In vivo marker of peroxynitrite, *Methods Enzymol.* 269, 185–194, 1996.

218. Luo, Y., Li, J., Zhang, N. et al., Identification of nitration sites by peroxynitrite on P16 protein, *Protein J.* 31, 393–400, 2012.

219. Berton, P., Domínguez-Romero, J.C., Wuilloud, R.G. et al., Determination of nitrotyrosine in *Arabidopsis thaliana* cell cultures with a mixed-mode solid-phase extraction cleanup followed by liquid chromatography time-of-flight mass spectrometry, *Anal. Bioanal. Chem.* 404, 1495–1503, 2012.

220. Thomsom, L., Tenopoulou, M., Lightfoot, R. et al., Immunoglobulins against tyrosine-nitrated epitopes in coronary artery disease, *Circulation* 126, 2392–2401, 2012.

221. Wheland, G.W., Brownell, R.M., and Mayo, E.C., The steric inhibition of resonance. III. Acid strengths of some nitro- and cyanophenols, *J. Am. Chem. Soc.* 70, 2492–2495, 1948.

222. Wheland, G.W., *Resonance in Organic Chemistry*, John Wiley & Sons, New York, 1955.

223. Vartak, D.G. and Menon, N.G., Solution stability constants of complexes of 4-nitro-2-aminophneol with some divalent metal ions, *J. Inorg. Nucl. Chem.* 28, 2911–2917, 1966.

224. Richards, P.G., Walton, D.J., and Heptinstall, J., The effects of tyrosine nitration on the structure and function of hen egg-white lysozyme, *Biochem. J.* 315, 473–479, 1996.

225. Morag, E., Bayer, E.A., and Wilchek, M., Reversibility of biotin-binding by selective modification of tyrosine in avidin, *Biochem. J.* 316, 193–199, 1996.
226. Morag, E., Bayer, E.A., and Wilchek, M., Immobilized nitro-avidin and nitro-streptavidin as reusable affinity matrices for application in avidin-biotin technology, *Anal. Biochem.* 243, 257–263, 1996.
227. Bolivar, J.G., Soper, S.A., and Carley, R.L., Nitroavidin as a ligand for the surface capture and release of biotinylated proteins, *Anal. Biochem.* 80, 9336–9342, 2008.
228. Tawfik, D.S. et al., pH on-off switching of antibody—Hapten binding by site-specific chemical modification of proteins, *Protein Eng.* 7, 431–434, 1994.
229. Riordan, J.F., Sokolovsky, M., and Vallee, B.L., Environmentally sensitive tyrosyl residues. Nitration with tetranitromethane, *Biochemistry* 6, 3582–3589, 1967.
230. Haas, J.A., Frederick, M.A., and Fox, B.G., Chemical and post-translational modifications of *Escherichia coli* acyl carrier protein for preparation of dansyl carrier protein, *Protein Expr. Purif.* 20, 274–284, 2000.
231. Haas, J.A. and Fox, B.G., Fluorescence anisotropy studies of enzyme-substrate complex formation in stearoyl-ACP-desaturase, *Biochemistry* 41, 14472–14481, 2002.
232. Nikov, G., Bhat, V., Wishnok, J.S., and Tannenbaum, S.R., Analysis of nitrated proteins by nitrotyrosine-specific affinity probes and mass spectrometry, *Anal. Biochem.* 320, 214–222, 2003.
233. Zhang, Q., Qian, W.J., Kuyushko, T.V. et al., A method for selective enrichment and analysis of nitrotyrosine-containing peptides in complex proteome samples, *J. Proteome Res.* 6, 2257–2268, 2007.
234. Sharov, V.S., Dremina, E.S., Pennington, J. et al., Selective fluorogenic derivatization of 3-nitrotyrosine and 3,4-dihydroxyphenylalanine in peptides: A method designed for quantitative proteomic analysis, *Methods Enzymol.* 441, 19–32, 2008.
235. Abello, N., Barroso, B., Kerstjens, H.A.M. et al., Chemical labeling and enrichment of nitrotyrosine-containing peptides, *Talanta* 80, 1503–1512, 2010.
236. Riordan, J.F., Sokolovsky, M., and Vallee, B.L., The functional tyrosyl residues of carboxypeptidase A. Nitration with tetranitromethane, *Biochemistry* 6, 3609–3617, 1967.
237. Burr, M. and Koshland, D.E., Jr., Use of a reporter groups in structure-function studies of proteins, *Proc. Natl. Acad. Sci. USA* 52, 1017–1024, 1964.
238. Muller, W.E. and Wollert, U., Spectroscopic studies on the complex formation of suramin with bovine and human serum albumin, *Biochem. Biophys. Acta* 427, 465–480, 1976.
239. Ajtai, K., Peyser, Y.M., Park, S. et al., Trinitrophenylated reactive lysine residues in myosin detects lever arm movement during the consecutive steps of ATP hydrolysis, *Biochemistry* 38, 6428–6440, 1999.
240. Pezzementi, L., Shi, J., Johnson, D.A. et al., Ligand-induced conformational changes in residues flanking the active site gorge of acetylcholinesterase, *Chem. Biol. Interact.* 157–158, 413–414, 2005.
241. Dempski, R.E., Friedrich, T., and Bamberg, E., Voltage clamp fluorometry: Combining fluorescence and electrophysiological methods to examine the structure-function of the Na$^+$K$^+$-ATPase, *Biochem. Biophys. Acta* 1787, 714–720, 2009.
242. Garel, J.R., Evidence for involvement of proline *cis-trans* isomerization in the slow unfolding of RNase A, *Proc. Natl. Acad. Sci. USA* 77, 795–798, 1980.
243. Parker, D.M., Jeckel, D., and Holbrook, J.J., Slow structural changes shown by the 3-nitrotyrosine-237 residue in pig heart [Tyr(3-NO$_2$)237] lactate dehydrogenase, *Biochem. J.* 201, 465–471, 1982.
244. Rischel, C. and Poulsen, F.M., Modification of a specific tyrosine enables tracing of the end-to-end distance during apomyoglobin, *FEBS Lett.* 374, 105–109, 1995.
245. De Fillippis, V., Draghi, A., Frasson, R. et al., *o*-Nitrotyrosine and *p*-iodophenylalanine as spectroscopic probes for structural characteristics of SH3 complexes, *Protein Sci.* 16, 1257–1265, 2007.
246. Hartings, M.R., Gray, H.B., and Winkler, J.R., Probing melittin helix-coil equilibria in solutions and vesicles, *J. Phys. Chem. B* 112, 3202–3207, 2008.
247. Digambaranath, J.L., Dang, L., Dembinska, M. et al., Conformations within soluble oligomers and insoluble aggregates revealed by resonance energy transfer, *Biopolymers* 93, 299–317, 2010.
248. Herz, J.M., Hrabeta, E., and Packer, L., Evidence for a carboxyl group in the vicinity of the retinal chromophore of bacteriorhodopsin, *Biochem. Biophys. Res. Commun.* 114, 872–881, 1983.
249. Lide, D.R. (ed.), *Handbook of Chemistry and Physics*, 82nd edn., pp. 3-207, CRC Press, Boca Raton, FL (CAS Registry Number 509-14-8).
250. Liang, P., Tetranitromethane, *Org. Synth. Coll.* 3, 803, 1955.
251. Fendler, J.H. and Liechti, R.R., Micellar catalysis of the reaction of hydroxide ion with tetranitromethane, *J. Chem. Soc. Perkins Trans.* 2 9, 1041–1043, 1972.
252. Walters, S.L. and Bruice, T.C., Reactions of tetranitromethane. II. Kinetics and products for the reactions of tetranitromethane with inorganic ions and alcohols, *J. Am. Chem. Soc.* 93, 2269–2282, 1971.

253. Selassie, C.D., Mekapati, S.B., and Verma, R.P., QSAR: Then and now, *Curr. Top. Med. Chem.* 2, 1357–1379, 2002.
254. Eberson, L. and Hartshorn, M.P., The formation and reactions of adducts from the photochemical reactions of aromatic compounds with tetranitromethane and other X-NO$_2$ reagents, *Aust. J. Chem.* 51, 1061–1081, 1998.
255. Capellos, C., Iyer, S., Liang, Y., and Gamms, L.A., Transient species and product formation from electronically excited tetranitromethane, *J. Chem. Soc. Faraday Trans.2* 82, 2195–2206, 1989.
256. Hodges, G.R., Young, M.J., Paul, T., and Ingold, K.U., How should xanthine oxidase-generated superoxide yields be measured?, *Free Radic. Biol. Med.* 29, 434–441, 2000.
257. Liochev, S.I. and Fridovich, I., Reversal of the superoxide dismutase reaction revisited, *Free Radic. Biol. Med.* 34, 908–910, 2003.
258. Sutton, H.C., Reactions of the hydroperoxyl radical (HO$_2$) with nitrogen dioxide and tetranitromethane in aqueous solutions, *J. Chem. Soc. Faraday Trans. 1* 71, 2142–2147, 1975.
259. Naqvi, K.R. and Melø, T.B., Reduction of tetranitromethane by electronically excited aromatics in acetonitrile: Spectra and molar absorption coefficients of radical cations of anthracene, phenanthrene and pyrene, *Chem. Phys. Lett.* 428, 83–87, 2006.
260. Butts, C.P., Eberson, L., Hartshorn, M.P. et al., Photochemical nitration by tetranitromethane. Part XXXIX. The photolysis of tetranitromethane with 2,8-dimethyl and 1,3,7,9-tetramethyl-dibenzofuran, *Acta Chem. Scand.* 51, 476–482, 1997.
261. Rasmusseon, M. et al., Ultrafast formation of trinitromethanide (C(NO$_2$)$_3$$^-$) by photo-induced dissociative electron transfer and subsequent ion pair coupling reaction in acetonitrile and dichloromethane, *J. Phys. Chem. B* 105, 2027–2035, 2001.
262. Scherrer, P. and Stoeckenius, W., Selective nitration of tyrosine-26 and -64 in bacteriorhodopsin with tetranitromethane, *Biochemistry* 23, 6195–6202, 1984.
263. Brown, L.S., Reconciling crystallography and mutagenesis: A synthetic approach to the creation of a comprehensive model for proton pumping by bacteriorhodopsin, *Biochem. Biophys. Acta* 1460, 49–59, 2000.
264. Scherrer, P. and Stoeckenius, W., Effects of tyrosine-26 and tyrosine-64 nitration on the photoreactions of bacteriorhodopsin, *Biochemistry* 24, 7733–7740, 1985.
265. Mayo, K.H., Schussheim, M., Visserr, G.W. et al., Mobility and solvent exposure of aromatic residues in bacteriorhodopsin investigated by ^{1}H-NMR and photo-CIDNP-NMR spectroscopy, *FEBS Lett.* 235, 163–168, 1988.
266. Renthal, R., Cothran, M., Dawson, N., and Harris, G.J., Light activates the reaction of bacteriorhodopsin aspartic acid-115 with dicyclohexylcarbodiimide, *Biochemistry* 24, 4275–4279, 1985.
267. Capellos, C., Iyer, S., Liang, S., and Gamms, L.A., Transient species and product formation from electronically excited tetranitromethane, *J. Chem. Soc. Faraday Trans. 2*, 82, 2195–2226, 1986.
268. Eberson, L., Hartshorn, M.P., and Persson, O., The reactions of some dienes with tetranitromethane, *Acta Chem. Scand.* 52, 450–452, 1998.
269. Fieser, M. and Fieser, L., *Reagents for Organic Synthesis*, Vol. 1, 1147pp., Wiley Interscience, New York, 1967.
270. Jewett, S.W. and Bruice, T.C., Reactions of tetranitromethane. Mechanisms of the reaction of tetranitromethane with pseudo acids, *Biochemistry* 11, 3338–3350, 1972.
271. Strosberg, A.D., Van Hoeck, B., and Kanarek, L., Immunochemical studies on hen's egg white lysozyme. Effect of selective nitration of the three tyrosine residues, *Eur. J. Biochem.* 19, 36–41, 1971.
272. Richards, P.G., Walton, D.J., and Heptinstall, J., The effects of nitration on the structure and function of hen egg-white lysozyme, *Biochem. J.* 315, 473–479, 1996.
273. Richards, P.G., Coles, B., Heptinstall, J., and Walton, D.J., Electrochemical modification of lysozyme: Anodic reaction of tyrosine residues, *Enzyme Microb. Technol.* 16, 795–801, 1994.
274. Skawinski, W.J., Adebodun, F., Cheng, J.T. et al., Labeling of tyrosines in proteins with [^{15}N]tetranitromethane, a new NMR reporter for nitrotyrosines, *Biochem. Biophys. Acta* 1162, 297–308, 1993.
275. Šantrůček, J., Strohalm, M., Kadlčk, V. et al., Tyrosine residues modification studied by MALDI-TOF mass spectrometry, *Biochem. Biophys. Res. Commun.* 323, 1151–1156, 2004.
276. Lecomte, J.T.J. and Llinás, M., Characterization of the aromatic proton magnetic resonance spectrum of crambin, *Biochemistry* 23, 4799–4807, 1984.
277. Lee, S.-B., Inouye, K., and Tomura, B., The states of tyrosyl residues in thermolysin as examined by nitration and pH-dependent ionization, *J. Biochem.* 121, 231–237, 1997.
278. Inouye, K., Lee, S.-B., and Tonomura, B., Effects of nitration and amination of tyrosyl residues in thermolysin on its hydrolytic activity and its remarkable activation by salts, *J. Biochem.* 124, 72–78, 1998.

279. Muta, Y., Oneda, H., and Inouye, K., Anomalous pH-dependence of the activity of human matrilysin (matrix metalloproteinase-7) as revealed by nitration and amination of its tyrosine residues, *J. Biochem.* 386, 263–270, 2005.

280. Christen, P. and Riordan, J.F., Syncatalytic modification of a functional tyrosyl residue in aspartate aminotransferase, *Biochemistry* 9, 3025–3024, 1970.

281. Betcher-Lange, S.L., Couglan, M.P., and Rajagopalan, K.V., Syncatalytic modification of chicken liver xanthine dehydrogenase by hydrogen peroxide. The nature of the reaction, *J. Biol. Chem.* 254, 8825–8829, 1979.

282. Usanov, S.A., Pikuleva, I.A., Chashchin, V.L., and Akhrem, A.A., Chemical modification of adrenocortical cytochrome P-450$_{scc}$ with tetranitromethane, *Biochem. Biophys. Acta* 790, 259–267, 1984.

283. Heyduk, T., Michalczyk, R., and Kochman, M., Long-range effects and conformational flexibility of aldolase, *J. Biol. Chem.* 266, 15650–15655, 1991.

284. Malhotra, O.P. and Dwivedi, U.N., Formation of enzyme-bound carbanion intermediate in the isocitrate lyase-catalyzed reaction: Enzyme reaction of tetranitromethane with substrates and its dependence on effector, pH, and metal ions, *Arch. Biochem. Biophys.* 250, 236–248, 1986.

285. Medda, R., Padiglia, A., Pedersen, J.Z., and Floria, G., Evidence for α-proton abstraction and carbanion formation involving a functional histidine residue in lentil seedling amine oxidase, *Biochem. Biophys. Res. Commun.* 196, 1349–1355, 1993.

286. Mekhanik, M.L. and Torchinskii, Y.M., Carbanion detection during the decarboxylation of quasisubstrates by glutamate dehydrogenase, *Biokhimiya* 37, 1308–1311, 1972.

287. Zappacosta, F., Ingallinella, P., Scaloni, P. et al., Surface topology of minibody by selective chemical modifications and mass spectrometry, *Protein Sci.* 6, 1901–1909, 1997.

288. Leite, J.F. and Cascio, M., Probing the topology of the glycine receptor by chemical modification couple to mass spectrometry, *Biochemistry* 41, 6140–6148, 2002.

289. Cascio, M., Glycine receptors: Lessons on topology and structural effects of the lipid bilayer, *Biopolymers* 66, 359–368, 2002.

290. D'Ambrosio, C., Talamo, F., Vitale, R.N. et al., Probing the dimeric structure of porcine aminoacylase 1 by mass spectrometric and modeling procedures, *Biochemistry* 42, 4430–4443, 2003.

291. Kurihara, K., Horinishi, H., and Shibata, K., Reaction of cyanuric halides with proteins. I. Bound tyrosine residues of insulin and lysozyme as identified with cyanuric fluoride, *Biochim. Biophys. Acta* 74, 678–687, 1963.

292. Gorbunoff, M.J., Cyanuration, *Methods Enzymol.* 25, 506–514, 1972.

293. Gorbunoff, M.J., The pH dependence of tyrosine cyanuration in proteins, *Biopolymers* 11, 2233–2240, 1972.

294. Gorbunoff, M.J. and Timasheff, S.N., The role of tyrosines in elastase, *Arch. Biochem. Biophys.* 152, 413–422, 1972.

295. Coffe, G. and Pudles, J., Chemical reactivity of the tyrosyl residues in yeast hexokinase. Properties of the nitroenzyme, *Biochim. Biophys. Acta* 484, 322–335, 1977.

296. Larionova, N.L., Kazanskaya, N.F., and Sakharov, I.Y., Soluble-high-molecular-weight derivatives of the pancreatic inhibitor of trypsin. Isolation and properties of the dextran-bound pancreatic inhibitor, *Biokhimiya* 42, 1237–1243, 1977.

297. Moller, M., Halogen cyanides. Molecular weight, stability and basic hydrolysis of cyanogen bromide, *Kgl. Danske Videnskab. Math. fys. Medd.* 12, 17, 1934.

298. Perret, A. and Perrot, R., Polymerization of cyanogen bromide. Preparation of pure cyanogen bromide, *Bull. Soc. Chim.* 7, 743–750, 1940.

299. Ferguson, S.J., Lloyd, W.J., Lyons, M.H., and Radda, G.K., The mitochondrial ATPase. Evidence for a single essential tyrosine residue, *Eur. J. Biochem.* 54, 117–126, 1975.

300. Ferguson, S.J., Lloyd, W.J., and Radda, G.K., The mitochondrial ATPase. Selective modification of a nitrogen residue in the β-subunit, *Eur. J. Biochem.* 54, 127–133, 1975.

301. Sesaki, H., Wong, E.F., and Siu, C.H., The cell adhesion molecule DdCAD-1 in *Dictyostelium* is targeted to the cell surface by a nonclassical transport pathway involving contractile vacuoles, *J. Cell Biol.* 138, 939–951, 1997.

302. Denu, J.M. and Tanner, K.G., Specific and reversible inactivation of protein tyrosine phosphatases by hydrogen peroxide: Evidence for a sulfenic acid intermediate and implications for redox regulation, *Biochemistry* 37, 5633–5642, 1998.

303. Nieslanik, B.S. and Atkins, W.M., The catalytic Tyr-9 of glutathione S-transferase A1–1 controls the dynamics of the C terminus, *J. Biol. Chem.* 275, 17447–17451, 2000.

304. Brudecki, L.E., Grindstaff, J.J., and Ahmad, Z., Role of αPhe-291 residue in the phosphate-binding sub-domain of catalytic sites of *Escherichia coli* ATP synthase, *Arch. Biochem. Biophys.* 471, 168–175, 2008.

305. Toyo'oka, T., Mantani, T., and Kato, M., Reversible labeling of tyrosine residues in peptide using 4-fluoro-7-nitro-2,1,3-benzoxadiazole and *N*-acetyl-L-cysteine, *Anal. Sci.* 19, 341–346, 2003.

306. Carballal, S., Radi, R., Kirk, M.C. et al., Sulfenic acid formation in human serum albumin by hydrogen peroxide and peroxynitrite, *Biochemistry* 42, 9906–9914, 2003.

307. Shetty, V., Spellman, D.S., and Neubert, T.A., Characterization by tandem mass spectrometry of stable cysteine sulfenic acid in a cysteine switch peptide of matrix metalloproteinases, *J. Am. Soc. Mass Spectrom.* 18, 1544–1551, 2007.

308. Carballal, S., Radi, R., Kirk, M.C. et al., Sulfenic acid formation in human serum albumin by hydrogen peroxide and peroxynitrite, *Biochemistry* 42, 9906–9914, 2003.

309. Lavergne, S.N., Wang, H., Callan, H.E. et al., "Danger" conditions increase sulfamethoxazole-protein adduct formation in human antigen-presenting cells, *J. Pharmacol. Exp. Ther.* 331, 372–381, 2009.

310. de Almeida Olivera, M.G., Rogana, E., Roas, J.C. et al., Tyrosine 151 is part of the substrate activation binding site of bovine trypsin. Identification by covalent labeling with *p*-diazonium-benzamidine and kinetic characterization of TYR-151-(*p*-benzamidine)-azo-β-trypsin, *J. Biol. Chem.* 268, 26893–26903, 1993.

311. Landsteiner, K., *The Specificity of Serological Reactions,* Harvard University Press, Cambridge, U.K., 1945.

312. Fraenkel-Conrat, H., Bean, R.S., and Lineweaver, H., Essential groups for the interaction of ovomucoid (egg white trypsin inhibitor) and trypsin, and for tryptic activity, *J. Biol. Chem.* 177, 385–403, 1949.

313. Riordan, J.F. and Vallee, B.L., Diazonium salts as specific reagents and probes of protein conformation, *Methods Enzymol.* 25, 521–531, 1972.

314. Tabachnick, M. and Sobotka, H., Azoproteins. I. Spectrophotometric studies of amino acid azo derivatives, *J. Biol. Chem.* 234, 1726–1730, 1959.

315. Tabachnick, M. and Sobotka, H., Azoproteins. II. A spectrophotometric study of the coupling of diazotized arsanilic acid with proteins, *J. Biol. Chem.* 235, 1051–1054, 1960.

316. Vallee, B.L. and Fairclough, G.F., Jr., Arsanilazochymotrypsinogen. The extrinsic Cotton effects of an arsanilazotyrosyl chromophore as a conformation probe of zymogen activation, *Biochemistry* 10, 2470–2477, 1971.

317. Gorecki, M., Wilchek, M., and Blumberg, S., Modulation of the catalytic properties of α-chymotrypsin by chemical modification at Tyr 146, *Biochim. Biophys. Acta* 535, 90–99, 1978.

318. Robinson, G.W., Reaction of a specific tryptophan residue in streptococcal proteinase with 2-hydroxy-5-nitrobenzyl bromide, *J. Biol. Chem.* 245, 4832–4841, 1970.

319. Gorecki, M. and Wilchek, M., Modification of a specific tyrosine residue of ribonuclease A with a diazonium inhibitor analog, *Biochim. Biophys. Acta* 532, 81–91, 1978.

320. Johansen, J.T., Livingston, D.M., and Vallee, B.L., Chemical modification of carboxypeptidase A crystals. Azo coupling with tyrosine-248, *Biochemistry* 11, 2584–2588, 1972.

321. Harrison, L.W. and Vallee, B.L., Kinetics of substrate and product interactions with arsanilazotyrosine-248 carboxypeptidase A, *Biochemistry* 17, 4359–4363, 1978.

322. Cueni, L. and Riordan, J.F., Functional tyrosyl residues of carboxypeptidase A. The effect of protein structure on the reactivity of tyrosine-198, *Biochemistry* 17, 1834–1842, 1978.

323. Dahoumane, S.A., Nguyen, M.N., Thorel, A. et al., Protein-functionalized hairy diamond nanoparticles, *Langmuir* 25, 9633–9638, 2009.

324. Flavel, B.S., Gross, A.J., Garrett, D.J. et al., A simple approach to patterned protein immobilization on silicon via electrografting from diazonium salt solutions, *ACS Appl. Mater. Interfaces* 2, 1184–1190, 2010.

325. Lerner, M.B., D'Souza, J., Pazina, T. et al., Hybrids of a genetically engineered antibody and a carbon nanotube transistor for detection of prostate cancer biomarkers, *ACS Nano* 6, 5143–5149, 2012.

326. Schlick, T.L., Ding, Z., Kovacs, E.W. et al., Dual-surface modification of the tobacco mosaic virus, *J. Am. Chem. Soc.* 127, 3718–3723, 2005.

327. Bruckman, M.A., Kaur, G., Lee, L.A. et al., Surface modification of tobacco mosaic virus with "click" chemistry, *ChemBioChem* 9, 519–523, 2008.

328. Gavrilyuk, J., Ban, H., Nagano, M. et al., Formylbenzene diazonium hexafluorophosphate reagent for tyrosine-selective modification of proteins and the introduction of a bioorthogonal aldehyde, *Bioconjug. Chem.* 23, 2321–2328, 2012.

329. Jones, M.W., Mantovani, G., Blindauer, C.A. et al., Direct peptide bioconjugation/PEGylation at tyrosine with linear and branched polymeric diazonium salts, *J. Am. Chem. Soc.* 134, 7406–7413, 2012.

330. Liao, T.-H., Ting, R.S., and Young, J.E., Reactivity of tyrosine in bovine pancreatic deoxyribonuclease with *p*-nitrobenzenesulfonyl fluoride, *J. Biol. Chem.* 257, 5637–5644, 1982.

331. Chen, B., Costantino, H.R., Liu, J. et al., Influence of calcium ions on the structure and stability of recombinant human deoxyribonuclease I in the aqueous and lyophilized states, *J. Pharm. Sci.* 88, 477–482, 1999.

332. Haniu, M., Yuan, H., Chen, S. et al., Structure-function relationship of NAD(P)H:quinone reductase: Characterization of NH$_2$-terminal blocking group and essential tyrosine and lysine residues, *Biochemistry* 27, 6877–6883, 1988.

333. Kulanthaivel, P., Leibach, F.H., Mahesh, V.B., and Ganapathy, V., Tyrosine residues are essential for the activity of the human placental taurine transporter, *Biochim. Biophys. Acta* 985, 139–146, 1989.

334. Gitlin, G., Bayer, E.A., and Wilchek, M., Studies on the biotin-binding sites of avidin and streptavidin—Tyrosine residues are involved in the binding-site, *Biochem. J.* 269, 527–530, 1990.

335. Suzuki, H., Higashi, Y., Naitoh, N. et al., Chemical modification of a catalytic antibody that accelerates the hydrolysis of carbonate esters, *J. Protein Chem.* 19, 419–424, 2000.

336. Chang, L.-S. and Lin, S.-R., Modification of tyrosine-3 (63) and lysine 6 of Taiwan cobra phospholipase A$_2$ affects its ability to enhance 8-anilinonapthaline-1-sulfonate fluorescence, *Biochem. Mol. Biol. Int.* 40, 235–241, 1996.

337. Chang, L.S., Lin, S.-R., and Chang, C.-C., The essentiality of calcium ion in the enzymic activity of Taiwan cobra phospholipase A$_2$, *J. Protein Chem.* 15, 701–707, 1996.

338. Soares, A.M. and Giglio, J.R., Chemical modifications of phospholipase A(2) from snake venom: Effects on catalytic and pharmacologic properties, *Toxicon* 42, 855–868, 2003.

339. Kresge, N., Simoni, R.D., and Hill, R.L., Unraveling the enzymology of oxidative phosphorylation: The work of Efraim Racker, *J. Biol. Chem.* 281, e4–e6, 2006.

340. Andrews, W.W. and Allison, W.S., 1-Fluoro-2,4-dinitrobenzene modifies a tyrosine residue when it inactivates the bovine mitochondrial F$_1$-ATPase, *Biochem. Biophys. Res. Commun.* 99, 813–819, 1981.

341. De Vendittis, E., Ursby, T., Rullo, R. et al., Phenylmethylsulfonyl fluoride inactivates an archaeal superoxide dismutase by chemical modification of a specific tyrosine residue. Cloning, sequencing and expression of the gene coding for *Sulfolobus solfataricus* superoxide dismutase, *Eur. J. Biochem.* 268, 1794–1801, 2001.

342. Murachi, T., Inagami, T., and Yasui, M., Evidence for alkyl phosphorylation of tyrosyl residues of stem bromelain by diisopropylphosphorofluoridate, *Biochemistry* 4, 2815–2825, 1965.

343. Means, G.E. and Wu, H.L., The reactive tyrosine residue of human serum albumin: Characterization of its reaction with diisopropylfluorophosphate, *Arch. Biochem. Biophys.* 194, 526–530, 1979.

345. Williams, N.H., Harrison, J.M., Read, R.W., and Black, R.M., Phosphorylated tyrosine in albumin as a biomarker of exposure to organophosphorous nerve agents, *Arch. Toxicol.* 81, 627–639, 2007.

346. Ding, S.J., Carr, J., Carlson, J.E. et al., Five tyrosine and two serines in human albumin are labeled by the organophosphorous agent FP-biotin, *Chem. Res. Toxicol.* 21, 1787–1794, 2008.

347. Tarhoni, M.H., Lister, T., Ray, D.E., and Carter, W.G., Albumin binding as a potential biomarker of exposure to moderately low levels of organophosphorous pesticides, *Biomarkers* 13, 343–363, 2008.

12 Modification of Tryptophan

Tryptophan is an electron-rich heterocyclic aromatic amino acid (Figure 12.1). The average content of tryptophan in proteins is 1.3%–1.5%.[1,2] Based on proteomic analysis of several microorganisms, tryptophan is present in 90% of the proteins.[3] Tryptophan is considered to be a hydrophobic amino acid,[3–6] but the meaning of that definition depends on the solvent.[7] Despite the hydrophobic properties, tryptophan residues can be found on the protein surface[8–10] and available for chemical modification.[11,12] The concept of *buried* and *surface* residues is discussed in Chapter 1. Tryptophan has been shown to be present at binding sites for antigen in antibody complementarity-determining regions (CDR) (paratopes)[13,14] and interface regions on proteins important for protein–protein interactions and protein/peptide–membrane interactions.[15–19] The relatively low prevalence combined with presence in a majority of proteins makes tryptophan a useful target for specific modification. However, the specific chemical modification of tryptophan in protein is one of the more challenging problems in protein chemistry as the solvent conditions for providing specificity of modification are, in general, somewhat harsh and there is considerable possibility of either the concomitant or separate modification of other amino acid residues. The problem of specific modification is driven by the chemistry of tryptophan. The pKa of the indole nitrogen is above 15, making it essentially unreactive. However, modification of N_1 of the indole ring of tryptophanyl residues in peptides was accomplished with malondialdehyde in acid (Figure 12.2); the adduct could be cleaved in base (pyrrolidine, pH > 8).[20] The aldehyde derivative can be further reacted with a hydrazide or hydrazone. This chemistry was taken further by this group[21] for coupling a tryptophan-containing peptide/protein, which could be used as *bait* in pull-down technology[22,23] to purify a specific protein/proteins; the resulting complex can be dissociated from the matrix by pyrrolidine.

Tryptophan is responsible for most of characteristic protein UV absorbance at 280 nm; tryptophan is also responsible for most of the intrinsic fluorescence of proteins. Tryptophan is also a target for photooxidation mediated by sensitizers such as methylene blue. There is further discussion of photooxidation in Chapter 5. For the purpose of the current discussion, photolysis is defined as the decomposition of substances on exposure to light, while photooxidation is the oxidation of a substance on exposure to light.[24] Irradiation of tryptophan with UV light results in the formation of tryptophan radical and a hydrated electron[25–30]; the formation of hydrogen atoms from the photolysis of tryptophan has been reported.[6,31] While the photolysis of tryptophan is still poorly understood, a mechanism where a hydrogen is removed from the indole nitrogen followed by rearrangement to yield a radical at C_8 or C_9 is consistent with the observations of Léonard and coworkers.[32] One of the practical consequences of the photolysis of tryptophan is the action of tryptophan as a photosensitizer,[33] which can modify other amino residues[12,34] that contributed to the oxidation of a methionine residue in a monoclonal antibody.[12] This oxidation is mediated by tryptophan radical–mediated formation of reactive oxygen species[35] by the tryptophan radical, which forms hydrogen peroxide.[36] Earlier work by Sidorkina and coworkers[37] reported on the modification of *E. coli* Fpg protein by UVB light (280–315 nm) in 50 mM sodium phosphate, pH 7.4, containing 2 mM 2-mercaptoethanol and 10% glycerol which resulted in polymerization of the protein. Polymerization of the protein was observed as a result of UVB irradiation. Three of the five tryptophanyl residues were modified in an air-saturated solution, while only one tryptophanyl residue was modified in an argon-saturated solution. Peptide bond cleavage is observed with the appearance of a fragment with a structure consistent with that of the first 32 or 33 amino acids in the N-terminal region. The peptide bond cleavage is observed within either the air-saturated medium or the argon-saturated medium. It is suggested the cleavage results from photolysis of Trp34. Subsequent work

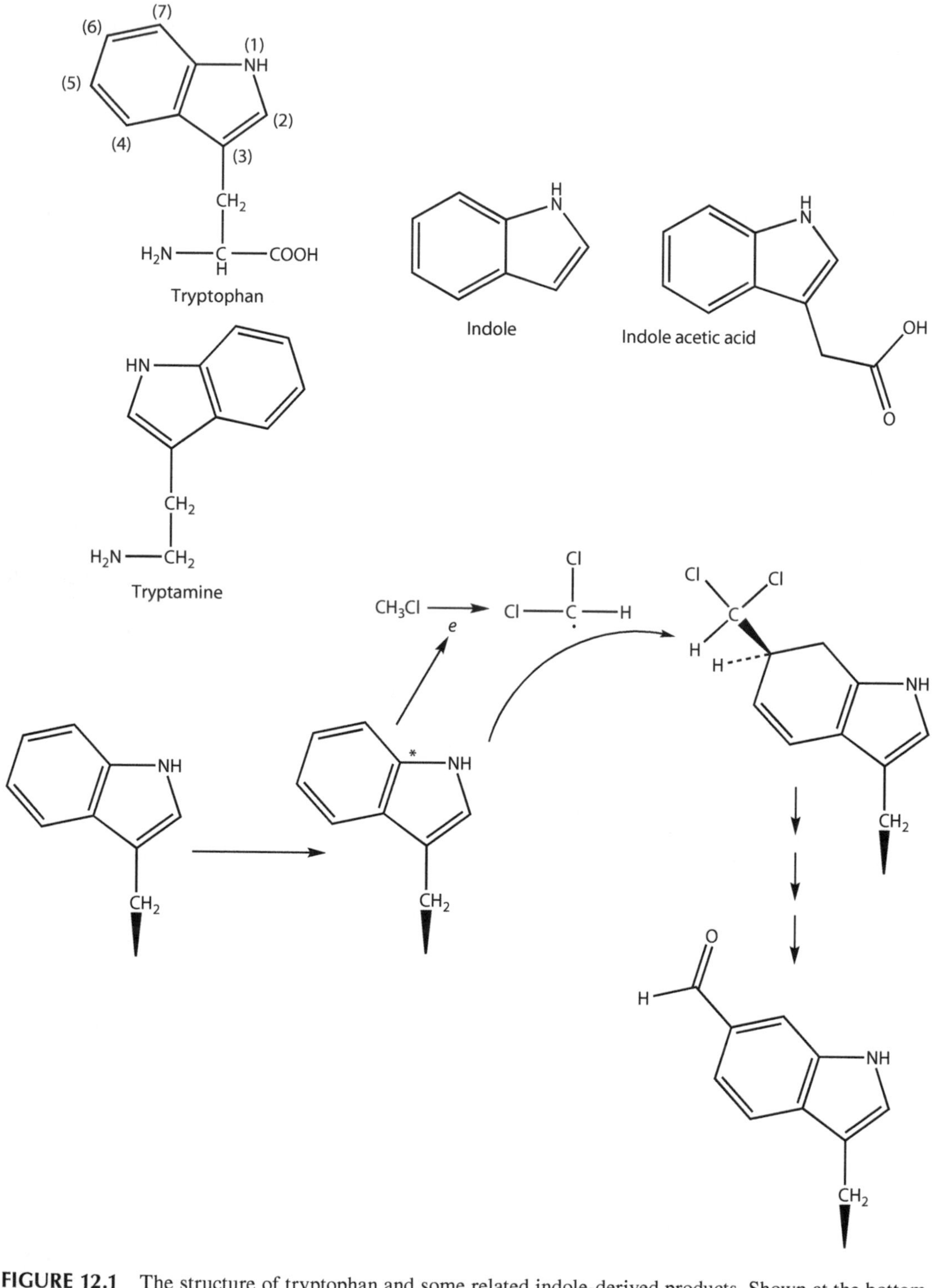

FIGURE 12.1 The structure of tryptophan and some related indole-derived products. Shown at the bottom is the photolytic reaction of tryptophan with chloroform to yield a formyltryptophan derivative (one of several possible isomers is shown). The reaction initially involved the formation of a tryptophan radical decaying with expulsion of a solvated electron that reacts with chloroform to yield a free radical that reacts with tryptophan decaying to yield the formyl derivative (Edwards, R.A., Jickling, G., and Turner, R.J., The light-induced reactions of tryptophan with halocompounds, *Photochem. Photobiol.* 75, 362–368, 2002).

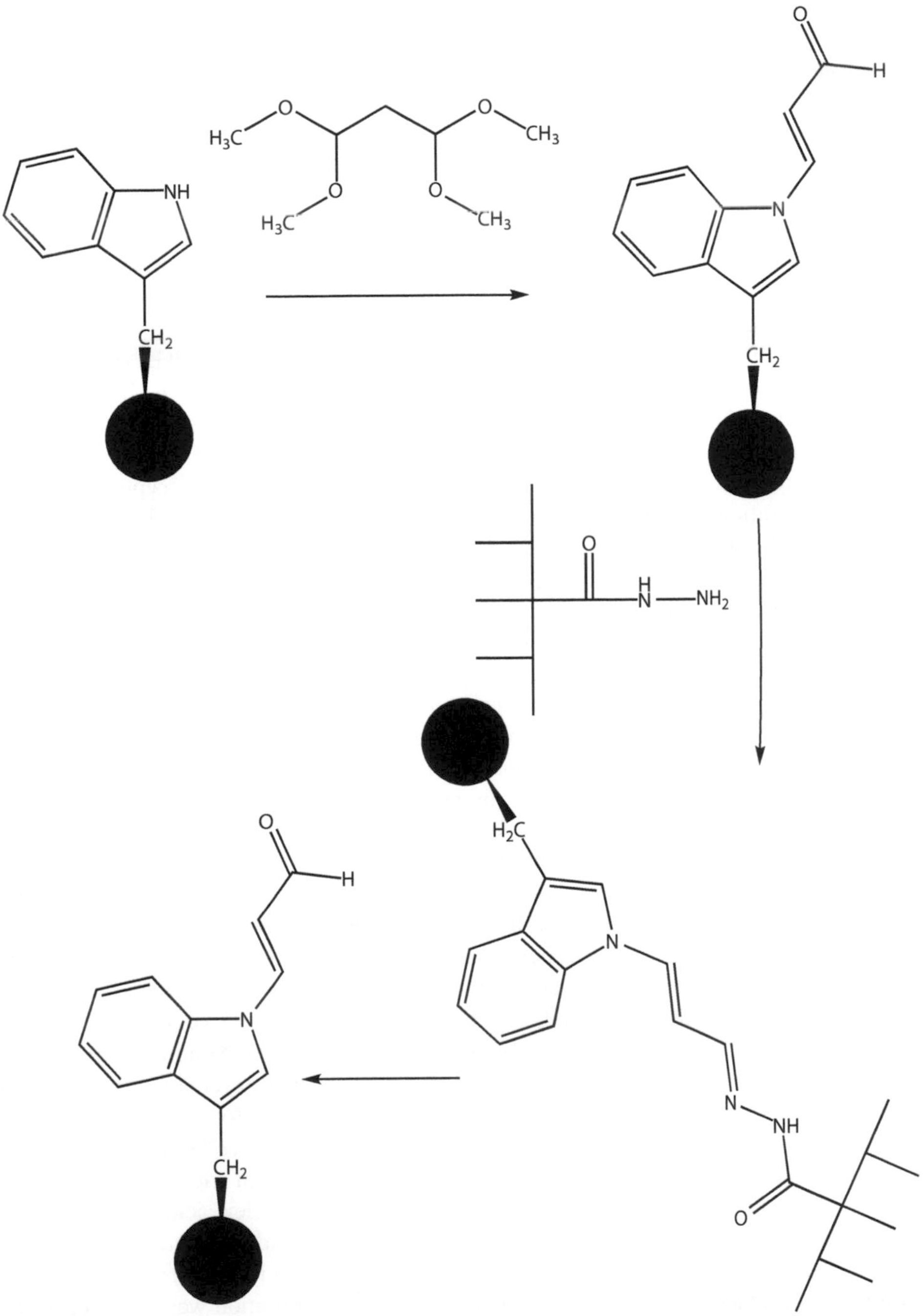

FIGURE 12.2 The reaction of malonaldehyde with tryptophan. Shown is the reaction of malonaldehyde bis(dimethylacetal) (1,1,3,3-tetramethoxypropane) with tryptophan in acid to yield a substituted acrolein with a free aldehyde, which can be coupled to hydrazide; the tryptophan residue can be regenerated by reaction with mild base (pyrrolidone, pH 8.0) (Foettinger, A., Melmer, M., Leitner, A., and Lindner, W., Reaction of the indole group with malonaldehyde: Application for the derivatization of tryptophan residues in peptides, *Bioconjug. Chem.* 18, 1678–1683, 2007). The example shown used a hydrazide function bound to a bead, and the tryptophanyl residue represents a *bait* protein/peptide (Sturm, M., Leitner, A., and Lindner, W., Development of an indole-based chemically cleavable linker concept for immobilizing bait compounds for protein pull-down experiments, *Bioconjug. Chem.* 22, 211–217, 2011).

by van der Kemp and coworkers[38] reported degradation of human OGG1 in either air-saturated or argon-saturated media. Human OGG1 contains 10 tryptophan residues, which are stated to be buried. Irradiation with UVB light results in a decrease in fluorescence and tryptophan content (at estimated by second-derivative spectroscopy). Loss of tryptophan was greater in argon-saturated buffer (50 mM potassium phosphate, pH 7.6, with 2 mM EDTA and 250 mM NaCl) than in air-saturated buffer. Modification was prevented by the presence of a 34 mer duplex (8-oxoG-C or G-C). Irradiation was associated with polymerization including the development of insoluble material; there were also several peptide fragments indicating multiple peptide bond cleavage sites consistent with the presence of multiple tryptophan residues in the human OGG1 protein. The reactions in proteins secondary to the photolysis of a tryptophanyl residue[39,40] are considered to reflect the action of the solvated electron ejected from the indole ring with the formation of the tryptophan free radical.[41–43] In their work on the effect of UVB irradiation of human OGG1, van der Kemp and coworkers[38] observed that the photolysis of tryptophan was more rapid in the OGG1 protein than with the free amino acid. These and other observations by van der Kemp and coworkers[38] support the suggestion that surrounding amino acid residues influence the rate of photolysis of tryptophanyl residues in proteins. Other studies suggest that the photodegradation of tryptophan in proteins depends on the location of the residue in a 3D structure of the proteins (solvent exposure).[44–46] Pigault and Gerard[44] studied the photolysis of four proteins, each with a single tryptophan with different degree of solvent exposure. Tryptophan loss was measured by amino acid analysis, UV spectroscopy, and fluorescence. The degree of tryptophan loss was related to exposure to solvent and surrounding amino acid; the type of photochemical product was also dependent on residue location. Studies with glucagon, staphylococcal nuclease, ribonuclease T_1, and melittin suggested that the *exposed* tryptophanyl residues are more susceptible to photodegradation, while reactivity appeared to be reduced in 4 M guanidine hydrochloride. These investigators concluded that the strongest correlation is between surface exposure and photodegradation.

Rao and coworkers[45] studied photolysis in native and random coil forms of melittin and β-lactoglobulin. The results support the concept that protein conformation is a major factor in photodegradation of tryptophan in proteins. Tallmadge and Borkman[46] measured the rate of photodegradation of the four tryptophanyl residues in γ-II crystallin. The rates of degradation of the individual tryptophanyl residues do reflect the local microenvironment.

The photolysis of tryptophan has also been used for the determination of tryptophan accessibility.[47] Ladner and coworkers[47] used the observations of Edwards and coworkers[48] to label surface tryptophan residues with halogenated compounds such as chloroform (Figure 12.1). As an example, Ladner and coworkers[47] irradiated a tripeptide, Lys-Trp-Lys, at 280 nm in 50 mM ammonium bicarbonate, pH 8.0, at 20°C in the presence of trichloroethanol obtaining the hydroxyethanone adduct. Reaction at tyrosine can also occur, but careful selection of the wavelength for irradiation can enhance specificity modification of tryptophan.[49] It is noted that Privat and Charlier[50] were the first to describe the photochemical modification of tryptophan in wheat-germ agglutinin with trichloroethanol.

Photolysis of tryptophan in the UVB range in the presence of nucleic acid resulted in a formation of a covalent complex between tryptophan and the polynucleotide.[51] There was dependence of product formation on the composition of polynucleotide with polyuridylic acid (polyrU) yielding the greatest amount of adduct, while the lowest amount of adduct formation was observed with polyguanylic acid (polyrG). There is a review containing information of the photochemical reaction of amino acids and nucleic acid components.[52] These two studies have focused on the reaction on free tryptophan (not in peptide linkage) and nucleic acids. There would appear wavelength dependence on the nature of the reaction of tryptophan with nucleic acid components. The use of UVB radiation promotes the formation of a covalent adduct between tryptophan and nucleic acid components. Tryptophan appears to protect DNA from photolytic damage with UVC light (100–280 nm) with little formation of a tryptophan–nucleic acid adduct.[53]

The determination of the extent of the modification of tryptophanyl residues in proteins independent of reagent is accomplished by MS.[54–64] Earlier work used amino acid analysis but special hydrolysis conditions were required.[65–69] Modification with alkylating reagents such as HNB and oxidation can be determined by UV–VIS spectroscopy. Oxidation has been used as technique to modify tryptophan in proteins for some 60 years.[70–79] A method based on the measurement of fluorescence of proteins in 6 M guanidine hydrochloride[80] has also been used to measure the oxidation of tryptophan in proteins.[81,82]

Treatment of tryptophan with various oxidizing agents results in the modification of the indole ring resulting a multiplicity of products (Figure 12.3).[54,60,78,83] NBS is a reagent used for the somewhat specific oxidation of tryptophan residues in proteins and is discussed in some detail in the following as are N-chlorosuccinimide and BPNS-skatole (2-(2′-nitrophenylsulfenyl)3-methyl-3-bromoindolenine).

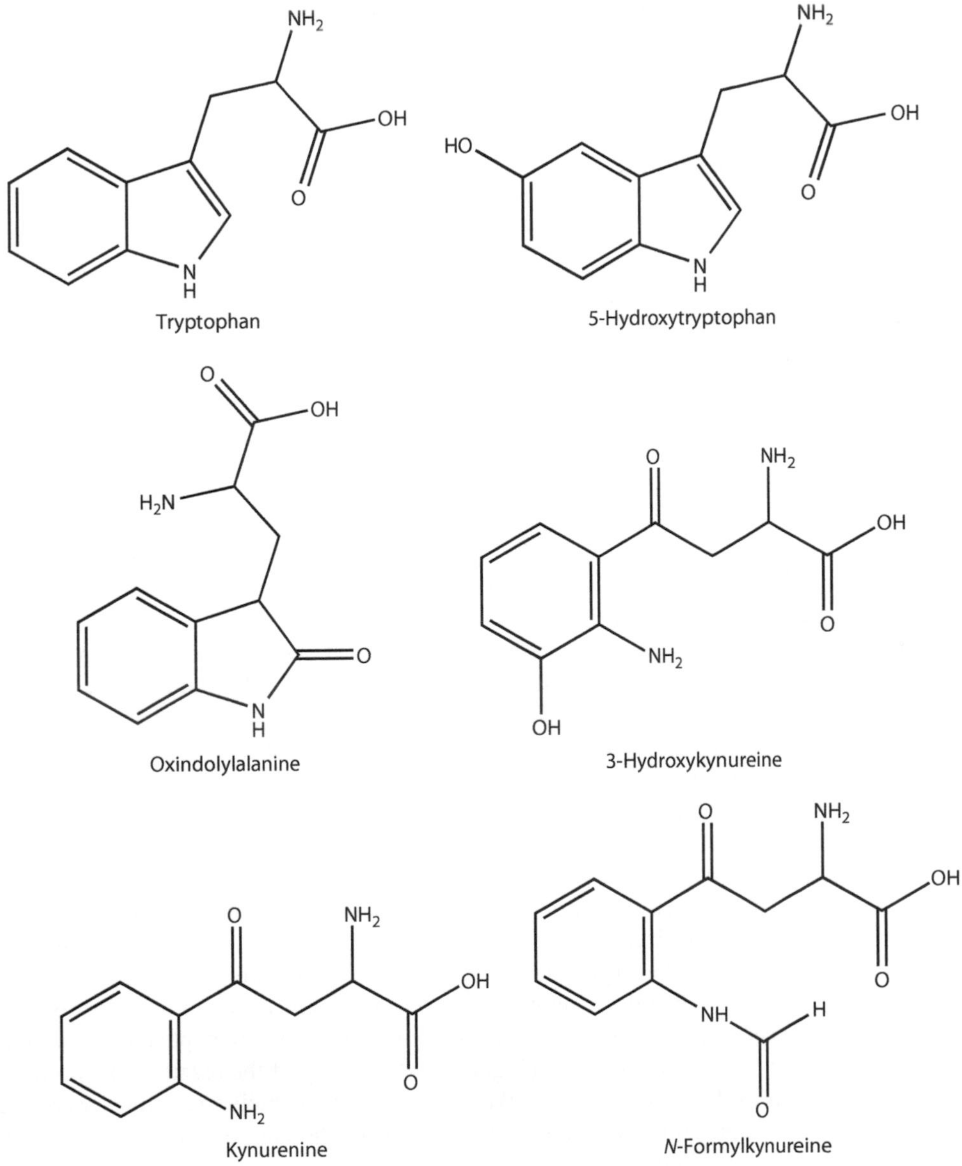

FIGURE 12.3 The structure of tryptophan and various tryptophan oxidation products (see Simat, T.J. and Steinhart, H., Oxidation of free tryptophan and tryptophan residues in peptides and proteins, *J. Agric. Food Chem.* 46, 490–498, 1998; Finley, E.L., Dillon, J., Crouch, R.K., and Schey, K.L., Identification of tryptophan oxidation products in bovine α-crystallin, *Protein Sci.* 7, 3291–2397, 1998).

Other reagents that have been demonstrated to oxidize tryptophan residues in proteins include iodine;[75,84–90] hydrogen peroxide in 0.5 M bicarbonate, pH 8.4 with 10% dioxane[76]; hydrogen peroxide/Fenton reagent ($FeCl_3$ + EDTA)[91,92]; chlorine dioxide[93]; tert-butylhydroperoxide[11]; hypohalous acids and derived haloamines[94,95]; and ferrous/EDTA/ascorbate.[96] It would appear that the use of the Fenton reagent provides the greatest extent of oxidation[92], but the modification of other amino acid such as methionine and cysteine is also subject to oxidation under these conditions. Oxidation by iodine of tryptophanyl residues in lysozyme provided important information, but there has been little use of iodine for the modification of tryptophan since 1974. It has been observed that the oxidation of tryptophanyl residues with iodine can result in peptide bond cleavage.[85,89]. The effect of hypohalous acids (hypochlorous acid, hypobromous acid) is quite variable with modest oxidation of tryptophan to hydroxytryptophan in one study[92] and a greater effect in another study.[94] This study[94] showed that the product of the reaction between hypobromous acid and taurine, taurine monobromamine, was much more potent than hypobromous acid in oxidizing tryptophan residues in hen egg white lysozyme. The potency of taurine monobromoamine as an oxidizing agent for tryptophan was demonstrated with human albumin in an earlier study.[95]

The oxidation of tryptophanyl residues in equine myoglobin by hypochlorous acid has been studied in some detail. Szuchman-Sapir and coworkers[97] observed that methionine and tryptophan were oxidized as low ratios (1–5 mol per mol protein) with methionine more susceptible to modification than tryptophan. Initial analysis of myoglobin modification by hypochlorous acid (prepared in solvent treated with Chelex® to remove any metal ions) was performed by amino acid analysis (hydrolysis in methanesulfonic acid). The identification of specific amino acid residues modified by hypochlorous acid was performed by LC/MS/MS. Met55 was oxidized before Met131; the modification of both methionyl residues preceded oxidation of either Trp7 or Trp14. Oxidation with PMA-activated neutrophils resulted in oxidation of the methionine residues but not of the tryptophanyl residues. It is suggested that the presence of other protein targets in the cell-based system resulted in the *consumption* of oxidant. A subsequent study by Szuchman-Sapir and coworkers[98] extended these observations with studies where the reaction of hypochlorous with an engineered recombinant human myoglobin where the cysteinyl residue at position 10 is replaced with alanine (C10A) is compared with the results with native human myoglobin. In the reaction of hypochlorous acid with the native human protein, cysteine was the most susceptible to oxidation by hypochlorous acid forming either cystine (a dimeric protein) or cysteic acid; cysteine sulfenic acid and/or cysteine sulfinic acid were not found as products. There was also oxidation of methionine; as with the equine protein, susceptibility of individual methionine residues varied somewhat on location of the methionine residue in the primary structure. Tryptophanyl residues in human myoglobin was far less susceptible to oxidation by hypochlorous acid than methionine; susceptibility of both methionine and tryptophan increased in the C110A variant. Under these reaction conditions, the order of susceptibility to oxidation with hypochlorous acid was Cys > Met > Trp; chlorination of tyrosine was observed after 24 h of reaction.

Hydrogen peroxide is known better for the conversion of methionine to methionine sulfoxide.[99–101] However, at alkaline pH, the oxidation of tryptophan by hydrogen peroxide is a favored reaction.[102] Cysteine is also oxidized by hydrogen peroxide resulting in the formation of cysteine sulfenic acid and cystine.[103–105] Peroxynitrite (see Chapter 4) is also known to oxidize tryptophanyl residues in proteins.[60,106] Hypothiocyanous acid is formed by the action of myeloperoxidase on thiocyanate and hydrogen peroxide. Hypothiocyanous acid reacts with tryptophan residues to form dioxindolylalanine and N-formylkynurenine (Figure 12.4). A study of the reaction of hypothiocyanous acid with protein by Hawkins and coworkers[107] showed that thiols in proteins were preferentially oxidized to unstable sulfenyl thiocyanate derivatives following oxidation of tryptophanyl residues. Hypothiocyanous acid oxidation of tryptophan in myoglobin occurred in the presence of glutathione (approximately fivefold molar excess) or ascorbate (25-fold molar excess). Hadfield and coworkers[108] reported that the action of hypothiocyanous acid on apolipoprotein A-1 was more specific than hypochlorous acid with only modification of tryptophanyl residues. There is modification of methionine with hypochlorous acid but not with hypothiocyanous acid. There is some difference

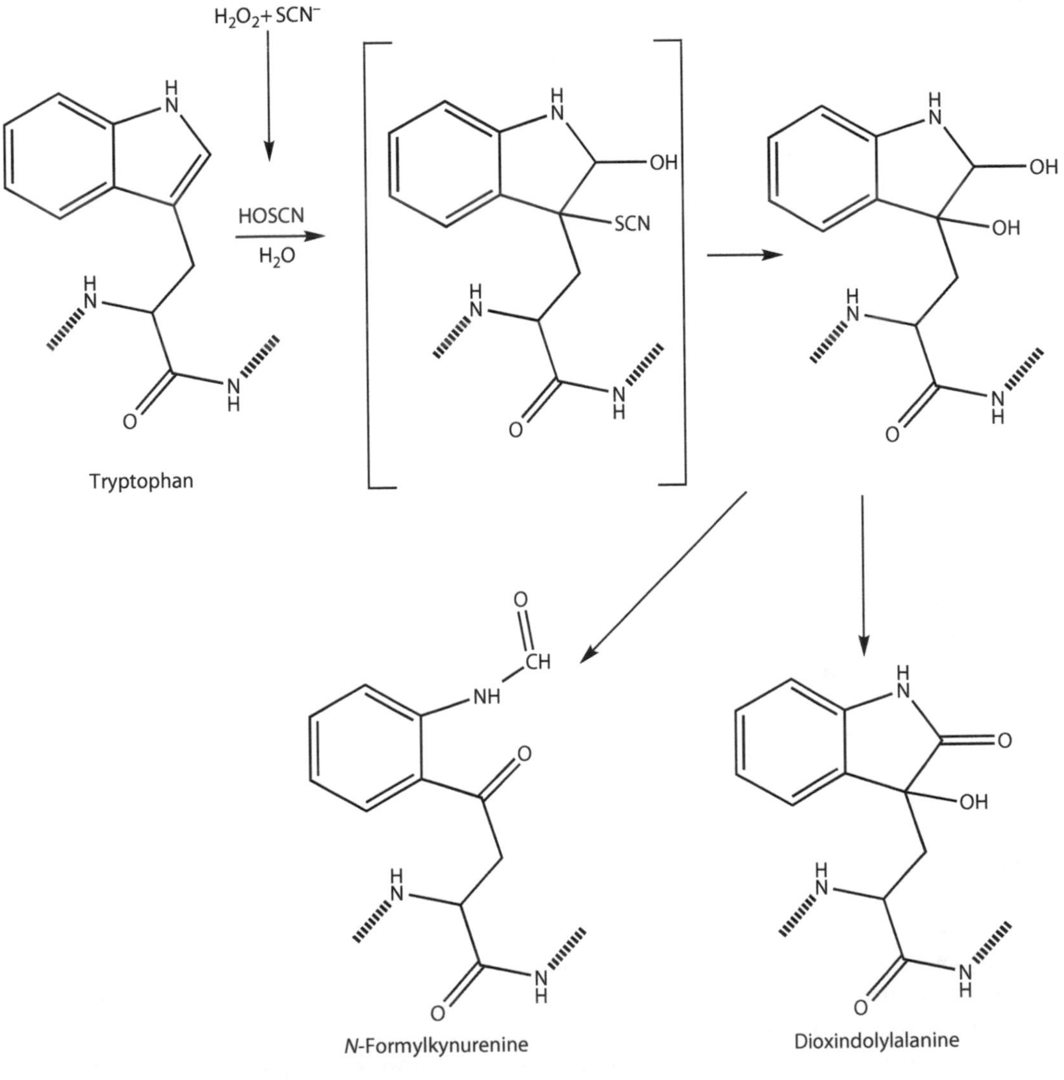

FIGURE 12.4 The reaction of hypothiocyanous acid with tryptophan. Hypothiocyanous is derived from the action of myeloperoxidase on hydrogen peroxide in the presence of chloride (Pattison, D.I., Davies, M.J., and Hawkins, C.L., Reaction and reactivity of myeloperoxidase-derived oxidants: Differential biological effects of hypochlorous and hypothiocyanous acids, *Free Radic. Res.* 46, 975–995, 2012). Shown is the reaction of hypothiocyanous acid with tryptophan to yield oxidized products (see Hawkins, C.L., Pattison, D.I., Stanley, N.R., and Davies, M.J., Tryptophan residues are targets in hypothiocyanous acid-mediated protein oxidation, *Biochem. J.* 416, 441–452, 2008).

in the susceptibility of the individual tryptophanyl residues (there are four tryptophan residues in apolipoprotein A-1) to modification with hypothiocyanous acid as compared to hypochlorous acid. Hadfield and coworkers[108] did observe that cyanate, a decomposition product from hypothiocyanous acid, reacts with the ε-amino group of lysine to form homocitrulline (Chapter 13).

The loss of tryptophan resulting from oxidation can be monitored by the methods described earlier including decrease in absorbance at 282 nm[76,79]; the difference in the molar extinction coefficient between tryptophan and the fully oxidized derivative is 3490 M^{-1} cm^{-1}. Reubsaet and colleagues[109] have reviewed the methods for the qualitative and quantitative analyses of tryptophan oxidation in peptides and proteins including UV spectroscopy, fluorescence, and HPLC analysis.

HPLC analysis of tryptophan oxidation products has been described.[83] Detail is provided for the separation of kynurenine, 5-hydroxytryptophan, tryptophan, and dioxindolealanine on a C_{18} column. The reader is directed to an excellent study by Mach and coworkers[110] for the extinction coefficients for tryptophan, tyrosine, and cystine in proteins. Fluorescence has also been used to measure tryptophan modification in proteins.[111] Reshetanyak and colleagues have reviewed the fluorescence properties of tryptophan in proteins.[112] While the previously mentioned studies have used hydrogen peroxide or other oxidizing agents, these reagents tend to be somewhat nonspecific. However, the power of MS for analysis makes hydrogen peroxide quite useful; also care needs to be taken with termination of reaction as, for example, modification continues in the frozen state.[113] Oxidation of proteins is a problem in the pharmaceutical industry, and various approaches are being developed to protect tryptophan, methionine, and histidine from oxidation.[114] Ji and coworkers[114] found that while free tryptophan protected the tryptophan in parathyroid hormone from oxidation with 2,2′-azobis(2-amidinopropane) dihydrochloride (a surrogate for polysorbate for generation of alkyl peroxides), only a mixture of methionine and tryptophan protected the tryptophan from modification with hydrogen peroxide or hydrogen peroxide plus ferric ions (Fenton oxidation) in addition to the alkyl peroxides.

Photooxidation is a problem for protein biologicals[115] that has been a subject for increased study.[116–121] In some of these studies, the observed photooxidation may reflect the action of a solvated electron from the photolysis of tryptophan as described earlier.[12,33] In other studies on biopharmaceutical products, tryptophan was the target for by photooxidation mediated by photosensitizers derived from histidine.[122] While unrelated to photooxidation, a recent study[123] showed that an organic peroxide derived from polysorbate 20 was responsible for the oxidation of a specific tryptophanyl residue in a Fab fragment in reaction involving His31. The histidine residue was suggested to bind a metal ion that catalyzed the reaction of the organic peroxide with the tryptophanyl residue; the tryptophanyl residue was refractory to oxidation by hydrogen peroxide.

The reaction of NBS with tryptophan (Figure 12.5) is the most common modification of tryptophanyl residues in proteins. This reagent was used earlier for quantitative analysis of tryptophan in proteins, but the development of hydrolysis methods obviated this necessity. NBS has been used for the modification of tryptophan residues in proteins. Work in this area prior to 2004 has been previously reviewed[124] and will only be discussed with respect to technique as relevant to current applications. The reaction of NBS with tryptophan yields the oxindole or dioxindole derivative, while other oxidizing agents yield kynurenine and other oxidation products.[125,126] Tryptophan oxidation to oxindolylalanine is also accomplished with hydrochloric acid in DMSO.[127–129] Huang and coworkers[128] demonstrated that it was possible to obtain cleavage of the oxidized peptides with cyanogen bromide; success does require technical expertise. Ohnishi and coworkers[130] provided a rigorous evaluation of the reaction of NBS with model tryptophanyl and tyrosyl compounds. At ratios of NBS to N-acetyltryptophan ethyl ester of greater than two in acetate buffer at pH 4.5, there is an apparent reversal of the decrease in absorbance at 280 nm. The maximal decrease in absorbance occurs at a ratio of NBS to tryptophan of two. If the data are obtained by stopped-flow spectroscopy, the molar excess of NBS does not have an effect on the maximum decrease observed, but when the spectrum is obtained 5 min after the initiation of the reaction, there is a decrease in the observed magnitude of change in absorbance at 280 nm. The evaluation of spectral changes in a protein can be further complicated by the reaction of NBS with tyrosine; however, the rate of tryptophan modification by NBS is approximately 10^3 faster than the modification of tyrosine. Daniel and Trowbridge[131] found that (at pH 4.0) the reaction of NBS with acetyl-L-tryptophan ethyl ester required 1.5 mol of NBS per mole of the acetyl-L-tryptophan ethyl ester, while trypsinogen required 2.0–2.3 mol NBS per mole of tryptophan oxidized, and trypsin required 1.5–2.0 mol NBS per mole of tryptophan oxidized. Sartin and coworkers[132] observed that the amount of NBS consumed in the modification of tryptophan in pancreatic DNAse is similar for the first two residues modified at either pH 4.0 or pH 5.5; above pH 5.5, more NBS is required for the modification of the third residue. At pH 5.5, the third residue modified is the most critical for enzyme activity, while at pH 4.0, the second residue modified is the most critical. These investigators also observed a discrepancy in the estimation of

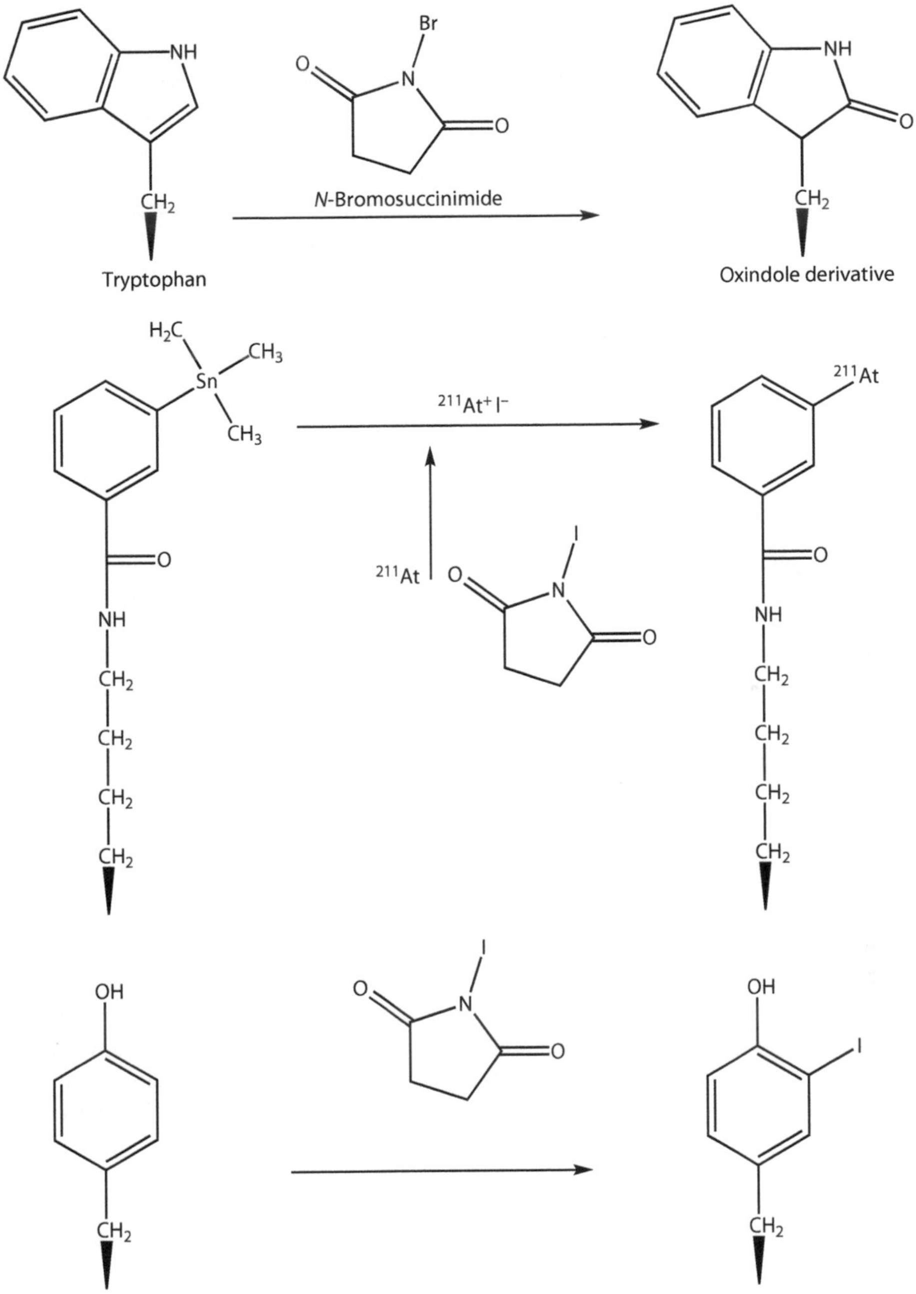

FIGURE 12.5 The reaction of NBS with tryptophan. Shown at the top is the reaction of NBS with tryptophan to produce the oxindole derivative (Spande, T.F. and Witkop, B., Determination of the tryptophan content of proteins with *N*-bromosuccinimide, *Methods Enzymol.* 11, 498–506, 1967). There is a decrease in absorbance at 278 nm associated with the conversion to the indole derivative. Shown below is the use of *N*-iodosuccinimide to oxidize astatine, which then substitutes an aromatic trimethylstannyl derivative to provide a radiolabeled derivative (Lindegren, S., Frost, S., Bäck, T. et al., Direct procedure for the production of 211At-labeled antibodies with ε-lysyl-3-(trimethylstannyl)benzamide immunoconjugate, *J. Nucl. Med.* 49, 1537–1545, 2008). Iodination of the 3(trimethylstannyl)benzamide immunoconjugate is accomplished by the use of NBS instead of *N*-iodosuccinimide. Shown at the bottom is the use of *N*-iodosuccinimide to iodiate tyrosyl residues in proteins (Liou, Y.-C., Davies, P.L., and Jia, Z., Crystallization and preliminary x-ray analysis of insect antifreeze protein from the beetle *Tenebrio molitor*, *Acta Crystallogr. D* 56, 354–356, 2000).

tryptophan modified between spectral analysis and amino acid analysis. However, other investigators[133] reported good agreement between spectral analysis and amino acid analysis (hydrolysis in 3.0 M toluenesulfonic acid) on the extent of modification of NBS modification of galactose oxidase. Freisheim and Huennekens[134] observed that the only tryptophan in dihydrofolate reductase reacts with NBS at pH 4.0, while at pH 6.0, a sulfhydryl group apparently is preferentially oxidized by the reagent prior to the reaction of tryptophan. Oxidation of sulfhydryl groups resulted in an increase in activity, while modification of the tryptophan residues leads to inactivation. Ohnishi and coworkers[135] followed up on their earlier study and used stopped-flow technology to study the modification of Trp82 in lysozyme with NBS. Modification occurred below pH 6 but not above pH 7. A similar study on the effect of pH on the modification of N-acetyl-L-tryptophan ethyl ester did not show dependence. These results suggest that, as with other functional groups in proteins, modification of tryptophan residues with NBS does depend on local electrostatic environment. It can be concluded that usually a twofold to fourfold molar excess of reagent is required for modification and specificity of modification may increase at mild acid pH (4.0).[136,137] Spande and coworkers[138] evaluated the effect of buffer, pH, and buffer concentration on the reaction of NBS with several proteins. In general, the oxidation reaction proceeds more effectively at lower pH (4.0 vs. 7.0) and at lower buffer concentration. Phosphate buffer was not as useful as acetate and later work by O'Gorman and Matthews[137] demonstrated an interaction with phosphate. It is noted that there is at least one example of the necessity for a large molar excess of NBS for the modification of proteins. Xue and coworkers[139] studied the reaction of NBS with tryptophanyl-tRNA synthetase (25 mM sodium acetate, pH 4.5, 22°C). This is an extremely interesting study as a large molar excess of NBS to enzyme (1500-fold) is required for complete inactivation. An even larger molar excess (10^5) of BPNS-skatole (3-bromo-3-methyl-3-(2-nitromercapto)-3H-indole) was required for inactivation. The stoichiometry of these reactions was not reported; however, intrinsic fluorescence measurement suggested modification of tryptophan residues by NBS. Tryptophan is, with tyrosine and phenylalanine, frequently described as an aromatic amino acid. While it is possible to modify tyrosine residues in proteins, the modification of phenylalanine is a lot more challenging. Human cytidine deaminase contains two conserved phenylalanine residues (Phe366 and Phe137) and a single tryptophan residue (Trp113). Vincenzetti and coworkers[140] prepared three mutant forms F36W/W113F, W113F, and F137W/W113F, each possessing a single Trp residue. Reaction with NBS suggested that Phe1377W was a surface residue as 1 mol of Trp was oxidized per mole of protein in the F137/W113F mutant; 0.4 mol of Trp per mole of protein was oxidized in the wild-type protein. There was no oxidation of Trp in the W113F mutant in either native (0.1 M citrate, pH 6.0) or denatured (0.1 M acetate, pH 4.0, with 8 M urea) protein; 1.14 mol of Trp per mole of protein was oxidized in the native protein under denaturing conditions, while 0.81 mol of Trp per mole of protein was oxidized in the F36W/W113F mutant protein. Munagala and coworkers[141] prepared a mutant form of a hypoxanthine–guanine–xanthine phosphoribosyltransferase (*Tritrichomonas foetus*) where Tyr74 was replaced with a tryptophan residue (T74W). These investigators were able to use the mutant as a probe for conformational change (intrinsic fluorescence). Reaction of the T74W mutant with NBS resulted in the loss of catalytic activity. These data allowed the investigators to suggest the importance of a flexible loop in catalysis by this enzyme. Tyagi and coworkers[142] introduced Trp residues into a region of human sodium/D-glucose transporter (hsGLT1) suggested to be important for biological function. These investigators had previously prepared a mutant of hsGLT1 where all of the tryptophanyl residues had been removed and replaced with phenylalanine.[143] Tyagi and coworkers[142] showed the Trp residues introduced at 457 and 460 are in contact with solvent based on NBS modification and intrinsic fluorescence measurements. Here, the insertion of tryptophanyl residues is used to map solvent exposure or accessibility in a manner similar to the use of cysteine[144,145] in the SCAM as described in Chapter 6. There is an earlier study by Takita and cowokers[114] that used mutation of tryptophanyl residues and modification with NBS to establish which Trp residue in a lysyl-tRNA synthetase from *Bacillus stearothermophilus* was buried (Trp314) and which was exposed to surface (Trp332). Of note in this study was the use of NBS in 0.1 M Tris, pH 8.0.

Some other selected examples of the application of NBS for the modification of tryptophanyl residues in proteins are presented in Table 12.1. In addition to some studies cited previously, there are several examples listed in Table 12.1 regarding the successful use of NBS for the selective modification of tryptophanyl residues in protein at or near neutral pH. Most investigators are able to obtain full modification of tryptophanyl residues in proteins at a molar ratio of 2–5 (NBS to protein). There are examples of the use of higher amounts of NBS as well as examples of tryptophanyl residues unreactive except in the presence of chaotropic agents.[146–151] There are also data[152] to suggest that the oxidation of *accessible* tryptophan residues with NBS results in denaturation of the protein with the concomitant oxidation of *buried* tryptophan residues by NBS. NBS is remarkably selective for tryptophan[153] although reaction with other amino acid residues has been reported. In the use of NBS for the determination of tryptophan concentration in proteins, possible reaction with tyrosine can present some difficulty; also proteins with substantial amount of tyrosine can provide challenges in the use of decreased absorbance at 278 nm for determination of the extent of modification.[154] Modification may occur at other amino acid residues including cysteine,[153,155] histidine,[153,155–158] tyrosine,[153,159,160] and methionine.[153,155,157,161,162]

There are two other *N*-halosuccinimide derivatives of interest: *N*-chlorosuccinimide and *N*-iodosuccinimide. Both *N*-chlorosuccinimide[163] and *N*-iodosuccinimide[164] have been used for the oxidation of thiol groups to disulfide bonds. *N*-Iodosuccinimide is also used for the oxidation of astatine for labeling antibody via electrophilic substitution of 3-(trimethylstannyl)benzoate.[165,166]

Peptide bond cleavage has been reported to occur during the reaction of NBS with tryptophanyl residues in proteins.[167] The peptide bond cleavage occurs by further oxidation of the oxindole derivative resulting in peptide bond cleavage (Figure 12.6). The example shown in Figure 12.6 depicts the use of *N*-chlorosuccinimide for the cleavage of tryptophanyl peptide bond as it may be more specific.[168] The cleavage of tyrosyl peptide bonds by NBS has also been reported.[169,170] *O*-Acetylation of tyrosine blocks the cleavage reagent, and this property has permitted the use of NBS for mapping of surface tyrosyl residues in ribonuclease.[171] The peptide bond cleavage by the *N*-halosuccinimides occurs at acid pH (0–5) and has not seen extensive citation in the literature. It is acknowledged that recombinant DNA technology and MS have largely supplanted the classical methods for the study of protein covalent structure. There has been somewhat more use of *N*-chlorosuccinimide[172–179] than either NBS[180] or *N*-iodosuccinimide.[168] Peptide bond cleavage is mentioned as potential side reaction when *N*-iodosuccinimide is used for protein iodination.[181] Cleavage of tryptophan peptide bonds following protein iodination has been described earlier.[85,89] The lactone derivative arising from the *N*-halosuccinimide cleavage of peptide bonds can be coupled to a solid-phase matrix (Figure 12.6).[182]

2-(2′-Nitrophenylsulfenyl)-3-methyl-3-bromoindolenine (BNPS-skatole; Figure 12.7) was developed by Omenn and coworkers[183] in 1970 as a reagent for the modification of the single tryptophan residues in staphylococcal nuclease. Studies with amino acids show that BNPS-skatole oxidized methionine to methionine sulfoxide, while NBS oxidized methionine to methionine sulfone; thioglycolic acid could reduce the methionine sulfoxide to methionine with amino acids but was less effective with intact protein. The reaction of staphylococcal nuclease with BPNS-skatole (ninefold molar excess, 80% acetic acid) resulted in the modification of the single tryptophan residue and oxidation of the four methionine residues; it had been demonstrated that a fivefold molar excess of BPNS-skatole was required for the oxidation of free tryptophan in 50% acetic acid at 23°C. Tyrosine and histidine were not modified by BNPS-skatole under these reaction conditions. Free cysteine was converted to cystine together with a small amount of cysteic acid (<10% conversion from cysteine). As with NBS, the oxidation of tryptophan with BNPS-skatole results in the oxindole with the decrease in absorbance at 278 nm. A 10-fold increase (50% acetic acid, 28 h, 23°C) in BNPS-skatole results in cleavage of the tryptophanyl peptide bond; the reaction of BNPS-skatole with bovine pancreatic ribonuclease, which does not contain tryptophan under the same conditions, did not result in peptide bond cleavage. There has been limited use of BNPS-skatole for the cleavage of proteins.[184–188] As an aside, skatole is considered to be the primary factor in the odor problems associated with pig farms.[189]

TABLE 12.1

Some Examples of the Use of NBS for the Modification of Proteins

Protein	Reaction Conditions and Comment	Molar Excess[a]	Reference
Trypsinogen	pH 4.0, pH-stat (KOH). A 1:1 ratio of NBS to trypsinogen resulted in less than 1 mol Trp modified per mole trypsinogen with no effect on activation; a 4:1 ratio of NBS to trypsinogen resulted in the modification of 2 mol Trp per mole trypsinogen with loss of ability to form trypsin.	1–4	12.1.1
Dihydrofolate reductase	0.1 M sodium phosphate, pH 6.0, 2 mol Trp oxidized per mole of protein, 0.1 M acetate, pH 4.0, 2.7 mol Trp oxidized per mole of protein, 0.13 M acetate–formate, pH 4.0 with 5.3 M urea (12-fold molar excess of NBS), 3.8 mol Trp oxidized per mole of protein. At pH 7.0 (0.1 M phosphate), the activation of enzyme activity was observed with a fivefold molar excess of reagent with inactivation at higher mole ratios of NBS. The activation of enzyme activity was shown to be due to the modification of a sulfhydryl group and activation was reduced/absent at lower pH.	15	12.1.2
Bovine pancreatic DNase	0.1 M sodium acetate, pH 3.0/4.0 containing 0.033 M $CaCl_2$, 25°C. Complete inactivation was observed at sixfold molar excess. Inactivation of 20%–30% was observed on the modification of a tryptophan residue with HNB.	6	12.1.3
Bovine pancreatic DNase	pH 4.0, 0.010 M $CaCl_2$. NBS was shown to modify three Trp residues, one of which is critical for enzymatic activity. At pH 4.0, the second residue modified as a function of NBS concentration is critical, while the third residue modified is critical for enzymatic activity. These investigators found that the decrease in absorbance at 280 nm is not an accurate measure of Trp modification.[b] It is noted that peptide bond cleavage was observed in this study.	1–6	12.1.4
Galactose oxidase	5 or 100 mM sodium acetate, pH 4.15. With the holoenzyme, inactivation is obtained in 5 mM acetate with modification of 2 of the 18 Trp residues in the protein. The modification of Trp residues was accomplished with a 3.5 M excess of NBS to Trp modified. An additional residue is available for modification in the holoenzyme in 100 mM sodium acetate buffer. There is 50% loss of activity in the apoenzyme with the modification of four Trp residues.	7	12.1.5
Bovine thrombin	0.1 M sodium acetate, pH 4.0, 23°C. One mole of tryptophan was modified with a twofold molar excess of reagent. Modification with NBS decreased subsequent reaction with HNB. A 50-fold molar excess resulted in the modification of 5 of the 8 tryptophan residues.	2	12.1.6
Papain	50 mM sodium acetate, pH 4.75, 25°C. A sixfold molar excess of NBS to protein resulted in the modification of one Trp and two to three Tyr residues. The active-site cysteine residue was protected with a mixed disulfide prior to reaction with NB.	6	12.1.7
Lac repressor protein	Reaction in 0.2 M potassium phosphate, pH 7.8, resulted in precipitation on the addition of NBS. Reaction of *lac* repressor protein in 1.0 M Tris HCl, pH 7.8, did not result in a precipitate but there was an increase in absorbance at 278 nM with increasing NBS concentration.[b] There was also modification of cysteine, tyrosine, and methionine.	5–10	12.1.8

TABLE 12.1 (continued)
Some Examples of the Use of NBS for the Modification of Proteins

Protein	Reaction Conditions and Comment	Molar Excess[a]	Reference
α-Amylase (*Bacillus subtilis*)	10 mM sodium phosphate, pH 7.0, 25°C. One mole of Trp residue per mole of protein is modified at a 10-fold excess of NBS to protein with four Trp residues (of a total of 11) at a 50-fold excess of reagent. Stopped-flow spectral analysis identified one fast-reacting residue. This study is somewhat unusual in that the modification reaction was performed at pH 7.0.	1–50	12.1.9
Dihydrofolate reductase	0.015 M bis tris HCl, pH 6.5 with 0.5 M KCl. 0°C. Inactivation of the enzyme is observed with an approximately sixfold molar excess of NBS; inactivation was associated with the modification of 1.1 mol Trp per mole protein. As NBS concentration was increased, turbidity was observed with an increase in absorbance at 278 nm.	1–6	12.1.10
Human serum vitamin D–binding protein	20 mM sodium acetate, pH 4.0, 25°C. Loss of binding activity (25-hydroxy vitamin D_3) at molar ratio of 4:1 (NBS to protein). Loss of tryptophan and binding activity on reaction with NBS was prevented by the presence of 25-hydroxy vitamin D_3 ($K_d \approx 10^{-10}$ M).	1–10	12.1.11
Ferredoxin–NADP$^+$ oxidoreductase	250 mM potassium phosphate, pH 7.7, dark, 4°C. Approximately 1 mol of Trp per mole of protein resulted in loss of 85% activity (Ferredoxin–NADP$^+$ oxidoreductase contains six Trp residues). A second mole of Trp was modified with a 100-fold molar excess of NBS. The reaction was time dependent with completion after 3 h of incubation. No loss of tyrosine or cysteine was detected under these reaction conditions (the cysteine residues are *buried* and available only on denaturation).	5	12.1.12
Alpha-amylase inhibitor (*Phaseolus vulgaris*)	100 mM sodium acetate, 50 mM NaCl, 1.0 mM $CaCl_2$, pH 5.0. Two moles of Trp was modified at a molar excess of 8:1 (NBS to protein). The modified residues were identified by MS.[c]		12.1.13
E. coli α-hemolysin	20 mM Tris HCl, pH 6.5. Measurement of intrinsic fluorescence shows the modification of 1 mol of Trp (there are four Trp residues in the α-hemolysin). It is noted that the reaction is performed near to neutral pH with a low molar excess of NBS.	0.25–2	12.1.14
Adenosine deaminase (Bovine spleen and bovine brain gray matter)	10 mM phosphate, pH 7.4. Four moles of Trp is modified with a 50-fold molar excess of reagent. There was no observed modification of tyrosine or cysteine under these reaction conditions.	0–50	12.1.15

[a] Molar ratio of reagent to protein unless otherwise indicated.

[b] It has been suggested that the use of decreased absorbance at 280 on modification with NBS is complication by high amounts of tyrosine and phenylalanine in the modified protein (Divita, G., Jauli, J.-M., Gautheron, D.C., and Di Pietro, A., Chemical modification of α-subunit tryptophan residues in *Schizosaccharomyces pombe* mitochondrial F_1 adenosine 5'-triphosphatase: Differential reactivity and role in activity, *Biochemistry* 32, 1017–1024, 1993).

[c] MS has added to the various analytical methods described in this table (Mendoza, V.L. and Vachet, R.W., Probing protein structure by amino acid-specific covalent labeling and mass spectrometry, *Mass Spectrom. Rev.* 28, 785–815, 2009).

(*continued*)

TABLE 12.1 (continued)

Some Examples of the Use of NBS for the Modification of Proteins

References to Table 12.1

12.1.1.   Daniel, V.W., III and Trowbridge, C.G., The effect of *N*-bromosuccinimide upon trypsinogen activation and trypsin catalysis, *Arch. Biochem. Biophys.* 134, 506–514, 1969.

12.1.2.   Freisheim, J.H. and Huennekens, F.M., Effect of *N*-bromosuccinimide on dihydrofolate reductase, *Biochemistry* 8, 2271–2276, 1969.

12.1.3.   Poulos, T.L. and Price, P.A., The identification of a tryptophan residue essential to the catalytic activity of bovine pancreatic deoxyribonuclease, *J. Biol. Chem.* 246, 4041–4044, 1971.

12.1.4.   Sartin, J.L., Hugli, T.E., and Liao, T.-H., Reactivity of the tryptophan residues in bovine pancreatic deoxyribonuclease with *N*-bromosuccinimide, *J. Biol. Chem.* 255, 8633–8637, 1980.

12.1.5.   Kosman, D.J., Ettinger, M.J., Bereman, R.D., and Giordano, R.S., Role of tryptophan in the spectral and catalytic properties of the copper enzyme, galactose oxidase, *Biochemistry* 16, 1597–1607, 1977.

12.1.6.   Uhteg, L.C. and Lundblad, R.L., The modification of tryptophan in bovine thrombin, *Biochim. Biophys. Acta* 491, 551–557, 1977.

12.1.7.   Glick, B.R. and Brubacher, L.S., The chemical and kinetic consequences of the modification of papain by *N*-bromosuccinimide, *Can. J. Biochem.* 55, 424–432, 1977.

12.1.8.   O'Gorman, R.B. and Matthews, K.S., *N*-bromosuccinimide modification of *lac* repressor protein, *J. Biol. Chem.* 252, 3565–3571, 1977.

12.1.9.   Fujimori, H., Ohnishi, M., and Hiromi, K., Tryptophan residues of saccharifying α-amylase from *Bacillus subtilis*. A kinetic discrimination of states of tryptophan residues using *N*-bromosuccinimide, *J. Biochem.* 83, 1503–1510, 1978.

12.1.10.   Thomson, J.W., Roberts, G.C.K., and Burgen, A.S.V., The effects of modification with *N*-bromosuccinimide on the binding of ligands to dihydrofolate reductase, *Biochem. J.* 187, 501–506, 1980.

12.1.11.   Swamy, N., Brisson, M., and Ray, R., Trp-145 is essential for binding of 25-hydroxyvitamin D to human serum vitamin D-binding protein. *J. Biol. Chem.* 270, 2636–2639, 1995.

12.1.12.   Hirasawa, M., Kleis-SanFrancisco, S., Proske, P.A., and Knaff, D.B., The effects of *N*-bromosuccinimide on ferredoxin: NADP+ oxidoreductase, *Arch. Biochem. Biophys.* 320, 280–289, 1995.

12.1.13.   Takahashi, T. et al., Identification of essential amino acid residues of an alpha-amylase inhibitor from *Phaseolus vulgaris* white kidney beans, *J. Biochem.* 126, 838–944, 1999.

12.1.14.   Verza, G. and Bakás, L., Location of tryptophan residues in free and membrane bound *Escherichia coli* α-hemolysin and their role on the lytic membrane properties, *Biochim. Biophys. Acta* 1464, 27–34, 2000.

12.1.15.   Mordanyan, S., Sharoyan, S., Antonyan, A. et al., Tryptophan environment in adenosine deaminase. I. Enzyme modification with *N*-bromosuccinimide in the presence of adenosine and EHNA analogues. *Biochim. Biophys. Acta* 1546, 185–195, 2001.

The conversion of tryptophanyl residues to 1-formyltryptophanyl (*N*-formyl) residues (Figure 12.8) has been reported. The reaction conditions are somewhat harsh (HCl in formic acid), but the procedure is reversible (in mild base) and it may be more useful for small peptides and as a protecting group in peptide synthesis.[190] Coletti-Previero and coworkers[191] have successfully applied this procedure to bovine pancreatic trypsin. Trypsin was dissolved in formic acid saturated with HCl at a concentration of 2.5 mg mL^{-1} at 20°C. The formylation reaction is associated with an increase in absorbance at 298 nm.[192] Therefore, it is possible to follow the reaction spectrophotometrically. The reaction is judged complete when there is no further increase in absorbance at 298 nm. The aforementioned reaction with trypsin was complete after an incubation period of 1 h. The solvent was partially removed in vacuo over KOH pellets followed by lyophilization. The formyltryptophan derivative is unstable at alkaline pH. At pH 9.5 (pH-stat), conversion back to tryptophan is complete after 200 min incubation at 20°C. Cooper and coworkers[193] reported that the *N*-formylation of tryptophanyl residues in proteins in cytochrome c. The single tryptophanyl residue was formylated with formic acid saturated with HCl. The modified protein had markedly reduced affinity for a monoclonal antibody resulting from local conformational change. Also shown in Figure 12.8 is the acylation or alkylation of N^1 position of

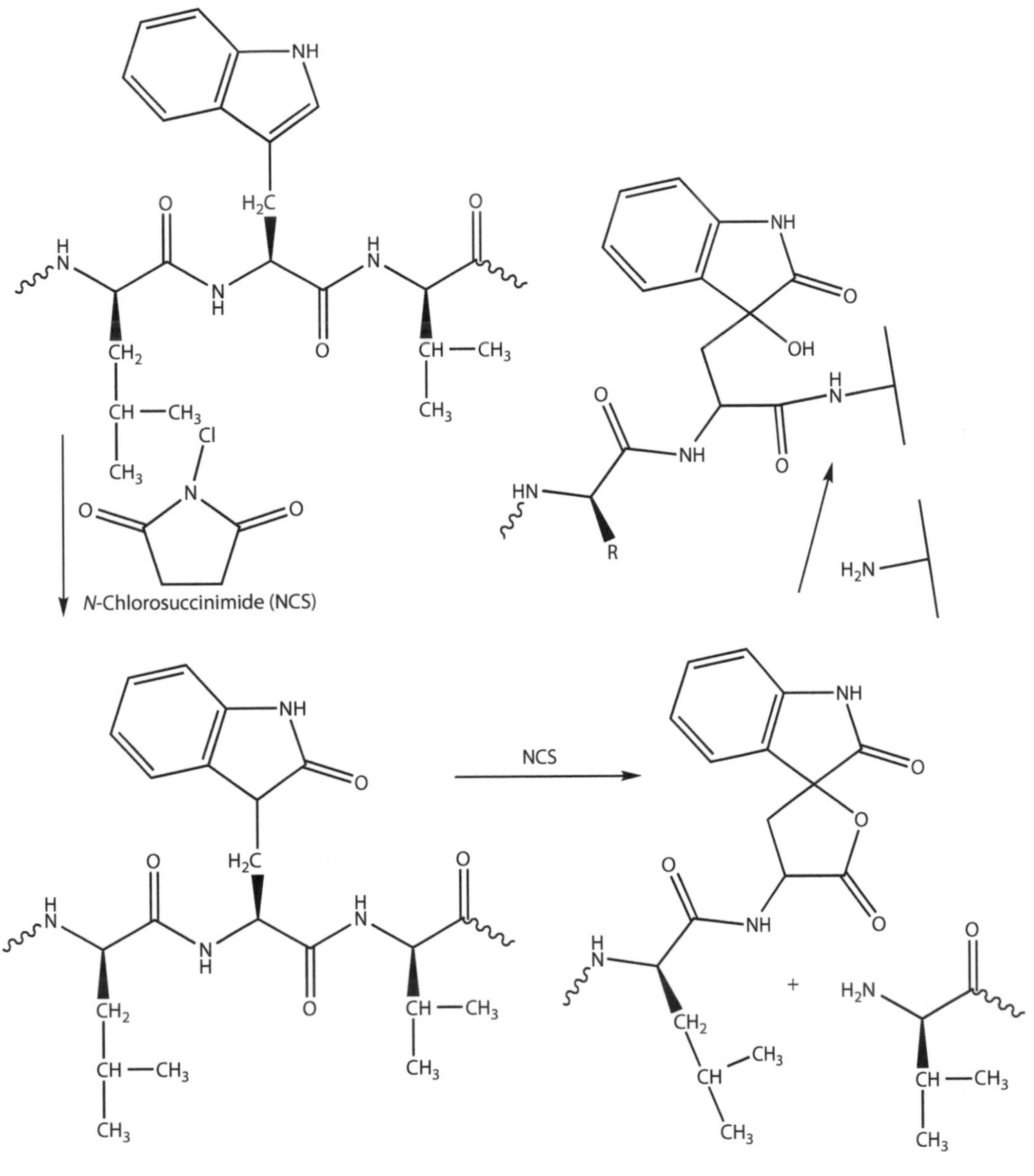

FIGURE 12.6 Peptide bond cleavage with *N*-halosuccinimides. Shown is the cleavage of tripeptide, Leu-Trp-Val by *N*-chlorosuccinimide with the formation of a lactone derivative (Lischwe, M.A. and Sung, M.T., Use of *N*-chlorosuccinimide/urea for the selective cleavage of tryptophanyl peptide bonds in proteins, *J. Biol. Chem.* 242, 4976–4980, 1972). Shown on right is the coupling of the lactone derivative to a matrix-attached amino function (Wachter, E. and Werhahn, R., Attachment of tryptophanyl peptides to 3-aminopropyl-glass suited for subsequent solid-phase Edman degradation, *Anal. Biochem.* 97, 56–64, 1979).

the tryptophan indole. As with formylation, these reactions occur under drastic (for proteins) conditions and are unlikely to be of importance for the modification of proteins. Nevertheless, these reactions do raise the potential of such modifications occurring in proteins.

HNB (Figure 12.9), frequently referred to as Koshland's reagent, was introduced by Koshland and coworkers[194,195] for the analysis and modification of tryptophanyl residues in proteins. Under appropriate reaction conditions (pH 4.0 or below), the reagent appears to be highly specific for reaction with

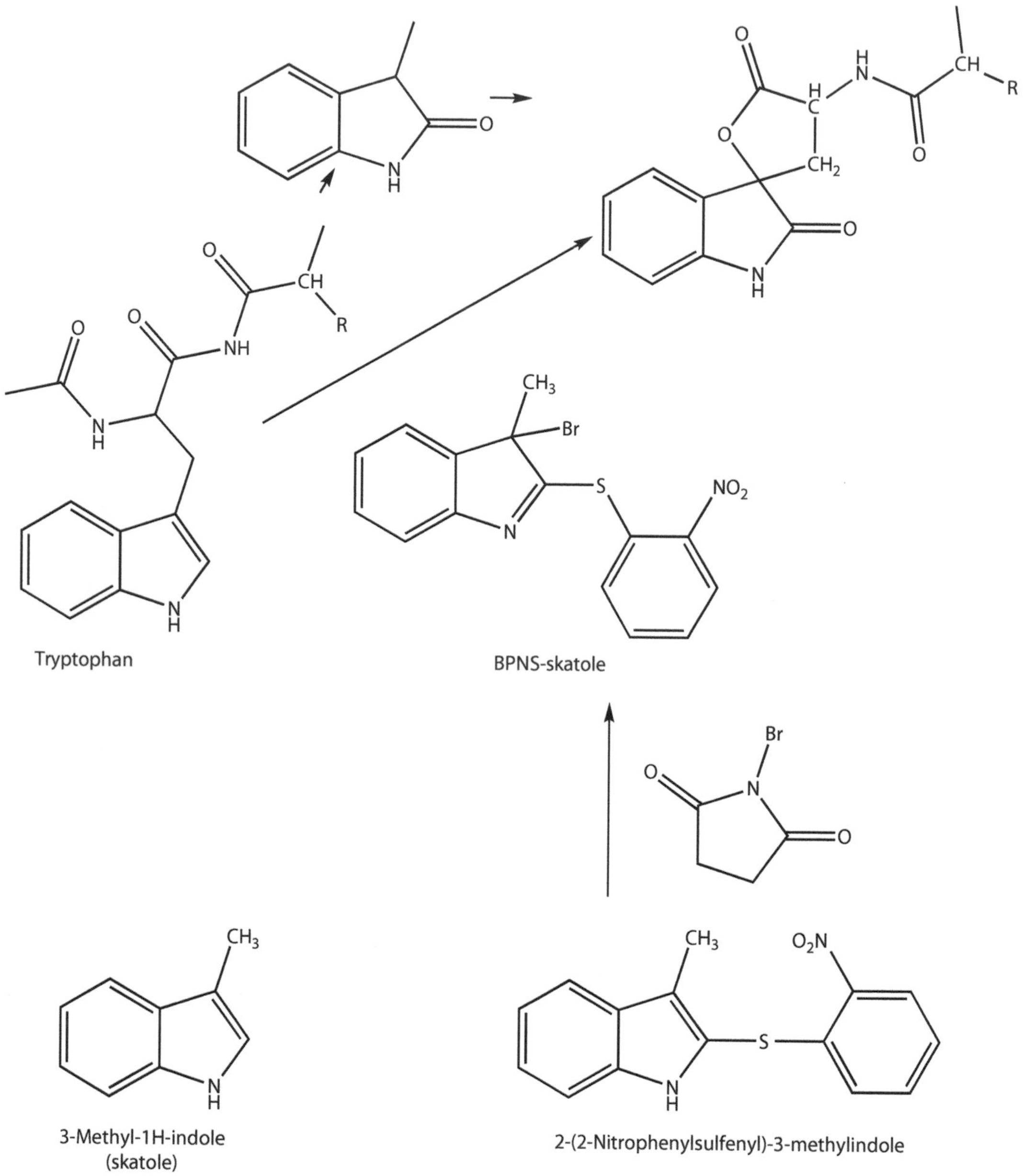

Tryptophan

BPNS-skatole

3-Methyl-1H-indole
(skatole)

2-(2-Nitrophenylsulfenyl)-3-methylindole

FIGURE 12.7   The reaction of BPNS-skatole with tryptophanyl residues in proteins. BPNS-skatole can be used to cleave peptide bonds in proteins (Halliday, J.A., Bell, K., and Shaw, D.C., The complete amino acid sequence of feline β-lactoglobulin II and a partial revision of the equine β-lactoglobulin II sequence, *Biochim. Biophys. Acta* 1077, 25–30, 1991). Also shown is the synthesis of BPNS-skatole (Fontana, A., Modification of tryptophan with BPNS-skatole (2-(2-nitrophenylsulfenyl)-3-methyl-3-bromoindolenine), *Methods Enzymol.* 45, 419–423, 1972).

tryptophanyl residues in proteins; this reagent also has the advantage of being a *reporter* group in the sense that the spectrum of the hydroxynitrobenzyl derivative is sensitive to changes in the microenvironment. This decrease observed in absorbance at 410 nm associated with an increase in absorbance at 320 nm upon the addition of dioxane is similar to that seen with acidification and reflects the increase in the pKa of the phenolic hydroxyl group. Titration curves of oxidized and reduced laccase[196]

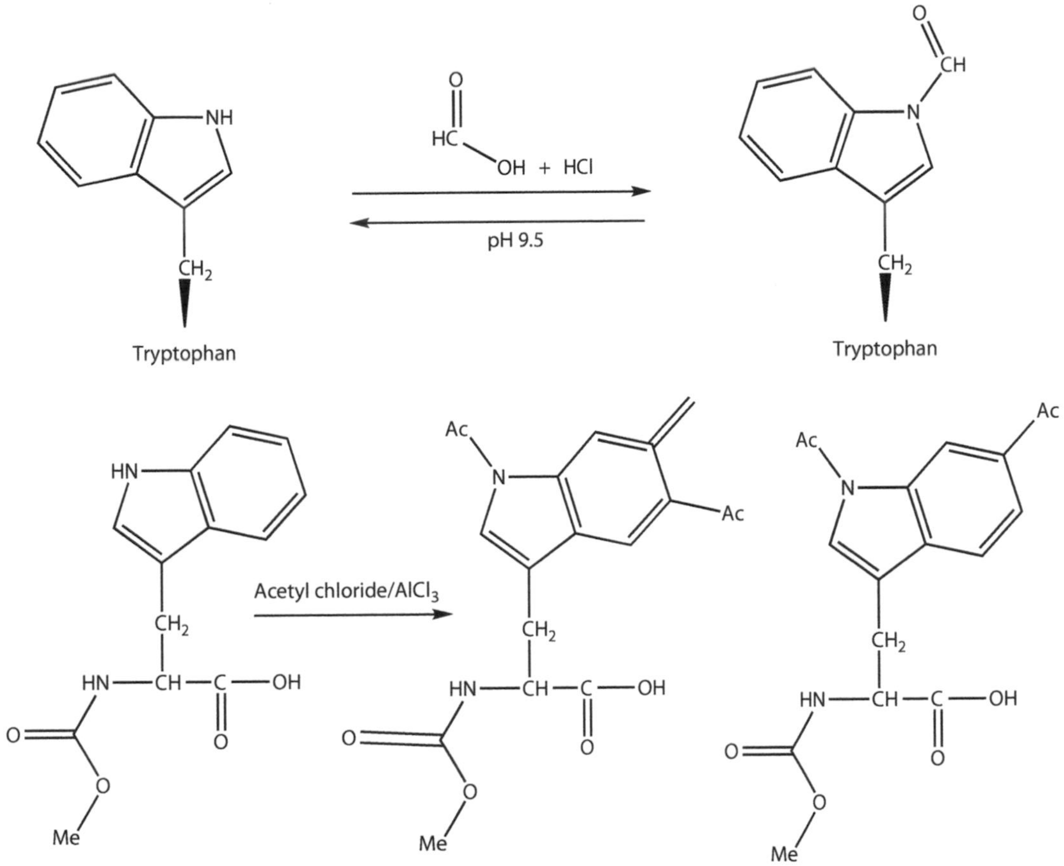

FIGURE 12.8 The formylation of tryptophan with formic acid. Shown is the formylation of tryptophan with HCl/HCOOH resulting in the formation of *N*-formyl tryptophan (Previero, A., Coletti-Previero, M.A., and Cavadore, J.C., Reversible chemical modification of the tryptophan residue, *Biochim. Biophys. Acta* 147, 453–461, 1967). The *N*-formyl group is removed by mild base (Odagami, T., Tsuda, Y., Kogami, Y. et al., Deprotection of the indole (*N*ind)-formyl group on tryptophan employing a new reagent, *N,N'*-dimethylethylenediamine (DMEDA) in an aqueous solution, *Chem. Pharm. Bull.* 57, 211–213, 2009). Shown below is the reaction of acetyl chloride with a tryptophan derivative to yield an *N*-acetyl derivative (and five or six acetyl products via Friedel–Crafts reaction) (Jiang, Y. and Ma, D., Regioselective acylation at the 5- or 6-position of L-tryptophan derivatives, *Tetrahedron Lett.* 43, 7013–7015, 2002). The selective acylation or alkylation of the *N*¹-position of tryptophan is also accomplished with alkanoic acid chloride (prepared by reaction with thionyl chloride) in the presence of 4-dimethylaminopyridine in methylene chloride (Beemelmanns, C., Lentz, D., and Reissig, H.-U., Samarium diiodide induced cyclizations of γ-, δ-, ε-indoyl ketones reductive coupling, intermolecular trapping, and subsequent transformations of indolines, *Chemistry* 17, 9720–9730, 2011).

that had been modified with HNB suggested that the residues modified with HNB are in an essentially aqueous microenvironment. This study provides titration curves for free HNB and HNB bound to laccase. Titration curves were based on A_{410} and provided a pKa of 6.83 for free HNB and 7.2 in laccase with 0.38 mol HNB, suggesting the microenvironment was slightly more hydrophobic. Figure 12.9 would suggest that there are a limited number of products obtained from the reaction of HNB with tryptophan in proteins but this is likely incorrect. Disubstitution on the indole ring occurs and is seen as a sudden *break* in the plot of extent of modification versus reagent excess.[197] These investigators used a 20-fold molar excess of reagent for the determination of tryptophan concentration in proteins as this concentration corresponded to the *breakpoint* on the plot. The reaction of HNB with protein occurs rapidly (less than a minute). Horton and Koshland[195] did observe the modification of cysteine

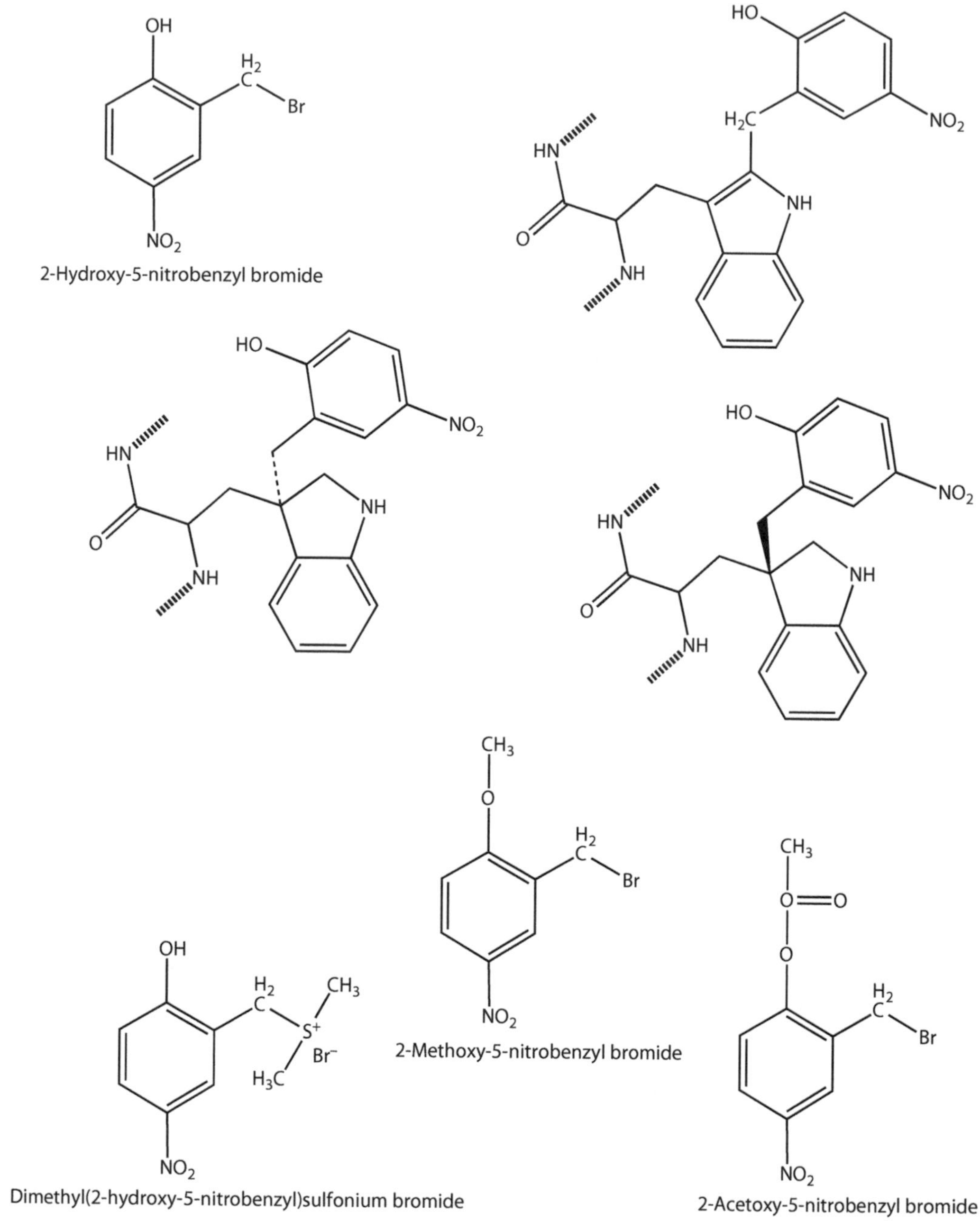

2-Hydroxy-5-nitrobenzyl bromide

Dimethyl(2-hydroxy-5-nitrobenzyl)sulfonium bromide

2-Methoxy-5-nitrobenzyl bromide

2-Acetoxy-5-nitrobenzyl bromide

FIGURE 12.9 The reaction of tryptophan with HNB. A variety of products can be obtained including substitution at the 2-carbon and diastereoisomers obtained from substitution at the 1-carbon (see Loudon, G.M. and Koshland, D.E., Jr., The chemistry of a reporter group 2-hydroxy-5-nitrobenzyl bromide, *J. Biol. Chem.* 245, 2247–2254, 1975; Strohalm, N., Kodíček, N., and Pechar, M., Tryptophan modification by 2-hydroxy-5-nitrobenzyl bromide studied by MALDI-TOF mass spectrometry, *Biochem. Biophys. Res. Commun.* 312, 811–816, 2003). Derivative forms include dimethyl(2-hydroxy-5-nitrobenzyl)sulfonium chloride (Horton, H.R. and Tucker, W.P., Dimethyl (2-hydroxy-5-nitrobenzyl)sulfonium salts, *J. Biol. Chem.* 245, 3397–3401, 1970), 2-methoxy-5-nitrobenzyl bromide (Horton, H.R., Kelly, H., and Koshland, D.E., Jr., Environmentally sensitive protein reagents. 2-Methoxy-5-nitrobenzyl bromide, *J. Biol. Chem.* 240, 722–724, 1965), and 2-acetoxy-5-nitrobenzyl chloride (Horton, H.R. and Young, G., 2-Acetoxy-5-nitrobenzyl chloride. A reagent designed to introduce a reporter group near the active site of chymotrypsin, *Biochim. Biophys. Acta* 194, 272–278, 1969).

with HNB with a rate considerably slower than tryptophan. There are mixed data as to whether HNB reacts with cysteine residues in proteins. Novak and colleagues[198] reported the inactivation of *E. coli* core RNA polymerase with HNB with data suggesting the modification of both tryptophan and cysteine. Baracca and coworkers[199] did not observe the inactivation of mitochondrial F1-ATPase with cysteine residues when inactivation was accomplished with HNB. Modification at serine or threonine was not observed by Barman and Koshland[197] nor was there reaction with carbohydrate. Modification at histidine by HNB was reported by Barman.[200] Lundblad and Noyes[201] analyzed the product of the reaction of HNB with a synthetic peptide (EAE peptide, FSWGAEGQR) that contained a single tryptophanyl residue. Chromatography on a C_8 demonstrated the presence of several products, some of which resulted from multiple substitution on the indole rings and others that were suggested to isomers and diastereoisomers. Diastereoisomers (Figure 12.9) resulting from the reaction of HNB with tryptophan compounds have been described by Loudon and Koshland.[202]

Strohalm and coworkers[203] used MALDI-TOF MS to characterize the products derived from the modification of tryptophan in a model peptide (GEGKGWGEGK) with HNB. A total of five products were obtained that reflected qualitative and quantitative differences in the substitution at the single tryptophan residue. The effect of HNB was evaluated over a 1–200 M excess range. The degree of modification increased as the concentration of reagent increased. It is observed that a disubstituted product could be obtained at an equimolar excess of reagent, while at a 200-fold molar excess, the most abundant product was a trisubstituted derivative; five different products were detected. The extent of modification increased with increasing pH with no major change in product distribution. This group[57] reported the use of HNB as reagent for determining the surface accessibility of tryptophan in proteins. Studies with lysozyme, cytochrome *c*, and myoglobin demonstrated the HNB modified *surface-accessible* residues and not *buried* residues. Denaturation of the proteins permitted modification of all tryptophanyl residues in the three proteins.

The issue of solution stability was mentioned previously; this difficulty is avoided and the characteristics of the reaction are preserved by the use of the dimethyl sulfonium salts (Figure 12.9). Dimethyl(2-hydroxy-5-nitrobenzyl)sulfonium chloride was obtained from the reaction of 2-hydroxy-5-nitrobenzyl chloride with dimethyl sulfide.[204,205] This water-soluble sulfonium salt derivative was used to modify tryptophan in rabbit skeletal myosin subfragment-1.[206] Purification of peptides containing modified tryptophanyl residues was achieved by immunoaffinity chromatography using rabbit antibody to bovine serum albumin previously modified with dimethyl-(2-hydroxy-5-nitrobenzyl) sulfonium chloride. This allowed the identification of the most rapidly reacting residue. 2-Methoxy-5-nitrobenzyl bromide (Figure 12.9) was synthesized by Horton and coworkers.[207] As would be expected, the spectra of this derivative were not sensitive to pH but were sensitive to solvent polarity. 2-Methoxy-5-nitrobenzyl bromide modified both methionine and tryptophan at similar rates, both much slower than the reaction of tryptophan with HNB. While HNB reacts with tryptophan in seconds, the reaction with tryptophan was 74% complete after 80 h of reaction (pH 5, 25°C, 10-fold molar excess of reagent). Some modification of cysteine was detected, but analysis was complicated by the competing oxidation of cysteine to cystine; the modification of other amino acids was not observed by these investigators. The lack of reactivity of the 2-methoxy-5-nitrobenzyl bromide has been used to develop a derivative that generates the active form of HNB at the enzyme active site. Horton and Young[208] prepared 2-acetoxy-5-nitrobenzyl bromide (Figure 12.9). This derivative, like the methoxy derivative, is essentially unreactive. There is a considerable structural identity between 2-acetoxy-5-nitrobenzyl bromide and *p*-nitrophenyl acetate, which is a nonspecific substrate for chymotrypsin. α-Chymotrypsin removes the acetyl group from 2-acetoxy-5-nitrobenzyl bromide, thus generating HNB at the active site, which then either rapidly reacts with a neighboring nucleophile or undergoes hydrolysis. Uhteg and Lundblad[209] have used both the acetoxy and butyroxy derivatives in the study of thrombin. Some selected examples of the use of HNB are shown in Table 12.2.

Reagents with reaction characteristics similar to HNB are the several nitrophenylsulfenyl derivatives[210] (Figure 12.10). The reaction product resulting from the sulfonylation of lysozyme[211] with 2-nitrophenylsulfenyl chloride (40-fold molar excess, pH 3.5, 0.1 M sodium acetate) has spectral

TABLE 12.2

Some Examples of the Use of HNB and Related Derivatives to Modify Proteins

Protein	Reaction Conditions and Comment	HNB/P[a]	Reference
Pepsin	0.1 M NaCl.[b] The modification of native pepsin at pH 3.5 resulted in the modification of approximately 2 mol Trp per mole protein; approximately 3 mol Trp per mole protein was modified in alkaline urea-denatured pepsin, while approximately 4 mol Trp modified per mole protein was found in reduced, carboxymethylated pepsin. Three moles of Trp per mole protein was modified in alkaline-denatured pepsinogen with 4 mol Trp modified in reduced, carboxymethylated pepsinogen. As a note, the modified pepsin retained 70% of enzymatic activity.	300[c]	12.2.1
Streptococcal proteinase	0.46 M sodium phosphate, pH 3.1, 23°C. The zymogen form of streptococcal proteinase was not modified with HNB. The modification of the active form of the enzyme with HNB resulted in the incorporation of 2 mol of reagent per mole protein. Analysis showed that this was the result of disubstitution of a single Trp residue.[d]	200	12.2.2
Pancreatic deoxyribonuclease	0.050 M CaCl$_2$, 23°C.[e] One mole of HNB is incorporated into DNase under these reaction conditions; a 200 M excess of reagent did not increase the amount of reagent incorporated into the protein; there was a maximum of 30% loss of enzyme activity resulted from reaction with HNB. Reaction of the modified protein with NBS resulted in the modification of a second mole of Trp with total loss of enzymatic activity.	100	12.2.3
Bovine and human carbonic anhydrase	0.1 M phosphate, pH 6.8, 23°C, 10 min.[f] There was little modification of the native bovine enzyme (0.3 HNB incorporated per mole protein); 5.4 mol HNB per mole of protein was incorporated in the presence of 8 M urea. The reaction of human carbonic anhydrase B with a 100-fold molar excess of reagent resulted in the incorporation of 0.8 mol HNB per mole protein; the addition of a second 100-fold excess of reagent resulted in the incorporation of 1.5 mol HNB per mole of protein. Human carbonic anhydrase C showed behavior similar to the bovine protein.	100	12.2.4
Glyceraldehyde-3-phosphate dehydrogenase	pH 6.75 (maintained by addition of 0.1 M NaOH), 23°C. The active-site sulfhydryl residue was protected by modification with 5,5′-dithiobis-2-(dinitrobenzoate) prior to reaction with dimethyl(2-hydroxynitrobenzyl)sulfonium bromide.[g] TRP193 was established as the site of modification with mono- and disubstitution.	30	12.2.5
Thrombin	0.2 M acetate, pH 4.0, 23°C. One mole of HNB incorporated per mole of protein. Reaction with 2-acetoxy-5-nitrobenzyl bromide[h] resulted in modification similar to that observed with HNB.	100	12.2.6
Laccase (*Rhus vernicifera*)	pH 3.30[i] (maintained by titration with NaOH), 23°C.	110	12.2.7
Laccase (*Rhus vernicifera*)	pH 3.30[i] (maintained by titration with NaOH), 33°C.	110	12.2.7
Human serum albumin	10 M urea, pH 4.4, 23°C. 1.1 mol HNB per mole of protein was observed at a 1000-fold molar excess; 1.3 mol HNB per mole of protein was observed at a 3000-fold molar excess. Lower urea concentrations and higher pH resulted in less specific modification of albumin.	1000[j]	12.2.8

TABLE 12.2 (continued)
Some Examples of the Use of HNB and Related Derivatives to Modify Proteins

Protein	Reaction Conditions and Comment	HNB/P[a]	Reference
Winged bean basic lectin (*Psophocarpus tetragonolobus*)	0.1 M sodium citrate, pH 3.1, 23°C, 30 min.[k] 0.9 mol HNB were incorporated into a 100-fold molar excess of reagent, 1.7 mol of HNB at 200-fold excess and 1.8 mol HNB at a 400-fold molar excess. There are four Trp residues in winged bean lectin. A 20-fold molar excess of NBS oxidized 2 mol of Trp per mole of protein in 0.1 M citrate, pH 6.1; all four Trp residues were oxidized by NBS in 0.1 M HOAc/8 M urea.	100–400	12.2.9

[a] Ratio of HNB to protein (P).

[b] pH adjusted to 3.5 for modification of pepsin with 50% acetic acid; alkaline samples are in 0.1 M NaOH. The reduced and carboxymethylated proteins were prepared in the conventional manner (Crestfield, A.M., Moore, S., and Stein, W.H., The preparation and enzymatic hydrolysis of reduced and *S*-carboxymethylated proteins, *J. Biol. Chem.* 238, 622–627, 1963).

[c] The reagent is subject to rapid hydrolysis in water with a $t_{1/2} < 30$ s (Horton, H.R. and Koshland, D.E., Jr., A highly colored reagent with selectivity for the tryptophan residue in proteins. 2-Hydroxy-5-nitrobenzyl bromide, *J. Am. Chem. Soc.* 87, 1126–1132, 1965). However, it is suggested that the rate of reaction of HNB with tryptophan is 10,000-fold faster than the reaction with water (Loudon, G.M. and Koshland, D.E., Jr., The chemistry of a reporter group: 2-Hydroxy-5-nitrobenzyl bromide, *J. Biol. Chem.* 245, 2247–2254, 1970).

[d] Multiple substitution at the indole ring has been observed with HNB (Lundblad, R.L. and Noyes, C.M., Observations on the reaction of 2-hydroxy-5-nitrobenzyl bromide with a peptide-bound tryptophanyl residue, *Anal. Biochem.* 136, 93–100, 1984).

[e] The pH of the reaction mixture remained between 4.0 and 4.5 in the absence of buffer and without addition of base to compensate for the hydrolysis of reagent.

[f] A yellow precipitate of 2-hydroxy-5-nitrobenzyl alcohol was removed by centrifugation before gel filtration of the reaction mixture.

[g] Dimethyl(2-hydroxy-5-nitrobenzyl)sulfonium bromide is a water-soluble derivative of HNB (Heinrish, C.P., Adam, S., and Arnold, W., The reaction of dimethyl(2-hydroxy-5-nitrobenzyl)sulfonium bromide with *N*-acetyl-L-tryptophan amide, *FEBS Lett.* 33, 181–183, 1973).

[h] 2-Acetoxy-5-nitrobenzyl bromide is inactive until the acetyl function is removed by hydrolysis at the enzyme active site (Horton, H.R. and Young, G., 2-Acetoxy-5-nitrobenzyl chloride. A reagent designed to introduce a reporter group near the active site of chymotrypsin, *Biochim. Biophys. Acta* 194, 272–278, 1969).

[i] The incorporation of HNB was evaluated over a pH range from 6.95 to 3.00 and showed increasing modification with decreasing pH (0.3 mol HNB per mole of protein at pH 6.95, 30 min reaction; 0.9 mol HNB per mole protein at pH 3.30 or pH 3.50, 60 min reaction). The modified proteins were separated from the reaction mixture by chromatography on CM Sephadex C-50 (column equilibrated with 0.01 M potassium phosphate, pH 6, and the modified protein eluted with 0.2 M potassium phosphate, pH 6). The modified protein was eluted in a broach band. The protein modified with a 100-fold molar excess of HNB at pH 3.3/33°C for 90 min was partially resolved into two peaks, one of which contained 0.91 mol HNB per mole protein and the second 2.41 mol per mole protein. The protein modified with a 110-fold molar excess of HNB at pH 3.3/23°C for 90 min was eluted as a single peak containing 2.39 mol HNB per mole protein.

[j] Modification of human serum albumin has been achieved at a lower molar ratio of HNB to albumin (albumin was incubated overnight in 8.0 M urea, pH 2.7 at 37°C) (Vallner, J.J. and Perrin, J.H., Circular dichroic examination of the interaction of some planar acidic drugs with tryptophan-modified human serum albumin, *J. Pharm. Phamacol.* 33, 697–700, 1981). The oxidation of the single tryptophanyl residue in human serum albumin with NBS does not require the presence of chaotropic agents (Sakamoto, H., Nagata, I., Kiruchi, K., *N*-bromosuccinimide-oxidized human serum albumin as a tool for the determination of drug binding sites of human serum albumin, *Chem. Pharm. Bull.* 31, 971–978, 1983).

[k] In the dark with stirring (most reactions with HNB are performed in the dark).

(continued)

TABLE 12.2 (continued)

Some Examples of the Use of HNB and Related Derivatives to Modify Proteins

References to Table 12.2

12.2.1. Dopheide, T.A.A. and Jones, W.M., Studies on the tryptophan residues in porcine pepsin, *J. Biol. Chem.* 243, 3906–3911, 1968.

12.2.2. Robinson, G.W., Reaction of a specific tryptophan residue in streptococcal proteinase with 2-hydroxy-5-nitrobenzyl bromide, *J. Biol. Chem.* 245, 4832–4841, 1970.

12.2.3. Poulos, T.L. and Price, P.A., The identification of a tryptophan residue essential to the catalytic activity of bovine pancreatic deoxyribonuclease, *J. Biol. Chem.* 246, 4041–4045, 1971.

12.2.4. Lindskog, S. and Nilsson, A., The location of tryptophanyl groups in human and bovine carbonic anhydrases. Ultraviolet difference spectra and chemical modification, *Biochim. Biophys. Acta* 295, 117–130, 1973.

12.2.5. Heilman, H.D. and Pfleiderer, G., On the role of tryptophan residues in the mechanism of action of glyceraldehyde-3-phosphate dehydrogenase as tested by specific modification, *Biochim. Biophys. Acta* 384, 331–341, 1975.

12.2.6. Uhteg, L.C. and Lundblad, R.L., The modification of tryptophan in bovine thrombin, *Biochim. Biophys. Acta* 491, 551–557, 1977.

12.2.7. Clemmer, J.D., Carr, J., Knaff, D.B., and Holwerda, R.A., Modification of laccase tryptophan residues with 2-hydroxy-5-nitrobenzyl bromide, *FEBS Lett.* 91, 346–350, 1978.

12.2.8. Fehske, K.J., Müller, W.F., and Wollert, U., The modification of the lone tryptophan residue in human serum albumin by 2-hydroxy-5-nitrobenzyl bromide. Characterization of the modified protein and the binding of L-tryptophan and benzodiazepines to the tryptophan-modified albumin, *Hoppe Seyler's Z. Physiol. Chem.* 359, 709–717, 1978.

12.2.9. Higuchi, M., Inoue, K., and Iwai, K., A tryptophan residue is essential to the sugar-binding site of winged bean basis lectin, *Biochim. Biophys. Acta* 829, 51–57, 1985.

characteristics that can be used to determine the extent of reagent incorporation (at 365 nm, $\varepsilon = 4 \times 10^3$ M^{-1} cm^{-1}). These reagents show considerable specificity for the modification of tryptophan at pH $\leq$ 4.0. Possible side reactions with other nucleophiles such as amino groups need to be considered. In the case of human chorionic somatomammotropin and human pituitary growth hormone,[212] reaction with o-nitrophenyl-sulfenyl chloride (2-nitrophenylsulfenyl chloride) was achieved in 50% acetic acid, but not in 0.1 sodium acetate, pH 4.0. Wilchek and Miron[213] have reported on the reaction of 2,4-dinitrophenylsulfenyl chloride with tryptophan in peptides and protein and subsequent conversion of the modified tryptophan to 2-thiotryptophan by reaction with 2-mercaptoethanol at pH 8.0. The thiolysis of the modified tryptophan is responsible for changes in the spectral properties of the derivative. The characteristics of the modified tryptophan have resulted in the development of a facile purification scheme for peptides containing the modified tryptophan residues.[214,215] Mollier and coworkers[216] examined the reaction of o-nitrophenylsulfenyl chloride (2-nitrophenylsulfenyl chloride) with notexin (a phospholipase obtained from *Notechis scutatus scutatus* venom that contains two tryptophanyl residues). Reactions with 2-nitrophenylsulfenyl chloride (twofold molar excess) in 50% (v/v) acetic acid resulted in two derivative proteins on HPLC analysis. One derivative contained two modified tryptophanyl residues (20 and 110), while the other derivative was modified only at position 20. There are several applications of this modification, which are of importance for investigators working in proteomics. Kuyama and coworkers[217] have developed a heavy (^{13}C) form of 2-nitrobenzenesulfonyl chloride for the differential labeling of tryptophan residues in protein mixtures. The application of the ICAT strategy[218] to tryptophanyl residues has significant advantage in that tryptophan is one of the least abundant residues in proteins. Interest in the 2-nitrophenylsulfenyl chloride in proteomic research has continued.[219,220] While both tryptophanyl and cysteinyl residues are labeled with 2-nitrobenzenesulfenyl chloride, the mixed disulfide product obtained on reaction with cysteine[221] is removed during reduction and alkylation prior to proteomic analysis. 2-Nitrobenzenesulfenyl chloride (Figure 12.10) is a relatively specific reagent for the modification of tryptophan residues in proteins. The reaction product undergoes extensive fragmentation

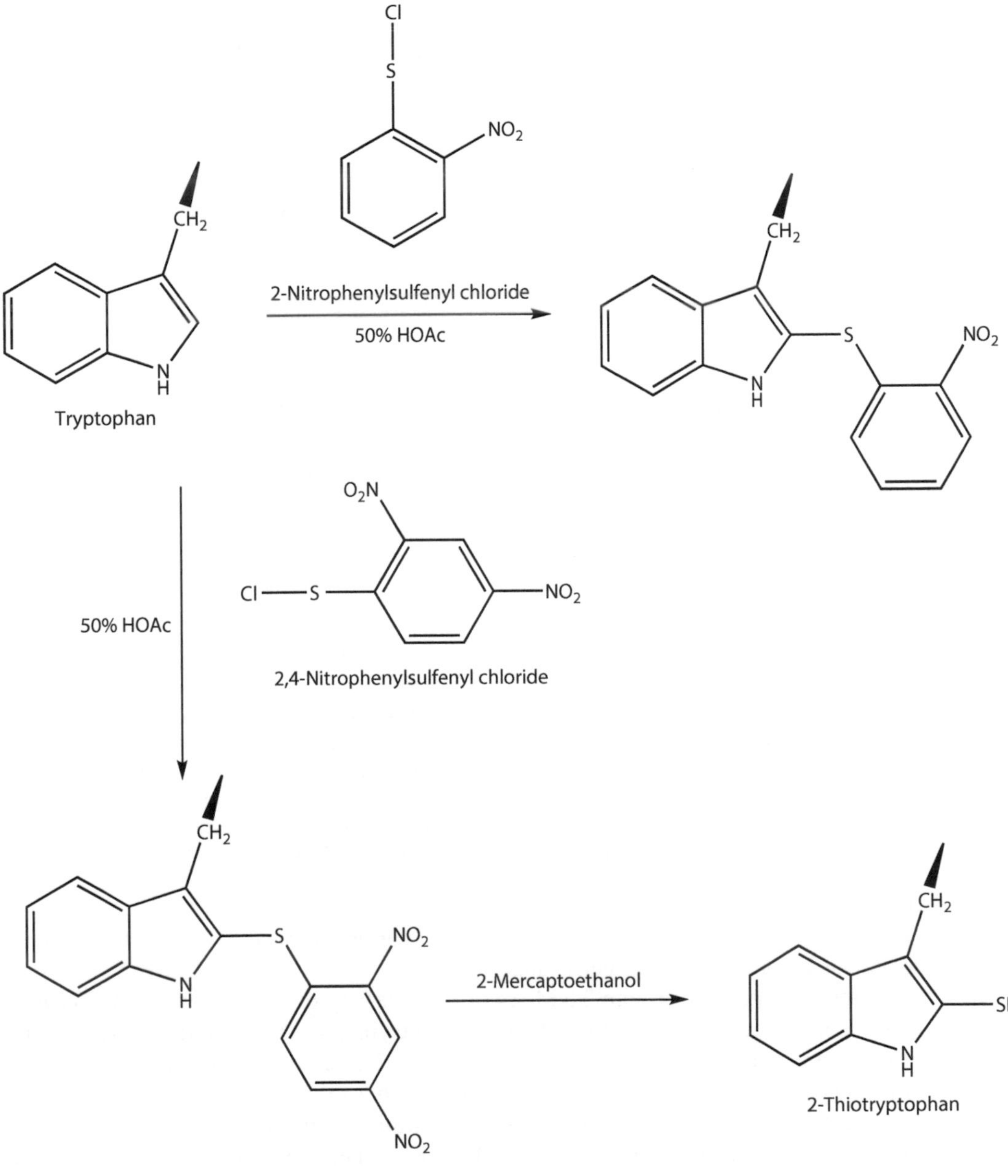

FIGURE 12.10 The reaction of nitrophenylsulfenyl compounds with tryptophan. Shown at the top is the reaction between 2-nitrophenylsulfenyl chloride and tryptophan to yield 2-nitrophenylsulfenyltryptophan (Scoffone, E., Fontana, A., and Rocchi, R., Sulfenyl halides as modifying reagents for polypeptides and proteins. I. Modification of tryptophan residues, *Biochemistry* 7, 971–979, 1968; Matsuo, E., Watanabe, M., Kuyama, H., and Nishimura, O., A new strategy for protein biomarker discovery utilizing 2-nitrobenzenesulfenyl (NBS) reagent and its application in clinical samples, *J. Chromatogr. B* 877, 2607–2614, 2009). Shown at the bottom is the conversion of 2,4-dinitrophenysulfenyltryptophan to 2-thiol tryptophan (Wilchek, M. and Miron, T., The conversion of tryptophan to 2-thioltryptophan, *Biochem. Biophys. Res. Commun.* 47, 1015–1020, 1972). The introduction of a thiol group permits the attachment of a fluorescent label (Heithier, H., Ward, L.D., Cantrill, R.C. et al., Fluorescent glucagon derivatives. I. Synthesis and characterization of fluorescent glucagon derivatives, *Biochim. Biophys. Acta* 971, 298–306, 1988).

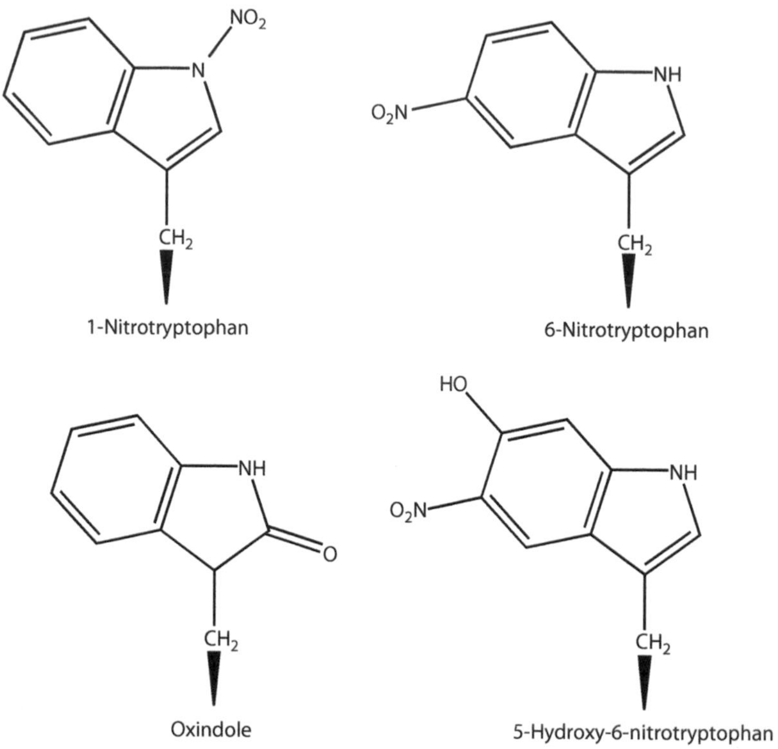

1-Nitrotryptophan

6-Nitrotryptophan

Oxindole

5-Hydroxy-6-nitrotryptophan

FIGURE 12.11 The reaction of TNM or peroxynitrite with tryptophan. The reaction of TNM or peroxynitrite with tryptophan can provide a variety of substitution and oxidation products. Most of the products have been identified as reaction products from the action of peroxynitrite or other RNS (Nuriel, T., Hansler, A., and Gross, S.S., Protein nitrotryptophan: Formation, significance and identification, *J. Proteomics* 74, 2300–2312, 2012; Uda, M., Kawasaki, H., Shigenaga, A. et al., Proteomic analysis of endogenous nitrotryptophan-containing proteins in the rat hippocampus and cerebellum, *Biosci. Rep.* 32, 521–530, 2012).

during MALDI MS. 2-(Trifluoromethyl)benzenesulfenyl chloride has been developed to provide a more stable derivative for mass spectrometric analysis.[222]

The modification of tryptophan with TNM results in a variety of products.[223–227] Sokolovsky and colleagues[223] have studied the reaction of TNM with tryptophan in some detail, suggesting that both nitration and oxidation of the indole ring can occur. These investigators also suggest that the tendency of tryptophan to be buried in proteins may either inhibit or facilitate modification (see preceding text for reagent partitioning argument). Nitration of tryptophan is associated with a modest increase in absorbance in the range of 340–360.[224] There has been more recent work on the reaction of peroxynitrite with tryptophan[228–233] with the 5- and 6-nitro derivatives having been identified by Padmaja and coworkers.[234] Padmaja and coworkers[234] also reported a second-order rate constant of 130 M^{-1} s^{-1} at 25°C for the reaction. Nuriel and coworkers[235] have recently reviewed the various products derived from the reaction of peroxynitrite with tryptophan, some of which are shown in Figure 12.11.

REFERENCES

1. Cornish-Bowden, A., The amino acid compositions of proteins are correlated with their molecular sizes, *Biochem. J.* 213, 271–274, 1983.
2. Gromiha, M.M. and Suvwa, M., A simple statistical method for discriminating outer membrane proteins with better accuracy, *Comput. Biol. Chem.* 29, 136–142, 2005.

3. Gevaert, K., Van Damme, P., Martens, L., and Vandekerckhove, J., Diagonal reverse-phase chromatography applications in peptide-centric proteomics: Ahead of catalogue-omics? *Anal. Biochem.* 345, 18–29, 2005.
4. Nozaki, Y. and Tanford, C., The solubility of amino acids and related compounds in aqueous urea solutions, *J. Biol. Chem.* 238, 4074–4081, 1963.
5. Beers, W.H. and Reich, E., Isolation and characterization of *Clostridium botulinum* type B toxin, *J. Biol. Chem.* 244, 4473–4479, 1969.
6. Kim, Y.S. and Brophy, E.J., Effect of amino acids on purified rat intestinal brush border membrane aminooligopeptidase, *Gastroenterology* 76, 82–87, 1979.
7. Wolfenden, R. and Radzicka, A., How hydrophilic is tryptophan? *Trend. Biochem. Sci.* 11, 69–70, 1986.
8. Catanzano, E., Graziano, G., Fusi, P. et al., Differential scanning calorimetry study of the thermodynamic stability of some mutants of Sso7d from *Sulfolobus solfataricus*, *Biochemistry* 37, 10493–10498, 1998.
9. Abbott, D.W. and Borasion, A., Structural analysis of a putative family 32 carbohydrate-binding module from the *Streptococcus pneumoniae* enzyme Endo D, *Acta Crystallogr. Sect. F: Struct. Biol. Cryst. Commun.* 67, 429–433, 2011.
10. Bromley, K.M., Kiss, A.S., Lokappa, S.B. et al., Dissecting amelogenin protein nanospheres: Characterization of metastable oligomers, *J. Biol. Chem.* 286, 34643–34653, 2011.
11. Hensel, M., Steurer, R., Fichtl, J. et al., Identification of potential sites for tryptophan oxidation in recombinant antibodies using *tert*-butylhydroperoxide and quantitative LC-MS, *PLoS One* 6, 17708, 2011.
12. Sreedhara, A., Lau, K., Li, C. et al., Role of surface exposed tryptophan as substrate generators for the antibody catalyzed water oxidation pathway, *Mol. Pharm.* 10, 278–288, 2013.
13. Painter, R.G., Sage, H.J., and Tanford, C., Contributions of heavy and light chains of rabbit immunoglobulin G to antibody activity. I. Binding studies on isolated heavy and light chains, *Biochemistry* 11, 1327–1337, 1972.
14. Brenke, R., Hall, D.R., Chuang, G.Y. et al., Application of asymmetric statistical potentials to antibody-protein docking, *Bioinformatics* 28, 2608–2614, 2012.
15. Thompson, R.B. and Lakowicz, J.R., Effect of pressure on the self-association of melittin, *Biochemistry* 23, 3411–3417, 1984.
16. Schibli, D.J., Epand, R.F., Vogel, H.J., and Epand, R.M., Tryptophan-rich antimicrobial peptides: Comparative properties and membrane interactions, *Biochem. Cell Biol.* 80, 667–677, 2002.
17. Al-Abdul-Wahid, M.S., Demill, C.M., Serwin, M.B. et al., Effect of juxtamembrane tryptophans on the immersion depth of Synaptobrevin, an integral vesicle membrane proteins, *Biochim. Biophys. Acta* 1818, 2994–2999, 2012.
18. Swartz, D.J., Weber, J., and Urbatsch, I.L., *P*-Glycoprotein is fully active after multiple tryptophan substitutions, *Biochim. Biophys. Acta* 1828, 1159–1168, 2013.
19. Herbst, D.A., Boll, B., Zocher, G. et al., Structural basis of the interaction of MbtH-like proteins, putative regulators of nonribosomal peptide biosynthesis, with adenylating enzymes, *J. Biol. Chem.* 288, 1991–2003, 2013.
20. Foettinger, A., Melmer, M., Leitner, A., and Lindner, W., Reaction of the indole group with malonaldehyde: Applications for the derivatization of tryptophan residues in proteins, *Bioconjug. Chem.* 18, 1678–1683, 2007.
21. Sturm, M., Leitner, A., and Lindner, W., Development of an indole-based chemically cleavable linker concept for immobilizing bait compounds for protein pull-down experiments, *Bioconjug. Chem.* 22, 211–217, 2011.
22. Craig, T.J., Ciufo, L.F., and Morgan, A., A protein-protein binding assay using coated microtitre plates: Increased throughput, reproducibility and speed compared to bead-based assays, *J. Biochem. Biophys. Methods* 60, 49–60, 2004.
23. Uhlen, M. and Ponten, F., Antibody-based proteomics for human tissue profiling, *Mol. Cell. Proteomics* 4, 384–393, 2005.
24. Shart, D.W.A., *Dictionary of Chemistry*, 2nd edn., Penguin Books, London, U.K., 1990.
25. Hase, H., The behavior of photo-induced trapped in solutions of tryptophan at 77°C, *J. Phys. Soc. Jpn.* 24, 223, 1968.
26. Santus, R., Hélène, C., and Ptak, M., Etude per resonance paramagnetique electronique et par spectrophotometrie d'absorption des processus primaires dans la photochimie de solutions aqueuses congelles A 77°K d'acides amines aromatiques e de polypeptides, *Photochem. Photobiol.* 7, 341–360, 1968.
27. Meybeck, A. and Windle, J.J., An E.P.R. study of peptides after U.V. irradiation, *Photochem. Photobiol.* 10, 1–12, 1969.

28. Santus, R. and Grossweiner, L.I., Primary products in the flash photolysis of tryptophan, *Photochem. Photobiol.* 15, 101–105, 1972.
29. Tassin, J.D. and Borkman, R.F., The photolysis rates of some di- and tripeptides of tryptophan, *Photochem. Photobiol.* 32, 577–583, 1980.
30. Szajdzinska-Pietek, E., Bednarek, J., and Plonka, A., Electron spin resonance studies on tryptophan photolysis in frozen micellar systems of anionic surfactants, *J. Photochem. Photobiol. A: Chem.* 75, 131–136, 1993.
31. Angiolillo, P.J. and Vanderkoo, J.M., Hydrogen atoms are produced when tryptophan within a protein is irradiated with ultraviolet light, *Photochem. Photobiol.* 64, 492–495, 1996.
32. Léonard, J., Sharma, D., Szafarowicz, B. et al., Formation dynamics and nature of tryptophan's primary photoproduct in aqueous solution, *Phys. Chem. Chem. Phys.* 12, 15744–15750, 2010.
33. Babu, V. and Joshi, P.C., Tryptophan as an endogenous photosensitizer to elicit harmful effects of ultraviolet B, *Ind. J. Biochem. Biophys.* 29, 296–298, 1992.
34. Barnes, S., Shonsey, E.M., Eliuk, S.M. et al., High-resolution mass spectrometry analysis of protein oxidations and resultant loss of function, *Biochem. Soc. Trans.* 36, 1037–1044, 2008.
35. Freinbichler, W., Colivicchi, M.A., Stefanini, C. et al., Highly reactive oxygen species: Detection, formation, and possible function, *Cell. Mol. Life Sci.* 68, 2067–2079, 2011.
36. Ananthaswamy, H.N. and Eisenstark, A., Near-UV-induced breaks in phage DNA: Sensitization by hydrogen peroxide (a tryptophan photoproduct), *Photochem. Photobiol.* 24, 439–442, 1976.
37. Sidorkina, O.M., Kuznetsov, S.V., Blais, J.C. et al., Ultraviolet-B-induced damage to *Escherichia coli* Fpg protein, *Photochem. Photobiol.* 69, 658–663, 1999.
38. van der Kemp, P.A., Blais, J.-C., Bazin, M. et al., Ultraviolet-B-induced inactivation of human OGG1, the repair enzyme for removal of 8-oxoguanine in DNA, *Photochem. Photobiol.* 76, 640–648, 2002.
39. Blum, A. and Grossweiner, L.I., Laser flash photolysis and photoinactivation of subtilisin BPN′, *Photochem. Photobiol.* 36, 617–622, 1982.
40. Neves-Peterson, M.T., Klitgaard, S., Pascher, T. et al., Flash photolysis of cutinase: Identification and decay kinetics of transient intermediates formed upon UV excitation of aromatic residues, *Biophys. J.* 97, 211–226, 2009.
41. Mialocq, J.C., Amouyai, E., Bernas, A., and Grand, D., Picosecond laser photolysis of aqueous indole and tryptophan, *J. Phys. Chem.* 86, 3173–3177, 1982.
42. Hirata, Y., Murata, N., Tanioka, Y., and Mataga, N., Dynamic behavior of solvated electrons produced by6 photoionization of indole and tryptophan in several polar solvents, *J. Phys. Chem.* 93, 4527–4530, 1989.
43. Stevenson, K.L., Papadantonakis, G.A., and LeBreton, P.R., Nanosecond UV laser photoionization of aqueous tryptophan: Temperature dependence of quantum yield, mechanism, and kinetics of hydrated electron decay, *J. Photochem. Photobiol. A: Chem.* 133, 159–167, 2000.
44. Pigault, C. and Gerard, D., Influence of the location of tryptophanyl residues in proteins on their photosensitivity, *Photochem. Photobiol.* 40, 291–296, 1984.
45. Rao, S.C., Ran, C.M., and Balasubramanian, D., The conformational status of a protein influences the aerobic photolysis of its tryptophan residues: Melittin, β-lactoglobulin, and the crystallins, *Photochem. Photobiol.* 41, 357–362, 1990.
46. Tallmadge, D.H. and Borkman, R.E., The rates of photolysis of the four individual tryptophan residues in UV exposed calf γII crystallin, *Photochem. Photobiol.* 51, 363–368, 1990.
47. Ladner, C.L., Turner, R.J., and Edwards, R.A., Development of indole chemistry to label tryptophan residues in protein for determination of tryptophan surface accessibility, *Protein Sci.* 16, 1204–1213, 2007.
48. Edwards, R.A., Jickling, G., and Turner, R.J., The light-induced reactions of tryptophan with halocompounds, *Photochem. Photobiol.* 75, 362–368, 2002.
49. Casas-Finet, J.R., Wilson, S.H., and Karpel, R.L., Selective photochemical modification by trichloroethanol of tryptophan residues in proteins with a high tyrosine-to-tryptophan ratio, *Anal. Biochem.* 205, 27–35, 1992.
50. Privat, J.-P. and Charlier, M., Photochemical modifications of the tryptophan residues of wheat-germ agglutinin in the presence of trichloroethanol, *Eur. J. Biochem.* 84, 79–85, 1978.
51. Reeve, A.E. and Hopkins, T.R., Photochemical reactions between tryptophan and polynucleotides, *Photochem. Photobiol.* 31, 413–415, 1980.
52. Cadet, J., Voituriez, L., Grand, A. et al., Recent aspects of the photochemistry of nucleic acids and related model compounds, *Biochimie* 67, 277–292, 1985.
53. Oladepo, S.A. and Loppnow, G.R., The effect of tryptophan on UV-induced DNA photodamage, *Photochem. Photobiol.* 86, 844–851, 2010.

54. Hara, I., Ueno, T., Ozaki, S.-i. et al., Oxidative modification of tryptophan 43 in the heme vicinity of the F43W/H64L myoglobin mutant, *J. Biol. Chem.* 276, 36087–36090, 2001.
55. Steen, H. and Mann, M., Analysis of bromotryptophan and hydroxyproline modifications by high-resolution, high-accuracy precursor ion scanning utilizing fragment ions with mass-deficient mass tags, *Anal. Chem.* 74, 6230–6236, 2002.
56. Strohalm, M., Santrusek, J., Hynek, R., and Kodicek, M., Analysis of tryptophan surface accessibility in proteins by MALDI-TOF mass spectrometry, *Biochem. Biophys. Res. Commun.* 32, 1134–1138, 2004.
57. Dyer, J.M., Bringans, S.D., and Bryson, W.G., Determination of photo-oxidation products within photoyellowed bleached wool proteins, *Photochem. Photobiol.* 82, 551–557, 2006.
58. Mouls, L., Silajdzic, E., Haroune, N. et al., Development of novel mass spectrometric methods for identifying HOCl-induced modifications of proteins, *Proteomics* 9, 1617–1631, 2009.
59. Madian, A.G. and Regnier, F.E., Profiling carbonylated proteins in human plasma, *J. Proteome Res.* 9, 1330–1343, 2010.
60. Perdivaral, I., Deterding, L.J., Przybylski, M., and Tomer, K.B., Mass spectrometric identification of oxidative modifications of tryptophan residues in proteins: Chemical artifact or post-translational modification? *J. Am. Soc. Mass Spectrom.* 21, 1114–1117, 2010.
61. Kawasaki, H., Ikeda, K., Shigenaga, A. et al., Mass spectrometric identification of tryptophan nitration sites on proteins in peroxynitrite-treated lysates from PC12 cells, *Free Radic. Biol. Med.* 50, 419–427, 2011.
62. Kawasaki, H., Shigenaga, A., Uda, M. et al., Nitration of tryptophan in ribosomal proteins is a novel post-translational modification of differentiated and naïve PC12 cells, *Nitric Oxide* 25, 176–182, 2011.
63. Todorovski, T., Federova, M., and Hoffmann, R., Identification of isomeric 5-hydroxytryptophan- and oxindolylalanine-containing peptides by mass spectrometry, *J. Mass Spectrom.* 47, 453–459, 2012.
64. Triquigneaux, M.M., Ehrenshaft, M., Roth, E. et al., Targeted oxidation of *Torpedo californica* acetylcholinesterase by singlet oxygen: Identification of *N*-formylkynurenine tryptophan derivatives within the active-site gorge of its complex with the photosensitizer methylene blue, *Biochem. J.* 448, 83–91, 2012.
65. Liu, T.-Y. and Chang, Y.H., Hydrolysis of proteins with *p*-toluenesulfonic acid. Determination of tryptophan, *J. Biol. Chem.* 246, 2842–2848, 1971.
66. Simpson, R.J., Neuberger, M.R., and Liu, T.-Y., Complete amino acid analysis of proteins from a single hydrolysate, *J. Biol. Chem.* 251, 1936–1940, 1976.
67. Molnár, I., Tryptophan analysis of peptides and proteins, mainly by liquid chromatography, *J. Chromatogr.* 763, 1–10, 1997.
68. Aiken, A. and Learmonth, M., Quantitation of tryptophan in proteins, in *Protein Protocols Handbook*, 2nd edn., ed. J.M. Walker, pp. 41–44, Humana Press, Totowa, NJ, 2002.
69. Friedman, M., Applications of the ninhydrin reaction for analysis of amino acids, peptides, and proteins in agricultural and biomedical sciences, *J. Agric. Food Chem.* 52, 385–406, 2004.
70. Drake, B.B. and Smythe, C.V., The oxidation of the aromatic amino acids, tyrosine, tryptophan and phenylalanine, *Arch. Biochem.* 4, 255–263, 1944.
71. Weil, L., Gordon, W.G., and Buchert, A.R., Photoöxidation of amino acids in the presence of methylene blue, *Arch. Biochem. Biophys.* 33, 90–105, 1951.
72. Yoshida, Z. and Kato, M., Photoöxidation products of tryptophan, *J. Am. Chem. Soc.* 76, 311–312, 1954.
73. Roberi, B., Prudomme, R.O., and Grabar, P., Mechanism of action of ultrasonics. I. Oxidation of aromatic nuclei, *Bull. Soc. Chim. Biol.* 37, 897–910, 1955.
74. Hayashi, K., Imoto, T., Funatsu, G., and Funatsu, M., The position of the active tryptophan residue in lysozyme, *J. Biochem.* 58, 227–235, 1965.
75. Hartdegan, F.J. and Rupley, J.A., The oxidation by iodine of tryptophan 108 in lysozyme, *J. Am. Chem. Soc.* 89, 1743–1745, 1967.
76. Hachimori, Y., Horinishi, H., Kurihara, K., and Shibata, K., States of amino residues in proteins. V. Different reactivities with H_2O_2 of tryptophan residues in lysozyme, proteinases and zymogens, *Biochim. Biophys. Acta* 93, 346–360, 1964.
77. Finley, E.L., Dillon, J., Crouch, R.K., and Schey, K.L. Identification of tryptophan oxidation products in bovine α-crystallin, *Protein Sci.* 7, 2391–2387, 1998.
78. Simat, T.J. and Steinhart, H., Oxidation of free tryptophan residues in peptides and proteins, *J. Agric. Food Chem.* 46, 490–498, 1998.
79. Matsuhima, A., Takiuchi, H., Salto, Y., and Inada, Y., Significance of tryptophan residues in the D-domain of the fibrin molecule in fibrin polymer formation, *Biochim. Biophys. Acta* 625, 230–236, 1980.
80. Pajot, P., Fluorescence of proteins in 6-M guanidine hydrochloride. A method for the quantitative determination of tryptophan, *Eur. J. Biochem.* 63, 263–269, 1976.

81. Clottes, E. and Vial, C. Discrimination between the four tryptophan residues of MM-creatine kinase on the basis of the effect of *N*-bromosuccinimide on activity and spectral properties, *Arch. Biochem. Biophys.* 329, 97–103, 1996.

82. Rodacka, A., Serafin, E., Bubinski, M. et al., The influence of oxygen on radiation-induced structural and functional changes in glyceraldehyde-3-phosphate dehydrogenase and lactate dehydrogenase, *Radiat. Phys. Chem.* 81, 807–815, 2012.

83. Simat, T., Meyer, K., and Steinhart, H., Syntheses and analysis of oxidation and carbonyl condensation compounds of tryptophan, *J. Chromatogr. A* 661, 93–99, 1994.

84. Hartdegen, F.J. and Rupley, J.A., Inactivation of lysozymes by iodine oxidation of a single tryptophan, *Biochim. Biophys. Acta* 92, 625–627, 1964.

85. Alexander, N.M., Oxidation and oxidative cleavage of tryptophanyl peptide bonds during iodination, *Biochem. Biophys. Res. Commun.* 54, 614–621, 1973.

86. Hartdegan, F.J. and Rupley, J.A., Oxidation of lysozyme by iodine: Isolation of an inactive product and its conversion to an oxindoleamine-lysozyme, *J. Mol. Biol.* 80, 637–648, 1973.

87. Hartdegen, F.J. and Rupley, J.A., Oxidation of lysozyme by iodine: Identification of oxindolealanine 108, *J. Mol. Biol.* 80, 649–656, 1973.

88. Imoto, T. and Rupley J.A., Oxidation of lysozyme by iodine: Identification and properties of an oxindolyl ester intermediate: Evidence for participation of glutamic acid 35 in catalysis, *J. Mol. Biol.* 80, 657–667, 1973.

89. Alexander, N.M., Oxidative cleavage of tryptophanyl peptide bonds during chemical- and peroxidase-catalyzed iodinations, *J. Biol. Chem.* 249, 1946–1952, 1974.

90. Romero, J.R., Martinez, R., Fresnedo, O., and Ochoa, B., Comparison of two methods for radioiodination on the oxidizability properties of low density lipoproteins, *J. Physiol. Biochem.* 57, 291–301, 2001.

91. Silva, A.M., Marcel, S.L., Vitorino, R. et al., Characterization of in vitro protein oxidation using mass spectrometry: A time course study of oxidized alpha-amylase, *Arch. Biochem. Biophys.* 530, 23–31, 2013.

92. Manzanares, D., Rodriguez-Capote, K., Liu, S. et al., Modification of tryptophan and methionine residues is implicated in the oxidative inactivation of surfactant protein B, *Biochemistry* 46, 5604–5615, 2007.

93. Stewart, D.J., Napolitano, M.J., Bakhmutova-Albert, E.V., and Margerum, D.W., Kinetics and mechanisms of chlorine dioxide oxidation of tryptophan, *Inorg. Chem.* 47, 1639–1647, 2008.

94. Petrônio, M.S. and Ximenes, V.F., Effects of oxidation of lysozyme by hypohalous acids and haloamines on enzymic activity and aggregation, *Biochim. Biophys. Acta* 1824, 1090–1096, 2012.

95. Ximenes, V.F., de Fonseca, L.M., and de Almeida, A.C., Taurine bromamine: A potent oxidant of tryptophan residues in albumin, *Arch. Biochem. Biophys.* 507, 315–322, 2011.

96. Chen, H.-H., Chen, C.-Y., Chow, L.-P. et al., Iron-catalyzed oxidation of Trp residues in low-density lipoprotein, *Biol. Chem.* 392, 859–897, 2011.

97. Szuchman-Sapir, A.J., Pattison, D.I., Ellis, N.A. et al., Hypochlorous acid oxidizes methionine and tryptophan residues in myoglobin, *Free Radic. Biol. Med.* 45, 789–798, 2008.

98. Szuchman-Sapir, A.J., Pattison, D.I., Davies, M.J., and Witting, P.K., Site-specific hypochlorous acid-induced oxidation of recombinant human myoglobin affects specific amino acid residues and the rate of cytochrome b_5-mediated heme reduction, *Free Radic. Biol. Med.* 48, 35–46, 2010.

99. Dixon, G.H. and Schachter, H., The chemical modification of chymotrypsin, *Can. J. Biochem.* 42, 695–714, 1964.

100. Neumann, N.P., Oxidation with hydrogen peroxide, *Methods Enzymol.* 25, 393–400, 1972.

101. Yin, J., Chu, J.W., Ricci, M.S. et al., Effects of excipients on the hydrogen peroxide-induced oxidation of methionine residues in granulocyte colony-stimulating factor, *Pharm. Res.* 22, 141–147, 2005.

102. Kleppe, K., The effect of hydrogen peroxide on glucose oxidase from *Aspergillus niger*, *Biochemistry* 5, 139–143, 1966.

103. Pirie, N.W., The oxidation of sulphydryl compounds by hydrogen peroxide: Catalysis of oxidation of cysteine by thiocarbamides and thiolglycoxalines, *Biochem. J.* 27, 1181–1188, 1933.

104. Madian, A.G., Hinupur, J., Hullerman, J.D. et al., Effect of single amino acid substitution on oxidative modifications of the Parkinson's disease-related protein, DJ-1, *Mol. Cell. Proteomics* 11, M111.010892, 2012.

105. Torres, M.J., Turell, L., Botti, H. et al., Modulation of the reactivity of the thiol of human serum albumin and its sulfenic derivative by fatty acids, *Arch. Biochem. Biophys.* 521, 102–110, 2012.

106. Kato, Y., Kawakishi, S., Aoki, T. et al., Oxidative modification of tryptophan residues exposed to peroxynitrite, *Biochem. Biophys. Res. Commun.* 234, 82–84, 1997.

107. Hawkins, C.L., Pattison, D.L., Stanley, N.R., and Davies, M.J., Tryptophan residues are targets in hypothiocyanous acid-mediated protein oxidation, *Biochem. J.* 416, 441–452, 2008.

108. Hadfield, K.A., Pattison, D.I., Brown, B.E. et al., Myeloperoxidase-derived oxidants modify apolipo-protein A-1 and generate dysfunctional high-density lipoproteins: Comparison of hypothiocyanous acid (HOSCN) with hypochlorous acid (HOCl), *Biochem. J.* 449, 531–542, 2013.

109. Reubsaet, J.L.E. et al., Analytical techniques used to study the degradation of proteins and peptides: Chemical instability. *J. Pharm. Biomed. Anal.* 17, 955–978, 1998.

110. Mach, H., Middaugh, C.R., and Lewis, R.V., Statistical determination of the average values of the extinc-tion coefficients of tryptophan and tyrosine in native proteins. *Anal. Biochem.* 200, 74–80, 1992.

111. Takita, T., Nakagoshi, M., Inouye, K., and Tonomura, B., Lysyl-tRNA synthetase from *Bacillus stearothermophilus*: The tryp312 residue is shielded in a non-polar environment and is responsible for the fluorescence changes observed in the amino acid activation reaction, *J. Mol. Biol.* 325, 677–695, 2003.

112. Reshetnyak, Y.K., Koshevnik, Y., and Burstein, E.A., Decomposition of protein tryptophan fluorescence spectra into log-normal components. III. Correlation between fluorescence and microenvironmental parameters of individual tryptophan residues, *Biophys. J.* 81, 1735–1758, 2001.

113. Hambly, D.M., and Gross, M.L., Cold chemical oxidation of proteins, *Anal. Chem.* 81, 7235–7242, 2008.

114. Ji, J.A., Zhang, B., Cheng, W., and Wang, J., Methionine, tryptophan, and histidine oxidation in a model protein, PTH: Mechanisms and stabilization, *J. Pharm. Sci.* 98, 4585–4900, 2009.

115. Kerwin, B.A. and Remmele, R.L., Jr., Protect from light: Photodegradation and protein biologics, *J. Pharm. Sci.* 96, 1468–1479, 2007.

116. Qi, P., Volkin, D.B., Zhao, H. et al., Characterization of the photodegradation of a human IgG1 monoclonal antibody formulated as a high-concentration liquid dosage form, *J. Pharm. Sci.* 98, 3117–3130, 2009.

117. Maity, H., O'Dell, C., Srivastava, A., and Goldstein, J., Effects of arginine on the photostability and ther-mal stability of IgG1 monoclonal antibodies, *Curr. Pharm. Biotechnol.* 10, 761–766, 2009.

118. Lau, H., Pace, D., Yan, B.X. et al., Investigation of degradation processes in IgG1 monoclonal antibod-ies by limited proteolysis coupled with weak cation-exchange HPLC, *J. Chromatogr. B Anal. Technol. Biomed. Life Sci.* 878, 868–876, 2010.

119. Grosvenor, A.J., Morton, J.D., and Dyer, J.M., Profiling of residue-level photo-oxidative damage in peptides, *Amino Acids* 39, 285–296, 2010.

120. Hawe, A., Wiggenhorn, M., van de Weert, M. et al., Forced degradation of therapeutic proteins, *J. Pharm. Sci.* 101, 895–913, 2012.

121. Mason, B.I., Schoneich, C., and Kerwin, B.A., Effect of light and pH on aggregation and conformation of an IgG1 mAB, *Mol. Pharm.* 9, 774–790, 2012.

122. Stroop, S.D., Conca, D.M., Lundgard, R.P. et al., Photosensitizers form in histidine buffer and mediate the photodegradation of a monoclonal antibody, *J. Pharm. Sci.* 100, 5142–5155, 2011.

123. Lam, X.M., Lai, W.G., Chan, E.K. et al., Site-specific tryptophan oxidation induced by autocatalytic reaction of polysorbate 20 in protein formulations, *Pharm. Res.* 28, 2543–2555, 2011.

124. Lundblad, R.L., *Chemical Reagents for Protein Modification*, 3rd edn., CRC Press, Boca Raton, FL, 2004.

125. Nagasaka, T. and Ohki, S., Indoles. II. Oxidation of *N*-phthaloyl-1-acetyltryptophan with chromium trioxide, *Chem. Pharm. Bull.* 19, 603–611, 1971.

126. Boccu, E., Veronese, F.M., Fontana, A., and Benassi, A., Studies on the function of tryptophan-108 on lysozyme, *Acta Vitaminol. Enzymol.* 29, 266–269, 1975.

127. Savige, W.E. and Fontana, A., Oxidation of tryptophan to oxindolylalanine by dimethyl sulfoxide-hydrochloric acid. Selective modification of tryptophan containing peptides, *Int. J. Pept. Protein Res.* 15, 285–297, 1980.

128. Huang, H.V., Bond, N.W., Hunkapiller, M.W., and Hood, L.W., Cleavage at tryptophanyl residues with dimethyl sulfoxide-hydrochloride acid and cyanogen bromide, *Methods Enzymol.* 91, 318–324, 1983.

129. Wagner, R.M. and Fraser, B.A., Analysis of peptides containing oxidizing methionine and/or tryptophan by fast atom bombardment mass spectrometry. *Biomed. Environ. Mass Spectrom.* 14, 69–72, 1987.

130. Ohnishi, M., Kawagishi, T., Abe, T., and Hiromi, K., Stopped-flow studies on the chemical modification with *N*-bromosuccinimide of model compounds of tryptophan residues, *J. Biochem.* 87, 273–279, 1980.

131. Daniel, V.W., III and Trowbridge, C.G., The effect of *N*-bromosuccinimide upon trypsinogen activation and trypsin catalysis, *Arch. Biochem. Biophys.* 134, 506–514, 1969.

132. Sartin, J.L., Hugli, T.E., and Liao, T.-H., Reactivity of the tryptophan residues in bovine pancreatic deoxyribonuclease with *N*-bromosuccinimide, *J. Biol. Chem.* 255, 8633–8637, 1980.

133. Kosman, D.J., Ettinger, M.J., Bereman, R.D., and Giordano, R.S., Role of tryptophan in the spectral and catalytic properties of the copper enzyme, galactose oxidase, *Biochemistry* 16, 1597–1601, 1977.

134. Freisheim, J.H. and Huennekens, F.M., Effect of *N*-bromosuccinimide on dihydrofolate reductase, *Biochemistry* 8, 2271–2776, 1969.

135. Ohnishi, M., Kawagishi, T., and Hiromi, K., Stopped-flow chemical modification with *N*-bromosuccinimide: A good probe for changes in the microenvironment of the Trp 62 residue of chicken egg white lysozyme, *Arch. Biochem. Biophys.* 272, 46–51, 1989.
136. Inokuchi, N., Takahashi, T., Yoshimoto, A., and Irie, M., *N*-Bromosuccinimide oxidation of a glucoamylase from *Aspergillus saitoi*, *J. Biochem.* 91, 1661–1668, 1982.
137. O'Gorman, R.B. and Matthews, K.S., *N*-Bromosuccinimide modification of lac repressor protein, *J. Biol. Chem.* 252, 3565–3571, 1977.
138. Spande, T.F., Green, N.M., and Witkop, B., The reactivity toward *N*-bromosuccinimide of tryptophan in enzymes, zymogens, and inhibited enzymes, *Biochemistry* 5, 1926–1933, 1966.
139. Xue, H., Xue, Y., Doublié, S., and Carter, C.W., Jr., Chemical modification of *Bacillus subtilis* tryptophanyl-tRNA synthetase, *Biochem. Cell. Biol.* 75, 709–715, 1997.
140. Vincenzitti, S., Cambi, A., Maury, G. et al., Possible role of two phenylalanine residues in the active site of human cytidine deaminase, *Protein Eng.* 13, 791–799, 2000.
141. Munagala, N., Basus, V.J., and Wang, C.C., Role of the flexible loop of hypoxanthine-guanine-xanthine phosphoribosyltransferase from *Tritrichomonas foetus* in enzyme catalysis, *Biochemistry* 40, 4303–4311, 2001.
142. Tyagi, N.K., Kumar, A., Goyal, P. et al., D-Glucose-recognition and phlorizin-binding sites in human sodium/D-glucose cotransporter 1: A tryptophan scanning study, *Biochemistry* 46, 13616–13628, 2007.
143. Kumar, A., Tyagi, N.K., Goyal, P. et al., Sodium-independent low-affinity D-glucose transport by human sodium/D-glucose cotransporter 1: Critical role of tryptophan 561, *Biochemistry* 46, 2758–2766, 2007.
144. Zhu, Q., Liu, W., Kao, L. et al., Topology of NBCe1 protein transmembrane segment 1 and structural effect of proximal renal tubular acidosis (pRTA)S427L mutation, *J. Biol. Chem.* 288, 7894–7906, 2013.
145. Rathman, D., Pedragosa-Badia, X., and Beck-Sickinger, A.G., In vitro modification of substituted cysteines as tool to study receptor functionality and structure-activity relationships, *Anal. Biochem.* 439, 173–183, 2013.
146. Spande, T.F. and Witkop, B., Reactivity toward *N*-bromosuccinimide as a criterion for buried and exposed tryptophan residues in proteins, *Methods Enzymol.* 11, 528–532, 1967.
147. Fairhead, S.M., Steel, J.S., Wreford, L.J., and Walker, I.O., Properties of tryptophan and tyrosine side chains in tobacco mosaic virus and tobacco mosaic virus protein, *Biochim. Biophys. Acta* 194, 584–593, 1969.
148. Masachika, I., State of tryptophan residue in ribonuclease T1 and carboxymethyl ribonuclease T1, *J. Biochem.* 68, 31–37, 1970.
149. Beveridge, H.J.T. and Nakai, S., Effects of chemical modification with 2-phenyl-1,4-dibromoacetoin on αs1-casein, *J. Dairy Sci.* 53, 1532–1539, 1970.
150. Masachika, I., Harada, M., and Sawada, F., State of tryptophan residues in ribonuclease from *Aspergillus saitoi*, *J. Biochem.* 72, 1351–1359, 1972.
151. Beattie, B.K., Prentice, G.A., and Merrill, A.R., Investigation into the catalytic role for the tryptophan residues within domain III of *Pseudomonas aeruginosa* exotoxin A, *Biochemistry* 35, 15134–15142, 1996.
152. Bray, M.R., Johnson, P.E., Gilkes, N.R. et al., Probing the role of tryptophan residues in a cellulose-binding domain by chemical modification, *Protein Sci.* 5, 2311–2318, 1996.
153. Spande, T.F. and Witkop, B., Determination of tryptophan content of proteins with *N*-bromosuccinimide, *Methods Enzymol.* 11, 498–506, 1967.
154. Divita, G., Jauli, J.-M., Gautheron, D.C., and Di Pietro, A., Chemical modification of α-subunit tryptophan residues in *Schizosaccharomyces pombe* mitochondrial F_1 adenosine 5′-triphosphatase: Differential reactivity and role in activity, *Biochemistry* 32, 1017–1024, 1993.
155. Manly, S.P. and Matthews, K.S., Activity changes in *lac* repressor with cysteine oxidation, *J. Biol. Chem.* 254, 3341–3347, 1979.
156. Brand, L. and Shatiel, S., Modification of histidine residues leading to the appearance of visible fluorescence, *Biochim. Biophys. Acta* 88, 338–351, 1964.
157. Williams, M.N., Effect of *N*-bromosuccinimide modification on dihydrofolate reductase from a methotrexate-resistant strain of *Escherichia coli*. Activity, spectrophotometric, fluorometric, and circular dichroism studies, *J. Biol. Chem.* 250, 322–330, 1975.
158. Skeberdis, V.A., Rimkute, L., Skeberdyte, A. et al., pH-dependent modulation of connexin-based gap junctional uncouplers, *J. Physiol.* 589, 3495–3506, 2011.
159. Kumar, T.K. and Ramachandran, L.K., Effect of modification of tyrosine residues in cytotoxin-1 from Indian cobra venom (*Naja naja naja*), *Biochem. Int.* 20, 879–885, 1990.
160. Bray, M.R., Carriere, A.D., and Clarke, A.J., Quantitation of tryptophan and tyrosine residues in proteins by fourth-derivative spectroscopy, *Anal. Biochem.* 221, 278–284, 1994.

161. Myer, Y.P., Thalium, K., Pande, J., and Verma, B.C., Selectivity of oxidase and reductase activity of horse heart cytochrome c, *Biochem. Biophys. Res. Commun.* 94, 1106–1112, 1980.

162. Ybarra, U., Prasad, A.R.S., and Nishimura, J.S., Chemical modification of tryptophan residues in *Escherichia coli* succinyl-CoA synthetase. Effect on structure and enzyme activity, *Biochemistry* 25, 7174–7178, 1986.

163. Postma, T.M. and Albericio, F., *N*-Chlorosuccinimide, an efficient reagent for on-resin disulfide formation in solid-phase peptide synthesis, *Org. Lett.* 15, 616–619, 2013.

164. Shih, H., New approaches to the synthesis of cystine peptides using *N*-iodoacetamide in the construction of disulfide bridges, *J. Org. Chem.* 58, 3003–3008, 1993.

165. Lindegren, S., Andersson, H., Bäck, T. et al., High-efficiency astatination of antibodies using *N*-iodosuccinimide as the oxidizing agent in labelling of *N*-succinimidyl 3-(trimethylstannyl)benzoate, *Nucl. Med. Biol.* 28, 35–39, 2001.

166. Lindegren, S., Frost, S., Bäck, T. et al., Direct procedure for the production of [211]At-labeled antibodies with an ε-lysyl-3-(trimethylstannyl)benzamide immunoconjugate, *J. Nucl. Med.* 49, 1537–1545, 2008.

167. Patchornik, A., Lawson, W.B., and Witkop, B., Selective cleavage of peptide bonds. II. The tryptophanyl peptide bond and the cleavage of glucagon, *J. Am. Chem. Soc.* 80, 4747–4748, 1958.

168. Lischwe, M.A. and Sung, M.T., Use of *N*-chlorosuccinimide/urea for the selective cleavage of tryptophanyl peptide bonds in proteins. Cytochrome c, *J. Biol. Chem.* 252, 4976–4980, 1972.

169. Schmidt, G.L., Cohen, L.A., and Witkop, B., The oxidative cleavage of tyrosyl peptide bonds. I. Cleavage of dipeptides and some properties of the resulting spirodienone-lactones, *J. Am. Chem. Soc.* 81, 2228–2233, 1959.

170. Ramachandran, L.K. and Witkop, B., *N*-Bromosuccinimide cleavage of peptides, *Methods Enzymol.* 11, 282–299, 1967.

171. Burstein, Y. and Patchornik, A., Use of a nonenzymatic cleavage reaction for identification of exposed tyrosine residues bovine pancreatic ribonuclease, *Biochemistry* 16, 2939–2944, 1972.

172. Welch, A.R., McNally, L.M., and Gibson, W., Cytomegalovirus assembly protein nested gene family: Four 3′-coterminal transcripts encode four in-frame overlapping proteins, *J. Virol.* 65, 4091–4100, 1991.

173. Schliess, F. and Stoffel, W., Evolution of the myelin integral membrane proteins of the central nervous system, *Biol. Chem. Hoppe-Seyler* 372, 865–874, 1991.

174. Chiang, C.S., Mitsis, P.G., and Lehman, I.R., DNA polymerase δ from embryos of *Drosophila melanogaster*, *Proc. Natl. Acad. Sci. USA* 90, 9105–9109, 1993.

175. Douady, D., Rousseau, B., and Caron, L., Fucoxanthin-chlorophyll a/c light-harvesting complexes of *Laminaria saccharina*. Partial amino acid sequences and arrangement in thylakoid membranes, *Biochemistry* 33, 3165–3170, 1994.

176. Wu, F., Bui, K.C., Buckley, S., and Warburton, D., Cell cycle-dependent expression of cyclin D1 and a 45 kD protein in human A549 lung carcinoma cells, *Am. J. Respir. Cell Mol. Biol.* 10, 437–447, 1994.

177. Zeng, F.-Y., Kratzin, H., and Gabius, H.-J., Migration inhibitory factor-binding sarcolectin from human placenta is indistinguishable from a subfraction of human serum albumin, *Biol. Chem. Hoppe-Seyler* 375, 393–399, 1994.

178. Kooi, C., Hodges, R.S., and Sokul, P.A., Identification of neutralizing epitopes on *Pseudomonas aeruginosa* elastase and effects of cross-reactions on other thermolysin-like proteases, *Infect. Immun.* 65, 472–477, 1997.

179. Smith, B.J., Chemical cleavage of proteins at tryptophan residues, in *Protein Protocols Handbook*, ed. J.M. Walker, Humana Press, Totowa, NJ, 1996.

180. Kowalski, A., Palyga, J., and Gormicka-Michalska, E., Two polymorphic linker histone loci in Guinea fowl erythrocytes, *Comp. Rend. Biol.* 334, 6–12, 2011.

181. Junek, H., Kirk, K.L., and Cohen, L.A., Oxidative cleavage of tyrosyl-peptide bonds during iodination, *Biochemistry* 252, 4976–4980, 1977.

182. Wachter, E. and Werhahn, R., Attachment of tryptophanyl peptides to 3-aminopropyl-glass suited for subsequent solid phase Edman degradation, *Anal. Biochem.* 97, 56–64, 1979.

183. Omenn, G.S., Fontana, A., and Anfinsen, C.B., Modification of the single tryptophan residue in staphylococcal nuclease by a new mild oxidizing agent, *J. Biol. Chem.* 245, 1895–1902, 1970.

184. Burnett, P.R. and Eylar, E.H., Allergic encephalomyelitis. Oxidation and cleavage of the single tryptophan residue of the A1 protein from bovine and human myelin, *J. Biol. Chem.* 246, 3425–3430, 1971.

185. Martenson, R.E. and Deibler, G.E., Partial characterization of basic proteins of chicken, turtle, and frog central nervous system myelin, *J. Neurochem.* 24, 79–88, 1975.

186. Hunziker, P.E., Hughes, G.J., and Wilson, K.J., Peptide fragmentation suitable for solid-phase microse-quencing. Use of *N*-bromosuccinimide and BNPS-skatole (3-bromo-3-methyl-2-[(2-nitrophenyl)thio]-3H-indole), *Biochem. J.* 187, 515–519, 1980.

187. Spycher, S.E., Nick, H., and Rickli, E.E., Human complement component C1s. Partial sequence determination of the heavy chain and identification of the peptide bond cleaved during activation, *Eur. J. Biochem.* 156, 49–57, 1986.

188. Rahali, V. and Gueguen, J., Chemical cleavage of bovine β-lactoglobulin by BNPS-skatole for preparative purposes: Comparative study of hydrolytic procedures and peptide characterization, *J. Protein Chem.* 18, 1–12, 1999.

189. Willig, S., Lösel, D., and Claus, R., Effects of resistant potato starch on odor emission from feces in swine production units, *J. Agric. Food Chem.* 53, 1173–1178, 2005.

190. Odagami, T., Tsudo, Y., Kogami, Y. et al., Deprotection of the indole (N^{ind})-formyl (For) group on trypto-phan employing a new reagent, *N,N'*-dimethylethylenediamine (DMEDA) in an aqueous solution, *Chem. Pharm. Bull.* 57, 211–213, 2009.

191. Coletti-Previero, M.-A., Previero, A., and Zuckerkandl, E., Separation of the proteolytic and esteratic activities of trypsin by reversible structural modification, *J. Mol. Biol.* 39, 493–501, 1969.

192. Previero, A., Coletti-Previero, M.-A., and Cavadore, J.C., A reversible chemical modification of the tryp-tophan residue, *Biochim. Biophys. Acta* 147, 453–461, 1967.

193. Cooper, H.M., Jemmerson, R., Hunt, D.F. et al., Site-directed chemical modification of horse cytochrome c results in changes in antigenicity due to local and long-range conformation perturbations, *J. Biol. Chem.* 262, 11591–11597, 1987.

194. Koshland, D.E., Jr., Karkhanis, Y.D., and Latham, H.G., An environmentally-sensitive reagent with selectivity for the tryptophan residue in proteins, *J. Am. Chem. Soc.* 86, 1448–1450, 1964.

195. Horton, H.R. and Koshland, D.E., Jr., A highly reactive colored reagent with selectivity for the tryptophanyl residue in proteins, 2-hydroxy-5-nitrobenzyl bromide, *J. Am. Chem. Soc.* 87, 1126–1132, 1965.

196. Clemmer, J.D., Carr, J., Knaff, D.B., and Holwerda, R.A., Modification of laccase tryptophan residues with 2-hydroxy-5-nitrobenzyl bromide, *FEBS Lett.* 91, 346–350, 1978.

197. Barman, T.E. and Koshland, D.E., Jr., A colorimetric procedure for the quantitative determination of tryptophan residues in proteins, *J. Biol. Chem.* 242, 5771–5776, 1967.

198. Novak, R.L., Banerjee, K., Dohnal, J. et al., Inhibition of RNA initiation in *E. coli* core RNA polymerase with 2-hydroxy-5-nitrobenzyl bromide, *Biochem. Biophys. Res. Commun.* 60, 833–837, 1974.

199. Baracca, A., Menegatti, D., Parenti Castelli, G. et al., Does 2-hydroxy-5-nitrobenzyl bromide react with the ε-subunit of the mitochondrial F1-ATPase? *Biochem. Int.* 21, 1135–1142, 1990.

200. Barman, T.E., The chemistry of the reaction of 2-hydroxy-5-nitrobenzyl bromide with his-32 of α-lactalbumin, *Eur. J. Biochem.* 83, 465–471, 1978.

201. Lundblad, R.L. and Noyes, C.M., Observations of the reaction of 2-hydroxy-5-nitrobenzyl bromide with a peptide-bound tryptophanyl residue, *Anal. Biochem.* 136, 93–100, 1985.

202. Loudon, G.M. and Koshland, D.E., Jr., The chemistry of a reporter group: 2-Hydroxy-5-nitrobenzyl bromide, *J. Biol. Chem.* 245, 2247–2254, 1970.

203. Strohalm, M., Kodíček, M., and Pechar, M., Tryptophan modification by 2-hydroxy-5-nitrobenzyl bromide studied by MALDI-TOF mass spectrometry. *Biochem. Biophys. Res. Commun.* 312, 811–816, 2003.

204. Horton, H.R. and Tucker, W.P., Dimethyl (2-hydroxy-5-nitrobenzyl) sulfonium salts. Water-soluble environmentally sensitive protein reagents, *J. Biol. Chem.* 245, 3397–3401, 1970.

205. Tucker, W.P., Wang, J., and Horton, H.R., The reaction of dimethyl (2-hydroxy-5-nitrobenzyl) sulfonium salts with tryptophan ethyl ester, *Arch. Biochem. Biophys.* 144, 730–733, 1971.

206. Peyser, Y.M., Muhlrad, A., and Werber, M.M., Tryptophan-130 is the most reactive tryptophan residue in rabbit skeletal myosin subfragment-1, *FEBS Lett.* 259, 346–348, 1990.

207. Horton, H.R., Kelly, H., and Koshland, D.E., Jr., Environmentally sensitive protein reagents. 2-Methoxy-5-nitrobenzyl bromide, *J. Biol. Chem.* 240, 722–724, 1965.

208. Horton, H.R. and Young, G., 2-Acetoxy-5-nitrobenzyl chloride. A reagent designed to introduce a reporter group near the active site of chymotrypsin, *Biochim. Biophys. Acta* 194, 272–278, 1969.

209. Uhteg, L.C. and Lundblad, R.L., The modification of tryptophan in bovine thrombin, *Biochim. Biophys. Acta* 491, 551–557, 1977.

210. Fontana, A. and Scoffone, E., Sulfenyl halides as modifying reagents for polypeptides and proteins, *Methods Enzymol.* 25, 482–494, 1972.

211. Shechter, Y., Burstein, Y., and Patchornik, A., Sulfenylation of tryptophan-62 in hen egg-white lysozyme, *Biochemistry* 11, 653–660, 1972.

212. Bewley, T.A., Kawauchi, H., and Li, C.H., Comparative studies of the single tryptophan residue in human chorionic somatomammotropin and human pituitary growth hormone, *Biochemistry* 11, 4179–4187, 1972.
213. Wilchek, M. and Miron, T., The conversion of tryptophan to 2-thioltryptophan in peptides and proteins, *Biochem. Biophys. Res. Commun.* 47, 1015–1020, 1972.
214. Chersi, A. and Zito, R., Isolation of tryptophan-containing peptides by adsorption chromatography, *Anal. Biochem.* 73, 471–476, 1976.
215. Rubinstein, M., Schechter, Y., and Patchornik, A., Covalent chromatography—The isolation of tryptophanyl containing peptides by novel polymeric reagents, *Biochem. Biophys. Res. Commun.* 70, 1257–1263, 1976.
216. Mollier, P., Chwetzoff, S., Bouet, F., Harvey, A.L., and Ménez, A., Tryptophan 110, a residue involved in the toxic activity but not in the enzymatic activity of notexin, *Eur. J. Biochem.* 185, 263–270, 1989.
217. Kuyama, H., Watanabe, M., Toda, C. et al., An approach to quantitative proteome analysis by labeling tryptophan residues, *Rapid Commun. Mass Spectrom.* 17, 1642–1650, 2003.
218. Ou, K., Kesuma, D., Ganesan, K. et al., Quantitative profiling of drug-associated proteomic alterations by combined 2-nitrobenzenesulfenyl chloride (NBS) isotope labeling and 2DE/MS identification, *J. Proteome Res.* 5, 2194–2206, 2006.
219. Matsuo, E., Watanabe, M., Kuyama, H., and Nishimura, O., A new strategy for protein biomarker discovery utilizing 2-nitrobenzenesulfenyl (NBS) reagent and its applications to clinical samples, *J. Chromatogr. B* 877, 2607–2614, 2009.
220. Fontana, A., Scoffone, E., and Benassi, C.A., Sulfenyl halides as modifying reagents for polypeptides and proteins II. Modification of cysteinyl residues, *Biochemistry* 7, 980–986, 1972.
221. Li, C., Gawandi, V., Protos, A. et al., A matrix-assisted laser desorption/ionization compatible reagent for tagging tryptophan residues, *Eur. J. Mass Spectrom.* 12, 213–221, 2006.
222. Sokolovsky, M., Fuchs, M., and Riordan, J.F., Reaction of tetranitromethane with tryptophan and related compounds, *FEBS Lett.* 7, 167–170, 1970.
223. Teuwissen, G., Masson, P.L., Osinski, P., and Heremans, J.F., Metal-combining properties of human lactoferrin. The effect of nitration of lactoferrin with tetranitromethane, *Eur. J. Biochem.* 35, 366–371, 1973.
224. Spande, T.F., Fontana, A., and Witkop, B., An unusual reaction of skatole with tetranitromethane, *J. Am. Chem. Soc.* 91, 6169–6170, 1969.
225. Riggle, W.L., Long, J.A., and Borders, C.L., Jr., Reaction of turkey egg-white lysozyme with tetranitromethane. Modification of tyrosine and tryptophan, *Can. J. Biochem.* 51, 1433–1439, 1973.
226. Katsura, T., Lam, E., Packer, L., and Seltzer, S., Light dependent modification of bacteriorhodopsin by tetranitromethane. Interaction of a tyrosine and a tryptophan residue with bound retinal, *Biochem. Int.* 5, 445–456, 1982.
227. Haddad, I.Y., Zhu, S., Ischiropoulos, H. et al., Nitration of a surfactant protein A results in decreased ability to aggregate lipids, *Am. J. Physiol.* 270, L281–L288, 1996.
228. Zhu, S., Haddad, I.Y., and Matalon, S., Nitration of surfactant protein A (SP-A) tyrosine residues results in decreased mannose binding ability, *Arch. Biochem. Biophys.* 333, 283–290, 1996.
229. Alverez, B., Rubbio, H., Kirk, M. et al., Peroxynitrite-dependent tryptophan nitration, *Chem. Res. Toxicol.* 9, 390–396, 1996.
230. Pollet, E., Martinez, A., Metha, B. et al., Role of tryptophan oxidation in peroxynitrite-dependent protein chemiluminescence, *Arch. Biochem. Biophys.* 349, 74–80, 1996.
231. Lehnig, M. and Kirsch, M., ^{15}N-CIDNP investigations during tryptophan, *N*-acetyl-L-tryptophan, and melatonin nitration with reactive nitrogen species, *Free Radic. Res.* 41, 523–535, 2007.
232. Rebrin, I., Bregere, C., Gallaher, T.K., and Sohal, R.S., Detection and characterization of peroxynitrite-induced modification of tyrosine, tryptophan, and methionine residues by tandem mass spectrometry, *Methods Enzymol.* 441, 284–294, 2008.
233. Padmaja, S., Ramazerian, M.S., Bounds, P.L., and Koppenal, W.H., Reaction of peroxynitrite with L-tryptophan, *Redox Rep.* 2, 173–177, 1996.
234. Nuriel, T., Hansler, A., and Gross, S.W., Protein nitrotryptophan. Formation, significance and identification, *J. Proteomics* 74, 2300–2312, 2011.

13 Modification of Amino Groups

The ε-amino group of lysine and the α-amino group at the N-terminus of polypeptide chains are the primary amine groups normally found in proteins. Alkaline conditions are usually required for the reaction of amino groups in proteins as amino groups must be unprotonated to function as satisfactory nucleophiles. As will be shown, the pKa of the ε-ammonium group of lysine is very sensitive to microenvironmental factors, and pKa value of 7 and below is not unknown. As shown in Table 1.1 in Chapter 1, the average pKa value for the epsilon-amino group of lysine is 10.79 and 9.68 for an α-amino group. The range for lysine is 5.7–12.1 (average 10.5 ± 1.1), while the range for the N-terminal amino group is 6.8–9.1 (average 7.2 ± 0.5).[1] Other protein nucleophiles, including thiol group of cysteine, the thioether function of methionine, the imidazole nitrogens of histidine, and the phenolic hydroxyl group of tyrosine, have the potential to react with some of the reagents used to modify primary amines in proteins. It is therefore necessary to characterize the products of chemical modification. Fortunately, the advent of MS has markedly increased the ability to characterize chemically modified proteins.[2–6] Modification of amino groups is usually performed at pH ≥ 8.0; it is possible to obtain modification at a lower pH. As with other functional groups in proteins, amino group reactivity is subject to microenvironmental influences. Schmidt and Westheimer[7] showed that the pH dependence of the acylation of the amino group at the active site of acetoacetate decarboxylase by 2,4-dinitrophenyl propionate was consistent with a pKa of 5.9, which is some 4 pKa units less than that of an *ordinary* ε-amino group of lysine. Notwithstanding the difficulties in the interpretation of such data,[8] it is clear that the pH dependence of the reaction is consistent with a difference in lysine group reactivity created by the local environment. Mukouyama and coworkers[9] showed that a sarcosine oxidase from *Corynebacterium* was inactivated with iodoacetamide. It was surprising to find that the inactivation was associated with the alkylation of two lysine residues; the pKa values for the two lysine residues were estimated to be 8.5 and 6.7. The unusual reactivity and abnormal pKa values were attributed to a unique microenvironment. There is discussion of microenvironment factors on functional group reactivity in Chapter 1.

A listing of some reagents commonly used to modify lysine residues in proteins is presented in Table 13.1. It is also useful to consult Chapters 2 (Alkylation) and 3 (Acylation) for more information on the chemistry of the reagents. The reagents do vary extensively on reaction rate and specificity of reaction; for example, the reaction rate of acetic anhydride is very rapid but the modification is relatively nonspecific. The reader is directed to a rather nice discussion by Cohen[10] for clarification of selectivity versus reactivity.

Acylation of amino groups in proteins with acid anhydrides (Figure 13.1) is one of the older modification procedures in protein chemistry. Modification most often occurs at the ε-amino group of lysine, somewhat less frequently at α-amino groups. Reaction can also occur at other nucleophilic functional groups including sulfhydryl, phenolic hydroxyl, and histidine imidazole nitrogens and at aspartic or glutamic via mixed acid anhydride formation. Most of these modifications are either exceedingly transient or labile under conditions (mild base) where N-acyl groups are stable. Acylation can be performed with a variety of acid anhydrides including citraconic anhydride,[11,12] maleic anhydride,[13–15] succinic anhydride,[16–18] 3-hydroxyphthalic anhydride,[19] trimellitic anhydride,[19] methyltetrahydrophthalic anhydride,[20,21] *cis*-aconitic anhydride,[17,18] fatty acid anhydrides,[22] hexahydrophthalic anhydride,[21,23] and phthalic anhydride.[24] The in vivo acylation of lysine has been reviewed[25] and this subject is outside the primary emphasis of the current work. However, I would be remiss if I did not point out that the nonenzymatic formylation of the ε-amino group has been suggested to occur via the mixed anhydride of formic acid and phosphoric acid.[26] This is an

TABLE 13.1

Some Chemical Reagents Used for Modification of Lysine in Proteins[a]

Reagent	Comment	References
Acetic anhydride	Modification also occurs at α-amino groups: tyrosine,[b] histidine,[b] and cysteine.[b] Specificity of modification at the α-amino group is increased at pH 7 and below; reaction at pH 6.0 was used to selectively (98% yield) modify an α-amino group with iodoacetic anhydride;[13.1.1] the resulting derivative could be used to prepare a bioconjugate or could be coupled to a matrix. As the rate of reaction is more rapid than many other reagents, acetic anhydride has been used for the identification of exposed amino acid residues in the study of protein conformation.[13.1.2–13.1.11] Reaction with acetic anhydride has been used to determine the effective charge on a protein.[13.1.12,13.1.13] Reaction with acetic anhydride can be accomplished in the gas phase;[13.1.14] α-chymotrypsin was used as the model protein for reaction with acetic anhydride (75°C after lyophilization from pH 9.0).[c] There was 25% modification of ε-amino groups and almost 90% modification of α-amino groups (compared to aqueous phase at pH 9.0). Formation of a mixed anhydride with protein carboxyl groups was also reported under these reaction conditions. It was the reagent used for the development of competitive labeling[13.1.15] as method for determining lysine residue reactivity. Differential labeling using reaction of lysine residues with acetic anhydride was developed to evaluate differences in reactivity occurring on interaction with other proteins or other macromolecules.[13.1.16] There is continued use for study of protein conformation.[13.1.17] Modification of amino groups with acetic anhydride does result in charge neutralization.	13.1.1–13.1.17
Organic acid chlorides	Modification occurs at α-amino groups with possible modification at histidine and tyrosine. Acetyl chloride has been used for C-terminal amino acid sequencing.[13.1.18] Oxalyl chloride has been used to block amino groups prior to coupling to a matrix.[13.1.19] Fatty acid chlorides such as myristoyl chloride and palmitoyl chloride have been used to modify protein properties including the ability to pass across membranes.[13.1.22]	13.1.18–13.1.22
Methyl acetyl phosphate	Modification can also occur at α-amino groups. Originally developed as a substrate analog,[13.1.23,13.1.24] it has received most use for the modification of anion-binding sites in proteins.[13.1.25–13.1.32]	13.1.23–13.1.32
Imidoesters	Modification can also occur at α-amino groups. Imidoesters such as methyl acetimidate modify amino groups with retention of charge.[13.1.33] Membrane-impermeable derivatives such as isethionyl acetimidate[13.1.34] and S-sulfethylthioacetimidate have been developed.[13.1.34,13.1.35] Bisimidoesters have extensive use as cross-linking reagents.[13.1.36–13.1.38] Modification with imidoesters preserves the charge of the amino group.	13.1.33–13.1.38
Pyridoxal-5′-phosphate	Modification can also occur at α-amino groups. The phosphate group in pyridoxal-5′-phosphate (PLPL) is thought to interact with anion-binding sites in proteins enhancing reactivity of PLP since, for example, pyridoxal is considerably less reactive than PLP with bovine pancreatic ribonuclease[13.1.39] and glutamate dehydrogenase.[13.1.40] The majority of studies suggest preferential reaction at the ε-amino group of lysine, but there is at least one report of reaction with an α-amino group.[13.1.41] In this work with deoxyhemoglobin, Val1 is, together with Lys82 that was also modified by PLP, in the polyphosphate binding site. The fluorescence properties[13.1.42] and UV spectra[13.1.43] of the PLP moiety have been used to measure conformational change. The initial reaction	13.1.39–13.1.48

TABLE 13.1 (continued)
Some Chemical Reagents Used for Modification of Lysine in Proteins[a]

Reagent	Comment	References
	product is a Schiff base that can be stabilized by reduction with sodium borohydride or sodium cyanoborohydride. There may be small differences in product depending on the reduced agent.[13.1.44] There are a number of studies in this area and a limited number are cited in the current work.[13.1.45–13.1.48]	
Reductive alkylation	Modification can also occur at α-amino groups. Reductive alkylation is the process of an alkyl aldehyde such as formaldehyde or butyraldehyde reacting with amino groups to form a Schiff base that is reduced to an imine by a reagent such as sodium borohydride or sodium cyanoborohydride.[13.1.49–13.1.52] The modification results in the preservation of positive charge on the amino group and differs from products obtained with monocarboxylic acid, which is charge neutral, or dicarboxylic acid anhydride such as succinic anhydride, which is charge reversal. This modification occurs without major change in protein structure as evidenced by the crystallography studies cited in the following text. The use of formaldehyde permits dimethylation, while aldehydes such as acetaldehyde and butyraldehyde that have larger alkyl groups tend to be restricted to monosubstitution. Formaldehyde and, to a lesser extent, larger alkyl aldehydes such as acetaldehyde will also modify N-terminal α-amino groups. Access to N-terminal α-amino groups can be restricted. However, it has been possible to use reductive alkylation to couple a butyraldehyde-poly (ethylene)glycol to a recombinant fibroblast growth factor bound to a heparin–agarose column.[13.1.53] The same technology was used for the improving pharmacokinetic properties of antagonists of neonatal Fc receptor.[13.1.54] Reducing sugars also can form Schiff base products with protein amines, and this reaction can be used for neoglycosylation.[13.1.55] Reductive methylation has been used for the incorporation of ^{2}H[13.1.56] of ^{13}C[13.1.57–13.1.59] into proteins and peptides for NMR studies. Reductive methylation has proved valuable in the preparation of samples for x-ray crystallography.[13.1.60–13.1.64]	13.1.49–13.1.64
TNBS	α-Amino groups and sulfhydryl groups. There is preference for reaction with amino groups,[13.1.65–13.1.67] and the product obtained with sulfhydryl groups is subject to amminolysis.[13.1.66] The extinction coefficient for the product obtained with the sulfhydryl group of cysteine is approximately 10% of that obtained with amino groups.[13.1.67] There is also reaction with ammonia.[13.1.68,13.1.69] The specificity for the modification of amino groups in the absence of cysteine has permitted the use of TNBS for modification of amino groups, mainly lysine, in enzymes, to determine function.[13.1.70–13.1.75] TNBS crosses a lipid membrane with difficulty[13.1.76] and had been used to modify surface lysine residues.[13.1.77] As a note of caution, TNBS has been demonstrated to pass the erythrocyte membrane via an inorganic anion-exchange pathway.[13.1.78] A chromogenic product allows the reagent to be used for measurement of amino groups in proteins[13.1.79,13.1.80] and allows the determination of amino group modification in proteins.[13.1.81–13.1.83] For example, TNBS was used to determine lysine modification in lactose repressor protein after reaction with diethylpyrocarbonate.[13.1.81] TNBS has been used to assess change in protein conformation.[13.1.84–13.1.86] The great majority of recent work has used TNBS to induce experimental colitis.[13.1.87,13.1.88]	13.1.65–13.1.82

(continued)

TABLE 13.1 (continued)
Some Chemical Reagents Used for Modification of Lysine in Proteins[a]

Reagent	Comment	References
NHS esters	Specific for lysine in solution; however, a gas-phase reaction with arginine has been reported.[13.1.89] The gas-phase reaction allows the arginine guanidino group to be unprotonated unlike the situation in solution where the NHS esters are too unstable at the alkaline pH (11–12) where the guanidino group is unprotonated. Previous work had shown preferential modification of the ε-amino group of lysine residues in peptides in gas phase.[13.1.90] There was little modification of α-amino groups. The reaction was facilitated by electrostatic binding of the reagent to the peptide by either the S-sulfo-NHS esters binding to protonated sites or NHS trimethylammonium butyrate esters. Preferential reaction at the α-amino group tumor necrosis factor-α was enhanced by reaction at pH 5.8, but there was still significant labeling at the ε-amino group of lysine.[13.1.91] The reaction of an α-amino group with the NHS ester of biotin appears to be sterically blocked in *Aphysia* egg-laying hormone.[13.1.92] Hydroxysuccinimide chemistry is used frequently for the biotinylation of proteins, and there are studies on the stability of the NHS reagent[13.1.93] and biotinylated proteins in buffer and plasma.[13.1.94] As a side observation, there was an increase in the hydrophobicity of biotinylated compared to unlabeled proteins as evaluated by hydrophobic interaction chromatography.[13.1.95] The sulfoderivative (i.e., sulfo-NHS acetate, acetic acid sulfo-NHS ester) is frequently used because of improved solubility in aqueous solution.[13.1.96,13.1.97] Sulfo-NHS esters interfere with the bicinchoninic acid assay for protein concentration.[13.1.98] Carbodiimide chemistry has been used to create NHS esters on agarose[13.1.99] and on proteins modified with succinic anhydride.[13.1.100] Carbodiimide chemistry is used with NHS that is used for zero-length cross-linking in proteins. An *N*-hydroxysuccinimide ester is formed with a carboxylic acid group on a protein in a reaction mediated by a carbodiimide. The *N*-hydroxysuccinimide ester then forms a peptide bond with a proximal amino group (usually the ε-amino group of a lysine residue) establishing a zero-length crosslink.[13.1.101] A more recent application of the combination of NHS and a carbodiimide has been in the preparation of protein-based scaffolds.[13.1.102,13.1.103]	13.1.89–13.1.103

[a] Information on the chemistry and reactivity of the various reagents is contained in Chapter 2 (Alkylation) and Chapter 3 (Acylation). The cited references direct the reader to studies using the reagent for the modification of lysine residues in proteins. As indicated in the table, most reagents also react with α-amino groups as well as the ε-amino groups. Emphasis is placed on the reaction of ε-amino groups of lysine.

[b] These modifications are either very transient or, as in the case of tyrosine, easily reversed by dilute base or alkaline hydroxylamine. The modification of tyrosine and histidine are also associated with change in UV spectra (see Chapter 11 for tyrosine and Chapter 14).

[c] pH memory is defined as the ability of a functional group to retain original ionization state in the lyophilized product (Govindarajan, R., Chatterjee, K., Gatlin, L. et al., Impact of freeze-drying on ionization of sulfonephthalein probe molecules in trehalose-citrate systems, *J. Pharm. Sci.* 95, 1498–1510, 2006). Proteins may exhibit pH memory after lyophilization (Vakos, H.T., Kaplan, H., and Black, B., Use of the pH memory in lyophilized proteins to achieve preferential methylation of α-amino groups, *J. Prot. Chem.* 19, 231–237, 2000).

References to Table 13.1

13.1.1.   Wetzel, R., Halualani, R., Stults, J.T., and Quan, C.A., A general method for highly selective cross-linking of unprotected polypeptides via pH-controlled modification of N-terminal α-amino groups, *Bioconjug. Chem.* 1, 114–122, 1990.

TABLE 13.1 (continued)
Some Chemical Reagents Used for Modification of Lysine in Proteins[a]

13.1.2. Suckau, D., Mak, M., and Przybylski, M., Protein surface topology-probing by selective chemical modification and mass spectrometric peptide mapping, *Proc. Natl. Acad. Sci. USA* 89, 5630–5634, 1992.

13.1.3. Cervenansky, C., Engstrom, A., and Karlsson, E., Study of structure-activity relationship of fasciculin by acetylation of amino groups. *Biochim. Biophys. Acta* 1199, 1–5, 1994.

13.1.4. Ohguro, H., Palczewski, K., Walsh, K.A., and Johnson, R.S., Topographical study of arrestin using differential chemical modifications and hydrogen deuterium exchange, *Protein Sci.* 3, 2428–2434, 1994.

13.1.5. Zappacosta, F., Ingallinella, P., Scaloni, A. et al., Surface topology of Minibody by selective chemical modification and mass spectrometry. *Protein Sci.* 6, 1901–1909, 1997.

13.1.6. Hochleitner, E.O., Borchers, C., Parker, C. et al., Characterization of a discontinuous epitope of the human immunodeficiency virus (HIV) core protein p24 by epitope excision and differential chemical modification follow by mass spectrometric peptide mapping analysis, *Protein Sci.* 9, 487–496, 2000.

13.1.7. Hlavica, P., Schulze, J., and Lewis, D.F.V., Function interaction of cytochrome P450 with its redox partners: A critical assessment and update of the topology of the predicted contact regions, *J. Inorg. Biochem.* 96, 279–297, 2003.

13.1.8. Calvete, J.J., Campanero-Rhodes, M.A., Raida, M., and Sanz, L., Characterisation of the conformational and quaternary structure-dependent heparin-binding region of bovine seminal plasma protein PDC-109, *FEBS Letters* 444, 260–264, 1999.

13.1.9. Smith, C.M., Gafken, P.R., Zhang, Z. et al., Mass spectrometric quantification of acetylation at specific lysines within the amino-terminal tail of histone H4, *Anal. Biochem.* 316, 23–33, 2003.

13.1.10. Yadev, S.P., Brew, K., and Puett, D., Holoprotein formation of human chorionic gonadotropin: Differential labeling with acetic anhydride, *Mol. Endocrinol.* 8, 1547–1558, 1994.

13.1.11. Hitchcock-DeGregori, S.E., Lewis, S.F., and Chou, T.M., Tropomyosin lysine reactivities and relationship to coiled-coil structure, *Biochemistry* 18, 3305–3314, 1985.

13.1.12. Gao, J., Gomez, F.A., Härter, R., and Whitesides, G.M., Determination of the effective charge of a protein in solution by capillary electrophoresis, *Proc. Nat. Acad. Sci. USA* 91, 12027–12030, 1994.

13.1.13. Gao, J. and Whitesides, G.M., Using protein charge ladders to estimate the effective charges and molecular weights of proteins in solution, *Anal. Chem.* 69, 575–580, 1997.

13.1.14. Taralp, A. and Kaplan, H., Chemical modification of lyophilized proteins in non-aqueous environments, *J. Prot. Chem.* 16, 183–193, 1997.

13.1.15. Kaplan, H., Stevenson, K.J., and Hartley, B.S., Competitive labelling, a method for determining the reactivity of individual groups in proteins. The amine groups of porcine elastase, *Biochem. J.* 124, 289–299, 1971.

13.1.16. Richardson, R.H. and Brew, K., Lactose synthase. An investigation of the interaction of α-lactalbumin for galactosyltransferase by differential kinetic labeling, *J. Biol. Chem.* 255, 3377–3385, 1980.

13.1.17. Gong, B., Ramos, A., Vázquez-Fernández, E. et al., Probing structural differences between PfP(C) and PrP(Sc) by surface nitration and acetylation: Evidence of conformational change in the C-terminus, *Biochemistry* 50, 4963–4972, 2011.

13.1.18. Shenoy, N.R., Shively, J.E., and Bailey, J.M., Studies in C-terminal sequencing: New reagents for the synthesis of peptidylthiohydantoins, *J. Protein Chem.* 12, 195–205, 1993.

13.1.19. Gautam, S. and Loh, K.C., Immobilization of hydrophobic peptidic ligands to hydrophilic chromatographic matrix: A preconcentration approach, *Anal. Biochem.* 423, 202–209, 2012.

13.1.20. Chopineau, J., Robert, S., Fénert, L. et al., Monoacylation of ribonuclease A enables its transport across an in vitro model of the blood-brain barrier, *J. Control. Release* 56, 231–237, 1998.

13.1.21. Batrakova, E.V., Vinogradov, S.V., Robinson, S.M. et al., Polypeptide point modifications with fatty acid and amphiphilic block copolymers for enhanced brain delivery, *Bioconjug. Chem.* 16, 793–802, 2005.

13.1.22. Kocevar, N., Obermajer, N., and Kreft, S., Membrane permeability of acylated cystatin depends on the fatty acyl chain length, *Chem. Biol. Drug Des.* 72, 217–224, 2008.

13.1.23. Kluger, R. and Tsui, W.-C., Methyl acetyl phosphate. A small anionic acetylating agent, *J. Org. Chem.* 45, 2723, 1980.

13.1.24. Kluger, R. and Tsui, W.C., Reaction of the anionic acetylation agent methyl acetyl phosphate with D-3-3-hydroxybutyrate dehydrogenase, *Biochem. Cell Biol.* 64, 434–440, 1986.

13.1.25. Ueno, H., Pospischil, M.A., Manning, J.M., and Kluger, R., Site-specific modification of hemoglobin by methyl acetyl phosphate, *Arch. Biochem. Biophys.* 244, 795–800, 1986.

(continued)

TABLE 13.1 (continued)

Some Chemical Reagents Used for Modification of Lysine in Proteins[a]

13.1.26. Ueno, H., Pospischil, M.A., and Manning, J.M., Methyl acetyl phosphate as a covalent probe for anion-binding sites in human and bovine hemoglobins, *J. Biol. Chem.* 264, 12344–12351, 1989.

13.1.27. Ueno, H. and Manning, J.M., The functional, oxygen-linked chloride binding sites of hemoglobin are contiguous within a channel in the central cavity, *J. Prot. Chem.* 11, 177–185, 1992.

13.1.28. Ueno, H., Popowicz, A.M., and Manning, J.M., Random chemical modification of the oxygen-linked chloride-binding sites of hemoglobin: Those in the central dyan axis may influence the transition between deoxy- and oxy-hemoglobin, *J. Protein Chem.* 12, 561–570, 1993.

13.1.29. Manning, J.M., Preparation of hemoglobin derivatives selectively or randomly modified at amino groups, *Methods Enzymol.* 231, 225–246, 1994.

13.1.30. Kataoka, K., Tanizawa, K., Fukui, T. et al., Identification of active site lysyl residues of phenylalanine dehydrogenases by chemical modification with methyl acetylphosphate combined with site-directed mutagenesis, *J. Biochem.* 116, 1370–1376, 1994.

13.1.31. Xu, A.S., Labotka, R.J. and London, R.E., Acetylation by human hemoglobin by methyl acetylphosphate; evidence of broad regio-selectivity revealed by NMR studies, *J. Biol. Chem.* 274, 26629–26632, 1999.

13.1.32. Raibekas, A.A., Bures, E.J., Siska, C.C. et al., Anion binding and controlled aggregation of human interleukin-1 receptor antagonist, *Biochemistry* 44, 9871–9879, 2005.

13.1.33. Makoff, A.J. and Malcolm, A.D.B., Properties of methyl acetimidate and its use as a protein-modifying reagent, *Biochem. J.* 193, 245–249, 1981.

13.1.34. Rack, M., Effects of chemical modification of amino groups by two different imidoesters on voltage-clamped nerve fibres of the frog, *Pflügers Arch.* 404, 120–130, 1983.

13.1.35. Jaffe, E.G., Lauber, M.A., Running, W.E., and Reilly, J.P., In vitro and in vivo chemical labeling of ribosomal proteins: A quantitative comparison, *Anal. Chem.* 84, 9335–9561, 2012.

13.1.36. Drews, G. and Rack, M., Modification of sodium and gating currents by amino group specific cross-linking and monofunctional reagents, *Biophys. J.* 54, 383–391, 1988.

13.1.37. Dodson, M.S., Dimethyl suberimidate cross-linking of oligo(dT) to DNA-binding proteins, *Bioconjug. Chem.* 11, 876–879, 2000.

13.1.38. Poucková, P., Morbio, M., Vottariello, F. et al., Cytotoxicity of polyspermine-ribonuclease A and polyspermine-dimeric ribonuclease A, *Bioconjug. Chem.* 18, 1946–1955, 2007.

13.1.39. Means, G.E. and Feeney, R.E., Affinity labeling of pancreatic ribonuclease, *J. Biol. Chem.* 246, 5532–5533, 1971.

13.1.40. Anderson, B.M., Anderson, C.D., and Churchich, J.E., Inhibition of glutamate dehydrogenase by pyridoxal-5′-phosphate, *Biochemistry* 5, 2893–2900, 1966.

13.1.41. Benesch, R., Benesch, R.E, Kwong, S. et al., Labeling of hemoglobin with pyridoxal phosphate, *J. Biol. Chem.* 257, 1320–1324, 1982.

13.1.42. Xiao, G.S. and Zhou, J.M., Conformational changes at the active site of bovine pancreatic RNase A at low concentrations of guanidine hydrochloride probed by pyridoxal 5′-phosphate, *Biochim, Biophys. Acta* 1294, 1–7, 1996.

13.1.43. Ahmed, S.A., McPhie, P., and Miles, E.W., A thermally induced reversible conformational transition of the tryptophan synthase β₂ subunit probed by the spectroscopic properties of pyridoxal phosphate and by enzymatic activity, *J. Biol. Chem.* 271, 8612–8617, 1996.

13.1.44. Fishman, R.M., Moore, G.L. Zegna, A., and Marini, M.A., High-performance liquid chromatographic evaluation of pyridoxal 5′-phosphate hemoglobin derivatives produced by different reduction procedures, *J. Chromatogr.* 532, 55–64, 1990.

13.1.45. Basu, S., Basu, A., and Modak, M.J., Pyridoxal 5′-phosphate mediated inactivation of *Escherichia coli* DNA polymerase I: Identification of lysine-635 as an essential residue for the processive mode of DNA synthesis, *Biochemistry* 27, 6710–6716, 1988.

13.1.46. Anderberg, S.J., Topological disposition of lysine 943 in native Na+/K(+)-transporter ATPase, *Biochemistry* 34, 9508–9516, 1995.

13.1.47. Lapko, A.G. and Ruckpaul, K., Discrimination between conformational states of mitochondrial cytochrome P-450scc by selective modification with pyridoxal-5-phosphate, *Biochemistry (Moscow)* 63, 568–572, 1998.

13.1.48. Vermeersch, J.J., Christmann-Franck, S., Kasrabashyan, S. et al., Pyridoxal 5′-phosphate inactivates DNA topoisomerase 1B by modifying the lysine general acid, *Nucleic Acids Res.* 32, 5649–5657, 2004.

13.1.49. Means, G.E., Reductive alkylation of amino groups, *Methods Enzymol.* 47, 469–478, 1977.

TABLE 13.1 (continued)
Some Chemical Reagents Used for Modification of Lysine in Proteins[a]

13.1.50. Means, G.E. and Feeney, R.E., Reductive alkylation of amino groups in proteins, *Biochemistry* 7, 2192–2201, 1968.

13.1.51. Means, G.E. and Feeney, F.E., Reductive alkylation of proteins, *Anal. Biochem.* 224, 1–16, 1995.

13.1.52. Kovanich, D., Cappadona, S., Raijmakers, R. et al., Applications of stable isotope dimethyl labeling in quantitative proteomics, *Bioanal. Chem.* 404, 991–1009, 2012.

13.1.53. Huang, Z., Ye, C., Liu, Z. et al., Solid-phase N-terminus PEGylation of recombinant human fibroblast growth factor 2 on heparin-sepharose column, *Bioconug. Chem.* 23, 740–750, 2012.

13.1.54. Mezo, A.R., Low, S.C., Hoehn, T. et al., PEGylation enhances the therapeutic potential of peptide antagonists of the neonatal Fc receptor, FcRn, *Bioorg. Med. Chem. Lett.* 21, 6332–6335, 2011.

13.1.55. Srimathi, S. and Jayaraman, G., Effect of glycosylation on the catalytic and conformation stability of homologous α-amylases, *Protein J.* 24, 79–88, 2005.

13.1.56. Brown, E.M., Pfeffer, P.E., Kumosinski, T.F., and Greenberg, R., Accessibility and mobility of lysine residues in β-lactoglobulin, *Biochemistry* 27, 5601–5610, 1988.

13.1.57. Zhang, M., Thulin, E., and Vogel, H.J., Reductive methylation and pKa determination of the lysine side chains in calbindin D9k, *J. Protein Chem.* 13, 527–535, 1994.

13.1.58. Larda, S.T., Bokoch, M.P., Evanics, F., and Prosser, R.S., Lysine methylation strategies for characterizing protein conformations by NMR, *J. Biomol. NMR* 54, 199–208, 2012.

13.1.59. Hattori, Y., Furuita, K., Ohki, I. et al., Utilization of lysine ^{13}C-methylation NMR for protein-protein interaction studies, *J. Biomol. NMR* 55, 19–31, 2013.

13.1.60. Rypniewski, W.R., Holden, H.M., and Rayment, I., Structural consequences of reductive methylation of lysine residues in hen egg white lysozyme: An X-ray analysis at 1.8 Å resolution, *Biochemistry* 32, 9851–9858, 1993.

13.1.61. Rayment, I., Reductive alkylation of lysine residues to alter crystallization properties of proteins, *Methods Enzymol.* 276, 171–179, 1997.

13.1.62. Kurinov, I.V. et al., X-ray crystallographic analysis of pokeweed antiviral protein-II after reductive methylation of lysine residues. *Biochem. Biophys. Res. Commun.* 275, 549–552, 2000.

13.1.63. Lara-González, S., Birktoft, J.J., and Lawson, C.L., Structure of the *Escherichia coli* RNA polymerase α subunit C-terminal domain, *Acta Crystallogr. D Biol. Crystallogr.* 66, 806–812, 2010.

13.1.64. Claesson, M., Siitonen, V., Dobritzsch, D. et al., Crystal structure of the glycosyltransferase SnogD from the biosynthetic pathway of nogalamycin in *Streptomyces nogalater*, *FEBS J.* 279, 3251–3263, 2012.

13.1.65. Satake, K., Okuyama, T., Ohashi, M., and Shinoda, T., The spectrophotometric determination of amine, amino acid and peptide with 2,4,6-trinitrobenzene 1-sulfonic acid, *J. Biochem.* 47, 654–660, 1960.

13.1.66. Kotaki, A., Harada, M., and Yagi, K., Reaction between sulfhydryl compounds and 2,4,6-trinitrobenzene-1-sulfonic acid, *J. Biochem.* 55, 553–561, 1964.

13.1.67. Fields, R., The rapid determination of amino groups, *Methods Enzymol.* 25, 464–468, 1972.

13.1.68. Whitaker, J.R., Granum, P.E., and Aasen, G., Reaction of ammonia with trinitrobenzene sulfonic acid, *Anal. Biochem.* 108, 73–75, 1980.

13.1.69. Means, G.E., Congdon, W.I., and Bender, M.L., Reactions of 2,4,6-trinitrobenzenesulfonate ion with amines and hydroxide ion. *Biochemistry* 11, 3564–3571, 1972.

13.1.70. Coffee, C.J., Bradshaw, R.A., Goldin, B.R., and Frieden, C., Identification of the sites of modification of bovine liver glutamate dehydrogenase reacted with trinitrobenzenesulfonate, *Biochemistry* 10, 3516–3526, 1971.

13.1.71. Whitson, P.A., Burgumn, A.A., and Matthews, K.S., Trinitrobenzenesulfonate modification of the lysine residues in lactose repressor protein, *Biochemistry* 23, 6046–6052, 1984.

13.1.72. Xia, C., Meyer, D.J., Chen, H. et al., Chemical modification of GSH transferase P1-1 confirms the presence of Arg-13, Lys-44 and one carboxylate group in the GSH-binding domain of the active site, *Biochem. J.* 293, 357–362, 1993.

13.1.73. Kurono, Y., Ichioka, K., Mori, S., and Ikeda, K., Kinetic-study on rapid reaction of trinitrobenzenesulfonate with human serum albumin, *J. Pharm. Sci.* 70, 1297–1298, 1981.

13.1.74. Kurono, Y., Ichioka, K., and Ikeda, K., Kinetics of the rapid modification of human serum albumin with trinitrobenzenesulfonate and localization of its site, *J. Pharm. Sci.* 72, 432–435, 1983.

(continued)

TABLE 13.1 (continued)
Some Chemical Reagents Used for Modification of Lysine in Proteins[a]

13.1.75. Haniu, M., Yuan, H., Chen, S.A. et al., Structure-function relationship of NAD(P)H:quinone reductase: Characterization of NH$_2$-terminal blocking group and essential tyrosine and lysine residues, *Biochemistry* 27, 6877–6883, 1988.

13.1.76. Wolf, D.F., Determination of the siddedness of carbocyanine dye labeling of membranes, *Biochemistry* 24, 582–586, 1985.

13.1.77. Cahalan, M.D. and Pappone, P.A., Chemical modification of potassium channel gating in frog myelinated nerve by trinitrobenzene sulfonic acid, *J. Physiol.* 342, 119–143, 1983.

13.1.78. Haest, C.W., Kamp, D., and Deutiche, B., Penetration of 2,4,6-trinitrobenzenesulfonte into human erythrocytes. Consequences for studies on phospholipid asymmetry, *Biochim. Biophys. Acta* 640, 535–543, 1981.

13.1.79. Sashidhar, R.B., Capoor, A.K., and Ramana, D., Quantitation of ε-amino group using amino acids as reference standards by trinitrobenzene sulfonic acid. A simple spectrophotometric method for the estimation of hapten to carrier protein ratio, *J. Immunol. Methods* 167, 121–127, 1994.

13.1.80. Cayot, P. and Tainturier, G., The quantification of protein amino groups by the trinitrobenzenesulfonic acid method: A reexamination, *Anal. Biochem.* 249, 184–200, 1997.

13.1.81. Sams, C.F. and Matthews, K.S., Diethyl pyrocarbonate reaction with the lactose repressor protein affects both inducer and DNA binding, *Biochemistry* 27, 2277, 1988.

13.1.82. Cayot, P., Roullier, L., and Tainturier, G., Electrochemical modifications of proteins. 1. Glycitolation, *J. Agric. Food Chem.* 47, 1915–1923, 1999.

13.1.83. Adamczyk, M., Buko, A., Chen, Y.Y. et al., Characterization of protein-hapten conjugates. 1. Matrix-assisted laser desorption ionization mass spectrometry of immuno BSA-hapten conjugates and comparison with other characterization methods, *Bioconjug. Chem.* 5, 631–635, 1994.

13.1.84. Yang, C.C. and Chang, L.S., Studies on the status of lysine residues in phospholipase A2 from *Naja naja atra* (Taiwan cobra) snake venom, *Biochem. J.* 262, 855–860, 1989.

13.1.85. Komatsu, H. Emoto, Y., and Tawada, K., Half-stoichiometric trinitrophenylation of myosin subfragment 1 in the presence of pyrophosphate or adenosine diphosphate, *J. Biol. Chem.* 268, 7789–7808, 1993.

13.1.86. Chang, L.S., Lin, S.R., and Chang, C.C., Probing calcium ion-induced conformational changes of Taiwan cobra phospholipase A2 by trinitrophenylation of lysine residues, *J. Protein Chem.* 16, 51–57, 1997.

13.1.87. te Velde, A.A., Verstege, M.I., and Hommes, D.W., Critical appraisal of the current practice in murine TNBS-induced colitis, *Inflamm. Bowel Dis.* 12, 995–999, 2006.

13.1.88. Guan, Q., Weiss, C.R., Qing, G. et al., An IL-17 peptide-based and virus-like particle vaccine enhances the bioactivity of IL-17 in vitro and *in vivo*, *Immunotherapy* 4, 1799–1807, 2012.

13.1.89. McGee, W.M., Mentinova, M., and McLuckey, S.A., Gas-phase conjugation to arginine residues in polypeptide ions via *N*-hydroxysuccinimide ester-based reagent ions, *J. Am. Chem. Soc.* 134, 11412–11414, 2012.

13.1.90. Mentinova, M. and McLuckey, S.A., Covalent modification of gaseous peptide ions with *N*-hydroxysuccinimide ester reagent ions, *J. Am. Chem. Soc.* 132, 18248–18257, 2010.

13.1.91. Magni, F., Curnis, F., Marazzine, L. et al., Biotinylation sites of tumor necrosis factor-α determined by liquid chromatography-mass spectrometry, *Anal. Biochem.* 298, 181–188, 2001.

13.1.92. Knock, S.L., Miller, B.T., Blankenship, J.E. et al., N-acylation of *Aphysia* egg-laying hormone with biotin. Characterization of bioactive and inactive derivatives, *J. Biol. Chem.* 266, 24413–24419, 1991.

13.1.93. Grumbach, I.M. and Veh, R.W., Sulpho-N-hydroxysuccinimide activated long chain biotin. A new microtitre plate assay for the determination of its stability at different pH values and its reaction rate with protein bound amino groups, *J. Immunol. Methods* 140, 205–210, 1991.

13.1.94. Bogusiewicz, A., Mock, N.I., and Mock, D.M., Instability of the biotin-protein bond in human plasma, *Anal. Biochem.* 327, 156–161, 2004.

13.1.95. Storm, D., Loos, M., and Kaul, M., Biotinylation of proteins via amino groups can induced binding to U937, HL-60 cells, monocytes, and granulocytes, *J. Immunol. Methods* 199, 87–99, 1996.

13.1.96. Hill, D.E., Fetterer, R.H., and Urban, J.F., Jr., Biotin as a probe of the surface of *Ascaris suum* developmental stages, *Mol. Biochem. Parasitol.* 41, 45–52, 1990.

13.1.97. Hurley, W.L. and Finkelstein, E., Identification of leukocyte surface proteins, *Methods Enzymol.* 184, 429–433, 1990.

13.1.98. Vashist, S.K., Zhang, B., Zheng, D. et al., Sulfo-N-hydroxysuccinimide interferes with bicinchoninic acid protein assay, *Anal. Biochem.* 417, 156–158, 2011.

TABLE 13.1 (continued)
Some Chemical Reagents Used for Modification of Lysine in Proteins[a]

13.1.99. Cuatrecasas, P. and Parikh, I., Adsorbents for affinity chromatography. Use of N-hydroxysuccinimide esters of agarose, *Biochemistry* 11, 2291–2299, 1972.

13.1.100. Kishida, Y., Olsen, B.R., Berg, R.A., and Prockop, D.J., Two improved methods for preparing ferritin-protein conjugates for electron microscopy, *J. Cell. Biol.* 64, 331–339, 1975.

13.1.101. Kobayashi, T., Grabarek, Z., Gergely, J., and Collins, J.H., Extensive interactions between troponins C and I. Zero-length cross-linking of troponin I and acetylated troponin C, *Biochemistry* 34, 10946–10952, 1995.

13.1.102. Lai, J.Y., Corneal stromal cell growth on gelatin/chondroitin sulfate scaffolds modified at different NHS/EDC molar ratios, *Int. J. Mol. Sci.* 14, 2036–2055, 2013.

13.1.103. Krishnamoorthy, G., Sehgal, P.K., Mandal, A.B., and Sadulla, S., Development of D-lysine-assisted 1-ethyl-3-(3-dimethylaminopropyl)-carbodiimide/N-hydroxysuccinimide-initiated cross linking of collagen matrix for design of scaffold, *J. Biomed. Mater. Res. A* 101, 1173–1183, 2013.

interesting piece of chemistry, and the N^ε-formyl lysine has been described as a relatively common posttranslational modification comparable to glycation.[27]

Modification of lysine residues with succinic anhydride[28] results in charge reversal (Figure 13.1). Reaction with succinic anhydride frequently results in the dissociation of multimeric proteins and has also been used to *solubilize* insoluble proteins. Meighen and coworkers[29] have produced a *variant* form of bacterial luciferase through reaction with succinic anhydride. The succinylated protein retained the dimeric subunit structure of the native enzyme. By complementation experiments involving the mixing/hybridization of the modified and native enzymes, it was determined that succinylation of bacterial luciferase resulted in the inactivation of the β-subunit without markedly affecting the function of the α-subunit. Shetty and Rao[30] used succinic anhydride to dissociate subunits of arachin. The modification reaction was performed in 0.1 M sodium phosphate, pH 7.8 with the pH maintained over the course of the reaction by the addition of 2.0 M NaOH. The extent of modification was determined by reaction of the unmodified primary amino groups on the protein with trinitrobenzenesulfonic acid. With a 200:1 molar excess of succinic anhydride, 82% of the available amino groups were succinylated with concomitant dissociation of the subunits of this protein. Succinic anhydride has been used to study other oligomeric proteins.[31–33]

Chymotrypsinogen and α-chymotrypsin were modified with succinic anhydride in 0.05 M sodium phosphate, pH 7.5; pH was maintained at 7.4 with the addition of 1.0 M NaOH.[34] Both of these studies emphasize the necessity of aggressive pH control in the modification of protein with organic acid anhydrides. Chymotrypsinogen (1 g) was dissolved in the sodium phosphate buffer, and 50 mg of succinic anhydride was added over a 30 min period at 23°C. Homogeneous products were obtained with both chymotrypsinogen and α-chymotrypsin with 8 or 14 lysine residues modified. As would be expected, the electrophoretic properties were changed as a result of the modification reactions. Friemuth and coworkers[35] compared the modification of lysine with acetic anhydride, succinic anhydride, and O-methylisourea. Reaction of O-methylisourea with lysine residues results in guanidation, which is useful in MS studies.[36,37] Friemuth and workers[35] did partially characterize the products from the modification reactions and observed that the isoelectric points of the acylated proteins moved into the acidic region, while the isoelectric points of the guanidated products became more basic. More recent work has used succinic anhydride for isobaric peptide termini labeling.[38] Succinylation of γ-interferon has been reported to increase in vitro affinity for cell receptors (Hep 2c cells) and had increased antiviral activity.[39] Succinylated β-lactoglobulin has been suggested as a tablet excipient for the delivery of probiotic bacteria to the colon.[40] Succinylation of streptavidin has reduced the renal localization of this protein in chemotherapy based on antibody pretargeting.[41,42] Pretargeting therapy involves the initial uses of an antibody or antibody fragment such as a Fab' labeled with biotin. The antibody or Fab' fragment binds to the target cell. Streptavidin is used to

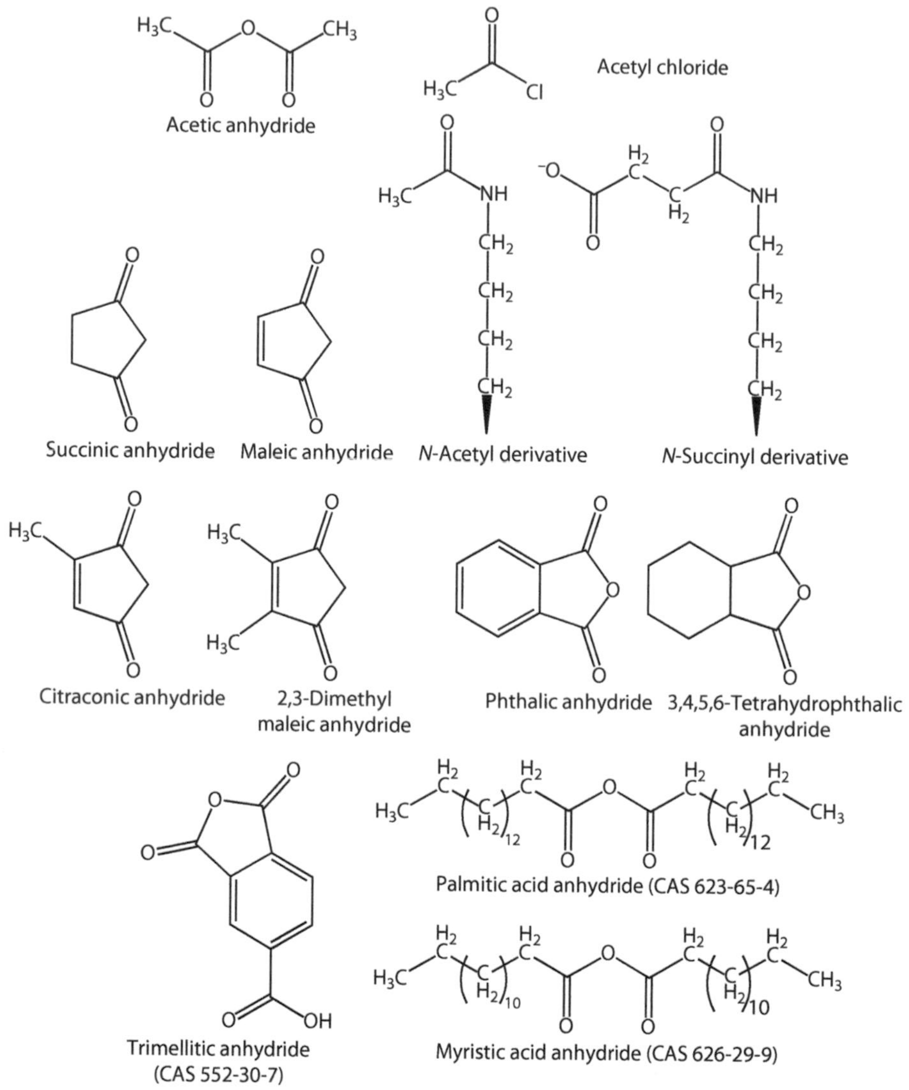

FIGURE 13.1 Reagents used for the acylation of amino groups in proteins. Shown at the top are acetic anhydride and acetyl chloride, both of which result in the N-acylation of proteins. Acetylation of amino groups results in charge neutralization in which reaction with a dicarboxylic anhydride such as succinic anhydride results in a charge reversal with a negative charge replacing a positive charge; the charge is preserved with reaction with imidoesters as shown in Figure 13.4. Acetyl chloride is highly reactive and somewhat unselective. Fatty acid chloride such as palmitoyl chloride or myristoyl chloride is used for the modification of protein and improving protein transport quality (Kocevar, N., Obermajer, N., Strukelj, B. et al., Improved acylation method enables efficient delivery of functional palmitoylated cystatin into epithelial cells, *Chem. Biol. Drug. Dev.* 69, 124–131, 2007). The modification with succinylation can be reversed by acid hydrolysis, while mild acid is sufficient to reverse citraconylation; the product obtained from the reaction of 2,3-dimethylmaleic anhydride is even less stable (Dixon, H.B.F. and Perham, R.N., Reversible blocking of amino groups with citraconic anhydride, *Biochem. J.* 109, 312–314, 1968). Shown at the bottom is trimellitic anhydride that has been infrequently used for protein modification (Neurath, A.R., Debnath, A.K., Strick, N. et al., Blocking of CD4 cell receptors for the human immunodeficiency virus type 1 (HIV-1) by chemically modified bovine milk proteins: Potential for AIDS prophylaxis, *J. Mol. Recognit.* 8, 304–316, 1995). Also shown at the bottom are palmitic acid anhydride and myristic acid anhydride (Pool, C.T. and Thompson, T.E., Methods for dual, site-specific derivatization of bovine pancreatic trypsin inhibitor: Trypsin protection of lysine-15 and attachment of fatty acids or hydrophobic peptides at the N-terminus, *Bioconjug. Chem.* 10, 221–230, 1999).

deliver the chemotherapeutic agent, such a short-lived, high-emitting isotope. The problem is to avoid the tendency of streptavidin to localize in the kidney. Succinylation of the streptavidin did not have a marked effect on the ability to bind avidin but did reduce renal binding.

Modification of proteins with citraconic anhydride (Figure 13.1) is a reversible reaction (see Chapter 3). The early work and development of citraconic anhydride for use in protein chemistry were reviewed by Atassi and Habeeb.[43] Reaction conditions for the modification of lysine residues in proteins with citraconic anhydride are similar to those described for succinic anhydride.[44] The difference is the stability of the product. Dixon and Perham[45] provided the initial description of the use of citraconic anhydride in 1968; the product obtained with maleic anhydride was too stable to be useful, while the product obtained with 2,3-dimethylmaleic anhydride was too labile to be useful. The citraconylated product was stable under conditions of tryptic digestion but was removed at pH 3.5/20°C. Barber and Warthesen[44] modified wheat gluten with succinic anhydride or citraconic anhydride in an attempt to improve the function in food products. The succinylated product was stable at pH 3.0 (stomach pH), while modification of the citraconylated product was fully reversed after 24 h. Habeeb and Atassi[46] evaluated several reagents including citraconic anhydride for the reversible blocking of amino groups. All of the primary amino groups in lysozyme were modified by multiple additions of reagent at pH 8.2 (the pH of the reaction mixture was maintained with a pH-stat) at room temperature. The product of the reaction was heterogeneous as judged by polyacrylamide gel electrophoresis. Citraconyl groups bound as esters (phenolic hydroxyl) could be removed by treatment with 1.0 M hydroxylamine, pH 10.0. Complete removal of the citraconyl groups was achieved by incubation at pH 4.2 for 3 h at 40°C yielding a mostly homogeneous protein with full enzymatic activity. While citraconylation is more stable at neutral to slightly basic pH, there is a finite loss of citraconyl groups. Reaction with citraconic anhydride has been used to dissociate nucleoprotein complexes.[47] Modification of the lysine residues with citraconic anhydride (pH 8.0–9.0 maintained with pH-stat) resulted in a marked change in the charge relationship between the α-amino groups of lysine and the phosphate backbone of the nucleic acid, allowing subsequent separation of protein from nucleic acid. The citraconyl groups were subsequently removed from this protein by incubation at pH 3.0–4.0 at 30°C for 3 h. As noted in Chapter 3, the majority of recent use of citraconic anhydride has been in antigen retrieval following formaldehyde fixation.[48] While citraconic anhydride is effective in antigen retrieval, it is not clear that the mechanism is understood. Antigen retrieval can be accomplished with other solvent conditions. Leong and Haffajee[48] used 0.1 M citrate, pH 6.0, or 0.05% (0.045 M) citraconic anhydride in H_2O (pH adjusted to 7.4 with 0.1 M NaOH). Tissue samples were subjected to microwave irradiation (98°) for 10 min and evaluated for reaction with a panel of commercial antibodies used in immunohistocytochemistry. Earlier work by Namimatsu and coworkers[49] used an electric kitchen pot for the heat step. This work appears to be the first use of citraconic anhydride. In principle, the reaction of formaldehyde with primary amines such as ε-amino group of lysine forms a Schiff base, which is unstable and can revert to the parent amine.[50]

Ketene (ethenone) was described early as reagents for acetylating amino and hydroxyl groups in proteins.[51] Diketene (4-methylene-2-oxetanone) is a component of various commercial organic chemical products[52] and is described as a potent acylating agent.[53] Both ketene and diketene are highly reactive materials that present challenges for use in that they are hazardous chemicals and present stability issues in aqueous solution. As such, these reagents have been used only occasionally in protein chemistry. More information on the chemistry of ketene and diketene is contained in Chapter 3. It is noted that an interesting ketene derivative has proved useful for the selective modification of an N-terminal α-amino group.[54] The acetoacetylation of amino groups by diketene can be reversed by neutral hydroxyl amine.[55] Mahley and coworkers have prepared the acetoacetyl derivatives of lipoproteins by reaction with diketene in 0.3 borate, pH 8.5.[56,57] The modification of tyrosyl and seryl residues also can occur under these conditions, but the O-acetoacetyl groups can be removed by dialysis against a mild alkaline buffer such as bicarbonate. The extent of modification was determined by subsequent titration with fluorodinitrobenzene (FDNB). The modification at lysyl residues was reversed by 0.5 M hydroxylamine, pH 7.0 at 37°C. Modification of 30%–60% of the total lysine residues of human low-density lipoprotein markedly increases the rate of clearance in rats; the clearance of human high-density lipoprotein was also increased by acetoacetylation. In contrast, acetoacetylation of rat or dog HDL retarded clearance in the rat model. The use of diketene has become a well-accepted procedure for the modification of lysine residues in apolipo-proteins.[58–64] Diketene has only been used on several other occasions.[65–68]

Urabe and coworkers[69] prepared various mixed carboxylic acid anhydrides of tetradecanoic acid and oxa derivatives, which varied in their *hydrophobicity*. This represented an attempt to change the surface properties of the enzyme molecule, in this case, thermolysin. The carboxylic acid anhy-drides were formed in situ from the corresponding acid and ethylchloroformate in dioxane with triethylamine. The modification reaction was performed in 0.013 M barbital, 0.013 M $CaCl_2$, pH 8.5 containing 39% (v/v) dioxane and was terminated with neutral hydroxylamine, which also served to remove *O*-acyl derivatives. The extent of reaction was determined by titration with trinitroben-zenesulfonic acid. Derivatives obtained with tetradecanoic acid and 4-oxatetradecanoic acid were insoluble. Derivatives obtained with 4,7,10-trioxatertradecanoic acid and 4,7,10,13-tetraoxotetra-decanoic acid both had approximately seven amino groups modified per mole of enzyme, showed little if any loss in either proteinase or esterase activity, and possessed enhanced thermal stability. Pool and Thompson[22,70] selectively attached long-chain (C_{12}–C_{16}) fatty acids to the amino-terminal amino acid of chemically modified (guanidinated) bovine pancreatic trypsin inhibitor. Fatty acid anhydrides were used to selectively acylate the α-amino group (1.5 μmol guanidinated bovine pan-creatic trypsin inhibitor was dissolved in 1.0 mL 100 mM phosphate, pH 7.8, and 100 μmol acid anhydride in 1.0 mL tetrahydrofuran added with stirring at 20°C–50°C [temperature depended on the physical nature of the anhydride solid or liquid] for 1 h. The modified protein was precipitated by the addition of tetrahydrofuran and purified by gel filtration followed by reverse-phase HPLC [C_3]). Fatty acid chlorides can also be used for the acylation reaction.[71] While there is considerable interest in the in vivo *S*- and *N*-acylation of proteins,[72] there is little interest in the in vitro modifica-tion of lysine residues with long-chain acyl compounds.

Howlett and Wardrop[73] were able to dissociate the components of human erythrocyte membrane by the use of 3,4,5,6-tetrahydrophthalic anhydride (Figure 13.1). The reaction was performed in 0.02 M Tricine, pH 8.5. The 3,4,5,6-tetrahydrophthalic anhydride was introduced into the reac-tion mixture as a dioxane solution (a maximum of 0.10 mL/5 mL reaction mixture). The pH was maintained at pH 8.0–9.0 with 1.0 M NaOH. The reaction was considered complete when no further change in pH was observed. The extent of modification was determined by titration with trinitro-benzenesulfonic acid. The reaction could be reversed by incubation for 24–48 h at ambient tempera-ture following the addition of an equal volume of 0.1 M potassium phosphate, pH 5.4 (the final pH of the reaction mixture was 6.0). More detail on the use of this reagent is contained in Chapter 3. It is noted that there has been limited use of this reagent over the past 20 years.[74–76]

Modification of amino groups with acetic anhydride (Figure 13.1) is one of the oldest[77] and still most extensively used approaches for the chemical modification of proteins. The reaction is preformed in half-saturated sodium acetate.[77,78] Acetic anhydride has the potential to modify pri-mary amino groups, aliphatic and aromatic hydroxyl groups, and the imidazole ring. Performing the modification reaction under these latter conditions (half-saturated sodium acetate) results in increased specificity since *O*-acetyl tyrosine is unstable in sodium acetate. However, more recent work as described in the following text is accomplished with either buffer or pH-stat control.[79,80] However, pH change during the reaction of an anhydride with a protein is always a concern because of charge neutralization and formation of a carboxylic acid.

Porcine pancreatic elastase was modified with acetic anhydride.[81] The reaction was carried out in a pH-stat or in various buffers. Inactivation of elastase at neutral pH (7.6, pH-stat) with acetic anhydride was reversible and activity recovered when acetic anhydride was no longer added to the reaction mixture; at pH 9.5, reactivation of elastase was complete within one minute. The revers-ibility of reaction suggests that there is *O*-acetylation of tyrosine and perhaps serine or threonine at neutral pH. Acetylation of tyrosine, serine, threonine, and histidine is readily reversed.[82] There is evidence to suggest that the reversal of serine and histidine in proteins is more readily reversed than model compounds suggesting a role for catalysis.[83] The inactivation of elastase by acetic anhydride was irreversible at pH 10.5 or at 9.0–9.5 in 4 M urea. The amino-terminal valine was not available for modification at pH 7.4 but could be modified at pH 11.0. Modification of this residue could be achieved in the presence of urea (4.0 M) at a lower pH (9.0). It was suggested that the reversible inactivation was likely due to the transient acetylation of a histidine residue (see Chapter 14) and that the irreversible inactivation was a reflection of the modification of the N-terminal valine residue. Subsequent work on the effect of pressure on the Raman spectrum of elastase showed that elastase was more stable than trypsin.[84] Trypsin showed a change in the amide I band with pressure and activity was lost, while no such change was seen with elastase; the lost of activity is suggested to reflect disruption of the salt bridge between the amino-terminal isoleucine and aspartic acid in the charge relay system. The results suggest that changes in amino acid residues at elastase active site stabilize the salt bridge between the amino-terminal valine and the aspartic acid in the elastase charge relay system.

In a study, which to my surprise has been largely ignored, Smit[85] used native electrophoresis to study the modification of human interleukin-3 with acetic anhydride. Changes in electrophoretic mobility correlated well with the amino group modification as measured by reaction with trinitrobenzenesulfonic acid.[86] This study is also quite useful in that it provides a study of the effect of pH on the modification of amino groups with acetic anhydride. Human interleukin-3 has eight amino groups for potential modification. Approximately two groups were available for modification at pH 5.0 (acetate), three at pH 6.0 (MES), and approximately eight at pH 9.0 (borate). Gao and Whitesidess[87] developed protein charge ladders for the purpose of determining the effective charge and molecular weight of a protein under nondenaturing conditions. Fourteen proteins differing in charge, size, and shape were modified with acetic anhydride to yield protein charge ladders differing in integral units of charge but minimum change in electrophoretic drag. The small size of the substitution (acetyl) assures that there would be minimal change in hydrodynamic radius and thus hydrodynamic drag. Careful management of reagent concentration, reaction pH, and time permitted the formation of a charge ladder on capillary electrophoresis. The charge ladder consists of rungs, each of which contains a discrete number of amino groups modified. In a related study, Anderson and coworkers[88] used acetyl-NHS instead of acetic anhydride for the acetylation of lysine residues in bovine carbonic anhydrase. These investigators noted that as modification increases, the reactivity of the remaining lysine residues decreases. As an example, the second-order rate constant (assumed 23°C; temperature not given) decreased from approximately 4.2 M^{-1} s^{-1} at zero modification to approximately 0.8 M^{-1} s^{-1} at 15 mol of lysine-modified/mole protein at pH 9.09. It could be argued that the decrease in modification reflects an increase in the relative number of protonated lysine residues. The preceding work using charge ladders focuses on application in electrophoretic systems. Chung and coworkers[89] extended the use of protein charge ladders to ion-exchange chromatography. A protein charge ladder was prepared by reaction of acetic anhydride with lysozyme and evaluated by chromatography on a sulfopropyl matrix. Effluent fractions were analyzed by capillary electrophoresis and MS. Some effluent fractions contained several variants differing in net charge. There was also a variant with more positive charge eluting before a variant with less positive charge. It is suggested that charge patches are important in retention on this cation-exchange matrix.

The use of acetic anhydride to modify amino groups in proteins evolved into a much more sophisticated approach for the study of protein conformation as a result of advances in analytical technology. In particular, MS has become extremely useful for the study of chemical modification in proteins.[90–92] Suckau and colleagues[93] used modification with acetic anhydride to study the surface topology of hen egg white lysozyme. Modification was performed in 0.5 M NH_4CO_3, pH 7.0 (maintained by the addition of NH_4OH [30% NH_c] with a 10–10,000 molar excess [to amino groups]), or acetic anhydride for 30 min at 20°C. Analysis by MS before and after tryptic hydrolysis permitted the identification of modified residues and the assignment of relative reactivity of the individual amino groups. This is an excellent paper on the modification of surface amino acid residues, which has been extensively cited since its publication in 1992.[3,94–96]

S-Ethylthiotrifluoroacetate was used to modify cytochrome c in 0.14 M sodium phosphate, pH 8.0.[97–99] The pH was maintained at 8.0 using a pH-stat. Singly substituted derivatives of cytochrome c can be separated by chromatography on anion-exchange resin (Bio Rex 70)[97] and carboxymethylcellulose.[99] It is critical to avoid lyophilization during the preparation of the various derivatives. These derivatives have been subjected to further investigation,[100,101] including the use of ^{19}F-containing derivatives for NMR probes.[102] It was possible to isolate 11 single-modified derivatives of cytochrome c by reaction with ethylthiotrifluoroacetate,[100–102] which were of considerable value in early studies of interaction of cytochrome c with other proteins including sulfite oxidase[100] and cytochrome c peroxidase.[102] There has been little use of S-ethylthiotrifluoroacetate in recent years. Kamo and Tsugita[103] demonstrated the cleavage of peptide chains at the amino groups serine and threonine in peptide chains. The cleavage was performed in the vapor phase (50°C for 6–24 h or 30°C for 24 h); reaction in solution resulted in cleavage at glycine. The reaction could also be performed on proteins blotted to a membrane. S-Ethyltrifluoroacetate has been demonstrated to potentiate the immune response to halothane in a rodent model.[104]

Competitive labeling (trace labeling, differential labeling) is a technique for determining intrinsic reactivity (accessibility and reactivity) of individual amino groups in a protein.[105] The method is based on the hypothesis that the individual amino groups will compete for a trace amount of radiolabeled reagent (the reagent is selected on the basis of nonselective reactivity with amino groups; with most studies, acetic anhydride has been the reagent of choice). The extent of radiolabel incorporation into the protein at a given site will then be a function of the pK_a, microenvironment, and inherent nucleophilicity of that particular amino group.[105] After the reaction with the radiolabeled reagent is complete, the protein is denatured, and complete modification at each amino group is achieved by the addition of an excess of unlabeled reagent. A reproducible digestion method (i.e., tryptic or chymotryptic hydrolysis) is used to obtain peptides from the completely modified protein. The peptides are separated by a chromatographic technique, and the extent of radiolabel at each site is determined. The extent of radiolabel incorporation at a given site is a function of the reactivity of that individual amino group under the reaction conditions used at the radiolabel step. The inclusion of a competing nucleophile such as phenylalanine enables the calculation of pK_a values for the various lysine residues in the protein. Xu[106] used competitive labeling with tritiated acetic anhydride on the α-polypeptide of Na^+, K^+ ion-activated ATPase. Lys501 was found to have a normal pK_a of 10.4 and second-order rate constant of 400 M^{-1} s^{-1} at 10°C. This rate is approximately 30% of that expected for a normal amine (see Table 3.2); the decreased rate of reaction is suggested to reflect the steric accessibility of Lys501. An alternative approach[107,108] involves a *trace* labeling step with tritiated acetic anhydride followed by complete modification with unlabeled acetic anhydride under denaturing conditions. This modified protein is then mixed with a preparation of the same protein that has been uniformly labeled with the ^{14}C-labeled acetic anhydride. Digestion and separation of peptide are performed by conventional techniques (see above), and the extent of radiolabeling is determined. The ratio of ^{3}H to ^{14}C in peptides containing amino groups is an indication of functional group reactivity. This method is somewhat more sensitive than the original method. Reductive methylation has also been used.[56] Rieder and Boshaard[109] used competitive (differential) labeling of lysine residues to determine binding sites on cytochrome c. These investigators used both radiolabeled acetic anhydride and formaldehyde (reductive methylation) to identify the binding site for cytochrome c oxidase on cytochrome c. Similar results were obtained with acetylation or reductive methylation. As an experimental control, cytochrome c oxidase did not have an effect on the reaction of acetic anhydride with ovalbumin supporting the specificity of competitive labeling.

Although competitive (differential) labeling was a laborious technique, the data obtained are excellent and provide considerable insight into the solution structure of proteins. There was use of this technique for the study of troponin-T,[110] troponin-C,[111] troponin-I,[112] calmodulin,[113–115] and tropomyosin.[116] Studies[115,116] that have used this technique to assess conformational change in solution have been useful.

The development of MS for measuring chemical modification permitted the direct determination of amino group reactivity without the use of radiolabeled material as described earlier. Ohguro and colleagues[117] used reaction with acetic anhydride to study conformational changes in arrestin as well as the association of arrestin with P-Rho. The effect of light/dark cycles on reactivity with acetic anhydride was evaluated as well as on the interaction of arrestin with P-Rho. The initial modification was performed with low levels (1–10 mM) of deuterated acetic anhydride at 0°C in 100 mM sodium borate, pH 8.5. This was followed by modification with higher concentrations of acetic anhydride (20 mM) in 100 mM sodium borate, pH 9.0 containing 6.0 M guanidine hydrochloride. The ratio of deuterated to protiated modification permitted the identification of residues *protected* from modification by interaction with P-Rho as well three lysine residues that were more reactive as a result of the interaction with P-Rho. A similar approach was used by Scaloni and coworkers[118] to study the interaction of thyroid transcription factor 1 homeodomain with DNA. Thyroid transcription factor 1 is a regulatory protein responsible for transcriptional activation and binds to DNA via a homeodomain region. A 1–10-fold molar excess of acetic anhydride was added to free or DNA-complexed thyroid transcription factor 1 homeodomain in 20 mM Tris-HCl–75 mM KCl, pH 7.5 at 25°C for 10 min. The acetylated samples were subjected to cyanogen bromide cleavage in 70% formic acid at room temperature for 18 h in the dark. The fractions were separated by HPLC and subjected to analysis by MS. Limited proteolysis of the free or DNA–protein complex was also used to define regions of DNA–protein interaction. The relatively small size of the homeodomain (61 amino acids) compared to the parent thyroid transcription factor 1 permitted NMR analysis. The 3-D structure obtained from NMR analysis was in good agreement with the surface topology model obtained from acetylation and limited proteolysis. D'Ambrosio and coworkers[119] used the combination of limited proteolysis and chemical modification to study the dimeric structure of porcine aminoacylase 1. Reaction with acetic anhydride was performed in 10 mM NH_4CO_3, 1 mM DTT, and 1 MM $ZnCl_2$, pH 7.5 at 25°C for 10 min using a 100–5000-fold molar excess of reagent. Modification occurred readily at the amino terminus and at 8 of the 17 lysine residues. These investigators also used TNM, cross-linking with *bis*(sulfosuccinimidyl)suberate, and limited proteolysis with several proteases to obtain a model for this multimeric protein. Zappacosta and coworkers[120] measured amino reactivity on minobody, a small de novo designed β-protein, by reaction with a low concentration of acetic anhydride (twofold molar excess to total amino groups). The modification reaction was performed in 50 mM NH_4CO_c, pH 7.5, at 25°C for 10 min. Two lysine residues were highly reactive, one lysine and the α-amino group were less reactive, and one lysine residue was not modified. MS was used to evaluate modification reactions. Chemical modification with TNM, NAI, and 1,2-cyclohexanedione was also used to measure functional group reactivity. Turner and coworkers[121] used acetic anhydride to evaluate solvent accessibility in factor XIII as a function of activation; cysteine residue reactivity was evaluated by reaction with an *N*-methylmaleimide or NEM (there was no difference observed between these two reagents). The acetylation of Lys73 and Lys211 was observed on activation of factor XIII with thrombin; acetylation of Lys677/Lys678 that was observed in the zymogen form was not observed in the activated enzyme. The relationship of functional group reactivity and exposure is problematic and discussed in some detail in Chapter 1.

Calvete and coworkers[122] used a clever approach to identify the heparin-binding domain of bovine seminal plasma protein PDC-109. The PDC-109 protein was bound to heparin–agarose in 16.6 mM Tris, 50 mM NaCl, and 1.6 mM EDTA, 0.025% NaN_3, pH 7.4. After washing the column to remove protein not bound to the matrix, the column was recycled at room temperature with the application buffer containing acetic anhydride (25–1600-fold molar excess over protein lysine). A similar experiment was performed with 1,2-cyclohexanedione (0.4–6 molar excess over arginine residues) to study arginine modification. Hydrolysis of acetic anhydride is a significant issue (Table 13.2), and a large molar excess of reagent is required compared to 1,2-cyclohexanedione, which is relatively stable in aqueous solution. Six basic residues (three lysine residues and three arginine residues) were

TABLE 13.2

Rate Constants for the Reaction of 2,4,6-Trinitrobenzene Sulfonic Acid (TNBS) with Model Compounds and Proteins

Reaction	Rate ($M^{-1} s^{-1}$)	Reference
N-acetyl-L-cysteine/50 mM borate, pH 9.5/23°C[a]	6.8×10^1	13.2.1
Glycine/50 mM borate, pH 9.5/23°C[a]	4.2	13.2.1
Glycylglycine/50 mM borate, pH 9.5/23°C[a]	2.1	13.2.1
N^α-acetyl-L-lysine/50 mM borate, pH 9.5/23°C[a]	1.3	13.2.1
Glycyl-L-lysine/50 mM borate, pH 9.5/23°C[a]	2.7^b	13.2.1
Glycyl-L-lysine/50 mM borate, pH 9.5/23°C[a]	0.6^c	13.2.1
Glycylglycine/25.4°C/I=0.49 sodium phosphate, pH 12.6	2.34	13.2.2
Glycine/25.4°C/I=0.49 sodium phosphate, pH 12.6	11.85	13.2.2
Alanine/25.4°C/I=0.49 sodium phosphate, pH 12.6	6.17	13.2.2
ε-Aminocaproic acid/25.4°C/I=0.49 sodium phosphate, pH 12.6	15.5	13.2.2
Diethylamine/25.4°C/I=0.49 sodium phosphate, pH 12.6	7.8×10^{-2}	13.2.2
Tris(hydroxymethyl)amino methane/25.4°C, at pK_1 corrected for degree of ionization	1.5×10^{-2}	13.2.2
Imidazole/25.4°C, at pK_1 corrected for degree of ionization	2.8×10^{-3}	13.2.2
Hydroxide ion/25.4°C/I=0.49 sodium phosphate, pH 12.6 (data extrapolated to zero ionic strength)	1.8×10^{-2}	13.2.2
N^α-acetyllysine[d]/25°C, pH 8.0 (50 mM Bicine,[e] 66 mM NaHCO$_3$, 10 mM MgCl$_2$)	3.3×10^{-2}	13.2.3
Lysine-166 of activated *Rhodospirillum rubrum* ribulose-P$_2$-carboxylase/oxygenase/25°C, pH 8.0 (50 mM Bicine, 66 mM NaHCO$_3$, 10 mM MgCl$_2$)	6.1	13.2.3
Lysine-334 of activated spinach ribulose-P$_2$-carboxylase/oxygenase/25°C, pH 8.0 (50 mM Bicine, 66 mM NaHCO$_3$, 10 mM MgCl$_2$)	6.8^f	13.2.3
Lysine-334 of deactivated spinach ribulose-P$_2$-carboxylase/oxygenase/25°C, pH 8.0 (50 mM Bicine, 0.1 mM EDTA)	6.8^f	13.2.3
Salicylate hydroxylase from *Pseudomonas putida* S-1/25°C, 100 mM potassium phosphate, pH 8.5 (in the dark)	2^g	13.2.4
Glycine/0.1 M phosphate, pH 7.4, 25°C	0.07 (11.3[h])	13.2.5
Tyrosine/0.1 M phosphate, pH 7.4, 25°C	0.16 (8.5[h])	13.2.5
α-N-acetyl-L-lysine amide/0.1 M phosphate, pH 7.4, 25°C	~0.7 (~83[h])	13.2.5
N-acetyl-DL-cysteine/0.1 M phosphate, pH 7.4, 25°C	0.4 (~167)	13.2.5

[a] Temperature not provided, assumed to be 23°C.

[b] α-Amino group.

[c] ε-Amino group.

[d] Presumed to be the L isomer.

[e] Bicine, N,N'-*bis*(2-hydroxyethyl)glycine.

[f] The pH-independent rate constants were higher. For example, the pH-independent second-order rate constant for reaction of Lys334 of the activated spinach enzyme was determined to be 75 $M^{-1} s^{-1}$. With the deactivated enzyme, the pH-independent rate constant was determined to be 433 $M^{-1} s^{-1}$.

[g] Determined from rate of inactivation.

[h] The pH-independent rate constant (intrinsic rate constant) is determined by dividing the observed second-order rate constant by fraction of amino acid or derivative with unprotonated amino group.

TABLE 13.2 (continued)
Rate Constants for the Reaction of 2,4,6-Trinitrobenzene Sulfonic Acid (TNBS) with Model Compounds and Proteins

References to Table 13.2

13.2.1. Fields, R., The measurement of amino groups in proteins and peptides, *Biochem. J.* 124, 581–590, 1971.

13.2.2. Means, G.E., Congdon, W.I., and Bender, M.L., Reactions of 2,4,6-trinitrobenzene-sulfonate ion with amines and hydroxide ion, *Biochemistry* 11, 3564–3571, 1972.

13.2.3. Hartman, F.C., Milanez, S., and Lee, E.H., Ionization constants of two active-site lysyl ε-amino groups of ribulosebisphosphate carboxylase/oxygenase, *J. Biol. Chem.* 260, 13968–13975, 1985.

13.2.4. Suzuki, K., Mizuguchi, M., Gomi, T., and Itagaki, E., Identification of a lysine residue in the NADH-binding site of salicylate hydroxylase from *Pseudomonas putida* S-1, *J. Biochem.* 117, 579–585, 1995.

13.2.5. Freedman, R.B. and Radda, G.K., The reaction of 2,4,6-trinitrobenzenesulphonic acid with amino acids, peptides and proteins, *Biochem. J.* 108, 383–391, 1968.

protected from modification by binding to the heparin matrix. Using a related approach, Hochleitner and coworkers[123] used reaction with acetic anhydride to define a discontinuous epitope in human immunodeficiency virus core protein p24. The protein was bound to the immobilized antibody and digested with endoproteinase Lys-C. Modification was then accomplished with acetic anhydride (10,000-fold molar excess of reagent, pH 7.8, 50 mM NH_4CO_c 20 min) followed by a 100,000-fold excess of the hexadeuterated reagents under the same conditions; pH was maintained by the addition of 10% NH_4OH.

Taralp and Kaplan[124] examined the reaction of acetic anhydride with lyophilized α-chymotrypsin in vacuo. α-Chymotrypsin was lyophilized from an unbuffered solution at pH 9.0 in one chamber in a reaction vessel. ^3H-acetic anhydride was added to another compartment in the reaction vessel. The reaction vessel was evacuated and placed in an oven at 75°C. Several reaction vessels were used and removed at various time intervals for analysis. The proteins were then modified with ^{14}C-acetic anhydride and the ratio of ^3H to ^{14}C was used to determine the extent of modification. While complete modification of amino groups is achieved at pH 9.0 in aqueous solution, in the nonaqueous system, only 25% of the ε-amino groups and 90% of the α-groups were modified. It also appeared that mixed anhydrides formed with carboxyl groups on the protein surface.

The studies by Calvete and coworkers, Hochleitner and coworkers, and Taralp and Kaplan illustrate the utility of chemical modification in the analysis of protein conformation with application to biopharmaceutical products.[125] The ability to use vapor-phase modification reaction presents the opportunity to modify lyophilized proteins.

Smith and coworkers[126] used reaction with acetic anhydride to determine the extent of posttranslational acetylation in histone H4. Histone H4 was modified with deuterated acetic anhydride (dried histone samples were suspected in deuterated glacial acid acetic and deuterated acetic anhydride added; the reaction mixture was allowed to stand for 6 h at ambient temperature). The modified samples were subjected to mass spectrometric analysis. The extent of endogenous acetylation was determined from the ratio of protiated to deuterated fragmentation ions. The extent of reaction with acetic anhydride can be measured by radioisotope incorporation or a decrease in reaction with a chromophoric reagent such as trinitrobenzenesulfonic acid. Radiolabeled acetic anhydride continues to be used to determine posttranslational modification in histones.[127] It should be mentioned that there is an antibody available for the determination of acetylation in proteins.[128–132]

Lysine residues can be modified by reaction with α-ketoalkyl halides (Chapter 2) such as iodoacetic acid.[133] Acylation can occur at pH > 7.0, but the rate of reaction is much slower than that of reaction with cysteinyl residues. Both the mono- and disubstituted derivatives of lysine have been reported. The monosubstituted derivative migrates close to methionine on amino acid analysis, while the disubstituted derivative migrates near aspartic acid. It should be noted that reaction with α-ketoalkyl halides is not considered particularly useful for the modification of primary amino groups. However, this is a reaction that has been demonstrated to occur during the reduction and carboxamidomethylation of proteins.[134] Nielsen and coworkers[134] showed that reaction with iodo-acetamide created an artifact resulting from dialkylation of lysine resembling ubiquitination; the modification was not observed with chloroacetamide. The reaction of iodoacetamide with polyly-sine, N^α-acetyllysine, and N^ε-acetyllysine has been described.[135] Iodoacetamide has been reported to alkylate a variety of nucleophiles including lysine and α-amino acids; disubstitution of an amino-terminal α-amino group was observed.[136] Heinrikson observed the alkylation of Lys41 in pancreatic RNase with bromoacetate.[137]

Reaction with aryl halides such as dinitrofluorobenzene (FDNB) (Figure 13.2) has been of considerable value in protein chemistry since Sanger and Tuppy's work on the structure of insu-lin.[138,139] A combination of site-specific mutagenesis and site-specific chemical modification with 2,4-dinitrofluorobenzene was used to study lysine residues in angiogenin.[140] Carty and Hirs[141] devel-oped the use of 4-sulfonyloxy-2-nitrofluorobenzene for the modification of amino groups in pancre-atic ribonuclease. The lysine residue at position 41 is the site of major substitution, which is a reflection of the lower pKa for the ε-amino group of this residue. The use of this compound did not present the solubility and reactivity problems posed by the fluoronitrobenzene compounds. It was possible to qualitatively determine the classes of amino groups in ribonuclease; these were the α-amino group, nine *normal* amino groups, and lysine 41. The reactivity of lysine 41 was influenced by the neighbor-ing functional groups. This effect was lost at pH > 11 or on thermal denaturation of the protein. The pH-independent second-order rate constant for the modification of Lys41 in pancreatic RNase with 4-sulfonoxy-2-nitrofluorobenzene was determined to be 7.2×10^{-3} M^{-1} s^{-1}.[142] Notwithstanding the improved physical properties, there has been no further use of 4-sulfonyloxy-2-nitrofluorobenzene.

1-Dimethylaminonaphthalene-5-sulfonyl chloride (dansyl chloride) (Figure 13.2) has been use-ful for the modification of both α-amino groups and ε-amino groups in proteins. Dansyl chlo-ride replaced FDNB in the determination of amino-terminal amino acids in proteins and peptides reflecting greater sensitivity.[143,144] In one study,[145] dansyl chloride (in acetone) is added to a solution of trypsin in 0.1 M phosphate, pH 8.0. The reaction is terminated after 24 h at 25°C by acidifica-tion to pH 3.0 with 1.0 M HCl. Insoluble material is removed by centrifugation, and the supernatant fraction is placed in dialysis. These investigators reported modification of the amino-terminal iso-leucine and one lysine residue. The extent of modification was determined by absorbance at 336 nm ($\varepsilon_m = 3.4 \times 10^4$ M^{-1} cm^{-1}). Reaction of dansyl chloride with phosphoenolpyruvate carboxylase has been used to introduce a fluorescent probe into this protein.[146] Eight lysine residues were modified by dansyl chloride at pH 8.0 in the tetrameric enzyme; Tsou plots (Chapter 1) showed that four of these lysine residues were essential for catalytic activity. The lysine residues essential for catalytic activity were modified at five times the rate of the nonessential residues. The extent of modification was determined by spectral analysis at 355 nm using an extinction coefficient of 3400 M^{-1} cm^{-1}. The dansylated enzyme was modified with eosin thiocyanate. The double-modified enzyme was used for fluorescence energy transfer experiments where the dansyl group was the donor and eosin the acceptor group. Despite early enthusiasm, there has been only limited use of dansyl chloride for the modification of proteins[147,148] including a dansylated antibody used for immunoblotting.[149] An anti-body to the dansyl group can also be used to identify modified proteins in electron microscopy.[150] There has been substantial use of dansyl chloride in analytical chemistry for amino acids[151–153] and other amines.[154,155]

4-Chloro-3,5-dinitrobenzoate (CDNB)[156] and 2-chloro-3,5-dinitrobenzoic acid[157] (Figure 13.2) react with amino groups and sulfhydryl groups in proteins forming a chromophoric negatively charged derivative. A dichloroderivative, 6-carboxy-2,4-dinitro-1,3-dichlorobenzene was also pre-pared for potential use as a cross-linking agent.[157] The reaction of CDNB with lysine results in

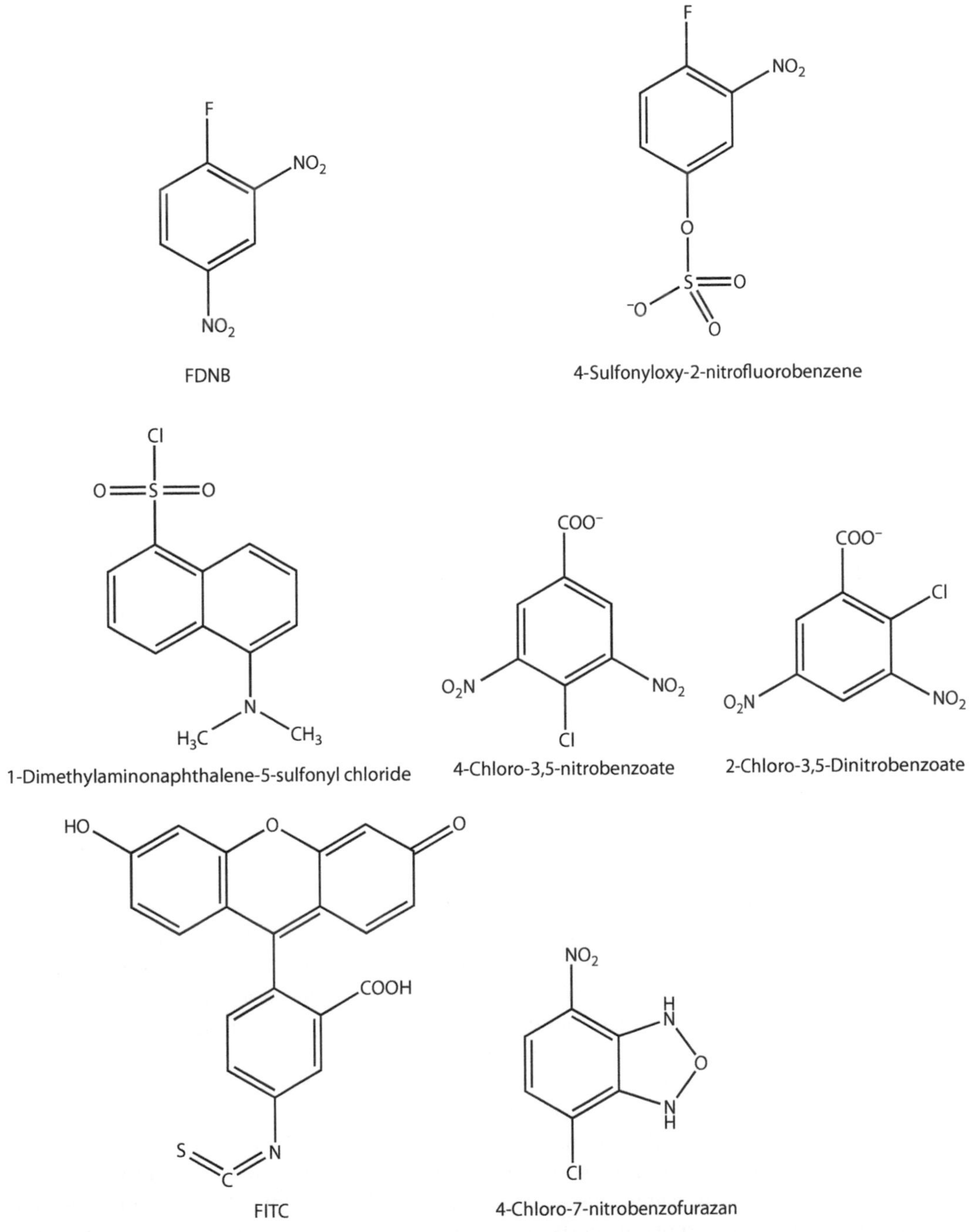

FIGURE 13.2 The structures of some aryl derivatives used for the modification of amino groups in proteins and serving as chromophores and fluorophores. Shown is the structure of FDNB, which has the longest history of use dating back to Sanger's work on insulin in 1949 (Sanger, F., Terminal peptides of insulin, *Biochem. J* 45, 563–574, 1949), and other derivatives have limited use and are discussed in the text. 1-Dimethylaminonapthaline-5-sulfonyl chloride (DNS chloride) and FITC are both chromophores and fluorophores and have been used in fluorescence energy transfer with dansyl as donor and fluorescein as acceptor (Dergunov, A.D., Vorotnikova, Y.Y., De Pauw, M., and Rosseneu, M., Apolipoprotein E self-association in solution studied by non-radiative energy transfer, *J. Biochem. Biophys. Methods* 29, 259–267, 1994).

the formation of the 4-carboxy-2,6-dinitrophenyl derivative.[156] Hall and coworkers[158] used CDNB to modify specific lysine residues in cytochrome c_2; derivatives with individual lysine residues were separated by cation-exchange chromatography (sulfoxyethyl cellulose). Subsequent work by Long and coworkers[159] used these derivatives with unique modified lysine residues to study electron transfer. The modification reaction (approximately sixfold molar excess of reagent) was performed in 0.2 M sodium bicarbonate, pH 9.0 at ambient temperature for 24 h.[156,158] The extent of modification was determined as described by Brautigan and coworkers.[156] The absorbance maximum of derivatives formed with various alkylamines was 436 nm with an extinction coefficient of 6.9×10^3 M^{-1} cm^{-1}. While CDNB appears to be useful for the site-specific modification of lysine residues in proteins, there has only been limited use of this reagent in protein chemistry.[160–164]

Hiratsuka and Uchida[165] examined the reaction of N-methyl-2-anilino-6-naphthalenesulfonyl chloride with lysyl residues in cardiac myosin. There was a difference in the nature of the reaction in the presence and absence of a divalent cation. N-Methyl-2-anilino-6-naphthalenesulfonyl chloride has been suggested for use as a fluorescent probe for hydrophobic regions of protein molecules.[166–168] The extent of incorporation of the N-methyl-2-anilino-6-naphthalenesulfonyl moiety into protein can be determined by spectral analysis at 327 nm ($\Delta\varepsilon = 2.0 \times 10^4$ M^{-1} cm^{-1}).[166,167] The use of N-methyl-2-anilino-6-naphthalenesulfonyl chloride appears to be restricted to studies on myosin.[169]

Modification of protein amino groups with isothiocyanate derivatives of various dyes has proved to be an effective means of introducing structural probes into proteins at specific sites.[168] Fluorescein isothiocyanate (FITC, Figure 13.2) has been used to modify cytochrome P-450 (reaction performed in 30 mM Tris, pH 8.0 containing 0.1% Tween 80; 2 h at 0°C in the dark),[170] actin (2 mM borate, pH 8.5; 3 h at ambient temperature and then at 4°C for 16 h),[171] and ricin (pH 8.1, 6°C for 4 h).[172] The extent of modification with FITC can be determined by spectroscopy using an extinction coefficient of 80,000 M^{-1} cm^{-1} at 495 nm (1% SDS with 0.1 M NaOH)[170] or 74,500 M^{-1} cm^{-1} (0.1 M Tris, pH 8.0).[171] Antibodies labeled with fluorescein have been used as targeted phototoxic agents.[173] In this approach, the fluorescein moiety is iodinated, resulting in a photodynamic sensitizer. The reader is directed to a study on the effect of microenvironment on the fluorescence of arylaminophthalenesulfonates.[174]

The reaction of primary amino groups in proteins with cyanate (Figure 13.3) has been a useful chemical modification procedure for several decades[175] and a technical challenge at one time.[176,177] More recently, there has been interest in the in vivo carbamylation of proteins by endogenously produced cyanate.[178–181] It is interesting to see solution chemistry modifications of proteins such as cyanylation and nitration that appeared to have disappeared, gaining much more prominence as in vivo event. Stark and coworkers[175] pursued the observation that ribonuclease was inactivated by urea in a time-dependent reaction. It was established that this inactivation was a reflection of the content of cyanate in the urea preparation. This observation was subsequently developed into a method for the quantitative determination of amino-terminal residues in peptides and proteins.[182] The reaction of cyanate with several amino acid residues was described by Stark.[183] The ε-amino group of lysine (ε-amino caproic acid) is the least reactive ($k = 2.0 \times 10^{-3}$ M^{-1} min^{-1}) compared to the α-amino group of glycylglycine ($k = 1.4 \times 10^{-1}$ M^{-1} min^{-1}). The rate of reaction with the thiol group of cysteine was much faster ($k = 4$ M^{-1} min^{-1}), while the rate of reaction with histidine was determined to be 1.8×10^{-1} M^{-1} min^{-1}; these measurements were made at 25°C. The carbamyl derivative of histidine is quite unstable as is the corresponding derivative of cysteine. There is also the possibility of modification at uniquely reactive residues. For example, the reaction of chymotrypsin with cyanate results in loss of catalytic activity associated with the carbamylation of the active-site serine residue.[184]

Manning and coworkers[185–188] established that the modification of sickle cell hemoglobin with cyanate increased the oxygen affinity of this protein. It was established that the amino-terminal value of hemoglobin is more reactive with cyanate in deoxygenated blood than in partially deoxygenated blood. At pH 7.4, the amino-terminal valyl residues of oxyhemoglobin S are carbamylated 50 to 100 times faster than lysyl residues.[188] The same laboratory has examined the carbamylation of α- and β-chains in some detail.[187] With the deoxy protein, the ratio of radiolabel from ^{14}C-cyanate

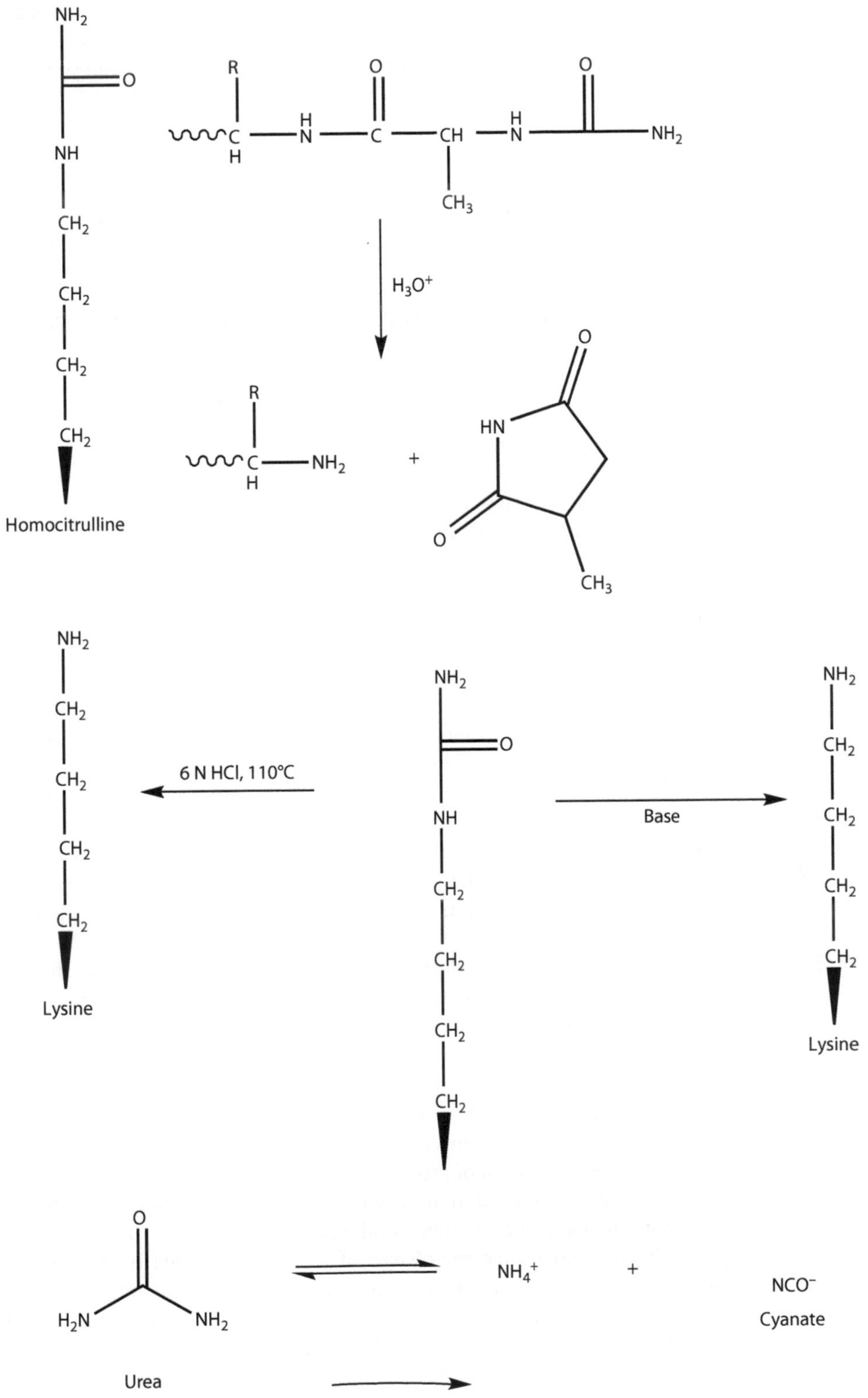

FIGURE 13.3

(*continued*)

FIGURE 13.3 (continued)   The reaction of cyanate with amino groups in proteins. The products of the reaction of cyanate with amino groups in proteins are shown. The rate of reaction of cyanate with the α-amino group is approximately 100 times faster than that with the ε-amino group of lysine at neutral pH (Stark, G.R., Modification of proteins with cyanate, *Methods Enzymol.* 11, 590–594, 1963). Homocitrulline, the product of the reaction of cyanate with the ε-amino group of lysine, slowly reverts to lysine under conditions of acid hydrolysis for amino acid analysis. The ε-carbamyl group is stable in dilute base at 23°C but can be quantitatively converted back to lysine in 0.2 M NaOH at 110°C for 16 h. The α-carbamyl product is converted to the hydantoin derivative in acid serving as the basis for use of cyanate for determining amino-terminal amino acids (Stark, G.R., Use of cyanate for determining NH$_2$-terminal residues in proteins, *Methods Enzymol.* 11, 125–138, 1963). Shown at the bottom is the conversion of urea to cyanate, which is a continuing problem in proteomics (Kollipara, L. and Zahedi, R.P., Protein carbamylation: In vivo modification or in vitro artifact? *Proteomics* 13, 941–944, 2013) and is an *in vivo* modification (Qin, W., Smith, J.B., and Smith, D.L., Rates of carbamylation of specific lysine residues in bovine α-crystallins, *J. Biol. Chem.* 267, 26128–26133, 1992; Koeth, R.A., Kalantar-Zadeh, K., Wang, Z. et al., Protein carbamylation predicts mortality in ESRD, *J. Am. Soc. Nephrol.* 24, 853–861, 2013).

on the α-chain as compared to the β-chain is 1.7:1.0, while it is 1:1 with the oxy protein. The carbamylation of the amino-terminal valine residues of hemoglobin is approximately 2.5-fold greater in partially deoxygenated media as compared to fully oxygenated media. Thus, it would appear that the reactivity of the amino-terminal valine is a sensitive index of conformational change.[188] The removal of Arg141 (α) with carboxypeptidase B abolishes the enhancement of carbamylation observed with the removal of oxygen from hemoglobin.

Weisgraber and coworkers[55] used carbamylation and reaction with diketene to explore the role of lysyl residues in the binding of plasma lipoprotein to fibroblasts. The reaction was performed in 0.3 M sodium borate, pH 8.0. The extent of modification was determined in two ways. In the first, the modified protein was subjected to acid hydrolysis. The amount of homocitrulline, the product of the reaction of the ε-amino group of lysine with cyanate, was considered equivalent to the number of lysine residues modified. However, homocitrulline is partially degraded on acid hydrolysis to produce lysine (17%–30%). In order to obviate this difficulty, these investigators used reaction with 2,4-dinitrofluorobenzene under denaturing conditions to measure unmodified amino groups. Modification of 15% of the lysine residues in low density lipoprotein (LDL) by carbamylation or 20% with diketene inhibited binding to LDL cell surface receptors. A similar effect was observed on reaction with; reversal of the acetoacetyl derivatives with hydroxylamine restored the ability of lipoprotein to bind to the LDL surface receptors. Plapp and coworkers[189] studied the modification of lysyl residues in bovine pancreatic deoxyribonuclease A with several different reagents including cyanate. The modification with cyanate was performed at 37°C in 1.0 M triethanolamine hydrochloride, pH 8.0. The extent of modification was determined by analysis for homocitrulline following acid hydrolysis. A time course of hydrolysis was utilized to provide for the accurate determination of homocitrulline, since this amino acid slowly decomposes to form lysine during acid hydrolysis (see above). There is further discussion of the carbamylation of DNase in the following text with imidoesters. I could find no reference to recent use of cyanate for the modification of lysine residues in proteins. The in vivo carbamylation of proteins (see above) has allowed the development of mass spectrometric methods for the determination of protein carbamylation.[180,190]

Imidoesters (Figure 13.4) have been used to modify amino groups in proteins with the majority of most use with reagents such as dimethyl suberimidate that are used for cross-linking. These reagents have the particular advantage that the charge of the lysine residue is maintained during the modification. The modification of a glutamine synthetase from *B. stearothermophilus* with ethyl acetimidate has been studied by Sekiguchi and coworkers.[191] The modification was performed at pH 9.5 with 0.2 M phosphate for 1 h at 35°C and terminated by dialysis at pH 7.2. The extent of modification was determined by titration of the modified protein with trinitrobenzenesulfonic acid (see below). As these investigators suggest, consideration must be given to the possibility of cross-linking occurring with this reagent under the conditions used.[192] Monneron and d'Alayer[193] examined the reaction of either methyl acetamidate or dimethyl suberimidate with

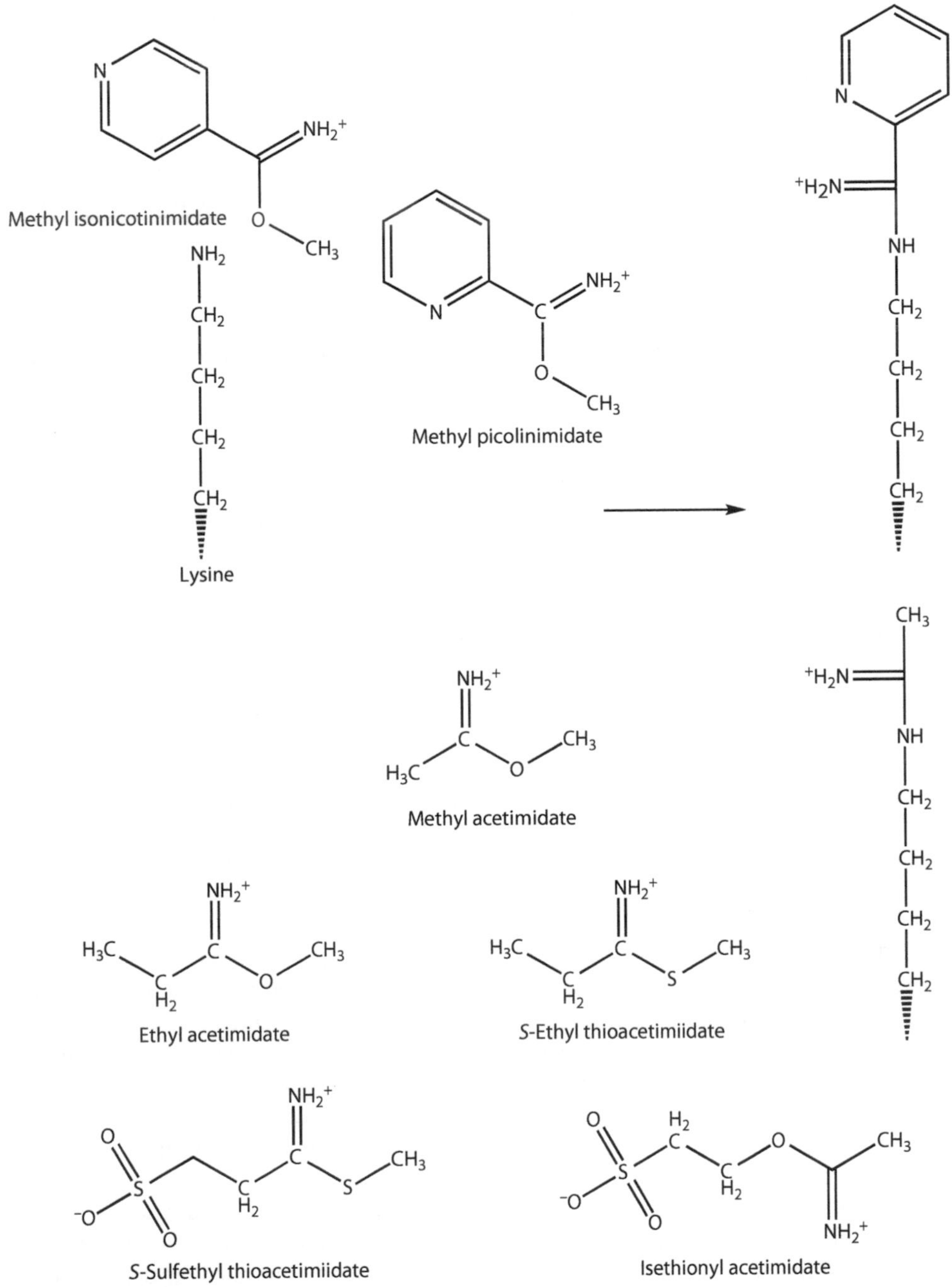

FIGURE 13.4 Structure of some imidoesters used for the modification of amino groups in proteins. Shown is the structure of methyl isonicotinimidate (Zoltobrocki, M., Kim, J.C., and Plapp, B.V., Activity of liver alcohol dehydrogenase with various substituents on the amino groups, *Biochemistry* 13, 899–903, 1974). Shown also are the structures of thioacetimidates (Thumm, M., Hoenes, J., and Pfleiderer, G., *S*-Methylthioacetimidate is a new reagent for the amidination of proteins at low pH, *Biochim. Biophys. Acta* 923, 263–267, 1987) and *S*-sulfethylthioacetimidate (Jaffee, E.G., Lauber, M.A., Running, W.E., and Reilly, J.P., In vitro and in vivo chemical labeling of ribosomal proteins: A quantitative comparison, *Anal. Chem.* 84, 9355–9361, 2012). Also at the bottom is isethionyl acetimidate, which, as with *S*-sulfethylthioacetimidate, is a membrane-impermeable reagent (Rack, M., Effects of chemical modification of amino groups by two different imidoesters on voltage-clamped nerve fibres of the frog, *Pflugers Arch.* 404, 126–130, 1985).

particulate adenylate cyclase. The reaction was performed in 0.05 M triethanolamine, 10% (w/v) sucrose, and 0.005 M $MgCl_2$ at pH 8.1.

Methyl picolinimidate (Figure 13.4) is an imidoester that reacts with the primary amino groups in proteins. Plapp and coworkers[189] examined the reaction of methyl picolinimidate with pancreatic deoxyribonuclease. The reaction was performed in 0.5 M triethanolamine hydrochloride, pH 8.0 containing 1 mM $CaCl_2$ with 0.1 M methyl picolinimidate for 22 h at 25°C, and then with 0.2 M methyl picolinimidate for an additional 8 h. Under these conditions, essentially all of the primary amino groups in deoxyribonuclease (nine lysine groups and one amino-terminal amino group) were modified, but there was no change in biological activity. Likewise, guanidation of the enzyme with O-methylisourea resulted in an active enzyme. Both of these modifications preserve the positive charge on the various amino groups. Carbamylation (see above) of the α-amino group and seven of the nine lysine residues resulted in approximately 50% loss of DNase activity; however, activity in the presence of calcium ions was equivalent to the native enzyme. Carbamylation of the last two lysine residues resulted in loss of activity. Trinitrophenylation was similar to carbamylation in that 4–5 mol of lysine per mole of protein could be modified without loss of activity. This is a useful study that compares the effect of different reagents on the modification of a well-understood protein. It has been extensively cited both for the chemistry of the modification reactions and for analytical methodology. The determination of the crystal structure of deoxyribonuclease I[194] established that lysine residues were not involved in the binding of substrate; the lack of involvement in catalysis had been inferred by Plapp and coworkers in their studies.[189] Chemical modification studies had shown the importance of tyrosine in DNase activity,[195] which was confirmed by the crystallographic results. Plapp has also studied the reaction of methyl picolinimidate with horse liver alcohol dehydrogenase (LADH).[196] This study was somewhat unique in that modification of the enzyme resulted in enhanced catalytic activity, reflecting more rapid dissociation of the enzyme–coenzyme complex. It was observed that the derivatized lysine reverts to lysine (60% yield) under the normal conditions of acid hydrolysis. This work on LADH was extended to the reaction of other imidoesters generated in situ.[197] It was shown that chlorobutyramide cyclizes to form 2-iminotetrahydrofuran, which modified the majority of amino groups in the protein. Plapp and coworkers also showed that modification of LADH with methyl isonicotinamide (Figure 13.4)[198] subsequently obtained the crystal structure of the isonicotinimidylated enzyme.[199] A variety of imidates have been shown to modify LADH with an observed enhancement of activity, while succinylation resulted in complete inhibition, while carbamylation had little effect on activity.[198] The apparent activation of LADH created a considerable amount of interest when Professor Plapp presented this observation at the annual FASEB meeting in Atlantic City, New Jersey, in 1969.[200] There was great disappointment when Professor Plapp informed the listeners that this effect was not observed in vivo.

Recent work has used thioimidates such as S-methylthioacetimidate,[96,201] which are suggested to have superior characteristics to oxyimidates such as methylacetimidate. S-Methylthioacetimidate was developed by Thumm and coworkers[202] and has the advantage reaction at pH as low as 5 with cross-linking or other side reactions. These investigators reported modification of 30 lysine residues in bovine serum albumin at pH 6.0 with a single addition of reagent. The half-life of S-methylthioacetimidate at neutral pH is approximately one hour.[202] The rate of hydrolysis of methyl acetimidate is more complex; at 20°C, the half-life is 29 min increasing to 55 min at pH 8.8.[203] The rate of hydrolysis is sensitive to temperature with the half-life in 0.1 M NaCl (corrected to pH 8.0) and decreases from 36 min at 20°C to 19 min at 30°C. Differing from the rate of hydrolysis of methyl acetimidate, the rate of reaction with denatured aldolase increases with increasing pH (0.84 M^{-1} min^{-1} at pH 6.8 to 2.32 M^{-1} min^{-1} at pH 8.8). There can be variability in the structure and stability of the aminidated product.[204,205] Isethionyl acetimidate (Figure 13.4) was developed as a membrane-impermeable reagent for the study of cell membranes.[206] The reaction of isethionyl acetimidate was compared with that of ethyl acetimidate.[206] This reagent has seen limited continued use[207,208] including the use for labeling amino groups in complex lipids.[209,210] S-Sulfethylthioacetimidate has been advanced as a membrane-impermeable derivative of S-methylthioacetimidate.[96] There is a relationship between the thioacetimidates and 2-iminothiolane,[211,212] which allows the addition of a thiol group to a lysine residue in a protein.

A variety of aldehydes react with amino groups in proteins, initially forming a Schiff base (imine) that can be reduced to form a carbon–nitrogen bond. The same chemistry is involved in the reaction of a reducing sugar such as glucose with ε-amino groups of lysine. A number of investigators have used pyridoxal-5'-phosphate (PLP) (Figure 13.5) to modify lysyl residues in proteins. PLP is the cofactor form of vitamin B_6 and plays an important role in biological catalysis.[213] PLP is useful for the modification of lysine because of the selectivity of reaction, spectral properties of the modified residue, and the potential reversibility of reaction. The Schiff base product can be stabilized by reduction with sodium borohydride, and it is possible to incorporate radiolabel by the use of radiolabeled sodium borohydride (sodium borotritide).[214] PLP can react with both ε-amino groups of lysine and the amino-terminal α-amino group in a protein. The phosphate group of PLP is important in binding to anion-binding sites in proteins as pyridoxal is much less effective.[215–221] It is of interest that Tris buffer, as a competing amine, attenuated the inactivation of RNA polymerase with PLP[215] but did not affect PLP inactivation of DNA polymerase.[216] Ohsawa and Gualerzi[218] studied the modification of E. coli initiation factor by pyridoxal phosphate in 0.020 M triethanolamine, and 0.03 M KCl, pH 7.8. In the course of the studies, it was observed that pyridoxal phosphate will not react with poly(AUG). These investigators also reported the preparation of N^6-pyridoxal lysine by reaction of pyridoxal phosphate with polylysine in 0.01 M sodium phosphate, pH 7.2 at 37°C followed by reduction with $NaBH_4$. The reduction was terminated by the addition of acetic acid. Acid hydrolysis (6 N HCl, 110°C, 22 h) yielded N^6-pyridoxal-L-lysine. Cake and coworkers[219] have demonstrated that modification of activated hepatic glucocorticoid receptor with PLP obviated the binding of the receptor to DNA. Greatly reduced inhibition was seen with pyridoxamine-5'-phosphate, pyridoxamine, or pyridoxine. Inhibition could be reversed by gel filtration or treatment with DTT, while treatment with $NaBH_4$ resulted in irreversible inhibition of DNA binding. These investigators used 0.2 M borate, 0.25 M sucrose, and 0.003 M $MgCl_2$ (pH 8.0) as the solvent for reaction with PLP. Shapiro and coworkers investigated the reaction of PLP with rabbit muscle aldolase.[220] The initial reaction produced a species with an absorbance maximum at 430–435 nm, reflecting the protonated Schiff base form of the PLP–protein complex. After reduction with sodium borohydride, the absorbance maximum was at 325 nm characteristic of the reduced Schiff base. This is a quite useful study in that the difference in reactivity between pyridoxal and PLP is demonstrated as is the reversible nature of the initial complex between the protein and PLP until stabilized by reduction. Schnackerz and Noltmann[221] compared the reaction of PLP and other aldehydes with rabbit muscle phosphoglucose at pH 8.0. PLP (0.19 mM) resulted in 82% inactivation, while the following results were obtained with other aldehydes: pyridoxal (8.4 mM), 16% inactivation; acetaldehyde (75 mM), 75% inactivation; and acetone (75 mM), 31% inactivation. This last reaction is of interest as many investigators are unaware that acetone can react with amino groups in proteins. The reaction of acetone with primary amino groups has been known for some time[222] and is discussed in further detail later within the topic of reductive alkylation. PLP can be considered to be an affinity label and saturation kinetics are observed on the reaction of lysine residues in proteins.[223–225]

The reaction of ribulose 1,5-bisphosphate carboxylase/oxygenase with PLP resulted in inactivation with or without reduction with sodium borohydride.[226] This reaction was performed in 0.1 M Bicine (N,N-(2-hydroxyethyl) glycine), 0.010 M $MgCl_2$, 0.2 mM EDTA, and 0.001 M DTT. The reaction demonstrated an optimum at pH 8.4. Spectral studies showed the formation of a species absorbing at 432 nm. As is the characteristic for the Schiff base derivative, this peak disappears on reduction to yield a species with an optimum at 325 nm ($\Delta\varepsilon = 4800$ M^{-1} cm^{-1}). This supports the suggestion that the loss of activity is observed on reaction with PLP due to the formation of a Schiff base that can be reduced with $NaBH_4$ to form a stable derivative as opposed to the formation of a 2-azolidine ring with a second nucleophile as has been observed by other investigators.[227–229] I would note that I found only one reference to the formation of an X-azolidine ring on reaction with PLP since the work in 1975.[230]

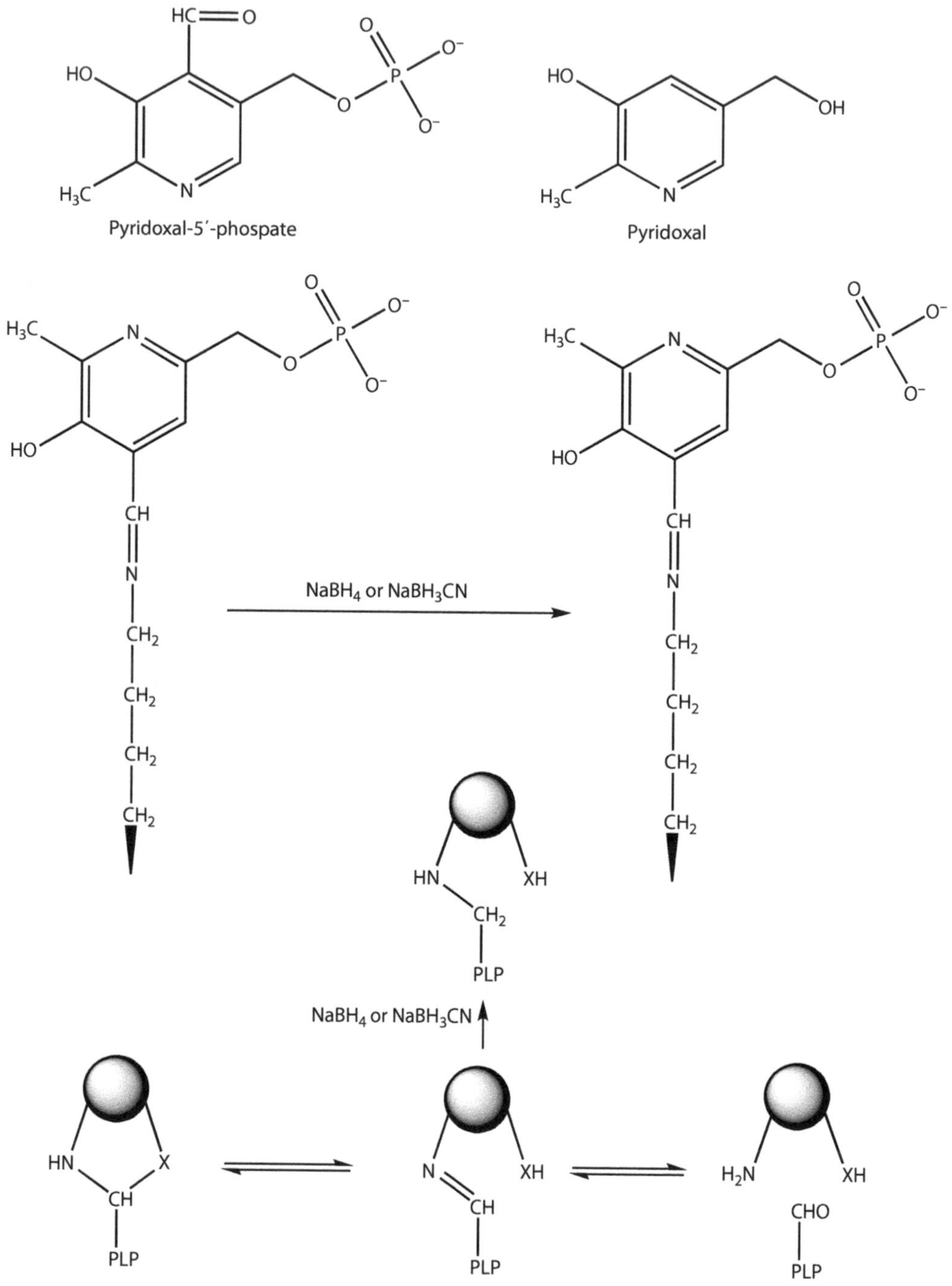

FIGURE 13.5   The reaction of PLP with protein amino groups. Shown is the reaction of PLP with the ε-amino group of lysine. The initial formation of the Schiff base derivative is reversible but can be reduced to form a stable secondary amine. Shown also is the structure of pyridoxal, which reacts more slowly than PLP, suggesting the importance of the phosphate group in binding to the protein (Cole, S.C. and Yon, R.J., Active-site-directed inactivation of wheat germ aspartate transcarbamylase by pyridoxal 5'-phosphate, *Biochem. J.* 248, 403–408, 1987). There is also support for the role of the phosphate group in forming the Schiff base (Huang, T.C., Chen, M.H., and Ho, C.T., Effect of phosphate on stability of pyridoxal in the presence of lysine, *J. Agric. Food Chem.* 49, 1559–1563, 2001). Shown at the bottom is the formation of an X-azolidine derivative (Kent, S.B., Krebs, E.G., and Fischer, E.H., Properties of crystalline phosphorylase b, *J. Biol. Chem.* 237, 549–558, 1958).

Cortijo and coworkers[231] have suggested the use of the ratio of absorbance at 415 and 335 nm of enzyme-bound PLP as an indication of the polarity of the medium. This group subsequently further evaluated the effect of solvent on the spectral properties of PLP and pyridoxal in organic solvents[232,233] and subsequently the stability of the Schiff base formed between PLP and hexylamine in dioxane.[234] June and coworkers used spectra of PLP to measure conformational changes in tryptophanase.[235] Slebe and Martinez-Carrion[236] have introduced the use of phosphopyridoxal trifluoroethylamine as a probe for pyridoxal phosphate binding sites in enzymes. Nishigori and Toft[237] studied the reaction of PLP with the avian progesterone receptor. The modification reaction of PLP with the avian progesterone receptor was performed in 0.02 M barbital, 10% (v/v) glycerol, pLP0.005 mM dithiothreitol, 0.010 M KCl, pH 8.0, 4°C. The reaction product, a Schiff base, required stabilization with $NaBH_4$. Consistent with earlier observations by Martial and coworkers,[215] Nishigori and Toft[237] noted that the modification was readily reversed in Tris buffer unless stabilized by $NaBH_4$. Moldoon and Cidlowski[238] demonstrate that 0.1 M Tris, pH 7.4, markedly interfered with the modification of rat uterine estrogen receptor with PLP. These investigators also noted that, as in the other studies, 0.05 M lysine would block the modification reaction and could also reverse the modification if the Schiff base had not been reduced. The action of Tris hydroxymethyl aminomethane (Tris) as a nucleophile is occasionally not fully appreciated and can yield very interesting results.[239,240] Sugiyama and Mukohata[241] observed that PLP modification of the lysine residue in chloroplast coupling factor using 0.020 M Tricine, 0.001 M EDTA, and 0.010 M $MgCl_2$ at pH 8.0 resulted in complete inactivation of the ATPase activity. Peters and coworkers[242] reported on the inactivation of the ATPase activity in a bacterial coupling factor by reaction with PLP. The modification was performed in 0.050 M morpholinosulfonic acid, pH 7.5. The inhibition was readily reversed by dilution or by 0.01 M lysine and was, as expected, stabilized by $NaBH_4$. Gould and Engel[243] reported on the reaction of mouse testicular lactate dehydrogenase with PLP in 0.050 M sodium pyrophosphate, pH 8.7 at 25°C. This reaction resulted in the inactivation of the dehydrogenase activity. The inactivation was reversed by cysteine and stabilized by $NaBH_4$. These investigators reported that the observed absorption coefficient at 325 nm may be decreased as much as 50% with protein-bound pyridoxal phosphate. Thus, estimation of the number of lysine residues modified using the absorption coefficient obtained with model compounds might provide only a minimum value. Ogawa and Fujioka[244] studied the reaction of PLP with saccharopine dehydrogenase in 0.1 M potassium phosphate, pH 6.8 at ambient temperature in the dark. Both spectral analysis and tritium incorporation from sodium borohydride reduction were consistent with the modification of one lysine residue per mole of enzyme being responsible for the loss of enzyme activity. A value of 1×10^4 M^{-1} cm^{-1} for the extinction coefficient at 325 nm[245] was used in this study. The concentrations of pyridoxal and PLP were determined spectrophotometrically in 0.1 M NaOH using an extinction coefficient of 5.8×10^3 M^{-1} cm^{-1} at 300 nm and 6.6×10^3 M^{-1} cm^{-1} at 388 nm, respectively.[246] Bürger and Görisch[247] reported the inactivation of histidinol dehydrogenase upon reaction with pyridoxal phosphate in 0.02 M Tris, pH 7.6. This modification could be reversed by dialysis unless the putative Schiff base was stabilized by reduction with $NaBH_4$ (n-octyl alcohol added to prevent foaming). These investigators used a $\Delta\varepsilon$ for ε-amino pyridoxal lysine of 1×10^4 M^{-1} cm^{-1} at 325 nm. A novel affinity label (pyridoxal-5'-diphospho-5'-adenosine) utilizing PLP chemistry has been used to study the adenine nucleotide binding sites in yeast hexokinase.[248] There has been limited use of PLP for the modification of proteins.[249–252] One novel observation is the oxidation of amino-terminal amino acids by PLP,[253] which can be used for coupling via oxime chemistry.[254]

Methyl acetyl phosphate (Figure 13.8) was originally developed as an affinity label for D-3-hydroxybutyrate dehydrogenase.[255] Manning and coworkers have examined the chemistry of the reaction of methyl acetyl phosphate with hemoglobin in some detail.[256,257] It appears to be an affinity label for the 2,3-diphosphoglycerate binding site.[256] More recent work suggests that this reagent may be a useful probe for other anion-binding sites in proteins.[258,259]

Reductive alkylation of amino groups in proteins (Figure 13.6) has the advantage that the basic charge properties of the modified residue are preserved. The early work on this modification has

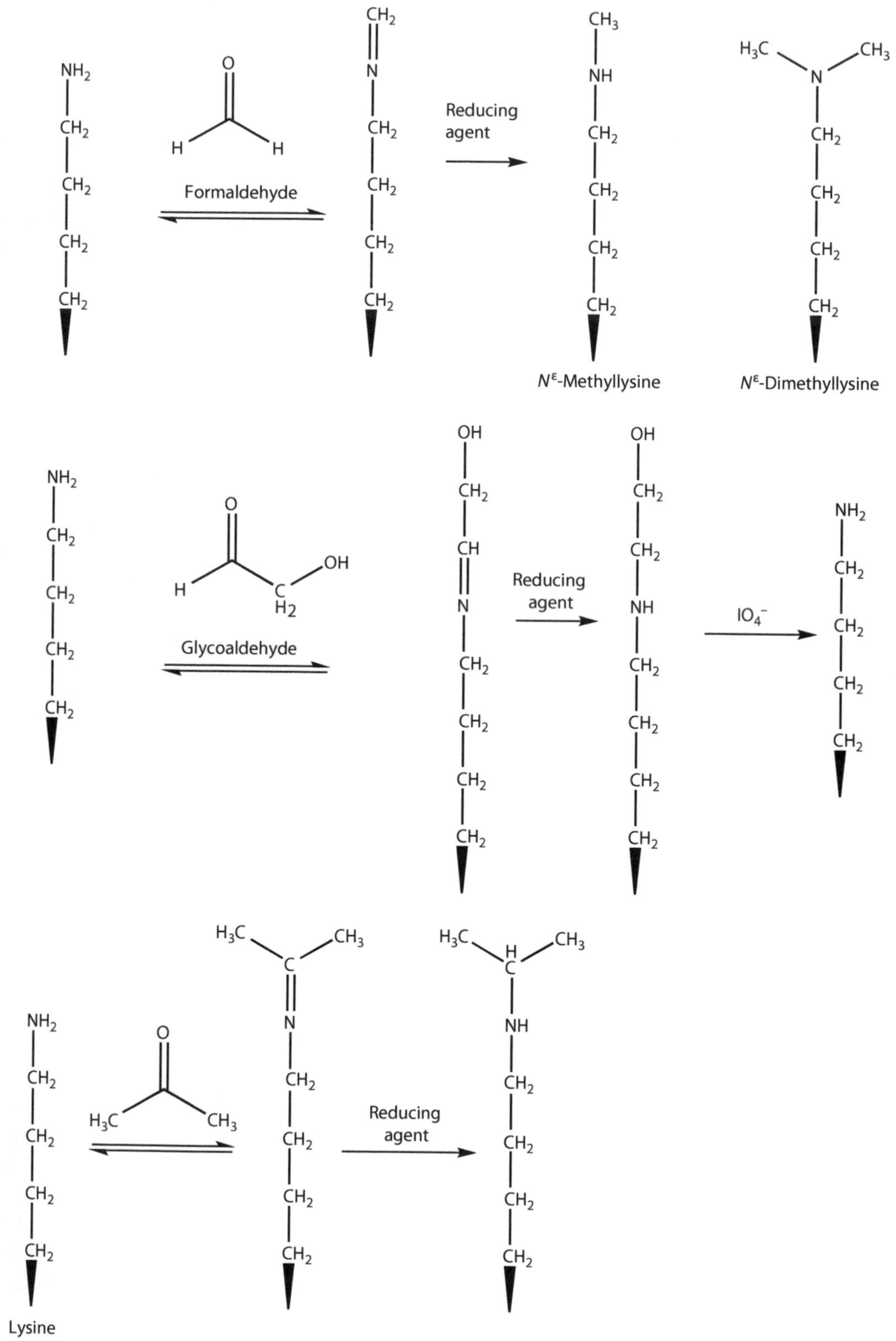

FIGURE 13.6 (continued)

been reviewed by Means.[260] Both monosubstituted and disubstituted derivatives can be prepared depending upon reaction conditions and the nature of the carbonyl compound. The use of ^{13}C formaldehyde enables the production of an NMR probe.[261–264] Other work has used deuterium-labeled carbonyl compounds with reductive alkylation for insertion of an NMR probe.[265] Brown and coworkers[265] prepared the [D]isopropyllysyl derivative by reductive alkylation with deuterated acetone. Dempsey and coworkers[266] used reductive methylation with $NaBD_3CN$ to place an NMR probe in melittin; a similar approach has been used to place deuterated probe in cytochrome c.[267] Rice and coworkers[268] reported the stabilization of trypsin by reductive methylation. This reaction utilized formaldehyde/sodium borohydride in 0.2 M sodium borate, pH 9.2 in the cold. Unsubstituted amino groups were present after the reaction as demonstrated by titration with trinitrobenzenesulfonic acid. The amino-terminal isoleucine residue was not modified under these conditions. Reduction of the Schiff base formed with lysine and formaldehyde can be reduced with sodium borotritide to incorporate a radiolabel.[269] The use of sodium cyanoborohydride as a reducing agent for reductive methylation provides advantages over sodium borohydride. Sodium cyanoborohydride is stable in aqueous solution at pH 7.0, and unlike sodium borohydride that can reduce aldehydes and disulfide bonds, sodium cyanoborohydride only reduces the Schiff base formed in the initial process of reductive alkylation. Dottavio-Martin and Ravel reported the radiolabeling of proteins using ^{14}C-formaldehyde and sodium cyanoborohydride.[270] The modification was performed in 0.04 M phosphate, pH 7.0 at 25°C. The modification can be performed equally well at 0°C, but as would be expected, it takes a longer period of time; there is no effect on the extent of the modification. In this regard, these authors estimated that the same extent of modification obtained in 1 h at 37°C could be achieved in 4 to 6 h at 25°C or 24 h at 0°C. Although the majority of experiments in this study were performed in phosphate buffer at pH 7.0, equivalent results can be obtained in Tris or HEPES buffer at pH 7.0. A greater extent of modification was observed with sodium cyanoborohydride at pH 7.0 than with sodium borohydride at pH 9.0. This work has been extensively cited on a consistent basis since its publication in 1978. Another highly cited article concerned with the methodology of reductive methylation was published by Jentoft and Dearborn in 1979.[271] While of less concern some 30+ years later, it was emphasized that the preparation of sodium cyanoborohydride is critical. At that time, recrystallization of sodium borohydride was considered to be critical. This reflected the presence of cyanide that limited the extent of the reductive alkylation (see below). Optimal reductive methylation was obtained at pH values greater than 8.0 during a short-term (10 min) incubation. The effect of pH is much less pronounced at longer periods of incubation (1 to 2 h), with optimal reductive methylation occurring between pH 7.0 and 8.0. These investigators also noted that Tris, 2-mercaptoethanol, DTT, ammonium ions (as ammonium sulfate), and guanidine (5 M) inhibited the reductive alkylation of albumin by formaldehyde and sodium cyanoborohydride in

FIGURE 13.6 (continued)   The reductive alkylation of amino groups in proteins. Shown, for example, are lysine residues although reaction can also occur at α-amino groups. At the top is the reaction of formaldehyde to form a Schiff base, which is then reduced to form N^ϵ-aminomethyllysine. The dimethyl derivative can also be formed. Although sodium borohydride is shown, other reducing agent such as sodium cyanoborohydride or amino boranes may also be used (see Geoghegan, K.F., Cabacungan, J.C., Dixon, H.B., and Feeney, R.E., Alternative reducing agents for reductive methylation of amino groups in proteins, *Int. J. Pept. Protein Res.* 17, 345–352, 1981; Means, G.E. and Feeney, R.E., Reductive alkylation of proteins, *Anal. Biochem.* 224, 1–16, 1995; Rayment, I., Reductive alkylation of lysine residues to alter crystallization properties of proteins, *Methods Enzymol.* 276, 171–179, 1997). Shown below is the reductive alkylation of lysine with glycoaldehyde followed by regeneration of lysine with periodate oxidation (Geoghegan, K.F., Ybarra, D.M., and Feeney, R.E., Reversible reductive alkylation of amino groups in proteins, *Biochemistry* 18, 5392–5399, 1979). Shown at the bottom is the reaction of lysine with acetone yielding a Schiff base (Yamashiro, D., Havran, R.T., Aanning, H.L., and Du Vigneaud, V., Inactivation of lysine vasopressin by acetone, *Proc. Natl. Acad. Sci. USA* 57, 1058–1061, 1967), which can be reduced to yield the isopropyl derivatives (Zhai, J., Liu, X., Huang, Z. et al., RABA (reductive alkylation by acetone): A novel stable isotope labeling approach for quantitative proteomics, *J. Am. Soc. Mass Spectrom.* 20, 1366–1377, 2009).

0.1 M HEPES, pH 7.5. Jentoft and Dearborn[272] characterized the inhibition by cyanide of reductive alkylation with sodium cyanoborohydride. This is of some importance since cyanide is a product of reductive alkylation with sodium cyanoborohydride. Inhibition by cyanide can be blocked by nickel (II) or cobalt (III). The observation that nickel (II) can preclude the inhibition of reductive alkylation by cyanide was shown to obviate the previously observed necessity for recrystallization of the sodium cyanoborohydride. More recent work has suggested other reducing agents such as boranes, which may be more useful than sodium cyanoborohydride.[273–276] Geoghegan and coworkers[273] compared sodium cyanoborohydride, dimethylamine borane, and trimethylamine borane with respect to effectiveness in reductive alkylation. Reduction at disulfide bonds was not observed with any of the three reagents. Dimethylamine borane was only slightly less effective than sodium cyanoborohydride, while trimethylamine borane was much less effective. This decrease in effectiveness in reductive alkylation is balanced by the absence of toxic by-products such as cyanide evolving during the reaction. Quantitative reductive methylation (equal to or greater than one methyl group per lysyl residue) is achieved at 10 mM formaldehyde with dimethylamine borane and at 50 mM formaldehyde with trimethylamine borane. It should be noted that a similar extent of modification is obtained with 5 mM formaldehyde using sodium cyanoborohydride. Consideration of a recent paper on reductive methylation[277] only mentions the use of freshly prepared sodium cyanoborohydride.

The effect of carbonyl compounds of different size on the extent of reductive alkylation has been examined by Fretheim and coworkers.[278] The extent of modification is more a reflection of the type of alkylating agent and reaction conditions than an intrinsic property of the protein under study. For example, nearly 100% disubstitution can be obtained with formaldehyde and approximately 35% disubstitution with n-butanol, while only monosubstitution can be obtained with acetone, cyclopentanone, cyclohexanone, and benzaldehyde. While most of the products of reductive alkylation retained solubility, the reaction products obtained with cyclohexanone and benzaldehyde tended to precipitate. Examination of the reductive alkylation of ovomucoid, lysozyme, and ovotransferrin with different aldehydes suggests that such modification occurs without major conformational change as judged by circular dichroism measurements.[279] The same study also examined the stability of the modified proteins by scanning differential calorimetry; the extensive modification of amino groups decreased thermal stability. The destabilization effect increases with increasing size (and hydrophobicity) of the modifying aldehyde. In another study, Geoghegan and coworkers[280] reported on the reversible reductive alkylation of proteins with either glycolaldehyde or acetol (Figure 13.6). The 2-hydroxyethyl derivative can be cleaved with periodate under mild basic condition to yield the free amine. Treatment of 30.0 mg lysozyme in 6.0 mL 0.2 M sodium borate, pH 9.0, with 60 mg glycolaldehyde and 10 mg sodium borohydride at ambient temperature resulted in 60% 2-hydroxyethylation. Treatment of 20 mg ovomucoid in 2.0 mL 0.2 M sodium borate, pH 9.0, with 10% acetol and 30 mg sodium borohydride (added in portions) resulted in 55% hydroxyisopropylation. In both situations, the reaction was terminated by adjustment of the pH to 5 with glacial acetic acid. The extent of modification was determined either by titration with trinitrobenzenesulfonic acid or by amino acid analysis after acid hydrolysis. Periodate oxidation could be accomplished with 0.015 M sodium periodate, pH 7.9, for 30 min at ambient temperature. Geoghegan and coworkers[280] also reported that a second molecule of glycolaldehyde can react forming a tertiary amine that is not cleaved by periodate; the disubstitution was favored by the use of sodium cyanoborohydride.

More recent work on reductive methylation has focused on the value of this process on the crystallization of proteins.[281–285] The use of reductive methylation for protein crystallization was introduced by Rayment and coworkers[286] for the study of the structure of myosin subfragment 1. It had not been possible to obtain crystals of myosin subfragment 1, frustrating structural analysis of this protein. Reductive methylation of myosin subfragment 1 did provide a form that could be crystallized and subjected to x-ray diffraction analysis. Reductive methylation of myosin subfragment 1 was accomplished with formaldehyde and dimethylamino borane at 4°C in the dark. Characterization of the modified protein by measurement of enzymatic activity suggests that there were no large conformational changes secondary to reductive methylation. While there was no suggestion of a gross conformational change in myosin subfragment 1, Rypniewski and coworkers[287] evaluated the effect of reductive methylation of the crystallographic properties of HEWL. It was necessary to use different conditions for crystal growth with the modified HEWL. Analysis of the modified and native protein revealed only small changes in the crystal structure; the positions of the α-carbons excluding the surface loops superimpose with a root-mean-square value of 0.4 Å. It was concluded that reductive methylation produces little structural perturbations but does markedly influence crystallization properties of a protein. Other workers[288] later showed that the radius of gyration of myosin subfragment 1 was not changed on reductive methylation.

Another class of aldehydes that reacts with protein to give interesting products is simple monosaccharides, which exist in solution in enol and keto forms (Figure 13.7). Wilson[289] showed that bovine pancreatic ribonuclease dimer would react with lactose in the presence of sodium cyanoborohydride to yield an active derivative that shows selectivity in uptake by the liver during in vivo experiments. The modification of ribonuclease dimer was performed in 0.2 M potassium phosphate, pH 7.4 (phosphate buffer was used to protect Lys41 from modification) at 37°C for 5 days with lactose and sodium cyanoborohydride. Under these conditions, 80% of the amino groups were modified. Reductive lactosamination of L-asparaginase had previously been shown to decrease circulating half-life.[290] The decreased circulatory half-life of both proteins is a reflection of increased uptake by hepatocytes in a process involving a specific surface lectin.[291] The clearance by binding to galactose lectin on the hepatocyte surface is well understood today but was an evolving concept[292] when these studies were performed. The importance of exposed galactosyl residues in accelerated clearance had been shown by Baynes and Wold[293] with the various pancreatic ribonucleases a year prior to the Marsh study.[290] Baynes and Wold[293] had shown that it was not the specific loss of glycan that was responsible for increased clearance but rather the specific exposure of a galactosyl moiety. The galactose lectin receptor is frequently referred to as the asialoglycoprotein receptor.[294] Lactoaminated albumin is used as a drug carrier for anticancer drugs such as doxorubicin.[295,296] Lactosamination has also been used for the hepatic targeting of peptide nucleic acids,[297] poly-L-lysine for gene therapy,[298] and a dendrimer/cyclodextrin conjugate to be used for siRNA delivery.[299]

Bunn and Higgins[300] have explored the reaction of monosaccharides with protein amino groups in the presence of sodium cyanoborohydride in some detail. These investigators studied the reaction of hemoglobin with various monosaccharides in Krebs–Ringer phosphate buffer, pH 7.3. The extent of modification was determined using tritiated sodium cyanoborohydride. The rate of modification was demonstrated to be a direct function of the amount of each sugar in the carbonyl (or keto) form. Thus, the k_1 ($\times 10^{-3}$ mM^{-1} h^{-1}) for D-glucose is 0.6 with 0.002% in the carbonyl form, while the k_1 for D-ribose is 10.0 with 0.05% in the carbonyl form. Watkins and coworkers[301] studied the reaction of RNase with glucose. As would be expected, the reaction is slow; with 0.4 M glucose, there was the incorporation of approximately one mole of glucose per mole of RNase after three days of reaction at 37°C in 0.1 M phosphate at pH 7.4. The rate of loss of catalytic activity was greater than the rate of lysine modification, which the authors suggest indicating preferential modification of active-site lysine residue(s). The rate of loss of enzyme activity and lysine modification that was markedly increased with sodium cyanoborohydride was included with glucose. Modification of the α-amino group of lysine-1 was the primary site of initial Schiff base formation stabilized with sodium cyanoborohydride, while formation of the Amadori product (ketoamine) was larger at Lys7 and Lys41. Subsequent work from this laboratory[302] suggested a specific role for phosphate in the inactivation of RNase. Reaction in the presence of MOPS [3-(N-morpholino)propanesulfonic acid] or TAPSO[3-(N-tris(hydroxylmethyl)methylamino)-2-hydroxypropanesulfonic acid] reduced the rate and extent of lysine modification (ketamine formation); there was no loss of enzyme activity with either of these two buffers. Gil and coworkers[303] reported that phosphate buffer enhanced the rate of glycation of gamma globulin but had no effect on the glycation of ovalbumin or human serum albumin. This observation supported the catalytic role of phosphate in protein glycation suggesting that the phosphate is the base removing the proton during the Amadori rearrangement. Histidine has also been suggested to have importance in catalyzing the Amadori rearrangment.[304]

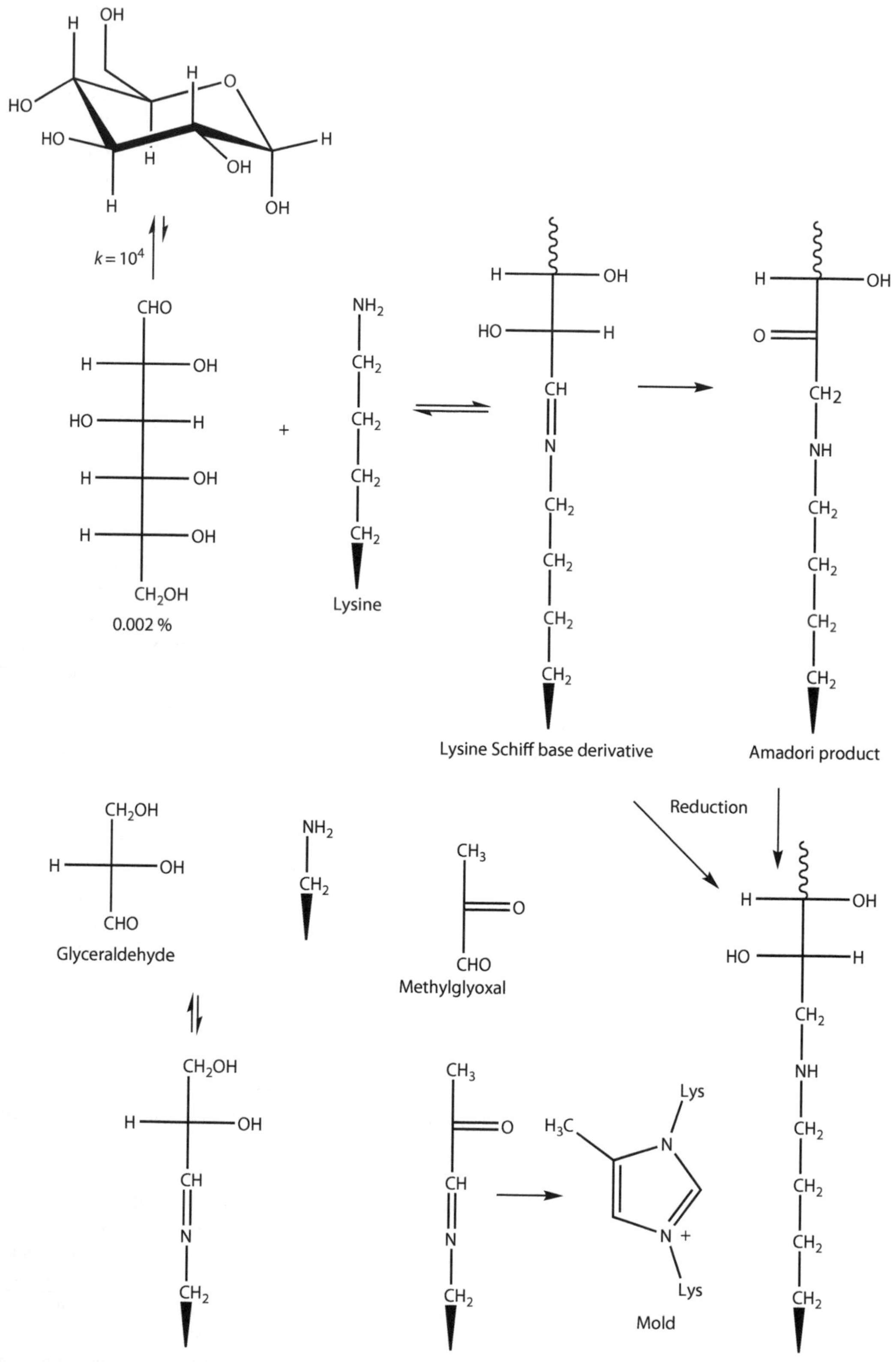

FIGURE 13.7 (continued)

The adjacent amino acid also has a profound effect on lysine glycation with reactivity increased by the presence of hydrophobic amino acid residues.[305] However, the effect of the specific residues also depended on the reducing sugar. For example, with Lys-Ile, there was 90% reaction with glucose after 30 min in anhydrous methanol at 64°C, while there was 40% reaction with lactose. Another interesting observation was the observed enhanced reaction of Lys-Arg with glucose (90% modification) compared to Lys-Lys (30% modification). While the process of glycation/Maillard reaction is of considerable importance in processes from cooking to the formation of AGEs in various disease processes,[306–312] the reaction of proteins with reducing sugars does not appear to be a useful chemical modification. There has, however, been increased interest in the glycation of monoclonal antibodies during cell culture manufacture.[313–315] Dehydroascorbic acid can also form a glycation product with lysine,[316] and cross-link formation with dehydroascorbic acid between lysine and arginine has been reported.[317]

The reaction of glyceraldehyde with carbonmonoxyhemoglobin S has been explored by Acharya and Manning.[318] This reaction was performed with 0.010 M glyceraldehyde in phosphate-buffered saline, pH 7.4, and the resultant Schiff bases were stabilized by reduction with sodium borohydride. Using radiolabeled glyceraldehyde, these investigators were able to obtain support for the concept that there is selectivity in the reaction of sugar aldehydes with hemoglobin. The reaction product between glyceraldehyde and hemoglobin S did have stability properties without reduction that were not consistent with only Schiff base products. These investigators suggested that the glyceraldehyde–hemoglobin Schiff base could undergo an Amadori rearrangement to form a stable ketoamine adduct (Figure 13.7), which could be reduced with sodium borohydride to form a product identical to that obtained by direct reduction of the Schiff base. In a subsequent study, these investigators did show that the glyceraldehyde–hemoglobin S Schiff base could rearrange to form a ketamine via an Amadori rearrangement.[319] These investigators were able to use reaction with phenylhydrazine to detect the protein-bound ketamine adduct. The reaction of glyceraldehyde with proteins continues to be of interest in the study of protein glycation and AGEs.[320–323] Glyceraldehyde has also been reported to react with the active cysteine residue in proteases such as papain and cathepsin B.[324] Methylglyoxal (Figure 13.7) is used as model for protein glycation.[325–330]

FIGURE 13.7 (continued)   The reduction of reducing sugars with proteins. Shown is the reaction of glucose with a lysine residue initially forming a Schiff base with subsequent conversion to an Amadori product. The reaction to form the Schiff base and subsequent Amadori product with lysine (Day, J.F., Thorpe, S.R., and Baynes, J.W., Nonenzymatically glucosylated albumin. in vitro preparation and isolation from normal human serum, *J. Biol. Chem.* 254, 595–597, 1979) is dependent on the presence of the carbonyl form of the reducing sugar. As shown for glucose, the carbonyl form is present at the level of 0.002% (Bunn, H.F. and Higgins, P.J., Reaction of monosaccharides with proteins: Possible evolutionary significance, *Science* 213, 722–724, 1981). The glucose Schiff base can be reduced to form a stable amine linkage. In studies with bovine pancreatic ribonuclease, Watkins and coworkers showed that the reaction of glucose in the presence of sodium borohydride shows Schiff base at amino-terminal amino acids (α-amino group), while in the absence, predominant reaction was at Lys7 and Lys41 to yield an Amadori product (Watkins, N.G., Thorpe, S.R., and Baynes, J.W., Glycation of amino groups in protein. Studies on the specificity of modification of RNase by glucose, *J. Biol. Chem.* 260, 10629–10636, 1985). Shown at the bottom left is the reaction of glyceraldehyde with a protein amino group to yield a Schiff base, which can go on to form more complex products (Acharya, A.S., Sussman, L.G., and Manning, J.M., Schiff base adducts of glyceraldehyde with hemoglobin. Differences in the Amadori rearrangement at the α-amino groups, *J. Biol. Chem.* 258, 2296–2302, 1983). Shown on the right is the structure of methylglyoxal (from either endogenous sources from dihydroxyacetone or an exogenous model compound for study of glycation), which reacts with amines to initially form a Schiff base (McLaughlin, J.A., Pethig, R., and Szent-Györgi, A., Spectroscopic studies of the protein-methylglyoxal adduct, *Proc. Natl. Acad. Sci. USA* 77, 949–951, 1980) but rapidly moves to more complex products such as *N,N*-di(*N*ε-lysino)-4-methyl-imidazolium (methylglyoxal-lysine dimer, MOLD) (Saralva, M.A., Borges, C.M., and Florêncio, M.H., Reactions of a modified lysine with aldehydic and diketonic dicarbonyl compounds: An electrospray mass spectrometry structure/activity study, *J. Mass. Spectrom.* 41, 216–228, 2006).

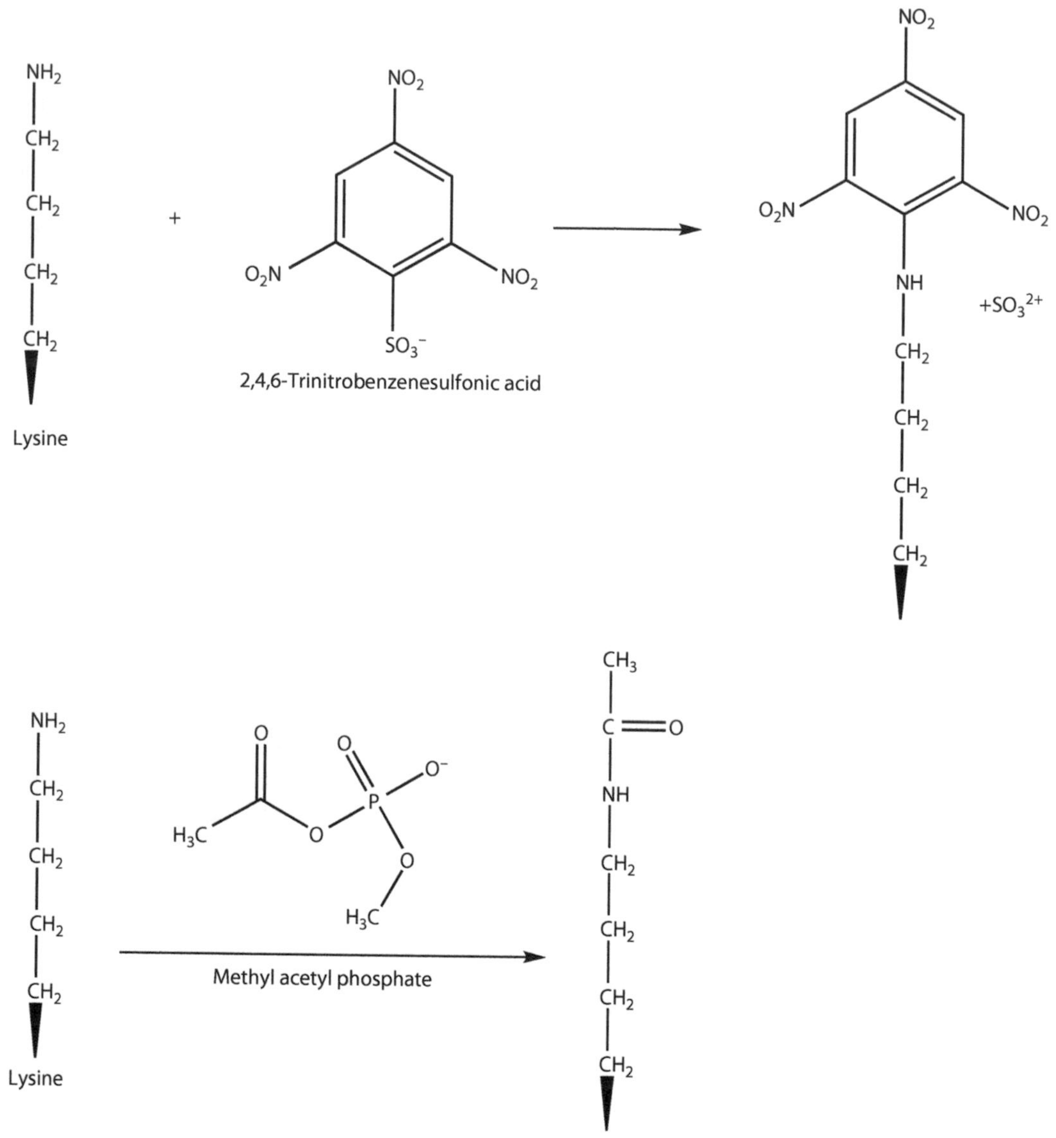

FIGURE 13.8   The reaction of TNBS and methyl acetyl phosphate with amino groups in proteins. Shown at the top is the reaction of TNBS with an amine (lysine) to form a derivatives with absorbance at 420 nm (Fields, R., The measurement of amino groups in proteins and peptides, *Biochem. J.* 124, 581–590, 1971; Brown, M.D., Schätzlein, A., Browlie, A et al., Preliminary characterization of novel amino acid based polymeric vesicles as gene and drug delivery agents, *Bioconjug. Chem.* 11, 880–891, 2000). Shown below is the reaction of methyl acetyl phosphate with lysine residues to yield the acetylated derivative (Kataoka, K., Tanizawa, K. Fukui, T. et al., Identification of active-site lysyl residues of phenylalanine dehydrogenase by chemical modification with methyl acetyl phosphate combined with site-directed mutagenesis, *J. Biochem.* 116, 1370–1376, 1994).

The reaction of TNBS with amino groups (Figure 13.8)[331–333] is reasonably specific and easy to measure by spectral analysis. In the presence of an excess of sulfite, absorbance at 420 nm is the most sensitive index, having $\varepsilon = 2.0 \times 10^4$ M^{-1} cm^{-1}. Absorbance at 420 nm is dependent upon the ability of the reaction product to form a complex with sulfite: the spectrum of a trinitrobenzyl amino compound has an isosbestic point at 367 nm with $\varepsilon = 1.05 \times 10^4$ M^{-1} cm^{-1}, which also can be useful. The reader is directed to an early review by Fields[334] for a wealth of technical information concerning the use of TNBS for reaction with amines. The derivatives of α- and ε-amino groups have similar spectra with the exception that α-amino derivatives have a slightly higher extinction coefficient at 420 nm ($\varepsilon = 2.20 \times 10^4$ M^{-1} cm^{-1}) than ε-amino groups ($\varepsilon = 1.92 \times 10^4$ M^{-1} cm^{-1}). Both of these derivatives have much higher extinction coefficients than the derivative obtained by reaction of TNBS with cysteinyl residues ($\varepsilon = 2.25 \times 10^3$ M^{-1} cm^{-1}). The α- and ε-amino derivatives can be differentiated by their stability to acid or base hydrolysis. The α-amino derivatives are unstable to acid hydrolysis (8 h at 110°C) or base hydrolysis.[335]

The reaction of TNBS with simple amines and hydroxide ions was studied in some detail by Gary Means and coworkers.[336] In general, reactivity of TNBS with amines increases with increasing basicity except that secondary amines and t-alkylamines are comparatively unreactive. The specific binding of TNBS to proteins, as with any chemical modification reagent, should be considered in the study of the reaction of this compound with proteins. These investigators advanced the following considerations regarding the reaction of trinitrobenzenesulfonic acid with proteins. Reactivity can be a sensitive measure of the basicity of an amino group. Adjacent charged groups have an influence on the rate of reaction with an increase observed with a positively charged group and a decrease with a negatively charged group. Proximity to surface hydrophobic regions that can bind TNBS can increase the observed reactivity of a particular amino group. Some second-order rate constants for the reaction of TNBS with model amines and proteins are shown in Table 13.2.

Cayot and Tainturier[337] extended our understanding of the reaction of TNBS with particular emphasis on reaction with glycated proteins. This paper contains a wealth of technical information regarding factors influencing the reaction of TNBS with amino groups in proteins. They observed that while raising the temperature of the reaction only slightly increases the rate of reaction with amino groups, the rate of the hydrolysis of the reagent is substantially increased; an increase in temperature may or may not affect exposure of lysine residues in the protein analyte. The presence of a chaotropic agent such as urea or SDS inhibits the reaction of TNBS with proteins; in the case of 4.0 M urea, the reaction is blocked and it was suggested this might be due to the presence of cyanate. They recommend reaction at pH 10 (0.1 M borate) with a 100-fold molar excess (to protein amino groups). Reaction is usually complete in 15 min and was measured by absorbance at 420 nm. This information is of considerable value considering that TNBS is used more for measurement of amino groups in polymers following modification reactions rather than of the study of amine functional groups.[338,339]

While there have been few studies in the past decade on the use of reaction of TNBS to define proteins function, there are earlier studies of interest. Bovine liver glutamate dehydrogenase provides an interesting example. Glutamate dehydrogenase is a hexamer with interesting allosteric properties,[340–346] which are beyond the scope of this work. The monomer form consists of six polypeptide chains of approximately 56 kDa molecular weight that form the active hexamer; the hexamer can undergo polymerization that can confound the chemical modification studies. A tentative sequence for the component polypeptide chain was published by Emil Smith and coworkers at UCLA in 1970[342] with corrections published several years later.[347]

This later work of great value since residue placement in the amino acid sequence did change. For example, the lysine residue modified by PLP was originally designated as Lys97 corrected to Lys126. It is assumed that Lys425 and Lys428 in the original sequence are Lys419 and Lys422 in the revised sequence. Freedman and Radda[348] reported that the intrinsic rate of reaction of TNBS with the ε-amino group of lysine was much more rapid than α-amino groups, which was in turn much slower than the rate of reaction with cysteine (Table 13.2). The trinitrophenyl derivative of the thiol group is unstable at alkaline pH and has a much lower extinction coefficient than the amine derivatives.[335] Freedman and Radda[348] also studied the reaction of TNBS with several proteins including bovine liver glutamate dehydrogenase. It was possible to differentiate lysine groups as the basis of reactivity, and two different sets of lysine residues were identified in glutamate dehydrogenase; the number of reactive groups decreased with aggregation at higher protein concentration. Later work by Freedman and Radda[349] showed that at pH 7.6 (0.1 M phosphate, 25°C), TNBS rapidly modified

($1.7–3.4$ M^{-1} s^{-1}) approximately one lysine residue per component polypeptide chain with 80% loss of activity; there was no detectable modification of cysteine under these reaction conditions. Clark and Yielding[350] also demonstrated that TNBS inactivated bovine glutamate dehydrogenase by modification of one mole of lysine/mole polypeptide chain with 20% loss of activity. This differed from Freedman and Radda who used a higher ratio of TNBS to protein. Glutamate dehydrogenase has a central role in nitrogen metabolism and, in the *forward* direction, uses α-ketoglutarate, ammonium, and reduced nicotinamide adenine dinucleotide (NADH) to synthesize glutamate.[351] It was not unreasonable to examine the effect of substrates and cofactors on the modification of enzymes by chemical reagents. It was, however, a surprise to see that TNBS was reduced to trinitrobenzene with NADH and glutamate dehydrogenase.[350,352] Subsequent work by Brown and Fisher[353] showed that the uncatalyzed rate of reduction of TNBS by NADPH (glutamate dehydrogenase can use either NADH or NADPH) was significant and that glutamate dehydrogenase might function only to increase the local concentrations of TNBS and NADH. Other work by Carlberg and Mannervick[354] showed that glutathione reductase catalyzed the NADH-dependent reduction of TNBS, but in the presence of oxygen, the reduced TNBS is oxidized with the formation of hydrogen peroxide. In the case of glutamate dehydrogenase, there is protection by *substrate*, but it has nothing to do with protection from modification but rather neutralization of the reagent. Coffee and coworkers[355] showed that the modification of glutamate dehydrogenase by TNBS resulted in the modification of Lys425 (Lys419 in the revised sequence; see above) and Lys428 (Lys422 in the revised sequence). Reaction was most rapid at Lys425 with modification of three of the six component polypeptide chains; reaction was slower at Lys428 with modification on the other three component polypeptide chains. There are other minor sites of reaction. Goldin and Frieden[356] characterized the TNBS reaction in greater detail suggesting that the residue modified by TNBS is primarily involved in the regulation of the enzyme. In any event, the story ends here as I could find no further work on the reaction of TNBS with glutamate dehydrogenase. There was one study that used TNBS to determine the extent of modification of glutamate dehydrogenase with dimethyladipimidate.[357] Reaction of PLP with Lys97 (Lys126) was shown to inactivate glutamate dehydrogenase.[358–360] Goldin and Frieden[361] demonstrated that prior modification of Lys428 (Lys422) did not affect modification by PLP at Lys97 (Lys126). Thus, we have a situation where two different reagents considered *specific* for lysine react with different lysine residues without overlap. While maleimide derivatives are considered to be reasonably specific for cysteine in proteins (Chapter 7), there is evidence of reaction with other functional groups. One such reaction appears to occur in bovine liver glutamate dehydrogenase with reaction at a lysine residue. While the modification of a lysine by a maleimide in glutamate dehydrogenase was initially discounted,[359] subsequent work by Blumenthal and Smith[363] showed that NEM reacted with the same lysine residue in *Neurospora* glutamate dehydrogenase that is modified by PLP. While the reaction of maleimide with lysyl residues in proteins is unusual, such reactions have been reported.[364] See Chapter 6 for a more thorough discussion of the reactions of NEM with proteins.

Flügge and Heldt have explored the labeling of a specific membrane component with TNBS[365] and PLP.[366] The modification of the phosphate translocation protein in spinach chloroplasts with TNBS was performed in 0.050 M HEPES, 0.33 M sorbitol, 0.001 M $MgCl_2$, 0.001 M $MnCl_2$, and 0.002 M EDTA at pH 7.6 at 4°C for periods of time up to 15 min, at which point tritiated sodium borohydride was added to both terminate the reaction and radiolabel the trinitrophenyl derivatives.[367] It is possible to label components on the surface of membranes with TNBS, as the sulfonate moiety does not permit membrane penetration. The same is true for PLP. The use of trinitrobenzenesulfonate in the selective modification of membrane surface components has been explored by Salem and coworkers.[368] This study involved the modification of intact cells with the TNBS (dissolved in methyl alcohol) diluted to a 1% methanolic solution. As mentioned previously, trinitrobenzenesulfonate does not pass across (or into) membranes, being more hydrophilic than, for example, FDNB.

Haniu and coworkers[369] have examined the reaction of lysine residues in NAD(P)H/quinone reductase with TNBS as compared to the reaction of tyrosine residues with *p*-nitrobenzenesulfonyl fluoride. Isolation and characterization of the peptides containing the modified residues showed that the modified tyrosyl residues are in hydrophobic regions of the protein, while the modified lysine residues are in hydrophilic regions.

Recent work with TNBS has focused on the use of this substance for the induction of inflammatory bowel disease in animal model systems.[370–375] The use of TNBS to measure proteolysis remains a valuable procedure in the food industry.[376–380] Reaction with TNBS is also used to determine the extent of amino group modification by other reagents.[381,382]

NHS esters (Figure 13.9) were developed by Anderson and coworkers[383] for use in peptide synthesis and used first for protein modification by Blumberg and Vallee.[384] In this work, the modification was at sites other than lysine and is discussed in more detail in the following text. NHS chemistry has not been extensively used for the modification of proteins for the study of function, but the chemistry is used for the attachment of probes and binding of protein to matrices. NHS chemistry is also of value for the preparation of protein conjugates and for the cross-linkage of proteins and manufacture of 3-D matrices such as hydrogels.

The Bolton–Hunter reagent (Figure 13.10)[385] for radiolabeling proteins is based on NHS chemistry and is used for the preparation of radiolabeled proteins for pharmacokinetic studies.[386–390] Sulfo-N-hydroxysuccinidimidyl esters were developed by Staros[391] as reagents with NHS chemistry but were more soluble and membrane impermeant compared to NHS esters. Harmon and coworkers[392] developed the use of fatty acid esters of sulfo-NHS for modification of plasma membrane proteins in rat adipocytes. Inhibition of fatty acid transport was observed as a result of this chemical modification. Inhibition was shown to be the result of reaction of the sulfo-NHS ester with fatty acid translocase/CD36. More recent studies[393] have demonstrated that sulfo-N-succinimidyl oleate is an inhibitor of the mitochondrial respiratory chain. The sulfo-NHS esters of fatty acids have been used to study the uptake and metabolism of long-chain fatty acids.[394] There has been continued use of sulfo-N-succinimidyl oleate to modify CD36, and this reagent can be considered a signature reagent for modification of CD36.[395,396]

NHS and sulfo-NHS esters have been used to label proteins with biotin. Yem and coworkers[397] used sulfo-NHS chemistry to introduce biotin into recombinant interleukin-1-α. It was determined that there were three monosubstituted products of the modification: the amino-terminal alanine, Lys93, and Lys94. The lysine residues are in a tripeptide sequence of three lysine residues. There were few other single-modified products despite the presence of other lysine residues. Yem and coworkers did use the sulfoderivative in their studies. It was possible to separate the amino-terminal-modified product from the product modified at the lysine residues, but it was not possible to separate the two lysine-modified species; the presence of singly modified species was inferred from isoelectric point measurement. These investigators used both sulfo-N-hydroxysuccinimidyl biotin and sulfo-N-hydroxysuccinimidyl hexanoyl biotin. Both reagents labeled the N-terminal alanine and Lys93 or Lys94; reaction at Lys103 was observed with sulfo-N-hydroxysuccinimidyl biotin but not with sulfo-N-hydroxysuccinimidyl hexanoyl biotin. Subsequent studies from this laboratory[398] showed modification of Lys103 with acrylodan without modification of Lys93 or Lys94. It is suggested that the hydrophobicity of the environment around Lys103 promotes interaction with the hydrophobic reagent acrylodan. In related work, Chang and coworkers[399] showed preferential reaction Lys103 in recombinant human interleukin-1β with 4-(N,N-dimethylamino)-4′-isothiocyanatoazobenzene-2′-sulfonic acid and suggested the importance of local microenvironmental factors. Given the results obtained with acetoacetate decarboxylase,[7] it is likely that the apolar environment decreases the pKa of Lys103 increasing its reactivity.

Wilchek and Bayer[400] reviewed the use of NHS derivatives of biotin for the labeling of proteins. An assay for protein biotinylation was also reviewed by Bayer and Wilchek in 1990.[401] Advances in analysis include commercial assays and MS.[402] NHS-biotin has been used to label surface lysine residues in proteins.[403] Smith has developed a kinetic model for the modification of protein with NHS-biotin.[404] N-hydroxysuccinimidyl biotin has been used to label protein to define binding sites.[403,405,406] The biotin label permits the facile purification of peptides, which are not protected from modification. Labeling of erythrocytes with biotin via reaction with N-hydroxysuccinimidyl

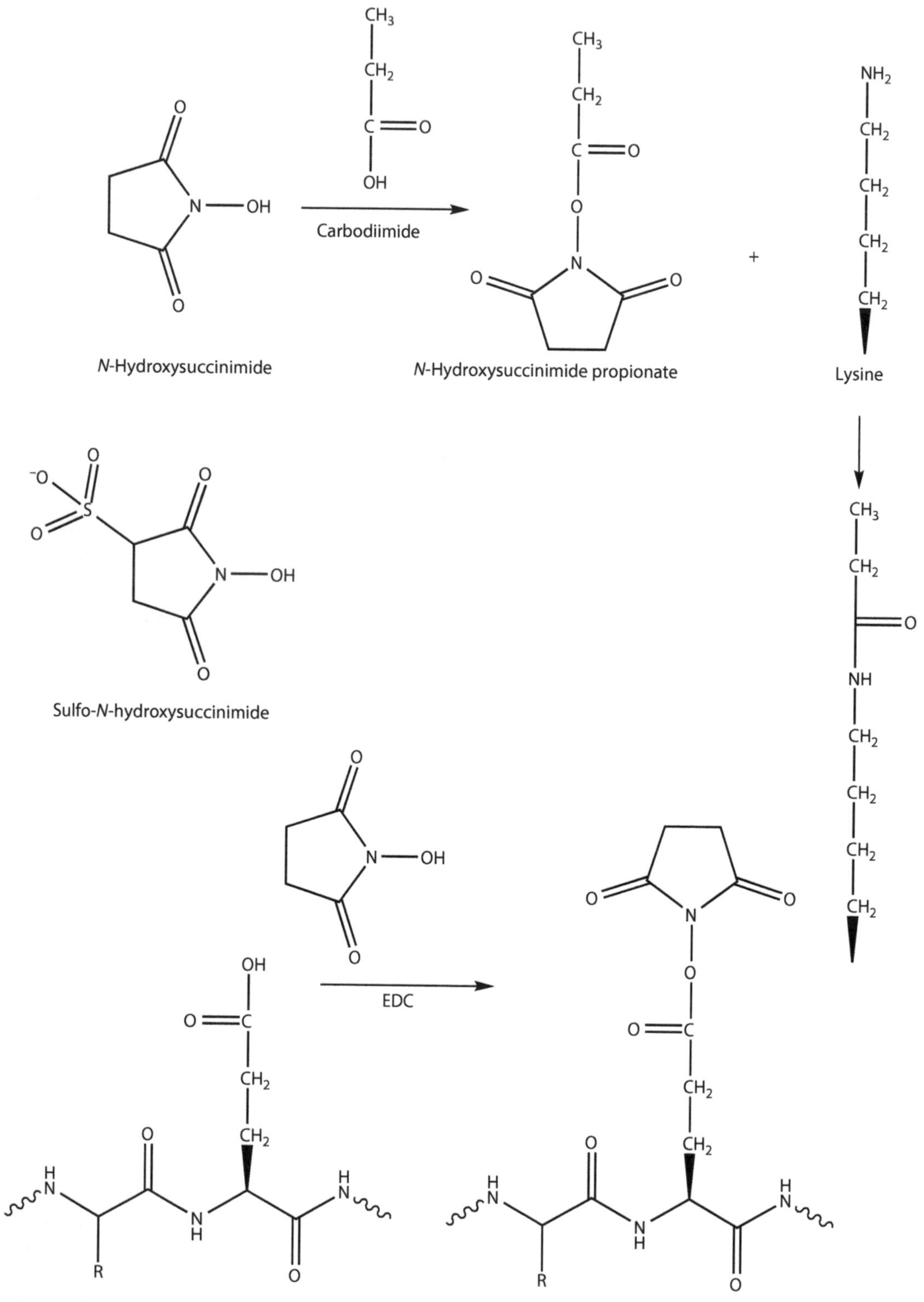

FIGURE 13.9 (continued)

biotin permits determination of half-life.[407–409] Biotin labeling has also been used to measure plate-let half-life.[410–412]

Fölsch has prepared the NHS ester of chloromercuriacetic acid.[413] This reagent was used for the preparation of mercury derivatives of proteins and nucleic acids for crystallographic analysis. Several years later, Levy prepared *p*-nitrophenyl-*p*-chloromercuribenzoate as a reagent to attach the *p*-chloromercuribenzoyl group to lysine residues in proteins.[414] NHS derivatives of agarose are used to prepare affinity matrices by providing a mechanism for the immobilization of proteins and other amino-containing molecules.[415] Kim and coworkers[416] have used NHS chemistry to immobilize human serum albumin on silica columns. Handké and coworkers[417] used an amphiphilic block copolymer containing *N*-acrylsuccinimide-co-*N*-vinylpyrrolidone to modify the surface of polylactide nanoparticles; a peptide, KKKVGBEESNDK, was then coupled to polylactide nanoparticles.

NHS esters will react with functional groups in proteins other than lysine including serine, threonine, and tyrosine as seen with NHS-based cross-linking agents.[418,] NHS has also been reported to interfere with the bicinchoninic assay for protein concentration.[419] Some of the earliest work on NHS esters involved the modification of a tyrosine residue in thermolysin with amino acid NHS esters.[384] This work resulted from the earlier observation of the *superactivation* of thermolysin with diethylpyrocarbonate in the presence of β-phenylpropionyl-L-phenylalanine.[420] Similar results were obtained with a mixed acid anhydride of β-phenylpropionyl-L-phenylalanine leading to the studies with the amino acid NHS esters.[420] The effect was reversed with neutral hydroxylamine. Superactivation of other neutral proteases was observed with amino acid NHS esters, and this effect was also reversed with hydroxylamine.[421] The reversibility with neutral hydroxylamine is consistent with the modification of tyrosine. Subsequent work by Blumberg[422] establishes Tyr110 in thermolysin as the site of modification with the amino acid NHS esters.

The recent work using NHS esters has focused on the preparation of conjugates and complex protein structures. Examples on conjugates include the preparation of single-chain VEGF (scVEGF) conjugates for PET imaging of VEGF receptors in angiogenic vasculature,[423] the PEGylation of recombinant nonglycosylated erythropoietin,[424] and an adenovirus vector to a cell-penetrating peptide.[425]

Teixeira and coworkers[426] used NHS chemistry to prepare a heparinized hydroxyapatite/collagen matrix. The reader is also directed to studies by Strehin and coworkers[427] on a *smart* hydrogel based on chondroitin sulfate and PEG and Deng and coworkers on a collagen/glycopolymer hydrogel.[428] The use of NHS chemistry to immobilize proteins for affinity chromatography was mentioned earlier,[413] and this chemistry continues to see active use in chromatographic applications.[429–432] NHS chemistry is also used for the immobilization of proteins for other solid-phase assay applications.[433–436] NHS is combined with EDC to form a peptide bond between a carboxyl group and an

FIGURE 13.9 (continued) The reaction of NHS esters with proteins. The reaction of NHS esters can acylate either α-amino group or ε-amino groups; reaction with the ε-amino group of lysine is shown in this figure. At the top is shown the synthesis of NHS propionic acid ester from propionic acid and NHS in the presence of a carbodiimide such as *N,N'*-dicyclohexylcarbodiimide (NHS can be synthesized from succinic anhydride and hydroxylamine) (Andeson, G.W., Zimmerman, J.E., and Callahan, F.M., The use of esters of *N*-hydroxysuccinimide in peptide synthesis, *J. Am. Chem. Soc.* 86, 1839–1842, 1964). Below is shown *N*-hydroxysulfosuccinimide (synthesized from *N*-hydroxymaleimide and sodium metabisulfite), which has superior solubility characteristics in aqueous solvents than NHS (Staros, J.V., *N*-Hydroxysulfosuccinimide active esters: *Bis*(*N*-hydroxysulfosuccinimide) esters of two dicarboxylic acids are hydrophilic, membrane-impermeant, protein cross-linkers, *Biochemistry* 21, 3950–3955, 1982). NHS esters are acylating agents and the formation of N^ε-propionyllysine is shown on the right-hand side of the figure (Smith, G.P., Kinetics of amine modification of proteins, *Bioconjug. Chem.* 17, 501–506, 2006). At the bottom is shown the conversion of a glutamic acid residue in an NHS ester by reaction with NHS in the presence of EDC (Graberek, Z. and Gergely, J., Zero-length crosslinking procedure with the use of active esters, *Anal. Biochem.* 185, 131–135, 1990); the active ester can be used for zero-length cross-linking and attachment of proteins to matrices or for the preparation of bioconjugates.

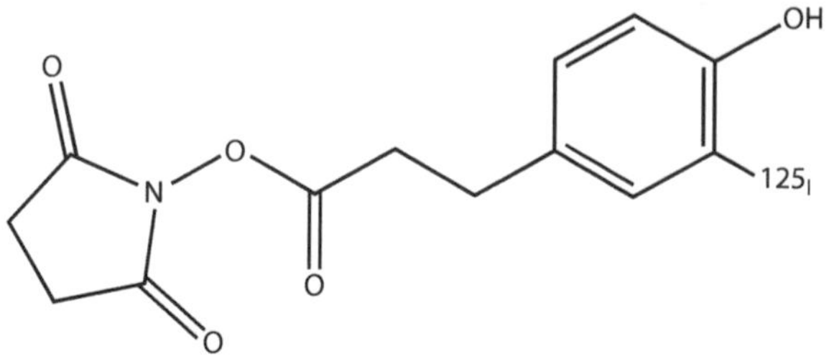

Succinimidyl-3-(3[^{125}I],4-hydroxyphenyl)propionate
Boulton-Hunter reagent

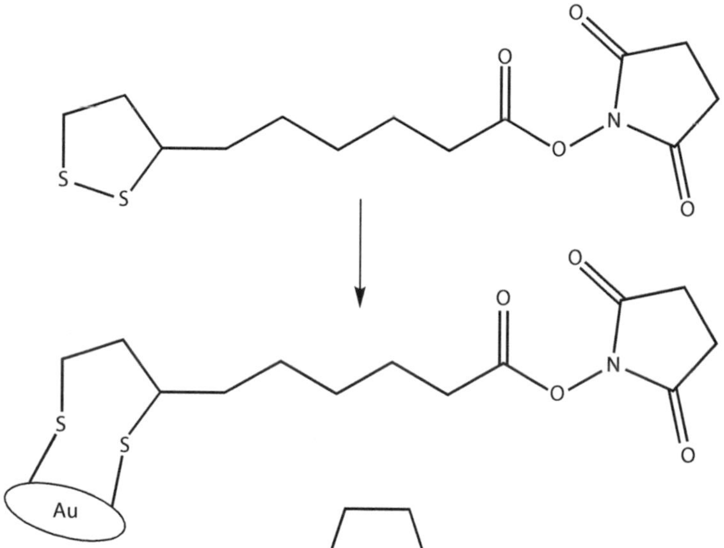

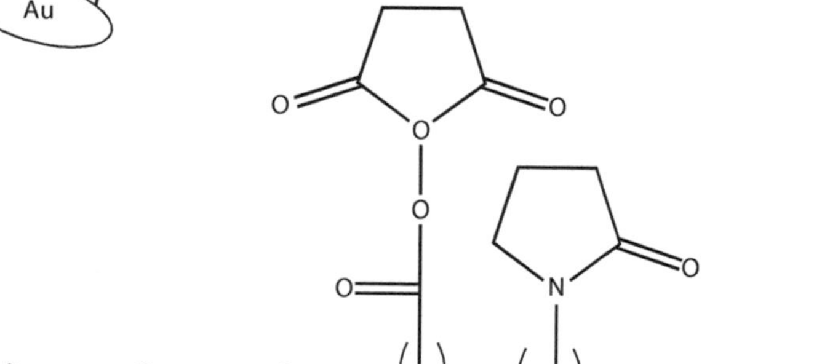

FIGURE 13.10   Some reagents based on NHS. Shown is the structure of the Bolton–Hunter reagent, which is used to radiolabel proteins (Boulton, A.E. and Hunter, W.M., The labelling of proteins to high specific radio-activities by conjugation to a ^{125}I-containing acylating agent, *Biochem. J.* 133, 529–539, 1973; Coutts, S.M. and Reid, D.M., Purification of small peptides labeled with Bolton–Hunter reagent, *Anal. Biochem.* 91, 446–450, 1978). A sulfoderivative with enhanced solubility in aqueous system is also available (Portolés, P., Rojo, J.M., and Janeway, C.A., Jr., A simple method for the radioactive iodination of CD4 molecules, *J. Immunol. Methods* 129, 105–109, 1990). Shown below is the structure of succinimidyl thioacetate (ester of NHS with thioctic acid) and binding to a gold surface (Scaramuzzo, F.A., Salvati, R., Paci, B. et al., Nanoscale in situ morphological study of protein immobilized on gold thin films, *J. Phys. Chem. B.* 113, 15895–15899, 2009). Below is a representation of a N-acryloxysuccinimide-co-*N*-vinylpyrrolidone amphiphilic copolymer for coating a polylactide copolymer, which is used to prepare a peptide-functionalized biodegradable carrier (Handké, N., Ficheux, D., Roller, M. et al., Lysine-tagged peptide coupling onto polylactide nanoparticles coated with activated ester-based amphiphilic copolymer: A route to highly peptide-functionalized biodegradable carriers, *Colloids Surf. B. Biointerfaces* 103, 298–303, 2013).

amino group (Figure 13.9) developed for zero-length cross-linking in proteins and the preparation of matrices.[437–444]

It is possible to selectively modify the α-amino groups of proteins by chemical transamination with glyoxylate at a slightly acid pH.[445,446] This modification has been applied to *Euglena* cytochrome C-552. This reaction was performed in 2.0 M sodium acetate, 0.10 M acetic acid, 0.005 M nickel sulfate, and 0.2 M sodium glyoxylate and resulted in the complete loss of the amino-terminal residue. Snake venom phospholipase A_2 has been subjected to chemical transamination.[446] This reaction was performed in 2.0 M sodium acetate, 0.4 M acetic acid, 0.010 M cupric ions, and 0.1 M glyoxylic acid, pH 5.5. More recent work by Gilmore and coworkers[447] and later by Scheck and coworkers[253] showed effective transamination of *N*-terminal amino acids with pyridoxal phosphate generating an aldehyde which can then be coupled with an amine or hydrazide to yield a bioconjugate.[254,448–449]

REFERENCES

1. Pace, C.N., Grimsley, G.R., and Scholtz, J.M., Protein ionizable groups: p*K* values and their contribution to protein stability and solubility, *J. Biol. Chem.* 284, 13285–13289, 2009.
2. Hopkins, C.E., O'Connor, P.B., Allen, K.N. et al., Chemical-modification rescue assessed by mass spectrometry demonstrates that γ-thialysine yields the same activity as lysine in aldolase, *Protein Sci.* 11, 1591–1599, 2000.
3. Hager-Braun, C., Hochleitner, E.O., Gorny, M.K. et al., Characterization of a discontinuous epitope of the HIV envelope protein gp120 recognized by a human monoclonal antibody using chemical modification and mass spectrometric analysis, *J. Am. Soc. Mass Spectrom.* 21, 1687–1698, 2010.
4. Koehler, C.J., Arntzen, M.Ø., Strozynski, M. et al., Isobaric peptide termini labeling utilizing site-specific *N*-terminal succinylation, *Anal. Chem.* 83, 4775–4781, 2011.
5. Pottiez, G. and Ciborowski, P., Elucidating protein inter- and intramolecular interacting domains using chemical cross-linking and matrix-assisted laser desorption ionization-time of flight/time of flight mass spectrometry, *Anal. Biochem.* 421, 712–718, 2012.
6. Maeshima, T., Honda, K., Chikazawa, M. et al., Quantitative analysis of acrolein-specific adducts generated during lipid peroxidation-modification of proteins in vitro: Identification of N^τ-(3-propanal)histidine as the major adduct, *Chem. Res. Toxicol.* 25, 1384–1392, 2012.
7. Schmidt, D.F., Jr. and Westheimer, F.H., pKa of the lysine amino group of acetoacetate decarboxylase, *Biochemistry* 10, 1249–1253, 1971.
8. Knowles, J.R., The intrinsic pKa values of functional groups in enzymes: Improper deductions from the pH-dependence of steady-state parameters, *CRC Crit. Rev. Biochem.* 4, 165–173, 1976.
9. Mukouyama, E.B., Oguchi, M., Kodera, Y. et al., Low pKa lysine residues at the active site of sarcosine oxidase from *Corynebacterium* sp. U-96, *Biochem. Biophys. Res. Commun.* 120, 846–851, 2004.
10. Cohen, S., Biological reactions of carbonyl halides, in *Chemistry of the Acyl Halides*, ed. S. Patel, Chapter 10, pp. 313–340, Interscience/Wiley, London, U.K., 1972.
11. Gould, A.R. and Norton, R.S., Chemical modification of cationic groups in the polypeptide cardiac stimulant anthopleurin-A, *Toxicon* 33, 187–199, 1995.
12. Becker, L., McLeod, R.S., Marcovina, S.M. et al., Identification of a critical lysine residue in apolipoprotein B-100 tat mediates noncovalent interaction with apolipoprotein (a), *J. Biol. Chem.* 276, 36155–36162, 2001.
13. Wink, M.R., Buffon, A., Bonan, C.D. et al., Effect of protein-modifying reagents on ecto-apyrase from rat brain, *Int. J. Biochem. Cell. Biol.* 32, 105–113, 2000.
14. Ehrhard, B., Misselwitz, R., Welfle, K. et al., Chemical modification of recombinant HIV-1 capsid protein p24 leads to the release of a hidden epitope prior to changes of the overall folding of the protein, *Biochemistry* 35, 9097–9105, 1996.
15. Paetzel, M., Strynadka, N.C., Tschantz, W.R. et al., Use of site-directed chemical modification to study an essential lysine in *Escherichia coli* leader peptidase, *J. Biol. Chem.* 272, 994–10003, 1997.
16. Alcalde, M., Plou, F.J. Teresa Martín, M. et al., Succinylation of cyclodextrin glycosyltransferase from *Thermoanaerobacter* s501 enhances its transferase activity using starch as a donor, *J. Biotechnol.* 86, 71–80, 2001.
17. Swart, P.J., Kuipers, M.E., Smit, C. et al., Antiviral effects of milk proteins: Acylation results in polyanionic compounds with potent activity against human immunodeficiency virus types 1 and 2 *in vitro*, *AIDS Res. Hum. Retroviruses* 12, 769–775, 1996.

18. Swart, P.J., Kuipers, E.M., Smit, C. et al., Lactoferrin. Antiviral activity of lactoferrin. *Adv. Exp. Med. Biol.* 443, 205–213, 1998.

19. Neurath, A.R., Debnath, A.K., Strick, N. et al., Blocking of CD4 cell receptors for the human immunodeficiency virus type 1 (HIV-1) by chemically modified milk proteins: Potential for AIDS prophylaxis, *J. Mol. Recognit.* 8, 204–216, 1995.

20. Jönsson, B.A., Wishnok, J.S. Skipper, P.L. et al., Lysine adducts between methyltetrahydrophthalic anhydride and collagen in guinea pig lung, *Toxicol. Appl. Pharmacol.* 135, 156–167, 1995.

21. Lindh, C.H. and Jonsson, B.A., Human hemoglobin adducts following exposure to hexahydrophthalic anhydride and methylhexahydrophthalic anhydride, *Toxicol. Appl. Pharmacol.* 153, 152–160, 1998.

22. Pool, C.T. and Thompson, T.E., Methods for dual, site-specific derivatization of bovine pancreatic trypsin inhibitor: Trypsin protection of lysine-15 and attachment of fatty acids or hydrophobic peptides at the N-terminus, *Bioconjug. Chem.* 10, 221–230, 1999.

23. Kristiansson, M.H., Jonsson, B.A., and Lindh, C.H., Mass spectrometric characterization of human hemoglobin adducts formed in vivo by hexahydrophthalic anhydride, *Chem. Res. Toxicol.* 15, 562–569, 2002.

24. O'Brien, A.M., Smith, H.T., and O'Fagain, C., Effects of phthalic anhydride modification on horse radish peroxidase stability and activity, *Biotechnol. Bioeng.* 81, 233–240, 2003.

25. Lin, H., Su, X., and He, B., Protein lysine acetylation and cysteine succination by intermediate of energy metabolism, *ACS Chem. Biol.* 7, 947–960, 2012.

26. Jiang, T., Zhou, X., Taghizadeh, K. et al., N-formylation of lysine in histone proteins as a secondary modification arising from oxidative DNA damage, *Proc. Natl. Acad. Sci. USA* 104, 60–65, 2007.

27. Wisniewski, J.R., Zougman, A., and Mann, M., N^ε-formylation of lysine is a widespread post-translational modification of nuclear proteins occurring at residues involved in regulation of chromatin function, *Nucleic Acids Res.* 36, 570–577, 2008.

28. Klotz, I.M., Succinylation, *Methods Enzymol.* 11, 576–580, 1967.

29. Meighen, E.A., Nicolim, M.Z., and Hustings, J.W., Hybridization of bacterial luciferase with a variant produced by chemical modification, *Biochemistry* 10, 4062–4068, 1971.

30. Shetty, K.J. and Rao, M.S.N., Effect of succinylation on the oligomeric structure of arachin, *Int. J. Pept. Protein Res.* 11, 305–313, 1978.

31. Rossi, A., Menezes, L.C., and Pudles, J., Yeast hexokinase A. Succinylation and properties of the active subunit, *Eur. J. Biochem.* 59, 423–432, 1975.

32. Nagel, G.M. and Schachman, H.K., Cooperative interactions in hybrids of aspartate transcarbamylase containing succinylated regulatory polypeptide chains, *Biochemistry* 14, 3195–3203, 1975.

33. Schwenke, K.D., Zirwer, D., Gast, K. et al., Changes in the oligomeric structure of legumin from pea (*Pisum sativum* L.) after succinylation, *Eur. J. Biochem.* 194, 621–627, 1990.

34. Shiao, D.D.F., Lumry, R., and Rajender, S., Modification of protein properties by change in charge. Succinylated chymotrypsinogen, *Eur. J. Biochem.* 29, 377–385, 1972.

35. Friemuth, U., Tebling, F., Boehme, F., and Ludwig, E., Studies on chemically modified proteins. Part I. Preparation and characterization, *Nahrung* 23, 215–222, 1979.

36. Bereszczak, J.Z., Brancia, F.L., Rojas Quijana, F.A., and Goux, W.J., Relative quantification of tau-related peptides using guanidino-labeling derivatization (GLaD) with online-LC on a hybrid ion trap (IT) time-of-flight (ToF) mass spectrometer, *J. Am. Soc. Mass Spectrom.* 18, 201–207, 2007.

37. King, G.J., Jones, A., Kobe, B. et al., Identification of disulfide-containing chemical cross-links in proteins using MALDI-TOF/TOF-mass spectrometry, *Anal. Chem.* 80, 5036–5043, 2008.

38. Koehler, C.J., Arntzen, M.O., Treumann, A., and Thiede, B., A rapid approach for isobaric peptide termini labeling, *Methods Mol. Biol.* 893, 129–141, 2012.

39. Martynov, A., Farber, B., and Farber, S., Quasi-life self-organizing systems: Based on ensembles of succinylated derivatives of interferon-gamma, *Curr. Med. Chem.* 18, 3431–3436, 2011.

40. Poulin, J.F., Caillard, R., and Subirade, M., β-Lactoglobulin tablets as a suitable vehicle for protection and intestinal delivery of probiotic bacteria, *Int. J. Pharm.* 405, 47–54, 2011.

41. Wilbur, D.S., Hamlin, D.K., Meyer, D.L. et al., Streptavidin in antibody pretargeting. 3. Comparison of biotin binding and tissue localization of 1,2-cyclohexanedione and succinic anhydride modified recombinant streptavidin, *Bioconjug. Chem.* 13, 611–620, 2002.

42. Wilbur, D.S., Hamlin, D.K., Meyer, D.L. et al., Streptavidin in antibody pretargeting. 5. Chemical modification of recombinant streptavidin for labeling with α-particle-emitting radionucleotides [213]Bi and [211]AT, *Bioconjug. Chem.* 19, 158–170, 2008.

43. Atassi, M.Z. and Habeeb, A.F.S.A., Reaction of protein with citraconic anhydride, *Methods Enzymol.* 25, 546–553, 1972.

44. Barber, K.J. and Warthesen, J.J., Some functional properties of acylated wheat gluten, *J. Agric. Food Chem.* 30, 930–934, 1982.

45. Dixon, H.B.F. and Perham, R.N., Reversible blocking of amino groups with citraconic anhydride, *Biochem. J.* 109, 312–314, 1968.

46. Habeeb, A.F.S.A. and Atassi, M.Z., Enzymic and immunochemical properties of lysozyme. Evaluation of several amino group reversible blocking reagents, *Biochemistry* 9, 4939–4944, 1970.

47. Shetty, J.K. and Kinsella, J.E., Ready separation of proteins from nucleoprotein complexes by reversible modification of lysine residues, *Biochem. J.* 191, 269–272, 1980.

48. Leong, A.S.-Y. and Haffajee, Z., Citraconic anhydride: A new antigen retrieval solution, *Pathology* 42, 77–81, 2010.

49. Namimatsu, S., Ghazizadeh, M., and Sugisaki, Y., Reversing the effects of formalin fixation with citraconic anhydride and heat: A universal antigen retrieval method, *J. Histochem. Cytochem.* 53, 3–11, 2005.

50. Holloway, C.J., Novel method for the isotachophoretic separation of amino acids at low to intermediate pH after derivatization with citraconic anhydride, *J. Chromatogr.* 390, 97–100, 1987.

51. Porter, R.R., The unreactive amino groups of proteins, *Biochem. Biophys. Acta* 2, 105–112, 1948.

52. Clemens, R.J., Diketene, *Chem. Rev.* 86, 241–318, 1986.

53. Gómez-Bombarelli, R., González, M., Pérez-Prior, M.T. et al., Chemical reactivity and biological activity of ketene, *Chem. Res. Toxicol.* 21, 1964–1969, 2008.

54. Chan, A.O., Ho, C.M., Chong, H.C. et al., Modification of N-terminal α-amino groups of peptides and proteins using ketenes, *J. Am. Chem. Soc.* 134, 2589–2598, 2012.

55. Marzotto, A., Pajetta, P., Galzigna, L., and Scoffone, E., Reversible acetoacetylation of amino groups in proteins, *Biochim. Biophys. Acta* 154, 450–456, 1968.

56. Weisgraber, K.H., Innerarity, T.L., and Mahley, R.W., Role of the lysine residues of plasma lipoproteins in high affinity binding to cell surface receptors on human fibroblasts, *J. Biol. Chem.*, 253, 9053–9062, 1978.

57. Mahley, R.W., Weisgraber, K.H., Innerarity, T.L., and Windmueller, H.G., Accelerated clearance of low-density and high-density lipoproteins and retarded clearance of E apoprotein-containing lipoproteins from the plasma of rats after modification of lysine residues, *Proc. Natl. Acad. Sci. USA* 76, 1746–1750, 1979.

58. Mahley, R.W., Innerarity, T.L., Weisgraber, K.B., and Oh, S.Y., Altered metabolism (*in vivo* and *in vitro*) of plasma lipoproteins after selective chemical modification of lysine residues of the apolipoproteins, *J. Clin. Invest.* 64, 743–750, 1979.

59. Jonas, A., Covinsky, K.E., and Sweeny, S.A., Effects of amino group modification in discoidal apolipoprotein A-I-egg phosphatidylcholine-cholesterol complexes on their reactions with lecithin:cholesterol acyltransferase, *Biochemistry* 24, 3508–3513, 1985.

60. Dodds, P.F., Lopez-Johnston, A., Welch, V.A., and Gurr, M.I., The effects of chemically modifying serum apolipoproteins on their ability to activate lipoprotein lipase, *Biochem. J.* 242, 471–478, 1987.

61. Saxena, U., Nagpurkar, A., Dolphin, P.J., and Mooerjea, S., A study on the selective binding of apoprotein b- and E-containing human plasma lipoproteins with lipoprotein, *J. Biol. Chem.* 262, 3011–3016, 1987.

62. Ye, S.Q., Trieu, V.N., Stiers, D.L., and McConathy, W.J., Interactions of low density lipoprotein 2 and other apolipoprotein B-containing lipoproteins with lipoprotein(a), *J. Biol. Chem.* 263, 6337–6343, 1988.

63. Bakillah, A., Jamil, H., and Hussain, M.M., Lysine and arginine residues in the N-terminal 18% of apolipoprotein B are critical for its binding to microsomal triglyceride transfer protein, *Biochemistry* 37, 3727–3734, 1998.

64. Brubaker, G., Peng, D.Q., Somerlot, B. et al., Apolipoprotein A-1 lysine modification: Effects on helical content, lipid binding and cholesterol acceptor activity, *Biochim. Biophys. Acta* 1761, 64–72, 2006.

65. Mueller, U., Reisman, R., Elliott, W. et al., Studies of chemically modified honeybee venom. I. Biochemical, toxicologic and immunologic characterization, *Int. Arch. Allergy Appl. Immunol.* 68, 312–319, 1982.

66. Mueller, U., Reisman, R., Wypych, J., and Arbesman, C., Studies of chemically modified honeybee venom. II. Immunogenicity and suppression of reaginic antibody formation, *Int. Arch. Allergy Appl. Immunol.* 68, 320–325, 1982.

67. Scandurra, R., Consalvi, V., Politi, L., and Gallina, C., Functional residues at the active site of horse liver phosphopantothenoylcysteine decarboxylase, *FEBS Lett.* 231, 192–196, 1988.

68. Seguro, K., Tamiya, T., Tsuchiya, T., and Matsumoto, J.J., Effect of chemical modifications on freeze denaturation of lactate dehydrogenase, *Cyrobiology* 26, 154–161, 1989.

69. Urabe, I., Yamamoto, M., Yamada, Y., and Okada, H., Effect of hydrophobicity of acyl groups on the activity and stability of acylated thermolysin, *Biochim. Biophys Acta*, 524, 435–441, 1978.

70. Pool, C.T. and Thompson, T.E., Chain length ad temperature dependence of the reversible association of model acylated proteins with lipid bilayers, *Biochemistry* 37, 10246–10255, 1998.

71. Béven, L., Adenier, H., Kichenama, R. et al., Ca2+-myristoyl switch and membrane binding of chemical acylated neurocalcins, *Biochemistry* 40, 8152–8160, 2001.

72. Aicart-Ramos, C., Valero, R.A., and Rodriguez-Crespo, I., Protein palmitoylation and subcellular trafficking, *Biochim. Biophys. Acta* 1808, 2981–2994, 2011.

73. Howlett, G.J. and Wardrop, A.J., Dissociation and reconstitution of human erythrocyte membrane proteins using 3,4,5,6-tetrahydrophthalic anhydride, *Arch. Biochem. Biophys.* 188, 429–437, 1978.

74. O'Connell, L.I., Bell, E.T., and Bell, J.E., 3,4,5,6-Tetrahydrophalic anhydride modification of gluta-mate dehydrogenase: The construction and activity of heterohexamers, *Arch. Biochem. Biophys.* 263, 315–322, 1988.

75. Wearne, S.J., Factor Xa cleavage of fusion proteins. Elimination of non-specific cleavage by reversible acylation, *FEBS Lett.* 263, 23–26, 1990.

76. Kimura, Y., Zaitsu, K., Motomura, Y., and Ohkura, Y., A practical reagent for reversible amino-protection of insulin, 3,4,5,6-tetrahydrophthalic anhydride, *Biol. Pharm. Bull.* 17, 881–885, 1994.

77. Fraenkel-Conrat, H., Methods for investigating the essential groups for enzyme activity, *Methods Enzymol.* 4, 247–269, 1957.

78. Riordan, J.F. and Vallee, B.L., Acetylation, *Methods Enzymol.* 11, 565–570, 1967.

79. Merle, M. et al., Acylation of porcine and bovine calcitonin: Effects on hypocalcemic activity in the rat, *Biochem. Biophys. Res. Commun.* 79, 1071, 1977.

80. Aviram, I., The role of lysines in Euglena cytochrome C-552. Chemical modification studies, *Arch. Biochem. Biophys.* 181, 199, 1977.

81. Karibian, D., Jones, C., Gertler, A. et al., On the reaction of acetic and maleic anhydrides with elastase. Evidence for a role of the NH_2-terminal valine, *Biochemistry* 13, 2891–2897, 1974.

82. Miyazaki, K. and Tsugita, A., C-terminal sequencing method for proteins in polyacrylamide gel by the reaction of acetic anhydride, *Proteomics* 6, 2026–2033, 2006.

83. Moore, G.J., Kinetics of acetylation-deacetylation of angiotensin II. Intramolecular interactions of the tyrosine and histidine side-chains, *Int. J. Pept. Protein Res.* 26, 469–481, 1985.

84. Heremans, L. and Heremans, K., Pressure effects on the Raman spectrum of proteins: Stability of the salt bridge in trypsin and elastase, *J. Mol. Struct.* 214, 305–314, 1989.

85. Smit, V., Native electrophoresis to monitor chemical modification of human intgerleukin-3, *Electrophoresis* 15, 251–254, 1994.

86. Claasen, E., Kors, N., and Van, R.N., Influence of carriers on the development and localization of anti-trinitrophenyl antibody forming cells in the murine spleen, *Eur. J. Immunol.* 16, 271–276, 1986.

87. Gao, J. and Whitesides, G.M., Using protein charge ladders to estimate the effective charges and molecu-lar weights of proteins in solution, *Anal. Chem.* 69, 575–580, 1997.

88. Anderson, J.R., Cherniayskaya, O., Gitlin, I. et al., Analysis by capillary electrophoresis of the kinetics of charge ladder formation for bovine carbonic anhydrase, *Anal. Chem.* 74, 1870–1878, 2002.

89. Chung, W.K., Evans, S.T., Freed, A.S. et al., Utilization of lysozyme charge ladders to examine the effects of protein surface charge distribution on binding affinity in ion exchange systems, *Langmuir* 26, 759–768, 2010.

90. Fligge, T.A., Kast, J., Bruns, K., and Przybylski, M., Direct monitoring of protein-chemical reactions utilizing nanoelectrospray mass spectrometry. *J. Am. Soc. Mass Spectrom.* 10, 112–118, 1999.

91. Bennett, K.L., Smith, S.V., Lambrecht, R.M. et al. characterization of chemically-modified proteins by electrospray mass spectrometry, *Bioconjug. Chem.* 7, 16–22, 1996.

92. Jahn, O., Hoffman, B., Brauns, O. et al., The use of multiple ion chromatograms in on-line HPLC-MS for the characterization of post-translational and chemical modifications of proteins, *Int. J. Mass Spectrom.* 214, 37–51, 2002.

93. Suckau, D., Mak, M., and Przybylski, M., Protein surface topology probing by selective chemical modi-fication and mass spectrometric peptide mapping, *Proc. Natl. Acad. Sci. USA* 89, 5630–5634, 1992.

94. Yeboah, F.K., Alli, I., Yaylayan, V.A. et al., Effect of limited solid-state glycation on the conformation of lysozyme by ESI-MSMS peptide mapping and molecular modeling, *Bioconjug. Chem.* 15, 27–34, 2004.

95. Brier, S., Pflieger, D., Le Mignon, M. et al., Mapping surface accessibility of the C1r/C1s tetramer by chemical modification and mass spectrometry provides new insights into assembly of the human C1 complex, *J. Biol. Chem.* 285, 32251–32263, 2010.

96. Jaffee, E.G., Lauber, M.A., Running, W.E., and Reilly, J.P., In vitro and in vivo chemical labeling of ribosomal proteins: A quantitative comparison, *Anal. Chem.* 84, 9355–9361, 2012.

97. Staudenmayer, N., Smith, M.B., Smith, H.T. et al., An enzyme kinetics and 19F nuclear magnetic resonance study of selectively trifluoroacetylated cytochrome *c* derivatives, *Biochemistry* 15, 3198–3205, 1976.

98. Staudenmayer, N., Ng, S., Smith, M.B., and Millett, F., Effects of specific trifluoroacetylation of individual cytochrome *c*: Lysines on the reaction with cytochrome oxidase, *Biochemistry* 16, 600–604, 1977.

99. Smith, M.B., Stonehuerner, J., Ahmed, A.J. et al., Use of specific trifluoroacetylation of lysine residues in cytochrome C to study the reaction with cytochrome b5, cytochrome, c1 and cytochrome oxidase, *Biochim. Biophys. Acta* 592, 303–313, 1980.

100. Webb, M., Stonehuerner, J., and Millett, F., The use of specific lysine modifications to locate the reaction site of cytochrome C with sulfite oxidase, *Biochim. Biophys. Acta* 593, 290–298, 1980.

101. Ahmed, A.J. and Millett, F., Use of specific lysine modifications to identify the site of reaction between cytochrome C and ferricyanide, *J. Biol. Chem.* 256, 1611–1615, 1981.

102. Smith, M.B. and Millett, F., A ^{19}F nuclear magnetic resonance study of the interaction between cytochrome C and cytochrome C peroxidase, *Biochim. Biophys. Acta* 626, 64–72, 1980.

103. Kamo, M. and Tsugita, A., Specific cleavage of amino side chains of serine and threonine in peptides and proteins with *S*-ethyltrifluorothioacetate vapor, *Eur. J. Biochem.* 255, 162–171, 1998.

104. Hastings, K.L., Thomas, C., Brown, A.P., and Gandolfi, A.J., Trifluoroacetylation potentiates the humoral immune response to halothane in the guinea pig, *Immunopharmacol. Immunotoxicol.* 17, 201–213, 1995.

105. Kaplan, H., Stevenson, K.J., and Hartley, B.S., Competitive labeling, a method for determining the reactivity of individual groups in proteins. The amino groups of porcine elastin, *Biochem. J.* 124, 289–299, 1971.

106. Xu, K., Acid dissociation constant and apparent nucleophilicity of lysine-501 of the α-polypeptide of sodium and potassium ion activated adenosinetriphosphatase, *Biochemistry* 28, 6894–6899, 1989.

107. Bosshard, H.R., Koch, G.L.E., and Hartley, B.S., The aminoacyl tRNA synthetase-tRNA complex: Detection by differential labeling of lysine residues involved in complex formation, *J. Mol. Biol.* 119, 377–389, 1978.

108. Richardson, R.H. and Brew, K., Lactose synthase. An investigation of the interaction site of alpha-lactalbumin for galactosyltransferase by differential kinetic labeling, *J. Biol. Chem.* 255, 3377–3385, 1980.

109. Rieder, R. and Bosshard, H.R., The cytochrome c oxidase binding site on cytochrome c. Differential chemical modification of lysine residues in free and oxidase-bound cytochrome c, *J. Biol. Chem.* 253, 6045–6053, 1978.

110. Hitchcock, S.E., Zimmerman, C.J., and Smalley, C., Study of the structure of troponin-T by measuring the relative reactivities of lysines with acetic anhydride, *J. Mol. Biol.* 147, 125–151, 1981.

111. Hitchcock, S.E., Study of the structure of troponin-C by measuring the relative reactivities of lysines with acetic anhydride, *J. Mol. Biol.* 147, 153–173, 1981.

112. Hitchcock-De Gregori, S.E., Study of the structure of troponin-I by measuring the relative reactivities of lysine with acetic anhydride, *J. Biol. Chem.* 257, 7372–7280, 1982.

113. Giedroc, D.P., Sinha, S.K., Brew, K., and Puett, D., Differential trace labeling of calmodulin: Investigation of binding sites and conformational states by individual lysine reactivities. Effects of beta-endorphin, trifluoperazine, and ethylene glycol bis(beta-aminoethyl ether)-N,N,N'N'-tetraacetic acid, *J. Biol. Chem.* 260, 13406–13413, 1985.

114. Wei, Q., Jackson, A.E., Pervaiz, S. et al., Effects of interactions of with calcineurin of the reactivities of calmodulin lysines, *J. Biol. Chem.* 263, 19541–19544, 1988.

115. Winkler, M.A., Fried, V.A., Merat, D.L., and Cheung, W.Y., Differential reactivities of lysines in calmodulin complexed to phosphatase, *J. Biol. Chem.* 262, 15466–15471, 1987.

116. Hitchcock-De Gregori, S.E., Lewis, S.F., and Mistrik, M., Lysine reactivities of tropomyosin complexed with troponin, *Arch. Biochem. Biophys.* 264, 410–416, 1988.

117. Ohguro, H., Palczewski, K., Walsh, K.A., and Johnson, R.S., Topographic study of arrestin using differential chemical modification and hydrogen/deuterium exchange, *Protein Sci.* 3, 2428–2434, 1994.

118. Scaloni, A., Monti, M., Acquaviva, R. et al., Topology of the thyroid transcription factor 1 homeodomain-DNA complex. *Biochemistry* 38, 64–72, 1999.

119. D'Ambrosio, C., Talamo, F., Vitale, R.M. et al., Probing the dimeric structure of porcine aminoacylase 1 by mass spectrometry and modeling procedures. *Biochemistry* 42, 4430–4443, 2003.

120. Zappacosta, F., Ingallinella, P., Scaloni, A. et al., Surface topology of Minibody by selective chemical modification and mass spectrometry, *Protein Sci.* 6, 1901–1909, 1997.

121. Turner, B.T., Sabo, T.M., Wilding, D., and Maurer, M.C., Mapping of factor XIII solvent accessibility as a function of activation state using chemical modification methods, *Biochemistry* 43, 9755–9765, 2004.

122. Calvete, J.J., Campanero-Rhodes, M.A., Raida, M., and Sanz, L., Characterisation of the conformation and quaternary structure-dependent heparin-binding region of bovine seminal plasma protein PDC-109, *FEBS Lett.* 444, 260–264, 1999.

123. Hochleitner, E.O. Borchers, C., Parker, C. et al., Characterization of a discontinuous epitope of the human immunodeficiency virus (HIV) core protein p24 by epitope excision and differential chemical modification followed by mass spectrometric peptide mapping analysis. *Protein Sci.* 9, 487–496, 2000.

124. Taralp, A. and Kaplan, H., Chemical modification of lyophilized proteins in nonaqueous environments. *J. Protein Chem.* 16, 183–193, 1997.

125. Lundblad, R.L., *Approaches to the Conformational Analysis of Biopharmaceuticals*, Chapter 18, pp. 251–297, CRC Press/Taylor & Francis, Boca Raton, FL, 2010.

126. Smith, C.M., Gafken, P.R., Zhang, Z. et al., Mass spectrometric quantification of acetylation at specific lysines within the amino-terminal tail of histone H4, *Anal. Biochem.* 316, 23–33, 2003.

127. Mackeen, M.M., Kramer, H.B., Chang, K.H. et al., Small-molecule-based inhibition of histone demethylation in cells assessed by quantitative mass spectrometry, *J. Proteome Res.* 9, 4082–4090, 2010.

128. Peng, L., Yuan, Z., Ling, H. et al., SIRT1 deacetylated the DNA methyltransferase 1 (DNMT1) protein and alters its activities, *Mol. Cell. Biol.* 31, 4720–4734, 2011.

129. Yang, L., Vaitheesvaran, B., Hartil, K. et al., The fasted/fed mouse metabolic acetylome: N^6-acetylation differences suggest acetylation coordinates organ-specific fuel switching, *J. Proteome Res.* 10, 4134–4149, 2011.

130. Price, J.V., Tangsombatvisit, S., Xu, G. et al., On silico peptide microarrays for high-resolution mapping of antibody epitopes and diverse protein-protein interactions, *Nat. Med.* 18, 1434–1440, 2012.

131. Chuang, C. and Yu-Lee, L.Y., Identifying acetylated proteins in mitosis, *Methods Mol. Biol.* 909. 181–204, 2012.

132. Nagaraj, R.H., Nahomi, R.B., Shanthakumar, S. et al., Acetylation of αA-crystallin in the human lens: Effects on structure and chaperone function, *Biochim. Biophys. Acta* 1822, 120–129, 2012.

133. Gurd, F.R.N., Carboxymethylation, *Methods Enzymol.* 11, 532–541, 1967.

134. Nielsen, M.L., Vermeulen, M., Bonaldi, T. et al., Iodoacetamide-induced artifact mimics ubiquitination in mass spectrometry, *Nat. Methods* 5, 459–460, 2008.

135. Parker, R.C., Stanley, S., and Kristol, D.S., Reaction of iodoacetamide with poly-l-lysine, α-acetyllysine, and ε-acetyllysine, *Int. J. Biochem.* 6, 863–866, 1975.

136. Yang, Z. and Attgalle, A.B., LC/MS characterization of undesired products formed during iodoacetamide derivatization of sulfhydryl groups of peptides, *J. Mass Spectrom.* 42, 233–243, 2007.

137. Heinrikson, R.L., Alkylation of amino acid residues at the active site of ribonuclease, *J. Biol. Chem.* 241, 1393–1405, 1966.

138. Sanger, F. and Tuppy, H., The amino acid sequence in the phenylalanyl chain of insulin. 1. The identification of lower peptides from partial hydrolysates, *Biochem. J.* 49, 463–481, 1951.

139. Sanger, F. and Tuppy, H., The amino acid sequence in the phenylalanyl chain of insulin. 2. The identification of peptides from enzymatic hydrolysates, *Biochem. J.* 49, 481–490, 1951.

140. Shapiro, R., Fox, E.A., and Riordan, J.F., Role of lysines in human angiogenin: Chemical modification and site-directed mutagenesis, *Biochemistry* 28, 1726–1732, 1989.

141. Carty, R.P. and Hirs, C.H.W., Modification of bovine pancreatic ribonuclease A with 4-sulfonoxy-2-nitrofluorobenzene. Isolation and identification of modified proteins, *J. Biol. Chem.* 243, 5244–5253, 1968.

142. Carty, R.P. and Hirs, C.H.W., Modification of bovine pancreatic ribonuclease A with 4-sulfonyloxy-2-nitrofluorobenzene. Effect of pH and temperature on the rate of reaction, *J. Biol. Chem.* 243, 5254–5265, 1968.

143. Gray, W.R., End-group analysis using dansyl chloride, *Methods Enzymol.* 25, 121–128, 1972.

144. Flengsrud, R., Detection of amino-terminal tryptophan in peptides and proteins using dansyl chloride, *Anal. Biochem.* 76, 547–550, 1976.

145. Franklin, J.G. and Leslie, J., Some enzymatic properties of trypsin after reaction with 1-dimethylaminonaphthalene-5-sulfonyl chloride, *Can. J. Biochem.* 49, 516–521, 1971.

146. Wagner, R., Podestá, F.E., González, D.H., and Andreo, C.S., Proximity between fluorescent probes attached to four essential lysyl residues in phosphoenolpyruvate carboxylase—A resonance energy transfer study, *Eur. J. Biochem.*, 173, 561–568, 1988.

147. Arduini, A., Di Paola, M., Di Ilio, C., and Federici, G., Effect of lipid peroxidation on the accessibility of dansyl chloride labeling to lipids and proteins of bovine myelin, *Free Radic. Res. Commun.* 1, 129–135, 1985.

148. Reichardt, P., Schreiber, A., Wichmann, G. et al., Identification and quantification of an in vitro adduct formed between protein reactive xenobiotics and a lysine-containing model peptide, *Environ. Toxicol.* 18, 29–36, 2003.

149. Abuharfeil, N.M., Atmeh, R.F., Abo-Shehada, M.N., and El-Sukhon, S.N., Detection of proteins after immunoblotting on nitrocellulose using fluorescent antibodies, *Electrophoresis* 12, 683–684, 1991.

150. Bergmann, U. and Wittmann-Liebold, B., Use of a hapten-specific anti-dansyl antibody for the localization of ribosomal proteins by immuno electron microscopy, *Biochem. Int.* 21, 741–751, 1990.

151. Jia, S., Kang, Y.P., Park, J.H. et al., Simultaneous determination of 23 amino acids and 7 biogenic amines in fermented food samples by liquid chromatography/quadrupole time-of-flight mass spectrometry, *J. Chromatogr. A.* 1218, 9174–9182, 2011.

152. Rebane, R., Oldekop, M.L., and Herodes, K., Comparison of amino acid derivatization reagents for LC-ESI-MS analysis. Introducing a novel phosphazene-based derivatization reagent, *J. Chromatogr. B. Anal. Technol. Biomed. Life Sci.* 904, 99–106, 2012.

153. Mazzoitti, F., Benabdelkamel, H., Di Donna, L. et al., Light and heavy dansyl reporter groups in food chemistry: Amino acid assay in beverages, *J. Mass Spectrom.* 47, 932–939, 2012.

154. Ulu, S.T., Sensitive spectrofluorimetric determination of tizanidine in pharmaceutical preparations, human plasma, and urine through derivatization with dansyl chloride, *Luminescence* 27, 426–430, 2012.

155. Nirogi, R., Komarneni, P., Kandikere, V. et al., A sensitive and selective quantification of catecholamine neurotransmitters in rat microdialysates by pre-column dansyl chloride derivatization using liquid chromatography-tandem mass spectrometry, *J. Chromatogr. B Anal. Technol. Biomed. Life Sci.* 913–914, 41–47, 2012.

156. Brautigan, D.L., Ferguson-Miller, S., and Margoliuash, E., Definition of cytochrome c binding domains by chemical modification. I. Reaction with 4-chloro-3,5-dinitrobenzoate and chromatographic separation of singly substituted derivatives, *J. Biol. Chem.* 253, 130–139, 1978.

157. Bello, J., Iijima, H., and Kartha, G., A new arylating agent, 2-carboxy-4,6-dinitrochlorobenzene. Reaction with model compounds and bovine pancreatic ribonuclease, *Int. J. Pept. Protein Res.*, 14, 199–212, 1979.

158. Hall, J., Zhu, X., Yu, L. et al., Role of specific lysine residues in the reaction of *Rhodobacter sphaeroides* cytochrome c2 with the cytochrome bc1 complex, *Biochemistry* 28, 2568–2571, 1989.

159. Long, J.E., Durham, B., Okamura, M., and Millett, F., Role of specific lysine residues in binding cytochrome c2 to the *Rhodobacter sphaeroides* reaction center in optimal orientation for rapid electron transfer, *Biochemistry* 28, 6970–6974, 1989.

160. Concar, D.W., Hill, H.A., Moore, G.R. et al., The modulation of cytochrome c electron self-exchange by site-specific chemical modification and anion binding, *FEBS Lett.* 206, 15–19, 1986.

161. Jacobs, A.A., van den Berg, P.A., Bak, H.J., and de Graaf, F.K., Localization of lysine residues in the binding domain of the K99 fibrillar subunit of enterotoxigenic *Escherichia coli*, *Biochim. Biophys. Acta* 872, 92–97, 1986.

162. Gross, E.L., Curtiss, A., Durell, S.R., and White, D., Chemical modification of spinach plastocyanin using 4-chloro-3,5-dinitrobenzoic acid: Characterization of four singly-modified forms, *Biochim. Biophys. Acta* 1016, 107–114, 1990.

163. Berger, S., Karamanos, Y., Schoentgen, F., and Julien, R., Characterization and use of biotinylated *Escherichia coli* K99 lectin, *Biochim. Biophys. Acta* 1206, 197–202, 1994.

164. Chang, L. and Lin, S., Modification of tyrosine-3(63) and lysine-6 of Taiwan cobra phospholipase A2 affects its ability to enhance 8-anilinonaphalthalene-1-sulfontate fluorescence, *Biochem. Mol. Biol. Int.* 40, 235–241, 1996.

165. Hiratsuka, T. and Uchida, K., Lysyl residues of cardiac myosin accessible to labeling with a fluorescent reagent, N-methyl-2-anilino-6-naphthalenesulfonyl chloride, *J. Biochem.* 88, 1437–1448, 1980.

166. Onondera, M., Shiokawa, H., and Takagi, T., Fluorescent probes for antibody active sites. I. Production of antibodies specific to the N-methyl-2-anilinonaphthalene-6-sulfonate group in rabbits and some fluorescent properties of the hapten bound to the antibodies, *J. Biochem.* 79, 195–202, 1976.

167. Cory, R.P., Becker, R.R., Rosenbluth, R., and Isenberg, I., Synthesis and fluorescent properties of some N-methyl-2-anilino-6-naphthalensulfonyl derivatives, *J. Am. Chem. Soc.* 90, 1643–1647, 1968.

168. Haugland, R.P., *Molecular Probes: Handbook of Fluorescent Probes and Research Chemicals*, Molecular Probes, Inc., Eugene, OR, 1989.

169. Kekic, M., Huang, W., Moens, P.D. et al., Distance measurements near the myosin head-rod junction using fluorescence spectroscopy, *Biophys. J.* 71, 40–47, 1996.

170. Tuls, J., Geren, L., and Millett, F., Fluorescein isothiocyanate specifically modifies lysine 338 of cytochrome P-450scc and inhibits adrenodoxin binding, *J. Biol. Chem.*, 264, 16421–16425, 1989.

171. Miki, M., Interaction of Lys-61 labeled actin with myosin subfragment-1 and the regulatory proteins, *J. Biochem.* 106, 651–55, 1989.

172. Bellelli, A., Ippoliti, R., Brunori, M. et al., Binding and internalization of ricin labelled with fluorescein isothiocyanate, *Biochem. Biophys. Res. Commun.*169, 602–609, 1990.

173. Devanathan, S., Dahl, T.A., Midden, W.R. et al., Readily available fluorescein isothiocyanate-conjugated antibodies can be easily converted into targeted phototoxic agents for antibacterial, antiviral, and anticancer therapy, *Proc. Natl. Acad. Sci. USA*, 87, 2980–2984, 1990.

174. Turner, D.C. and Brand, L., Quantitative estimation of protein binding site polarity. Fluorescence of N-arylaminonaphthalenesulfonates, *Biochemistry* 7, 3381–3390, 1968.

175. Stark, G.R., Stein, W.H., and Moore, S., Reaction of the cyanate present in aqueous urea with amino acids and proteins, *J. Biol. Chem.* 235, 3177–3181, 1960.

176. McCarthy, J., Hopwood, F., Oxley, D. et al., Carbamylation of proteins in 2-D electrophoresis—Myth or reality? *J. Proteome Res.* 2, 239–242, 2003.

177. Ruth, M.C., Old, W.M., Emrick, M.A. et al., Analysis of membrane proteins from human chronic myelogenous leukemia cells: Comparison of extraction methods for multidimensional LC-MS/MS, *J. Proteome Res.* 5, 709–729, 2006.

178. Kraus, L.M., Miyamura, S., Pecha, B.R. et al., Carbamoylation of hemoglobin in uremic patients determined by antibody specific for homocitrulline (carbamoylated ε-N-lysine), *Mol. Immunol.* 28, 459–463, 1991.

179. Wang, Z., Nicholls, S.J., Rodriguez, E.R. et al., Protein carbamylation links inflammation, smoking, uremia and atherogenesis, *Nat. Med.* 13, 1176–1184, 2007.

180. Holzer, M., Gauster, M., Pfeifer, T. et al., Protein carbamylation renders high-density lipoprotein dysfunctional, *Antioxid. Redox. Signal.* 14, 2337–2346, 2011.

181. Schreier, S.M., Hollaus, M., Hermann, M. et al., Carbamoylated free amino acids in uremia: HOCl generates volatile protein modifying and cytotoxic oxidant species from *N*-carbamoyl-threonine but not threonine, *Biochimie* 94, 2441–2447, 2012.

182. Stark, G.R. and Smyth, D.G., The use of cyanate for the determination of NH_2-terminal residues in proteins, *J. Biol. Chem.* 238, 214–226, 1963.

183. Stark, G.R., Modification of proteins with cyanate, *Methods Enzymol.* 25, 579–584, 1972.

184. Shaw, D.C., Stein, W.H., and Moore, S., Inactivation of chymotrypsin by cyanate, *J. Biol. Chem.* 239, 671, 1964.

185. Cerami, A. and Manning, J.M., Potassium cyanate as an inhibitor of the sickling of erythrocytes in vitro, *Proc. Natl. Acad. Sci. USA*, 68, 1180–1183, 1971.

186. Lee, C.K. and Manning, J.M., Kinetics of the carbamylation of the amino groups of sickle cell hemoglobin by cyanate, *J. Biol. Chem.* 248, 5861–5865, 1973.

187. Njikam, N., Jones, W.M., Nigen, A.M. et al., Carbamylation of the chains of hemoglobin S by cyanate in vitro and in vivo, *J. Biol. Chem.* 248, 8052–8056, 1973.

188. Nigen, A.M., Bass, B.D., and Manning, J.M., Reactivity of cyanate with valine-1 (α) of hemoglobin. A probe of conformational change and anion binding, *J. Biol. Chem.* 251, 7638–7643, 1976.

189. Plapp, B.V., Moore, S., and Stein, W.H., Activity of bovine pancreatic deoxyribonuclease A with modified amino groups, *J. Biol. Chem.* 246, 939–945, 1971.

190. Zhang, J., Yan, H, Harding, J.J. et al., Identification of the primary targets of carbamylation in bovine lens proteins by mass spectrometry, *Curr. Eye Res.* 33, 963–976, 2008.

191. Sekiguchi, T. et al., Chemical modification of ε-amino groups in glutamine synthetase from *Bacillus stearothermophilus* with ethyl acetimidate, *J. Biochem.*, 85, 75–78, 1979.

192. Browne, D.J. and Kent, S.B.H., Formation of non-amidine products in the reaction of primary amines with imido esters, *Biochem. Biophys. Res. Commun.*, 67, 126–132, 1975.

193. Monneron, A. and d'Alayer, J., Effects of imido-esters on membrane-bound adenylate cyclase, *FEBS Lett.* 122, 241–246, 1980.

194. Suck, D., Lahm, A., and Oefner, C., Structure refined to 2Å of a nicked DNA octanucleotide complex with DNase I, *Nature* 332, 464–468, 1988.

195. Hugli, T.E. and Stein, W.H., Involvement of a tyrosine residue in activity of bovine pancreatic deoxyribonuclease A, *J. Biol. Chem.* 246, 7191–7200, 1971.

196. Plapp, B.V., Enhancement of the activity of horse liver alcohol dehydrogenase by modification of amino groups at the active sites, *J. Biol. Chem.* 245, 1727–1735, 1970.

197. Fries, R.W., Bohlken, D.P., Blakley, R.T., and Plapp, B.V., Activation of liver alcohol dehydrogenases by imidoesters generated in solution, *Biochemistry* 14, 5233–5238, 1975.

198. Zoltobrocki, M., Kim, J.C., and Plapp, B.V., Activity of liver alcohol dehydrogenase with various substituents on the amino groups, *Biochemistry* 13, 899–903, 1974.

199. Plapp, B.V., Eklund, H., Jones, T.A., and Bränden, C.-I., Three-dimensional structure of isonicotinimidylated liver alcohol dehydrogenase, *J. Biol. Chem.* 258, 5537–5547, 1983.

200. Plapp, B.V., Amino groups at the active site of liver alcohol dehydrogenase, FASEB, Atlantic City, April 1969, *Fed. Proc.* 28, 601, 1969.

201. Running, W.E. and Reilly, J.P., Ribosomal proteins of *Deinococcus radiodurans*: Their solvent accessibility and reactivity, *J. Proteome Res.* 8, 1228–1246, 2009.

202. Thumm, M., Hoenes, J., and Pfleiderer, G., *S*-Methylthioacetimidate is a new reagent for the amidination of proteins at low pH, *Biochim. Biophys. Acta* 923, 263–267, 1987.

203. Makoff, A.J. and Malcolm, D.B., Properties of methyl acetimidate and its use as a protein-modifying reagent, *Biochem. J.* 193, 245–249, 1981.

204. Browne, D.T. and Kent, S.B., Formation of non-amidine products in the chemical modification of horse liver alcohol dehydrogenase with imido esters, *Biochem. Biophys. Res. Commun.* 67, 133–138, 1975.

205. Nureddin, A. and Inagami, T., Chemical modification of amino groups and guanidino groups of trypsin. Preparation of stable and soluble derivatives, *Biochem. J.* 147, 71–81, 1975.

206. Whitely, N.M. and Berg, H.C., Amidination of the outer and inner surfaces of the human erythrocyte membrane, *J. Mol. Biol.* 87, 541–561, 1974.

207. Drews, G. and Rack, M., Modification of sodium and gating currents by amino group specific cross-linking and monofunctional reagents, *Biophys. J.* 54, 383–391, 1988.

208. Rack, M. and Drews, G., Effects of chemical modification on Na channel function, *J. Protein Chem.* 8, 394–397, 1989.

209. Ruiz-Albusac, J.M., Zueco, J.A., Velásquez, E., and Blázquez, E., Isolation of a glycosyl phosphatidylinositol in rat fetal hepatocytes, *Diabetes* 42, 1262–1272, 1993.

210. Velázquez, E., Ruiz-Albusac, J.M., Carrion, M., and Blázquez, E., Isolation of a glycosyl-phosphatidylinositol (GPI) from rat brain, *Neuroreport* 5, 261–264, 1993.

211. Traut, R.R., Bollen, A., Sun, T.T. et al., 4-Mercaptobutyrimidate as a cleavable cross-linking reagent and its application to the *Escherichia coli* 30S ribosome, *Biochemistry* 12, 3266–3273, 1973.

212. Jue, R., Lambert, J.M., Pierce, L.R., and Traut, R.R., Addition of sulfhydryl groups to *Escherichia coli* ribosomes by protein modification with 2-iminothiolane (methyl-4-mercaptobutyrimidate), *Biochemistry* 17, 5399–5406, 1978.

213. Dunathan, H.C., Stereochemical aspects of pyridoxal phosphate catalysis, *Adv. Enzymol.* 35, 79–134, 1971.

214. Garadi, R. and Babitch, J.A., Externally disposed polypeptides of chick synaptosomal plasma membrane. Identification with pyridoxal phosphate and sodium borotritide, *Biochim. Biophys. Acta* 595, 31–40, 1980.

215. Martial, J., Zaldivar, J., Bull, P. et al., Inactivation of rat liver RNA polymerases I and II and yeast RNA polymerase I by pyridoxal-5′-phosphate. Evidence for the participation of lysyl residues at the active site, *Biochemistry* 14, 4907–4911, 1974.

216. Modak, M.J., Observations on the pyridoxal-5′-phosphate inhibition of DNA polymerases, *Biochemistry* 15, 3620–3626, 1976.

217. Ozyhar, A., Kiltz, H.H., and Pongs, O., Pyridoxal phosphate inhibits the DNA-printing activity of the ecdysteroid receptor, *Eur. J. Biochem.* 192, 167–174, 1990.

218. Ohsawa, H. and Gualerzi, C., Structure-function relationship in Escherichia coli inhibition factors. Identification of a lysine residue in the ribosomal binding site of initiation factor by site-specific chemical modification with pyridoxal phosphate, *J. Biol. Chem.* 256, 4905–4912, 1981.

219. Cake, M.A., DiSorbo, D.M., and Litwack, G., Effect of pyridoxal phosphate on the DNA binding site of activated hepatic glucocorticoid receptor, *J. Biol. Chem.*, 253, 4886–4891, 1978.

220. Shapiro, S., Enser, M., Pugh, E., and Horecker, B.L., The effect of pyridoxal phosphate on rabbit muscle aldolase, *Arch. Biochem. Biophys.* 128, 554–562, 1968.

221. Schnackerz, K.D. and Noltmann, E.A., Pyridoxal-5′-phosphate as a site-specific protein reagent for a catalytically critical lysine residue in rabbit muscle phosphoglucose isomerase, *Biochemistry* 10, 4837–4843, 1971.

222. Havran, R.T. and du Vigneaud, V., The structure of acetone-lysine vasopressin as established through its synthesis from the acetone derivative of S-benzyl-l-cysteinyl-l-tyrosine, *J. Am. Chem. Soc.* 91, 2696–2698, 1969.

223. Bull, P., Zaldivar, J., Venegas, A. et al., Inactivation of *Escherichia coli* RNA polymerase by pyridoxal 5′-phosphate. Identification of a low pKa lysine as the modified residue, *Biochem. Biophys. Res. Commun.* 64, 1152–1159, 1975.

224. Quang, K.H., Kishore, G.M., and Bild, G.S., 5-Enolpyruvyl shikimate 3-phosphate synthase from *Escherichia coli*. Identification of Lys-33 as a potential active site residue, *J. Biol. Chem.* 263, 735–739, 1988.

225. Lilley, K.S. and Engel, P.C., The essential active-site lysines of clostridial glutamate dehydrogenase. A study with pyridoxal-5′-phosphate, *Eur. J. Biochem.* 207, 533–540, 1992.

226. Paech, C. and Tolbert, N.E., Active site studies of ribulose-1,5-bisphosphate carboxylase/oxygenase with pyridoxal-5′-phosphate, *J. Biol. Chem.*, 253, 7864–7873, 1978.

227. Kent, A.B., Krebs, E.G., and Fischer, E.H., Properties of crystalline phosphorylase b, *J. Biol. Chem.* 232, 549–558, 1958.

228. Wimmer, M.J., Mo, T., Sawyers, D.L., and Harrison, J.H., Biphasic inactivation of porcine heart mito-chondrial malate dehydrogenase by pyridoxal-5′-phosphate, *J. Biol. Chem.* 250, 710–715, 1975.

229. Bleile, D.M., Jameson, J.L., and Harrison, J.H., Inactivation of porcine heart cytoplasmic malate dehy-drogenase by pyridoxal-5′-phosphate, *J. Biol. Chem.* 251, 6304–6307, 1976.

230. Bang, S.K. and Kim, Y.S., Chemical modification of *Pseudomonas fluorescens* malonyl-CoA synthetase with pyridoxal-5′-phosphate, *Han'guk Saenghwa Hakhoechi* 23, 276–280, 1990.

231. Cortijo, M., Jimenez, J.S., and Lior, J., Criteria to recognize the structure and micropolarity of pyridoxal-5′-phosphate binding sites in protein, *Biochem. J.* 171, 497–500, 1978.

232. Llor, J., Lopez-Cantarero, E., and Coritjo, M., Polarography of pyridoxal 5′-phosphate in aqueous and non-aqueous solvents, *Bioelectrochem. Bioenerg.* 5, 276–284, 1978.

233. Llor, J., Sanchez Ruiz, J.M., and Cortijo, M., Thermodynamic-equilibrium constants for pyridoxal and pyridoxal-5′-phosphate in dioxane water mixtures, *J. Chem. Soc. Perkin Trans. 2* (2), 951–955, 1988.

234. Delpino, I.M.P. and Sanchez Ruiz, J.M., A potentiometric study into the stability of the Schiff-base formed between pyridoxal 5′-phosphate and hexylamine in water dioxane mixtures, *J. Chem. Soc. Perkins Trans. 2* (3), 573–579, 1993.

235. June, D.S., Suelter, C.H., and Dye, J.L., Kinetics of pH-dependent interconversion of trypto-phanase spectral forms studied by scanning stopped-flow spectrophotometry, *Biochemistry* 20, 2707–2713, 1981.

236. Slebe, J.C. and Martinez-Carrion, M., Selective chemical modification and 19F NMR in the assignment of a pK-value to the active site lysyl residue in aspartate transaminase, *J. Biol. Chem.* 253, 2093–2097, 1978.

237. Nishigori, H. and Toft, D., Chemical modification of the avian progesterone receptor by pyridoxal-5′-phosphate, *J. Biol Chem.* 254, 9155–9161, 1979.

238. Moldoon, T.G. and Cidlowski, J.A., Specific modification of rat uterine estrogen receptor by pyridoxal-5′-phosphate, *J. Biol. Chem.* 255, 3100–3107, 1980.

239. Acharya, A.S., Roy, R.P., and Dorai, B., Aldimine to ketoamine isomerization (Amadori rearrange-ment) potential at the individual nonenzymic glycation sites of hemoglobin A: Preferential inhibition of glycation by nucleophiles at sites of low isomerization potential, *J. Protein Chem.* 10, 345–358, 1991.

240. Mattson, A. Boutelje, J., Csöregh, I. et al., Enhanced stereoselectivity in pig liver esterase catalyzed dies-ter hydrolysis. The role of a competitive nucleophile, *Bioorg. Med. Chem.* 2, 501–508, 1994.

241. Sugiyama, Y. and Mukohata, Y., Modification of one lysine by pyridoxal phosphate completely inacti-vates chloroplast coupling factor 1 ATPase, *FEBS Lett.*, 98, 276, 1979.

242. Peters, H., Risi, S., and Dose, K., Evidence for essential primary amino groups in a bacterial coupling factor F1 ATPase, *Biochem. Biophys. Res. Commun.*, 97, 1215, 1980.

243. Gould, K.G. and Engel, P.C., Modification of mouse testicular lactate dehydrogenase by pyridoxal 5′-phosphate, *Biochem. J.*, 191, 365, 1980.

244. Ogawa, H. and Fujioka, M., The reaction of pyridoxal-5′-phosphate with an essential lysine residue of saccharopine dehydrogenase (L-lysine-forming), *J. Biol. Chem.* 255, 7420, 1980.

245. Forrey, A.W. et al., Synthesis and properties of α- and ε-pyridoxyl lysines and their phosphorylated derivatives, *Biochimie* 53, 269, 1971.

246. Sober, H.A., *Handbook of Biochemistry*, 2nd edn., The Chemical Rubber Company, Cleveland, OH, 1970.

247. Bürger, E. and Görisch, H., Evidence for an essential lysine at the active site of l-histidinol:NAD+ oxidoreductase, a bifunctional dehydrogenase, *Eur. J. Biochem.* 118, 125, 1981.

248. Tamura, J.K., LaDine, J.R., and Cross, R.L., The adenine nucleotide binding site on yeast hexoki-nase PII. Affinity labeling of Lys-111 by pyridoxal 5′-diphospho-5′-adenosine, *J. Biol. Chem.* 263, 7907–7912, 1988.

249. Masuda, T., Ide, N., and Kitabatake, N., Effects of chemical modification of lysine residues on the sweet-ness of lysozyme, *Chem. Senses* 30, 253–264, 2005.

250. Chang, L.S., Cheng, Y.C., and Chen, C.P., Modification of Lys-6 and Lys-65 affects the structural stabil-ity of Taiwan cobra phospholipase A2, *Protein J.* 25, 127–134, 2006.

251. Schoenhofen, I.C., Lunin, V.V., Julien, J.P. et al., Structural and functional characterization of PseC, an aminotransferase involved in the biosynthesis of pseudaminic acid, as essential flagellar modification in *Helicobacter pylori*, *J. Biol. Chem.* 281, 8907–8916, 2006.

252. Grin, I.R., Rieger, R.A., and Zharkov, D.O., Inactivation of NEIL2 DNA glycosylation by pyridoxal phosphate reveals a loop important for substrate binding, *Biochem. Biophys. Res. Commun.* 394, 100–105, 2010.

253. Scheck, R.A., Dedeo, M.T., Iavarone, A.T., and Francis, M.B., Optimization of a biomimetic transamination reaction, *J. Am. Chem. Soc.* 130, 11762–11770, 2008.

254. Witus, L.S., Moore, T., Thuronyi, B.W. et al., Identification of highly reactive sequences for PLP-mediated bioconjugation using a combinatorial peptide library, *J. Am. Chem. Soc.* 132, 16812–16817, 2010.

255. Kluger, R., Methyl acetyl phosphate. A small anionic acetylating agent, *J. Org. Chem.*, 45, 2733, 1980.

256. Ueno, H., Pospischil, M.A., Manning, J.M., and Kluger, R., Site-specific modification of hemoglobin by methyl acetyl phosphate, *Arch. Biochem. Biophys.*, 244, 795, 1986.

257. Ueno, H., Pospischil, M.A., and Manning, J.M., Methyl acetyl phosphate as a covalent probe for anion-binding sites in human and bovine hemoglobins, *J. Biol. Chem.*, 264, 12344, 1989.

258. Xu, A.S., Labotka, R.J., and London, R.E., Acetylation of human hemoglobin by methyl acetylphosphate. Evidence of broad regio-selectivity revealed by NMR studies, *J. Biol. Chem.* 274, 26629–26632, 1999.

259. Raibekas, A.A., Bures, E.J., Siska, C.C. et al., Anion binding and controlled aggregation of human interleukin-1 receptor antagonist, *Biochemistry* 44, 9871–9879, 2005.

260. Means, G.E., Reductive alkylation of amino groups, *Methods Enzymol.* 47, 469–478, 1977.

261. Jentoft, J.E., Jentoft, N., Gerken, T.A., and Dearborn, D.G., 13C NMR studies of ribonuclease A methylated with [13C] formaldehyde, *J. Biol. Chem.* 254, 4366, 1979.

262. Dick, L.R., Geraldes, C.F.G.C., Sherry, A.D., Gray, C.W., and Gray, D.M., ^{13}C NMR of methylated lysines of fd gene 5 protein: Evidence for a conformational change involving lysine 24 upon binding of a negatively charged lanthanide chelate, *Biochemistry* 28, 7896, 1989.

263. Larda, S.T., Bokoch, M.P., Evanics, F., and Prosser, R.S., Lysine methylation strategies for characterization protein conformation by NMR, *J. Biomol. NMR* 54, 199–209, 2012.

264. Hattori, Y., Furuita, K., Ohki, I. et al., Utilization of lysine ^{13}C-methylation NMR for protein-protein interactions studies, *J. Biomol. NMR*, 55, 19–31, 2013.

265. Brown, E.M., Pfeffer, P.E., Kumosinski, T.F., and Greenberg, R., Accessibility and mobility of lysine residues in β-lactoglobulin, *Biochemistry* 27, 5601–5610, 1988.

266. Dempsey, C.E., Cryer, G.D., and Watts, A., The interaction of amino-deuteromethylated melittin with phospholipid membranes studied by deuterium NMR, *FEBS Lett.* 218, 173–177, 1987.

267. Spooner, P.J. and Watts, A., Reversible unfolding of cytochrome c upon interaction with cardiolipin bilayers. 1. Evidence from deuterium NMR measurements, *Biochemistry* 30, 3871–3879, 1991.

268. Rice, R.H., Means, G.E., and Brown, W.D., Stabilization of bovine trypsin by reductive methylation, *Biochim. Biophys. Acta* 492, 316–321, 1977.

269. Heller, M., Loomes, K.M., and Cooper, G.J., Synthesis of biologically active tritiated amylin and salmon calcitonin analogues, *Anal. Biochem.* 285, 100–104, 2000.

270. Dottavio-Martin, D. and Ravel, J.M., Radiolabeling of proteins by reductive alkylation with [14C]formaldehyde and sodium cyanoborohydride, *Anal. Biochem.* 87, 562–565, 1978.

271. Jentoft, N. and Dearborn, D.G., Labeling of proteins by reductive methylation using sodium cyanoborohydride, *J. Biol. Chem.* 254, 4359–4365, 1979.

272. Jentoft, N. and Dearborn, D.G., Protein labeling by reductive methylation with sodium cyanoborohydride effect of cyanide and metal ions on the reaction, *Anal. Biochem.* 106, 186–190, 1980.

273. Geoghegan, K.F., Cabacungan, J.C., Dixon, H.B., and Feeney, R.E., Alternative reducing agents for reductive methylation of amino groups in proteins, *Int. J. Pept. Protein Res.* 17, 345–352, 1981.

274. Wong, W.S., Osuga, D.T., and Feeney, R.E., Pyridine borane as a reducing agent for proteins, *Anal. Biochem.* 139, 58–67, 1984.

275. Rauert, W., Eddine, A.N., Kaufmann, S.H. et al., Reductive methylation to improve crystallization of the putative oxidoreductase Rv0765c from *Mycobacterium tuberculosis*, *Acta Crystallogr. Sect. F. Struct. Biol. Cryst. Commun.* 63, 507–511, 2007.

276. Unterieser, I. and Mischnick, P., Labeling of oligosaccharides for quantitative mass spectrometry, *Carbohydr. Res.* 346, 68–75, 2011.

277. Boersema, P.J., Foong, L.Y., Ding, V.M.Y. et al., In-depth qualitative and quantitative profiling of tyrosine phosphorylation using a combination of phosphopeptide immunoaffinity purification and stable isotope dimethyl labeling, *Mol. Cell. Proteomics* 9, 84–99, 2010.

278. Fretheim, K., Iwai, S., and Feeney, R.F., Extensive modification of protein amino groups by reductive addition of different sized substituents, *Int. J. Pept. Protein Res.* 14, 451–456, 1979.

279. Fretheim, K., Edelandsdal, B., and Harbitz, O., Effect of alkylation with different size substituents on the conformation of ovomucoid, lysozyme and ovotransferrin, *Int. J. Pept. Protein Res.* 25, 601–607, 1985.

280. Geoghegan, K.F., Ybarra, D.M., and Feeney, R.E., Reversible reductive alkylation of amino groups in proteins, *Biochemistry* 18, 5392–5399, 1979.

281. Rayment, I., Reductive alkylation of lysine residues to alter crystallization properties of proteins, *Methods Enzymol.* 276, 171–179, 1997.

282. Walter, T.S. and Grimes, J.M., Lysine methylation as a routine rescue strategy for protein crystallization, *Structure* 14, 1617–1622, 2006.

283. Sledz, P., Zheng, H. Murzyn, K. et al. New surface contacts formed upon reductive lysine methylation: Improving the probability of protein crystallization, *Protein Sci.* 19, 1395–1404, 2010.

284. Fan, Y. and Jochimiak, A., Enhanced crystal packing due to solvent reorganization through reductive methylation of lysine residues in oxidoreductase from *Streptococcus pneumoniae*, *J. Struct. Funct. Genomics* 11, 101–111, 2010.

285. Kim, Y., Quartey, P., Li, H. et al., Large-scale evaluation of protein reductive methylation for improving protein crystallization, *Nat. Methods* 5, 853–854, 2008.

286. Rayment, I., Rypniewski, W.R., Schmidt-Börn, K. et al., Three-dimensional structure of myosin subfragment-1: A molecular motor, *Science* 261, 50–58, 1993.

287. Rypniewski, W.R., Holden, H.M., and Rayment, I., Structural consequences of reductive methylation of lysine residues in hen egg white lysozyme: An x-ray analysis at 1.8 Å resolution, *Biochemistry* 32, 9851–9858, 1993.

288. Stone, D.B., Schneider, D.K., Huang, Z., and Mendelson, R.A., The radius of gyration of native and reductive methylated myosin subfragment-1 from neutron scattering, *Biophys. J.* 69, 767–776, 1995.

289. Wilson, G., Effect of reductive lactosamination on the hepatic uptake of bovine pancreatic ribonuclease A dimer, *J. Biol. Chem.* 253, 2070–2072, 1978.

290. Marsh, J.W., Denis, J., and Wriston, J.C., Glycosylation of *Escherichia coli* L-asparaginase, *J. Biol. Chem.* 252, 7673–7684, 1977.

291. McMahon, A., O'Neill, M.J., Gomez, E. et al., Targeted gene delivery to hepatocytes with galactosylated amphiphilic cyclodextrins, *J. Pharm. Pharmacol.* 64, 1063–1073, 2012.

292. Connolly, D.T., Townsend, R.R., Kawaguchi, K. et al., Binding and endocytosis of glycoproteins and neoglycoproteins by isolated rabbit hepatocytes, *Biochem. J.* 214, 421–431, 1983.

293. Baynes, J.W. and Wold, F., Effect of glycosylation on the in vivo half-life of ribonuclease, *J. Biol. Chem.* 251, 6016–6024, 1976.

294. Hashida, M., Nishikawa, M., Yamashita, F., and Takakura, Y., Cell-specific delivery of genes with glycosylated carriers, *Adv. Drug. Deliv. Rev.* 52, 187–196, 2001.

295. Fiume, L., Baglioni, M., Bolondi, L. et al., Doxorubicin coupled to lactosaminated human albumin: A hepatocellular carcinoma targeted drug, *Drug. Discov. Today* 13, 1002–1009, 2008.

296. Fiume, L. and Di Stefano, G., Lactosaminated human albumin, a hepatotropic carrier of drugs, *Eur. J. Pharm. Sci.* 40, 253–262, 2010.

297. Zhang, X., Simmons, C.G., and Corey, D.R., Liver cell specific targeting of peptide nucleic acid oligomers, *Bioorg. Med. Chem. Lett.* 11, 1269–1272, 2001.

298. Choi, Y.H., Liu, F., Park, J.S. and Kim, S.W., Lactose-poly(ethylene glycol)-grafted poly-L-lysine as hepatoma cell-targeted gene carrier, *Bioconjug. Chem.* 9, 708–718, 1998.

299. Hayashi, Y., Mori, Y., Yamashita, S. et al., Potential use of lactosaminated dendrimer (G3)/α-cyclodextrin conjugates as hepatocyte-specific siRNA carriers for the treatment of familial amyloidotic polyneuropathy, *Mol. Pharm.* 9, 1645–1653, 2012.

300. Bunn, H.F. and Higgins, P.J., Reaction of monosaccharides with proteins: Possible evolutionary significance, *Science* 213, 222–224, 1981.

301. Watkins, N.G., Thorpe, S.R., and Baynes, J.W., Glycation of amino groups in protein. Studies on the specificity of modification of RNase by glucose, *J. Biol. Chem.* 260, 10629–10636, 1985.

302. Watkins, N.G., Neglia-Fisher, C.I. Dyer, D.G. et al., Effect of phosphate on the kinetics and specificity of glycation of proteins. *J. Biol. Chem.* 262, 7207–7212, 1987.

303. Gil, H., Salcedo, D., and Romero, R., Effect of phosphate buffer on the kinetics of glycation of proteins, *J. Phys. Org. Chem.* 18, 183–186, 2005.

304. Ito, S., Nakahari, T., and Yamamoto, D., The structural feature surrounding glycated lysine residues in human hemoglobin, *Biomed. Res.* 32, 217–223, 2011.

305. Mennella, C., Visciano, M., Napolitano, A. et al., Glycation of lysine-containing dipeptides, *J. Pept. Sci.* 12, 291–296, 2006.

306. Ruderman, N. and Williamson, J.R., eds., *Hyperglycemia, Diabetes, and Vascular Disease*, Oxford University Press, New York, 1992.

307. Dalle-Donne, I. and Scaloni, A., *Redox Proteomics: From Protein Modifications to Cellular Dysfunction and Diseases*, Wiley-Interscience, Hoboken, NJ, 2006.

308. Baynes, J.W. (ed), The maillard reaction: Chemistry at the interface of nutrition, aging, and disease, *Annals N.Y. Acad. Sci* 1043, 2005.

309. Pietropaoli, D., Monaco, A., Del Pinto, R. et al., Advanced glycation end products: Possible link between metabolic syndrome and periodontal diseases, *Int. J. Immunopathol. Pharmacol.* 25, 9–17, 2012.

310. Yap, F.Y., Kantharidis, P., Couglan, M.T. et al., Advanced glycation end products as environmental risk factors for the development of type 1 diabetes, *Curr. Drug Targets* 13, 526–540, 2012.

311. Li, J., Liu, D., Sun, L. et al., Advanced glycation end products and neurodegenerative diseases: Mechanisms and perspectives, *J. Neurol. Sci.* 317, 1–5, 2012.

312. Prasad, A., Bekker, P., and Tsimikas, S., Advanced glycation end products and diabetic cardiovascular disease, *Cardiol. Rev.* 20, 177–183, 2012.

313. Quan, C., Alcala, E., Petkovska, I. et al., A study in glycation of a therapeutic recombinant humanized monoclonal antibody: Were it is, how it got there, and how it affects charge-based behavior, *Anal. Biochem.* 373, 179–191, 2008.

314. Zhang, B., Yang, Y., Yuk, I. et al., Unveiling a glycation hot spot in a recombinant humanized recombinant monoclonal antibody, *Anal. Chem.* 80, 2379–2390, 2008.

315. Yuk, I.H., Zhang, B.Y., Yang, Y. et al., Controlling glycation of recombinant antibody in fed-batch cell culture, *Biotechnol. Bioeng.* 108, 2600–2610, 2011.

316. Argirov, O.K., Lin, B., Olesen, P., and Ortwerth, B.J., Isolation and characterization of a new advanced glycation end product of dehydroascorbic acid and lysine, *Biochim. Biophys. Acta* 1620, 235–244, 2003.

317. Reihl, O., Lederer, M.O., and Schwack, W., Characterization and detection of lysine-arginine cross-links derived from dehydroascorbic acid, *Carbohydr. Res.* 339, 483–491, 2004.

318. Acharya, A.S. and Manning, J.M., Reactivity of the amino groups of carbonmonoxyhemoglobin S with glyceraldehyde, *J. Biol. Chem.* 255, 1406–1412, 1980.

319. Acharya, A.S. and Manning, J.M., Amadori rearrangement of glyceraldehyde-hemoglobin Schiff base adducts. A new procedure for the determination of ketoamine adducts in proteins, *J. Biol. Chem.* 255, 7218–7224, 1980.

320. Seneviratne, C., Dombi, G.W., Liu, W., and Dain, J.A., The in vitro glycation of human serum albumin in the presence of Zn(II), *J. Inorg. Biochem.* 105, 1548–1554, 2011.

321. Hwang, Y.J., Granelli, J., and Lyubovitsky, J.G., Multiphoton optical image guided spectroscopy method for characterization of collagen-based materials modified by glycation, *Anal. Chem.* 83, 200–206, 2011.

322. Seneviratne, C., Narayanan, R., Liu, W., and Dain, J.A., The in vitro inhibition effect of 2 nm gold nanoparticles on non-enzymatic glycation of human serum albumin, *Biochem. Biophys. Res. Commun.* 422, 447–454, 2012.

323. Ko, S.Y., Lin, I.H., Shieh, T.M. et al., Cell hypertrophy and MEK/ERK phosphorylation are regulated by glyceraldehyde-derived AGEs in cardiomyocyte H9c2 cells, *Cell. Biochem. Biophys.*, 66, 537–544, 2013.

324. Zeng, J., Dunlop, R.A., Rodgers, K.J., and Davies, M.J., Evidence for inactivation of cysteine proteases by reactive carbonyls via glycation of active site thiol, *Biochem. J.* 398, 197–206, 2006.

325. Bose, T., Bhattacherjee, A., Banerjee, S., and Chakraborti, A.S., Methylglyoxal-induced modifications of hemoglobin: Structural and functional characteristics, *Arch. Biochem. Biphys.* 529, 99–104, 2013.

326. Nahomi, R.B., Oya-Ito, T., and Nagaraj, R.H., The combined effect of acetylation and glycation on the chaperone and anti-apoptotic functions of human α-crystallin, *Biochim. Biophys. Acta* 1832, 195–203, 2013.

327. Liu, W., Cohenford, M.A., Frost, L. et al., Non-enzymatic glycation of melamine with sugars and sugar like compounds, *Bioorg. Chem.*, 46, 1–9, 2013.

328. Oliveira, L.M., Gomes, R.A., Yang, D. et al., Insights into the molecular mechanism of protein native-like aggregation upon glycation, *Biochim. Biophys. Acta*, 1834, 1010–1022, 2013.

329. Gomes, R.A., Vicentie Miranda, H., Silva, M.S. et al., Yeast protein glycation in vivo by methylglyoxal. Molecular modification of glycolytic enzymes and heat shock proteins, *FEBS J.* 272, 5273–5287, 2006.

330. Lip, Z., Yang, K., Macallister, S.L., and O'Brien, P.J., Glyoxal and methylglyoxal: Autoxidation from dihydroxyacetone and polyphenol cytoprotective antioxidant mechanisms, *Chem. Biol. Interact.*, 202, 267–274, 2013.

331. Goldfarb, A.R., A kinetic study of the reactions of amino acids and peptides with trinitrobenzenesulfonic acid, *Biochemistry*, 5, 2570, 1966.

332. Goldfarb, A.R., Heterogeneity of amino groups in proteins. I. Human serum albumin, *Biochemistry*, 5, 2574, 1966.

333. Habeeb, A.F.S.A., Determination of free amino groups in proteins by trinitrobenzenesulfonic acid, *Anal. Biochem.*, 14, 328, 1966.

334. Fields, R., The rapid determination of amino groups with TNBS, *Methods Enzymol.* 25, 464–468, 1972.

335. Kotaki, A. and Satake, K., Acid and alkaline degradation of the TNP-amino acids and peptides, *J. Biochem.* 56, 299–307, 1964.

336. Means, G.E., Congdon, W.I., and Bender, M.L., Reactions of 2,4,6-trinitrobenzenesulfonate ion with amines and hydroxide ion, *Biochemistry* 11, 3564–3571, 1972.

337. Cayot, P. and Tainturier, G., The quantification of protein amino groups by the trinitrobenzenesulfonic acid method: A reexamination, *Anal. Biochem.* 249, 184–200, 1997.

338. Liu, Y., Li, M., Wang, D., PolyPEGylation of protein using semitelechelic and mid-functional poly (PEGMA) is synthesized by RAFT polymerization, *Aust. J. Chem.* 64, 1602–1610, 2011.

339. Sathe, M., Derveni, M., Broadbent, G. et al., Synthesis and characterization of immunogens for the production of antibodies against small hydrophobic molecules with biosignature properties, *Anal. Chim. Acta* 708, 97–106, 2011.

340. Chen, S.S. and Engel, P.C., The allosteric mechanism of bovine liver glutamate dehydrogenase. Evidence from circular-dichroism studies for a conformational change in the ternary complex enzyme -(oxidized nicotinamide-adenine dinucleotide)-glutarate, *Biochem. J.* 163, 297–302, 1977.

341. Pal, P.K. and Colman, R.F., Affinity labeling of an allosteric GTP site of bovine liver glutamate dehydrogenase by 5′-*p*-fluorosulfonylbenzoylguanosine, *Biochemistry* 18, 838–845, 1979.

342. Smith, E.L., Landon, M., Piszkiewicz, D. et al., Bovine liver glutamate dehydrogenase: Tentative amino acid sequence; identification of a reactive lysine; nitration of a specific tyrosine and loss of allosteric inhibition by guanosine triphosphate, *Proc. Natl. Acad. Sci. USA*, 67, 724–730, 1970.

343. Madhusoodanan, K.S. and Colman, R.F., Adenosine 5′-O-[S-(4-succinimidyl-benzophenone) thiophosphate]: A new photoaffinity label of the allosteric ADP site of bovine liver glutamate dehydrogenase, *Biochemistry* 40, 1577–1586, 2001.

344. Smith, T.J. and Stanley, C.A., Untangling the glutamate dehydrogenase allosteric nightmare, *Trends Biochem. Sci.* 33, 557–564, 2008.

345. Engel, P.C., A marriage full of surprises; forty-five years living glutamate dehydrogenase, *Neurochem. Int.* 59, 489–494, 2011.

346. Li, M., Li, C., Allen, A. et al., The structure and allosteric regulation of mammalian glutamate dehydrogenase, *Arch. Biochem. Biophys.* 519, 69–80, 2012.

347. Moon, K., Piszkiewicz, D., and Smith, E.L., Glutamate dehydrogenase: Amino acid sequence of the bovine enzyme and comparison with that from chicken liver, *Proc. Natl. Acad. Sci. USA* 69, 1380–1383, 1972.

348. Freedman, R.B. and Radda, G.K., The reaction of 2,4,6-trinitrobenzenesulphonic acid with amino acids, peptides and proteins, *Biochem. J.* 108, 383–391, 1968.

349. Freedman, R.B. and Radda, G.K., Chemical modification of glutamate dehydrogenase by 2,4,6-trinitrobenzenesulphonic acid, *Biochem. J.* 114, 611–619, 1969.

350. Clark, C.E. and Yielding, K.L., Chemical modification of the catalytic, regulatory, and physical properties of glutamate dehydrogenase by trinitrobenzene sulfonate, *Arch. Biochem. Biophys.* 143, 158–165, 1971.

351. Frieden, C., A lifetime of kinetics, *J. Biol. Chem.* 283, 19873–19878, 2008.

352. Bates, D.J., Goldin, B.R., and Frieden, C., A new reaction of glutamate dehydrogenase: The enzyme-catalyzed formation of trinitrobenzene from TNBS in the presence of reduced coenzyme, *Biochem. Biophys. Res. Commun.* 39, 502–507, 1970.

353. Brown, A. and Fisher, H.F., A comparison of the glutamate dehydrogenase catalyzed oxidation of NADPH by trinitrobenzenesulfonate with the uncatalyzed reaction, *J. Am. Chem. Soc.* 98, 5682–5688, 1976.

354. Carlberg, I. and Mannervik, B., Reduction of 2,4,6-trinitrobenzenesulfonate by glutathione reductase and the effect of NADP+ on the electron transfer, *J. Biol. Chem.* 261, 1629–1635, 1986.

355. Coffee, C.J., Bradshaw, R.A., Goldin, B.R., and Frieden, C., Identification of the sites of modification of bovine liver glutamate dehydrogenase reacted with trinitrobenzene-sulfonate, *Biochemistry* 10, 3516–3522, 1971.

356. Goldin, B.R. and Frieden, C., Effects of trinitrophenylation of specific lysyl residues on the catalytic, regulatory and molecular properties of bovine liver glutamate dehydrogenase, *Biochemistry*, 10, 3527, 1971.

357. Rasched, I.R., Bohn, A., and Sund, H., Studies of glutamate dehydrogenase. Analysis of the quaternary structure and contact areas between the polypeptide chains, *Eur. J. Biochem.* 74, 365–377, 1977.

358. Anderson, B.M., Anderson, C.D., and Churchich, J.E., Inhibition of glutamic dehydrogenase by pyridoxal 5′-phosphate, *Biochemistry* 5, 2893–2900, 1966.

359. Piszkiewicz, D., Landon, M., and Smith, E.L., Bovine liver glutamate dehydrogenase. Sequence of a hexadecapeptide containing a lysyl residue reactive with pyridoxal 5′-phosphate, *J. Biol. Chem.* 245, 2622–2626, 1970.

360. Chen, S.-S. and Engel, P.C., Ox liver glutamate dehydrogenase. The role of lysine-126 in the light of studies of inhibition and inactivation by pyridoxal 5′-phosphate, *Biochem. J.* 149, 619–626, 1975.

361. Goldin, B.R. and Frieden, C., The effect of pyridoxal phosphate modification on the catalytic and regulatory properties of bovine liver glutamate dehydrogenase, *J. Biol. Chem.* 247, 2139–2144, 1972.

362. Holbrook, J.J. and Jeckel, R., A peptide containing a reactive lysyl group from ox liver glutamate dehydrogenase, *Biochem. J.* 111, 689–694, 1969.

363. Blumenthal, K.M. and Smith, E.L., Nicotinamide adenine dinucleotide phosphate-specific glutamate dehydrogenase of *Neurospora*. 1. Isolation, subunits, amino acid composition, sulfhydryl groups, and identification of a lysine residue reactive with pyridoxal phosphate and *N*-ethylmaleimide, *J. Biol. Chem.* 248, 6002–6008, 1973.

364. Aliverti, A., Gadda, G., Ronchi, S., and Zanetti, G., Identification of Lys116 as the target of *N*-ethylmaleimide inactivation of ferredoxin NADP+ oxidoreductase, *Eur. J. Biochem.* 198, 21–24, 1991.

365. Flügge, U.I. and Heldt, H.W., Specific labelling of the active site of the phosphate translocator in spinach chloroplasts by 2,4,6-trinitrobenzene sulfonate, *Biochem. Biophys. Res. Commun.* 84, 37–44, 1978.

366. Flügge, U.I. and Heldt, H.W., Specific labelling of a protein involved in phosphate transport of chloroplasts by pyridoxal-5′-phosphate, *FEBS Lett.* 82, 29–33, 1977.

367. Parrott, C.L. and Shifrin, S., A spectrophotometric study of the reaction of borohydride with trinitrophenyl derivatives of amino acids and proteins, *Biochim. Biophys Acta* 491, 114–120, 1977.

368. Salem, N., Jr., Lauter, C.J., and Trams, E.G., Selective chemical modification of plasma membrane ectoenzymes, *Biochim. Biophys. Acta* 641, 366–376, 1981.

369. Haniu, M., Yuan, H., Chen, S.A. et al., Structure-function relationship of NAD(P)H:quinone reductase: Characterization of NH$_2$-terminal blocking group and essential tyrosine and lysine residues, *Biochemistry* 27, 6877–6883, 1988.

370. Whittle, B.J., Cavicchi, M., and Lamarque, D., Assessment of anticolitic drugs in the trinitrobenzene sulfonic acid (TNBS) rat model of inflammatory bowel disease, *Methods Mol. Biol.* 225, 209–222, 2003.

371. te Velde, A.A., Verstege, M.I., and Hommes, D.W., Critical appraisal of the current practice in murine TNBS-induced colitis, *Inflamm. Bowel Dis.* 12, 995–999, 2006.

372. Kawada, M., Arihiro, A., and Mizoguchi, E., Insights from advances in research of chemically induced experimental models of human inflammatory bowel disease, *World J. Gastroenterol.* 13, 5581–5593, 2007.

373. Qin, H.Y., Wu, J.C., Tong, X.D. et al., Systemic review of animal models of post-infections/post-inflammatory irritable bowel syndrome, *J. Gastroenterol.* 46, 164–174, 2011.

374. Takagi, T., Naito, Y., Uchiyama, K., and Yoshikawa, T., The role of heme oxygenase and carbon monoxide in inflammatory bowel disease, *Redox Rep.* 15, 193–201, 2010.

375. Bautzová, T., Rabišková, M., and Lamprecht, A., Multiparticulate systems containing 5-aminosalicylic acid for the treatment of inflammatory bowel disease, *Drug Dev. Ind. Pharm.* 37, 1100–1109, 2011.

376. Polychroniadou, A., A simple procedure using trinitrobenzenesulfonic acid for monitoring proteolysis in cheese, *J. Dairy Res.* 55, 585–596, 1988.

377. Boulten, Y. and Grappin, R., Measurement of proteolysis in cheese: Relationship between phosphotungstic acid-soluble N fraction by Kjeldahl and 2,4,6-trinitrobenzenesulfonic acid-reactive groups in water-soluble N, *J. Dairy Res.* 61, 437–440, 1994.

378. Hernández-Herrero, M.M., Roig-Sagués, A.X., López-Sabater, E.I. et al., Protein hydrolysis and proteinase activity during the ripening of salted anchovy (*Engraulis encrasicholus* I.). A microassay method for determining protein hydrolysis, *J. Agric. Food Chem.* 47, 3319–3324, 1999.

379. Bu, G., Luo, Y., Zhang, Y., and Chen, F., Effects of fermentation by lactic acid bacteria on the antigenicity of bovine whey proteins, *J. Sci. Food Agric.* 90, 2015–2020, 2010.

380. Chove, L.M., Grandison, A.S., and Lewis, M.J., Comparison of methods for analysis of proteolysis by plasmin in milk, *J. Dairy Res.* 17, 1–7, 2011.

381. Everaerts, F., Torrianni, M., Hendriks, M., and Feijen, J., Quantification of carboxyl groups in carbodiimide cross-linked collagen sponges, *J. Biomed. Mater. Res. A* 83, 1176–1183, 2007.

382. Castaneda, L., Valle, J., Yang, N. et al., Collagen cross-linking with Au nanoparticles, *Biomacromolecules* 9, 3383–3388, 2008.

383. Anderson, G.W., Zimmerman, J.E., and Callahan, F.M., The use of esters of *N*-hydroxysuccinimide in peptide synthesis, *J. Am. Chem. Soc.* 86, 1839–1842, 1984.
384. Blumberg, S. and Vallee, B.L., Superactivation of thermolysin by acylation with amino acid *N*-hydroxysuccinimide esters, *Biochemistry* 14, 2410–2419, 1975.
385. Boulton, A.E. and Hunter, W.M., The labelling of proteins to high specific radioactivities by conjugation to a ^{125}I-containing acylating agent, *Biochem. J.* 133, 529–539, 1973.
386. Song, S.H., Jung, K.H., Paik, J.Y. et al., Distribution and pharmacokinetic analysis of angiostatin radio-iodine labeled with high stability, *Nucl. Med. Biol.* 32, 845–850, 2005.
387. Mougin-Degraef, M., Jestin, E., Bruel, D. et al., High-activity radio-iodine labeling of conventional and stealth liposomes, *J. Liposome Res.* 16, 91–102, 2006.
388. Patel, Z.S., Yamamoto, M., Ueda, H. et al., Biodegradable gelatin microparticles as delivery systems for the controlled release of bone morphogenetic protein-2, *Acta Biomater.* 4, 1126–1138, 2008.
389. Boado, R.J., Zhou, Q.H., Lu, J.X. et al., Pharmacokinetics and brain uptake of a genetically engineered bifunctional fusion antibody targeting the mouse transferrin receptor, *Mol. Pharm.* 7, 237–244, 2010.
390. Welp, A., Manz, B., and Peschke, E., Development and validation of a high throughput direct radioimmuno-assay for the quantitative determination of serum and plasma melatonin (*N*-acetyl-5-methoxytryptamine) in mice, *J. Immunol. Methods* 358, 1–8, 2010.
391. Staros, J.V., *N*-Hydroxysulfosuccinimidyl active esters: Bis(*N*-hydroxysulfosuccinimide) esters of two dicarboxylic acids are hydrophilic, membrane-impermeant, protein cross-linkers, *Biochemistry* 21, 3950–3955, 1982.
392. Harmon, C.M., Luce, P., Beth, A.H., and Abumrad, N.A., Labeling of adipocyte membrane by sulfo-*N*-succinimidyl derivatives of long-chain fatty acids: Inhibition of fatty acid transport, *J. Membr. Biol.* 121, 261–268, 1991.
393. Drahota, Z., Vrbacký, M., Nůsková, H. et al., Succinimidyl oleate, established inhibitor of CD36/FAT translocase inhibits complex III of mitochondrial respiratory chain, *Biochem. Biophys. Res. Commun.* 391, 1348–1351, 2010.
394. Bonen, A., Luiken, J.J., Liu, S. et al., Palmitate transport and fatty acid transporters in red and white muscles, *Am. J. Physiol.* 275, E471–E478, 1998.
395. Coort, S.L., Willems, J., Coumans, W.A. et al., Sulfo-*N*-succinimidyl esters of long chain fatty acids specifically inhibit fatty acid translocase (FAS/CD36)-mediated cellular fatty acid uptake, *Mol. Cell. Biochem.* 239, 213–219, 2002.
396. Baines, R.J., Chana, R.S., Hall, M. et al., CD36 mediates proximal tubular binding and uptake of albumin and is upregulated in proteinuric neuropathies, *Am. J. Physiol. Renal Physiol.* 303, F1006–F1014, 2012.
397. Yem, A.W., Zurcher-Neely, H.A., Richard, K.A. et al., Biotinylation of reactive amino groups in native recombinant human interleukin-1β, *J. Biol Chem.* 264, 17691–17697, 1989.
398. Yem, A.W., Epps, D.E., Mathews, W.R. et al., Site-specific chemical modification of interleukin 1β by acrylodan at cysteine 8 and lysine 103, *J. Biol. Chem.* 267, 3122–3128, 1992.
399. Chang, J.-Y., Ngai, P.K., Priestle, J.P. et al., Identification of a reactive lysyl residue (Lys[103]) of recombinant human interleukin-1β. Mechanism of its reactivity and implication of its functional role in receptor binding, *Biochemistry* 31, 2874–2878, 1992.
400. Wilchek, M. and Bayer, E., Biotin-containing reagents, *Methods Enzymol.* 184, 123, 1990.
401. Bayer, E. and Wilchek, M., Protein biotinylation, *Methods Enzymol.* 184, 148, 1990.
402. Blazer, L.L. and Boyle, M.D., Use of protein chip mass spectrometry to monitor biotinylation reactions, *Appl. Microbiol. Biotechnol.* 74, 717–722, 2007.
403. Dreger, M., Leung, B.W., Brownlee, G.G., and Deng, T., A quantitative strategy to detect changes in accessibility of protein regions to chemical modification on heterodimerization, *Protein Sci.* 18, 1448–1458, 2009.
404. Smith, G.P., Kinetics of amine modification of proteins, *Bioconjug. Chem.* 17, 501–506, 2006.
405. Gabant, G., Augier, J., and Armengaud, J., Assessment of solvent residues accessibility using three sulfo-NHS-biotin reagents in parallel: Application to footprint changes of a methyltransferase upon binding its substrate, *J. Mass Spectrom.* 43, 360–370, 2008.
406. Ori, A., Free, P., Courty, J. et al., Identification of heparin-binding sites in proteins by selective labeling, *Mol. Cell. Proteomics* 8, 2256–2265, 2009.
407. Owens, S.D., Johns, I.L., Walker, N.J. et al., Use of an in vitro biotinylation technique for determining posttransfusion survival of fresh and stored autologous red blood cells in Thoroughbreds, *Am. J. Vet. Res.* 71, 960–966, 2010.
408. Garon, C.L., Cohn, L.A., and Scott, M.A., Erythrocyte survival time in Greyhounds as assessed by use of in vivo biotinylation, *Am. J. Vet. Res.* 71, 1033–1038, 2010.

409. Mudge, M.C., Walker, N.J., Borjesson, D.L. et al., Post-transfusion survival of biotin-labeled allogeneic RBCs in adult horses, *Vet. Clin. Pathol.* 41, 56–62, 2012.

410. Ault, K.A. and Knowles, C., In vivo biotinylation demonstrates that reticulated platelets are the youngest platelets in circulation, *Exp. Hematol.* 23, 996–1001, 1995.

411. Valeri, C.R., Macgregor, H., Giorgio, A., and Ragno, G., Comparison of radioisotope methods and a non-radioisotope method to measure platelet survival in the baboon, *Transfus. Apher. Sci.* 32, 275–281, 2005.

412. Dowling, M.R., Josefsson, E.C., Henley, K.J. et al., Platelet senescence is regulated by an internal timer, not damage inflicted by hits, *Blood* 116, 1776–1778, 2010.

413. Fölsch, G., *N*-Hydroxysuccinimide ester of chloromercuriacetic acid, a new reagent for preparing mercury derivatives of amino acids, proteins and aminoacyl transfer ribonucleic acids, *Acta Chem. Scand.* 24, 1115–1117, 1970.

414. Levy, D., The selective chloromercuration of insulin, *Biochim. Biophys. Acta* 317, 473–481, 1973.

415. Cuatrecasas, P. and Parikh, I., Adsorbents for affinity chromatography. Use of *N*-hydroxysuccinimide esters of agarose, *Biochemistry* 11, 2291–2299, 1972.

416. Kim, H.S., Kye, Y.S., and Hage, D.S., Development and evaluation of *N*-hydroxysuccinimide-activated silica for immobilizing human serum albumin in liquid chromatography columns, *J. Chromatogr. A.* 1049, 51–61, 2004.

417. Handké, N., Ficheux, D., Rollet, M. et al., Lysine-tagged peptide coupling onto polylactide nanoparticles coated with activated ester-based amphiphilic copolymer: A route to highly peptide-functionalized biodegradable carriers, *Colloids Surf. B. Biointerfaces* 103, 298–303, 2013.

418. Kalkof, S. and Sinz, A., Chances and pitfalls of chemical cross-linking with amino-reactive *N*-hydroxysuccinimide esters, *Anal. Bioanal. Chem.* 392, 305–312, 2008.

419. Vashist, S.K. and Dixit, C.K., Interference of N-hydroxysuccinimide with bicinchoninic acid protein assay, *Biochem. Biophys. Res. Commun.* 411, 455–457, 2011.

420. Blomberg, S., Holmquist, B., and Vallee, B.L., Reversible inactivation and superactivation by covalent modification of thermolysin, *Biochem. Biophys. Res. Commun.* 51, 987–992, 1974.

421. Holmquist, B., Blumberg, S., and Vallee, B.L., Superactivation of neutral proteases with *N*-hydroxysuccinimide esters, *Biochemistry* 15, 4675–4680, 1976.

422. Blomberg, S., Amino acid residue modified during superactivation of neutral protease: Tyrosine 110 of thermolysin, *Biochemistry* 18, 2815–2820, 1979.

423. Eder, M., Krivoshein, A.V., Backer, M. et al., ScVEGF-PEG-HBED-CC and ScVEGF-PEG-NOTA conjugates: Comparison of easy-to-label recombinant proteins for ^{68}Ga PET imaging of VEGF receptors in angiogenic vasculature, *Nucl. Med. Biol.* 37, 405–412, 2010.

424. Wang, Y.J., Hao, S.J., Liu, Y.D. et al., PEGylation markedly enhanced the in vivo potency of recombinant human non-glycosylated erythropoietin: A comparison with glycosylated erythropoietin, *J. Control. Release*, 145, 306–313, 2010.

425. Kida, S., Eto, Y. Yoshioka, Y. et al., Evaluation of synthetic cell-penetrating peptides, Pro-rich peptide and octaargine derivatives, as adenovirus vector carriers, *Protein Pept. Lett.* 17, 164–167, 2010.

426. Teixeira, S., Yang, L., Dijkstra, P.J. et al., Heparinized hydroxyapatite/collagen three-dimensional scaffolds for tissue engineering, *J. Mater. Sci. Mater. Med.*, 21, 2385–2392, 2010.

427. Strehin, I., Nahas, Z., Arora, K. et al., A versatile pH sensitive chondroitin sulfate-PEG tissue adhesive and hydrogel, *Biomaterials* 31, 2788–2797, 2010.

428. Deng, C., Li, F., Hackett, J.M. et al., Collagen and glycopolymer based hydrogel for potential corneal application, *Acta Biomater.* 6, 187–194, 2010.

429. Kim, H.S., Mallik, R., and Hage, D.S., Chromatographic analysis of carbamazepine binding to human serum albumin. II. Comparison of the Schiff base and *N*-hydroxysuccinimide immobilization methods, *J. Chromatogr. B. Anal. Technol. Biomed. Life Sci.* 837, 138–146, 2006.

430. Johannsesson, G.A., Kristiansson, M.H., Jönsson, B.A. et al., Evaluation of an immunoaffinity extraction column for enrichment of adducts between human serum albumin and hexahydrophthalic anhydride in plasma, *Biomed. Chromatogr.* 22, 327–332, 2008.

431. Ramírez, A.R., Guerra, Y., Otero, A. et al., Generation of an affinity matrix useful in the purification of natural inhibitors of plasmepsin II, an antimalarial-drug target, *Biotechnol. Appl. Biochem.* 52, 149–157, 2009.

432. Maeno, K., Hirayama, A., Sakuma, K., and Miyazawa, K., An activated medium with high durability and low nonspecific adsorption: Application to protein A chromatography, *Anal. Biochem.* 409, 123–129, 2011.

433. Nelson, G.W., Perry, M., He, S.M. et al., Characterization of covalently bonded proteins on poly(methyl methacrylate) by X-ray photoelectron spectroscopy, *Colloids Surf. B. Biointerfaces* 78, 61–68, 2010.

434. Xu, Z., Du, W., Zhang, P. et al., Development of a protein biochip to identify 6 monoclonal antibodies against subtypes of recombinant human interferons, *Assay Drug. Dev. Technol.* 8, 212–218, 2010.
435. Kim, J. Cho, J., Seidler, P.M. et al., Investigations of chemical modifications of amino-terminated organic films on silicon substrates and controlled protein immobilization, *Langmuir* 26, 2599–2608, 2010.
436. El Khoury, G., Laurenceau, E., Chevolot, Y. et al., Development of miniaturized immunoassay: Influence of surface chemistry and comparison with enzyme-linked immunosorbent assay and Western blot, *Anal. Biochem.* 400, 10–18, 2010.
437. Calderon, L., Collin, E., Velasco-Bayon, D. et al., Type II collagen-hyaluronan hydrogel—A step towards a scaffold for intervertebral disc tissue engineering, *Eur. Cell. Mater.* 20, 134–148, 2010.
438. Wang, C., Yan, Q., Liu, H.B. et al., Different EDC/NHS activation mechanisms between PAA and PMAA brushes and the following amidation reactions, *Langmuir* 27, 12058–12068, 2011.
439. Hsieh, S.C., Tang, C.M., Huang, W.T. et al., Comparison between two different methods of immobilizing NGF in poly (DL-lactic acid-co-glycolic acid) conduit for peripheral nerve regeneration by EDC/NHS/MES and genipin, *J. Biomed. Mater. Res. A.* 99, 576–585, 2011.
440. Gorgieva, S. and Kokol, V., Preparation, characterization, and in vitro enzymatic degradation of chitosan-gelatine hydrogel scaffolds as potential biomaterials, *J. Biomed. Mater. Res. A.* 100, 1655–1667, 2012.
441. Zhang, Y., Xu, J.L., Yuan, Z.H. et al., Artificial intelligence techniques to optimize the EDC/NHS-mediated immobilization of cellulase on Eudragit L-100, *Int. J. Mol. Sci.* 13, 7952–7962, 2012.
442. Lai, J.Y., Corneal stromal cell growth on gelatin/chondroitin sulfate scaffolds modified at different NHS/EDC molar ratios, *Int. J. Mol. Sci.* 14, 2036–2055, 2013.
443. Krishnamoorthy, G., Sehgal, P.K., Mandal, A.B., and Sadulla, S., Development of D-lysine-assisted 1-ethyl-3-(3-dimethylaminopropyl)-carbodiimide/N-hydroxysuccinimide-initiated cross linking of collage matrix for design of scaffold, *J. Biomed. Mater. Res. A.*, 101, 1173–1183, 2013.
444. Bou-Akl, T., Banglmaier, R., Miller, R., and Vandevord, P., Effect of crosslinking on the mechanical properties of mineralized and non-mineralized collagen fibers, *J. Biomed. Mater. Res. A*, 101, 2507–2514, 2013.
445. Dixon, H.B.F. and Fields, R., Specific modification of NH_2-terminal residues by transamination, *Methods Enzymol.* 25, 409–419, 1972.
446. Verheij, H.M., Egmond, M.R., and de Haas, G.H., Chemical modification of the α-amino group in snake venom phospholipases A_2. A comparison of the interaction of pancreatic and venom phospholipases with lipid-water interfaces, *Biochemistry* 20, 94–99, 1981.
447. Gilmore, J.M., Scheck, R.A., Esser-Kahn, A.P., Joshi, N.S., and Francis, M.B., N-terminal protein modification through a biomimetic transamination reaction, *Angew. Chem. Int. Ed.* 45, 5307–5311, 2006.
448. Carrico, Z.M., Farkas, M.E., Zhou, Y. et al., N-terminal labeling of filamentous phage to create cancer marker imaging agents, *ACS Nano* 6, 6675–6680, 2012.
449. Netirojjanakui, C., Witus, L.S., Behrens, C.B. et al., Synthetically modified Fc domains as building blocks for immunotherapy applications, *Chem. Sci.* 4, 266–272, 2013.
450. Agarwal, P., van der Weijden, J., Sletten, E.M. et al., A Pictet-Spengler ligation for protein chemical modification, *Proc. Natl. Acad. Sci. USA* 110, 46–51, 2013.

14 Modification of Histidine

Histidine is an amino acid which is present at many enzyme active sites and contains two nitrogen atoms on the imidazole ring which can function as is considered to have unique characteristics as an acid–base catalyst.[1] The pKa of histidine residues in peptide and proteins is commonly between 6 and 7 permitting function at physiological pH, but there can be considerable variability.[2] Histidine is known for its ability to bind metal ions,[3] and the binding of metal ions such as zinc ions affects observed reactivity with chemical reagents such as iodoacetate[4] and diethylpyrocarbonate.[5] The ability to interact with metal ions provides the basis for the use of the hexahistidine *tail* binding to nitrilotriacetic matrix[6] (Figure 14.1) in the purification of recombinant proteins.[7–9] The hexahistidine is used for other purposes such as the design of sensors.[10,11] Dasa and coworkers[12] used a hexahistidine tail to concentrate copper ions in bulk solutions. Zaitouna and Lai[13] described a self-assembled imidazole monolayer on gold for the immobilization of a histidine–methylene blue probe (Figure 14.1). The imidazole monolayer uses nickel ions in a manner similar to that of the immobilized metal ion affinity matrix that used nitrilotriacetic acid. Histidine is also important in binding sites on proteins for macromolecules, heme groups, phosphate groups, and metal ions.[14–121] Nyarko and coworkers[23] demonstrated that the acid-induced dissociation of the LC8 subunit of cytoplasmic dynein is due to the protonation of a specific histidine residue (His55). Histidine has been described as a ligand (pseudoaffinity[24]) for the affinity chromatography of immunoglobulins.[25–27] Watanabe and coworkers[15] inserted histidine residues into streptococcal protein G B1 that is used to purify IgG; the inserted histidine residues increased the pH sensitivity of the interaction between the streptococcal protein and the IgG. Anderson and colleagues[28] established the importance of a histidine residue in the neonatal Fc receptor for binding albumin. This binding is important for the recycling of albumin. Batalla and coworkers[17] describe the importance of the histidine residues in the Fc domain for the metal-ion affinity chromatography of IgG.[29–32]

A substantial number of the studies on the modification of histidine have been directed at the importance of this amino acid residue in the catalytic mechanism of enzymes and somewhat less in the function of protein binding sites. Thus, many of the modifying reagents have been directed at the enzyme active site and are considered to be affinity labels[33] such as tosyl-phenylalanine chloromethyl ketone[34] or 2′(3′)-*O*-bromoacetyluridine.[35] TPCK reacts with active-site histidine (His57) in chymotrypsin, while 2′(3′)-*O*-bromoacetyluridine reacts with His12 in bovine pancreatic ribonuclease. The reaction rate of 2′(3′)-*O*-bromoacetyluridine with RNAse is approximately 3000 times than with free histidine, and the rate of inactivation of RNAse is 4.5 times faster than that observed with bromoacetate. The development of 1-chloro-3-tosylamido-7-amino-2-heptanone (TLCK, *N*-α-*p*-tosyl-L-lysine chloromethyl ketone)[36,37] followed the development of TPCK by Elliott Shaw and coworkers.[34] It is my understanding that the synthesis of the TLCK presented somewhat more of a challenge than TPCK. This material has the somewhat interesting characteristic of smelling like stale popcorn. The rate of trypsin inactivation by TLCK at pH 7.0 was determined to be 12.6 M^{-1} s^{-1}, the rate of inactivation by L-lysine chloromethyl ketone was determined to be 6.1 M^{-1} s^{-1}, and D-TLCK is inactive.[37] There is some other work that occurred at the same time as the elegant work of Shaw and coworkers, which deserves mention. Inagami[38] observed that iodoacetamide was a poor inhibitor of trypsin (3.3×10^{-2} M^{-1} min^{-1} at pH 7.0); the inclusion of methylguanidine increased the rate inactivation to 2×10^{-1} M^{-1} s^{-1}. It would seem that binding of the methylguanidine changes the conformation of trypsin increasing reactivity of the active-site histidine. This is supported by the work showing a conformational change in trypsin on the binding of N^{α}-benzoyl-L-arginine ethyl ester.[39] It would be predicted that the rate of reaction of chloroacetamide with the active-site histidine in trypsin would

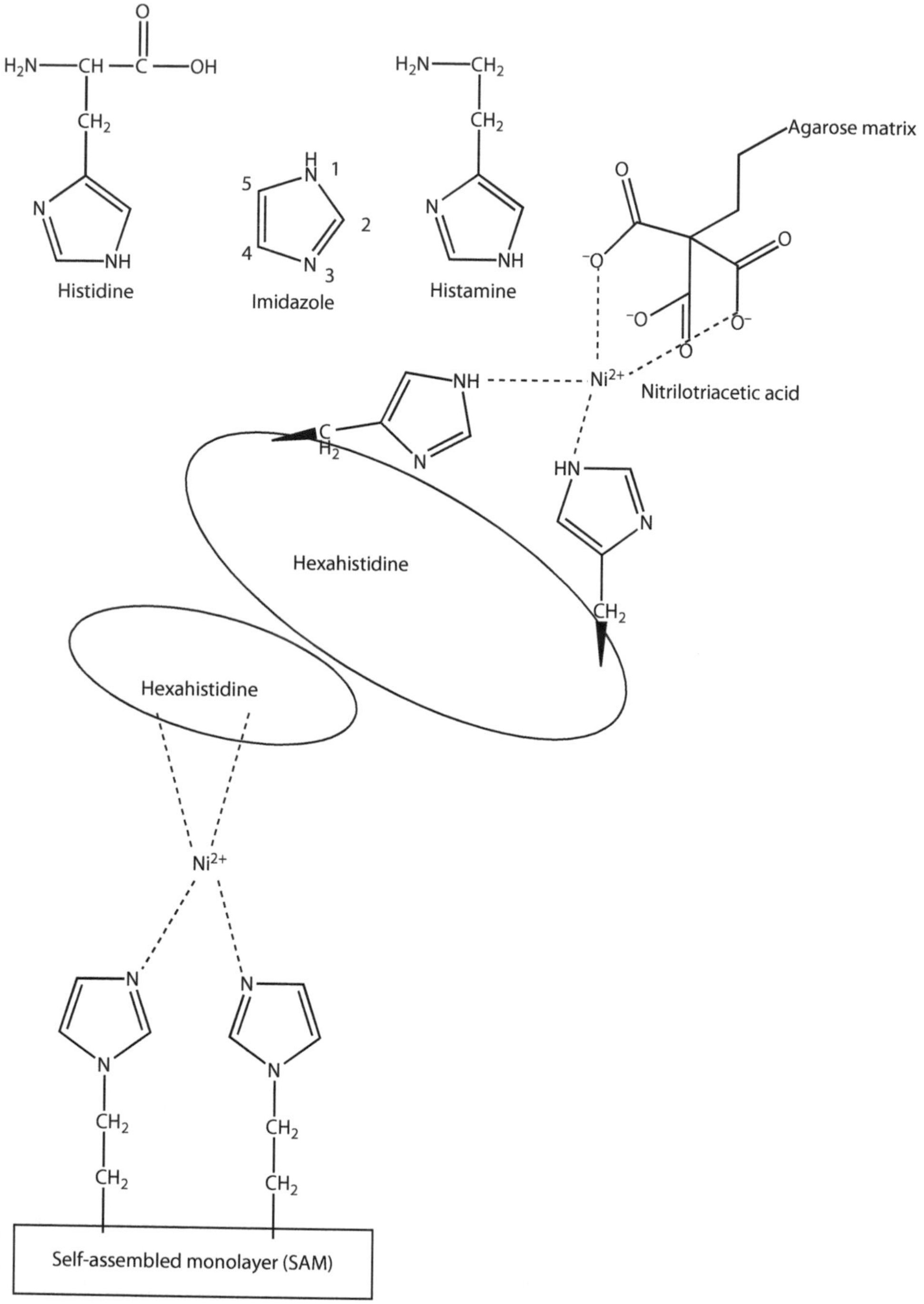

FIGURE 14.1 Structure of histidine and some derivative forms. Shown is the structure of histidine, imidazole, and histamine. In the following is a schematic of the interaction of a hexahistidine tag with a nitrilotriacetic acid matrix (Adapted from Hellman, L.M., Zhao, C., Melikishvili, M. et al., Histidine-tag-directed chromophores for tracer analyses in the analytical ultracentrifuge, *Methods* 54, 31–38, 2011). Shown at the bottom is a schematic drawing of imidazole SAM for the metal-ion immobilization of histidine-tagged peptides (Zaltouna, A.J. and Lai, R.Y., Design and characterization of a metal ion-imidazole self-assembled monolayer for reversible immobilization of histidine-tagged peptides, *Chem. Commun.* 47, 12391–12393, 2011).

be much less than that observed with iodoacetamide (see Chapter 2) as mentioned earlier emphasizing the power of the specific binding of TLCK.[37] I could find no work on the reaction of chloroacetamide or chloroacetic acid with trypsin. As discussed in Chapter 7, reagents such as TLCK or TPCK do react with enzymes other than serine proteases. One example is the work of Kupfer and coworkers[40] that demonstrated that TLCK inhibits a cyclic AMP-dependent protein kinase by reaction with a sulfhydryl group; inactivation was not observed with either chloroacetic acid or chloroacetamide. The modification of histidine residues in thiol proteases such a streptococcal protease is a complicated presence of a more potent nucleophile such as cysteine. It was necessary to block the sulfhydryl group with tetrathionate before it possibly modifies the active-site histidine. It was not possible to use iodoacetic acid or iodoacetamide, but modification was accomplished with N^α-bromoacetylarginine methyl ester.[41] The positive charge of histidine residue at the active site can serve to orient a reagent such as chloroacetic acid in the reaction with the active-site thiol.[42] Dahl and McKinley-McKee[43] have observed that imidazole accelerates the rate of reaction of haloacids and derivatives such as iodoacetate and iodoacetamide with the sulfhydryl group at the active site of liver alcohol dehydrogenase. Bloxham[44] methylated the active-site cysteine residue in lactate dehydrogenase with MMTS and observed that the rate of reaction of diethylpyrocarbonate (10°C, 0.1 M sodium pyrophosphate, pH 7.2) with histidine-195 decreased from 173 to 8.7 M^{-1} s^{-1}.

While there has been substantial interest in the modification of histidine residues in enzymes, a relatively small number of reagents have been studied for the group-specific modification of this amino acid in proteins (Figure 14.2). Some data on the rate of reaction of these reagents with histidine and histidyl residues in proteins are present in Table 14.1. This is a very limited selection, but it is apparent that reaction with diethylpyrocarbonate is much more rapid than haloacetyl derivatives but, with some proteins, comparable to that with p-bromophenacyl bromide. There is considerable variation with diethylpyrocarbonate that likely reflects increased reactivity of a histidine residue not unlike that seen in the reaction of ribonuclease with bromoacetate compared to the reaction of bromoacetate with histidine or histidine hydantoin (Table 14.1).

The imidazole ring of histidine contains two nitrogens which can be potential nucleophilic sites for modification. There are two tautomeric forms of histidine (Figure 14.2), which in the current work will be designated as N-1(N1, *pros*, $N^{\delta 1}$,N^τ) and N-3(N3 *tele*, $N^{\epsilon 2}$, N^π).[45–48] Results with histidine[45,49] and histamine[49] suggest a preference for protonation of the N3 nitrogen, but location of the histidine residue is important in the determination of tautomeric form.[48] Calleman and Poirier[47] observed that the rate of reaction at the N3 position in N^α-acetylhistidine methyl amide was slightly faster (pH 7.4, 37°C) than at the N1 position with MMTS (1.91×10^{-4} M^{-1} s^{-1} compared to 1.36×10^{-4} M^{-1} s^{-1}), while the reaction with iodoacetate was slightly slower (1.8×10^{-5} M^{-1} s^{-1} compared to 3.6×10^{-5} M^{-1} s^{-1}). In earlier work with the cupric ion complex of L-histidine, Weighardt and Goren[46] had shown that the rate of reaction of N3 (*tele*) with bromoacetate was three times the rate of reaction at N1 (*pros*); the rate of a second carboxymethylation reaction is six to seven times at either N1 or N3 than the first reaction.

Early work on the modification of histidine in proteins focused on photooxidation[50] and reagents such as FDNB, mustard gas, iodine, and diazonium compounds[51] but was hampered by the lack of analytical capability. In an excellent review in 1973, Glazer[52] stated that there were no specific reagents for histidine; the situation is a bit more promising some 40 years later on the basis of work with diethylpyrocarbonate (ethoxyformic anhydride) and, at least with phospholipase A2, p-bromophenacyl bromide. There is less use of methyl p-nitrobenzenesulfonate and haloalkyl compounds such as iodoacetate and other alkylating agents. That said, it is noted that a large amount of very interesting information has been obtained through the use of haloacid derivatives such as iodoacetate or iodoacetamide. The modification of histidine was observed in keratin (wool) on reaction with methyl iodide at pH 9.0 in phosphate buffer.[53] Ebert[53] also reported that the methylation of cysteine under these reaction conditions was associated with the formation of substantial amounts of dehydroalanine that subsequently reacted via Michael addition reaction with cystine and lysine. It was stated that the reaction was heterogeneous reflecting the insolubility of wool and the low

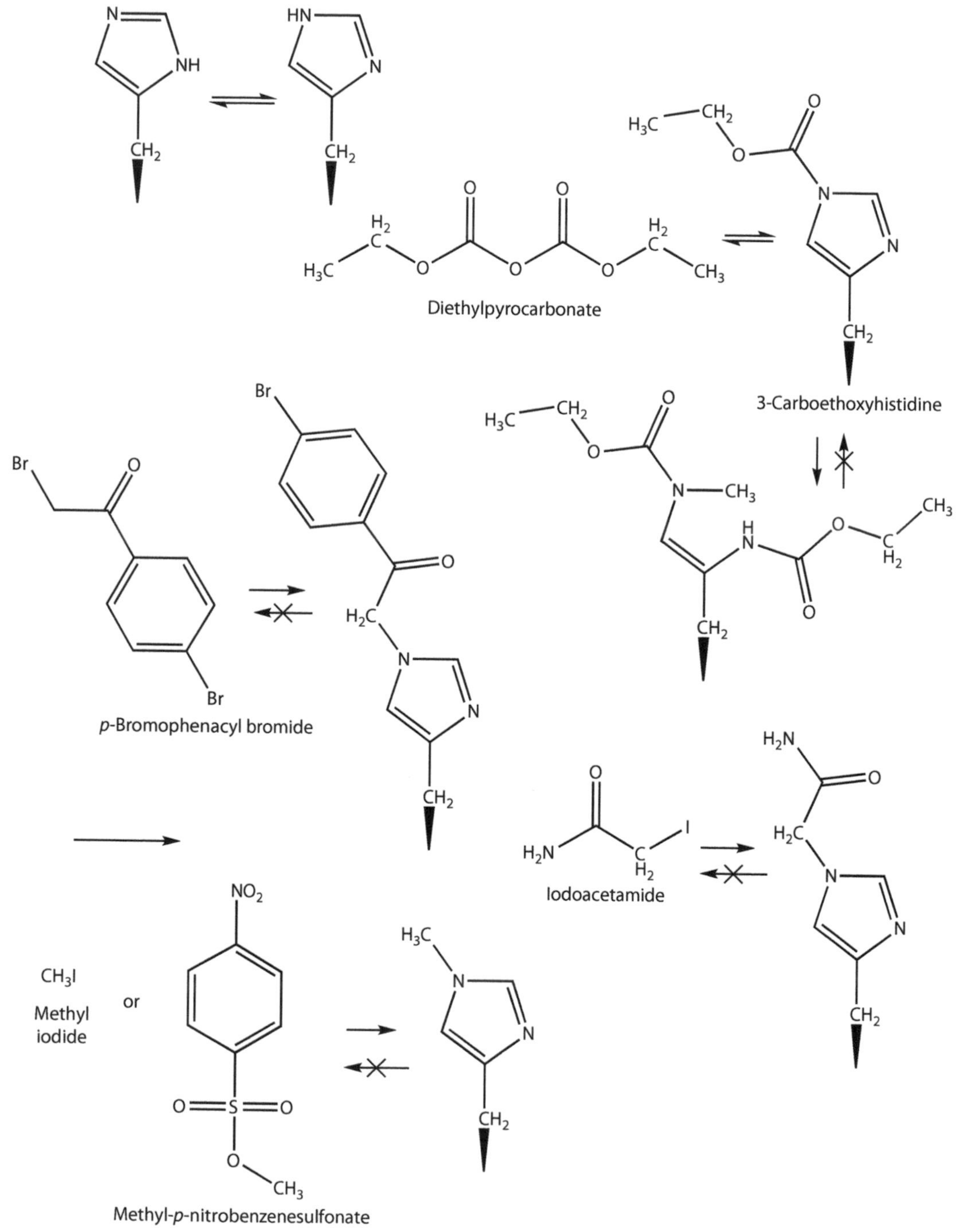

Diethylpyrocarbonate

3-Carboethoxyhistidine

p-Bromophenacyl bromide

Iodoacetamide

CH₃I
Methyl
iodide   or

Methyl-p-nitrobenzenesulfonate

FIGURE 14.2 (continued)

solubility of methyl iodide in aqueous media. The volatility of methyl iodide permits reaction in the gas phase[54] and was used to suggest molecular imprinting of α-chymotrypsin on lyophilization.[55] In these studies, the modification of the active-site histidine was enhanced in α-chymotrypsin lyophilized in the presence of sorbitol, while protein lyophilized in the presence of citrate showed a decrease in the rate of modification. Histidine is subject to glycation and subsequent Maillard reaction (Figure 14.3).[56] There have been several reports[57,58] on the modification of histidine with N-ethylmaleimide; this reaction is most likely a Michael addition that should be considered when this reagent is used for the modification of cysteine.

Photooxidation can be a relatively specific modification method for the modification of histidine in proteins but also can result in the modification of cysteine, methionine, tryptophan, and, to as lesser extent, cystine and tyrosine.[59] Studies on the photooxidation of proteins date to at least 1926 to the studies of Harris on blood,[60] plasma,[61] and proteins/amino acids.[62] The work on the photooxidation of plasma[61] was a seminal work in establishing the importance of a sensitizer with the observation in the velocity between clear plasma and plasma with some hemolysis. Hematoporphyrin, a degradation product of heme, was identified as a material derived in hemolysis that could function as a photosensitizer in the studies in plasma. Subsequent work by Harris on proteins and amino acids[62] established the importance of proteins in blood/plasma as the targets of photooxidation. This work used hematoporphyrin as a photosensitizer for the photooxidation of egg white, isolated ovalbumin, and amino acids. Tryptophan and tyrosine were susceptible to photooxidation in the presence of hematoporphyrin. A variety of other amino acids including histidine were not susceptible to photooxidation under these conditions. It is interesting to note that Harris mentions the theory relating chemical composition to light absorption referring back to the work of Witt[63] in 1876 on defining chromophores and chromogens. Studies on the photooxidation of isolated plasma proteins appeared in 1941[64] and continue today with the use of photodynamic virus inactivation in the manufacture of blood plasma products.[65,66]

Photooxidation is discussed in further detail in Chapter 5. This short section will focus on the effect of singlet oxygen generated with a photosensitizer such as methylene blue (photooxidation type II); there will be only limited mention of the role of radicals such as superoxide anion in photooxidation (photooxidation type I) (Figure 14.4). Tsai and coworkers[67] studied the photooxidation of lipoamide dehydrogenase and lysozyme with several dyes including rose bengal, methylene blue, and eosin Y. The dye bound to the protein prior to the process of photooxidation; the dissociation constant for rose bengal was determined to be 48 mM, while the value for methylene blue was 22 mM. Histidine was the target in lipoamide dehydrogenase and tryptophan was the target in lysozyme. The binding of methylene blue is important in the photooxidation of *Torpedo californica*

FIGURE 14.2 (continued)   The structure of histidine and some reagents modifying histidine. Shown at the top left is the tautomerism of histidine. The left form shows protonation on the N-1 position (*pros*), while the right form shows protonation on N-3 position (*tele*). It is estimated that the majority of histidine shows protonation at the N-3 position (Reynolds, W.F., Peat, I.R., Freedman, M.H., and Lyerla, J.R., Jr., Determination of the tautomeric form of the imidazole ring of L-histidine in basic solution by carbon-13 magnetic resonance spectroscopy, *J. Am. Chem. Soc.* 95, 328–331, 1973) although the existence of a specific tautomeric form in protein depends on location (Vila, J.A., Arnautova, Y.A., Vorobjev, Y., and Scheraga, H.A., Assessing the fractions of tautomeric forms of the imidazole ring of histidine in proteins as a function of pH, *Proc. Natl. Acad. Sci. USA* 108, 5602–5607, 2011). Shown is one product of the reaction of diethylpyrocarbonate with histidine. This reaction is reversible but is enhanced in the presence of hydroxylamine. The monosubstituted derivative is subject to a second reaction to yield the disubstituted derivative which is subject to degradation in an irreversible reaction (Miles, E.W., Modification of histidyl residues in proteins in diethylpyrocarbonate, *Methods Enzymol.* 47, 431–442, 1977). Also shown is the product of the reaction of p-bromophenacyl bromide [2-bromo-1-(4-bromophenyl)ethanone], the N3-methyl histidine that arises from the reaction with iodomethane or methyl p-nitrobenzenesulfonate, and the product arising from reaction with iodoacetate. The products shown are those where modification occurred at the N3 position ($N^{\varepsilon 2}$, *tele*); reaction can also occur at the N1 ($N^{\delta 1}$, *pros*) position.

TABLE 14.1
Modification of Histidine Residues in Proteins

Amino Acid, Peptide, or Protein and Reaction Conditions	$M^{-1}s^{-1}$	Reference
Histidine/bromoacetate/I=0.1/0.1 M acetate, pH 5.5/25°C	8.6×10^{-6}	14.1.1
Histidine hydantoin/bromoacetate/25°/pH > pKa	$1.1 \times 10^{-4,a}$	14.1.2
Ribonuclease/iodoacetate/I=0.1/0.1 M acetate, pH 5.5/25°C	5.8×10^{-3}	14.1.2
Ribonuclease/bromoacetate/I=0.1/0.1 M acetate, pH 5.5/25°C	2.1×10^{-2}	14.1.1
Ribonuclease/chloroacetate/I=0.1/0.1 M acetate, pH 5.5/25°C	2×10^{-4}	14.1.1
Ribonuclease/iodoacetate/0.2 M MES, pH 5.5/37°C	5×10^{-2}	14.1.3
Ribonuclease/bromoacetate/0.2 M MES, pH 5.5/37°C	9×10^{-2}	14.1.3
Ribonuclease/chloroacetate/0.2 M MES, pH 5.5/37°C	3×10^{-3}	14.1.3
Deoxyribonuclease/bromoacetate/0.05 M Tris—4 mM CuCl$_2$/25°C	0.1	14.1.3
Porcine pancreatic phospholipase A$_2$/p-bromophenacyl bromide, 0.1 M sodium cacodylate and 0.1 M NaCl, pH 6.1/30°C	2.1	14.1.4
Phospholipase A$_2$/p-bromophenacyl bromide/0.04 MES, pH 7.0/25°C	2.2	14.1.5
Imidazole/diethylpyrocarbonate/18°C–22°C	20[a]	14.1.6
Lactate dehydrogenase/diethylpyrocarbonate/18°C–22°C	216[a]	14.1.6
Lactate dehydrogenase/diethylpyrocarbonate/10°C	173	14.1.7
Thiomethylated lactate Dehydrogenase/diethylpyrocarbonate/10°C	8.7	14.1.7
Bacterial luciferase/diethylpyrocarbonate 0.1 M phosphate, pH 6.1 at 0°C	2.4	14.1.8
Prostatic acid phosphatase/diethylpyrocarbonate 0.025 M sodium barbital, 0.15 M NaCl, pH 6.9 at 25°C	0.12	14.1.9
Pyridoxamine-5'-phosphate oxidase/diethylpyrocarbonate, 0.1 M phosphate, pH 7.0, 25°C	12.5[a]	14.1.10
L-α-Hydroxy acid oxidase; 0.020 M MES, pH 7.0 at 25°C	14	14.1.11
Ribulose bisphosphate carboxylase/oxygenase (*Rhodospirillum rubrum*); 0.050 M Tris, 0.001 M EDTA, 0.020 M MgCl$_2$, pH 7.0 at 21°C	39	14.1.12
Ribulose bisphosphate carboxylase/oxygenase (spinach); 0.050 M Tris, 0.001 M EDTA, 0.020 M MgCl$_2$, pH 7.0 at 21°C	14	14.1.12
Rat liver arginase/diethylpyrocarbonate/0.050 M HEPES, 5 mM MnCl$_2$, pH 7.0/25°C	113	14.1.13
Yeast alcohol dehydrogenase/diethylpyrocarbonate[b]/phosphate (I=0.1), pH 7.0/23°C	100	14.1.14

[a] pH-independent rate constant.

[b] This study contains information on the half-life of diethylpyrocarbonate at different pH values. At pH 7.0, the $t_{1/2}$ is 13.8 min and decreases only slightly with increasing pH $t_{1/2}$=13.2 min at pH 8.75.

References to Table 14.1

14.1.1. Heinrikson, R.L., Stein, W.H., Crestfield, A.M., and Moore, S., The reactivities of the histidine residues at the active site of ribonuclease toward halo acids of different structures, *J. Biol. Chem.* 240, 2921–2934, 1965.

14.1.2. Lennette, E.P. and Plapp, B.V., Kinetics of carboxymethylation of histidine hydantoin, *Biochemistry* 18, 3933–3938, 1979.

14.1.3. Plapp, B.V., Mechanisms of carboxymethylation of bovine pancreatic nucleases by haloacetates and tosylglycolate, *J. Biol. Chem.* 248, 4896–4900, 1973.

14.1.4. Volwerk, J.J., Pieterson, W.A., and de Haas, G.H., Histidine at the active site of phospholipase A$_2$, *Biochemistry* 13, 1446–1454, 1974.

14.1.5. Roberts, M.F., Deems, R.A., Mincey, T.C., and Dennis, E.A., Chemical modification of the histidine residue in phospholipase A$_2$ (*Naja naja naja*), *J. Biol. Chem.* 252, 2405–2411, 1977.

14.1.6. Holbrook, J.J. and Ingram, V.A., Ionic properties of an essential histidine residue in pig heart lactate dehydrogenase, *Biochem. J.* 131, 729–738, 1973.

14.1.7. Bloxham, D.P., The chemical reactivity of the histidine-195 residue in lactate dehydrogenase thiomethylated at the cysteine-165 residue, *Biochem. J.* 193, 93–97, 1981.

14.1.8. Cousineau, J. and Meighen, E., Chemical modification of bacterial luciferase with ethoxyformic anhydride: Evidence for an essential histidyl residue, *Biochemistry* 15, 4992–5000, 1976.

TABLE 14.1 (continued)

Modification of Histidine Residues in Proteins

14.1.9. McTigue, J.J. and van Etten, R.L., An essential active-site histidine residue in human prostatic acid phosphatase. Ethoxyformylation by diethylpyrocarbonate and phosphorylation by a substrate, *Biochim. Biophys. Acta* 523, 407–421, 1978.

14.1.10. Horiike, K., Tsuge, H., and McCormick, D.B., Evidence for an essential histidyl residue at the active site of pyridoxamine (pyridoxine)-5′-phosphate oxidase from rabbit liver, *J. Biol. Chem.* 254, 6638–6643, 1979.

14.1.11. Meyer, S.E. and Cromartie, T.H., Role of essential histidine residues in l-α-hydroxy acid oxidase from rat kidney, *Biochemistry* 19, 1874–1881, 1980.

14.1.12. Saluja, A.K. and McFadden, B.A., Modification of the active site histidine in ribulosebisphosphate carboxylase/oxygenase, *Biochemistry* 21, 89–95, 1982.

14.1.13. Colleluori, D.M., Reszkowski, R.S., Emig, F.A. et al., Probing the role of the hyper-reactive histidine residue of arginase, *Arch. Biochem. Biophys.* 444, 15–26, 2005.

14.1.14. Dickenson, C.J. and Dickinson, F.M., The role of an essential histidine residue of yeast alcohol dehydrogenase, *Eur. J. Biochem.* 52, 595–603, 1975.

acetylcholinesterase in the presence of methylene blue.[68,69] Tryptophan was the target in the *T. californica* acetylcholinesterase. The work of Tsai and coworker showed that both type I photooxidation and type II photooxidation were occurring in their system, but these investigators suggested that singlet oxygen (photooxidation type II) was dominant in the modification of histidine. It is possible to differentiate between type I and type II photooxidation through the use of quenchers such as azide (type II)[70] and ferricyanide (type I)[71]; the effect of singlet oxygen is extended in ^2H_2O providing another test for type II photooxidation.[72]

There has been steady use of photooxidation to identify functional histidine residues in enzymes[73–87] and some other proteins[88] including fibrinogen.[89–91] The observations on fibrinogen are of interest as it resulted in an innovative approach when methylene blue is used for viral inactivation.[65,66] While the use of photooxidation to identify functional histidine residues is valuable, the work on cross-linking and aggregation is also of considerable importance as it has value in understanding biopharmaceutical stability. Tomita and coworkers[92] presented results on the photooxidation of histidine in 1969 and proposed a mechanism that results in a histidine dimer (Figure 14.4). Balasubramanian and coworkers[93] reported on the reaction of singlet oxygen (photooxidation type II) with crystallins. The oxidation of tryptophan with the formation of *N*-formylkynurenine did not result in the formation of aggregates, while the oxidation of histidine was necessary for protein aggregation. Shen and coworkers[94] used photooxidation in the presence of rose bengal to cross-link *N*-(2-hydroxypropyl) methylacrylamide with ε-aminocaproic acid side chains terminating in histidine or lysine. Cross-links were formed between histidine residues and between histidine and lysine. Subsequent work from this group studied the cross-linking of N^α-benzoyl-L-histidine.[95] Agon and coworkers[96] studied the photooxidation of histidine and other imidazole derivatives in the presence of rose bengal and other photosensitizers and demonstrated the formation of peroxides (Figure 14.4). Finally, photodegradation of a monoclonal antibody in a histidine-containing buffer has been reported.[97]

Histidine residues can be modified by α-halo carboxylic acids and amides (i.e., bromoacetate and bromoacetamide) (Figure 14.5), but there has been little current use of these reagents as other amino acids, most notably cysteine, are modified more rapidly than histidine.[98] As an example, Rahimi and coworkers[99] could modify the cysteine in a far-red fluorescent protein with iodoacetamide and the histidine residue with diethylpyrocarbonate (see below). The advent of diethylpyrocarbonate and, to a lesser extent, *p*-bromophenacyl bromide has decreased the use of reagents such as iodoacetate and iodoacetamide for the modification of histidine residues. It should be noted that, for all practical purposes, the reaction of histidine with a reagent such as iodoacetate or iodoacetamide outside of an enzyme active site is quite slow.[4,38] Nevertheless, there are individual applications where a histidine

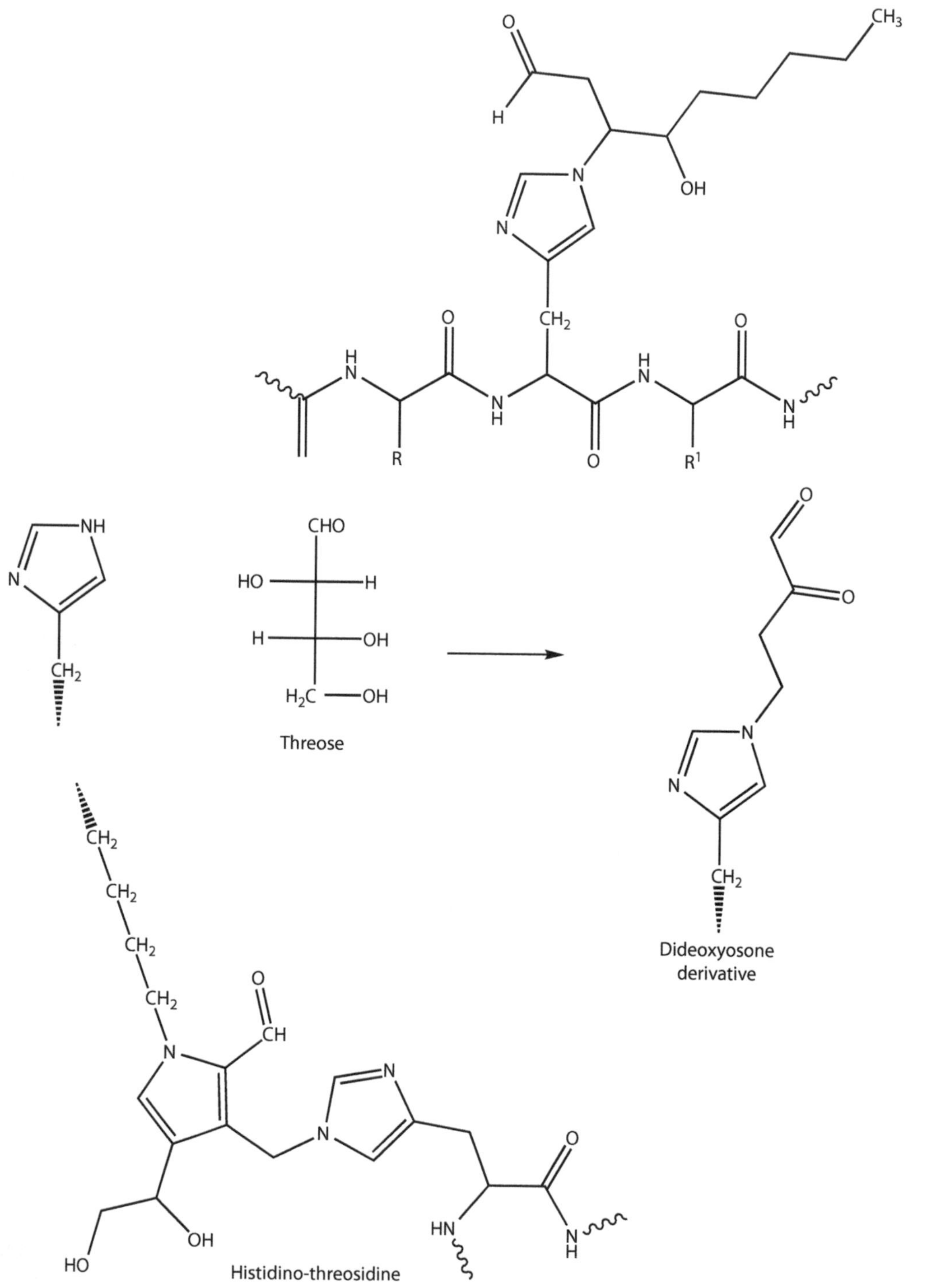

FIGURE 14.3 The glycation reactions of histidine. Shown is a glycation reaction (Puttaiah, S., Zhang, Y., Pilch, H.A. et al., Detection of dideoxyosone intermediates of glycation using a monoclonal antibody: Characterization of major epitope structures, *Arch. Biochem. Biophys.* 446, 186–196, 2006) and a Maillard reaction product (Dai, Z., Nemet, I., Shen, W., and Monnier, V.M., Isolation, purification and characterization of histidino-threosidine, a novel Maillard reaction protein crosslink from threose, lysine and histidine, *Arch. Biochem. Biophys.* 463, 78–88, 2007).

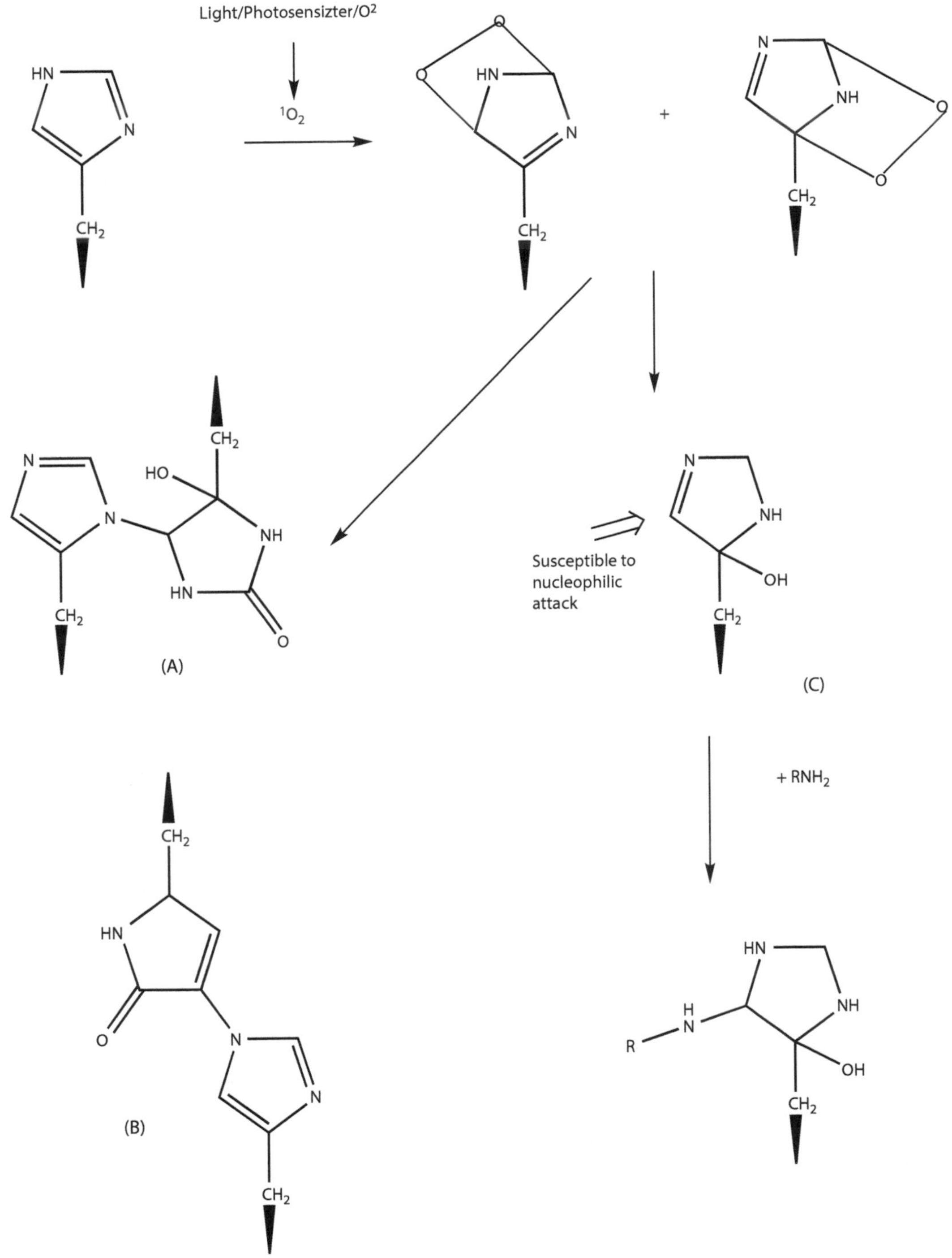

Light/Photosensizter/O^2

$^1O_2$

+

Susceptible to
nucleophilic
attack

(A)

(B)

(C)

+ RNH$_2$

FIGURE 14.4

(*continued*)

FIGURE 14.4 (continued)   Some possible reactions from the type II photooxidation of histidine in proteins. Shown is the reaction of singlet oxygen with histidine to yield endoperoxide intermediates. One potential cross-linked product (A) is shown (Tomita, M., Irie, M., and Ukita, T., Sensitized photooxidation of histidine and its derivatives. Products and mechanism of reaction, *Biochemistry* 8, 5149–5160, 1969). Subsequent work suggests that at least six dimers are formed with one such structure shown (B). (Shen, H.-R., Spikes, J.D., Smith, C.J., and Kopeček, J., Photodynamic cross-linking of proteins. IV. Nature of the His-His bond(s) formed in the rose bengal-photosensitized cross-linking of *N*-benzoyl-L-histidine, *J. Photochem. Photobiol. A Chem.* 130, 1–6, 2000). Shown is one product (B) and earlier product suggested by another group. Shown at (C) is a product derived from the photosensitized oxidation of histidine as suggested by Agon (Agon, V.V., Bubb, W.A., and Wright, A., Sensitizer-mediated photooxidation of histidine residues. Evidence for the formation of reactive side-chain peroxides, *Free Radic. Biol. Med.* 40, 698–710, 2006). The arrow indicates a site for nucleophilic attack and the possible product of reaction with an amine-like lysine is shown.

residue has been modified in the presence of a more potent nucleophile.[100,101] Gregory[100] showed that bovine heart mitochondrial malate dehydrogenase was inactivated with iodoacetamide but not iodoacetic acid, while a sulfhydryl group is modified with some difficulty by *N*-ethylmaleimide. Likewise, Anderton and Rabin[101] reported that pig heart malate dehydrogenase was inactivated by iodoacetamide but not by iodoacetate. Thus, here is a situation in contradiction to the preceding statement on the preference for alkylation with cysteine residues. In the case of the porcine heart mitochondrial malate dehydrogenase, the cysteine residues are not readily available for modification with PCMB, but the modification of all cysteine residues does result in the loss of activity.[102] Cai and coworkers[103] studied the reaction of several peptides containing cysteine, histidine, and lysine residues as well as free amino terminal with acrolein. Cysteine was the most reactive amino acid via the Michael addition with acrolein (see Chapter 6) with much slower reaction at histidine and lysine. If the cysteine residues were blocked by carboxamidomethylation, higher concentrations of acrolein result in adduct formation with histidine and the amino terminal but no reaction at lysine. Malinowski and Fridovich[104] studied the reaction of bromoacetate with bovine erythrocyte superoxide dismutase under native and denaturing conditions. Both the sulfhydryl group of the cysteine and position 6 and histidine residues were unreactive with bromoacetic acid under native or in the presence of 8.0 M urea. Both the cysteine residue and histidine residues were modified with bromoacetic acid in the presence of 6.0 M guanidine hydrochloride with 1.0 M EDTA. Thomas and coworkers[105,106] used affinity labels to modify histidine and cysteine residues at the active site of human placental 17β,20α-hydroxysteroid dehydrogenase. Reaction of human placental 17β,20α-hydroxysteroid dehydrogenase with estrone 3-(bromoacetate) in the presence of NADH resulted in total inactivation; recovery of activity was accomplished with 2-mercaptoethanol at alkaline pH (Figure 14.6). Enzyme inactivated in the absence of NADH showed only modest recovery of activity. The reaction of either preparation with another affinity reagent, 16α-(bromoacetoxy) estradiol 3-(methyl ether), resulted in inactivation with the formation of 1,3-dicarboxymethyl histidine in the reactivated enzyme and alkylation of a sulfhydryl group at the active site.

It is possible to modify a histidine residue in the presence of cysteine residue by blocking the cysteine residue. Liu[38] found it necessary to modify the cysteine residue at the active site of streptococcal proteinase with sodium tetrathionate in order to alkylate the active-site histidine residue. The alkylation of the active-site histidine was accomplished by using a positively charged reagent, α-*N*-bromoacetylarginine methyl ester, as the histidine was resistant to modification with iodoacetic acid, iodoacetamide, *N*-chloroacetyltryptophan, or *N*-chloroacetylglycyleucine. Gleisner and Liener used a similar approach to modify the histidine residue at the active site of ficin.[107] The modification of active-site cysteine with tetrathionate and oxidation of the methionine residue with sodium periodate permitted the modification of the histidine at the enzyme active site with bromoacetone in the presence of 2 M urea.

Bromoacetic acid and iodoacetic acid were first used as metabolic inhibitors and were important in the understanding of glycolysis.[108] Concomitant with this early research, it was observed

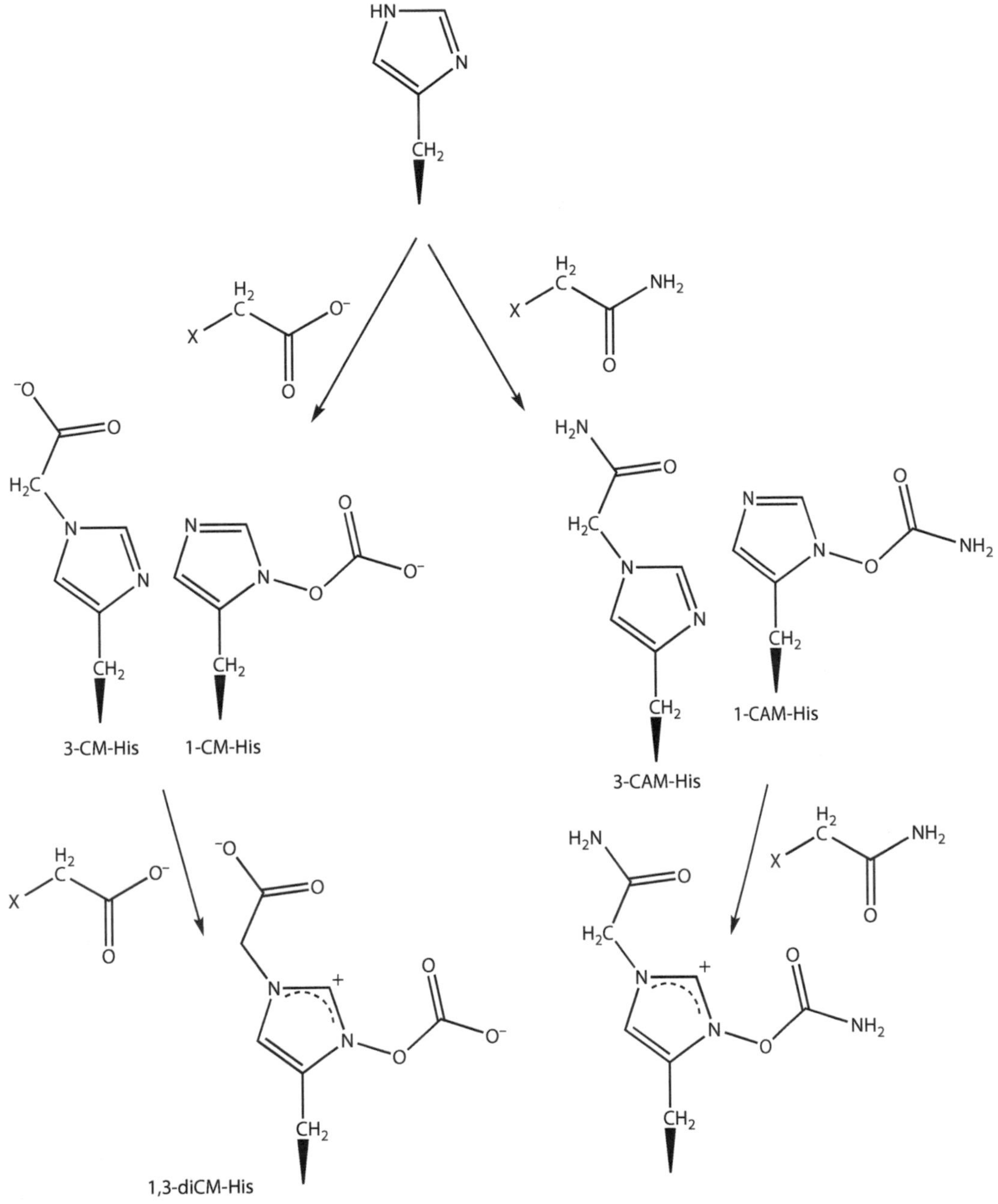

FIGURE 14.5 The reaction of histidine with haloacetate or haloacetamide. Shown on the left is the reaction of haloacetate with histidine to yield the 1-carboxymethyl(1-CM), 3-carboxymethyl (3-CM), and 1,3-dicarboxymethyl (1,3-diCM) derivatives. Shown on the right is the reaction of haloacetamide to yield the 1-carboxamidomethyl (1-CAM), 3-carboxamidomethyl (3-CAM), and the 1,3-dicarboxamidomethyl (1,3-diCAM) derivatives.

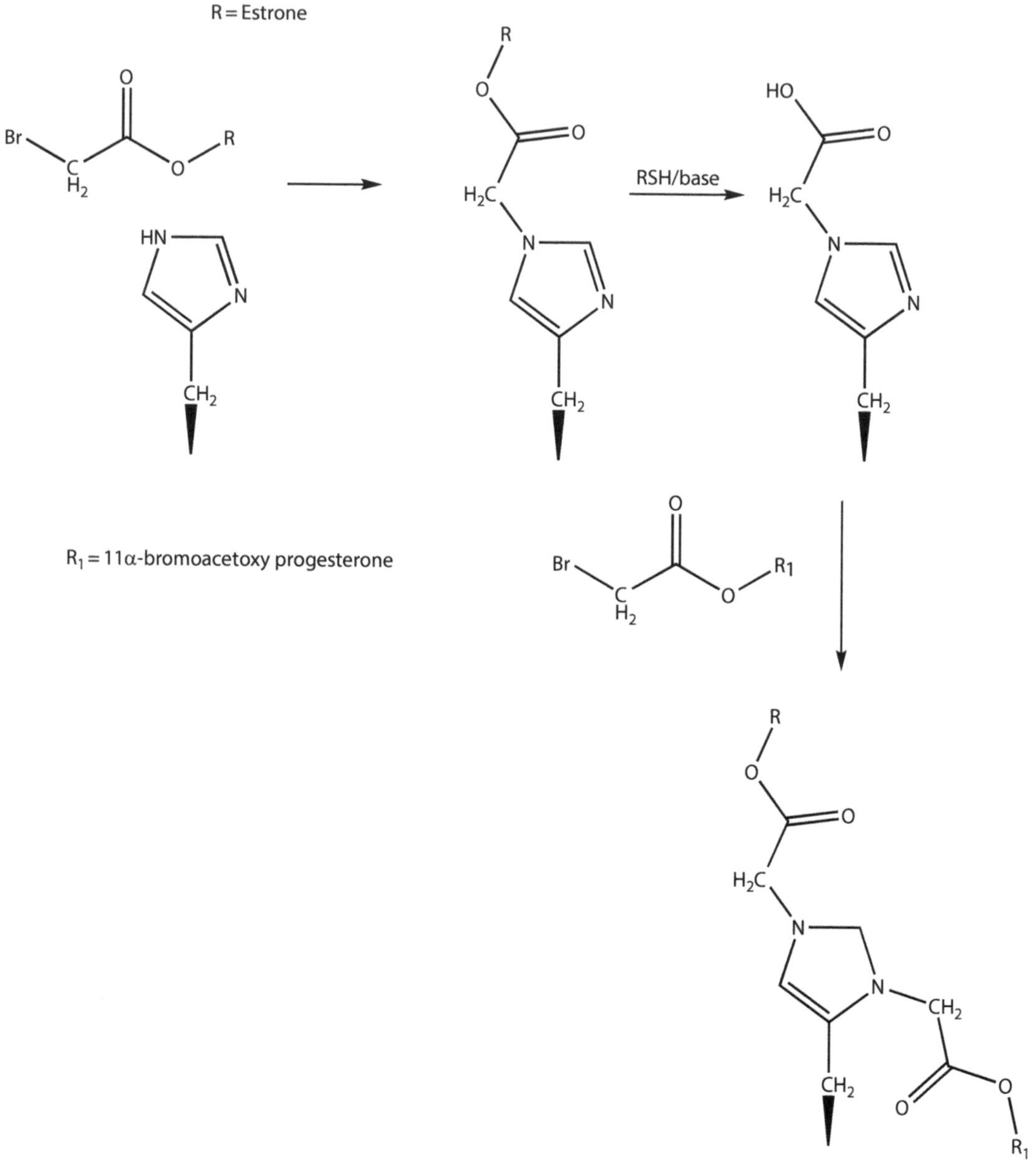

FIGURE 14.6 The reaction of human placental 17β,20α-hydroxysteroid dehydrogenase. Shown is the reaction of estrone-3-bromoacetate with a histidine residue in human placental 17β,20α-hydroxysteroid dehydrogenase resulting in modification of the *tele*-nitrogen on the imidazole ring. Mild base removes estrone. If this reaction is performed in the presence of NADH, the reactivated enzyme can be modified with 11α-bromoacetoxyprogesterone resulting in alkylation at the *pros* nitrogen. Not shown is the alkylation of a cysteine residue, which occurs with estrone-3-bromoacetate in the absence of NADH (see Thomas, J.L., LaRochelle, M.C., Asibey-Berko, E., and Stricker, R.C., Reactivation of human placental 17β,20α-hydroxysteroid dehydrogenase affinity alkylated by estrone-3-(bromoacetate): Topographic studies with 16α-(bromoacetoxy)estradiol-1-(methyl ether), *Biochemistry* 24, 5361–5367, 1985; Thomas, J.L., Asibey-Berko, E., and Stricker, R.C., The affinity alkylators, 11α-bromoacetoxyprogesterone and estrone-3-bromoacetate modify a common histidyl residue in the active site of human placental 17β, 20α-hydroxysteroid dehydrogenase, *J. Steroid Biochem.* 25, 103–108, 1986).

that iodoform and related compounds reacted with thiol groups in biological fluids. Studies on the inhibition of glyoxalase by iodoacetate in the early 1930s demonstrated the importance of thiol groups in enzymes.[65] This early work resulted in the identification of iodoacetate and bromoacetate (and related compounds) as specific reagents for the modification of sulfhydryl groups in proteins, and the potential for reaction at other functional groups was not fully appreciated.[109] It is thus not surprising that Zittle[110] in 1946 ascribed the inhibition of RNAse (ribonucleinase*) observed with iodoacetate to the modification of a cysteinyl residue in the enzyme. Considering later observations, it is of interest that iodoacetamide was much less effective than iodoacetic acid. Determination of the primary structure of RNAse by Werner Hirs working with Stanford Moore and William H. Stein at the Rockefeller Institute revealed the absence of free cysteine residues.[111,112] The Rockefeller group proceeded to examine the reaction of iodoacetate with RNAse in light of this information. Gundlach, Moore, and Stein[113] observed that the rate of inactivation of RNAse by iodoacetate was most rapid at pH 5.5–6.0 and at pH 2.2; the rate of inactivation was less rapid both at intermediate pH values and at more alkaline pH. Amino acid analysis demonstrated modification of histidine at pH 5.5–6, modification of methionine at pH 2.2, and modification of the ε-amino group of lysine at alkaline pH. Chromatography of the modified protein on IRC-50 yielded three fractions, two of which were inactive. These investigators also reported that the reaction of iodoacetate with N-acetylhistidine yielded two products. George Stark[114] then proceeded to show that the modification of histidine residues was dependent on the native conformation of the protein. No modification of histidine occurred at pH 5.5 in the presence of 8 M urea and 4 M guanidine or after reduction with sodium borohydride. These investigators did report the modification of methionine with iodoacetic acid under these reaction conditions. The lack of reaction of iodoacetamide was also confirmed.

Another research group with a different W. Stein at King's College in London studied the reaction of bromoacetate with RNase[115] reporting that the inactivation was associated with the modification of a single histidine residue. The reaction proceeded optimally at pH 5.5–7.0 and the enzyme was protected by cytidylic acid. The rate of reaction of bromoacetate with RNAse was estimated to be some 30 times more rapid than the rate of reaction of bromoacetate with free histidine[116] or bovine serum albumin.[117] Korman and Clarke[116] also demonstrated the rate of reaction of bromoacetate with histidine was more rapid than the rate of reaction with chloroacetate; iodoacetate was similar to bromoacetate. The reaction of iodoacetate is more rapid with amino groups or phenolic hydroxyl groups than the reaction of these functional groups with bromoacetate. Subsequent structural analysis by the King's College group[118] showed that the modification of RNAse with bromoacetate occurred at His119. The early work of Korman and Clarke[117] on the carboxymethylation of albumin and trypsin deserves further comment. The reaction of albumin (500 mg) with bromoacetate (6 mmol; an approximately 1000-fold excess of reagent) with MgO in 20 mL H_2O at 35°C resulted in the progressive denaturation of the protein; reaction with trypsin resulted in the loss of enzyme activity. Serum albumin contains a free sulfhydryl group that is readily modified by iodoacetic acid,[119] and other investigators have noted structural change on modification with iodoacetamide[120] or iodoacetic acid.[121] In summary then, at this point in time, it had been demonstrated that the inactivation of RNAse by iodoacetate at neutral pH resulted in the modification of a histidine residue(s), the reaction yielded two inactive products, and the reaction of histidine in RNAse with iodoacetate, but not methionine, was dependent on the native conformation of the protein. The crystal structure of the protein was not yet available, and RNAse, together with lysozyme and, to a lesser extent, albumin, had become model proteins to develop basic information on protein chemistry.

---

* It is not clear as to origin of the term ribonucleinase. The study activity of ribonuclease dates to 1915 (Jones, W. and Richards, D.E., Simpler nucleotides from yeast nucleic acid, *J. Biol. Chem.* 20, 25–3, 1915). The activity was described as possible ferment in 1920 (Jones, W., The actions of boiled pancreas extract on yeast nucleic acid, *Am. J. Physiol.* 52, 203–207, 1920) and later as a polynucleotidase in 1937 (Dubos, R.J. The decomposition of yeast nucleic acid by a heat-resistant enzyme, *Science* 85, 549–550, 1937). The protein was crystallized in 1940 and described as ribonuclease (Kunitz, M., Crystalline ribonuclease, *J. Gen. Physiol.* 24, 15–32, 1940).

The author joined the laboratory of Stanford Moore and William H. Stein in 1966. I was settling in after returning from the Federation Meeting in Atlantic City when the late Rachael Fruchter introduced herself and asked me to sign a chromatographic chart (see Figure 1 in Ref. [44]). This chart showed the separation of the reaction products of iodoacetate and RNAse on IRC-50 and was taken from a paper by Art Crestfield and coworkers[122] published in 1963. The work showed that the reaction of iodoacetate with RNAse yielded a major product (1-carboxymethyl-His119-RNase) and a minor product (3-carboxymethyl-His12-RNAse). The 119-derivative was inactive with either cyclic cytidylic acid or RNA as substrate, while the 12-derivative had 7% activity with cyclic cytidylic acid and less than 5% activity with RNA. Polymorphism was excluded when the same results were obtained by modification of RNAse derived from a single cow. As an aside, the author was informed by Stanford Moore that all of the early work on bovine pancreatic ribonuclease was performed on protein derived from a Kosher abattoir. The issue of polymorphism in protein isolated from multiple animals was an issue with early work on bovine carboxypeptidase A.[123] It was determined that the heterogeneity seen in protein obtained from multiple animals was also seen in protein isolated from the pancreas of a single cow. A subsequent study from the Rockefeller group[124] showed the modification of one of the two histidine residues precluded modification at the other residue. The alkylation of histidine also affected the ability of RNAse to bind phosphate; chromatography on IRC-50 in the presence of 0.2 M NaCl instead of 0.2 M sodium phosphate (both at pH 6.47) changes the elution position of the products of the reaction; in phosphate, the native protein is eluted first followed by the 119-derivative and then the 12-derivative, while in the presence of 0.2 M NaCl, the 199-derivative is eluted first followed by the 12-derivative and then the native protein. This work and studies from other laboratories cited in Ref. [44] showed that the enzymatic activity of RNAse was provided by both histidine-12 and histidine-119 and a dimer formed from the 12-derivative and 119-derivative would have activity as, because both modified forms are inactive, each derivative as an unmodified residue that could contribute an active-site residue. Crestfield and colleagues[124] prepared a dimer from equimolar quantities of the 12-derivatrive and 119-derivative by lyophilization from 50% acetic acid. Assay showed that the hybrid dimer has 45% of the specific activity of the native enzyme; the majority of this activity was lost on heating under conditions where there was dissociation of the dimer. These studies and other studies permitted the suggestion that the two active-site residues are approximately 5 Å apart in the native protein. In later work, Fruchter and Crestfield[125] subsequently demonstrated that pancreatic ribonuclease formed two dimer species on lyophilization from 50% acetic acid and that the dimers could be inactivated by iodoacetate at a rate comparable to the native monomer form; alkylation of methionine also occurred under these reaction conditions. The dimers could be dissociated by heating at 65°C. The analysis of the product of the reaction of dimer with iodoacetate showed the presence of the 3-CM-119RNAse, the 1-CM12 RNAse, native protein, and 1-CM12, 3-CMRNase; a disubstituted histidine derivative was not observed. This work and later studies[126,127] suggested that mutual displacement of the amino-terminal region in RNAse occurred under conditions of lyophilization in the presence of 50% acetic acid.

Crestfield and colleagues in their 1963 papers[122,124] had suggested the mutually exclusive modification of the histidine residues with preferential reaction at His119 reflected orientation of the negatively charged reagent by a positive charge in the protein and that a protonated His12 could supply that charge. There is a suggestion that such a charge could be supplied by a lysine residue, but Michael Lin and coworkers[128] were able to confirm this original suggestion by eliminating the possible role of either Lys7 or Lys41 in fulfilling this orientation function. This work is providing the understanding for the reaction of iodoacetate with RNAse. Implicit in this consideration would be the necessity of a higher pKa for His12 compared to His119. This value was supported by NMR measurements[129] but not necessarily by other physical techniques.[130] The relative ineffectiveness of iodoacetamide in the inactivation of RNAse was mentioned previously; differences in the relative reactivity of iodoacetate and iodoacetamide with proteins are not uncommon and usually reflect the difference between a charged reagent and a neutral compound. Fruchter and Crestfield[131] studied the reaction of iodoacetamide with RNAse in more detail and observed that the rate of inactivation was quite slow compared to iodoacetic acid but still some 10–100 times faster than the reaction with free histidine. Two products were obtained from the reaction of iodoacetamide with RNAse: 3-carboxyamidomethyl-His12 and fully active S-carboxamidomethyl-Met. It is possible that the lower pKa value for His12 is responsible for the increased reactivity of this residue. The issue of the roles of His12 in the modification of His119 in RNAse was further investigated by Bryce Plapp in a work published in 1973.[132] The study showed that a careful examination of the effect of concentration on reaction rate showed a Michaelis–Menten relationship. This work also studied the effect of various leaving groups (chloride, bromide, iodine, and tosylate) on reaction rate. The data thus obtained support the importance of the binding of iodoacetate to RNAse prior to the alkylation of His119. This study also reported the reaction of tosylglycolate with RNAse with a rate similar to the various halo compounds but with a somewhat increased specificity for reaction at His119. The reader is referred to the Nobel comment by Stanford Moore and William H. Stein[133] and the review[134] by Garland Marshall and others for further discussion of these studies. The work by Marshall emphasizes the work of the Merrifield group at the Rockefeller in the chemical synthesis of bovine pancreatic ribonuclease.

Hirs and colleagues noted in their publication on the primary structure of RNAse[111] that much had been learned in that and previous studies, which made the work much easier. The same can be said for the aforementioned story that started with Zittle's observations in 1946,[110] which were followed by the studies of Gerd Gundlach in 1959[113] and then by a series of publications that resulted in a considerable increase in our understanding of basic protein chemistry. As with Hirs' observation mentioned earlier, the answers to the RNAse problem would have come more rapidly with today's instrumentation, but our understanding of solution chemistry would not be rich. While today's protein therapeutics are efficiently prepared by recombinant DNA technology rather than being obtained from tissue sources and purified by *black box* technologies, there is knowledge that is critical to the understanding of the protein going from an active pharmaceutical ingredient to a final drug product.

The chemistry of histidine alkylation with α-halo carboxylic acids and amides provided the basis for the development of peptide chloromethylketones (Figure 14.7) for the affinity labeling of proteolytic enzymes.[135–138] These reagents were mentioned earlier and the reader is directed to the review by Bryce Plapp[33] for the discussion of affinity labeling technology. Specificity can be enhanced by increasing the complexity of the reagents as demonstrated by comparing the inactivation of urokinase by Pro-Gly-Arg-chloromethyl ketone to N^α-tosyl-L-lysine chloromethyl ketone.[139] Inactivation with the tripeptide chloromethylketone occurred at micromolar concentrations, while millimolar concentrations of TLCK provided only slow inactivation. Care must be taken with the use of peptide chloromethyl ketones in complex systems as these materials react with sulfhydryl groups via an alkylation mechanism.[37,140]

There has been little use of carboxymethylation or carboxamidomethylation for the modification of histidine in the last decade, but there are several studies that merit some mention. Carboxymethyl poly(L-histidine) was prepared by the reaction of iodoacetate with poly(L-histidine) at pH 5.5 at 23°C for 3 days; after dialysis, the product was subjected to a second carboxymethylation cycle.[141] The resulting polymer, carboxymethyl poly(L-histidine), is a novel pH-sensitive material, which has been used for gene delivery.[142–144]

One of the few recent studies used bromoacetate and iodoacetamide to alkylate the solvent-accessible histidine residues in sperm whale myoglobin.[145] This is a slow reaction with either iodoacetamide or bromoacetate requiring 5–7 days in 0.1 M sodium phosphate, pH 7.0/23°C. Reaction of iodoacetamide or bromoacetate with cysteine or lysine would be a complication with most proteins, but sperm whale myoglobin (and other myoglobins) does not contain cysteine and reaction at pH 7.0 or below minimizes reaction at amino groups. The carboxylation of histidine residues in myoglobin dates at least to 1963 with studies by Banaszak and coworkers[146] and Banaszak and Gurd.[147] The work of Banaszak and Gurd[147] showed that with two adjacent histidine residues, one is converted to a dicarboxymethyl derivative (His81) and the other is unreactive (His82). These earlier studies were also characterized by long reaction times of 7–10 days with one study extended to 30 days. Eight of the twelve histidine residues were modified under these reaction conditions. Nigen and coworkers[148]

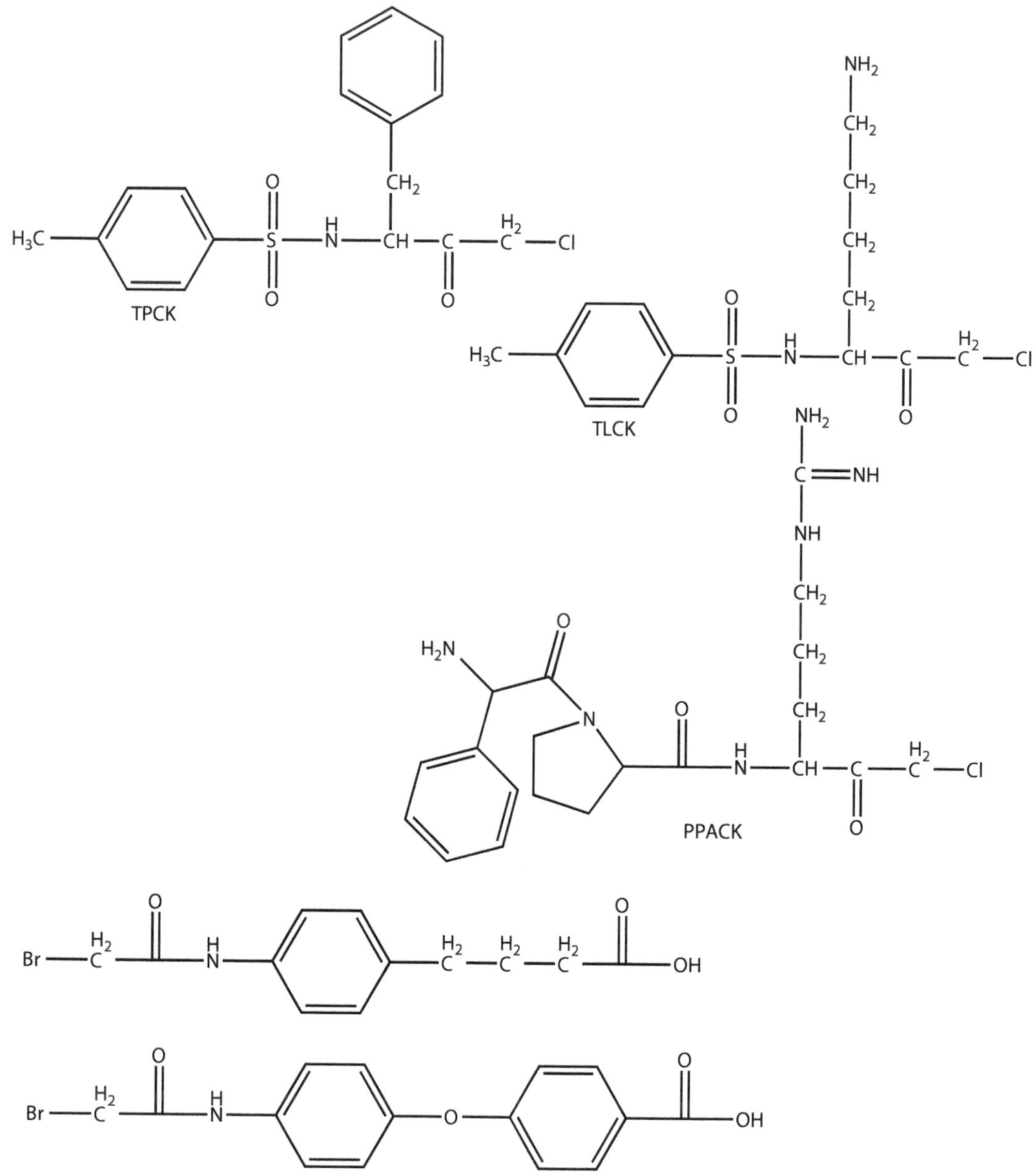

FIGURE 14.7 Some halomethylketones used for affinity labeling of enzymes. Shown is the structure of N^α-tosyl-L-phenylalanine chloromethyl ketone, which was developed for chymotrypsin-like proteinases; N^α-tosyl-L-lysine chloromethyl ketone, which was developed for tryptic-like proteinases; and a tripeptide chloromethyl ketone, D-phenylalanine-proline-arginine chloromethyl ketone (PPACK), which was developed for the inhibition of thrombin (Kettner, C. and Shaw, E., Inactivation of trypsin-like enzymes with peptides or arginine chloromethyl ketone, *Methods Enzymol.* 80, 826–842, 1981). Also shown are the structures of several bromoacetamido derivatives used as affinity labels for horse liver alcohol dehydrogenase (Chen, W.-S. and Plapp, B.V., Affinity labeling with ω-bromoacetamido fatty acids and analogs, *Methods Enzymol.* 72, 587–591, 1981).

prepared the ^{13}C-labeled carboxymethyl derivative of myoglobin for NMR studies. Sperm whale myoglobin (1.1 mM) was placed under nitrogen with 0.2 M sodium bromoacetate at pH 6.8 (pH Stat)/25°C for 8 days. For comparison in the same study, pancreatic ribonuclease (0.9 mM) was placed with 25 mM sodium acetate at pH 5.6/35°C for 50 min resulting in the modification 0.9 histidine residues. Dautrevaux and coworkers[149] compared the reaction of myoglobin with TPCK at pH 7.0 (0.07 M phosphate) for 6 h at 37°C, with iodoacetate at pH 7.0 (0.05 M Tris) for 6 h at 37°C, and with bromoacetate at pH 7.0 (1 M phosphate) for 6–7 days at 25°C. No modification of histidine was observed with TPCK, one histidine residue was modified at pH 5.5, two histidine residues were modified in denatured myoglobin at pH 5.5, six histidine residues were modified in native myoglobin with bromoacetate, while nine residues of histidine were modified at pH 7 in denatured myoglobin. In a subsequent study, Delacourte and coworkers[150] studied the reaction of myoglobin with iodoacetate at pH 5.5 (8 h, 37°C) and obtained a derivative of horse muscle myoglobin with the majority of modification at His113 (1-carboxymethyl derivative) with minor modification at other histidine residues and alkylation of two methionine residues. Banaszak and Gurd[148] used 0.2 M bromoacetate with sperm whale myoglobin at pH 7.0 (1 M phosphate) at 23°C for 4 days; additional bromoacetate was added and the reaction continued for 4 days. Analysis showed that there are 3.7 unmodified histidine residues, 0.6 3-carboxymethyl histidine residues, and 7.4 1,3-dicarboxymethyl histidine residues per mole of protein. Two other studies from the Indiana group on the reaction of bromoacetate with sperm whale myoglobin deserve some additional mention. Hugli and Gurd reported on the reaction of bromoacetate with crystalline sperm whale myoglobin[151] and in solution.[152] Reaction with the crystalline protein was performed in concentrated ammonium sulfate at pH 6.8 for 10 days at 25°C, while reaction in solution was performed at pH 6.8 (pH Stat) at 22°C–24°C with 0.2 sodium bromoacetate. There was considerable similarity between the products obtained under the two reaction conditions; however, histidine-36 was not modified in the solution reaction. Hugli and Gurd also confirmed the formation of dicarboxymethylhistidine at His81 and lack of reaction at His82.

Histidine residues in myoglobin have been modified with diethylpyrocarbonate for the purpose of surface mapping,[153] but the stability of modification presents challenges.

A compound related to α-halo carboxylic acids is p-bromophenacyl bromide [2-bromo-1-(4-bromophenyl)ethanone] (Figure 14.8), which has been used to modify histidyl and carboxyl residues in proteins. This reagent modifies a histidine residue in the active site of phospholipase A$_2$,[154] and this modification is used as discriminating factor in identifying the physiological impact of phospholipase A$_2$ in complex biological systems.[155–169] A study reported by Shaw and coworkers[165] illustrates the value of p-bromophenacyl bromide for the inactivation of phospholipase A$_2$ in studies with cobra venom factor, an anticomplement protein obtained from Naja naja cobra venom. These investigators note that the hydrophobic properties of phospholipase A$_2$ complicated separation from cobra venom factor during purification but it was possible to inhibit the phospholipase A$_2$ with p-bromophenacyl bromide permitting clear interpretation of data obtained from whole-animal studies. There are calcium-independent phospholipase A$_2$ enzymes that are not inhibited by p-bromophenacyl bromide[170–172] but apparently one from nervous tissues that is inhibited by p-bromophenacyl bromide.[173]

p-Bromophenacyl bromide was developed for the preparation of esters of organic acids[174] as the bromophenacyl derivative absorbs light at 254 nm.[175,176] Bernard Erlanger and colleagues at Columbia[177] observed the p-bromophenacyl bromide inactivated pepsin via the modification of a carboxyl group at the enzyme active site; this observation was confirmed by Gross and Morrell.[178] It should be noted that iodoacetate has also been shown to react with a glutamic acid residue in ribonuclease T1.[179] p-Bromophenacyl bromide would be expected to react with sulfhydryl groups in proteins, but there is a limited literature on this possibility.[180] While most of the subsequent work on p-bromophenacyl bromide was on phospholipase A$_2$, there is a work that I have overlooked (and found by accident) by Jeng and Fraekel-Conrat[181] on the reaction of p-bromophenacyl bromide with crotoxin. The crotoxin complex was unreactive with p-bromophenacyl bromide as was the A component; there was the modification of the B component that was associated with a decrease in lethality. Modification with diethylpyrocarbonate did not modify histidine residues but rather resulted in

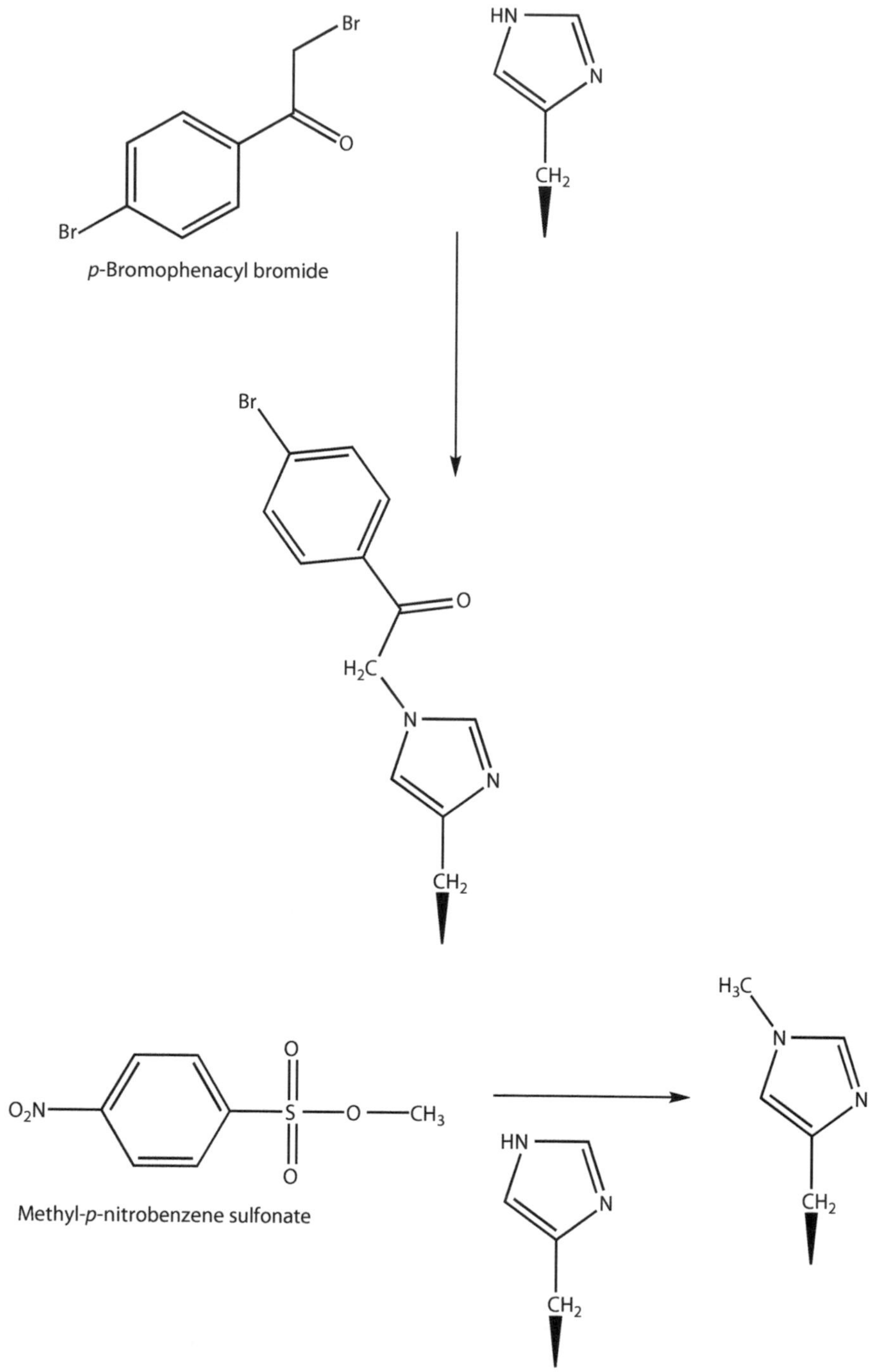

FIGURE 14.8 *p*-Bromophenacyl bromide [2-bromo-1-(4-bromophenyl)ethanone] and methyl *p*-nitrobenzene sulfonate. Shown is the reaction of *p*-bromophenacyl bromide with histidine. Alkylation at the N-3 position is shown although there are no data to support that assignment. Shown also is the formation of *N*-3 methyl histidine (see Nakagawa, Y. and Bender, M.L., Methylation of histidine-57 in α-chymotrypsin by methyl *p*-nitrobenzenesulfonate. A new approach to enzyme modification, *Biochemistry* 9, 259–267, 1970). Shown at the bottom is the reaction of a tertiary amine with methyl *p*-nitrobenzenesulfonate as a model for nucleophilic substitution (Ford, J.W., Janakat, M.E., Lu, J. et al., Local polarity in CO$_2$-expanded acetonitrile: A nucleophilic substitution reaction and solvatochromic probes, *J. Org. Chem.* 73, 3364–3368, 2008).

lysine modification, which also resulted in a loss of lethality. The reader might well question (as the author did) how did *p*-bromophenacyl bromide become a *signature* reagent for phospholipase A2.* There are precious few reagents that get to that status. A colleague of mine recalls a quote from Efraim Racker that goes something like "a reagent is never as specific as when it is first discovered." The first citation that the author could readily locate on the reaction of *p*-bromophenacyl bromide with phospholipase A$_2$ appeared in 1974.[182] This is rather a good study that showed the modification of a single histidine at the active site; a minor amount of modification occurred at another histidine residue. The enzyme was protected from inactivation by substrate and cofactors. Phospholipase A$_2$ occurs as a zymogen and Volwerk and colleagues[183] showed that modification of this histidine residue with *p*-bromophenacyl bromide also occurred in the zymogen form of the enzyme. Volwerk and colleagues do reference an earlier report[184] from their laboratory that describes the initial experiments with a reference to Erlanger and colleagues[177] so that it is presumed that *p*-bromophenacyl bromide was one of the several reagents used to characterized the phospholipase A$_2$, and Bunsen and coworkers[184] were astute enough to recognize that inactivation reflected the modification of histidine, not carboxyl groups. A decade later, a review by Ed Dennis,[185] one of the more distinguished investigators in lipid biochemistry, describes *p*-bromophenacyl bromide as a specific inhibitor of phospholipase A$_2$. Snake venoms exhibit a spectrum of physiological responses,[186] and *p*-bromophenacyl bromides continue to be used, as mentioned previously, as an inhibitor of phospholipase A$_2$ in the study of these complex biological mixtures.[164] The crystal structure of dimeric Lys49-phospholipase A$_2$ complexed with *p*-bromophenacyl bromide has been reported[187] Lys49-phospholipase A$_2$ is a naturally occurring inactive form of phospholipase A$_2$ where Asp49 is substituted with a lysine residue (D49K). Modification of the active-site histidine with *p*-bromophenacyl bromide results in loss of myotoxic activity suggesting an association between the active site and the C-terminal domain in this activity, which is separate from catalytic activity. The structure also shows that the *p*-bromophenacyl group is bound to the N^1 in the imidazole ring of His48 making hydrophobic interactions with a cystine residue (there are seven disulfide bonds in this relatively small [14.5 kDa] protein) and lys49 together with other nonpolar interactions. It seems reasonable that the hydrophobic nature of *p*-bromophenacyl bromide contributes to the usual specificity of this reagent for phospholipase A$_2$. Prior studies[188,189] on the reaction of *p*-bromophenacyl bromide with phospholipase A$_2$ from several sources suggested that complex formation occurs between inhibitor and enzyme prior to the modification of the histidine residue. A cursory consideration of the rate constants for the inactivation of phospholipase A$_2$ by *p*-bromophenacyl bromide suggests that there are only small differences between the enzymes from various sources. Fujii and coworkers[189] also showed the phospholipase was inactivated by *p*-bromophenacyl bromide at about one-half that of the active enzyme.

The following studies have been selected as being, at least to the author, some examples of the more useful studies on the *p*-bromophenacyl bromide modification of proteins. Taipoxin is an extremely lethal neurotoxin isolated from the venom of *Oxyuranus scutellatus* (Australian taipan).[190] While taipoxin is considered a neurotoxic phospholipase A2, the ability to block synaptic transmission is more complex than the enzymatic activity and involves the affinity of binding to the presynaptic membrane.[191] *p*-Bromophenacyl bromide modifies a single histidine residue in taipoxin with a 350-fold decrease in neurotoxicity.[192] The extent of modification was assessed by both amino acid analysis (loss of histidine) and spectral analysis ($\Delta\varepsilon$ 271 = 17,000 M^{-1} cm^{-1}).[193] Two of seven histidine residues are modified (1 mol/mol in α-subunit; 1 mol/mol in β-subunit) under these reaction conditions. This study shows the use of spectral properties of *p*-bromophenacyl bromide for the determination of

---

* The last decade has seen a marked increase in our understanding of phospholipase A2, which is now considered to be a large family of enzymes, only some of which are inhibited by *p*-bromophenacyl bromide (Kim, T.S., Sundaresh, C.S., Feinstein, S.T. et al., Identification of a human cDNA clone for lysosomal type Ca^{2+}-independent phospholipase A2 and properties of the expressed protein, *J. Biol. Chem.* 272, 2542–2550, 1997; Hui, D.Y., Phospholipase A$_2$ enzymes in metabolic and cardiovascular diseases, *Curr. Opin. Lipidol.* 23, 235–240, 2012; Cao, J., Burke, J.E., and Dennis, E.A., Using hydrogen/deuterium exchange mass spectrometry to define the specific interactions of the phospholipase A2 superfamily with lipid substrates, *J. Biol. Chem.* 288, 1806–1813, 2013).

the extent of modification. As cited earlier, *p*-bromophenacyl bromide is used for the derivatization of organic acids prior to analysis. Verheij and coworkers[194] reported on the rate of inactivation of porcine pancreatic phospholipase A$_2$ by *p*-bromophenacyl bromide and other reagents that modified His48 at the enzyme active site. The second-order rate constants (0.1 M cacodyate–0.1 M NaCl, pH 6.0, 30°C) for the inactivation of porcine pancreatic phospholipase A$_2$ by *p*-bromophenacyl bromide are 125 M^{-1} min^{-1} as compared to 79 M^{-1} min^{-1} for phenacyl bromide and 75 M^{-1} min^{-1} for 1-bromooctan-2-one; the inactivation of porcine pancreatic phospholipase A$_2$ with iodoacetamide was not detected under these reaction conditions. Roberts and coworkers[195] reported half-site reactivity in phospholipase A$_2$ from *Naja naja naja* with the histidine in one subunit reactive toward *p*-bromophenacyl bromide, while the histidine residue in other subunit is modified with diethylpyrocarbonate; reaction with *p*-bromophenacyl bromide results in almost complete loss of enzyme activity, while reaction with diethylpyrocarbonate resulted in a product with 15% residual activity. There was 3% loss of activity with iodoacetamide and no loss of activity with iodoacetic acid. These investigators also demonstrated that the presence of calcium ions inhibited the rate of enzyme inactivation by *p*-bromophenacyl bromide. Of interest was the effect of Triton X-100, which also inhibited the rate of enzyme inactivation. These investigators speculated that the *p*-bromophenacyl bromide was sequestered into the Triton X-100 micelles, thus lowering the effective reagent concentration.

Other than the pepsin papers cited previously, there are only several studies of the reaction of *p*-bromophenacyl bromide with proteins other than phospholipase A$_2$. Battaglia and Radomiska-Pandya[196] studied the functional role of histidine in the UDP-glucuronic acid carrier by measuring the effect of chemical modification on the uptake of radiolabeled UDP-glucuronic acid in rat liver endoplasmic reticulum. Inhibition of uptake was more pronounced with either *p*-bromophenacyl bromide or diethylpyrocarbonate (both hydrophobic reagents) than with *p*-nitrobenzenesulfonic acid and methyl ester, a hydrophilic reagent. De Vet and van den Bosch[197] examined the role of histidine in recombinant guinea pig alkyl-dihydroxyacetonephosphate synthase by modification with *p*-bromophenacyl bromide and oligonucleotide-directed mutagenesis. Modification with *p*-bromophenacyl bromide was performed in 10 mM Tris-HCl and 0.15% Triton X-100, pH 7.4, at room temperature. Inactivation was observed showing pseudo-first-order kinetics and the enzyme was protected by substrate. A remarkable increase in the rate of inactivation was observed at pH ≥ 8.0 (50 mM Tris-HCl). Replacement of His617 with alanine also eliminated catalytic activity. Dorsch and coworkers[198] have shown that the phospholipase A2–like activity of the VP1 region of parvovirus B19 is inactivated by *p*-bromophenacyl bromide. Parvovirus is the causative agent of erythema infection and perhaps other diseases. Primary structure analysis showed the presence of a phospholipase A motif. A new analytical method demonstrated the presence of enzyme activity in the VP1 region that was Ca^{2+}-dependent and inhibited by *p*-bromophenacyl bromide.

Methyl *p*-nitrobenzenesulfonate (Figure 14.8) is a reagent that has been used infrequently to modify histidine residues in proteins. Nakagawa and Bender[199,200] used methyl *p*-nitrobenzenesulfonate as a substrate analog to modify the active-site histidine in α-chymotrypsin. Modification of trypsin or subtilisin is not observed under these reaction conditions; reaction was not observed with free imidazole, *N*-acetyl-L-serinamide, or *N*-acetyl-L-methioninamide. Subsequently, Glick used methyl *p*-nitrobenzenesulfonate to methylate histidine residue(s) in ribosomal peptidyl transferase.[201] The author suggests only histidine residues are modified, but definitive evidence on this point is absent. Marcus and Dekker[202] examined the effect of methyl *p*-nitrobenzenesulfonate on the activity of *E. coli* L-threonine dehydrogenase. The reaction was performed in 100 mM potassium phosphate, pH 7.0 at 25°C; early reactions in this study were performed at pH 8.4, but it was observed that the reagent was more specific for the modification of histidine at the lower pH. Examination of the effect of reagent concentration on reaction rate demonstrated saturation kinetics with a limit value for the rate of inactivation of 0.01 min^{-1} suggesting the binding of inhibitor prior to the inactivation reaction. Analysis of the modified protein showed that His90 had been methylated at the N^3 position. Subsequent studies[203] with oligonucleotide-directed mutagenesis confirmed the importance of this residue in enzyme function. Verheij and coworkers[194] reported the inactivation of equine, porcine, and bovine pancreatic phospholipase A_2 with methyl p-nitrobenzenesulfonate. Reagent stability precluded determination of rate constants; it was possible to show that the equine enzyme was inactivated more rapidly than either the porcine or bovine enzyme; the rate of inactivation was an order of magnitude slower with methyl p-toluenesulfonate. While there were minor sites of modification on the protein, the inactivation of phospholipase A_2 with methyl p-nitrobenzenesulfonate is associated with methylation at the N^1 position on the imidazole ring of His48. The majority of the recent work with methyl p-nitrobenzenesulfonate has focused on the use of methyl p-nitrobenzenesulfonate as model compound in nucleophilic substitution reactions.[204–207] Histidine residues in proteins may also be modified with methyl iodide, but the reaction is not specific.[208,209] While most of the studies are based on the increased reactivity of histidyl residues at enzyme active sites, the study by Taralp and Kaplan[51] used methyl iodide to study the reactivity of lyophilized proteins in a nonaqueous (n-octane) environment.

Diethylpyrocarbonate (pyrocarbonic acid, diethyl ester, dicarbonic acid, diethyl ester, oxydiformic acid diethyl ester, ethoxyformic anhydride; DEPC) was used as a pesticide and antifungal agent but its use was banned in food and food products in 1972.[210] There has been recent interest in the use of a related compound, dimethyl dicarbonate (DMDC), for decontamination of grape musts.[211] It has received attention as nuclease inhibitor for the isolation of undegraded nucleic acid from tissues.[212] The use of DEPC as a nuclease inhibitor for isolation of undegraded nucleic acids resulted in the observation that DEPC modified adenine residues in nucleic acids.[213] The utility of DEPC for the preparation of undegraded nucleic acids has resulted in the use of DEPC water for processing of nucleic acids although degradation is still possible. While useful for the initial extraction, the value of long-term use of DEPC in aqueous solvent is questionable.[214]

Diethylpyrocarbonate is the most extensively used reagent for the specific modification of histidine in proteins (Figure 14.9). In the pH range from 5.5 to 7.5, diethylpyrocarbonate is reasonably specific, but not totally specific, for reaction with histidyl residues. The reaction can be complex, and the modification of other functional groups such as tyrosine, the ε-amino group of lysine, and cysteine occurs. With the exception of the modification of the ε-amino group of lysine and secondary products derived from histidine, the modifications are reversible in mild base and, in some cases, the modification of histidine can be quite transient. The rate of reaction is reasonably rapid (Table 14.1).

The use of diethylpyrocarbonate for the modification of histidine in proteins dates back to work by Fedorcsák and colleagues at the Royal University of Stockholm in the 1960s.[215–217] This work was performed to obtain understanding of the mechanism of action of cold sterilization.[164] The work suggested modification of tryptophan and lysine resulting in irreversible structural changes in protein upon reaction with DEPC. A variety of other studies suggest other mechanisms of DEPC modification of proteins.[218–221] Further support was provided for reaction at tryptophan,[219] and a cross-linking mechanism was proposed involving isopeptide bond formation between lysine and either aspartic or glutamic acid.[221] One of the more interesting observations was protection of RNAse from inactivation by DEPC by albumin.[220] As other studies were performed,[222,223] it appeared that it was possible to selectively modify histidine residues, but consideration of modification at other sites was clearly necessary as frequent modification does occur at lysine.[224] Modification also can occur at tyrosine and less frequently at serine and threonine.[225] This latter study by Mendoza and Vachet[225] noted that lysine modification occurs slowly at neutral pH. The modification of arginine and cysteine by DEPC has also been reported,[226] but, as with tryptophan, there is no recent documentation of the modification of arginine residues in proteins with DEPC. The modification of cysteine by diethylpyrocarbonate would expect to yield an unstable thioester derivative. Garrison and Himes[227] did report an unusual reaction between DEPC and cysteine in carboxylate (succinate or acetate) buffers where a thioester is formed between buffer and cysteine. The spectral properties of this derivative could pose problems in the spectrometric analysis of the reaction of DEPC with histidine. It is noted that Rua and coworkers[228] reported the modification of cysteinyl residues in *E. coli* isocitrate lyase with DEPC as determined by analysis with Ellman's reagent. The early work on diethylpyrocarbonate has been reviewed by Miles in 1967.[219]

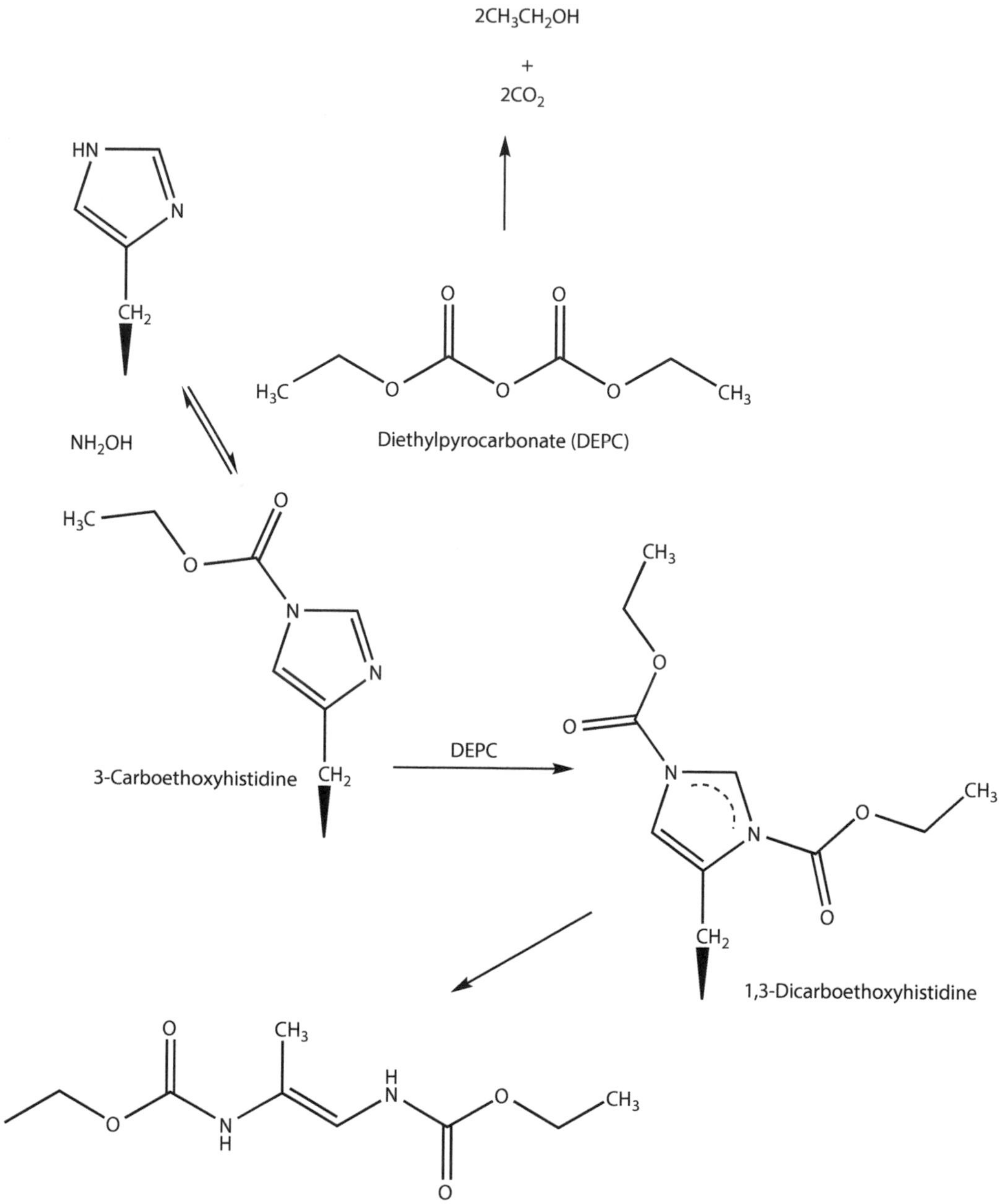

FIGURE 14.9 The chemistry of the reaction of diethylpyrocarbonate with proteins. Shown is the reaction of diethylpyrocarbonate (ethoxyformic anhydride) with histidine to yield the monosubstituted derivative. The increase in absorbance at 240 nm (3200 M^{-1} cm^{-1}) of the monosubstituted derivative can be used to determine stoichiometry of modification. This reaction is reversed by hydroxylamine (Miles, E.W., Modification of histidyl residues in proteins by diethylpyrocarbonate, *Methods Enzymol.* 47, 431–442, 1977). As shown at the top, diethylpyrocarbonate undergoes hydrolysis and kinetic analysis should account for a decrease in the concentration of diethylpyrocarbonate (Topham, C.M. and Dalziel, K., Chemical modification of sheep-liver 6-phosphogluconate dehydrogenase by diethylpyrocarbonate. Evidence for an essential histidine residue, *Eur. J. Biochem.* 155, 87–94, 1986). The monocarboxylated histidine may be modified at the other nitrogen on the imidazole ring, usually the *prox*, yielding the dicarboethoxy derivative that is associated with a further increase in absorbance at 242 nm. Treatment of the dicarboethoxy derivative with hydroxylamine or base results in ring cleavage. The formation of the dicarboethoxy derivative and subsequent ring cleavage is irreversible.

Reaction of diethylpyrocarbonate with histidine residues at a *moderate excess* of diethylpyrocarbonate at neutral pH results in substitution at one of the nitrogen positions on the imidazole ring. This reaction is associated with an increase in absorbance at 240 nm ($\Delta\varepsilon = 3200$ M^{-1} cm^{-1}). Monosubstitution is readily reversed at alkaline pH and, in particular, in the presence of nucleophiles such as hydroxylamine. Tris and other nucleophilic buffers can also reverse the modification and their use should be avoided with diethylpyrocarbonate. Generally, treatment with neutral hydroxylamine (0.1–1.0 M, pH 7.0) is used to regenerate histidine. As with the deacylation of *O*-acetyl tyrosine by neutral hydroxylamine, the higher the concentration of hydroxylamine, the more rapid the process of decarboxymethylation. Carboxymethylation at both N_1 and N_3 (disubstitution) results in a derivative with altered spectral properties compared to the monosubstituted derivative; this derivative does not regenerate histidine, and treatment with neutral hydroxylamine or a base likely results in scission of the imidazole ring similar to that suggested for diacetylation of histidine with acetic anhydride. Loss of histidine is detected by amino acid analysis after acid hydrolysis; sequence analysis using Edman degradation chemistry also shows the absence of histidine with the presence of a disubstituted derivative.[230] In these studies, a PTH derivative eluting near PTH-glycine was observed, and the structure was verified by MS. The monosubstituted derivative is unstable under conditions of acid hydrolysis and yields free histidine. As with the modification of other amino acid residues in proteins, MS is the method of choice in the analysis of the chemical modification of histidine in proteins including carboxymethylated histidine.[231–237] Glocker and coworkers[231] were able to demonstrate the modification of lysine and tyrosine in addition to histidine modification; these investigators were also able to distinguish between the monosubstitution and disubstitution of histidine with DEPC. Dage and coworkers[233] used MS to determine the extent of ethoxyformylation of two histidine residues and a lysine residue using MS of peptides isolated from α_1-acid glycoprotein. The modification at the histidine residues showed modest pH dependence, while the modification at lysine increased markedly with increased pH. Tyrosine modification was also demonstrated in this study. Krell and coworkers[234] used MS to measure protein modification with diethylpyrocarbonate. These investigators reported the use of MS to study protein modification with phenylglyoxal, TNM, and trinitrobenzenesulfonic acid. Analysis was only limited by the stability of the derivatives. These investigators did find it difficult to locate specific histidine residues modified by DEPC because of multiple sites of modification. It was possible to obtain some information on the presence of histidine binding sites from protection experiments. Protection studies were also used by Qin and coworkers[235] in the identification of histidine residues present in copper binding sites on prion proteins using MALDI-TOF mass spectrometric footprinting. Ginotra and Kulkarni[236] used reaction of histidine with DEPC to provide support for a neutral five-coordinate structure for complex of histidine and cupric ions. Willard and Kinter[237] used modification of histidine with diethylpyrocarbonate to improve product ion spectra with improved sequence information. More recent studies have used mass spectrometric analysis of diethylpyrocarbonate modification to characterize the heme-coordinating histidine residues in cytochrome b_5.[238] Nakanishi and coworkers[239] used reaction with diethylpyrocarbonate combined with EPR spectroscopy to characterize the heme-coordinating residues of a myoglobin mutant. UV spectroscopy and MALDI-TOF MS were used to characterize the histidine residues involved in heme binding. Konkle and coworkers[240] used modification with diethylpyrocarbonate to study the histidine residues in Rieske protein. An elegant study[241] has examined the reaction of diethylpyrocarbonate with histidyl residues in cytochrome b_5. Using (NMR) spectroscopy with this well-characterized protein, it has been possible to identify factors influencing histidine modification with this reagent; three major factors include (1) the pKa of the individual histidine residue, (2) solvent exposure of the residue, and (3) hydrogen bonding of the imidazolium ring. NMR spectroscopy enabled these investigators to assign the relative reactivities of the four histidine residues modified in cytochrome b_5 by diethylpyrocarbonate.

The use of diethylpyrocarbonate for the modification of histidine residues at enzyme active sites has been previously reviewed by this author.[242] This is still a useful technique provided that sufficient care is made to assure specificity of modification. It is, however, the sense of the author that the value of diethylpyrocarbonate for protein modification is in the use for the measurement of surface-exposed residues.[224,225] Given the ability to determine the extent and site of modification with MS and with the appropriate protein, NMR, reaction with diethylpyrocarbonate is a useful method for studying protein–protein interaction, ligand binding, and conformational change.

The introductory statement to this chapter noted the peculiar character of the imidazole ring of the histidine. It would appear that the nucleophilic character is maximum in the range of 5.5–7.0 reflecting the ionization of the protonated nitrogen. While the reactivity of histidine in the active sites of serine proteases is fairly well understood, the reactivity of histidine residues outside the active site is somewhat more difficult to understand. Likewise, the apparently random reactivity of histidine in proteins is equally difficult to understand. The author recalls a conversation with Stanford Moore at the Rockefeller Institute many years ago that went something like this: "Bill and I thought that when we had the final sequence of RNAse, we would know how it worked—no such luck—then when Fred Richard at Yale brought the model derived from crystallographic analysis, we were again optimistic but again disappointed." Given our lack of understanding of molecular interactions, it is not unreasonable that there are histidine residues that one would expect to modify but are not as well as unexpected modifications of histidine residues in proteins. As with other amino acids, location of the histidine residue in a protein profoundly influences reaction. Consider, for example, the carboxymethylation of histidine residues in myoglobin discussed earlier. Thus, we have examples of residues that are not modified and residues that react with reagents not usually associated with the modification of histidine.

The modification of histidine in interferon provides an example of an atypical modification. Succinimidyl carbonate PEG (Figure 14.10) was developed to modify lysine residues in protein to yield a more effective biopharmaceutical product.[243] This reagent uses NHS chemistry as described in Chapter 2. The succinimidyl carbonate derivative was developed to replace succinimidyl succinate–PEG. It was thought that the reagent would be specific for the modification of amino groups although care was taken to examine the possibility of tyrosine modification. The succinimidyl carbonate PEG forms a carbamate derivative[244] with protein amino groups (Figure 14.10).

Wang and coworkers[147] reported that the reaction of interferon α-2b with succinimidyl carbonate PEG at pH 6.5 (phosphate) yielded a mixture of PEGylated products; 47% of the product was a carbamate histidine (His34) PEG derivative of interferon α-2b (Figure 14.10). Subsequently, Wiley and coworkers[246] studied the PEGylation of interferon α-2b with succinimidyl carbonate PEG and observed the pH-dependent formation of a PEG–histidine derivative. At mildly acidic pH (pH 5.4–6.5), there were several products; the major product was modified at His34. At more alkaline pH, there was more extensive modification of lysine. Both groups noted that the PEGylation of histidine was reversed under mild basic conditions consistent the stability of N acetyl histidine or ethoxyformylated histidine. Both studies argue that solvent exposure of His34 is important for PEGylation; however, to the best of the author's knowledge, this reaction has not been reported with the use of succinimidyl carbonate in other proteins[247,248] and was not observed with the solid-phase modification of interferon α-2a with succinimidyl carbonate PEG[249]. In this study, PEGylation occurred at the amino-terminal amino acid. The study of Wang and coworkers[245] is instructive in this regard. These investigators compared the stability of the His34-PEGylated interferon α-2b with products prepared from the reaction of a peptide, Ac-βDRH(PEG)DFGFPQ with succinimidyl carbonate PEG, or from the PEG-His peptide isolated from the subtilisin digestion of the modified interferon. The His34-PEGylated derivative was much more stable than either peptide derivative. In a work not cited in any of the aforementioned studies, Borukhov and Ya.Strongin[250] did not see any effect of diethylpyrocarbonate or TLCK with either αA-interferon or β-interferon; the inhibition of β-interferon was observed with TPCK. This study lacks data on the extent of modification in the several proteins. TPCK was developed as an active-site-directed reagent that reacts with an active-site histidine as an affinity reagent,[7] but reaction at other sites such as sulfhydryl groups has been reported.[251]

Another example of the modification of histidine by reagents that, in general, react more avidly with residues other than histidine is the reaction with dansyl chloride (Figure 14.11). Hartley and

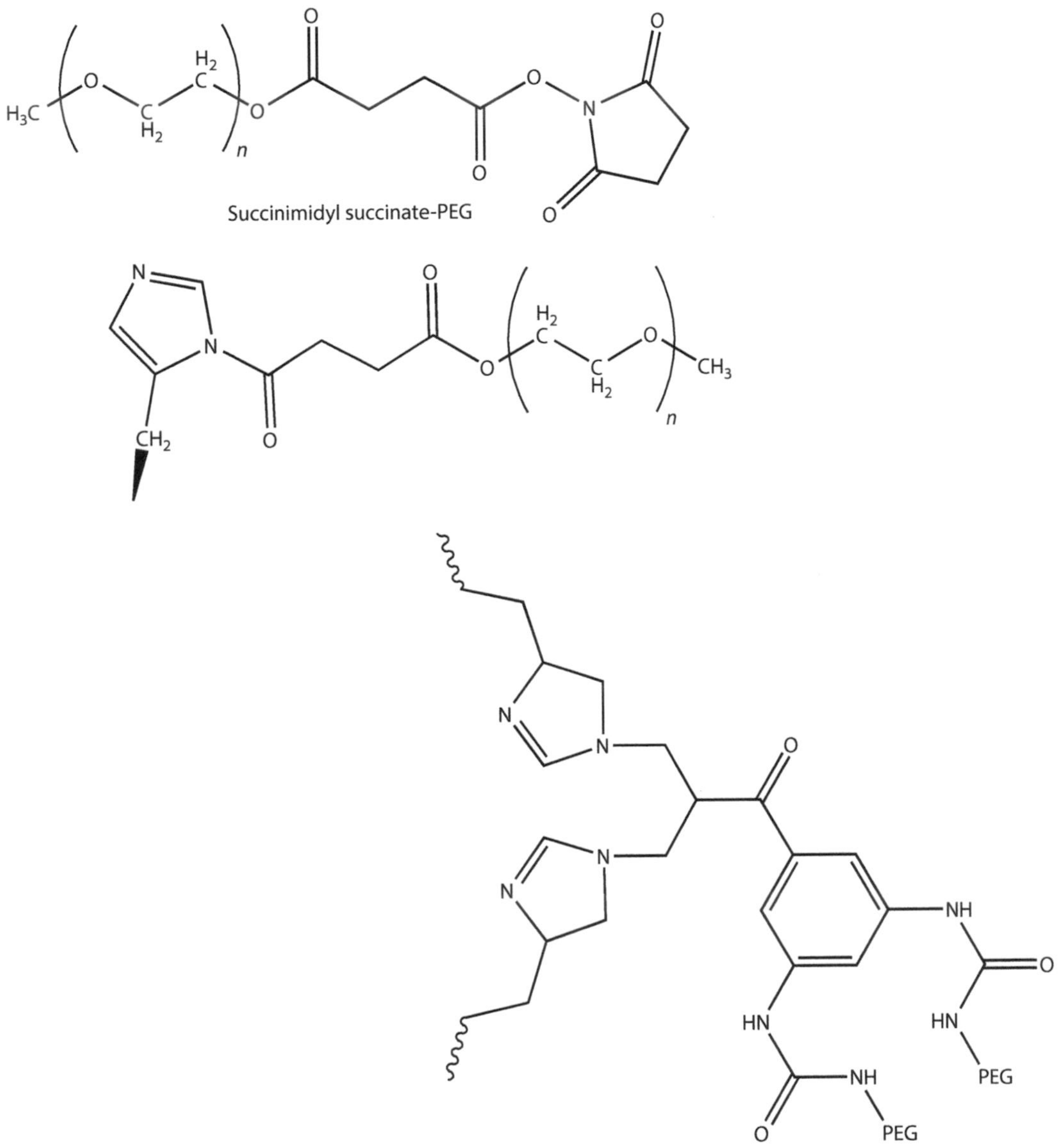

Succinimidyl succinate-PEG

FIGURE 14.10 Acylation reactions of histidine. Shown is the reaction of a succinimidyl carbonate derivative with histidine (Wang, Y.S., Youngster, S., Bausch, J. et al., Identification of the major positional isomer
of pegylated interferon α-2B, *Biochemistry* 39, 10634–10640, 2000; Wang, M., Basu, A., Palm, T. et al.,
Engineering an arginine catabolizing bioconjugate: Biochemical and pharmacological characterization of
PEGylated derivatives of arginine deiminase from *Mycoplasma arthritidis, Bioconjug. Chem.* 17, 1447–1459,
2005). This derivative is labile at alkaline pH or in the presence of hydroxylamine. Shown at the bottom is the
suggested structure for the site-specific PEGylation of a polyhistidine tag (Cong, Y., Pawlisz, E., Bryant, P.
et al., Site-specific PEGylation at histidine tags, *Bioconjug. Chem.* 23, 248–263, 2012).

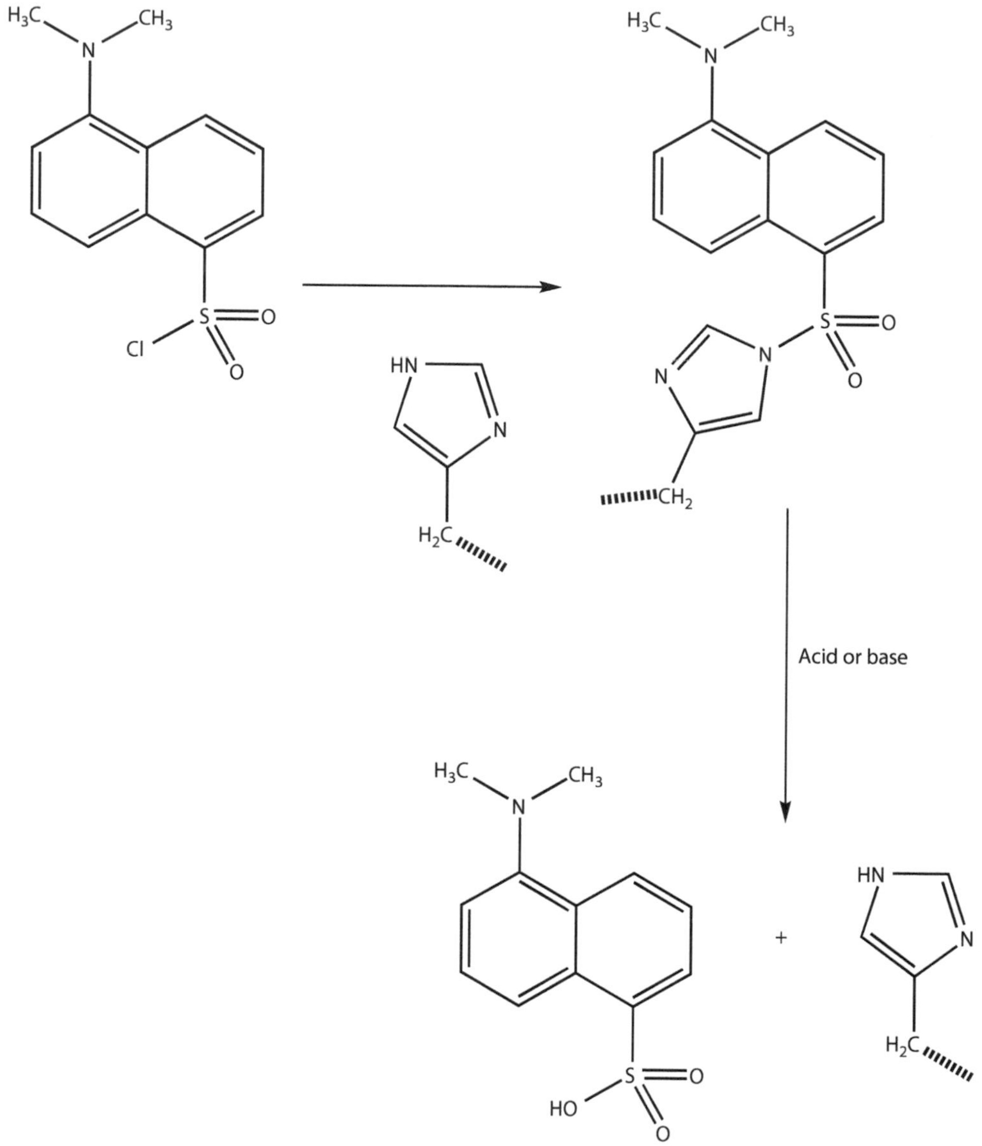

FIGURE 14.11 The reaction of dansyl chloride with histidine residues in proteins. The product of the reaction is not stable in acid or base or in the presence of hydroxylamine. See Hartley, B.S. and Massey, V., Active centre of chymotrypsin I. Labeling with a fluorescent dye, *Biochim. Biophys. Acta* 21, 58–70, 1956; Nishino, T., Massey, V., and Williams, C.H., Jr., Chemical modification of D-amino acid oxidase. Evidence for active site histidine tyrosine and arginine residues, *J. Biol. Chem.* 255, 3610–3615, 1980; Gadda, G., Beretta, G.L., and Pilone, M.S., Reactivity of histidyl residues in D-amino acid oxidase from *Rhodotorula gracilis*, *FEBS Lett.* 363, 307–310, 1995. Dansyl histidine is unstable in either acid or base (Horton, H.R. and Koshland, D.E., Jr., Environmentally sensitive groups attached to proteins, *Methods Enzymol.* 11, 856–870, 1967). The product obtained by reaction with primary amines is stable.

Massey[252] described the inactivation of chymotrypsin by dansyl chloride. The inactivated protein was stable in the range of pH 4–6; enzyme activity was recovered on incubation in either more acidic or more basic conditions. A comparison of the spectral property of the modified chymotrypsin and dansylated histidine suggested that the modification of chymotrypsin occurs by a modification of histidine at the enzyme active site. Tamura and others[253] observed that dansyl chloride reacted with the α-amino group and the imidazole ring (specific nitrogen not identified). The dansyl imidazole derivative is a potent dansylation reagent and is therefore not stable. Nishino and coworkers[254] reported on the inactivation of D-amino acid oxidase with dansyl chloride. Substantially complete reactivation occurred with 0.5 M hydroxylamine (NH_2OH) at pH 6.6. This reactivation excluded reaction with primary amino functional groups such as lysine, and amino acid analysis suggested the reaction had not occurred with an oxygen nucleophile such as tyrosine. Treatment of the enzyme with diethyl pyrocarbonate also resulted in the loss of catalytic activity and reduced the amount of dansyl groups incorporated in a subsequent reaction, suggesting that dansyl chloride reacts with the same functional group that reacted with diethylpyrocarbonate. Subsequent structural analysis from this group[255] showed the modification of His217 in the active center of D-amino acid oxidase. Gadda and coworkers[256] observed the inactivation of D-amino acid oxidase from *Rhodotorula gracilis* by dansyl chloride. The reaction was performed in 50 mM phosphate, pH 6.6 containing 10% glycerol at 18°C in the dark with a 300-fold molar excess of dansyl chloride. The enzyme was protected from inactivation by benzoate. Nonlinearity was observed in the time course of inactivation reflecting the hydrolysis of dansyl chloride during the modification reaction. The modified enzyme retained activity with altered substrate specificity.

The cyanation of histidine residues in myoglobin using an equimolar ratio of cyanogen bromide and protein at pH 7.0 has been reported.[257–259] This derivative is somewhat unstable, but it has proved useful in spectral studies (NMR, IR, UV–VIS) of this protein. It was not possible to find application of cyanogen bromide for the modification of histidine in any protein except myoglobin. A search for cyanation (cyanylation) of histidine with other cyanylation reagents such as 2-nitro-5-thiocyanobenzoic acid was unsuccessful.

Competitive labeling of proteins is a method for determining residue accessibility as a measure of protein conformation and protein–protein interaction.[260] Competitive labeling of the amino-terminal histidine residue in secretin with FDNB has been used to study the reactivity of this residue versus other nucleophiles.[261] The amino-terminal functional group has a pKa of 8.83 and fivefold greater reactivity than the model compound (histidyl-glycine), while the imidazolium ring has a pKa value of 8.24 and a 26-fold greater reactivity than the model compound. These results were interpreted as reflecting a conformational state where the histidine is interacting with a carboxylate function. Kaplan and Oda[262] reported on the use of FDNB for the selective isolation of free and blocked amino-terminal peptides. The dinitrophenyl group on histidine imidazole ring (Figure 14.12) can be removed by thiolysis.[263] Thiolysis will also remove the dinitrophenyl group from tyrosine or cysteine.

4-Hydroxy-2-nonenal (HNE) and 4-oxo-2-nonenal are aldehydes derived from the oxidation of lipids[264] that can react with proteins to form a variety of products.[265–269] MS can be used to identify the various products including the Michael addition product with histidine (Figure 14.13).[270,271] If either sodium cyanoborohydride or sodium borohydride were included with HNE and the oxidized B-chain of insulin, reaction occurs with the ε-amino group of lysine via a Schiff mechanism with the aldehyde.[272] In the absence of reducing agent, HNE modification of histidine in the oxidized B-chain of insulin proceeded via a Michael addition as the predominant reaction. Monoclonal antibodies to the HNE adduct with histidine have been developed.[273–276]

The formation of *N*-acetylhistidine in proteins (Figure 14.14) is not unusual, but the establishment of the presence of such a derivative is difficult because of product stability.[277,278] This product stability issue is also observed with other protein modifications such as the ethoxyformylation of

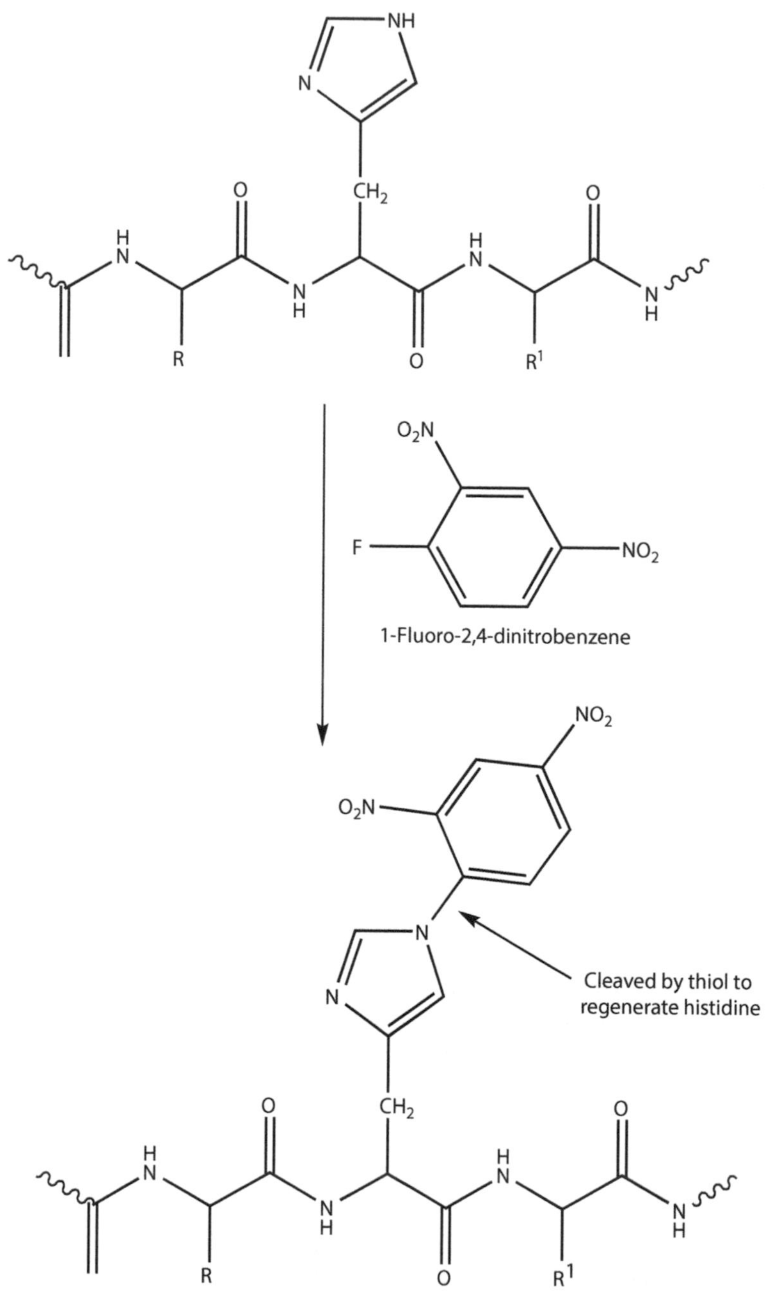

FIGURE 14.12 The reaction of histidyl residues with 1-fluoro-2,4-dinitrofluorobenzene to yield the N^{im}-dinitrophenyl derivative. Histidine can be regenerated by thiolysis to yield free histidine and the corresponding 1-thio-2,4-dintrobenzyl derivative. (See Shatiel, S., Thiolysis of some dinitrophenyl derivatives of amino acids, *Biochem. Biophys. Res. Commun.* 29, 178–183, 1967; Kaplan, H. and Oda, G., Selective isolation of free and blocked amino-terminal peptides from enzymatic digestion of proteins, *Analyt. Biochem.* 132, 384–388, 1983). Histidine is less reactive than lysine and specific modification of lysine can be achieved with low levels of reagent (Chen, X., Anderson, V.E., and Chen, Y.H., Isotope edited product ion assignment by α-N labeling of peptides with [^{2}H$_3$(50%)]2,4-dinitrofluorobenzene, *J. Am. Soc. Mass Spectrom.* 10, 448–452, 1999).

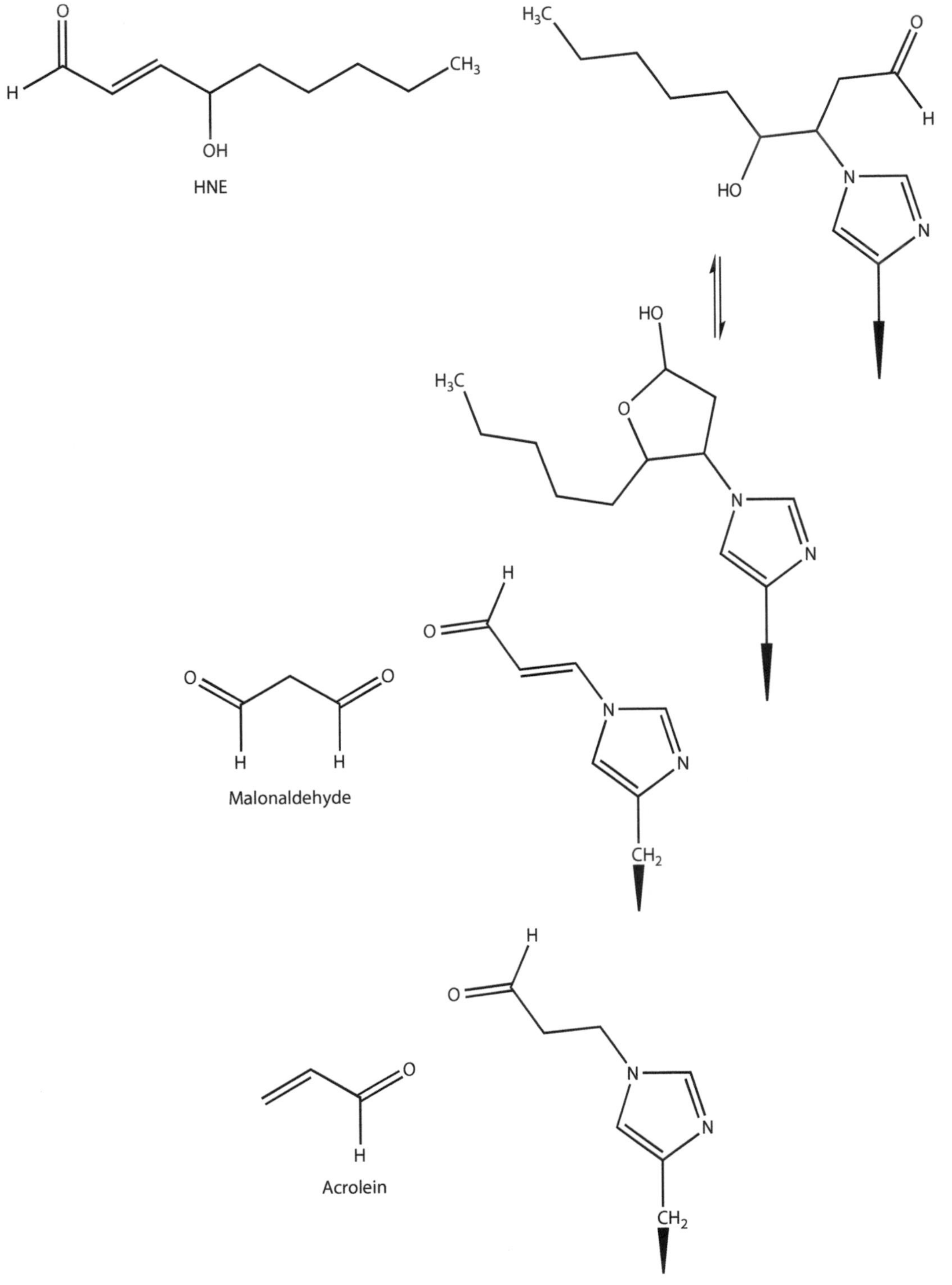

FIGURE 14.13

(continued)

FIGURE 14.13 (continued) The modification of histidine by HNE and other biologically important alde-hydes. Shown are the reaction products obtained with HNE, malonaldehyde, and acrolein. HNE and acrolein add to the imidazole ring of histidine by a Michael addition while malonaldehyde reacts with the imidazole ring forming a Schiff base. See LoPachin, R.M., Gavin, T., Petersen, D.R., and Barber, D.S., Molecular mecha-nisms of 4-hydroxy-2-nonenal and acrolein toxicity: Nucleophilic targets and adduct formation, *Chem. Res. Toxicol.* 22, 1499–1508, 2009; Rauniyar, N., Prokai-Tatral, K., and Prokal, L., Identification of carbonylation sites in apomyoglobin after exposure to 4-hydroxy-2-nonenal by solid-phase enrichment and liquid chroma-tography-electrospray ionization tandem mass spectrometry, *J. Mass Spectrom.* 45, 398–410, 2010; Zhao, J., Chen, J., Zhu, H., and Xiong, Y.L., Mass spectrometric evidence of malonaldehyde and 4-hydroxynonenal adductions to radical-scavenging soy peptides, *J. Agric. Food Chem.* 60, 9727–9736, 2012; Maeshima, T., Honda, K., Chikazawa, M. et al., Quantitative analysis of acrolein-specific adducts generated during lipid peroxidation-modification of proteins in vitro: Identification of Nτ-(3-propanal)histidine as the major adduct, *Chem. Res. Toxicol.* 25, 1384–1392, 2012; Xie, Z., Baba, S.P., Sweeney, B.R., and Barski, O.A., Detoxification of aldehydes by histidine-containing dipeptides: From chemistry to clinical implications, *Chem. Biol. Interact.* 202, 288–297, 2013.

histidine, the formation of cysteine sulfenic acid, and the acylation of serine or threonine. There are, however, several examples where a stable modification of a protein by acetic anhydride has been reported. MacDonald and coworkers[278] reported that NMR spectroscopy showed the transient acetylation of histidine followed by the formation of a stable *N*-acetyllysine derivative. These inves-tigators also observed the participation of a histidine residue in the acetylation of lysine by acetyl salicylate. Moore[279] reported the acetylation by acetic anhydride of the imidazole ring of histidine in angiotensin II. It was observed that the imidazole side chains were acetylated and deacetylated at a more rapid rate than the free amino acid. Welsch and Nelsestuen[230] observed the diacety-lation of a histidine residue with the reaction of prothrombin fragment 1 with acetic anhydride. The diacetylation of histidine was associated with the opening of the imidazole ring and the loss of the C2 carbon; this modification was not reversed by hydroxylamine. This reaction is similar to the disubstitution of histidine with diethylpyrocarbonate. Kinnunen and colleagues[280] reported the inactivation of acyl-CoA/cholesterol *O*-acyltransferase by diethylpyrocarbonate or acetic anhydride. The inactivation by either reagent is reversed by hydroxylamine. There was a marked difference between the liver enzyme and the enzyme derived from aorta.

Several groups have reported the modification of histidine with Woodward's reagent K (*N*-ethyl-5-phenylisoxaxolium-3-sulfonate).[281–283] This reagent is usually considered specific for carboxyl groups (see Chapter 3). One group[283] observed saturation kinetics in the inactivation of an acylphos-phatase with Woodward's reagent K suggesting the formation of a reversible complex prior to the inactivation reaction.

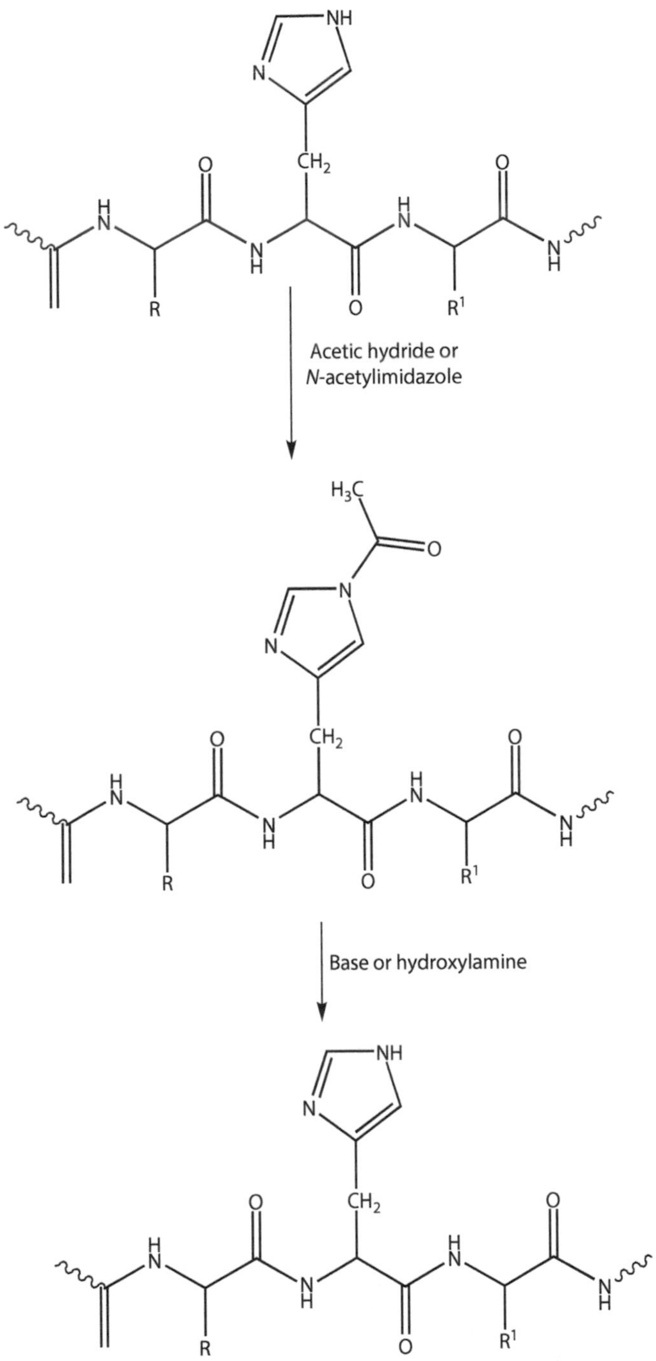

FIGURE 14.14 The acetylation of histidyl residues in proteins. Histidyl residues can be modified with various acetylating agents such as acetic anhydride and *N*-acetylimidazole. See Macdonald, J.M., Haas, A.L., and London, R.E., Novel mechanisms of surface catalysis of protein adduct formation. NMR studies of the acetylation of ubiquitin, *J. Biol. Chem.* 275, 31908–31923, 2000; Houston, L.L. and Walsh, K.A., The transient inactivation of trypsin by mild acetylation with *N*-acetylimidazole, *Biochemistry* 9, 156–166, 1970; Lundblad, R.L., The reaction of bovine thrombin with *N*-butyrylimidazole. Two different reactions resulting in the inhibition of catalytic activity, *Biochemistry* 14, 1033–1037, 1975.

REFERENCES

1. Hoffman, K.W., Romei, M.G., and Londergan, C.H., A new Raman spectroscopic probe of both the protonation state and non-covalent interactions of histidine residues, *J. Phys. Chem. A* 117, 5987–5996, 2013.

2. Edgcomb, S.P. and Murphy, K.P., Variability in the pKa of histidine side-chains correlates with burial within proteins, *Proteins* 49, 1–6, 2002.

3. Sundberg, R.J. and Martin, R.B., Interactions of histidine and other imidazole derivatives with transition metal ions in chemical and biological systems, *Chem. Rev.* 74, 471–517, 1974.

4. Covelli, I., Frati, L., and Wolff, J., Carboxymethylation of the histidyl of insulin, *Biochemistry* 12, 1043–1047, 1973.

5. Røjkaer, R. and Schousboe, I., Partial identification of the Zn^{2+}-binding sites in factor XII and its activation derivatives, *Eur. J. Biochem.* 247, 491–496, 1997.

6. Block, H., Maertens, B., Spriestersbach, A. et al., Immobilized-metal affinity chromatography, *Methods Enzymol.* 463, 439–473, 2009.

7. Fagerlund, R.D., Ooi, P.L., and Wilbanks, S.M., Soluble expression and tumor suppressor WT1 and its zinc finger domain, *Protein Expr. Purif.* 85, 165–172, 2012.

8. Schlager, B., Straessle, A., and Hafen, E., Use of anionic denaturing detergents to purify insoluble proteins after overexpression, *BMC Biotechnol.* 12, 95, 2012.

9. Honjo, T., Hoe, K., Tabayashi, S. et al., Preparation of affinity membranes using thermally induced phase separation for one-step purification of recombinant proteins, *Anal. Biochem.* 434, 269–274, 2013.

10. Boeneman Gemmill, K., Deschamps, J.R., Delehanty, J.B. et al., Optimizing protein coordination to quantum dots with designer peptidyl linkers, *Bioconjug. Chem.* 24, 269–281, 2013.

11. Conzuelo, F., Gamella, M., Campuzano, S. et al., Integrated amperometric affinity biosensors using Co^{2+}-tetradentate nitrilotriacetic acid modified disposable carbon electrodes: Application to the determination of β-lactam antibiotics, *Anal. Chem.* 85, 3246–3254, 2013.

12. Dasa, S.S., Jin, Q., Chen, C.T., and Chen, L., Target-specific copper hybrid t7 phage particles, *Langmuir* 28, 17372–17380, 2012.

13. Zaitouna, A.J. and Lai, R.Y., Design and characterization of a metal ion-imidazole self-assembled monolayer for reversible immobilization of histidine-tagged peptides, *Chem. Commun.* 47, 12391–12393, 2011.

14. Sebollela, A., Cagliari, T.D., Limaverde, G.S. et al., Heparin-binding sites in granulocyte-macrophage colony-stimulating factor. Localization and regulation by histidine ionization, *J. Biol. Chem.* 280, 31949–31956, 2005.

15. Watanabe, H., Matsumaru, H., Ooishi, A. et al., Optimizing pH response of affinity between protein G and IgG Fc: How electrostatic modulations affect protein-protein interactions, *J. Biol. Chem.* 284, 12373–12383, 2009.

16. Gera, N., Hill, A.B., White, D.P. et al., Design of pH sensitive binding proteins from the hyperthermophilic Sso7d scaffold, *PLoS One* 7(11), e48928, 2012.

17. Batalla, P., Bolívar, J.M., Lopez-Gallego, F., and Guisan, J.M., Oriented covalent immobilization of antibodies onto heterofunctional agarose supports: A highly efficient immuno-affinity chromatography platform, *J. Chromatogr. A* 1262, 56–63, 2012.

18. Engleset, M., Xia, S., Okada, C. et al., Structural and mechanistic insights into guanylylation of RNA-splicing ligase RtcB joining RNA between 3'-terminal phosphate and 5'-OH, *Proc. Natl. Acad. Sci. USA* 109, 15235–15240, 2012.

19. Tadwal, V.S., Sundararaman, L., Manimekalai, M.S. et al., Relevance of the conserved histidine and asparagine residues in the phosphate-binding loop of the nucleotide binding subunit B of A_1A_0 ATP synthases, *J. Struct. Biol.* 180, 509–518, 2012.

20. Higgins, K.A., Hu, H.Q., Chivers, P.T., and Maroney, M.J., Effects of select histidine to cysteine mutations on transcriptional regulation by *Escherichia coli* RcnR, *Biochemistry* 52, 84–97, 2013.

21. Zhang, L., Andersen, E.M., Khajo, A. et al., Dynamic factors affecting gaseous ligand binding in an artificial oxygen transport protein, *Biochemistry* 52, 447–455, 2013.

22. Bálint, E.E., Petres, J., Szabó, M. et al., Fluorescence of a histidine-modified enhanced green fluorescent protein (EGFP) effectively quenched by copper (II) ions, *J. Fluoresc.* 23, 273–281, 2013.

23. Nyarko, A., Cochran, L., Norwood, S. et al., Ionization of His 55 at the dimer interface of dynein light chain LC8 is coupled to dimer dissociation, *Biochemistry* 44, 14248–14255, 2005.

24. Fitton, V., Verdoni, N., Sanchez, J., and Santarelli, X., Penicillin acylase purification with the aid of pseudo-affinity chromatography, *J. Biochem. Biophys. Methods* 49, 553–560, 2001.

25. Mandjiny, S. and Vijayalakhmi, M.A., Quantitation of adsorption capacity of immunoglobulin G on histidine-aminohexyl sepharose and determination of affinity constant, *J. Chromatogr.* 616, 189–195, 1993.

26. Todorova-Balvay, D., Pitiot, O., Bourhim, M. et al., Immobilized metal-ion affinity chromatography of human antibodies and their proteolytic fragments, *J. Chromatogr. B Analyt. Technol. Biomed. Life Sci.* 808, 57–62, 2004.

27. Hu, X., Li, G., and Huang, E., Click chemistry: A route to designing and preparing pseudo-biospecific immunoadsorbent for IgG adsorption, *J. Chromatogr. B Analyt. Technol. Biomed. Life Sci.* 899, 96–102, 2012.

28. Andersen, J.T., Dee Qian, J., and Sandlie, I., The conserved histidine 166 residue of the human neonatal Fc receptor heavy chain is critical for the pH-dependent binding to albumin, *Eur. J. Immunol.* 36, 3044–3051, 2006.

29. Belew, M., Yip, T.T., Andersson, L., and Ehrnström, R., High-performance analytical applications of immobilized metal ion affinity chromatography, *Anal. Biochem.* 164, 457–465, 1987.

30. Hari, P.R., Paul, W., and Sharma, C.P., Adsorption of human IgG on Cu^{2+}-immobilized cellulose affinity membrane: Preliminary study, *J. Biomed. Mater. Res.* 50, 110–113, 2000.

31. Jain, S. and Gupta, M.N., Purification of goat immunoglobulin G by immobilized metal-ion affinity using cross-linked alginate beads, *Biotechnol. Appl. Biochem.* 39, 319–322, 2004.

32. Bresolin, I.T., Borsoi-Ribeiro, M., Tamashiro, W.M. et al., Evaluation of immobilized metal-ion affinity chromatography (IMAC) as a technique for IgG(1) monoclonal antibodies purification: The effect of chelating ligand and support, *Appl. Biochem. Biotechnol.* 160, 2148–2165, 2010.

33. Plapp, B.V., Application of affinity labeling for studying structure and function of enzymes, *Methods Enzymol.* 87, 469–499, 1982.

34. Schoellmann, G. and Shaw, E., Direct evidence for the presence of histidine in the active center of chymotrypsin, *Biochemistry* 2, 252–255, 1963.

35. Pincus, M., Thi, L.L., and Carty, R.P., The kinetics and specificity of the reaction of 2′(3′)-O-bromoacetyluridine with bovine pancreatic ribonuclease A, *Biochemistry* 14, 3653–3661, 1975.

36. Shaw, E., Mares-Guia, M., and Cohen, W., Evidence for an active-center histidine in trypsin through use of a specific reagent, 1-chloro-3-tosylamido-7-amino-2-heptanone, the chloromethyl ketone derived from N^α-tosyl-L-lysine, *Biochemistry* 4, 2219–2224, 1965.

37. Shaw, E. and Glover, G., Further observations on substrate-derived chloromethyl ketones that inactivate trypsin, *Arch. Biochem. Biophys.* 139, 298–305, 1970.

38. Inagami, T., The alkylation of the active site of trypsin with iodoacetamide in the presence of alkylguanidines, *J. Biol. Chem.* 240, PC3453–PC3455, 1965.

39. Royer, G.P. and Uy, R., Evidence for the induction of a conformational change of bovine trypsin by a specific substrate at pH 8.0, *J. Biol. Chem.* 248, 2627–2629, 1973.

40. Kupfer, A., Gani, V., Jiménez, J.S., and Shaltiel, S., Affinity labeling of the catalytic subunit of cyclic AMP-dependent protein kinase by N^α-tosyl-L-lysine chloromethyl ketone, *Proc. Natl. Acad. Sci. USA* 76, 3073–3077, 1979.

41. Liu, T.-Y., Demonstration of the presence of a histidine residue at the active site of streptococcal proteinase, *J. Biol. Chem.* 242, 4029–4032, 1967.

42. Gerwin, B.I., Properties of the single sulfhydryl group of streptococcal proteinase. A comparison of the rates of alkylation by chloroacetic acid and chloroacetamide, *J. Biol. Chem.* 242, 451–456, 1967.

43. Dahl, K.H. and McKinley-McKee, J.S., The imidazole-promoted inactivation of horse-liver alcohol dehydrogenase, *Eur. J. Biochem.* 120, 451–459, 1981.

44. Bloxham, D.P., The chemical reactivity of the histidine-195 residue in lactate dehydrogenase thiomethylated at the cysteine-165 residue, *Biochem. J.* 93, 93–97, 1981.

45. Reynolds, W.F., Peat, I.R., Freedman, M.H., and Lyerla, J.R., Jr., Determination of the tautomeric form of the imidazole ring of L-histidine in basic solution by carbon-13 magnetic resonance spectroscopy, *J. Am. Chem. Soc.* 95, 328–331, 1973.

46. Weighardt, T. and Goren, H.J., The reactivity of imidazole nitrogens in histidine to alkylation, *Bioorg. Chem.* 4, 30–40, 1975.

47. Calleman, C.J. and Poirier, V., The nucleophilic reactivity and tautomerism of the imidazole nitrogens of N^2-acetyl-histidine methylamide and N^2-acetyl-histidine, *Acta Chem. Scand. B* 37, 809–815, 1983.

48. Vila, J.A., Arnautova, Y.A., Vorobjev, Y., and Scheraga, H.A., Assessing the fractions of tautomeric forms of the imidazole ring of histidine in proteins as a function of pH, *Proc. Natl. Acad. Sci. USA* 108, 5602–5607, 2011.

49. Reynolds, W.F. and Tzeng, C.W., Determination of the preferred tautomeric forms of histamine by ^{13}C nmr spectroscopy, *Can. J. Biochem.* 55, 576–578, 1977.

50. Webb, J.L., Interaction of inhibitors with enzymes, in *Enzyme and Metabolic Inhibitors*, Vol. 1, Chapter 6, pp. 193–318, Academic Press, New York, 1963.

51. Putnam, F.W., The chemical modification of proteins, in *The Proteins Chemistry, Biological Activity and Methods*, Vol. 1, Part B, Chapter 10, eds. H. Neurath and K. Bailey, pp. 893–972, Academic Press, New York, 1953.

52. Glazer, A.N., Chemical modification of proteins by group-specific and site-specific reagents, in *The Proteins*, 3rd edn., Vol. II, Chapter 1, eds. H. Neurath, R.L. Hill, and C.-L. Boeder, pp. 1–103, Academic Press, New York, 1973.

53. Ebert, C., Über die chemische Modifizierung von Keratinen bei der Methylierung von Aminosäureseitengruppen, *Colloid Interface Sci.* 252, 100–116, 1974.

54. Taralp, A. and Kaplan, H., Chemical modification of lyophilized proteins in nonaqueous environments, *J. Protein Chem.* 16, 183–193, 1997.

55. Stewart, N.A.S., Taralp, A., and Kaplan, H. Imprinting of lyophilized α-chymotrypsin affects reactivity of the active-site imidazole, *Biochem. Biophys. Res. Commun.* 240, 27–31, 1997.

56. Dai, Z., Nemet, I., Shen, W., and Monnier, V.M., Isolation, purification and characterization of histidino-threosidine, a novel Maillard reaction protein crosslink from threose, lysine, and histidine, *Arch. Biochem. Biophys.* 463, 78–88, 2007.

57. Smyth, D.G., Nagamatsu, A., and Fruton, J.S., Reactions of *N*-ethylmaleimide, *J. Am. Chem. Soc.* 82, 4600–4604, 1960.

58. Papini, A., Rudolph, S., Sigmueller, G. et al., Alkylation of histidine with maleimido-compounds, *Int. J. Pept. Protein Res.* 39, 348–355, 1992.

59. Ray, W.J., Jr., Photochemical oxidation, *Methods Enzymol.* 11, 490–497, 1967.

60. Harris, D.T., XXXVI. The action of light on blood, *Biochem. J.* 20, 271–279, 1926.

61. Harris, D.T., XXXVII. Photo-oxidation of plasma. With a note on its sensitization, *Biochem. J.* 20, 280–287, 1926.

62. Harris, D.T., Observations on the velocity of the photo-oxidation of proteins and amino acids, *Biochem. J.* 20, 288–292, 1926.

63. Witt, O., Verhältniss der chemischern Constitution eines aromatischen Körpers zu seiner färbenden Kraft, *Berichte der Deutschen chemischen Gesellschaft zu Berlin*, 950, 1876.

64. Smetana, H. and Shemin, D., Studies on the photo-oxidation of antigen and antibodies, *J. Exp. Med.* 73, 223–242, 1941.

65. Suontaka, A.-M., Blombäck, M., and Chapman, J., Changes in functional activities of plasma fibrinogen after treatment with methylene blue and red light, *Transfusion* 43, 568–573, 2003.

66. Steinmann, E., Gravemann, U., Friesland, M. et al., Two pathogen reduction technologies-methylene blue plus light and shortwave ultraviolet light-effectively inactivates hepatitis C virus in blood products, *Transfusion*, 53, 1010–1018, 2013.

67. Tsai, C.S., Godin, J.R.P., and Ward, A.J., Dye-sensitized photo-oxidation of enzymes, *Biochem. J.* 225, 203–208, 1985.

68. Paz, A., Roth, E., Ashani, Y. et al., Structural and functional characterization of the interaction of the photosensitizing probe methylene blue with *Torpedo californica* acetylcholinesterase, *Protein Sci.* 21, 1138–1152, 2012.

69. Triquigeneaux, M.M., Ehrenshaft, M., Roth, E. et al., Targeted oxidation of *Torpedo californica* acetyl-cholinesterase by singlet oxygen: Identification of *N*-formylkynurenine tryptophan derivatives within the active-site gorge of its complex with the photosensitizer methylene blue, *Biochem. J.* 448, 83–91, 2012.

70. Buchko, G.W., Wagner, J.R, Cadet, J. et al., Methylene blue-mediated photooxidation of 7.8-dihydro-8-oxo-2'-deoxyguanosine, *Biochim. Biophys. Acta* 1263, 17–24, 1995.

71. Andley, U.P. and Clark, B.A., Conformational changes of β H-crystallin in riboflavin-sensitized photo-oxidation, *Exp. Eye Res.* 47, 1–15, 1988.

72. Pecci, L., Costa, M., Antonucci, A. et al., Methylene blue photosensitized oxidation of cysteine sulfinic acid and other sulfinates: The involvement of a singlet oxygen and the azide paradox, *Biochem. Biophys. Res. Commun.* 270, 782–786, 2000.

73. Waku, K. and Nakazawa, Y., Photooxidation of ribonuclease T1 in the presence of substrate analog, *J. Biochem.* 68, 63–67, 1970.

74. Martinez-Carrion, M., Kuczenski, R., Tiemeier, D.C., and Peterson, D.L., The structure and enzyme-coenzyme relationship for supernatant aspartate transaminase after dye sensitized photooxidation, *J. Biol. Chem.* 245, 799–805, 1970.

75. Tsurushiin, S., Hiramatsu, S., Inamasu, M., and Yasunobo, K.T., The essential histidine residues of bovine amine oxidase, *Biochim. Biophys. Acta* 400, 451–460, 1975.

76. Okumura, K. and Murachi, T., Photooxidation of histidine and tryptophan residues of papain in the presence of methylene blue, *J. Biochem.* 77, 913–918, 1975.
77. Oara, A., Fujimoto, S., Kanazawa, H., and Nakagawa, T., Studies on the active site of papain. V. Photooxidation of histidine residues, *Chem. Pharm. Bull. (Tokyo)* 23, 967–970, 1975.
78. Kandal, M., Gornall, A.G., Lam, L.K., and Kandel, S.I., Photooxidation of dinitrophenylhistidine-200 human carbonic anhydrase B, *Can. J. Biochem.* 53, 599–608, 1975.
79. Murachi, T., Tsudzuki, T., and Okumura, K., Photosensitized inactivation of stem bromelain. Oxidation of histidine, methionine, and tryptophan residues, *Biochemistry* 14, 249–255, 1975.
80. Murachi, T. and Okumura, K., Normal apparent pKa value for the ionization of the histidine residue of papain and stem bromelain as determined by photooxidation reaction, *FEBS Lett.* 40, 127–129, 1974.
81. Hiramatsu, A., Tsurushiin, S., and Yasunobu, K.T., Evidence for essential histidine residues in bovine-liver mitochondrial monoamine oxidase, *Eur. J. Biochem.* 57, 587–593, 1975.
82. Silva, E. and Barrera, M., The riboflavin-sensitized photooxidation of horseradish apoperoxidase, *Radiat. Environ. Biophys.* 24, 57–61, 1985.
83. Pfiffner, E. and Lerch, K., Histidine at the active site of *Neurospora* tyrosinase, *Biochemistry* 20, 6029–6035, 1981.
84. Kuno, S., Fukui, S., and Toraya, T., Essential histidine residues in coenzyme B12-dependent diol dehydrase: Dye-sensitized photooxidation and ethoxycarbonylation, *Arch. Biochem. Biophys.* 277, 211–217, 1990.
85. Shiroya, Y. and Samejima, T., The specific modification of histidyl residues of inorganic pyrophosphatase from *Bacillus stearothermophilus* by photooxidation, *J. Biochem.* 98, 333–339, 1985.
86. Amutha, B., Khire, J.M., and Khan, M.I., Active site characterization of the exo-*N*-acetyl-β-D-glucosaminidase from thermotolerant *Bacillus* sp. NCIM 5120: Involvement of tryptophan, histidine and carboxylate residues in catalytic activity, *Biochim. Biophys. Acta* 1427, 121–132, 1999.
87. Shinoda, M., Hara, A., Nakayama, T. et al., Modification of pig liver dimeric dihydrodiol dehydrogenase with diethylpyrocarbonate and by rose bengal-sensitized photooxidation: Evidence for an active-site histidine residue, *J. Biochem.* 112, 834–839, 1992.
88. Stuart, J., Pessah, I.N., Favero, T.G., and Abramson, J.J., Photooxidation of skeletal muscle sarcoplasmic reticulum induces rapid calcium release, *Arch. Biochem. Biophys.* 292, 512–521, 1992.
89. Inada, Y., Hessel, B., and Blombäck, B., Photooxidation of fibrinogen in the presence of methylene blue and its effect on polymerization, *Biochim. Biophys. Acta* 532, 161–170, 1978.
90. Shimizu, A., Saito, Y., Matsushima, A., and Inada, Y., Identification of an essential histidine residue for fibrin polymerization. Essential role of histidine 16 of the Bβ-chain, *J. Biol. Chem.* 258, 7915–7917, 1983.
91. Shimizu, A., Saito, Y., and Inada, Y., Distinctive role of histidine-16 of the Bβ chain of fibrinogen in the end-to-end association of fibrin, *Proc. Natl. Acad. Sci. USA* 83, 591–593, 1986.
92. Tomita, M., Irie, M., and Ukita, T., Sensitized photooxidation of histidine and its derivatives. Products and mechanism of the reaction, *Biochemistry* 8, 5149–5160, 1969.
93. Balasubramanian, D., Du, X., and Zigler, J.S., Jr., The reaction of singlet oxygen with proteins with special reference to crystallins, *Photochem. Photobiol.* 52, 761–768, 1990.
94. Shen, H.-R., Spikes, J.D., Kopečeková, P., and Kopeček, J., Photodynamic crosslinking of proteins. I. Model studies using histidine- and lysine-containing *N*-(2-hydroxypropyl) methacrylamide copolymer, *J. Photochem. Photobiol. B Biol.* 34, 203–210, 1996.
95. Shen, H.-R., Spikes, J.D., Smith, C.J., and Kopeček, J., Photodynamic cross-linking of proteins IV. Nature of the his-his bond(s) formed in the Rose Bengal-photosensitized cross-linking of *N*-benzoyl-L-histidine, *J. Photochem. Photobiol. A Chem.* 130, 1–6, 2000.
96. Agon, V.V., Bubb, W.A., Wright, A. et al., Sensitizer-mediated photooxidation of histidine residues: Evidence for the formation of reactive side-chain peroxides, *Free Radic. Biol. Med.* 40, 698–710, 2006.
97. Stroop, S.D., Conca, D.M., Lundgard, R.P. et al., Photosensitizers form in histidine buffer and mediate the photodegradation of a monoclonal antibody, *J. Pharm. Sci.* 100, 5142–5155, 2011.
98. Gurd, F.R., Carboxymethylation, *Methods Enzymol.* 11, 532–541, 1967.
99. Rahimi, Y., Shrestha, S., Banerjee, T., and Deo, S.K., Copper sensing based on the far-red fluorescent protein, HcRed, from *Heteractis crispa*, *Anal. Biochem.* 370, 60–67, 2007.
100. Gregory, E.M., Chemical modification of bovine heart mitochondrial malate dehydrogenase. Selective modification of cysteine and histidine, *J. Biol. Chem.* 250, 5470–5474, 1975.
101. Anderton, B.H. and Rabin, B.R., Alkylation studies in a reactive histidine in pig heart malate dehydrogenase, *Eur. J. Biochem.* 15, 568–573, 1970.
102. Anderton, B.H., Identification of an essential, reactive histidine in pig heart mitochondrial malate dehydrogenase, *Eur. J. Biochem.* 15, 562–567, 1970.

103. Cai, J., Bhatnagar, A., and Pierce, W.M., Jr., Protein modification by acrolein: Formation and stability of cysteine adducts, *Chem. Res. Toxicol.* 22, 708–716, 2009.
104. Malinowski, D.P. and Fridovich, I., Subunit association and side-chain reactivities of bovine erythrocyte superoxide dismutase in denaturing solvents, *Biochemistry* 18, 5055–5060, 1979.
105. Thomas, J.L., LaRochelle, M.C., Asibey-Berko, E., and Strickler, R.C., Reactivation of human placental 17β,20α-hydroxysteroid dehydrogenase affinity alkylated by estrone 3-(bromoacetate): Topographic studies with 16α-(bromoacetoxy)estradiol 3-(methyl ether), *Biochemistry* 24, 5361–5367, 1985.
106. Thomas, J.L., Asibey-Berko, E., and Strickler, R.C., The affinity alkylators, 11α-bromoacetoxy-progesterone and estrone 3-bromoacetate, modify a common histidyl residue in the active site of human placental 17β,20α-hydroxysteroid dehydrogenase, *J. Steroid Biochem.* 25, 103–108, 1986.
107. Gleisner, J.M. and Liener, I.E., Chemical modification of the histidine residue located at the active site of ficin, *Biochim. Biophys. Acta* 317, 482–491, 1973.
108. Webb, J.L., Iodoacetate and iodoacetamide, in *Enzyme and Metabolic Inhibitors*, Vol. III, Chapter 1, pp. 1–283, Academic Press, New York, 1966.
109. Dudley, H.W., LIII. Intermediary carbohydrate metabolism. The effect of sodium iodoacetate on glyoxalase, *Biochem. J.* 25, 439–445, 1931.
110. Zittle, C.A., Ribonucleinase III. The behavior of copper and calcium in the purification of nucleic acid and the effect of these and other reagents on the activity of ribonucleinase, *J. Biol. Chem.* 163, 111–117, 1946.
111. Hirs, C.H., Moore, S., and Stein, W.H., The sequence of the amino acid residues in performic acid-oxidized ribonuclease, *J. Biol. Chem.* 235, 633–647, 1960.
112. Spackman, D.H., Stein, W.H., and Moore, S., The disulfide bonds of ribonuclease, *J. Biol. Chem.* 235, 648–659, 1960.
113. Gundlach, H.G., Stein, W.H., and Moore, S., The nature of the amino acid residues modified on the inactivation of ribonuclease by iodoacetate, *J. Biol. Chem.* 234, 1754–1760, 1959.
114. Stark, G.R., Stein, W.H., and Moore, S., Relationship between the conformation of ribonuclease and its reactivity with iodoacetate, *J. Biol. Chem.* 236, 436–442, 1961.
115. Barnard, E.A. and Stein, W.D., The histidine residue in the active centre of ribonuclease. I. A specific reaction with bromoacetic acid, *J. Mol. Biol.* 1, 333–349, 1959.
116. Korman, S. and Clarke, H.T., Carboxymethylamino acids and peptides, *J. Biol. Chem.* 221, 113–131, 1956.
117. Korman, S. and Clarke, H.T., Carboxymethyl proteins, *J. Biol. Chem.* 221, 133–141, 1956.
118. Stein, W.D. and Barnard, E.A., The histidine residue in the active centre of ribonuclease. II. The position of this residue in the primary protein chain, *J. Mol. Biol.* 1, 350–358, 1959.
119. Cha, M.-K. and Kim, I.-H., Glutathione-linked thiol peroxidase activity of human serum albumin: A possible antioxidant role of serum albumin in blood plasma, *Biochem. Biophys. Res. Commun.* 222, 619–625, 1996.
120. Huggins, C. and Jenson, E.V., Thermal coagulation of serum proteins I. The effects of iodoacetate, iodoacetamide, and thiol compounds on coagulation, *J. Biol. Chem.* 179, 845–854, 1949.
121. Batra, P.P., Sasa, K., Ueki, T., and Takeda, K., Circular dichroic study of the conformational stability of sulfhydryl-blocked bovine serum albumin, *Int. J. Biochem.* 21, 857–862, 1989.
122. Crestfield, A.M., Stein, W.H., and Moore, S., Alkylation and identification of the histidine residues at the active site of ribonuclease, *J. Biol. Chem.* 238, 2413–2420, 1963.
123. Kumar, K.S., Walsh, K.A., and Neurath, H., Chemical characterization of bovine carboxypeptidase A isolated from a single pancreas, *Biochemistry* 3, 1726–1727, 1964.
124. Crestfield, A.M., Stein, W.H., and Moore, S., Properties and conformation of the histidine residues at the active site of ribonuclease, *J. Biol. Chem.* 238, 2421–2428, 1963.
125. Fruchter, R.G. and Crestfield, A.M., Preparation and properties of two active forms of ribonuclease dimer, *J. Biol. Chem.* 240, 3868–3874, 1965.
126. Fruchter, R.G. and Crestfield, A.M., On the structure of ribonuclease dimer. Isolation and identification of monomers derived from inactive carboxymethylated dimers, *J. Biol. Chem.* 240, 3875–3882, 1965.
127. Crestfield, A.M. and Fruchter, R.G., The homologous and hybrid dimers of ribonuclease A and the carboxymethylhistidine derivatives, *J. Biol. Chem.* 242, 3279–3284, 1967.
128. Lin, M.C., Stein, W.H., and Moore, S., Further studies on the alkylation of the histidine residues of pancreatic ribonuclease, *J. Biol. Chem.* 243, 6167–6170, 1968.
129. Meadows, D.H., Jardetsky, O., Epand, R.M. et al., Assignment of histidine peaks in the nuclear magnetic resonance spectroscopy spectrum of ribonuclease, *Proc. Natl. Acad. Sci. USA* 60, 766–772, 1968.

130. Miyagi, M. and Nakazawa, T., Determination of pKa values of individual histidine residues in proteins using mass spectrometry, *Anal. Chem.* 80, 6481–6487, 2008.

131. Fruchter, R.G. and Crestfield, A.M., The specific alkylation by iodoacetamide of histidine 12 in the active site of ribonuclease, *J. Biol. Chem.* 242, 5807–5812, 1967.

132. Plapp, B.V., Mechanisms of carboxymethylation of bovine pancreatic nucleases by haloacetates and tosylglycolate, *J. Biol. Chem.* 248, 4896–4900, 1973.

133. Moore, S. and Stein, W.H., Chemical structures of pancreatic ribonuclease and deoxyribonuclease. *Science* 180, 458–464, 1973.

134. Marshall, G.R., Fend, J.A., and Kuster, D.J., Back to the future: Ribonuclease A, *Biopolymers* 90, 259–277, 2008.

135. Kettner, C. and Shaw, E., Inactivation of trypsin-like enzymes with peptides of arginine chloromethyl ketone, *Methods Enzymol.* 80, 826–842, 1981.

136. Bock, P.E., Active site selective labeling of serine proteases with spectroscopic probes using thioester peptide chloromethyl ketones: Demonstration of thrombin labeling using N$^\alpha$ [T-[(acetylthio)acetyl]-d-Phe-Pro-Arg-Ch2Cl, *Biochemistry* 27, 6633–6639, 1988.

137. Bock, P.E., Active-site-selective labeling of blood coagulation proteinases with fluorescence probes by the use of thioester peptide chloromethyl ketones. II. Properties of thrombin derivatives as reporters of prothrombin fragment 2 binding and specificity of the labeling approach for other proteinases, *J. Biol. Chem.* 267, 14974–14981, 1992.

138. Williams, E.B., Krishnaswamy, S., and Mann, K.G., Zymogen/enzyme discrimination using peptide chloromethyl ketones, *J. Biol. Chem.* 264, 7536–7545, 1989.

139. Kettner, C. and Shaw, E., The susceptibility of urokinase to affinity labeling by peptides of arginine chloromethyl ketone, *Biochim. Biophys. Acta* 569, 31–40, 1979.

140. Perez-G, M., Cortes, J.R., and Rivas, M.D., Treatment of cells with N-α-tosyl-L-phenylalanine-chloromethyl ketone induces the proteolytic loss of STAT6 transcription factor, *Mol. Immunol.* 45, 3896–3901, 2008.

141. Asayama, S., Kato, H., Kawakami, H., and Nagaoka, S., Carboxymethyl poly(L-histidine) as a new pH-sensitive polypeptide at endosomal/lysosomal pH, *Polym. Adv. Technol.* 18, 329–333, 2007.

142. Asayama, S., Sudo, M., Nagaoka, S., and Kawakami, H., Carboxymethyl poly (L-histidine) as a new pH-sensitive polypeptide to enhance polyplex gene delivery, *Mol. Pharm.* 5, 898–901, 2008.

143. Park, J.K., Singha, K., Arote, R.B. et al., pH-responsive polymers as gene carriers, *Macromol. Rapid Commun.* 31, 1122–1133, 2010.

144. Gu, J.J., Wang, X., Jiang, X.Y. et al., Self-assembled carboxymethyl poly (L-histidine) coated poly (β-amino ester) DNA complexes for gene transfection, *Biomaterials* 33, 644–658, 2012.

145. Postnikova, G.B., Moiseeva, S.A., and Shekhovisova, E.A., The main role of inner histidines in the molecular mechanism of myoglobin oxidation catalyzed by copper compounds, *Inorg. Chem.* 49, 1347–1354, 2010.

146. Banaszak, L.J., Andrews, P.A., Burgner, J.W. et al., Carboxymethylation of sperm whale metmyoglobin, *J. Biol. Chem.* 238, 3307–3314, 1963.

147. Banaszak, L.J. and Gurd, F.R., Carboxymethylation of sperm whale metmyoglobin. Reactivity of the adjacent histidine residues, *J. Biol. Chem.* 239, 1836–1838, 1964.

148. Nigen, A.M., Keim, P., Marshall, R.C. et al., Carbon 13 nuclear magnetic resonance spectroscopy of myoglobins and ribonuclease A carboxymethylated with enriched (2–^{13}C)bromoacetate, *J. Biol. Chem.* 247, 4100–4102, 1972.

149. Dautrevaux, M., Han, K., Boulanger, Y., and Biserte, G., Alkylation des radicaux d'histidine du centre actif de la myoglobine de Cheval, *Bull. Soc. Chim. Biol.* 47, 2178–2182, 1965.

150. Delacourte, A., Han, K.-K., and Dautrevaux, M., Alkylation des résidus d'histidine de la myoglobine de chevel. Etude de la myoglobine 1-CM His-113, *Biochemie* 55, 869–876, 1973.

151. Hugli, T.E. and Gurd, F.R.N., Carboxymethylation of sperm whale myoglobin in the crystalline state, *J. Biol. Chem.* 245, 1930–1938, 1970.

152. Hugli, T.E. and Gurd, F.R.N., Carboxymethylation of sperm whale myoglobin in the dissolved state, *J. Biol. Chem.* 245, 1939–1946, 1970.

153. Zhou, Y. and Vachet, R.W., Increased protein structural resolution from diethylpyrocarbonate-based covalent labeling and mass spectrometric detection, *J. Am. Soc. Mass Spectrom.* 23, 708–717, 2012.

154. Viljoen, C.C., Visser, L., and Botes, D.P., Histidine and lysine residues and the activity of phospholipase A2 from the venom of *Bitis gabonica*, *Biochim. Biophys. Acta* 483, 107–120, 1977.

155. Jiménez, M., Cabanes, J., Gandía-Herrero, F. et al., A continuous spectrophotometric assay for phospholipase A$_2$ activity, *Anal. Biochem.* 319, 131–137, 2003.

156. Ram, A., Das, M., Gangal, S.V., and Ghosh, B., p-Bromophenacyl bromide alleviates airway hyperre-sponsiveness and modulates cytokines, IgE and eosinophil levels in ovalbumin-sensitized and-challenged mice, *Int. Immunopharmacol.* 4, 1697–1707, 2004.

157. Tariq, M., Elfaki, I., Khan, H.A. et al., Bromophenacyl bromide, a phospholipase A2 inhibitor attenuates chemically induced gastroduodenal ulcers in rats, *World J. Gastroenterol.* 12, 5798–5804, 2006.

158. Merchant, M., Heard, R., and Monroe, C., Characterization of phospholipase A_2 activity in the serum of the American alligator (*Alligator mississippiensis*), *J. Exp. Zool. A Ecol. Genet. Physiol.* 311, 662–666, 2009.

159. Zychar, B.C., Dale, C.S., Demarchi, D.S., and Goncalves, L.R., Contribution of metalloproteases, ser-ine proteases and phospholipases A2 to the inflammatory reaction induced by *Bothrops jararaca* crude venom in mice, *Toxicon* 55, 227–234, 2010.

160. Romero-Vargas, F.F., Ponce-Soto, L.A., Martins-de-Souza, D., and Marangoni, S., Biological and biochemical characterization of two new PLA2 isoforms Cdc-9 and Cdc-10 from *Crotalus durissus cumanensis* snake venom, *Comp. Biochem. Physiol. C Toxicol. Pharmacol.* 151, 66–74, 2010.

161. Ravindran, S., Lodoen, M.B., Verhelst, S.H. et al., 4-Bromophenacyl bromide specifically inhibits rhoptry secretion during Toxoplasma invasion, *PLoS One* 4(12), e8143, 2009.

162. Fonseca, F.V., Baldissera, L., Jr., Carmargo, E.A. et al., Effect of the synthetic coumarin, ethyl 2-oxo-2H-chromene-3-carboxylate, on activity *Crotalus durissus ruruima* sPLA2 as well as on edema and platelet aggregation induced by this factor, *Toxicon* 55, 1527–1530, 2010.

163. Berger, M., Reck, J., Jr., Terra, R.M. et al., *Lonomia obliqua* venomous secretion induces human platelet adhesion and aggregation, *J. Thromb. Thrombolysis* 30, 300–310, 2010.

164. Blacklow, B., Escoubas, P., and Nicholson, G.M., Characterization of the heterotrimeric presynaptic phospholipase A_2 neurotoxin complex from the venom of the common death adder (*Acanthophis ant-arcticus*), *Biochem. Pharmacol.* 80, 277–287, 2010.

165. Shaw, J.O., Roberts, M.F., Ulevitch, R.J. et al., Phospholipase A_2 contamination of cobra venom fac-tor preparations. Biologic role in complement-dependent in vivo reactions and inactivation with p-bromophenacyl bromide, *Am. J. Pathol.* 91, 517–530, 1978.

166. Diz Filho, E.B., Maragoni, S.,Toyama, D.O. et al., Enzymatic and structural characterization of new PLA2 isoform isolated from white venom of Crotalus *durissus ruruima*, *Toxicon* 53, 104–114, 2009.

167. Ximenes, R.M., Alves, R.S., Pereira, T.P. et al., Harpalycin 2 inhibits the enzymatic and platelet aggrega-tion activities of PrTX-III, a D49 phospholipase A2 from *Bothrops pirajai* venom, *BMC Complement. Altern. Med.* 12, 139, 2012.

168. Galkina, S.I., Fedorova, N.V., Serebryakova, M.V. et al., Proteome analysis identified human neutrophil membrane tubulovesicular extensions (cytonemes, membrane tethers) as bactericide trafficking, *Biochim. Biophys. Acta* 1820, 1705–1714, 2012.

169. Kumar, R. and Atreja, S.K., Effect of incorporation of additives in tris-based egg yolk extender on buffalo (*Bubalus bubalis*) sperm tyrosine phosphorylation during cryopreservation, *Reprod. Domest. Anim.* 47, 485–490, 2012.

170. Wang, R., Dodia, C.R., Jain, M.K., and Fisher, A.B., Purification and characterization of a calcium-independent acidic phospholipase A2 from rat lung, *Biochem. J.* 304, 131–137, 1994.

171. Akiba, S., Dodia, C., Chen, X., and Fisher, A.B., Characterization of a acidic Ca^{2+}-independent phospho-lipase A_2 of bovine lung, *Comp. Biochem. Physiol. B Biochem. Mol. Biol.* 120, 393–404, 1998.

172. Jacob, M., Weech, P.K., and Salesse, C., Phospholipase A_2 of rod outer segment-free bovine retinae are different from well-known phospholipases A_2, *Biochim. Biophys. Acta* 1391, 169–180, 1998.

173. Brustovetsky, T., Antonsson, B., Jemmerson, R. et al., Activation of calcium-independent phospholi-pase A (iPLA) in brain mitochondria and release of apoptogenic factors by BAX and truncated BID, *J. Neurochem.* 94, 980–994, 2005.

174. Berger, J., Identification of organic compounds. I. Preparation of p-bromophenacyl esters of carboxylic acids, *Acta Chem. Scand.* 10, 638–642, 1956.

175. Durst, H.D., Milano, M., Kikta, E.J. et al., Phenacyl esters of fatty acids via crown ether catalysts for enhanced ultraviolet detection in liquid chromatography, *Anal. Chem.* 47, 1797–1801, 1975.

176. Zamir, I., Derivatization of saturated long-chain fatty acids with phenacyl bromide in ionic micelles, *J. Chromatogr.* 586, 347–350, 1991.

177. Erlanger, B.F., Vratsanos, S.M., Wasserman, N., and Cooper, A.G., Chemical investigation of the active center of pepsin, *Biochem. Biophys. Res. Commun.* 23, 245–245, 1966.

178. Gross, E. and Morrell, J.L., Evidence for an active carboxyl group in pepsin, *J. Biol. Chem.* 241, 3638–3639, 1966.

179. Takahashi, K., Stein, W.H., and Moore, S., The identification of a glutamic acid residue as part of the active site of ribonuclease T₁, *J. Biol. Chem.* 242, 4682–4690, 1967.
180. Kyger, E.M. and Franson, R.C., Nonspecific inhibition of enzymes by *p*-bromophenacyl bromide, *Biochim. Biophys. Acta* 794, 96–103, 1984.
181. Jeng, T.-W. and Fraenkel-Conrat, H., Chemical modification of histidine and lysine residues of crotoxin, *FEBS Lett.* 87, 291–296, 1978.
182. Volwerk, J.J., Pieterson, W.A., and de Haas, G.H., Histidine at the active site of phospholipase A₂, *Biochemistry* 13, 1446–1454, 1974.
183. Abita, J.P., Lazdunski, M., Bonsen, P.P. et al., Zymogen-enzyme transformations. On the mechanism of activation of prophospholipase A, *Eur. J. Biochem.* 30, 37–47, 1972.
184. Bonsen, P.P.M., Pieterson, W.A., Volwerk, J.J., and de Haas, G.H., Phospholipase A and its zymogen from porcine pancreas, in *Current Trends in the Biochemistry of Lipids (Biochemical Society Symposium 35)*, eds. P. Ganguly and R.M.S. Smellie, pp. 189–200, Academic Press, London, U.K., 1971.
185. Dennis, E.A., Phospholipases, in *The Enzymes*, 3rd edn., Vol. 16, Chapter 9, ed. P.D. Boyer, Academic Press, New York, pp. 307–353, 1983.
186. Mackessey, S.P., Evolutionary trends in venom composition in the western rattlesnakes (*Crotalus viridis sensu lato*): Toxicity vs. tenderness, *Toxicon* 55, 1463–1474, 2010.
187. Marchi-Salvador, D.P., Fernandes, C.A.H., Silveira, L.B. et al., Crystal structure of a phospholipase A₂ homolog complexed with *p*-bromophenacyl bromide reveals important structural changes associated with the inhibition of myotoxic activity, *Biochim. Biophys. Acta* 1794, 1583–1590, 2009.
188. Miyake, T., Inoue, S., and Ikeda, K., pH Dependence of the reaction rate of His 48 with *p*-bromophenacyl bromide and the binding constant to Ca²⁺ of the monomeric forms of intact and α-NH₂ modified phospholipase A₂ from *Trimeresurus flavoviridis*, *J. Biochem.* 105, 565–572, 1989.
189. Fujii, S., Meida, M., Tani, T. et al., pH Dependence of the reaction rate of *p*-bromophenacyl bromide and of the binding constants of Ca²⁺ and an amide-type substrate analog to bovine pancreatic phospholipase A₂, *Arch. Biochem. Biophys.* 354, 73–82, 1998.
190. Fohlman, J., Eaker, D., Karlsoon, E., and Theslff, S., Taipoxin, an extremely potent presynaptic neurotoxin from the venom of the australian snake taipan (*Oxyuranus s. scutellatus*). Isolation, characterization, quaternary structure and pharmacological properties, *Eur. J. Biochem.* 68, 457–469, 1976.
191. Tzeng, M.C., Yen, C.H., Hseu, M.J. et al., Binding proteins on synaptic membranes for crotoxin and taipoxin, two phospholipases A2 with neurotoxicity, *Toxicon* 33, 451–457, 1995.
192. Fohlman, J., Eaker, D., Dowdall, M.J., Lüllmann-Rauch, R., Sjödin, T., and Leander, S., Chemical modification of taipoxin and the consequences for phospholipase activity, pathophysiology, and inhibition of high-affinity choline uptake, *Eur. J. Biochem.* 94, 531–540, 1979.
193. Halpert, J., Eaker, D., and Karlsson, E., The role of phospholipase activity in the action of a presynaptic neurotoxin of *Notechis scutatus scutatus* (Australian Tiger Snake), *FEBS Lett.* 61, 72–76, 1976.
194. Verheij, H.M., Volwerk, J.J., Jansen, E.H. et al., Methylation of histidine-48 in pancreatic phospholipase A2. Role of histidine and calcium ion in the catalytic mechanism, *Biochemistry* 19, 743–750, 1980.
195. Roberts, M.F., Deems, R.A., Mincey, T.C., and Dennis, E.A., Chemical modification of the histidine in phospholipase A₂ (*Naja naja naja*), *J. Biol. Chem.* 252, 2405–2411, 1977.
196. Battaglia, E. and Radominska-Pandya, A., A functional role for histidyl residues of the UDP-glucuronic acid carrier in rate liver endoplasmic reticulum membranes, *Biochemistry* 37, 258–263, 1998.
197. de Vet, E.C. and van den Bosch, H., Characterization of recombinant guinea pig alkyl-dihydroxyacetonephosphate synthase expressed in *Escherichia coli*: Kinetics, chemical modification, and mutagenesis, *Biochim. Biophys. Acta* 1436, 299–306, 1999.
198. Dorsch, S., Liebisch, G., Kaufmann, B. et al., The VP1 unique region of parvovirus B19 and its constituent phospholipase A2-like activity, *J. Virol.* 76, 2014–2018, 2002.
199. Nakagawa, Y. and Bender, M.L., Modification of α-chymotrypsin by methyl *p*-nitrobenzenesulfonate, *J. Am. Chem. Soc.* 91, 1566–1567, 1967.
200. Nakagawa, Y. and Bender, M.L., Methylation of histidine-57 in α-chymotrypsin by methyl *p*-nitrobenzenesulfonate. A new approach to enzyme modification, *Biochemistry* 9, 259–267, 1970.
201. Glick, B.R., The chemical modification of *Escherichia coli* ribosomes with methyl p-nitrobenzenesulfonate. Evidence for the involvement of a histidine residue in the functioning of the ribosomal peptidyl transferase, *Can. J. Biochem.* 58, 1345–1347, 1980.
202. Marcus, J.P. and Dekker, E.E., Identification of a second active site residue in *Escherichia coli* L-threonine dehydrogenase: Methylation of histidine-90 with methyl-*p*-nitrobenzenesulfonate, *Arch. Biochem. Biophys.* 316, 413–420, 1995.

203. Johnson, A.R. and Dekker, E.E., Site-directed mutagenesis of histidine-90 in *Escherichia coli* L-threonine dehydrogenase alters its substrate specificity, *Arch. Biochem. Biophys.* 351, 8–16, 1998.

204. Lancaster, N.L. and Welton, T., Nucleophilicity in ionic liquids. 3. Anion effects on halide nucleophilicity in a series of 1-butyl-3-methylimidazolium ionic liquids, *J. Org. Chem.* 80, 5986–5992, 2004.

205. Crowhurst, L., Lancaster, N.L., Perez Arlandis, J.M., and Welton, T., Manipulating solute nucleophilicity with room temperature ionic liquids, *J. Am. Chem. Soc.* 126, 11549–11555, 2004.

206. Ford, J.W., Janaka, M.E., Lu, J. et al., Local polarity in CO_2-expanded acetonitrile: A nucleophilic substitution reaction and solvatochromic probes, *J. Org. Chem.* 73, 3364–3368, 2008.

207. Hayaki, S., Kido, K., Sato, H., and Sakaki, S., *Ab initio* study on SN_2 reaction of methyl *p*-nitrobenzene-sulfonate and chloride anion in [mmim][PF6], *Phys. Chem. Chem. Phys.* 12, 1822–1826, 2010.

208. Edmondson, D.E., Kenney, W.C., and Singer, T.P., Structural elucidation and properties of 8α-(N[1]-histidyl)riboflavin: The flavin component of thiamine dehydrogenase and β-cyclopiazonate oxidocyclase, *Biochemistry* 15, 2937–2945, 1976.

209. Kamińska, J., Wiśniewska, A., and Kościelak, J., Chemical modifications of α 1,6-fucosyltransferase define amino acid residues of catalytic importance, *Biochimie* 85, 303–310, 2003.

210. 21 CFR 189.150, Diethyl pyrocarbonate (DEPC), August 2, 1972.

211. Delfini, C., Gaia, P., Schellino, R. et al., Fermentability of grape must after inhibition with dimethyl dicarbonate, *J. Agric. Food Chem.* 50, 5601–5611, 2002.

212. Solymosy, F., Fedorcsák, I., Gulyás, A. et al., A new method based on the use of diethyl pyrocarbonate as a nuclease inhibitor for the extraction of undegraded nucleic acid, *Eur. J. Biochem.* 5, 520–527, 1968.

213. Leonard, N.J., McDonald, J.J., and Reichmann, M.E., Reaction of diethyl pyrocarbonate with nucleic acid components. I. Adenine, *Proc. Natl. Acad. Sci. USA* 67, 93–98, 1970.

214. Huang, Y.H., Leblanc, P., Apostolou, V. et al., Comparison of Milli-Q PF plus water with DEPC-treated water in the preparation and analysis of RNA, *Nucleic Acids Symp. Ser.* (33), 129–133, 1995.

215. Hullán, L., Szontagh, T., Turtóczky, I., and Fedorcsák, I., The inactivation of trypsin by diethyl pyrocarbonate, *Acta Chem. Scand.* 19, 2440–2441, 1965.

216. Fedorcsák, I. and Ehrenberg, L., Effects of diethyl pyrocarbonate and methyl methanesulfonate on nucleic acids and nucleases, *Acta Chem. Scand.* 20, 107–112, 1966.

217. Rosén, C.G. and Fedorcsák, I., Studies on the action of diethyl pyrocarbonate on proteins, *Biochim. Biophys. Acta* 130, 401–405, 1966.

218. Ovádi, J. and Keleti, T., Effect of diethyl pyrocarbonate on the conformation and enzymatic activity of d-glyceraldehyde-3-phosphate dehydrogenase, *Acta Biochim. Biophys. Sci. Hung.* 4, 365–378, 1969.

219. Rosén, C.G., Gejvall, T., and Andersson, L.O., Reaction of diethyl pyrocarbonate with indole derivatives with special reference to the reaction with tryptophan residues in a protein, *Biochim. Biophys. Acta* 221, 207–213, 1970.

220. Wiener, S.L., Wiener, R., Urivetzky, M., and Meilman, E., Inactivation of ribonuclease by diethyl pyrocarbonate and other methods, *Biochim. Biophys. Acta* 259, 378–385, 1972.

221. Wolf, B., Lesnaw, J.A., and Reichmann, M.E., A mechanism of irreversible inactivation of bovine pancreatic ribonuclease by diethylpyrocarbonate. A general reaction of diethyl pyrocarbonate with proteins, *Eur. J. Biochem.* 13, 519–525, 1970.

222. Morris, D.L. and McKinley-McKee, J.S., The histidines in liver alcohol dehydrogenase. Chemical modification with diethylpyrocarbonate, *Eur. J. Biochem.* 29, 515–520, 1972.

223. Holbrook, J.J., Lodola, A., and Illesley, N.P., Histidine residues and the enzyme activity of pig heart supernatant malate dehydrogenase, *Biochem. J.* 139, 797–800,1974.

224. Hnízda, A., Šantrůcek, J., Šanda, M. et al., Reactivity of histidine and lysine side-chains with diethyl pyrocarbonate—A method to identify surface exposed residues in proteins, *J. Biochem. Biophys. Methods* 70, 1091–1097, 2008.

225. Mendoza, V.L. and Vachet, R.W., Protein surface mapping using diethyl pyrocarbonate with mass spectrometric detection, *Anal. Chem.* 81, 2895–2901, 2008.

226. Muhirad, A., Hegyi, G., and Toth, G., Effect of diethyl pyrocarbonate on proteins. I. Reaction of diethyl pyrocarbonate with amino acids, *Acta Biochim. Biophys. Acad. Sci. Hung.* 2, 19–29, 1967.

227. Garrison, C.K. and Himes, R.H., The reaction between diethyl pyrocarbonate and sulfhydryl groups in carboxylate buffers, *Biochem. Biophys. Res. Commun.* 67, 1251–1255, 1975.

228. Rua, J., Robertson, A.G.S., and Nimmo, H.G., Identification of the histidine residues in *Escherichia coli* isocitrate lyase that reacts with diethylpyrocarbonate, *Biochim. Biophys. Acta* 1122, 212–218, 1992.

229. Miles, E.W., Modification of histidyl residues in proteins by diethylpyrocarbonate, *Methods Enzymol.* 47, 431–442, 1977.

230. Welsch, D.J. and Nelsestuen, G.L., Irreversible degradation of histidine-96 of prothrombin fragment 1 during protein acetylation: Another unusually reactive site in the kringle, *Biochemistry* 27, 7513–7518, 1988.
231. Glocker, M.O., Kalkum, M., Yamamoto, R., and Schreurs, J., Selective biochemical modification of function residues in recombinant human macrophage colony-stimulating factor beta (rhM-CSFbeta): Identification by mass spectrometry, *Biochemistry* 35, 14625–14633, 1996.
232. Kalkum, M., Prxybylski, M., and Glocker, M.O., Structural characterization of functional histidine residues and carbethoxylated derivatives in peptides and proteins by mass spectrometry, *Bioconjug. Chem.* 9, 226–235, 1998.
233. Dage, J.L., Sun, H., and Halsall, H.B., Determination of diethyl pyrocarbonate-modified amino acid residues in alpha-1-acid glycoprotein by high-performance liquid chromatography electrospray ionization mass spectrometry and matrix-assisted laser desorption/ionization time-of-flight mass spectrometry, *Anal. Biochem.* 257, 176–185, 1998.
234. Krell, T., Chackrawarthy, S., Pitt, A.R. et al., Chemical modification monitored by electrospray mass spectrometry: A rapid and simple method for identifying and studying functional residues in enzymes, *J. Pept. Res.* 51, 201–209, 1998.
235. Qin, K.,Yang, Y., Mastrangelo, P., and Westaway, D., Mapping Cu(II) binding sites in prion protein by diethyl pyrocarbonate modification of matrix-assisted laser desorption time-of-flight (MALDI-TOF) mass spectrometric footprinting. *J. Biol. Chem.* 277, 1981–1990, 2002.
236. Ginotra, Y.P. and Kulkarni, P.P., Solution structure of physiological Cu(His)$_2$: Novel considerations into imidazole coordination, *Inorg. Chem.* 48, 7000–7002, 2009.
237. Willard, B.B. and Kintes, M., Effects of internal histidine residues on the collision-induced fragmentation of triply protonated tryptic peptides, *J. Am. Soc. Mass Spectrom.* 12, 1262–1271, 2001.
238. Nakanishi, N., Takeuchi, F., Okamoto, H. et al., Characterization of heme-coordinating histidyl residues of cytochrome b_5 based on the reactivity with diethylpyrocarbonate: A mechanism for the opening of axial imidazole rings, *J. Biochem.* 140, 561–571, 2006.
239. Nakanishi, N., Takeuchi, F., Park, S.Y. et al., Characterization of heme-coordinating histidyl residues of an engineered six-coordinated myoglobin mutant based on the reactivity with diethylpyrocarbonate, mass spectrometry and electron paramagnetic resonance spectroscopy, *J. Biosci. Bioeng.* 105, 604–613, 2008.
240. Konkle, M.E., Eisenheimer, K.N., Hakala, K. et al., Chemical modification of the Rieske protein from *Thermus thermophilus* using diethyl pyrocarbonate modifies ligating histidine 154 and reduces the [2Fe-2S] cluster, *Biochemistry* 49, 7272–7281, 2010.
241. Altman, J., Lipka, J.J., Kuntz, I., and Waskell, L., Identification by proton nuclear magnetic resonance of the histidine in cytochrome b_5 modified by diethyl pyrocarbonate, *Biochemistry* 28, 7516–7523, 1989.
242. Lundblad, R.L., *Chemical Reagents for Protein Modification*, 3rd edn., CRC Press, Boca Raton, FL, 2005.
243. Zalipsky, S., Seltzer, R., and Menon-Rudolph, S., Evaluation of a new reagent for covalent attachment of polyethylene glycol to proteins, *Biotechnol. Appl. Biochem.* 15, 100–114, 1992.
244. Sumiyoshi, H., Shimizu, T., Katoh, M. et al., Solution-phase parallel synthesis of carbamates using polymer-bound N—Hydroxysuccinimide, *Org. Lett.* 4, 3923–3926, 2002.
245. Wang, Y.-S., Youngster, S., Bausch, J. et al., Identification of the major positional isomers of pegylated interferon Alpha-2b, *Biochemistry* 39, 10634–10640, 2000.
246. Wylie, D.C., Voloch, M., Lee, S. et al., Carboxyalkylated histidine is a pH dependent product of pegylation with SC-PEG, *Pharm. Res.* 18, 1354–1360, 2001.
247. Miron, T. and Wilchek, M., A simplified method for the preparation of succinimidyl carbonate polyethylene glycol for coupling to proteins, *Bioconjug. Chem.* 4, 580–589, 1993.
248. Wang, M., Basu, A., Palm, T. et al., Engineering an arginine catabolizing bioconjugate: Biochemical and pharmaceutical characterization of PEGylated derivatives of arginine deiminase from *Mycoplasma arthritidis*, *Bioconjug. Chem.* 17, 1447–1459, 2005.
249. Lee, B.K., Kwon, J.S., Kim, H.J. et al., Solid-phase PEGylation of recombinant interferon α-2b for site-specific modification: Process performance, characterization, and in vitro bioactivity, *Bioconjug. Chem.* 18, 1728–1734, 2007.
250. Borukhov, S.I. and Strongin, A.Ya., Chemical modification of the recombinant human αA- and β-interferons, *Biochem. Biophys. Res. Commun.* 167, 74–80, 1990.
251. Tsan, M.F., Inhibition of neutrophil sulfhydryl groups by chloromethyl ketones. A mechanism for their inhibition of superoxide production, *Biochem. Biophys. Res. Commun.* 112, 671–677, 1983.
252. Hartley, B.S. and Massey, V., Active centre of chymotrypsin I. Labeling with a fluorescent dye, *Biochim. Biophys. Acta* 21, 58–70, 1956.

253. Tamura, Z., Nakajima, T., Nakayama, T. et al., Identification of peptides with 5-dimethylaminonaphthalenesulfonyl chloride, *Anal. Biochem.* 52, 595–606, 1973.

254. Nishino, T., Massey, V., and Williams, C.H., Jr., Chemical modifications of d-amino acid oxidase. Evidence for active site histidine, tyrosine, and arginine residues, *J. Biol. Chem.* 255, 3610–3615, 1980.

255. Swenson, R.P., Williams, C.R., Jr., and Massey, V., Identification of the histidine residue in D-amino acid oxidase that is covalently modified during inactivation with 5-dimethylaminonapthalene-1-sulfonyl chloride, *J. Biol. Chem.* 258, 497–502, 1983.

256. Gadda, G., Beretta, G.L., and Pilone, M.S., Reactivity of histidyl residues in D-amino acid oxidase from *Rhodotorula gracilis*, *FEBS Lett.* 363, 307–310, 1995.

257. Morishima, I., Shiro, Y., Adachi, S., Yano, Y., and Orii, Y., Effect of the distal histidine modification (cyanation) of myoglobin on the ligand binding kinetics and the heme environmental structures, *Biochemistry* 28, 7582–7586, 1989.

258. Shiro, Y. and Morishima, I., Modification of the heme distal side chain in myoglobin by cyanogen bromide. Heme environmental structures and ligand binding properties of the modified myoglobin, *Biochemistry* 23, 4879–4884, 1984.

259. Tanguchi, I., Sonoda, K., and Mie, Y., Electroanalytical chemistry of myoglobin with modification of distal histidines by cyanated imidazole, *J. Electronanalyt. Chem.* 468, 9–16, 1999.

260. Oomwn, R.P. and Kaplan, H., Competitive labeling as an approach to defining the binding surfaces of proteins: Binding of monomeric insulin to lipid bilayers, *Biochemistry* 26, 303–308, 1987.

261. Hefford, M.A. and Kaplan, H., Chemical properties of the histidine residue of secretin: Evidence for a specific intramolecular interaction, *Biochim. Biophys. Acta*, 998, 262–270, 1989.

262. Kaplan, H. and Oda, G., Selective isolation of free and blocked amino-terminal peptides from enzymatic digestion of proteins, *Anal. Biochem.* 132, 384–388, 1983.

263. Shatiel, S., Thiolysis of some dinitrophenyl derivatives of amino acids, *Biochem. Biophys. Res. Commun.* 29, 178–183, 1967.

264. Uchida, K., 4-Hydroxy-2-nonenal: A product and mediator of oxidative stress, *Prog. Lipid Res.* 42, 318–343, 2003.

265. Hidalgo, F.J., Alaiz, M., and Zamero, R., A spectrophotometric method for the determination of proteins damaged by oxidized lipids, *Anal. Biochem.* 262, 129–136, 1998.

266. Refsgaard, H.H.F., Tsai, L., and Stadtman, E.R., Modification of proteins by polyunsaturated fatty acid peroxidation products, *Proc. Natl. Acad. Sci. USA* 97, 611–616, 2000.

267. Oe, T. et al., A novel lipid peroxide-derived cyclic covalent modification to histone H4. *J. Biol. Chem.* 278, 42098–42105, 2003.

268. Zhu, X., Anderson, V.E., and Sayre, L.M., Charge-derivatized amino acids facilitate model studies on protein side-chain modifications by matrix-assisted laser desorption/ionization time-of-flight mass spectrometry, *Rapid Commun. Mass Spectrom.* 23, 2113–2124, 2009.

269. Grimsrud, P.A., Xie, H., Griffin, T.J., and Bernlohn, D.A., Oxidative stress and covalent modification of protein with bioactive aldehydes, *J. Biol. Chem.* 283, 21837–21841, 2008.

270. Fenaiile, F., Tabet, J.C., and Guy, P.A., Identification of 4-hydroxy-2-nonenal-modified peptides within unfractionated digests using matrix-assisted laser desorption/ionization time-of-flight mass spectrometry, *Anal. Chem.* 76, 867–873, 2004.

271. Raunlyar, N., Prokal-Tatrai, K., and Prokai, L., Identification of carbonylation sites in apomyoglobin after exposure to 4-hydroy-2-nonenal by solid-phase enrichment and liquid chromatography-electrospray ionization tandem mass spectrometry, *J. Mass Spectrom.* 45, 398–410, 2010.

272. Fenaille, F., Guy, P.A., and Tabet, J.-C., Study of protein modification by 4-hydroxy-2-nonenal and other short chain aldehydes analyzed by electrospray ionization tandem mass spectrometry, *J. Am. Soc. Mass Spectrom.* 14, 215–226, 2003.

273. Toyokuni, S., Miyake, N., Hiai, H. et al., The monoclonal antibody specific for the 4-hydroxy-2-nonenal histidine adduct, *FEBS Lett.* 359, 189–191, 1995.

274. Uchida, K., Itakura, K., Kawakishi, S. et al., Characterization of epitopes recognized by 4-hydroxy-2-nonenal specific antibodies, *Arch. Biochem. Biophys.* 324, 241–248, 1995.

275. Ozeki, M., Miyagawa-Hayshino, A., Akatsuka, S. et al., Susceptibility of actin to modification by 4-hydroxy-2-nonenal, *J. Chromatogr. B* 827, 119–126, 2005.

276. Waeg, G., Dimsity, G., and Esterhaven, H., Monoclonal antibodies for detection of 4-hydoxy-nonenal modified proteins, *Free Radic. Res.* 25, 149–159, 1996.

277. Jerfy, A. and Roy, A.B., Sulfatase of ox liver. XII. Effects of tyrosine and histidine reagents on the activity of sulfatase A, *Biochim. Biophys. Acta* 175, 355–364, 1969.

278. MacDonald, J.M., Hass, A.L., and London, R.K., Novel mechanism of surface catalysis of protein adduct formation. NMR studies of the acetylation of ubiquitin, *J. Biol. Chem.* 275, 31908–31913, 2000.
279. Moore, G.J., Kinetics of acetylation-deacetylation of angiotensin II. Intramolecular interactions of the tyrosine and histidine side-chains, *Int. J. Pept. Protein Res.* 26, 469–481, 1985.
280. Kinnunen, P.M., DeMichele, A., and Lange, L.G., Chemical modification of acyl-CoA:cholesterol *O*-acyltransferase. 1. Identification of acyl-CoA:cholesterol *O*-acyltransferase subtypes by differential diethyl pyrocarbonate sensitivity, *Biochemistry* 27, 7344–7350, 1988.
281. Johnson, A.R. and Dekker, E.E., Woodward's Reagent K inactivation of *Escherichia coli* L-threonine dehydrogenase: Increased absorbance at 340–350 nm is due to modification of cysteine and histidine residues, not aspartate or glutamate carboxyl groups, *Protein Sci.* 5, 382–390, 1996.
282. Bustos, P. et al., Woodward's Reagent K reacts with histidine and cysteine residues in *Escherichia coli* and *Saccharomyces cerevisiae* phosphoenolpyruvate carboxykinase, *J. Protein Chem.* 15, 467–472, 1996.
283. Paoli, P. et al., Mechanism of acylphosphatase inactivation by Woodward's Reagent K, *Biochem. J.* 328, 855–861, 1997.

15 Modification of Arginine

Arginyl residues account for approximately 5% of the amino acid composition in proteins[1,2] with the percentage higher in cationic proteins.[3,4] Arginine is an essential amino acid to the young but not adult animals[5] but is used extensively in nutritional supplements[6-8] and appears to be useful in wound healing.[9,10] Arginine peptides have antimicrobial activity[11-13] and are also useful for transfection and the delivery of RNA-based therapeutic products.[14-17] Arginine is a precursor of nitric oxide,[18,19] and arginine-containing peptides are precursors of citrulline-containing peptides, which are biomarkers in inflammatory arthritis.[20,21] I have not found any evidence for the actual participation of arginine in catalysis, but it is critical in binding sites.[22,23] The best-known binding function is in the RGD (arginine–glycine–glutamic acid) sequence,[24-27] and the presence of arginine is recognized in anion-binding sites in proteins.[28-32]

The chemical modification of arginyl residues in proteins presents a considerable challenge. The average pKa for the guanidino group of arginine is 12.48 (Table 1.1), making it the most basic group in a protein and, as such, a poor nucleophile. As with other amino acids in proteins with functional groups, arginyl residues are subject to posttranslational modification. The methylation of arginine resulting in three different derivative forms (Figure 15.1)[33,34] is of considerable importance in the control or protein–protein interactions[35] and protein–nucleic acid interactions.[36] Arginyl residues are subject to glycation (Figure 15.1) with metabolites such as glyoxal[37] and methylglyoxal.[38] Chetyrkin and coworkers[39] observed that the nonoxidative addition of glucose to proteins (lysozyme and the RDG–α3NCI domain of collagen IV; see Chapter 13 for the discussion of the reaction of glucose with amino groups in proteins) did not affect activity, while reaction with glucose under oxidative conditions, which produces glyoxal and methylglyoxal, did result in inactivation. Arginine undergoes reversible phosphorylation (Figure 15.1) in process considered to be an important regulatory event.[40,41] There is early work on the nitration of arginine residues in proteins with nitric acid in fuming sulfuric acid,[42] and nitro group was used to protect the guanidine group during peptide synthesis,[43] but other derivatives have been the subject of recent study.[44] Nitroarginine is an inhibitor of nitric oxide synthesis[45] and, as the methyl esters (L-NAME), has in vivo physiological effect.[46] Kethoxal is a chemical used as a footprinting reagent for RNA[47-50] where it modifies guanine residues.[49-51] The antiviral activity of kethoxal dates at least to 1959,[52] and there was some early clinical use of this chemical.[53,54] Delihas and coworkers[55] demonstrated the kethoxal formed a stable complex with proteins in *E. coli* ribosomes. These investigators also demonstrated reaction with bovine serum albumin. Subsequent work from this laboratory[56] showed that kethoxal reacted with arginyl residues. Somewhat later, Iijima and coworkers[57] reported the reaction of kethoxal with amino acids and pancreatic ribonuclease. The primary site of reaction was arginine; there was slow reaction with primary amines including the ε-amino group of lysine. Reaction with N^α-arginine resulted in multiple products, which were stable at pH 7 and reactive with periodate; 15% of the modified arginine reverts to arginine on acid hydrolysis (6 N HCl/110°C/22 h). The extent of modification depended on solvent with the greatest extent of modification in water with decreasing modification in cacodylate and acetate with the lowest degree of modification in phosphate buffer. The degree of loss of activity also varied with solvent and pH of assay. Aggregation of RNase was also observed on reaction with kethoxal. In the most recent work, Akinsiku and coworkers[58] used MS to determine the structure of the adduct formed between kethoxal and arginine and the borate complex (Figure 15.2). I could find no current use of kethoxal for the modification of proteins.

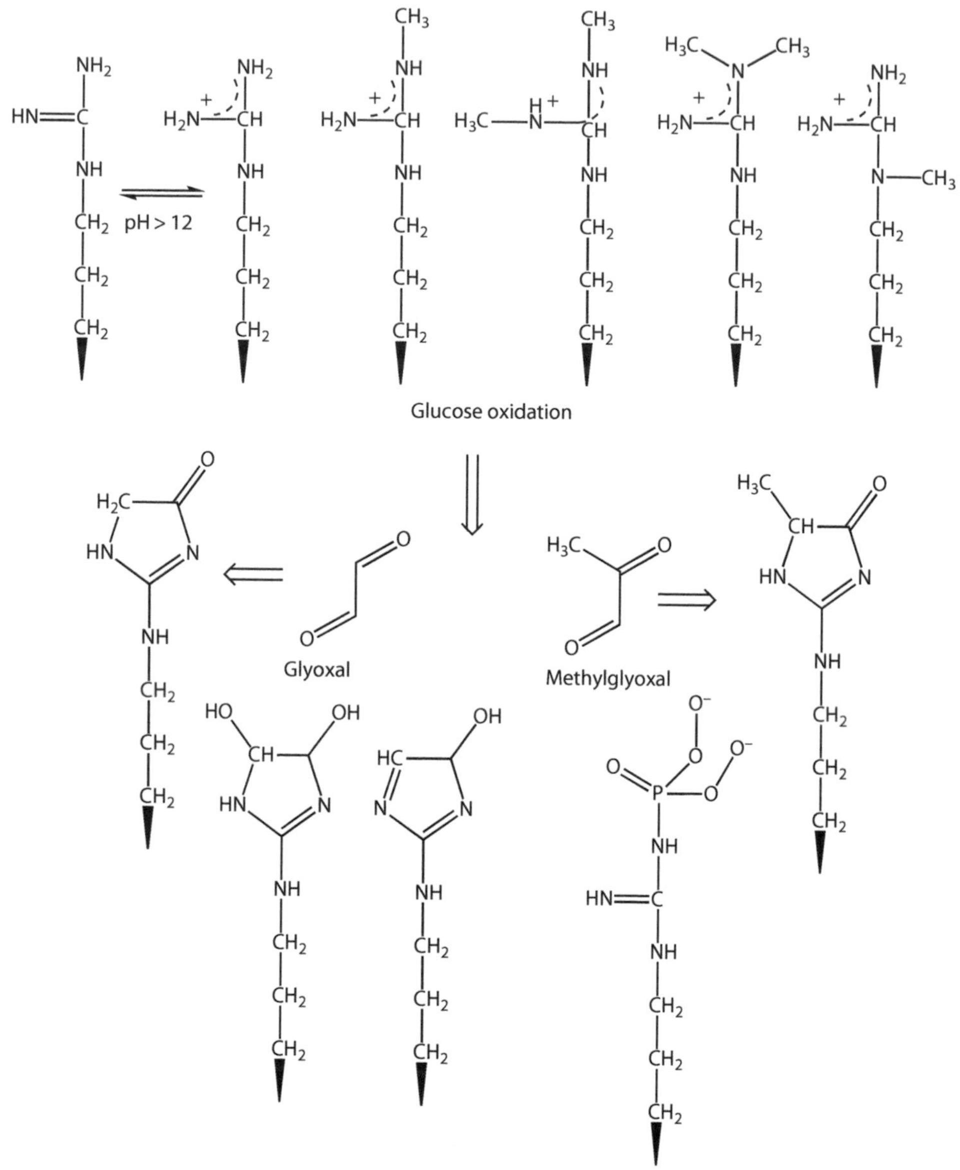

FIGURE 15.1 Some posttranslational modifications of arginine. Shown at the top right are the tautomeric forms of arginine. Given the high pKa value, the unprotonated form is rare. At the top right are the various forms of methylated arginine including monomethylarginine (ω), symmetrical dimethylarginine, and unsymmetrical dimethylarginine. Also shown is the other monomethyl derivative (δ), which is not as common (Bedford, M.T. and Clarke, S.G., Protein Arginine methylation in mammals: Who, what, and why, *Mol. Cell.* 33, 1–13, 2009; Le, D.D., Cortesi, A.T., Myers, S.A. et al., Site-specific and regiospecific installation of methylarginine analogues into recombinant histones and insights into effector protein binding, *J. Am. Chem. Soc.*, 135, 2879–2882, 2013). Shown below are the products from the glycation of arginine with glyoxal and methylglyoxal, which arise from the oxidation of glucose (Chetyrkin, S., Mathis, M., Pedchenko, V. et al., Glucose autoxidation induces functional damage to proteins via modification of critical arginine residues, *Biochemistry* 50, 6102–6112, 2011). The hydrated forms are also shown (Lopez-Clavijo, A.F., Barrow, M.P., Rabbani, N. et al., Determination of types and binding sites of advanced glycation end products for substance P, *Anal. Chem.* 84, 10568–10575, 2012) as is the structure of phosphorylated arginine (Ilg, T. and Werr, M., Arginine kinase of the sheep blowfish *Lucilia cuprina*: Gene identification and characterization of the native and recombinant enzyme, *Pestic. Biochem. Physiol.* 102, 115–123, 2012).

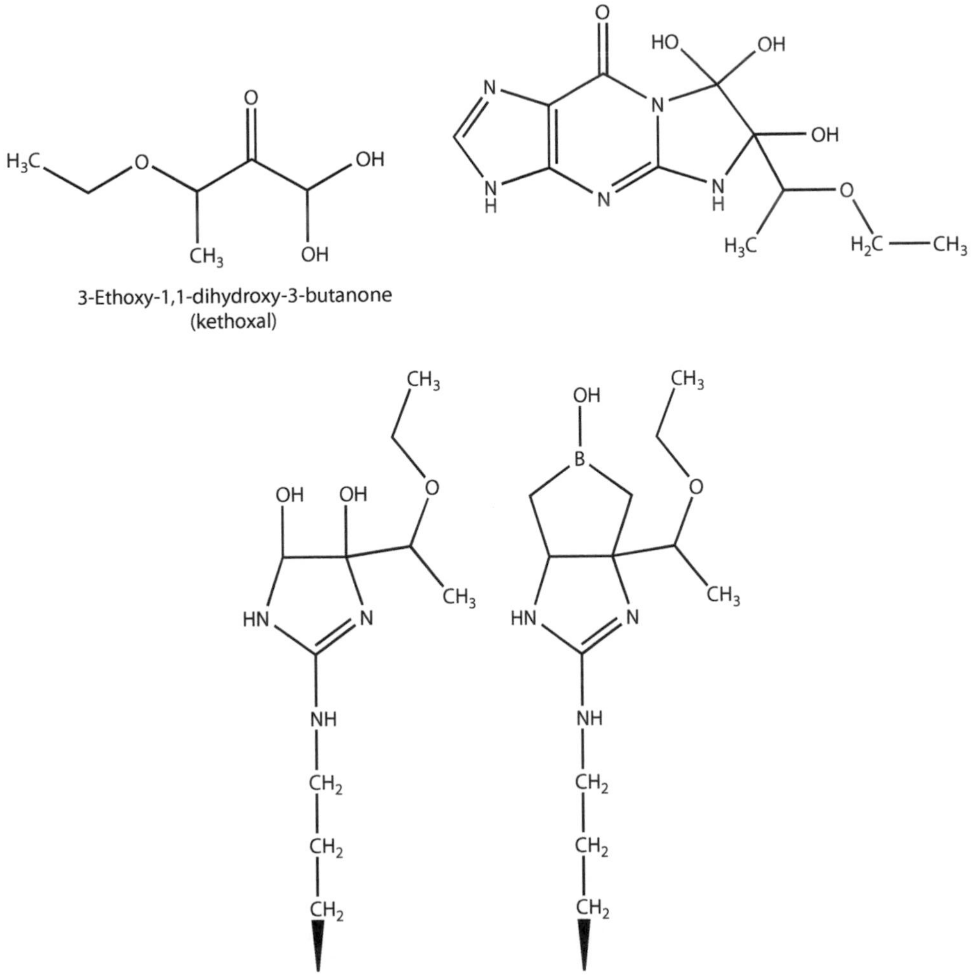

3-Ethoxy-1,1-dihydroxy-3-butanone
(kethoxal)

FIGURE 15.2 The reaction of kethoxal with guanine and arginine. Shown is the structure of kethoxal (3-ethoxy-1,1-dihydroxy-3-butanone; β-ethoxy-α-ketobutyraldehyde); the hydrate form is shown. Also shown is the structure of the product obtained between the guanine base and kethoxal (Shapiro, R., Cohen, B.I., Shiuey, S.J., and Maurer, H., On the reaction of guanine with glyoxal, pyruvaldehyde, and kethoxal, and the structure of the acylguanines. A new synthesis of N^2-alkylguanines, *Biochemistry* 8, 238–245, 1969). At the bottom is the suggested structure of the adduct formed between kethoxal and arginine and the borate complex (Akinsiku, O.T., Yu, E.T., and Fabria, D., Mass spectrometric investigation of protein alkylation by the RNA footprinting probe kethoxal, *J. Mass Spectrom.* 40, 1372–1381, 2005).

There has been limited work on the modification of arginine in proteins. The majority of past approaches to the site-specific modification of arginyl residues in proteins have used three reagents: phenylglyoxal (and derivatives such as *p*-hydroxyphenylglyoxal),[59] 2,3-butanedione,[60] and 1,2-cyclohexanedione.[61] The determination of the extent of arginine modification can be determined by amino acid analysis after acid hydrolysis, but conditions generally need to be modified to prevent loss of the arginine derivative.[61] While amino acid analysis is mentioned in several papers, details are not presented regarding the results. However, more frequently, such analysis is totally ignored in more recent publications. Many of the studies are qualitative in nature, and such data are not required. Chromogenic reagents such as *p*-hydroxyphenylglyoxal and 4-hydroxy-3-nitrophenylglyoxal are described in the following. MS has been of value in identifying the modification of arginine in proteins.[62–68]

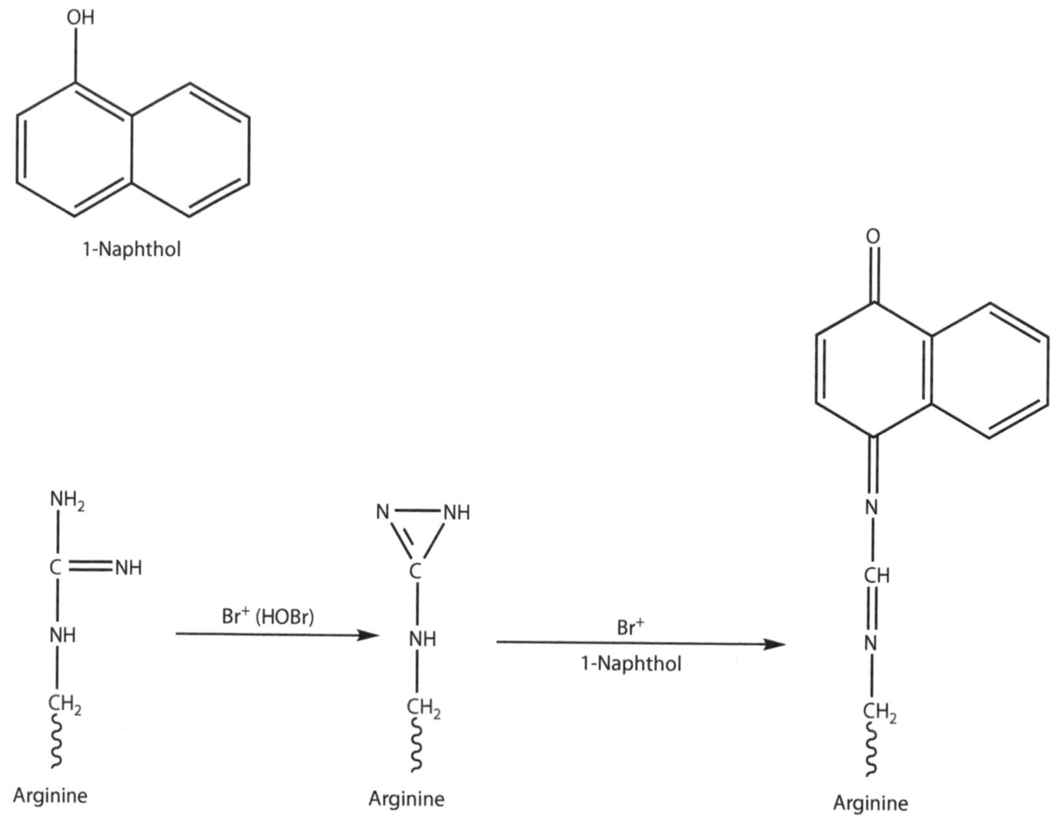

FIGURE 15.3 The Sakaguchi reaction for determination of arginine. The Sakaguchi reaction used reaction with hypobromite followed by addition of a chromogen such as 1-naphthol. The color product is not stable, and a suggested structure is shown in this figure (Casdebaig, F., Mesnard, P., and Devaux, G., Mise au point. La réaction de sakaguchi, *Bull. Soc. Pharm. Bordeaux* 114, 82–94, 1976; Dupin, J.-P., Casadebaig, F., Pometan, J.-P. et al., La réaction de Sakaguchi, *Anns. Pharmaceut. Francoises* 37, 523–530, 1979).

There is one method for the accurate determination of arginine, which is of both historical interest and current practical value. The Sakaguchi reaction (Figure 15.3) is based on the reaction of α-naphthol with guanidino compounds in the presence of hypohalite to yield a red product.[69] The Sakaguchi reaction[70,71] has been used, after acid hydrolysis, to determine the extent of arginine modification by 2,3-butanedione and 1,2-cyclohexanedione in rat liver microsomal stearoyl-coenzyme A desaturase[72] and the extent of arginine modification by 4-hydroxybenzil in gelatin.[73] This modification is part of a program for the development of novel protein–polymer grafts.[74] It is noted that the modification occurs under rigorous (at least for proteins) conditions (70% ethanolic NaOH at 23°C). It has also been used for the assay of arginine in food products.[75,76] It has been suggested[77] that the enzymatic assay for arginine is more valuable than the Sakaguchi reaction. Other applications of Sakaguchi chemistry have been in histochemical studies[78,79] and for the assay of guanidine compounds (guanidinoacetic acid) in screening for hepatic guanidinoacetate-methyltransferase activity.[80,81] There are two fundamental problems with the Sakaguchi reaction. The first is the short life of the reaction product, which can be addressed, in part, by the substitution of hypobromite for hypochlorite and the addition of urea to remove excess hypobromite.[82] Linearity is another issue[83] complicated by a lack of complete understanding of the reaction. A fluorometric method for the determination of arginine using 9,10-phenanthrenequinone[84] has been described. This method is some 1000-fold more sensitive than the Sakaguchi reaction, but some concern remains concerning the absolute accuracy of the reagent for the determination of arginine in peptide linkage. This reagent has also been used to study the modification of arginine in proteins by methylglyoxal,[85,86] glycation,[86,87] and oxidation.[86]

The use of phenylglyoxal (Figure 15.4) was developed by Takahashi[59] in 1968 and has since been applied to the study of the role of arginyl residues in proteins as shown in Table 15.1. As with other modifying agents, there has been an increased use of phenylglyoxal to study complex systems.[62,88–107]

It is extremely difficult to consider these studies to be useful for protein chemistry as there is usually insufficient analytical information to support the conclusions. However, with cautious interpretation, these studies can be useful. It is possible that the use of multiple arginine reagents might prove instructive. Many of these studies fail to recognize that phenylglyoxal, like glyoxal, will react with ε-amino groups at an appreciable rate and polymerization was noted in a sample of bovine pancreatic ribonuclease incubated with phenylglyoxal for 21 h (pH 8.0/25°C).[59] As established by Takahashi, the stoichiometry of the reaction involves the reaction of 2 mol of phenylglyoxal with 1 mol of arginine. The amino-terminal lysine residue was also rapidly modified under these conditions yielding the α-keto acid. The possible effect of light on the reaction of phenylglyoxal with arginine, as has been reported for 2,3-butanedione,[108–112] has not been studied. Phenylglyoxal is a substrate for liver alcohol dehydrogenase yielding α-hydroxyacetophenone providing a method for assaying for the purity of phenylglyoxal.[113]

Borders and coworkers[114] have reported the synthesis of a chromophoric derivative, 4-hydroxy-3-nitrophenylglyoxal (Figure 15.4). The adduct between 4-hydroxy-3-nitrophenylgyloxal and arginine absorbs light at 316 nm ($\varepsilon = 1.09 \times 10^4\,M^{-1}\,cm^{-1}$). The derivative is unstable to acid hydrolysis (6 N HCL, 110°C, 24 h) but can be stabilized by the inclusion of thioglycolic acid. Fresh solutions of the reagent should be used to avoid problems associated with the decomposition of the reagent. This same group subsequently used this reagent to identify the reactive arginine in yeast Cu, Zn superoxide dismutase.[115] The reaction of 4-hydroxy-3-nitrophenylglyoxal (50 mM bicine, 100 mM $NaHCO_3$, pH 8.3) with yeast Cu, Zn superoxide dismutase is slower ($0.57\,M^{-1}\,min^{-1}$) than that observed with phenylglyoxal ($28\,M^{-1}\,min^{-1}$). A similar difference in the rate of reaction with the two reagents was observed with creatinine kinase.[114,116] 4-Hydroxy-3-nitrophenylglyoxal has been used to measure creatinine.[117] This reagent is used infrequently but is useful.[118–124]

There are a number of examples of the use of phenylglyoxal for a variety of proteins presented in Table 15.1. A consideration of these various studies clearly indicates that the reaction of phenylglyoxal with proteins is more complex than the model presented by Kenji Takahashi in 1968. I had the good fortune of working in the same laboratory with Kenji in the 1960s at Rockefeller University. Following several major contributions in New York, Kenji returned to a distinguished academic career in Japan. The model proposed by Takahashi involved the sequential addition of two molecules to one molecule of arginine (Figure 15.4) in a reversible reaction. The various studies described in Table 15.1 suggest that (1) there is a product formed with one phenylglyoxal and one arginine, which may not be reversible, and (2) low concentrations of borate may stabilize the product of the reaction in a manner similar to that observed for 2,3-butanedione. While there are no solid data, there are studies that can be interpreted in differences in the reaction phenylglyoxal and 2,3-butanedione with arginyl residues in proteins; steric issues may preclude the reaction of phenylglyoxal, especially in reactions where the second molecule of phenylglyoxal is required for a stable product.

p-Hydroxyphenylglyoxal (Figure 15.4) was developed by Yamasaki and colleagues[125] for the detection of available arginine residues in proteins. As with phenylglyoxal, it reacts with arginine under mild conditions (pH 7–9, 25°C, 30–60 min). The concentration of the resulting adduct (2:1 stoichiometry) with arginine can be determined at 340 nm ($\varepsilon = 1.83 \times 10^4\,M^{-1}\,cm^{-1}$). The modification is slowly reversed under basic conditions. The general characteristics of the reaction of *p*-hydroxyphenylglyoxal are similar to those described for phenylglyoxal. Most studies on the modification of arginine in proteins have used multiple reagents, but there are a few studies that have used on *p*-hydroxyphenylglyoxal. These are presented in Table 15.2.

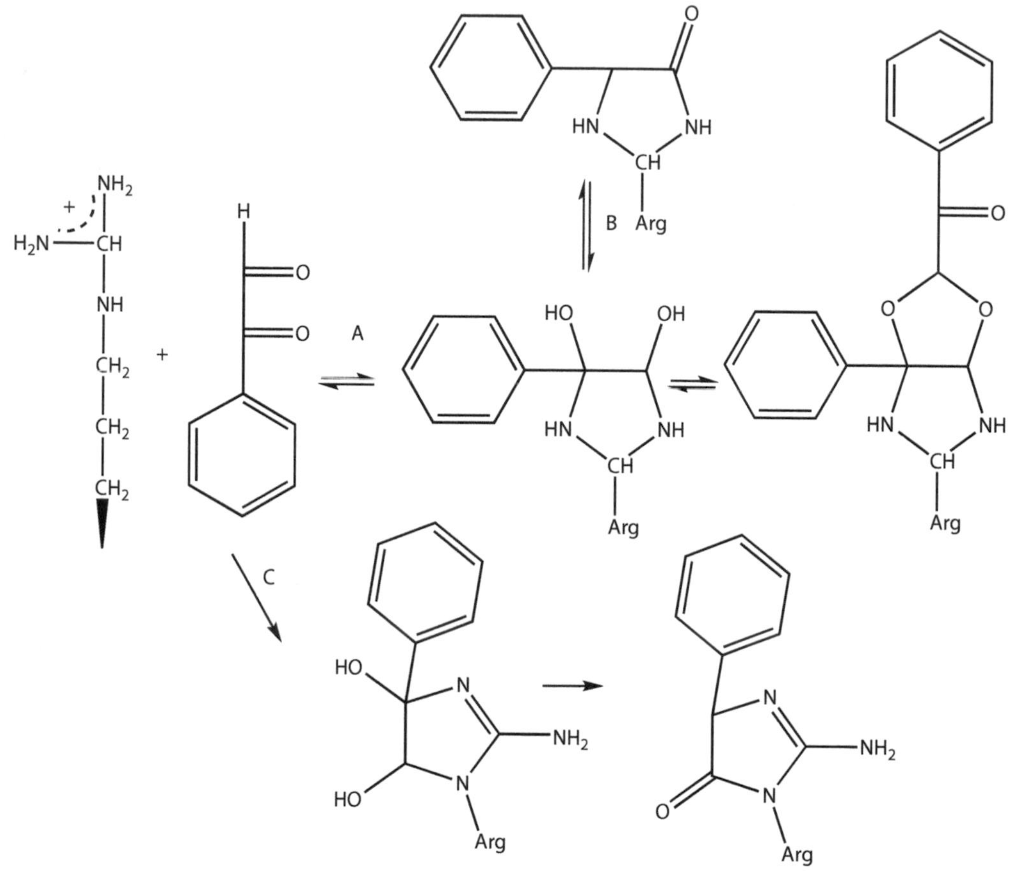

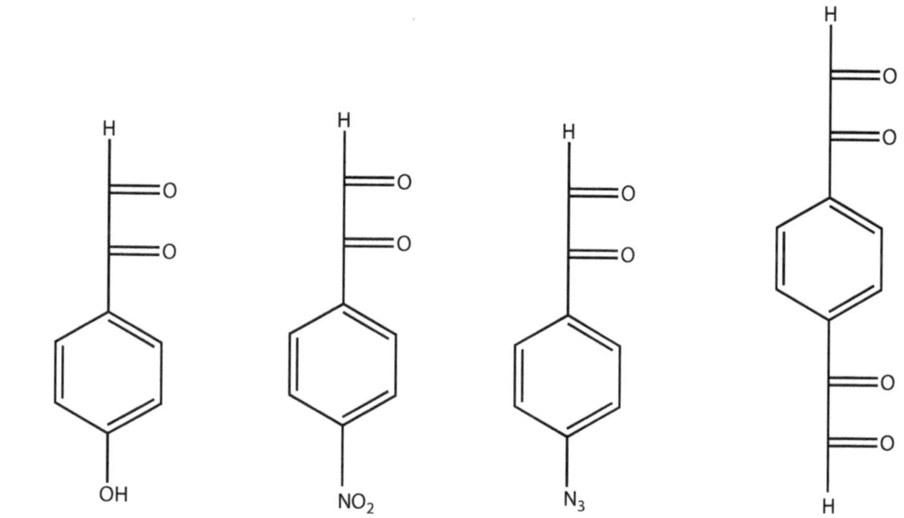

FIGURE 15.4 (continued)

There have been several studies that have compared p-hydroxyphenylglyoxal and phenylglyoxal. It can be argued that p-hydroxyphenylglyoxal is more hydrophilic than phenylglyoxal. Béliveau and coworkers[126,127] showed that the rate of inactivation of phosphate transport by rat kidney brush-border membranes was more rapid ($k_{obsvd}=0.052$ s^{-1}) with phenylglyoxal than with p-hydroxyphenylglyoxal ($k_{obsvd}=0.012$ s^{-1}). The rate of inactivation of glucose uptake by phenylglyoxal was slower than that for phosphate. Reaction order suggested that the inactivation is associated with a modification of a single arginine residue. Mukouyama and colleagues[128] have compared the modification of an L-phenylalanine oxidase from *Pseudomonas* sp. P-501 by phenylglyoxal and p-hydroxyphenylglyoxal. The rate for the phenylglyoxal modification of a single essential arginine residue was 10.6 M^{-1} min^{-1}, while the rate observed for the modification by 4-hydroxy-3-nitrophenylglyoxal was 15.1 M^{-1} min^{-1}. The most remarkable observation on differences between p-hydroxyphenylglyoxal and phenylglyoxal comes from studies by Linder and coworkers.[89] These investigators observed that the treatment of mitochondria with phenylglyoxal (10 mM HEPES, 250 mM sucrose, 10 mM succinate, 100 μM EGTA, 3 μM rotenone, pH 8.0) results in the closing of the permeability pore, while reaction with p-hydroxyphenylglyoxal results in pore opening.

The reaction of arginine with phenylglyoxal is greatly accelerated in bicarbonate–carbonate buffer systems.[129] The reaction of methylglyoxal with arginine is also enhanced by bicarbonate, while a similar effect is not seen with either glyoxal or 2,3-butanedione. The molecular basis for this specific buffer effect is not clear at this time, nor is it known whether reaction with α-amino functional groups occurs at a different rate than with other solvent systems used for this modification of arginine with phenylglyoxal. Yamasaki and coworkers[132] reported that p-nitrophenylglyoxal reacts with arginine in 0.10 M sodium pyrophosphate—0.15 M sodium ascorbate, pH 9.0 (30 min at 30°C), to yield a derivative that absorbs at 475 nm. Reaction also occurs with histidine under the same reaction conditions providing a product with approximately 20% of the absorbance of arginine.

FIGURE 15.4 (continued) Phenylglyoxal and analogs for modification of arginine residues in proteins. The reaction of phenylglyoxal with arginine is shown at the top. Pathway A shows the original description of the reaction of phenylglyoxal with arginine as a two-step process with a reversible reaction of one molecule of phenylglyoxal to complex the guanidino function of arginine followed by the reversible addition of a second molecule of phenylglyoxal to yield a product that is stable at mild acid pH but decomposes at neutral/basic pH with the recovery of free arginine. (Takahashi, K., The reaction of phenylglyoxal with arginine residues in proteins, *J. Biol. Chem.* 243, 6171–6179, 1968; see also Riordan, J.F., Arginyl residues and anion binding sites in proteins, *Mol. Cell. Biochem.* 26, 71–92, 1979). Modification of arginine in proteins shows variable recovery of arginine suggesting that there may other products with different stability characteristics (Branlant, G., Tritsch, D., and Biellman, J.-F., Evidence for the presence of anion-recognition sites in pig-liver aldehyde reductase. Modification by phenyl glyoxal and p-carboxyphenyl glyoxal of an arginyl residue located close to the substrate-binding site, *Eur. J. Biochem.* 116, 505–512, 1981). Single substituted products resulting from the reaction of phenylglyoxal and arginine have been described, and pathway B shows a pathway suggested on the basis of mass spectrometric data (Krell, T., Pitt, A.R., and Coggins, J.R., The use of electrospray mass spectrometry to identify an essential arginine residue in type II dehydroquinases, *FEBS Lett.* 360, 93–96, 1995). Pathway C shows a potential pathway yielding a different initial product; this pathway is based on work with methylglyoxal and arginine (Klöpfer, A., Spanneberg, R., and Glomb, M.A., Formation of arginine modifications in a model system of N^α-tert-butyloxycarbonyl (Boc)-arginine with methylglyoxal, *J. Agric. Food Chem.* 59, 394–401, 2011). 4-Hydroxy-3-nitrophenylglyoxal is a chromophoric derivative of phenylglyoxal (Borders, C.L., Jr., Pearson, L.J., McLaughlin, A.E. et al., 4-Hydroxy-3-nitrophenylglyoxal. A chromophoric reagent for arginyl residues in proteins, *Biochim. Biophys. Acta* 568, 491–495, 1979) as is p-hydroxyphenylglyoxal (Eur, H.-M. and Miles, E.W., Reaction of phenylglyoxal and (p-hydroxyphenyl)glyoxal with arginine and cysteines in the α subunit of tryptophan synthase, *Biochemistry* 23, 6484–6491, 1984). The azidophenyl derivative (Politz, S.M., Noller, H.F., and McWhirter, P.D., Ribonucleic acid-protein cross-linking in *Escherichia coli* ribosomes: (4-azidophenyl)glyoxal, a novel heterobifunctional reagent, *Biochemistry* 20, 372–378, 1981) and 1,4-phenyldiglyoxal (Zhang, Q., Crosland, E., and Fabris, D., Nested arg-specific bifunctional crosslinkers for MS structural analysis of proteins and protein assemblies, *Anal. Chim. Acta* 627, 117–128, 2008) are cross-linking agents based on phenylglyoxal.

TABLE 15.1
Reaction of Phenylglyoxal with Proteins

Protein	Solvent	Reagent Excess[a]	Residues Modified	References
Pancreatic RNase	0.1 M N-ethylmorpholine acetate, pH 8.0, 25°C	—	2–3/4[b,c]	15.1.1
Porcine carboxypeptidase B	0.3 M borate, pH 7.9, 37°C, 1 h	200[d]	1[e]	15.1.2
Aspartate transcarbamylase	0.125 M potassium bicarbonate, pH 8.3, or 0.1 M N-ethylmorpholine, pH 8.3 at 25° for 1 h	—	2.2[f]	15.1.3
Pyruvate kinase	0.1 M triethanolamine, pH 7.0, 37°C	—	3–6[g]	15.1.4
Horse liver alcohol dehydrogenase	0.125 M sodium bicarbonate, pH 7.9, 25°C	—	2 per subunit[h]	15.1.5, 15.1.6
Mitochondrial ATPase	0.097 M sodium borate, 0.097 M EDTA, pH 8.0, 30°C	111	4[i]	15.1.7
Porcine heart adenylate kinase	0.1 M triethanolamine·HCl, pH 7.0, 37°C	—	1/11[j]	15.1.8, 15.1.9
Rhodospirillum rubrum chromatophores	0.05 M borate, pH 7.8, 25°C	—	—[k]	15.1.10
Erythrocyte (Ca^{2+}+ Mg^{2+}) ATPase	20 mM imidazole (pH 7.4) with 3.0 mM MgCl$_2$ and 100 μM CaCl$_2$ (final pH of reaction mixture was 7.17 at 37°C)	—	—[l]	15.1.11
Ribulose bisphosphate carboxylase (*Rhodospirillum rubrum* and spinach enzymes)	0.066 M sodium carbonate, 0.050 M bicine, 0.1 M EDTA, pH 8.0[m], 30°C (*R. rubrum*), and 21°C (spinach)	—	2–3[n]	15.1.12
Yeast hexokinase	0.035 M veronal, pH 7.5	—	1/18[o]	15.1.13
Superoxide dismutase	0.125 M sodium bicarbonate, pH 8.0, 25°C	—	1/4[p]	15.1.14
Myosin (subfragment 1)	0.1 M potassium bicarbonate, pH 8.0, 15°C or 25°C	—	1.7/35[q]	15.1.15
Thymidylate synthetase	0.125 M bicarbonate, pH 8.0, 25°C[r]	—	1.8/12	15.1.16
Glutamate decarboxylase	0.125 M sodium bicarbonate, pH 7.5, 23°C	300	1/23[s]	15.1.17
Cardiac myosin S1	0.1 M N-ethylmorpholine acetate, pH 7.6, 25°C	—	2.8/42[t]	15.1.18
Fatty acid synthetase	0.1 M sodium phosphate, 0.0005 M dithioerythritol, 0.001 M EDTA, pH 7.6, 30°C	—	4/106[u]	15.1.19
Yeast inorganic pyrophosphatase	0.08 M N-ethylmorpholine acetate, pH 7.0, 35°C	—	1/6[v]	15.1.20, 15.1.21
Porcine phospholipase A	0.125 M potassium bicarbonate, pH 8.5	—	1.4/4[w]	15.1.22
Superoxide dismutase[x]	0.100 M sodium bicarbonate, pH 8.3, 25°C	50–100	0.88/4.0[y]	15.1.23, 15.1.24[z]
p-Hydroxybenzoate hydroxylase	0.050 M potassium phosphate, pH 8.0, 25°C	250	2–3/24[aa]	15.1.25

TABLE 15.1 (continued)
Reaction of Phenylglyoxal with Proteins

Protein	Solvent	Reagent Excess[a]	Residues Modified	References
Thymidylate synthetase	0.200 M N-ethylmorpholine, pH 7.4, 25°C[bb]	65	2/12[cc]	15.1.26
Acetylcholine esterase	0.025 M borate, 0.005 phosphate, 0.050 M NaCl, pH 7.0, 25°C	—	3/31[dd]	15.1.27
D-β-Hydroxybutyrate dehydrogenase	0.05 M HEPES, pH 7.5, 25°C	—	—[ee]	15.1.28
Ornithine transcarbamylase	0.05 M bicine, 0.1 M KCl, 0.0001 M EDTA, pH 8.05, 25°C	—	—[ff]	15.1.29
Coenzyme B_{12}-dependent diol dehydrase	0.05 M borate, pH 8.0, 25°C	—	—[gg]	15.1.30
Transketolase	0.125 M sodium bicarbonate, pH 7.6, 22°C	—	4/34[hh]	15.1.31
Pig-liver aldehyde reductase	20 mM sodium phosphate, pH 7.0, 30°C	—	0.6/16[ii]	15.1.32
ATP citrate lyase	0.050 M HEPES, pH 8.0, 30°C[jj]	—	8.5/40	15.1.33
Rat liver S-adenosyl-homocysteinase	50 mM potassium pyrophosphate, pH 8.5, 25°C	—	2/14[kk]	15.1.34
Pyridoxamine-5'-phosphate oxidase	0.1 M potassium phosphate, pH 8.0, containing 5% ETOH, 25°C	—	4/40[ll]	15.1.35
Acetate kinase	0.050 M triethanolamine, pH 7.6, 25°C	—	—[mm]	15.1.36
Pancreatic phospholipase A_2	0.2 M N-ethylmorpholine, pH 8.0, 23°C	30	1.0–1.2[nn]	15.1.37
L-Lactate dehydrogenase	50 mM lutidine, pH 8.4, 23°C	—	1[oo]	15.1.38
Choline acetyltransferase	50 mM HEPES, pH 7.8, 25°C	—	—[pp]	15.1.39
ADP-glucose synthetase	0.05 M potassium phosphate, 0.25 mM EDTA, pH 7.5, 25°C	110	1[qq]	15.1.40
Pyruvate oxidase (E. coli)	0.1 M sodium phosphate, 0.010 M magnesium chloride, pH 7.8 (temperature not provided)	—	2.5/5[rr]	15.1.41
Carbon monoxide	20 mM sodium phosphate, pH 8.2 with 4 mM dithiothreitol, 25°C	—	6/68[ss]	15.1.42
Phosphomannose isomerase (Candida albicans)	50 mM HEPES, pH 8.5, 37°C	—	1.2[tt]	15.1.43
1,4,5-trisphosphate 3-kinase	50 mM HEPES, 0.1 M EDTA, 12 mM 2-mercaptoethanol, 10 mM $MgCl_2$, pH 7.4, 23°C	—	1.2[uu]	15.1.44
Type II dehydroquinases	100 mM sodium bicarbonate, pH 9.4, 25°C	—	1[vv]	15.1.45

(continued)

TABLE 15.1 (continued)

Reaction of Phenylglyoxal with Proteins

Protein	Solvent	Reagent Excess[a]	Residues Modified	References
Maize branching enzymes	50 mM triethanolamine, pH 8.5, 25°C	—	~1[ww]	15.1.46
Nitrate reductase in whole cells of *Paracoccus denitrificans*	0.1 M sodium phosphate, pH 7.3 of 0.1 M sodium borate, pH 7.3	—	1–2[xx]	15.1.47
K+-dependent leucine transport in brush-border membrane	10 mM HEPES–Tris, pH 7.4 with 100 mM mannitol	—	—[yy]	15.1.48

[a]   Reagent/protein.

[b]   After 3 h at 25°C.

[c]   There was also modification of α-amino groups and lysine residues in pancreatic ribonuclease under these reaction conditions. Deamination of the α-amino group was observed as a general side reaction, while reaction at the ε-amino group of lysine (assessed by loss of lysine on amino acid analysis following acid hydrolysis) was observed after a long period (21 h) of reaction.

[d]   Mole of reagent per mole of arginine.

[e]   Loss of activity was associated with modification of 1 mol of arginine per mole of protein (amino acid analysis and ^{14}C-labeled reagent incorporation). It was observed that while modification with phenylglyoxal resulted in the loss of peptidase activity, there was a stimulation of esterase activity to a maximum at 200-fold molar excess of reagent; esterase activity decreased at higher phenylglyoxal concentration.

[f]   Arginine residues modified per 50 kDa (one regulatory chain and one catalytic chain) after 3 h at 25°C. Approximately one-third of the modification is on the catalytic chain and two-thirds on the regulatory chain. The extent of modification was assessed by both amino acid analysis and incorporation of ^{14}C-labeled reagent.

[g]   The extent of modification depends on reagent concentration. ATP protects one arginine from modification; the modified protein is not as soluble as the native protein. It is noted that phenylglyoxal is considered to be a hydrophobic reagent.

[h]   Inactivation was also observed with 2,3-butanedione in the presence of borate. The derivative with 2,3-butane is stable in the presence of borate but not stable for the isolation of modified peptide. The availability of ^{14}C-labeled phenylglyoxal permitted the identification of one modified arginine residue (Arg84); the second arginine derivative was not sufficiently stable for isolation of the modified peptide.

[i]   Thirty minutes of reaction at 30°C; the presence of efrapeptin, a low-molecular-weight antibiotic, which is a potent inhibitor of oxidative phosphorylation, prevented the modification of one *fast-reacting* arginyl residue.

[j]   A double-logarithmic plot of reagent concentration versus $t_{1/2}$ yielded a straight line consistent with the modification of a single arginine residue responsible for the loss of catalytic activity (Levy, H.M., Leber, P.D., and Ryan, E.M., Inactivation of myosin by 2,4-dinitrophenol and protection by adenosine triphosphate and other phosphate compounds, *J. Biol. Chem.* 238, 3654–3659, 1963). There is a bit of problem with this conclusion as the product of the reaction of phenylglyoxal with arginine involves 2 mol of phenylglyoxal.[15.1.1] The reaction of phenylglyoxal with arginine is a two-step process with the initial formation of a stoichiometric adduct between the guanidino group of arginine and phenylglyoxal resulting in blocking arginine function. This comment is applicable to other references using the Levy plot. Subsequent work on the reaction of phenylglyoxal on adenylate cyclase showed similar reaction characteristics for adenylate cyclase from rabbit muscle and pig skeletal muscle. It also shown that Arg97 was the residue modified by phenylglyoxal in pig heart adenylate cyclase.

[k]   Work was performed with chromatophores isolated from *Rhodospirillum rubrum* cells. The rate of reaction with either phenylglyoxal or 2,3-butanedione was determined either by loss of photophosphorylation activity or loss of Mg-ATPase activity. The second-order rate constant for loss of photophosphorylation (3.4 M^{-1} min^{-1} for phenylglyoxal; 1.2 M^{-1} min^{-1} for 2,3-butanedione) was similar to that observed for loss of Mg-ATPase activity (1.6 M^{-1} min^{-1} for phenylglyoxal; 0.6 M^{-1} min^{-1} for 2,3-butanedione). A double-logarithmic plot of reagent concentration versus first-order rate constant provided a line with a slope near unity for either reagent and either measured activity consistent with the modification of a single arginine residue as responsible for the loss of activity (Levy, H.M., Leber, P.D., and Ryan, E.M., Inactivation of myosin by 2,4-dinitrophenol and protection by adenosine triphosphate and other phosphate compounds, *J. Biol. Chem.* 238, 3654–3659, 1963).

TABLE 15.1 (continued)
Reaction of Phenylglyoxal with Proteins

l   The rate of reaction of phenylglyoxal with human erythrocyte plasma membrane was measured by loss of $Ca^{2+}+Mg^{2+}$-ATPase. A double-logarithmic plot of reagent concentration versus the reciprocal of reaction half-life yielded a line with a slope close to unity consistent with the modification of a single arginine residue responsible for the loss of enzymatic activity (Levy, H.M., Leber, P.D., and Ryan, E.M., Inactivation of myosin by 2,4-dinitrophenol and protection by adenosine triphosphate and other phosphate compounds, *J. Biol. Chem.* 238, 3654–3659, 1963). Reaction with intact cells did not result in the loss of enzymatic activity suggesting that phenylglyoxal does not cross the membrane or, since ATP prevents inactivation by phenylglyoxal, intracellular ATP concentrations are high enough to block the reaction in intact erythrocytes.

m   Solvent made metal free using Bio-Rad Chelex; reaction performed with and without $MgCl_2$.

n   There are 35 arginine residues in spinach enzyme protomer and 29 arginine residues in the *R. rubrum* enzyme protomer. Analysis showed the modification of two to three arginine residues per protomer for either enzyme showing that a small proportion of the total arginine residues are modified with the loss of activity. The second-order rate constant for phenylglyoxal with the *R. rubrum* enzyme is 0.31 M^{-1} s^{-1} at 21°C and 2.5 M^{-1} s^{-1} for the spinach enzyme; 5.1 M^{-1} s^{-1} at 30°C. A double-logarithmic plot of reagent concentration versus first-order rate constant yielded a line with a slope close to unity consistent with the modification of a single arginine residue as responsible for the loss of activity (Levy, H.M., Leber, P.D., and Ryan, E.M., Inactivation of myosin by 2,4-dinitrophenol and protection by adenosine triphosphate and other phosphate compounds, *J. Biol. Chem.* 238, 3654–3659, 1963). Amino acid analysis demonstrated that phenylglyoxal is specific for the modification of arginine.

o   Partial reactivation of modified enzyme was observed reflecting lability of modified arginine residues. Reaction also shows saturation kinetics reflecting *specific* affinity of reagent for enzyme possibly from hydrophobic interaction. These authors suggest that this phenomenon is observed with the reaction of other hydrophobic reagents with this enzyme. A similar phenomenon has been observed with a number of other reagents used for the chemical modification of proteins.

p   Modification with phenylglyoxal caused a loss of Cu^{2+}. The second-order rate constant for reaction with phenylglyoxal was 2.5 M^{-1} min (0.125 M sodium bicarbonate) and 4.0 M^{-1} min^{-1} (50 mM sodium borate, pH 9.0), both reactions at 25°C. There was some loss of the phenylglyoxal label on reduction and carboxymethylation of the modified enzyme; these investigators pursued the isolation of a peptide obtained from the 2,3-butanedione/borate modification and demonstrated modification of Arg141. While glyoxal resulted in the modification of both lysine and arginine, reaction of superoxide dismutase with phenylglyoxal, 2,3-butanedione, and 1,2-cyclohexanedione resulted in modification of only arginine.

q   The rate of reaction of phenylglyoxal with the S1 fragment of myosin measured by loss of K-ATPase activity. Saturation kinetics were observed consistent with affinity labeling as has been with phenylglyoxal and other proteins (see footnote o). The rate of reaction is biphasic consistent with the modification of two arginyl residues at different rates. There is an initial stimulation of Ca^{2+}-ATPase and Mg^{2+}-ATPase on reaction with phenylglyoxal.

r   Rates of enzyme inactivation were dependent upon buffer; at 5.9 mM phenylglyoxal, the following data were obtained: bicarbonate ($t_{1/2}=6.0$ min), MOPS ($t_{1/2}=11.5$ min), borate ($t_{1/2}=34.0$ min), and phosphate ($t_{1/2}=48.0$ min).

s   This study demonstrated the pronounced effect of bicarbonate on the reaction of phenylglyoxal with arginyl residues in proteins. The reaction of the holoenzyme was slow with either phenylglyoxal (0.59 M^{-1} min^{-1}, 125 mM bicarbonate, pH 7.5 at 23°C) or 2,3-butanedione (0.09 M^{-1} min^{-1} in 50 mM sodium borate, pH 7.5 at 23°C). The rate of reaction was much faster (108 M^{-1} min^{-1} for phenylglyoxal; 0.7 M^{-1} min^{-1} for 2,3-butanedione) with the apoenzyme under the same reaction conditions. The rate of reaction of phenylglyoxal was most rapid with bicarbonate buffer; progressively slower rates were found with *N*-ethylmorpholine, borate, and veronal.

t   Initial studies with 5 mM phenylglyoxal in 100 mM potassium bicarbonate, 0.5 M KCl, pH 7.8, at 22°C showed modification of four arginine residues in 15 s without loss of activity; modification of 12 arginine residues resulted in 50% loss of activity. Reaction with 5.0 mM phenylglyoxal in 0.1 M *N*-ethylmorpholine acetate, pH 7.8, with 0.2 mM dithiothreitol and 1 mM EDTA at 25°C for 10 min resulted in the modification of two to three arginine residues. It is suggested that the modification of four arginine residues would result in total inactivation; modification of more than 70% of the arginine residues resulted in gelation of the myosin fragment.

u   Complete inactivation was obtained with 5 mM phenylglyoxal, but the reaction did not follow first-order kinetics. Since it was possible that the inactivation of fatty acid synthetase by phenylglyoxal reflected modification of a binding site for NADPH, it was reasonable to see if NADPH would protect from inactivation. This was not possible as the NADPH concentration was continually changing reflecting the reduction of phenylglyoxal by the enzyme. The activity of horse liver

(continued)

TABLE 15.1 (continued)
Reaction of Phenylglyoxal with Proteins

alcohol dehydrogenase in reducing the α-ketoaldehydes is used as a spectrophotometric assay for the α-ketoaldehydes (Yang, C.F. and Brush, E.J., A spectrophotometric assay for α-ketoaldehydes using horse liver alcohol dehydrogenase, *Anal. Biochem.* 214, 124–127, 1993). The reaction of fatty acid synthetase with phenylglyoxal results in the initial modification of approximately two arginyl residues with no loss of activity, while the subsequent modification of two additional arginyl residues results in the complete loss (by extrapolation) of fatty acid synthase activity, ketoacyl reductase activity, and enoyl reductase activity. It is remarkable that only 4 of 104 arginyl residues are modified by phenylglyoxal in the fatty acid synthase from goose uropygial gland or in fatty acid synthase from chicken liver (Vernon, C.M. and Hsu, R.Y., The presence of essential arginine residues at the NADPH-binding sites of β-ketoacyl reductase and enoyl reductase domains of the multifunctional fatty acid synthetase in chicken liver, *Biochim. Biophys. Acta* 788, 124–131, 1984). It is suggested that the modification of arginine by phenylglyoxal inhibits the binding of the 2′-phosphate of NADPH.

[v] The second-order rate constant for the inactivation of inorganic pyrophosphatase by phenylglyoxal was found to be 0.82 M^{-1} min^{-1} at pH 7.0 (25°C), which is faster than the value of 0.15 M^{-1} min^{-1} found for the reaction of phenylglyoxal with free arginine or ribonuclease (Takahashi, K., The reaction of phenylglyoxal with arginine residues in proteins, *J. Biol. Chem.* 243, 6171–6179, 1968). The enhanced reactivity of arginyl residues with dicarbonyl compounds has been suggested to reflect a decreased pKa for the guanidino function (see Patthy, L. and Thész, J., Origin of the selectivity of α-dicarbonyl reagents for arginyl residues of anion-binding sites, *Eur. J. Biochem.* 105, 387–393, 1980). Investigations into the stability of phenylglyoxal–arginine adducts showed markedly increased stability at pH < 5. With this information, a peptide containing the modified arginine residue was isolated, and the site of modification established as Arg77.

[w] Determined at 99% inactivation (25°C) of phospholipase activity (release of fatty acid from egg yolk in water with 3 mM $CaCl_2$ and 1.4 mM sodium deoxycholate). These investigators did examine the possibility of amino-terminal alanine modification; no loss of alanine was observed with 75% inactivation (0.9 mol Arg modified per mole of protein), while enzyme samples with a greater extent of inactivation did have some loss of amino-terminal alanine (quantity not given). These investigators did examine the pH dependence of enzyme inactivation by phenylglyoxal (presumably, a direct measure of the rate of arginine modification) and reported the following second-order rate constants (M^{-1} min^{-1}): pH 6.5, 0.3; pH 7.5, 1.5; pH 8.5, 3.3; and pH 9.5, 3.9. These investigators also showed that phenylglyoxal ($t_{1/2}$ = 1 min) was more effective than 2,3-butanedione ($t_{1/2}$ = 20 min) and 1,2-cyclohexanedione ($t_{1/2}$ = 120 min).

[x] Cu, Zn superoxide dismutase from *Saccharomyces cerevisiae*.

[y] Determined at 80% loss of enzymatic activity using reaction of the modified enzyme with 9,10-phenanthrenequinone. This value corresponded to that determined by the incorporation of radiolabeled phenylglyoxal assuming 2:1 adduct. Amino acid analysis with samples prepared using normal hydrolytic conditions (6 N HCl, 110°C, 20 h) suggested only approximately 50% of this extent of arginine modification. When thioglycolic acid was included during the hydrolysis, values for the extent of arginine modification approached those determined by the fluorescence technique and radiolabel incorporation.

[z] The study uses reaction with 4-hydroxy-3-nitrophenylglyoxal, a chromophoric derivative of phenylglyoxal, to identify the specific arginine residue modified. It is of some interest that the rate of reaction with this derivative is approximately sixfold less than that with the parent phenylglyoxal.

[aa] Reaction was performed at 25°C for 60–120 min. Loss of lysine residues was not observed under these reaction conditions. Amino acid analysis (hydrolysis in 6 N HCl, 110°C, 24 h) correlated well with radiolabeled phenylglyoxal incorporation assuming 2:1 stoichiometry (i.e., amino acid analysis gave 3.6 mol Arg loss per mole of enzyme, while 7.54 mol radiolabel was incorporated).

[bb] Second-order rate constant of 32 M^{-1} min^{-1}. An approximate 100-fold increase in the rate of inactivation is observed at pH 8.4. Reaction at pH > 8.0 was characterized by a rapid first phase and a slower second phase, while reaction at pH 7.4 was characterized by a pseudo-first-order process.

[cc] The presence of substrate, 2′-deoxyuridylate, prevents the modification of 1 mol of arginine per mole of enzyme. Amino acid analysis showed only modification of arginine residues in the protein. The presence of a chaotropic agent, sodium perchlorate (2.4 M), resulted in a 20% loss of activity and an increase in the extent of arginine modification at 50% inactivation.

[dd] The modification with phenylglyoxal is associated with an approximate 15% loss of enzyme activity. Treatment with 2,3-butanedione under similar reaction conditions results in the modification of approximately one more mole of arginine per mole enzyme with an approximate 75% loss of catalytic activity.

[ee] Stoichiometry was not established, but the data are consistent with the loss of activity resulting from the modification of a single arginine residue. Submitochondrial vesicles were used as the source of enzyme in these studies. A second-order rate constant of 1.03 M^{-1} min^{-1} was obtained from the reaction with phenylglyoxal. A value of 0.8 M^{-1} min^{-1} was obtained

TABLE 15.1 (continued)
Reaction of Phenylglyoxal with Proteins

for reaction with 1,2-cyclohexanedione (0.050 M borate, pH 7.5), while a value of 4.6 $M^{-1} min^{-1}$ was obtained for 2,3-butanedione in the borate buffer system.

[ff] See more complete discussion of this study under Table 15.2. For inactivation by phenylglyoxal, a second-order rate constant of 56 $M^{-1} min^{-1}$ was obtained at pH 8.04. The reactions were performed in the dark.

[gg] A double-logarithmic plot of first-order rate constant versus phenylglyoxal concentration suggests that the inactivation is the result of modification of a single arginine residue. These investigators obtained similar results with 2,3-butanedione (50 mM potassium borate, pH 8.5, 25°C). The enzyme was protected by phosphate from inactivation by phenylglyoxal but not from inactivation by 2,3-butanedione.

[hh] The extent of modification was assessed by incorporation of radiolabeled phenylglyoxal and amino acid analysis. A lower limit of four modified arginyl residues was established. Amino acid analysis also showed that arginine was the only amino acid modified with phenylglyoxal. The use of a Tsou plot (see Wang, H.R., Bai, J.H., Zheng, S.Y. et al., Ascertaining the number of essential thiol groups for the folding of creatine kinase, *Biochem. Biophys. Res. Commun.* 221, 174–180, 1996) that uses the relationship between extent of modification and loss of activity to determine the number of essential groups in the modified protein suggests the importance of a single arginine residue. The inactivation of transketolase by phenylglyoxal was irreversible; there is example of reversibility of arginine modification by phenylglyoxal under mild conditions.

[ii] A second-order rate constant of $k = 2.6 \ M^{-1} min^{-1}$ was found for the reaction of phenylglyoxal with an arginine residue in pig liver aldehyde reductase assuming that the loss of activity seen with phenylglyoxal directly reflects the loss of an arginine residue. A double-logarithmic plot of first-order rate constant versus phenylglyoxal concentration provided a line with a slope of 0.93 suggesting the modification of a single arginine residue is responsible for the loss of enzymatic activity. These investigators also used *p*-carboxyphenyl glyoxal, which demonstrated saturation kinetics. The extent of modification was assessed by both amino acid analysis and radiolabel incorporation.

[jj] Most studies were performed in this solvent at 30°C with a second-order rate constant of 0.33 $M^{-1} s^{-1}$. The rate was reduced in potassium phosphate ($k = 0.25 \ M^{-1} s^{-1}$) and borate ($k = 0.078 \ M^{-1} s^{-1}$). These investigators also used 2,3-butanedione in 50 mM sodium borate, pH 8.0 (30°C), and obtained a second-order rate constant of 0.042 $M^{-1} s^{-1}$.

[kk] A plot of loss of activity versus the square of phenylglyoxal concentration provided a straight line. The use of ^{14}C-labeled phenylglyoxal showed the incorporation of 4.2 mol/mol protein (2 mol/mol subunit). The reaction was readily reversed at 50% inactivation to 95% of the original activity (pH 8/25°C/3 h).

[ll] The rate of inactivation at 25°C for the apoenzyme was determined to be 3.7 and 11.1 $M^{-1} min^{-1}$ for the holoenzyme.

[mm] Saturation kinetics are observed with phenylglyoxal, suggesting the formation of an enzyme–inhibitor complex prior to reaction with an arginine residue. Inactivation was associated with the reaction of 1 mol phenylglyoxal per mole of protein. These investigators note that the initial reaction of phenylglyoxal with arginine to form the glyoxaline ring is sufficient to inactivate the enzyme.

[nn] This study shows that this level of arginine modification is associated with 80% loss of amino-terminal alanine. It was necessary to protect the α-amino group of the amino-terminal alanine with a *t*-butyloxycarbonyl group to avoid modification with phenylglyoxal. Reaction of phenylglyoxal with the amino-terminal-blocked enzyme resulted in the incorporation of 2 mol of phenylglyoxal (^{14}C-labeled reagent) at Arg6. Removal of the amino-terminal blocking agent yielded a modified enzyme with 30%–38% of original enzymatic activity. Reaction of the enzyme with 1,2-cyclohexanedione also modified Arg6 but with little loss of enzymatic activity.

[oo] One arginine is modified (total of 26 arginine residues in L-lactate dehydrogenase) with the incorporation of 2 mol of phenylglyoxal. A double-logarithmic plot of rate of inactivation versus phenylglyoxal concentration yielded a straight line with a slope of 1.9 consistent with the reaction of 2 mol of phenylglyoxal per mole of protein. The inactivation by phenylglyoxal was not reversed by removal of phenylglyoxal by gel filtration.

[pp] Phenylglyoxal was much more effective than 2,3-butanedione or camphorquinone-10-sulfonic acid.

[qq] Phenylglyoxal is much more effective than 1,2-cyclohexanedione and also enhanced specificity of modification. Determination of stoichiometry from inactivation kinetics suggested reaction of 1 mol phenylglyoxal per mole protein; determination of stoichiometry by incorporation of ^{14}C-labeled reagent showed incorporation of 2 mol of phenylglyoxal per mole of protein. These investigators also note that the addition of one molecule of phenylglyoxal with arginine forming a glyoxaline ring is sufficient for inactivation with a second molecule of phenylglyoxal adding to form the final phenylglyoxal–arginine complex but not contributing to the observed effect on enzyme activity.

[rr] Five phenylglyoxal molecules are incorporated per subunit in the initial rapid phase with a slower phase with the incorporation of another five molecules per subunit. In the presence of thiamine pyrophosphate, labeling is reduced

(continued)

TABLE 15.1 (continued)
Reaction of Phenylglyoxal with Proteins

by 0.7 phenylglyoxal molecules per subunit. Kinetic evidence suggests a 1:1 stoichiometry for enzyme inactivation. There are clearly at least two classes of reactive arginine residue. When the reaction is performed at pH 6.0, inactivation with phenylglyoxal can be partially reversed on dilution in pH 6.0 buffer. The rate of reaction at pH 6.0 is much slower than the reaction at 7.8. The reaction at pH 7.8 shows two phases demonstrated by both loss of activity and incorporation of ^{14}C-labeled reagent. It is suggested that there are two independent mechanisms of phenylglyoxal inactivation of enzyme activity and that inactivation is a result of the formation of a 1:1 adduct of phenylglyoxal and arginine. The inactivation with phenylglyoxal at pH 6.0 is partially reversed after a 16-fold dilution in 0.1 M phosphate buffer, pH 6.0, containing magnesium chloride, sodium pyruvate, and thiamine pyrophosphate.

[ss] Two arginine residues are protected from modification by the presence of coenzyme A. All 68 residues are modified in a tryptic digest of the protein. The extent of arginine modification was determined by the use of *p*-nitrophenylglyoxal (Yamasaki, R.B., Shimer, D.A., and Feeney, R.E., Colorimetric determination of arginine residues in proteins by *p*-nitrophenylglyoxal, *Anal. Biochem.* 111, 220–226, 1981).

[tt] The pH optimum for the reaction of phenylglyoxal with phosphomannose isomerase was 9.1. The second-order rate constant for the reaction of phenylglyoxal with phosphomannose isomerase at pH 8.5 (50 mM HEPES) at 37°C was determined to be 22 M^{-1} min^{-1}. Partial recovery of activity (50%) was obtained by transfer into 100 mM Tris, pH 8.0, or 100 mM ammonium bicarbonate (4 h at 37°C). Arginine 304 was identified as the site of modification.

[uuu] Phenylglyoxal of 1.2 mol is incorporated into a mole of protein at 100% inactivation. It is suggested that the reaction of phenylglyoxal with the arginine residue in 1,3,5-trisphosphate 3-kinase is a two-step process involving binding of phenylglyoxal prior to reaction with an arginyl residue. Structural analysis demonstrated the modification of Arg317.

[vv] A second-order rate constant of 89 M^{-1} min^{-1} was derived from the loss of activity in the *Streptomyces coelicolor*, while a second-order rate constant of 150 M^{-1} min^{-1} was obtained for the enzyme from *Aspergillus nidulans* (both in 100 mM sodium bicarbonate, pH 9.4, 25°C). The extent of modification was determined by MS. A peptide with a mass gain of +116 was obtained consistent with a 1:1 adduct of phenylglyoxal and arginine together with minor fragments, one of which contained the disubstituted arginine (Figure 15.1).

[ww] Data consistent with the modification of a single arginyl residue were obtained for the several enzymes in this study. There are two observations that are of interest. The first is the lack of an effect of sodium bicarbonate on the rate inactivation of branching enzyme. The other is that there was no effect of light on the reaction of phenylglyoxal with branching enzyme. The inactivation of the enzymes by phenylglyoxal was irreversible.

[xx] The slope of line obtained from double-logarithmic plot of reaction rate versus phenylglyoxal concentration gives value dependent on solvent. A slope of 1.40 was obtained with 0.1 M sodium phosphate, pH 7.3, while a slope of 0.87 is obtained with 0.1 sodium borate, pH 7.3. While interpretation of the double-logarithmic plot is not absolute, it has been useful with assessing stoichiometry. In addition, this is a whole-cell system, and the Racker principle ("don't waste clean thinking on dirty enzymes") is operative. That said, it is tempting to speculate that the borate buffer drives formation of a 1:1 phenylglyoxal–arginine complex and a 2:1 phenylglyoxal complex in the presence of sodium phosphate buffer. There is at least one example of a positive effect of borate on the reaction of phenylglyoxal with arginine (Parenti, P., Hanozet, G.M., Villa, M., and Giordana, B., Effect of arginine modification on K⁺-dependent leucine uptake in brush-border membrane vesicles from the midgut of *Philosamia cynthia* larvae, *Biochim. Biophys. Acta* 1191, 27–32, 1994).

[yy] The rate of inactivation of K⁺-dependent leucine transport by phenylglyoxal in 10 mM HEPES–Tris, pH 7.3, at 25°C was biphasic (2 M^{-1} min^{-1} for the slow fast and 9.1 M^{-1} min^{-1} for the fast phase). Substitution of a borate buffer (100 mM sodium borate, pH 7.5) resulted in a single-phase reaction with a rate said to be similar to that of the fast phase in the HEPES–Tris buffer.

References to Table 15.1

15.1.1. Takahashi, K., The reaction of phenylglyoxal with arginine residues in proteins, *J. Biol. Chem.* 243, 6171–6179, 1968.
15.1.2. Werber, M.M. and Sokolovsky, M., Chemical evidence for a functional arginine residue in carboxypeptidase B, *Biochem. Biophys. Res. Commun.* 48, 384–390, 1972.
15.1.3. Kantrowitz, E.R. and Lipscomb, W.N., Functionally important arginine residues of aspartate transcarbamylase, *J. Biol. Chem.* 252, 2873–2880, 1977.
15.1.4. Berghäuser, J., Modifizierung von argininresten in pyruvat-kinase, *Hoppe-Seyler's Physiol. Chem.* 358, 1565–1572, 1977.
15.1.5. Lange, L.G., III, Riordan, J.F., and Vallee, B.L., Functional argininyl residues as NADH binding sites of alcohol dehydrogenases, *Biochemistry* 13, 4361–4370, 1974.

TABLE 15.1 (continued)
Reaction of Phenylglyoxal with Proteins

15.1.6. Jörnvall, H., Lange, L.G., III, Riordan, J.F., and Vallee, B.L., Identification of a reactive arginyl residue in horse liver alcohol dehydrogenase, *Biochem. Biophys. Res. Commun.* 77, 73–78, 1977.

15.1.7. Kohlbrenner, W.E. and Cross, R.L., Efrapeptin prevents modification by phenylglyoxal of an essential arginyl residue in mitochondrial adenosine triphosphatase, *J. Biol. Chem.* 253, 7609–7611, 1978.

15.1.8. Berghäuser, J., A reactive arginine in adenylate kinase, *Biochim. Biophys. Acta* 397, 370–376, 1975.

15.1.9. Berghäuser, J. and Schirmer, R.H., Properties of adenylate kinase after modification of Arg-97 by phenylglyoxal, *Biochim. Biophys. Acta* 537, 428–435, 1978.

15.1.10. Vallejos, R.H., Lescano, W.I.M., and Lucero, H.A., Involvement of an essential arginyl residue in the coupling activity of *Rhodospirillum rubrum* chromatophores, *Arch. Biochem. Biophys.* 190, 578–584, 1978.

15.1.11. Raess, B.U., Record, D.M., and Tunnicliff, G., Interaction of phenylglyoxal with the erythrocyte $(Ca^{2+}+Mg^{2+})$ ATPase, *Mol. Pharmacol.* 27, 444–450, 1985.

15.1.12. Schloss, J.V., Norton, I.L., Stringer, C.D., and Hartman, F.C., Inactivation of ribulosebisphosphate carboxylase by modification of arginyl residues with phenylglyoxal, *Biochemistry* 17, 5626–5631, 1978.

15.1.13. Philips, M., Pho, D.B., and Pradel, L.-A., An essential arginyl residue in yeast hexokinase, *Biochim. Biophys. Acta* 566, 296–304, 1979.

15.1.14. Malinowski, D.P. and Fridovich, I., Chemical modification of arginine at the active site of the bovine erythrocyte superoxide dismutase, *Biochemistry* 18, 5909–5917, 1979.

15.1.15. Mornet, D., Pantel, P., Audemard, E., and Kassab, R., Involvement of an arginyl residue in the catalytic activity of myosin heads, *Eur. J. Biochem.* 100, 421–431, 1979.

15.1.16. Cipollo, K.L. and Dunlap, R.B., Essential arginyl residues in thymidylate synthetase from amethopterin-resistant *Lactobacillus casei*, *Biochemistry* 18, 5537–5541, 1979.

15.1.17. Cheung, S.-T. and Fonda, M.L., Kinetics of the inactivation of *Escherichia coli* glutamate apodecarboxylase by phenylglyoxal, *Arch. Biochem. Biophys.* 198, 541–547, 1979.

15.1.18. Morkin, E., Flink, I.L., and Banerjee, S.K., Phenylglyoxal modification of cardiac myosin S-1. Evidence for essential arginine residues at the active site, *J. Biol. Chem.* 254, 12647, 1979.

15.1.19. Poulose, A.J. and Kolattukudy, P.E., Presence of one essential arginine that specifically binds the 2′-phosphate of NADPH on each of the ketoacyl reductase and enoyl reductase active sites of fatty acid synthetase, *Arch. Biochem. Biophys.* 199, 457–464, 1980.

15.1.20. Cooperman, B.S. and Chiu, N.Y., Yeast inorganic pyrophosphatase. III. Active-site mapping by electrophilic reagents and binding measurements, *Biochemistry* 12, 1676–1682, 1973.

15.1.21. Bond, M.W., Chiu, N.Y., and Cooperman, B.S., Identification of an arginine residue important for enzymatic activity within the covalent structure of yeast inorganic pyrophosphatase, *Biochemistry* 19, 94, 1980.

15.1.22. Vensel, L.A. and Kantrowitz, E.R., An essential arginine residue in porcine phospholipase A2, *J. Biol. Chem.* 255, 7306–7310, 1980.

15.1.23. Borders, C.L., Jr. and Johansen, J.T., Essential arginyl residues in Cu, Zn superoxide dismutase from *Saccharomyces cerevisiae*, *Carlsberg Res. Commun.* 45, 185–194, 1980.

15.1.24. Borders, C.L., Jr. and Johansen, J.T., Identification of Arg-143 as the essential arginyl residue in yeast Cu, Zn superoxide dismutase by the use of a chromophoric arginine reagent, *Biochem. Biophys. Res. Commun.* 96, 1071–1078, 1980.

15.1.25. Shoun, H., Beppu, T., and Arima, K., An essential arginine residue at the substrate-binding site of p-hydroxybenzoate hydroxylase, *J. Biol. Chem.* 255, 9319–9324, 1980.

15.1.26. Belfort, M., Maley, G.F., and Maley, F., A single functional arginyl residue involved in the catalysis promoted by *Lactobacillus casei* thymidylate synthetase, *Arch. Biochem. Biophys.* 204, 340–349, 1980.

15.1.27. Müllner, H. and Sund, H., Essential arginine residue in acetylcholinesterase from *Torpedo californica*, *FEBS Lett.* 119, 283–286, 1980.

15.1.28. El Kebbaj, M.S., Latruffe, N., and Gaudemer, Y., Presence of an essential arginine residue in d-β-hydroxybutyrate dehydrogenase from mitochondrial inner membrane, *Biochem. Biophys. Res. Commun.* 96, 1569–1578, 1980.

15.1.29. Marshall, M. and Cohen, P.P., Evidence for an exceptionally reactive arginyl residue at the binding site for carbamyl phosphate in bovine ornithine transcarbamylase, *J. Biol. Chem.* 255, 7301–7305, 1980.

15.1.30. Kuno, S., Toraya, T., and Fukui, S., Coenzyme B12-dependent diol dehydrase: Chemical modification with 2,3-butanedione and phenylglyoxal, *Arch. Biochem. Biophys.* 205, 240–245, 1980.

(continued)

TABLE 15.1 (continued)

Reaction of Phenylglyoxal with Proteins

15.1.31. Kremer, A.B., Egan, R.M., and Sable, H.Z., The active site of transketolase. Two arginine residues are essential for activity, *J. Biol. Chem.* 255, 2405–2410, 1980.

15.1.32. Branlant, G., Tritsch, D., and Biellmann, J.-F., Evidence for the presence of anion-recognition sites in pig-liver aldehyde reductase. Modification by phenyl glyoxal and *p*-carboxyphenyl glyoxal of an arginine located close to the substrate binding site, *Eur. J. Biochem.* 116, 505–512, 1981.

15.1.33. Ramakrishna, S. and Benjamin, W.B., Evidence for an essential arginine residue at the active site of ATP citrate lyase from rat liver, *Biochem. J.* 95, 735–743, 1981.

15.1.34. Takata, Y. and Fujioka, M., Chemical modification of arginine residues of rat liver *S*-adenosylhomocysteinase, *J. Biol. Chem.* 258, 7374–7378, 1983.

15.1.35. Choi, J.-D. and McCormick, D.B., Roles of arginyl residues in pyridoxamine-5′-phosphate oxidase from rabbit liver, *Biochemistry* 20, 5722–5728, 1981.

15.1.36. Wong, S.S. and Wong, L.-J., Evidence for an essential arginine residue at the active site of Escherichia coli acetate kinase, *Biochim. Biophys. Acta* 660, 142–147, 1981.

15.1.37. Fleer, E.A.M., Puijk, W.C., Slotboom, A.J., and DeHaas, G.H., Modification of arginine residues in porcine pancreatic phospholipase A2, *Eur. J. Biochem.* 116, 277–284, 1981.

15.1.38. Peters, R.G., Jones, W.C., and Cromartie, T.H., Inactivation of L-lactate monooxygenase with 2,3-butanedione and phenylglyoxal, *Biochemistry* 20, 2564–2571, 1981.

15.1.39. Mautner, H.G., Pakyla, A.A., and Merrill, R.E., Evidence for presence of an arginine residue in the coenzyme A binding site of choline acetyltransferase, *Proc. Natl. Acad. Sci. USA* 78, 7449–7452, 1981.

15.1.40. Carlson, C.A. and Preiss, J., Involvement of arginine residues in the allosteric activation of *Escherichia coli* ADP-glucose synthetase, *Biochemistry* 21, 1929–1934, 1982.

15.1.41. Koland, J.G., O'Brien, T.A., and Gennis, R.B., Role of arginine in the binding of thiamin pyrophosphate to *Escherichia coli* pyruvate oxidase, *Biochemistry* 21, 2656–2660, 1982.

15.1.42. Shanmugasundaram, T., Kumar, G.K., Shenoy, B.C., and Wood, H.G., Chemical modification of the functional arginine residues of carbon monoxide dehydrogenase from *Clostridium thermoaceticum*, *Biochemistry* 28, 7112–7116, 1989.

15.1.43. Wells, T.N.C., Scully, P., and Magnena, E., Arginine 304 is an active site residue in phosphomannose isomerase from *Candida albicans*, *Biochemistry* 35, 5777–5782, 1994.

15.1.44. Communi, D., Lecocq, R., Vanweyenberg, V., and Erneux, C., Active site of labelling of inositol 1,4,5-trisphosphate 3-kinase A by phenylglyoxal, *Biochem. J.* 310, 109–115, 1995.

15.1.45. Krell, T., Pitt, A.R., and Coggins, J.R., The use of electrospray mass spectrometry to identify an essential arginine residue in type II dehydroquinases, *FEBS Lett.* 360, 93–96, 1995.

15.1.46. Cao, H. and Preiss, J., Evidence for essential arginine residues at the active sites of maize branching enzymes, *J. Protein Chem.* 15, 291–304, 1996.

15.1.47. Kučera, I., Inhibition by phenylglyoxal of nitrate transport in *Paracoccus denitrificans*: A comparison with the effect of a protonophorous uncoupler, *Arch. Biochem. Biophys.* 409, 327–334, 2003.

15.1.48. Parenti, P., Hanozet, G.M., Villa, M., and Giordana, B., Effect of arginine modification on K+-dependent leucine uptake in brush-border membrane vesicles for the midgut of *Philosamia cynthia* larvae, *Biochim. Biophys. Acta* 1191, 27–32, 1994.

There is a slightly stronger response with 1-methylhistidine and no reaction with 3-methylhistidine. There is also reaction with the sulfhydryl group of cysteine or 2-mercaptoethanol yielding a product absorbing at 475 nm but a much weaker response (3% of the absorbance of arginine). Branlant and coworkers[131] have used *p*-carboxyphenyl glyoxal in bicarbonate buffer at pH 8.0 to modify aldehyde reductase. A second-order rate constant of 26 M^{-1} min^{-1} was observed in 80 mM bicarbonate and 2.9 M^{-1} min^{-1} in 20 mM sodium phosphate at pH 7.0. Saturation kinetics was observed with this reagent under certain conditions suggesting that a binding event occurs prior to actual chemical modification. Phenylglyoxal also inactivated this enzyme with a second-order rate constant of 2.6 M^{-1} min^{-1} in 20 mM sodium phosphate at pH 7.0. Inactivation is reversible on dialysis. Further work with *p*-carboxyphenyl glyoxal has not been reported.

TABLE 15.2

Use of *p*-Hydroxyphenylglyoxal to Modify Arginine Residues in Proteins

Protein	Conditions	References
Bacterial peptidyl dipeptiase-4	0.2 M *N*-ethylmorpholine, pH 8.0, 25°C	15.2.1
Recombinant soluble CD4[a]	25 mM NaHCO$_3$, 15 h, 25°C, dark	15.2.2
Amadoriase II from *Aspergillus* sp.	10 mM HEPES, pH 8.0, 23°C, dark	15.2.3
HLA-DR1	25 mM sodium bicarbonate, 25°C, 15 h[b]	15.2.4
Recombinant monoclonal antibody	50 mM HEPES, pH 7.5, 37°C, 2 h[c]	15.2.5
Human anion transporter band 3 protein (red cell ghosts)	Red cell ghosts (sealed and unsealed) in 10 mM sodium phosphate, pH 7.4, or 5 mM HEPES, pH 7.4. Buffers contained 200 mM sucrose, 25 mM gluconate, 25 mM citrate, and 1 mM sodium sulfate at 37°C[d]	15.2.6, 15.2.7

[a] Analysis of modification by MS.
[b] Analysis of modification by MS showed that three arginyl residues were modified in empty DR1 but not in peptide-loaded DR1.
[c] A positively charged exosite was identified using *p*-hydroxyphenylglyoxal as a probe. Differential modification was achieved by modification of the monoclonal antibody bound to a cation-exchange matrix.
[d] Modification was performed with both phenylglyoxal and hydroxyphenylglyoxal. Hydroxyphenylglyoxal is considered a more hydrophilic reagent. The specific modification of one arginine residue, Arg901, is reported for reaction with either reagent.

References to Table 15.2

15.2.1. Lanzillo, J.J., Dascrathy, Y., and Fanburg, B.L., Detection of essential arginine in bacterial peptidyl dipeptidase-4: Arginine is not the anion binding site, *Biochem. Biophys. Res. Commun.* 160, 243–249, 1989.

15.2.2. Hager-Braun, C. and Tomer, K.B., Characterization of the tertiary structure of soluble CD4 bound to glycosylated full-length HIVgp120 by chemical modification of arginine residues and mass spectrometric analysis, *Biochemistry* 41, 1759–1766, 2002.

15.2.3. Wu, X. et al., Alternation of substrate specificity through mutation of two arginine residues in the binding site of amadoriase II from *Aspergillus* sp., *Biochemistry* 41, 4453–5548, 2002.

15.2.4. Carven, G.J. and Stern, L.J., Probing the ligand-induced conformational change in HLA-DR1 by selective chemical modification and mass spectrometric mapping, *Biochemistry* 44, 13625–13637, 2005.

15.2.5. Zhang, L., Lilyestrom, W., Li, C. et al., Revealing a positive charge patch on a recombinant monoclonal antibody by chemical labeling and mass spectrometry, *Anal. Chem.* 83, 8501–8508, 2011.

15.2.6. Takazaki, S., Abe, Y., Kang, D. et al., The functional role of arginine 901 at the C-terminus of the human anion transporter band 3 protein, *J. Biochem.* 139, 903–912, 2006.

15.2.7. Takazaki, S., Abe, Y., Yamaguchi, T. et al., Arg 901 in the AE1 C-terminal tail is involved in conformational change but not in substrate binding, *Biochim. Biophys. Acta* 1818, 658–665, 2012.

Eun[132] has examined the effect of borate on the reaction of arginine with phenylglyoxal and *p*-hydroxyphenylglyoxal. The base buffer of these studies was 0.1 M sodium pyrophosphate at pH 9.0. Spectroscopy was used to follow the rate of arginine modification. The rate of modification of either free arginine or *N*-acetyl-L-arginine with phenylglyoxal was 10–15 times faster than that of *p*-hydroxyphenylglyoxal in the base buffer system. The inclusion of sodium borate (10–50 mM) markedly increased the rate of the reaction (approximately 20-fold) of *p*-hydroxyphenylglyoxal with either arginine or *N*-acetyl-L-arginine, while there was only a slight enhancement of the phenylglyoxal reaction. In a related study,[120] the effect of phenylglyoxal on sodium-channel gating in frog myelinated nerve was compared with that of *p*-hydroxyphenylglyoxal or *p*-nitrophenylglyoxal. Both *p*-hydroxyphenylglyoxal and *p*-nitrophenylglyoxal had less effect than phenylglyoxal in reduced sodium current. The results are discussed in terms of the differences in hydrophobicity of the reagents, but it is clear that the intrinsic difference in reagent effectiveness described by Eun may be responsible, in part, for the observed differences. Phenylglyoxal has been used to modify arginine residues in *Klebsiella aerogenes* urease.[133] Previous studies have shown Arg336 to be present at the enzyme active site. The R336Q variant shows greatly reduced enzyme activity (decreased k_{cat}, normal K_m). Modification of this variant with phenylglyoxal resulted in a further decrease of activity suggesting that the modification of nonactive site residues can eliminate activity.

2,3-Butanedione (Figure 15.5) is the second well-characterized reagent for the selective modification of arginyl residues in proteins. Yankeelov and coworkers introduced the use of this reagent.[60,134] There were problems with the specificity of the reaction and the time required for modification until the observation of Riordan[135] that borate had a significant effect on the nature of the reaction of 2,3-butanedione with arginyl residues in proteins. The ability of 2,3-butanedione to act as a photosensitizing agent for the destruction of amino acids and proteins in the presence of oxygen was emphasized in the work by Fliss and Viswanatha.[108] As would be expected from consideration of early photooxidation work, tryptophan and histidine are lost most rapidly with methionine; cystine and tyrosine are lost at a much slower rate. Loss is not seen on irradiation in the absence of 2,3-butanedione. Azide (10 mM), a singlet oxygen scavenger, greatly reduces the rate of loss of amino acids. The absence of oxygen also greatly reduces the rate of loss of sensitive amino acids. These observations have been confirmed and extended by other laboratories.[99,110]

There are several recent studies on the use of 2,3-butanedione to modify proteins that are worthy of some specific comment. Alkena and coworkers[136] have modified penicillin acylase (*E. coli*) with 2,3-butanedione (50 mM borate, pH 8.0, 25°C). The modified enzyme had decreased specificity (k_{cat}/K_m) for 2-nitro-5-[(phenylacetyl)amino]benzoic acid. Oligonucleotide-directed mutagenesis of two arginine residues yielded similar results although there is dependence on the replacement amino acid. The kinetic parameters of βR263K were similar to wild type and was inactivated by 2,3-butanedione. The βR263L variant showed a greater than 1000-fold decrease in k_{cat}/K_m. It is of interest that while phenylglyoxal also inhibited the enzyme, analysis was complicated by observation that phenylglyoxal was also a competitive inhibitor. Clark and Ensign[137] studied the inactivation of 2-[(R)-2-hydroxypropylthio]ethanesulfonate dehydrogenase with 2,3-butanedione. The reaction was performed in 50 mM sodium borate, pH 9.0, at 25°C. In experiments for amino acid analysis of the modified enzyme, reaction was performed at 30°C for 30 min and then terminated by the addition of sodium borohydride to trap the adducts. The modification was specific for arginine with a second-order rate constant of 0.031 M^{-1} s^{-1} for the loss of enzyme activity. Leitner and Linder[92,138] have developed an approach for the general labeling of guanidino groups in proteins via reaction with 2,3-butanedione in the presence of an arylboronic acid (e.g., phenylboronic acid, see Figure 15.5) under alkaline conditions (pH 8–10). The sample is then subjected to electrospray ionization MS without further processing. The complex with phenylboronic acid with the modified arginine residue is not stable under mild acidic condition. The differential stability of the butanedione–phenylboronic acid complex was subsequently used by Foettinger and coworkers[139] for the purification of peptide containing arginine (Figure 15.6). This process has been used to purify separate C-terminal arginine peptides from endopeptidase ArgC peptidase digests resulting in the purification of the C-terminal peptide from the parent protein.[140] This is an example of the use of negative selection for purification, which is used more often for cellular elements.[141,142] C-terminal arginine peptides bound to an agarose matrix have been used for the purification of trypsin-like enzymes,[143] and solid-phase anhydrotrypsin has been used for the purification of C-terminal arginine peptides.[144] The Vienna group extended the studies on phenylboronic acid to study the buried and available arginine residues in proteins[145] and for the identification of protein by peptide mapping.[146] This group has also studied the reaction of malondialdehyde with arginine (Figure 15.6).[147–149] There is preferential modification of tryptophanyl residues by malondialdehyde under acidic condition (see Chapter 11).[150] A list of some recent studies using 2,3-butanedione to modify arginyl residues in proteins is presented in Table 15.3. The reaction of 2,3-butanedione with arginyl residues in proteins is less complex than

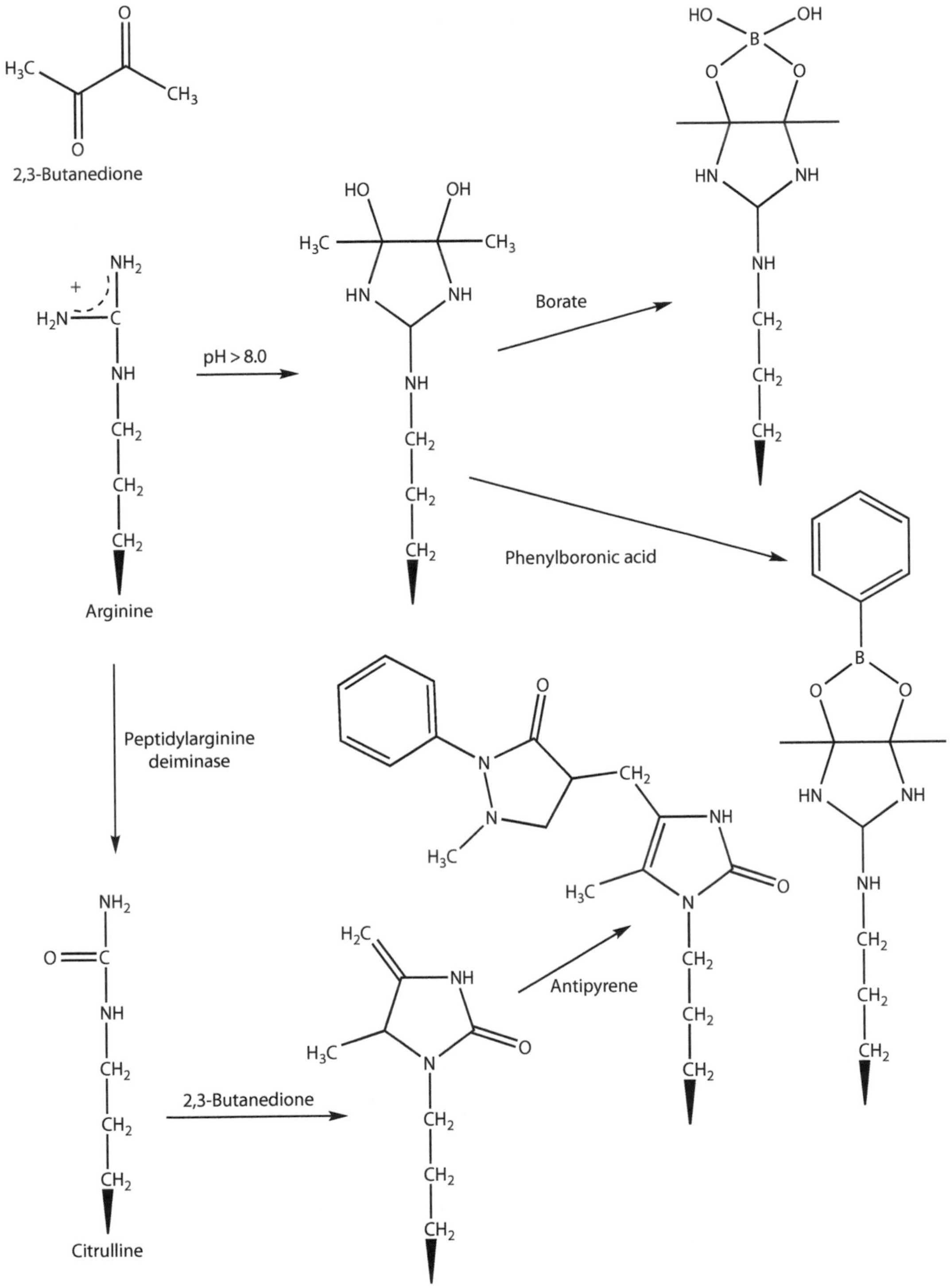

FIGURE 15.5

(*continued*)

FIGURE 15.5 (continued) 2,3-Butanedione modification of arginine residues in proteins. This is modification in the presence of borate (Riordan, J.F., Functional arginyl residues in carboxypeptidase A. Modification with butanedione, *Biochemistry* 12, 3915–2923, 1973). As shown, this reaction uses borate to stabilize the product of the reaction between 2,3-butanedione and arginine. It is also possible to use arylboronic acids to stabilize the product as shown with phenylboronic acid (Leitner, A. and Lindner, W., Probing of arginine residues in peptides and proteins using selective tagging and electrospray ionization mass spectrometry, *J. Mass Spectrom.* 38, 891–899, 2003). Also shown is the reaction of 2,3-butanedione with citrulline (Stensland, M., Holm, A., Kiehne, A., and Fleckenstein, B., Targeted analysis of protein citrullination using chemical modification and tandem mass spectrometry, *Rapid Commun. Mass Spectrom.* 23, 2754–2762, 2009; De Ceuleneer, M., De Wit, V., Van Steendam, K. et al., Modification of citrulline residues with 2,3-butanedione facilitates their detection by liquid chromatography/mass spectrometry, *Rapid Commun. Mass Spectrom.* 25, 1536–1542, 2011). Also shown is the reaction of antipyrine with citrulline as method for the determination of citrulline in proteins (Moelants, E.A.V., Van Damme, J., and Proost, P., Detection and quantification of citrullinated chemokines, *PLoS One* 6, e28976, 2011).

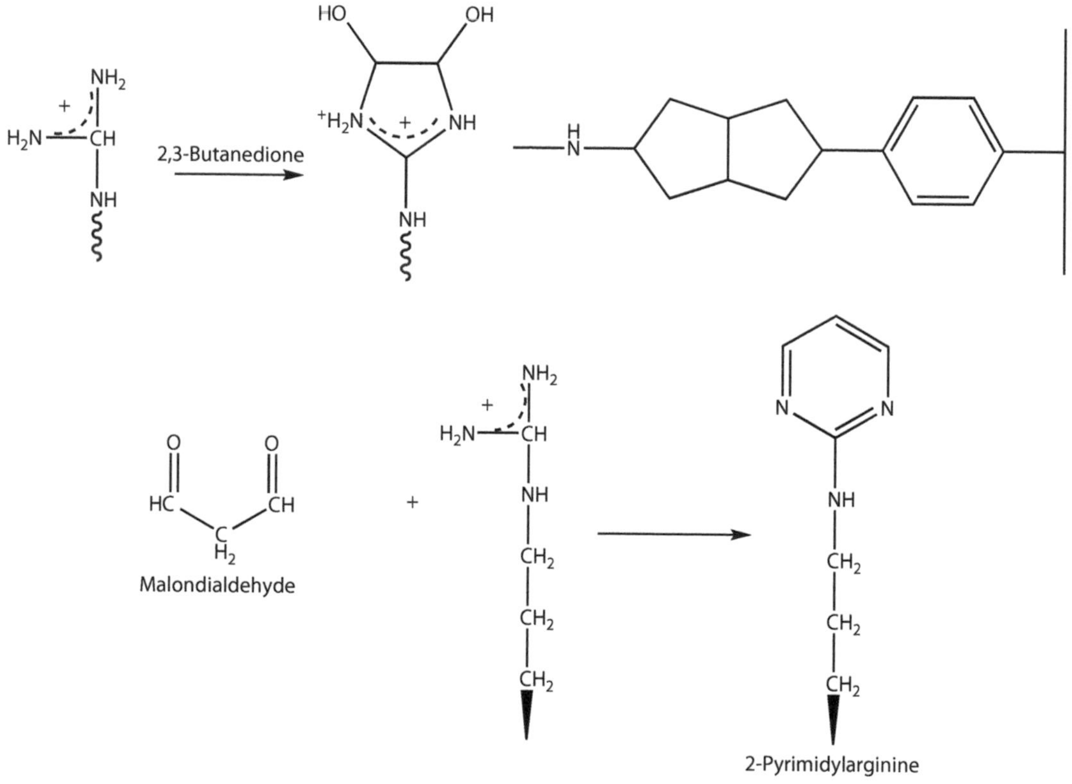

FIGURE 15.6 1,2-Cyclohexanedione modification of arginine. Shown is the reaction of 1,2-cyclohexanedione with arginine to form the N^7,N^8-(1,2-dihydroxycyclohex-1,2-ylene)-L-arginine derivative, which is stabilized by formation of a complex with borate at mild alkaline pH (Patthy, L. and Smith, E.L., Reversible modification of arginine residues. Application to sequence studies by restriction of tryptic hydrolysis to lysine residues, *J. Biol. Chem.* 250, 557–564, 1975). The modification is reversed by hydroxylamine resulting in the formation of the diox-ime, which can be detected as the nickel complex (Patthy, L. and Smith, E.L., Identification of functional arginine residues in ribonuclease A and lysozyme, *J. Biol. Chem.* 250, 565–569, 1975). For synthesis of 1,2-cyclohexane dioxime, see Hach, C.C., Banks, C.V., and Diehl, H., 1,2-Cyclohexane dioxime, *Org. Synth. Coll.* 4, 229, 1963. Reaction of 1,2-cyclohexanedione with arginine at pH > 12 (0.2 N NaOH) results in the irreversible formation of N^5-(4-oxo-1,3-diazaspiro[4,4]non-2-ylidene)-L-ornithine (Toi, K., Bynum, E., Norrie, E., and Itano, H.A., Studies on the chemical modification of arginine. I. The reaction of 1,2-cyclohexanedione with arginine and arginyl resi-dues of proteins, *J. Biol. Chem.* 242, 1036–1043, 1967). Also shown is the structure of benzil, which had early use for modification of arginine (Itano, H.A. and Gottlieb, A.J., Blocking of tryptic cleavage of arginyl bonds by the chemical modification of the guanido group with benzil, *Biochem. Biophys. Res. Commun.* 12, 405–408, 1963).

TABLE 15.3
Use of 2,3-Butanedione to Modify Arginine Residues in Proteins[a]

Protein	Solvent	Reagent Excess[b]	Stoichiometry	Reference
Carboxypeptidase A	0.05 M borate, 1.0 M NaCl, pH 7.5, 20°C	—	2/10[c]	15.3.1
Chymotrypsin	0.1 M phosphate, pH 6.0, 35°C	100[d]	1/3[e]	15.3.2
Thymidylate synthetase	0.050 M borate, pH 8.0, 25°C	—	—[f]	15.3.3
Prostatic acid phosphatase	0.050 M borate, pH 8.0, 30°C	—	—	15.3.4
Purine nucleoside phosphorylase	0.0165 M borate, pH 8.0, 23°C	—	—[g]	15.3.5
Yeast hexokinase PII	0.050 M borate, pH 8.3, 25°C	—	4.2/18[h]	15.3.6
Isocitrate dehydrogenase	0.05 M MES, pH 6.2, 20% glycerol, 0.0021 M MnSO$_4$, 30°C	—	1.6/13.4[i]	15.3.7
Stearoyl-coenzyme A desaturase	50 mM sodium borate, pH 8.1, 25°C	2500	2[j]	15.3.8
Superoxide dismutase	50 mM borate, pH 9.0, 25°C	—	1.3/4[k]	15.3.9
Enolase	50 mM borate, pH 8.3, 1.0 mM Mg (OAc)$_2$, 0.01 mM EDTA, 25°C	260	3/16[l]	15.3.10
NADPH-dependent aldehyde reductase	50 mM borate, pH 7.0, 37°C	—	1/18[m]	15.3.11
Arylsulfatase A	50 mM NaHCO$_3$, pH 8.0, 25°C	—	—[n]	15.3.12
Carbamate kinase	5 mM triethanolamine, 50 mM borate, pH 7.5, 25°C	2,000	1.2/3.0[o]	15.3.13
(K$^+$+H$^+$)-ATPase	0.125 M sodium borate, pH 7.0, 37°C	—	—[p]	15.3.14
Acetylcholinesterase	5 mM phosphate, 25 mM borate, 50 mM NaCl, pH 7.0, 25°C	—	4/31[q]	15.3.15
Coenzyme B$_{12}$–dependent diol dehydrase	50 mM borate, pH 8.5, 25°C	—	—[r]	15.3.16
Ornithine transcarbamylase	50 mM bicine,[s] 0.1 mM EDTA, 0.1 M KCl, pH 7.67, 25°C	—	0.88/11	15.3.17
Saccharopine dehydrogenase	8 mM HEPES, 0.2 M KCl, 0.01 M borate, pH 8.0, 25°C[t]	—	8/38	15.3.18
Testicular hyaluronidase	50 mM borate, pH 8.3, 20°C[u]	—	3.6/28	15.33.19
Stonustoxin (*Synanceia horrida*)	0.05 M borate, pH 8.1, 60 min, 23°C	25–100[v]	—[w]	15.3.20
ADP-glucose pyrophosphorylase (*Rhodobacter sphaerodies* 2.4.1)	50 mM HEPES, 20 mM potassium borate, 1 mM DTE, pH 7.5, dark, 25°C	—	—[x]	15.3.21
Helix stabilizing nucleoid protein (HSNP-C; *Sulfolobus acidocaldarius*)	20 mM triethanolamine, pH 8.0, 37°C, 30 min	—	—[y]	15.3.22
2-[(R)-2-hydroxypropylthio] ethanesulfonate dehydrogenase	50 mM sodium borate, pH 9.0, 25°C	—	8/15[z]	15.3.23

(*continued*)

TABLE 15.3

Use of 2,3-Butanedione to Modify Arginine Residues in Proteins[a]

[a] It must be recognized that 2,3-butanedione is a hazardous material and must be handled with care. Earlier studies described the purification of the reagent prior to use. A consideration of recent work using 2,3-butanedione suggests that commercial material is of sufficient purity to use without purification. (see Holm, A., Rise, F., Sessler, N. et al., Specific modification of peptide-bound citrulline residues, *Anal. Biochem.* 352, 68–72, 2006. Moelants, E., Van Damme, J., and Proost, P., Detection and quantification of citrullinated chemokines, *PLoS One* 6, e28976, 2011).

[b] Mole reagent per mole protein unless otherwise indicated.

[c] This study demonstrated that, in the presence of borate, there is essentially no difference in the reaction of 2,3-butanedione monomer and butadione trimer. In earlier work, it was necessary that the commercially available 2,3-butanedione should be distilled immediately prior to use. It is not clear that such a procedure is required for today.

[d] This study used 2,3-butanedione trimer prepared by allowing 2,3-butanedione (40 mL) to stand with 80 g untreated Permutit under dry air (after shaking to obtain an even dispersion of 2,3-butanedione in Permutit) for 4–6 weeks at ambient temperature. The mixture was extracted with anhydrous ether. The ether extract was taken to an oil with dry air. The oil was allowed to stand for 5–7 days to permit crystallization of the trimer. Early work had suggested the importance of the trimer as a reagent (Yankeelov, J.A., Jr., Mitchell, C.D., and Crawford, T.H., A simple trimerization of 2,3-butanedione yielding a selective reagent for the modification of arginine in proteins, *J. Am. Chem. Soc.* 90, 1664–1666, 1968).

[e] In the absence of light, there is also some loss of lysine; there is no loss of catalytic activity. In the presence of sunlight, there was rapid inactivation of the enzyme with loss of lysine, arginine (less than in the dark), and tyrosine. With the exception of tyrosine modification, the changes in amino acid composition in the reaction exposed to light were less than those for the dark reaction despite the more significant loss of activity. Study of the wavelength dependence demonstrates that light of 300 nm is most effective. 2,3-Butanedione monomer was not effective in this photoinactivation process.

[f] Stoichiometry of reaction was not established. Inactivation was reversed by gel filtration in 0.05 M Tris, 0.010 M 2-mercaptoethanol, pH 8.0.

[g] Ambient temperature. Calf spleen enzyme had 26 Arg modified at 98% loss of activity. Reaction with arginyl residues (as judged by loss of catalytic activity) was 50% as rapid with 2,3-butanedione in borate ($t_{1/2} = 40.3$ min) as with phenylglyoxal in Tris buffer ($t_{1/2} = 19.2$ min).

[h] Reaction was performed at 25°C, determined by amino acid analysis after acid hydrolysis (6 N HCl, 110°C, 18 h). MgATP (5 mM) did not protect against either modification or loss of enzymatic activity, but MgATP and glucose reduced extent of modification from 3.3 arginine residues per subunit (65% inactivation) to 2.1 residues per subunit (20% inactivation). Inactivation was also observed with phenylglyoxal in 0.050 M bicine, pH 8.3. Stoichiometry with this modification was not established.

[i] The reaction was termination by taking 1 M HCl at 0°C followed by dialysis against 1 M HCl at 4°C and taken to dryness over solid NaOH followed by hydrolysis in 6 N HCl. It was found that a maximum of 1.6 arginine residues (of an average value of 13.4 arginine residues/subunit) were modified at maximum inactivation. A second-order rate constant of 0.031 M^{-1} s^{-1} was determined for the inactivation of the enzyme by 2,3-butanedione in 20 mM MES, pH 6.2, containing 20% glycerol and 2.1 mM MnSO$_4$. The maximum value obtained is 1.6 residues modified out of an average of 13.4 arginyl residues per subunit.

[j] The modification was performed at 25°C. The presence of stearoyl-CoA greatly decreased the rate and extent of inactivation by 2,3-butanedione. When the modified enzyme is taken into 0.020 Tris (acetate), 0.100 M NaCl, pH 8.1, by gel filtration, there is the rapid recovery of activity and the concomitant decrease in the extent of arginine modification. A similar extent of modification and loss of catalytic activity was seen with 1,2-cyclohexanedione in 0.1 M sodium borate, pH 8.1.

[k] Inactivation occurred at a rate of 10.9 M^{-1} min^{-1} under these conditions (compared to 4.0 M^{-1} min^{-1} with phenylglyoxal in bicarbonate/carbonate and 6.6 M^{-1} min^{-1} with 1,2-cyclohexanedione in 0.050 M borate, pH 9.0). Inactivation with 2,3-butanedione is not observed in 0.05 M bicarbonate/carbonate, pH 9.0 at 25°C; however, there is reduced modification of arginine (0.4 residue per subunit as compared to 1.3 residues per subunit with 77% inactivation). The majority of arginine modification by 2,3-butanedione could be reversed by the removal of reagent and borate solvent by dialysis versus 0.05 M potassium phosphate, pH 7.8. Enzymatic activity was also recovered as a result of the dialysis procedure. These investigators were able to obtain evidence supporting the selective modification of Arg[141] by either 2,3-butanedione, 1,2-cyclohexandione, or phenylglyoxal.

[l] The extent of modification was determined by amino acid analysis after acid hydrolysis. The extent of modification reported was obtained after 75 min of reaction concomitant with 85% loss of activity. The presence of substrate, α-phosphoglycerate, reduced the extent of modification to 2 mol arginine per subunit with only 5% loss of catalytic activity.

TABLE 15.3 (continued)
Use of 2,3-Butanedione to Modify Arginine Residues in Proteins[a]

[m] A second-order rate constant of 0.0635 M^{-1} min^{-1} was obtained for the loss of enzymatic activity upon reaction with 2,3-butanedione in 0.050 M borate, pH 7.0, at 25°C. This presumably reflects the modification of a single arginine residue (see footnote p). The inactivation of the enzyme by 1,2-cyclohexanedione, methylglyoxal, and phenylglyoxal is compared with that by 2,3-butanedione (all at 10 mM in 0.05 M borate, pH 7.0). 2,3-Butanedione is clearly the most effective followed by phenylglyoxal, methylglyoxal, and 1,2-cyclohexanedione. The authors note that the enzyme under study aldehyde reductase can utilize methylglyoxal and phenylglyoxal as substrates, precluding their rigorous evaluation in this study. The extent of modification was determined by amino acid analysis after acid hydrolysis (6 N HCl, 110°C, 24 h). The control preparation yielded a value of 17.8 ± 1 Arg, while the modified enzyme yielded a value of 16.7 ± 1 Arg. The presence of cofactor yielded a preparation with 17.5 ± 1 Arg.

[n] Borate buffers could not be used since borate is a competitive inhibitor of the enzyme and prevents inactivation in bicarbonate buffer. Reaction with phenylglyoxal in the same solvent.

[o] Stoichiometry was established by amino acid analysis after acid hydrolysis (6 N HCl, 100°C, 20 h). Arginine is the only amino acid modified under these reaction conditions. These values were obtained at 80% inactivation. The presence of ADP reduced activity loss to 55%, with the extent of arginine modification reduced to 0.4–0.5 residues.

[p] The use of isolated *membrane fraction* prevented the establishment of stoichiometry in these studies. Analysis of the dependence of reaction rate on concentration of 2,3-butanedione is consistent with the modification of a single arginine residue. As expected, the stability of modification is dependent upon the presence of borate. Gel filtration into HEPES (0.125 M, pH 7.0) and subsequent inactivation at 37°C resulted in the recovery of a substantial amount of catalytic activity. Similar results were obtained with imidazole and Tris buffers under similar reaction conditions. This reactivation does not occur when the incubation following gel filtration is performed at 0°C instead of 37°C.

[q] Reactions were performed at 25°C. The modification of arginyl residues is associated with an approximate 70% loss of enzymatic activity. The presence of *N*-phenylpyridinium-2-aldoxine iodide reduces the extent of arginine modification by approximately 1 mol/mol of enzyme with concomitant protection of enzymatic activity. It should be noted that modification of this enzyme with phenylglyoxal results in the modification of 3 mol of arginine per mole of enzyme with 17% loss of enzymatic activity.

[r] Rigorous evaluation of the stoichiometry of the reaction is not available. Analysis of the dependence of first-order rate constants on reagent concentration (double-logarithmic relationships) is consistent with the modification of a single arginyl residue. The inactivation was reversed by 100-fold dilution into 0.05 M potassium phosphate, pH 8.5, at 25°C. The holoenzyme, prepared by incubation of the apoenzyme with adenosylcobalamin, has reduced sensitivity to 2,3-butanedione and is refractory to inactivation by phenylglyoxal. The protein modified with 2,3-butanedione showed increased anodic mobility on native gel electrophoresis. Increased anodic electrophoretic mobility has been observed for bovine serum albumin, which has been modified by methylglyoxal (Westwood, M.E. and Thornally, P.J., Molecular characteristics of methylglyoxal-modified bovine and human serum albumins. Comparison with glucose-derived advanced glycation endproduct-modified serum albumins, *J. Protein Chem.* 14, 359–372, 1995).

[s] The inactivation of ornithine transcarbamylase is readily reversible in this solvent; the presence of borate precludes reactivation observed on dilution of modified enzyme in solvent. A value of 179 M^{-1} min^{-1} for the second-order rate constant for reaction of 2,3-butanedione with ornithine transcarbamylase under these conditions was recorded.

[t] Determined by amino acid analysis on 95+% inactivated enzyme. Plotting loss of activity versus arginine residues modified suggests that inactivation is due to the modification of a single arginine residue. There was a small loss of cysteine on reaction with 2,3-butanedione (2.45 residues in the modified enzyme versus 2.85 residues in the control enzyme preparation as determined by reaction with Ellman's reagent). Other analysis suggests that the modification of cysteine is not responsible for the inactivation of enzyme activity with 2,3-butanedione. Second-order rate constant of $k = 7.5$ M^{-1} min^{-1} at 25°C (80 mM HEPES, pH 8.0, with 0.2 M KCl and 10 mM sodium borate) was obtained from the analysis of reaction rate data; a pH dependence study showed optimal rate of inactivation at pH 8.2.

[u] Second-order rate constant of $k = 13.57$ M^{-1} min^{-1} obtained at 20°C in 50 mM sodium borate, pH 8.3; inactivation is much less rapid in 0.050 M HEPES, pH 8.3. Bovine testicular hyaluronidase was protected from inactivation by 2,3-butanedione by D-glucuronamide in borate buffer but not HEPES; it is suggested the protection by D-glucuronamide in borate buffer reflects the binding of borate by D-glucuronamide.

[v] Mole ratio of 2,3-butanedione to arginine residues.

(*continued*)

TABLE 15.3 (continued)
Use of 2,3-Butanedione to Modify Arginine Residues in Proteins[a]

[w] Stonustoxin is a pore-forming toxin composed of two subunits. Modification of 37 arginine residues in stonustoxin with 2,3-butanedione (50:1, 2,3-butanedione:arginine, 60 min, 23°C) results in a change in net charge from +21 to –54 and a loss in hemolytic activity. The increase in negative charge is due to borate binding to the dihydroxyimidazoline complex. Circular dichroism measurements did not show a change suggesting maintenance of conformational integrity.

[x] Modification with 2,3-butanedione eliminates the allosteric interactions of phosphate, fructose 6-phosphate, and adenosine monophosphate; as would be expected, these allosteric modifiers protected the enzyme from modification by 2,3-butanedione.

[y] The modification reaction in triethanolamine was followed by dialysis against 100 mM sodium borate, pH 9.0, containing 2 mM 2-mercaptoethanol. The modified protein showed deceased affinity for an immobilized double-stranded DNA cellulose matrix.

[z] One arginine is protected by modification by the presence of substrate, 2-[(S)-2-hydroxypropylthio]ethanesulfonate. The rate of enzyme inactivation by 2,3-butanedione is 0.031 M^{-1} s^{-1} in 50 mM sodium borate, pH 9.0, 25°C.

References to Table 15.3

15.3.1. Riordan, J.F., Functional arginyl residues in carboxypeptidase A. Modification with butanedione, *Biochemistry* 12, 3915–3923, 1973.

15.3.2. Fliss, H., Tozer, N.M., and Viswanatha, T., The reaction of chymotrypsin with 2,3-butanedione trimer, *Can. J. Biochem.* 53, 275–283, 1975.

15.3.3. Cipollo, K.L. and Dunlap, R.B., Essential arginyl residues in thymidylate synthetase, *Biochem. Biophys. Res. Commun.* 81, 1139–1144, 1978.

15.3.4. McTigue, J.J. and Van Etten, R.L., An essential arginine residue in human prostatic acid phosphatase, *Biochim. Biophys. Acta* 523, 422–429, 1978.

15.3.5. Jordan, F. and Wu, A., Inactivation of purine nucleoside phosphorylase by modification of arginine residues, *Arch. Biochem. Biophys.* 190, 699–704, 1978.

15.3.6. Borders, C.L., Jr., Cipollo, K.L., Jordasky, J.F., and Neet, K.E., Role of arginyl residues in yeast hexokinase PII, *Biochemistry* 17, 2654–2658, 1978.

15.3.7. Hayman, S. and Colman, R.F., Effect of arginine modification on the catalytic activity and allosteric activation by adenosine diphosphate of the diphosphopyridine nucleotide specific isocitrate dehydrogenase of pig heart, *Biochemistry* 17, 4161–4168, 1978.

15.3.8. Enoch, H.G. and Strittmatter, P., Role of tyrosyl and arginyl residues in rat liver microsomal stearyl-coenzyme A desaturase, *Biochemistry* 17, 4927–4932, 1978.

15.3.9. Malinowski, D.P. and Fridovich, I., Chemical modification of arginine at the active site of the bovine erythrocyte superoxide dismutase, *Biochemistry* 18, 5909–5917, 1979.

15.3.10. Borders, C.L., Jr. and Zurcher, J.A., Rabbit muscle enolase also has essential argininyl residues, *FEBS Lett.* 108, 415–418, 1979.

15.3.11. Davidson, W.S. and Flynn, T.G., A functional arginine residue in NADPH-dependent aldehyde reductase from pig kidney, *J. Biol. Chem.* 254, 3724–3729, 1979.

15.3.12. James, G.T., Essential arginine residues in human liver arylsulfatase A, *Arch. Biochem. Biophys.* 197, 57–62, 1979.

15.3.13. Pillai, R.P., Marshall, M., and Villafranca, J.J., Modification of an essential arginine of carbamate kinase, *Arch. Biochem. Biophys.* 199, 16–20, 1980.

15.3.14. Schrijen, J.J., Luyben, W.A.H.M., DePont, J.J.H.M., and Bonting, S.L., Studies on (K++H+)-ATPase. I. Essential arginine residue in its substrate binding center, *Biochim. Biophys. Acta* 597, 331–344, 1980.

15.3.15. Müllner, H. and Sund, H., Essential arginine residue in acetylcholinesterase from *Torpedo californica*, *FEBS Lett.* 119, 283–286, 1980.

15.3.16. Kuno, S., Toraya, T., and Fukui, S., Coenzyme B12-dependent diol dehydrase: Chemical modification with 2,3-butanedione and phenylglyoxal, *Arch. Biochem. Biophys.* 205, 240–245, 1980.

15.3.18. Fujioka, M. and Takata, Y., Role of arginine residue in saccharopine dehydrogenase (l-lysine forming) from baker's yeast, *Biochemistry* 20, 468–472, 1981.

15.3.19. Gacesa, P., Savitsky, M.J., Dodgson, K.S., and Olavesen, A.H., Modification of functional arginine residues in purified bovine testicular hyaluronidase with butane-2,3-dione, *Biochim. Biophys. Acta* 661, 205–212, 1981.

TABLE 15.3 (continued)
Use of 2,3-Butanedione to Modify Arginine Residues in Proteins[a]

15.3.20. Chen, D., Haemolytic activity of stonustoxin from stonefish (*Synanceia horrida*) venom: Pore formation and the role of cationic amino acid residues, *Biochem. J.* 375, 685–691. 1997.

15.3.21. Meyer, C.R., Characterization of ADP-glucose pyrophosphorylase from *Rhodobacter sphaeroides* 2.4.1.: Evidence for the involvement of arginine in allosteric regulation, *Arch. Biochem. Biophys.* 372, 178–188, 1999.

15.3.22. Celestina, F. and Suryanarayana, T., Biochemical characterization and helix stabilizing properties of HSNP-C′ from the thermoacidophilic archeon *Sulfolobus acidocaldarius*, *Biochem. Biophys. Res. Commun.* 267, 614–618, 2000.

15.3.23. Clark, D.D. and Ensign, S.A., Characterization of the 2-[(R)-2-hydroxypropyl]ethanesulfonate dehydrogenase from *Xanthobacter* strain PY2: Product inhibition, pH dependence of kinetic parameters, site-directed mutagenesis, rapid equilibrium inhibition, and chemical modification, *Biochemistry* 41, 2727–2740, 2002.

that observed with phenylglyoxal. The reversibility of 2,3-butanedione modification is more likely than phenylglyoxal; the reaction proceeds quite slowly in the absence of borate. While there are some notable exceptions, the reaction of 2,3-butanedione with arginyl residues is slower than that observed with phenylglyoxal.

1,2-Cyclohexanedione (Figure 15.7) was shown to modify arginyl residues under very basic conditions (0.2 N NaOH) in 1967.[151] This work was based on earlier observations on the reaction of benzil with arginine in proteins.[152] However, it was not until Patthy and Smith[61] reported on the reaction of 1,2-cyclohexanedione in borate with arginyl residues in proteins that the use of this reagent became practical. These investigators reported that 1,2-cyclohexanedione reacted with arginyl residues in 0.2 M borate at pH 9.0. At alkaline pH, reaction of 1,2-cyclohexanedione with arginine forms N^5-(4-oxo-1,3-diazaspiro[4,4]non-2-yliodene)-L-ornithine (CHD-arginine), a reaction that cannot be reversed. Between pH 7.0 and 9.0, a compound is formed from arginine and 1,2-cyclohexanedione, N^7-N^8-(1,2-dihydroxycyclohex-1,2-ylene)-L-arginine (DHCH-arginine). This compound is stabilized by the presence of borate and is unstable in the presence of buffers such as Tris. This compound is readily converted back to free arginine in 0.5 M hydroxylamine at pH 7.0.

Patthy and Smith subsequently used this reagent to identify functional residues in bovine pancreatic ribonuclease A and egg white lysozyme.[153] Extent of modification of arginine residues in protein by 1,2-cyclohexanedione is generally assessed by amino acid analysis after acid hydrolysis. Under the conditions normally used for acid hydrolysis (6 N HCl, 110°C, 24 h), the borate-stabilized reaction product between arginine and 1,2-cyclohexanedione is unstable, and there is partial regeneration of arginine and the formation of unknown degradation products.[61] Acid hydrolysis in the presence of an excess of mercaptoacetic acid prevents the destruction of DHCD-arginine. Modification with cyclohexanedione is reversed by dialysis against mildly alkaline Tris buffer.[154] Ullah and Sethumadhaven[155] have demonstrated differences in the susceptibility of two phytases from *Aspergillus ficuum* to modification of arginine. Phytase A was rapidly inactivated by either 1,2-cyclohexanedione (borate, pH 9.0) or phenylglyoxal (NaHCO₃, pH 7.5). Phytase B was resistant to inactivation by 1,2-cyclohexanedione and less susceptible than phytase A to inactivation with phenylglyoxal. Calvete and colleagues[156] used a novel approach to the modification of arginine residues in bovine seminal plasma protein PDC-109. The protein was bound to a heparin–agarose column and the 1,2-cyclohexanedione (in 16 mM Tris, 50 mM NaCl, 1.6 mM EDTA, 0.025% NaN₃, pH 7.4) circulated through the column overnight at room temperature. The modified protein was eluted with 1.0 M NaCl. Residues shielded from modification were presumed to be the heparin-binding site. Table 15.4 lists some of the enzymes in which structure–function relationships have been studied by reaction with 1,2-cyclohexanedione.

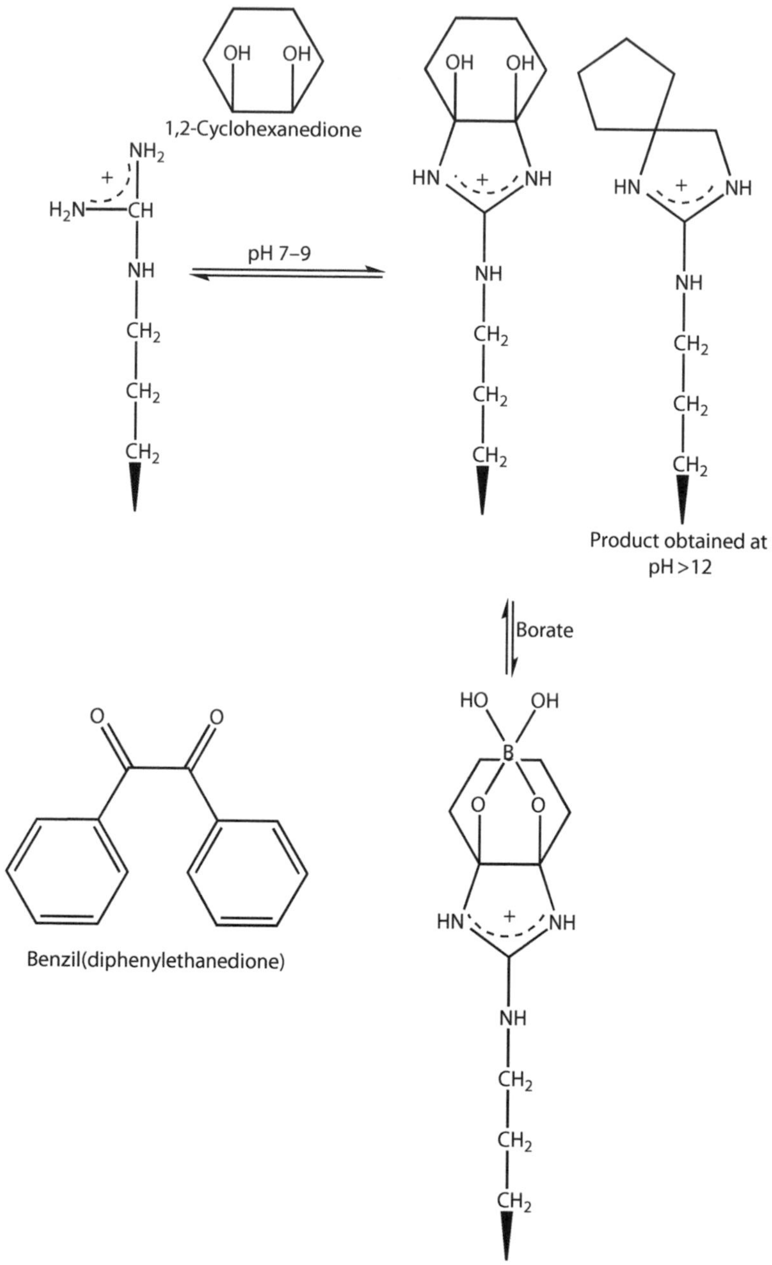

FIGURE 15.7 Oxidation of arginine and reaction of oxidation products with lysine. Shown at the top is the oxidation (metal-catalyzed oxidation [MCO]) of arginine to γ-glutamyl semialdehyde (Amici, A., Levine, R.L., Tsai, L., and Stadtman, E.R., Conversion of amino acid residues in proteins and amino acid homopolymers to carbonyl derivatives by metal-catalyzed oxidation reactions, *J. Biol. Chem.* 264, 3341–3346, 1989; Climent, I. and Levine, R.L., Oxidation of the active site of glutamine synthetase: Conversion of arginine-344 to γ-glutamyl semialdehyde, *Arch. Biochem. Biophys.* 289, 371–375, 1991). Shown below is the oxidation of arginine to citrulline with the formation of nitric oxide (Gorren, A.C. and Mayer, B., The versatile and complex enzymology of nitric oxide synthase, *Biochemistry (Moscow)* 63, 734–743, 1998; Tennyson, A.G. and Lippard, S.J., Generation, translocation, and action of nitric oxide in living systems, *Chem. Biol.* 18, 1211–1220, 2011). Shown at the bottom is some structure of products obtained from the reaction of methylglyoxal with arginine (Klöpper, A., Spannberg, R., and Glomb, M.A., Formation of arginine modifications in a model system of *N*$^\alpha$-*tert*-butoxycarbonyl (Boc)-arginine with methylglyoxal, *J. Agric. Food Chem.* 59, 394–401, 2011).

TABLE 15.4

Use of 1,2-Cyclohexanedione to Modify Arginine Residues in Proteins

Protein	Solvent	Reagent Excess[a]	Residues Modified	Reference
Ribonuclease A	0.2 M sodium borate, pH 9.0	50,000	3/4	15.41
Lysozyme	0.2 M sodium borate, pH 9.0	50,000	11/11	15.4.1
Kunitz bovine trypsin inhibitor	0.2 M sodium borate, pH 9.0	—	5.5/6	15.4.2
Threonine dehydrogenase	50 mM triethanolamine, pH 8.4, containing 5 mM 2-mercaptoethanol and 25 μM sodium borate/37°C	—	—[b]	15.4.3
Phosphoenolpyruvate carboxykinase	65 mM Tris-Cl, pH 7.4	—	—[c]	15.4.4
Recombinant streptavidin	Phosphate-buffered saline with NaOH to 0.2 N NaOH final concentration (pH 12), 23°C	6–200[d]	1.6[e]	15.4.5
Notechis scutatus scutatus notexin	Notexin (0.56 mg/mL; ~43 μM[f]) in 0.1 M sodium borate, pH 8.5 (assume 23°C)	400–2,000	1[g]	15.4.6
Hen egg white lysozyme	0.14 mM lysozyme in 0.2 M sodium borate, pH 9.0/37°C/120 min	33, 100[h]	4	15.4.7
Hen egg white lysozyme	0.14 mM lysozyme in 0.2 M sodium borate, pH 9.0/37°C/120 min	33, 100[i]	2–3/11[j]	15.4.8
Antifreeze protein from *Dendroides canadensis*	0.1 M borate, pH 9.0/37°C/dark/1–3 h[k]	10[h]	1/4[l]	15.4.9

[a] Mole reagent/mole protein.

[b] Rate of inactivation with 1,2-cyclohexanedione is less than that observed with corresponding molar excesses of either phenylglyoxal or 2,3-butanedione; analysis of the rates of inactivation with any of the three reagents is consistent with the modification of one arginine residue per catalytic site of the enzymes.

[c] Rate constant for inactivation of 0.313 M^{-1} min^{-1}, pH 7.4, at 22°C.

[d] Reagent added as solid.

[e] Residues/subunit; the reaction mixture was observed to turn slightly yellow.

[f] Assuming a molecular weight of 13,000 (Frances, B., Schmidt, J., Yang, Y. et al., Anions and the anomalous gel filtration behavior of notexin and scutoxin, *Toxicon* 33, 779–789, 1995).

[g] These investigators also studied the reaction of notexin with phenylglyoxal. The rate of reaction as judged by loss of enzymatic activity was more rapid with phenylglyoxal (5.71 M^{-1} min^{-1}) than with 1,2-cyclohexanedione (0.06 M^{-1} min^{-1}). A plot of log of first-order rate constant versus log reagent concentration yields a line with a slope approaching unity consistent with the loss of activity resulting from the modification of one residue of arginine. It is therefore of interest to see that modification had occurred at both Arg43 and Arg79.

[h] Per mole of arginine.

[i] These investigators stated that they used the conditions of reactions given in Ref. [15.4.7].

[j] Two or three arginine residues modified. The modification did not affect the *sweetness* of lysozyme.

[k] Reaction was shown to be complete at 1 h.

[l] The modification of 1 mol of arginine/mole of protein was associated with a loss of antifreeze properties; the antifreeze properties were regained on treatment of the modified protein with hydroxylamine. The ability of citrate to enhance the antifreeze activity of the beetle protein was lost with modification with 1,2-cyclohexanedione. It is noted that citrate affects the gel filtration behavior of notexin and the possibility binding to arginine is discussed (Francis, B., Schmidt, J., Yang, Y. et al., Anions and the anomalous gel filtration behavior of notexin and scutoxin, *Toxicon* 33, 779–789, 1995).

(continued)

TABLE 15.4 (continued)
Use of 1,2-Cyclohexanedione to Modify Arginine Residues in Proteins

References to Table 15.4

15.4.1. Patthy, L. and Smith, E.L., Identification of functional arginine residues in ribonuclease A and lysozyme, *J. Biol. Chem.* 250, 565, 1975.

15.4.2. Menegatti, E., Ferroni, R., Benassi, C.A., and Rocchi, R., Arginine modification in Kunitz bovine trypsin inhibitor through 1,2-cyclohexanedione, *Int. J. Pept. Protein Res.* 10, 146, 1977.

15.4.3. Epperly, B.R. and Dekker, E.E., Inactivation of *Escherichia coli* l-threonine dehydrogenase by 2,3-butanedione. Evidence for a catalytically essential arginine residue, *J. Biol. Chem.* 264, 18296–18301, 1989.

15.4.4. Cheng, K.-C. and Nowak, T., Arginine residues at the active site of avian liver phosphoenolpyruvate carboxykinase, *J. Biol. Chem.* 264, 3317, 1989.

15.4.5. Wilbur, D.S., Hamlin, D.K., Meyer, D.L. et al., Streptavidin in antibody pretargeting. 3. Comparison of biotin binding and tissue localization of 1,2-cyclohexanedione and succinic anhydride modified recombinant streptavidin, *Bioconjug. Chem.* 13, 611–620, 2002.

15.4.6. Chang, L.-S., Wu, P.-F., Liou, J.-C. et al., Chemical modification of arginine residues of *Notechis scutatus scutatus* notexin, *Toxicon* 44, 491–497, 2004.

15.4.7. Suckau, D., Mak, M., and Przybylski, M., Protein surface topology-probing by selective chemical modification and mass spectrometric peptide mapping, *Proc. Natl. Acad. Sci. USA* 89, 5630–5634, 1992.

15.4.8. Masuda, T., Ide, N., and Kitabatake, N., Structure-sweetness relationship in egg white lysozyme: Role of lysine and arginine residues on the elicitation of lysozyme sweetness, *Chem. Senses* 30, 667–681, 2005.

15.4.9. Wang, S., Amornwittawat, N., Juwita, V. et al., Arginine, a key residue for the enhancing ability of an antifreeze protein of the beetle *Dendroides canadcensis*, *Biochemistry* 48, 9696-9703, 2009.

Patthy and Thész[157] advanced a hypothesis for the specificity of dicarbonyl compounds such as 1,2-cyclohexanedione and 2,3-butanedione for arginyl residues in anion-binding sites. These investigators suggest that there is a preference for reaction with the unprotonated guanidino group and suggest that the pKa is lower for arginine residues in anion-binding sites due to the strong electrostatic potential at anion-binding sites.

Arginine is susceptible to oxidation and reaction with products of the oxidation of lipids and carbohydrates to form a variety of products.[158–161] An excellent review of the methods for the analysis of the various oxidation products is available.[159] Methylglyoxal is formed during aerobic glycolysis and reacts with arginine to form imidazolium adducts (Figure 15.8).[162–164] Ascorbic acid has been shown to react with N^α-acetyl-L-arginine to form N^α-acetyl-N^δ-[4-(1,2-dihydroxy-3-propyliden)-3-imidazolin-5-on-2-yl]-L-ornithine.[165]

Ninhydrin was developed as a reagent for the modification of arginine residues (Figure 15.9) in proteins by Takahashi.[166] The modification reaction was performed in 0.1 M morpholine acetate, pH 8.0, at 25°C in the dark. Modification of both arginine and lysine occurs under these reaction conditions. Specific modification of arginine residues may be accomplished by first modifying the lysine residues with citraconic anhydride. Verri and coworkers[101] observed that thiamin transport in renal brush-border membranes was decreased by treatment with phenylglyoxal but not with ninhydrin. Earlier studies[167] on an enterokinase from kidney beans (*Phaseolus vulgaris*) showed inactivation with either 1,2-cyclohexanedione or ninhydrin. Use of ninhydrin under basic conditions (0.4 M NaOH, alkaline ninhydrin) provides a specific method for the detection of arginine peptides.[168,169] For the purpose of historical accuracy, it is noted that Ruehmann reported on the reaction between ninhydrin (triketohydrindene hydrate) and guanidine in 1910[170]; this investigator also noted the relationship between ninhydrin and phenylglyoxal. As reported by Retinger,[171] ninhydrin is said to have been discovered by Aderhalden for use in his test for the presence of proteolytic enzymes in serum.[172–174]

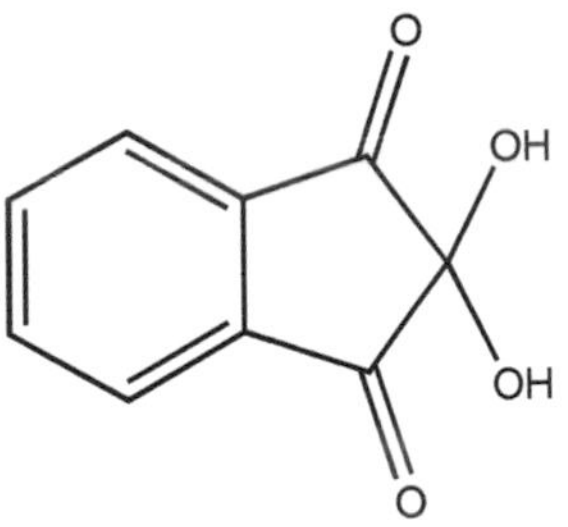

Ninhydrin

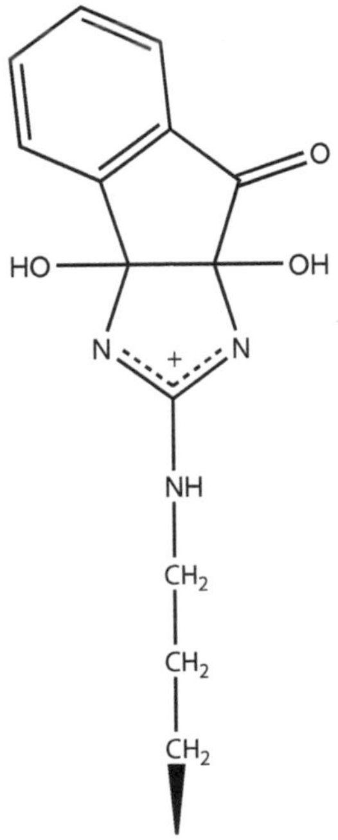

FIGURE 15.8 The reaction of ninhydrin with the guanidino group of arginine. Shown are the structure of ninhydrin and the proposed structure for the adduct of ninhydrin and the guanidino group of arginine (Takakashi, K., Specific modification of arginine residues in proteins with ninhydrin, *J. Biochem.* 80, 1173–1176, 1976; Chaplin, M.E., The use of ninhydrin as a reagent for the reversible modification of arginine residues in proteins, *Biochem. J.* 155, 457–459, 1976).

REFERENCES

1. Gromiha, M.M. and Suwa, M., A simple statistical method for discriminating outer membrane proteins with better accuracy, *Comput. Biol. Chem.* 29, 136–142, 2005.
2. Grimsley, G.R., Scholtz, J.M., and Pace, C.N., A summary of the measurement pK values of the ionizable groups in folded proteins, *Protein Sci.* 18, 247–251, 2009.
3. Olsson, I., Venge, P., Spitznagel, J.K., and Lehrer, R.I., Arginine-rich cationic proteins of human eosinophil granules: Comparison of the composition of the constituent of eosinophilic and neutrophilic leukocytes, *Lab. Invest.* 36, 493–500, 1977.
4. Patterson-Delafield, J., Szklarek, D., Martinez, R.J., and Lehrer, R.I., Microbicidal cationic proteins of rabbit alveolar macrophages: Amino acid composition and functional attributes, *Infect. Immun.* 31, 723–731, 1981.
5. Wu, G., Jaeger, L.A., Bazer, F.W. et al., Arginine deficiency in preterm infants: Biochemical mechanisms and nutritional implications, *J. Nutr. Biochem.* 15, 442–451, 2004.

6. Shao, A. and Hathcock, J.N., Risk assessment for the amino acids taurine, L-glutamine and L-arginine, *Regul. Toxicol. Pharmacol.* 50, 376–399, 2008.
7. Sureda, A. and Pons, A., Arginine and citrulline supplementation in sports and exercise: Ergogenic nutrients? *Med. Sport. Sci.* 59, 18–28, 2012.
8. Coburn, L.A., Gong, X., Singh, K. et al., L-Arginine supplementation improves responses to injury and inflammation in dextran sulfate sodium colitis, *PLoS One* 7(3), e33546, 2012.
9. Leigh, B., Desneves, K., Rafferty, J. et al., The effect of different doses of an arginine-containing supplement on the healing of pressure ulcers, *J. Wound Care* 21, 150–156, 2012.
10. Dort, J., Sirois, A., Leblanc, N. et al., Beneficial effects of cod protein on skeletal muscle repair following injury, *Appl. Physiol. Nutr. Metab.* 37, 489–498, 2012.
11. Arnusch, C.J., Pieters, R.J., and Breukink, E., Enhanced membrane pore formation through high-affinity targeted antimicrobial peptides, *PLoS One* 7(6), 39768, 2012.
12. Wang, A., Chen, F., Wang, Y. et al., Enhancement of antiviral activity of human α-defensin 5 against herpes simplex virus 2 by arginine mutagenesis at adaptive evolution sites, *J. Virol.* 87, 2835–2845, 2013.
13. Deslouches, B., Steckbeck, J.D., Craigo, J.K. et al., Rational design of engineered cationic antimicrobial peptides consisting exclusively of arginine and tryptophan: WR eCAP activity against multidrug-resistant pathogens, *Antimicrob. Agents Chemother.* 57(6), 2511–2521, 2013.
14. Liu, B.R., Lin, M.D., Chiang, H.J., and Lee, H.J., Arginine-rich cell-penetrating peptides deliver gene into living human cells, *Gene* 505, 37–45, 2012.
15. Naik, R.J., Chatterjee, A., and Ganguli, M., Different roles of cell surface and exogenous glycosaminoglycans in controlling gene delivery by arginine-rich peptides with varied distribution of arginines, *Biochim. Biophys. Acta* 1828, 1484–1493, 2013.
16. Lee, S.K., Siefert, A., Beloor, J. et al., Cell-specific siRNA delivery by peptides and antibodies, *Methods Enzymol.* 502, 91–122, 2012.
17. Nakase, I., Akita, H., Kogure, K. et al., Efficient intracellular delivery of nucleic acid pharmaceuticals using cell-penetrating peptides, *Acc. Chem. Res.* 45, 1132–1139, 2012.
18. Palmer, R.M., Rees, D.D., Ashton, D.S., and Moncada, S., L-Arginine is the physiological precursor for the formation of nitric oxide in endothelium-dependent relaxation, *Biochem. Biophys. Res. Commun.* 153, 1251–1256, 1988.
19. Davids, M., Peters, J.M., Jong, S.D., and Teerlink, T., Measurement of nitric oxide-related amino acids in serum and plasma: Effects of blood clotting and type of anticoagulant, *Clin. Chim. Acta* 421, 164–167, 2013.
20. Kinloch, A., Lundberg, K., Wait, R. et al., Synovial fluids is a site of citrullination of autoantigens in inflammatory arthritis, *Arthritis Rheum.* 58, 2287–2295, 2008.
21. Luban, S. and Li, Z.B., Citrullinated peptide and its relevance to rheumatoid arthritis: An update, *Int. J. Rheum. Dis.* 13, 284–287, 2010.
22. Schug, K.A. and Lindner, W., Noncovalent binding between guanidinium and anionic groups: Focus on biological- and synthetic-based arginine/guanidinium interactions with phosph[on]ate and sulf[on]ate residues, *Chem. Rev.* 105, 67–114, 2005.
23. Blondeau, P., Segura, M., Pérez-Fernández, R. et al., Molecular recognition of oxoanions based on guanidinium receptors, *Chem. Soc. Rev.* 36, 198–210, 2007.
24. Ruoslahti, E., RGD and other recognition sequences for integrins, *Annu. Rev. Cell Dev. Biol.* 12, 697–715, 1996.
25. Takagi, J., Structural basis for ligand recognition by RGD (Asp-Gly-Asp)-dependent integrins, *Biochem. Soc. Trans.* 32, 403–406, 2004.
26. Mavropoulos, E., Hausen, M., Costa, A.M. et al., The impact of the RGD peptide on osteoblast adhesion and spreading on zinc-substituted hydroxyapatite surface, *J. Mater. Sci. Med.* 24, 1271–1283, 2013.
27. Wu, C., Chen, M., Skelton, A.A. et al., Adsorption of arginine-glycine-aspartate tripeptide onto negatively charged Rutile (110) mediated cations: The effect of surface hydroxylation, *ACS Appl. Mater. Interfaces* 5, 2567–2579, 2013.
28. Jonas, A. and Weber, G., Presence of arginine residues at the strong, hydrophobic anion binding sites of bovine serum albumin, *Biochemistry* 10, 1335–1339, 1971.
29. Riordan, J.F., Arginyl residues and anion binding sites in proteins, *Mol. Cell. Biochem.* 26, 71–92, 1979.
30. Anderson, B.F., Baker, H.M., Norris, G.E. et al., Structure of human lactoferrin: Crystallographic structure analysis and refinement at 2.8 Å resolution, *J. Mol. Biol.* 209, 711–734, 1989.
31. Zak, O., Ikuta, K., and Aisen, P., The synergistic anion-binding sites of human transferrin: Chemical and physiological effects of site-directed mutagenesis, *Biochemistry* 41, 7416–7416, 2002.

32. Sato, M., Kubo, M., Aizawa, T. et al., Role of putative anion-binding sites in cytoplasmic and extracellular channels of *Natronomonas pharaonis* halorhodopsin, *Biochemistry* 44, 4775–4784, 2005.

33. Bedford, M.T. and Clarke, S.G., Protein arginine methylation in mammals: Who, what, and why, *Mol. Cell* 33, 1–13, 2009.

34. Le, D.D., Cortesi, A.T., Myers, S.A. et al., Site-specific and regiospecific installation of methylarginine analogues into recombinant histones and insights into effector protein binding, *J. Am. Chem. Soc.* 135, 2879–2882, 2013.

35. Boisvert, F.M., Chénard, C.A., and Richard, S., Protein interfaces in signaling regulated by arginine methylation. *Sci. STKE* 2005(271), re2, 2005.

36. Migliori, V., Phalke, S., Bezzi, M. et al., Arginine/lysine-methyl/methyl switches: Biochemical role of histone arginine methylation in transcriptional regulation, *Epigenomics* 2, 119–137, 2010.

37. Lopez-Clavijo, A.F., Barrow, M.P., Rabbani, N. et al., Determination of types and binding site for advanced glycation end products for substance P, *Anal. Chem.* 84, 10568–10575, 2012.

38. Wang, T., Kartika, R., and Spiegel, D.A., Exploring post-translational arginine modifications using chemically synthesized methylglyoxal hydroimidazolones, *J. Am. Chem. Soc.* 134, 8958–8967, 2012.

39. Chetyrkin, S., Mathis, M., Pedchenko, V. et al., Glucose autoxidation induced functional damage to proteins via modification of critical arginine residues, *Biochemistry* 50, 6102–6112, 2011.

40. Berger, S.L., Cell signalling and transcriptional regulation via histone phosphorylation, *Cold Spring Harb. Sym. Quant. Biol.* 75, 23–26, 2010.

41. Elsholz, A.K.W., Turgay, K., Michalik, S. et al., Global impact of protein arginine phosphorylation on the physiology of *Bacillus subtilis*, *Proc. Natl. Acad. Sci. USA* 109, 7451–7456, 2012.

42. Roche, J. and Mourgue, M., Nitration des proteines et réactivité des groupements guanidiques de l'arginine, *Comt. Rend. Hebd. Séances Acad. Sci.* 228, 1848–1850, 1948.

43. Hayakawa, T., Fujiwara, Y., and Noguchi, J., A new method of reducing nitroarginine-peptide into arginine-peptide, with reference to the synthesis of poly-L-arginine hydrochloride, *Bull. Chem. Soc. Jpn.* 40, 1205–1208, 1967.

44. Martin, N.I. and Liskamp, R.M., Preparation of N^G-substituted L-arginine analogues suitable for solid phase peptide synthesis, *J. Org. Chem.* 73, 7849–7851, 2008.

45. Holzer, P., Wachter, C., Jocic, M., and Heinemann, A., Vascular bed-dependent roles of the peptide CGRP and nitric oxide in acid-evoked hyperaemia of the rat stomach, *J. Physiol.* 480, 575–585, 1994.

46. Kuo, P., Gentilcore, D., Nair, N. et al., The nitric oxide synthase inhibitor, N^g-nitro-L-arginine methyl ester, attenuates the delay in gastric emptying induced by hyperglycaemia in healthy humans, *Neurogastroenenterol. Motil.* 21, 1175–e103, 2009.

47. Hogan, J.J., Gutell, R.R., and Noller, H.F., Probing the conformation of 18S rRNA in yeast 40S ribosomal subunits with kethoxal, *Biochemistry* 23, 3322–3330, 1984.

48. Litt, M. and Hancock, V., Kethoxal—A potentially useful reagent for the determination of nucleotide sequences in single-stranded regions of transfer ribonucleic acid, *Biochemistry* 6, 1848–1854, 1967.

49. Litt, M., Structural studies on transfer ribonucleic acid. I. Labeling of exposed guanine sites in yeast phenylalanine transfer ribonucleic acid with kethoxal, *Biochemistry* 8, 3249–3253, 1969.

50. Quarrier, S., Martin, J.S., Davis-Neulander, L. et al., Evaluation of the information content of RNA structure mapping data for secondary structure prediction, *RNA* 16, 1108–1117, 2010.

51. Miura, K., Tsuda, S., Ueda, T. et al., Chemical modification of guanine residues of mouse 5S ribosomal RNA with kethoxal, *Biochim. Biophys. Acta* 739, 281–285, 1983.

52. Underwood, G.E., Siem, R.A., Gerpheide, S.A., and Hunter, J.H., Binding of an antiviral agent (kethoxal) by various metabolites, *Proc. Soc. Exp. Biol. Med.* 100, 312–315, 1959.

53. Underwood, G.E., Kethoxal for treatment of cutaneous herpes simplex, *Proc. Soc. Exp. Biol. Med.* 129, 235–239, 1968.

54. Underwood, G.E. and Nichol, F.R., Clinical evaluation of kethoxal against cutaneous herpes simplex, *Appl. Microbiol.* 22, 588–592, 1971.

55. Delihas, N., Zorn, G.A., and Strobel, E., The reaction of *Escherichia coli* ribosomes with kethoxal, *Biochimie* 55, 1227–1234, 1973.

56. Steinberg, B., Dodt, S., and Delihas, N., unpublished data (cited in Benkov, K. and Delihas, N., Analysis of kethoxal bound to ribosomal proteins from *Escherichia coli* 70S reacted ribosomes, *Biochem. Biophys. Res. Commun.* 60, 901–908, 1974).

57. IIjima, H., Patrzyc, H., and Bello, J., Modification of amino acids and bovine pancreatic ribonuclease A by kethoxal, *Biochim. Biophys. Acta* 491, 305–316, 1977.

58. Akinsiku, O.T., Yu, E.T., and Fabris, D., Mass spectrometric investigation of protein alkylation by the RNA footprinting probe kethoxal, *J. Mass Spectrom.* 40, 1372–1381, 2005.

59. Takahashi, K., The reaction of phenylglyoxal with arginine residues in proteins, *J. Biol. Chem.* 243, 6171–6179, 1968.
60. Yankeelov, J.A., Jr., Mitchell, C.D., and Crawford, T.H., A simple trimerization of 2,3-butanedione yielding a selective reagent for the modification of arginine in proteins, *J. Am. Chem. Soc.* 90, 1664–1666, 1968.
61. Patthy, L. and Smith, E.L., Reversible modification of arginine residues. Application to sequence studies by restriction of tryptic hydrolysis to lysine residues, *J. Biol. Chem.* 250, 557–564, 1975.
62. Hager-Braun, C. and Tomer, K.M., Characterization of the tertiary structure of soluble CD4 bound to glycosylated full-length HIV gp120 by chemical modification of arginine residues and mass spectrometric analysis, *Biochemistry* 41, 1759–1766, 2002.
63. Biswas, A., Lewis, S., Wang, B. et al., Chemical modulation of the chaperone function of human αA-crystalline, *J. Biochem.* 144, 21–32, 2008.
64. Hermansson, M., Artemenko, K., Ossipova, E. et al., MS analysis of rheumatoid arthritic synovial tissue identifies specific citrullination sites on fibrinogen, *Proteomics Clin. Appl.* 4, 511–518, 2010.
65. Moelants, E.W., Van Damme, J., and Proost, P., Detection and quantification of citrullinated chemokines, *PLoS One* 6, e28976, 2011.
66. Hart-Smith, G., Low, J.K., Erce, M.A., and Wilkins, M.R., Enhanced methylarginine characterization by post-translational modification-specific targeted data acquisition and electron-transfer dissociation mass spectrometry, *J. Am. Soc. Mass Spectrom.* 23, 1376–1389, 2012.
67. Stutzman, J.R. and McLuckey, S.A., Ion/ion reactions of MALDI-derived peptide ions: Increased sequence coverage via covalent and electrostatic modification upon charge inversion, *Anal. Chem.* 84, 10679–10685, 2012.
68. Afjehi-Sadat, L. and Garcia, B.A., Comprehending dynamic protein methylation with mass spectrometry, *Curr. Opin. Chem. Biol.* 17, 12–19, 2013.
69. Greenstein, J.P. and Winitz, M., *Chemistry of the Amino Acids*, Chapter 22, pp. 1841–1855, John Wiley & Sons, New York, 1961.
70. Sakaguchi, S., Über eine neue farbenreaktion von protein und arginin, *J. Biochem.* 5, 25–31, 1925.
71. Izumi, Y., New Sakaguchi reaction, *Anal. Biochem.* 10, 218–226, 1965.
72. Enoch, H.G. and Strittmatter, P., Role of tyrosyl and arginyl residues in rat liver microsomal stearyl-coenzyme A desaturase, *Biochemistry* 17, 4927–4932, 1978.
73. Kaleem, K., Chertok, F., and Erhan, S., Protein-polymer grafts III A. Modification of amino groups Ib.i Modification of protein-bound arginine with 4-hydroxybenzil, *J. Biol. Phys.* 15, 71–74, 1987.
74. Kaleem, K., Chertok, F., and Erhan, S., Novel materials form protein-polymer grafts, *Nature* 325, 328–329, 1987.
75. Young, C.T., Automated colorimetric measurement of free arginine in peanuts as a means to evaluated maturity and flavor, *J. Agric. Food Chem.* 21, 556–558, 1973.
76. Wang, H., Liang, X.-h., Zhao, R. et al., Spectrophotometric determination of arginine in grape juice using 8-hydroquinoline, *Agric. Sci. China* 7, 1210–1215, 2008.
77. Mira de Orduna, R., Quantitative determination of L-arginine by enzymatic end-point analysis, *J. Agric. Food Chem.* 49, 549–552, 2001.
78. Murakami, T., Kosaka, M., Sato, H. et al., The intensely positively charged perioneuronal net in the adult rat brain, with special reference to its reactions to oxine, chondroitinase ABC, hyaluonidase, and collagenase, *Arch. Histol. Cytol.* 64, 313–318, 2001.
79. Mahmoud, L.H. and el-Alfy, N.M., Electron microscopy and histochemical studies on four Egyptian helminthes eggs of medical importance, *J. Egypt Soc. Paristol.* 33, 229–243, 2003.
80. Schulze, A., Mayatepek, E., Rating, D. et al., Sakaguchi reaction: A useful method for screening guanidinoacetate-methyltransferase deficiency, *J. Inhert. Metab. Dis.* 19, 706, 1996.
81. Schulze, A., Hess, T., Wevers, R. et al., Creatinine deficiency syndrome caused by guanidinoacetate methyl transferase deficiency: Diagnostic tools for a new inborn error of metabolism, *J. Pediatr.* 131, 626–631, 1997.
82. Weber, C.J., A modification of Sakaguchi's reaction for the determination of arginine, *J. Biol. Chem.* 86, 217–222, 1930.
83. Brand, E. and Kassell, B., Photometric determination of arginine, *J. Biol. Chem.* 145, 359–264, 1942.
84. Smith, R.E. and MacQuarrie, R., A sensitive fluorometric method for the determination of arginine using 9,10-phenanthrenequinone, *Anal. Biochem.* 90, 246, 1978.
85. Fan, X., Subramaniam, R., Weiss, M.F., and Monnier, V.M., Methylglyoxal-bovine serum albumin stimulates tumor necrosis factor alpha secretion in RAW 264.7 cells though activation of mitogen-activating protein kinase, nuclear factor κB and intracellular reactive oxygen species formation, *Arch. Biochem. Biophys.* 409, 274–285, 2002.

657

86. Knott, H.M., Brown, B.E., Davies, M.J., and Dean, R.T., Glycation and glycoxidation of low-density lipoproteins by glucose and low-molecular mass aldehydes. Formation of modified and oxidized proteins, *Eur. J. Biochem.* 270, 3572–3582, 2003.
87. Miele, C., Riboulet, A., Maitan, M.A. et al., Human glycated albumin affects glucose metabolism in L6 skeletal muscle cells by impairing insulin-induced insulin receptor substrate (IRS) signaling through a protein kinase Cα-mediated mechanism, *J. Biol. Chem.* 278, 47376–47387, 2003.
88. Schepens, I., Ruelland, E., Miginiac-Maslow, M. et al., The role of active site arginines of sorghum NADP-malate dehydrogenase in thioredoxin-dependent activation and activity, *J. Biol. Chem.* 275, 35792–35798, 2000.
89. Linder, M.D., Morkunaite-Haimi, S., Kinnunen, P.K. et al., Ligand-selective modulation of the permeability transition pore by arginine modification. Opposing effects of *p*-hydroxyphenylglyoxal and phenylglyoxal, *J. Biol. Chem.* 277, 937–942, 2002.
90. Wu, X., Chen, S.G., Petrash, J.M., and Monnier, V.M., Alteration of substrate specificity through mutation of two arginine residues in the binding site of amadoriase II from *Aspergillus* sp., *Biochemistry* 41, 4453–4458, 2002.
91. Xu, G., Takamoto, K., and Chance, M.R., Radiolytic modification of basic amino acid residues in peptides: Probes for examining protein-protein interactions, *Anal. Chem.* 75, 6995–7007, 2003.
92. Leitner, A. and Linder, W., Probing of arginine residues in peptides and proteins using selective tagging and electrospray ionization mass spectrometry, *J. Mass Spectrom.* 38, 891–899, 2003.
93. Eriksson, O., Fontaine, E., Petronilli, V. et al., Inhibition of the mitochondrial cyclosporine A-sensitive permeability pores by the arginine reagent phenylglyoxal, *FEBS Lett.* 409, 361–364, 1997.
94. Eriksson, O., Fontaine, E., and Bernardi, P., Chemical modification of arginines by 2,3-butanedione and phenylglyoxal causes closure of the mitochondrial permeability transition pore, *J. Biol. Chem.* 273, 12664–12674, 1998.
95. Skysgaard, J.M., Modification of Cl⁻ transport in skeletal muscle of *Rana temporaria* with the arginine-binding reagent phenylglyoxal, *J. Physiol.* 510, 591–604, 1998.
96. Cook, L.J., Davies, J., Yates, A.P. et al., Effects of methylglyoxal on rat pancreatic β-cells, *Biochem. Pharmacol.* 55, 1361–1367, 1999.
97. Emmons, C., Transport characterization of the apical anion exchanger of rabbit cortical collecting duct beta-cells, *Am. J. Physiol.* 276, F635–F643, 1999.
98. Tamai, I., Ogihara, T., Takanaga, H. et al., Anion antiport mechanism is involved in transport of lactic acid across intestinal epithelial brush-border membrane, *Biochim. Biophys. Acta* 1468, 285–292, 2000.
99. Raw, P.E. and Gray, J.C., The effect of amino acid-modifying reagents on chloroplast protein import and the formation of early import intermediates, *J. Exp. Bot.* 52, 57–66, 2001.
100. Cremaschi, D., Vallin, P., Sironi, C., and Porta, C., Inhibitors of the Cl⁻/HCO₃⁻ exchanger activate an anion channel with similar features in the epithelial cells of rabbit gall bladder: Analysis in the intact epithelium, *Pfuegers Arch.* 441, 456–466, 2001.
101. Verri, A., Laforenza, U., Gastaldi, G. et al., Molecular characteristics of small intestinal and renal brush border thiamin transporters in rats, *Biochim. Biophys. Acta* 1558, 187–197, 2002.
102. Forster, I.E., Köhler, K., Stange, G. et al., Modulation of renal type IIa Na⁺/Pᵢ cotransporter kinetics by the arginine modifier phenylglyoxal, *J. Membr. Biol.* 187, 85–96, 2002.
103. Hermann, A., Varga, V., Oja, S.S. et al., Involvement of amino acid side chains of membrane proteins in the binding of glutathione to pig cerebral cortical membranes, *Neurochem. Res.* 27, 389–394, 2002.
104. Castagna, M., Vincenti, S., Marciani, P., and Sacchi, V.F., Inhibitors of the lepidopteran amino acid co-transporter KAAT1 by phenylglyoxal: Role of Arg76, *Insect Mol. Biol.* 11, 389–394, 2002.
105. Kucera, I., Inhibition by phenylglyoxal of nitrate transport in *Paracoccus denitrificans*: A comparison with the effect of a protonophorous uncoupler, *Arch. Biochem. Biophys.* 409, 327–334, 2003.
106. Sacchi, V.F., Castagna, M., Mari, S.A. et al., Glutamate 59 is critical for transport function of the amino acid cotransporter KAAT1, *Am. J. Physiol.* 285, C623–C632, 2003.
107. Winters, C.J. and Adreoli, T.E., Chloride channels is basolateral TAL membranes XVIII. Phenylglyoxal induces functional mcCIC-Ka activity in basolateral MTAL membranes, *J. Membr. Biol.* 195, 63–71, 2003.
108. Fliss, H. and Viswanatha, T., 2,3-Butanedione as a photosensitizing agent: Application to α-amino acids and α-chymotrypsin, *Can. J. Biochem.* 57, 1267–1272, 1979.
109. Gripon, J.-C. and Hofmann, T., Inactivation of aspartyl proteinases by butane-2,3-dione. Modification of tryptophan and tyrosine residues and evidence against reaction of arginine residues, *Biochem. J.* 193, 55–65, 1981.
110. Mäkinen, K.K., Mäkinen, P.-L., Wilkes, S.H., Bayliss, M.E., and Prescott, J.M., Photochemical inactivation of Aeromonas aminopeptidase by 2,3-butanedione, *J. Biol. Chem.* 257, 1765–1772, 1982.

111. Willassen, N.P. and Little, C., Effect of 2,3-butanedione on human myeloperoxidase, *Int. J. Biochem.* 21, 755–759, 1989.

112. Inano, H., Ohba, H., and Tamaoki, B., Photochemical inactivation of human placental estradiol 17 β-dehydrogenase in the presence of 2,3-butanedione, *J. Steroid Biochem.* 19, 1617–1622, 1983.

113. Yang, C.F. and Brush, E.J., A spectrophotometric assay for α-ketoaldehydes using horse liver alcohol dehydrogenase, *Anal. Biochem.* 214, 124–127, 1993.

114. Borders, C.L., Jr., Pearson, L.J., McLaughlin, A.E., Gustafson, M.E., Vasiloff, J., An, F.Y., and Morgan, D.J., 4-Hydroxy-3-nitrophenylglyoxal. A chromophoric reagent for arginyl residues in proteins, *Biochim. Biophys. Acta* 568, 491–495, 1979.

115. Borders, C.L., Jr. and Johansen, J.T., Identification of Arg-143 as the essential arginine residue in yeast Cu, Zn superoxide dismutase by the use of a chromophoric arginine reagent, *Biochem. Biophys. Res. Commun.* 96, 1071–1078, 1980.

116. Borders, C.L., Jr. and Riordan, J.F., An essential arginyl residue at the nucleotide binding site of creatinine kinase, *Biochemistry* 14, 4699–4704, 1975.

117. Aminlari, M. and Vaseghi, T., A new colorimetric method for determination of creatinine phosphokinase, *Anal. Biochem.* 164, 397–404, 1987.

118. Wijnands, R.A., Müller, F., and Visser, A.J.W., Chemical modification of arginine residues in *p*-hydroxybenzoate hydroxylase from *Pseudomonas fluorescens*: A kinetic and fluorescence study, *Eur. J. Biochem.* 163, 535–544, 1987.

119. Julian, T. and Zaki, L., Studies on the inactivation of anion transport in human blood red cells by reversibly and irreversibly acting arginine-specific reagents, *J. Membr. Biol.* 102, 217–224, 1988.

120. Meves, H., Ruby, N., and Staempfil, R., The action of arginine-specific reagents on ionic and gating currents in frog myelinated nerve, *Biochim. Biophys. Acta* 943, 1–12, 1988.

121. Betakis, E., Fritzsch, G., and Zaki, L., Inhibition of anion transport in the human red blood cell membrane with *p*- and *m*-methoxyphenylglyoxal, *Biochim. Biophys. Acta* 1110, 75–80, 1992.

122. Boehm, R. and Zaki, L., Toward the localization of the essential arginine residues in the band 3 protein of human red blood cell membranes, *Biochim. Biophys. Acta* 2390, 238–242, 1996.

123. Zaki, L., Boehm, R., and Merckel, M., Chemical labeling of arginyl-residues involved in anion transport mediated by human band 3 protein and some aspects of its location in the peptide chain, *Cell. Mol. Biol.* 42, 1053–1063, 1996.

124. Belousova, L.V. and Muizhnek, E.L., Kinetics of chemical modification of arginine residues in mitochondrial creatine kinase from bovine heart: Evidence for negative cooperativity, *Biochemistry (Moscow)* 69, 455–461, 2004.

125. Yamasaki, R.B., Vega, A., and Feeney, R.E., Modification of available arginine residues in proteins by *p*-hydroxyphenylglyoxal, *Anal. Biochem.* 109, 32–40, 1980.

126. Béliveau, R., Bernier, M., Giroux, S., and Bates, D., Inhibition by phenylglyoxal of the sodium-coupled fluxes of glucose and phosphate in renal brush-border membrane, *Biochem. Cell. Biol.* 66, 1005–1012, 1988.

127. Strevey, J., Vachon V., Beaumier, B. et al., Characterization of essential arginine residues implicated in the renal transport of phosphate and glucose, *Biochim. Biophys. Acta* 1106, 110–116, 1992.

128. Mukouyama, E.B., Hirose, T., and Suzuki, H., Chemical modification of L-phenylalanine oxidase from *Pseudomonas* sp. 5012 by phenylglyoxal. Identification of a single arginine residue, *J. Biochem.* 123, 1097–1103, 1998.

129. Cheung, S.-T. and Fonda, M.L., Reaction of phenylglyoxal with arginine. The effect of buffers and pH, *Biochem. Biophys. Res. Commun.* 90, 940–947, 1979.

130. Yamasaki, R.B., Shimer, D.A., and Feeney, R.E., Colorimetric determination of arginine residues in proteins by p-nitrophenylglyoxal, *Anal. Biochem.* 111, 220–226, 1981.

131. Branlant, G., Tritsch, D., and Biellmann, J.-F., Evidence for the presence of anion-recognition sites in pig-liver aldehyde reductase. Modification by phenylglyoxal and p-carboxyphenyl glyoxal of an arginyl residue located close to the substrate-binding site, *Eur. J. Biochem.* 116, 505–512, 1981.

132. Eun, H.-M., Arginine modification by phenylglyoxal and (p-hydroxyphenyl)glyoxal: Reaction rates and intermediates, *Biochem. Int.* 17, 719–727, 1988.

133. Pearson, M.A., Park, I.S., Schaller, R.A. et al., Kinetic and structural characterization of urease active site variants, *Biochemistry* 39, 8575–8584, 2000.

134. Yankeelov, J.A., Jr., Modification of arginine in proteins by oligomers of 2,3-butanedione, *Biochemistry* 9, 2433–2399, 1970.

135. Riordan, J.F., Functional arginyl residues in carboxypeptidase A. Modification with butanedione, *Biochemistry* 12, 3915–2923, 1973.

136. Alkema, W.B.L., Prins, A.K., de Vries, E., and Janssen, D.B., Role of αArg[145] and βArg[263] in the active site of penicillin acylase of *Escherichia coli*, *Biochem. J.* 365, 303–309, 2003.

137. Clark, D.D. and Ensign, S.A., Characterization of the 2-[(R)-2-hydroxypropyl]ethanesulfonate dehydrogenase from *Xanthobacter* strain Py2: Product inhibition, pH dependence of kinetic parameters, site-directed mutagenesis, rapid equilibrium inhibition, and chemical modification, *Biochemistry* 41, 2727–2740, 2002.

138. Leitner, A. and Lindner, W., Effects of an arginine-selective tagging procedure on the fragmentation of peptides studied by electrospray ionization tandem mass spectrometry (ESI-MS/MS). *Anal. Chim. Acta* 528, 165–171, 2005.

139. Foettinger, A., Leitner, A., and Lindner, W., Solid-phase capture and release of arginine peptides by selective tagging and boronate affinity chromatography, *J. Chromatogr. A* 1079, 187–196, 2005.

140. Kuyama, H., Nakajima, C., and Tanaka, K., Enriching C-terminal peptide from endopeptidase ArgC digest for protein C-terminal analysis, *Bioorg. Med. Chem. Lett.* 22, 7163–7168, 2012.

141. Landry, F.J. and Findlay, S.R., Purification of human basophils by negative selection, *J. Immunol. Methods* 63, 329–336, 1983.

142. Hansel, T.T., Pound, J.D, Pilling, D. et al., Purification of human blood eosinophils by negative selection using immunomagnetic beads, *J. Immunol. Methods* 122, 97–103, 1989.

143. Kasai, K. and Ishii, S., Affinity chromatography of trypsin using a Sepharose derivative coupled with peptides containing L-arginine in carboxyl termini, *J. Biochem.* 71, 363–366, 1972.

144. Ohta, T., Inoue, Y., Fukumoto, Y., and Takitani, S., Preparation of anhydrotrypsin-immobilized diol silica as a selective adsorbent for high-performance affinity chromatography of peptides containing arginine or lysine at the C-termini, *Chromatographia* 30, 410–413, 1990.

145. Leitner, A. and Lindner, W., Functional probing of arginine residues in proteins using mass spectrometry and an arginine-specific covalent tagging concept, *Anal. Chem.* 77, 4481–4488, 2005.

146. Leitner, A., Amon, S., Rizzi, A., and Lindner, W., Use of the arginine-specific butanedione/phenylboronic acid tag for analysis of peptides and protein digests using matrix-assisted laser desorption/ionization mass spectrometry, *Rapid Commun. Mass Spectrom.* 21, 1321–1330, 2007.

147. Foettinger, A., Leitner, A., and Lindner, W., Derivatisation of arginine residues with malondialdehyde for the analysis of peptides and protein digests by LC-ESI-MS/MS, *J. Mass Spectrom.* 41, 623–632, 2006.

148. Leitner, A., Foettinger, A., and Lindner, W. Improving fragmentation of poorly fragmenting peptides and phosphopeptides during collision-induced dissociation by malonaldehyde modification of arginine residues, *J. Mass Spectrom.* 42, 950–957, 2007.

149. Onofrejova, L., Leitner, A., and Lindner, W., Malondialdehyde tagging improves the analysis of arginine oligomers and arginine-containing dendrimers by HPLC-MS, *J. Sep. Sci.* 31, 499–506, 2008.

150. Foettinger, A., Melmer, M., Leitner, A., and Lindner, W., Reaction of the indole group with malondialdehyde: Application for the derivatization of tryptophanyl residues in peptides, *Bioconjug. Chem.* 18, 1678–1683, 2007.

151. Toi, K., Bynum, E., Norris, E., and Itano, H.A., Studies on the chemical modification of arginine. I. The reaction of 1,2-cyclohexanedione with arginine and arginyl residues of proteins, *J. Biol. Chem.* 242, 1036–1043, 1967.

152. Itano, N.A. and Gottlieb, A.J., Blocking of tryptic cleavage of arginyl bonds by the chemical modification of the guanido group with benzil, *Biochem. Biophys. Res. Commun.* 12, 405–408, 1963.

153. Patthy, L. and Smith, E.L., Identification of functional arginine residues in ribonuclease A and lysozyme, *J. Biol. Chem.* 250, 565–569, 1975.

154. Samy, T.S., Kappen, L.S., and Goldberg, I.H., Reversible modification of arginine residues in neocarzinostatin. Isolation of a biologically active 89-residue fragment from tryptic hydrolysate, *J. Biol. Chem.* 255, 3420–3426, 1980.

155. Ullah, A.H.J. and Sethumadhaven, K., Differences in the active site environment of *Aspergillus ficuum* phytases, *Biochem. Biophys. Res. Commun.* 243, 458–462, 1998.

156. Calvete, J.J., Campanero-Rhodes, M.A., Raida, M., and Sanz, L., Characterization of the conformation and quaternary structure-dependent heparin-binding region of bovine seminal plasma protein PDC-109, *FEBS Lett.* 444, 260–264, 1999.

157. Patthy, L. and Thész, J., Origin of the selectivity of α-dicarbonyl reagents for arginyl residues of anion-binding sites, *Eur. J. Biochem.* 105, 387–393, 1980.

158. Baynes, J.W. and Thorpe, S.R., Role of oxidative stress in diabetic complications: A new perspective on an old paradigm, *Diabetes* 48, 1–9, 1995.

159. Reubsaet, J.L.E., Beijnen, J.H., Bult, A. et al., Analytical techniques used to study the degradation of proteins and peptides: Chemical instability, *J. Pharm. Biomed. Anal.* 17, 955–978, 1998.

160. Thornalley, P.J., Glutathione-dependent detoxification of alpha-oxoaldehyde by the glyoxalase system: Involvement in disease mechanisms and antiproliferative activity of glyoxalase I inhibitors, *Chem. Biol. Interact.* 111–112, 137–151, 1998.

161. Deyl, Z. and Miksik, I., Post-translational non-enzymatic modification of proteins. I. Chromatography of marker adducts with special emphasis on glycation reactions, *J. Chromatogr. B* 699, 287–209, 1997.

162. Westwood, M.E., Argirov, O.K., Abordo, E.A., and Thornalley, P.J., Methylglyoxal-modified arginine residues—A signal for receptor-mediated endocytosis and degradation of proteins by monocytic THP-1 cells, *Biochim. Biophys. Acta* 1356, 84–94, 1997.

163. Degenhardt, T.P., Thorpe, S.R., and Baynes, J.W., Chemical modification of proteins by methylglyoxal, *Cell. Mol. Biol.* 44, 1139–1145, 1998.

164. Klöpfer, A., Spannberg, R., and Glomb, M.A., Formation of arginine modifications in a model system of N^α-*tert*-butoxycarbonyl (Boc)-arginine with methylglyoxal, *J. Agric. Food Chem.* 59, 394–401, 2011.

165. Pischetsrieder, M., Reaction of L-ascorbic acid with L-arginine derivatives, *J. Agric. Food Chem.* 44, 2081–2085, 1996.

166. Takahashi, K., Specific modification of arginine residues in proteins with ninhydrin, *J. Biochem.* 80, 1173–1176, 1976.

167. Jacob, R.T., Bhat, P.G., and Pattabiraman, T.N., Isolation and characterization of a specific enterokinase inhibitor from kidney bean (*Phaseolus vulgaris*), *Biochem. J.* 209, 91–97, 1983.

168. Rhodes, G.R. and Boppana, V.K., High-performance liquid chromatographic analysis of arginine-containing peptides in biological fluids by means of a selective post-column reaction with fluorescence detection, *J. Chromatogr.* 444, 123–131, 1988.

169. Boppana, V.K. and Rhodes, G.R., High-performance liquid chromatographic determination of an arginine-containing octapeptide antagonist of vasopressin in human plasma by means of a selective post-column reaction with fluorescence detection, *J. Chromatogr.* 507, 79–84, 1990.

170. Ruhemann, S., Triketohydrindene hydrate, *J. Chem. Soc. Trans.* 97, 2025–2031, 1910.

171. Retinger, J.M., The mechanism of the ninhydrin reaction. A contribution to the theory of color of salts of alloxantine-like compounds, *J. Am. Chem. Soc.* 39, 1059–1066, 1917.

172. Bronfenbrenner, J., The mechanism of the Abderhalden reaction: Studies on immunity I. *J. Exp. Med.* 21, 221–238, 1915.

173. Abderhalden, E. and Schmidt, H., Über die Verwendung von Triketohydrindenthydrat[1]) zum Nachweis von Eiweszstoffen under deren Abbaustufen, *Hoppe-Seyler's Z. Physiol. Chem.* 72, 37–43, 1913

174. Jobling, J.W., Eggstein, A.A., and Petersen, W., Serum proteases and the mechanism of the Abderhalden reaction: Studies on ferment action. XX, *J. Exp. Med.* 21, 239–249, 1915.

Index